Materials for Sustainable Energy

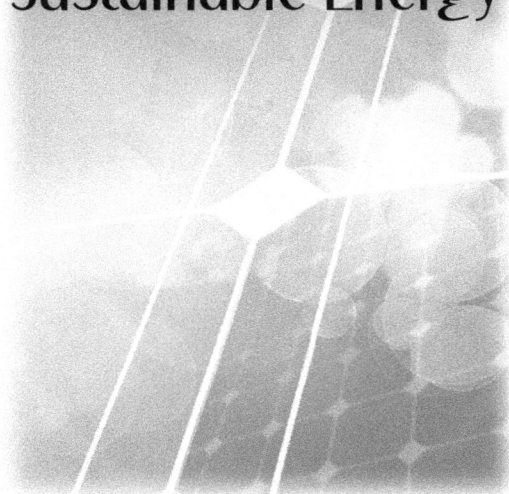

A Collection of Peer-Reviewed Research and Review Articles
from Nature Publishing Group

Materials for Sustainable Energy

A Collection of Peer-Reviewed Research and Review Articles
from Nature Publishing Group

Edited by

Vincent Dusastre

With an Introduction by
Jean-Marie Tarascon and Michael Grätzel

World Scientific

nature publishing group **npg**

Published by

Macmillan Publishers Ltd
(trading as Nature Publishing Group)
4–6 Crinan Street
London N1 9XW
United Kingdom

and

World Scientific Publishing Co. Pte. Ltd.

5 Toh Tuck Link, Singapore 596224
USA office: 27 Warren Street, Suite 401-402, Hackensack, NJ 07601
UK office: 57 Shelton Street, Covent Garden, London WC2H 9HE

Library of Congress Cataloging-in-Publication Data
Materials for sustainable energy : a collection of peer-reviewed research and review articles from nature
   publishing group / edited by Vincent Dusastre ; with an Introduction by Jean-Marie Tarascon and
   Michael Grätzel.
      p. cm.
   Includes bibliographical references.
   ISBN-13: 978-981-4317-64-1 (hardcover : alk. paper)
   ISBN-10: 981-4317-64-0 (hardcover : alk. paper)
   ISBN-13: 978-981-4317-65-8 (paperback : alk. paper)
   ISBN-10: 981-4317-65-9 (paperback : alk. paper)
   1. Energy conservation--Equipment and supplies. 2. Sustainable engineering--Materials.
I. Dusastre, Vincent.
   TJ163.3.M385 2010
   621.042--dc22

                                                           2010033305

British Library Cataloguing-in-Publication Data
A catalogue record for this book is available from the British Library.

Printed in Singapore.

CONTENTS

THERMOELECTRIC CONVERTERS

BATTERIES AND SUPERCAPACITORS

PREFACE

The availability of affordable energy has so far led to spectacular industrialization and development, but with growth accelerating in developing countries, demands on energy infrastructure and sources are being stretched to new limits. The recent realization that the observed rise in global temperature since pre-industrial time is most likely due to the increase in the concentration of greenhouse gases, most crucially carbon dioxide, together with the uneven distribution of energy sources around the world, must change the way we view the generation and supply of energy. If nothing is done to mitigate the carbon dioxide problem, the consequences for human life and our well-being could be catastrophic[1,2].

To deal with these rapid changes we basically have three options: improve the efficiency with which energy is used through better technology; push up the price of emitting carbon through regulation; and reduce the costs of generating energy without fossil fuels, most notably from renewable sources such as light, wind, water and plant growth. The usual tendency among governments and traditional utilities to see renewable energy sources as add-ons is shortsighted. These sources are, alongside nuclear power[3], what we need.

Many new energy-saving technologies such as electric cars, light-emitting diodes and high-voltage direct-current transmission of electric power are emerging. None is without its problems, and the renewable resources tend to be generally diffuse and sporadic. Yet there is a great opportunity for technological progress in many of these fields, most notably in solar power[4] and in energy storage.

The need for abundant energy sources that do not rely on fossil fuels is the great technological challenge of the twenty-first century — fundamental to further economic development and, to some degree, of climate stability. Fossil-fuel depletion and carbon dioxide emission is driving the quest for renewable energy resources and the search for cleaner, cheaper, smaller and more efficient energy technology is to a large extent driven by the development of new materials[5,6].

The aim of this collection of articles is therefore to focus on what materials-based solutions can offer, and show how the rational design and improvement of their physical and chemical properties can lead to energy-production alternatives that have the potential to compete with existing technologies. This collection of 48 peer-reviewed articles including reviews and research papers published over the past 10 years in *Nature* and *Nature Materials* looks at the technologies and science-base needed to meet the challenge of clean energy on a global scale.

We hope this book can serve as an introduction to researchers who may be thinking of moving into a new field related to energy and for others (students, politicians) who simply want to learn about how we can tackle the energy challenge. In terms of alternative means to generate electricity that utilize renewable energy sources, the most dramatic breakthroughs for both mobile and stationary applications are taking place in the fields of photovoltaic cells and thermoelectric converters. And from an energy-storage perspective, exciting developments have been emerging from the fields of rechargeable batteries, supercapacitors, fuel cells, hydrogen storage and superconductors.

The articles are introduced in a new essay by Jean-Marie Tarascon (CNRS/Université de Picardie Jules Verne) and Michael Grätzel (Ecole Polytechnique Fédérale de Lausanne), which explores the challenges and opportunities that are facing researchers for developing new materials and technologies capable of quenching the world's increasing thirst for energy. Focusing on energy conversion and storage, the article outlines the fundamental and practical challenges associated with each technology and discusses ways of moving towards renewable energy.

This collection mostly relates to fundamental research related to materials discovery and sustainable technological solutions, but the global landscape of materials issues associated with energy is much broader, and all existing energy technologies should be considered as they all have an environmental footprint. For example, considering present reserves, coal will continue to be a critical component of the energy portfolio for this century. Generating energy from coal efficiently and with limited environmental damage presents notable challenges for developing higher-performance materials. Technologies exist to transform a fossil fuel to other uses, such as coal to gas or liquid. New materials that increase the efficiency of the transformation and lower its cost are welcomed, and materials should also be evaluated in terms of their entire life cycle to establish which will be the most efficient. This principle should also be applied to other technologies such as nuclear power, carbon capture and storage, and biofuels.

Another challenge is how to scale up these new technologies for a global economic market. A significant part of the answer is to invest in focused research and development. However, there is also a role for regulation and subsidy. The types of subsidy that have worked so far are expensive even at the megawatt scale. At the scale of hundreds of gigawatts they are likely to be unsustainable. Another risk is that poorly designed subsidies will damage financial investments and thwart the economic and industrial development they seek to encourage. That said, there are strong reasons to encourage technologies in a smart and flexible way and not to leave the market forces to decide.

All in all there is still a great lack of public appreciation of the scientific and technological difficulties in supplying energy once the sources of fossil fuels are used up. Many difficulties remain in making the substantial improvements to materials that are needed to increase efficiency in the generation, conversion, transmission and use of energy. As we all share the atmosphere of our planet, international cooperation is crucial not only on a scientific level, but more importantly on a political level. The recent failure in Copenhagen to reach a political agreement on carbon emission should serve as a wake-up call[7]. The challenge is not just how to preserve our environment most efficiently, but how to save our planet.

Vincent Dusastre
Chief Editor
Nature Materials

References

1. A pledge for immediate action. *Nature Mater.* **8,** 81 (2009).
2. http://www.ipcc.ch/pdf/assessment-report/ar4/syr/ar4_syr_spm.pdf
3. 'Nuclear' is not the question. *Nature Mater.* **7,** 679 (2008).
4. Let the Sun shine. *Nature Mater.* **7,** 825 (2008).
5. Dusastre, V. Materials for clean energy. *Nature* **414,** 331 (2001).
6. Harnessing materials for energy. *Mater. Res. Soc. Bull.* **33** (Special issue), 261–477 (2008).
7. Copenhagen no more. *Nature Mater.* **9,** 89 (2010).

Acknowledgements

For the introductory pieces, thanks to Jane Morris for copy editing and to Karen Moore for preparing the artwork. Thanks also to Cecilia Högström, Angelina Storti, Hollie Gill, Laura Graham-Clare and Theresa Tuson for obtaining permission to reproduce figures in the book; to Emma Green and Jason Wilde for project management; to colleagues on various NPG journals for suggesting articles to include and, finally, to all the authors.

When citing any of these articles, please cite the original version. Full citation details are available on the contents pages of each chapter.

RESEARCH INITIATIVES TOWARDS SUSTAINABLE HANDLING OF ENERGY

Jean-Marie Tarascon* & Michael Grätzel[†]

The world's hunger for energy is ever increasing but fossil fuel reserves are limited. As the planet's energy needs will at least double within the next 20 years, a major energy shortage is unavoidable unless renewable energy can take over from fossil fuels. Public concern over energy resources has increased as a result of the disastrous environmental pollution from oil spills and the frightening potential climatic consequences of global warming from the combustion of fossil fuels. Researchers are assessing various alternative storage and conversion technologies to solve this equation. Here, a few that rely on innovations in sustainable materials-composites are reviewed and their technological impact considered.

Energy is the lifeblood of modern societies with electricity being the most used vector carrier. Global warming, finite stocks of fossil fuel and increasing city pollution show how important it is to move towards intense and efficient use of renewable energies, and to find innovative solutions to ease the progressive transition from thermal to electric vehicles. Both the intermittency of renewable energies, and the need to provide enough power on board electric vehicles to ensure they can travel long distances without needing refuelling, make the development of new technologies for electrochemical storage of energy and energy-conversion the greatest challenge this century. We are dependent on the electrical chain (electricity generation, distribution and storage) and we must generate electricity when and where people need it. Meeting such requirements necessitates, among others: efficient photovoltaics or thermo-electric devices to convert light or heat, respectively, into electricity; high-electronic-conducting media such as superconductors to minimize the joule effect; storage media such as batteries or supercapacitors to hold energy in its chemical form and restore it as electricity on demand. Such production, transportation and storage issues, although requiring different strategies, will also require consideration of alternative vector carriers, such as hydrogen.

Progress in the fields of energy conversion or storage has been slow compared with progress in memory storage (Fig. 1). Breakthroughs are needed to depart from the weakly sloping curve, and public awareness of such needs is a clear impetus for spurring research activities dealing with energy. Although the hunt for high-performance energy-storage and -conversion devices remains the main objective, the cost associated with producing these devices, and the dependence on the abundance of materials and fabrication processes, are now becoming other overriding factors. Notions of sustainability, renewability and green chemistry must be taken into consideration when selecting electrode materials for the next generation of batteries and photovoltaic cells, especially when high-volume

*RCS-UMR 6007, Université de Picardie Jules Verne, 80039 Amiens, France.
[†]Laboratory of Photonics and Interfaces, Station 6, Institute of Chemical Science and Engineering Faculty of Basic Science, Ecole Polytechnique Federale de Lausanne, CH-1015 Lausanne, Switzerland.

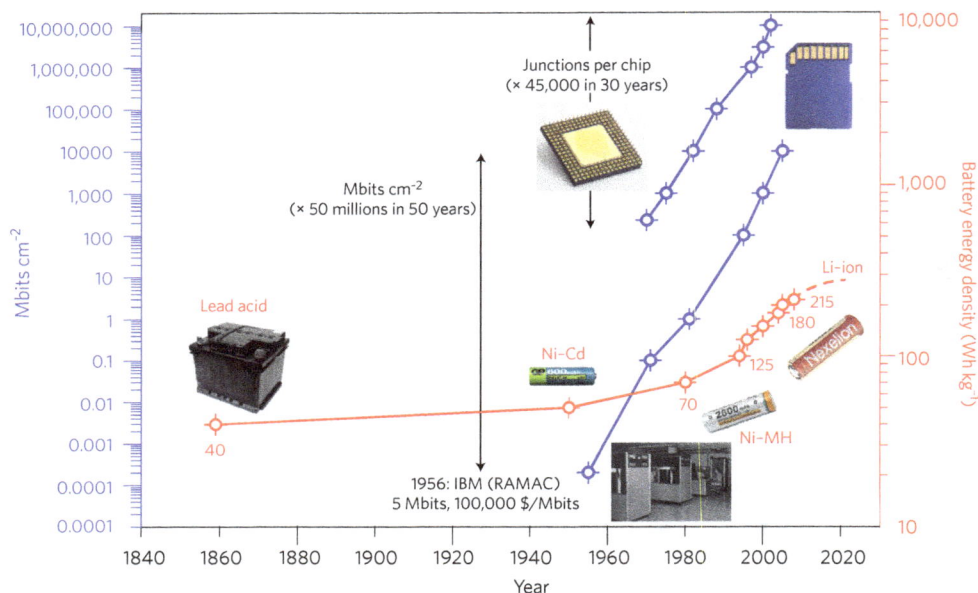

Figure 1 Comparative evolution of memory storage versus electrochemical storage. This shows that progress in battery storage capacity is far from following Moore's law (doubling of capacity every 18 months) as it has only increased by a factor of five over 200 years. Such slow progress is mainly linked to materials limitations. On an optimistic note, with the advent of Li-ion technology, the energy density has been boosted by a factor of two in twenty years. Whether such a rate can be maintained over the next century remains an open question. It should be noted that the curve starts to level off, calling for a paradigm shift in battery research. (Figure courtesy of D. Larcher; credits: battery, ©iStockphoto/Difydave; chip, ©iStockphoto/Gecko753; memory card, ©iStockphoto/Trebuchet.)

applications are considered. Such new requirements are adding constraints while providing conceptual opportunities (beyond the use of nanomaterials), relying on the rapid ingress of organic components and multifunctional devices (for example, associated with storage and conversion functions within the same device).

Many years after microelectronics, the field of energy and conversion storage is going through its 'nanomania' with the arrival of nanoporous materials, nanotextured and nanostructured electrodes and nano-architectured current collectors. Such a paradigm shift from bulk to nanomaterial electrodes has benefited fuel cells and thermoelectric devices, enabling great progress in photovoltaics and has boosted fuel cell, supercapacitor and battery performances.

Photovoltaic cells

Ever since the French scientist Henri Becquerel discovered the photoelectric effect in 1839, researchers and engineers have been infatuated with the idea of converting light into electric power. There is a magical aspect to solar cells that attracts a wide range of people, from amateurs to highly trained professionals, from penniless entrepreneurs to wealthy sponsors. Their common dream is to collect the energy that is freely available from sunlight and turn it into the valuable and strategically important asset that is electric power. The world will face a huge power-supply gap of 14 terawatts by the year 2050, equal to today's entire consumption, threatening to create a planetary emergency of gigantic dimensions. As the sun provides about 120,000 terawatts to the Earth's surface, covering only 0.1% of it with

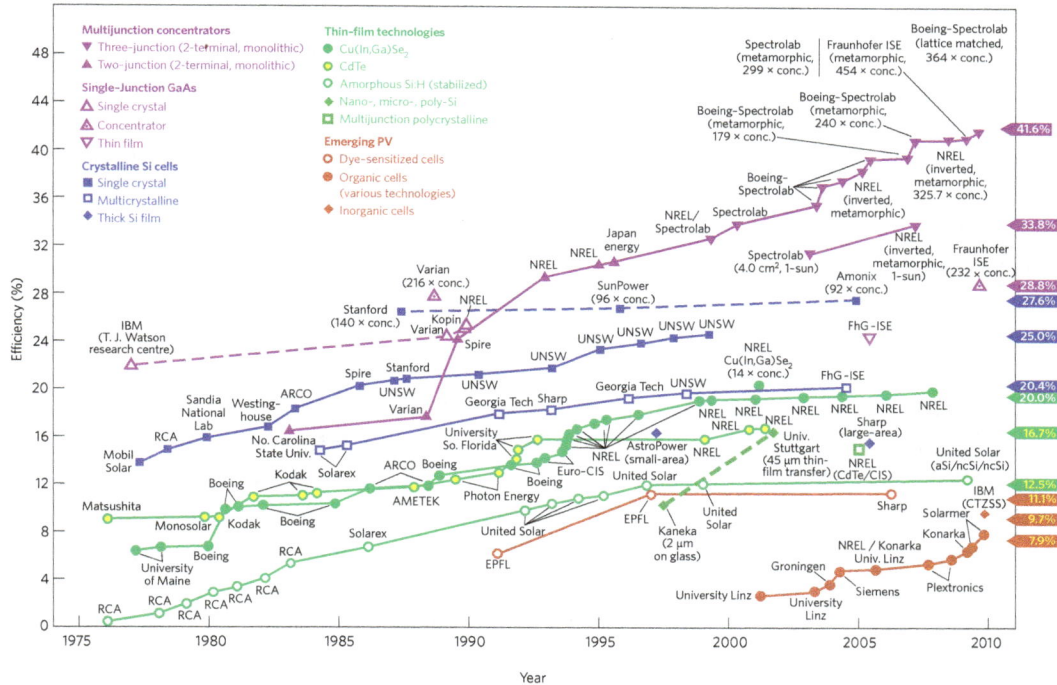

Figure 2 Evolution of the best laboratory cell efficiencies for different solar cell technologies. The plot distinguishes three generations of solar cells according to their state of development. First-generation cells are made of single-layer p–n junction diodes of GaAs or silicon wafers. The second-generation cells are thin films of materials such as amorphous silicon, CdTe and CuInSe$_2$ (CIGS). Third-generation cells are emerging PV technologies such as dye-sensitized cells and various organic cell technologies. Multijunction concentrator cells are a separate PV category that currently achieves efficiencies over 40%. (Figure courtesy of L. Kamarski, NERL USA, April 2010.)

solar cells having an efficiency of 10% would fulfil these needs. To tap into the Sun's huge energy reservoir remains, nevertheless, a major challenge.

Figure 2 gives an overview of the different families of solar cells and the evolution of their power-conversion efficiency over the past few decades. Note that the power-conversion efficiencies refer to small laboratory cells, values for commercial modules being 20% to 50% lower than those quoted here. First-generation silicon wafer cells are flat, single-layer, p–n junction diodes, which remain the dominant technology in the commercial production of solar cells, accounting for at least 85% of the solar market. However, owing to the high production cost in terms of energy and finance — payback time is about three years — doubts have been expressed as to whether this technology would be suitable for delivering electric power on the terrawatt scale[1]. The same holds for second-generation inorganic thin-film photovoltaic (PV) cells based on CdTe or CuIn(Ga)Se$_2$ (CIGS) containing toxic elements and/or those of low abundance[2]. Quarternary compounds, such as Cu$_2$ZnSn(Se,S)$_4$ (CZTS) are promising thin-film PV materials[3] as they avoid the use of In, Cd and Te. Small laboratory CZTS cells have recently attained an efficiency of 9.6% (ref. 4), which is respectable, albeit still half that of CIGS cells.

Thanks to government-imposed cost incentives, annual production of solar cells has increased 30 times over the past 10 years, reaching 7.9 gigawatts of peak power worldwide in 2008. However, the set-back in sales in 2009 revealed the great sensitivity of the PV market to changes in these subsidies; the cost of PV power being at present still 5–10 times too high to compete with conventional energy suppliers on the wholesale level. Clearly, disruptive breakthroughs are required for PVs to meet the cost target of 0.5$ per peak watt, corresponding to a kilowatt-hour price of around 5 cents, which is required for PVs to become an unsubsidized market.

The impressive advances made by the first two generations of PV cells over the past two decades are now being increasingly challenged by third-generation photovoltaic cells. Rather than relying on traditional flat p–n junctions to separate photo-generated charge carriers, most of these cells use films of mesoscopic morphology for the conversion of solar to electric power. Most papers on PV cells selected for this collection belong to this category, giving a vivid demonstration of how research has been booming over the past decade in this rapidly evolving field. This new paradigm was introduced with the advent of dye-sensitized solar cells (DSSCs), which used for the first time a three-dimensional interpenetrating network junction based on nanocrystalline metal oxide particles sensitized by a monolayer of dye molecules[5]. The role of the dye is to absorb sunlight and generate charge carriers by electron injection in the oxide particles; the positive charge left on the sensitizer is injected in the electrolyte or 'hole conductor' present in the pores of the film. Thus, in analogy to the natural photosynthetic process, the DSSC separates the two functions of light harvesting and charge-carrier transport, greatly increasing the choice for the absorber material. Research on this new type of bulk heterojunction has been extended to blends of polymers with C60 derivatives[6] as well as several other embodiments extending from pure organic to hybrid and inorganic materials. These offer the prospect of cheap fabrication together with other attractive features, such as flexibility. The phenomenal recent progress in fabricating and characterizing nanocrystalline materials has opened up whole new vistas of opportunity. The photoactive materials come in different shapes, ranging, for example, from nanoparticles to nanotubes and nanorods. Owing to their small feature size the diffusion path of electron–hole pairs photogenerated in these structures is only a few nanometres, giving the carriers very short diffusion times to reach the junction where they are separated. Thus the new devices can reach high conversion efficiencies, competing with those of conventional solar cells even though they have much higher levels of impurities or defects than solar-grade semiconductor materials. This strategy has also been successfully applied to photoelectrochemical cells that decompose water into hydrogen and oxygen[7]. Thus, Fe_2O_3 (haematite) the main component of rust, is notorious for its very inefficient performance as a photo-anode for the production of oxygen from water. Photogenerated electron–hole pairs live for only a few picoseconds in haematite, and their mobility is so low that the charge carriers diffuse only a few nanometres before they recombine. Despite this huge predicament, the recent introduction of mesoscopic haematite films having a feature size in the nanometre range has made Fe_2O_3 the most efficient oxide photo-anode for solar water-splitting known so far[8,9].

Apart from tolerance to defects and impurities, high-aspect-ratio structures, such as silicon wire arrays, show light-containment effects that render them very effective in capturing sunlight using less material than the flat p–n junction cells[10]. Metallic nanostructures, which support surface plasmons, also provide very promising opportunities for light trapping, by reducing the semiconductor layer thickness required to absorb sunlight[11–13]. Apart from their effective scattering action these structures act as antennas, capturing incoming light and transferring the excitation energy to a nearby absorber. This near-field effect also entails a significant local concentration of the light that may be exploited for photon up-conversion and concentrator cells[14].

Implementation of photonic crystals in thin-film solar cells provides another attractive means of improving their light-harvesting capacity[15,16]. In the DSSC this effect can be exploited to enhance dye absorbance on the longer wavelength side of the photonic

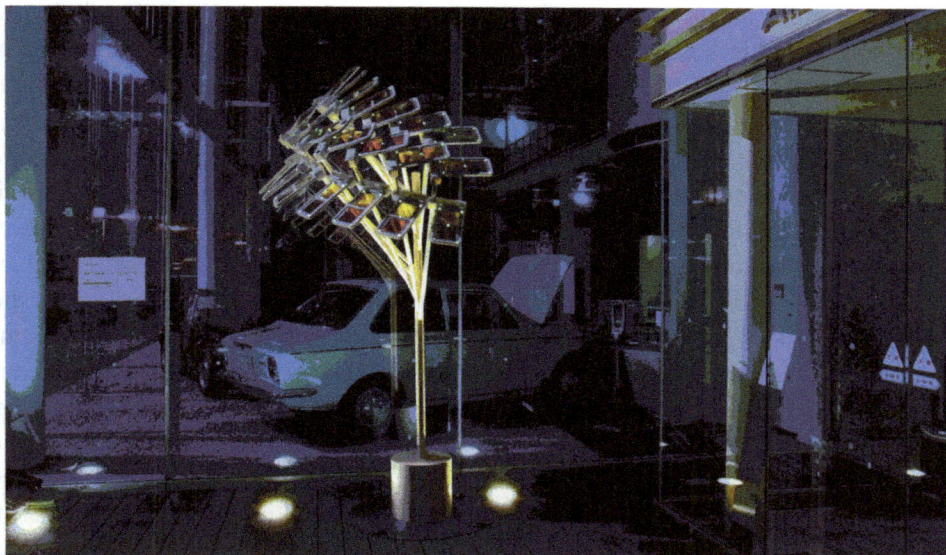

Figure 3   Light-producing tree for street lamps. (Image courtesy of Aisin Seiki, Japan.)

bandgap, leading to an increase in the efficiency of the conversion of the incident photon into current. Importantly, enhanced absorption due to the slow group-velocity of light with wavelengths near to the band-edge and the localization of light in nanocrystalline dye-sensitized TiO_2 films has now been observed[16].

Great excitement has followed the reports by Klimov and Nozik[17,18] that several excitons can be produced from the absorption of a single photon by very small semiconductor particles — called quantum dots because their electronic properties are different from those of bulk materials due to the confinement of the electron–hole pairs produced by optical excitation. This multi-exciton generation (MEG) effect occurs if the energy of the absorbed photon is several times higher than the semiconductor bandgap, and was reported to be much more efficient in the quantum dots than in bulk size materials[17,18]. However, whether a significant enhancement of the MEG phenomenon by the quantum size effect really occurs is still under debate. Current research aims to find ways of collecting the excitons as electric charges before they recombine. A calculation shows that the maximum conversion efficiency of a single junction cell could theoretically be increased from 31% to 44% by fully exploiting MEG effects[18].

The intermittent nature of solar radiation renders the development of energy-storage systems very important. At present much of the electricity derived from PVs is injected into the grid. However, this option is limited by the low capacity of the grid and problems with the management and storage of the extra solar power. Future systems will therefore integrate the solar panels into systems that combine them with a rechargeable battery, and optionally other elements such as a light-emitting diode. Figure 3 shows a light-producing tree that uses transparent DSSCs in different colours to charge a battery during the day. Below the DSSC is a light-emitting diode that produces light during the night.

Thermoelectric converters

Thermoelectric converters provide a means of converting heat directly into electricity without resorting to any mechanical or chemical processes. Geothermal, solar or simply waste heat can serve as an energy source for thermoelectric generators that are emission-free during operation, noiseless and extremely durable, providing a sustainable energy supply. Thermoelectric materials are ranked based on a figure of merit ZT, where T is temperature, and Z is proportional to the Seebeck coefficient (S) according to the formula $Z = S\sigma^2/\kappa$, with σ and κ being electrical and thermal conductivity, respectively. To be of practical interest a value of ZT larger than 1 would be desirable. Several strategies are being applied to tackle this goal, and they have been discussed in a recent review by Heath[19]. The two reports on thermoelectric conversion by silicon nanowires (Boukai et al.[20] and Hochbaum et al.[21]) that are included in this volume are landmark papers in as much as they show for the first time how, by endowing the silicon nanowires with a rough surface structure, their thermal conductivity can be reduced by a factor of over 100, without significantly decreasing their electrical conductance. As a result, ZT values of between 0.6 and 1 have been achieved for the first time with silicon-based systems. At the same time, complex bulk materials such as skutterudites, clathrates, Zintl and semi-Heusler phases have been explored, and high efficiencies could occasionally be achieved with, for instance, a ZT value of 1.4 at 525 K for $Zn_{3.2}Cd_{0.8}Sb_3$. In contrast, research on cobaltites ($NaCoO_3$, $Ca_3Co_4O_9$), despite early excitement, has been somewhat disappointing as measured ZT values still remain lower than 1 even for a temperature greater than 1,000 K. Overall, it seems that the most promising way of achieving greater ZT values is nested in combining both innovative crystal chemistry and nanostructuring approaches so as to obtain materials/composites with high conducting properties that could manage phonons at all length scales[22].

Batteries

To cope with the intermittent aspects of the PVs, storage of electricity is needed, hence batteries. Within the field of batteries, moving from bulk to nanosized particles[23,24] has, for instance, enabled (1) the previously disregarded and poorly conducting $LiFePO_4$ insertion electrode to be turned into the most praised electrode material for future applications for electric vehicles[25]; (2) the use of metal-based (tin or silicon) negative electrodes based on alloy reactions, such that tin-based lithium-ion battery technologies have already reached the market place (NEXELION) and that silicon-based ones are on the verge of doing so[26]; (3) new lithium reaction mechanisms to be triggered with the advent of conversion reaction electrodes[27] showing huge capacity gains, although with poor energy efficiency.

The possibility of bypassing the limitations of charge transport in materials via particle downsizing and carbon-coating approaches[28,29], coupled with high-throughput computational searches of new chemistries and structures to maximize energy-density and rate capabilities, has opened the door to fertile materials research. Fluorine-based electrode materials known for their poor conducting properties but higher redox potential than their oxides are attracting growing interest. Many fluorine-based compounds ($LiVPO_4F$ (ref. 30), Na_2FePO_4F (ref. 31), and more recently $LiFeSO_4F$ (ref. 32)) have been developed to propel the next-generation Li-ion batteries.

Despite the new lease of life recently brought by the arrival of nanomaterials, the present Li-ion technology falls short of powering vehicles for long distances ($\sim 500\,km$). One of the few approaches with the potential to achieve this is the Li-air battery, which can provide up to a 10-fold increase in gravimetric energy density compared with the Li-ion technology[33]. As a result, this battery is receiving much attention, and its reversibility, for at least a few tens of cycles and with the help of a carbon/a-MnO_2 nanotextured oxygen electrode, has been demonstrated[34]. However, there are many problems to be addressed before Li-air can even be considered as a potential technology, despite the worldwide excitement and colossal financial investment. In parallel, the Li-sulphur battery is coming back to the scene with the use of a highly ordered nanostructured carbon–sulphur cathode to minimize the polysulphide $(S–S)_n$ solubility. This has been the main aim of this system until now because it enables more sustainable capacity retention[35].

Other long-term demands, linked to the large-volume application for Li-ion technology, deal with materials resources and abundance, as well as with their synthesis and recycling processes, which have a heavy energy cost. This is the reason why notions of materials sustainability and materials with a minimum footprint made via eco-efficient processes are integrated in new research endeavours towards the next generation of sustainable and 'greener' Li-ion batteries. The recently reported $Li_xC_6O_6$ organic electrodes[36] seem to be ideal candidates as they can provide environmental benefits: they can be synthesized via 'green chemistry' from natural organic sources, are biodegradable, not resource limited and show gravimetric–volumetric energy densities that compare favourably with today's most common electrodes. Mindful of these requirements, bio-inspired or biomimetic approaches towards materials sustainability and renewability issues are also appealing. We are therefore witnessing increasing numbers of studies in the biomimetic field of electrode materials for Li-ion batteries. Belcher's recent work[37] on the use of engineered viruses as templates to make high-power $FePO_4$/single-walled carbon nanotube electrodes at room temperature illustrates what can be done. It is likely that these biological principles will be more intensely implemented in the rational design and assembly of battery components.

Looking to the future, there is no doubt that the huge increase in hybrid energy vehicles will strain the world's supply of lithium. In anticipation of such a scenario, new chemistries must be created[38], and among the most appealing alternatives is the use of sodium instead of lithium. There are several reasons: sodium has similar intercalation chemistry; its resources are in principle unlimited and it is very easy to recuperate; and finally it does not alloy with aluminium, therefore aluminium current collectors can be used as the negative electrode. All of which leads to a technology for which the kWh could be cheaper than for lithium, although a penalty in terms of gravimetric energy density will have to be paid as sodium is both heavier and less electropositive than lithium. Room-temperature Na-ion cells that are cost effective, sustainable and environmentally friendly could be developed, as we could use the knowledge gained by developing Li-ion insertion electrodes to design new sodium intercalation compounds. One problem is that sodium metal is somewhat less prone to developing a solid–electrolyte interface, which is vital to the proper functioning of Li-ion cells. Other abundant metals worth considering are magnesium and aluminium, because repeated plating of the latter is also possible, however only in very few electrolytes[39,40]. Owing to the high voltage penalty of using magnesium and aluminium anodes — a drop in voltage of 0.8 V and 1.3 V, respectively, compared with lithium — such negative electrodes need to be associated with high-capacity/high-voltage positive electrodes to make this system attractive, which is so far a poorly explored research field. Furthermore, far less is known on the kinetics of electrode reactions involving the motion of multivalent species than on single-valent species such as lithium and sodium. These are not insurmountable tasks, and overall, it can only be a good thing to search for alternative chemistries to decrease our dependence on lithium.

Supercapacitors

Unfortunately, Li-ion batteries have a somewhat slow power delivery/uptake, therefore for applications requiring faster, higher-power energy-storage systems, electrochemical capacitors (ECs) have been used. These can complement or replace batteries in electrical

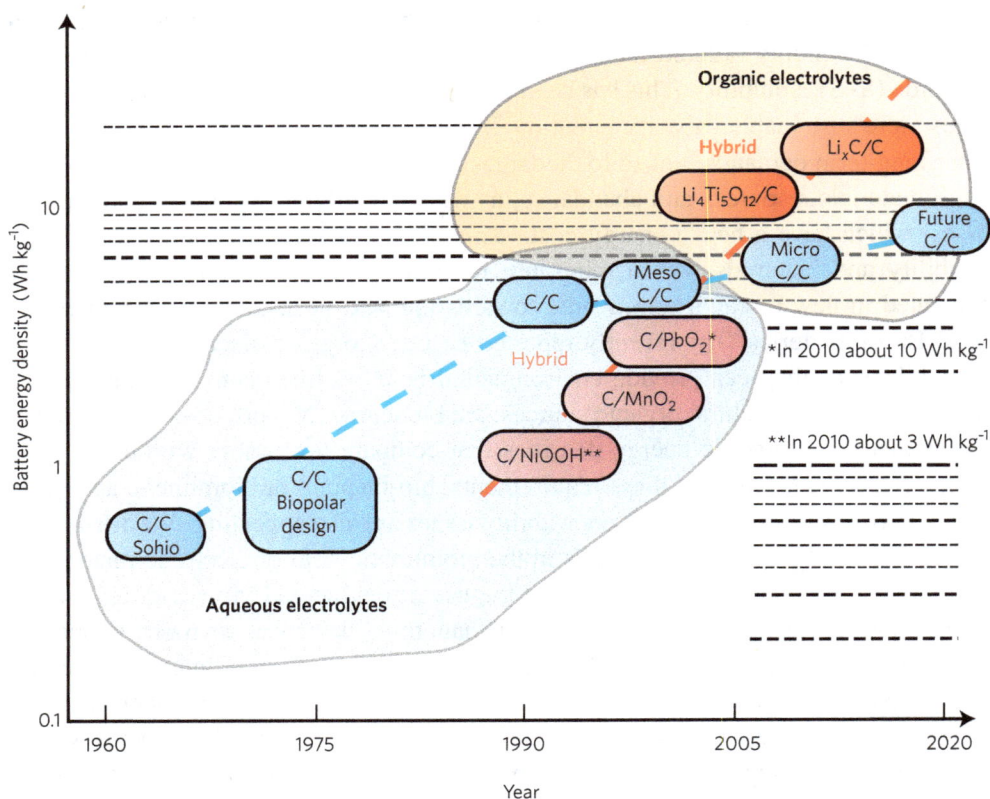

Figure 4 Evolution in performance of electrochemical capacitors over the past half-century, represented by distinguishing aqueous (blue) electrolytes from non-aqueous ones (pink). Step changes were obtained by moving from aqueous to non-aqueous electrolytes as the voltage could be increased from 0.9 V to 2.5 V, and by subsequently decreasing the pore size (for example, from mesoporous to microporous carbons). A greater increase in slope is provided by hybrid technology (see text) regardless of the type of aqueous electrolyte; however, a price has to be paid in terms of cycling performance. For reasons of clarity, redox-based electrochemical capacitors (for example, RuO_2-based) are not reported. (Figure courtesy of P. Simon.)

energy storage and harvesting applications when high-power delivery or uptake is needed for a few seconds. One possible example can be found in hybrid vehicles, where ECs could allow for fast acceleration, and for recovering the energy produced by braking, while increasing the battery life by diminishing the strain caused by these high-rate activities. Like batteries, the development of advanced nanostructured materials has had a considerable impact on EC performance (Fig. 4). The discovery that ion desolvation occurs in smaller pores than solvated ions[41] has opened a new research avenue to increase the capacitance for electrochemical double-layer capacitors by designing a variety of carbon electrodes having subnanometre pores. It is within this context that carbon with subnanometre pores tuned to fit the size of the ions of the electrolyte with ångström accuracy, known as carbide-derived carbons (CDC), was developed[42]. Notable capacitance increases were also achieved by the fabrication of nanotextured electrodes based on novel pseudo-capacitive materials including oxides and nitrides, and the development of hybrid systems that combine a battery-like electrode with a capacitor-like electrode[43]. However, this gain is achieved at the expense of the life of the device. The performance of supercapacitors can potentially be further increased by designing multifunctional materials combining capacitive and pseudo-capacitive charge storage — this is achieved by decorating high-surface-area nanoporous carbons with nanometric oxide particles — and by using ionic liquids resistant to oxidation to high voltages like those for batteries[44] — however, this is achieved at the cost of the power for temperature applications below ambient.

Cost, environmental and sustainability issues (although the latter is less problematic for supercapacitors than for batteries because of their longer lifetime) are receiving an increasing amount of attention. Two such examples are the development of (1) a low-cost carbon–MnO_2 hybrid system that combines a high capacitance in neutral aqueous electrolytes with high cell voltages (1.6 V)[45], making it a green alternative to electric double-layer capacitors (which use acetonitrile-based solvents and fluorinated salts); and (2) a low-cost carbon–PbO_2 hybrid system using a H_2SO_4 electrolyte solution and operating at 2.1 V. Another noteworthy trend in this area is the colossal amount of work currently being performed to develop highly capacitive carbons from biomass.

Fuel cells

Fuel cells constitute another category of electrochemical storage devices that can convert chemical energy directly into electrical energy with high efficiency and low emission of pollutants, and are therefore an appealing solution for powering future generations of 'green' vehicles[46]. Their positive attributes are linked to the use of hydrogen as a fuel, which can be made from renewable energy sources. When hydrogen fuel is burned (that is, reacts with oxygen from the air) to produce electricity, the exhaust is pure water with no net increase in carbon dioxide emissions. Out of the several different types of fuel cells currently under development, polymer electrolyte membrane fuel cells (PEMFCs) are gaining momentum for applications in transportation, and solid oxide fuel cells (SOFCs) are more promising for large stationary applications[47]. At present, research into the use of nanostructured materials and processing methods for low-temperature fuel cells (usually PEMFCs) is focused on the development and dispersion of precious or non-precious catalysts for efficient hydrogen reduction, and the elaboration of perfluoro sulphonic membranes containing nanoceramic fillers (SiO_2) in the polymer electrolyte network to minimize problems with dehydration. The use of nanomaterials in fuel-cell research is not limited to PEMFCs; nanostructured electroceramic materials are increasingly also used in SOFCs, as they significantly reduce the firing temperature during the membrane-formation step in the cell-fabrication process.

Although attractive, to become commercially viable PEMFCs must reduce the costs associated with the use of platinum and platinum-based catalysts. This has caused research to tend towards non-platinum catalysts such as cobalt–polypyrrole–carbon composites and cobaxolamines[48], while waiting for the realization of biomimetic solutions inspired by the ability of metalloenzymes to reversibly catalyse the H_2/H^+ interconversion. Similarly, for SOFCs the most important technical barrier relates to the anode, as we want this electrode to operate efficiently with readily available fuels in addition to H_2, whose use is limited by production and storage difficulties. The best anode material so far, Ni-YSZ (yttria-stabilized zirconia) has problems related to a low tolerance of sulphur and cracking reactions that can lead to carbon deposition. These issues have led to the development of alternative anode materials, relying on the formulation of new single-phase anode perovskites, such as LSTMG ($La_4Sr_8Ti_{11}Mn_{0.5}Ga_{0.5}O_{3.75}$), which are highly catalytically active for the oxidation of methane[49].

Fuel cells, supercapacitors and batteries share the objective of producing electricity on demand and have all benefited from recent advances in the design of nanocomposite electrode–electrolyte components, the development of either new renewable electrodes or new catalysts via 'green' or biomimetic approaches, and the formulation of specific ionic-liquid-based electrolytes. An increase in cross-fertilization between these different domains is therefore necessary in the future if we are to rapidly meet clean-energy imperatives.

Hydrogen storage

The most essential component for the operation of fuel cells is hydrogen, which for the past decade has been considered a promising energy carrier. However, such promises have failed to be realized because of major problems associated with the efficient, safe and economic production and storage of this gas. Reforming hydrogen-containing compounds is still the basis for production of the gas, but processes such as the electrolysis of water have to be developed for clean hydrogen production especially in light of the growing

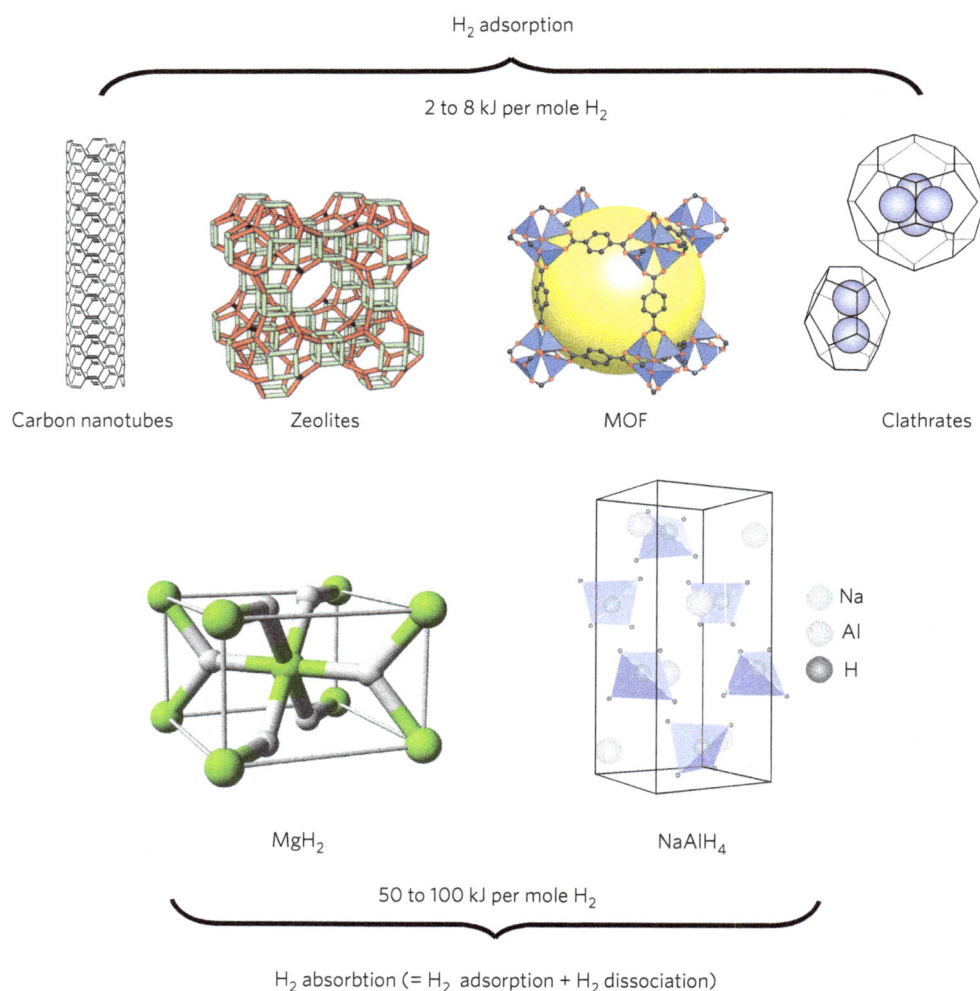

Figure 5 Hydrogen-storage materials can be split into two classes. Materials that trap molecular H_2 by adsorption onto their surfaces and/or into their pores, which include materials such as carbon nanotubes, zeolites, metal–organic frameworks and clathrates (top). Materials such as hydrides (MgH_2) or complex hydrides ($NaAlH_4$), which capture atomic hydrogen via a two-step process first by the adsorption of H_2 and then by the dissociation and diffusion of H within the bulk of the material (bottom). (Figure courtesy of R. Janot; credits: zeolite, ref. 58 ©2002 NPG; MOF, ref. 59 ©2003 NPG; MgH_2, ©Ben Mills.)

interest in high-temperature electrolysis, which could use some aspects of future nuclear reactors. Regarding storage[50], there is currently a debate on the choice between the use of energetically costly high-pressure cryogenic liquids; inefficient high-surface-area adsorbents such as, for instance, tetrahydrofuran-containing binary-clathrate hydrates, which can reversibly take up 2.1 wt% of H_2 at room temperature and under modest ($\sim$ 50 bars) pressures[51]; or storage in metal–organic frameworks, which can adsorb up to 7 wt% of hydrogen at a very low temperature (77 K)[52]; and finally (Fig. 5) chemical storage in metal hydrides and complex hydrides with the greatest hope being that MgH_2 could absorb and release its 7.6 wt% hydrogen content at both a low temperature (100 °C) and pressure (1 bar), conditions required by electric vehicle applications — although at present there is little knowledge on how to master the material aspects of this hydride. As pursuing these storage techniques most likely leads to a dead end, researchers are currently exploring alternative compounds such as alanates[53] and amides–imides[54] to reversibly store hydrogen, with some promising results. However, the chemistry of these materials is somewhat complex and questions remain regarding the feasibility of implementing them into large-scale production. In summary, currently we are still far from reaching targets defined by the US Department of Energy (5.5 wt% for an on-board storage system in 2015) and other national institutions; however, this fact is not alarming as some vehicle manufacturers are currently making good use of AB_2-type intermetallic compounds, which only reversibly absorb 2 wt% of hydrogen at near ambient pressure–temperature conditions. Although the hydrogen storage capacity of these materials is not great, the lack of an infrastructure for hydrogen distribution seems to be the overriding limitation for its application in vehicles.

Superconductors

As important as the storage of energy is the need to transport it to the end user with minimum loss. The use of superconducting wires, which are able to transport electricity without joule losses below their superconducting transition temperature, is still a very appealing solution. Even though they are far from replacing copper wires on the grid, some recent progress has to be considered[55]. First of all, there are a few but conclusive experiments on replacing copper wires by high-temperature superconductors in dense urban areas, as in Long Island (New York) for example, where superconducting cables carry three times the energy previously carried by regular wires in the same pipe. Second, a third generation of cables, based on YBCO deposited on metallic tapes, is coming, with better performances in terms of critical current and the ability to be shaped into useful forms such as coils. Third, new materials appear, such as MgB_2, with a high critical temperature and high critical fields. MgB_2 can be shaped into cables and coils by means of rather standard techniques used, for example, for A15 superconducting wires (those formed from intermetallic compounds such as Nb_3Sn that crystallize in a cubic unit cell). Recently, a complete magnetic-resonance imaging system with a MgB_2 superconducting magnet has been operated successfully. Finally, the newly discovered pnictide-based materials[56,57] of the general formula La–Fe–As–O, with critical temperatures of up to 50 K, enlarges the possibilities by adding a new class of materials to the superconducting world. This is a scientific breakthrough, because Fe was considered to be detrimental to achieving superconductivity; however, this advance may not be significant from an application perspective owing to the poisonous nature of As.

Outlook

New life has been given to the energy storage and conversion field because of recent advances in analytical and computational techniques. Such techniques include the development of *in situ* experiments (for example, magnetic-resonance imaging, transmission electron microscopy and others) that enable the inside of the energy device to be investigated passively, and provide precious information about the mechanistics of electrochemical reaction, the evolution of the interfaces and the dynamic processes with good time and spatial resolution. These aspects, along with recent advances in modelling, computational high-throughput screening of electrode, electrolyte and electrocatalytic materials, and energy-storage management, must be eagerly pursued.

In addition to the development of novel functionalized materials and processes, innovative analytical and modelling techniques, and future innovations in the harvesting and redistribution of energy will be driven by research aimed at integrating different functions within the same device. This combination of storage and conversion functions could result in self-sustained solar cells or thermoelectric devices integrating photovoltaic, thermoelectric and battery components. For such advances to occur, more integrated research is needed. Newcomers are joining the field of energy storage and conversion, and this is a good thing as it is the source of original ideas stemming from experience in other fields, but it could be a wasted effort if a real knowledge of the basic background and challenges in the field is lacking, leading to numerous repetitive studies. Besides, it should be realized that the ways of storing and harvesting energy are not competitive but rather complementary, and future cars will not simply rely on a single technology (Fig. 6) but will

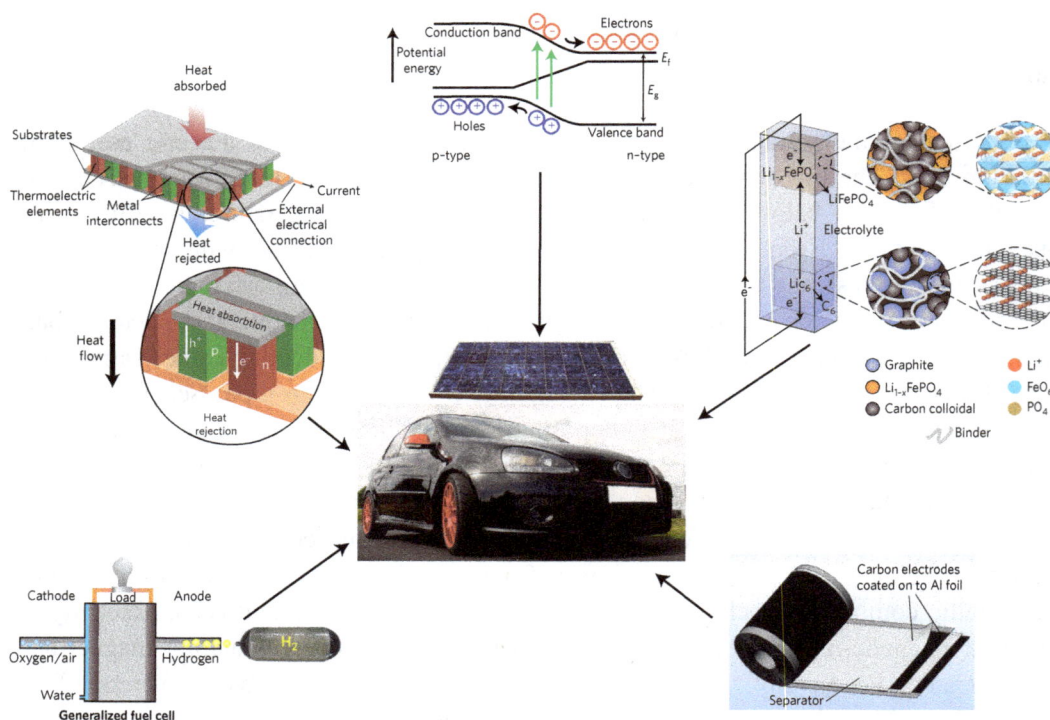

Figure 6 Storage and conversion devices used to power electric vehicles. The state-of-the-art of these devices is described in this chapter, and discussed in detail within the articles collected in this book. (Figure courtesy of D. Larcher; credits: thermoelectric module, ref. 22 ©2008 NPG; car, ©Niklas Ramberg/123RF; Li-ion battery, ©2009 La Recherche/Philippe Mouche; double-layer capacitor, ref. 42 ©2008 NPG.)

most likely contain joint technologies. Success will probably depend on how efficiently we could couple them.

Issues associated with the generation and storage of energy are a worldwide problem. Nevertheless, while sharing similar objectives and road maps, various countries tend to favour national over worldwide programmes. Time is limited, and it is necessary for governments to find means and infrastructures to enhance the distribution of information between national programmes dealing with energy-related matters at both the European and international levels. Concrete action must be rapidly taken if we want to secure a bright future for the generations to come and for our planet as a whole.

References

1. Tao, M. Inorganic photovoltaic solar cells: silicon and beyond. *Interface* **4,** 30–35 (The Electrochemical Society, Winter 2008).
2. Green, M. A. Thin-film solar cells: review of materials, technologies and commercial status. *J. Mater. Sci. Mater. Electron.* **18** (Suppl. 1), S15–S19 (2007).
3. Katagiri, H. *et al.* Development of CZTS based thin film solar cells. *Thin Solid Films* **517,** 2455–2460 (2007).
4. Todorov, T. K., Reuter, K. B. & Mitzi, D. B. High-efficiency solar cell with earth-abundant liquid-processed absorber. *Adv. Mater.* **22,** E156–E159 (2010).
5. O'Regan, B. & Grätzel, M. A low cost, high efficiency solar cell based on the sensitization of colloidal titania particles. *Nature* **335,** 7377–7382 (1991).
6. Yu, G., Gao, J., Hummelen, J. C., Wudl, F. & Heeger, A. J. Polymer photovoltaic cells, enhanced efficiencies via a network of internal donor-acceptor heterojunctions. *Science* **270,** 1789–1791 (1995).
7. Grätzel, M. Photoelectrochemical cells. *Nature* **414,** 338–344 (2001).
8. Brillet, J. *et al.* Examining architectures of photoanode-photovoltaic tandem cells for solar water splitting. *J. Mater. Res.* **25,** 17–24 (2010).
9. Kay, A., Cesar, I. & Grätzel, M. New benchmark for water photo-oxidation by nanostructured a-Fe_2O_3 films. *J. Am. Chem. Soc.* **128,** 15714–15721 (2006).
10. Kelzenberg, M. D. *et al.* Enhanced absorption and carrier collection in Si wire arrays for photovoltaic applications. *Nature Mater.* **9,** 239–244 (2010).
11. Atwater, H. A. & Polman, A. Plasmonics for improved photovoltaic devices. *Nature Mater.* **9,** 205–213 (2010).
12. Hägglund, C., Zäch, M. & Kasemo, B. Enhanced charge carrier generation in dye sensitized solar cells by nanoparticle plasmons. *Appl. Phys. Lett.* **92,** 013113 (2008).
13. Hägglund, C., Zäch, M., Petersson, G. & Kasemo, B. Electromagnetic coupling of light into a silicon solar cell by nanodisk plasmons. *Appl. Phys. Lett.* **92,** 053110 (2008).
14. Schuller, J. A. *et al.* Plasmonics for extreme light concentration and manipulation. *Nature Mater.* **9,** 193–204 (2010).
15. Yip, C. H., Chiang, Y. M. & Wong, C. C. Dielectric band edge enhancement of energy conversion efficiency in photonic crystal dye-sensitized solar cell. *J. Phys. Chem.* **112,** 8735–8740 (2008).
16. Colodrero, S. *et al.* Porous one-dimensional photonic crystals improve the power-conversion efficiency of dye-sensitized solar cells. *Adv. Mater.* **21,** 764–770 (2009).
17. Schaller, R. D., Sykora, M., Pietryga, J. M. & Klimov, V. I. Seven excitons at a cost of one: redefining the limits for conversion efficiency of photons into charge carriers. *Nano Lett.* **6,** 424–429 (2006).
18. Nozik, A. J. Multiple exciton generation in semiconductor quantum dots. *Chem. Phys. Lett.* **457,** 3–11 (2008).
19. Heath, J. R. in *Nanoscience and Technology* (ed. Rodgers, P.) xi–xx (World Scientific, 2009).
20. Boukai, A. I. *et al.* Silicon nanowires as efficient thermoelectric materials. *Nature* **451,** 168–171 (2008).
21. Hochbaum, A. I. *et al.* Enhanced thermoelectric performance of rough silicon nanowires. *Nature* **451,** 163–167 (2008).

22. Snyder, J. & Toberer, E. S. Complex thermoelectric materials. *Nature Mater.* **7**, 105–114 (2008).

23. Aricò, A. S., Bruce, P., Scrosati, B., Tarascon, J.-M. & van Schalkwijk, W. Nanostructured materials for advanced energy conversion and storage devices. *Nature Mater.* **4**, 366–377 (2005).

24. Maier, J. Nanoionics: ion transport and electrochemical storage in confined systems. *Nature Mater.* **4**, 805–815 (2005).

25. Yamada, A., Chung, S. C. & Hinokuna, K. Optimized $LiFePO_4$ for lithium battery cathodes. *J. Electrochem. Soc.* **148**, A224–A229 (2001).

26. Obrovac, M. N. & Christensen, L. Structural changes in silicon anodes during lithium insertion/extraction. *Electrochem. Solid-State Lett.* **7**, A93–A96 (2004).

27. Poizot, P., Laruelle, S., Grugeon, S., Dupont, L. & Tarascon, J.-M. Nanosized transition metal oxides as negative electrode materials for lithium-ion batteries. *Nature* **407**, 496–499 (2000).

28. Ravet, N. *et al.* in *PriME 2008* Abstract 127 (ECS, 1999).

29. Subramanya Herle, P., Ellis, B., Coombs, N. & Nazar, L. F. Nano-network electronic conduction in iron and nickel olivine phosphates. *Nature Mater.* **3**, 147–152 (2004).

30. Barker, J., Saidi, M. Y. & Swoyer, J. L. A Comparative investigation of the Li insertion properties of the novel fluorophosphate phases, $NaVPO_4F$ and $LiVPO_4F$. *J. Electrochem. Soc.* **151**, A1670–A1677 (2004).

31. Ellis, B. L., Makahnouk, W. R. M., Makimura, Y., Toghill, K. & Nazar, L. F. A multifunctional 3.5 V iron-based phosphate cathode for rechargeable batteries. *Nature Mater.* **6**, 749–753 (2007).

32. Recham, N. *et al.* A 3.6 V Li-based fluorosulphate insertion positive electrode for Li-ion batteries. *Nature Mater.* **9**, 68–74 (2010).

33. Abraham, K. M. & Jiang, Z. A Polymer electrolyte-based rechargeable lithium/oxygen battery. *J. Electrochem. Soc.* **143**, 271–275 (1996).

34. Ogasawara, T., Debart, A., Holzapfel, M., Novak, P. & Bruce, P. Rechargeable Li_2O_2 electrode for lithium batteries. *J. Am. Chem. Soc.* **128**, 1390–1393 (2006).

35. Ji, X., Lee, K. T. & Nazar, L. F. A highly ordered nanostructured carbon–sulphur cathode for lithium-sulphur batteries. *Nature Mater.* **8**, 500–506 (2009).

36. Chen, H., Poizot, P., Dolhem, M., Armand, M. & Tarascon, J.-M. From biomass to the first example of a renewable $Li_xC_6O_6$ organic electrode for Li-ion batteries. *ChemSusChem* **1**, 348–355 (2008).

37. Nam, K. T. *et al.* Virus-enabled synthesis and assembly of nanowires for lithium ion battery electrodes. *Science* **312**, 885–888 (2006).

38. Armand, M. & Tarascon, J.-M. Building better batteries. *Nature* **451**, 652–657 (2008).

39. Aurbach, D. *et al.* Electrolyte solutions for rechargeable magnesium batteries based on organomagnesium chloroaluminate complexes. *J. Electrochem. Soc.* **149**, A115–A121 (2002).

40. Kamavaram, V. & Reddy, R. G. *Electrochemical Studies of Aluminum Deposition in Ionic Liquids at Ambient Temperatures. Light Metals* (Warrendale, 2002).

41. Salitra, G., Soffer, A., Eliad, L., Cohen, Y. & Aurbach, D. Carbon electrodes for double layer capacitors. I. Relations between ions and pore dimensions. *J. Electrochem. Soc.* **147**, 2486–2493 (2000).

42. Simon, P. & Gogotsi, Y. Materials for electrochemical capacitors. *Nature Mater.* **7**, 845–854 (2008).

43. Amatucci, G. G., Badway, F. & DuPasquier, A. in *Intercalation Compounds for Battery Materials* (ECS Proc. Vol. 99) 344–359 (ECS, 2000).

44. Armand, M., Endres, F., MacFarlane, D. R., Ohno, H. & Scrosati, B. Ionic-liquid materials for the electrochemical challenges of the future. *Nature Mater.* **8**, 621–629 (2009).

45. Fisher, A. E., Pettigrew, K. A., Rolinson, D. R., Stroud, R. M. & Long, J. W. Incorporation of homogeneous nanoscale MnO_2 with untraporous carbon structures via self-limiting electroless deposition: implications for electrochemical capacitors. *Nano Lett.* **7**, 281–286 (2007).

46. Steele, C. H. & Heinzel, A. Materials for fuel-cell technologies. *Nature* **414**, 345–352 (2001).

47. Atkinson, A. *et al.* Advanced anodes for high-temperature fuel cells. *Nature Mater.* **3**, 17–27 (2004).

48. Bashyam, R. & Zelenay, P. A class of non-precious metal composite catalysts for fuel cells. *Nature* **443**, 63–66 (2006).

49. Ruiz-Morales, J. C. *et al.* Disruption of extended defects in solid oxide fuel cells for methane oxidation. *Nature* **439,** 568–571 (2006).

50. Eberle, U., Felderhoff, M. & Schüth, F. Chemical and physical solutions for hydrogen storage. *Angew. Chem. Int. Ed.* **48,** 6608–6630 (2009).

51. Lee, H. *et al.* Tuning clathrate hydrates for hydrogen storage. *Nature* **434,** 743–746 (2005).

52. Chae, H. K. *et al.* A route to high surface area, porosity and inclusion of large molecules in crystals. *Nature* **427,** 523–527 (2004).

53. Xiong, Z. *et al.* High-capacity hydrogen storage in lithium and sodium amidoboranes. *Nature Mater.* **7,** 138–141 (2007).

54. Chen, P., Xiong, Z., Luo, J., Lin, J. & Tan, K. L. Interaction of hydrogen with metal nitrides and imides. *Nature* **420,** 302–304 (2002).

55. Larbalestier, D., Gurevich, A., Feldmann, D. M. & Polyanski, A. High-T_c superconducting materials for electric power applications. *Nature* **414,** 368–377 (2001).

56. Kamihara, Y., Watanabe, T., Hirani, M. & Hosono, H. Iron-based layered superconductor La[$O_{1-x}F_x$]FeAs ($x = 0.05$–012] with $T_c = 26$ K. *J. Am. Chem. Soc.* **130,** 3296–3297 (2008).

57. Day, C. Iron-based superconductors. *Phys. Today* 36–40 (August 2009).

58. Corma, A., Díaz-Cabañas, M. J., Martínez-Triguero, J., Rey, F. & Rius, J. A large-cavity zeolite with wide pore windows and potential as an oil refining catalyst. *Nature* **418,** 514–517 (2002).

59. Yaghi, O. M. *et al.* Reticular synthesis and the design of new materials. *Nature* **423,** 705–714 (2003).

PHOTOVOLTAIC CELLS

Plasmonics for improved photovoltaic devices

Harry A. Atwater[1]* and Albert Polman[2]*

The emerging field of plasmonics has yielded methods for guiding and localizing light at the nanoscale, well below the scale of the wavelength of light in free space. Now plasmonics researchers are turning their attention to photovoltaics, where design approaches based on plasmonics can be used to improve absorption in photovoltaic devices, permitting a considerable reduction in the physical thickness of solar photovoltaic absorber layers, and yielding new options for solar-cell design. In this review, we survey recent advances at the intersection of plasmonics and photovoltaics and offer an outlook on the future of solar cells based on these principles.

Photovoltaics, the conversion of sunlight to electricity, is a promising technology that may allow the generation of electrical power on a very large scale. Worldwide photovoltaic production was more than 5 GW in 2008, and is expected to rise above 20 GW by 2015. Photovoltaics could thus make a considerable contribution to solving the energy problem that our society faces in the next generation. To make power from photovoltaics competitive with fossil-fuel technologies, the cost needs to be reduced by a factor of 2–5. At present most of the solar-cell market is based on crystalline silicon wafers with thicknesses between 180–300 μm, and most of the price of solar cells is due to the costs of silicon materials and processing. Because of this, there is great interest in thin-film solar cells, with film thicknesses in the range 1–2 μm, that can be deposited on cheap module-sized substrates such as glass, plastic or stainless steel. Thin-film solar cells are made from a variety of semiconductors including amorphous and polycrystalline Si, GaAs, CdTe and CuInSe$_2$, as well as organic semiconductors. A limitation in all thin-film solar-cell technologies is that the absorbance of near-bandgap light is small, in particular for the indirect-bandgap semiconductor Si. Therefore, structuring the thin-film solar cell so that light is trapped inside to increase the absorbance is very important. A significant reduction in thin-film solar-cell thickness would also allow the large-scale use of scarce semiconductor materials such as In and Te that are available in the Earth's crust in only small quantities.

In conventional thick Si solar cells, light trapping is typically achieved using a pyramidal surface texture that causes scattering of light into the solar cell over a large angular range, thereby increasing the effective path length in the cell[1–3]. Such large-scale geometries are not suitable for thin-film cells, for geometrical reasons (as the surface roughness would exceed the film thickness) and because the greater surface area increases minority carrier recombination in the surface and junction regions.

A new method for achieving light trapping in thin-film solar cells is the use of metallic nanostructures that support surface plasmons: excitations of the conduction electrons at the interface between a metal and a dielectric. By proper engineering of these metallodielectric structures, light can be concentrated and 'folded' into a thin semiconductor layer, thereby increasing the absorption. Both localized surface plasmons excited in metal nanoparticles and surface plasmon polaritons (SPPs) propagating at the metal/semiconductor interface are of interest.

In the past few years, the field of plasmonics has emerged as a rapidly expanding new area for materials and device research[4]. This is a result of the large array of tools that have become available for nanoscale fabrication and nanophotonics characterization, and also because of the availability of powerful electromagnetic simulation methods that are critical to understanding and harnessing plasmon excitations. Studies are just starting to appear illustrating the coupling of plasmons to optical emitters[5–7]; plasmon focusing[8,9]; hybridized plasmonic modes in nanoscale metal shells[10]; nanoscale waveguiding[11–14]; nanoscale optical antennas[15]; plasmonic integrated circuits[16,17]; nanoscale switches[18]; plasmonic lasers[19–21]; surface-plasmon-enhanced light-emitting diodes[22]; imaging below the diffraction limit[23]; and materials with negative refractive index[24–26]. Despite all these exciting opportunities, until recently little systematic thought has been given to the question of how plasmon excitation and light localization might be used advantageously in high-efficiency photovoltaics.

Plasmonics for photovoltaics

Conventionally, photovoltaic absorbers must be 'optically thick' to allow near-complete light absorption and photocarrier current collection. Figure 1a shows the standard AM1.5 solar spectrum together with a graph that illustrates what fraction of the solar spectrum is absorbed on a single pass through a 2-μm-thick crystalline Si film. Clearly, a large fraction of the solar spectrum, in particular in the intense 600–1,100 nm spectral range, is poorly absorbed. This is the reason that, for example, conventional wafer-based crystalline Si solar cells have a much larger thickness of typically 180–300 μm. But high-efficiency solar cells must have minority carrier diffusion lengths several times the material thickness for all photocarriers to be collected (see Fig. 1b), a requirement that is most easily met for thin cells. Solar-cell design and materials-synthesis considerations are strongly dictated by these opposing requirements for optical absorption thickness and carrier collection length.

Plasmonic structures can offer at least three ways of reducing the physical thickness of the photovoltaic absorber layers while keeping their optical thickness constant. First, metallic nanoparticles can be used as subwavelength scattering elements to couple and trap freely propagating plane waves from the Sun into an absorbing semiconductor thin film, by folding the light into a thin absorber layer (Fig. 2a). Second, metallic nanoparticles can be used as subwavelength antennas in which the plasmonic near-field is coupled to the semiconductor, increasing its effective absorption cross-section (Fig. 2b). Third, a corrugated metallic film on the back surface of a thin photovoltaic absorber layer can couple sunlight into SPP modes supported at the metal/semiconductor interface as well as

[1]Caltech Center for Sustainable Energy Research and Thomas J. Watson Laboratories of Applied Physics, California Institute of Technology, Pasadena, California 91125, USA. [2]Center for Nanophotonics, FOM Institute AMOLF, Science Park 104, 1098 XG Amsterdam, The Netherlands. *e-mail: haa@caltech.edu; polman@amolf.nl

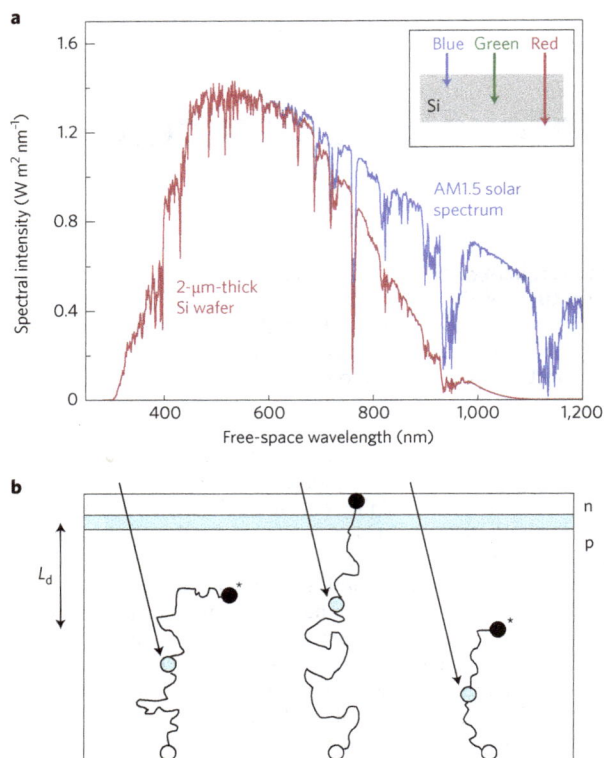

Figure 1 | Optical absorption and carrier diffusion requirements in a solar cell. a, AM1.5 solar spectrum, together with a graph that indicates the solar energy absorbed in a 2-μm-thick crystalline Si film (assuming single-pass absorption and no reflection). Clearly, a large fraction of the incident light in the spectral range 600–1,100 nm is not absorbed in a thin crystalline Si solar cell. **b**, Schematic indicating carrier diffusion from the region where photocarriers are generated to the p–n junction. Charge carriers generated far away (more than the diffusion length L_d) from the p–n junction are not effectively collected, owing to bulk recombination (indicated by the asterisk).

guided modes in the semiconductor slab, whereupon the light is converted to photocarriers in the semiconductor (Fig. 2c).

As will be discussed in detail in the next section, these three light-trapping techniques may allow considerable shrinkage (possibly 10- to 100-fold) of the photovoltaic layer thickness, while keeping the optical absorption (and thus efficiency) constant.

Various additional ways of using plasmonic nanostructures to increase photovoltaic energy conversion are described in the section on other plasmonic solar-cell designs.

Plasmonic light trapping in thin-film solar cells

Light scattering using particle plasmons. Light scattering from a small metal nanoparticle embedded in a homogeneous medium is nearly symmetric in the forward and reverse directions[27,28]. This situation changes when the particle is placed close to the interface between two dielectrics, in which case light will scatter preferentially into the dielectric with the larger permittivity[29]. The scattered light will then acquire an angular spread in the dielectric that effectively increases the optical path length (see Fig. 2a). Moreover, light scattered at an angle beyond the critical angle for reflection (16° for the Si/air interface) will remain trapped in the cell. In addition, if the cell has a reflecting metal back contact, light reflected towards the surface will couple to the nanoparticles and be partly reradiated into the wafer by the same scattering mechanism. As a result, the incident light will pass several times through the semiconductor film, increasing the effective path length.

The enhanced incoupling of light into semiconductor thin films by scattering from plasmonic nanoparticles was first recognized by Stuart and Hall, who used dense nanoparticle arrays as resonant scatterers to couple light into Si-on-insulator photodetector structures[30,31]. They observed a roughly 20-fold increase in the infrared photocurrent in such a structure. This research field then remained relatively dormant for many years, until applications in thin-film solar cells emerged, with papers published on enhanced light coupling into single-crystalline Si (ref. 32), amorphous Si (refs 33,34), Si-on-insulator[35], quantum well[36] and GaAs (ref. 37) solar cells covered with metal nanoparticles.

Although there is now considerable experimental evidence that light scattering from metal nanoparticle arrays increases the photocurrent spectral response of thin-film solar cells, many of the underlying physical mechanisms and their interplay have not been studied systematically. The full potential of the particle scattering concept, taking into account integration with optimized anti-reflection coatings, is being studied by several research groups. In recent papers[38,39], we reported that both shape and size of metal nanoparticles are key factors determining the incoupling efficiency. This is illustrated in Fig. 3a, which shows that smaller particles, with their effective dipole moment located closer to the semiconductor layer, couple a larger fraction of the incident light into the underlying semiconductor because of enhanced near-field coupling. Indeed, in the limit of a point dipole very near to a silicon substrate, 96% of the incident light is scattered into the substrate, demonstrating the power of

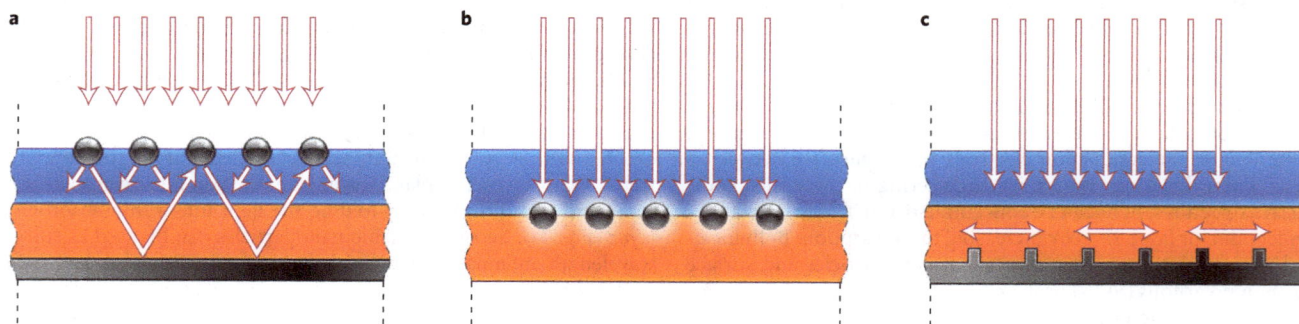

Figure 2 | Plasmonic light-trapping geometries for thin-film solar cells. a, Light trapping by scattering from metal nanoparticles at the surface of the solar cell. Light is preferentially scattered and trapped into the semiconductor thin film by multiple and high-angle scattering, causing an increase in the effective optical path length in the cell. **b**, Light trapping by the excitation of localized surface plasmons in metal nanoparticles embedded in the semiconductor. The excited particles' near-field causes the creation of electron–hole pairs in the semiconductor. **c**, Light trapping by the excitation of surface plasmon polaritons at the metal/semiconductor interface. A corrugated metal back surface couples light to surface plasmon polariton or photonic modes that propagate in the plane of the semiconductor layer.

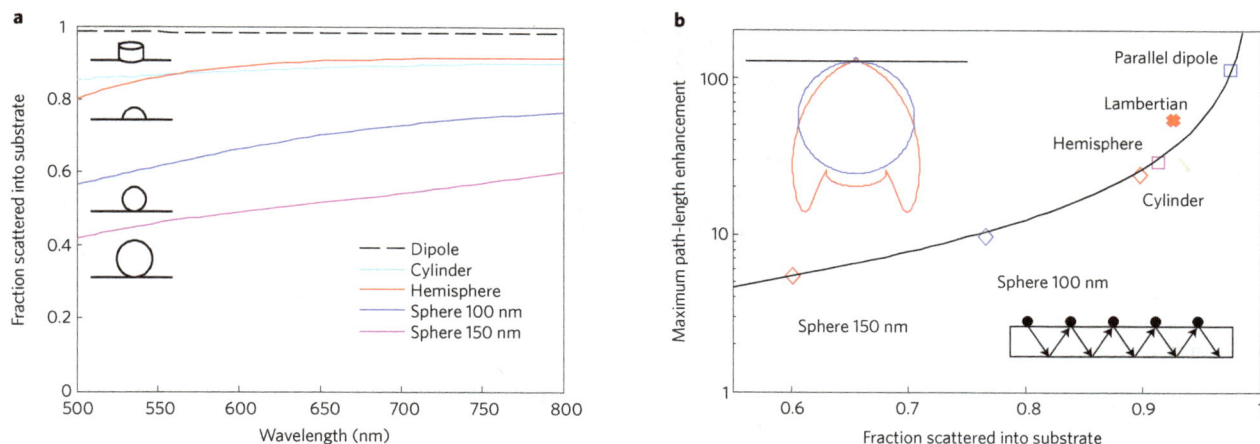

Figure 3 | Light scattering and trapping is very sensitive to particle shape. a, Fraction of light scattered into the substrate, divided by total scattered power, for different sizes and shapes of Ag particles on Si. Also plotted is the scattered fraction for a parallel electric dipole that is 10 nm from a Si substrate. **b**, Maximum path-length enhancement, according to a first-order geometrical model, for the same geometries as in **a** at a wavelength of 800 nm. Absorption within the particles is neglected for these calculations, and an ideal rear reflector is assumed. The line is a guide for the eye. Insets: (top left) angular distribution of scattered power for a parallel electric dipole that is 10 nm above a Si surface (red) and a Lambertian scatterer (blue); (lower right) geometry considered for calculating the path-length enhancement. Figure reproduced with permission: © 2008 AIP.

the particle scattering technique. Figure 3b shows the path-length enhancement in the solar cell derived from Fig. 3a using a simple first-order scattering model. For 100-nm-diameter Ag hemispheres on Si, a 30-fold enhancement is found. These light-trapping effects are most pronounced at the peak of the plasmon resonance spectrum, which can be tuned by engineering the dielectric constant of the surrounding matrix. For example, small Ag or Au particles in air have plasmon resonances at 350 nm and 480 nm respectively; they can be redshifted in a controlled way over the entire 500–1,500 nm spectral range by (partially) embedding them in SiO_2, Si_3N_4 or Si (refs 40–42; Fig. 4a), which are all standard materials in solar-cell manufacturing. A dielectric spacer layer between the nanoparticles and the semiconductor also causes electrical isolation, thereby avoiding additional surface recombination owing to the presence of the metal. The scattering cross-sections for metal nanoparticle scattering can be as high as ten times the geometrical area[28,38], and a ~10% coverage of the solar cell would thus suffice to capture most of the incident sunlight into plasmon excitations.

Optimization of plasmonic light trapping in a solar cell is a balancing act in which several physical parameters must be taken into account. First of all, Fig. 3 demonstrates the advantage of using small particles to create the forward scattering anisotropy. But very small particles suffer from significant ohmic losses, which scale with volume v, whereas scattering scales with v^2, so that using larger particles is advantageous to increase the scattering rate. For example, a 150-nm-diameter Ag particle in air has an albedo (fraction of light emitted as radiation) as large as 95%. Interestingly, the effective scattering cross-section can be increased by spacing the particles further away from the substrate, as this avoids destructive interference effects between the incident and reflected fields, albeit at the price of reduced near-field coupling[38,39]. For frequencies above the plasmon resonance, Fano resonance effects can cause destructive interference between scattered and unscattered light beams, and thus can cause reflection rather than enhanced incoupling[42,43]. One way to overcome the latter problem is by using a geometry in which particles are placed on the rear of the solar cell[42,44]. In this case, blue and green light is directly absorbed in the cell, whereas poorly absorbed (infra-)red light is scattered and trapped using metal nanoparticles. Similarly, light may be coupled in by using arrays of metal strips placed on the solar-cell surface. Calculations show that using this geometry can enhance the short-circuit current by 45% compared with a solar cell

with a flat back contact[45]. Finally, in designing optimized plasmonic light-trapping arrays, we must take into account coupling between the nanoparticles, ohmic damping, grating diffraction effects[46] and the coupling to waveguide modes[47–49].

Light concentration using particle plasmons. An alternative use of resonant plasmon excitation in thin-film solar cells is to take advantage of the strong local field enhancement around the metal nanoparticles to increase absorption in a surrounding semiconductor material. The nanoparticles then act as an effective 'antenna' for the incident sunlight that stores the incident energy in a localized surface plasmon mode (Fig. 4b). This works particularly well for small (5–20 nm diameter) particles for which the albedo is low. These antennas are particularly useful in materials where the carrier diffusion lengths are small, and photocarriers must thus be generated close to the collection junction area. For these antenna energy conversion effects to be efficient, the absorption rate in the semiconductor must be larger than the reciprocal of the typical plasmon decay time (lifetime ~10–50 fs), as otherwise the absorbed energy is dissipated into ohmic damping in the metal. Such high absorption rates are achievable in many organic and direct-bandgap inorganic semiconductors.

Several examples of this concept have recently appeared that demonstrate enhanced photocurrents owing to the plasmonic near-field coupling. Enhanced efficiencies have been demonstrated for ultrathin-film organic solar cells doped with very small (5 nm diameter) Ag nanoparticles[50]. Plasmon-enhanced organic solar cells have also been demonstrated using electrodeposited Ag (ref. 51). An increase in efficiency by a factor of 1.7 has been shown for organic bulk heterojunction solar cells[52,53]. Dye-sensitized solar cells can also be enhanced by embedding small metal nanoparticles, as reported elsewhere[54–56]. Also, inorganic solar cells have shown increased photocurrents owing to near-field effects, such as CdSe/Si hetero-structures[57]. Work on increased light absorption by metal nano-particles embedded in Si has also been reported[58,59]. More recently, the coupling between plasmons in arrays of metal nanoparticles has been used to engineer the field enhancement to overlap with a selected area of the junction[53]. The dynamics (and optimization) of the coupling between plasmons, excitons and phonons in metal–semiconductor nanostructures is a rich field of research that so far has not received much attention with photovoltaics in mind.

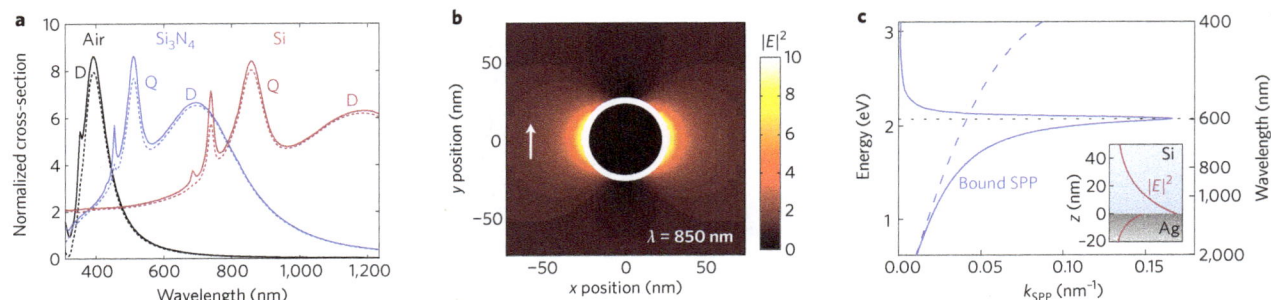

Figure 4 | Characteristics of surface plasmons. a, Metal nanoparticles scatter light over a broad spectral range that can be tuned by the surrounding dielectric. The plots show the scattering cross-section spectrum for a 100-nm-diameter Ag particle embedded in three different dielectrics (air, Si_3N_4 and Si). Dipole (D) and quadrupole (Q) modes are indicated. The cross-section is normalized to the geometrical cross-section of the particle. **b,** Metal nanoparticles show an intense near-field close to the surface. Intensity enhancement around a 25-nm-diameter Au particle embedded in a medium with index $n = 1.5$ (plasmon resonance peak at 500 nm). Light with a wavelength $\lambda = 850$ nm is incident with a polarization indicated by the vertical arrow. The magnitude of the enhanced electric-field intensity E is indicated by the colour scale. **c,** SPPs are bound waves at the interface between a semiconductor and a dielectric. This dispersion diagram, plotting the relationship between frequency and wavevector ($2\pi/\lambda$) for SPPs on a Ag/Si interface. The 'bound' SPP mode occurs at energies below the surface plasmon resonance energy of 2.07 eV (600 nm). The dispersion of light in Si is indicated by the dashed line. The inset shows a schematic of the SPP mode profile along the Si/Ag interface, at a free-space wavelength of 785 nm.

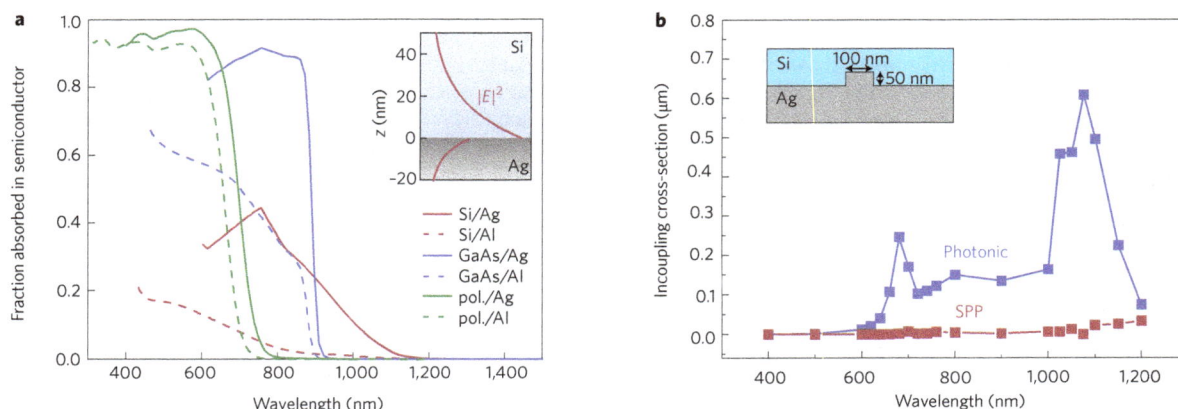

Figure 5 | Light scattering into SPP and photonic modes in thin semiconductor films. a, Fraction of light absorbed into the semiconductor for SPPs propagating along interfaces between semi-infinite layers of GaAs, Si and an organic semiconductor film made of a PF10TBT:[C60]PCBM polymer blend (termed 'pol.' in the legend), in contact with either Ag or Al. Graphs are plotted for wavelength down to the surface plasmon resonance. The inset shows the SPP field intensity near the interface. **b,** Two-dimensional calculation of the incoupling cross-section for SPP and photonic modes as a function of wavelength for a 200-nm-thick Si slab on an optically thick Ag substrate with a 100-nm-wide, 50-nm-tall Ag ridge, as shown in the inset. Figure reproduced with permission: **b,** © 2008 ACS.

Light trapping using SPPs. In a third plasmonic light-trapping geometry, light is converted into SPPs, which are electromagnetic waves that travel along the interface between a metal back contact and the semiconductor absorber layer (see Fig. 4c)[60]. Near the plasmon resonance frequency, the evanescent electromagnetic SPP fields are confined near the interface at dimensions much smaller than the wavelength. SPPs excited at the metal/semiconductor interface can efficiently trap and guide light in the semiconductor layer. In this geometry the incident solar flux is effectively turned by 90°, and light is absorbed along the lateral direction of the solar cell, which has dimensions that are orders of magnitude larger than the optical absorption length. As metal contacts are a standard element in the solar-cell design, this plasmonic coupling concept can be integrated in a natural way.

At frequencies near the plasmon resonance frequency (typically in the 350–700 nm spectral range, depending on metal and dielectric) SPPs suffer from relatively high losses. Further into the infrared, however, propagation lengths are substantial. For example, for a semi-infinite Ag/SiO$_2$ geometry, SPP propagation lengths range

from 10 to 100 μm in the 800–1,500 nm spectral range. By using a thin-film metal geometry the plasmon dispersion can be further engineered[61–64]. Increased propagation length comes at the expense of reduced optical confinement and optimum metal-film design thus depends on the desired solar-cell geometry. Detailed accounts of plasmon dispersion and loss in metal–dielectric geometries are found in refs 61–64.

The SPP coupling mechanism is beneficial for efficient light absorption if absorption of the SPP in the semiconductor is stronger than in the metal. This is illustrated in Fig. 5a, which shows a calculation of the fraction of light that is absorbed in a silicon or GaAs film in contact with either Ag or Al. The inset shows the mode-field distribution at the Si/Ag interface at a wavelength of 850 nm. As can be seen, for SPPs at a GaAs/Ag interface the semiconductor absorption fraction is high in the spectral range from the GaAs/Ag SPP resonance (600 nm) to the bandgap of GaAs (870 nm). For a Si/Ag interface, with smaller optical absorption in Si owing to the indirect bandgap, plasmon losses dominate over the entire spectral range, although absorption in the 700–1,150 nm spectral range is still higher than single-pass absorption through a 1-μm-thick Si film. In this case, it is more beneficial to use a geometry in which a thin metal film is embedded in the semiconductor, as in this case plasmon absorption is much smaller owing to the smaller mode overlap with the metal[63,64]. Figure 5a also shows a calculation of the absorption of an organic semiconductor (PF10TBT:[C60]PCBM) (ref. 65) in contact with Ag or Al. Optical constants for the polymer are taken from ref. 65. Here the absorption efficiency is high over the entire spectral range below 650 nm because the organic semiconductor has high absorption, and the low dielectric constant of the organic material causes a small overlap of the modal field with the metal and thus lower ohmic losses.

Owing to the momentum mismatch between the incident light and the in-plane SPPs (see Fig. 4c) a light-incoupling structure must be integrated in the metal/dielectric interface. Figure 5b plots the results of full-field electrodynamics simulations (taking into account in- and outcoupling) of the scattering from an incident plane wave into SPPs by a ridge fabricated in the Ag/Si interface[66]. The Si layer is 200 nm thick and the Ag ridge 50 nm tall. The simulations show that light is coupled into an SPP mode as well as a photonic mode that propagates inside the Si waveguide, and that the strength of coupling to each mode can be controlled by the height of the scattering object[66]. The photonic modes are particularly interesting as they suffer from only very small losses in the metal. The fraction of light coupled into both modes increases with increasing wavelength. This is mainly because at shorter wavelengths the incoming light beam is directly absorbed in the Si layer. The data demonstrate that light with $\lambda > 800$ nm, which would not be well absorbed by normal incidence on the Si layer (see Fig. 1a), can now be efficiently absorbed by conversion into the in-plane SPP and photonic modes. Although this example shows coupling from a single, isolated ridge, the shape, height and interparticle arrangements of incoupling structures can all be optimized for preferential coupling to particular modes. In the ultimate, ultrathin Si solar cell (thickness <100 nm), no photonic modes exist and all scattered light is converted into SPPs. Further enhancements are expected for three-dimensional scattering structures integrated in the back contact, and more research is required to investigate this. Most recently, we have reported experiments on amorphous Si thin-film solar cells deposited on a textured metal back reflector, which show a 26% enhancement in short-circuit current, with the primary photocurrent enhancement in the near-infrared[67].

In relation to the plasmonic coupling effect, a similar conversion mechanism into surface polariton waves has recently been demonstrated using lossy dielectrics rather than metals[68]. Here too, light is efficiently coupled into a two-dimensional wave that can then be absorbed in a semiconductor layer. Several other reports on the integration of SPP geometries with thin-film solar-cell geometries are now appearing, including organic solar cells[69–71].

The data in Fig. 5b were calculated for light under normal incidence, with a polarization perpendicular to the ridge. The next challenge is to engineer coupling structures that depend weakly on frequency, polarization and angle of incidence. A two-dimensional mesh structure with features much smaller than the wavelength may serve such a purpose.

It is clear that with these scattering structures made on the rear side of the cell, the concepts of plasmonic scattering, concentration and coupling from this section and the previous two sections become closely integrated. Indeed, any solar cell with a non-planar metal back contact will have geometric scattering and higher local fields as well as scattering into photonic and plasmonic modes, and all these effects must be carefully engineered. We note that there is at present considerable activity on thin-film solar cells with textured metal back reflectors for light trapping[72–77].

Advantages of reduced semiconductor absorber layer thickness.
The plasmonic light-trapping concepts described in the previous sections lead to new solar-cell designs with smaller semiconductor thicknesses and thus lower material costs. This has important technical and strategic consequences as photovoltaics scales-up in manufacturing capacity from its present level of ~8 GW in 2009 to >50 GW by 2020, and eventually to the terawatt scale. For this to happen, the materials used in solar cells must be sufficiently abundant in the Earth's crust and amenable to the formation of efficient photovoltaic devices. For single-junction devices, Si has proved to be a near-ideal photovoltaic material. It is abundant, with a nearly optimal bandgap, excellent junction formation characteristics, high minority carrier diffusion length and effective methods for surface passivation to reduce unwanted carrier recombination. Its only shortcoming — low spectral absorption of the terrestrial solar spectrum — means that it requires cell thicknesses of over 100 μm, which gives rise to a relatively high material cost per output power. Plasmonic light trapping makes it possible to design crystalline Si thin-film cells with spectral quantum efficiency approaching that seen in thick cells, even for absorber layers a few micrometres in thickness.

Materials resources are a significant limitation for large-scale production of two of the most common thin-film solar-cell materials: CdTe and CuInSe$_2$. Manufacturing costs for these cells have fallen, and solar-cell production using these semiconductors is expanding rapidly. Table 1 lists the (projected) annual solar-cell production per year, as well as the materials feedstock required for the production of the corresponding solar-cell volume using Si, CdTe or CuInSe$_2$. As can be seen, the materials feedstock required in 2020 exceeds the present annual world production of Te and In, and in the case of In is even close to the total reserve base. If it were possible to reduce the cell thickness for such compound semiconductor cells by 10–100 times as a result of plasmon-enhanced light absorption, this could considerably extend the reach of these compound semiconductor thin-film solar cells towards the terawatt scale. Earth-abundance considerations will also influence plasmonic cell designs at large-scale production: although Ag and Au have been the metals of choice in most plasmonic designs and experiments, they are relatively scarce materials, so scalable designs will need to focus on abundant metals such as Al and Cu.

Reducing the active-layer thickness by plasmonic light trapping not only reduces costs but also improves the electrical characteristics of the solar cell[78]. First of all, reducing the cell thickness reduces the dark current (I_{dark}), causing the open-circuit voltage V_{oc} to increase, as $V_{oc} = (k_B T/q) \ln(I_{photo}/I_{dark} + 1)$, where k_B is the Boltzmann constant, T is temperature, q is the charge and I_{photo} is the photocurrent. Consequently, the cell efficiency rises in logarithmic proportion to the decrease in thickness, and is ultimately

Table 1 | Photovoltaic resource requirements: materials by production and reserve.

Year	Annual solar-cell production*	Material		
		Si (c-Si)	Te (CdTe)	In (CuInSe$_2$)
2000	0.3 GW$_p$	4	0.03	0.03
2005	1.5 GW$_p$	15	0.15	0.15
2020	50 GW$_p$	150	5	5
World production (1,000 tonnes per year)		1,000	0.3	0.5
Reserve base (1,000 tonnes)		Abundant	47	6

*W$_p$ = peak output power under full solar illumination.
The annual solar-cell production and the required materials feedstock are indicated, assuming cells are made of Si, CdTe or CuInSe$_2$. The assumed material use for Si is 13 g W$_p^{-1}$ in 2000, 10 g W$_p^{-1}$ in 2005, 3 g W$_p^{-1}$ in 2020 (projected); it is 0.1 g W$_p^{-1}$ for Te and In. The two bottom rows indicate the world reserve base data for Si, Te and In (that is, resources that are economic at present or marginally economic and some that are at present subeconomic).
Sources: refs 93, 94 and G. Willeke (Fraunhofer Institute for Solar Energy Systems), personal communication.

limited by surface recombination. Second, in a thin-film geometry, carrier recombination is reduced as carriers need to travel only a small distance before being collected at the junction. This leads to a higher photocurrent. Greatly reducing the semiconductor layer thickness allows the use of semiconductor materials with low minority carrier diffusion lengths, such as polycrystalline semiconductors, quantum-dot layers or organic semiconductors. Also, this could render useful abundant and potentially inexpensive semiconductors with significant impurity and defect densities, such as Cu_2O, Zn_3P_2 or SiC, for which the state of electronic materials development is not as advanced as it is for Si.

Other new plasmonic solar-cell designs

The previous section has focused on the use of plasmonic scattering and coupling concepts to improve the efficiency of single-junction planar thin-film solar cells, but many other cell designs can benefit from the increased light confinement and scattering from metal nanostructures. First of all, plasmonic 'tandem' geometries may be made, in which semiconductors with different bandgaps are stacked on top of each other, separated by a metal contact layer with a plasmonic nanostructure that couples different spectral bands in the solar spectrum into the corresponding semiconductor layer (see Fig. 6a)[79]. Coupling sunlight into SPPs could also solve the problem of light absorption in quantum-dot solar cells (see Fig. 6b). Although such cells offer potentially large benefits because of the flexibility in engineering the semiconductor bandgap by particle size, effective light absorption requires thick quantum-dot layers,

through which carrier transport is problematic. As we have recently demonstrated[80], a 20-nm-thick layer of CdSe semiconductor quantum dots deposited on a Ag film can absorb light confined into SPPs within a decay length of 1.2 μm at an incident photon energy above the CdSe quantum-dot bandgap at 2.3 eV. The reverse geometry, in which quantum dots are electrically excited to generate plasmons, has also recently been demonstrated[81]. We note that the plasmon light-trapping concepts described in the previous section rely on scattering using localized modes, and are thus relatively insensitive to angle of incidence[66,82]. This is an advantage for solar-cell designs made for areas where incident sunlight is mostly diffuse rather than direct.

In a recent example of nanoscale plasmonic solar-cell engineering, an organic photovoltaic light absorber was integrated in the gap between the arms of plasmonic antennas arranged in arrays (see Fig. 6c)[83]. Other examples of nanoscale antennas are coaxial holes fabricated in a metal film, which show localized plasmonic modes owing to Fabry–Perot resonances (see Fig. 6d)[84–86]. Such nanostructures, with field enhancements up to a factor of about 50, could be used in entirely new solar-cell designs, in which an inexpensive semiconductor with low minority carrier lifetime is embedded inside the plasmonic cavity. Similarly, quantum-dot solar cells based on multiple-exciton generation[87], or cells with solar upconverters or downconverters based on multiphoton absorption effects, could benefit from such plasmonic field concentration. In general, field concentration in plasmonic nanostructures is likely to be useful in any type of solar cell where light concentration is

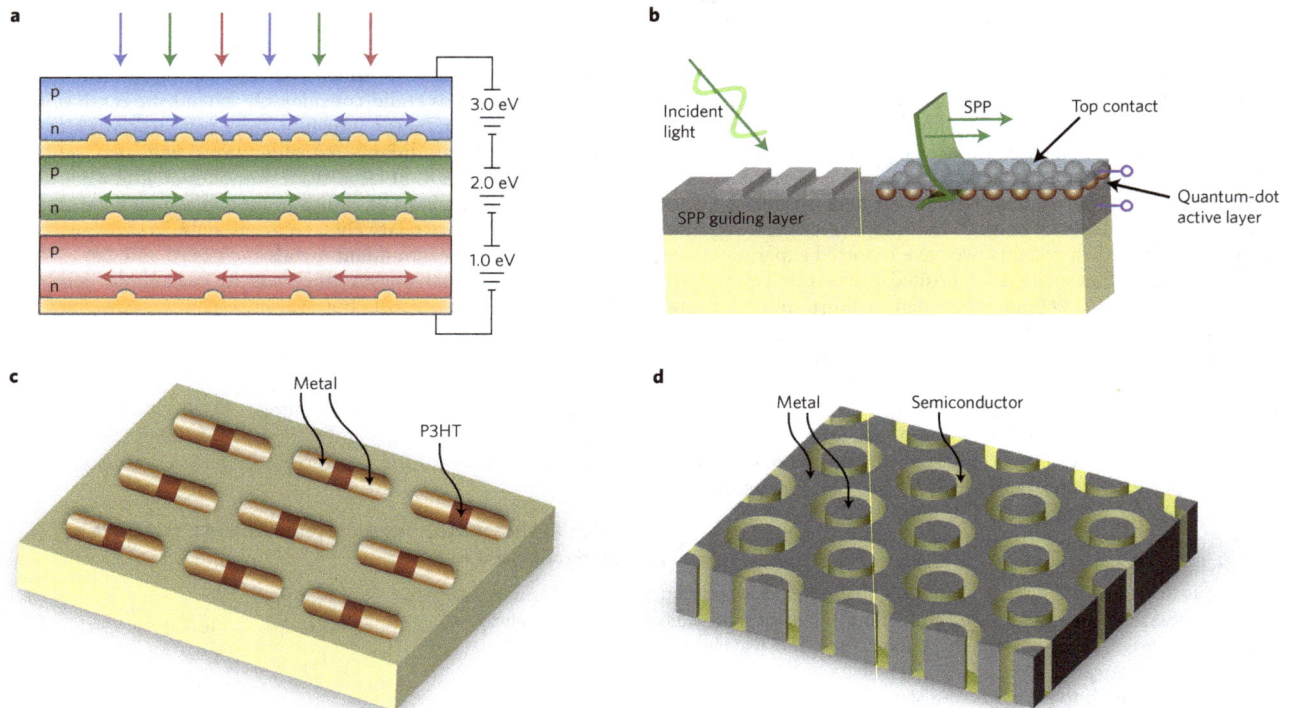

Figure 6 | New plasmonic solar-cell designs. a, Plasmonic tandem solar-cell geometry. Semiconductors with different bandgaps are stacked on top of each other, separated by a metal contact layer with a plasmonic nanostructure that couples different spectral bands of the solar spectrum into the corresponding semiconductor layer. **b,** Plasmonic quantum-dot solar cell designed for enhanced photoabsorption in ultrathin quantum-dot layers mediated by coupling to SPP modes propagating in the plane of the interface between Ag and the quantum-dot layer. Semiconductor quantum dots are embedded in a metal/insulator/metal SPP waveguide. **c,** Optical antenna array made from an axial heterostructure of metal and poly(3-hexylthiophene) (P3HT). Light is concentrated in the nanoscale gap between the two antenna arms, and photocurrent is generated in the P3HT semiconductor[83]. **d,** Array of coaxial holes in a metal film that support localized Fabry–Perot plasmon modes. The coaxial holes are filled with an inexpensive semiconductor with low minority carrier lifetime, and carriers are collected by the metal on the inner and outer sides of the coaxial structure. Field enhancements up to a factor of about 50 are possible and may serve to enhance nonlinear photovoltaic conversion effects[85].

beneficial, such as a high-efficiency multijunction solar cells[88,89]. More details on plasmon light concentration are found in another review in this issue[90].

Large-area fabrication of plasmonic solar-cell structures

The plasmonic coupling effects described in this review all require integration of dense arrays of metal nanostructures with control over dimension tolerances at the nanometre scale. Although systematic laboratory experiments can be performed using cleanroom techniques such as electron-beam lithography or focused ion-beam milling, true application in large-area photovoltaic module production requires inexpensive and scalable techniques for fabricating metal nanopatterns in a controlled way. Several research groups have studied these techniques and have shown their feasibility.

The simplest way to form metal nanoparticles on a substrate is by thermal evaporation of a thin (10–20 nm) metal film, which is then heated at a moderate temperature (200–300 °C), to cause agglomeration by surface tension of the metal film into a random array of nanoparticles. As has been shown[35,42], this leads to the formation of random arrays of Ag nanoparticles with a diameter (100–150 nm) and hemispherical shape that is well suited for light trapping[38]. More control over the Ag nanoparticle size, aspect ratio and density can be achieved using deposition through a porous alumina template, as was demonstrated for example in ref. 37. Here, the nanoparticle shape was further tuned by thermal annealing at 200 °C, leading to hemispherical particles as well. More recently, substrate conformal imprint lithography[91] has been developed, in which a sol–gel mask is defined by soft lithography using a rubber stamp, followed by Ag evaporation and lift-off. This process leads to <0.1 nm control over particle pitch. Examples of nanopatterns fabricated using these three techniques are shown in Fig. 7a–c. Figure 7d shows a photograph of a 6-inch-diameter Si wafer structured using soft lithography with a regular array of metal nanoparticles.

Integration of metallic nanostructures on the surface of a solar cell can also reduce the cell surface sheet resistance, which leads to greater output power as well[37]. Indeed, in an optimized plasmonic solar-cell design, the metal nanostructures will be integrated with the metal-finger structure that conventionally collects current from the cell. In further advanced designs, a percolating metal nanostructure may itself be used as a means to collect the photocurrent. An inverse structure using metal hole arrays for light concentration and collection has also been demonstrated[92]. Using nanoscale metal contacts could potentially reduce the optical reflection loss that typically results from the coverage of a solar cell by macroscopic metal fingers.

Summary and outlook

The ability to construct optically thick but physically very thin photovoltaic absorbers could revolutionize high-efficiency photovoltaic device designs. This becomes possible by using light trapping through the resonant scattering and concentration of light in arrays of metal nanoparticles, or by coupling light into surface plasmon polaritons and photonic modes that propagate in the plane of the semiconductor layer. In this way extremely thin photovoltaic absorber layers (tens to hundreds of nanometres thick) may absorb the full solar spectrum.

Because of the small required thickness, relatively impure and defective semiconductor absorbers (for example polycrystalline films of Si, Cu_2O, Zn_3P_2 or SiC) may be used, without suffering from volume recombination losses. The use of thin absorber layers also avoids some of the problems of materials scarcity for compound semiconductors.

Guiding and concentration of light using plasmonic geometries allows entirely new solar-cell designs in which light is fully absorbed in a single quantum well, or in a single layer of quantum dots or molecular chromophores. This would also open up new approaches

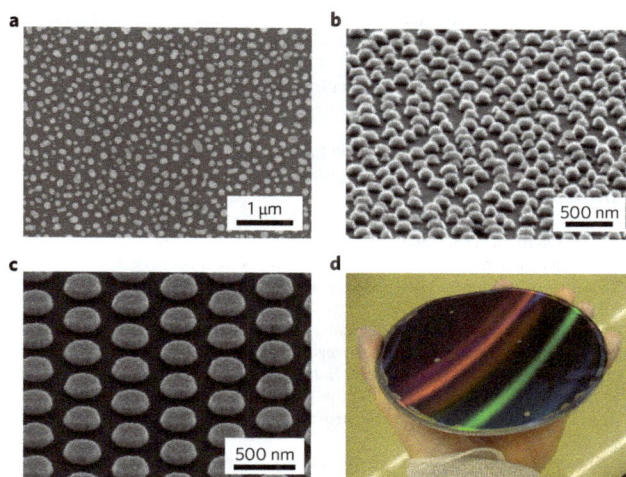

Figure 7 | Large-area metal nanopatterns for plasmonic solar cells.
a–c, Scanning electron micrographs of arrays of Ag nanoparticles.
a, A 14-nm-thick Ag film was deposited onto a thermally oxidized Si wafer (SiO_2 thickness 10 nm) by thermal evaporation and annealed at 200 °C in a N_2 ambient atmosphere for 60 minutes in forming gas. Silver nanoparticles form by surface-tension-induced agglomeration. The average particle diameter is 135 nm, and surface coverage is 26%. **b**, Silver nanoparticles evaporated through a porous alumina template, annealed at 200 °C. The average particle diameter is 135 nm. **c**, Hexagonal array of Ag nanoparticles deposited using substrate-conformal imprint lithography using the SCIL technique. The particle diameter is 300 nm. From M. A. Verschuuren, P. Gerlach, A. Polman and H. A. van Sprang, manuscript in preparation. **d**, Wafer-scale metal nanopatterns made using the SCIL technique. From M. A. Verschuuren, Y. Ni, M. Megens, H. A. van Sprang and A. Polman, manuscript in preparation. Figures reproduced with permission: **a**, © 2007 AIP; **b**, © 2008 AIP.

to carrier collection from quantum-well and quantum-dot solar cells that do not rely on inter-well or inter-dot carrier transfer before photocarrier collection at the device contacts. The extreme confinement of light that can be achieved using plasmonic nanostructures can also enhance nonlinear effects such as up- and down-conversion or multiple carrier generation in nanoscale solar-cell geometries.

References

1. Green, M. A. *Solar Cells: Operating Principles, Technology and System Applications* (Univ. New South Wales, 1998).
2. Yablonovitch, E. & Cody, G. D. Intensity enhancement in textured optical sheets for solar cells. *IEEE Trans. Electr. Dev.* **29**, 300–305 (1982).
3. Deckman, H. W., Roxlo, C. B. & Yablonovitch, E. Maximum statistical increase of optical absorption in textured semiconductor films. *Opt. Lett.* **8**, 491–493 (1983).
4. Polman, A. Plasmonics applied. *Science* **322**, 868 (2008).
5. Andrew, P. & Barnes, W. L. Energy transfer across a metal film mediated by surface plasmon polaritons. *Science* **306**, 1002–1005 (2004).
6. Mertens, H., Biteen, J. S., Atwater, H. A. & Polman, A. Polarization-selective plasmon-enhanced Si quantum dot luminescence. *Nano Lett.* **6**, 2622–2625 (2006).
7. Kühn, S., Hakanson, U., Rogobete, L. & Sandoghdar, V. Enhancement of single molecule fluorescence using a gold nanoparticle as an optical nano-antenna. *Phys. Rev. Lett.* **97**, 017402 (2006).
8. Stockman, M. I. Nanofocusing of optical energy in tapered plasmonic waveguides. *Phys. Rev. Lett.* **93**, 137404 (2004).
9. Verhagen, E., Spasenović, M., Polman, A. & Kuipers, L. Nanowire plasmon excitation by adiabatic mode transformation. *Phys. Rev. Lett.* **102**, 203904 (2008).
10. Prodan, E., Radloff, C., Halas, N. J. & Nordlander, P. A hybridization model for the plasmon response of complex nanostructures. *Science* **302**, 419–422 (2003).
11. Quinten, M., Leitner, A., Krenn, J. R. & Aussenegg, F. R. Electromagnetic energy transport via linear chains of silver nanoparticles. *Opt. Lett.* **23**, 1331–1333 (1998).

12. Maier, S. A. *et al.* Local detection of electromagnetic energy transport below the diffraction limit in metal nanoparticle plasmon waveguides. *Nature Mater.* **2,** 229–232 (2003).

13. Koenderink, A. F. & Polman, A. Complex response and polariton-like dispersion splitting in periodic metal nanoparticle chains. *Phys. Rev. B* **74,** 033402 (2006).

14. Zia, R., Selker, M. D., Catrysse, P. B. & Brongersma, M. L. Geometries and materials for subwavelength surface plasmon modes. *J. Opt. Soc. Am. A* **21,** 2442–2446 (2004).

15. Mühlschlegel, P., Eisler, H. J., Martin, O. J. F., Hecht, B. & Pohl, D. W. Resonant optical antennas. *Science* **308,** 1607–1609 (2005).

16. Ditlbacher, H., Krenn, J. R., Schider, G., Leitner, A. & Aussenegg, F. R. Two-dimensional optics with surface plasmon polaritons. *Appl. Phys. Lett.* **81,** 1762–1764 (2002).

17. Bozhevolnyi, S. I., Volkov, V. S., Devaux, E., Laluet, J. Y. & Ebbesen, T. W. Channel plasmon subwavelength waveguide components including interferometers and ring resonators. *Nature* **440,** 508–511 (2006).

18. Krasavin, A. V. & Zheludev, N. I. Active plasmonics: Controlling signals in Au/Ga waveguide using nanoscale structural transformations. *Appl. Phys. Lett.* **84,** 1416–1418 (2004).

19. Colombelli, R. *et al.* Quantum cascade surface-emitting photonic crystal laser. *Science* **302,** 1374–1377 (2003).

20. Hill, M. T. *et al.* Lasing in metallic-coated nanocavities. *Nature Photon.* **1,** 589–594 (2007).

21. Oulton, R. F. *et al.* Plasmon lasers at deep subwavelength scale. *Nature* **461,** 629–632 (2009).

22. Okamoto, K. *et al.* Surface-plasmon-enhanced light emitters based on InGaN quantum wells. *Nature Mater.* **3,** 601–605 (2004).

23. Pendry, J. B. Negative refraction makes a perfect lens. *Phys. Rev. Lett.* **85,** 3966–3968 (2000).

24. Shelby, R. A., Smith, D. R. & Schultz, S. Experimental verification of a negative index of refraction. *Science* **292,** 77–79 (2001).

25. Linden, S. *et al.* Magnetic response of metamaterials at 100 terahertz. *Science* **306,** 1351–1353 (2004).

26. Fang, N., Lee, H., Sun, C. & Zhang, X. Sub-diffraction-limited optical imaging with a silver superlens. *Science* **308,** 534–537 (2005).

27. Kreibig, U. & Vollmer, M. *Optical Properties of Metal Clusters* (Springer, 1995).

28. Bohren, C. F. & Huffman, D. R. *Absorption and Scattering of Light by Small Particles* (Wiley, 2008).

29. Mertz, J. Radiative absorption, fluorescence, and scattering of a classical dipole near a lossless interface: a unified description. *J. Opt. Soc. Am. B* **17,** 1906–1913 (2000).

30. Stuart, H. R. & Hall, D. G. Absorption enhancement in silicon-on-insulator waveguides using metal island films. *Appl. Phys. Lett.* **69,** 2327–2329 (1996).

31. Stuart, H. R. & Hall, D. G. Island size effects in nanoparticle-enhanced photodetectors. *Appl. Phys. Lett.* **73,** 3815–3817 (1998).

32. Schaadt, D. M., Feng, B. & Yu, E. T. Enhanced semiconductor optical absorption via surface plasmon excitation in metal nanoparticles. *Appl. Phys. Lett.* **86,** 063106 (2005).

33. Derkacs, D., Lim, S. H., Matheu, P., Mar, W. & Yu, E. T. Improved performance of amorphous silicon solar cells via scattering from surface plasmon polaritons in nearby metallic nanoparticles. *Appl. Phys. Lett.* **89,** 093103 (2006).

34. Matheu, P., Lim, S. H., Derkacs, D., McPheeters, C. & Yu, E. T. Metal and dielectric nanoparticle scattering for improved optical absorption in photovoltaic devices. *Appl. Phys. Lett.* **93,** 113108 (2008).

35. Pillai, S., Catchpole, K. R., Trupke, T. & Green, M. A. Surface plasmon enhanced silicon solar cells. *J. Appl. Phys.* **101,** 093105 (2007).

36. Derkacs, D. *et al.* Nanoparticle-induced light scattering for improved performance of quantum-well solar cells. *Appl. Phys. Lett.* **93,** 091107 (2008).

37. Nakayama, K., Tanabe, K. & Atwater, H. A. Plasmonic nanoparticle enhanced light absorption in GaAs solar cells. *Appl. Phys. Lett.* **93,** 121904 (2008).

38. Catchpole, K. R. & Polman, A. Design principles for particle plasmon enhanced solar cells. *Appl. Phys. Lett.* **93,** 191113 (2008).

39. Catchpole, K. R. & Polman, A. Plasmonic solar cells. *Opt. Express* **16,** 21793–21800 (2008).

40. Xu, G., Tazawa, M., Jin, P., Nakao, S. & Yoshimura, K. Wavelength tuning of surface plasmon resonance using dielectric layers on silver island films. *Appl. Phys. Lett.* **82,** 3811–3813 (2003).

41. Mertens, H., Verhoeven, J., Polman, A. & Tichelaar, F. D. Infrared surface plasmons in two-dimensional silver nanoparticle arrays in silicon. *Appl. Phys. Lett.* **85,** 1317–1319 (2004).

42. Beck, F. J., Polman, A. & Catchpole, K. R. Tunable light trapping for solar cells using localized surface plasmons. *J. Appl. Phys.* **105,** 114310 (2009).

43. Lim, S. H., Mar, W., Matheu, P., Derkacs, D. & Yu, E. T. Photocurrent spectroscopy of optical absorption enhancement in silicon photodiodes via scattering from surface plasmon polaritons in gold nanoparticles. *J. Appl. Phys.* **101,** 104309 (2007).

44. Beck, F. J., Mokkapati, S., Polman, A. & Catchpole, K. R. Asymmetry in light-trapping by plasmonic nanoparticle arrays located on the front or on the rear of solar cells. *Appl. Phys. Lett.* (in the press).

45. Pala, R. A., White, J., Barnard, E., Liu, J. & Brongersma, M. L. Design of plasmonic thin-film solar cells with broadband absorption enhancements. *Adv. Mater.* **21,** 3504–3509 (2009).

46. Mokkapati, S., Beck, F. J., Polman, A. & Catchpole, K. R. Designing periodic arrays of metal nanoparticles for light trapping applications in solar cells. *Appl. Phys. Lett.* **95,** 53115 (2009).

47. Stuart, H. R. & Hall, D. G. Thermodynamic limit to light trapping in thin planar structures. *J. Opt. Soc. Am. A* **14,** 3001–3008 (1997).

48. Stuart, H. R. & Hall, D. G. Enhanced dipole–dipole interaction between elementary radiators near a surface. *Phys. Rev. Lett.* **80,** 5663–5668 (1998).

49. Catchpole, K. R. & Pillai, S. Absorption enhancement due to scattering by dipoles into silicon waveguides. *J. Appl. Phys.* **100,** 044504 (2006).

50. Rand, B. P., Peumans, P. & Forrest, S. R. Long-range absorption enhancement in organic tandem thin-film solar cells containing silver nanoclusters. *J. Appl. Phys.* **96,** 7519–7526 (2004).

51. Kim, S. S., Na, S.-I., Jo, J., Kim, D. Y. & Nah, Y.-C. Plasmon enhanced performance of organic solar cells using electrodeposited Ag nanoparticles. *Appl. Phys. Lett.* **93,** 073307 (2008).

52. Morfa, A. J., Rowlen, K. L., Reilly, T. H., Romero, M. J. & Van de Lagemaat, J. Plasmon-enhanced solar energy conversion in organic bulk heterojunction photovoltaics. *Appl. Phys. Lett.* **92,** 013504 (2008).

53. Lindquist, N. C., Luhman, W. A., Oh, S. H. & Holmes, R. J. Plasmonic nanocavity arrays for enhanced efficiency in organic photovoltaic cells. *Appl. Phys. Lett.* **93,** 123308 (2008).

54. Kume, T., Hayashi, S., Ohkuma, H., Yamamoto, K. Enhancement of photoelectric conversion efficiency in copper phthalocyanine solar cell: white light excitation of surface plasmon polaritons. *Jpn. J. Appl. Phys.* **34,** 6448–6451 (1995).

55. Westphalen, M., Kreibig, U., Rostalski, J., Lüth, H. & Meissner, D. Metal cluster enhanced organic solar cells. *Sol. Energy Mater. Sol. C.* **61,** 97–105 (2000).

56. Hägglund, C., Zäch, M. & Kasemo, B. Enhanced charge carrier generation in dye sensitized solar cells by nanoparticle plasmons. *Appl. Phys. Lett.* **92,** 013113 (2008).

57. Konda, R. B. *et al.* Surface plasmon excitation via Au nanoparticles in n-CdSe/ p-Si heterojunction diodes. *Appl. Phys. Lett.* **91,** 191111 (2007).

58. Hägglund, C., Zäch, M., Petersson, G. & Kasemo, B. Electromagnetic coupling of light into a silicon solar cell by nanodisk plasmons. *Appl. Phys. Lett.* **92,** 053110 (2008).

59. Kirkengena, M., Bergli, J. & Galperin, Y. M. Direct generation of charge carriers in c-Si solar cells due to embedded nanoparticles. *J. Appl. Phys.* **102,** 093713 (2007).

60. Raether, H. *Surface Plasmons on Smooth and Rough Surfaces and on Gratings.* (Springer Tracts in Modern Physics III, Springer, 1988).

61. Berini, P. Plasmon–polariton waves guided by thin lossy metal films of finite width: bound modes of symmetric structures. *Phys. Rev. B* **61,** 10484–10503 (2000).

62. Berini, P. Plasmon–polariton waves guided by thin lossy metal films of finite width: bound modes of asymmetric structures. *Phys. Rev. B* **63,** 125417 (2001).

63. Dionne, J. A., Sweatlock, L., Atwater, H. A. & Polman, A. Planar plasmon metal waveguides: frequency-dependent dispersion, propagation, localization, and loss beyond the free electron model. *Phys. Rev. B* **72,** 075405 (2005).

64. Dionne, J. A., Sweatlock, L., Atwater, H. A. & Polman, A. Plasmon slot waveguides: towards chip-scale propagation with subwavelength-scale localization. *Phys. Rev. B* **73,** 035407 (2006).

65. Slooff, L. H. *et al.* Determining the internal quantum efficiency of highly efficient polymer solar cells through optical modeling. *Appl. Phys. Lett.* **90,** 143506 (2007).

66. Ferry, V. E., Sweatlock, L. A., Pacifici, D. & Atwater, H. A. Plasmonic nanostructure design for efficient light coupling into solar cells. *Nano Lett.* **8,** 4391–4397 (2008).

67. Ferry, V. *et al.* Improved red-response in thin film a-Si:H solar cells with nanostructured plasmonic back reflectors. *Appl. Phys. Lett.* **95,** 183503 (2009).

68. Giannini, V., Zhang, Y., Forcales, M. & Gómez Rivas, J. Long-range surface plasmon polaritons in ultra-thin films of silicon. *Opt. Express* **16,** 19674–19685 (2008).

69. Mapel, J. K., Singh, M., Baldo, M. A. & Celebi, K. Plasmonic excitation of organic double heterostructure solar cells. *Appl. Phys. Lett.* **90,** 121102 (2007).

70. Tvingstedt, K., Persson, N. K., Ingan, O., Rahachou, A. & Zozoulenko, I. V. Surface plasmon increase absorption in polymer photovoltaic cells. *Appl. Phys. Lett.* **91,** 113514 (2007).

71. Heidel, T. D., Mapel, J. K., Singh, M., Celebi, K. & Baldo, M. A. Surface plasmon polariton mediated energy transfer in organic photovoltaic devices. *Appl. Phys. Lett.* **91,** 093506 (2007).

72. Haug, F. J., Söderström, T., Cubero, O., Terrazzoni-Daudrix, V. & Ballif, C. Plasmonic absorption in textured silver back reflectors of thin film solar cells. *J. Appl. Phys.* **104,** 064509 (2008).

73. Tvingstedt, K., Person, N. K., Inganäs, O., Rahachou, A. & Zozoulenko, I. V. Surface plasmon increase absorption in polymer photovoltaic cells. *Appl. Phys. Lett.* **91,** 113514 (2007).

74. Franken, R. H. *et al.* Understanding light trapping by light-scattering textured back electrodes in thin-film n–i–p silicon solar cells. *J. Appl. Phys.* **102,** 014503 (2007).

75. Schropp, R. E. I. *et al.* Nanostructured thin films for multibandgap silicon triple junction solar cells. *Sol. Energy Mater. Sol. C.* **93,** 1129–1133 (2009).

76. Rockstuhl, C., Fahr, S. & Lederer, F. Absorption enhancement in solar cells by localized plasmon polaritons. *J. Appl. Phys.* **104,** 123102 (2008).

77. Keevers, M. J., Young, T. L., Schubert, U. & Green, M. A. 10% efficient CSG minimodules. *Proc. 22nd Eur. Photovoltaic Solar Energy Conf.* Milan, Italy, 3–7 September 2007.

78. Green, M. A., Zhao, J., Wang, A. & Wenham, S. R. Very high efficiency silicon solar cells: science and technology. *IEEE Trans. Electr. Dev.* **46,** 1940–1947 (1999).

79. Fahr, S., Rockstuhl, C. & Lederer, F. Metallic nanoparticles as intermediate reflectors in tandem solar cells. *Appl. Phys. Lett.* **95,** 121105 (2009).

80. Pacifici, D., Lezec, H. & Atwater, H. A. All-optical modulation by plasmonic excitation of CdSe quantum dots. *Nature Photon.* **1,** 402–406 (2007).

81. Walters, R. J., van Loon, R. V. A., Brunets, I., Schmitz, J. & Polman, A. A silicon-based electrical source of surface plasmon polaritons. *Nature Mater.* **9,** 21–25 (2010).

82. Verhagen, E., Kuipers, L. & Polman, A. Field enhancement in metallic subwavelength aperture arrays probed by erbium upconversion luminescence. *Opt. Express* **17,** 14586–14597 (2009).

83. O'Carroll, D., Hofmann, C. E. & Atwater, H. A. Conjugated polymer/metal nanowire heterostructure plasmonic antennas. *Adv. Mater.* doi:10.1002/adma.200902024 (2009).

84. Labeke, D. V., Gerard, D., Guizal, B., Baida, F. I. & Li, L. An angle-independent frequency selective surface in the optical range. *Opt. Express* **14,** 11945–11951 (2006).

85. De Waele, R., Burgos, S. P., Polman, A. & Atwater, H. A. Plasmon dispersion in coaxial waveguides from single-cavity optical transmission measurements. *Nano Lett.* **9,** 2832–2837 (2009).

86. Kroekenstoel, E. J. A., Verhagen, E., Walters, R. J., Kuipers, L. & Polman, A. Enhanced spontaneous emission rate in annular plasmonic nanocavities. *Appl. Phys. Lett.* **95,** 263106 (2009).

87. Pijpers, J. J. H. *et al.* Assessment of carrier-multiplication efficiency in bulk PbSe and PbS. *Nature Phys.* **5,** 811–814 (2009).

88. King, R. R. *et al.* 40% efficient metamorphic GaInP/GaInAs/Ge multijunction solar cells. *Appl. Phys. Lett.* **90,** 183516 (2007).

89. Coutts, T. J. *et al.* Critical issues in the design of polycrystalline, thin-film tandem solar cells. *Prog. Photovolt. Res. Appl.* **11,** 359–375 (2003).

90. Schuller, J. A. *et al.* Plasmonics for extreme light concentration and manipulation. *Nature Mater.* **9,** 193–204 (2010).

91. Verschuuren, M. A. & van Sprang, H. A. 3D photonic structures by sol-gel imprint lithography. *Mater. Res. Soc. Symp. Proc.* **1002,** 1002-N03-05 (2007).

92. Reilly, T., van de Lagemaat, J., Tenent, R. C., Morfa, A. J. & Rowlen, K. L. Surface-plasmon enhanced transparent electrodes in organic photovoltaics. *Appl. Phys. Lett.* **92,** 243304 (2008).

93. US Geological Survey 2004 <http://minerals.usgs.gov/minerals/pubs/mcs/>.

94. Green, M. A. Improved estimates for Te and Se availability from Cu anode slimes and recent price trends. *Prog. Phot.* **14,** 743 (2006).

Acknowledgements

We thank M. Bonn, K. Catchpole, V. E. Ferry, J. Gomez Rivas, M. Hebbink, J. N. Munday, P. Saeta, W. C. Sinke, R. E. I. Schropp, K. Tanabe, E. Verhagen, M. A. Verschuuren, R. de Waele and E. T. Yu for discussions. This work is supported by the Global Climate and Energy Project. The FOM portion of this work is part of the research programme of FOM and of the Joint Solar Panel programme, which are both financially supported by the Netherlands Organisation for Scientific Research (NWO). The Caltech portion of this work was supported by the Department of Energy under grant number DOE DE-FG02-07ER46405.

Additional information

The authors declare no competing financial interests.

Excitons in nanoscale systems

Nanoscale systems are forecast to be a means of integrating desirable attributes of molecular and bulk regimes into easily processed materials. Notable examples include plastic light-emitting devices and organic solar cells, the operation of which hinge on the formation of electronic excited states, excitons, in complex nanostructured materials. The spectroscopy of nanoscale materials reveals details of their collective excited states, characterized by atoms or molecules working together to capture and redistribute excitation. What is special about excitons in nanometre-sized materials? Here we present a cross-disciplinary review of the essential characteristics of excitons in nanoscience. Topics covered include confinement effects, localization versus delocalization, exciton binding energy, exchange interactions and exciton fine structure, exciton–vibration coupling and dynamics of excitons. Important examples are presented in a commentary that overviews the present understanding of excitons in quantum dots, conjugated polymers, carbon nanotubes and photosynthetic light-harvesting antenna complexes.

GREGORY D. SCHOLES*[1] AND GARRY RUMBLES[2]

[1]Department of Chemistry 80 St George Street, Institute for Optical Sciences, and Centre for Quantum Information and Quantum Control, University of Toronto, Toronto, Ontario M5S 3H6, Canada
[2]National Renewable Energy Laboratory, Chemical and Biosciences Center, MS3216, 1617 Cole Boulevard, Golden, Colorado 80401-3393, USA
*e-mail: gscholes@chem.utoronto.ca

An exciting aspect of nanoscience is that relationships between structure and electronic properties are being revealed through a combination of synthesis, structural characterization, chemical physics and theory. Spectroscopy is a tool that is sensitive to aspects of the electronic structure of materials. As such, spectroscopic studies of nanoscale systems often provide insights into the collective absorption and redistribution of excitation energy. This vibrant field is the study of excitons in nanoscale systems. These electronic excitations differ from excitons in bulk materials, and represent a distinct class of exciton that can be thought of either as a confined bulk-type exciton or a molecular excitation. In this review for the non-specialist, our aim is to capture the essence of the field and highlight areas of current interest. For the expert, we aim to provoke cross-disciplinary discussion and emphasize some of the challenges at the leading edge of the field. Overall we aim to identify and discuss characteristic features of excitons in materials with nanometre-size dimensions

or organization. Part of our emphasis is to highlight topics at the forefront of the field, and to suggest questions that should be asked in future work.

Excitons can be formed by association of electrical charge units (free carriers) or by direct photoexcitation. The exciton luminescence that can follow this charge association underlies organic light-emitting diode technologies[1]. Conversely, excitons that are formed readily by photoexcitation can dissociate into free carriers (unbound electrons and holes) and thus play a central role in photovoltaic and solar cell devices[2]. Excitons that remain bound and emit light can be used as laser media[3]. These observations help to illustrate the concept of excitons compared with free carriers in bulk materials[4]. Photoexcitation creates an electron in the conduction band, leaving a 'hole' in the valence band. If the interaction between the electron and hole is assumed to be negligible—justified when each wavefunction is spread over an expanse of atoms—then the pair are free carriers. On the other hand, an attractive coulombic interaction between the electron and hole 'quasi-particles' binds them into an exciton. A distinguishing feature of excitons is that the spatial extent of an electronic excited state is increased through coherent sharing of the excitation among subunits of the material. That characteristic is determined by electronic coupling among the repeat units that make up the material[5-8].

Excitons in nanoscale systems are formed by light absorption in molecules such as polyacenes and polyenes, conjugated polymers, quantum dots, molecular aggregates, carbon nanotubes and so on. The new aspect of excitons that is prevalent in, or

When citing this article, please cite the original version as shown on the contents page of this chapter.

even defines, nanoscience is that the physical size and shape of the material strongly influences the nature and dynamics of the electronic excitation. Therefore, a deciding property of excitons in nanoscale systems is that the exciton size is dictated not by the electron–hole Coulomb interaction, but by the physical dimensions of the material or the arrangement of distinct building blocks. That interests the chemist because excitons can be engineered in a material according to structure. It is of interest to physicists because the spatial confinement of the exciton accentuates many of its interesting physical properties, exposing them for examination. The theorist may conclude that perhaps we should not try to force-fit existing theories to model excitations in nanostructures as they often carry with them assumptions that we have forgotten, or that we erroneously ignore. Nanoscale materials thus provide both a testbed and an inspiration for new quantum mechanical approaches to the calculation of electronic properties of large systems. Through the study of excitons in nanoscale systems we are learning some general rules defining how size and shape tunes electronic structure and dynamics. In Fig. 1 a summary of excitons in bulk materials (adapted from ref. 6) is extended to indicate how excitons in nanoscale systems

differ. An overview and introduction to elements of excitons in nanoscale systems is given in Box 1.

A challenge in this field is that the materials under investigation are often complex; they contain many atoms, they can be structurally disordered or influenced by surface effects, and samples often have an inhomogeneous composition in terms of structural disorder and size polydispersity. We attempt below to provide some insights into how such hurdles have been overcome. Questions being addressed include: can the properties of confined excitons be used to change the way that solar cells or lasers operate? How should we transfer the language of solid-state physics to organic systems[9], and vice versa?

A unifying theme in nanoscience is that size and shape are important. How best can we understand and compare nanomaterials and the implications of size-tunable properties for excitons? A powerful approach is to combine experiment and theory. However, owing to the intermediate length scales characteristic of many nanomaterials, unique challenges for theory are presented. For example, the huge number of electrons presents difficulties for molecular-type electronic structure calculations,

Box 1: An introduction to excitons in nanoscale systems

1. Bands and molecular orbitals for extended systems
Periodic, strongly coupled molecular or atomic systems are characterized by a high density of delocalized orbitals that form the valence (VB) and conduction (CB) bands. Those bands can be represented in real space as shown on the left side of the diagram, or equivalently, the energy of each band can be plotted against its wave vector k, as in the right schematic. Excitation of an electron from VB to CB creates free carriers. The minimum energy required is that of the bandgap, E_g.

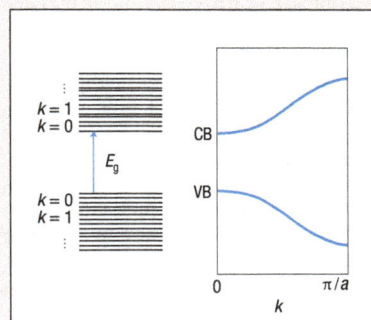

2. Excitons in extended systems: bound electron–hole pairs
The language that describes the features of the quasi-particle approach is nicely intuitive. An exciton in a spatially entended system is described as a neutral excitation particle: an electron–hole pair. In an exciton the electron and hole are bound by the electron-hole Coulomb interaction. In the chemist's electron–electron language, the 'attraction' derives from decreased electron–electron repulsion integrals in the excited state configuration compared with the ground state, thus lowering the energy of that configuration relative to the continuum by the exciton binding energy, $E_b = E_g - E_x$. Related electron–hole exchange interactions mix the Hartree–Fock (single particle) configurations, so that the exciton states are finally obtained through a configuration interaction (CI-singles) calculation. Correct spin eigenstates can only be obtained by antisymmetrizing wavefunctions and mixing configurations through exchange interactions.

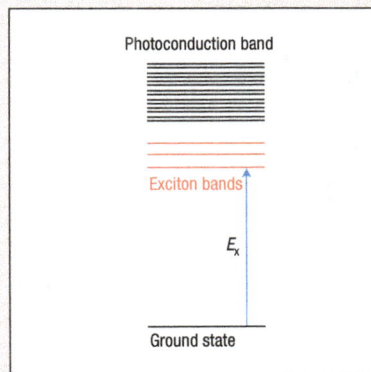

3. Molecular orbital picture for confined systems

Occupied (in the ground state) and virtual molecular orbitals for a series of polyacenes. As the conjugation size increases, the bandgap decreases and the density of states increases. These quantum size effects derive from changes in electron–electron interactions that are proportional to molecular orbital wavefunction delocalization. A single-excitation configuration is constructed by excitation of an electron from an occupied to a virtual orbital. That configuration is delocalized over the extent of the molecule. However, to be consistent with the definition of an exciton in an extended system, the change in electron–electron interactions after excitation needs to be considered to describe the exciton. The single-excitation configurations are mixed by exchange interactions so that each excited state — exciton state — is a linear combination of the single-excitation configurations. A manifold of singlet and triplet excitons are thus obtained.

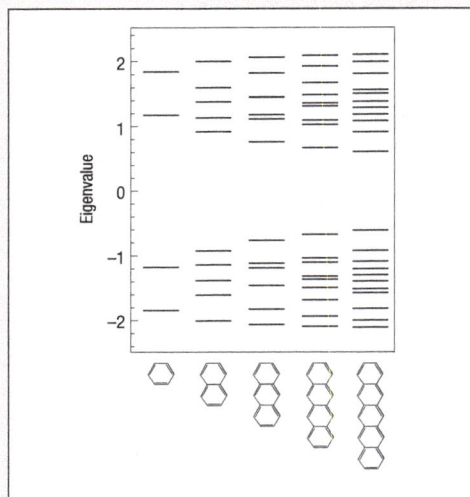

4. Spectroscopy of molecular excitons

Absorption spectra of a series of naphthyl dimer molecules[117] compared with a monomeric model chromophore (dashed line). The dimer spectra consist of two bands, with a splitting equal to the electronic coupling H_{LR} between the naphthyl groups on the left (L) and right (R), indicating that the excitation is shared between the two chromophores: a molecular exciton. A schematic diagram of the relative energies and wavefunctions of the napthyl moieties compared to the dimer is given below. The redshift of the centre of gravity of the exciton bands is due to orbital mixing effects. These interactions induce the electron and hole to delocalize over the subunits. Model dimers such as shown here allow for detailed examination of interchromophore interactions, including realistic treatment of solvation[118]. (Spectra reprinted with permission from ref. 117. Copyright (1993) American Chemical Society).

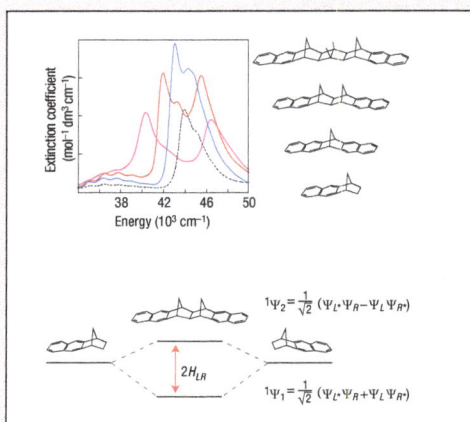

which scale steeply with the number of electrons, yet the importance of atomic-scale details needs to be captured. Great progress has been made in solving such problems, as has been reviewed elsewhere[10–13]. In this review we will cover four broad topics of importance in the spectroscopy of excitons in nanoscale systems. To illustrate the ways in which excitons in nanoscale systems are unique or interesting, and to formulate some general observations that may stimulate cross-disciplinary discussion, we will describe examples pertaining to quantum dots, conjugated polymers, carbon nanotubes and photosynthetic light-harvesting antenna complexes within each of these subject areas. We begin by providing some background on each material.

OVERVIEW OF THE MATERIALS

PHOTOSYNTHETIC LIGHT-HARVESTING COMPLEXES

Proteins organized in the photosynthetic membranes of higher plants, photosynthetic bacteria and algae accomplish the capture of incident photons (light-harvesting), subsequent spatial redistribution of that excitation energy, and the ultimate trapping of the energy via photo-induced electron transfer, with remarkably high efficiencies[14]. Suspensions of surfactant-isolated light-harvesting complexes (known as LH2) from purple bacteria provide a good example. They show two absorption bands (Fig. 2), attributed to two distinct arrangements of bacteriochlorophyll-a (Bchl) molecules. The B800 absorption band is due to

Bulk materials   Nanomaterials

Electron,hole
Bloch functions   *Confinement*
delocalized bands }

High

Wannier–Mott
exciton model

* Discrete optical
transitions
* Increased exciton
binding energy
* Enhanced singlet–triplet
exchange interaction

Dielectric constant

A general
exciton model

* Increased absorption
coefficient and nonlinear optical
properties
* Decreased exciton
binding energy
* Diminished singlet–triplet
exchange interaction

Frenkel
exciton model

Low

Atomic or
molecular basis
localized excitations }   *Delocalization*

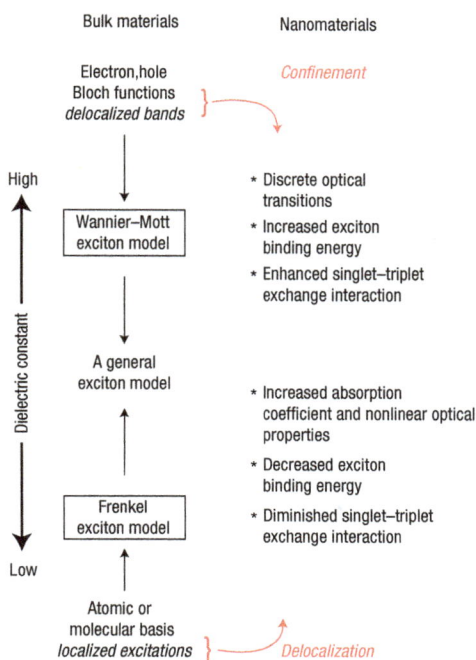

Figure 1 Excitons in nanoscale systems. The diagram relates the Wannier and Frenkel exciton models and sketches the basis for describing excitons. The description may proceed from delocalized bands, wherein the electron and hole may move freely throughout the material. Such a model can be used to model excitons in crystalline materials with high dielectric constant, such that electron–hole Coulomb interactions are small compared with the bandgap of the material. One may also develop a model starting from a localized basis set. The tight-binding model is an example of this approach. In their usual implementation, either approach contains approximations or limitations that restrict its application. A general exciton model overcomes such limitations, and contains all the key features necessary to describe the excited states of a molecule as well as an infinite system. Nanoscale excitons can be described starting from the point of view of either a delocalized or a localized excitation. In a delocalized representation of the problem, nanoscale excitons are subject to quantum confinement, and their spectroscopy becomes richer. With respect to a model based on localized functions, nanoscale materials can sustain more delocalized excitations. The Frenkel model uses the point of view, particularly useful for molecular aggregates, that the wavefunction is separable into distinct electron groups tightly associated with each repeat unit. Such an approach is applied to molecular excitons, such as photosynthetic excitons or J-aggregates.

a ring of essentially non-interacting Bchl molecules (nearest-neighbour electronic coupling ~30 cm^{-1}). The B850 band, however, is the absorption into the optically bright exciton states characteristic of the ring of strongly interacting Bchl molecules (nearest-neighbour electronic coupling ~300 cm^{-1}). Here is an example of electronic coupling among molecules that creates a new chromophore: a Frenkel exciton. The excitons that are of interest in these systems include Frenkel excitons, but are primarily weakly coupled localized excitations[15].

Relatively weak electronic coupling amongst the many hundreds of chromophores that are bound in light-harvesting proteins enables the excitation energy to be funnelled efficiently through space to reaction centre proteins[16].

In Fig. 2 a simple light-harvesting antenna protein, PE545, is shown together with a sketch of the organization of the photosynthetic membrane of a type of cryptophyte algae[17]. Each PE545 antenna protein binds eight chromophores, as seen in the structural model. Such model systems, where we know the precise arrangement of light-absorbing molecules, have provided superb model systems for learning about molecular excitons (also known as Frenkel excitons; see Fig. 1) and electronic (resonance) energy transfer. Electronic energy transfer allows excitation energy absorbed by an antenna protein to migrate rapidly (over a few picoseconds) to the peripheral chromophores, circled in the figure, then transfer into the core antenna of one of the membrane-bound photosystems, PS I or PS II (50–100 ps). In most organisms the antenna protein is also membrane-bound. The funnelling of excitation energy to the photosynthetic reaction centres involves many tens of hops, occurring on timescales of 100 fs to 100 ps. These fast, light-initiated processes occurring in proteins isolated from photosynthetic organisms have been studied in great detail over the past years[14,17]. It has become evident that the organization of light-absorbing molecules (usually chlorophylls and carotenoids) on the nanometre length scale provides important—sometimes subtle and ingenious—optimizations that control light capture and funnelling[18].

CONJUGATED POLYMERS

Conjugated polymers are highly conjugated linear macromolecules that are of great current interest because of their semiconductor-like properties and ease of processing. Excitons in conjugated polymers such as functionalized poly(phenylene vinylene) (for example MEH-PPV, poly[2-methoxy-5-(2′-ethyl)hexyloxy-1,4-phenylene vinylene]) and alkyl polyfluorenes have been intensively studied from both experimental and theoretical viewpoints to understand their nature and dynamics. Two basic types of exciton have been identified: intrachain and interchain. The former are formed by the extended π-conjugation along sections of the polymer backbone. Interchain excitons form when two chain segments couple through space, either because the chains are near to each other in a solid film, or because the chain is folded back on itself. Thus optical properties depend strongly on the aggregation state of the polymer. Details can be found in recent reviews[19–22].

Conjugated polymer excitons play important roles in electroluminescent and photovoltaic devices, and some intriguing properties relevant to device optimization have been reported. For example, the observation of wavelength-dependent photocurrent generation in MEH-PPV was explained by analysis of quantum chemical calculations that revealed significant changes in mixing of configuration wavefunctions of the polymer[23]. That is, the electronic make-up of higher-energy excitons differs from excitons at the optical gap in such a way that mixing with extended charge-separated states is increased. A challenge now is to elucidate a microscopic picture that explains observations of interfacial excitons in heterojunction polymer blends[24]. Evidence suggests that such excitons can reversibly dissociate into an exciplex (an excited-state hetero-complex)[25], in which the electron and hole become spatially separated. In that case the Frenkel exciton model breaks down because in the excited state electron density is transferred from one molecule to the other. Dissociation of an exciplex produces geminate ion pairs that might play important intermediate roles in electron–hole dissociation or capture processes[26].

SEMICONDUCTING SINGLE-WALL CARBON NANOTUBES

The spectroscopy of semiconducting carbon nanotubes (CNTs)[27] has been of great interest in the last few years. CNTs are formed in a great variety of sizes and types, and typically aggregate into macroscopic bundles. Until recently, the accepted description of their electronic states and spectroscopy was a simple model that neglected interactions between electrons. However, that viewpoint changed as a result of the report of bandgap fluorescence for CNTs[28], made possible by surfactant isolation of individual tubes or small bundles of tubes, thus providing samples without large bundles and ropes in which fluorescence is quenched. Although it was not immediately recognized, that observation provided clear evidence for excitons in CNTs, supporting the earlier predictions of Ando[29].

Not surprisingly, the key advance underpinning the observation of CNT exciton fluorescence was material preparation. Nonetheless, optical spectra are highly inhomogeneously broadened, as shown in Fig. 3, by a sum of contributions from a series of CNT macromolecules. A plot of excitation energy versus fluorescence emission, Fig. 3b reveals distinct peaks corresponding to a series of individual tube types[30], which in combination with simple theory has allowed assignment of the spectral features to the various chemical types and sizes of CNTs. Nanotube types are designated by the chiral index (n,m), according to how the parent graphene sheet is rolled up to form a tube[31]. The fluorescence data[30] made it possible to assign the optical transitions and elucidate their size-evolution. An interesting observation from inspection of the fluorescence spectra is that spectral linewidths are narrow, being ~30 meV at room temperature[32]; this is probably a result of exchange narrowing[33], whereby the linewidth is reduced because the delocalized exciton states average over disorder in the site energies of primitive cells.

SEMICONDUCTOR QUANTUM DOTS

Semiconductor quantum dots (QDs), or nanocrystals, are nanometre-sized crystallites that can be prepared with great tunability of size and shape through

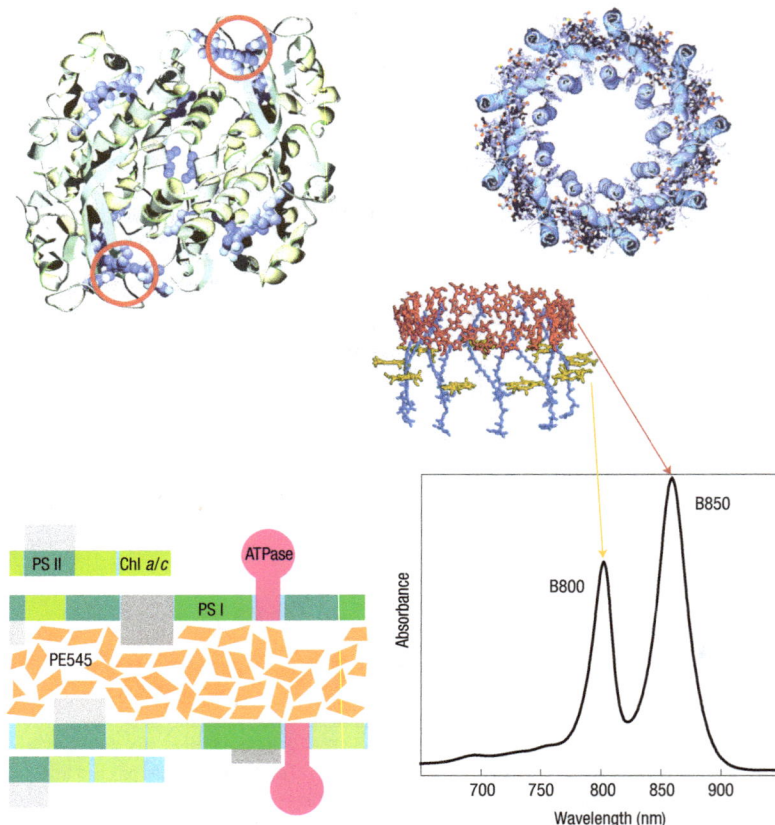

Figure 2 Frenkel excitons in photosynthesis. Left, the structure of a simple light-harvesting protein PE545 isolated from a marine cryptophyte algae. It binds eight light-absorbing molecules (blue). A cascade of electronic energy transfer steps equilibrates excitation, over timescales of tens of femtoseconds to a few tens of picoseconds, to the periphery of the protein, onto the encircled chromophores[17]. Ultimately that excitation is transferred into core antenna complexes associated with the reaction centres, photosystems I and II (PS I and PS II), organized in the thylakoid membrane as indicated schematically. (Bottom left image adapted from C. D. de Witt *et al.*, submitted to *J. Phys. Chem B*.)PE545, unlike the peripheral light-harvesting complexes of higher plants (LHC-II) or purple bacteria (LH2), is not membrane-bound. Right, the light-harvesting antenna of purple bacteria, LH2, is an example of a system whose spectroscopy is characteristic of a molecular exciton. The structural model shown at the top emphasizes the α-helices serving as a scaffold binding the light-absorbing molecules. These molecules are shown in the simplified structural model below, indicating the B850 ring (red), B800 ring (yellow) and cartenoids (blue). The absorption spectrum clearly shows the marked distinction between the B800 absorption band, arising from essentially 'monomeric' bacteriochlorophyll-*a* (Bchl) molecules, and the redshifted B850 band that is attributed to the optically bright lower exciton states of the 18 electronically coupled Bchl molecules.

colloidal routes[34–38]. Self-assembled QDs can be formed spontaneously in a semiconductor heterostructure grown using molecular beam epitaxy[39,40]. Typically, self-assembled QDs are lens-shaped, having a diameter of tens of nanometres and a height of a few nanometres, whereas colloidal QDs are usually spherical with radii of 1 to 4 nm and contain around 200 to thousands of atoms. The most common materials studied with respect to excitons include CuCl, CdSe, CdS, CdTe, InP, PbSe and PbS. In the early 1980s it was predicted that QDs should possess discrete electronic energy levels, more like molecules than bulk semiconductors[41,42].

Figure 3 Excitons in single-wall carbon nanotubes.
a, The absorption spectrum of an aqueous suspension of CNTs made by high-pressure CO conversion (HiPco), showing inhomogeneous line broadening. A single-wall CNT structural model is inset. **b**, CNT size and 'wrapping' determine the exciton energies. Samples contain many different kinds of tubes, as evident from the absorption spectrum. However, by scanning excitation wavelengths and recording a map of fluorescence spectra[30], the emission bands from various different CNTs can be discerned (see text). Assignments of the major emission peaks for this sample are indicated (T. McDonald, M. Heben & M. Jones, unpublished work; cf. ref. 32). **c**, The index (n,m) is used to define the vector $\mathbf{C}_n = n\mathbf{a}_1 + m\mathbf{a}_2$, which in turn determines the direction that the graphene sheet is rolled to form a particular CNT. The sheet is rolled so that a particular lattice point (n,m) is superimposed with $(0,0)$.

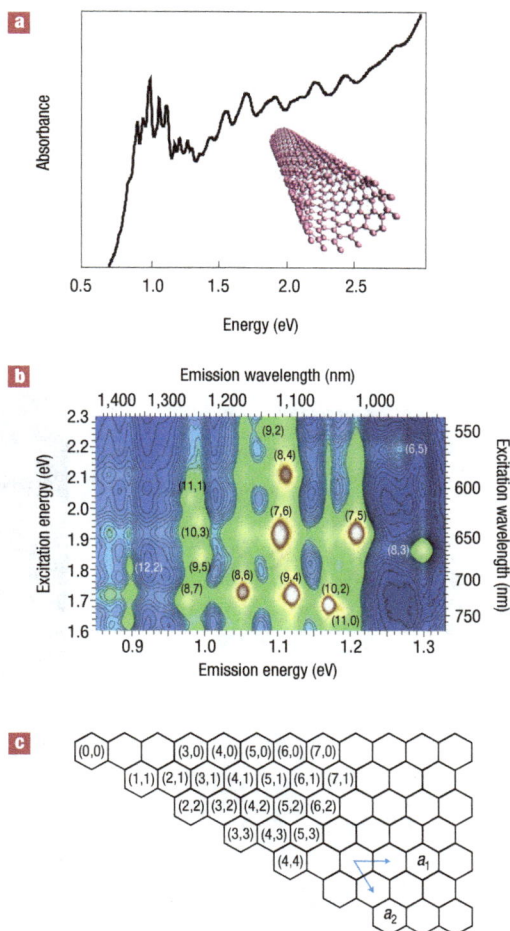

That idea was confirmed through an accumulation of experimental work[34]. QDs are clearly differentiated from molecules in that many properties have analogues in those of bulk semiconductors, spin–orbit coupling can be significant, and they have large dielectric constants. An important consideration in deciding confinement of the exciton to the QD, in particular the size of the wavefunction compared to that of the QD, is the dielectric medium surrounding the QD[40,43]. In the coming years, shape-dependent photophysics of QD excitons will be topical, driven by recent synthetic breakthroughs[44]. Electron micrograph images of a range of CdSe QD shapes are shown in Fig. 4c–f. Nanoscale 'exciton engineering' — epiaxial growth of one semiconductor over another — opens further possibilities for improving surface passivation, changing excited state dynamics of the exciton, or manipulating wavefunctions, thereby increasing the diversity of exciton photophysics that can be examined[36].

SIZE-TUNABLE SPECTROSCOPY AND EXCITONS

What are the unique aspects of excitons in nanometre-sized materials? Certainly the prevalence of size-dependent properties is notable. A myriad of electronic properties are governed by size, including the bandgap, the exciton binding

energy, and the exchange interactions. These quantities determine the spectroscopic states of the material and its (photo-)electrochemical properties. The goal of this section is to establish the spectroscopic signatures of excitons in nanoscale systems and to suggest notable aspects of their spectroscopy.

Size-tunable spectroscopic properties have been of great interest for over half a century[45], although recent interest has centred on QDs. The Bohr radius of a semiconductor exciton provides a reference as to the exciton size in the bulk. Size-tuning of properties in QDs is attributed to confinement of the exciton in a nanocrystal significantly smaller than the bulk exciton. For example, the exciton Bohr radius of PbS is ~20 nm and its bulk bandgap is 0.41 eV. Absorption spectra for PbS QDs[46] of radii ranging from ~1.3 to ~3.5 nm are shown in Fig. 5, revealing QD exciton energies in the range 0.7 to 1.5 eV.

The spectroscopic properties of molecular materials that scale in size can be elucidated in great detail, thus providing a foundation for a deeper understanding of spectroscopic measures of excitons in nanoscale materials. The size dependence of exciton transition energies for a range of organic[30,45] and QD[46–49] materials is plotted in Fig. 5. Measurement of the exciton energy does not directly reveal information that allows us to learn generally about excitons by comparing different materials. For example, size-tunable optical properties derive primarily from the bandgap, not the exciton. The bandgap is the energy difference between occupied and unoccupied orbitals. That provides only a starting point for establishing the excitation energy because, first, the electronic excited-state wavefunction is not accurately written as a one-electron excitation between these orbitals. For example, to develop a theory for calculating electronic spectra Pariser and Parr[50] decided that a method involving antisymmetrized product wavefunctions and configuration interaction was essential. Second, the interactions between electrons in the excited state differ from those for ground-state configurations. There is lower repulsion between electrons in an excited state than in a closed-shell ground state because a pair of electrons can occupy different orbitals. That decreased electron–electron repulsion gives the exciton binding energy. In addition, an associated quantum mechanical effect arises, known as the exchange interaction. Both of these effects are intimately tied to the size of the exciton because that parameter dictates the average electron–electron separation.

The exchange interaction plays a central role in determining the ordering of electronic spin states in any material. An exchange interaction raises the energy of each singlet state and lowers the energy of the three corresponding triplet spin eigenstates. Such a distinction between singlet and multiplet states is a traditional backbone to atomic and molecular spectroscopy. Similarly, the exchange interaction mixes the single-excitation configurations in a QD, evident as a splitting between the bright and dark exciton states at the band edge[47,51,52] (Fig. 6). QDs with wurtzite structure are complicated by crystal field effects, otherwise the exciton fine structure splitting is determined by the same kind of exchange interaction that splits singlet and triplet states of molecules. As an exciton is spatially compressed by confinement, exchange interactions are increased, as evidenced by larger singlet–triplet splitting. The distinct manifold of states provides an opportunity in spectroscopy to learn about the excited states and electronic structure of nanoscale systems. The challenge, however, for materials such as QDs is that the spectrum is obscured by inhomogeneous line broadening, so specialized techniques are required to probe these states.

The singlet–triplet splittings for the conjugated oligomers[53–56] are closely comparable to those for the aromatic[57] molecule and polyene series[58]. The long-range character of the exchange interaction is evident in the attenuation of the exchange interaction in the polyenes, polyacenes and conjugated oligomers plotted in Fig. 6. If only one-centre exchange integrals contributed to the singlet–triplet splitting, it would diminish as the inverse of the length L of these quasi-one-dimensional excitons (that is the approximation often used for solid state materials). It turns out that the splitting falls off approximately as $L^{1/3}$ for each of these materials, which is a result of long-range exchange interactions, clearly significant over 3 nm. The polyacenes and conjugated oligomers are 'less linear' than the polyenes: in other words the exciton is about twice as wide (compare the chemical structures). That extra width is manifest in the delocalization of the exciton, through molecular orbital coefficients, hence reducing the magnitude of the exchange interaction for any length by approximately a factor of two compared with the corresponding polyene. One can see, therefore, how the exchange interaction (singlet–triplet splitting) is a good measure of exciton size, regardless of the chemical structure of a material. It is in that way that one should think about the CNT data[59] plotted in Fig. 6. The exciton is very long, but an increase in CNT diameter further delocalizes the exciton around the tube, reducing the exchange interaction.

We see that nanoscale materials provide a fascinating intermediate ground between molecular and bulk materials. Interesting spectroscopic properties of excitons, which are often minuscule in bulk materials, are greatly accentuated in nanometre-sized materials and molecules. For example, the electron–hole exchange interaction is of the order of millielectronvolts in bulk semiconductor materials, but is increased 1,000-fold in QDs[51,60–65]. That is still several hundred times smaller than exchange interactions in organic materials that are <10 nm in size (Fig. 6). However, once an exciton in an organic material is as large as that in a CNT, the exchange interaction is reduced to tens of millielectronvolts.

As yet, a number of aspects of the photophysics of CNTs have not been satisfactorily resolved. For example, the reasons underlying the low quantum yield (~10^{-4}) and the fluorescence decay behaviour (~10–200 ps) of the band edge exciton in CNTs are unclear[66], but a closely located triplet state is considered to be a possible contributor. The singlet–triplet splitting for the CNTs is calculated to be of the order of 10 to 40 meV, and is predicted[59], according

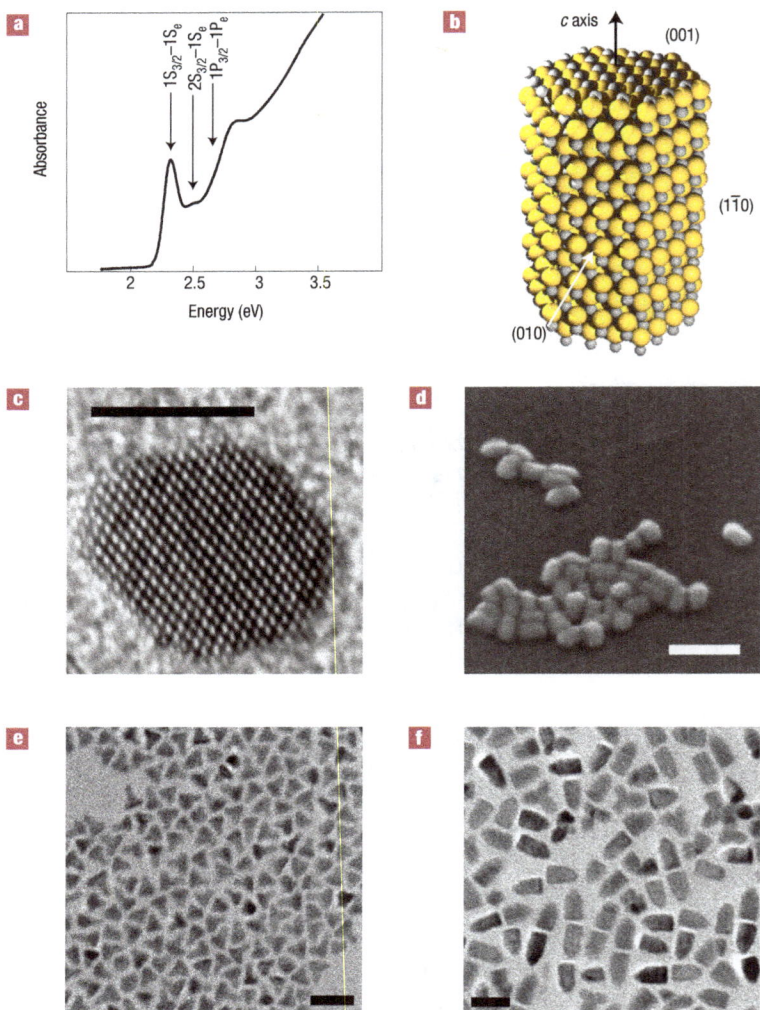

Figure 4 Colloidal CdSe quantum dots. a, An absorption spectrum recorded at 4 K of CdSe QDs with 4.0 nm mean diameter. Hole and electron states associated with excitons seen as the prominent absorption features are labelled according to the assignments of ref. 42. That notation[34] specifies a hole or electron state as $n*(l, l + 2)F$, where $n*$ labels ground and excited states, l is the minimum orbital quantum number and $F = l + j$ is the total angular momentum quantum number (orbital plus spin), containing F_z from $-F$ to $+F$. Usually just the major contributor out of l or $l + 2$, or alternatively just l, is written as S, P, D and so on. That quantum number is associated with the envelope function, usually modelled mathematically as a spherical Bessel function $j_l(x)$. The quantum number j is associated with the Bloch function. For example, the upper valence band of a zinc blende structure has $j = 3/2$ and $j_z = -3/2, -1/2, 1/2, 3/2$. **b**, A structural model of a small, idealized wurtzite nanorod with some crystallographic planes and the direction of the c-axis indicated. **c**, A high-resolution transmission electron micrograph of an individual CdSe wurtzite nanorod, looking down the nanocrystal c-axis. Scale bar corresponds to 5.2 nm. Reprinted from ref. 119. Copyright (2006), with permission from Elsevier. **d**, A scanning-mode electron micrograph of CdSe nanorods. Scale bar corresponds to 50 nm. **e,f**, By varying growth conditions a myriad of other shapes can be produced by selective growth of different crystal faces. The scale bars each correspond to 20 nm. Note that the ends of each rod are not identical, as there is no centre of inversion or reflection symmetry in a plane normal to the c axis of the crystal, so asymmetric rods like the bullet-shaped colloids seen in **f** can be grown (P. S. Nair, K. P. Fritz & G. D. Scholes, unpublished data).

to calculations based on a tight-binding Hamiltonian, to depend on the inverse of the tube diameter, as shown in Fig. 6. Considering the trend for the other

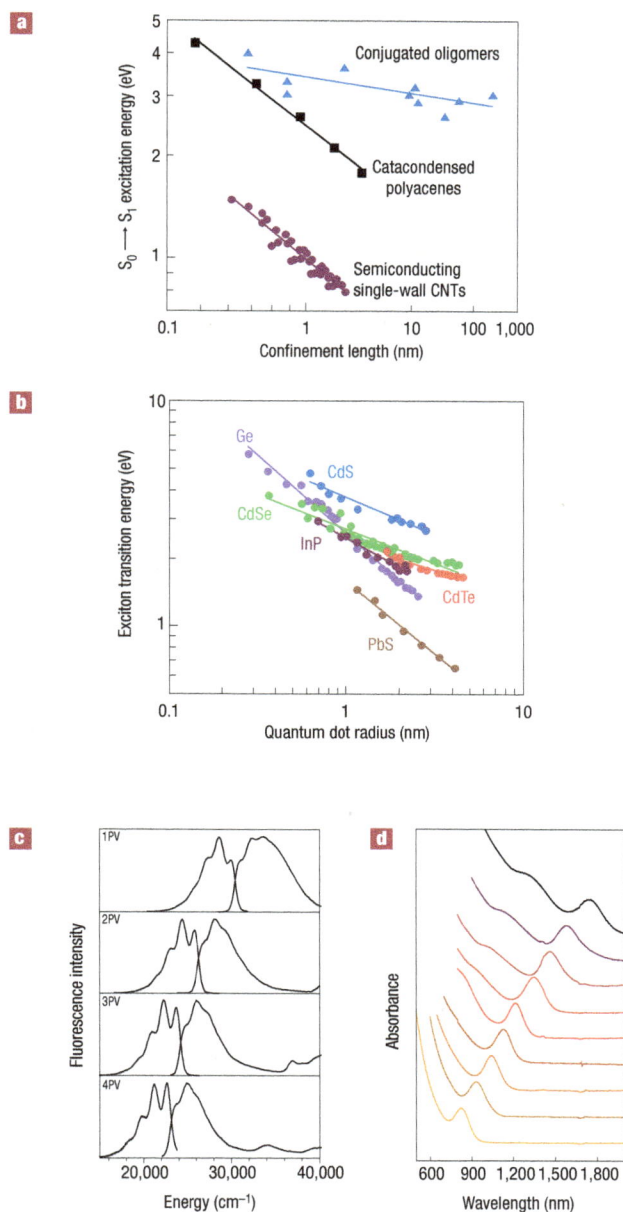

Figure 5 Size dependence of exciton transition energies. Log–log plots of excitation energy versus confinement length (dimension) of the ground and first excited states. For CNTs, the confinement dimension is taken to be the tube diameter. For the other organic materials it is the approximate conjugation length from end to end. **a**, Organic materials. Squares show data for the series of aromatic molecules[45] naphthalene, anthracene, naphthacene, pentacene and hexacene. Triangles indicate data for some oligomers that serve as models for conjugated polymers: methyl-substituted ladder-type poly(p-phenylene) (MeLPPP), poly(2,7-(9,9-bis(2-ethylhexyl)fluorene)) (PF2/6)[53,54], oligofluorenes[56] and oligothiophenes[55]. Data were typically recorded for samples at 80 K. The filled circles show CNT data recorded at ambient temperature[30]. **b**, Size-tuning of excitation energies for various QD materials: Ge[49], calculated using *ab initio* density functional theory using ΔSCF, CdSe, CdS and CdTe (ref. 48), measured at 250 K, PbS (ref. 46), measured at 293 K, and InP (refs 47,61) measured at 10 K. **c**, The size-scaling of conjugated molecules, for example phenylenevinylene (PV) oligomers[120], derives from size-limited delocalization of the molecular orbitals. Absorption and fluorescence spectra are shown as a function of the number of PV repeat units. **d**, In semiconductor QDs, the small size of the nanocrystal compared with the exciton in the bulk material confines the exciton. A result is size-dependent optical properties, as shown for the absorption spectra of PbS quantum dots[46].

organic materials, long-range exchange interactions may modify that predicted size-dependence and magnitude. Experimental observation of this splitting seems to be an important next step.

The basic size-scaling of the dark state/bright state splitting for the range of QDs follows a similar trend to that seen for the organic materials, but the magnitudes of the splitting are substantially smaller. The main reason for that is that the high dielectric constant of QDs attenuates the long-range electron–electron repulsion compared with these organic materials. It is striking that the exchange interactions over the range of materials plotted in Fig. 6 are similar in magnitude for each QD size. Confinement of the QD exciton is in three dimensions, therefore the exchange is predicted to scale as R^{-3}, where R is the QD radius. That is the case for some of the data plotted here, which include a range of experimental observations and good quality calculations. A scaling of $\sim R^{-3}$ is found for bright–dark splitting of CdSe, CdTe and GaAs QDs. Silicon QDs are calculated to show $R^{-2.6}$, whereas the splitting for AgI, InAs and InP scales as R^{-2}, or slightly more weakly. Future work may expose more details of the spectroscopic signatures of the exciton from beneath the inhomogeneously broadened absorption band. An example was reported recently where the dynamics associated with flipping between two QD fine-structure states was measured[67].

A concluding message is that nanoscale materials of vastly different structure and composition can be directly compared through spectroscopic signatures of exciton delocalization. Any effect that is determined by a two-electron integral involving orbitals that constitute the exciton is appropriate, although the exchange interaction seems the obvious choice. Finally we note that in this section we have focused on strongly delocalized excitons. The discussion becomes more complex for consideration of Frenkel excitons.

EXCITON BINDING ENERGY

The exciton binding energy in a quantum confined system can be taken to be the energy difference between the exciton transition energy (optical gap) and the electronic bandgap, as shown in Box 1. The electronic bandgap can be written as the difference between the ionization potential and electron affinity, assuming no structural relaxation of the material or its surroundings[68]. This Coulomb energy, thought of as electron–hole attraction, assumes marked significance in nanoscale materials. In this section we discuss the binding energy of excitons in various materials, aiming to understand how to think about confinement in terms of the prevalence of excitons as elementary excitations. We give further examples that illustrate how the size and binding energy of an exciton can be dynamic, changing considerably from optical absorption to equilibration prior to photoluminescence emission.

In a bulk semiconductor material with high dielectric constant, the exciton binding energy is typically small: 27 meV for CdS, 15 meV for CdSe, 5.1 meV for InP and 4.9 meV for GaAs. Excitons are therefore not a distinctive feature in the spectroscopy of such materials at room temperature, making them ideal for photovoltaic applications. On the other hand, in molecular materials, the electron–hole Coulomb interaction is substantial—a few electronvolts. In nanoscale materials we find a middle ground where exciton binding energies are significant in magnitude— that is, excitons are important—and they are size-tunable. As a consequence of the size-dependence of the binding energy, delocalization versus localization of the exciton complicates matters. For example, exciton self-trapping (see the following section) in organic materials, such as conjugated polymers, leads to a strongly bound (hundreds of millielectronvolts) exciton. Hence the need to dope conjugated polymer solar cells with electron acceptors, such as other polymers[69], fullerenes[70] or quantum dots[71], to promote charge separation. In a strongly delocalized exciton, free carriers could be photogenerated directly upon excitation of conjugated polymers. The extent to which that occurs is still a matter for debate.

The exciton binding energy in QDs of radii $R \approx 1–2$ nm is in the range 200–50 meV, scaling approximately as $1/R$, according to the size-dependence of the electron–hole Coulomb interaction[60]. A key difference between these inorganic materials and organic materials is dielectric constant. Dielectric constant is central in determining the exciton binding energy because it shields electron–electron repulsions, or equivalently electron–hole interactions. Interestingly, it has been reported that dielectric constant diminishes with size for QDs, further increasing the size effect of the exciton binding energy[60].

The importance of electron–electron interactions in determining the properties of excitons in CNTs has been revealed by quantum chemical calculations[72–74]. The most striking result is the large binding energy, found to be between ~0.3 and 1.0 eV, depending on the chiral index and computational methodology. The remarkable finding is that they are of the order of half the bandgap. Zhou and Mazumdar[74] report the scaling of the binding energy for various CNT types, indicating that the binding energy is greater for smaller tube diameters, as expected. Compelling experimental evidence for a significant exciton binding energy in CNTs has been recently communicated[75,76]. In each case, experimental studies suggested that the exciton binding energy in (8,3) CNTs is ~0.4 eV. Evidence for the existence of excitons in CNTs can also be inferred from a report of phonon sidebands in fluorescence data[32]. In the work reported by Wang et al.[75] the energy of the one-photon allowed exciton absorption was compared with the two-photon allowed absorption into a second exciton state and nearby continuum absorption. That direct observation of the exciton transition and the continuum provides a clean measure of the exciton binding energy. In the report of Ma et al.[76] the one-exciton state and continuum were detected using ultrafast transient absorption spectroscopy, similarly enabling the exciton binding energy to be deduced.

The situation becomes less clear when the material is disordered or when there is strong coupling between electronic and nuclear degrees of freedom. Organic materials typically exhibit structural disorder and tend to contain torsional

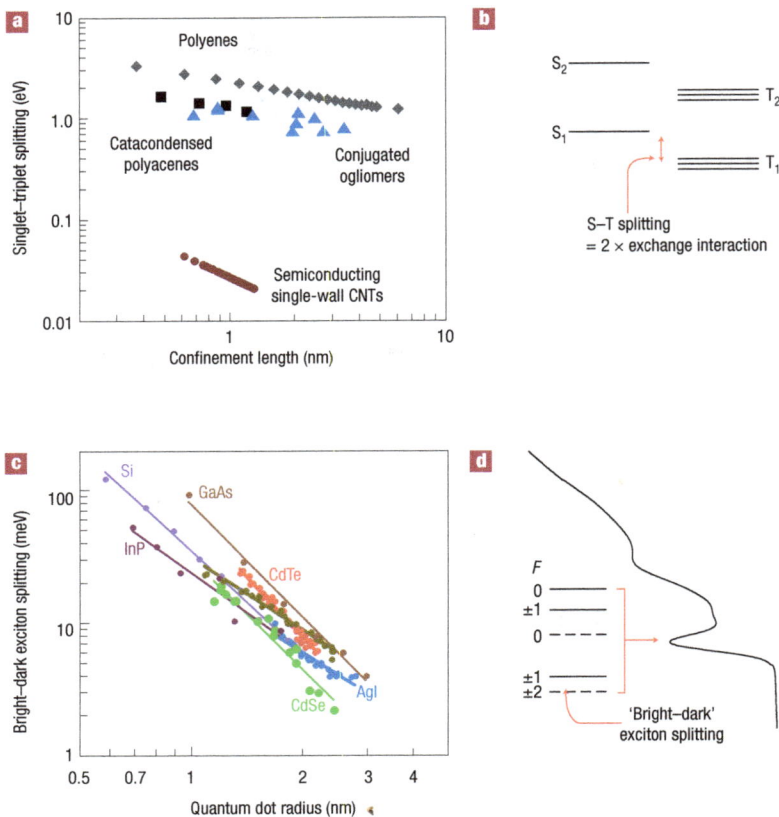

Figure 6 Size dependence of exchange interactions. Log–log plots of singlet–triplet splitting energy versus size. As in Fig. 5, for CNTs the confinement dimension is taken to be the tube diameter; for other organic materials it is the approximate conjugation length from end to end. **a**,The organic materials from Fig. 5, together with calculated singlet–triplet splitting for a series of all-trans polyenes. The diamonds show the results of these time-dependent density functional calculations of polyene excited states[58]. **b**, A typical Jablonskii diagram, indicating that the singlet–triplet splitting is determined by an exchange interaction. That diagram is compared to that for a CdSe quantum dot, shown in **d**, where one considers a manifold of eight states—the exciton fine structure—rather than just four. The splitting between the lowest bright and dark exciton fine structure states is analogous to the singlet–triplet splitting of the organic materials when crystal field splitting is negligible compared with the electron–hole exchange interaction. That is usually not the case for materials with a wurtzite structure, such as CdSe. **c**, Data showing reported bright–dark exciton splitting for a selection of quantum dots: Si (ref. 62; tight-binding calculations), InP (ref. 47; fluorescence line narrowing, measured at 10 K), GaAs (ref. 60; pseudopotential calculations of rectangular nanocrystals), CdTe (ref. 64; fluorescence line narrowing, 10 K, of colloids in glass), AgI (ref. 65; fluorescence line narrowing, 2 K), InAs (ref. 63; fluorescence line narrowing, 10 K), and CdSe(ref. 51; fluorescence line narrowing, 10 K). These data are a combination of experimental results and calculations.

motions and related low-frequency vibrations that couple to the exciton. Such a combination of static disorder and electron–vibration coupling disrupts delocalization. That is the trade-off for the great structural diversity in both repeat units and extended bonding patterns found in organic supramolecular assemblies and macromolecules. These ideas have been elucidated in detail for J-aggregates[77], which are well known linear assemblies of dye molecules.

It has taken many years to develop an acceptable understanding of excitons in conjugated polymers that

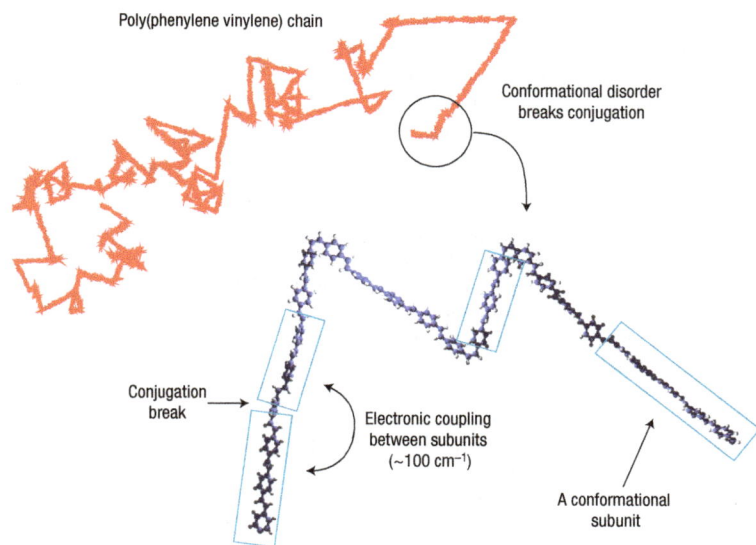

Figure 7 Conformational subunits of conjugated polymers. A typical 'defect cylinder' conformation of a conjugated polymer, poly(phenylene vinylene), is shown in red. Zooming in on a small part of the chain, it is evident that the π-bonds extending over the phenylene vinylene repeat units are frequently disrupted by rotations of sufficient angle to break the conjugation. The result is a series of conformational subunits, outlined conceptually by boxes. This ensemble of chromophores makes up the zero-order picture of the conjugated chain exciton. Long-range Coulomb interactions couple the conformational subunits, as indicated. After light-absorption, energy migration transfers the excitation to a narrow distribution of conformational subunits, from which fluorescence emission is observed.

explains molecular and semiconducting aspects of the materials in the context of their highly disordered structure[20]. The binding energy and precise nature of the excited state (exciton versus polaron-exciton) are still a matter for discussion. We note that coupling to phonons, exciton self-trapping and energy migration can influence the observed binding energy. Electronic structure calculations in conjunction with experiment have proven central for elucidating the properties of excitons in conjugated polymers[78]. Moses et al.[79] report that the exciton binding energy can be as low as 60 meV for excitons in poly(phenylene vinylene). The report of Silva et al.[80] on poly-6,6′,12,12′-tetraalkyl-2,8-indenofluorene and that of Arkhipov and Bässler[81] should be considered in conjunction with those results. Many other experimental and theoretical reports conclude that the exciton binding energy is 0.3 to 0.6 eV for a range of conjugated polymers. Films differ from isolated polymers owing to interchain interaction effects. The reader is referred to the article by Brédas et al.[82] for detailed discussion of the binding energy in conjugated polymers. An important message of that paper is that an interplay between electron–electron and electron–lattice interactions is important in determining a description of conjugated polymer excitons, and hence their binding energies.

The binding energy of the exciton in the B850 ring of LH2 is apparently lowered by 90 meV relative to that in monomeric Bchl, as can be inferred from the energy difference between the 800-nm and 850-nm absorption bands (Fig. 2). Note that B850 is a Frenkel exciton, so the electron and hole are closely associated on each molecule in the aggregate. If the molecules in B850 were sufficiently closely spaced, then charge transfer between Bchl molecules would become important— that is, electron–hole separation. The consequence would be a substantially increased exciton binding energy for essentially the same-sized aggregate.

DISORDER AND COUPLING TO VIBRATIONS

The purely 'electronic' models for describing excitons are modified by the coupling between the exciton and the bath of nuclear motions. That coupling is manifest in spectroscopy as line broadening, Stokes shift and vibronic structure in absorption and photoluminescence spectra[83,84]. The aim of this section is to categorize the important elements of line broadening in nanomaterials, highlight their size dependences and provide some specific examples that give an overall picture of the diverse consequences of disorder and exciton–bath coupling.

Any model for the eigenstates and energies of excitons contains the energy at each repeat unit (molecule, unit cell, atom), known as the site energy, and information on how those repeat units are coupled. In the most general model, each site can accommodate an excitation, an electron or a hole[85]. Clearly both the composition of the eigenstates and the corresponding energies will be affected by energetic disorder in the site energies or couplings. Such disorder is caused by random fluctuations in the positions of atoms[84]. The timescale of those fluctuations is important in delimiting the effects as static (on the timescale of the measurement) or fluctuating. The former gives rise to inhomogeneous line broadening, the latter to homogeneous line broadening[86]. Static disorder is seen as a temperature-independent gaussian line shape in absorption or photoluminescence, whereas homogeneous line broadening can play an important role in trapping an exciton on a fast timescale[87] (femtoseconds to picoseconds).

The interaction between excitons and nuclear motions can introduce time-dependent confinement effects, known as exciton self-trapping, where the exciton becomes trapped in a lattice deformation[87]. Those nuclear motions are contributed by intramolecular vibrations and the environment. Exciton self-trapping occurs through a local collective structural change, connected to random nuclear fluctuations by the fluctuation-dissipation theorem. The resultant exciton is also called a polaron-exciton. Hence the size and electronic make-up of an organic exciton can change markedly on short timescales after photoexcitation (tens of femtoseconds)[13,88]. That is understood as the tendency of molecules to change their equilibrium geometry in the excited state compared with the ground state, which is observed, together with solvation[86], as spectral diffusion. The associated reorganization energy is equal to half the Stokes shift. A very small change in the equilibrium coordinates of many vibrational modes can inspire surprisingly rapid localization of the exciton. In other words, exciton self-trapping is induced by the amplitude of fluctuations as well as their characteristic timescales. The precise excited state dynamics are dictated by competition between the delocalizing effect of electronic coupling and the localizing influence of electron–phonon coupling.

The LH2 antenna protein (Fig. 2) from purple bacteria provides a clear example of the implications of static disorder at each site in a Frenkel exciton. To understand the optical properties, one must consider the importance of static disorder in the transition energy gaps from ground state to excited state of each Bchl molecule and/or in the electronic coupling[89–91]. The result is that each LH2 complex in an ensemble has a unique absorption spectrum, as shown strikingly in the low-temperature fluorescence excitation spectra of individual LH2 complexes reported by van Oijen et al.[92]. Implications are that ensemble measurements must be carefully unravelled to expose the properties and dynamics characteristic of individual light-harvesting complexes. Although static disorder is important, particularly in determining the temperature dependence of spectroscopic data, dynamic electron–vibration coupling also needs to be considered to predict electronic energy transfer rates and to understand ultrafast exciton dynamics. Some insight into the role of electron–vibration coupling can be gathered from the observation that the entire ring of Bchl molecules needs to be considered in a theoretical model to predict correctly the absorption or circular dichroism spectra of B850[17]. Nonetheless, excitation is self-trapped in just tens of femtoseconds so that the exciton size determined by experiments such as pump-probe[93] and time-resolved fluorescence spectroscopy[94] is small enough to cover just small part of the ring, typically three to four Bchl molecules only. The size of an exciton therefore can depend on the timescale on which it is probed.

Conjugated polymers provide another good example where static disorder is significant. By considering single-molecule studies and simulations of the polymer chain it was established that MEH-PPV adopts a fairly ordered defect cylinder chain conformation[95], illustrated in Fig. 7. That work provides an important 'low-resolution' picture of chain disorder. The π-electron system of conjugated polymers is disrupted on shorter length scales by conformational disorder[20]. It has been thought that this disorder impacts the spectroscopy by disrupting the π-electron conjugation along the backbone of the polymer, giving a chain of chromophores known as conformational subunits (Fig. 7). Thus, the optical properties of conjugated polymers are highly dependent on the polymer chain conformation[20,22,96]. Single-molecule studies and site-selective fluorescence studies together provide clear evidence that energy absorbed by a polymer chain is efficiently funnelled to low-energy chromophores along the chain by electronic energy transfer[97].

Recent work has aimed to understand better how to think about conformational subunits and how they define conjugated polymer excitons. Simple considerations of electronic coupling (for example, dipole–dipole coupling) suggest that conformational subunits should be electronically coupled, meaning that a delocalized exciton might be formed by photoexcitation[98–100]. Careful quantum

chemical investigations furthermore reveal that it is difficult to define conformational subunits with respect to torsional angles: there is no clear point when the strong interactions that depend on π–π overlap 'switch off' and, even when weak, those interactions are important for defining the nature of the excited states. Beenken and Pullerits[101] conclude that conformational subunits arise concomitantly with dynamic localization of the excitation[102] (exciton self-trapping). An interesting contrast is contributed by the highly ordered polydiacetylene chains investigated by Schott and co-workers[103]. Those molecules resemble an organic semiconductor quantum wire in many respects, therefore sustaining a highly delocalized one-dimensional exciton. Key properties that differentiate these chains from many other conjugated polymers are their structural order on all length scales, their rigid, crystalline environment, and weaker electron–phonon coupling.

QD excitons couple relatively weakly to nuclear motions and the homogeneous line broadening is consequently narrow compared with organic materials, suggesting possible applications of QDs as single-photon sources and 'qubits' for quantum computation. That is perhaps because QDs, and indeed also CNTs, are considerably more rigid and ordered than the flexible structures of organic materials. There are two characteristic types of vibrations in a QD[104]. The higher-frequency mode is that of the longitudinal optical (LO) phonon, which has a frequency that is typically very similar to the frequency known for the corresponding bulk material. For CdSe the LO-phonon mode frequency is 207 cm^{-1}. The radial breathing modes of CNTs[105] are conceptually similar kinds of vibrations. The frequencies of radial breathing modes, however, scale with the CNT diameter d (in nm), according to the empirical relation[106] $\omega = (214.4/d) + 18.7$ cm^{-1}. Coupling of the exciton to vibrations with frequencies greater than thermal energies leads to phonon sidebands in frequency-resolved spectroscopies and quantum beats (oscillations) in ultrafast time domain experiments.

The principal vibrations that contribute to QD line shape are acoustic phonons. In the bulk these are analogous to sound waves travelling through the crystal. Owing to the small size of QDs, the acoustic phonon modes are quantized torsional and spheroidal modes—the motions of an elastic sphere that leave its volume constant. The frequencies of these phonon modes lie in the range 5 to 40 cm^{-1}, depending on the QD size. The exciton–phonon coupling scales as $1/R^2$, where R is the QD radius[104]. Models for phonons in bulk crystals assume the displacements of the modes are small. That makes sense for an infinite-sized crystal, in which coupling between many oscillators limits the amplitude of motion. On the other hand, for nanoscale materials such as QDs, exciton–phonon coupling may be better described by considering aspects of theories developed for molecules.

DYNAMICS OF EXCITONS

The dynamics of excitons have been of interest for many years. Here we give only a glimpse of some areas topical in nanoscience that characterize each particular material. Photosynthetic proteins provide wonderful model materials for the measurement and investigation of electronic energy transfer. Energy migration in conjugated polymers has been extensively studied, particularly with respect to elucidating the essential differences between dilute solutions and films[97]. Exciton dynamics in CNTs have recently been exposed, so this area is only now emerging. Relaxation dynamics—radiationless as well as radiative relaxation—of highly excited excitons and multiexcitons are a topic of continued interest in QDs.

As introduced in Fig. 2, the light-harvesting complexes in photosynthesis gather incident solar energy and transfer that excitation energy to reaction centres using a series of electronic energy transfer steps. The B850 unit plays an important role in the intra-complex energy funnel for purple bacteria. Examination of B800–B850 resonance energy transfer revealed that excitons are unique in the way they trap excitation energy[17,18]. A surprisingly large spatial and spectral cross-section for donor quenching is achieved through the physical arrangement of the acceptor molecules, together with the fact that the full extent of the exciton density of states can be harnessed to optimize spectral overlap between donor 'emission' and acceptor 'absorption', regardless of the distribution of oscillator strength. This second point is rather subtle, as it is not captured by the dipole approximation for donor–acceptor electronic couplings. The way that a delocalized exciton acts as a unique acceptor in resonance energy transfer may be understood in terms of coupling between exciton transition densities[16-18].

Exciton states are used in photosynthetic light harvesting and redistribution because they substantially increase the absorption cross-section for incident light without the cost of dissipative processes that might result from diffusive trapping of excitation energy absorbed by a weakly coupled aggregate of molecules. Once captured by the exciton manifold, excitation is rapidly trapped in the lowest state. Recent work reveals that such exciton trapping and relaxation is not a simple cascading relaxation process, but can actually involve spatial redistribution, a new picture that was discovered through two-dimensional photon echo spectroscopy[107]. This method opens the possibility of new insights into mechanisms underlying ultrafast dynamics of excitons, because information about electronic couplings between exciton states can be captured in addition to populations.

Similar to conjugated polymers, CNT excitons may migrate along the length of the tube. A possible implication of energy migration is that trap and defect sites along the tube may quench excitation, contributing to the low fluorescence yield. At higher excitation intensities, two excitons on the same CNT may encounter each other and annihilate[108]. Thus multiexciton states are typically short-lived (<100 fs) in organic materials because excitons resident on proximate but spatially distinct parts of a macromolecule or aggregate annihilate efficaciously by a resonance energy transfer mechanism, forming a localized higher electronic state that relaxes rapidly through internal conversion[109,110].

Translational dynamics of excitons are obviously not important in QDs. However, radiationless relaxation processes from higher excited states in QDs to the lowest exciton state have been widely investigated[36]. The electron and hole levels in small QDs, typical of colloidal QDs, are widely spaced compared with the LO-phonon frequency. It was therefore expected that relaxation of the electron and hole, and hence exciton, would be impeded: the phonon bottleneck. In fact, it has been found that the relaxation is very fast in CdSe[111,112]. For CdSe the valence band consists of more closely spaced levels than the conduction band. Bearing this in mind, the fast exciton relaxation has been explained on the basis of an Auger-like process whereby the highly excited electron interacts with the hole through an electronic (coulombic) interaction that enables the electron to relax to its lowest level by scattering the hole deeper into the valence band[111]. Because the valence band has a high density of states, phonon-mediated relaxation of the hole is possible. However, that rationale does not satisfactorily explain the fast exciton relaxation observed in IV–VI QDs, for which the electron and hole states are thought to be approximately equally spaced. Guyot-Sionnest et al. have further suggested that surface ligands may play a role in assisting exciton relaxation[113]. Whether independent consideration of the electron and hole relaxations is a good approximation may also be questioned given the significant Coulomb and exchange energies that bind the exciton.

As opposed to most organic materials, QDs can support multiple exciton populations, biexcitons, triexcitons and so on, for times significant compared with the exciton lifetime. Biexcitons in CdSe QDs have recombination times of the order of tens of picoseconds, depending on the QD size[3,112]. Nozik proposed that the unique properties of multiexcitons in QDs could be harnessed to increase the energy conversion efficiency of solar cells[114]. Recently that prediction was confirmed for PbS and PbSe QDs by the striking observations of band-edge exciton yields of >300% when the pump energy is tuned to greater than three times the bandgap[115,116].

OUTLOOK

We emphasize the great strides that have been gained from elucidating size-dependent properties within each class of materials we have reviewed. What are the signatures of nanoscale exciton spectroscopy? As an example we looked at singlet–triplet splitting and its relationship to exciton delocalization (or confinement). The properties of quantum dots stood out from those of the other materials examined; the relatively small magnitude of properties such as the exchange interaction highlights the role of dielectric constant in screening long-range interactions. It is clear, however, that size plays an important role here also, as can be concluded by comparing the singlet–triplet splitting of the majority of organic materials surveyed in Fig. 6 with those calculated for CNTs.

We conclude that an opportunity for nanoscience is that solid-state materials acquire molecular-like spectroscopic features: discrete transitions, spin state manifolds and so on, allowing for a more detailed examination of the excited states. That, in turn, will provide inspiration for theory, which in the past has focused mostly on bandgaps. Future work will examine more closely how spectroscopy can reveal properties unique to excitons in nanoscale systems. It is likely that our understanding of excited electronic states characteristic of nanomaterials will evolve in a similar manner to our present understanding of the excited states of molecules. Standard approximations and concepts relevant to bulk materials may need to be reconsidered in light of the significant exciton binding energies, exchange interactions, and exciton–vibration couplings apparent in the spectroscopy of nanoscale excitons. Perhaps we need to look back before we go small?

doi: 10.1038/nmat1710

References

1. Friend, R. H. *et al.* Electroluminescence in conjugated polymers. *Nature* **397**, 121–128 (1999).
2. Gregg, B. A. Excitonic solar cells. *J. Phys. Chem. B* **107**, 4688–4698 (2003).
3. Klimov, V. I. *et al.* Optical gain and stimulated emission in nanocrystal quantum dots. *Science* **290**, 314–317 (2000).
4. Koch, S. W., Kira, M., Khitrova, G. & Gibbs, H. M. Semiconductor excitons in a new light. *Nature Mater.* **5**, 523–531 (2006).
5. Elliott, R. J. in *Polarons and Excitons* (eds Kuper, C. G. & Whitfield, G. D.) 269–293 (Plenum, New York, 1962).
6. Knox, R. S. in *Collective Excitations in Solids* (ed. Bartolo, B. D.) 183–245 (Plenum, New York, 1981).
7. McRae, E. G. & Kasha, M. in *Physical Processes in Radiation Biology*, 23–42 (Academic, New York, 1964).
8. Slater, J. C. & Shockley, W. Optical absorption by the alkali halides. *Phys. Rev.* **50**, 705–719 (1936).
9. Hoffmann, R. How chemistry meets physics in the solid state. *Angew. Chem. Int. Edn. Engl.* **26**, 846–878 (1987).
10. Cohen, M. L. Nanotubes, nanoscience, and nanotechnology. *Mater. Sci. Eng. C.* **15**, 1–11 (2001).
11. Cohen, M. L. The theory of real materials. *Annu. Rev. Mater. Sci.* **30**, 1–26 (2000).
12. Tretiak, S. & Mukamel, S. Density matrix analysis and simulation of electronic excitations in conjugated and aggregated molecules. *Chem. Rev.* **102**, 3171–3212 (2002).
13. Brédas, J. L., Beljonne, D., Coropceanu, V. & Cornil, J. Charge-transfer and energy-transfer processes in pi-conjugated oligomers and polymers: a molecular picture. *Chem. Rev.* **104**, 4971–5003 (2004).
14. Sundström, V., Pullerits, T. & van Grondelle, R. Photosynthetic light-harvesting: reconciling dynamics and structure of purple bacterial LH2 reveals function of photosynthetic unit. *J. Phys. Chem. B* **103**, 2327–2346 (1999).
15. Scholes, G. D. & Fleming, G. R. Energy transfer and photosynthetic light harvesting. *Adv. Chem. Phys.* **132**, 57–130 (2005).
16. Scholes, G. D. Long-range resonance energy transfer in molecular systems. *Annu. Rev. Phys. Chem.* **54**, 57–87 (2003).
17. Doust, A. B., Wilk, K. E., Curmi, P. M. G. & Scholes, G. D. The photophysics of cryptophyte light harvesting. *J. Photochem. Photobiol. A. Chem.* (in the press); doi:10.1016/j.jphotochem.2006.06.006.
18. Scholes, G. D., Jordanides, X. J. & Fleming, G. R. Adapting the Förster theory of energy transfer for modeling dynamics in aggregated molecular assemblies. *J. Phys. Chem. B* **105**, 1640–1651 (2001).
19. Rothberg, L. J. *et al.* Photophysics of phenylenevinylene polymers. *Synth. Met.* **80**, 41–58 (1996).
20. Bässler, H. & Schweitzer, B. Site-selective fluorescence spectroscopy of conjugated polymers and oligomers. *Acc. Chem. Res.* **32**, 173–182 (1999).
21. Sariciftci, N. S. (ed.) *Primary Excitations in Conjugated Polymers: Molecular Exciton versus Semiconductor Band Model* (World Scientific, Singapore, 1997).
22. Barbara, P. F., Gesquiere, A. J., Park, S.-J. & Lee, Y. J. Single-molecule spectroscopy of conjugated polymers. *Acc. Chem. Res.* **38**, 602–610 (2005).
23. Köhler, A. *et al.* Charge separation in localized and delocalized electronic states in polymeric semiconductors. *Nature* **392**, 903–906 (1998).
24. Morteani, A. C. *et al.* Barrier-free electron–hole capture in polymer blend heterojunction light-emitting diodes. *Adv. Mater.* **15**, 1708–1712 (2003).
25. Morteani, A. C., Sreearunothai, P., Hertz, L. M., Friend, R. H. & Silva, C. Exciton regeneration at polymeric semiconductor heterojunctions. *Phys. Rev. Lett.* **92**, 247402 (2004).
26. Knibbe, H., Rȯllig, K., Schäfer, F. P. & Weller, A. Charge-transfer complex and solvent-shared ion pair in fluorescence quenching. *J. Chem. Phys.* **47**, 1184–1185 (1967).
27. Terrones, M., Hsu, W. K., Kroto, H. W. & Walton, D. R. M. Nanotubes: a revolution in materials science and electronics. *Top. Curr. Chem.* **199**, 189–234 (1999).
28. O'Connell, M. J. *et al.* Bandgap fluorescence from individual single-walled carbon nanotubes. *Science* **297**, 593–596 (2002).
29. Ando, T. Excitons in carbon nanotubes. *J. Phys. Soc. Japan* **66**, 1066–1073 (1997).
30. Bachilo, S. M. *et al.* Structure-assigned optical spectra of single-walled carbon nanotubes. *Science* **298**, 2361–2366 (2002).
31. Reich, S., Thomsen, C. & Maultzsch, J. Carbon nanotubes: basic concepts and physical properties (Wiley, New York, 2004).
32. Jones, M. *et al.* Analysis of photoluminescence from solubilized single-walled carbon nanotubes. *Phys. Rev. B* **71**, 115426 (2005).
33. Didraga, C. & Knoester, J. Exchange narrowing in circular and cylindrical molecular aggregates: degenerate versus nondegenerate states. *Chem. Phys.* **275**, 307–318 (2002).
34. Gaponenko, S. V. *Optical Properties of Semiconductor Nanocrystals* (Cambridge Univ. Press, Cambridge, 1998).
35. Alivisatos, A. P. Perspectives on the physical chemistry of semiconductor nanocrystals. *J. Phys. Chem.* **100**, 13226–13239 (1996).
36. Burda, C., Chen, X. B., Narayanan, R. & El-Sayed, M. A. Chemistry and properties of nanocrystals of different shapes. *Chem. Rev.* **105**, 1025–1102 (2005).
37. Weller, H. Colloidal semiconductor Q-particles: chemistry in the transition region between solid state and molecules. *Angew. Chem. Int. Edn Engl.* **32**, 41–53 (1993).
38. Klimov, V. I. (ed.) Semiconductor and metal nanocrystals: synthesis and electronic and optical properties (Marcel Dekker, New York, 2004).
39. Bimberg, D., Grundman, M. & Ledentsov, N. N. *Quantum Dot Heterostructures* (Wiley, Chichester, 1999).
40. Yoffe, A. D. Semiconductor quantum dots and related systems: electronic, optical, luminescence and related properties of low dimensional systems. *Adv. Phys.* **50**, 1–208 (2001).
41. Efros, A. L. & Efros, A. L. Interband absorption of light in a semiconductor sphere. *Sov. Phys. Semicond.* **16**, 772–775 (1982).
42. Ekimov, A. I. *et al.* Absorption and intensity-dependent photoluminescence measurement on CdSe quantum dots: assignment of the first electronic transitions. *J. Opt. Soc. Am. B* **10**, 100 (1993).
43. Rabani, E., Hetenyi, B., Berne, B. J. & Brus, L. E. Electronic properties of CdSe nanocrystals in the absence and presence of a dielectric medium. *J. Chem. Phys.* **110**, 5355–5369 (1999).
44. Yin, Y. & Alivisatos, A. P. Colloidal nanocrystal synthesis and the organic-inorganic interface. *Nature* **437**, 664–670 (2005).
45. Klevens, H. B. & Platt, J. R. Spectral resemblances of cata-condensed hydrocarbons. *J. Chem. Phys.* **17**, 470–481 (1949).
46. Hines, M. A. & Scholes, G. D. Colloidal PbS nanocrystals with size-tunable near-infrared emission: observation of post-synthesis self-narrowing of the particle size distribution. *Adv. Mater.* **15**, 1844–1849 (2003).
47. Micic, O. I. *et al.* Size-dependent spectroscopy of InP quantum dots. *J. Phys. Chem. B* **101**, 4904 (1997).
48. Yu, W. W., Qu, L. H., Guo, W. Z. & Peng, X. G. Experimental determination of the extinction coefficient of CdTe, CdSe, and CdS nanocrystals. *Chem. Mater.* **15**, 2854–2860 (2003).
49. Nesher, G., Kronik, L. & Chelikowsky, J. R. Ab initio absorption spectra of Ge nanocrystals. *Phys. Rev. B* **71**, 35344 (2005).
50. Pariser, R. & Parr, R. G. A semi-empirical theory of the electronic spectra and electronic structure of complex unsaturated molecules. I. *J. Chem. Phys.* **21**, 466–471 (1953).
51. Efros, A. L. *et al.* Band-edge exciton in quantum dots of semiconductors with a degenerate valence band: dark and bright exciton states. *Phys. Rev. B* **54**, 4843–4856 (1996).
52. Leung, K., Pokrant, S. & Whaley, K. B. Exciton fine structure in CdSe nanoclusters. *Phys. Rev. B* **57**, 12291–12301 (1998).
53. Hertel, D. *et al.* Phosphorescence in conjugated poly(para-phenylene)-derivatives. *Adv. Mater.* **13**, 65–70 (2001).
54. Köhler, A. & Beljonne, D. The singlet–triplet exchange energy in conjugated polymers. *Adv. Funct. Mater.* **14**, 11–18 (2004).
55. Wasserberg, D., Marsal, P., Meskers, S. C. J., Janssen, R. A. J. & Beljonne, D. Phosphorescence and triplet state energies of oligothiophenes. *J. Phys. Chem. B* **109**, 4410–4415 (2005).
56. Chi, C. Y., Im, C. & Wegner, G. Lifetime determination of fluorescence and phosphorescence of a series of oligofluorenes. *J. Chem. Phys.* **124**, 24907 (2006).
57. McGlynn, S. P., Azumi, T. & Kinoshita, M. *Molecular Spectroscopy of the Triplet State* (Prentice-Hall, Englewood Cliffs, NJ, 1969).
58. Catalán, J. & de Paz, J. L. G. On the ordering of the first two excited electronic states in all-trans linear polyenes. *J. Chem. Phys.* **120**, 1864–1872 (2004).
59. Perebeinos, V., Tersoff, J. & Avouris, P. Radiative lifetime of excitons in carbon nanotubes. *Nanoletters* **5**, 2495–2499 (2005).
60. Franceschetti, A. & Zunger, A. Direct pseudopotential calculation of exciton coulomb and exchange energies in semiconductor quantum dots. *Phys. Rev. Lett.* **78**, 915–918 (1997).

61. Fu, H. X. & Zunger, A. InP quantum dots: electronic structure, surface effects, and the redshifted emission. *Phys. Rev. B* **56**, 1496–1508 (1997).

62. Leung, K. & Whaley, K. B. Electron-hole interactions in silicon nanocrystals. *Phys. Rev. B* **56**, 7455–7468 (1997).

63. Banin, U., Lee, J. C., Guzelian, A. A., Kadavanich, A. V. & Alivisatos, A. P. Exchange interaction in InAs nanocrystal quantum dots. *Superlattices Microstruct.* **22**, 559–567 (1997).

64. Lavallard, P. *et al.* Exchange interaction and acoustical phonon modes in CdTe nanocrystals. *Solid State Commun.* **127**, 439–442 (2003).

65. Gogolin, O., Mshvelidze, G., Tsitishvili, E., Djanelidze, R. & Klingshirn, C. Exchange interaction in argentum iodide nanocrystals. *J. Lumin.* **102**, 414–416 (2003).

66. Wang, F., Dukovic, G., Brus, L. E. & Heinz, T. F. Time-resolved fluorescence of carbon nanotubes and its implication for radiative lifetimes. *Phys. Rev. Lett.* **92**, 177401 (2004).

67. Huxter, V. M., Kovalevskij, V. & Scholes, G. D. Dynamics within the exciton fine structure of colloidal CdSe quantum dots. *J. Phys. Chem. B* **109**, 20060–20063 (2005).

68. Yaron, D., Moore, E. E., Shuai, Z. & Brédas, J. L. Comparison of density matrix renormalization group calculations with electron–hole models of exciton binding in conjugated polymers. *J. Chem. Phys.* **108**, 7451–7458 (1998).

69. Halls, J. J. M. *et al.* Efficient photodiodes from interpenetrating polymer networks. *Nature* **376**, 498–500 (1995).

70. Sariciftci, N. S., Smilowitz, L., Heeger, A. J. & Wudl, F. Photoinduced electron transfer from a conducting polymer to buckminsterfullerene. *Science* **258**, 1474–1476 (1992).

71. Huynh, W. U., Dittmer, J. J. & Alivisatos, A. P. Hybrid nanorod–polymer solar cells. *Science* **295**, 2425–2427 (2002).

72. Spataru, C. D., Ismail-Beigi, S., Benedict, L. X. & Louie, S. G. Excitonic effects and optical spectra of single-walled carbon nanotubes. *Phys. Rev. Lett.* **92**, 77402 (2004).

73. Chang, E., Bussi, G., Ruini, A. & Molinari, E. Excitons in carbon nanotubes: an ab initio symmetry-based approach. *Phys. Rev. Lett.* **92**, 196401 (2004).

74. Zhou, H. & Mazumdar, S. Electron–electron interaction effects on the optical excitations of semiconducting single-walled carbon nanotubes. *Phys. Rev. Lett.* **93**, 157402 (2004).

75. Wang, F., Dukovic, G., Brus, L. E. & Heinz, T. F. The optical resonances in carbon nanotubes arise from excitons. *Science* **308**, 838–841 (2005).

76. Ma, Y.-Z., Valkunas, L., Bachilo, S. M. & Fleming, G. R. Exciton binding energy in semiconducting single-walled carbon nanotubes. *J. Phys. Chem. B* **109**, 15671–15674 (2005).

77. Heijs, D. J., Malyshev, V. A. & Knoester, J. Decoherence of excitons in multichromophoric systems: thermal line broadening and destruction of superradiant emission. *Phys. Rev. Lett.* **95**, 177402 (2005).

78. Brédas, J.-L., Cornil, J., Beljonne, D., dos Santos, D. A. & Shuai, Z. Excited-state electronic structure of conjugated oligomers and polymers: a quantum-chemical approach to optical phenomena. *Acc. Chem. Res.* **32**, 267–276 (1999).

79. Moses, D., Wang, J., Heeger, A. J., Kirova, N. & Brazovski, S. Singlet exciton binding energy in poly(phenylene vinylene). *Proc. Natl Acad. Sci.* **98**, 13496–13500 (2001).

80. Silva, C. *et al.* Exciton and polaron dynamics in a step-ladder polymeric semiconductor: the influence of interchain order. *J. Phys.: Condens. Matter* **14**, 9803–9824 (2002).

81. Arkhipov, V. I. & Bässler, H. Exciton dissociation and charge photogeneration in pristine and doped conjugated polymers. *Phys. Status Solidi A* **201**, 1152–1187 (2004).

82. Brédas, J. L., Cornil, J. & Heeger, A. J. The exciton binding energy in luminescent conjugated polymers. *Adv. Mater.* **8**, 447–452 (1996).

83. Lax, M. The Franck–Condon principle and its application to crystals. *J. Chem. Phys.* **20**, 1752–1760 (1952).

84. Sumi, H. Exciton–lattice interaction and the line shape of exciton absorption in molecular crystals. *J. Chem. Phys.* **67**, 2943–2954 (1977).

85. Scholes, G. D., Harcourt, R. D. & Ghiggino, K. P. Rate expressions for excitation transfer. III. An ab initio study of electronic factors in excitation transfer and exciton resonance interactions. *J. Chem. Phys.* **102**, 9574–9581 (1995).

86. Fleming, G. R., Passino, S. A. & Nagasawa, Y. The interaction of solutes with their environments. *Phil. Trans. R. Soc. London A* **356**, 389–404 (1998).

87. Nakajima, S. *The Physics of Elementary Excitations* (Springer, New York, 1980).

88. Franco, I. & Tretiak, S. Electron-vibrational dynamics of photoexcited polyfluorenes. *J. Am. Chem. Soc.* **126**, 12130–12140 (2004).

89. Jimenez, R., Dikshit, S. N., Bradforth, S. E. & Fleming, G. R. Electronic excitation transfer in the LH2 complex of *Rhodobacter sphaeroides*. *J. Phys. Chem.* **100**, 6825–6834 (1996).

90. Alden, R. G. *et al.* Calculations of spectroscopic properties of the LH2 bacteriochlorophyll-protein antenna complex from *Rhodopseudomonas acidophila*. *J. Phys. Chem. B* **101**, 4667–4680 (1997).

91. Jang, S., Dempster, S. E. & Silbey, R. J. Characterization of the static disorder in the B850 band of LH2. *J. Phys. Chem. B* **105**, 6655–6665 (2001).

92. Van Oijen, A. M., Ketelaars, M., Köhler, J., Aartsma, T. J. & Schmidt, J. Unraveling the electronic structure of individual photosynthetic pigment-protein complexes. *Science* **285**, 400–402 (1999).

93. Pullerits, T., Chachisvilis, M. & Sundström, V. Exciton delocalization length in the B850 antenna of *Rhodobacter sphaeroides*. *J. Phys. Chem.* **100**, 10787–10792 (1996).

94. Monshouwer, R., Abrahamsson, M., van Mourik, F. & van Grondelle, R. Superradiance and exciton delocalization in bacterial photosynthetic light-harvesting systems. *J. Phys. Chem. B* **101**, 7241–7248 (1997).

95. Hu, D. *et al.* Collapse of stiff conjugated polymers with chemical defects into ordered, cylindrical conformations. *Nature* **405**, 1030–1033 (2000).

96. Schindler, F. *et al.* Counting chromophores in conjugated polymers. *Angew. Chem. Int. Edn Engl.* **44**, 1520–1525 (2005).

97. Beljonne, D. *et al.* Interchain vs. intrachain energy transfer in acceptor-capped conjugated polymers. *Proc. Natl Acad. Sci.* **99**, 10982–10987 (2002).

98. Chang, R., Hayashi, M., Lin, S. H., Hsu, J.-H. & Fann, W. S. Ultrafast dynamics of excitations in conjugated polymers: a spectroscopic study. *J. Chem. Phys.* **115**, 4339–4348 (2001).

99. Wang, X., Dykstra, T. E. & Scholes, G. D. Photon-echo studies of collective absorption and dynamic localization of excitation in conjugated polymers and oligomers. *Phys. Rev. B* **71**, 45203 (2005).

100. Ruseckas, A. *et al.* Ultrafast depolarization of the fluorescence in a conjugated polymer. *Phys. Rev. B* **72**, 115214 (2005).

101. Beenken, W. J. D. & Pullerits, T. Spectroscopic units in conjugated polymers: a quantum chemically founded concept? *J. Phys. Chem. B* **108**, 6164–6169 (2004).

102. Tretiak, S., Saxena, A., Martin, R. L. & Bishop, A. R. Conformational dynamics of photoexcited conjugated molecules. *Phys. Rev. Lett.* **89**, 97402 (2002). **[author: this is your ref 105, which was cited out of order]**

103. Lécuiller, R. *et al.* Fluorescence yield and lifetime of isolated polydiacetylene chains: evidence for a one-dimensional exciton band in a conjugated polymer. *Phys. Rev. B* **66**, 125205 (2002).

104. Takagahara, T. Electron–phonon interactions and excitonic dephasing in semiconductor nanocrystals. *Phys. Rev. Lett.* **71**, 3577–3580 (1993).

105. Plentz, F., Ribeiro, H. B., Jorio, A., Strano, M. S. & Pimenta, M. A. Direct experimental evidence of exciton–phonon bound states in carbon nanotubes. *Phys. Rev. Lett.* **95**, 247401 (2005).

106. Telg, H., Maultzsch, J., Reich, S., Hennrich, F. & Thomsen, C. Chirality distribution and transition energies of carbon nanotubes. *Phys. Rev. Lett.* **93**, 177401 (2004).

107. Brixner, T. *et al.* Two-dimensional spectroscopy of electronic couplings in photosynthesis. *Nature* **434**, 625–628 (2005).

108. Valkunas, L., Ma, Y.-Z. & Fleming, G. R. Exciton–exciton annihilation in single-walled carbon nanotubes. *Phys. Rev. B* **73**, 115432 (2006).

109. De Schryver, F. C. *et al.* Energy dissipation in multichromophoric single dendrimers. *Acc. Chem. Res.* **38**, 514–522 (2005).

110. Sternlicht, H., Nieman, G. C. & Robinson, G. W. Triplet–triplet annihilation and delayed fluorescence in molecular aggregates. *J. Chem. Phys.* **38**, 1326–1335 (1963).

111. Klimov, V. I., McBranch, D. W., Leatherdale, C. A. & Bawendi, M. G. Electron and hole relaxation pathways in semiconductor quantum dots. *Phys. Rev. B* **60**, 13740–13749 (1999).

112. Klimov, V. I., Mikhailovsky, A. A., McBranch, D. W., Leatherdale, C. A. & Bawendi, M. G. Quantization of multiparticle Auger rates in semiconductor quantum dots. *Science* **287**, 1011–1013 (2000).

113. Guyot-Sionnest, P., Wehrenberg, B. & Yu, D. Intraband relaxation in CdSe nanocrystals and the strong influence of the surface ligands. *J. Chem. Phys.* **123**, 74709 (2005).

114. Nozik, A. J. Quantum dot solar cells. *Physica E* **14**, 115–120 (2002).

115. Schaller, R. D. & Klimov, V. I. High efficiency carrier multiplication in PbSe nanocrystals: Implications for solar energy conversion. *Phys. Rev. Lett.* **92**, 186601 (2004).

116. Ellingson, R. J. *et al.* Highly efficient multiple exciton generation in colloidal PbSe and PbS quantum dots. *Nanoletters* **5**, 865–871 (2005).

117. Scholes, G. D., Ghiggino, K. P., Oliver, A. M. & Paddon-Row, M. N. Through-space and through-bond effects on exciton interactions in rigidly linked dinaphthyl molecules. *J. Am. Chem. Soc.* **115**, 4345–4349 (1993).

118. Curutchet, C. & Mennucci, B. Toward a molecular scale interpretation of excitation energy transfer in solvated bichromophoric systems. *J. Am. Chem. Soc.* **127**, 16733–16744 (2005).

119. Fritz, K. P. *et al.* Structural characterization of CdSe nanorods. *J. Cryst. Growth* **293**, 203–208 (2006).

120. Gierschner, J., Mack, H.-G., Lüer, L. & Oelkrug, D. Fluorescence and absorption spectra of oligophenylenevinylenes: vibronic coupling, band shapes, and solvatochromism. *J. Chem. Phys.* **116**, 8596–8609 (2002).

Acknowledgements

G.D.S. thanks G. J. Wilson for introducing him to excitons. He acknowledges funding from the Natural Sciences and Engineering Research Council of Canada and the A. P. Sloan Foundation. G.R. acknowledges funding through the Photochemistry and Radiation research program of the US Department of Energy, Office of Science, Office of Basic Energy Sciences, Chemical Sciences, Geosciences and Biosciences (DoE contract DE-AC36-99G010337).
Correspondence should be addressed to G.D.S.

Photoelectrochemical cells

Michael Grätzel

Institute of Photonics and Interfaces, Swiss Federal Institute of Technology, CH-1015, Lausanne, Switzerland (e-mail michael.graetzel@epfl.ch)

Until now, photovoltaics — the conversion of sunlight to electrical power — has been dominated by solid-state junction devices, often made of silicon. But this dominance is now being challenged by the emergence of a new generation of photovoltaic cells, based, for example, on nanocrystalline materials and conducting polymer films. These offer the prospect of cheap fabrication together with other attractive features, such as flexibility. The phenomenal recent progress in fabricating and characterizing nanocrystalline materials has opened up whole new vistas of opportunity. Contrary to expectation, some of the new devices have strikingly high conversion efficiencies, which compete with those of conventional devices. Here I look into the historical background, and present status and development prospects for this new generation of photoelectrochemical cells.

Ever since the French scientist Edmond Becquerel[1] discovered the photoelectric effect, researchers and engineers have been infatuated with the idea of converting light into electric power or chemical fuels. Their common dream is to capture the energy that is freely available from sunlight and turn it into the valuable and strategically important asset that is electric power, or use it to generate fuels such as hydrogen. Photovoltaics takes advantage of the fact that photons falling on a semiconductor can create

electron–hole pairs, and at a junction between two different materials, this effect can set up an electric potential difference across the interface. So far, the science of solar cells has been dominated by devices in which the junction is between inorganic solid-state materials, usually doped forms of crystalline or amorphous silicon, and profiting from the experience and material availability resulting from the semiconductor industry. Recently, we have seen more use of devices made from compound semiconductors — the III/V compounds for high-efficiency aerospace components and the copper–indium–sulphide/selenide materials for thin-film, low-cost terrestrial cells. But the dominance of the field by inorganic solid-state junction devices faces new challenges in the coming years. Increasingly, there is an awareness of the possible advantages of nanocrystalline and conducting polymer devices, for example, which are relatively cheap to fabricate (the expensive and energy-intensive high-temperature and high-vacuum processes needed for the traditional devices can be avoided), can be used on flexible substrates, and can be shaped or tinted to suit domestic devices or architectural or decorative applications. It is now even possible to depart completely from the classical solid-state junction device, by replacing the phase in contact with the semiconductor by an electrolyte (liquid, gel or organic solid), thereby forming a photoelectrochemical device.

The development of these new types of solar cells is promoted by increasing public awareness that the Earth's oil reserves could run out during this century. As the energy needs of the planet are likely to double within the next 50 years, the stage is set for a major energy shortage, unless renewable energy can cover the substantial deficit left by fossil fuels. Public concern has been heightened by the disastrous environmental pollution arising from all-too-frequent oil spills and the frightening climatic consequences of the greenhouse effect caused by fossil fuel combustion. Fortunately the supply of energy from the Sun to the Earth is gigantic: 3×10^{24} joules a year, or about 10,000 times more than the global population currently consumes. In other words, covering 0.1% of the Earth's surface with solar cells with an efficiency of 10% would satisfy our present needs. But to tap into this huge energy reservoir remains an enormous challenge.

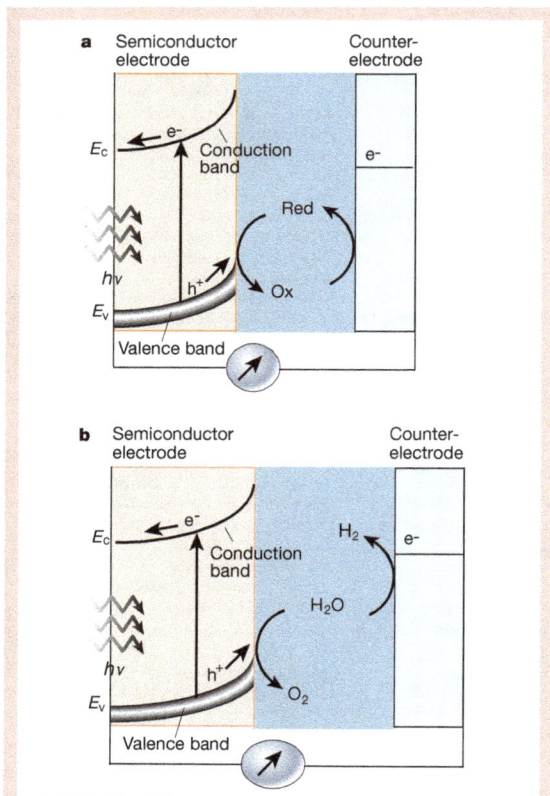

Figure 1 Principle of operation of photoelectrochemical cells based on *n*-type semiconductors. **a**, Regenerative-type cell producing electric current from sunlight; **b**, a cell that generates a chemical fuel, hydrogen, through the photo-cleavage of water.

Historical background

Becquerel's pioneering photoelectric experiments in 1839 were done with liquid not solid-state devices — a fact that is often ignored. His research, in which illumination of

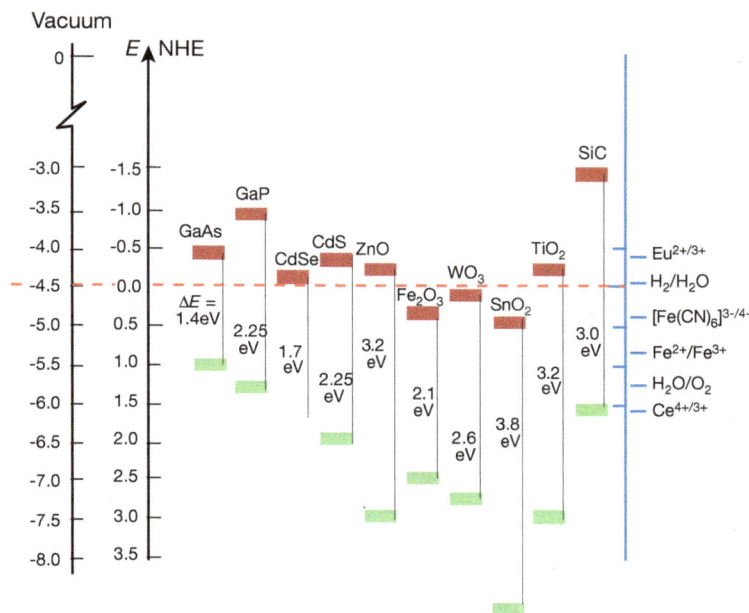

Figure 2 Band positions of several semiconductors in contact with aqueous electrolyte at pH 1. The lower edge of the conduction band (red colour) and upper edge of the valence band (green colour) are presented along with the band gap in electron volts. The energy scale is indicated in electron volts using either the normal hydrogen electrode (NHE) or the vacuum level as a reference. Note that the ordinate presents internal and not free energy. The free energy of an electron–hole pair is smaller than the band gap energy due to the translational entropy of the electrons and holes in the conduction and valence band, respectively. On the right side the standard potentials of several redox couples are presented against the standard hydrogen electrode potential.

solutions containing a metal halide salt produced a current between two platinum electrodes immersed in the electrolyte, was motivated by photography. Daguerre had made the first photographic images in 1837, and Fox Talbot followed with the silver halide process in 1839. Although the art of formulating emulsions only became a science with the theoretical analysis of the process by Gurney and Mott[2] in 1938, there was constant empirical progress in the sensitivity of photographic films. Initially, films were particularly insensitive to mid-spectrum and red light. This is now recognized as being due to the semiconductor nature of the silver halide grains: they have a band gap (a gap in the allowed electronic energy levels) which ranges from 2.7 to 3.2 electron volts (eV) and negligible light absorption at wavelengths longer than 460 nm. Vogel's discovery[3] in 1883 that silver halide emulsions could be sensitized by adding a dye extended the photosensitivity to longer wavelengths. Four years later, the concept of dye enhancement was carried over by Moser[4] from photography to photoelectrochemical cells using the dye erythrosine on silver halide electrodes. This parallel between photography and photoelectrochemistry comes as a surprise to many chemists[5]. That the same dyes were particularly effective for both processes was recognized by Namba and Hishiki at the 1964 International Conference on Photosensitization of Solids in Chicago, which was a seminal event in the history of sensitization[6]. It was also recognized that the dye should be adsorbed on the semiconductor electrodes in a closely packed monolayer for maximum efficiency[7]. At this stage it was still debated whether sensitization occurred by transfer of electrons or of energy from the dye to the semiconductor[8]. Subsequent studies, notably by Hauffe[9], Tributsch and Gerischer[10], showed electron transfer to be the prevalent mechanism both for photographic and for photoelectrochemical sensitization processes.

Photosynthetic and regenerative cells

The foundation of modern photoelectrochemistry, marking its change from a mere support of photography to a thriving research direction on its own, was laid down by the work of Brattain and Garret[11] and subsequently Gerischer[12] who undertook the first detailed electrochemical and photoelectrochemical studies of the semiconductor–electrolyte interface. Research on photoelectrochemical cells went through a

frantic period after the oil crisis in 1973, which stimulated a worldwide quest for alternative energy sources. Within a few years well over a thousand publications appeared (see ref. 13 for a list).

Investigations focused on two types of cells whose principle of operation is shown in Fig. 1. The first type is the regenerative cell, which converts light to electric power leaving no net chemical change behind. Photons of energy exceeding that of the band gap generate electron–hole pairs, which are separated by the electric field present in the space-charge layer (see Box 1 for an explanation). The negative charge carriers move through the bulk of the semiconductor to the current collector and the external circuit. The positive holes are driven to the surface where they are scavenged by the reduced form of the redox relay molecule (R), oxidizing it: $h^+ + R \rightarrow O$. The oxidized form O is reduced back to R by the electrons that re-enter the cell from the external circuit. Much of the work on regenerative cells has focused on electron-doped (n-type) II/VI or III/V semiconductors using electrolytes based on sulphide/polysulphide, vanadium(II)/vanadium(III) or I_2/I^- redox couples. Conversion efficiencies of up to 19.6% have been reported for multijunction regenerative cells[14].

The second type, photosynthetic cells, operate on a similar principle except that there are two redox systems: one reacting with the holes at the surface of the semiconductor electrode and the second reacting with the electrons entering the counter-electrode. In the example shown, water is oxidized to oxygen at the semiconductor photoanode and reduced to hydrogen at the cathode. The overall reaction is the cleavage of water by sunlight. Titanium dioxide has been the favoured semiconductor for these studies, following its use by Fujishima and Honda for water photolysis[15]. Unfortunately, because of its large band gap (3–3.2 eV, as shown in Fig. 2), TiO_2 absorbs only the ultraviolet part of the solar emission and so has low conversion efficiencies. Numerous attempts to shift the spectral response of TiO_2 into the visible, or to develop alternative oxides affording water cleavage by visible light, have so far failed.

In view of these prolonged efforts, disillusionment has grown about the prospects of photoelectrochemical cells being able to give rise to competitive photovoltaic devices, as those semiconductors with band gaps narrow enough for efficient absorption of visible light

Box 1
The semiconductor–electrolyte interface

When a semiconductor is placed in contact with an electrolyte, electric current initially flows across the junction until electronic equilibrium is reached, where the Fermi energy of the electrons in the solid (E_F) is equal to the redox potential of the electrolyte (E_{redox}), as shown in the figure. The transfer of electric charge produces a region on each side of the junction where the charge distribution differs from the bulk material, and this is known as the space-charge layer. On the electrolyte side, this corresponds to the familiar electrolytic double layer, that is, the compact (Helmholtz) layer followed by the diffuse (Gouy–Chapman) layer. On the semiconductor side of the junction the nature of the band bending depends on the position of the Fermi level in the solid. If the Fermi level of the electrode is equal to the flat band potential, there is no excess charge on either side of the junction and the bands are flat. If electrons accumulate at the semiconductor side one obtains an accumulation layer. If, however, they deplete from the solid into the solution, a depletion layer is formed, leaving behind a positive excess charge formed by immobile ionized donor states. Finally, electron depletion can go so far that their concentration at the interface falls below the intrinsic level. As a consequence, the semiconductor is p-type at the surface and n-type in the bulk, corresponding to an inversion layer. The illustration in the figure refers to n-type materials where electrons are the mobile charge carriers. For p-type semiconductors, analogous considerations apply. Positive holes are the mobile charge carriers and the immobile negatively charged states of the acceptor dopant form the excess space charge within the depletion layer.

The flat band potential is a very useful quantity in photoelectrochemistry as it facilitates location of the energetic position of the valence and conduction band edge of a given semiconductor material. It is obtained by measuring the capacity of the semiconductor–electrolyte junction. The semiconductor is subjected to reverse bias — that is, a voltage is applied to increase the potential step across the junction — and the differential capacity is determined as a function of the applied potential, V. The space-charge capacity of the semiconductor (C_{sc}) is in series with that of the Helmholtz layer (C_H) present at the electrolyte side of the interface. In the depletion regime the condition $C_H > C_{sc}$ applies, so the measured capacity is that of the space-charge layer. This depends on the applied bias voltage according to the Mott–Schottky equation: $1/(C_{sc})^2 = 2 (\Delta\phi_{sc}RT/F)/(\varepsilon_o\varepsilon^1 N)$, where $\Delta\phi_{sc} = V - V_{fb}$ is the voltage drop in the space-charge layer, R is the gas constant, F the Faraday number, ε the dielectric constant of the semiconductor, ε_o the permittivity of vacuum and ^1N the ionized donor dopant concentration. A plot of the square of the reciprocal capacity against the applied voltage gives a straight line and this is extrapolated to $1/(C_{sc})^2 = 0$ to derive the flat band potential V_{fb}.

Flat band potentials have been determined for a large number of materials[49] and some representative examples are shown in Fig. 2. Apart from the type of semiconductor they depend on the nature and composition of the electrolyte. In aqueous solution the flat band potentials of most oxide semiconductors shifts by 0.059 V when the pH is changed by one unit. This is a consequence of the fact that protons are potential-determining ions for these solids.

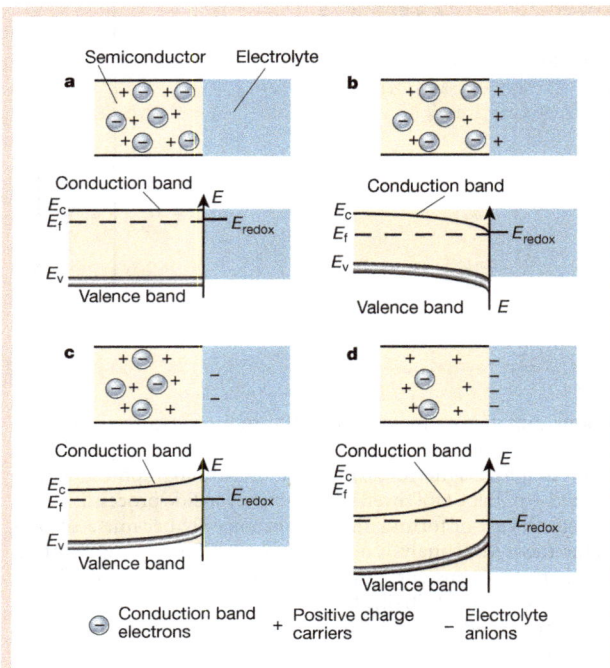

Box 1 Figure Schematic showing the electronic energy levels at the interface between an n-type semiconductor and an electrolyte containing a redox couple. The four cases indicated are: **a**, flat band potential, where no space-charge layer exists in the semiconductor; **b**, accumulation layer, where excess electrons have been injected into the solid producing a downward bending of the conduction and valence band towards the interface; **c**, depletion layer, where electrons have moved from the semiconductor to the electrolyte, producing an upward bending of the bands; and **d**, inversion layer where the electrons have been depleted below their intrinsic level, enhancing the upward band bending and rendering the semiconductor p-type at the surface.

are unstable against photocorrosion. The width of the band gap is a measure of the chemical bond strength. Semiconductors stable under illumination, typically oxides of metals such as titanium or niobium, therefore have a wide band gap, an absorption edge towards the ultraviolet and consequently an insensitivity to the visible spectrum.

The resolution of this dilemma came in the separation of the optical absorption and charge-generating functions, using an electron transfer sensitizer absorbing in the visible to inject charge carriers across the semiconductor–electrolyte junction into a substrate with a wide band gap, and therefore stable. Figure 3 shows the operational principle of such a device.

Nanocrystalline junctions and interpenetrating networks
The need for dye-sensitized solar cells to absorb far more of the incident light was the driving force for the development of mesoscopic semiconductor materials[16] — minutely structured materials with an enormous internal surface area — which have attracted great attention during recent years. Mesoporous oxide films are made up of arrays of tiny crystals measuring a few nanometres across. Oxides such as TiO_2, ZnO, SnO_2 and Nb_2O_5, or chalcogenides such as $CdSe$, are the preferred compounds. These are interconnected to allow electronic conduction to take place. Between the particles are mesoscopic pores filled with a semiconducting or a conducting medium, such as a p-type semiconductor, a polymer, a hole transmitter or an electrolyte. The net result is a junction of extremely large contact area between two interpenetrating, individually continuous networks. Particularly intriguing is the ease with which charge carriers percolate across the mesoscopic particle network, making the huge internal surface area electronically addressable. Charge transport in such mesoporous

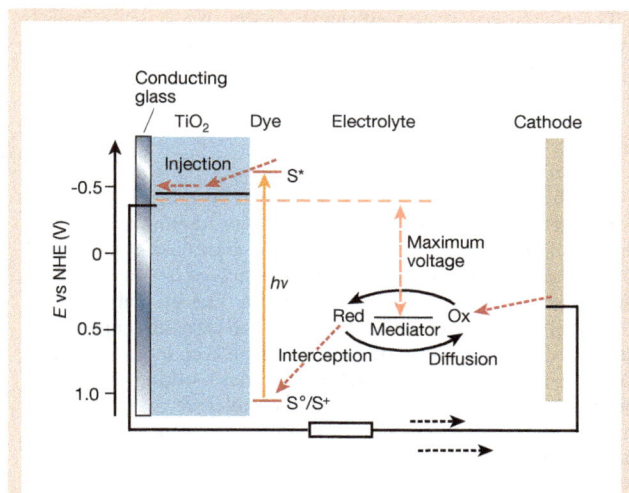

Figure 3 Schematic of operation of the dye-sensitized electrochemical photovoltaic cell. The photoanode, made of a mesoporous dye-sensitized semiconductor, receives electrons from the photo-excited dye which is thereby oxidized, and which in turn oxidizes the mediator, a redox species dissolved in the electrolyte. The mediator is regenerated by reduction at the cathode by the electrons circulated through the external circuit. Figure courtesy of P. Bonhôte/EPFL-LPI.

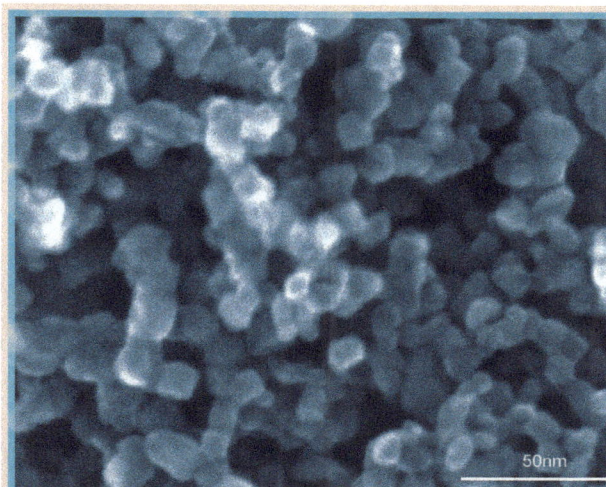

Figure 4 Scanning electron micrograph of the surface of a mesoporous anatase film prepared from a hydrothermally processed TiO_2 colloid. The exposed surface planes have mainly {101} orientation.

systems is under intense investigation today[17,18] and is best described by a random walk model[19].

The semiconductor structure, typically 10 μm thick and with a porosity of 50%, has a surface area available for dye chemisorption over a thousand times that of a flat, unstructured electrode of the same size. If the dye is chemisorbed as a monomolecular layer, enough can be retained on a given area of electrode to provide absorption of essentially all the incident light. Figure 4 shows an electron micrograph of a nanocrystalline TiO_2 film with a grain size in the range of 10–80 nm. (For details of the hydrothermal deposition procedure, starting with hydrolysis of a titanium isopropoxide precursor and terminating with screen printing and firing of the semiconductor layer on a conductive transparent substrate, see ref. 20.)

The nanostructuring of the semiconductor introduces profound changes in its photoelectrochemical properties. Of great importance is the fact that a depletion layer (see Box 1) cannot be formed in the solid — the particles are simply too small. The voltage drop within the nanocrystals remains small under reverse bias, typically a few millivolts. As a consequence there is no significant local electric field present to assist in the separation of photogenerated electron–hole pairs. The photoresponse of the electrode is determined by the rate of reaction of the positive and negative charge carriers with the redox couple present in the electrolyte. If the transfer of electrons to the electrolyte is faster than that of holes, then a cathodic photocurrent will flow, like in a p-type semiconductor/liquid junction. In contrast, if hole transfer to the electrolyte is faster, then anodic photocurrent will flow, as in n-type semiconductor photoelectrochemical cells. Striking confirmation of the importance of these kinetic effects came with the demonstration[21] that the same nanocrystalline film could show alternatively n- or p-type behaviour, depending on the nature of the hole or electron scavenger present in the electrolyte phase. This came as a great surprise to a field where the traditional thinking was to link the photoresponse to formation of a charge-depletion layer at the semiconductor–electrolyte interface.

What, then, is the true origin of the photovoltage in dye-sensitized solar cells? In the conventional picture, the photovoltage of photoelectrochemical cells does not exceed the potential drop in the space-charge layer (Box 1 Figure). But nanocrystalline cells can develop photovoltages close to 1 V even though the junction potential is in the millivolt range. It has been suggested that a built-in potential difference at the back contact of the nanocrystalline film with the conducting glass is responsible for the observed photovoltage[22]. Other evidence[23] suggests that under illumination, electron injection from the sensitizer increases the electron concentration in the nanocrystalline electrode, raising the Fermi level of the oxide and thus shifting its potential[24]. From recent electrical impedance studies[25], it seems that both changes — the potential drop across the back contact and the Fermi level shift of the TiO_2 nanoparticles — contribute to the photovoltage of dye-sensitized solar cells.

Accumulation layers (part b of Box 1 Figure) can be produced in the nanocrystals under forward bias when majority carriers are injected, rendering the film highly conductive. Under reverse bias the carriers are withdrawn, turning it into an insulator. Thus, by changing the applied potential, the film can be switched back and forth from a conducting to an insulating state. Space-charge limitation of the current (arising from limitation of the density of charge carriers because they are repelled by each other's electric field) is not observed as the injected majority carriers are efficiently screened by the electrolyte present in the pores surrounding the nanoparticles. The factors controlling the rate of charge carrier percolation across the nanocrystalline film are under intense scrutiny[17]. A technique known as intensity-modulated impedance spectroscopy has proved to be an elegant and powerful tool[25,26] for addressing these and other important questions related to the characteristic time constants for charge carrier transport and reaction dynamics.

An interesting feature specific to nanocrystalline electrodes is the appearance of quantum confinement effects. These appear when the films are made up of small quantum dots, such as 8-nm-sized CdTe particles[27]. Such layers have a larger band gap than the bulk material, the band edge position being shifted with respect to the positions indicated in Fig. 2 for macroscopic materials. The conduction band redox potential is lowered and that of the valence band is increased. As a consequence, electrons and holes can perform reduction and oxidation reactions that cannot proceed on bulk semiconductors.

The astounding photoelectrochemical performance of nanocrystalline semiconductor junctions is illustrated in Fig. 5 where we

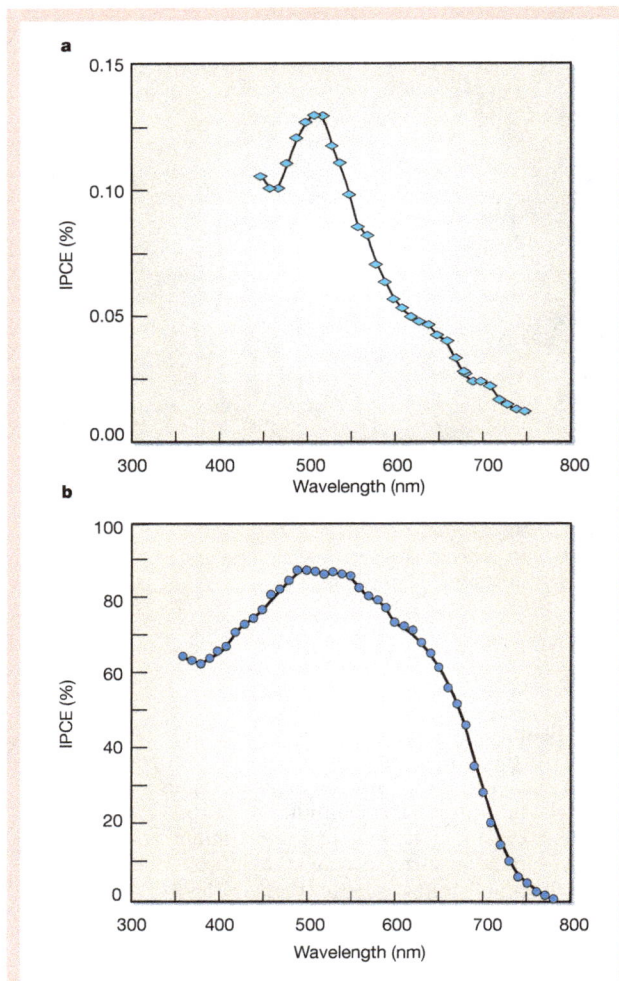

Figure 5 The nanocrystalline effect in dye-sensitized solar cells. In both cases, TiO$_2$ electrodes are sensitized by the surface-anchored ruthenium complex *cis*-RuL$_2$(SCN)$_2$. The incident-photon-to-current conversion efficiency is plotted as a function of the excitation wavelength. **a**, Single-crystal anatase cut in the (101) plane. **b**, Nanocrystalline anatase film. The electrolyte consisted of a solution of 0.3M LiI and 0.03M I$_2$ in acetonitrile.

The overall conversion efficiency of the dye-sensitized cell is determined by the photocurrent density measured at short circuit (i_{ph}), the open-circuit photo-voltage (V_{oc}), the fill factor of the cell (ff) and the intensity of the incident light (I_s)

$$\eta_{global} = i_{ph} V_{oc} (ff/I_s)$$

Under full sunlight, short-circuit photocurrents ranging from 16 to 22 mA cm^{-2} are reached with state-of-the-art ruthenium sensitizers, while V_{oc} is 0.7–0.8 V and the fill factor values are 0.65–0.75. A certified overall power conversion efficiency of 10.4% has been attained at the US National Renewable Energy Laboratory[30]. Although this efficiency makes dye-sensitized cells fully competitive with the better amorphous silicon devices, an even more significant parameter is the dye lifetime achieved under working conditions. For credible system performance, a dye molecule must sustain at least 10^8 redox cycles of photo-excitation, electron injection and regeneration, to give a device service life of 20 years. The use of solvents such as valeronitrile, or γ-butyrolactone, appropriately purified, in the electrolyte formulation provide a system able to pass the standard stability qualification tests for outdoor applications, including thermal stress for 1,000 h at 85 °C, and this has been verified independently[31].

Tandem cells for water cleavage by visible light

The advent of nanocrystalline semiconductor systems has rekindled interest in tandem cells for water cleavage by visible light, which remains a highly prized goal of photoelectrochemical research. The 'brute force' approach to this goal is to use a set of four silicon photovoltaic cells connected in series to generate electricity that is subsequently passed into a commercial-type water electrolyser. Solar-to-chemical conversion efficiencies obtained are about 7%. Much higher efficiencies in the range of 12–20% have been reported for tandem cells based on III/V semiconductors[14,32], but these single-crystal materials cost too much for large-scale terrestrial applications.

A low-cost tandem device that achieves direct cleavage of water into hydrogen and oxygen by visible light was developed recently[33]. This is based on two photosystems connected in series as shown in the electron flow diagram of Fig. 6. A thin film of nanocrystalline tungsten trioxide, WO$_3$ (ref. 34), or Fe$_2$O$_3$ (ref. 35) serves as the top electrode absorbing the blue part of the solar spectrum. The valence-band holes (h$^+$) created by band-gap excitation of the film oxidize water to oxygen:

$$4h^+ + H_2O \Rightarrow O_2 + 4H^+$$

compare the photoresponse of an electrode made of single-crystal anatase, one of the crystal forms of TiO$_2$, with that of a mesoporous TiO$_2$ film. Both electrodes are sensitized by the ruthenium complex *cis*-RuL$_2$(SCN)$_2$ (L is 2,2′-bipyridyl-4-4′-dicarboxylate), which is adsorbed as a monomolecular film at the titania surface. The incident-photon-to-current conversion efficiency (IPCE) is plotted as a function of wavelength. The IPCE value obtained with the single-crystal electrode is only 0.13% near 530 nm, where the sensitizer has an absorption maximum, whereas it reaches 88% with the nanocrystalline electrode[28] — more than 600 times as great. The photocurrent in standard sunlight augments 10^3–10^4 times when passing from a single crystal to a nanocrystalline electrode (standard, or full, sunlight is defined as having a global intensity (I_s) of 1,000 W m^{-2}, air mass 1.5; air mass is the path length of the solar light relative to a vertical position of the Sun above the terrestrial absorber). This striking improvement is due largely to the far better light harvesting of the dye-sensitized nanocrystalline film as compared with a flat single-crystal electrode, but is also due, at least in part, to the mesoscopic film texture favouring photogeneration and collection of charge carriers[29].

and the conduction-band electrons are fed into the second photosystem consisting of the dye-sensitized nanocrystalline TiO$_2$ cell discussed above. The latter is placed directly under the WO$_3$ film, capturing the green and red part of the solar spectrum that is transmitted through the top electrode. The photovoltage generated by the second photosystem enables hydrogen to be generated by the conduction-band electrons.

$$4H^+ + 4e^- \Rightarrow 2H_2$$

The overall reaction corresponds to the splitting of water by visible light. There is close analogy to the 'Z-scheme' (named for the shape of the flow diagram) that operates in photosynthesis. In green plants, there are also two photosystems connected in series, one that oxidizes water to oxygen and the other generating the compound NADPH used in fixation of carbon dioxide. As discussed above, the advantage of the tandem approach is that higher efficiencies can be reached than with single junction cells if the two photosystems absorb complementary parts of the solar spectrum. At present, the overall conversion efficiency from standard solar light to chemical energy achieved with this device stands at 4.5%, and further improvements are being sought.

Dye-sensitized solid heterojunctions and ETA cells

Interest is growing in devices in which both the electron- and hole-carrying materials are solids, but are grown as interpenetrating networks forming a heterojunction of large contact area. From conventional wisdom one would have predicted that solar cells of this kind would work very poorly, if at all. The disordered character of the junction and the presence of the huge interface are features one tries to avoid in conventional photovoltaic cells, because the disruption of the crystal symmetry at the surface produces electronic states in the band gap of the semiconductor, enhancing the recombination of photogenerated carriers. The fact that molecular photovoltaic cells based on the sensitization of nanocrystalline TiO_2 were able to achieve overall conversion efficiencies from solar to electric power of over 10% encouraged work on solid-state analogues, that is, dye-sensitized heterojunctions. The first devices of this type used inorganic p-type semiconductors, for example CuI (ref. 36) or CuSCN (ref. 37), as hole conductors replacing the redox electrolyte. Respectable conversion efficiencies exceeding 1% have been reached with such cells. But the lack of photostability of the Cu(I) compounds and the difficulty of realizing a good contact between the two mesoscopic inorganic materials still present considerable practical challenges.

Organic charge-transport materials have advantages in this respect. An amorphous hole conductor can be introduced into the mesoporous TiO_2 film by a simple spin-coating process and readily adapts its form to the highly corrugated oxide surface. Cells based on a spirobisfluorene-connected arylamine hole transmitter[38], which fills the pores of a dye-sensitized nanocrystalline TiO_2 film, have reached a conversion efficiency of 2.56% at full sunlight[39]. The high open-circuit voltage of these devices, exceeding 900 mV, is particularly noteworthy and promising for further substantial improvements in performance. In general, dye-sensitized heterojunction cells offer great flexibility because the light absorber and charge-transport material can be selected independently to obtain optimal solar energy harvesting and high photovoltaic output. The great advantage of such a configuration is that the charge carriers are generated by the dye precisely at the site of the junction where the electric field is greatest, enhancing charge separation.

Extremely thin absorber (ETA) solar cells are conceptually close to dye-sensitized heterojunctions. The molecular dye is replaced by an extremely thin (2–3 nm) layer of a small-band-gap semiconductor, such as $CuInS_2$. A hole conductor such as CuSCN is placed on top of the absorber, producing a junction of the PIN type (p-type semiconductor/insulator/n-type semiconductor)[40]. The structure has the advantage of enhanced light harvesting due to the surface enlargement and multiple scattering. Because photo-induced charge separation occurs on a length scale of a few nanometres, higher levels of defects and impurities can be tolerated than in flat thin-film devices, where the minority carriers are required to diffuse several microns. On the other hand, making PIN-junctions of such high contact area is difficult and this has hampered the performance of these cells. Their conversion efficiency so far has remained below 5%, which is less than one-third of the yield obtained with similar semiconductor materials in a flat junction configuration.

Soft junctions and organic solar cells

Organic materials have the advantage of being cheap and easy to process. They can be deposited on flexible substrates, bending where their inorganic competitors would crack. The choice of materials is practically unlimited, and specific parts of the solar spectrum can be selectively absorbed. Although organic cells are still considerably less efficient than single-crystal gallium arsenide or silicon, progress has been impressive over the past few years. In particular, solar cells based on interpenetrating polymer networks[41], polymer/fullerene blends[42],

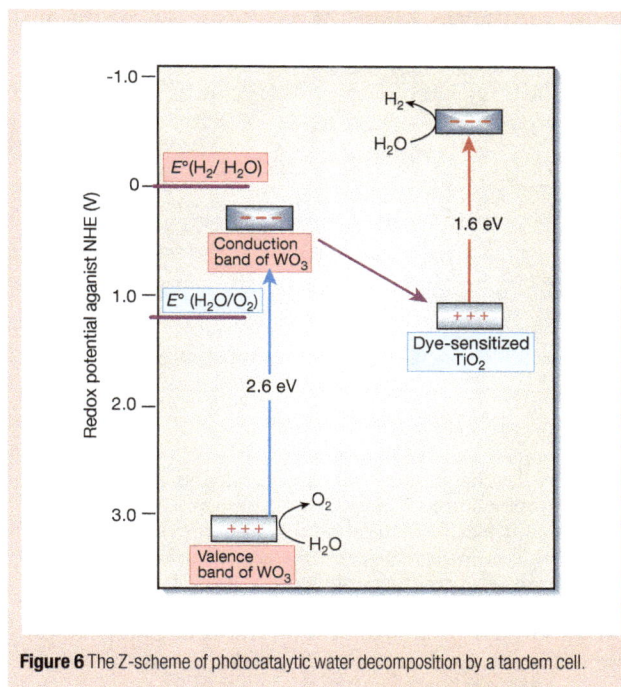

Figure 6 The Z-scheme of photocatalytic water decomposition by a tandem cell.

halogen-doped organic crystals[43] and the solid-state dye-sensitized devices mentioned above[38] have shown surprisingly high solar conversion efficiencies, currently reaching values of 2–3%.

Conducting polymers, for example poly-(phenylenevinylene) (PPV) derivatives or C_{60} particles, are attracting great interest as photovoltaic materials[44,49]. Bulk donor–acceptor heterojunctions are formed simply by blending the two organic materials serving as electron donor (p-type conductor) and electron acceptor (n-type conductor). The advantage of these new structures over the flat-junction organic solar cells investigated earlier[45] is the interpenetration of the two materials that conduct positive and negative charge carriers, reducing the size of the individual phase domains to the nanometre range. This overcomes one of the problems of the first generation of organic photovoltaic cells: the unfavourable ratio of exciton diffusion length to optical absorption length. An exciton is a bound electron–hole pair produced by absorption of light; to be useful, this pair must reach the junction and there dissociate into two free charge carriers — but excitons typically diffuse only a few nanometres before recombining. Light is absorbed (and generates excitons) throughout the composite material. But in the composite, the distance the exciton has to travel before reaching the interface is at most a few nanometres, which is commensurate with its diffusion length. Hence photo-induced charge separation can occur very efficiently. A conversion efficiency from incident photons to current of over 50% has been achieved with a blend containing PPV and methanofullerene derivatives[46]. The overall conversion efficiency from solar to electric power under full sunlight achieved with this cell was 2.5%. Although these results are impressive, the performance of the cell declined rapidly within hours of exposure to sunlight[47]. In contrast, the output of dye-sensitized solar cells is remarkably stable even under light soaking for more than 10,000 h. Similar long-term stability will be required for large-scale application of polymer solar cells.

Summary

Photovoltaic devices based on interpenetrating mesoscopic networks have emerged as a credible alternative to conventional solar cells. Common to all these cells is an ultrafast initial charge separation step, occurring in femtoseconds, and a much slower back-reaction.

Table 1 Performance of photovoltaic and photoelectrochemical solar cells

Type of cell	Efficiency (%)*		Research and technology needs
	Cell	Module	
Crystalline silicon	24	10–15	Higher production yields, lowering of cost and energy content
Multicrystalline silicon	18	9–12	Lower manufacturing cost and complexity
Amorphous silicon	13	7	Lower production costs, increase production volume and stability
CuInSe₂	19	12	Replace indium (too expensive and limited supply), replace CdS window layer, scale up production
Dye-sensitized nanostructured materials	10–11	7	Improve efficiency and high-temperature stability, scale up production
Bipolar AlGaAs/Si photoelectrochemical cells	19–20	—	Reduce materials cost, scale up
Organic solar cells	2–3	—	Improve stability and efficiency

*Efficiency defined as conversion efficiency from solar to electrical power.

This allows the charge carriers to be collected as electric current before recombination takes place. Table 1 compares the performance of the new photoelectrochemical systems with conventional devices. Although still of lower efficiency, the nanostructured cells offer several advantages over their competitors. They can be produced more cheaply and at less of a cost in energy than silicon cells, for which 5 GJ have to be spent to make 1 m² of collector area. Unlike silicon, their efficiency increases with temperature, narrowing the efficiency gap under normal operating conditions. They usually have a bifacial configuration, allowing them to capture light from all angles. Transparent versions of different colour can readily be made that could serve as electric power-producing windows in buildings. These and other attractive features[48] justify the present excitement about these cells and should aid their entry into a tough market. □

1. Bequerel, E. Recherches sur les effets de la radiation chimique de la lumière solaire, au moyen des courants électriques. *C.R. Acad. Sci.* **9**, 145–149 (1839).
2. Gurney, R. W. & Mott, N. F. Theory of the photolysis of silver bromide and the photographic latent image. *Proc. R. Soc. Lond. A* **164**, 151–167 (1938).
3. West, W. First hundred years of spectral sensitization. *Proc. Vogel Cent. Symp. Photogr. Sci. Eng.* **18**, 35–48 (1974).
4. Moser, J. Notiz über die Verstärkung photoelectrischer Ströme durch optische Sensibilisierung. *Monatsh. Chem.* **8**, 373 (1887).
5. Spitler, M. T. Dye photo-oxidation at semiconductor electrodes—a corollary to spectral sensitization in photography. *J. Chem. Educ.* **60**, 330–332 (1983).
6. Namba, S. & Hishiki, Y. Color sensitization of zinc oxide with cyanide dyes. *J. Phys. Chem.* **69**, 774–779 (1965).
7. Nelson, R. C. Minority carrier trapping and dye sensitization. *J. Phys. Chem.* **69**, 714–718 (1965).
8. Boudon, J. Spectral sensitization of chemical effects in solids. *J. Phys. Chem.* **69**, 705–713 (1965).
9. Gerischer, H. & Tributsch, H. Electrochemische Untersuchungen zur spectraleu sensibilisierung von ZnO-Einkristallen. *Ber. Bunsenges. Phys. Chem.* **72**, 437–445 (1968).
10. Hauffe, K., Danzmann, H. J., Pusch, H., Range, J. & Volz, H. New Experiments on the sensitization of zinc oxide by means of the electrochemical cell technique. *J. Electrochem. Soc.* **117**, 993–999 (1970).
11. Brattain, W. H. & Garrett, C. G. B. Experiments on the interface between germanium and an electrolyte. *Bell Syst. Tech. J.* **34**, 129–176 (1955).
12. Gerischer, H. Electrochemical behavior of semiconductors under illumination. *J. Electrochem. Soc.* **113**, 1174–1182 (1966).
13. Kalyanasundaram; K. Photoelectrochemcial cell studies with semiconductor electrodes: a classified bibliography (1975-1983). *Solar Cells* **15**, 93–156 (1985).
14. Licht, S. Multiple band gap semiconductor/electrolyte solar energy conversion. *J. Phys. Chem.* **105**, 6281–6294 (2001).
15. Fujishima. A. & Honda, K. Electrochemical photolysis of water at a semiconductor electrode. *Nature* **238**, 37–38 (1972).
16. O'Regan, B. & Grätzel, M. A low-cost, high efficiency solar cell based on dye-sensitized colloidal TiO₂ films. *Nature* **353**, 737–740 (1991).
17. Hagfeldt, A. & Grätzel, M. Molecular photovoltaics. *Acc. Chem. Res.* **33**, 269–277 (2000).
18. Hilgendorff, M., Spanhel, L., Rothenhäusler, Ch. & Müller, G. From ZnO colloids to nanocrystalline highly conductive films. *J. Electrochem. Soc.* **145**, 3632–3637 (1998).
19. Nelson, J. Continuous time random walk model of electron transport in nanocrystalline TiO₂ electrodes. *Phys. Rev. B* **59**, 15374–15380 (1999).
20. Barbé, Ch. J. et al. Nanocrystalline titanium oxide electrodes for photovoltaic applications. *J Am. Ceram. Soc.* **80**, 3157–3171 (1997).
21. Hodes, G., Howell, I. D. J. & Peter, L. M. Nanocrystalline photoelectrochemical cells: a new concept in photovoltaic cells. *J. Electrochem. Soc.* **139**, 3136–3140 (1992).
22. Schwarzburg, K. & Willig, F. Origin of photovoltage in dye sensitized elelectrochemical solar cells. *J. Phys. Chem. B* **103**, 5743–5746 (1999).
23. Pichot, F. & Gregg, B. A. The photovoltage-determining mechanism in dye-sensitized solar cells. *J. Phys. Chem. B* **104**, 6–10 (1999).
24. Cahen, D., Hodes, G., Grätzel, M. Guillemoles, J. F. & Riess, I. Nature of photovoltaic action in dye-sensitized solar cells. *J. Phys. Chem. B* **104**, 2053–2059 (2000).
25. Van de Lagemaat, J., Park, N.-G. & Frank, A. J. Influence of electrical potential distribution, charge transport, and recombination on the photopotential and photocurrent conversion efficiency of dye-sensitized nanocrystalline TiO₂ solar cells: a study by electrical impedance and optical modulation techniques. *J. Phys. Chem. B* **104**, 2044–2052 (2000).
26. Dloczik, L. et al. Dynamic response of dye-sensitized nanocrystalline solar cells: characterization by intensity-modulated photocurrent spectroscopy. *J. Phys. Chem. B* **101**, 10281–10289 (1997).
27. Masatai, Y. & Hodes, G. Size quantization in electrodeposited CdTe nanocrystalline films. *J. Phys. Chem. B* **101**, 2685–2690 (1997).
28. Nazeeruddin, M. K. et al. Conversion of light to electricity by cis X2-Bis(2,2′-bipyridyl-4,4′-dicarboxalate) ruthernium(II) charge transfer sensitizers. *J. Am. Chem. Soc.* **115**, 6382–6390 (1993).
39. Kavan, L., Grätzel, M., Gilbert, S. E., Klemenz, C. & Scheel, H. J. Electrochemical and photoelectrochemical investigations of single-crystal anatase. *J. Am. Chem. Soc.* **118**, 6716–6723 (1996).
30. Nazeeruddin, M. K. et al. Engineering of efficient panchromatic sensitizers for nanocrystalline TiO₂-based solar cells. *J. Am. Chem. Soc.* **123**, 1613–1624 (2001).
31. Hinsch, A. et al. in *Proc. 16th Eur. PV Solar Energy Conf., Glasgow, May 2000* (eds Scheel, H. et al.) 32 (James & James, London, 2000).
32. Khaselev, O. & Turner, J. A. A monolithic photovoltaic-photoelectrochemical device for hydrogen production via water splitting. *Science* **280**, 425–427 (1998).
33. Grätzel, M. The artificial leaf, bio-mimetic photocatalysis. *Cattech* **3**, 3–17 (1999).
34. Santato, C., Ulmann, M. & Augustynski, J. Photoelectrochemical properties of nanostructured tungsten trioxide films. *J. Phys. Chem. B* **105**, 936–940 (2001).
35. Khan, S. U. M. & Akikusa, J. Photoelectrochemical splitting of water at nanocrystalline n-Fe₂O₃ thin-film electrodes. *J. Phys. Chem. B* **103**, 7184–7189 (1999).
36. Tennakone, K., Kumara, G. R. R. A., Kumarasinghe, A. R., Wijayantha, K. G. U. & Sirimanne, P. M. A dye-sensitized nano-porous solid-state photovoltaic cell. *Semicond. Sci. Technol.* **10**, 1689–1693 (1995).
37. O'Regan, B. & Schwarz, D. T. Large enhancement in photocurrent efficiency caused by UV illumination of dye sensitized heterojunctions. *Chem. Mater.* **10**, 1501–1509 (1998).
38. Bach, U. et al. Solid-state dye-sensitized mesoporous TiO₂ solar cells with high photon-to-electron conversion efficiencies. *Nature* **395**, 583–585 (1998).
39. Krüger, J., Bach, U. & Grätzel, M. High efficiency solid-state photovoltaic device due to inhibition of interface charge recombination. *Appl. Phys. Lett.* **79**, 2085–2087 (2001).
40. Kaiser, I. et al. The eta-solar cell with CuInS₂: a photovoltaic cell concept using an extremely thin absorber. *Sol. Energy Mater. Sol. Cells* **67**, 89–96 (2001).
41. Halls, J. J. M., Pickler, K., Friend, R. H., Morati, S. C. & Holmes, A. B. Efficient photodiodes from interpenetrating polymer networks. *Nature* **376**, 498–500 (1995).
42. Yu, G., Gao, J., Hummelen, J. C., Wudl, F. & Heeger, A. J. Polymer photovoltaic cells: enhanced efficiencies via a network of internal donor acceptor heterojunctions. *Science* **270**, 1789–1791 (1995).
43. Schön, J. H., Kloc, Ch., Bucher, E. & Batlogg, B. Efficient organic photovoltaic diodes based on doped pentacene. *Nature* **403**, 408–410 (2000).
44. Brabec, C. J. & Sariciftci, N. S. Polymeric photovoltaic devices. *Mater. Today* 3–8 (2000).
45. Wöhrle,.D. & Meissner D. Organic solar cells. *Adv. Mat.* **3**, 129–138 (1991).
46. Shaheen, S. E. et al. 2.5% efficient orga-nic plastic solar cells. *Appl. Phys. Lett.* **78**, 841–843 (2001).
47. Tuladhar, D. et al. Abstract, Int. Workshop Nanostruct. Photovoltaics, Dresden, Germany <http://www.mpipks-dresden.mpg.de> (2001).
48. Grätzel, M. Perspectives for dye-sensitized nanocrystalline solar cells. *Prog. Photovoltaic Res. Applic.* **8**, 171–185 (2000).
49. Savenije, T. J., Warman, J. M. & Goosens, A. Visible light sensitization of titanium dioxide using a phenylene vinylene polymer. *Chem. Phys. Lett.* **278**, 148–153 (1998).

Acknowledgements

I thank the members of the Swiss Federal Institute of Technology (EPFL) electrochemical photovoltaics development team, some of whose work is referenced; the industrial organizations whose interest in this PV system has induced them to license the concept and thereby support the research; EPFL; and OFEN (Swiss Federal Office of Energy) for past encouragement and support.

High-performance dye-sensitized solar cells based on solvent-free electrolytes produced from eutectic melts

YU BAI[1]*, YIMING CAO[1]*, JING ZHANG[1], MINGKUI WANG[2], RENZHI LI[1], PENG WANG[1†], SHAIK M. ZAKEERUDDIN[2] AND MICHAEL GRÄTZEL[2†]

[1] State Key Laboratory of Polymer Physics and Chemistry, Changchun Institute of Applied Chemistry, Chinese Academy of Sciences, Changchun 130022, China
[2] Laboratory for Photonics and Interfaces, Swiss Federal Institute of Technology, CH 1015, Lausanne, Switzerland
*These authors contributed equally to this work
† e-mail: peng.wang@ciac.jl.cn; michael.graetzel@epfl.ch

Published online: 29 June 2008; doi:10.1038/nmat2224

Low-cost excitonic solar cells based on organic optoelectronic materials are receiving an ever-increasing amount of attention as potential alternatives to traditional inorganic photovoltaic devices. In this rapidly developing field, the dye-sensitized solar cell[1] (DSC) has achieved so far the highest validated efficiency of 11.1% (ref. 2) and remarkable stability[3]. However, the cells with the best performance use volatile solvents in their electrolytes, which may be prohibitive for outdoor solar panels in view of the need for robust encapsulation. Solvent-free room-temperature ionic liquids[4–11] have been pursued as an attractive solution to this dilemma, and device efficiencies of over 7% were achieved by using some low-viscosity formulations containing 1-ethyl-3-methylimidazolium thiocyanate[8], selenocyanate[9], tricyanomethide[10] or tetracyanoborate[11]. Unfortunately, apart from tetracyanoborate, all of these low-viscosity melts proved to be unstable under prolonged thermal stress and light soaking. Here, we introduce the concept of using eutectic melts to produce solvent-free liquid redox electrolytes. Using a ternary melt in conjunction with a nanocrystalline titania film and the amphiphilic heteroleptic ruthenium complex Z907Na (ref. 10) as a sensitizer, we reach excellent stability and an unprecedented efficiency of 8.2% under air-mass 1.5 global illumination. Our results are of importance to realize large-scale outdoor applications of mesoscopic DSCs.

The ionic liquid formulations used in the dye-sensitized solar cell (DSC) as solvent-free electrolytes use iodide melts as their major component, because high concentrations of iodide are required to intercept quantitatively the geminate recombination between the electrons injected by the photo-excited sensitizer in the nanocrystalline titania film and its oxidized form[10]. The viscosity of the iodide melts should be as low as possible to avoid mass transport limitation of the photocurrent and loss of fill factor under cell operation in full sunlight. Amongst the iodide salts that form room-temperature ionic liquids, 1-propyl-3-methylimidazolium iodide (PMII) has the lowest viscosity. PMII has therefore also been the candidate of choice for binary ionic liquids that have so far achieved the highest efficiency in solar electricity generation by solvent-free DSCs[10].

Figure 1 shows plots of the specific conductivity versus temperature for several imidazolium salts and their mixtures. The conductivity of pure iodide melts increases in the order 1-hexyl-3-methylimidazolium iodide < 1-butyl-3-methylimidazolium iodide < PMII, mirroring the behaviour of their fluidity, which follows the same trend. The viscous behaviour of ionic liquids is dictated by the interplay of coulombic and van der Waals interactions as well as hydrogen bond formation[12]. Increasing the alkyl chain length decreases the electrostatic attraction between cations and iodide but the van der Waals interaction between the imidazolium cations increases. 1-ethyl-3-methylimidazolium iodide (EMII) and 1,3-dimethylimidazolium iodide (DMII) are solids at ambient temperature on probable account of their high lattice Gibbs energies due to the conformational rigidity of small and symmetric cations. Importantly, Fig. 1 shows that above their melting temperatures, both salts are more conductive than PMII. This surprising observation encouraged us to mix EMII and DMII at a molar ratio of 1:1, resulting in a highly conductive, low-melting binary salt with a eutectic temperature of 47.5 °C (see Fig. 1 and Supplementary Information, Fig. S1). The decrease of melting temperature is due to the increase in entropy[13] of the components in the eutectic mixture.

We further observed that 1-allyl-3-methylimidazolium iodide (AMII), which has a melting point of 60 °C, also shows a higher conductivity than PMII in the liquid state. We therefore reasoned that a mixture formed from the three solids EMII, DMII and AMII could provide a room-temperature ionic liquid with superior fluidity and ionic conductivity compared with the binary melt. This expectation is borne out by curve h in Fig. 1 referring to the mixture of AMII, DMII and EMII at a molar ratio of 1:1:1. The resulting ternary melt has a melting point below 0 °C and a strikingly high room-temperature conductivity of 1.68 mS cm⁻¹, exceeding that of PMII (0.58 mS cm⁻¹) by almost a factor of three. We attribute this remarkable increase in the conductivity over the pure PMII melt to enhanced fluidity along with higher ion concentrations and the smaller size of the cations in the melt.

In light of this result, we prepared four melts to assess the potential advantages of eutectic-based melts relative

When citing this article, please cite the original version as shown on the contents page of this chapter.

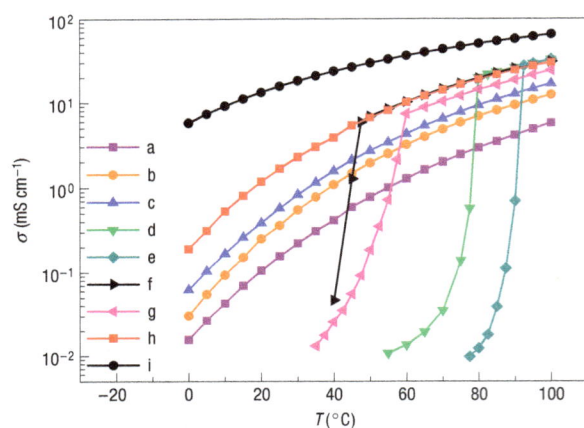

Figure 1 Plots of specific conductivity (σ) of imidazolium melts versus temperature (T). (a) 1-hexyl-3-methylimidazolium iodide. (b) < 1-butyl-3-methylimidazolium iodide. (c) PMII. (d) EMII. (e) DMII. (f) DMII/EMII (1:/1). (g) AMII. (h) DMII/EMII/AMII (1:1:1). (i) EMITCB. Before measurements in a sealed tube, all of the samples were dried at 80 °C under a vacuum of ~3 torr for 8 h.

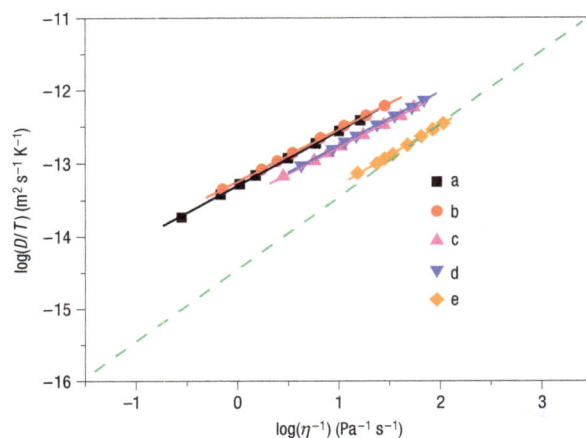

Figure 2 Stokes–Einstein plots of diffusion coefficient (D) versus fluidity (η^{-1}). (a)–(e), Melts I–V. The dashed line is calculated from the Stokes–Einstein relation using a hydrodynamic radius of 2.1 Å for triiodide.

to state-of-the-art systems[6,11] used in DSCs. The following compositions were tested for comparison (the molar ratio of the components in the melt is given in parenthesis): melt I, PMII/I_2 (24:1); melt II, DMII/EMII/AMII/I_2 (8:8:8:1); melt III, PMII/1-ethyl-3-methylimidazolium tetracyanoborate(EMITCB)/I_2 (24:16:1.67); melt IV, DMII/EMII/EMITCB/I_2 (12:12:16:1.67). We measured the effect of temperature (T) on the density (d), viscosity (η), specific conductivity (σ), molar conductivity (Λ) and apparent triiodide diffusion coefficient (D) of melts I–IV after drying them at 80 °C under a vacuum of ~3 torr for 8 h. Even under these severe conditions, we did not detect any volatility of iodine by the simple iodine–starch colourimetric method, suggesting that the complexation of iodine by iodide to form triiodide ions prevents any significant evaporative iodine loss. This removes one major concern for long-term outdoor application of these melts, where loss of iodine would adversely affect the stability of photovoltaic operation.

Compared with the PMII-containing reference melts I and III, formulations II and IV based on the eutectic of DMII and EMII show much improved properties (see Supplementary Information, Figs S3–S6). They exhibit lower viscosities, higher conductivities and faster triiodide transport. These features are highly desirable to enhance the photovoltaic performance in solvent-free DSCs. The electrolyte viscosity can be well fitted by the Vogel–Fulcher–Tammann equation. Recent work by Angell and co-workers[14] has shown that the empirical Walden's rule applies rather well to pure ionic liquids. As shown in the Supplementary Information, Fig. S7, log(Λ) of electrolytes I–IV increases linearly with log(η^{-1}). The relationship can be expressed as $\Lambda \eta^\alpha$ = constant, with α being the slope of the line in the Walden plot and reflecting the degree of ion dissociation. The slopes of the fitted lines corresponding to our four melts and pure EMITCB are all slightly less than unity predicted by the 'ideal' Walden rule, indicating progressive augmentation in the population of less conductive ion-pairs with increasing temperature. However, melts I–IV exhibit anomalous high molar conductivities in view of their high viscosities. This will be further elaborated below

by analysing the triiodide diffusion coefficients measured with ultramicroelectrode voltammetry.

In Fig. 2, the temperature-dependent apparent triiodide diffusion coefficients (D) are plotted versus fluidity (η^{-1}) in the Stokes–Einstein coordinate. Whereas log(D/T) increases linearly with log(η^{-1}) for melts I–IV, the fitted slopes (0.74, 0.71, 0.71, and 0.73) are less than unity, departing considerably from the description of the Stokes–Einstein relation, equation (1)

$$\log(D/T) = \log(k_B/6\pi r_H) + \log(1/\eta). \qquad (1)$$

Moreover, the effective hydrodynamic radii (r_H) of triiodide derived from the fitted intercepts are unrealistically small, being only several tenths of an ångström as compared with the expected value of 2.1 Å (ref. 15). Previously[4,6,10,16], this anomalous transport behaviour has been qualitatively explained by the Grotthus-like exchange mechanism and rationalized by the Dahms–Ruff equation ($D = D_{phys} + D_{ex} = D_{phys} + k_{ex}\delta^2 c/6$, where D_{phys} and D_{ex} are the physical and exchange diffusion coefficients, respectively, k_{ex} is the rate constant of the iodide–triiodide bond exchange, and c and δ are the iodide concentration and average centre-to-centre distances between iodide and triiodide in the encounter complex).

Similar to proton transfer in water, the transport of triiodide by the Grotthus mechanism occurs by bond exchange. The triiodide approaches iodide from one end forming an encounter complex, from which triiodide is released at the other end. In this fashion, the triiodide is displaced by the length of one I–I bond, ~2.9 Å (ref. 17), without having to cross that distance. The bond exchange occurs immediately on formation of the encounter complex, rendering the process diffusion-controlled. Thus, the viscosity-dependent transport of triiodide in ionic liquid electrolytes with high iodide concentration can be described by physical diffusion coupled to the Grotthus bond exchange, which augments the diffusional displacement by ~2.9 Å for each diffusional encounter with iodide. This manifests itself by an apparent acceleration of the triiodide diffusion, which however can only be felt at high iodide concentration where the average distance between two adjacent iodide ions is small.

Melt V containing 1-ethyl-3-methylimidazolium bis(trifluoromethanesulphonyl)imide as an inert salt has a low iodide concentration, DMII/EMII/1-ethyl-3-methylimidazolium

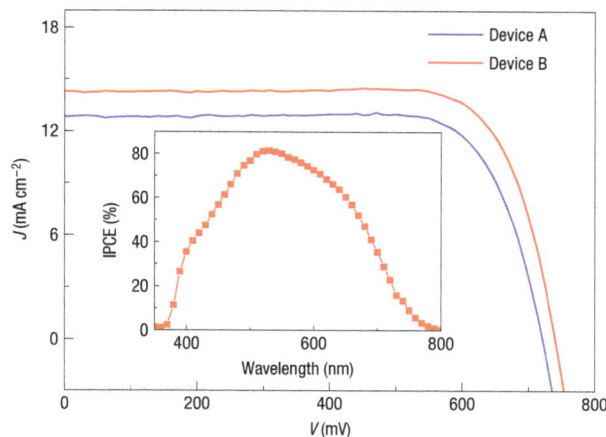

Figure 3 Photocurrent density–voltage (*J–V*) characteristics of devices A and B under AM 1.5G illumination (100 mW cm^{-2}). The inset shows the IPCE of device B. Cells were tested using a metal mask with an aperture area of 0.158 cm^2.

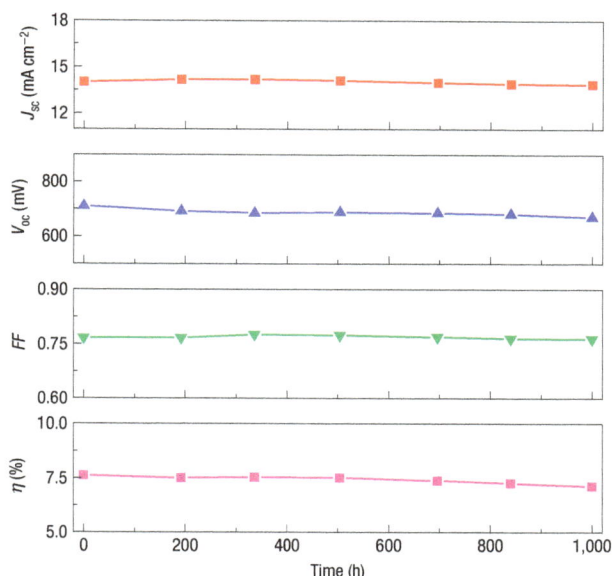

Figure 4 Detailed photovoltaic parameters of a cell measured under the irradiance of AM 1.5G sunlight during successive full-sun visible-light soaking at 60 °C. J_{sc}: short-circuit photocurrent density; V_{oc}: open-circuit voltage; *FF*: fill factor; η: power conversion efficiency.

bis(trifluoromethanesulphonyl)imide/I_2 (6:6:800:0.83). Hence, it serves as a reference where the Grotthus contribution to the triiodide transport is negligible. The enhancement of the diffusion coefficient due to the Grotthus contribution can be derived from the ratio of the D/T values for melts I–IV to that for melt V. For example, at a $\log(\eta^{-1})$ of 1.5, owing to the Grotthus contribution the D/T value for the ternary melt II is about five times larger than that for melt V. The fact that the straight lines for the five melts in Fig. 2 are parallel to each other indicates that the enhancement of the diffusion coefficient due to the Grotthus bond exchange mechanism is similar over the whole investigated viscosity range. Closer inspection of Fig. 2 also shows that the acceleration of the triiodide transport due to the Grotthus mechanism increases with the iodide packing densities in the melts. Melt II has the highest iodide concentration and shows the largest Grotthus effect. As expected, the Grotthus acceleration is significantly smaller in melts III and IV, which both have lower iodide packing densities than melts I and II.

Taking advantage of the properties of eutectic-based melts, we used the routine additives guanidinium thiocyanate (GNCS) and *N*-butylbenzoimidazole (NBB) to formulate two practical electrolytes for photovoltaic device evaluation. The compositions tested were as follows. Electrolyte A in device A: DMII/EMII/AMII/I_2/NBB/GNCS (8:8:8:1:2:0.4); electrolyte B in device B: DMII/EMII/EMITCB/I_2/NBB/GNCS (12:12:16:1.67:3.33:0.67). Detailed fabrication procedures for the mesoporous TiO_2 electrode and sealed cell have been described elsewhere[18]. A 7-µm-thick film of 20-nm-sized TiO_2 particles was first printed on the fluorine-doped SnO_2 conducting glass electrode and further coated by a 5-µm-thick second layer of 400-nm-sized light-scattering anatase particles. The sintered TiO_2 electrodes were immersed at room temperature for 12 h into a solution containing 300 µM Z907Na dye and 300 µM 3-phenylpropionic acid in acetonitrile and *t*-butyl alcohol (1:1, v/v).

Figure 3 shows the photocurrent density–voltage (*J–V*) characteristics of devices A and B using the above-mentioned molten salts measured under air-mass 1.5 global (AM 1.5G) sunlight. The short-circuit photocurrent density (J_{sc}), open-circuit voltage (V_{oc}) and fill factor (*FF*) of device A with the ternary iodides are 12.82 mA cm^{-2}, 721 mV and 0.768, respectively,

yielding an overall power conversion efficiency (η) of 7.1%, which is much higher than that of 6.0% previously reported[6] for the corresponding device with a solvent-free, PMII-based ionic liquid electrolyte. Note that this ternary electrolyte composed of simple imidazolium iodides has shown a comparable efficiency to previously reported solvent-free electrolytes[8–11] containing relatively precious ionic liquids. The photovoltaic parameters (J_{sc}, V_{oc}, *FF* and η) of device B with mixed imidazolium iodides and tetracyanoborate are 14.26 mA cm^{-2}, 741 mV, 0.774 and 8.2%, respectively. For the first time, such a high efficiency under AM 1.5G sunlight is obtained for DSCs with solvent-free electrolytes. It is worth noting that simply replacing the PMII used in our previous work[11] by the eutectic melt of DMII and EMII without changing the sensitizer or titania film has resulted in a remarkable 17% photovoltaic performance enhancement. The photocurrent action spectrum of device B is shown in the inset of Fig. 3. The incident photon-to-current conversion efficiencies (IPCEs) exceed 60% in a broad spectral range from 460 to 650 nm, reaching its maximum of about 81% at 540 nm. From the overlap integral of this curve with the standard global AM 1.5 solar emission spectrum, a short-circuit photocurrent density (J_{sc}) of 14.30 mA cm^{-2} is calculated, which is in excellent agreement with the measured photocurrent. This confirms that there is no mismatch between the simulator used and AM 1.5 solar emission, showing that the conversion efficiency value of 8.2% truly refers to the standard reporting condition. To test the stability of the eutectic-based electrolyte B for DSCs, photovoltaic cells without 3-phenylpropionic acid as a coadsorbent were subjected for 1,000 h to light soaking with full solar intensity (1,000 W m^{-2}) at 60 °C under a Suntest lamp (Hanau corporation). As shown in Fig. 4, the cell retained over 93% of its initial conversion efficiency over this period, confirming the very high stability of the device. With respect to a reference cell using the best previously reported ionic liquid, a DSC based on this new electrolyte B shows a remarkable enhancement in device

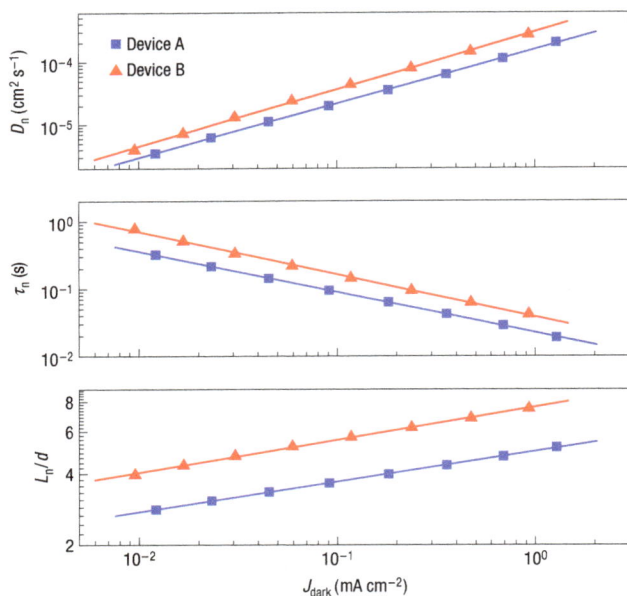

Figure 5 Plots of effective electron diffusion coefficient (D_n), lifetime (τ_n) and normalized diffusion length (L_n/d) versus dark current density (J_{dark}).

efficiency while maintaining a similar stability (see Supplementary Information, Fig. S8).

The higher V_{oc} of device B compared with A can be explained by the more positive Nernst potential of its electrolyte due to a lower iodide concentration. In addition, the higher J_{sc} of device B produces an increase in V_{oc}[19–21] and electrochemical impedance[22] measurements were carried out to determine the origin of the difference in the photocurrent. As shown in Supplementary Information, Fig. S9, under the same extracted charge densities, device B always has slower rates of charge recombination than A due to a lower triiodide concentration in the electrolyte. This has been further confirmed by electrochemical impedance measurements of the electron lifetime (τ_n) shown in Fig. 5. Note also that the less viscous electrolyte for device B gives larger effective electron diffusion coefficients (D_n) than A at the same dark current densities. This is in keeping with the notion[23] that the diffusion coefficient is in fact ambipolar, reflecting apart from the electron motion also the mobility of cations that screen the photoinjected electrons in the mesoporous titania film[24]. The normalized electron diffusion length (L_n/d, where d is the titania film thickness) is significantly larger for device B than for A, indicating that the increase in J_{sc} is caused by a higher charge collection yield. As shown in the Supplementary Information, Fig. S10, the enhanced photovoltaic performance of electrolyte B compared with the previously reported ionic liquid[11] can also be rationalized in terms of an increased effective electron diffusion length.

In summary, using mixtures of solid salts has led to the discovery of solvent-free electrolytes for DSCs showing unprecedented efficiency and excellent stability. An efficiency of 8.2% achieved in full sunlight sets a benchmark for solvent-free DSCs. Importantly, their performance now matches that of a low-volatility, 3-methoxypropionitrile-based electrolyte[21], rendering the use of such solvents obsolete. This is expected to have important practical consequences as the need for using organic-solvent-based electrolytes to achieve high conversion efficiencies has impaired large-scale production and outdoor application of the DSC. Organic solvents not only present challenges for sealing but also permeate across plastics, excluding their use in flexible cells. In contrast, the vapour pressure of molten salts and their permeation rate across plastics are negligible at the typical operating temperature of photovoltaic converters. We believe that our findings on eutectic-based melts will provide useful clues for further improvement of solvent-free electrolytes on the basis of rational design of their constituents, greatly facilitating the large-scale practical application of light-weight, flexible dye-sensitized thin-film cells.

METHODS

The viscosity measurements were carried out using a Brookfield DV-II+Pro viscometer. Densities were determined with an Anton Paar DMA 35N density meter. A Radiometer CDM210 conductivity meter was used to measure conductivities. The Radiometer CDC749 conductivity cell with a nominal cell constant of 1.70 cm^{-1} was calibrated with 0.1 M KCl aqueous solution before the experiments. A two-electrode electrochemical cell, consisting of a 5.0-μm-radius Pt ultramicroelectrode as the working electrode and a Pt foil as the counter electrode, was used for the measurements of triiodide diffusion coefficients in combination with a CHI 660C electrochemical workstation. A heating–cooling cycle pump was used to control the sample temperatures. J–V and IPCE measurements were carried out as reported previously[3]. A white-light-emitting diode array powered by varied driving voltages was used to supply different steady-state light intensities in the transient photoelectrical experiments. A red-light-emitting diode array controlled with a fast solid-state switch was used to generate a perturbation pulse with a width of 200 ms. Electrical impedance experiments were carried out with an Echo Chemie Autolab electrochemical workstation, with a frequency range of 0.01–10^6 Hz and a potential modulation of 5 mV. The obtained impedance spectra were fitted with the Z-view software (v2.8b, Scribner Associates) in terms of appropriate equivalent circuits[22].

Received 16 March 2008; accepted 27 May 2008; published 29 June 2008.

References
1. Grätzel, M. Photoelectrochemical cells. *Nature* **414**, 338–344 (2001).
2. Chiba, Y. *et al.* Dye-sensitized solar cells with conversion efficiency of 11.1%. *Japan. J. Appl. Phys. 2* **45**, L638–L640 (2006).
3. Wang, P. *et al.* A stable quasi-solid-state dye-sensitized solar cell with an amphiphilic ruthenium sensitizer and polymer gel electrolyte. *Nature Mater.* **2**, 402–407 (2003).
4. Papageorgiou, N. *et al.* The performance and stability of ambient temperature molten salts for solar cell applications. *J. Electrochem. Soc.* **143**, 3099–3108 (1996).
5. Kubo, W. *et al.* Quasi-solid-state dye-sensitized solar cells using room temperature molten salts and a low molecular weight gelator. *Chem. Commun.* 374–375 (2002).
6. Wang, P. *et al.* Gelation of ionic liquid-based electrolytes with silica nanoparticles for quasi-solid-state dye-sensitized solar cells. *J. Am. Chem. Soc.* **125**, 1166–1167 (2003).
7. Wang, P. *et al.* A new ionic liquid electrolyte enhances the conversion efficiency of dye-sensitized solar cells. *J. Phys. Chem. B* **107**, 13280–13285 (2003).
8. Wang, P. *et al.* A binary ionic liquid electrolyte to achieve ≥7% power conversion efficiencies in dye-sensitized solar cells. *Chem. Mater.* **16**, 2694–2696 (2004).
9. Wang, P. *et al.* A solvent-free, SeCN$^-$/(SeCN)$_3^-$ based ionic liquid electrolyte for high-efficiency dye-sensitized nanocrystalline solar cells. *J. Am. Chem. Soc.* **126**, 7164–7165 (2004).
10. Wang, P. *et al.* Charge separation and efficient light energy conversion in sensitized mesoscopic solar cells based on binary ionic liquids. *J. Am. Chem. Soc.* **127**, 6850–6856 (2005).
11. Kuang, D. *et al.* Stable mesoscopic dye-sensitized solar cells based on tetracyanoborate ionic liquid electrolyte. *J. Am. Chem. Soc.* **128**, 7732–7733 (2006).
12. Wasserscheid, P. & Keim, W. Ionic liquids-new 'solutions' for transition metal catalysis. *Angew. Chem. Int. Ed.* **39**, 3773–3789 (2000).
13. Belieres, J.-P., Gervasio, D. & Angell, C. A. Binary inorganic salt mixtures as high conductivity liquid electrolytes for >100 °C fuel cells. *Chem. Commun.* 4799–4801 (2006).
14. Xu, W., Cooper, E. I. & Angell, C. A. Ionic liquids: Ion mobilities, glass temperature, and fragilities. *J. Phys. Chem. B* **107**, 6170–6178 (2003).
15. Spiro, M. & Creeth, A. M. Tracer diffusion coefficients of I$^-$, I$_3^-$, Fe^{2+} and Fe^{3+} at low temperatures. *J. Chem. Soc. Faraday Trans.* **86**, 3573–3576 (1990).
16. Kawano, R. & Watanabe, M. Anomaly of charge transport of an iodide/tri-iodide redox couple in an ionic liquid and its importance in dye-sensitized solar cells. *Chem. Commun.* 2107–2109 (2005).
17. Said, F. F. *et al.* Hydrogen bonding motifs of N, N′, N″-trisubstituted guanidinium cations with spherical and rodlike monoanions: Synthesis and structures of I$^-$, I$_3^-$, and SCN$^-$ salts. *Cryst. Growth Des.* **6**, 258–266 (2006).
18. Wang, P. *et al.* Enhance the performance of dye-sensitized solar cells by co-grafting amphiphilic sensitizer and hexadecylmalonic acid on TiO$_2$ nanocrystals.. *J. Phys. Chem. B* **107**, 14336–14341 (2003).
19. O'Regan, B. C. & Lenzmann, F. Charge transport and recombination in a nanoscale interpenetrating network of n-type and p-type semiconductors: Transient photocurrent and photovoltage studies of TiO$_2$/dye/CuSCN photovoltaic cells. *J. Phys. Chem. B* **108**, 4342–4350 (2004).
20. Walker, A. B. *et al.* Analysis of photovoltage decay transients in dye-sensitized solar cells. *J. Phys. Chem. B* **110**, 25504–25507 (2006).

21. Zhang, Z. *et al.* Effects of ω-guanidinoalkyl acids as coadsorbents in dye-sensitized solar cells. *J. Phys. Chem. C* **111**, 398–403 (2007).

22. Bisquert, J. Theory of the impedance of electron diffusion and recombination in a thin layer. *J. Phys. Chem. B* **106**, 325–333 (2002).

23. Kopidakis, N. *et al.* Ambipolar diffusion of photocarriers in electrolyte-filled, nanoporous TiO_2. *J. Phys. Chem. B* **104**, 3930–3936 (2000).

24. Lanning, O. J. & Madden, P. A. Screening at a charged surface by a molten salt. *J. Phys. Chem. B* **108**, 11069–11072 (2004).

Supplementary Information accompanies this paper on www.nature.com/naturematerials.

Acknowledgements

The National Key Scientific Program-Nanoscience and Nanotechnology (No. 2007CB936700), the National Science Foundation of China (50773078) and the '100-Talent Program' of the Chinese Academy of Sciences have supported this work. M.W., S.M.Z. and M.G. thank G. Rothenberger for helpful discussions and the Swiss National Science Foundation for financial support. P.W. thanks EPFL for a visiting professorship.

Author information

Reprints and permission information is available online at http://npg.nature.com/reprintsandpermissions. Correspondence and requests for materials should be addressed to P.W., S.M.Z. or M.G.

Ultrathin silicon solar microcells for semitransparent, mechanically flexible and microconcentrator module designs

JONGSEUNG YOON[1,2,3]*, ALFRED J. BACA[3,4]*, SANG-IL PARK[1,2,3], PAULIUS ELVIKIS[5],
JOSEPH B. GEDDES III[2], LANFANG LI[1,4], RAK HWAN KIM[1,2,3], JIANLIANG XIAO[6], SHUODAO WANG[6],
TAE-HO KIM[1,2,3], MICHAEL J. MOTALA[3,4], BOK YEOP AHN[1,3], ERIC B. DUOSS[1,3], JENNIFER A. LEWIS[1,3],
RALPH G. NUZZO[1,3,4], PLACID M. FERREIRA[5], YONGGANG HUANG[6,7], ANGUS ROCKETT[1]
AND JOHN A. ROGERS[1,2,3,4,5]†

[1]Department of Materials Science and Engineering, University of Illinois at Urbana-Champaign, Illinois 61801, USA
[2]Beckman Institute for Advanced Science and Technology, University of Illinois at Urbana-Champaign, Illinois 61801, USA
[3]Frederick Seitz Materials Research Laboratory, University of Illinois at Urbana-Champaign, Illinois 61801, USA
[4]Department of Chemistry, University of Illinois at Urbana-Champaign, Illinois 61801, USA
[5]Department of Mechanical Science and Engineering, University of Illinois at Urbana-Champaign, Illinois 61801, USA
[6]Department of Mechanical Engineering, Northwestern University, Evanston, Illinois 60208, USA
[7]Department of Civil and Environmental Engineering, Northwestern University, Evanston, Illinois 60208, USA
*These authors contributed equally to this work
†e-mail: jrogers@uiuc.edu

Published online: 5 October 2008; doi:10.1038/nmat2287

The high natural abundance of silicon, together with its excellent reliability and good efficiency in solar cells, suggest its continued use in production of solar energy, on massive scales, for the foreseeable future. Although organics, nanocrystals, nanowires and other new materials hold significant promise, many opportunities continue to exist for research into unconventional means of exploiting silicon in advanced photovoltaic systems. Here, we describe modules that use large-scale arrays of silicon solar microcells created from bulk wafers and integrated in diverse spatial layouts on foreign substrates by transfer printing. The resulting devices can offer useful features, including high degrees of mechanical flexibility, user-definable transparency and ultrathin-form-factor microconcentrator designs. Detailed studies of the processes for creating and manipulating such microcells, together with theoretical and experimental investigations of the electrical, mechanical and optical characteristics of several types of module that incorporate them, illuminate the key aspects.

Research in silicon photovoltaics represents a robust and diverse effort, with foci that seek to improve performance, cost and capabilities of these systems, ranging from structures for light trapping[1–3] to advanced doping techniques[4–7], innovative spherical[8–10], rectangular[11–13] and ultrathin[14–16] cell designs and advanced manufacturing techniques[17,18]. The results presented here contribute to this progress by introducing practical means to create and manipulate monocrystalline Si solar cells that are much thinner (down to ~100 nm, or limited only by junction depth) and smaller (down to a few micrometres) than those possible with other process technologies[19–21]. The small sizes of the cells and the room-temperature schemes for integrating them into modules enable the use of thin, lightweight flexible substrates for ease of transport and installation. The ability to define the spacings between cells in sparse arrays provides a route to modules with engineered levels of transparency, thereby creating opportunities for use in windows and other locations that benefit from this feature. Alternatively, such layouts of cells can be combined with moulded micro-optic concentrators to increase the power output and provide an unusual appearance with some aesthetic appeal. Such design attributes, together with the thin geometries of the microcells (μ-cells), are also advantageous because they can optimally balance optical absorption and carrier separation/collection efficiency with materials usage and purity requirements to reduce system cost. The following describes these aspects, beginning with the materials and integration strategies, and following with characteristics of the μ-cells and various different modules that incorporate them.

Figure 1a schematically illustrates the steps for fabricating ultrathin, monocrystalline silicon solar μ-cells along with methods for integrating them into interconnected modules (Fig. 1b). The process, which builds on our recent work in single-crystalline silicon for flexible electronics[22–24], begins with delineation of the lateral dimensions of microbar (μ-bar) structures on a Si(111) p-type, boron-doped, single-crystalline Czochralski wafer with a resistivity of 10–20 Ω cm, which we refer to as the source wafer,

When citing this article, please cite the original version as shown on the contents page of this chapter.

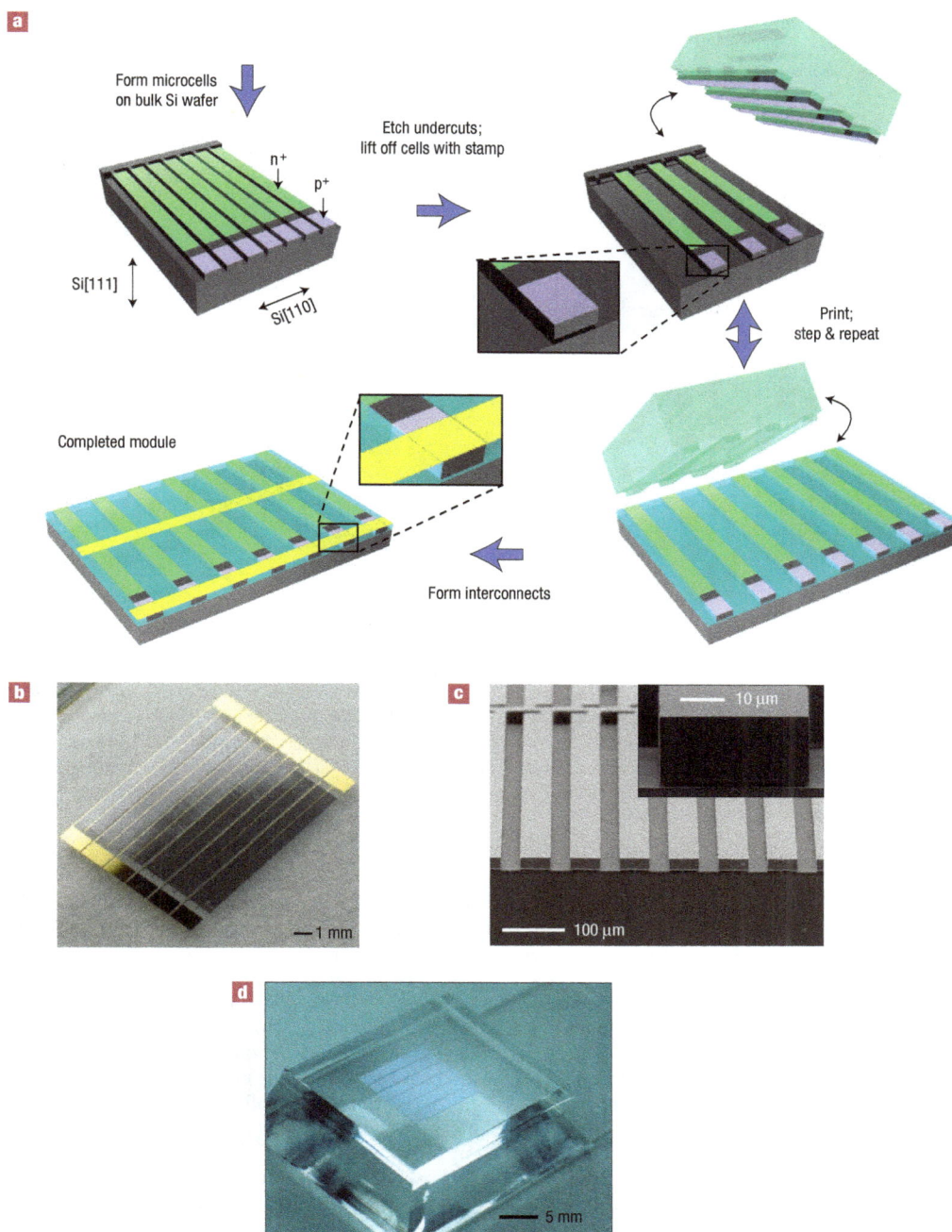

Figure 1 Schematic illustrations, scanning electron microscopy (SEM) image and optical images of key steps in the fabrication of monocrystalline silicon photovoltaic modules that incorporate arrays of microscale solar cells (μ-cells). **a**, Schematic illustration of steps for fabricating ultrathin μ-cells from a bulk wafer, printing them onto a target substrate and forming electrical interconnections to complete a module. **b**, Optical image of a completed module consisting of printed μ-cell arrays, interconnected by metal grid lines (Cr–Au, width ~80 μm, thickness ~0.6 μm) that each connect 130 μ-cells. **c**, SEM image of an array of μ-cells on a source wafer, ready for printing, after doping and KOH undercut. The inset shows a magnified cross-sectional SEM image of a typical μ-cell, with thickness of ~20 μm. **d**, Optical image of an array of μ-cells on a flat elastomeric poly(dimethylsiloxane) (PDMS) stamp, immediately after retrieval from a source wafer.

by etching through a patterned mask. Aligning the lengths of these structures perpendicular to the Si(1̄10) direction of the wafer places their long axes along the preferential {110} etching plane for anisotropic, undercut etching with KOH. Regions of narrowed widths at the ends of the μ-bars serve as anchors to retain

their lithographically defined positions throughout the processing. Maintaining sharp-angled corners at the positions of these anchors leads to stress focusing for controlled fracture[25] in the printing step, as described below. After etching, selective-area diffusion of boron (p+) and phosphorus (n+) from solid doping sources

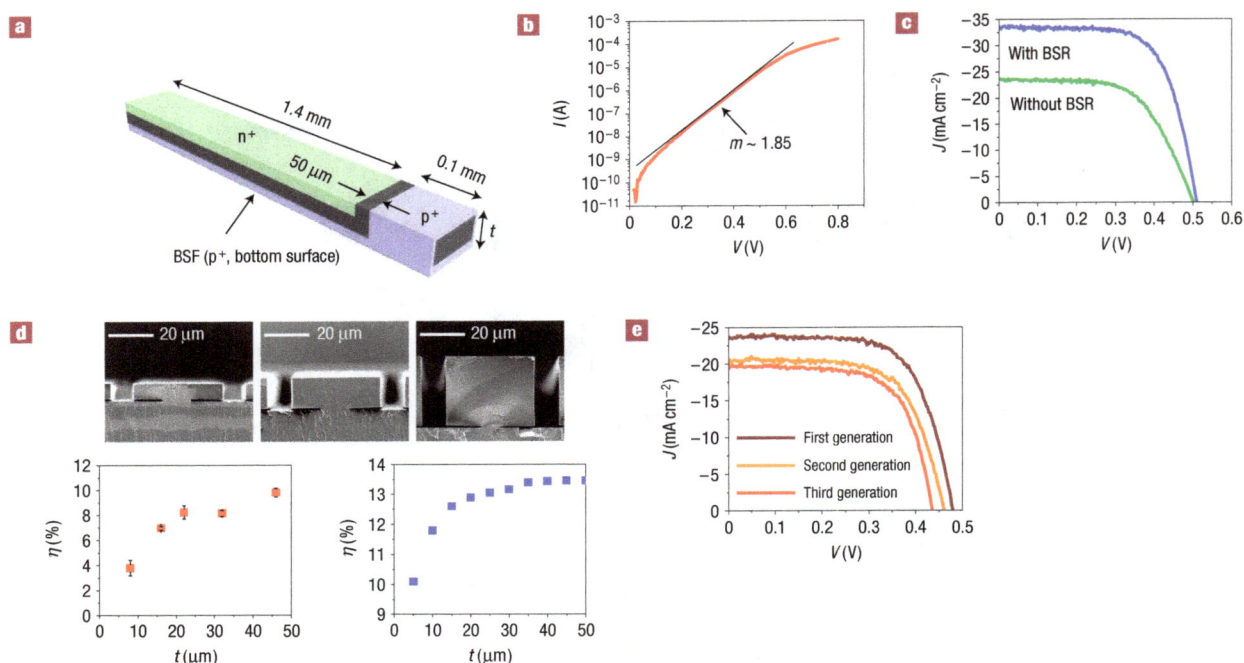

Figure 2 Doping layout and performance characteristics of individual μ-cells. a, Schematic illustration of a μ-cell, showing the dimensions and the doping profiles. **b,** Semilog plot of the forward-bias dark-current (I)–voltage (V) characteristics of an individual μ-cell. The linear fit corresponds to a diode ideality factor (m) of ~1.85. **c,** Representative current-density (J) and voltage (V) data from an individual μ-cell with thickness of ~15 μm under Air Mass 1.5 (AM 1.5) illumination of 1,000 W m^{-2}, with and without a white diffuse backside reflector (BSR). **d,** SEM images, experimental efficiency (η) data (with a metallic BSR) and PC-1D software modelling results corresponding to studies of the scaling properties with thicknesses between ~8 and ~45 μm. **e,** Light J–V curves of individual μ-cells corresponding to the first, second and third generations from a single source wafer, with thickness of ~15 μm.

through patterned diffusion barriers of SiO_2 creates rectifying pn junctions and top contacts. Deposition of etch masks (SiO_2–Si_3N_4, Cr–Au) on the top surfaces and sidewalls of the μ-bars followed by KOH etching releases them from the source wafer everywhere except at the positions of the anchors. Boron doping at the exposed bottom surfaces of the μ-bars, again using a solid doping source, creates a back-surface field to yield fully functional Si solar μ-cells. Figure 1c provides a scanning electron micrograph of a representative array of μ-cells on a source wafer where the bars have lengths (L), widths (W) and thicknesses (t) of 1.55 mm, 50 μm and 15 μm, respectively.

These μ-cells can be selectively retrieved, by controlled fracture at the anchors, with a soft, elastomeric stamp (Fig. 1d) and then printed onto a substrate, in a room-temperature process with overall yields of ~99.9% (ref. 26). Defining electrodes by an etch-back process after metal evaporation (Fig. 1b), by evaporation of metal through a shadow mask or by direct ink writing (see Supplementary Information, Fig. S1) interconnects the μ-cells and completes the fabrication process. Figure 1b shows a module that incorporates 130 μ-cells on a glass substrate, where a photo-cured polymer (NOA61, Norland Products Inc.) serves as a planarizing layer and as an adhesive for the printing process. This device was fabricated with a flat stamp to create a system with μ-cells in arrangements that match those on the source wafer. Stamps with appropriately designed relief features (see Supplementary Information, Figs S2–S3) can retrieve selected sets of μ-cells and print them in layouts (for example spacings between adjacent μ-cells) that are different from those on the source wafer[27]. For example, the μ-cells can be printed in sparse arrays, using a

step-and-repeat process to enable overall module sizes that are much larger than that of the source wafer. These layouts enable semitransparent and micro-optic concentrator module designs, as described below. The source wafer can be reprocessed (following a surface re-polishing step using KOH etching, see Supplementary Information) after all of the μ-cells are retrieved, to yield new generations of cells for additional rounds of printing. This process can be repeated until the entire wafer is consumed.

Figure 2a schematically illustrates the layout of a representative μ-cell design, highlighting the details of the doping profiles. An individual cell ($L = 1.55$ mm) consists of phosphorus-doped ($L_{n+} = 1.4$ mm), boron-doped ($L_{p+} = 0.1$ mm) and un-doped ($L_p = 0.05$ mm) regions, respectively. The thicknesses, t, can be selected by suitable processing to lie between tens of micrometres and hundreds of nanometres. The boron-doped region on the top of the cell connects to the back-surface field on the bottom through doping on the sidewalls, in a manner that enables access to both emitter (n^+) and base (p^+) contacts on the top surface. This configuration greatly simplifies the process of electrical interconnection to form modules, by providing both contacts on the same side of the device. Surface doping concentrations of n^+ (phosphorus), p^+ (boron) and back-surface field (boron) regions are ~1.2×10^{20} cm^{-3}, ~1.8×10^{20} cm^{-3} and ~5.8×10^{19} cm^{-3}, respectively, as measured by secondary-ion mass spectrometry[28] (see Supplementary Information). To fabricate interconnects with high yields, we identified two convenient means for planarizing the relief associated with the μ-cells and for ensuring electrical isolation of the emitter and the base. The first uses a photocurable polymer as both an adhesive and planarization

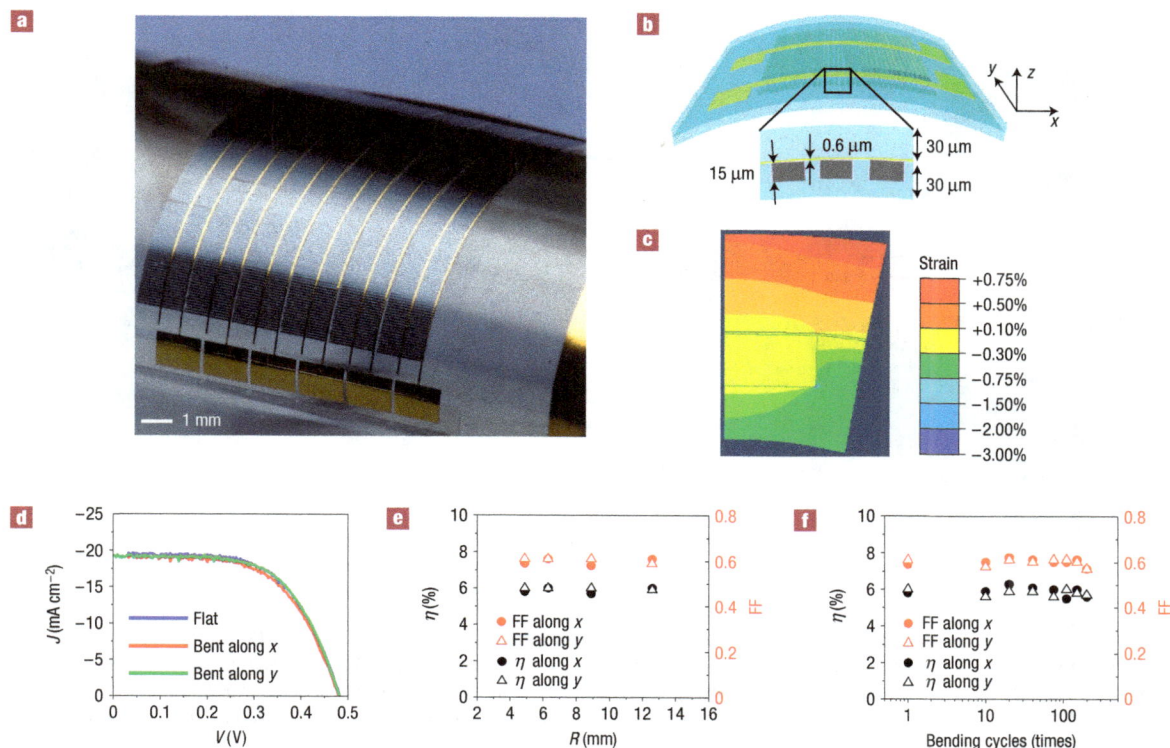

Figure 3 Optical image, schematic illustration, mechanics modelling and photovoltaic performance of mechanically flexible modules that incorporate arrays of interconnected μ-cells. **a**, Optical image of a module bent along a direction parallel to the widths of the μ-cells, to a bending radius (R) of 4.9 mm. **b**, Schematic illustration of an optimized design in which the neutral mechanical plane is positioned near the centre of the μ-cells (grey) through judicious choices of thickness for the polymer (blue) substrate and overcoat. **c**, Colour contour plot of calculated bending strains (ε_{xx}) through the cross-section of a mechanically flexible μ-cell module, bent along the cell width direction at R = 4.9 mm. The calculations use symmetry boundary conditions for evaluation of a single unit cell of the system. The black lines delineate the boundaries of the μ-cell and metal interconnect line (top). **d**, J–V data from a module under AM 1.5 illumination in a flat configuration and bent along the cell width (x) and length (y) directions, both for R = 4.9 mm. **e**, Plot of η and fill factor (FF) under AM 1.5 illumination for R = 12.6, 8.9, 6.3 and 4.9 mm. **f**, Plot of η and fill factor as a function of bending cycles up to 200 times at R = 4.9 mm.

medium, as described in the context of Fig. 1 (see Supplementary Information, Fig. S4), such that a single step accomplishes both printing and planarization. Here, arrays of μ-cells on the stamp press down into a liquid, photocurable polymer (NOA61) coated on the receiving substrate. The polymer fills the empty space between the μ-cells by capillary action. Curing by ultraviolet exposure through the transparent stamp and then removing the stamp completes the process. The flat surface of the stamp coincides precisely with the top surfaces of the μ-cells, to define the planarized surface of the module. Another approach (see Supplementary Information, Fig. S5), which is better suited to a step-and-repeat process, involves printing μ-cells on a substrate that is coated with a layer (~10 μm thick) of cured PDMS (Dow Corning) as a soft, elastomeric adhesive. Covering the printed μ-cells with thin layers of SiO$_2$ (~150 nm thick) and NOA61 (~30 μm thick), pressing a flat piece of PDMS on top of the structure and then ultraviolet curing through the stamp accomplishes planarization with a tolerance (less than 1 μm) similar to that achieved in the first approach. In this second method, a short oxygen reactive-ion etching step is often needed to remove the thin, residual layer of NOA61 that tends to coat partially the top surfaces of the μ-cells. In both approaches, the shallow junction depth (~0.3 μm) creates challenging demands on the extent of planarization. Extending the phosphorus doping down the sidewalls, to a distance of

~1/3 of the μ-cell thickness, as illustrated in Fig. 2a, relaxes the requirements on planarization. Direct ink writing and other approaches that form conformal electrodes provide further benefits in this sense.

I–V measurements of individual μ-cells and completed modules were made in the dark and in a simulated AM 1.5 illumination condition of 1,000 W m^{-2} at room temperature. Figure 2b shows a representative dark I–V curve recorded from an individual μ-cell under forward bias, indicating a diode ideality factor (m) of ~1.85 at room temperature. Figure 2c shows I–V curves from typical μ-cells with and without a backside reflector (BSR) under AM 1.5 illumination, evaluated without metal contacts or antireflection coatings. Without a BSR, this μ-cell, which has t ~ 15 μm, shows a short-circuit current density, J_{sc}, of 23.6 mA cm^{-2}, an open-circuit voltage, V_{oc}, of 503 mV, a fill factor of 0.61 and an overall solar-energy conversion efficiency (η) of 7.2%, where the calculations relied on the spatial dimensions of the μ-cells rather than the surface area of the p–n junction. We also do not explicitly account for contributions from light incident on the edges of the cells. The device-to-device variations in properties of the μ-cells of 15–20 μm thickness without BSR are typically in the range of 6–8% (10–13% with BSR) for η and 450–510 mV for V_{oc}.

In this ultrathin regime, the absorption length of monocrystalline Si for near-infrared and visible wavelengths is

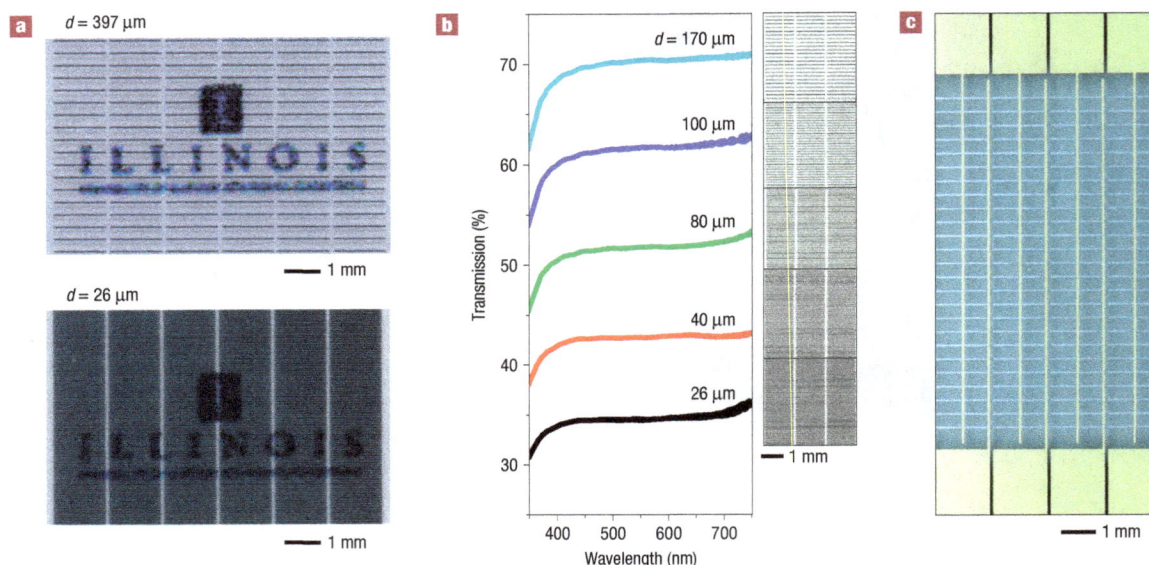

Figure 4 Optical images and transmission spectra of printed, semitransparent μ-cell arrays and interconnected modules. a, Optical images of printed μ-cell arrays on PDMS- (~10 μm thickness) coated polyethylene terephthalate substrates (~50 μm thickness) at inter-cell spacings (*d*) of 26 μm and 397 μm, respectively, resting on a piece of paper with text and logos to illustrate the differences in transparency. **b,** Transmission spectra recorded at normal incidence through printed semitransparent μ-cell arrays with *d* = 26 μm, 40 μm, 80 μm, 100 μm and 170 μm, respectively. Corresponding optical images of μ-cell arrays are also shown. **c,** Optical image of an interconnected semitransparent module with *d* = 397 μm.

greater than or comparable to *t* (refs 29,30). As a result, the efficiency can be improved significantly by adding structures for light-trapping and/or a BSR. The top curve in Fig. 2c shows the effects of a diffuse white BSR, where J_{sc} and η increase to 33.6 mA cm^{-2} (~42% increase) and 11.6% (61% increase), respectively. The J_{sc} value without a BSR in Fig. 2c is close to the theoretical maximum of ~26 mA cm^{-2} that would be expected on the basis of the solar spectrum and absorption coefficient of Si, suggesting that the surface and contact recombination in the device was modest under short-circuit conditions. With the BSR the gain in J_{sc} to 33.6 mA cm^{-2} is consistent with a 56 μm equivalent thickness (on the basis of the required thickness for sufficient absorption of light)[29]. The much higher optical path length shows that the BSR is working well.

To further examine the dependence of performance on thickness, we tested μ-cells with *t* between ~8 and ~45 μm and compared the measurements with numerical simulation of conventional cells using PC-1D software[31], in vertical-type (n$^+$–p–p$^+$) configurations (see Supplementary Information, Fig. S6 and Table S1). Figure 2d shows the results, which indicate sharp increases in efficiency with thickness up to ~15 μm, followed by a gradual saturation from 20 to 30 μm to a plateau above ~40 μm. Increases in efficiency with *t* are due mainly to increased absorption associated with the longer optical path lengths. For *t* above ~40 μm, however, the total absorption does not increase significantly, though the bulk recombination of minority carriers does. Although there are some quantitative differences between measurement and theory owing to non-ideal features of the μ-cells (such as edge surface recombination due to un-passivated surfaces), the qualitative trends are consistent. These observations highlight the value of ultrathin (that is, less than 40 μm) cell designs, both in optimizing materials usage and in minimizing sensitivity to impurities that can lead to trapping of carriers. As described previously, multiple generations of such ultrathin

cells can be created from a single wafer. Figure 2e shows results from first-, second- and third-generation devices produced from a single source wafer in conventional vertical-type (n$^+$–p–p$^+$) cell configurations. Only moderate changes, comparable to typical cell-to-cell variations in properties, are observed. Improved doping profiles, ohmic contacts, antireflection coatings, surface texturization, light trapping structures, surface passivation layers and other advanced designs for monocrystalline Si cells can all be implemented within the schemes described here; each has the potential to provide improvements over the performance indicated in Fig. 2.

The μ-cell designs and printing techniques enable new opportunities at the module level, with performance consistent with that of the individual cells. For example, the sequence in Fig. 1 separates high-temperature processing steps from the module substrate. As a result, integration of μ-cells on rollable, plastic sheets, for ease of transport and installation, is possible. High levels of bendability can be achieved by exploiting optimized mechanical designs. The example shown in Fig. 3a,b involves a composite structure consisting of a planarizing/adhesive layer (NOA61; thickness ~ 30 μm), which also serves as the substrate, arrays of μ-cells and metal interconnects, and a polymer encapsulation layer (NOA61; thickness ~30 μm). Spin-coating and then curing this encapsulation layer represents the final step in the fabrication sequence (see Supplementary Information, Fig. S7). Analytical modelling indicates that this module design places the neutral mechanical plane near the centre of the Si μ-cells, such that the maximum strains in the silicon and metal interconnects (see Supplementary Information, Figs S8,S9), by far the most brittle materials in the system, are less than 0.3% even for bend radii less than 5 mm, for bending in any direction (that is, inward or outward, along the lengths of the μ-cells or perpendicular to them) (see Supplementary Information, Figs S10,S11). Finite-element modelling, with representative results shown in Fig. 3c, confirms

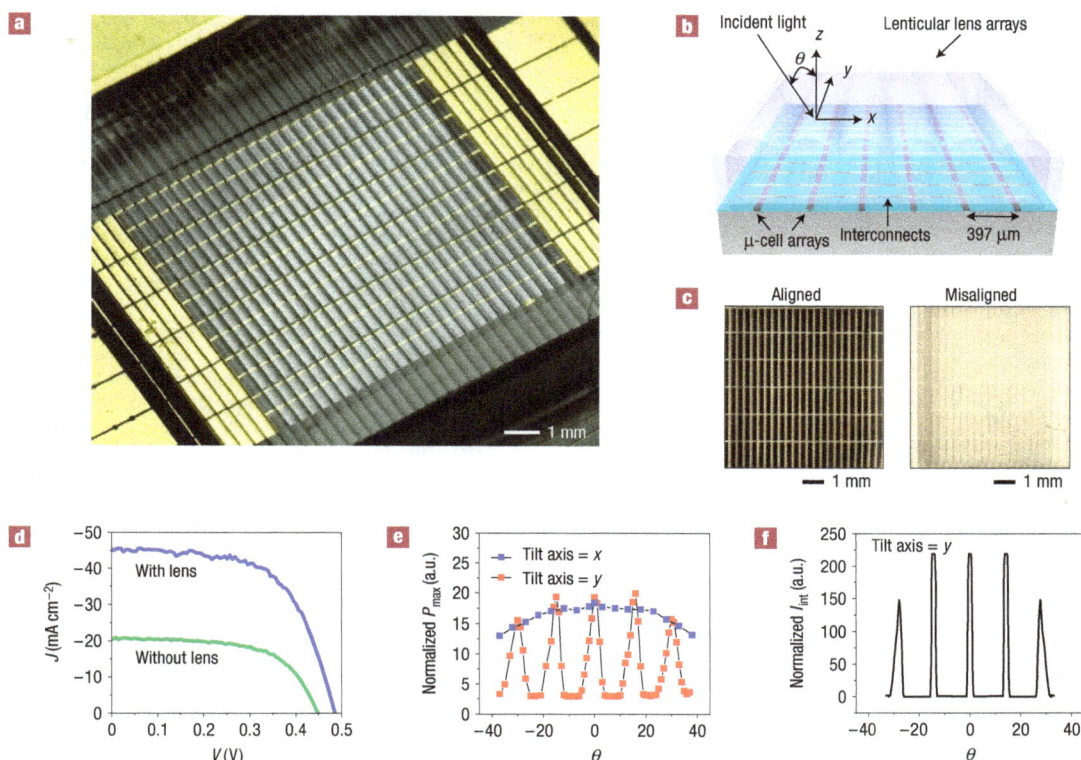

Figure 5 Optical images, schematic illustration and performance characteristics of μ-CPV modules. a, Optical image of a μ-CPV module that combines moulded lenticular lens arrays and printed μ-cells, viewed at an angle that corresponds to alignment of the focal positions of the lenses with the locations of the μ-cells. **b**, Schematic illustration of this type of device. **c**, Optical images of lenticular lens arrays aligned (left) and misaligned (right) to arrays of printed μ-cells, in the layout of the schematic illustration above. **d**, J–V curves of this μ-CPV module with and without the lenticular concentrator optics under AM 1.5 illumination. The lenses in these relatively low-concentration-ratio systems increase the current density and the maximum output power by ~2.5 times. **e**, Normalized output power (P_{out}) from a μ-CPV module as a function of incidence angle (θ) for tilt along the x and y axes. Zero degrees corresponds to normal incidence. **f**, Normalized computed intensity (I_{int}) integrated over the top surface of the μ-cells as a function of θ for tilting parallel to the cell length (with respect to the y axis). The periodicity observed results from focusing of light on μ-cells from neighbouring sets of lenses.

these predictions (see Supplementary Information, Figs S12,S13). Module performance, evaluated in outward bending along and perpendicular to the cell length under AM 1.5 illumination, shows behaviour consistent with expectation on the basis of mechanics analysis and relative insensitivity of the degree of illumination across the modest area of the module, for the bend radii examined here. For example, at bending radii of 12.6, 8.9, 6.3 and 4.9 mm, the module efficiency (~6.0%) and fill factor (~0.60) remain unchanged as summarized in Fig. 3d,e. Fatigue tests, with bending up to 200 cycles, also show little change in performance, as summarized in Fig. 3f. The slightly reduced module efficiency and fill factor compared with the individual cell performance can be partially attributed to the shadowing effect and resistive losses arising from metal interconnects.

Another feature of the module designs and fabrication processes introduced here is their ability to achieve definable levels of optical transparency, which can be valuable for applications in architectural or automotive glass and others. This outcome can be achieved either through the use of extremely thin μ-cells (see, for example, Fig. 2d) or sparse arrays, defined by etching procedures or step-and-repeat printing. This latter approach is particularly easy to implement, and offers a significant degree of control over visually uniform levels of greyscale (that is, individual μ-cells with dimensions reported here are not readily visible to the unaided eye).

Figure 4a shows printed text and logos viewed through arrays of μ-cells with high and low areal coverages, to demonstrate the effect. Automated printers (see Supplementary Information, Fig. S14) enable programmable selection of coverages and, therefore, levels of transparency, for any given arrangement of μ-cells on the source wafer. Figure 4b shows normal-incidence transmission spectra and optical micrographs for cases of cell spacings ranging from 170 to 26 μm (areal coverages from ~20% to 60%), corresponding to levels of transparency from ~70% to ~35%, all generated from arrays of μ-cells on a single source substrate. The transmittance in each case is constant throughout the visible range, and increases approximately linearly with areal coverage, as expected. Figure 4c provides an image of a completed module, with interconnects, consisting of μ-cells at a spacing of 397 μm.

For cells in such layouts, concentrator photovoltaic designs that use integrated micro-optic focusing elements for ultrathin-form-factor microconcentrator photovoltaic (μ-CPV) systems can improve the module's output power. Here, we demonstrate this possibility with moulded arrays of cylindrical lenses, for possible implementation with a single-axis tracker. These devices use arrays of μ-cells with spacings (~397 μm) that match the layouts of low-cost, commercially available lenticular lens arrays (Edmund Optics), from which we could form replicas by soft lithographic moulding of a composite silicone-based epoxy resin that was thermally matched to the photovoltaic module by filling with silica nanoparticles[32]. The radius of curvature of the commercial and replicated cylindrical microlenses was ~0.83 mm, corresponding to a focal length of ~2.2 mm. With collimated light, the widths of the focused lines of light (full-width at 90% maximum) were ~35 μm. We aligned the lens arrays to interconnected arrays of μ-cells using a thin PDMS film as a spacer, and a coupler on an XYZ and angle-controlled stage. Figure 5a,b shows an optical image and a schematic illustration of such a μ-CPV device. Figure 5c presents images corresponding to the cases when the lens arrays are aligned and misaligned to the μ-cells. In the aligned state, the module seems to incorporate silicon at a nearly full areal coverage. When misaligned, the system assumes the colour of the module substrate, and the silicon is invisible. The I–V characteristics of a module with and without aligned lens arrays, under AM 1.5 illumination, are shown in Fig. 5d. The maximum output power with the lenses is ~2.5 times larger than that without the lenses. This ratio is somewhat smaller than the expectation on the basis of simple estimates (see Supplementary Information), owing partly to the relatively large size of the light source in the solar simulator (91192-1000W, Oriel) and its close proximity to the module. These features result in a degree of collimation that is both non-ideal and substantially less than that of sunlight. However, the small area and ultrathin microdesigns presented here can in principle lead to consumption of less silicon material than conventional and related microspherical silicon concentrator modules[9]. Owing to the cylindrical geometry of the lenses and the bar shapes of the μ-cells, decreases in output power associated with angular tilting about the x-axis are minimal, as illustrated in Fig. 5e. Rotations about the y-axis cause dramatic changes, consistent with the nature of the optics and the images shown in Fig. 5c. The periodicity observed in this case results from focusing of light on μ-cells from neighbouring sets of lenses. The angular positions and relative values of the first, second and third peaks match well with simulated data from numerical ray-tracing calculations (Fig. 5f).

The types of module reported here may create new possibilities for monocrystalline silicon photovoltaics, particularly in applications that benefit from thin, lightweight construction, mechanical flexibility, semitransparency or the unusual optical properties of the μ-CPV designs. In most cases, we chose materials that have the potential for long lifetime and high reliability. The procedures themselves are compatible with substrates, encapsulation, adhesive and optical materials used in existing photovoltaic systems. Similarly, as noted previously, advanced monocrystalline silicon cell designs and enhancement techniques can also be incorporated for improved performance. Although the focus of the strategies presented here is on module capabilities and designs, rather than cost or performance, a notable feature of these approaches is that the ultrathin cell geometries and, for μ-CPV and semitransparent designs, the sparse coverages represent efficient ways to use silicon. The former aspect can also relax requirements on the purity of the silicon. An obvious consequence of these aspects is the potential to reduce the silicon component of the module cost. Such reductions are balanced, however, by increased processing costs associated with creating and interconnecting the μ-cells. Low-cost printing, doping and etching techniques suitable for high-performance μ-cell and module fabrication, together with other means to reduce cost or increase performance, are, therefore, important areas for further work.

METHODS

FABRICATING MICROCELLS

The fabrication process began with a p-type (111) Czochralski Si wafer (3 inch diameter, 10–20 Ω cm, 375 μm thickness, Montco Silicon Technology) that

was coated with a layer of SiO_2 (~ 600 nm) formed by plasma-enhanced chemical vapour deposition (PlasmaTherm SLR) at 250 °C. Spin casting, exposing (365 nm light, through a Karl Suss MJB mask aligner) and developing a layer of photoresist (AZ5214, Clariant; developer, AZ327MIF, Clariant) formed a pattern that defined the lateral dimensions and layouts of the μ-cells, in rectangular geometries (that is, μ-bars). The SiO_2 not protected by the resist was removed with buffered oxide etchant (6:1, Transene). Inductively coupled plasma reactive-ion etching (STS)[23,24] formed trench structures with typical depths of 15–20 μm in the regions of exposed silicon. The photoresist and remaining SiO_2 were then removed with acetone and hydrofluoric acid (HF, Fisher, 49% concentration), respectively. Selective area doping of top contacts was conducted using solid-state sources of boron (BN-1250, Saint Gobain) and phosphorus (PH-1000N, Saint Gobain) at 1,000 °C under N_2 atmosphere for 30 min (boron) and 10 min (phosphorus). A layer of SiO_2 (900 nm) deposited by plasma-enhanced chemical vapour deposition at 250 °C and patterned by photolithography (photoresist, AZ4620, Clariant; developer, deionized H_2O:AZ400K = 3:1 by volume, Clariant) and etching in buffered oxide etchant served as a doping mask. The doped wafer was then cleaned and coated with SiO_2 (100 nm) and Si_3N_4 (500 nm) by plasma-enhanced chemical vapour deposition at 250 °C, and subsequently with Cr (80 Å) and Au (800 Å) via directional deposition in an electron-beam evaporator (Temescal, FC 1800) at an angle of ±30° with respect to the wafer surface. Reactive-ion etching (PlasmaTherm 790 series) using CHF_3–O_2 (40-2 s.c.c.m., 50 mtorr, 150 W, 7 min) and SF_6 (40 s.c.c.m., 50 mtorr, 100 W, 1 min) exposed regions of Si at the bottoms of the trenches formed by inductively coupled plasma reactive-ion etching. Immersion in KOH (PSE-200, Transene) at 100 °C for ~30 min initiated anisotropic undercut etching at these locations to define the bottom surfaces of the μ-cells with overall yields of over 99%. After removing Au and Cr with commercial etchants (Transene), these bottom surfaces were doped with boron again using the solid-state doping source at 1,000 °C for 5 min. Cleaning of the resulting sample in Piranha solution (H_2SO_4:H_2O_2 = 3:1 by volume, 3 min) and HF completed the process.

FABRICATING ELASTOMERIC STAMPS

Simple, flat stamps for by-hand printing were prepared by curing a PDMS prepolymer and cross-linking agent (Sylgard 184, Dow Corning Corp.) mixed at 10:1 by volume at 75 °C for 2 h. Forming composite stamps suitable for use in our automated printer system involved several steps (see Supplementary Information, Fig. S3). First, the template that defined the geometry of relief on the stamp was prepared on a Si wafer (4 inch diameter) by optical lithography using a negative-tone photoresist (SU8-50, 100 μm thickness, Microchem) and a developer (SU-8 developer, Microchem). This substrate was then exposed to a vapour of (tridecafluoro-1,2,2-tetrahydrooctyl)-1-tricholorosilane (T2492-KG, United Chemical Technologies) for 3 h at room temperature. A 10:1 PDMS prepolymer mixture was poured onto the substrate, to a thickness of 100–200 μm, and then partially cured at 75 °C for 30 min. A thin glass disc (~0.1 mm thickness, 3 inch diameter, Corning Incorporated) was placed on top, to form a backing layer capable of reducing in-plane deformations during printing. As a final step, another layer of 10:1 PDMS prepolymer mixture was poured on top. The entire composite stamp was heated at 75 °C for 2 h, to complete the curing.

TRANSFER PRINTING MICROCELLS

Transfer printing used a custom-built, automated machine consisting of motion-controlled stages with 1 μm resolution and an optical microscope vision system with a zoom range of 4× to 26×. Vacuum chucks mount on manually controlled rotational stages with 6 arc seconds sensitivity to support the processed wafers and the target substrates and to align them with each other and the relief features of the stamp. These chucks rest on a computer-controlled stage capable of 8 inches of motion in the X and Y directions. A PDMS composite stamp bolts into a vertical printhead assembly that can move in the vertical (Z) direction up to 2 inches. The stamp mount has a square, 3 inch aperture enabling an optical microscope vision system to image through the transparent composite stamp onto the stages below. The steps for printing are as follows.

To ensure high yields, it is critical that all components of the system are properly aligned. The tilt of the PDMS composite stamp relative to the source wafer and target substrate was manually adjusted, with 20 arc seconds of sensitivity, using the vision system for guidance. The μ-cells on the source wafer were aligned to the corresponding relief features on the composite stamp using rotational stages on the XY stage. A two-point calibration must be carried out on the source wafer, target substrate and cleaning substrate (six points in total) to account for tilt in the Y direction as well as misalignment of the XY motion axes relative to the orientation of the stamp.

Unlike alignment, the printing itself was fully automated. XYZ calibration data were first entered into custom software along with the desired spacing and number of rows in the printed cell arrays. This software calculated XYZ data for each pickup, print and cleaning position. The stages used these data to guide the printing process in a step-and-repeat procedure. The cycle time for a single pickup and printing procedure was approximately three minutes. One minute was required for positioning, pickup, printing and cleaning. For two minutes the cells were allowed to rest on the PDMS substrate before printing to increase adhesion.

PLANARIZING MICROCELLS

In planarization method 1, a precleaned substrate (glass or polyethylene terephthalate) was exposed to ultraviolet-induced ozone for 10 min and then spin-coated with an ultraviolet-curable polymer (NOA61, Norland Products Inc.). Retrieved μ-cells on a flat PDMS stamp were placed against this substrate and then the entire system was exposed to an ultraviolet source for ∼30 min to cure the NOA. The PDMS stamp was then slowly peeled from the substrate, leaving planarized μ-cells in a NOA matrix.

In planarization method 2, after printing arrays of μ-cells on a substrate with a thin PDMS coating, SiO_2 (150 nm) was deposited by electron-beam evaporation (Temescal, FC 1800). Spin-coating a layer of NOA61 (∼30 μm) and then contacting a bare, flat PDMS element caused the NOA to flow to conform to and planarize the relief presented by the μ-cells. Curing the NOA by exposure to ultraviolet light followed by removal of the stamp and, sometimes, a brief exposure of the substrate to an oxygen reactive-ion etch (10 s.c.c.m., 50 mtorr, 150 W, 2–3 min) completed the process.

FABRICATION OF MICROCONCENTRATORS

A commercially available cylindrical lens array (Edmund Optics NT43-028) served as a 'master' for the formation of replica lenses by soft lithography. The process began with cleaning of the master in soapy water under ultrasonic vibration for 20 min, followed by the same process with deionized water, and finally blowing the structure dry with compressed nitrogen. This cleaned lens master was then exposed to a vapour of (tridecafluoro-1,2,2-tetrahydrooctyl)-1-tricholorosilane (T2492-KG, United Chemical Technologies) for 1 h. A glass spacer of 1 mm thickness placed between the lens master and a glass backing plate prepared the system for casting and curing of a 10:1 mixture of PDMS prepolymer and curing agent (Sylgard 184, Dow Corning Corp.) at room temperature for 48 h. Peeling away yielded a PDMS mould on a glass backing plate. A separate, optically flat PDMS slab on a glass backing plate was prepared in a similar way, with a flat silicon wafer instead of the lens master. The photocurable polymeric material from which lens array replicas were made was prepared using commercially available 9–15 nm silica nanoparticles (IPA-ST, Nissan Chemicals, Ltd), a silicone-based epoxy resin (PCB 35-54B, Polyset) and a coupling agent (3-glycidyloxypropyl)trimethoxysilane (Sigma-Aldrich) according to published procedures[32]. After exposure to a vapour of (tridecafluoro-1,2,2-tetrahydrooctyl)-1-tricholorosilane for 1 h, the negative mould and the flat PDMS surface were assembled with a ∼2.2 mm spacer. The photocurable polymer prepared as above was poured into the cavity and cured under ultraviolet (9 mW cm^{-2}) exposure for 10 min. Removing the flat PDMS and negative mould completed the fabrication of the replica lens array. The long-term stability and cost of the moulded lens arrays are expected to be reasonable for solar applications owing to enhanced thermal, optical and mechanical properties of the composite system and the inexpensive soft-lithography replication process.

ELECTRICAL AND OPTICAL MEASUREMENTS

Light and dark $I–V$ measurements of μ-cells were carried out at room temperature using a d.c. source meter (model 2400, Keithley) operated by LabVIEW5, and a 1,000 W full-spectrum solar simulator (model 91192, 4 × 4 inch source diameter, ±4° collimation, Oriel) equipped with AM 0 and AM 1.5 direct filters. The input power of light from the solar simulator was measured with a power meter (model 70260, Newport) and a broadband detector (model 70268, Newport) at the point where the sample's top surface was placed. A typical $I–V$ scan was conducted between −0.5 V and +0.6 V with a 2.2 mV increment (500 data points). $I–V$ measurements of individual

μ-cells were made after transfer printing, with cells in an array format on a glass substrate. The reported efficiencies are based on the top surface area of the μ-cell, without specifically accounting for sidewall coupling of light. For electrical characterization of the mechanically flexible μ-cell arrays, the completed module was attached to the outer surfaces of glass test tubes with various radii. The centre of the module was aligned for normal-incidence illumination (see Supplementary Information, Fig. S15). Owing to the small sample size (∼5.1 mm × 1.6 mm), the average flux of light into the module's active area in the flat state is only slightly larger (∼1.05 times) than that even in the most highly bent state studied here (that is, $R = 4.9$ mm). Light and dark $I–V$ measurements at various bending geometries (that is, outward bending, along the cell length and the cell width) and bending radii were then made at room temperature. For fatigue tests, one bending cycle was defined such that the module was bent, relaxed to a flat state and bent again over the test tube. $I–V$ measurements were conducted at bent states after a selected number of bending cycles. $I–V$ measurements of the μ-CPV module were made under the same experimental conditions as non-concentrated modules without an additional active cooling system, where the $I–V$ characteristics were maintained as stable as the non-concentrated case throughout the entire measurement process (a few hours).

Optical transmission spectra of semitransparent μ-cell modules were obtained at normal incidence of light using an ultraviolet–visible–near-infrared spectrophotometer (CARY 5G, Varian). Data collection was conducted with an aperture of ∼5 mm diameter in the wavelength range between 350 and 750 nm, where baseline corrections for 0% (no light) and 100% (air) transmission were made.

OPTICS SIMULATION

The calculation was carried out with a commercial ray-tracing package (Rayica 3.0, Optica Software). We assumed that the rays of light were incident at one angle and had a wavelength of 550 nm, that the lens array was infinite and that Fresnel reflections were negligible. The curved surface of the lens array was profiled experimentally and fitted to a parabola; the width of each lens was ∼0.4 mm and its centre thickness was ∼2.2 mm. The lens material was taken to be BK7 glass (for the purposes of this calculation). The lens array was positioned 0.1 mm from the top surface of the μ-cells, whose width was 0.05 mm. The integrated top surface intensity is an imperfect predictor of the power incident and absorbed by the μ-cells. However, the periodicity of the μ-cell response with incident angle is captured.

Received 16 June 2008; accepted 1 September 2008; published 5 October 2008.

References

1. Campbell, P. & Green, M. A. Light trapping properties of pyramidally textured surfaces. *J. Appl. Phys.* **62**, 243–249 (1987).
2. Heine, C. & Morf, R. H. Submicrometer gratings for solar-energy applications. *Appl. Opt.* **34**, 2476–2482 (1995).
3. Feng, N.-N. *et al.* Design of highly efficient light-trapping structures for thin-film crystalline silicon solar cells. *IEEE Trans. Electron Devices* **54**, 1926–1933 (2007).
4. Wenham, S. R., Honsberg, C. B. & Green, M. A. Buried contact silicon solar cells. *Sol. Energy Mater. Sol. Cells* **34**, 101–110 (1994).
5. Sinton, R. A., Kwark, Y., Gan, J. Y. & Swanson, R. M. 27.5-Percent silicon concentrator solar-cells. *IEEE Electron Devices Lett.* **7**, 567–569 (1986).
6. Kerschaver, E. V. & Beaucarne, G. Back-contact solar cells: A Review. *Prog. Photovolt.* **14**, 107–123 (2006).
7. Zhao, J. H., Wang, A. H. & Green, M. A. 24.5% efficiency silicon PERT cells on MCZ substrates and 24.7% efficiency PERL cells on FZ substrates. *Prog. Photovolt.* **7**, 471–474 (1999).
8. Biancardo, M. *et al.* Characterization of microspherical semi-transparent solar cells and modules. *Sol. Energy* **81**, 711–716 (2007).
9. Liu, Z. X. *et al.* A concentrator module of spherical Si solar cell. *Sol. Energy Mater. Sol. Cells* **91**, 1805–1810 (2007).
10. Minemoto, T. & Takakura, H. Fabrication of spherical silicon crystals by dropping method and their application to solar cells. *Jpn. J. Appl. Phys.* **46**, 4016–4020 (2007).
11. Taguchi, M. *et al.* HIT (TM) cells—high-efficiency crystalline Si cells with novel structure. *Prog. Photovolt.* **8**, 503–513 (2000).
12. Weber, K. J. *et al.* A novel low-cost, high-efficiency micromachined silicon solar cell. *IEEE Electron Devices Lett.* **25**, 37–39 (2004).
13. Verlinden, P. J. *et al.* Sliver (R) solar cells: A new thin-crystalline silicon photovoltaic technology. *Sol. Energy Mater. Sol. Cells* **90**, 3422–3430 (2006).
14. Brendel, R., Bergmann, R. B., Lolgen, P., Wolf, M. & Werner, J. H. Ultrathin crystalline silicon solar cells on glass substrates. *Appl. Phys. Lett.* **70**, 390–392 (1997).
15. Brendel, R. Review of layer transfer processes for crystalline thin-film silicon solar cells. *Jpn. J. Appl. Phys.* **40**, 4431–4439 (2001).
16. Tayanaka, H., Yamauchi, K. & Matsushita, T. Thin-film crystalline silicon solar cells obtained by separation of a porous silicon sacrificial layer. *Proc. 2nd World Conf. Photovolt. Sol. Energy Conv.* 1272–1275 (Institute of Electrical and Electronics Engineers (IEEE), 1998).
17. Yamamoto, K. *et al.* Thin-film poly-Si solar cells on glass substrate fabricated at low temperature. *Appl. Phys. A* **69**, 179–185 (1999).
18. Shah, A. *et al. Photovoltaic Specialists Conference, Conference Record of the Twenty-Sixth IEEE* 569–574 (Institute of Electrical and Electronics Engineers (IEEE), 1997).

19. Bergmann, R. B. Crystalline Si thin-film solar cells: A review. *Appl. Phys. A* **69**, 187–194 (1999).
20. Green, M. A. Crystalline and thin-film silicon solar cells: State of the art and future potential. *Sol. Energy* **74**, 181–192 (2003).
21. Kazmerski, L. L. Solar photovoltaics R&D at the tipping point: A 2005 technology overview. *J. Electron Spectrosc. Relat. Phenom.* **150**, 105–135 (2006).
22. Mack, S., Meitl, M. A., Baca, A. J., Zhu, Z. T. & Rogers, J. A. Mechanically flexible thin-film transistors that use ultrathin ribbons of silicon derived from bulk wafers. *Appl. Phys. Lett.* **88**, 213101 (2006).
23. Ko, H. C., Baca, A. J. & Rogers, J. A. Bulk quantities of single-crystal silicon micro-/nanoribbons generated from bulk wafers. *Nano Lett.* **6**, 2318–2324 (2006).
24. Baca, A. J. *et al.* Printable single-crystal silicon micro/nanoscale ribbons, platelets and bars generated from bulk wafers. *Adv. Funct. Mater.* **17**, 3051–3062 (2007).
25. Meitl, M. A. *et al.* Stress focusing for controlled fracture in microelectromechanical systems. *Appl. Phys. Lett.* **90**, 083110 (2007).
26. Meitl, M. A. *et al.* Transfer printing by kinetic control of adhesion to an elastomeric stamp. *Nature Mater.* **5**, 33–38 (2006).
27. Lee, K. J. *et al.* Large-area, selective transfer of microstructured silicon: A printing-based approach to high-performance thin-film transistors supported on flexible substrates. *Adv. Mater.* **17**, 2332–2336 (2005).
28. Wilson, R. G., Stevie, F. A. & Magee, C. W. *Secondary Ion Mass Spectrometry: A Practical Handbook for Depth Profiling and Bulk Impurity Analysis* (Wiley, 1989).
29. Hull, R. (ed.) *Properties of Crystalline Silicon* (The Institution of Electrical Engineers (IEE), 1999).
30. Budianu, E., Purica, M., Rusu, E., Manea, E. & Gavrila, R. *Semiconductor Conf. 2002. CAS 2002 Proc. Int.* Vol. 1 (Institute of Electrical and Electronics Engineers (IEEE), 2002).
31. Clugston, D. A. & Basore, P. A. *Photovoltaic Specialists Conf. 1997., Conf. Record of the Twenty-Sixth IEEE* 207–210 (Institute of Electrical and Electronics Engineers (IEEE), 1997).
32. Kunnavakkam, M. V. *et al.* Low-cost, low-loss microlens arrays fabricated by soft-lithography replication process. *Appl. Phys. Lett.* **82**, 1152–1154 (2003).

Supplementary Information accompanies the paper at www.nature.com/naturematerials.

Acknowledgements

We thank T. Banks, K. Colravy and D. Sievers for help with processing, T. Spila for assistance with secondary-ion mass spectrometry measurements and H. Kim and D. Stevenson for help with photography. The materials parts of this effort were supported by the US Department of Energy (DoE), Division of Materials Sciences, under award DE-FG02-07ER46471, through the Materials Research Laboratory (MRL). The general characterization facilities were provided through the MRL with support from the University of Illinois and from DoE grants DE-FG02-07ER46453 and DE-FG02-07ER46471. The mechanics theory and the transfer-printing systems were developed under support from the Center for Nanoscale Chemical Electrical Mechanical Manufacturing Systems at the University of Illinois (funded by the NSF under grant DMI-0328162). J.Y. and J.B.G. acknowledge support from a Beckman postdoctoral fellowship. A.J.B. acknowledges support from the Department of Defense Science, Mathematics and Research for Transformation (SMART) fellowship.

Author contributions

J.Y., J.A.L., R.G.N., P.M.F., Y.H., A.R. and J.A.R. designed the experiments. J.Y., A.J.B., S.-I.P., P.E., L.L., R.H.K., T.-H.K., M.J.M., B.Y.A., E.B.D., J.A.L., R.G.N., P.M.F., Y.H., A.R. and J.A.R. carried out experiments. J.Y., A.J.B., S.-I.P., J.A.L., R.G.N., P.M.F., Y.H., A.R. and J.A.R. analysed the data. P.E., J.Y., R.H.K., T.-H.K., J.A.L., R.G.N., P.M.F., Y.H., A.R. and J.A.R. contributed to transfer printing. J.X., S.W., Y.H., J.A.L., R.G.N., P.M.F., Y.H., A.R. and J.A.R. contributed to mechanics modelling and analysis. J.B.G., J.A.L., R.G.N., P.M.F., Y.H., A.R. and J.A.R. contributed to optics simulation. J.Y., A.J.B., J.A.L., R.G.N., P.M.F., Y.H., A.R. and J.A.R. wrote the paper.

Author information

Reprints and permissions information is available online at http://npg.nature.com/reprintsandpermissions. Correspondence and requests for materials should be addressed to J.A.R.

Efficiency enhancement in low-bandgap polymer solar cells by processing with alkane dithiols

J. PEET, J. Y. KIM, N. E. COATES, W. L. MA, D. MOSES, A. J. HEEGER* AND G. C. BAZAN*

Center for Polymers and Organic Solids, University of California at Santa Barbara, Santa Barbara, California 93106, USA
*e-mail: ajhe@physics.ucsb.edu; Bazan@chem.ucsb.edu

Published online 27 May 2007; Corrected 4 June 2007 (see details online); doi:10.1038/nmat1928

High charge-separation efficiency combined with the reduced fabrication costs associated with solution processing and the potential for implementation on flexible substrates make 'plastic' solar cells a compelling option for tomorrow's photovoltaics[1]. Attempts to control the donor/acceptor morphology in bulk heterojunction materials as required for achieving high power-conversion efficiency have, however, met with limited success[2–4]. By incorporating a few volume per cent of alkanedithiols in the solution used to spin-cast films comprising a low-bandgap polymer and a fullerene derivative, the power-conversion efficiency of photovoltaic cells (air-mass 1.5 global conditions) is increased from 2.8% to 5.5% through altering the bulk heterojunction morphology[5]. This discovery can potentially enable morphological control in bulk heterojunction materials where thermal annealing is either undesirable or ineffective.

Bulk heterojunction 'plastic' solar cells are based on phase-separated blends of polymer semiconductors and fullerene derivatives[6–8]. Because of self-assembly on the nanometre length scale, mobile carriers and excitons formed after absorption of solar irradiation diffuse to a heterojunction before recombination and are dissociated at the polymer/fullerene interface[6–8]. Ultrafast charge transfer from semiconducting polymers to fullerenes guarantees that the quantum efficiency for charge transfer at the interface approaches unity[9–11], with electrons on the fullerene network and holes on the polymer network. After breaking the symmetry by using different metals for the two electrodes, electrons migrate towards the lower work-function metal and holes migrate towards the higher work-function metal. Despite high charge-separation efficiency, a significant fraction of carriers recombine at donor/acceptor interfaces before extraction from the device owing, in part, to the inherently random interpenetrating network morphology formed after spin-casting. Carrier recombination before reaching the electrodes and low mobility limit both the device fill factor (FF) and the overall photon harvesting by reducing the optimum active-layer thickness[12]. The carrier lifetime is largely controlled by the phase morphology between the donor and acceptor materials[13]. Although significant advances in the performance of polymer-based photovoltaic devices have been made during the past few years[3,14–16], the ability to control the morphology of the donor/acceptor network is critical to optimizing efficiency.

While investigating the use of alkanethiol-stabilized gold nanoparticles for polymer solar-cell applications, we discovered that by incorporating small concentrations of alkanethiols

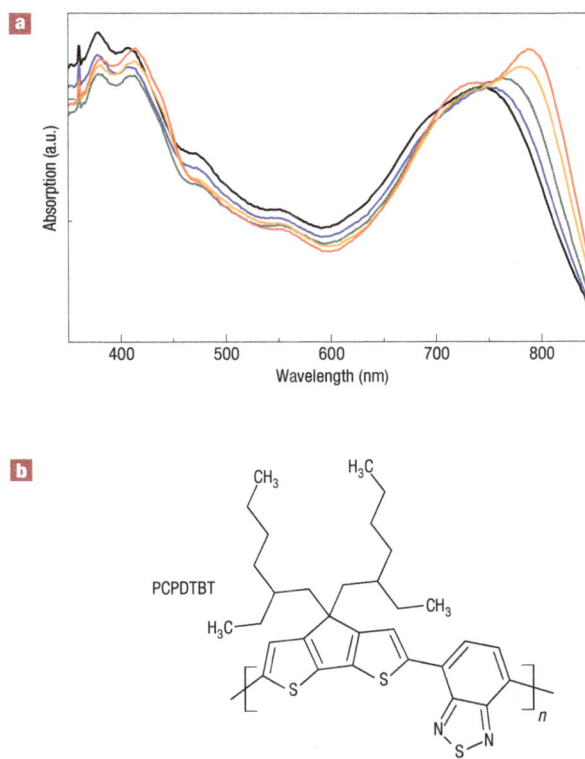

Figure 1 Ultraviolet–visible absorption. a, The spectral response indicates a red-shift from a pure PCPDTBT:C_{71}-PCBM film (black line) to films cast from chlorobenzene containing 24 mg ml^{-1} 1,3-propanedithiol (blue line), 1,4-butanedithiol (green line), 1,6-hexanedithiol (orange line) and 1,8-octanedithiol (red line). **b**, The molecular structure of PCPDTBT.

in the solutions from which the bulk heterojunction films are cast, the poly(3-hexylthiophene) (P3HT) aggregation and P3HT/fullerene phase separation can be modified[5]. This approach offers the potential to introduce morphology control to bulk heterojunction materials during device fabrication without the need for subsequent thermal annealing.

When citing this article, please cite the original version as shown on the contents page of this chapter.

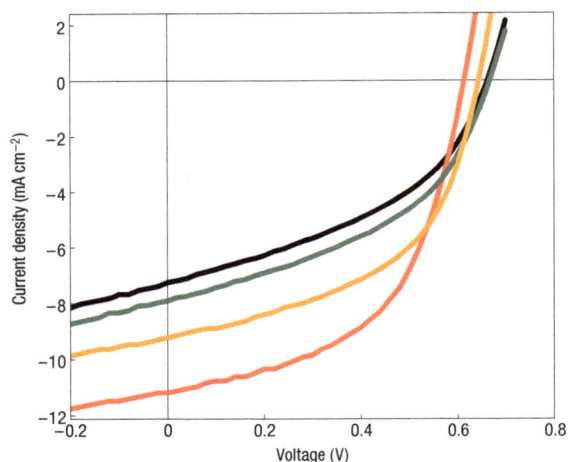

Figure 2 Device I–V characteristics. Current density versus voltage curves (under simulated air-mass 1.5 global, AM 1.5G, radiation at 80 mW cm^{-2}) for a series of PCPDTBT:C$_{71}$-PCBM solar cells. The PCPDTBT:C$_{71}$-PCBM films were cast at 1,200 r.p.m. from pristine chlorobenzene (black line) and chlorobenzene containing 24 mg ml^{-1} butanedithiol (green line), hexanedithiol (orange line) or octanedithiol (red line).

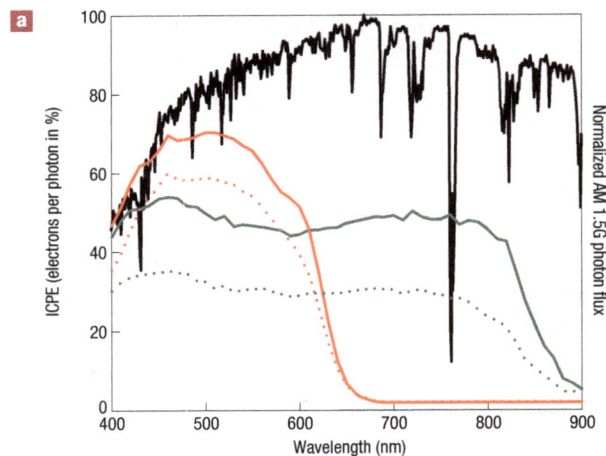

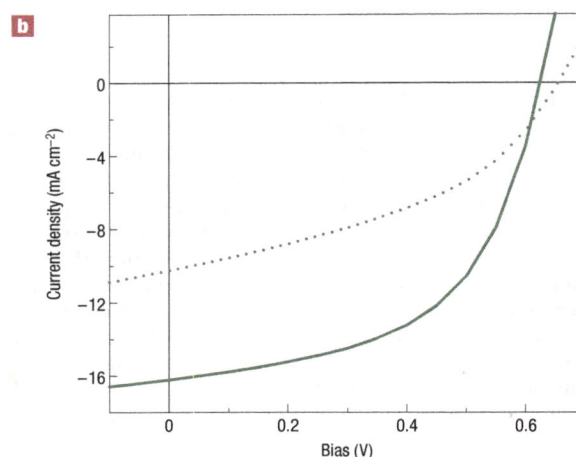

Figure 3 IPCE. a, IPCE spectra of polymer bulk heterojunction solar cells composed of P3HT:C$_{61}$-PCBM before (dotted red line) and after (solid red line) annealing, and PCPDTBT:C$_{71}$-PCBM with (solid green line) and without (dotted green line) the use of 1,8-octanedithiol. The AM 1.5G reference spectrum is shown for reference[17] (black line). **b**, Current–voltage characteristics of the same PCPDTBT devices used for the IPCE measurements, processed with (solid line) and without (dotted line) 1,8-octanedithiol under simulated 100 mW cm^{-2} AM 1.5G illumination; $I_{sc} = 16.2$ mA cm^{-2}, FF = 0.55 and $V_{oc} = 0.62$ V.

The most efficient bulk heterojunction devices so far have used the high-mobility semiconducting polymer, P3HT, as the photo-donor, and the soluble fullerene derivative, [6,6]-phenyl C$_{61}$-butyric acid methyl ester (C$_{61}$-PCBM), as the acceptor[3,14–16]. Despite considerable effort, however, power-conversion efficiencies obtained from P3HT:C$_{61}$-PCBM solar cells are limited to values of approximately 5% owing principally to the poor overlap between the P3HT absorption spectrum and the solar emission spectrum[1,17]. With improved light harvesting in the near-infrared region by a low-bandgap polymer, such as poly[2,6-(4,4-bis-(2-ethylhexyl)-4H-cyclopenta[2,1-b;3,4-b']-dithiophene)-alt-4,7-(2,1,3-benzothiadiazole)] (PCPDTBT), higher power-conversion efficiencies should be attainable[4,18]. The energy gap of PCPDTBT is nearly ideal for solar photovoltaic applications, $E_g = 1.46$ eV; the molecular structure of PCPDTBT is shown in Fig. 1b (ref. 1). The asymmetric C$_{70}$ fullerene was chosen because of its increased absorption in the visible region, which leads to better overlap with the solar spectrum relative to that obtained with the C$_{61}$ analogue[4,19].

Thermal annealing or controlled solvent evaporation after film casting have proved critical for optimizing charge separation and transport within the bulk heterojunction morphology[2,3,14–16]. Unfortunately, attempts to improve the performance of PCPDTBT:C$_{71}$-PCBM solar cells through similar methods have not been successful[4,18].

Figure 1a shows the shift in the film absorption caused by adding different alkanedithiols to the PCPDTBT:C$_{71}$-PCBM solution in chlorobenzene before spin-casting. The addition of 24 mg ml^{-1} of 1,8-octanedithiol into the chlorobenzene causes the largest change; the film absorption peak red-shifts 41 nm to 800 nm. Such a shift to lower energies and the emergence of a structure on the absorption peak associated with the π–π^* transition when films are processed with alkanedithiols indicates that the PCPDTBT chains interact more strongly and that there is improved local structural order compared with films processed from pure chlorobenzene[4,5].

Analysis by Fourier-transform infrared (FTIR) and Raman spectroscopies yielded no resolvable thiol signals after drying in vacuum ($\sim 10^{-3}$ torr) for 10 min at room temperature (when FTIR was measured on wet films before vacuum drying, a small thiol peak was observed). Because of the weak oscillator strength of the S–H stretch, infrared spectroscopy of the PCPDTBT:C$_{71}$–PCBM films is insensitive to residual dithiol (using 2-naphthalene thiol to calibrate the sensitivity, we concluded that less than 0.1 dithiol per PCPDTBT repeat unit in the dried films would not be detectable); the FTIR and Raman data are shown in Supplementary Information, Fig. S1. X-ray photoelectron spectroscopy (XPS) averaged over several samples with multiple scans per sample indicates no significant content of dithiol, certainly less than 0.1 dithiol per PCPDTBT repeat unit after thorough vacuum drying. The XPS data are shown in detail in Supplementary Information, Fig. S2.

The current–voltage characteristics obtained under 80 mW cm^{-2} for devices processed from chlorobenzene and from

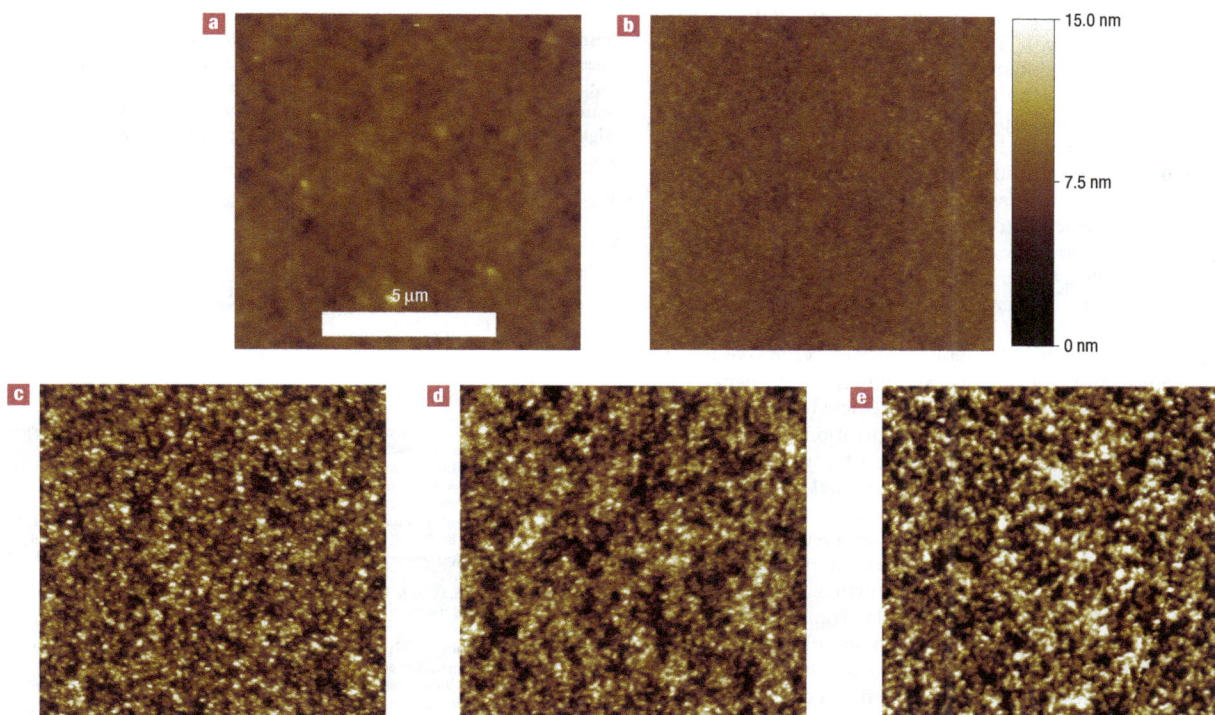

Figure 4 Morphology characterization. AFM topography images (10 μm × 10 μm) of films cast from chlorobenzene with 24 mg ml^{-1} of progressively longer alkyl chains. **a**, No additive. **b**, 1,4-butanedithiol. **c**, 1,6-hexanedithiol. **d**, 1,8-octanedithiol. **e**, 1,9-nonanedithiol. The height scale is 15 nm for all images.

chlorobenzene with 24 mg ml^{-1} by volume alkanedithiols with different chain lengths are shown in Fig. 2. From the I–V curves in Fig. 2, it is apparent that processing with 1,8-octanedithiol increases both I_{sc} and FF.

Device optimization involved over 1,000 devices made from over 250 independently prepared PCPDTBT:C$_{71}$-PCBM films; optimum photovoltaic efficiencies of between 5.2% and 5.8% were obtained. The most efficient devices comprised a polymer/fullerene ratio of between 1:2 and 1:3, a spin speed of between 1,200 and 1,600 r.p.m., a polymer concentration of between 0.8 and 1% by weight and a 1,8-octanedithiol concentration of between 17.5 and 25 mg ml^{-1}. The most repeatable series of high-efficiency devices had an average power-conversion efficiency of 5.5% under 100 mW cm^{-2}, with short-circuit current $I_{sc} = 16.2$ mA cm^{-2}, FF $= 0.55$ and open circuit voltage $V_{oc} = 0.62$ V. More than 40 devices gave efficiencies of over 5.2%. Nevertheless, as implied by the measured FF $= 0.55$, there are significant improvements to be made that could lead to more efficient solar-energy conversion.

As shown in Fig. 3a, the IPCE spectra of photovoltaic cells made with films cast from chlorobenzene compare very well with those previously reported for PCPDTBT:C$_{71}$-PCBM films formed by 'doctor blade' deposition from ortho-dichlorobenzene[4]. Devices processed with 1,8-octanedithiol increased the IPCE by a factor of approximately 1.6 between 400 and 800 nm compared with devices processed without the additive. The integrated IPCE and the I_{sc} measured on the same device, shown in Fig. 3b, agree to within approximately 5%. Note that as the IPCE for P3HT:C$_{61}$-PCBM is limited to a narrower fraction of the solar spectrum, higher efficiencies are anticipated for the PCPDTBT:C$_{71}$-PCBM devices.

To determine the effect of the alkanedithiol processing on carrier transport, we measured the steady-state and transient photoconductivity in films fabricated with and without processing with the alkanedithiol additive. The steady-state responsivity at $E = 20$ KV cm^{-1} (E is the applied electric field) and the magnitude of the current response in the transient photoconductivity from films processed using 1,8-octanedithiol are each larger than the values obtained from films processed without thiols by approximately a factor of two. The measured current in the transient data indicates an increase in the number of extracted mobile carriers, and the waveform indicates an additional increase in the carrier lifetime. As in the absorption, a red-shift is observed in the peak photoconductivity from 765 nm to 825 nm with the use of alkanedithiols[18].

When 1-alkanethiols were previously used to enhance the photoresponsivity of P3HT:C$_{61}$-PCBM blends, the primary effect of thiol processing was attributed to enhanced crystallization of the P3HT phase[5]. For P3HT:C$_{61}$-PCBM blends, an increase in the hole mobility by two orders of magnitude was demonstrated (as measured in field-effect transistors, FETs) on incorporation of 4.8 mg ml^{-1} 1-octanethiol in the toluene solution[5]. As shown in Supplementary Information, Fig. S3, FETs fabricated in the same way with PCPDTBT and C$_{71}$-PCBM show no increase in hole mobility when processed with octanedithiol. X-ray diffraction data shown in Supplementary Information, Fig. S4 show no indication of crystallization in the PCPDTBT:C$_{71}$-PCBM films either with or without dithiol processing. Therefore, the increased efficiencies must result from an increase in the mobile-carrier-generation efficiency and an increase in the mobile-carrier lifetime as a result of changes in the heterojunction morphology.

Changes in the surface topography of films cast from pure chlorobenzene and from chlorobenzene with 24 mg ml^{-1} of the alkanedithiols, as measured by atomic force microscopy

(AFM), are shown in Fig. 4. Comparison of the data obtained from films processed with 1,4-butanedithiol, 1,6-hexanedithiol, 1,8-octanedithiol or 1,9-nonanedithiol indicate that approximately six methylene units are required for the alkanedithiol to have an appreciable effect on the morphology (higher resolution AFM scans are shown in Supplementary Information, Fig. S5). Corresponding changes in the internal nanostructure are also seen by transmission electron microscopy, as shown in Supplementary Information, Fig. S6. The strong dependence of the absorption, the morphology and the device performance on the alkyl chain length implies that processing with dithiols influences the physical interactions between the polymer chains and/or between the polymer and fullerene phases.

The potential for PCPDTBT-based devices to yield significantly higher currents and fill factors should the interpenetrating network morphology be improved was predicted previously[4]. The approach described here provides an operationally simple and versatile tool available for the tailoring of the heterojunction solar-cell morphology in systems where thermal annealing is not effective. Note that our new approach works even on a system in which polymer crystallinity is not observed. On the basis of calculations by Brabec and co-workers on photovoltaic cells fabricated from PCPDTBT:C_{71}-PCBM, further optimization of morphology and equalization of ambipolar transport could lead to power-conversion efficiencies as high as 7% (ref. 4). This expectation is fully consistent with the incident-photon-conversion efficiency (ICPE) data shown in Fig. 3a; there is a clear opportunity to increase the IPCE.

METHODS

Photovoltaic cells were fabricated by spin-casting the active bulk heterojunction layer onto a 60 nm layer of H.C. Stark Baytron P poly(3,4-ethylenedioxythiophene):poly(styrenesulfonate) on Corning 1737 active-matrix liquid-crystal-display glass patterned with 140 nm of indium tin oxide by Thin Film Devices. A 100-nm-thick aluminium cathode was deposited (area 17 mm^2) by thermal evaporation with no heating of the sample either before or after deposition. Unless otherwise stated, the bulk heterojunction layer was spin-cast at 1,200 r.p.m. from a solution of 24 mg ml^{-1} octanedithiol in chlorobenzene containing 10 mg ml^{-1} PCPDTBT and 20 mg ml^{-1} C_{71}-PCBM. PCPDTBT was obtained from Konarka and the C_{71}-PCBM was purchased from Nano-C. AFM showed that the active layers were approximately 110 nm thick regardless of the alkanedithiol concentration in the solution.

Device efficiencies were measured with a 150 W Newport-Oriel AM 1.5G light source operating at 80 mW cm^{-2} and independently cross-checked using a 300 W AM 1.5G source operating at 100 mW cm^{-2} for verification. Solar-simulator illumination intensity was measured using a standard silicon photovoltaic with a protective KG5 filter calibrated by the National Renewable Energy Laboratory. Many of the most efficient devices were fabricated independently by different individuals (J.P. and J.Y.K.) in two separate laboratories and cross-checked under the two different illumination sources. IPCE spectra measurements were made with a 250 W Xe source, a McPherson EU-700-56 monochromater, optical chopper and lock-in amplifier, and a National Institute of Standards and Technology traceable silicon photodiode for monochromatic power-density calibration.

Photoconductive devices were fabricated by spin-casting on alumina substrates as described previously[5]. For steady-state photoconductivity measurements, the standard modulation technique was used[20]; for transient photoconductivity, an Auston switch configuration was used[21]. Thin-film transistor devices were fabricated and tested as described previously with a bottom contact geometry with gold electrodes[5]. AFM images were taken on a Veeco multimode AFM with a nanoscope IIIa controller. Ultraviolet–visible absorption spectroscopy was measured on a Shimadzu 2401 diode array spectrometer. XPS spectra were recorded on a Kratos Axis Ultra XPS system with a base pressure of 1×10^{-10} mbar using a monochromated Al Kα X-ray source. XPS survey scans were taken at a pass energy of 160 eV and high-resolution scans were taken at a pass energy of 10 eV. Data analysis was done with the CASA XPS software package.

Received 14 March 2007; accepted 3 May 2007; published 27 May 2007.

References

1. Scharber, M. C. et al. Design rules for donors in bulk-heterojunction solar cells—Towards 10% energy-conversion efficiency. Adv. Mater. 18, 789–794 (2006).
2. Hoppe, H. & Sariciftci, N. S. Morphology of polymer/fullerene bulk heterojunction solar cells. J. Mater. Chem. 16, 45–61 (2006).
3. Ma, W. L., Yang, C. Y., Gong, X., Lee, K. & Heeger, A. J. Thermally stable, efficient polymer solar cells with nanoscale control of the interpenetrating network morphology. Adv. Funct. Mater. 15, 1617–1622 (2005).
4. Muhlbacher, D. et al. High photovoltaic performance of a low-bandgap polymer. Adv. Mater. 18, 2884–2889 (2006).
5. Peet, J. et al. Method for increasing the photoconductive response in conjugated polymer/fullerene composites. Appl. Phys. Lett. 89, 252105 (2006).
6. Yu, G. & Heeger, A. J. Charge separation and photovoltaic conversion in polymer composites with internal donor/acceptor heterojunctions. J. Appl. Phys. 78, 4510–4515 (1995).
7. Yu, G., Gao, J., Hummelen, J. C., Wudl, F. & Heeger, A. J. Polymer photovoltaic cells: Enhanced efficiencies via a network of internal donor-acceptor hetrojunctions. Science 270, 1789–1791 (1995).
8. Halls, J. J. M., Pichler, K., Friend, R. H., Moratti, S. C. & Holmes, A. B. Exciton diffusion and dissociation in a poly(p-phenylenevinylene)/C-60 heterojunction photovoltaic cell. Appl. Phys. Lett. 68, 3120–3122 (1996).
9. Kraabel, B., McBranch, D., Sariciftci, N. S., Moses, D. & Heeger, A. J. Ultrafast spectroscopic studies of photoinduced electron transfer from semiconducting polymers to C_{60}. Phys. Rev. B 50, 18543–18552 (1994).
10. Kraabel, B. et al. Subpicosecond photoinduced electron transfer from conjugated polymers to functionalized fullerenes. J. Chem. Phys. 104, 4267–4273 (1996).
11. Brabec, C. J. et al. Tracing photoinduced electron transfer process in conjugated polymer/fullerene bulk heterojunctions in real time. Chem. Phys. Lett. 340, 232–236 (2001).
12. Sievers, D. W., Shrotriya, V. & Yang, Y. Modeling optical effects and thickness dependent current in polymer bulk-heterojunction solar cells. J. Appl. Phys. 100, 114509 (2006).
13. Dennler, G. et al. Charge carrier mobility and lifetime versus composition of conjugated polymer/fullerene bulk-heterojunction solar cells. Org. Electron. 7, 229–234 (2006).
14. Li, G. et al. High-efficiency solution processable polymer photovoltaic cells by self-organization of polymer blends. Nature Mater. 4, 864–868 (2005).
15. Reyes-Reyes, M., Kim, K. & Carroll, D. L. High-efficiency photovoltaic devices based on annealed poly(3-hexylthiophene) and 1-(3-methoxycarbonyl)-propyl-1-phenyl-(6,6)C-61 blends. Appl. Phys. Lett. 87, 083506 (2005).
16. Kim, Y. et al. A strong regioregularity effect in self-organizing conjugated polymer films and high-efficiency polythiophene: fullerene solar cells. Nature Mater. 5, 197–203 (2006).
17. ASTM G-173-03. Terrestrial Reference Spectra for Photovoltaic Performance Evaluation (American Society for Testing Materials (ASTM) International, West Conshohocken).
18. Soci, C. et al. Photoconductivity of a low-bandgap conjugated polymer. Adv. Funct. Mater. 200600267 (2007).
19. Wienk, M. M. et al. Efficient methano[70]fullerene/MDMO-PPV bulk heterojunction photovoltaic cells. Angew. Chem. Int. Edn 42, 3371–3375 (2003).
20. Lee, C. H. et al. Sensitization of the photoconductivity of conducting polymers by C-60-photoinduced electron-transfer. Phys. Rev. B 48, 15425–15433 (1993).
21. Auston, D. H. Impulse-response of photoconductors in transmission-lines. IEEE J. Quantum Electron. 19, 639–648 (1983).

Acknowledgements

The research was financially supported by grants from the Department of Energy (DE-FC26-04NT42277) (G.C.B.), the Office of Naval Research (N0014-0411) (G.C.B.), the Department of Energy (DE-FG02-06ER46324)(A.J.H.), Konarka Technologies (A.J.H.) and by the Ministry of Science & Technology of Korea under the International Cooperation Research Program Global Research Laboratory Program. J.P. thanks the NDSEG fellowship for support. The authors thank J. Yuen for thin-film transistor substrate preparation and Tom Mates for assistance with XPS measurement and analysis. We thank C. J. Brabec, Z. Zou, D. Waller and R. Gaudiana at Konarka Technologies for making the PCPDTBT available for our use. We thank C. Waldauf for important discussions.
Correspondence and requests for materials should be addressed to G.C.B. or A.J.H. Supplementary Information accompanies this paper on www.nature.com/naturematerials.

Author contributions

Devices were fabricated by J.P. and J.Y.K. Photoconductivity work was done by N.E.C. Transmission electron microscopy measurements were taken by W.L.M. AFM, FET, XPS, wide-angle X-ray diffraction, ultraviolet–visible, FTIR and Raman spectroscopy measurements were taken by J.P. D.M., A.J.H. and G.C.B. contributed to project planning, experimental design and manuscript preparation.

Metallated conjugated polymers as a new avenue towards high-efficiency polymer solar cells

WAI-YEUNG WONG[1,2]*, XING-ZHU WANG[1], ZE HE[1], ALEKSANDRA B. DJURIŠIĆ[3]*, CHO-TUNG YIP[3], KAI-YIN CHEUNG[3], HAI WANG[3], CHRIS S. K. MAK[4] AND WAI-KIN CHAN[4]

[1]Department of Chemistry, Hong Kong Baptist University, Waterloo Road, Kowloon Tong, Hong Kong, PR China
[2]Centre for Advanced Luminescence Materials, Hong Kong Baptist University, Waterloo Road, Kowloon Tong, Hong Kong, PR China
[3]Department of Physics, The University of Hong Kong, Pokfulam Road, Hong Kong, PR China
[4]Department of Chemistry, The University of Hong Kong, Pokfulam Road, Hong Kong, PR China
*e-mail: rwywong@hkbu.edu.hk; dalek@hkusua.hku.hk

Published online: 13 May 2007; doi:10.1038/nmat1909

Bulk heterojunction solar cells have been extensively studied owing to their great potential for cost-effective photovoltaic devices. Although recent advances resulted in the fabrication of poly(3-hexylthiophene) (P3HT)/fullerene derivative based solar cells with efficiencies in the range 4.4–5.0%, theoretical calculations predict that the development of novel donor materials with a lower bandgap is required to exceed the power-conversion efficiency of 10%. However, all of the lower bandgap polymers developed so far have failed to reach the efficiency of P3HT-based cells. To address this issue, we synthesized a soluble, intensely coloured platinum metallopolyyne with a low bandgap of 1.85 eV. The solar cells, containing metallopolyyne/fullerene derivative blends as the photoactive material, showed a power-conversion efficiency with an average of 4.1%, without annealing or the use of spacer layers needed to achieve comparable efficiency with P3HT. This clearly demonstrates the potential of metallated conjugated polymers for efficient photovoltaic devices.

Bulk heterojunctions fabricated by blending polymers with fullerene have resulted in great improvements in the polymer photovoltaic cell efficiencies. The most commonly used materials for the fabrication of bulk heterojunctions are poly(3-hexylthiophene) (P3HT) and [6,6]-phenyl C_{61}-butyric acid methyl ester (PCBM)[1–10]. With optimized device structure and fabrication conditions, P3HT/PCBM bulk heterojunction solar cells can reach efficiencies as high as 4.4–5.0% (refs 1–3). However, novel materials with lower energy gaps need to be developed to improve the coverage of the solar spectrum and consequently improve the efficiency[1]. Therefore, great effort has been made to optimize the spectral response of polymer photovoltaic cells by extending the absorption to longer wavelengths, as absorption of the active layer must cover most of the solar spectrum. The lowering of the bandgap of the polymer is predicted to result in efficiency exceeding 6%, whereas optimization of the energy levels and electron transfer to PCBM can lead to a further increase to ~8% (ref. 11). The ultimate efficiency limit for fully optimized polymer/fullerene solar cells is ~11% (ref. 11). Other reports also predict that for power-conversion efficiencies exceeding 10%, the donor polymer should have a bandgap below 1.74 eV (ref. 12). However, although several lower bandgap polymers have been proposed for photovoltaic applications[13–25], the highest reported power-conversion efficiency was merely ~2.2% (refs 17,18,20) to ~3.2% (refs 21,22), which is lower than that of P3HT/PCBM cells. There are a number of possible approaches to develop a polymer with electronic properties that have been identified as favourable by theoretical predictions. Yet limited success has been achieved

with polymers based on polyfluorene[12–16,18,23], benzothiadiazole[24] and thiophene derivatives[17,19,22,25].

One possible approach that has not been commonly explored involves the use of organometallic polymers. Although organometallic donor materials are commonly used in small-molecule solar cells[26,27], soluble π-conjugated organometallic polyyne polymers have rarely been used in high-performance polymer solar cells[28–30]. The maximum photocurrent quantum efficiencies in single-layer neat platinum polyyne cells were ~0.04–1% (refs 28,31), whereas quantum efficiencies of ~1–2% were achieved for these cells with the addition of 7 wt% C_{60} (ref. 29). More recently, an external quantum efficiency (EQE) of ~9% has been achieved in bulk heterojunction cells using the metallopolyyne/PCBM composite, resulting in a power-conversion efficiency of 0.27% (ref. 30). The polymer absorbed only in the blue–violet spectral region with a maximum efficiency at 400 nm, and consequently the efficiency was low owing to the low coverage of the solar spectrum, although the photoinduced charge separation in the blend involved the triplet state of the organometallic polymer[30]. Obviously, the photovoltaic application of this type of material is hampered by their wide bandgaps, and possibly unfavourable energy levels and charge-transport properties for forming efficient blends with PCBM. Although polymetallaynes of the type trans-$[-Pt(PR_3)_2-C≡C-R-C≡C-]_n$ (R = carbocycles, heterocycles, main group elements and so on) have aroused much attention for use in optoelectronic devices[28,31–33], most of these platinum(II) polyynes are usually characterized by relatively large bandgaps[34–37],

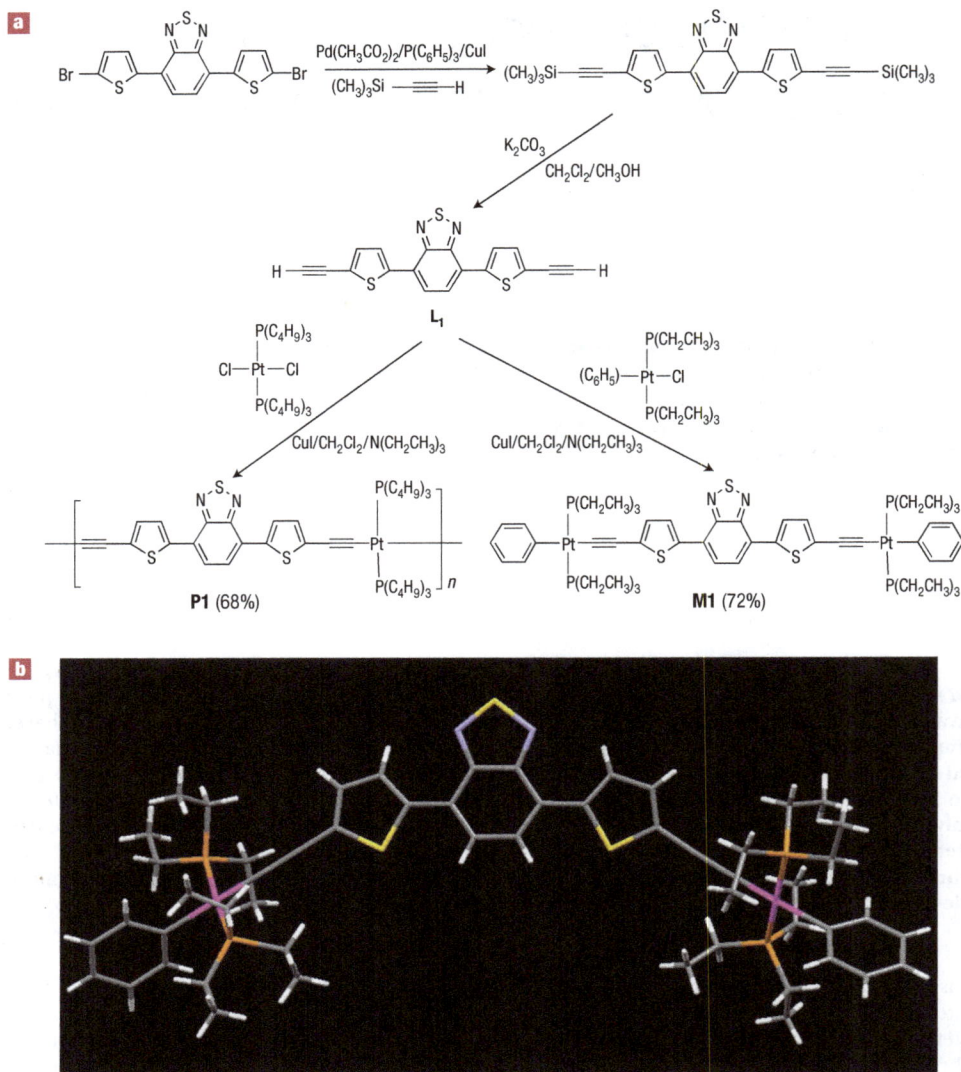

Figure 1 Synthesis scheme of low-bandgap organometallic diyne and polyyne and solid-state structure of M1. a, Synthesis scheme. **b,** A perspective view of the crystal structure of the model complex **M1**.

which compare unfavourably with those of some conjugated organic polymers comprising alternating electron donor (D) and acceptor (A) units (<1 eV) (refs 38,39). However, successful strategies for creating narrow-bandgap metallopolyynes involving the construction of D–A type systems are very scarce[31,40]. As a promising step towards our goal in designing metal polyynes with small bandgaps, we have exploited 4,7-di-2'-thienyl-2,1,3-benzothiadiazole as a core component to create a new π-conjugated system with extended absorption that features unique D–A characteristics. The novel platinum metallopolyyne is based on planar conjugated segments with internal D–A functions between electron-rich Pt–ethynyl groups to achieve a low-bandgap material (about 1.85 eV).

The synthetic route of the new platinum(II) polyyne polymer **P1** and its discrete model compound **M1** is shown in Fig. 1a. They were prepared by the Sonogashira-type dehydrohalogenation reaction between **L₁** and the platinum chloride precursors[37].

P1 was purified by silica column chromatography and repeated precipitation, leading to a dark purple powder of the polymer in good yield and high purity. Both metal compounds are thermally and air stable and soluble in common chlorinated hydrocarbons and toluene. Tough, free-standing thin films can be readily cast from **P1** solution. Gel-permeation chromatography on **P1** suggests a high-molecular-weight material with a degree of polymerization of ~22 (~66 heterocyclic rings in total). The structures were unequivocally characterized using elemental analyses, mass spectrometry and infrared and NMR spectroscopies. The rigid-rod nature of **P1** was also confirmed by X-ray crystal structure analysis of **M1** (Fig. 1b).

The absorption and emission spectra of **L₁**, **P1** and **M1** were measured in CH_2Cl_2 solutions at 298 K (Fig. 2). The absorption of a deep purple solution of **P1** is dominated by a strong band peaking at 554 nm, and the polymer exhibits an intense fluorescence band at 680 nm. The absorption and emission properties of **M1** are very similar to those of **P1**, with the absorption maximum at 548 nm. The apparent lack of an energy shift in the electronic transition between **M1** and **P1** indicates that the lowest singlet excited state is confined to a single repeat unit in the polymer chain. The optical bandgap, determined from the onset of absorption, is 1.85 eV for **P1**. Owing to the presence of a more extended π-electron delocalized system throughout the chain and the creation of an alternate D–A chromophore based on highly electron-accepting benzothiadiazole rings in combination with the electron-donating thiophene rings[41–45], the bandgap of **P1** is significantly lowered by about 0.35–0.70 eV relative to the purely electron-rich bithienyl (2.55 eV) (ref. 28) or electron-deficient benzothiadiazole (2.20 eV) linked counterpart[46], both of which appear yellow in the solid and solution phases. This is a clear indication of the role of strong D–A interactions in **P1**, that is, the bithienyl–benzothiadiazole hybrid here imparts gap-narrowing characteristics to **P1**. The idea behind the material design is that a conjugated polymer is formed by a regular alternation of D- and A-like moieties, possibly separated by neutral parts. If the D and A regions are extended through the Pt centre and alkynyl units we have a case analogous to the inorganic n–i–p–i superlattice quantum-well structures, resulting in a reduction of the bandgap[44]. The measured lifetimes for **P1** and **M1** are very short and lie in the nanosecond range. Unlike other Pt polyynes, the low-temperature emission for **P1** has no vibronic structure and only a weak temperature dependence (Fig. 2c). Such observations and the small Stokes shift (<0.40 eV) indicate that the emitting state is not a triplet but a singlet excited state instead. Again in contrast to Pt polyynes with higher bandgaps[28,46], we observed no emission from a triplet excited state over the measured spectral window between 1.2 and 3.1 eV. This is not surprising and is in accordance with the energy-gap law for Pt-containing conjugated polymers, in which the non-radiative decay rate from the triplet state increases exponentially with decreasing triplet–singlet energy gap[47]. There is always a trade-off between the bandgap and the rate of phosphorescence in such a system. In other words, a high-bandgap polymer with a high triplet-state energy will normally favour the observation of triplet emission, whereas the low-bandgap congener will probably not be phosphorescent even at low temperatures. A similar phenomenon has been observed for a low-bandgap Pt polyyne spaced by a thieno-pyrazine unit[31,42]. Therefore, it is not the triplet state but mostly a charge-transfer excited state that contributes to the efficient photoinduced charge separation in the energy conversion for **P1** (ref. 31), which is different from the Pt–thiophene polyyne-based blends where charge separation occurs via the triplet state of the polymer[30]. From these results, we attribute the localized state centred at 2.22 eV to a strong D–A charge-transfer type interaction. The charge-transfer nature of the transition was also supported by the solvent-dependent emission spectra of **P1** in solvents of different polarities. **P1** exhibited a marked positive solvatochromism with a red-shift of the emission maximum of ~40 nm from toluene (λ_em = 670 nm) to CHCl₃ (λ_em = 710 nm), which suggests a less polar excited state (see the Supplementary Information). We consider that the D–A interaction already present in the 4,7-di-2'-thienyl-2,1,3-benzothiadiazole unit is remarkably enhanced by the extensive π-delocalization in the polymer backbone containing the electron-rich platinum ion. This causes a charge-transfer type transition, which leads to a substantial lowering of the optical gap. **P1** shows an irreversible oxidation wave at +0.57 and a reversible reduction peak at −1.66 V in CH₂Cl₂. The oxidation (E_{ox}) and reduction (E_{red}) potentials were used to determine the highest occupied molecular orbital (HOMO) and lowest unoccupied molecular orbital (LUMO) energy levels using the equations $E_{HOMO} = -(E_{ox} + 4.8)$ eV and $E_{LUMO} = -(E_{red} + 4.8)$ eV, which were calculated using the internal

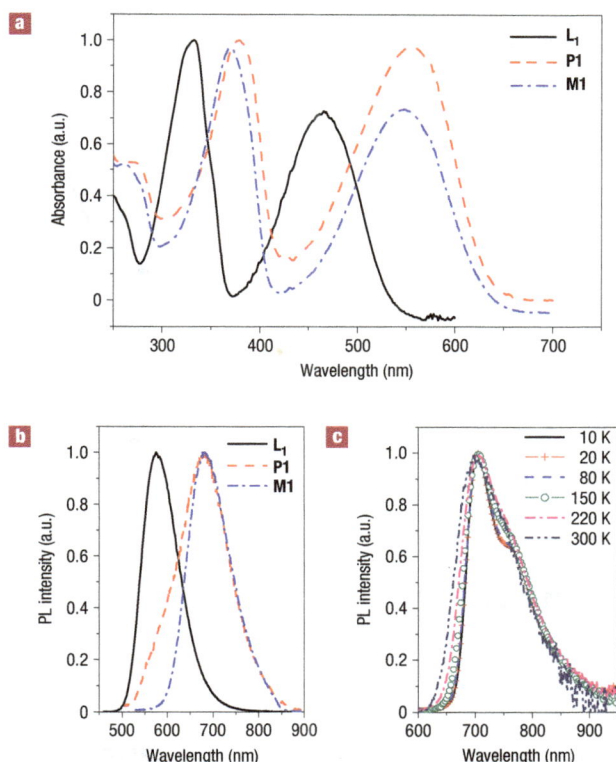

Figure 2 Optical properties of L₁, P1 and M1. a, Absorption spectra of **L₁**, **P1** and **M1**. **b,** Emission spectra of **L₁**, **P1** and **M1** in solution. **c,** Emission spectra of **P1** films at different temperatures.

ferrocene standard value of −4.8 eV with respect to the vacuum level[48]. Hence, the calculated HOMO and LUMO energies are −5.37 and −3.14 eV, respectively.

Polymer solar cells were fabricated by using **P1** as an electron donor and PCBM as an electron acceptor. The hole-collection electrode consisted of indium tin oxide (ITO) with a spin-coated poly(3,4-ethylene-dioxythiophene)/poly(styrene sulphonate) (PEDOT/PSS) layer, whereas Al served as the electron-collecting electrode. Different solvents, solution concentrations and spinning speeds were tested to identify optimal fabrication conditions for **P1**/PCBM blend cells. Toluene was identified as the solvent resulting in the best performance, followed by chlorobenzene, whereas cells prepared from chloroform and xylene solutions exhibited significantly worse performance. The optimal concentration and spinning speed resulted in a layer thickness of 70–75 nm, as verified by a step profiler and spectroscopic ellipsometry (JA Woollam V-VASE), whereas the thickness of PEDOT/PSS was ~27 nm. Solar cells with P3HT/PCBM active layers were also prepared for comparison. As the performance of P3HT/PCBM cells is known to be strongly dependent on the material properties and processing conditions[4], the same processing conditions (apart from solution preparation owing to solubility differences) were chosen for P3HT/PCBM solar cells. As P3HT/PCBM cells have been extensively studied, we have cited relevant literature for the best performance of P3HT/PCBM cells instead of optimizing the fabrication conditions for this type of cells. A considerable increase in the short-circuit current density and the power-conversion efficiency can be observed in

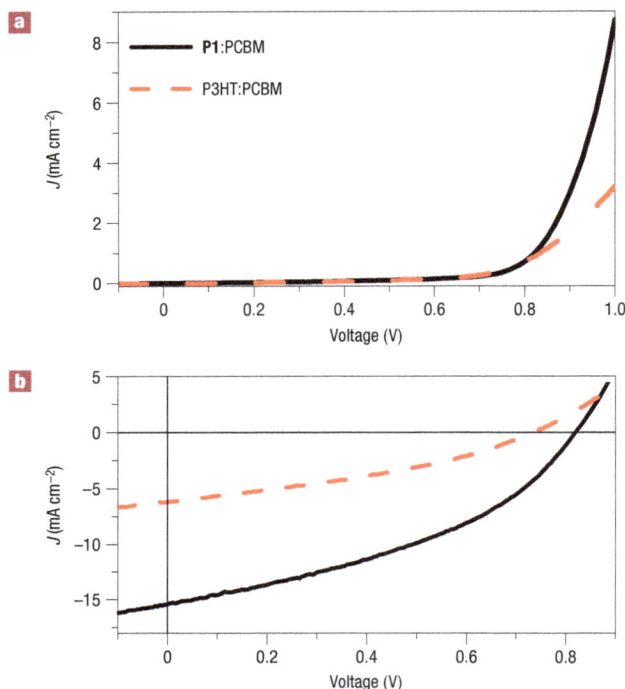

Figure 3 J–V curves of solar cells with P1/PCBM (1:4) and P3HT/PCBM (1:4) active layers. a, J–V curves in the dark. **b**, J–V curves under simulated AM1.5 solar irradiation.

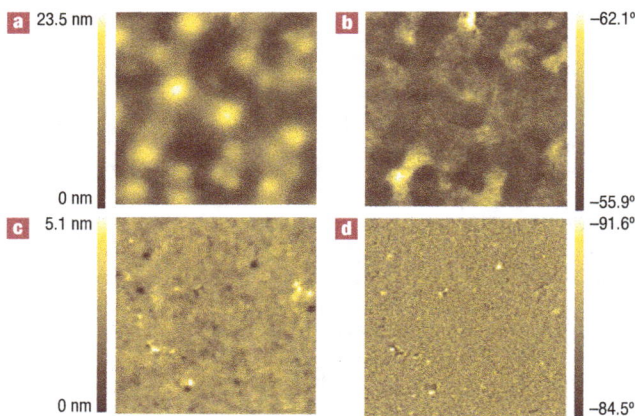

Figure 4 AFM images of P1/PCBM films for different blend compositions. Topography (left) and phase contrast (right) images of bulk heterojunction films. **a,b**, P1/PCBM 1:4. **c,d**, P1/PCBM 1:1. The area is $2\,\mu m \times 2\,\mu m$.

the case of **P1** (Fig. 3b). The open-circuit voltage obtained for the best cell was $V_{oc} = 0.82\,V$, the short-circuit current density was $J_{sc} = 15.43\,\text{mA cm}^{-2}$ and the fill factor was FF = 0.39, resulting in the power-conversion efficiency $\eta = 4.93\%$. The open-circuit voltage is also higher compared with P3HT/PCBM cells, which is probably due to the lower HOMO level of **P1** ($-5.37\,eV$, compared with $-5.20\,eV$ for P3HT[1]). The P3HT/PCBM cell exhibited $V_{oc} = 0.74\,V$, $J_{sc} = 6.22\,\text{mA cm}^{-2}$, FF = 0.35 and $\eta = 1.61\%$. It

should be noted that the efficiency of our P3HT/PCBM cells is lower than the best reported results in the literature[1–3]. However, such high efficiencies require either a more complicated device structure including a spacer layer[1] or annealing[2–6] to improve the cell performance.

It should also be noted that we use a different blend ratio (1:4 w/w) compared with that commonly used in P3HT/PCBM cells, which is typically 1:1 (refs 3–7) or 1:0.8 (ref. 1) as it has been shown that the optimal blend ratio is in the range 1:0.9–1:1 (ref. 7). However, optimal blend concentration is dependent on the polymer used. For **P1**/PCBM blends, cells with a 1:4 ratio resulted in the best performance, similar to the case of poly[2-methoxy-5-(3′,7′-dimethyloctyloxy)-1,4-phenylene vinylene] (MDMO-PPV)/PCBM blends[49,50], as well as poly(3-octylthiophene-2,5-diyl-*co*-3-decyloxythiophene-2,5-diyl)/PCBM blends[19]. The improved cell performance for 1:4 ratios in MDMO-PPV/PCBM blends was attributed to improved charge transport and improved dissociation efficiency[50]. Atomic force microscopy (AFM) was used to characterize the blends of **P1**/PCBM with ratios of 1:4 and 1:1, as shown in Fig. 4. It can be observed that, similar to (MDMO-PPV)/PCBM blends, phase separation occurs for films with a 1:4 blend ratio, whereas films with a 1:1 ratio are smooth ($\sim$3–5 nm for the 1:1 ratio versus $\sim$20 nm for the 1:4 ratio) and do not show significant phase separation or the formation of larger PCBM domains. Similar film roughening and the formation of larger PCBM aggregates is also observed in annealed P3HT/PCBM 1:1 blends, which result in high solar-cell efficiency[2]. Phase separation with the formation of PCBM-rich domains facilitates improved charge transport and carrier collection efficiency, which results in a reduction of recombination losses and an increase in short-circuit current density[49,51]. Consequently, the performance of **P1**/PCBM cells with a 1:4 blend ratio is improved compared with those with a 1:1 ratio.

To investigate the reasons for the improved efficiency, the EQE for solar cells with both active materials has been measured, as shown in Fig. 5a. The cells with a **P1**/PCBM active layer exhibit a higher maximum EQE (87% versus 67% for P3HT/PCBM) and a wider coverage of the visible spectral range, as expected from the absorption spectrum of the active layer (see Fig. 5a, inset). Charge transport in **P1**/PCBM and P3HT/PCBM layers was studied using the time-of-flight (TOF) technique and space-charge-limited current (SCLC) modelling. Experimental results for the electric-field dependences of the electron and hole mobilities are shown in Fig. 5b,c. For TOF measurements, the drop-casting technique had to be used instead of spin-coating to achieve sufficiently thick films, which involved very slow drying of the film in a solvent-saturated atmosphere. As a result, in addition to increased roughness of the films compared with spin-coated films, the slow drying inherent to the drop-casting process would also affect the mobility of the films, as it has been shown that slow drying of P3HT/PCBM films results in a significant enhancement of the hole mobility, leading to a more balanced charge transport[9,52]. It can be observed that electron and hole transport is more balanced in **P1**/PCBM blends, and both blend films exhibit small negative electric-field dependence, in agreement with previously reported results for P3HT/PCBM blends[52]. The TOF and SCLC data for **P1**/PCBM blends give comparable hole and electron mobilities, although the mobilities obtained from the TOF measurements are lower. However, the TOF measurements yield comparable hole and electron mobilities in P3HT/PCBM blends, whereas the SCLC model gives higher electron mobilities than hole mobilities, as expected from the blend film composition in agreement with previous results[53]. Thus, the mobilities obtained from the SCLC measurements can be considered more representative for describing the performance of solar cells prepared by the same

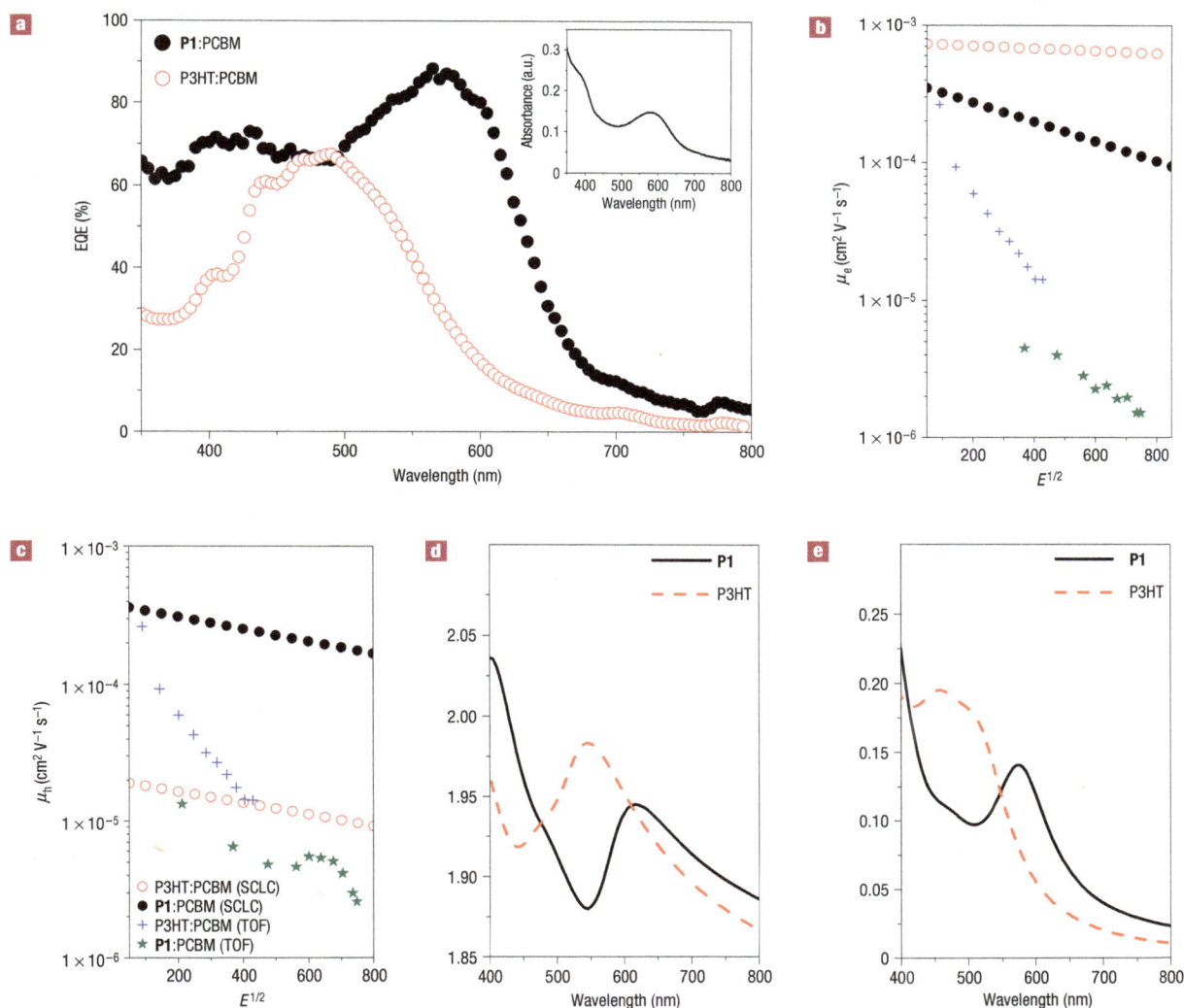

Figure 5 Comparison of P1/PCBM and P3HT/PCBM blends. a, EQE wavelength dependencies of solar cells with **P1**/PCBM (1:4) and P3HT/PCBM (1:4) active layers. Inset: The absorption of the **P1**/PCBM active layer. **b,c**, Electrical field dependence of electron (**b**) and hole (**c**) mobility of **P1**/PCBM (1:4) and P3HT/PCBM (1:4) blends. **d,e**, Real (**d**) and imaginary (**e**) parts of the refractive index of P1/PCBM and P3HT/PCBM films determined by spectroscopic ellipsometry.

fabrication method (spin-coating). Therefore, **P1**/PCBM blends exhibit balanced charge transport, as required for efficient solar-cell performance[53], and the mobility values are comparable to those measured in P3HT/PCBM cells with the optimal (1:1) blend ratio reported in the literature[52,53]. However, it should be noted that the fill factor of the devices is relatively low, which can be a sign of space-charge-limited photocurrent[10,50]. The photocurrent dependence on the light intensity for voltages approaching V_{oc} ($V_0 - V = 0.2$ V) shows a slope of ~3/4, which indicates the possibility of space-charge-limited device performance. Device performance is also very sensitive to atmosphere (oxygen and moisture) exposure, which probably results in the presence of traps and worsening of the fill factor. A comprehensive study of charge transport and the influence of traps is necessary to further improve the fill factor and device performance.

To investigate in more detail the reasons for the excellent performance of **P1**/PCBM cells, we carried out spectroscopic ellipsometry measurements of **P1**/PCBM and P3HT/PCBM blend films. The obtained real and imaginary parts of the index of refraction are shown in Fig. 5d,e. It can be clearly observed that the extinction coefficient of the **P1**/PCBM blend is higher above 550 nm, leading to improved absorption of the solar radiation. Typically, in the design of low-bandgap polymer solar cells, the main concerns are the HOMO and LUMO levels of the polymer[11,12], and charge transport in the cells[11]. However, the amount of absorbed light is dependent not only on the cut-off wavelength of the absorption, but also on how intense the absorption is. Therefore, the design of low-bandgap polymers for photovoltaic applications should consider not only lowering the gap but also increasing the absorption coefficient of the polymer.

The performance of the devices with optimal composition and active layer thickness was studied in detail. The $J-V$ curves under different excitation powers and the power dependence of the solar-cell parameters are shown in Fig. 6. The short-circuit current density exhibits a linear dependence on the optical power, whereas V_{oc} and fill factor show some increase and then saturate at higher

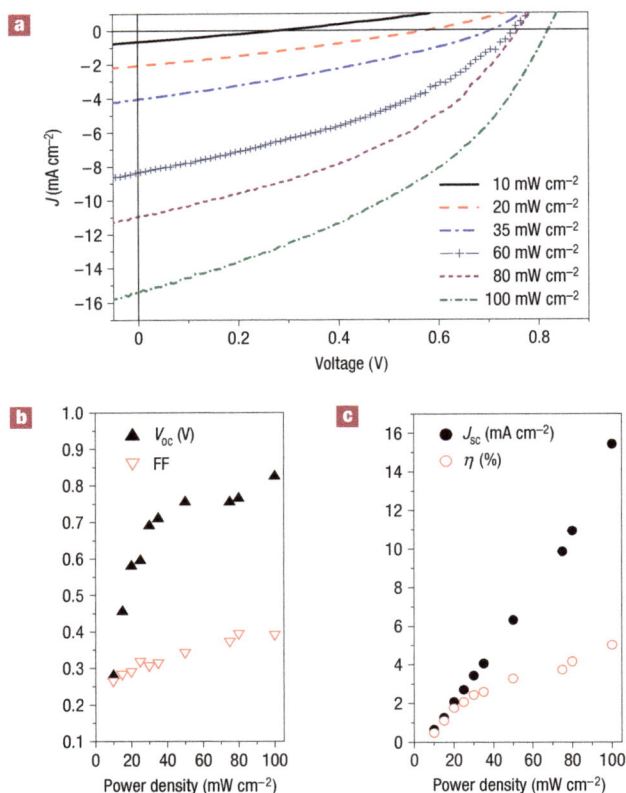

Figure 6 Performance of the P1/PCBM device for different illumination powers. a, *J–V* curves. **b,c,** Power dependences of open-circuit voltage and fill factor (**b**) and short-circuit current density and power-conversion efficiency (**c**).

intensity, as expected[26]. To test the reproducibility of the results obtained, a large number (90) of cells were prepared. The solar-cell performance parameters were (average value ± standard deviation) $V_{oc} = 0.82 \pm 0.03$ V, $J_{sc} = 13.1 \pm 3.2$ mA cm^{-2}, FF $= 0.37 \pm 0.03$ and $\eta = 4.1 \pm 0.9\%$. If we compare only the devices within the same batch (fabricated from the same solution, electrode evaporated at the same time, 6–8 cells), much smaller variations were obtained, with $V_{oc} = 0.80 \pm 0.02$ V, $J_{sc} = 15.1 \pm 2.2$ mA cm^{-2}, FF $= 0.40 \pm 0.04$ and $\eta = 4.8 \pm 0.4\%$. Overall, the reproducibility of the results was very good, and it is expected that it would be further improved if the fabrication was carried out in nitrogen instead of an air atmosphere. Although spectral mismatch correction has not been carried out[54], there is a good agreement between the measured short-circuit current density and the short-circuit current density estimated from EQE and the theoretical AM1.5 global solar spectrum, with on average 1.4 mA cm^{-2} difference between the measured J_{sc} and J_{sc} estimated from EQE, which is smaller than the standard deviation of the J_{sc} values of different cells. In addition, the measurements of the same devices in our laboratory and at CEA-INES, Laboratoire Composants Solaires, were in excellent agreement. We anticipate that the performance of these devices can be further improved by device optimization, as well as further chemical modifications of these metallized systems. It is expected that this demonstration will stimulate further work on the development of novel organometallic polymers for photovoltaic applications and bring bulk heterojunction polymer photovoltaic devices closer to commercialization.

METHODS

All chemicals for the syntheses were purchased from Aldrich or Acros Organics. PCBM and regioregular P3HT (M_w 20,000–50,000 g mol^{-1}) were purchased from American Dyes. PEDOT/PSS (Baytron P VPAI 4083) was purchased from H. C. Starck. Reactions and manipulations were carried out under an atmosphere of prepurified nitrogen using Schlenk techniques. The cyclic voltammetry experiment of the polymer film was carried out by casting the polymer on the glassy-carbon working electrode with a Pt wire as the reference electrode, at a scan rate of 100 mV s^{-1}. The solvent in all measurements was deoxygenated acetonitrile, and the supporting electrolyte was 0.1 M [($^nC_4H_9)_4$N][BF$_4$]. Ferrocene was added as an internal standard after each set of measurements, and all potentials are quoted relative to that of the ferrocene–ferrocenium couple.

SYNTHESIS OF PLATINUM POLYYNE POLYMER **P1**

Polymerization was carried out by mixing L_1 (50.0 mg, 0.143 mmol), *trans*-[PtCl$_2$(P(C$_4$H$_9)_3)_2$] (96.0 mg, 0.143 mmol) and CuI (5.0 mg) in N(CH$_2$CH$_3)_3$/CH$_2$Cl$_2$ (40 ml, 1:1, v/v). After stirring at room temperature overnight, the solution mixture was evaporated to dryness. The residue was redissolved in CH$_2$Cl$_2$ and filtered through a short silica column using the same eluent to remove ionic impurities and catalyst residues. After removal of the solvent, the crude product was purified by precipitating it in CH$_2$Cl$_2$ from methanol several times. Subsequent washing with *n*-hexane and drying *in vacuo* gave a dark purple solid of **P1** (92.0 mg, 68%).

SOLAR-CELL FABRICATION AND CHARACTERIZATION

ITO glass substrates (10 Ω □$^{-1}$) were cleaned by sonication in toluene, acetone, ethanol and deionized water, and dried in an oven. The samples were then cleaned with ultraviolet ozone for 300 s. As-received PEDOT/PSS solution was passed through a 0.45 μm filter and spin-coated on patterned ITO substrates at 5,000 r.p.m. for 3 min. The substrates were then baked in N$_2$ at 150 °C for 15 min. The **P1**/PCBM (1:4) active layer was prepared by spin-coating the toluene solution (4 mg ml^{-1} of **P1**, 16 mg ml^{-1} of PCBM) at 1,000 r.p.m. for 2 min. The substrates were dried in a vacuum oven at room temperature for 1 h, and then stored in high vacuum (10^{-5}–10^{-6} torr) overnight. Then, an Al electrode (100 nm) was evaporated through a shadow mask to define the active area of the devices (2 and 4 mm diameter circles). For comparison purposes, solar cells with a P3HT/PCBM (1:4) active layer were also prepared. The procedure for cell fabrication was the same as in the case of **P1**, except that P3HT/PCBM solution was prepared in chlorobenzene and stirred overnight in N$_2$ at 40 °C. All of the fabrication procedures were carried out in air, whereas the cells were measured in N$_2$ inside a sealed sample holder with a quartz window. The current–voltage characteristics were measured using a Keithley 2400 sourcemeter. For white-light efficiency measurements (at 100 mW cm^{-2}), an Oriel 66002 solar light simulator with an AM1.5 filter was used. The light intensity was measured by a Molectron Power Max 500D laser power meter. For the measurement of the EQE, a different wavelength of light was selected with a Thermo Oriel 257 monochromator. The photocurrent generated was measured using a Keithley 2400 sourcemeter. The light intensity was measured with a Newport 1830-C optical power meter equipped with an 818-UV detector probe. AFM images were obtained using a Seiko SPI 3800N Probe Station with a SPA 300 HV scanning probe microscope controller. The scanning speed was 0.1–0.3 Hz and the tip resonant frequency and force constant were 109 kHz and 16 N m^{-1}, respectively. The charge-carrier mobility was determined using the TOF method. The samples were prepared by drop-casting the blend on ITO glass. Toluene was used as a solvent for the **P1**/PCBM (1:4) blend, whereas chlorobenzene was used as a solvent for the P3HT/PCBM (1:4) blend. The film thickness was 1.11 μm for **P1**/PCBM and 2.45 μm for P3HT/PCBM. After drying the film in vacuum, an Al electrode (50 nm) was evaporated in high vacuum (10^{-6} torr). A nitrogen laser was used to generate pulsed excitation ($\lambda = 337.1$ nm). The transient photocurrent was monitored using an oscilloscope Tektronix TDS 3052 B. Because a different method was used to fabricate the films for the TOF measurements, spin-coated films were also prepared and the mobilities were determined from the SCLC modelling. For the hole mobility determination, the device structure used was ITO/PEDOT/PSS/blend film/Au, whereas for electron mobility determination, the device configuration used was ITO/Mg/blend film/LiF/Al. The mobilities were determined by fitting the measured *I–V* curves to the SCLC model[10]. The relative permittivity of the material was estimated from the modified Lorentz model fitting[55] of the ellipsometry data. Spectroscopic ellipsometry measurements were carried out using a JA Woollam V-VASE ellipsometer.

Received 1 February 2007; accepted 11 April 2007; published 13 May 2007.

References

1. Kim, J. Y. *et al.* New architecture for high-efficiency polymer photovoltaic cells using solution-based titanium oxide as an optical spacer. *Adv. Mater.* **18**, 572–576 (2006).
2. Reyes-Reyes, M., Kim, K. & Carroll, D. J. High-efficiency photovoltaic devices based on annealed poly(3-hexylthiophene) and 1-(3-methoxycarbonyl)-propyl-1-phenyl-(6,6)C₆₁ blends. *Appl. Phys. Lett.* **87**, 083506 (2005).
3. Kim, Y. *et al.* A strong regioregularity effect in self-organizing conjugated polymer films and high-efficiency polythiophene:fullerene solar cells. *Nature Mater.* **5**, 197–203 (2006).
4. Li, G. *et al.* High-efficiency solution processable polymer photovoltaic cells by self-organization of polymer blends. *Nature Mater.* **4**, 864–868 (2005).
5. Li, G., Shortriya, V., Yao, Y. & Yang, Y. Investigation of annealing effects and film thickness dependence of polymer solar cells based on poly(3-hexylthiophene). *J. Appl. Phys.* **98**, 043704 (2005).
6. Kim, Y. *et al.* Device annealing effect in organic solar cells with blends of regioregular poly(3-hexylthiophene) and soluble fullerene. *Appl. Phys. Lett.* **86**, 063502 (2006).
7. Chirvase, D., Parisi, J., Hummelen, J. C. & Dyakonov, V. Influence of nanomorphology on the photovoltaic action of polymer–fullerene composites. *Nanotechnology* **15**, 1317–1323 (2004).
8. De Bettignies, R., Leroy, J., Firon, M. & Sentein, C. Accelerated lifetime measurements of P3HT:PCBM solar cells. *Synth. Met.* **156**, 510–513 (2006).
9. Mihailetchi, V. D. *et al.* Origin of the enhanced performance in poly(3-hexylthiophene):[6,6]-phenyl C₆₁ butyric acid methyl ester solar cells upon slow drying of the active layer. *Appl. Phys. Lett.* **89**, 012107 (2006).
10. Mihailetchi, V. D., Xie, H., de Boer, B., Koster, L. J. A. & Blom, P. W. M. Charge transport and photocurrent generation in poly(3-hexylthiophene):methanofullerene bulk heterojunction solar cells. *Adv. Funct. Mater.* **16**, 699–708 (2006).
11. Koster, L. J. A., Mihailetchi, V. D. & Bloom, P. W. Ultimate efficiency of polymer/fullerene bulk heterojunction solar cells. *Appl. Phys. Lett.* **88**, 093511 (2006).
12. Scharber, M. C. *et al.* Design rules for donors in bulk-heterojunction solar cells—towards 10% energy-conversion efficiency. *Adv. Mater.* **18**, 789–794 (2006).
13. Wienk, M. M., Turbiez, M. G. R., Struijk, M. P., Fonrodona, M. & Janssen, R. Low-band gap poly(di-2-thienylthienopyrazine):fullerene solar cells. *Appl. Phys. Lett.* **88**, 153511–153513 (2006).
14. Wang, X. *et al.* Infrared photocurrent spectral response from plastic solar cell with low-band-gap polyfluorene and fullerene derivative. *Appl. Phys. Lett.* **85**, 5081–5083 (2004).
15. Wang, X. *et al.* Enhanced photocurrent spectral response in low-bandgap polyfluorene and C₇₀-derivative-based solar cells. *Adv. Funct. Mater.* **15**, 1665–1670 (2005).
16. Xia, Y. *et al.* Novel random low-band-gap fluorene-based copolymers for deep red/near infrared light-emitting diodes and bulk heterojunction photovoltaic cells. *Macromol. Chem. Phys.* **207**, 511–520 (2006).
17. Zhou, Q. *et al.* Fluorene-based low band-gap copolymers for high performance photovoltaic devices. *Appl. Phys. Lett.* **84**, 1653–1655 (2004).
18. Svensson, M. *et al.* High-performance polymer solar cells of an alternating polyfluorene copolymer and a fullerene derivative. *Adv. Mater.* **15**, 988–991 (2003).
19. Shi, C., Yao, Y., Yang, Y. & Pei, Q. Regioregular copolymers of 3-alkoxythiophene and their photovoltaic application. *J. Am. Chem. Soc.* **128**, 8980–8986 (2006).
20. Zhang, F. *et al.* Low-bandgap alternating fluorene copolymer/methanofullerene heterojunctions in efficient near-infrared polymer solar cells. *Adv. Mater.* **18**, 2169–2173 (2006).
21. Hou, J. *et al.* Synthesis and photovoltaic properties of two-dimensional conjugated polythiophenes with bi(thienylenevinylene) side chains. *J. Am. Chem. Soc.* **128**, 4911–4916 (2006).
22. Mühlbacher, D. *et al.* High photovoltaic performance of a low-bandgap polymer. *Adv. Mater.* **18**, 2884–2889 (2006).
23. Zhang, F. *et al.* Polymer solar cells based on a low-bandgap fluorene copolymer and a fullerene derivative with photocurrent extended to 850 nm. *Adv. Mater.* **18**, 2169–2173 (2006).
24. Brabec, C. J. *et al.* A low-bandgap semiconducting polymer for photovoltaic devices and infrared emitting diodes. *Adv. Funct. Mater.* **12**, 709–712 (2002).
25. Shaheen, S. A. *et al.* Low band-gap polymeric photovoltaic devices. *Synth. Met.* **121**, 1583–1584 (2001).
26. Xue, J., Uchida, S., Rand, B. P. & Forrest, S. R. 4.2% Efficient organic photovoltaic cells with low series resistances. *Appl. Phys. Lett.* **84**, 3013–3015 (2004).
27. Robertson, N. Optimizing dyes for dye-sensitized solar cells. *Angew. Chem. Int. Edn* **45**, 2338–2345 (2006).
28. Chawdhury, N. *et al.* Evolution of lowest singlet and triplet excited states with number of thienyl rings in platinum poly-ynes. *J. Chem. Phys.* **110**, 4963–4970 (1999).
29. Köhler, A., Wittman, H. F., Friend, R. H., Khan, M. S. & Lewis, J. Enhanced photocurrent response in photocells made with platinum-poly-yne/C₆₀ blends by photoinduced electron transfer. *Synth. Met.* **77**, 147–150 (1996).
30. Guo, F., Kim, Y.-G., Reynolds, J. R. & Schanze, K. S. Platinum-acetylide polymer based solar cells: Involvement of the triplet state for energy conversion. *Chem. Commun.* **2006**, 1887–1889 (2006).
31. Younus, M. *et al.* Synthesis, electrochemistry, and spectroscopy of blue platinum(II) polyynes and diynes. *Angew. Chem. Int. Edn* **37**, 3036–3039 (1998).
32. Wilson, J. S. *et al.* Spin-dependent exciton formation in π-conjugated compounds. *Nature* **413**, 828–831 (2001).
33. Manners, I. Putting metals into polymers. *Science* **294**, 1664–1666 (2001).
34. Long, N. J. & Williams, C. K. Metal alkynyl σ complexes: Synthesis and materials. *Angew. Chem. Int. Edn* **42**, 2586–2617 (2003).
35. Manners, I. *Synthetic Metal-Containing Polymers* (Wiley-VCH, Weinheim, 2004).
36. Köhler, A. & Beljonne, D. The singlet-triplet exchange energy in conjugated polymers. *Adv. Funct. Mater.* **14**, 11–18 (2004).
37. Wong, W.-Y. Recent advances in luminescent transition metal polyyne polymers. *J. Inorg. Organomet. Polym. Mater.* **15**, 197–219 (2005).
38. Karikomi, M., Kitamura, C., Tanaka, S. & Yamashita, Y. New narrow-bandgap polymer composed of benzobis(1,2,5-thiadiazole) and thiophenes. *J. Am. Chem. Soc.* **117**, 6791–6792 (1995).
39. Roncali, J. Synthetic principle for bandgap control in linear π-conjugated systems. *Chem. Rev.* **97**, 173–205 (1997).
40. Wong, W.-Y., Choi, K.-H., Lu, G.-L. & Shi, J.-X. Synthesis, redox and optical properties of low-bandgap platinum(II) polyynes with 9-dicyanomethylene-substituted fluorene acceptors. *Macromol. Rapid Commun.* **22**, 461–465 (2001).
41. Bredas, J. L. Theoretical design of polymeric conductors. *Synth. Met.* **17**, 115–121 (1987).
42. Köhler, A. *et al.* Donor-acceptor interactions in organometallic and organic poly-ynes. *Synth. Met.* **101**, 246–247 (1999).
43. Bredas, J. L., Heeger, A. J. & Wudi, F. Towards organic polymers with very small intrinsic band gaps. I. Electronic structure of polyisothianaphthene and derivatives. *J. Chem. Phys.* **85**, 4673–4678 (1986).
44. Havinga, E. E., Haeve, W. & Wynberg, H. Alternate donor-acceptor small-band-gap semiconducting polymers: Polysquaraines and polycroconaines. *Synth. Met.* **55–57**, 299–306 (1993).
45. Kertesz, M. & Lee, Y.-S. Electronic structures of small gap polymers. *Synth. Met.* **28**, C545–C552 (1989).
46. Wong, W.-Y. & Ho, C.-L. Di-, oligo- and polymetallaynes: Synthesis, photophysics, structures and applications. *Coord. Chem. Rev.* **250**, 2627–2690 (2006).
47. Wilson, J. S. *et al.* The energy gap law for triplet states in Pt-containing conjugated polymers and monomers. *J. Am. Chem. Soc.* **123**, 9412–9417 (2001).
48. Ashraf, R. S., Shahid, M., Klemm, E., Al-Ibrahim, M. & Sensfuss, S. Thienopyrazine-based low-bandgap poly(heteroaryleneethynylene)s for photovoltaic devices. *Macromol. Rapid Commun.* **27**, 1454–1459 (2006).
49. van Duren, J. K. J. *et al.* Relating the morphology of poly(*p*-phenylene vinylene)/methanofullerene blends to solar-cell performance. *Adv. Funct. Mater.* **14**, 425–434 (2004).
50. Mihailetchi, V. D. Compositional dependence of the performance of poly(*p*-phenylene vinylene):methanofullerene bulk-heterojunction solar cells. *Adv. Funct. Mater.* **15**, 795–801 (2005).
51. Shikler, R., Chiesa, M. & Friend, R. H. Photovoltaic performance and morphology of polyfluorene blends: The influence of phase separation evolution. *Macromolecules* **39**, 5393–5399 (2006).
52. Huang, J., Li, G. & Yang, Y. Influence of composition and heat-treatment on the charge transport properties of poly(3-hexylthiophene) and (6,6)-phenyl C₆₁–butyric acid methyl ester blends. *Appl. Phys. Lett.* **87**, 112105 (2005).
53. von Hauff, E., Parisi, J. & Dyakonov, V. Investigations of the effects of tempering and composition dependence on charge carrier field effect mobilities in polymer and fullerene films and blends. *J. Appl. Phys.* **100**, 043702 (2006).
54. Shrotriya, V. *et al.* Accurate measurement and characterization of organic solar cells. *Adv. Funct. Mater.* **16**, 2016–2023 (2006).
55. Liu, Z. T., Kwok, H. S. & Djurišić, A. B. The optical functions of metal phthalocyanines. *J. Phys. D* **37**, 678–688 (2004).

Acknowledgements

This work was supported by the Research Grants Council of The Hong Kong Special Administrative Region, China (Project Numbers HKBU 2024/04P, HKU 7008/04P and 7010/05P). Financial support from the Hong Kong Baptist University, the Strategic Research Theme, University Development Fund, a Seed Funding Grant and an Outstanding Young Researcher Award (administered by The University of Hong Kong) are also acknowledged. The authors would like to thank J. Gao from the University of Hong Kong for step-profiler thickness measurements, A. M. C. Ng for ellipsometry measurements and W. C. H. Choy and HKU-CAS Joint Laboratory on New Materials and C. M. Che for the use of a glove box for encapsulation. The authors would also like to thank R. de Bettignies and S. Guillerez from CEA-INES RDI, Laboratoire Composants Solaires, in Bourges Du Lac, France for independent verification of the solar-cell results.
Correspondence and requests for materials should be addressed to W.-Y.W. or A.B.D.
Supplementary Information accompanies this paper on www.nature.com/naturematerials.

Author contributions

X.-Z.W. and Z.H. were responsible for the synthesis and chemical analyses, C.-T.Y., K.-Y.C., H.W. and C.S.-K.M. carried out film and device fabrication and characterization, C.-T.Y., C.S.-K.M., A.B.D., W.-K.C. and W.-Y.W. were responsible for data analysis and interpretation, W.-Y.W., A.B.D. and W.-K.C. were responsible for project planning and experiment design and W.-Y.W. and A.B.D. were responsible for manuscript preparation.

Competing financial interests

The authors declare no competing financial interests.

Reprints and permission information is available online at http://npg.nature.com/reprintsandpermissions/

Coaxial silicon nanowires as solar cells and nanoelectronic power sources

Bozhi Tian[1]*, Xiaolin Zheng[1]*, Thomas J. Kempa[1], Ying Fang[1], Nanfang Yu[2], Guihua Yu[1], Jinlin Huang[1] & Charles M. Lieber[1,2]

Solar cells are attractive candidates for clean and renewable power[1,2]; with miniaturization, they might also serve as integrated power sources for nanoelectronic systems. The use of nanostructures or nanostructured materials represents a general approach to reduce both cost and size and to improve efficiency in photovoltaics[1–9]. Nanoparticles, nanorods and nanowires have been used to improve charge collection efficiency in polymer-blend[4] and dye-sensitized solar cells[5,6], to demonstrate carrier multiplication[7], and to enable low-temperature processing of photovoltaic devices[3–6]. Moreover, recent theoretical studies have indicated that coaxial nanowire structures could improve carrier collection and overall efficiency with respect to single-crystal bulk semiconductors of the same materials[8,9]. However, solar cells based on hybrid nanoarchitectures suffer from relatively low efficiencies and poor stabilities[1]. In addition, previous studies have not yet addressed their use as photovoltaic power elements in nanoelectronics. Here we report the realization of p-type/intrinsic/n-type (p-i-n) coaxial silicon nanowire solar cells. Under one solar equivalent (1-sun) illumination, the p-i-n silicon nanowire elements yield a maximum power output of up to 200 pW per nanowire device and an apparent energy conversion efficiency of up to 3.4 per cent, with stable and improved efficiencies achievable at high-flux illuminations. Furthermore, we show that individual and interconnected silicon nanowire photovoltaic elements can serve as robust power sources to drive functional nanoelectronic sensors and logic gates. These coaxial silicon nanowire photovoltaic elements provide a new nanoscale test bed for studies of photo-induced energy/charge transport and artificial photosynthesis[10], and might find general usage as elements for powering ultralow-power electronics[11] and diverse nanosystems[12,13].

We have focused on p-i-n coaxial silicon nanowire structures (Fig. 1a) consisting of a p-type silicon nanowire core capped with i- and n-type silicon shells. An advantage of this core/shell architecture is that carrier separation takes place in the radial versus the longer axial direction, with a carrier collection distance smaller or comparable to the minority carrier diffusion length[8]. Hence, photogenerated carriers can reach the p-i-n junction with high efficiency without substantial bulk recombination. An additional consequence of this geometry is that material quality can be lower than in a traditional p-n junction device without causing large bulk recombination[1].

Silicon nanowire p-cores were synthesized by means of a nanocluster-catalysed vapour–liquid–solid (VLS) method[14,15]. Silicon shells were then deposited at a higher temperature and lower pressure than for p-core growth (Fig. 1a, right panel) to inhibit axial elongation of the silicon nanowire core during the shell deposition, where phosphine was used as the n-type dopant in the outer shell[15]. The growth temperatures were sufficiently low to ensure that

minimal amounts of metal catalyst were incorporated into the silicon nanowire structure. Scanning electron microscopy (SEM) images of a typical p-i-n coaxial silicon nanowire recorded in the back-scattered electron mode (Fig. 1b) highlight several key features. First, the uniform contrast of the nanowire core is consistent with a single-crystalline structure expected for silicon nanowires obtained by the VLS method[14,15]. Second, contrast variation observed in the shells is indicative of a polycrystalline structure grain of the order of 30–80 nm. Third, the core/shell silicon nanowires have uniform

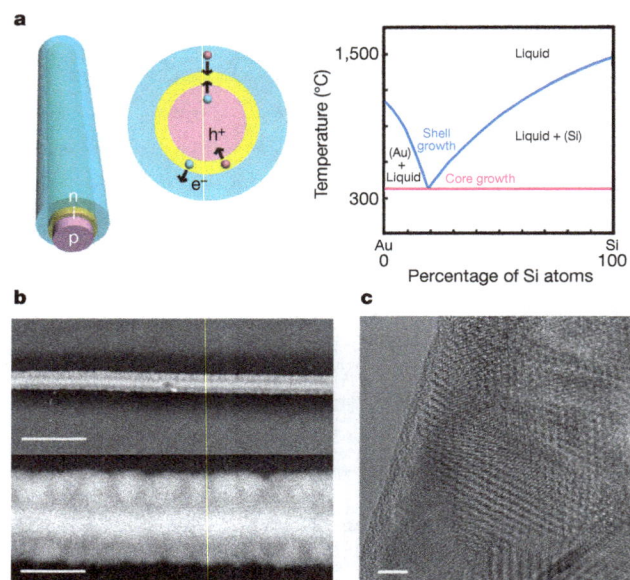

Figure 1 | Schematics and electron microscopy images of the p-i-n coaxial silicon nanowire. a, Illustrations of the core/shell silicon nanowire structure; its cross-sectional diagram shows that the photogenerated electrons (e^-) and holes (h^+) are swept into the n-shell and p-core, respectively, by the built-in electric field. The phase diagram of gold (Au)–silicon (Si) alloy on the right panel illustrates that the core is grown by means of the VLS mechanism, whereas the shells are deposited at higher temperature and lower pressure to inhibit further nanowire axial elongation. b, SEM images (back-scattered electron mode) of the p-i-n coaxial silicon nanowire at two different magnifications. Scale bar, 1 μm (top), 200 nm (bottom). The p-i-n silicon nanowire was grown with 100-nm-diameter gold catalyst, and with i- and n-shell growth times of 60 min and 30 min, respectively. The feeding ratios of silicon:boron and silicon:phosphorus are 500:1 and 200:1, respectively. c, High-resolution TEM image (spherical-aberration-corrected) of the p-i-n coaxial silicon nanowire. Scale bar, 5 nm.

---

[1]Department of Chemistry and Chemical Biology, [2]School of Engineering and Applied Sciences, Harvard University, Cambridge, Massachusetts 02138, USA.
*These authors contributed equally to this work.

Nature | Vol 449 | 18 October 2007 | doi:10.1038/nature06181 diameters of ~300 nm (280–360 nm for other nanowires), which is in agreement with independent transmission electron microscopy (TEM) and atomic force microscopy measurements. In addition, high-resolution TEM images (Fig. 1c) confirm that the nanowire shell is indeed polycrystalline. We note that this nanocrystalline shell structure could enhance light absorption in the nanowires (see below).

To characterize electrical transport through the p-i-n coaxial silicon nanowires, we fabricated metal contacts selectively to the inner p-core and outer n-shell (Fig. 2a). Briefly, core/shell silicon nanowires were etched selectively using potassium hydroxide (KOH) solution (see Methods) to expose the p-core in a lithographically defined region, and then metal contacts were made to the p-core and n-shell after a second lithographic patterning step, as shown in the SEM images of Fig. 2b. Dark current–voltage (I–V) curves obtained from devices fabricated in this way (Fig. 2c) exhibit several notable features. First, the linear I–V curves from core–core (p1-p2) and shell–shell (n1-n2) configurations indicate that ohmic contacts are made to both core and shell portions of the nanowires. Second, the I–V curve for the shell–shell contact reveals a shell

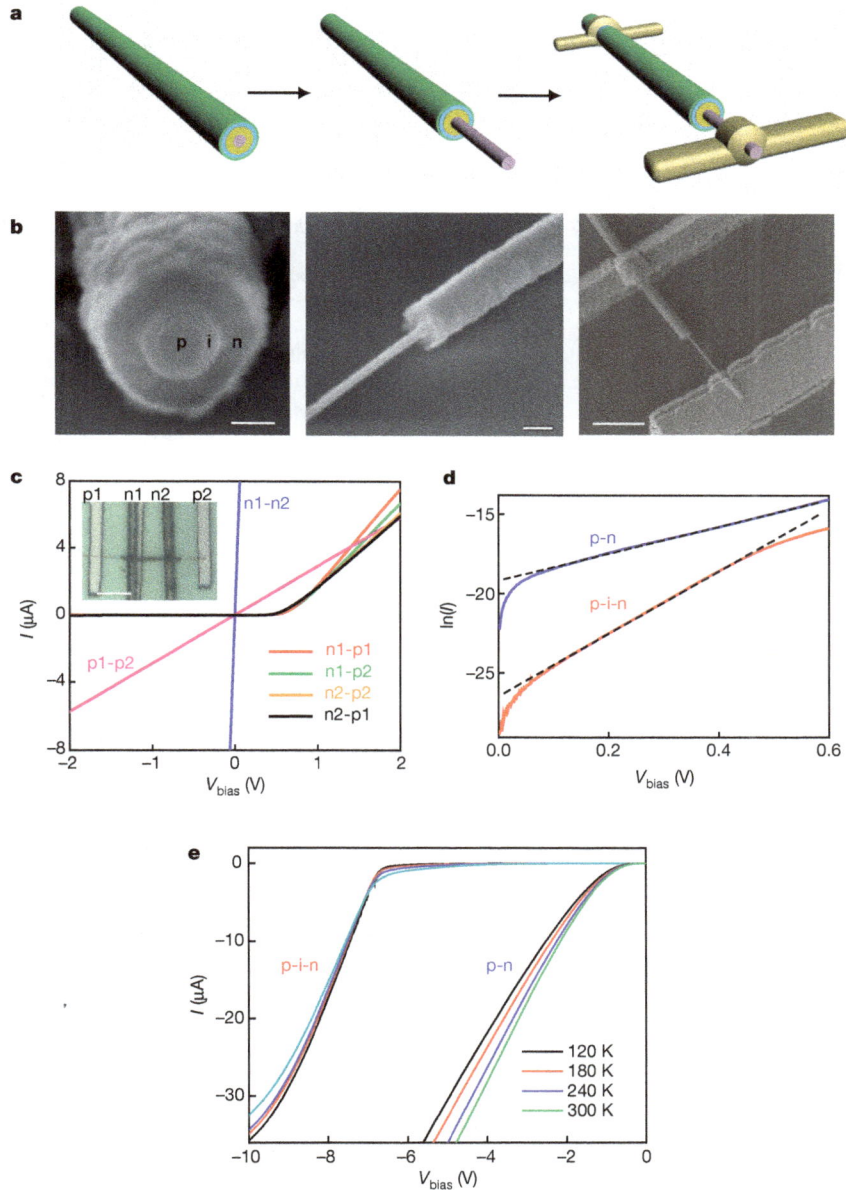

Figure 2 | Device fabrication and diode characterization. **a,** Schematics of device fabrication. Left, pink, yellow, cyan and green layers correspond to the p-core, i-shell, n-shell and PECVD-coated SiO_2, respectively. Middle, selective etching to expose the p-core. Right, metal contacts deposited on the p-core and n-shell. **b,** SEM images corresponding to schematics in **a**. Scale bars are 100 nm (left), 200 nm (middle) and 1.5 μm (right). **c,** Dark I–V curves of a p-i-n device with contacts on core–core, shell–shell and different core–shell combinations. V_{bias}, the applied bias voltage. Inset, optical microscope image of the device. Scale bar, 5 μm. **d,** Semi-log scale I–V curves of p-i-n and p-n diodes. The ideality factor N can be extrapolated (dashed lines) from the diode linear regimes (p-i-n diode, 0.12–0.50 V; p-n diode, 0.10–0.60 V), which are 1.96 and 4.52 for p-i-n and p-n diodes, respectively. To keep the total diameters of the p-n and p-i-n silicon nanowires approximately the same, the p-n silicon nanowire was grown with a 100-nm diameter gold catalyst and an n-shell growth time of 100 min. The SiH_4/dopants feeding ratios for p-n and p-i-n nanowires are the same (silicon:boron, 500:1; silicon:phosphorus, 200:1). **e,** Temperature-dependent I–V measurement of the p-n and p-i-n diodes in the reverse bias voltage regime.

conductance of 132 μS, higher than that of the core (3 μS); the calculated shell resistivity is within a factor of two of that measured for single-crystal n-type silicon nanowire prepared with a similar SiH_4:PH_3 ratio[15]. The highly conductive n-shell will reduce or eliminate potential drop along the shell, thereby enabling uniform radial carrier separation and collection when illuminated[8]. Third, I–V curves recorded from different core–shell contact geometries show rectifying behaviour, and demonstrate that the p-i-n coaxial silicon nanowires behave as well-defined diodes. The reproducibility of the selective etching and contact formation to p-cores and n-shells was further demonstrated by defining more complex 'AND' and 'OR' diode logic gates using single p-i-n coaxial silicon nanowires (Supplementary Fig. 1).

The core/shell silicon nanowire diodes were further characterized by analysing data recorded with and without the i-layer as a function of temperature. Fits to $\ln(I)$–V data recorded in forward bias from p-i-n and p-n coaxial structures (Fig. 2d) are linear, and yield diode ideality factors N of 1.96 and 4.52, respectively (see Methods). The N-values show that introduction of the i-layer yields much better quality diodes. Reverse bias measurements from p-i-n and p-n diodes (Fig. 2e) also show markedly different behaviour: the p-i-n diode breaks down at much larger reverse-bias voltage (approximately -7 V) than the p-n diode (approximately -1 V) for all temperatures studied. In addition, the reverse-bias breakdown voltage of the p-n diode increases with decreasing temperature, which is consistent with a Zener (tunnelling) breakdown mechanism, whereas the breakdown voltage of the p-i-n structures exhibits little temperature dependence, suggesting

contributions from tunnelling and avalanche mechanisms[16]. Overall, these results indicate that tunnelling or leakage currents are more significant in the p-n diode[17], and that the diode quality factor and breakdown behaviour are readily controlled during nanowire growth by the introduction of the i-layer as in planar structures[18,19].

The photovoltaic properties of the p-i-n coaxial silicon nanowire diodes were characterized under air mass 1.5 global (AM 1.5G) illumination. I–V data recorded from one of the better devices (Fig. 3a) yields an open-circuit voltage V_{oc} of 0.260 V, a short-circuit current I_{sc} of 0.503 nA and a fill factor F_{fill} of 55.0%. The maximum power output P_{max} for the silicon nanowire device at 1-sun (see Methods) is $\sim$72 pW. Notably, these values were constant for measurements made over a seven-month period, thus demonstrating excellent stability of our nanowire photovoltaic elements. In addition, I–V data recorded using contacts to the n-shell that were 5.9 μm (n1) and 13.3 μm (n2) from the p-core contact (Fig. 3b) exhibited essentially the same photovoltaic response, thus indicating that the n-shell is equipotential with radial carrier separation occurring uniformly along the entire length of the core/shell silicon nanowire device. Measurements of I_{sc} as a function of the p-i-n coaxial silicon nanowire (Fig. 3c) length show linear scaling with values of 1 nA silicon nanowire^{-1} readily achieved for lengths of 10 μm (1-sun), whereas V_{oc} is essentially independent of length. The linear scaling of I_{sc} with silicon nanowire length suggests that photogenerated carriers are collected uniformly along the length of these radial nanostructures, and that scattering of light by the metal contacts does not make a major contribution to the observed photocurrent.

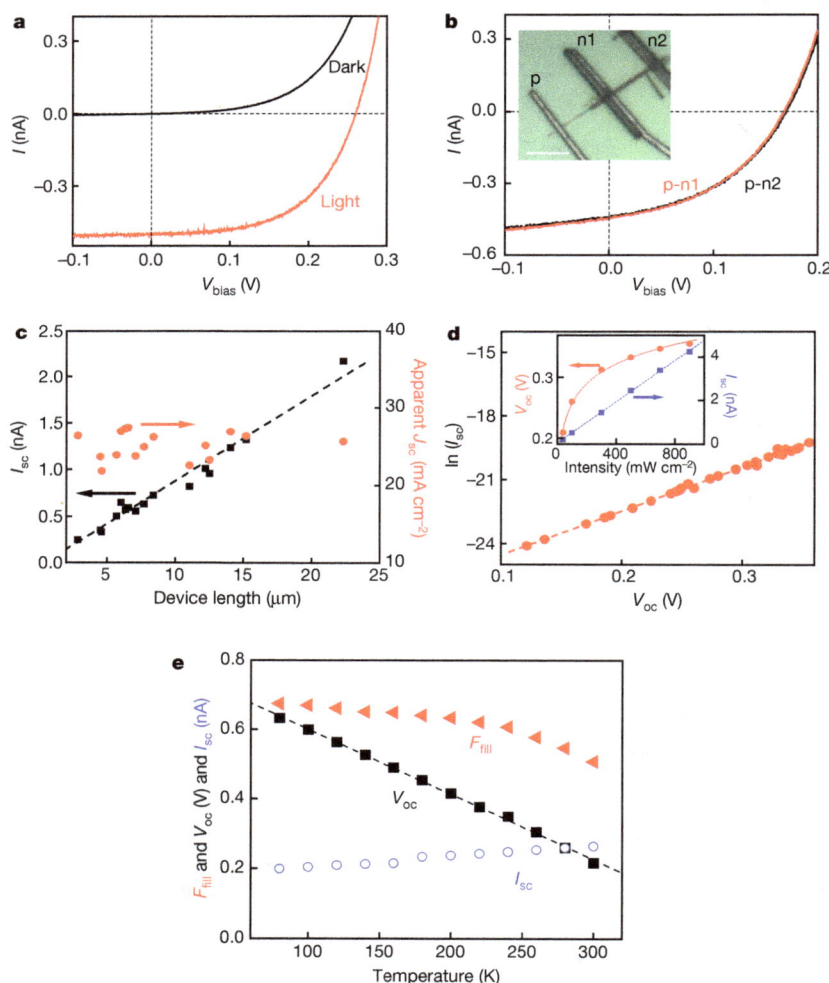

Figure 3 | Characterization of the p-i-n silicon nanowire photovoltaic device. a, Dark and light I–V curves. **b,** Light I–V curves for two different n-shell contact locations. Inset, optical microscopy image of the device. Scale bar, 5 μm. **c,** Device-length-dependent I_{sc} and J_{sc} (upper bound) plots. **d,** Plot of $\ln(I_{sc})$ versus V_{oc}; each point corresponds to a different light intensity. Inset, light-intensity-dependent I_{sc} and V_{oc} plots. **e,** Temperature-dependent measurement. The device was illuminated at 0.6-sun to reduce sample heating, which may cause temperature fluctuations. The red triangle, black square and blue circle correspond to F_{fill}, V_{oc} and I_{sc}, respectively. The same device was characterized in **a,** **d** and **e**. The p-i-n coaxial silicon nanowire was grown using conditions as in Fig. 2.

The apparent short-circuit current density J_{sc} calculated using the projected area of the core/shell nanowire structure was 23.9 ± 1.2 mA cm^{-2} (upper bound, excluding metal covered and exposed p-core areas) and 16.0 ± 0.8 mA cm^{-2} (lower bound, including metal-covered and exposed p-core areas) for the device in Fig. 3a. The use of projected area to estimate the apparent current density is consistent with the methodology used with other nanostructured photovoltaic devices[3–6] and our use of devices as nanoscale power sources (see below, Fig. 4). Control experiments were also carried out to investigate the principal area of light absorption by devices. For example, measurements made on devices with and without lithographic masks that block illumination of the nanowire (Supplementary Fig. 2), with and without external scattering centres, and as a function of incident angle of illumination verify that the reported photocurrents and large apparent photocurrent densities arise primarily from direct nanowire absorption and are not much enhanced by scattering and/or waveguiding of incident light remote from devices. We note that the large nanowire J_{sc} values (Fig. 3c) imply substantial absorption across the solar spectrum. Such absorption is consistent with the nanocrystalline shell structure of the nanowires and previous studies of microcrystalline thin films[18], although the detailed nature of absorption will require further investigation. The apparent photovoltaic efficiency η of this device is $3.4 \pm 0.2\%$ (upper bound) and $2.3 \pm 0.2\%$ (lower bound), but might be improved through increased understanding of absorption and better coupling of light into the devices, for example, by vertical integration[8] or multilayer stacking[20].

I_{sc} and V_{oc} depend linearly and logarithmically, respectively, on the light intensity incident on the chip (inset, Fig. 3d), consistent with systematic increase in photogenerated carriers[19,21]. We note that the apparent efficiency is substantially higher at multiple-sun illumination: $4.1 \pm 0.2\%$ and $4.5 \pm 0.3\%$ (upper bounds) under 3-sun and 5-sun conditions, respectively. Although this apparent efficiency enhancement is larger than that in a planar silicon solar cell[19], it is consistent with the larger ideality factor (N) and lower 1-sun V_{oc} of the nanowire devices[19]. Analysis of a plot of $\ln(I_{sc})$ versus V_{oc} (Fig. 3d) yields values of the diode ideality factor and saturation current of $N = 1.86$ and $I_0 = 2.72$ pA, respectively (see Methods). These values are similar to those extrapolated from the dark measurements ($N = 1.96$, $I_0 = 3.24$ pA), and thus demonstrate good consistency in the behaviour and analysis of these core/shell silicon nanowire diode devices.

In addition, the temperature dependences of I_{sc}, V_{oc} and F_{fill} were characterized to understand better the behaviour of the silicon nanowire photovoltaic devices (Fig. 3e). I_{sc} decreases slightly with decreasing temperature, and can be attributed to reduced light absorption due to increasing bandgap as temperature is reduced[22]. V_{oc} exhibits a substantial linear increase with decreasing temperature, where the slope (dV_{oc}/dT) of -1.9 mV K^{-1} is close to the value (-1.7 mV K^{-1}) calculated in single crystalline silicon solar cells[21]. The observed increase in V_{oc} can be attributed to a reduced recombination rate at lower temperature[21,22], and yields an apparent efficiency of 6.6% (upper bound) at 80 K (0.6-sun). The F_{fill} also increases with decreasing temperature (as expected from the negative dV_{oc}/dT)[22]. Taken together, these V_{oc} and F_{fill} results indicate that the silicon nanowire photovoltaic performance at room temperature (298 K in our experiments) can be significantly improved by reducing recombination processes, for example, by improving the crystalline structure of the shells and/or passivating the nanowire surface and grain boundaries[19,23].

Our core/shell silicon nanowire results can be compared to nanocrystal-based[4] and nanorod-based[5,6] photovoltaic devices. The best silicon nanowire device exhibits large apparent short-circuit current densities—23.9 mA cm^{-2} (upper bound) and 16.0 mA cm^{-2} (lower bound)—with upper limits that are comparable to the 24.4 mA cm^{-2} value for the best thin film nanocrystalline silicon solar cell[24], and substantially better than values reported for CdSe nanorod/poly-3-hexathiophene[4] and dye-sensitized ZnO nanorod[5,6] solar cells. The V_{oc} value, 0.260 V, is 2–2.8 times lower than reported in these previous

studies[4–6,24] and represents an area that should be addressed in future studies. However, the overall apparent efficiency of the p-i-n coaxial silicon nanowire photovoltaic elements—3.4% (upper bound) and 2.3% (lower bound)—exceeds reported nanorod/polymer and nanorod/dye systems[4–6], and could be increased substantially with improvements in V_{oc} by means of, for example, surface passivation. In addition, increasing the illumination intensity can yield stable improvements in the apparent efficiency of our p-i-n coaxial silicon nanowire photovoltaic elements in contrast to other nanostructured solar cells, which often exhibit degradation[4–6].

The ability of individual core/shell silicon nanowires to function as robust photovoltaic elements might indicate their potential as nanoscale power sources that might be integrated 'on-chip' with other

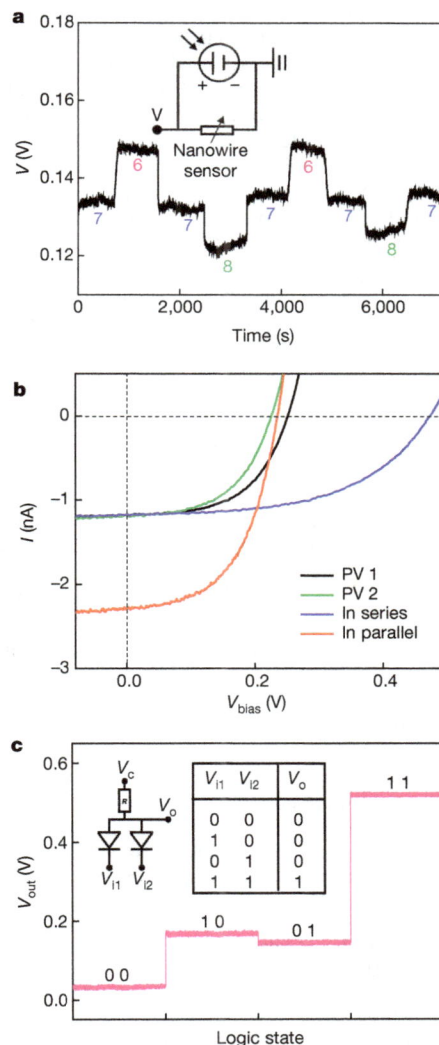

Figure 4 | Self-powered nanosystems. a, Real-time detection of the voltage drop across an aminopropyltriethoxysilane-modified silicon nanowire at different pH values. The silicon nanowire pH sensor is powered by a single silicon nanowire photovoltaic device operating under 8-sun illumination ($V_{oc} = 0.34$ V, $I_{sc} = 8.75$ nA). Inset, circuit schematics. **b,** Light I–V curves (1-sun, AM 1.5G) of two silicon nanowire photovoltaic devices (PV 1 and PV 2) individually and connected in series and in parallel. **c,** Nanowire AND logic gate powered by two silicon nanowire photovoltaic devices in series. Insets, circuit schematics and truth table for the AND gate. The resistance of CdSe nanowire is ~5 GΩ; the V_{oc} of two photovoltaic devices in series is 0.53 V. The large resistance of the CdSe nanowire and reverse-biased p-i-n diode makes V_c and V_i (HIGH) very close to V_{oc} of the photovoltaic device. To get V_i (LOW), the diode is simply grounded.

Nature | Vol 449 | 18 October 2007 | doi:10.1038/nature06181 semiconductor nanowire- and carbon-nanotube-based nanoelectronic elements, given that these elements require power as low as a few nanowatts[25–27]. Recent work addressing this key issue has involved the use of piezoelectric ZnO nanowires for mechanical-to-electrical conversion, although the direct current (d.c.) power developed by this nanogenerator[28,29], 1–4 fW per nanowire, is at present less than is needed to drive nanoelectronic devices. Silicon nanowire photovoltaic elements can produce 50–200 pW per nanowire at 1-sun illumination, and thus could function as nanoscale power supplies for nanoelectronics by either increasing the light intensity or using several coupled elements. For example, a single silicon nanowire photovoltaic device, operating under 8-sun illumination ($P_{max} = 1.86$ nW, $\eta = 4.8\%$) was used to drive a silicon nanowire pH sensor[25] without additional power (Fig. 4a). Measurements of the voltage drop across the p-type silicon nanowire sensor (powered solely by the silicon nanowire photovoltaic element) as a function of time (Fig. 4a) show reversible increase (or decrease) in voltage as the solution pH is decreased (or increased) that are consistent with the expected changes in resistance of the silicon nanowire with surface charge[25]. In addition, we note that the photovoltaic (under constant 8-sun illumination) and sensor devices both exhibited excellent stability over the approximately two-hour time of experiments.

Last, the core/shell silicon nanowire photovoltaic devices were interconnected in series and in parallel to demonstrate scaling of the output characteristics and to drive larger loads. $I–V$ data recorded from two illuminated silicon nanowire elements (Fig. 4b) show several important features. First, the individual elements exhibit very similar behaviour, highlighting the good reproducibility of our core/shell nanowire devices. Second, interconnection of the two elements in series and parallel yields V_{oc} and I_{sc} values, respectively, that are approximately the sum of two, as expected. Notably, we have used interconnected silicon nanowire photovoltaic elements as the sole power supply driving a nanowire-based AND logic gate (Fig. 4c), where V_c and the voltage inputs 1 and 2 V_{i1} (HIGH) and V_{i2} (HIGH) are provided by two nanowire photovoltaic devices in series at 2-sun illumination. (HIGH is the input state and V_{i1} (HIGH) and V_{i2} (HIGH) are close to V_{oc} of the PV devices.) A summary of the input/output results (right inset, Fig. 4c) shows correct AND logic. This work thus demonstrates the potential for self-powered nanowire-based logic circuits and, more generally, the possibility of self-powered functional nanoelectronic systems through, for example, the integration of multiple stacked silicon nanowire photovoltaic elements with nanoelectronic, photonic and biological sensing devices.

METHODS SUMMARY

Single-crystalline silicon nanowire p-cores were synthesized by means of a nanocluster-catalysed VLS method[14,15], and then chemical vapour deposition was used to deposit i- and n-type nanocrystalline silicon shells; shell growth was carried out at higher temperature and lower pressure than those used in core growth to inhibit axial elongation of the silicon nanowire core. After nanowire growth, SiO$_2$ was deposited conformally by means of plasma-enhanced chemical vapour deposition (PECVD). Standard electron beam lithography, silicon wet chemical etching (KOH etchant) and thermal evaporation were used to make coaxial nanowire devices, with selective contacts on the p-core and n-shell. A standard solar simulator (150 W, Newport Stratford) with an AM 1.5G filter was used to characterize the photovoltaic device response, where the average intensity was calibrated using a power meter. For multiple-sun illumination, an aspheric lens was placed between the light source and nanowire devices. All electrical measurements were made with a probe station (TTP-4, Desert Cryogenics). For self-powered pH sensing and AND logic gate experiments, a computer-controlled analogue-to-digital converter (6030E, National Instruments) was used to record the voltage drop or voltage output of the silicon nanowire devices.

Full Methods and any associated references are available in the online version of the paper at www.nature.com/nature.

Received 15 May; accepted 7 August 2007.

1. Lewis, N. S. Toward cost-effective solar energy use. *Science* 315, 798–801 (2007).

2. Lewis, N. S. & Crabtree, G. (eds) *Basic Research Needs for Solar Energy Utilization.* (Report of the Basic Energy Sciences Workshop on Solar Energy Utilization, US Department of Energy, Washington DC, 2005); <http://www.er.doe.gov/bes/reports/abstracts.html#SEU> (18–21 April, 2005).

3. Gratzel, M. Photoelectrochemical cells. *Nature* 414, 338–344 (2001).

4. Huynh, W. U., Dittmer, J. J. & Alivisatos, A. P. Hybrid nanorod-polymer solar cells. *Science* 295, 2425–2427 (2002).

5. Law, M., Greene, L. E., Johnson, J. C., Saykally, R. & Yang, P. Nanowire dye-sensitized solar cells. *Nature Mater.* 4, 455–459 (2005).

6. Baxter, J. B. & Aydil, E. S. Nanowire-based dye-sensitized solar cells. *Appl. Phys. Lett.* 86, 053114 (2005).

7. Luque, A., Marti, A. & Nozik, A. J. Solar cells based on quantum dots: multiple exciton generation and intermediate bands. *MRS Bull.* 32, 236–241 (2007).

8. Kayes, B. M., Atwater, H. A. & Lewis, N. S. Comparison of the device physics principles of planar and radial p-n junction nanorod solar cells. *J. Appl. Phys.* 97, 114302 (2005).

9. Zhang, Y., Wang, L. W. & Mascarenhas, A. Quantum coaxial cables. *Nano Lett.* 7, 1264–1269 (2007).

10. Gust, D., Moore, T. A. & Moore, A. L. Mimicking photosynthetic solar energy transduction. *Acc. Chem. Res.* 34, 40–48 (2001).

11. Klauk, H., Zschieschang, U., Pflaum, J. & Halik, M. Ultralow-power organic complementary circuits. *Nature* 445, 745–748 (2007).

12. Browne, W. R. & Feringa, B. L. Making molecular machines work. *Nature Nanotechnol.* 1, 25–35 (2006).

13. Avouris, P. & Chen, J. Nanotube electronics and optoelectronics. *Mater. Today* 9, 46–54 (2006).

14. Wagner, R. S. & Ellis, W. C. Vapor–liquid–solid mechanism of single crystal growth. *Appl. Phys. Lett.* 4, 89 (1964).

15. Zheng, G. F., Lu, W., Jin, S. & Lieber, C. M. Synthesis and fabrication of high-performance n-type silicon nanowire transistors. *Adv. Mater.* 16, 1890–1893 (2004).

16. Hayden, O., Agarwal, R. & Lieber, C. M. Nanoscale avalanche photodiodes for highly sensitive and spatially resolved photon detection. *Nature Mater.* 5, 352–356 (2006).

17. Karpov, V. G., Cooray, M. L. C. & Shvydka, D. Physics of ultrathin photovoltaics. *Appl. Phys. Lett.* 89, 163518 (2006).

18. Shah, A. V. *et al.* Thin-film silicon solar cell technology. *Prog. Photovolt. Res. Appl.* 12, 113–142 (2004).

19. Luque, A. & Hegedus, S. *Handbook of Photovoltaic Science and Engineering* (Wiley, Chichester, 2003).

20. Javey, A., Nam, S., Friedman, R. S., Yan, H. & Lieber, C. M. Layer-by-layer assembly of nanowires for three-dimensional, multifunctional electronics. *Nano Lett.* 7, 773–777 (2007).

21. Würfel, P. *Physics of Solar Cells, From Principles to New Concepts* (Wiley-VCH, Weinheim, 2005).

22. Green, M. A. General temperature dependence of solar cell performance and implications for device modeling. *Prog. Photovolt. Res. Appl.* 11, 333–340 (2003).

23. Aberle, A. G. Surface passivation of crystalline silicon solar cells: a review. *Prog. Photovolt. Res. Appl.* 8, 473–487 (2000).

24. Green, M. A., Emery, K., King, D. L., Hishikawa, Y. & Warta, W. Solar cell efficiency tables (version 29). *Photovolt. Res. Appl.* 15, 35–40 (2007).

25. Cui, Y., Wei, Q. Q., Park, H. K. & Lieber, C. M. Nanowire nanosensors for highly sensitive and selective detection of biological and chemical species. *Science* 293, 1289–1292 (2001).

26. Huang, Y. *et al.* Logic gates and computation from assembled nanowire building blocks. *Science* 294, 1313–1317 (2001).

27. Bachtold, A., Hadley, P., Nakanishi, T. & Dekker, C. Logic circuits with carbon nanotube transistors. *Science* 294, 1317–1320 (2001).

28. Wang, Z. L. & Song, J. Piezoelectric nanogenerators based on zinc oxide nanowire arrays. *Science* 312, 242–246 (2006).

29. Wang, X., Song, J., Liu, J. & Wang, Z. L. Direct-current nanogenerator driven by ultrasonic waves. *Science* 316, 102–105 (2007).

Supplementary Information is linked to the online version of the paper at www.nature.com/nature.

Acknowledgements We thank D. W. Pang, D. C. Bell, H. G. Park, H. S. Choe, H. Yan and P. Xie for help with experiment and data analysis. C.M.L. acknowledges support from the MITRE Corporation and the Air Force Office of Scientific Research, and T.J.K. acknowledges an NSF graduate fellowship.

Author Contributions C.M.L., B.T., X.Z. and T.J.K. designed the experiments. B.T., X.Z., T.J.K., Y.F., N.Y. and G.Y. performed experiments and analyses. C.M.L., B.T., X.Z. and T.J.K. wrote the paper. All authors discussed the results and commented on the manuscript.

Author Information Reprints and permissions information is available at www.nature.com/reprints. Correspondence and requests for materials should be addressed to C.M.L. (cml@cmliris.harvard.edu).

A strong regioregularity effect in self-organizing conjugated polymer films and high-efficiency polythiophene:fullerene solar cells

YOUNGKYOO KIM[1]*, STEFFAN COOK[2], SACHETAN M. TULADHAR[1], STELIOS A. CHOULIS[1], JENNY NELSON[1]*, JAMES R. DURRANT[2], DONAL D. C. BRADLEY[1]*, MARK GILES[3], IAIN MCCULLOCH[3], CHANG-SIK HA[4] AND MOONHOR REE[5]

[1] Department of Physics, Blackett Laboratory, Imperial College London, London SW7 2BW, UK
[2] Department of Chemistry, Imperial College London, London SW7 2AZ, UK
[3] Merck Chemicals, Chilworth Science Park, Southampton SO16 7QD, UK
[4] Department of Polymer Science and Engineering, Pusan National University, Pusan 609-735, South Korea
[5] Department of Chemistry & Pohang Accelerator Laboratory, Pohang University of Science and Technology, Pohang 790-784, South Korea
*e-mail: y.kim@ic.ac.uk; jenny.nelson@imperial.ac.uk; d.bradley@imperial.ac.uk

Published online: 5 February 2006; doi:10.1038/nmat1574

Low-cost photovoltaic energy conversion using conjugated molecular materials has become increasingly feasible through the development of organic 'bulk heterojunction (BHJ)' structures[1–7], where efficient light-induced charge separation is enabled by a large-area donor–acceptor interface[2,3]. The highest efficiencies have been achieved using blends of poly(3-hexylthiophene) (P3HT) and a fullerene derivative[8–12], but performance depends critically on the material properties and processing conditions. This variability is believed to be influenced by the self-organizing properties of P3HT, which means that both optical[13,14] and electronic[15,16] properties are sensitive to the molecular packing. However, the relationship between molecular nanostructure, optoelectronic properties of the blend material and device performance has not yet been demonstrated. Here we focus on the influence of polymer regioregularity (RR) on the molecular nanostructure, and hence on the resulting material properties and device performance. We find a strong influence of RR on solar-cell performance, which can be attributed to enhanced optical absorption and transport resulting from the organization of P3HT chains and domains. Further optimization of devices using the highest RR material resulted in a power conversion efficiency of 4.4%, even without optimization of electrodes[7].

Polymer/fullerene BHJ solar cells have a layered structure (Fig. 1a) in which the light is incident on the BHJ layer, here blend films of P3HT and 1-(3-methoxycarbonyl)-propyl-1-phenyl-(6,6) C_{61} (PCBM), through the indium/tin-oxide (ITO)-coated glass substrate[6,10]. Light absorption by this BHJ layer generates excitons that are dissociated into holes and electrons by charge separation between the electron-donor (P3HT) and the electron-acceptor (PCBM)[10]. Photocurrent generation by devices based on such blends is limited by the poor red-light absorption of the polymer and the low mobilities for charge transport to the electrodes[5]. Both the optical[13] and transport[15] properties of pristine P3HT polymer films are known to improve with the degree of RR, defined as the percentage of monomers adopting a head-to-tail configuration rather than head-to-head (Fig. 1). RR P3HT chains tend to stack into planar structures known as 'lamellae', which are normally oriented perpendicular to the substrate (see the schematic for lamella folding in Fig. 1b). Higher RR enables closer packing of these lamellae such that the optical[13] and charged[17] excitations adopt some interchain character. These interchain contributions are believed to be responsible for the distinct shoulder observed on the long-wavelength side of the absorption maximum[13] and for the larger field-effect mobility[15] in high RR P3HT films. In Fig. 1b we illustrate the influence of RR on chain packing using an energy-minimized molecular structure of 11 units of P3HT with RR of 100% and 90%. Even though the interchain spacing within the thiophene ring plane ($d_{a–a}$) (herein referred to as intraplane stacking) is found to be unaffected by this variation in RR, the interchain spacing perpendicular to the ring plane ($d_{b–b}$) (referred to as interplane stacking) is found to increase in the lower RR case, indicating that higher RR should indeed increase interchain interactions. Moreover, the tendency of lamellae to orient normal to the substrate in ordered films has been suggested to increase optical anisotropy leading to higher absorption for

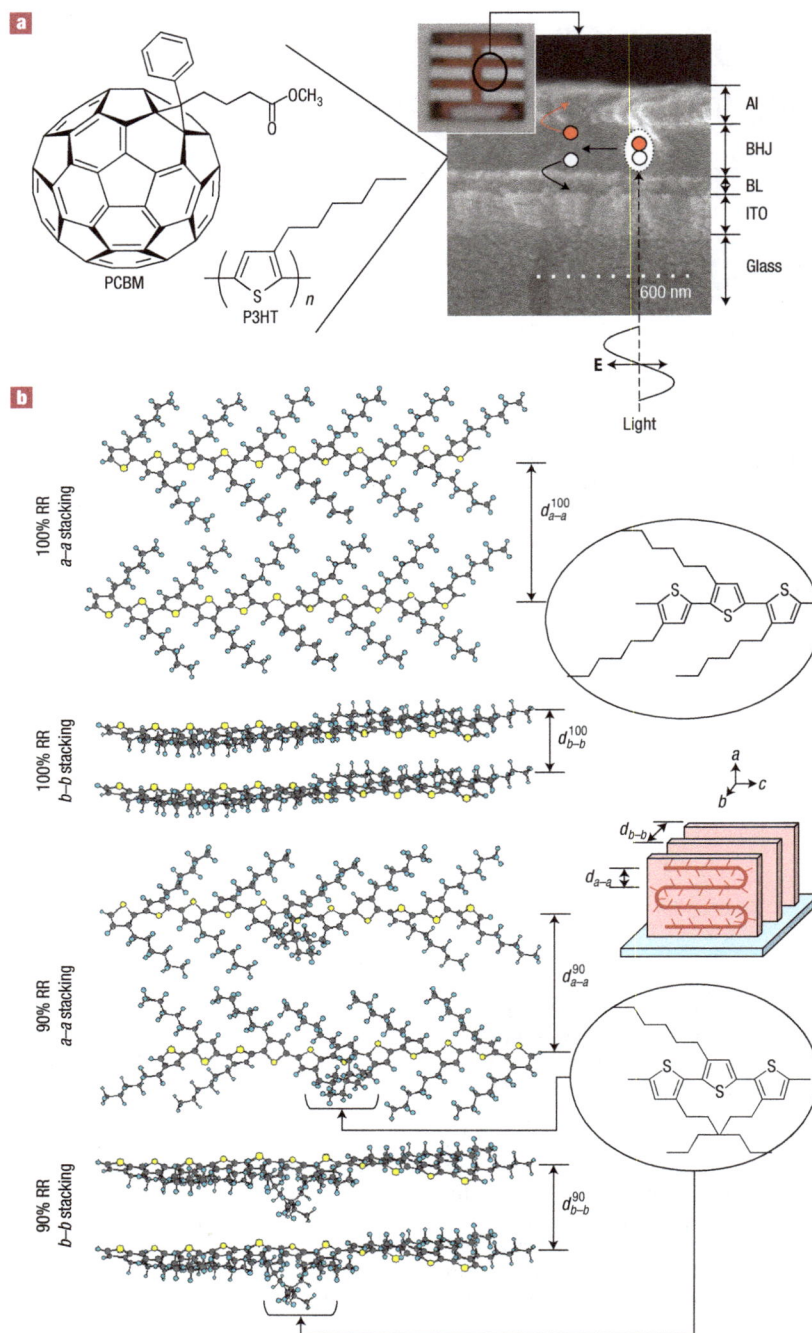

Figure 1 Schematic structure of polymer solar cells and energy-minimized structure of P3HTs. a, Cross-sectional structure (SEM, main image) of a polymer solar cell (inset photograph: top-view) with a BHJ layer (P3HT + PCBM) and a PEDOT:PSS layer (BL). The exciton generation and charge (electrons in red circles; holes in white circles) separation/collection under depolarized light illumination (**E** is the electric field vector) are shown schematically. **b**, The energy-minimized molecular structure of P3HTs (11 repeating units per chain) for 100% RR (d_{a-a}^{100} and d_{b-b}^{100} are the a-direction and b-direction chain-stacking spacings, respectively) and 90% RR (d_{a-a}^{90} and d_{b-b}^{90} are the corresponding spacings). The a-direction and b-direction are parallel and perpendicular to the thiophene ring plane, respectively (see the chemical structures within the ovals as well as the schematic illustration for lamella folding and ordering on a substrate).

light incident normal to the substrate[14]. Achieving similar ordering of P3HT chains within a BHJ structure is therefore desirable, as it promises higher optical absorption and improved transport properties, although in a BHJ the segregation of donor and acceptor phases must also be considered. Despite the recognized importance of RR in influencing chain packing and ordering,

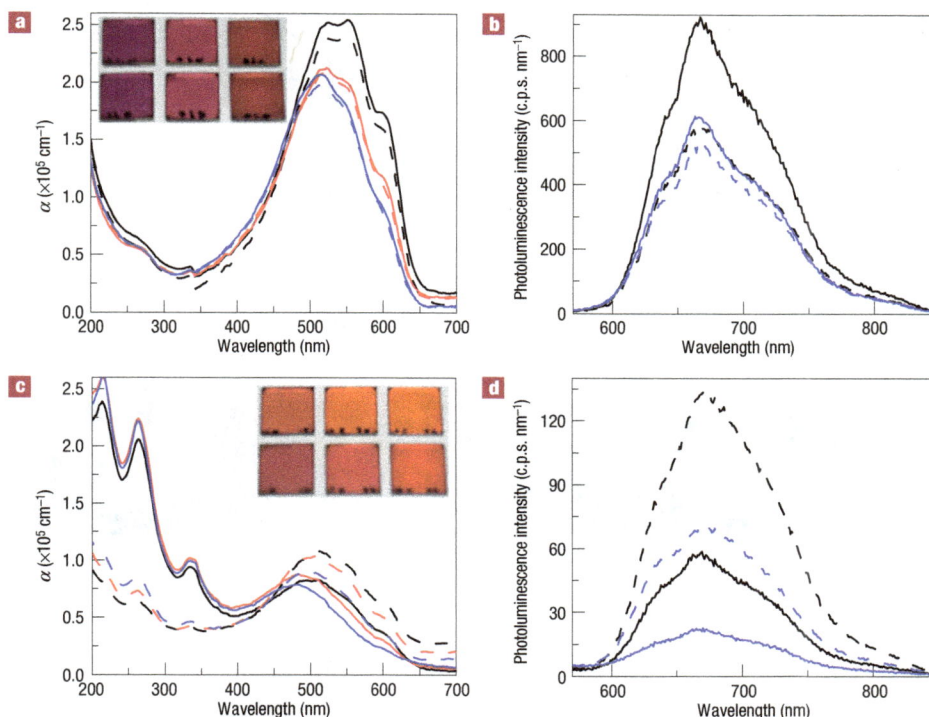

Figure 2 Optical spectra of pristine and blend films. a–d, Optical absorption and photoluminescence spectra (93% RR omitted) of pristine P3HT films (**a,b,** approximately 100 nm thick) and P3HT:PCBM (1:1) blend films (**c,d,** approximately 200 nm thick): black, red and blue colours represent 95.2%, 93% and 90.7% RR, respectively, and solid and dashed lines denote not-annealed and annealed (140 °C for 2 h) films, respectively. c.p.s. is counts per second. Insets to **a,c:** photographs of films, sample size 12 mm × 12 mm; from left 95.2%, 93% and 90.7% RR; top, not-annealed; bottom, annealed.

its influence on the performance of BHJ solar cells has not yet been demonstrated.

In this work we demonstrate that the performance of P3HT:PCBM solar cells is improved by both increasing RR and annealing, and that these improvements result from the beneficial impact of P3HT chain packing on the optical and transport properties of the blend. The effect of chain stacking on the optical properties of pristine and blend films made from P3HTs with three different degrees of RR is illustrated in Fig. 2. For pristine P3HT films, the absorption coefficient α increases with increasing RR, particularly on the red shoulder at 600 nm, which is assigned to interplane interactions[13] (see the colour changes on the photographs in Fig. 2a). No striking absorption coefficient changes result, however, from thermal annealing at 140 °C. From the intensity of photoluminescence spectra of the pristine P3HT (Fig. 2b) it is clear that higher RR tends to increase photoluminescence intensity, suggesting a reduction in non-radiative quenching pathways in the more ordered material. Annealing of the highest RR films invokes a slight reduction in photoluminescence intensity. For the P3HT:PCBM (1:1) blend films, the absorption coefficient trends are the same as for the pristine films with respect to RR, with a growth in the red shoulder attributed to interplane interactions, showing that self-organization of P3HT is still effective even in the presence of 50 wt% PCBM (Fig. 2c)[10]. After thermal annealing, all blend films show increased absorption coefficient and photoluminescence intensity (Fig. 2d)[10]. These effects are attributed to the segregation of ordered P3HT domains, with improved intraplane and interplane stacking, out of the blend on thermal annealing.

The influence of RR and annealing on intraplane chain stacking in the pristine and blend films is demonstrated with grazing-incidence X-ray-diffraction (GIXRD) measurements using a synchrotron X-ray beam (Fig. 3a). As shown in Fig. 3c (Fig. 3b for two-dimensional images), the out-of-plane (OOP) diffraction patterns (diffraction angle $2\theta = 5.3°$, $10.7°$ and $15.9°$ for the primary (100), secondary (200) and tertiary (300) peaks) indicate that all pristine P3HT films show a well-organized intraplane structure with lamellae oriented normal to the substrate (see the schematic for lamella folding in Fig. 1b)[15], with the degree of order, inferred from the peak intensity, increasing with RR. From the in-plane (IP) patterns (inset to Fig. 3c) the RR effect is clearly observed: the higher RR, the weaker the intraplane stacking in the plane of the substrate, indicating that intraplane stacking perpendicular to the substrate is most pronounced[15]. After thermal annealing, this contrast in the peak intensity for OOP and IP geometries becomes more pronounced, whilst the trend with RR is preserved. For the blend film with the 90.7% RR P3HT, the primary peak in both OOP and IP geometries is greatly reduced compared with the pristine polymer, indicating that intraplane stacking almost disappears on blending with PCBM molecules, whereas the blend films with higher RR P3HTs still show the intraplane stacking features. After thermal annealing, this intraplane stacking feature is enhanced for all the blend films, showing clear primary and secondary peaks in both OOP and IP geometries. This indicates an increase in the size or number of P3HT crystallites in the blend films on thermal annealing, supporting the increased photoluminescence intensity on annealing (see Fig. 2d). Therefore, these GIXRD results support the proposal that the thermal

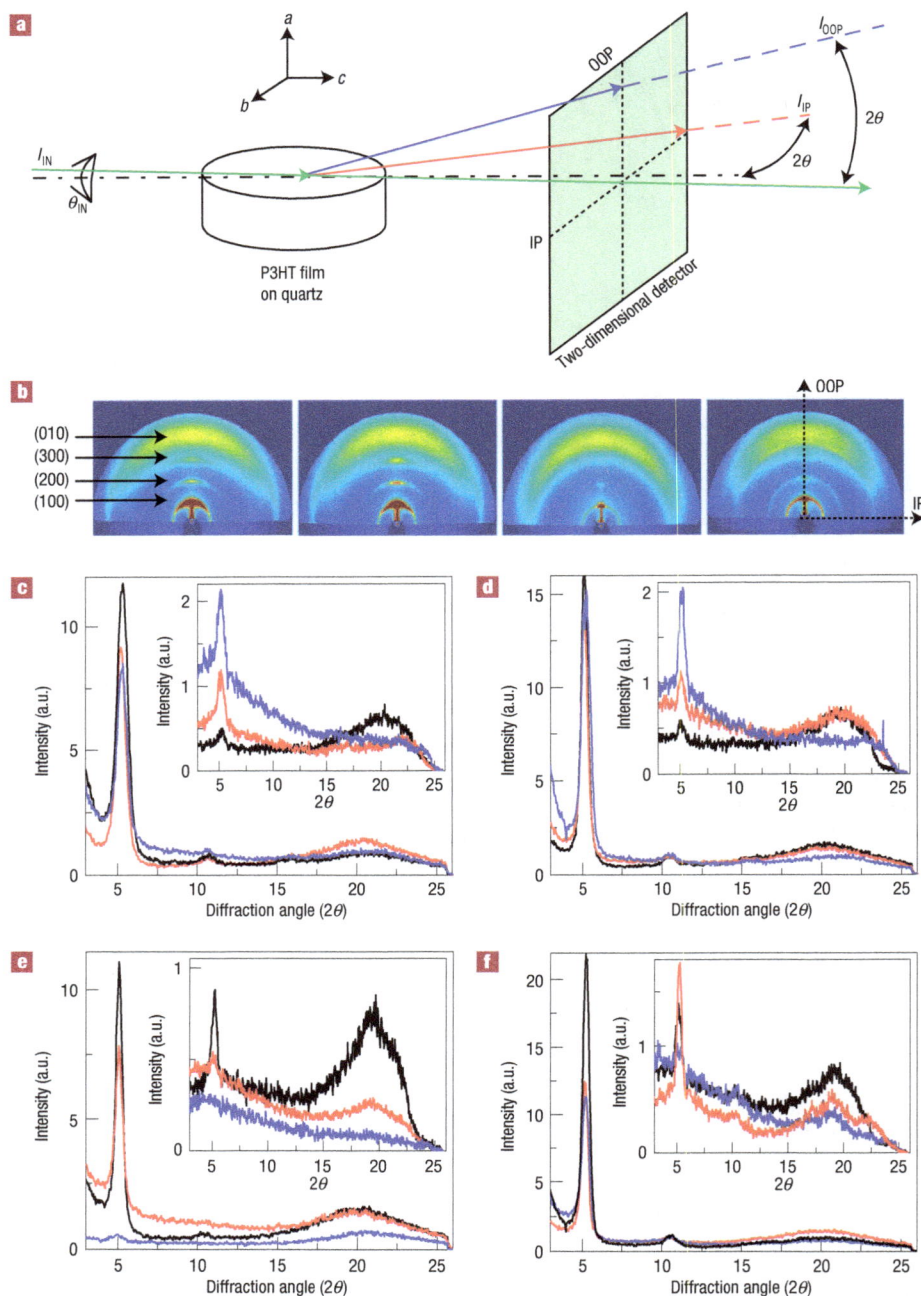

Figure 3 GIXRD setup and results for pristine and blend films. a,b, Schematic illustration (**a**) of the two-dimensional GIXRD measurement (incident X-ray angle $\theta_{IN} = 0.2°$) and representative two-dimensional images (**b**) of pristine P3HT film (not-annealed in first from left; annealed in second from left) and P3HT:PCBM (1:1) blend films (not-annealed in third from left; annealed in fourth from left). I_{IN}, incident X-ray intensity; I_{OOP}, diffracted X-ray intensity in the OOP direction; I_{IP}, diffracted X-ray intensity in the IP direction. **c–f,** OOP GIXRD diffractograms of pristine P3HT films (**c**, not-annealed; **d**, annealed) and P3HT:PCBM (1:1) blend films (**e**, not-annealed; **f**, annealed; approximately 200 nm thick): insets show corresponding IP GIXRD patterns. Black, red and blue colours represent 95.2%, 93% and 90.7% RR, respectively.

annealing of blend films at 140 °C strongly develops the intraplane chain stacking perpendicular to the film substrate and, in the case of low RR polymers, stimulates the segregation of crystalline P3HT domains out of the disordered blend.

The influence of these differences in chain packing with RR is examined for polymer solar cells made with the blend films.

The devices have the layer structure shown in Fig. 1a with a blend film thickness of ~200 nm. A comparison of current-density–voltage (J–V) characteristics under air-mass 1.5 simulated irradiance shows that the short-circuit current density (J_{SC}) and external quantum efficiency (EQE) (Fig. 4a,b) both increase with the RR of P3HT. This is attributed partly to the enhanced red-light

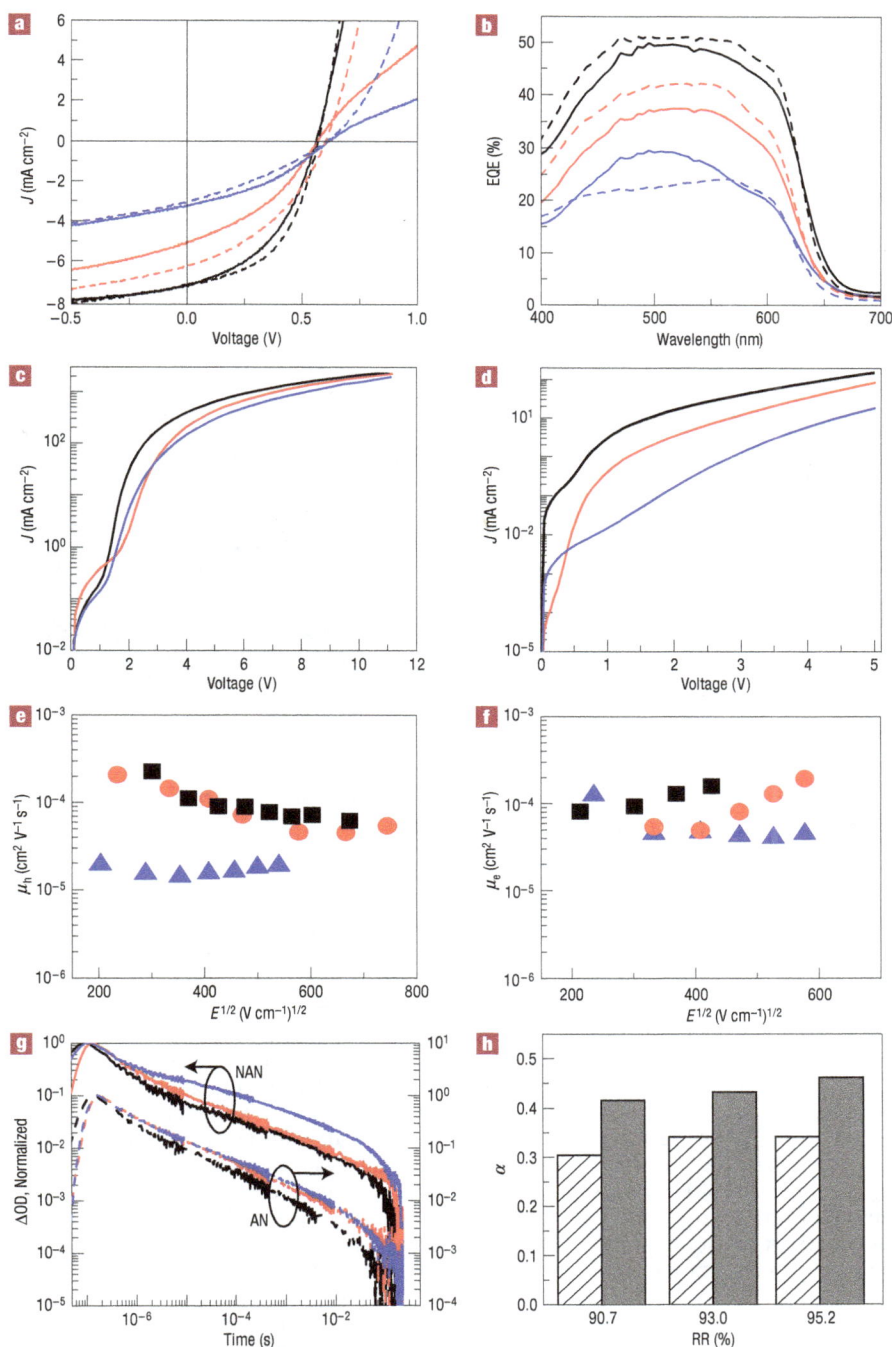

Figure 4 **Electrical properties and polaron dynamics of pristine and/or blend films. a,b,** Light J–V characteristics (air mass 1.5, incident-light power density $P_{IN} = 85$ mW cm^{-2}) (**a**) and EQE spectra (**b**) of the devices with P3HT:PCBM (1:1) blend films (approximately 200 nm thick): symbol and line colours are as in Fig. 2. Solar-cell characteristics are summarized as follows: for not-annealed device with 95.2% RR P3HT, $J_{SC} = 7.26$ mA cm^{-2}, open circuit voltage $V_{OC} = 0.56$ V, fill factor FF = 43.9%, PCE = 2.1%; for annealed device with 95.2% RR, $J_{SC} = 7.28$ mA cm^{-2}, $V_{OC} = 0.57$ V, FF = 49.7%, PCE = 2.4%; for not-annealed device with 93% RR, $J_{SC} = 5.1$ mA cm^{-2}, $V_{OC} = 0.57$ V, FF = 37.8%, PCE = 1.3%; for annealed device with 93% RR, $J_{SC} = 6.28$ mA cm^{-2}, $V_{OC} = 0.6$ V, FF = 39.4%, PCE = 1.8%; for not-annealed device with 90.7% RR, $J_{SC} = 3.27$ mA cm^{-2}, $V_{OC} = 0.62$ V, FF = 37.0%, PCE = 0.9%; for annealed device with 90.7% RR, $J_{SC} = 3.07$ mA cm^{-2}, $V_{OC} = 0.61$ V, FF = 32.1%, PCE = 0.7%. **c,d,** Dark J–V characteristics of not-annealed devices with pristine P3HT films (**c**) and P3HT:PCBM (1:1) blend films (**d**). **e,f,** Hole (**e**) and electron (**f**) mobilities (TOF) of P3HT:PCBM (1.2:1) blend films (approximately 1–2 μm thick) as a function of the square-rooted electric field ($E^{1/2}$): black squares, red circles and blue triangles, which were excited at 355 nm, denote the blend films prepared with 95.2%, 93% and 90.7% RR P3HT. **g,h,** Normalized $\triangle$OD transients (**g**, line colours as in Fig. 2) of not-annealed (NAN) and annealed (AN) blend films (approximately 200 nm thick), and resulting slope (α) changes with respect to RR (**h**, hatched bar for not-annealed; grey bar for annealed blend films).

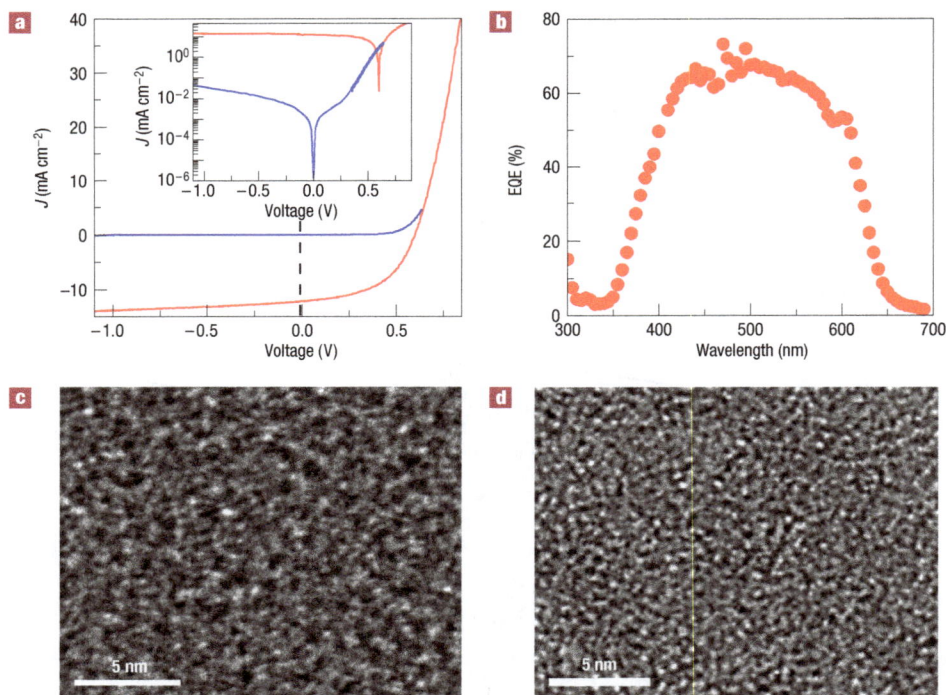

Figure 5 Best performance of polymer solar cells made in this work. a, Dark (blue line) and light (red line, air mass 1.5, $P_{IN} = 85$ mW cm^{-2}) J–V characteristics (inset with semi-logarithmic scale) of the annealed (140 °C for 2 h) device with P3HT:PCBM (1:1) blend films (approximately 175 nm thick). **b,** EQE spectrum ($P_{IN} = 33$ μW cm^{-2} at 520 nm) of the same device. **c,d,** FEG-TEM images of not-annealed (**c**) and annealed (**d**) blend films.

absorption by the higher RR P3HT owing to its more organized chain packing (see Fig. 1b). After thermal annealing, this trend was preserved. In addition, the fill factor was improved, partly owing to reduced series resistance, particularly for the devices with low RR P3HT. However, the effect of RR on absorption coefficient illustrated in Fig. 2c does not account for this large effect on J_{SC}: from the relative absorption coefficients of the 90.7% and 95.2% RR P3HT blend films, an increase in J_{SC} of only 10–20% can be expected on increasing RR from 90.7% to 95.2%, whilst the measured J_{SC} increases by more than 100% (see Supplementary Information). Therefore, the improvement in J_{SC} must arise from other factors such as improvement in charge transport with RR. For this reason, the charge-transport characteristics for blend films of different RR were investigated by measuring steady-state dark J–V characteristics and time-of-flight (TOF) mobilities[18].

The dark-current density of as-fabricated devices made with pristine P3HT increases with the RR of P3HT in the higher voltage regime (Fig. 4c). This trend is preserved after thermal annealing, and may be attributed to improved mobilities, improved injection or both. The increase in dark current is modest compared with increases observed in field-effect hole mobility when regiorandom P3HT is replaced with high RR P3HT[15]. This may be expected because the preferred direction for interplane interchain (π–π) stacking, revealed by GIXRD measurements above, is parallel to the substrate, and the enhanced interplane interactions available with high RR are therefore less likely to benefit vertical (such as J–V) than lateral (such as field-effect) transport. TOF hole mobilities in thick ($\sim$1 μm) films of P3HT showed negligible influence of RR on either electron or hole mobility, again indicating that interplane stacking does not strongly affect vertical transport in

the pristine polymer. In as-fabricated devices with blend films, the dark-current densities (Fig. 4d) and TOF mobilities (Fig. 4e,f) both show a stronger dependence on RR than for pristine films. The TOF hole mobility, μ_h, is about one order of magnitude lower for the blend film with 90.7% RR P3HT than for the higher RR P3HTs (Fig. 4e), which is consistent with the structural evidence for poor P3HT domain formation in that blend before annealing. The electron mobility, μ_e, also increases with RR, most probably owing to PCBM domain formation but possibly assisted by improved morphology for electron transport in the ambipolar P3HT[18]. From these J–V and TOF results we conclude that charge transport in blend films becomes faster with increasing RR of P3HT, and that the increase is correlated to improvements in P3HT chain packing and their influence on P3HT domain formation. The increased charge mobilities, together with the increase in light absorption, are capable of explaining the large increase in J_{SC} with RR.

Further evidence for this strong effect of RR on charge transport was obtained using transient absorption spectroscopy (TAS)[19,20], where the absorption ΔOD of photogenerated negative polarons[21] of PCBM in the blend films is monitored at 980 nm following low-intensity pulsed laser excitation (Fig. 4g). All blend films show a slow power-law decay, which is assigned to the diffusion-limited recombination of the remaining, separated charges. In the presence of deep traps for either charge, the slow decay dynamics reflect the detrapping of charges, which is the rate-limiting step for both recombination and charge transport. These dynamics are quantified by fitting the slow decay to a power law (ΔOD $\sim t^{-\alpha}$), where t is time, $\alpha = T/T_0$, T is temperature and T_0 represents the characteristic temperature of an exponential distribution of charge traps. As shown in Fig. 4g, these decay dynamics are faster (higher α) for the higher RR P3HT films (Fig. 4h), indicating fewer deep traps in the high RR blend films[19,20]. This is consistent with the faster hole transport in higher RR blend films demonstrated above. Annealed blend films show increased α regardless of RR, again indicating improved charge-carrier diffusion, which is ascribed to the improved ordering of P3HT chains on thermal annealing, as measured with GIXRD (see Fig. 3), and, moreover, indicating that some of the charge traps are of structural, rather than chemical, origin. Accordingly, these TAS results confirm that charge transport in these blend films is affected by the RR of P3HT as well as by thermal annealing, apparently through the effects of RR and thermal annealing on the self-organizing crystal structure of the P3HT.

On the basis of these results, we anticipate that the highest device efficiencies will be achieved with the highest RR P3HT. Previous studies show that thickness and annealing conditions influence device performance[10,22,23] and that they should be optimized for each batch of materials. Accordingly, we have achieved 4.4% power conversion efficiency (PCE) and 68–73% EQE (450–550 nm) using the P3HT polymer with the highest available RR (95.4%), after optimizing film thickness ($\sim$175 nm) and device annealing conditions (Fig. 5a,b). In particular, the optimized device showed very high dark rectification ratios of over 140 at $\pm$0.6 V (open-circuit condition) and 2.1×10^3 at $\pm$0.85 V (injection-limited regime), which indicates improved shunt and series resistances[24] by device annealing. This improved device performance is mainly ascribed to the enhanced film packing parallel to the substrate on thermal annealing as observed from the high-resolution transmission electron microscopy (TEM) images in Fig. 5c,d, which stress again the importance of interplane packing density.

METHODS

P3HTs with 90.7%, 93% and 95.2% RR were synthesized using the Merck synthetic method and were end-capped with hydrogen[25]. Weight (M_{w}) and number (M_{n}) average molecular weights, polydispersity indices (PDI), and melting points (T_{m}) were as follows: for 95.2% RR, $M_{\mathrm{w}} = 2.19 \times 10^4$, $M_{\mathrm{n}} = 1.42 \times 10^4$, PDI = 1.57, $T_{\mathrm{m}} = 212$ °C; for 93% RR, $M_{\mathrm{w}} = 3.19 \times 10^4$, $M_{\mathrm{n}} = 1.78 \times 10^4$, PDI = 1.79, $T_{\mathrm{m}} = 211$ °C; for 90.7% RR, $M_{\mathrm{w}} = 4.59 \times 10^4$, $M_{\mathrm{n}} = 2.37 \times 10^4$, PDI = 1.94, $T_{\mathrm{m}} = 203$ °C (here we note that the trend in molecular weights is in the wrong direction to explain the present RR data). For the 4.4% PCE devices a separate batch of P3HT (95.4% RR), synthesized in the same way, was used: this was characterized by $M_{\mathrm{w}} = 2.11 \times 10^4$, $M_{\mathrm{n}} = 1.16 \times 10^4$, PDI = 1.82, $T_{\mathrm{m}} = 222$ °C. All polymers were purified in the same manner to give typical metallic impurities Ni $<$ 2 µg g^{-1} and Mg $<$ 50 µg g^{-1}. PCBM was synthesized by the University of Groningen (The Netherlands) and used as received[26].

Pristine P3HT solutions were prepared using chlorobenzene (30 mg ml^{-1}). Blend solutions (P3HT:PCBM = 1:1 by weight)[10,22,23] for thin-film and device fabrication were prepared using chlorobenzene (60 mg ml^{-1}), whereas for TOF measurements concentrated blend solutions (P3HT:PCBM = 1.2:1 by weight) were prepared using chlorobenzene (120 mg ml^{-1}). For the characterization of film samples, the pristine and blend solutions were spin-coated onto quartz substrates (spectrosil B). It was found that the measured optical density of pristine P3HT films ($\sim$100 nm thick) was slightly higher or the same, whereas the thickness was increased by about 10–20% after thermal annealing. Polymer solar cells were fabricated in the same way as in ref. 10, and film and device annealing was carried out at 140 °C for 2 h inside a nitrogen-filled glove box[10]. The structures for TOF mobility measurements were fabricated in a similar way to those reported in refs 18,24.

The film and device measurements were performed as reported in previous studies, namely optical absorption as per ref. 24, solar-cell characteristics as per refs 10,24, TOF mobility as per refs 18,24 and photoluminescence measurements using an integrating sphere with excitation light at a wavelength $\lambda = 525$ nm as per ref. 24. For TAS measurements, the PCBM radical anion

decay dynamics were obtained by pulsed excitation at 525 nm (repetition rate 4 Hz, pulse duration $<$1 ns, excitation density $\sim$16 µJ cm^{-2}) monitoring the decay dynamics of the photoinduced absorption at 980 nm (0.8 eV) assigned primarily to PCBM radical anion absorption[19,21]. Scanning electron microscopy (SEM) images were measured using an SEM/energy dispersive X-ray system (Hitachi S4200). Two-dimensional GIXRD images were measured using a high-power X-ray beam (photon flux $\approx 10^{11}$ photons s^{-1} mrad^{-1} per 0.1%, beam size $\leq$0.5 mm^2) from a synchrotron radiation source (4C2 beamline, Pohang Accelerator Laboratory, South Korea) and a detection system equipped with a two-dimensional X-ray detector (PI-SCX4300-165/2, Princeton Instrument). The IP nanomorphology of blend films was measured using a high-resolution TEM equipped with a field emission gun (FEG-TEM, JEM-2100F, JEOL, Japan).

The energy-minimized molecular structures of 11 units of P3HT with RR of 100% and 90% were calculated using MM2 (CambridgeSoft).

Received 31 August 2005; accepted 1 December 2005; published 5 February 2006.

References

1. Tang, C. W. A two-layer organic solar cell. *Appl. Phys. Lett.* **48**, 183–185 (1986).
2. Sariciftci, N. S., Smilowitz, L., Heeger, A. J. & Wudl, F. Photoinduced electron transfer from a conducting polymer to buckminsterfullerene. *Science* **258**, 1474–1476 (1992).
3. Halls, J. J. M. *et al.* Efficient photodiodes from interpenetrating networks. *Nature* **376**, 498–500 (1995).
4. Yu, G., Gao, J., Hummelen, J. C., Wudl, F. & Heeger, A. J. Polymer photovoltaic cells: enhanced efficiencies via a network of internal donor–acceptor heterojunctions. *Science* **270**, 1789–1791 (1995).
5. Nelson, J. Organic photovoltaic films. *Mater. Today* **5**, 20–27 (2002).
6. Brabec, C. J., Sariciftci, N. S. & Hummelen, J. C. Plastic solar cells. *Adv. Funct. Mater.* **11**, 15–26 (2001).
7. Brabec, C. J., Shaheen, S. E., Winder, C., Sariciftci, N. S. & Denk, P. Effect of LiF/metal electrodes on the performance of plastic solar cells. *Appl. Phys. Lett.* **80**, 1288–1290 (2002).
8. Schilinsky, P., Waldauf, C. & Brabec, C. J. Recombination and loss analysis in polythiophene based bulk heterojunction photodetectors. *Appl. Phys. Lett.* **81**, 3885–3887 (2002).
9. Padinger, F., Ritterberger, R. S. & Sariciftci, N. S. Effects of postproduction treatment on plastic solar cells. *Adv. Funct. Mater.* **13**, 85–88 (2003).
10. Kim, Y. *et al.* Device annealing effect in organic solar cells with blends of regioregular poly(3-hexylthiophene) and soluble fullerene. *Appl. Phys. Lett.* **86**, 063502 (2005).
11. Reyes-Reyes, M., Kim, K. & Carroll, D. L. High-efficieny photovoltaic devices based on annealed poly(3-hexylthiophene) and 1-(3-methoxycarbonyl)-propyl-1-pheny-(6,6)C$_{61}$ blends. *Appl. Phys. Lett.* **87**, 083506 (2005).
12. Li, G. *et al.* High-efficiency solution processable polymer photovoltaic cells by self-organization of polymer blends. *Nature Mater.* **4**, 864–868 (2005).
13. Brown, P. J. *et al.* Effect of interchain interactions on the absorption and emission of poly(3-hexylthiophene). *Phys. Rev. B* **67**, 064203 (2003).
14. Zhokhavets, U., Gobsch, G., Hoppe, H. & Sariciftci, N. S. Anisotropic optical properties of thin poly(3-octylthiophene)-films as a function of preparation conditions. *Synth. Met.* **143**, 113–117 (2004).
15. Sirringhaus, H. *et al.* Two-dimensional charge transport in self-organized, high mobility conjugated polymers. *Nature* **401**, 685–688 (1999).
16. Kline, R. J. *et al.* Dependence of regioregular poly(3-hexylthiophene) film morphology and field-effect mobility on molecular weight. *Macromolecules* **38**, 3312–3319 (2005).
17. Österbacka, R., An, C. P., Jiang, X. M. & Vardeny, Z. V. Two-dimensional electronic excitations in self-assembled conjugated polymer nanocrystals. *Science* **287**, 839–842 (2000).
18. Choulis, S. A. *et al.* High ambipolar and balanced carrier mobility in regioregular poly(3-hexylthiophene). *Appl. Phys. Lett.* **85**, 3890–3892 (2004).
19. Montanari, I. *et al.* Transient optical studies of charge recombination dynamics in a polymer/fullerene composite at room temperature. *Appl. Phys. Lett.* **81**, 3001–3003 (2002).
20. Nelson, J. Diffusion-limited recombination in polymer-fullerene blends and its influence on photocurrent collection. *Phys. Rev. B* **67**, 155209 (2003).
21. Guldi, D. M., Hungerbühler, H. & Asmus, K.-D. Redox and excitation studies with C$_{60}$-substituted malonic acid diethyl esters. *J. Phys. Chem.* **99**, 9380–9385 (1995).
22. Kim, Y. *et al.* Composition and annealing effects in polythiophene/fullerene solar cells. *J. Mater. Sci.* **40**, 1371–1376 (2005).
23. Chirvase, D., Parisi, J., Hummelen, J. C. & Dyakonov, V. Influence of nanomorphology on the photovoltaic action of polymer-fullerene composites. *Nanotechnology* **15**, 1317–1323 (2004).
24. Kim, Y. *et al.* Organic photovoltaic devices based on blends of regioregular poly(3-hexylthiophene) and poly(9,9-dioctylfluorene-co-benzothiadiazole). *Chem. Mater.* **16**, 4812–4818 (2004).
25. Koller, G., Falk, B., Weller, C., Giles, M. & McCulloch, I. Process of preparing regioregular poly-(3-substituted) thiophenes. World patent WO2005/014691 (2005).
26. Hummelen, J. C. *et al.* Preparation and characterization of fulleroid and methanofullerene derivatives. *J. Org. Chem.* **60**, 532–538 (1995).

Acknowledgements

The authors thank Merck Chemicals for supplying the P3HT polymer and British Petroleum International for financial support through the OSCER project. Y.K. thanks G. Y. Heo (Postech, Korea), J. W. Park and H. J. Kim (PNU, Korea) for their help with the GIXRD and SEM measurements. Correspondence and requests for materials should be addressed to Y.K. or J.N. or D.D.C.B. Supplementary Information accompanies this paper on www.nature.com/naturematerials.

Competing financial interests

The authors declare that they have no competing financial interests.

Solution-processed PbS quantum dot infrared photodetectors and photovoltaics

STEVEN A. MCDONALD[1], GERASIMOS KONSTANTATOS[1], SHIGUO ZHANG[1], PAUL W. CYR[1,2], ETHAN J. D. KLEM[1], LARISSA LEVINA[1] AND EDWARD H. SARGENT[1]*

[1]Department of Electrical and Computer Engineering, University of Toronto, Toronto, Ontario, M5S 3G4, Canada
[2]Department of Chemistry, University of Toronto, Toronto, Ontario, M5S 3H6, Canada
*e-mail: ted.sargent@utoronto.ca

Published online: 9 January 2005; doi:10.1038/nmat1299

In contrast to traditional semiconductors, conjugated polymers provide ease of processing, low cost, physical flexibility and large area coverage[1]. These active optoelectronic materials produce and harvest light efficiently in the visible spectrum. The same functions are required in the infrared for telecommunications (1,300–1,600 nm), thermal imaging (1,500 nm and beyond), biological imaging (transparent tissue windows at 800 nm and 1,100 nm), thermal photovoltaics (>1,900 nm), and solar cells (800–2,000 nm). Photoconductive polymer devices have yet to demonstrate sensitivity beyond ~800 nm (refs 2,3). Sensitizing conjugated polymers with infrared-active nanocrystal quantum dots provides a spectrally tunable means of accessing the infrared while maintaining the advantageous properties of polymers. Here we use such a nanocomposite approach in which PbS nanocrystals tuned by the quantum size effect sensitize the conjugated polymer poly[2-methoxy-5-(2′-ethylhexyloxy-*p*-phenylenevinylene)] (MEH-PPV) into the infrared. We achieve, in a solution-processed device and with sensitivity far beyond 800 nm, harvesting of infrared-photogenerated carriers and the demonstration of an infrared photovoltaic effect. We also make use of the wavelength tunability afforded by the nanocrystals to show photocurrent spectra tailored to three different regions of the infrared spectrum.

Organic/nanocrystal composites have been demonstrated to enable a number of important optoelectronic devices operating in the visible region[4-8]. In the infrared, electroluminescence has been demonstrated from such materials[9-11]. In the area of infrared photodetection using nanocomposites there is one report with a low internal quantum efficiency of 10^{-5} at 5 V bias that necessitated the use of modulated illumination and a lock-in amplifier to observe the photocurrent[12]. Thus far, there has been no demonstration of an infrared photovoltaic effect from such a material system.

We demonstrate, using solution-processed materials, both a three-order-of-magnitude improvement in infrared photoconductive internal quantum efficiency compared with previous results[12], allowing observation of the photocurrent under continuous-wave illumination without reliance on lock-in techniques; and also the first observation of an infrared photovoltaic effect in such materials.

Under −5 V bias and illumination from a 975 nm laser, our detectors show an internal quantum efficiency of 3%, a ratio of photocurrent to dark current of 630, and a maximum responsivity of 3.1×10^{-3} A W^{-1}. The photovoltaic response under 975 nm excitation results in a maximum open-circuit voltage of 0.36 V, short-circuit current of 350 nA, and short-circuit internal quantum efficiency of 0.006%. We also demonstrate, by varying the size of the nanocrystals during processing, photocurrent spectra with peaks tailored to 980 nm, 1.200 μm and 1.355 μm.

Quantum dot nanocrystals of PbS were chosen for their ability to sensitize MEH-PPV, which on its own absorbs between ~400 nm and ~600 nm, into the infrared. Our nanocrystals have absorption peaks tunable from ~800 nm to ~2,000 nm (ref. 13). We show herein that our devices' photocurrent spectrum corresponds to the nanocrystals' absorption spectrum, indicating that the sensitivity of the nanocomposite could potentially be tuned across the 800–2,000 nm spectral region.

The selection of the semiconducting polymer is critical to achieving charge separation between the nanocrystal and polymer. Conjugated polymers typically have better hole than electron mobility. Thus, photoconductivity in polymer/nanocrystal composites requires a band alignment that favours transfer of the photogenerated hole to the polymer; that is, the ionization potential of the polymer should, ideally, lie closer to vacuum than that of the nanocrystal. The bulk ionization potential of PbS is ~4.95 eV, whereas most conjugated polymers have ionization potentials greater than ~5.3 eV (ref. 14). The low ionization potential of PbS relative to other semiconductor materials used in nanocrystal-based photoconductive devices, such as the cadmium chalcogenides (bulk ionization potentials between ~6.4 eV and ~7.3 eV), limits the number of readily available conjugated polymers that provide a favourable energy alignment. MEH-PPV was selected for its low ionization potential, variously reported[15,16] between ~4.9 eV and ~5.1 eV. It was not obvious at the outset that MEH-PPV/PbS would provide the type-II heterojunction needed for efficient photoconduction and for the observation of a photovoltaic effect. The vacuum-referenced band edge of the organic component is uncertain; it is possible that a dipole layer could be formed at the

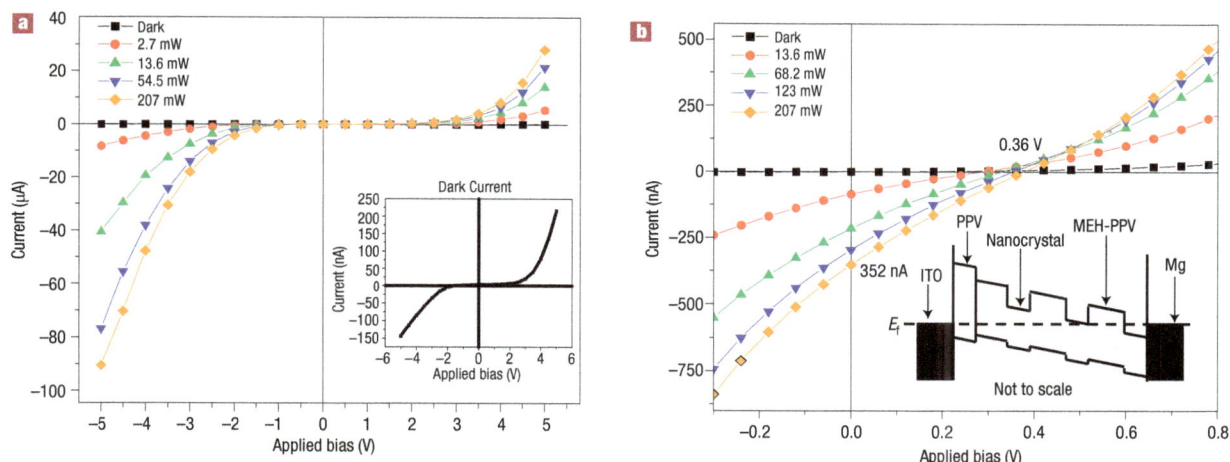

Figure 1 Dark current and photocurrent versus applied bias at the ITO electrode. The pump powers are shown in the key. **a**, Main panel: dark current and photocurrent results for a sample with ~90% by weight nanocrystals in the polymer/nanocrystal blend. Inset: dark current for the main panel. **b**, Main panel: dark current and photocurrent curves near zero bias, demonstrating the photovoltaic effect; these data were obtained from a different sample from that shown in **a** and represent the best results so far for short-circuit current and open-circuit voltage. Inset: proposed simplified band diagram depicting the relative energy alignments after the magnesium electrode has been deposited and the sample has reached equilibrium.

interface between materials, altering the effective band alignment; and the nanocrystal energy levels vary with size. However, MEH-PPV was one available conjugated polymer candidate to provide the correct heterostructure for this application.

The devices consist of a sandwich structure of glass, indium tin oxide (ITO), poly(p-phenylenevinylene) (PPV), MEH-PPV/PbS nanocrystal blend, and an upper magnesium contact. In addition to acting as a hole transport layer, the PPV layer provides a number of improvements over samples with the MEH-PPV/nanocrystal layer deposited directly on the ITO: it provides better electrical stability by forming a smooth and pinhole-free pre-layer on which the blend films can be cast, eliminating catastrophic shorts from the upper contact directly through to the ITO; it decreases the dark current by introducing an injection barrier at the ITO contact, allowing larger ratios of photocurrent to dark current; and it permits a higher bias to be applied to the samples before electrical breakdown, allowing the establishment of a higher internal field, more efficient photogenerated carrier extraction and consequently higher photocurrents. The PPV layer slightly reduces the photocurrent internal quantum efficiency because it also poses a barrier to the extraction of both photogenerated holes in the reverse bias and photogenerated electrons in the forward bias (it will be shown below that the barrier to extracting holes from the active region is less than that for electrons, resulting in higher photocurrent in the reverse bias). However, the PPV layer poses less of an extraction barrier than it does an injection barrier, which allows for the improved ratio of photocurrent to dark current. The slight reduction in efficiency was a compromise to obtain low dark current and to maximize the on/off ratio, of critical importance in detection and imaging applications.

The dark current and photocurrent are shown as a function of bias applied at the ITO electrode in Fig. 1a. The data were taken using an Agilent 4155C Semiconductor Parameter Analyzer and microprobe station. The optical excitation was provided by a 975 nm continuous-wave semiconductor laser, which allowed selective excitation of the nanocrystal phase. The dark current is 216 nA at a bias of 5 V and 144 nA at −5 V (inset, Fig. 1a). The photocurrent I–V curves show diode-like behaviour, with higher photocurrents in the

reverse bias. At a bias of −5 V, the photocurrent is 8.43 µA for 2.7 mW incident power and 90.61 µA for 207 mW incident power, which represents a ratio of photocurrent to dark current of 59 and 630, respectively. The above photocurrent under 2.7 mW illumination represents a responsivity of 3.1×10^{-3} A W^{-1}. When ITO is positively biased at 5 V, the photocurrent is reduced to 5.39 and 28.12 µA for incident powers of 2.7 and 207 mW, respectively.

The asymmetry of the photocurrent I–V curves can be ascribed to the work-function difference between ITO (~4.8 eV) and magnesium (3.7 eV) and to the energy levels of the PPV layer. The inset of Fig. 1b shows a possible band diagram for the structure after the magnesium contact has been deposited and the device has reached the equilibrium state. For this diagram, the lowest-energy absorption peak is assumed to be the first excitonic absorption of the PbS nanocrystals. This is used to estimate an increase in bandgap energy relative to bulk PbS, which has a bandgap of 0.41 eV; for the nanocrystals with absorption peaks centred at 955, 1,200 and 1,355 nm (depicted in Fig. 3), the effective bandgaps are 1.30, 1.03 and 0.92 eV, respectively. Because of the nearly equal effective masses for holes and electrons in PbS, it is also assumed that the confinement energy is shared equally in the conduction and valence bands so the bands move up and down, respectively, by equal energies. The barrier for electrons comes from the octylamine ligand, which passivates the nanocrystal surface, and/or the MEH-PPV (ionization energy ~4.9 eV and electron affinity ~2.9 eV; ref. 15), that surrounds the nanocrystal. To align the Fermi level in all layers, the magnesium side tilts down and the ITO side tilts up. (Similar band tilting in polymers[17] and in C$_{60}$-doped polymers sandwiched between two different conductors[18] have been discussed previously.) After an electron in the valence band of the nanocrystal is transferred to the conduction band by absorbing a photon, the hole in the valence band may transfer to the hole-conducting MEH-PPV, and the electron can either remain in the quantum dot or move through the nanocrystal network by hopping or tunnelling. Depending on the polarity of the built-in and/or applied field, the electron and hole can move towards the ITO or towards the magnesium. When the electron moves to the ITO side, it will see a higher barrier by the tilted band

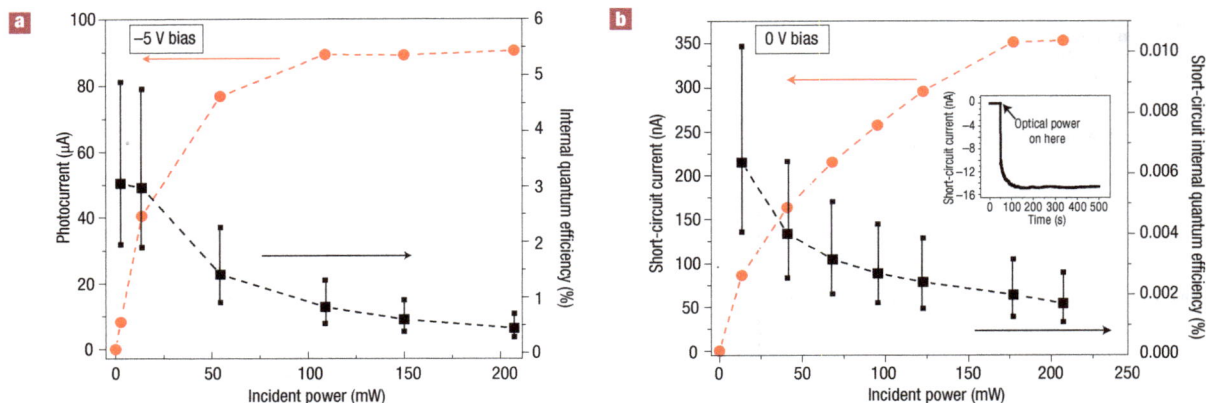

Figure 2 Photocurrent and internal quantum efficiency versus incident optical power. a, The photocurrent in red (circles) on the left axis and the internal quantum efficiency (squares) in black on the right axis are shown as a function of incident power at −5 V bias. **b,** Main panel: short-circuit current (circles) and corresponding internal quantum efficiency (squares) as a function of incident power. The lines are merely provided to guide the eye. Inset: stability of the short-circuit current as a function of time for a sample with much lower photovoltaic response than in Fig. 1b.

and PPV (ionization energy ~5.1 eV and electronic affinity ~2.7 eV; ref. 19) than when moving to the magnesium electrode. When the hole moves to the ITO, it also faces a barrier between MEH-PPV and PPV, and no barrier if it moves to the magnesium side. Reverse bias results in photogenerated holes being extracted through the ITO/PPV side of the sample, whereas forward bias results in electrons being extracted through that side. Thus, the higher photocurrent in the reverse bias suggests that the electron barrier posed by the PPV may be more severe than the hole barrier in carrier extraction.

Figure 1b shows dark and illuminated I–V curves for the region near 0 V, demonstrating the presence of a photovoltaic effect under continuous-wave illumination at 975 nm. The maximum short-circuit current is 350 nA for an incident power of 207 mW. The open-circuit voltage is 0.36 V. The photovoltaic effect was checked for hysteresis effects to see whether slow charge reorganization alone could be the cause. A sample demonstrating much lower short-circuit current (~15 nA) than shown in Fig. 1b was used to provide more convincing evidence that, even with very low short-circuit currents, the effect is not simply an artefact of hysteresis. The inset of Fig. 2b shows the result of this test where the sample was held at zero bias and the short-circuit current monitored over 500 s: the signal was stable over this time span. Further evidence that the effect was not hysteresis-based was provided by performing voltage scans in both directions (forward bias to reverse bias and vice versa); the direction of scan had negligible effect on the photovoltaic response. Although this photovoltaic response, which allows separation of an electron–hole pair at zero applied bias, could be indicative of a type-II heterostructure between the PbS nanocrystals and MEH-PPV, the built-in field in the device under zero bias is significant and could also allow charge separation with a marginal type-I heterostructure.

Figure 2a presents the dependence of photocurrent and internal quantum efficiency on the incident power at −5 V bias. The internal quantum efficiency is defined as the ratio of the number of collected charges to the number of absorbed photons at the pump wavelength. The calculation of internal quantum efficiency using absorption values obtained in reflection mode, the handling of optical interference effects and the range bars on these efficiency values are described in detail in the Methods. The percentage absorption at the 975 nm wavelength used to obtain the main efficiency points in Fig. 2a and b was 12.7%; the upper and lower range bars represent upper and lower bounds obtained by consideration of multiple

pass propagation through the active layer. From the figure it can be seen that the photocurrent does not increase linearly with incident power. Above ~50 mW, the photocurrent increases more slowly with increased power. In the low power region, the recombination of trapped electrons in the nanocrystal network with holes in the neighbouring polymer dominates. When more photons are absorbed at higher powers, bimolecular recombination between free holes and electrons occurs in addition to the recombination at electron trap centres[8]. The additional bimolecular recombination reduces the number of photo-excited carriers and, hence, lowers the internal quantum efficiency as shown in Fig. 2a. At an incident power of 2.7 mW the internal quantum efficiency is about 3% (external quantum efficiency of ~0.38%), whereas at 207 mW the internal quantum efficiency is reduced to about 0.4%. The short-circuit current and corresponding internal quantum efficiency is plotted in Fig. 2b, showing a maximum value of ~0.006% (short-circuit external quantum efficiency of ~0.0008%). These zero-bias internal quantum efficiencies are about 500 times lower than at −5 V and show similar signs of a roll-off caused by bimolecular recombination at higher powers. The short-circuit internal quantum efficiency is much lower than the best previous reports for CdSe nanocrystal-based systems where the trioctylphosphine oxide (TOPO) ligands were removed by treatment with pyridine; for samples with the TOPO ligands still present on the nanocrystal surface, these systems showed internal quantum efficiencies closer to, but still slightly higher than, that reported here[8,20]. Further efforts are required in the PbS system to remove the ligands, as this could markedly improve efficiencies (especially in photovoltaic mode) in our system.

The 3% internal quantum efficiency at −5 V is an increase of three orders of magnitude over that reported previously[12] and is attributed principally to an improvement in film quality across these large-area devices. The MEH-PPV in previous work was typically cast from toluene and was not ultrasonicated or filtered. In the present work, the MEH-PPV was cast from chloroform, ultrasonicated for 1 hour before casting the films, and both the polymer and nanocrystal solutions were independently filtered. The combination of the above treatments was shown using atomic force microscopy to provide smoother films with fewer defects and pinholes than the previous process. The films produced as described previously[12] showed large centres of aggregated material and many pinholes; the newer films show much smaller regions of aggregated, transport-impeding material

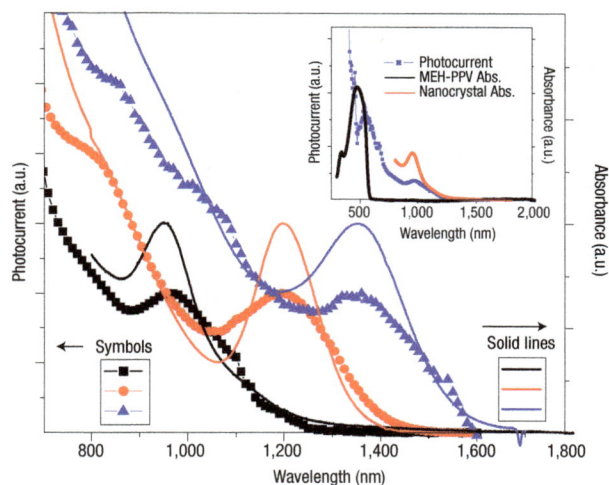

Figure 3 Photocurrent spectral responses and absorption spectra. Main panel: photocurrent spectral response (symbols) and the corresponding absorption spectra (solid line) for three different samples. The absorption peaks are tuned to 955 (black), 1,200 (red) and 1,355 nm (blue). Inset: extended spectral response for the sample centred at 955 nm, indicating the response in the region below ~600 nm where both the polymer and nanocrystal are excited. Also shown are the absorption spectra of the polymer and the nanocrystals.

and are pinhole-free. The improved surface of the films allows better interfacial contact with the upper metal electrode, resulting in better carrier extraction[21]. Films in the earlier work contained only ~60% nanocrystal by weight because higher loading gave films that suffered from excessive shorts. In this work, films containing 90% nanocrystal by weight were successfully cast by optimizing the concentration, and hence viscosity, of both the nanocrystal and polymer solutions. The increased nanocrystal loading probably resulted in improved electron transport. Pinholes previously[12] led to a photocurrent-to-dark-current ratio of ~10^{-4}, necessitating the use of lock-in techniques to detect the photocurrent signal. The present devices, with their orders-of-magnitude greater photocurrent-to-dark ratios and efficiencies, were readily studied using continuous-wave methods.

Figure 3 shows the absorbance spectrum of the nanocrystals (measured using a Varian CARY 500 Scan spectrophotometer) used in three different devices, each tuned to a different part of the infrared spectrum, and the measured photocurrent spectral response of each device. No bias was applied to the devices during measurement of the photocurrent spectrum, and the excitation was provided by narrow wavelength bands selected from a white light source by a monochromator. At wavelengths longer than 600 nm, the absorption of MEH-PPV is negligible; thus, all absorption at these wavelengths is assigned solely to the PbS nanocrystals. The absorption peaks at 955, 1,200 and 1,355 nm correspond to the first excitonic absorption features in the three different choices of PbS nanocrystals. The photocurrent spectra show peaks that match closely the absorption peaks associated with the nanocrystals. Along with demonstrating control over the spectral response, this result adds further evidence that the photocurrent is due to exciton formation in the nanocrystals followed by charge separation. The inset of Fig. 3 shows the spectral response of the device with the 955-nm response peak, including shorter wavelengths where the polymer is also excited. The photocurrent response peaks at a wavelength of 60 nm redshifted relative to the polymer absorption peak.

METHODS

PBS NANOCRYSTAL SYNTHESIS AND LIGAND EXCHANGE

The synthesis followed that used previously[13]. The as-prepared nanocrystals were capped with oleic acid ligands. A post-synthesis ligand exchange was performed to replace these with octylamine ligands. The original oleic-acid-capped nanocrystals were precipitated with methanol, dried and dispersed in an excess of octylamine. This solution was heated at 70 °C for ~16 hours. After heating, the octylamine capped nanocrystals were precipitated with N,N-dimethylformamide and redispersed in chloroform. The nanocrystals were then mixed with MEH-PPV to give a known weight fraction.

DEVICE PROCESSING

A 40 nm PPV hole transport layer was spin-coated on 2.5 × 2.5 cm² ITO-coated glass slide and annealed at 200 °C for 3 hours in vacuum to allow polymerization. A MEH-PPV/nanocrystal blend (90% nanocrystal by weight) dissolved in chloroform was spin-coated on the PPV layer to form a film of thickness 100–150 nm. Finally, the upper contact was deposited by vacuum evaporation forming a 3-mm² metal stack of 150 nm Mg/100 nm Ag/10 nm Au.

MEASUREMENT OF PHOTOCURRENT SPECTRAL RESPONSE

A bias of 0 V was applied to the sample connected in series with a load resistor of ~100 Ω, which was about three orders of magnitude smaller than the resistance of the sample. Illumination was provided by a white light source dispersed by a monochromator (Triax 320) and mechanically chopped at a frequency of ~250 Hz. Various filters were used to avoid overtones of the monochromator's grating from illuminating the sample. The potential drop across the load resistor was read by a lock-in amplifier (model SR803 DSP). The light intensity at each wavelength was separately measured. Then, the photocurrent at each wavelength was scaled to the same incident light intensity by assuming that the photocurrent was linearly proportional to the light intensity in the low-intensity region used.

CALCULATION OF INTERNAL QUANTUM EFFICIENCY

The percentage absorption used in the internal quantum efficiency calculation needs to account for the device structure, which creates multiple optical passes because of the mirror-like upper metallic contact. Hence, optical interference effects must be considered[22]. Two separate approaches were taken and the results compared. First, the device's absorption was directly measured in reflection mode (see Supplementary Information, Fig. S1). For the second method, the single-pass absorption was measured and scaled by a factor determined by the intensity increase created by interference. Using a multilayer program, it was determined that the maximum enhancement would be between 2 and 2.5 depending on the exact layer thickness and index of refraction of each layer. Because there is considerable uncertainty in these values we chose the more severe absorption enhancement factor of 2.5 to provide a conservative lower bound on efficiency. The single-pass absorption was used as the lower bound on absorption (upper bound on efficiency) to represent that case where negligible Fabry–Perot enhancement occurs. The intermediate efficiency points in Fig. 2 were obtained using the measured multi-pass absorption at 975 nm (12.7%), and the upper and lower bounds were obtained using the single-pass absorption at 975 nm (7.9%) and 2.5 enhancement absorption at 975 nm (19.8%), respectively.

Received 21 July 2004; accepted 19 November 2004; published 9 January 2005.

References

1. Forrest, S. R. The path to ubiquitous and low-cost organic electronic appliances on plastic. *Nature* **428**, 911–918 (2004).
2. Brabec, C. J. *et al.* A low-bandgap semiconducting polymer for photovoltaic devices and infrared diodes. *Adv. Funct. Mater.* **12**, 709–712 (2002).
3. Yoshino, K. *et al.* Near IR and UV enhanced photoresponse of C₆₀-doped semiconducting polymer photodiode. *Adv. Mater.* **11**, 1382–1385 (1999).
4. Huynh, W. U., Dittmer, J. J. & Alivisatos, A. P. Hybrid nanorod-polymer solar cells. *Science* **295**, 2425–2427 (2002).
5. Wang, Y. & Herron, N. Photoconductivity of CdS nanocluster-doped polymers. *Chem. Phys. Lett.* **200**, 71–75 (1992).
6. Dabbousi, B. O., Bawendi, M. G., Onitsuka, O. & Rubner, M. F. Electroluminescence from CdSe quantum-dot/polymer composites. *Appl. Phys. Lett.* **66**, 1316–1318 (1995).
7. Mattoussi, H. *et al.* Electroluminescence from heterostructures of poly(phenylene vinylene) and inorganic CdSe nanocrystals. *J. Appl. Phys.* **83**, 7965–7947 (1998).
8. Greenham, N. C., Peng, X. & Alivisatos, A. P. Charge separation and transport in conjugated-polymer/semiconductor-nanocrystal composites studied by photoluminescence quenching and photoconductivity. *Phys. Rev. B* **54**, 17628–17637 (1996).
9. Bakueva, L. *et al.* Size-tunable infrared (1000–1600 nm) electroluminescence from PbS quantum-dot nanocrystals in a semiconducting polymer. *Appl. Phys. Lett.* **82**, 2895–2897 (2003).
10. Tessler, N., Medvedev, V., Kazes, M., Kan, S. & Banin, U. Efficient near-infrared polymer nanocrystal light-emitting diodes. *Science* **295**, 1506–1508 (2002).
11. Steckel, J. S., Coe-Sullivan, S., Bulovic, V. & Bawendi, M. 1.3 μm to 1.55 μm tunable electroluminescence from PbSe quantum dots embedded within an organic device. *Adv. Mater.* **15**, 1862–1866 (2003).
12. McDonald, S. A., Cyr, P. W., Levina, L. & Sargent, E. H. Photoconductivity from PbS-nanocrystal/semiconducting polymer composites for solution-processible, quantum-size tunable infrared photodetectors. *Appl. Phys. Lett.* **85**, 2089–2091 (2004).
13. Hines, M. A. & Scholes, G. D. Colloidal PbS nanocrystals with size-tunable near-infrared emission: observation of post-synthesis self-narrowing of the particle size distribution. *Adv. Mater.* **15**, 1844–1849 (2003).
14. Skotheim, T. A. (ed.) *Handbook of Conducting Polymers* (Dekker, New York, 1986).
15. Greenwald, Y. *et al.* Polymer–polymer rectifying heterojunction based on poly(3,4-dicyanothiophene) and MEH-PPV. *J. Polym. Sci. A* **36**, 3115–3120 (1998).
16. Jin, S.-H. *et al.* Synthesis and characterization of highly luminescent asymmetric poly(p-phenylene vinylene) derivatives for light-emitting diodes. *Chem. Mater.* **14**, 643–650 (2002).
17. Greczynski, G., Kugler, Th. & Salaneck, W. R. Energy level alignment in organic-based three-layer structures studied by photoelectron spectroscopy. *J. Appl. Phys.* **88**, 7187–7191 (2000).

18. Brabec, C. J. *et al.* Origin of the open circuit voltage of plastic solar cells. *Adv. Funct. Mater.* **11**, 374–380 (2001).
19. Schlamp, M. C., Peng, X. & Alivisatos, A. P. Improved efficiencies in light emitting diodes made with CdSe(CdS) core/shell type nanocrystals and a semiconducting polymer. *J. Appl. Phys.* **82**, 5837–5842 (1997).
20. Ginger, D. S. & Greenham, N. C. Charge injection and transport in films of CdSe nanocrystals. *J. Appl. Phys.* **87**, 1361–1368 (2000).
21. Nguyen, T.-Q., Kwong, R. C., Thompson, M. E. & Schwartz, B. J. Improving the performance of conjugated polymer-based devices by control of interchain interactions and polymer film morphology. *Appl. Phys. Lett.* **76**, 2454–2456 (2000).
22. Peumans, P., Yakimov, A. & Forrest, S. R. Small molecular weight organic thin-film photodetectors and solar cells. *J. Appl. Phys.* **93**, 3693–3723 (2003).

Acknowledgements

We thank S. Hoogland for discussions and the following for support: the Government of Ontario through the Ontario Graduate Scholarships program (S.A.M.); Materials and Manufacturing Ontario, a division of the Ontario Centres of Excellence; the Natural Sciences and Engineering Research Council of Canada through its Collaborative Research and Development Program; Nortel Networks; the Canada Foundation for Innovation; the Ontario Innovation Trust; and the Canada Research Chairs Programme. Correspondence and requests for materials should be addressed to E.H.S.
Supplementary Information accompanies the paper on www.nature.com/naturematerials.

Competing financial interests

The authors declare that they have no competing financial interests.

Nanowire dye-sensitized solar cells

MATT LAW[1,2]*, LORI E. GREENE[1,2]*, JUSTIN C. JOHNSON[1], RICHARD SAYKALLY[1] AND PEIDONG YANG[1,2]†

[1]Department of Chemistry, University of California, Berkeley, California 94720, USA
[2]Materials Science Division, Lawrence Berkeley National Laboratory, Berkeley, California 94720, USA
*These authors contributed equally to this work.
†e-mail: p_yang@berkeley.edu

Published online: 15 May 2005; doi:10.1038/nmat1387

Excitonic solar cells[1]—including organic, hybrid organic–inorganic and dye-sensitized cells (DSCs)—are promising devices for inexpensive, large-scale solar energy conversion. The DSC is currently the most efficient[2] and stable[3] excitonic photocell. Central to this device is a thick nanoparticle film that provides a large surface area for the adsorption of light-harvesting molecules. However, nanoparticle DSCs rely on trap-limited diffusion for electron transport, a slow mechanism that can limit device efficiency, especially at longer wavelengths. Here we introduce a version of the dye-sensitized cell in which the traditional nanoparticle film is replaced by a dense array of oriented, crystalline ZnO nanowires. The nanowire anode is synthesized by mild aqueous chemistry and features a surface area up to one-fifth as large as a nanoparticle cell. The direct electrical pathways provided by the nanowires ensure the rapid collection of carriers generated throughout the device, and a full Sun efficiency of 1.5% is demonstrated, limited primarily by the surface area of the nanowire array.

The anodes of dye-sensitized solar cells are typically constructed using thick films (~10 μm) of TiO₂ or, less often, SnO₂ or ZnO nanoparticles[4–6] that are deposited as a paste and sintered to produce electrical continuity. The nanoparticle film provides a large internal surface area (characterized by a roughness factor, defined as the total film area per unit substrate area, of ~1,000) for the anchoring of sufficient chromophore (usually a ruthenium-based dye) to yield high light absorption in the 400–800 nm region, where much of the solar flux is incident. During operation, photons intercepted by the dye monolayer create excitons that are rapidly split at the nanoparticle surface, with electrons injected into the nanoparticle film and holes leaving the opposite side of the device by means of redox species (traditionally the I^-/I_3^- couple) in a liquid or solid-state[7] electrolyte. The classic report[8] of a 10% efficient TiO₂ DSC initiated a decade of research into the electrical transport physics of nanoparticle anodes, which were shown to collect injected electrons with high efficiency despite their disordered, polycrystalline topology.

The nature of electron transport in oxide nanoparticle films is fairly well understood. Time-resolved photocurrent and photovoltage measurements[9,10] and modelling studies[11,12] indicate that electron transport in wet, illuminated nanoparticle networks proceeds by a trap-limited diffusion process, in which photogenerated electrons repeatedly interact with a distribution of traps as they undertake a random walk through the film. Drift transport, a vital mechanism in most photovoltaic cells, is prevented in DSCs by ions in the electrolyte that screen macroscopic electric fields and couple strongly with the moving electrons, effectively rendering them neutral carriers (that is, there is ambipolar diffusion)[13]. Under full sunlight, an average injected electron may experience a million trapping events before either percolating to the collecting electrode or recombining with an oxidizing species, predominantly I_3^- in the electrolyte[14]. Transit times for electron escape from the film average in the milliseconds[15]. Yet despite the extremely slow nature of such trap-mediated transport (characterized by an electron diffusivity, $D_n \leq 10^{-4}$ cm² s⁻¹, several orders of magnitude smaller than in TiO₂ and ZnO single crystals[16,17]), electron collection remains favoured over recombination because of the even slower multi-electron kinetics of I_3^- reduction on oxide surfaces. Electron diffusion lengths of 7–30 μm have been reported for cells operating at light intensities up to 0.1 Sun[9,10,18]. This is strong evidence that electron collection is highly efficient for the 10-μm-thick nanoparticle films normally used in devices.

Insight into the factors that limit DSC performance is gained by comparing theoretical cell efficiencies with those of current state-of-the-art cells. The power conversion efficiency η of a solar cell is given as $\eta = (FF \times |J_{sc}| \times V_{oc})/P_{in}$, where FF is the fill factor, $|J_{sc}|$ is the absolute value of the current density at short circuit, V_{oc} is the photovoltage at open circuit and P_{in} is the incident light power density. In principle, the maximum J_{sc} of a DSC is determined by how well the absorption window of its dye overlaps the solar spectrum. Record cells achieve current densities (and overall efficiencies) that are between 55 and 75% of their theoretical maxima at full sunlight, depending on the exact dye used[19]. Much of the shortfall is due to the poor absorption of low-energy photons by available dyes. Considerable efforts have been made to develop dyes and dye mixtures that absorb better at long wavelengths[20,21], so far with little success. Another option for improving the absorption of red and near-infrared light—thickening the nanoparticle film to increase its optical density—is unsuccessful because the film thickness comes to exceed the electron diffusion length through the nanoparticle network.

One promising solution to this impasse is to increase the electron diffusion length in the anode by replacing the nanoparticle film with an array of oriented single-crystalline nanowires. Electron transport in crystalline wires is expected to be several orders of magnitude faster than percolation through a random polycrystalline network.

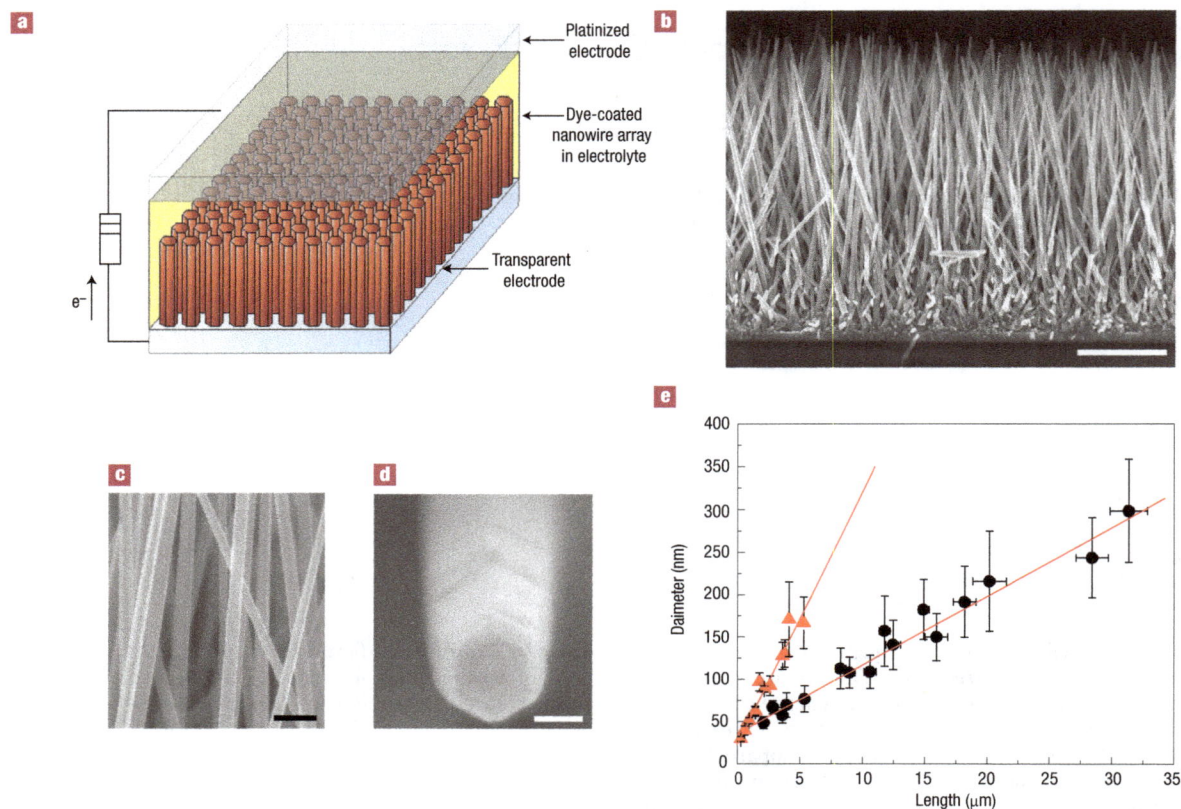

Figure 1 The nanowire dye-sensitized cell, based on a ZnO wire array. a, Schematic diagram of the cell. Light is incident through the bottom electrode. **b**, Typical scanning electron microscopy cross-section of a cleaved nanowire array on FTO. The wires are in direct contact with the substrate, with no intervening particle layer. Scale bar, 5 μm. **c**, Magnified view of the oriented wires. In this array, wire length and diameter vary from 16 to 17 μm and 130 to 200 nm, respectively. Scale bar, 500 nm. **d**, Typical top view of a single nanowire, showing its faceting, surface texture and a slight taper to its tip. Scale bar, 50 nm. **e**, Wire length against diameter with (circles) and without (triangles) PEI added to the growth bath. Lines are least-squares fits to the data, and error bars represent one standard deviation.

Using a sufficiently dense array of long, thin nanowires as a dye scaffold, it should be possible to increase the DSC dye loading (and so its absorption of red light) while simultaneously maintaining very efficient carrier collection. Moreover, the rapid transport provided by a nanowire anode would be particularly favourable for cell designs that use non-standard electrolytes, such as polymer gels or solid inorganic phases, in which recombination rates are high compared with the liquid electrolyte cell[22]. Here we present the first ordered nanowire DSC (Fig. 1a) and illustrate how this topology could improve the understanding and performance of DSCs and other types of excitonic solar cells.

A high-performance nanowire photoanode must foremost have a large surface area for dye adsorption, comparable to that of a nanoparticle film. We made ZnO nanowire arrays of high surface area in aqueous solution using a seeded growth process[23] that was modified to yield long wires. Briefly, a 10–15-nm-thick film of ZnO quantum dots was deposited onto F:SnO₂ conductive glass (FTO) substrates by dip coating, and wires were grown from these nuclei through the thermal decomposition of a zinc complex. This two-step process is a simple, low-temperature and environmentally benign route to forming dense arrays (up to 35 billion wires per square centimetre) on arbitrary substrates of any size. Solution-grown ZnO nanowire arrays reported previously have been limited to aspect ratios of less than 20, too small for efficient DSCs. We boosted the

aspect ratio of our nanowires above 125 by using polyethylenimine (PEI), a cationic polyelectrolyte, to hinder only the lateral growth of the nanowires in solution, while maintaining a relatively high nanowire density (Fig. 1b–d and Methods). The striking effect of this molecule is seen by plotting nanowire length against diameter at different growth times with and without PEI (Fig. 1e). The longest arrays presented here (20–25 μm) have one-fifth the active surface area of a nanoparticle anode.

The wire films are good electrical conductors along the direction of the wire axes. Two-point electrical measurements of dry arrays on FTO substrates gave linear current–voltage (I–V) traces (see Supplementary Information, Fig. S1), indicating barrier-free contacts between nanowire and substrate. Individual nanowires were extracted from the arrays, fashioned into field-effect transistors using standard electron-beam lithography procedures, and analysed to determine their resistivity, carrier concentration and mobility (Fig. S2). Measured resistivity values ranged from 0.3 to 2.0 Ω cm, with an electron concentration of 1–5 × 10¹⁸ cm⁻³ and mobility of 1–5 cm² V⁻¹ s⁻¹. Using the Einstein relation, $D = k_B T \mu / e$, we estimate an electron diffusivity $D_n = 0.05$–0.5 cm² s⁻¹ for single dry nanowires. This value is several hundred times larger than the highest reported diffusivity for TiO₂ or ZnO nanoparticle films in operating cells[15,24]. Moreover, the conductivity of the wire arrays increased by 5–20% when they were bathed in the standard DSC

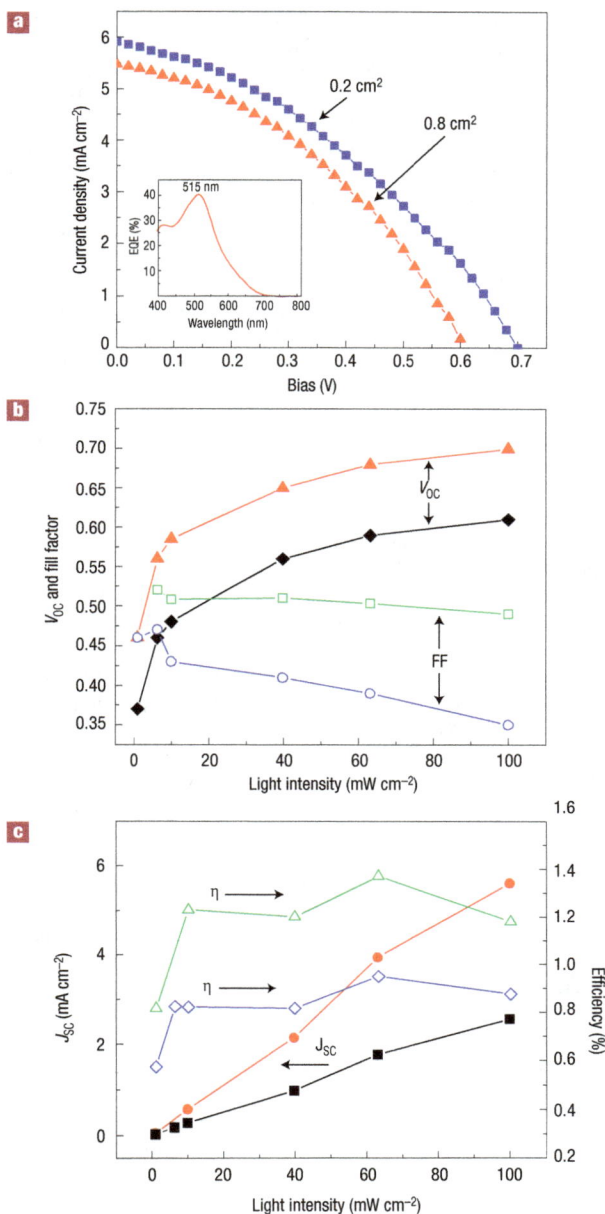

Figure 2 **Device performance under AM 1.5G illumination. a**, Traces of current density against voltage (J–V) for two cells with roughness factors of ~200. The small cell (active area: 0.2 cm²) shows a higher V_{oc} and J_{sc} than the large cell (0.8 cm²). The fill factor and efficiency are 0.37 and 1.51% and 0.38 and 1.26%, respectively. Inset, the external quantum efficiency against wavelength for the large cell. **b**, Open-circuit voltage and fill factor against light intensity, and **c**, short-circuit current density and efficiency against light intensity for cells with roughness factors from 75 to 200. Each of the four parameters is represented by data from two different devices in order to provide an estimate of the range of their variability. In general, cells of high roughness factor have low V_{oc} and fill factor, but high J_{sc} and efficiency (see Fig. S7). Note that each of the eight plots is taken from a different device. Active cell size: 0.8 cm².

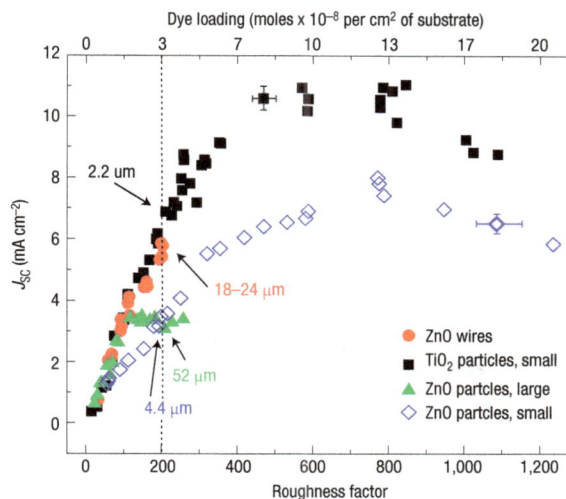

Figure 3 **Comparative performance of nanowire and nanoparticle cells.** Short-circuit current density versus roughness factor for cells based on ZnO wires, small TiO₂ particles, and large and small ZnO particles. The TiO₂ films show a higher maximum current than either of the ZnO films and a larger initial slope than the small ZnO particles, consistent with better transport through TiO₂ particle networks. The large ZnO particle cells attain a smaller maximum current than the small particles because the film thickness (and therefore the electron escape length) becomes larger than the electron diffusion length at a much lower roughness factor. The wire data fall on the TiO₂ line and significantly exceed the current output from both types of ZnO particle cells above a roughness factor of ~100. A slight sag of the wire data off the TiO₂ line at high roughness factor may be a sign of excessive scattering within the opaque wire films. Cell thickness is directly proportional to roughness factor and is labelled for each cell type at a roughness factor of 200. Error bars are provided on only two points to maximize figure clarity, and they are an estimate of the maximum range of the values. The error bars for cells with roughness factors below 250 are smaller than the size of the data points. Data points were made by measuring the roughness factor of dye-sensitized films through ultraviolet–visible spectroscopy of the desorbed dye in basic H₂O and then re-sensitizing the films for fabrication into cells. See also Fig. S6. Fig. S7 shows comparative trends in η, V_{oc} and fill factor. Cell size: 0.8 cm².

electrolyte (Fig. S3). Thus, facile transport through the nanowire array is retained in device-like environments and should result in faster carrier extraction in the nanowire cell.

Solar cells were constructed with wire arrays of various surface areas and tested in simulated sunlight. At a full Sun intensity of 100 ± 3 mW cm⁻², our highest-surface-area devices are characterized by $J_{sc} = 5.3–5.85$ mA cm⁻², $V_{oc} = 0.61–0.71$ V, FF = 0.36–0.38 and efficiency $\eta = 1.2–1.5\%$. (Fig. 2a) The external quantum efficiency of these cells peaks at 40–43% near the absorption maximum of the dye and is limited chiefly by the relatively low dye loadings of the nanowire films.

Figure 2b and c shows the effect of light intensity on the performance characteristics of the wire cells. The open-circuit voltage and short-circuit current depend logarithmically and linearly on light flux, respectively. The fill factors are low compared with nanoparticle cells, do not vary with cell size (see Fig. S4), and fall off with increasing light intensity owing to the development of a large photo-shunt of unknown origin. These poor fill factors halve the potential efficiency of our best nanowire cells, and they are robust with respect to changes in the nanowire electrical properties (Fig. S5), electrolyte concentration and choice of substrate (FTO or

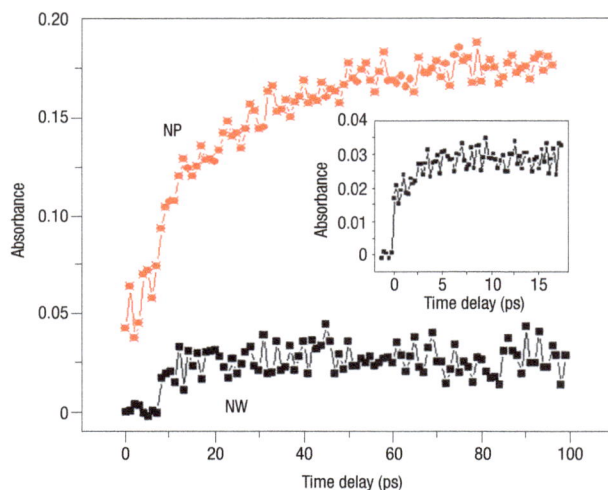

Figure 4 Transient mid-infrared absorption traces of dye-sensitized ZnO nanowire (NW) and ZnO nanoparticle (NP) films pumped at 400 nm. The large difference in injection amplitudes is due to the larger surface area of the particle film. Injection in wires is complete after ~5 ps but continues for ~100 ps in the particle case. A high-resolution trace (inset) shows the ultrafast step (<250 fs) and ~3 ps rise time for a nanowire sample. The slower time constant showed a weak dependence on pump wavelength (see Figs S8 and S9). Particles were synthesized[6] and films were prepared[25] (using dye N719) as described elsewhere. Films were deposited on Al_2O_3 substrates. Spectra are offset by ~0.05 absorbance units for clarity.

indium tin oxide). The efficiency of our devices is fairly flat above a power density of ~5 mW cm^{-2}.

To assess the relative efficiency with which carriers are extracted from the nanowire devices, we compare in Fig. 3 the short-circuit current densities of the wire cells to those of TiO_2 and ZnO nanoparticle cells as a function of the internal surface area (roughness factor). A hypothetical photoanode that maintained a near-unity carrier collection efficiency independent of roughness factor would trace out a line in this plot that gradually tapered off at high surface areas to a large J_{sc} value (>25 mA cm^{-2}). In contrast, the rapid saturation and subsequent decline of the current from cells built with 12-nm TiO_2 particles, 30-nm ZnO particles or 200-nm ZnO particles confirms that the transport efficiency of particle films falls off above a certain film thickness, as we argued above. Crucially, the nanowire films show a nearly linear increase in J_{sc} that maps almost directly onto the TiO_2 data. Because transport in the thin TiO_2 particle films is very efficient (with J_{sc} = 7.8–8.7 mA cm^{-2} at a roughness factor of 250), this is strong evidence of an equally high collection efficiency for nanowire films as thick as ~25 μm. In addition, the nanowire cells generate considerably higher currents than either of the ZnO particle cells over the accessible range of roughness factors (55–75% higher at a roughness of 200). This is direct confirmation of the superiority of the nanowire photoanode as a charge collector.

Better electron transport within the nanowire photoanode is a product of both its higher crystallinity and an internal electric field that can assist carrier collection by separating injected electrons from the surrounding electrolyte and sweeping them towards the collecting electrode. The Debye–Hückel screening length of ZnO (roughly one-third of the thickness of the space-charge layer in the semiconductor at the semiconductor–electrolyte junction) is about 4 nm for a carrier concentration of 10^{18} cm^{-3}, making our nanowires

thick enough to support the sort of radial electric field (depletion layer) that is impossible in smaller TiO_2 or ZnO nanoparticles with fewer carriers. This upward band bending at the nanowire surface should suppress recombination by corralling injected electrons within the wire cores. At the same time, an axial field along each nanowire encourages carrier motion towards the external circuit. These macroscopic fields should act synergistically to increase electron transport relative to nanoparticle cells, which lack such fields. Ambipolar diffusion is consequently a less dominant mechanism in the nanowire devices.

A switch from particles to wires also affects the kinetics of charge transfer at the dye–semiconductor interface, as particle and wire films have dissimilar surfaces onto which the sensitizing dye adsorbs. Whereas ZnO particles present an ensemble of surfaces having various bonding interactions with the dye, our wire arrays are dominated by a single crystal plane (the {100}) that accounts for over 95% of their total area. We used femtosecond transient absorption spectroscopy to measure the rate of electron injection from photoexcited ruthenium dyes into nanowire and nanoparticle films. Dye-sensitized samples were excited with 400-nm, 510-nm or 570-nm pulses and the free carrier concentration of the oxide was monitored with a mid-infrared probe (see Figs S8 and S9). The transient responses for wires and particles (Fig. 4) were considerably different. Injection in wires was characterized by bi-exponential kinetics with time constants of less than 250 fs and around 3 ps, whereas the particle response was tri-exponential and significantly slower (time constants: <250 fs, 20 ps, 200 ps). Our data on particle injection are in excellent agreement with published results[25], validating our evidence for faster electron injection in nanowires.

The nanowire dye-sensitized solar cell is an exciting variant of the most successful of the excitonic photovoltaic devices. As an ordered topology that increases the rate of electron transport, a nanowire electrode may provide a means to improve the quantum efficiency of DSCs in the red region of the spectrum, where their performance is currently limited. Important differences in transport, internal electric field distribution and light scattering should make comparative studies of wire and particle devices fruitful. Raising the efficiency of the nanowire cell to a competitive level depends on achieving higher dye loadings through an increase in surface area. We are now extending our synthetic strategy to design nanowire electrodes with much larger areas available for dye adsorption. The advantages of the nanowire geometry are even more compelling for other types of excitonic photocells, such as inorganic–polymer hybrid devices[26], in which an oriented, continuous and crystalline inorganic phase of the proper dimensions could greatly improve the collection of both electrons and holes.

METHODS

SYNTHESIS OF NANOWIRE ARRAYS

Arrays of ZnO nanowires were synthesized on FTO substrates (TEC-7, 7 Ω per square), Hartford Glass Co.) that were first cleaned thoroughly by acetone/ethanol sonication and then coated with a thin film of ZnO quantum dots, 3–4 nm in diameter, by dip-coating in a concentrated ethanol solution. Nanowires were grown by immersing seeded substrates in aqueous solutions containing 25 mM zinc nitrate hydrate, 25 mM hexamethylenetetramine and 5–7 mM polyethylenimine (branched, low molecular weight, Aldrich) at 92 °C for 2.5 hours. Because nanowire growth slowed after this period, substrates were repeatedly introduced into fresh solution baths in order to obtain long wire arrays (total reaction times of up to 50 hours). The arrays were then rinsed with deionized water and baked in air at 400 °C for 30 minutes to remove any residual organics and to optimize cell performance.

SOLAR CELL FABRICATION AND CHARACTERIZATION

Nanowire arrays were first sensitized in a solution (0.5 mmol l^{-1}) of (Bu$_4$N)$_2$Ru(dcbpyH)$_2$(NCS)$_2$ (N719 dye) in dry ethanol for one hour and then sandwiched together and bonded with thermally platinized FTO counter electrodes separated by 40-μm-thick hot-melt spacers (Bynel, Dupont). The internal space of the cell was filled with a liquid electrolyte (0.5 M LiI, 50 mM I$_2$, 0.5 M 4-tertbutylpyridine in 3-methoxypropionitrile (Fluka)) by capillary action. Cells were immediately tested under AM 1.5G simulated sunlight (300 W Model 91160, Oriel). Intensity measurements were made with a set of neutral density filters. External quantum efficiency (EQE) values (uncorrected for transmission and reflection losses) were obtained with a 150-W xenon lamp coupled to a monochromator, and calibrated with a silicon photodiode.

Identical procedures were used to build and test DSCs based on TiO$_2$ and ZnO particle films prepared by spin-coating or spreading pastes with a thin glass rod (doctor-blading). Films of TiO$_2$ made from a commercial paste of 10–15-nm anatase crystals (Ti-Nanoxide T, Solaronix) were sintered at 450 °C for 30 minutes, treated with a 0.2 M aqueous TiCl$_4$ solution for 12 hours as described previously[2], sintered a second time at 450 °C for 30 minutes and sensitized with dye for 24 hours. Pastes of small, spherical ZnO particles (30 ± 14 nm) and large, irregular ZnO particles (200 ± 75 nm) were formulated as described previously[27], and sintered and sensitized similarly to the nanowire cells. Film thickness was varied by using different spacers for doctor-blading and/or by diluting the pastes with water. All films were free of cracks. The small ZnO particles were synthesized by heating 0.8 g zinc acetate dihydrate and 50 ml ethanol in an autoclave at 125 °C for 2 hours. The large particles were obtained as a commercial powder (200 mesh, 99.999%, Cerac).

ELECTRICAL MEASUREMENTS

For the single wire studies, nanowires 8–10 μm long were dispersed from ethanol solution on oxidized silicon substrates (300 nm SiO$_2$) and fired in air at 400 °C for 30 minutes. Electron-beam lithography was used to pattern and deposit contacts (100 nm Ti) linking the wires to prefabricated electrode sets. Most devices showed ohmic I–V plots without annealing treatments. Measurements were made with a global back gate using a semiconductor parameter analyser (4145B, Hewlett-Packard). Samples for array transport studies were made by encapsulating fired arrays (grown on FTO) in a matrix of spin-cast poly(methylmethacrylate) (PMMA), exposing the wire tips by ultraviolet development and dissolution of the top portion of the PMMA film, and then depositing metal contacts by thermal evaporation. The insulating PMMA matrix prevented potential short circuits due to pinholes in the nanowire array and provided mechanical stability for the measurement.

MID-INFRARED TRANSIENT ABSORPTION MEASUREMENTS

Transient absorption measurements were made with a home-built Ti:sapphire oscillator (30 fs, 88 MHz) and commercial regenerative amplifier (Spitfire, Spectra-Physics) that operates at 810 nm and 1 kHz repetition rate. About 800 μJ of the beam was used to pump an optical parametric amplifier (TOPAS, Quantronix), while 80 J was retained and frequency-doubled in β-barium borate (BBO) for use as the 405-nm pump beam. This beam was delayed by a motorized stage and directed to the sample. The signal and idler beams from the optical parametric amplifier were combined in a AgGaS$_2$ crystal to create tuneable mid-infrared pulses (1,000–3,500 cm^{-1}). The residual 810-nm beam and the residual signal and idler beams were re-combined in a BBO crystal to create sum-frequency generation at 510 nm and 575 nm. The 510-nm beam was directed to a separate delay stage and then to the sample. The pump beams were focused to a spot size of roughly 200–300 μm, with typical pump energies of 0.5–2 μJ. The pump beams were mechanically chopped at 500 Hz (synchronous with the laser), and separate boxcar integrators were triggered by the rejected and passed beams, allowing for independent detection channels of probe with pump ('sample') and without pump ('reference'). The sample signal was subtracted from the reference signal, and the result was divided by the reference to give the differential transmittance, which was converted to effective absorbance. The probe beam, which was typically centred at 2,150 cm^{-1} with a bandwidth of 250 cm^{-1}, was focused with a CaF$_2$ lens to a size of roughly 100–200 μm. The probe beam was collected after transmission through the sample and directed through bandpass filters before being focused onto a single-element HgCdTe detector (IR Associates). An instrument response of 250–300 fs was determined by measuring the rise of free-electron absorption (less than 50 fs) in a thin silicon wafer after blue or green pump.

Each transient plot is an average of points taken on both forward and reverse scans, checked for reproducibility. Each point consists of about 500 averaged laser shots. Samples were translated after each scan to minimize probing dye photoproducts. They were not moved during the scan because small inhomogeneities caused changes in the amplitude of the transient signal, obscuring the true kinetics.

Received 4 March 2005; accepted 31 March 2005; published 15 May 2005.

References

1. Gregg, B. A. Excitonic solar cells. *J. Phys. Chem. B* **107**, 4688–4698 (2003).
2. Nazeeruddin, M. K. *et al.* Engineering of efficient panchromatic sensitizers for nanocrystalline TiO$_2$-based solar cells. *J. Am. Chem. Soc.* **123**, 1613–1624 (2001).
3. Wang, P. *et al.* A stable quasi-solid-state dye-sensitized solar cell with an amphiphilic ruthenium sensitzer and polymer gel electrolyte. *Nature Mater.* **2**, 402–407 (2003).
4. Rensmo, H. *et al.* High light-to-energy conversion efficiencies for solar cells based on nanostructured ZnO electrodes. *J. Phys. Chem. B* **101**, 2598–2601 (1997).
5. Tennakone, K., Kumara, G. R. R. A., Kottegoda, I. R. M. & Perera, V. P. S. An efficient dye-sensitized photoelectrochemical solar cell made from oxides of tin and zinc. *Chem. Commun.* 15–16 (1999).
6. Keis, K., Magnusson, E., Lindström, H., Lindquist, S.-E. & Hagfeldt, A. A 5% efficient photoelectrochemical solar cell based on nanostructured ZnO electrodes. *Sol. Energy Mater. Sol. Cells* **73**, 51–58 (2002).
7. Krüger, J., Plass, R., Grätzel, M., Cameron, P. J. & Peter, L. M. Charge transport and back reaction in solid-state dye-sensitized solar cells: a study using intensity-modulated photovoltage and photocurrent spectroscopy. *J. Phys. Chem. B* **107**, 7536–7539 (2003).
8. O'Regan, B. & Grätzel, M. A low-cost, high-efficiency solar cell based on dye-sensitized colloidal TiO$_2$ films. *Nature* **353**, 737–740 (1991).
9. Fisher, A. C., Peter, L. M., Ponomarev, E. A., Walker, A. B. & Wijayantha, K. G. U. Intensity dependence of the back reaction and transport of electrons in dye-sensitized nanocrystalline TiO$_2$ solar cells. *J. Phys. Chem. B* **104**, 949–958 (2000).
10. Oekermann, T., Zhang, D., Yoshida, T. & Minoura, H. Electron transport and back reaction in nanocrystalline TiO$_2$ films prepared by hydrothermal crystallization. *J. Phys. Chem. B* **108**, 2227–2235 (2004).
11. Nelson, J. Continuous-time random-walk model of electron transport in nanocrystalline TiO$_2$ electrodes. *Phys. Rev. B* **59**, 15374–15380 (1999).
12. van de Lagemaat, J. & Frank, A. J. Nonthermalized electron transport in dye-sensitized nanocrystalline TiO$_2$ films: transient photocurrent and random-walk modeling studies. *J. Phys. Chem. B* **105**, 11194–11205 (2001).
13. Kopidakis, N., Schiff, E. A., Park, N.-G., van de Lagemaat, J. & Frank, A. J. Ambipolar diffusion of photocarriers in electrolyte-filled, nanoporous TiO$_2$. *J. Phys. Chem. B* **104**, 3930–3936 (2000).
14. Benkstein, K. D., Kopidakis, N., van de Lagemaat, J. & Frank, A. J. Influence of the percolation network geometry on electron transport in dye-sensitized titanium dioxide solar cells. *J. Phys. Chem. B*. **107**, 7759–7767 (2003).
15. Kopidakis, N., Benkstein, K. D., van de Lagemaat, J. & Frank, A. J. Transport-limited recombination of photocarriers in dye-sensitized nanocrystalline TiO$_2$ solar cells. *J. Phys. Chem. B* **107**, 11307–11315 (2003).
16. Kavan, L., Grätzel, M., Gilbert, S. E., Klemenz, C. & Schell, H. J. Electrochemical and photoelectrochemical investigation of single-crystal anatase. *J. Am. Chem. Soc.* **118**, 6716–6723 (1996).
17. Wagner, P. & Helbig, R. Hall effect and anisotropy of the mobility of the electrons in zinc oxide. *J. Phys. Chem. Sol.* **35**, 327–335 (1974).
18. Nakade, S. *et al.* Dependence of TiO$_2$ nanoparticle preparation methods and annealing temperatures on the efficiency of dye-sensitized solar cells. *J. Phys. Chem. B* **106**, 10004–10010 (2002).
19. Frank, A. J., Kopidakis, N. & van de Lagemaat, J. Electrons in nanostructured TiO$_2$ solar cells: transport, recombination and photovoltaic properties. *Coord. Chem. Rev.* **248**, 1165–1179 (2004).
20. Renouard, T. *et al.* Novel ruthenium sensitizers containing functionalized hybrid tetradentate ligands: synthesis, characterization, and INDO/S analysis. *Inorg. Chem.* **41**, 367–378 (2002).
21. Hara, K. *et al.* Design of new coumarin dyes having thiophene moieties for highly efficient organic-dye-sensitized solar cells. *New J. Chem.* **27**, 783–785 (2003).
22. Kron, G., Egerter, T., Werner, J. H. & Rau, U. Electronic transport in dye-sensitized nanoporous TiO$_2$ solar cells—comparison of electrolyte and solid-state devices. *J. Phys. Chem. B* **107**, 3556–3564 (2003).
23. Greene, L. *et al.* Low-temperature wafer scale production of ZnO nanowire arrays. *Angew. Chem. Int. Edn Engl.* **42**, 3031–3034 (2003).
24. Noack, V., Weller, H. & Eychmüller, A. Electron transport in particulate ZnO electrodes: a simple approach. *J. Phys. Chem. B*. **106**, 8514–8523 (2002).
25. Anderson, N. A., Ai, X. & Lian, T. Electron injection dynamics from Ru polypyridyl complexes to ZnO nanocrystalline thin films. *J. Phys. Chem. B* **107**, 14414–14421 (2003).
26. Huynh, W. U., Dittmer, J. J. & Alivisatos, A. P. Hybrid nanorod–polymer solar cells. *Science* **295**, 2425–2427 (2002).
27. Park, N.-G. *et al.* Morphological and photoelectrochemical characterization of core–shell nanoparticle films for dye-sensitized solar cells: Zn-O type shell on SnO$_2$ and TiO$_2$ cores. *Langmuir* **20**, 4246–4253 (2004).

Acknowledgements

We thank M. Graetzel, A. P. Alivisatos, J. Frechet, B. O'Regan, E. Kadnikova, U. Bach, D. Milliron and I. Gur for discussions, T. Lavarone and S. Hamzehpour for technical assistance and A. P. Alivisatos for use of the solar simulator. This work was supported by the US Department of Energy, Office of Basic Sciences.

Correspondence and requests for materials should be addressed to P.Y.

Supplementary Information accompanies the paper on www.nature.com/naturematerials.

Competing financial interests

The authors declare that they have no competing financial interests.

High-efficiency solution processable polymer photovoltaic cells by self-organization of polymer blends

GANG LI[1], VISHAL SHROTRIYA[1], JINSONG HUANG[1], YAN YAO[1], TOM MORIARTY[2], KEITH EMERY[2] AND YANG YANG[1]*

[1]Department of Materials Science and Engineering, University of California Los Angeles, Los Angeles, California 90095, USA
[2]National Renewable Energy Laboratory, Golden, Colorado 80401, USA
*e-mail: yangy@ucla.edu

Published online: 9 October 2005; doi:10.1038/nmat1500

Converting solar energy into electricity provides a much-needed solution to the energy crisis the world is facing today. Polymer solar cells have shown potential to harness solar energy in a cost-effective way. Significant efforts are underway to improve their efficiency to the level of practical applications. Here, we report highly efficient polymer solar cells based on a bulk heterojunction of polymer poly(3-hexylthiophene) and methanofullerene. Controlling the active layer growth rate results in an increased hole mobility and balanced charge transport. Together with increased absorption in the active layer, this results in much-improved device performance, particularly in external quantum efficiency. The power-conversion efficiency of 4.4% achieved here is the highest published so far for polymer-based solar cells. The solution process involved ensures that the fabrication cost remains low and the processing is simple. The high efficiency achieved in this work brings these devices one step closer to commercialization.

Polymer solar cells have evolved as a promising cost-effective alternative to silicon-based solar cells[1–3]. Some of the important advantages of these so-called 'plastic' solar cells include low cost of fabrication, ease of processing, mechanical flexibility and versatility of chemical structure from advances in organic chemistry. However, low efficiency[4–6] of these plastic solar cells limits their feasibility for commercial use. The efficiencies of polymer photovoltaic (PV) cells got a major boost with the introduction of the bulk heterojunction (BHJ) concept[7,8] consisting of an interpenetrating network of electron donor and acceptor materials. This concept has also been demonstrated in small-molecular organic PVs[9]. It is argued that owing to the space–charge effects inherent in the BHJ structure, the fill factor (FF) is usually low and the disordered structure will be ultimately limited by high series resistance[10]. To achieve a highly efficient PV device, solar radiation needs to be efficiently absorbed, for which the device thickness needs to be increased. However, this will further increase the series resistance of the device. Here we demonstrate that the series resistance of the polymer BHJ PV cells can be significantly reduced by polymer self-organization (to values as low as 1.56 Ω cm^2) by controlling the growth rate of the active polymer layer from solution to solid state. The FF also increased to a value of more than 67%, which is among the highest values reported for polymer solar cells. As a result, we achieved device power conversion efficiency (PCE) of 4.4% under the standard AM1.5G 1 Sun test condition. The PV cells fabricated in this study are made by a solution-based process and large device area can be achieved by this process at relatively low cost.

The polymer PV device in this study consisted of an active layer of poly(3-hexylthiophene): [6,6]-phenyl-C$_{61}$-butyric acid methyl ester (P3HT/PCBM) sandwiched between metallic electrodes. The thickness of the active layer was $\sim$210−230 nm and the active device area was roughly 0.11 cm^2. The growth rate of the polymer layer was controlled by varying the film solidification time. The details on device fabrication process and characterization

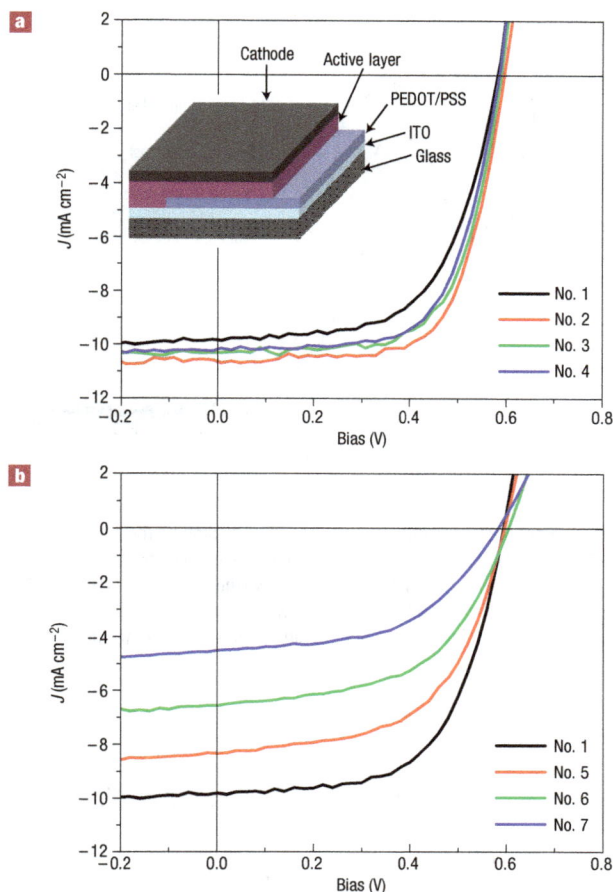

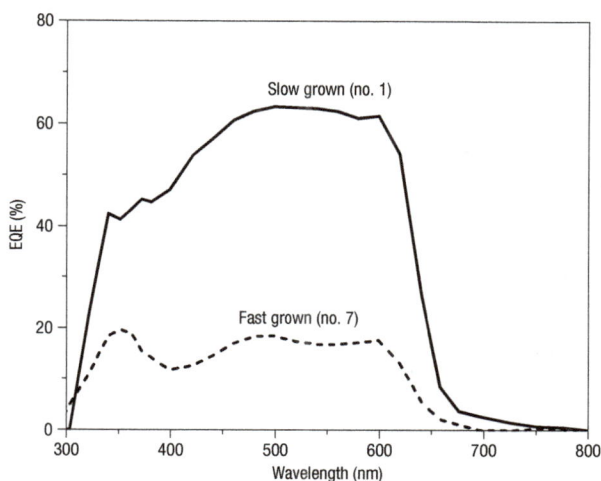

Figure 1 Effect of thermal annealing and film growth rate on the performance of the plastic solar cells. a, The different $J–V$ curves correspond to the devices with active layer before (no. 1) and after thermal annealing at 110 °C for 10 min (no. 2), 20 min (no. 3) and 30 min (no. 4). The active layer thickness was ∼210 nm and the film growth time was ∼20 min. **b**, $J–V$ characteristics under illumination for devices with different film growth rates by varying the solvent evaporation time, t_{evp}. The t_{evp} for different films were 20 min (no. 1), 3 min (no. 5), 40 s (no. 6) and 20 s (no. 7).

Figure 2 Effect of film growth rate on EQE of polymer solar cells. EQE measurements are plotted for P3HT/PCBM photovoltaic cells for two types of active layer: slow grown (no. 1) and fast grown (no. 7). The efficiency maximum for slow-grown film is ∼63% which is more than three times that of fast-grown film (∼19%).

techniques are given in the Methods section. Encapsulated devices were brought to the National Renewable Energy Laboratory (NREL), Colorado, for testing under the Standard Test Condition (STC). On the basis of the mismatch between the efficiency measurements at University of California Los Angeles (UCLA) and NREL, a correction factor was created by the UCLA team and applied to similar measurements done at UCLA (see Methods for details). The current–voltage ($J–V$) curves under illumination for four devices with annealing times (t_A) of 0 (device no. 1), 10 (no. 2), 20 (no. 3) and 30 min (no. 4) are shown in Fig. 1a. The annealing was performed at 110 °C. After annealing for 10 minutes (device no. 2), the short-circuit current (J_{SC}) increases slightly from 9.9 to 10.6 mA cm^{-2} and the FF increases from 60.3 to 67.4%, which is among the highest FFs in organic solar cells. The open-circuit voltage (V_{OC}) remains unchanged after annealing. As a result, the PCE improves from 3.5% (no. 1) to 4.4% (no. 2). Under the dark conditions, the rectification ratios are close to 10^7 at a bias of 2 V for all four devices. The reason behind the high FFs of our devices is believed to be the significantly large thickness of

the active layer. The thickness of the active layer makes it free of pinholes and microcracks and all devices show very high shunt resistance of 180–640 MΩ as derived from the $J–V$ characteristics measured under dark conditions. To our knowledge, these values are the highest reported for organic solar cells. The ultrahigh shunt resistance reduces the noise equivalent power (NEP) of the device and makes it ideal for photodetector applications. Moreover, the surface of the active layer becomes smoother on annealing, which enables a very good, defect-free contact with the metal cathode, thereby increasing the FF values. Based on 16 devices of the same kind (no. 2, t_A = 10 min), the efficiency variation is 4.2 ± 0.2%. Although the formerly highest reported PV cell efficiency is for a P3HT/PCBM system in the 1:2 wt/wt ratio[5], various independent studies have shown that the 1:1 wt/wt ratio should be superior[11,12]. It has also been reported that to absorb greater than 95% of the incident light over the wavelength range of 450–600 nm, a P3HT film of 240 nm thickness is needed[2]. With P3HT/PCBM of 1:1 wt/wt ratio and the reflective cathode, 210–230-nm-thick P3HT/PCBM film efficiently absorbs incident light. It is widely believed that the fundamental limitation of the photocurrent of polymer solar cells is because of the low mobility of the holes in donor polymer[3]. Compared with the devices with P3HT/PCBM of 1:2 wt/wt ratio, the donor/acceptor network transports holes more efficiently in 1:1 wt/wt ratio, resulting in a more balanced electron and hole transport. Time-of-flight (TOF) measurements on P3HT/PCBM blend films with different wt/wt ratios verified that only 1:1 wt/wt ratio film gives balanced, non-dispersive electron and hole transport. The much improved FF of 67.4% for our devices supports this argument.

The highly regular chain structure of poly(3-alkylthiophene)s (P3ATs) facilitates their self-organization into two-dimensional sheets by means of inter-chain stacking[13]. Self-organization has been shown to improve field-effect carrier mobility in RR-P3HT by more than a factor of 100 to 0.1 cm^2 V^{-1} s^{-1} (refs 14,15). The slow growth will assist the formation of self-organized ordered structure in the P3HT/PCBM blend system. However, the ordering in this blend system might be different from the kind of ordering

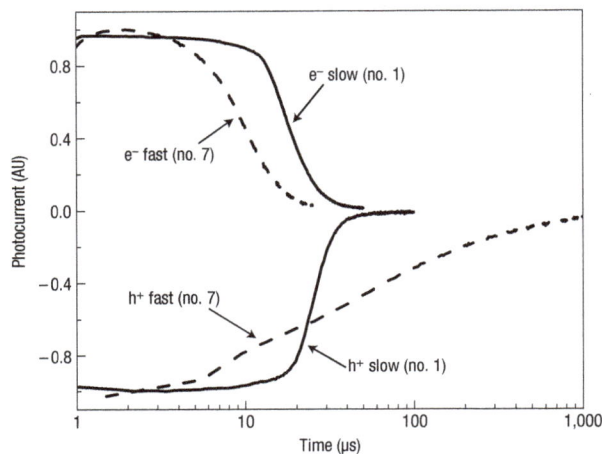

Figure 3 **Effect of film growth rate on the mobility of charge carriers in the active layer.** TOF signals of slow-grown (no. 1) and fast-grown (no. 7) films plotted in a semi-log scale. Note that μ_h in fast-grown film (no. 7) is significantly lower compared with μ_e in the film, as well as compared with μ_h and μ_e in the slow-grown film (no. 1). Also, the hole transport is dispersive in film no. 7 and the carrier transport is unbalanced. Films in TOF devices were prepared the same way as cells in this study and their film thickness is ~1 μm.

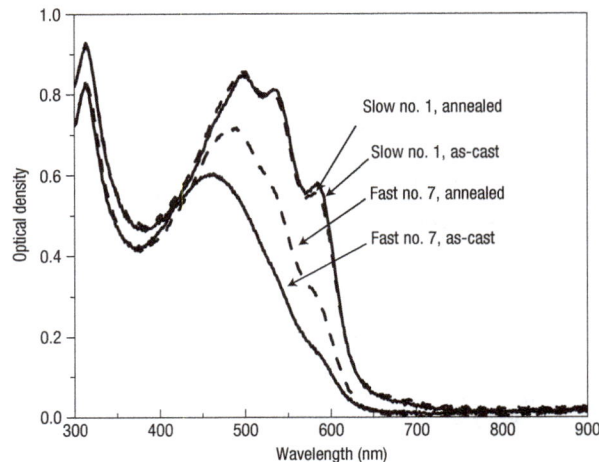

Figure 4 **Effect of film growth rate and thermal annealing on the absorbance of the P3HT/PCBM films.** Ultraviolet–visible absorption spectra for films of P3HT/PCBM (in 1:1 wt/wt ratio), for both slow-grown (no. 1) and fast-grown (no. 7) films, before (solid line) and after (dashed line) annealing. The films were spun cast at 600 r.p.m. for 60 s (film thickness ~ 210 nm) and the annealing was done at 110 °C for 20 min.

observed for pristine P3HT films, in the sense that the presence of PCBM molecules will have an effect on the stacking of the polymer chains[16]. The degree of self-organization can be varied by controlling the film growth rate or, in other words, by controlling the time it takes for the wet films to solidify. In Fig. 1b we compare the J–V characteristics of four devices with different solvent evaporation times (t_{evp}) after spin-coating, judging by visual inspection of the change in film colour when it solidifies from the liquid phase. Device no. 1 was covered in a glass petri dish while drying and had t_{evp} ~ 20 min, no. 5 was left open in N_2 ambient and had t_{evp} ~ 3 min, no. 6 and no. 7 were dried by putting on hot plate at 50 and 70 °C, respectively, and had t_{evp} ~ 40 and 20 s. The J_{SC} reduces from 9.9 to 8.3, 6.6 and 4.5 mA cm^{-2}, and the device series resistance, R_{SA}, increases from 2.4 to 4.5, 12.5 and 19.8 Ω cm^2 with reducing t_{evp}. Series and shunt resistances were derived from the slope of the J–V characteristic curve under dark conditions close to 2 and 0 V, respectively[17]. The FF also consistently decreases from 60.3 to 52.0%. The low R_{SA} of 2.4 Ω cm^2 (for device no. 1) achieved is comparable to that of much thinner devices[18] (~48 nm), underlining the effect of self-organization. The summary of device performance for various PV devices fabricated in this work (device nos 1–7) is provided in Table 1. Figure 2 shows the results of external quantum efficiency (EQE) measurements for two types of device, slow grown (no. 1) and fast grown (no. 7). The parameters for the highest efficiency device (no. 2) are shown in bold. The EQE for the device with fast-grown film shows a maximum of ~19% at a wavelength of 350 nm. On the other hand, for the device with slow-grown film, the EQE maximum increases by more than a factor of three to ~63% at 500 nm. The integral of the product of this absolute EQE and the global reference spectrum yields a J_{SC} of 9.47 mA cm^{-2}, which matches closely to the J_{SC} that we measured for this particular device. This increase in EQE over the wavelength range of 350–650 nm contributes to the increase in the PCE of our devices. We believe that this enhancement in EQE originates from two important contributions, an increase in the charge-carrier mobility and increased absorption in the active

Table 1 Summary of device performance for various PV devices in the work.

Device no.	J_{SC} (mA cm^{-2})	V_{OC} (V)	PCE (%)	FF (%)	R_{SA} (Ω cm^2)
1	9.86	0.59	3.52	60.3	2.4
2	**10.6**	**0.61**	**4.37**	**67.4**	**1.7**
3	10.3	0.60	4.05	65.5	1.6
4	10.3	0.60	3.98	64.7	1.6
5	8.33	0.60	2.80	56.5	4.9
6	6.56	0.60	2.10	53.2	12.5
7	4.50	0.58	1.36	52.0	19.8

layer. The discussion on the effect of slow growth rate on the charge-carrier mobility and the absorption spectra of the films follows.

A TOF study was conducted on slow-grown (no. 1) and fast-grown (no. 7) films at an electric field strength of E ~ 2 × 10^5 V cm^{-1}. The film-preparation method for TOF measurements was the same as that for device fabrication to maintain maximum similarity for reliable comparison. As clearly seen in Fig. 3, in the film prepared with conditions similar to device no. 1 (or the slow-grown film), both the electrons and holes transport non-dispersively with electron and hole mobilities of $\mu_e = 7.7 \times 10^{-5}$ and $\mu_h = 5.1 \times 10^{-5}$ cm^2 V^{-1} s^{-1}, whereas for film no. 7, fast growth leads to dispersive hole transport and a significant reduction in μ_h to 5.1×10^{-6} cm^2 V^{-1} s^{-1}. The electron mobility increases slightly to 1.1×10^{-4} cm^2 V^{-1} s^{-1}. The destruction of ordered structure during fast growth is believed to be the reason for this. For the slow-grown film, μ_h is similar to (or slightly higher than) the values reported earlier for poly(p-phenylenevinylene)/PCBM blends[19,20] and the ratio between electron and hole mobilities is close to unity (μ_e/μ_h ~ 1.5) resulting in balanced carrier transport in the active layer. When the charge transport in the device is unbalanced (for example, in the pristine P3HT/PCBM system where μ_h is significantly lower than μ_e) hole accumulation occurs in the device and the

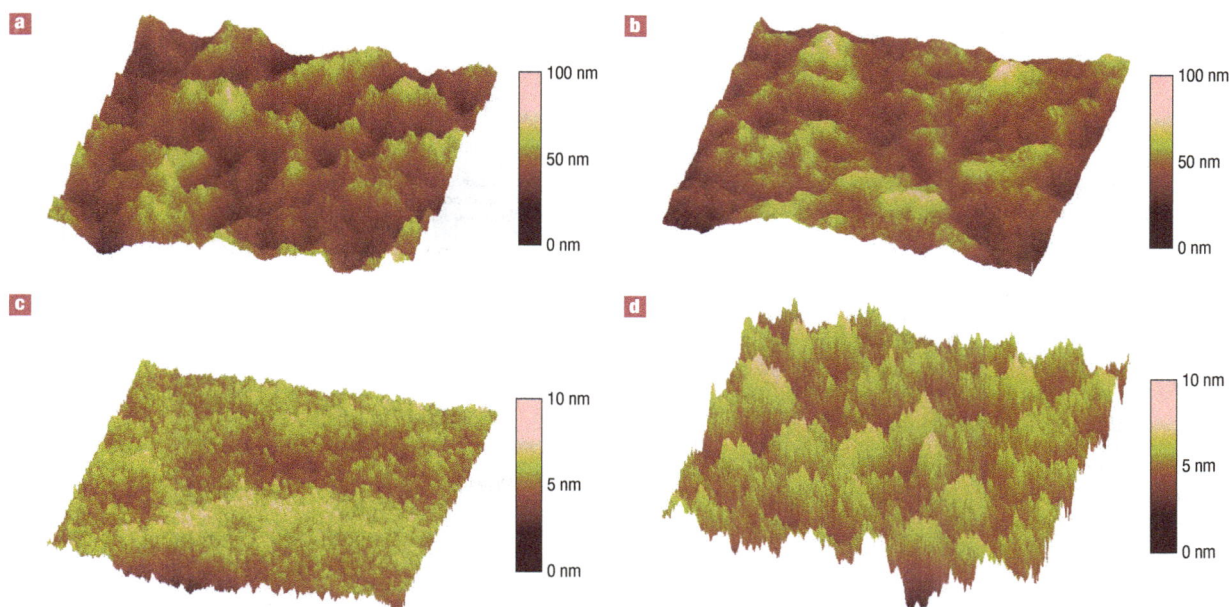

Figure 5 Effect of growth rate and thermal annealing on the morphology of the active layer. AFM height images of the P3HT/PCBM composite films (PCBM concentration = 50 wt%) showing a 5 μm × 5 μm surface area. **a**, Slow-grown film (no. 1) before thermal annealing. **b**, Slow-grown film (no. 1) after thermal annealing at 110 °C for 10 min. **c**, Fast-grown film (no. 7) before thermal annealing. **d**, Fast-grown film (no. 7) after thermal annealing at 110 °C for 20 min. Note that the colour scale for the films **a** and **b** is 0–100 nm, whereas for films **c** and **d** it is 0–10 nm.

photocurrent is space–charge limited[16]. The photocurrent is then governed by a square-root dependence on bias and an FF above 40% is unachievable[21]. However, when the carrier transport is more balanced, the current is not limited by space–charge effects and high FFs are possible. For example, blending PCBM into a poly(phenylene vinylene) derivative has shown to increase the hole mobility by a factor of 200 compared with the pristine polymer films, resulting in μ_h being lower than μ_e by only an order of magnitude[16]. As a result, FF values of up to 60% can be achieved for PV devices based on these polymer/PCBM blends[4]. For our devices μ_e/μ_h of only 1.5 results in FF values of more than 67%. For devices with active layer thickness larger than the mean carrier drift length of the charge carrier, the device performance is limited by the carrier with lower mean free path (holes)[22]. To maintain the electrical neutrality in the device, the unbalanced transport will result in a loss in efficiency by increased recombination. The more balanced transport of holes and electrons in slow-grown films will therefore result in better device performance. On the other hand, for the fast-grown film, the unbalanced electron and hole transport and the significantly reduced μ_h results in low photocurrent and poor FF.

The measured optical densities for slow-grown (no. 1) and fast-grown (no. 7) films are shown in Fig. 4, before and after thermal annealing at 110 °C for 20 min. Compared with the film dried at 70 °C (no. 7), the absorption in the red region of the slow grown film (no. 1) is much stronger. The three vibronic absorption shoulders (peaks) are much more pronounced in film no. 1, indicating a higher degree of ordering[23]. After annealing at 110 °C for 20 min, the absorbance of film no. 7 shows significant increase and the vibronic features become clearer, indicating a partial recovery of ordering. During the fast growth of the film, the orientation of P3HT supermolecules is forced by the short timescale and is not thermodynamically stable. On thermal annealing, the chains become mobile and self-organization can occur to form ordering. Significant red-shift appears in the more ordered films as high crystalline order involves an enhanced conjugation length and hence a shift of the absorption spectrum to lower energies[23]. The observed enhancement of conjugation length is consistent with our TOF results. As the film thickness is same for all four films, the change in absorption spectra is believed to be a result of increased inter-chain interaction and not because of an increase in film thickness. For the slow-grown film (no. 1), the absorption spectrum shows no significant difference before and after thermal annealing, further strengthening our argument that the slow-grown film already has a high degree of ordering. X-ray diffraction (XRD; Bede Microsource D1 Diffractometer) is a standard method for investigating crystalline structures. XRD studies on pure poly(3-alkylthiophene) films have been reported[13,24–26]. From the preliminary XRD measurements on the slow-grown P3HT/PCBM-blend films we could not clearly identify crystalline peaks. Although P3HT is among the polymers with highest degree of crystalline structure, P3HT film is still mostly amorphous. In the blend film, with 1:1 wt/wt ratio, P3HT and PCBM each act as heavily doped impurities of their counterpart and inhibit their formation of crystalline structure. Thus, the ordered structure indicated by absorption spectroscopy is at most nanocrystalline in structure in the film, and the detection is beyond the capability of our XRD system. Only indirect evidence for increased ordering can be seen from the increase in the roughness of the films (discussed in the following), increased inter-band absorption from absorption spectra and increased mobility from TOF measurements.

Figure 5 shows the atomic force microscopy (AFM) images of the slow-grown (no. 1) and fast-grown (no. 7) films, before and after thermal annealing. For film no. 1, the surface is very rough with r.m.s. roughness, σ, of 11.5 nm (Fig. 5a). The film after 10 min of annealing at 110 °C shows similar $\sigma \sim 9.5$ nm (Fig. 5b), but the texture is smoother than that of un-annealed film. For film no. 7, a very smooth surface with $\sigma \sim 0.87$ nm is observed (Fig. 5c). After heating at 110 °C for 20 min, the roughness increases with $\sigma \sim 1.9$ nm (Fig. 5d). By comparing these results with device performance, we first suspected that the rough surface may effectively reduce the charge-transport distance and increase the J_{SC}, at the same time providing nanoscaled texture that further enhances internal light scattering and light absorption. However, using the surface-area calculation function in our AFM program, the surface area in the roughest film was found to be only 0.4% more than that of an absolutely flat film. Thus, these mechanisms can only account for minor efficiency improvement at most. Instead, the rough surface is probably a signature of polymer (blend) self-organization, which in turn enhances ordered structure formation in the thin film. This assumption is strongly supported by (i) the very rough surface of slow-grown thick film, (ii) significant roughness increment after annealing of fast-grown thick film and (iii) roughness increment after annealing of thin films[18], all of which significantly enhance device efficiency. The peak to valley height of the slow-grown film is ~100 nm, corresponding to ~50% of the mean thickness. Thermal annealing of P3AT films has been shown to enhance crystallization and increase the μ_h in PV cells[5]. However, this might not be true for well-ordered thick films such as those in our case, as seen from the absorption spectra and no significant increase in J_{SC} on annealing. We observed that FFs tended to increase when devices were left in the vacuum for long times. Thus, we suspect annealing the slow-grown films might mainly help in removing the residue solvent, reducing the free volume and improving the interface with the electrode, instead of inducing further self-organization in already ordered films. As a result of annealing, the number of trapping sites is reduced for carrier transport and extraction. Another possible mechanism is the improvement of cathode contact from morphology change. After annealing, the surface of slow-grown film becomes smoother and this in turn improves the organic/cathode contact. Annealing at 110 °C reduces R_{SA} from 2.41 to 1.56 Ω cm^2, which is among the lowest values reported for similar device dimensions. Along with a better balanced charge transport this gives a very high FF of 67.4% and the PCE of 4.4%. Similar series resistance values have been reported earlier for devices based on P3HT[11,27], poly(*p*-phenylenevinylene) derivative[28] and copper phthalocyanine[29]; however, in our case the carrier transport is more balanced, as observed from the TOF measurements.

These results will help in better understanding the underlying relationship between the polymer morphology and the device performance and will subsequently help in enhancing the efficiencies of plastic solar cells to a level of practical applications.

METHODS

The polymer PV devices were fabricated by spin-coating a blend of P3HT/PCBM in 1:1 wt/wt ratio, sandwiched between a transparent anode and a cathode. The anode consisted of glass substrates pre-coated with indium tin oxide (ITO), modified by spin-coating polyethylenedioxythiophene/polystyrenesulphonate (PEDOT/PSS) layer (~25 nm), and the cathode consisted of Ca (~25 nm) capped with Al (~80 nm). Before device fabrication, the ITO-coated (~150 nm) glass substrates were cleaned by ultrasonic treatment in detergent, deionized water, acetone and isopropyl alcohol sequentially. A thin layer (~30 nm) of PEDOT/PSS (Baytron P VP A1 4083) was spin-coated to modify the ITO surface. After baking at 120 °C for 1 h, the substrates were transferred to a nitrogen-filled glove box (<0.1 p.p.m. O_2 and H_2O). P3HT (purchased from Rieke Metals, used as received) was first dissolved in 1,2-dichlorobenzene (DCB) to make 17 mg ml^{-1} solution, followed by blending with PCBM (purchased from Nano-C, used as received) in 50 wt%. The blend was stirred for ~14 h at 40 °C in the glove box. The active layer was obtained by spin-coating the blend at 600 r.p.m. for 60 s and the thickness of film was ~210–230 nm, as measured with a Dektek profilometer. The active device area was 0.11 cm^2. Spin-coating at 600 r.p.m. left the films wet, which were then dried in covered glass petri dishes. Before cathode deposition, the films were thermally annealed at 110 °C for various times. Testing was done in N_2 under simulated AM1.5G irradiation (100 mW cm^{-2}) using a xenon-lamp-based solar simulator (Oriel 96000 150W Solar Simulator). The films used in TOF measurements were spun-cast on ITO-coated glass substrates from P3HT/PCBM 1:1 wt/wt ratio solution (100 mg ml^{-1} total, DCB as solvent) and then Al was deposited as electrode. Immediately following the spin-coating, the films were either kept in covered petri dish for slow growth or were baked at 70 °C for ~30 s for fast growth. The thickness of the films for TOF measurements was ~1 μm. The absorption spectra were taken using a Varian Cary 50 Ultraviolet–Visible Spectrophotometer. To mimic the device fabrication conditions, all of the films were spun-cast on PEDOT/PSS-covered silica glass, also used as the absorption baseline. A Digital Instruments Multimode Scanning Probe Microscope was used to obtain the AFM images. The film preparation conditions for the AFM measurements were kept the same as device fabrication for accurate comparison.

Encapsulated devices were brought to NREL, Colorado, for testing of PCE and EQE under STC. NREL measured a set of devices, the best yielding 3.6% efficiency. There was a mismatch between the results for the same devices as measured at UCLA, probably the result of spectral mismatch primarily owing to a poorly matched reference cell. Although the open circuit voltages and FFs were the same, the currents and efficiencies as measured by NREL were only 84% of those measured in our lab. This 0.84 correction factor was applied to the larger set of devices measured at UCLA to obtain correct efficiency values. The efficiency values reported in this work are all after the correction. Although the efficiency values are calibrated, we did not consider three important factors: (i) device degradation during shipping to NREL owing to non-ideal encapsulation; (ii) reduction in effective device area on reaction with epoxy after encapsulation; and (iii) the optical losses owing to reflection by glass substrate and absorption from ITO.

Received 11 March 2005; accepted 19 August 2005; published 9 October 2005.

References

1. Brabec, C. J., Sariciftci, N. S. & Hummelen, J. C. Plastic solar cells. *Adv. Funct. Mater.* **11,** 15–26 (2001).
2. Coakley, K. M. & McGehee, M. D. Conjugated polymer photovoltaic cells. *Chem. Mater.* **16,** 4533–4542 (2004).
3. Brabec, C. J. Organic photovoltaics: technology and market. *Solar Energy Mater. Solar Cells* **83,** 273–292 (2004).
4. Shaheen, S. E. *et al.* 2.5% efficient organic plastic solar cells. *Appl. Phys. Lett.* **78,** 841–843 (2001).
5. Padinger, F., Rittberger, R. S. & Sariciftci, N. S. Effects of post production treatment on plastic solar cells. *Adv. Funct. Mater.* **13,** 85–88 (2003).
6. Walduf, C., Schilinsky, P., Hauch, J. & Brabec, C. J. Material and device concepts for organic photovoltaics: towards competitive efficiencies. *Thin Solid Films* **451–452,** 503–507 (2004).
7. Yu, G., Gao, J., Hummelen, J. C., Wudl, F. & Heeger, A. J. Polymer photovoltaic cells: enhanced efficiencies via a network of internal donor-acceptor heterojunctions. *Science* **270,** 1789–1791 (1995).
8. Saraciftci, N. S., Smilowitz, L., Heeger, A. J. & Wudl, F. Photoinduced electron transfer from a conducting polymer to buckminsterfullerene. *Science* **258,** 1474–1476 (1992).
9. Peumans, P., Uchida, S. & Forrest, S. R. Efficient bulk-heterojunction photovoltaic cells using small-molecular-weight organic thin films. *Nature* **425,** 158–162 (2003).
10. Yang, F., Shtein, M. & Forrest, S. R. Controlled growth of a molecular bulk heterojunction photovoltaic cell. *Nature Mater.* **4,** 37–41 (2005).
11. Chirvase, D., Parisi, J., Hummelen, J. C. & Dyakonov, V. Influence of nanomorphology on the photovoltaic action of polymer-fullerene composites. *Nanotechnology* **15,** 1317–1323 (2004).
12. Shrotriya, V., Ouyang, J., Tseng, R. J., Li, G. & Yang, Y. Absorption spectra modification in poly(3-hexylthiophene):methanofullerene blend thin films. *Chem. Phys. Lett.* **411,** 138–143 (2005).
13. Grevin, B., Rannou, P., Payerne, R., Pron, A. & Travers, J. P. Multi-scale scanning tunneling microscopy imaging of self-organized regioregular poly(3-hexylthiophene) films. *J. Chem. Phys.* **118,** 7097–7102 (2003).
14. Sirringhaus, H. *et al.* Two-dimensional charge transport in self-organized, high-mobility conjugated polymers. *Nature* **401,** 685–688 (1999).
15. Bao, Z., Dodabalapur, A. & Lovinger, A. J. Soluble and processable regioregular poly(3-hexylthiophene) for thin film field-effect transistor applications with high mobility. *Appl. Phys. Lett.* **69,** 4108–4110 (1996).
16. Melzer, C., Koop, E. J., Mihailetchi, V. D. & Blom, P. W. M. Hole transport in poly(phenylene vinylene)/methanofullerene bulk-heterojunction solar cells. *Adv. Funct. Mater.* **14,** 865–870 (2004).
17. Shirland, F. The history, design, fabrication and performance of CdS thin film solar cells. *Adv. Energy Conversion* **6,** 201–222 (1966).
18. Li, G., Shrotriya, V., Yao, Y. & Yang, Y. Investigation of annealing effects and film thickness dependence of polymer solar cells based on poly(3-hexylthiophene). *J. Appl. Phys.* **98,** 043704 (2005).
19. Choulis, S. A. *et al.* Investigation of transport properties in polymer/fullerene blends using time-of-flight photocurrent measurements. *Appl. Phys. Lett.* **83,** 3812–3814 (2003).
20. Mihailetchi, V. D. *et al.* Compositional dependence of the performance of poly(p-phenylenevinylene):methanofullerene bulk-heterojunction solar cells. *Adv. Funct. Mater.* **15,** 795–801 (2005).
21. Goodman, A. M. & Rose, A. Double extraction of uniformly generated electron–hole pairs from insulators with noninjecting contacts. *J. Appl. Phys.* **42,** 2823–2830 (1971).
22. Snaith, H. L., Greenham, N. C. & Friend, R. H. The origin of collected charge and open-circuit voltage in blended polyfluorene photovoltaic devices. *Adv. Mater.* **16,** 1640–1645 (2004).
23. Sunderberg, M., Inganas, O., Stafstrom, S., Gustafsson, G. & Sjogren, B. Optical absorption of poly(3-alkylthiophenes) at low temperatures. *Solid State Commun.* **71,** 435–439 (1989).
24. Prosa, T. J., Moulton, J., Heeger, A. J. & Winokur, M. J. Diffraction line-shape analysis of poly(3-docecylthiophene): A study of layer disorder through the liquid crystalline polmer transition. *Macromolecules* **32,** 4000–4009 (1999).
25. Aasmundtveit, K. E. *et al.* Structural anisotropy of poly(alkylthiophene) films. *Macromolecules* **33,** 3120–3127 (2000).
26. Samuelsen, E. J., Breiby, D. W., Konovalov, O., Struth, B. & Smilgies, D. -M. In situ studies of transition from solution to solid film of poly(octylthiophene). *Synth. Met.* **123,** 165–170 (2001).
27. Schilinsky, P., Waldauf, C., Hauch, J. & Brabec, C. J. Simulation of light intensity dependent current characteristics of polymer solar cells. *J. Appl. Phys.* **95,** 2816–2819 (2004).
28. Mozer, A. J. *et al.* Novel regiospecific MDMO-PPV copolymer with improved charge transport for bulk heterojunction solar cells. *J. Phys. Chem. B* **108,** 5235–5242 (2004).
29. Xue, J., Uchida, S., Rand, B. P. & Forrest, S. R. 4.2% efficient organic photovoltaic cells with low series resistance. *Appl. Phys. Lett.* **84,** 3013–3015 (2004).

Acknowledgements

We thank J. Ouyang for very helpful technical discussions. This research work is supported in part by the Office of Naval Research (grant no. N00014-01-1-0136, program manager P. Armistead), and the Air Force Office of Scientific Research (grant no. F49620-03-1-0101, program manager C. Lee). Correspondence and requests for materials should be addressed to Y.Y.

Competing financial interests

The authors declare that they have no competing financial interests.

A photovoltaic device structure based on internal electron emission

Eric W. McFarland* **& Jing Tang†**

* Department of Chemical Engineering, † Materials Department, University of California, Santa Barbara, California 93106-5080, USA

There has been an active search for cost-effective photovoltaic devices since the development of the first solar cells in the 1950s (refs 1–3). In conventional solid-state solar cells, electron–hole pairs are created by light absorption in a semiconductor, with charge separation and collection accomplished under the influence of electric fields within the semiconductor. Here we report a multilayer photovoltaic device structure in which photon absorption instead occurs in photoreceptors deposited on the surface of an ultrathin metal–semiconductor junction Schottky diode. Photoexcited electrons are transferred to the metal and travel ballistically to—and over—the Schottky barrier, so providing the photocurrent output. Low-energy (~1 eV) electrons have surprisingly long ballistic path lengths in noble metals[4,5], allowing a large fraction of the electrons to be collected. Unlike conventional cells, the semiconductor in this device serves only for majority charge transport and separation. Devices fabricated using a fluorescein photoreceptor on an Au/TiO$_2$/Ti multilayer structure had typical open-circuit photovoltages of 600–800 mV and short-circuit photocurrents of 10–18 μA cm^{-2} under 100 mW cm^{-2} visible band illumination: the internal quantum efficiency (electrons measured per photon absorbed) was 10 per cent. This alternative approach to photovoltaic energy conversion might provide the basis for durable low-cost solar cells using a variety of materials.

The device structure is a solid-state multilayer; it consists of a photoreceptor layer deposited on a ~10–50-nm Au film, which caps 200 nm of TiO$_2$ on an ohmic metal back contact (Fig. 1). Au and TiO$_2$ are chosen and prepared so that a Schottky barrier, ϕ, of approximately 0.9 V is formed at the metal–semiconductor interface, and the dye, merbromin (2,7-dibromo-5-(hydroxymercurio)-fluorescein), is selected such that the photoexcited donor level is energetically above the barrier. The photon-to-electron conversion in this device occurs in four steps. First, light absorption occurs in the surface-absorbed photoreceptors, giving rise to energetic electrons. Second, electrons from the photoreceptor excited state are injected into conduction levels of the adjacent conductor, where they travel ballistically through the metal at an energy, ε_e, above the Fermi energy, E_f. Third, provided that ε_e is greater than the Schottky barrier height, ϕ, and the carrier mean-free path is long compared to the metal thickness, the electrons will traverse the metal and enter conduction levels of the semiconductor (internal electron emission). The absorbed photon energy is preserved in the remaining excess electron free energy when it is collected at the back ohmic contact, giving rise to the photovoltage, V. The photo-oxidized dye is reduced by transfer of thermalized electrons from states near E_f in the adjacent metal. Like electrochemical dye-sensitized solar cells[6,7], this structure physically separates the photon absorption process from charge separation and transport; but it does not have the disadvantage of needing a reducing agent in an electrolyte for intermolecular charge transport.

In addition to the desired production of ballistic electrons in the metal, there are other processes available for energy dissipation by the electron in the excited photoreceptor. The efficiency of the device will depend upon favouring specific energy transfer pathways from the major competing processes: (1) radiative intramolecular de-excitation with photon emission (luminescence), (2) non-radiative intramolecular or intermolecular de-excitation with direct coupling to phonons, (3) non-radiative intermolecular de-excitation with coupling to the metal conduction electrons, and (4) electron transfer of the energetic electron from the photoexcited state into the unoccupied conduction states of the metal. Both pathways (3) and (4) may produce energetic electrons in metal conduction levels above the Fermi energy (hot electrons) that also have sufficient energy and momentum to traverse the Schottky barrier; it is these electrons that give rise to the primary photocurrent in the device. We believe that the dominant pathway is (4) in our device, but we cannot rule out contributions to the measured photocurrent from pathway (3). Pathway (2) as well as non-transmitted electrons from (3) and (4) constitute the principal competing 'quenching' processes whereby the electronic excitation energy is dissipated as heat, and neither appears as light nor electron free energy in the device current.

The current–voltage characteristics of a representative device measured in the dark and under 100 mW cm^{-2} visible band

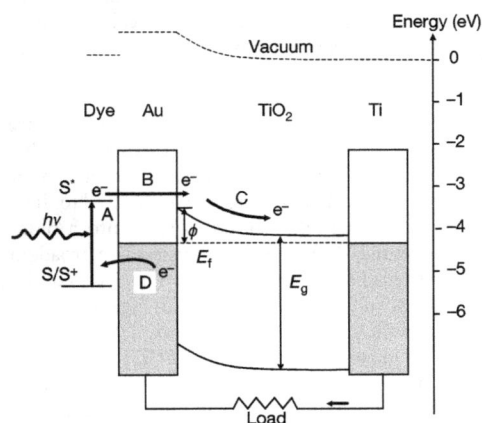

Figure 1 Electron transfer in the operating photovoltaic device. Process A, photon absorption and electron excitation from the chromophore ground state, S, to the excited state, S*; B, energetic electron transfer from S* into and (ballistically) through the conducting surface layer and over the potential energy barrier into the semiconductor; C, electron conduction as a majority carrier within the semiconductor to the ohmic back contact and through the load; and D, reduction of the oxidized chromophore, S$^+$, by a thermal electron from the conductor surface. Shown schematically are the relative energies of the electron levels within the device structures, the Schottky barrier, ϕ, the Fermi energy, E_f, and the semiconductor bandgap, E_g.

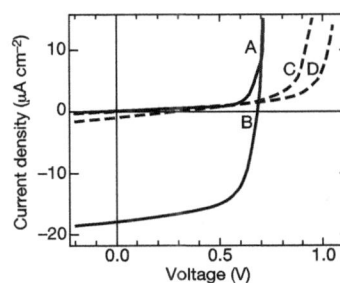

Figure 2 Current–voltage characteristics of the multilayer merbromin/Au/TiO$_2$/Ti photovoltaic device. The response is shown for dark conditions (curve A), and for 1,000 W m^{-2} broadband visible illumination (B). For comparison, curves obtained from the same device before adding merbromin are also shown for dark conditions (C), and under the same illumination (D).

When citing this article, please cite the original version as shown on the contents page of this chapter.

illumination before and after applying merbromin are shown in Fig. 2. After deposition of merbromin, the surface-activated device had an open-circuit photovoltage, V_{oc}, of 685 mV, and a short-circuit photocurrent, j_{sc}, of 18.0 μA cm^{-2}. The fill factor was determined to be 0.63 from the current–voltage characteristics under illumination. Before deposition of merbromin on the Au surface, excitation of defect levels in the thermally oxidized TiO$_2$ is responsible for a weak visible photoresponse ($V_{oc} = 325$ mV, $j_{sc} = 0.9$ μA cm^{-2}).

The photovoltage, V, is the difference, under illumination, between the Fermi level of the semiconductor and the Fermi level of the ultrathin metal film. For the surface-sensitized Schottky diode, the expression for the maximum value under open circuit conditions, V_{oc} is identical to a conventional Schottky solar cell where photon absorption occurs by bandgap excitation:

$$V_{oc} = \left(\frac{nkT}{e}\right)[\ln(j_\gamma/(T^2 A^{**})) + e\phi_B/kT] \qquad (1)$$

where n is the ideality factor ($\sim$1), k is Boltzmann's constant, e is the charge on the electron, A^{**} is the Richardson constant, and T is the device temperature[5]. The photocurrent density produced by photon absorption, j_γ, is balanced by the effective forward bias current from V_{oc} such that no net current is observed. The maximum photovoltages and photocurrents can be achieved by choice and preparation of the device materials to control and match the Schottky barrier height, ϕ, the semiconductor conduction band positions, and the chromophore donor/acceptor levels.

The photocurrent produced from a monochromatic photon flux, $F(\varepsilon_\gamma)$, is determined largely by the photon capture efficiency and the internal quantum efficiency, IQE, which is the number of hot electrons injected into the semiconductor (and thus detected) per absorbed photon:

$$j_\gamma(\varepsilon_\gamma) = F(\varepsilon_\gamma)\eta_\gamma(\varepsilon_\gamma)\text{IQE}(\varepsilon_\gamma) \qquad (2)$$

$$\text{IQE}(\varepsilon_\gamma) = \int_0^{\varepsilon_\gamma} \eta_{CM}(\varepsilon_\gamma, \varepsilon_e)\eta_{MS}(\varepsilon_e)\eta_S \, d\varepsilon_e \qquad (3)$$

where $\eta_\gamma(\varepsilon_\gamma)$ is the efficiency for absorption of a photon of energy ε_γ; $\eta_{CM}(\varepsilon_\gamma, \varepsilon_e)$ is the probability that absorption will result in the injection of an excited electron into the metal at an energy, ε_e, above the Fermi energy; $\eta_{MS}(\varepsilon_e)$ is the efficiency for charge transport across the metal film and into the semiconductor; and η_S is the charge collection efficiency in the semiconductor. Experimentally, the incident photon-to-electron conversion efficiency, IPCE(ε_γ) = $j(\varepsilon_\gamma)/F(\varepsilon_\gamma)$, is typically determined by measuring the photocurrent from a monochromatic photon source. The IQE is then calculated by correcting the IPCE for photon absorption in the dye as IQE(ε_γ) = IPCE(ε_γ)/$\eta_\gamma(\varepsilon_\gamma)$.

What makes the IQE for this device structure acceptable is the surprisingly high value of $\eta_{MS}(\varepsilon_e)$. At energies of $\sim$1 eV above the Fermi level, the ballistic mean free path for electrons in Au and other metals with low-lying or filled d bands has been measured to be extremely long, $\sim$20–150 nm (refs 4, 5). Thus, despite the competing processes of quasi-elastic scattering by phonons and inelastic scattering from other electrons, provided the metal conducting film is ultrathin, a significant fraction of the injected electrons will reach the Schottky barrier.

Although the relationship between V_{oc} and j_γ in our devices is identical to that of a conventional semiconductor Schottky solar cell, equation (1), the bulk semiconductor is not utilized for photon absorption; thus the bandgap and semiconductor thickness constraints are largely removed. The significance of the Schottky barrier height, however, is increased. The absorption process occurs in the photoreceptor, which is selected to balance high solar-spectrum absorbance with maximum photovoltage as well as chemical stability. In this design, several different chromophores could be used simultaneously to provide more efficient conversion (for example, dyes, quantum structures); however, in an idealized cell using a single dye, the optimal chromophore absorption maximum would be the same as the ideal bandgap of a conventional Schottky diode solar cell ($\sim$1.5 eV; refs 2, 8). Thus, the theoretical maximum power output of the Schottky device utilizing internal electron emission can be no greater than an idealized conventional cell for the solar spectrum, $\sim$25% (refs 2, 8).

The physical and electronic coupling of the chromophore to the metal conduction levels are crucial to the performance of the device. In aqueous solution, merbromin has an absorption maximum at 511 nm wavelength; but when merbromin is attached to the Au film, the primary absorption peak appears to split into a blue-shifted peak and a stronger red-shifted peak, giving rise to a broadening and red-shift of the overall spectrum (Fig. 3a). The wavelength-dependent photoresponse of the device is shown in Fig. 3b, where we plot the IPCE under short-circuit conditions, IPCE = $j(\varepsilon_\gamma)/F(\varepsilon_\gamma)$, equation (2). The general features of the IPCE response spectrum, and the broad maximum, are approximately the same as the surface-bound dye absorption; this is consistent with the mechanism of action described in Fig. 1, with both the associated red- and blue-shifted dye states coupling to the metal such that hot electrons are produced. The IQE is determined by correcting the IPCE for the fraction of incident photons absorbed in the dye (Fig. 3c). Although the broad absorption edge of the TiO$_2$ overlaps with the dye adsorption between 400 nm and 450 nm, above 500 nm the TiO$_2$ absorption is negligible and the IQE of approximately 10% reflects the efficiency of the internal emission process.

The main considerations in choosing the materials for the device are: (1) the relative energies of the donor/acceptor levels in the

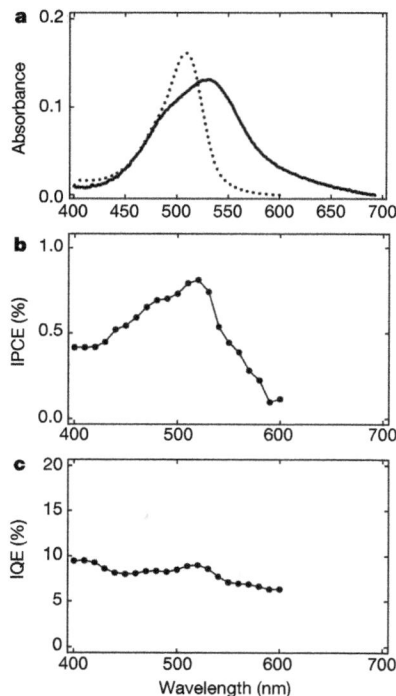

Figure 3 The wavelength-dependent photoresponse and absorbance of the photovoltaic device. **a**, The absorbance of merbromin adsorbed to the surface of the Au/TiO$_2$/Ti device (solid line). Shown for reference is the absorbance (in arbitrary units) of merbromin in water (dotted line). **b**, The incident photon-to-electron conversion efficiency (IPCE) of the dye-sensitized device under 0.4 mW cm^{-2} illumination. **c**, The internal quantum efficiency (IQE), determined by correcting the IPCE by the absorption efficiency of the affixed dye, reflects the absorbed photon-to-electron conversion efficiency of the dye.

photoreceptor, the conductor Fermi level, the barrier height, and the position of the semiconductor conduction band edges, (2) the ballistic mean free path of electrons, and (3) the physical and electronic coupling of the chromophore to the conductor for high absorbance and high electron transfer efficiency. The devices studied here represent only one of several different configurations of photovoltaics taking advantage of ballistic electron transport and internal electron emission in a photovoltaic device. For example, modifications of the structure to utilize hot hole injection (rather than hot electrons) in a p-type junction[9] would allow use of hole conducting polymers instead of inorganic semiconductors.

The IPCE and overall energy efficiency are limited by low dye coverage (8×10^{14} molecules cm^{-2}) and the resulting low photon absorption. Significant increases are expected with improved optical design (reduced surface reflection), decreased metal thickness, increased dye loading, and an engineered surface morphology with significantly higher surface area structured such that multiple passes through a dye-covered surface are possible for each photon. Although the ultimate efficiency of an optimized device based on the concept presented here is approximately the same as an ideal conventional semiconductor cell, there appear to be practical and economic advantages in terms of the wide choice of inexpensive, durable, and readily synthesized device materials that may be utilized. ☐

Methods

Device fabrication

Devices were fabricated on titanium foil substrates (Alfa Aesar), which served as ohmic back contacts. A 250-nm layer of titanium (99.9999%) was evaporated under vacuum onto the foil following cleaning and polishing (using 10 μm grit). A 200-nm layer of TiO$_2$ was grown on the substrate by thermal oxidation at 500 °C. The polycrystalline TiO$_2$ is predominantly rutile phase, with oxygen vacancies giving rise to n-type doping. Au films were electrodeposited onto the TiO$_2$ from a solution containing 0.2 M KCN and 0.1 M AuCN at pH 14. The TiO$_2$ served as the working electrode, with a Pt wire counter electrode. A 100-ms galvanostatic pulse at -200 mA cm^{-2} was used to nucleate Au uniformly on the surface, followed by a periodic galvanostatic pulse train of 5 ms at $+0.2$ mA cm^{-2} and 5 ms at -1.7 mA cm^{-2} for 10 s to form a film ~10–50 nm thick. Photoactive merbromin (2,7-dibromo-5-(hydroxymercurio)fluorescein disodium salt, 5 mM in water) was adsorbed onto the surface by immersion at room temperature for 10–12 h, followed by rinsing in water.

Characterization

Current–voltage (I–V) curves were measured using a voltage ramp rate of 0.05 V s^{-1} in the dark and under illumination from a 250-W tungsten lamp (Oriel, 6129), with intensity measured using a radiometer (IL1700, International Light). The fill factor was calculated at 1,000 W m^{-2} by dividing the maximum product of current and voltage from the illuminated I–V curve by the product of open-circuit voltage and short-circuit current at the same illumination. The spectral response was determined using a 150-W Xe lamp and monochromator (Oriel 7240). IPCE was calculated from the current density under short-circuit conditions and the photon flux as measured by the radiometer. The optical absorbance (and absorption efficiency, $\eta_\gamma(\varepsilon_\gamma)$) of the dye on the device surface and dye photon absorption was determined from the transmission and reflectance of a device fabricated on a transparent substrate before and after application of the dye, using an integrating sphere (LabSphere) and fibre-optic coupled monochromator (Ocean Optics). Free-solution dye absorbance was measured with an optical spectrometer (UV-1610, Shimadzu). Dye loading was determined by detaching the dye from the activated device surface in 1 mM NaOH solution, and determining the amount removed from the difference in optical absorbance at 511 nm of the NaOH solution.

Received 17 July; accepted 18 November 2002; doi:10.1038/nature01316.

1. Chapin, D. M., Fuller, C. S. & Pearson, G. L. A new silicon p-n junction photocell for converting solar radiation into electrical power. *J. Appl. Phys.* **25**, 676–677 (1954).
2. Archer, M. D. & Hill, R. (eds) *Clean Electricity from Photovoltaics* (Series on Photoconversion of Solar Energy, Vol. 1, Imperial College Press, London, 2001).
3. Goetzberger, A. & Hebling, C. Photovoltaic materials, past, present, future. *Sol. Energy Mater. Sol. Cells* **62**, 1–19 (2000).
4. Seah, M. P. & Dench, W. A. Quantitative electron spectroscopy of surfaces: a standard data base for electron inelastic mean free paths in solids. *Surf. Interf. Anal.* **1**, 2–11 (1979).
5. Frese, K. W. & Chen, C. Theoretical models of hot carrier effects at metal-semiconductor electrodes. *J. Electrochem. Soc.* **139**, 3234–3249 (1992).
6. Grätzel, M. Photoelectrochemical cells. *Nature* **414**, 338–344 (2001).
7. Green, M. A., Emery, K., King, D. L., Igari, S. & Warta, W. Solar cell efficiency tables (version 17). *Prog. Photovolt. Res. Applic.* **9**, 49–56 (2001).
8. Sze, S. M. *Physics of Semiconductor Devices* 2nd edn, Ch. 14 (Wiley & Sons, New York, 1981).
9. Nienhaus, H. *et al.* Electron-hole pair creation at Ag and Cu surfaces by adsorption of atomic hydrogen and deuterium. *Phys. Rev. Lett.* **82**, 446–448 (1999).

Acknowledgements We thank M. White, A. Tavakkoly, A. Kochhar, N. Shigeoka, G. Stucky and W. Siripala for technical assistance and discussions. The project was supported by Adrena Inc. Financial support for J.T. was provided by the NSF-MRSEC funded Materials Research Laboratory (UCSB).

Competing interests statement The authors declare competing financial interests: details accompany the paper on *Nature*'s website (▶ http://www.nature.com/nature).

Correspondence and requests for materials should be addressed to E.W.M. (e-mail: mcfar@engineering.ucsb.edu).

A stable quasi-solid-state dye-sensitized solar cell with an amphiphilic ruthenium sensitizer and polymer gel electrolyte

PENG WANG[1], SHAIK M. ZAKEERUDDIN*[1], JACQUES E. MOSER[1], MOHAMMAD K. NAZEERUDDIN[1], TAKASHI SEKIGUCHI[2] AND MICHAEL GRÄTZEL*[1]

[1]Laboratory for Photonics and Interfaces, Swiss Federal Institute of Technology, CH-1015 Lausanne, Switzerland
[2]Hitachi Maxell, 1-1-88, Ushitora, Ibaraki-Shi, Osaka 567-8567, Japan
*e-mail: shaik.zakeer@epfl.ch; michael.graetzel@epfl.ch

Published online: 18 May 2003; doi:10.1038/nmat904

Dye-sensitized nanocrystalline solar cells (DSC) have received considerable attention as a cost-effective alternative to conventional solar cells. One of the main factors that has hampered widespread practical use of DSC is the poor thermostability encountered so far with these devices. Here we show a DSC with unprecedented stable performance under both thermal stress and soaking with light, matching the durability criteria applied to silicon solar cells for outdoor applications. The cell uses the amphiphilic ruthenium sensitizer cis-RuLL'(SCN)$_2$ (L = 4,4'-dicarboxylic acid-2,2'-bipyridine, L' = 4,4'-dinonyl-2,2'-bipyridine) in conjunction with a quasi-solid-state polymer gel electrolyte, reaching an efficiency of >6% in full sunlight (air mass 1.5, 100 mW cm^{-2}). A convenient and versatile new route is reported for the synthesis of the heteroleptic ruthenium complex, which plays a key role in achieving the high-temperature stability. Ultramicroelectrode voltammetric measurements show that the triiodide/iodide couple can perform charge transport freely in the polymer gel. The cell sustained heating for 1,000 h at 80 °C, maintaining 94% of its initial performance. The device also showed excellent stability under light soaking at 55 °C for 1,000 h in a solar simulator (100 mW cm^{-2}) equipped with a ultraviolet filter. The present findings should foster widespread practical application of dye-sensitized solar cells.

Dye-sensitized nanocrystalline solar cells (DSC) provide an economically credible alternative to conventional inorganic photovoltaic devices. Owing to their high-energy conversion efficiency and low production cost, they have received considerable attention over the past decade[1–4]. The mesoscopic texture of the TiO$_2$ film in these cells significantly increases the cross-section of light harvesting by surface-anchored charge-transfer sensitizers, while maintaining a good contact with electrolytes. In these photovoltaic devices, ultrafast electron injection from a photoexcited dye into the conduction band of an oxide semiconductor, and subsequently dye regeneration and hole transportation to the counter electrode, are responsible for the efficient generation of electricity. Although a respectable 10.4% light-to-electricity conversion efficiency at air mass (AM) 1.5 solar irradiance has been obtained for photovoltaic devices with a panchromatic dye and a liquid electrolyte containing the triiodide/iodide couple[5], the achievement of long-term stability at temperatures of about 80–85 °C, which is an important requirement for outdoor application of the DSC, still remains a major challenge[1–11].

The leakage of liquid electrolyte from such modules, possible desorption of loosely attached dyes, and photodegradation in the desorbed state as well as corrosion of the Pt counter electrode by the triiodide/iodide couple have been suggested as some of the critical factors limiting the long-term performance of the DSC, especially at elevated temperature. Thus considerable efforts have been made to realize devices with high efficiencies that meet the stability criteria for outdoor use. In this context, new counter-electrode materials[12–14], alternative redox couples[15–17] and sensitizers[18–25] have been screened. Additionally, p-type semiconductor[26,27], hole conductor[28] and polymeric or gel materials incorporating triiodide/iodide as a redox couple[29–33] were introduced to substitute the liquid electrolytes by solid-state or quasi-solid-state materials. Numerous investigations of polymer gel electrolytes have been carried out with special emphasis on applications to lithium batteries[34]. Gelation of the liquid electrolytes mitigate the potential instability against solvent leakage under thermal stress. In this paper, a photochemically stable fluorine polymer, poly(vinylidenefluoride-co-hexafluoropropylene) (PVDF-HFP), was used to solidify a 3-methoxypropionitrile (MPN)-based liquid electrolyte to obtain a quasi-solid-state gel electrolyte. When used in combination with a new amphiphilic polypyridyl ruthenium dye, this new embodiment of a DSC, which reaches a conversion efficiency under full sunlight of over 6%, shows strikingly high stabilities under both thermal stress at 80 °C and prolonged soaking with light.

When citing this article, please cite the original version as shown on the contents page of this chapter.

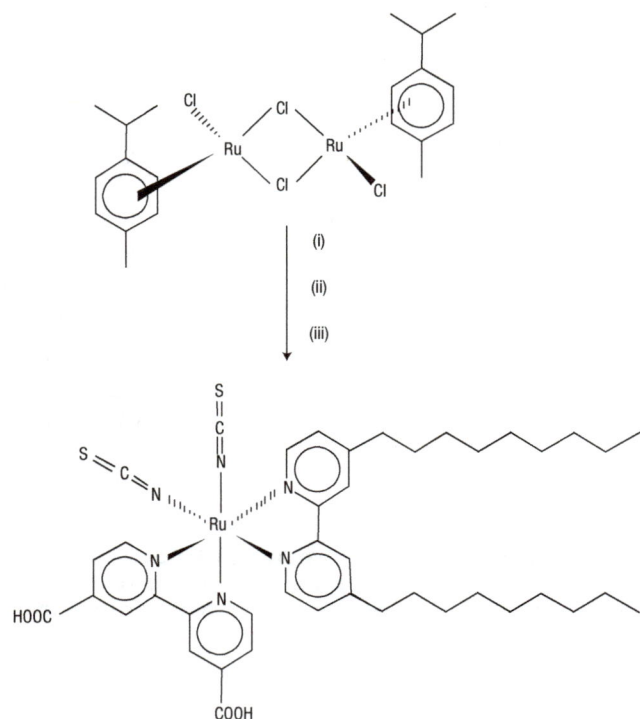

Scheme 1 One-pot synthetic route for Z-907 dye.
(i) DMF, dnbpy, 70 °C, N₂, 4 h.
(ii) H₂dcbpy, 150 °C, N₂, 4 h.
(iii) NH₄NCS, 150 °C, N₂, 4 h.

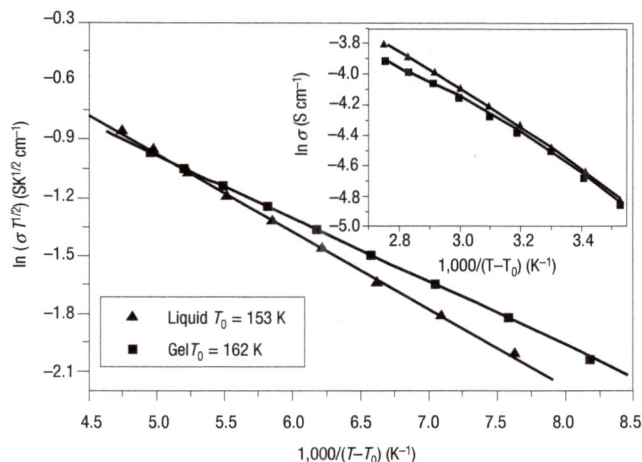

Figure 1 Plots of conductivity–temperature data in the VTF coordinates for the liquid and polymer gel electrolytes. Inset: Arrhenius plots of conductivity–temperature data. σ = specific conductivity; T = absolute temperature; T_0 = glass-transition temperature.

We have previously prepared a series of hydrophobic ruthenium polypyridyl sensitizers for low-power photoelectrochemical cells to increase their tolerance to water in the electrolytes[35]. The amphiphilic dye Z-907 (cis-Ru(H₂dcbpy)(dnbpy)(NCS)₂, where the ligand H₂dcbpy is 4,4'-dicarboxylic acid-2, 2'-bipyridine and dnbpy is 4,4'-dinonyl-2, 2'-bipyridine) was synthesized according to Scheme 1. A novel synthetic route starting from the RuCl₂(p-cymene)₂ complex, and using sequential addition of ligands to it at different time intervals, was designed for the synthesis of heteroleptic polypyridyl ruthenium complexes. This method is more facile and gives better yields than that previously reported[35].

The conductivity (σ) of the electrolyte used in this study was first examined. This provides information on the mobility of the ions, their interaction with the solvent and on ion-pairing phenomena, which are expected to affect the photovoltaic performance and in particular the fill factor of the solar cell. It is given by[36]:

$$\sigma(T) = \sum_i |Z_i| F c_i \, \mu_i = \sum_i \frac{|Z_i|^2 F c_i e D_i}{K_B T} \qquad (1)$$

where Z_i, c_i, μ_i and D_i are the charge, concentration, mobility and diffusion coefficient of the ith ion, e is the electronic charge, T is the absolute temperature and K_B and F are the Boltzmann and Faraday constants, respectively. As is apparent from the inset of Fig. 1, the Arrhenius equation[36] cannot be used to describe the conductivity–temperature behaviour of the liquid and polymer gel

electrolytes. A better fit to the data in Fig. 1 is obtained by the Vogel–Tammann–Fulcher (VTF) equation[37].

$$\sigma(T) = AT^{-1/2} \exp[-B/(T-T_0)] \qquad (2)$$

In equation (2), A and B are constants and T_0 is the glass-transition temperature. Surprisingly, whereas gelation resulted in a slight decrease in conductivities, for example, from 10.4 to 10.2 mS cm⁻¹ at room temperature, the steady-state voltammograms (Fig. 2) for a Pt ultramicroelectrode in the liquid and polymer gel electrolytes are practically identical. Thus, the triiodide/iodide redox couple can diffuse freely in the liquid domains entrapped by the three-dimensional network of the PVDF-HFP, despite the solid nature of the electrolyte. The apparent diffusion coefficients (D_{app}) of iodide and triiodide can be calculated from anodic and cathodic steady-state currents (I_{ss}) according to the following equation[38].

$$I_{ss} = 4ncaFD_{app} \qquad (3)$$

where n is the electron number per molecule, a is the microelectrode radius, F is the Faraday constant and c is the bulk concentration of electroactive species. The calculated diffusion coefficients of triiodide and iodide are 3.60×10^{-6} and 4.49×10^{-6} cm² s⁻¹, respectively. Furthermore, using equation (1) the diffusion coefficient of 1,2-dimethyl-propylimidazolium (DMPI) cations in the liquid and polymer gel electrolyte was derived to be 2.39×10^{-7} and

Table 1 Device efficiencies of cells with the liquid and polymer gel electrolytes.

	η (%) at different incident light intensities*			
	0.01 Sun	0.1 Sun	0.5 Sun	1.0 Sun
Liquid	7.5	7.4	6.9	6.2
Gel	7.6	7.3	6.8	6.1

*The spectral distribution of the lamp mimics air mass 1.5 solar light. 1.0 Sun corresponds to an intensity of 100 mW cm⁻².

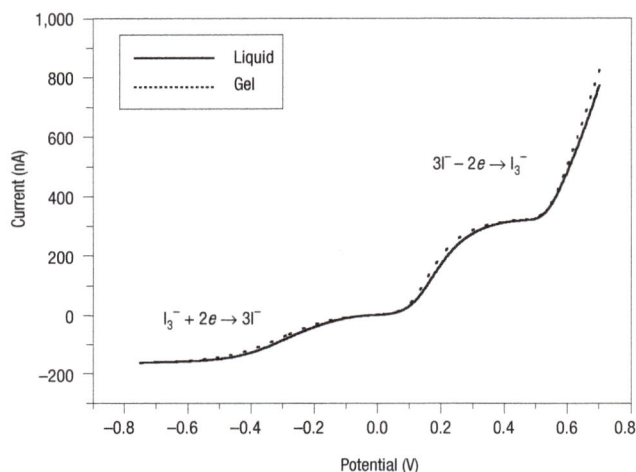

Figure 2 **Steady-state voltammograms of the liquid and polymer gel electrolytes with a Pt ultramicroelectrode.** Scan rate: 10 mV s^{-1}.

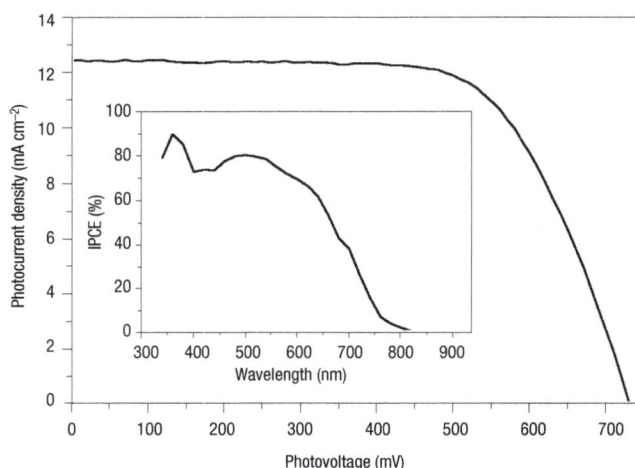

Figure 3 **Typical photocurrent density–voltage characteristic of photovoltaic cells.** The polymer gel electrolyte and Z-907 dye were at an irradiance of AM 1.5 sunlight (99.8 mW cm^{-2}). Cell area: 0.152 cm^2. The inset is the incident photon-to-current efficiency (IPCE) for the quasi-solid-state cells.

1.55×10^{-7} cm^2 s^{-1}. The lower value obtained in the gel indicates hydrogen-bonding interaction between DMPI cations and PVDF-HFP copolymer.

Figure 3 shows a typical photocurrent density–voltage curve for cells based on the Z-907 dye and the polymer gel electrolyte under AM 1.5 sunlight illumination. The short-circuit photocurrent density (J_{sc}), open-circuit voltage (V_{oc}), and fill factor (FF) are 12.5 mA cm^{-2}, 730 mV and 0.67, respectively, yielding an overall energy conversion efficiency (η) of 6.1%. The action spectrum of the photocurrent is shown in the inset of Fig. 3. The incident photon-to-current conversion efficiency (IPCE) reaches a maximum efficiency of 80% at 540 nm. The photovoltaic performance obtained with liquid and polymer gel electrolytes is almost identical (Table 1), indicating that gelation has no adverse effect on the conversion efficiency.

The use of the amphiphilic Z-907 dye in conjunction with the polymer gel electrolyte was found to result in remarkably stable device performance both under thermal stress and soaking with light. The high conversion efficiency of the cell was sustained even under heating for 1,000 h at 80 °C, maintaining 94% of its initial value after this time period as shown in Fig. 4a. The device using the liquid electrolyte retained only 88% of its initial performance under the same conditions. The difference may arise from a decrease in solvent permeation across the sealant in the case of the polymer gel electrolyte. The polymer gel electrolyte is quasi-solid at room temperature, but becomes a viscous liquid (viscosity: 4.34 mPa s) at 80 °C compared with the blank liquid electrolyte (viscosity: 0.91 mPa s). Tolerance of such a severe thermal stress by a DSC having over 6% efficiency is unprecedented. In the case of N-719 dye, the overall efficiency decreased almost 35% during the first week at 80 °C, which clearly reflects the effect of molecular structure of the sensitizer on the stability of DSC. The difference between N-719 and Z-907 is that one of the 4,4′-dicarboxylic acid-2, 2′-bipyridine is replaced with 4,4′-dinonyl-2, 2′-bipyridine to make the dye more hydrophobic. We believe that desorption of N-719 at high temperature resulted in the poor thermostability of related devices. So far, dye-sensitized solar cells have been plagued by performance degradation at temperatures between 80 and 85 °C. The best result obtained in previous studies[10] was a decline in conversion efficiency from initially 4.5 to 3% when the cell was maintained for 875 h at 85 °C.

Figure 5 presents the detailed behaviour of device parameters during the aging tests performed at 80 °C with the DSC containing polymer gel electrolyte. After the first week of aging, the efficiency was moderately enhanced due to an increase in the J_{sc} and FF values. Then a gradually small decrease in V_{oc}, without much variation in J_{sc} and FF, caused a decrease in the overall efficiency by 6%. This is well within the limit of thermal degradation accepted for silicon solar cells.

The device also showed excellent photostability when submitted to accelerated testing in a solar simulator at 100 mW cm^{-2} intensity. After 1,000 h of light-soaking at 55 °C the efficiency had dropped by less than 5% (Fig. 4b) for cells covered with an ultraviolet absorbing polymer film. The efficiency difference for devices tested with and without the polymer film was only 4% at AM 1.5 sunlight, indicating a very small sacrifice in efficiency due to ultraviolet filter.

Detailed studies are currently under way to explore the reasons for the remarkable high-temperature stability of solar cells based on the amphiphilic ruthenium sensitizer and the quasi-solid-state gel electrolyte. Here we report on nanosecond time-resolved laser experiments designed to scrutinize the dynamics of the recombination of the electrons injected in the conduction band of TiO$_2$ (e^-_{cb}) with the oxidized dye (S$^+$) and the dye regeneration reaction with iodide. Indeed, kinetic competition between these two interfacial charge-transfer processes controls, to a large extent, the photon-to-current conversion efficiency of the photovoltaic devices. Transient absorbance measurements shown in Fig. 6 monitor directly the concentration of oxidized ruthenium sensitizer following photoinduced electron injection from the dye into the conduction band of the semiconductor film[16,39]. In the absence of iodide (Fig. 6a, trace 1), the decay of the absorption signal reflects the dynamics of recombination of injected electrons with the oxidized dye (S$^+$). Because the pulsed laser intensity was kept at a very low level (fluence ≤ 40 mJ cm^{-2} per pulse at the sample), less than one e^-_{cb}/S$^+$ charge-separated pair was produced on the average per TiO$_2$ nanocrystal. This is comparable to the condition prevailing when the photovoltaic device functions under natural sunlight.

The kinetics of S$^+$ transient absorbance decay could be fitted by a stretched exponential with a typical half-reaction time ($t_{1/2}$) of the

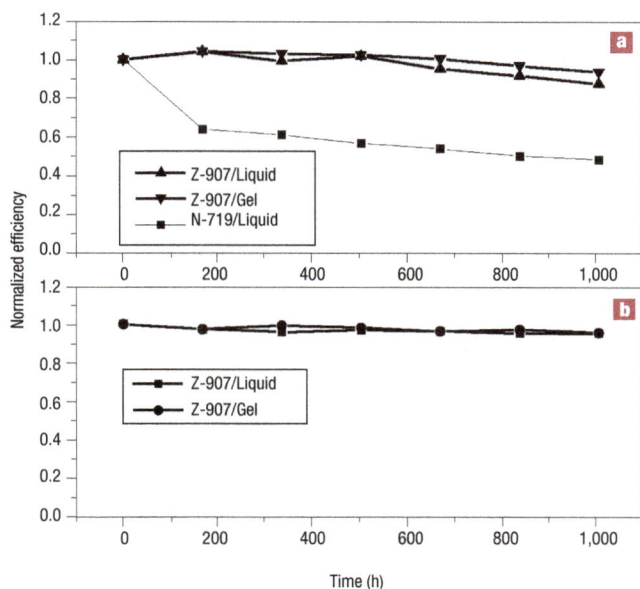

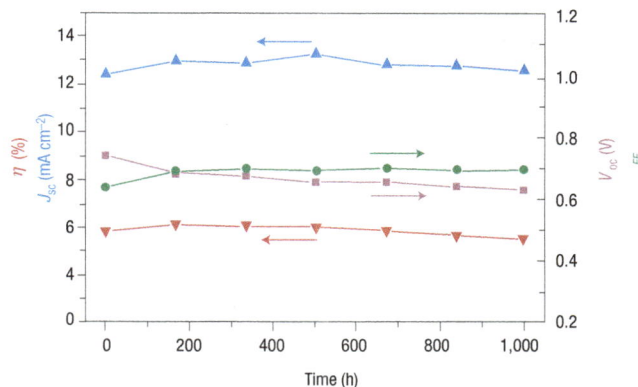

Figure 5 Detail of device parameter variations for cells with polymer gel electrolyte during accelerated aging at 80 °C. η: energy conversion efficiency ; J_{sc}: short-circuit photocurrent density; V_{oc}: open current voltage; FF: fill factor.

Figure 4 Normalized device efficiencies for cells with the liquid and polymer gel electrolytes. During **a**, accelerated aging at 80 °C and **b**, successive one sun visible-light soaking at 55 °C.

order of 180 µs, similar to what has been reported for the [cis-Ru(dcbpy)$_2$(NCS)$_2$] dye[40]. Thus the replacement of two carboxylate groups by the hydrocarbon chains does not appear to alter significantly the rate of the interfacial charge recombination process. In the presence of iodide, the decay of the oxidized dye signal is accelerated ($t_{1/2} \approx 30$ µs), showing that the mediator intercepts back-electron transfer. Hardly any difference is observed between the kinetics obtained with the liquid (Fig. 6a, trace 2) and polymer gel (trace 3) electrolytes. In both cases, however, the interception is rather slow compared with what is typically observed for the standard dye N-719 [cis-Ru(H$_2$dcbpy)(dcbpy)(NCS)$_2$](TBA)$_2$ under identical conditions[39] (Fig. 6b, trace 5, $t_{1/2} \approx 2$ µs). A significant amount of Z-907 cations survives for hundreds of microseconds indicating that the interception is incomplete.

Flash photolysis experiments were conducted under open-circuit conditions, which are in variance with those corresponding to the measurement of IPCE in a short-circuited photovoltaic device. However, the low light intensity used for both laser excitation and probing minimized the build-up of the concentration of conduction-band electrons in TiO$_2$. In the light of previously reported results[40], it can be inferred that observed back-electron-transfer kinetics corresponded to a very moderate potential of the nanocrystalline TiO$_2$ photoanode of ~0 mV versus Ag/AgCl. The same previous data show that the slow component of the back-transfer kinetics beyond 100 µs is only slightly affected at higher anodic biases. As a consequence, one can estimate that approximately 10–15% of the initial S$^+$ species population undergoes recombination with conduction-band electrons. This limits the quantum conversion efficiency and the maximum photocurrent density obtainable in practice. Figure 6b shows the effect of ageing at 80 °C on the kinetics of the interception process. Heat treatment for 72 h clearly produces an increase in the dye regeneration rate, with $t_{1/2}$ diminishing from 30 µs to about 5 µs (Fig. 6b, trace 3). No further change in the interception kinetics is obtained by prolonging the heat exposure of the samples (trace 4). The enhancement of the interception

reaction rate by a factor of six after the heat treatment easily accounts for the increase in overall conversion efficiency after one week of ageing at 80 °C: the photocurrent increases as more charge carriers escape from recombination. This effect is likely to be due to improved self-assembly of the Z-907 dye molecules on the surface of TiO$_2$ nanocrystals during annealing, and/or electrolyte penetration into cavities containing adsorbed sensitizer molecules where the electrolyte was originally not able to access.

Signals recorded at a probe wavelength of 630 nm are known to be quite sensitive to chemical degradation of Ru^{II}L$_2$(NCS)$_2$ dyes. This region of the transient spectrum between 610 and 900 nm is indeed dominated by a ligand-to-metal charge transfer transition of the NCS ligand to the Ru(III) metal centre of the oxidized dyes. Alterations in the coordination geometry of the complex result in a marked decrease[41] of the transient absorbance above 610 nm. Remarkably, no change of the transient absorbance signal amplitude was observed within experimental errors after ageing cells with the Z-907 dye for one week at 80 °C. This indicates that thermal aging does not affect significantly the chemical integrity of the dye molecules.

In summary, we have demonstrated that PVDF-HFP polymer can gel MPN-based liquid electrolyte without hampering charge transport of the triiodide/iodide couple inside the polymer network. Consequently, there is no difference in the conversion efficiencies of dye-sensitized solar cells with liquid and polymer gel electrolytes even at AM 1.5 sunlight. For the first time, long-term thermostable devices with higher than 6% energy conversion efficiency have been obtained by combining an amphiphilic polypyridyl ruthenium sensitizer with a polymer gel electrolyte. The extraordinary stabilities of the device under both thermal stress and soaking with light match the durability criteria applied to solar cells for outdoor use, rendering these devices viable for practical application.

METHODS

All organic solvents used were of puriss quality from Fluka, Switzerland. MPN (Fluka) was distilled before use. Other compounds used were: dnbpy (Aldrich, Switzerland), [RuCl$_2$(p-cymene)]$_2$ (Aldrich), N-methylbenzimidazole (NMBI) (Aldrich), NH$_4$NCS (Fluka), Sephadex LH-20 (Pharmacia, Sweden), poly(vinylidenefluoride-co-hexafluoropropylene (PVDF-HFP; Solvay, Belgium). 1,2-Dimethyl-3-propyl-limidazolium iodide (DMPII) was synthesized according to the method described in our previous work[42].

ONE-POT SYNTHESIS OF Z-907

In a typical one-pot synthesis of Z-907, [RuCl$_2$(p-cymene)]$_2$ (0.1 g, 0.16 mmol) was dissolved in DMF (50 ml) and dnbpy (0.133 g, 0.32 mmol) then added. The reaction mixture was heated to 60 °C under

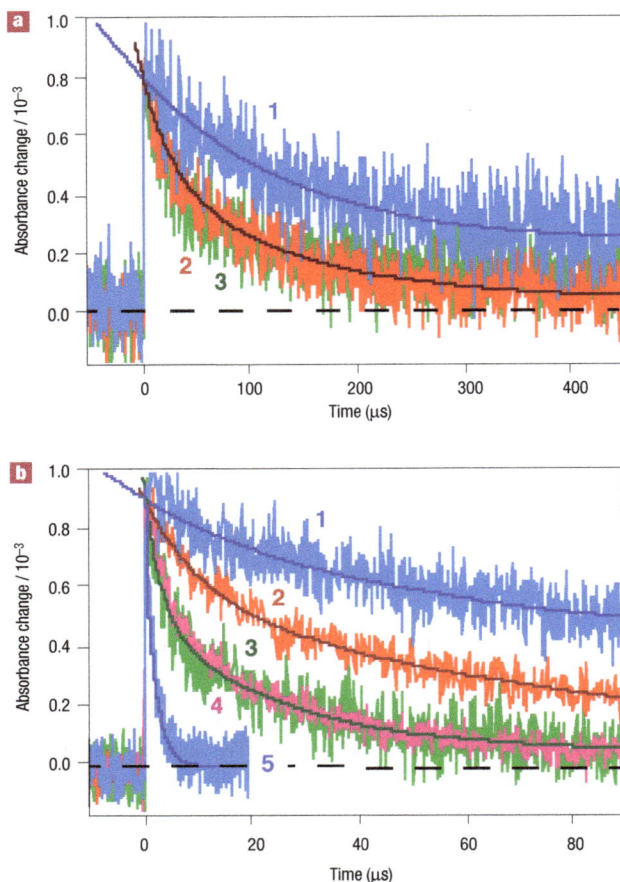

Figure 6 Transient absorbance decay kinetics of the oxidized state of Z-907 dye adsorbed on transparent TiO₂ nanocrystalline films. Measured at 630 nm on pulsed laser excitation at 510 nm (5 ns full-width at half-maximum pulse duration, 40 μJ cm⁻² pulse fluence). **a**, In the presence of pure solvent 3-methoxypropionitrile (1), the liquid electrolyte (2), and the polymer gel electrolyte (3). **b**, Cells containing pure 3-methoxypropionitrile (1), the liquid electrolyte before heat treatment (2), and after keeping at 80 °C for 72 h (3) and 280 h (4), respectively. Trace (5) shows, for direct comparison, the temporal behaviour of the standard N-719 dye under conditions similar to those of trace (2). Solid curves drawn on the top of experimental data are bi-exponential fits of decay kinetics.

nitrogen for 4 h with constant stirring. To this reaction flask H₂dcbpy (0.08 g, 0.32 mmol) was added and refluxed for 4 h. Finally, excess of NH₄NCS (13 mmol) was added to the reaction mixture and the reflux continued for another 4 h. The reaction mixture was cooled down to room temperature and the solvent was removed by using a rotary evaporator under vacuum. Water was added to the flask and the insoluble solid was collected on a sintered glass crucible by suction filtration. The solid was washed (5 × 20 ml) with pH 12 aqueous solution, distilled water and diethyl ether and dried. On a Sephadex LH-20 column the complex was further purified with methanol as an eluent. ¹H NMR (δ_H/p.p.m. in CD₃OD) 9.72 (d, 1H), 9.28 (d, 1H), 9.08 (s, 1H), 8.92 (s, 1H), 8.55 (s, 1H), 8.42 (s, 1H), 8.28 (d, 1H), 7.88 (t, 1H), 7.70 (t, 2H), 7.40 (d, 1H), 7.05 (d, 1H), 2.95 (t, 2H), 2.75 (t, 2H), 1.95 (m, 2H), 1.40 (m, 26H), 0.90 (t, 6H). Analytical calculation for RuC₄₂H₅₂N₆O₄S₂: C, 57.99; H, 5.98; N, 9.66%. Found: C, 57.90; H, 5.97; N, 9.61%.

FABRICATION OF SOLAR CELLS

A screen-printed double layer of TiO₂ particles was used as the photoanode. A 10-μm-thick film of 20-nm-diameter TiO₂ particles was first printed on the fluorine-doped SnO₂ conducting glass electrode and further coated by 4-μm-thick second layer of 400 nm light-scattering anatase particles (CCIC, Japan). After sintering at 500 °C and cooling down to 80 °C, the TiO₂ electrodes were dye-coated by immersing them into a 0.3 mmol l⁻¹ solution of Z-907 in acetonitrile and t-butanol (volume ratio of 1:1) at room temperature for 12 hours and then assembled with thermally platinized conducting glass electrodes. The electrodes were separated by a 35-μm-thick hot-melt ring (Bynel, DuPont) and sealed up by heating. PVDF-HFP (5 wt%) was mixed with the liquid electrolyte consisting of DMPII (0.6 mol l⁻¹),

iodine (0.1 mol l⁻¹), NMBI (0.5 mol l⁻¹) in MPN and heated until no solid was observed. The internal space of the cell was filled with the resulting hot solution using a vacuum pump. After cooling down to room temperature, a uniform motionless polymer gel layer was formed in the cells. The electrolyte-injecting hole made with a sand-ejecting drill on the counter electrode glass substrate was sealed with a Bynel sheet and a thin glass cover by heating. To have a good comparison with the polymer gel electrolyte, devices with the liquid electrolyte were also fabricated using the above procedure.

CONDUCTIVITY, VISCOSITY, VOLTAMMETRIC AND PHOTOELECTROCHEMICAL MEASUREMENTS

A CDM210 conductivity meter (Radiometer Analytical, SAS, France) was used to measure conductivities of the liquid and polymer gel electrolytes. The CDC749 conductivity cell (Radiometer Analytical, SAS, France) with a nominal cell constant of 1.70 cm⁻¹ was calibrated with 0.1 mol l⁻¹ KCl aqueous solution before the experiments. Dynamic viscosity measurements were carried out on a microviscosimeter (VT500, Haake, Germany). A DT Hetotherm cycle heat pump (Heto, Denmark) was used to control the temperature of electrolytes. Steady-state voltammograms were recorded on an Autolab P20 electrochemical workstation (Eco Chimie, Netherlands) at 25 °C. A two-electrode electrochemical cell was used, consisting of a Pt ultramicroelectrode with a radius of 5.0 μm as working electrode and a Pt foil as counter electrode. Photoelectrochemical data was obtained with a set-up as explained in our earlier publication[6].

THERMAL STRESS AND VISIBLE LIGHT-SOAKING TESTS

Hermetically sealed cells were used for long-term stability tests. For thermal stress, the cells were stored in the oven at 80 °C. In light-soaking tests the cells were covered with a 50-μm-thick of polyester film (Preservation Equipment, Norfolk, UK) as an ultraviolet cut-off filter (up to 400 nm) were irradiated at open circuit under a Suntest CPS plus lamp (ATLAS, Netherlands; 100 mW cm⁻², 55 °C).

LASER TRANSIENT ABSORBANCE MEASUREMENTS

Nanosecond pulsed laser excitation was applied using a GWU-355 (GWU-Lasertechnik, Erfstadt-Friesheim, Germany) broadband optical parametric oscillator (OPO) pumped by a Continuum Powerlite 7030 frequency-tripled Q-switched Nd:YAG laser (Continuum, Santa Clara, California, USA). The output of the OPO (30 Hz repetition rate, pulse-width at half-height of 5 ns) was tuned at a wavelength of 510 nm and attenuated by filters. The beam was expanded by a planoconcave lens to irradiate a large cross-section of the sample, whose surface was kept at a 30° angle to the excitation beam. The laser fluence on the sample was kept at a very low level (≤40 μJ cm⁻² per pulse) to ensure that, on average, less than one electron is injected per nanocrystalline TiO₂ particle on pulsed irradiation. The probe light, produced by a continuous wave xenon arc lamp, was passed through a first monochromator tuned at 630 nm, various optical elements, the sample, and a second monochromator, before being detected by a fast photomultiplier tube. Data waves were recorded on a DSA 602A digital signal analyser (Tektronix, Beaverton, Oregon, USA) Satisfactory signal-to-noise ratios were typically obtained by averaging over 3,000 laser shots. Except for normal glass and 8 μm transparent TiO₂ film, sealed cells were used for laser transient absorbance experiments.

Received 24 February 2003; accepted 25 March 2003; published 18 May 2003.

References

1. O'Regan, B. & Grätzel, M. A low cost, high efficiency solar cell based on dye-sensitized colloidal TiO₂ films. *Nature* **353**, 737–740 (1991).
2. Nazeeruddin, M. K. *et al.* Conversion of light to electricity by cis-X₂-bis(2,2'-bipyridyl-4,4'-dicarboxalate)ruthenium(II) charge transfer sensitizers (X=Cl⁻, Br⁻, I⁻, CN⁻, and SCN⁻) on nanocrystalline TiO₂ electrodes. *J. Am. Chem. Soc.* **115**, 6382–6390 (1993).
3. Hagfeldt, A. & Grätzel, M. Molecular photovoltaics. *Accounts Chem. Res.* **33**, 269–277 (2000).
4. Grätzel, M. Photoelectrochemical cells. *Nature* **414**, 338–344 (2001).
5. Nazeeruddin, M. K. *et al.* Engineering of efficient panchromatic sensitizers for nanocrystalline TiO₂-based solar cells. *J. Am. Chem. Soc.* **123**, 1613–1624 (2001).
6. Papageogiou, N. *et al.* The performance and stability of ambient temperature molten salts for solar cell applications. *J. Electrochem. Soc.* **143**, 3009–3108 (1996).
7. Kohle, O. *et al.* The photovoltaic stability of bis(isothiocyanato)ruthenium(II)-bis-2,2'-bipyridine-4,4'-dicarboxylic acid and related sensitizers. *Adv. Mater.* **9**, 904–906 (1997).
8. Pettersson, H. & Gruszecki, T. Long-term stability of low-power dye-sensitised solar cells prepared by industrial methods. *Solar Energy Mater. Solar Cells* **70**, 203–211 (2001).
9. Kern, R. *et al.* Long-term stability of dye-sensitized solar cells for large area power applications. *Opto-Electron. Rev.* **8**, 284–288 (2001).
10. Hinsch, A. *et al.* Long-term stability of dye-sensitised solar cells. *Prog. Photovoltaics* **9**, 425–438 (2001).
11. Pettersson, H. *et al.* Manufacturing method for monolithic dye-sensitized solar cells permitting long-term stable low-power modules. *Solar Energy Mater. Solar Cells* **77**, 405–413 (2003).
12. Kay, A. & Grätzel, M. Low cost photovoltaic modules based on dye sensitized nanocrystalline titanium dioxide and carbon powder. *Solar Energy Mater. Solar Cells* **44**, 99–117 (1996).
13. Saito, Y. *et al.* Application of poly(3,4-ethylenedioxythiophene) to counter electrode in dye-sensitized solar cells. *Chem. Lett.* 1060–1061 (2002).
14. Suzuki, K. *et al.* Application of carbon nanotubes to counter electrodes of dye-sensitized solar cells. *Chem. Lett.* **32**, 28–29 (2003).
15. Oskam, G. *et al.* Pseudohalogens for dye-sensitized TiO₂ photoelectrochemical cells. *J. Phys. Chem. B* **105**, 6867–6873 (2001).
16. Nusbaumer, H. *et al.* Co^II(dbbip)₂²⁺ complex rivals tri-iodide/iodide redox mediator in dye-sensitized photovoltaic cells. *J. Phys. Chem. B* **105**, 10461–10464 (2001).
17. Sapp, S. A. *et al.* Substituted polypyridine complexes of cobalt(II/III) as efficient electron transfer mediators in dye-sensitized solar cells. *J. Am. Chem. Soc.* **124**, 11215–11222 (2002).
18. Ferrere, S. & Gregg, B. A. Photosensitization of TiO₂ by [Fe^II(2,2'-bipyridine-4,4'-dicarboxylic acid)₂(CN)₂]: band selective electron injection from ultra-short-lived excited states. *J. Am. Chem. Soc.* **120**, 843–844 (1998).
19. Hou, Y.-J. *et al.* Influence of the attaching group and substituted position in the photosensitization behavior of ruthenium polypyridyl complexes. *Inorg. Chem.* **38**, 6320–6322 (1999).
20. Monat, J. E. & McCusker, J. K. Femtosecond excited dynamics of an iron(II) polypyridyl solar cell sensitizer model. *J. Am. Chem. Soc.* **122**, 4092–4097 (2000).
21. Sauvé, G. *et al.* Dye sensitization of nanocrystalline titanium dioxide with osmium and ruthenium polypyridyl complexes. *J. Phys. Chem. B* **104**, 6821–6836 (2000).

22. Yanagida, M. *et al.* Dye-sensitized solar cells based on nanocrystalline TiO₂ sensitized with a novel pyridylquinoline ruthenium(II) complex. *New J. Chem.* **26**, 963–965 (2002).

23. Li, X. *et al.* New peripherally-substituted naphthalocyanines: synthesis, characterization and evaluation in dye-sensitised photoelectrochemical solar cells. *New J. Chem.* **26**, 1076–1080 (2002).

24. He, J. *et al.* Modified phthalocyanines for efficient near-IR sensitisation of nanostructured TiO₂ electrode. *J. Am. Chem. Soc.* **124**, 4922–4932 (2002).

25. Hara, K. *et al.* Novel polyene dyes for highly efficient dye-sensitized solar cells. *Chem. Commun.* 252–253 (2003).

26. Kumara, G. R. A. *et al.* Fabrication of dye-sensitized solar cells using triethylamine hydrothiocyanate as a CuI crystal growth inhibitor. *Langmuir* **18**, 10493–10495 (2002).

27. O'Regan, B. *et al.* A solid-state dye-sensitized solar cell fabricated with pressure-treated P25-TiO₂ and CuSCN: analysis of pore filling and IV characteristics. *Chem. Mater.* **14**, 5023–5029 (2002).

28. Krüger, J. *et al.* Improvement of the photovoltaic performance of solid-state dye-sensitized device by silver complexation of the sensitizer cis-bis(4,4'-dicarboxy-2,2' bipyridine)-bis(isothiocyanato) ruthenium(II). *Appl. Phys. Lett.* **81**, 367–369 (2002).

29. Cao F., Oskam, G. & Searson, P. C. A solid-state, dye-sensitized photoelectrochemical cell. *J. Phys. Chem.* **99**, 17071–17073 (1995).

30. Nogueira, A. F., Durrant, J. R. & De Paoli, M. A. Dye-sensitized nanocrystalline solar cells employing a polymer electrolyte. *Adv. Mater.* **13**, 826–830 (2001).

31. Kubo, W. *et al.* Quasi-solid-state dye-sensitized solar cells using room temperature molten salts and a low molecular weight gelator. *Chem. Commun.* 374–375 (2002).

32. Wang, P. *et al.* High efficiency dye-sensitized nanocrystalline solar cells based on ionic liquid polymer gel electrolyte. *Chem. Commun.* 2972–2973 (2002).

33. Wang, P. *et al.* Gelation of ionic liquid-based electrolytes with silica nanoparticles for quasi-solid-state dye-sensitized solar cells. *J. Am. Chem. Soc.* **125**, 1166–1167 (2003).

34. Tarascon, J.-M. & Armand, M. Issues and challenges facing rechargeable lithium batteries. *Nature* **414**, 359–367 (2001).

35. Zakeeruddin, S. M. *et al.* Design, synthesis, and application of amphiphilic ruthenium polypyridyl photosensitizers in solar cells based on nanocrystalline TiO₂ films. *Langmuir* **18**, 952–954 (2002).

36. Kataoka, H. *et al.* Interactive effect of the polymer on carrier migration nature in the chemically cross-linked polymer gel electrolyte composed of poly(ethylene glycol) dimethacrylate. *J. Phys. Chem. B* **106**, 12084–12087 (2002).

37. Gu, G. Y. *et al.* 2-Methoxyethyl (methyl) carbonate-based electrolytes for Li-ion batteries. *Electrochim. Acta* **45**, 3127–3139 (2000).

38. Wightman, R. M & Wipf, D. O. in *Electroanalytical Chemistry* Vol. 15 (ed. Bard, A. J.) 283 (Marcel Dekker, New York, 1989).

39. Pelet, S., Moser, J. E & Grätzel, M. Cooperative effect of adsorbed cations and iodide on the interception of back electron transfer in the dye sensitization of nanocrystalline TiO₂. *J. Phys. Chem. B* **104**, 1791–1795 (2000).

40. Haque, S. *et al.* Parameters influencing charge recombination kinetics in dye-sensitized nanocrystalline titanium dioxide films. *J. Phys. Chem. B* **104**, 538–547 (2000).

41. Moser, J. E. *et al.* Comment on "Measurement of ultrafast photoinduced electron transfer from chemically anchored Ru-dye molecules into empty electronic states in a colloidal anatase TiO₂ film". *J. Phys. Chem. B* **102**, 3649–3650 (1998).

42. Bonhôte, P *et al.* Hydrophobic, highly conductive ambient-temperature molten salts. *Inorg. Chem.* **35**, 1168–1178 (1996).

Acknowledgements

We are grateful to P. Péchy, I. Exnar and P. Liska for fruitful discussions, T. Koyanagi (CCIC, Japan) for providing the 400 nm TiO₂ particles and P. Comte for TiO₂ film fabrication. The Swiss Science Foundation, Swiss Federal Office for Energy (OFEN) and the European Office of US Air Force under Contract No. F61775–00–C0003 have supported this work.

Correspondence and requests for materials should be addressed to S.M.Z. or M.G.

Competing financial interests

The authors declare that they have no competing financial interests.

blends of the donor and acceptor materials[3–5]: phase separation during spin-coating leads to a bulk heterojunction that removes the exciton diffusion bottleneck by creating an interpenetrating network of the donor and acceptor materials. The realization of bulk heterojunctions using mixtures of vacuum-deposited small-molecular-weight materials has, on the other hand, posed elusive: phase separation induced by elevating the substrate temperature inevitably leads to a significant roughening of the film surface and to short-circuited devices. Here, we demonstrate that the use of a metal cap to confine the organic materials during annealing prevents the formation of a rough surface morphology while allowing for the formation of an interpenetrating donor–acceptor network. This method results in a power conversion efficiency 50 per cent higher than the best values reported for comparable bilayer devices, suggesting that this strained annealing process could allow for the formation of low-cost and high-efficiency thin film organic solar cells based on vacuum-deposited small-molecular-weight organic materials.

The external quantum efficiency of a photovoltaic cell based on exciton dissociation at a donor–acceptor interface is[6] $\eta_{EQE} = \eta_A \times \eta_{ED} \times \eta_{CC}$. Here, η_A is the absorption efficiency. The exciton diffusion efficiency, η_{ED}, is the fraction of photogenerated excitons that reaches a donor–acceptor interface before recombining. The carrier collection efficiency, η_{CC}, is the probability that a free carrier, generated at a donor–acceptor interface by dissociation of an exciton, reaches its corresponding electrode. Typically, in bilayer donor–acceptor photovoltaic cells with a total thickness, L, of the order of the optical absorption length, L_A, we have $\eta_A = 1 - \exp(-L/L_A) > 50\%$ if optical interference effects are ignored, and $\eta_{CC} \approx 100\%$. However, since the exciton diffusion length (L_D) is typically an order of magnitude smaller than L_A, a large fraction of the photogenerated excitons remains unused for photocurrent generation[6] (Fig. 1a), limiting η_{EQE} and hence the power conversion efficiency, η_P, of this type of planar junction cell.

In polymer photovoltaic cells, the exciton diffusion bottleneck has been removed through the introduction of bulk heterojunctions[4,5] (Fig. 1b). In a bulk heterojunction, the donor–acceptor interface is highly folded such that photogenerated excitons find an interface within a distance L_D of their generation site. Currently, state-of-the-art bulk heterojunction polymer photovoltaic cells have power conversion efficiencies of up to 3.5% (refs 7–9). The bulk heterojunction is fabricated by spin-coating a mixture of the donor and acceptor materials. During spin coating and solvent evaporation, the donor and acceptor materials phase separate, creating an intricate interpenetrating network.

Attempts to achieve a bulk heterojunction through codeposition of the donor and acceptor materials yielded devices with η_P values falling short of those achievable in optimized bilayer cells using the same materials[10–16]. Strong quenching of the photoluminescence in mixed materials indicates that $\eta_{ED} \approx 100\%$. Therefore, the low efficiencies are attributed to poor charge transport, resulting in a low η_{CC} (Fig. 1c). If charge collection is assisted by the application of an external voltage, a high η_{EQE} can be obtained[17].

Growth of mixed layers at elevated substrate temperatures leads to phase separation and the appearance of crystalline domains. However, this increase in crystallinity and possibly larger L_D comes at the cost of an increased film roughness[13,18]. The high density of pinholes leads to short circuits and makes device fabrication impractical. The same problem occurs when mixed-layer films are annealed post-deposition to induce phase separation.

Here, we present a method for the fabrication of bulk heterojunctions in small-molecule systems based on annealing mixed-layer films in a confined geometry. In this case, the devices are completed with a ~1,000-Å-thick metal cathode, and then subsequently annealed. The metal cathode stresses the organic film during annealing, preventing morphological relaxation and the concomitant formation of a high density of pinholes, while

Efficient bulk heterojunction photovoltaic cells using small-molecular-weight organic thin films

Peter Peumans, Soichi Uchida & Stephen R. Forrest

Center for Photonics and Optoelectronic Materials (POEM), Department of Electrical Engineering and the Princeton Materials Institute, Princeton University, Princeton, New Jersey 08544, USA

The power conversion efficiency of small-molecular-weight and polymer organic photovoltaic cells has increased steadily over the past decade. This progress is chiefly attributable to the introduction of the donor–acceptor heterojunction[1,2] that functions as a dissociation site for the strongly bound photogenerated excitons. Further progress was realized in polymer devices through use of permitting phase separation to occur in the bulk of the organic film leading to the desired highly folded bulk heterojunction.

Annealing of completed devices was previously used to improve the lifetime[19] and luminous efficiency[20,21] of organic light-emitting diodes, and to improve the efficiency of polymer photovoltaic cells[7].

However, these effects were not attributed to changes in the interface morphology but to polymer chain alignment, an increase in the charge carrier mobility or the formation of an interfacial layer at the organic/cathode interface.

Scanning electron microscope (SEM) images of the film surfaces

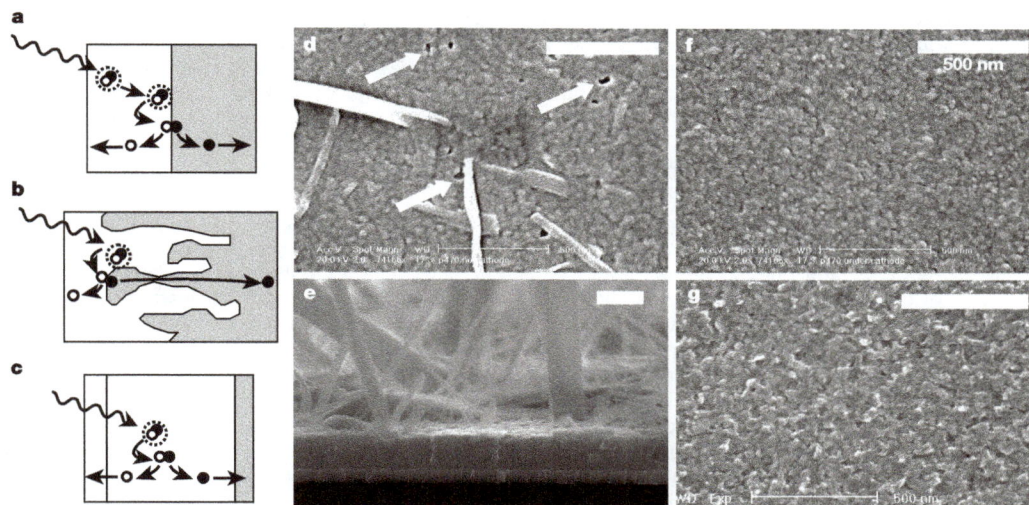

Figure 1 Diagrams of types of organic donor–acceptor photovoltaic cells and SEM images of the surface of a ~5,000-Å-thick CuPc:PTCBI film on ITO. **a–c**, Diagrams. **a**, Bilayer cell. **b**, Bulk heterojunction cell. **c**, Mixed-layer cell. In these diagrams, the incident light (wavy arrows), electrons (filled circles), holes (open circles) and excitons (paired circles) are indicated. **d–g**, SEM images. In **d** the film was annealed in the absence of a metal cap. White arrows indicate several pinholes. **e**, Cross-section of the same film obtained by cleaving the substrate. **f**, The film was capped by a 1,000-Å -thick film of Ag which was removed before imaging. For comparison, in **g** the organic surface of a non-annealed ITO/400 Å CuPc/400 Å PTCBI/1,000 Å Ag is shown after removal of the Ag cap. The features in this image correspond to crystalline domains of pure, planar-stacking PTCBI[18]. Scale bars, all 500 nm.

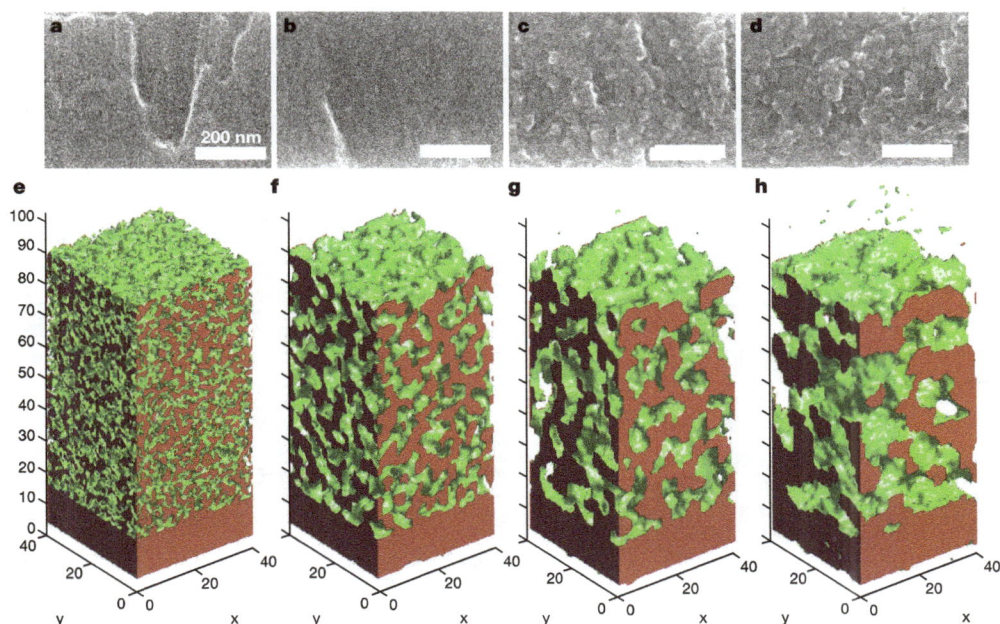

Figure 2 SEM images of cross-sections of a 5,000-Å-thick CuPc:PTCBI(4:1) film on ITO. **a**, Not annealed. **b–d**, Annealed for **b**, 15 min at 450 K, **c**, 500 K and **d**, 550 K. **e–h**, The simulated effects of annealing on the interface morphology of a mixed-layer photovoltaic cell. The initial configuration (**e**) is generated using a random number generator, and assumes a mixture composition of 1:1. This also assumes that no significant phase segregation occurs during deposition. The interface between CuPc and PTCBI is shown as a green surface. CuPc is shown in red and PTCBI is left 'transparent'. The as-grown, or initial, configuration is shown in **a**. The configurations after annealing at **f**, $T_{A1} = 0.067 E_{coh}/N_A k$, **g**, $T_{A1} = 0.13 E_{coh}/N_A k$ and **h**, $T_{A1} = 0.20 E_{coh}/N_A k$ are also shown. Note the resemblance between the structure in images **a–d** and the simulated structures **e–h**. The vertical axis represents film thickness, and hence is the direction of film growth.

in Fig. 1d, e show the effect of capping by a 1,000-Å-thick Ag film during annealing. The layer structure was indium tin oxide (ITO)/ 100 Å copper phthalocyanine (CuPc)/600 Å CuPc:PTCBI (3:4)/ 100 Å 3,4,9,10-perylene tetracarboxylic bis-benzimidazole (PTCBI). The concentration of CuPc to PTCBI in this case was 3:4, by weight, achieved through codeposition. The images show the organic surface morphology after annealing for 2 min at 560 K. In Fig. 1d, e, the film was uncapped by metal during the annealing process, resulting in a high density of pinholes ($\sim 8 \times 10^8 \, cm^{-2}$) and of large crystallites protruding from the film surface. In Fig. 1f, the organic layers were covered with a 1,000-Å-thick Ag cap during annealing that was subsequently peeled off using sticky tape. The resulting organic film is pinhole-free and lacks large ($\sim 1 \, \mu m$) crystalline domains, suggesting that the metal layer prevents morphological changes from occurring in the underlying film. For comparison, the surface morphology of a conventional non-annealed bilayer structure: ITO/400 Å CuPc/400 Å PTCBI/1,000 Å Ag after removing the Ag cap is shown in Fig. 1g.

While the presence of the Ag cap prevents the development of surface roughness, it does not prevent phase segregation within the bulk of the mixed layer itself. Evidence for phase segregation, leading to domains rich in CuPc or PTCBI, is provided in Fig. 2a–d. Here, SEM images of cross-sections of the layer structure: ITO/5,000 Å CuPc:PTCBI (4:1)/1,000 Å Ag are shown for an as-grown film (Fig. 2a), and for films annealed for 15 min at $T_{A1} = 450$ K (Fig. 2b), $T_{A1} = 500$ K (Fig. 2c), and $T_{A1} = 550$ K (Fig. 2d). The cross-section of the as-grown film (Fig. 2a) does not exhibit any morphological features other than artefacts of the cleaving process. Upon annealing, the cross-sections reveal domains whose size increases with increasing annealing temperature. At 550 K, domain sizes of ~ 20 nm are observed. This is corroborated by the X-ray diffraction data shown in Fig. 3. Upon annealing,

diffraction peaks corresponding to the β-CuPc phase emerge[22], and the broad amorphous background signal between $2\theta = 2.5°$ and $12.5°$ is reduced.

In Fig. 2e–h, we modelled the effect of the annealing temperature T_{A1} on the interface morphology of a mixed-layer device using the cellular automaton model discussed in the Methods. Annealing at $T_{A1} = 0.067 E_{coh}/N_A k$ (Fig. 2f), $T_{A1} = 0.13 E_{coh}/N_A k$ (Fig. 2g) and $T_{A1} = 0.20 E_{coh}/N_A k$ (Fig. 2h) has a dramatic influence on the morphology of the mixed-layer device, and the calculated morphology bears a remarkable resemblance to the observed cross-sections in Fig. 2a–d. Here E_{coh} is the cohesive energy of the crystalline organic solid per mole, N_A is Avogadro's number and k is Boltzmann's constant. Phase separation leads to the appearance of branches of pure material that grow increasingly thicker with increasing T_{A1}. The η_{ED} value is reduced in the thicker branches, but their presence improves η_{CC}.

By measuring the room-temperature η_{EQE} as a function of T_{A1}, the effect of this morphological change on exciton and charge transport can be inferred. In the inset of Fig. 4, the action spectrum of a mixed-layer device is shown as a function of T_{A1}. We observe a remarkable 30-fold increase in η_{EQE} at a wavelength of $\lambda = 690$ nm from 0.6% to 19%. The increase is uniform over the entire absorption spectrum of both CuPc and PTCBI, and cannot be identified with a single component. This confirms that the increase in η_{EQE} is associated with a change in morphology of the entire mixed layer.

In Fig. 4, η_{EQE} at $\lambda = 632$ nm as a function of T_{A1} is shown for a bilayer device (closed squares), and for mixed-layer devices (open symbols). Annealing a bilayer device does not significantly improve η_{EQE}, and annealing at $T_{A1} > 450$ K even results in its decrease. In contrast, for all mixed-layer devices, a dramatic increase in η_{EQE} is observed upon annealing at $T_{A1} > 450$ K, with an optimal annealing temperature of $T_{A1} = 540$ K. While the maximum attainable η_{EQE} clearly depends on the composition of the mixed layer, the η_{EQE} versus T_{A1} characteristics have a similar shape, independent of the mixture composition.

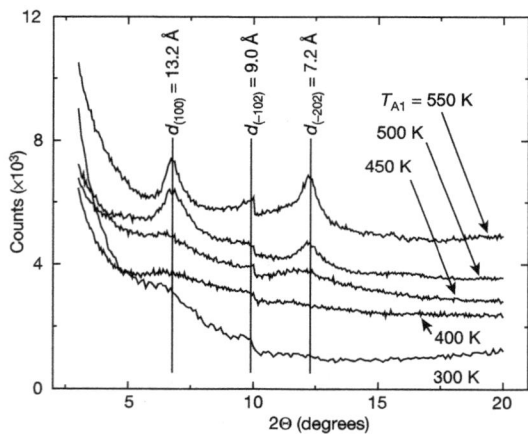

Figure 3 Bragg–Brentano X-ray diffractograms of a 5,000-Å-thick film on ITO using the Cu-Kα line. The film was covered with a 1,000-Å-thick cap of Ag and annealed at 300 K (not annealed), and $T_{A1} = 400$ K, 450 K, 500 K and 550 K. The Ag cap was removed before performing the scan. CuPc crystal indices are noted. The amorphous background is indicated by the broad curvature at low X-ray angles. The large width of the peaks suggests limited crystalline domain size. For the film annealed at 550 K, using the full-width at half-maximum of the peaks at angles $2\theta = 6.7°$ and $2\theta = 12.2°$, corresponding to interplanar spacings of $d = 13.2$ Å, 9.0 Å and 7.2 Å, we calculate a domain size of (12 ± 1) nm, which is consistent with our observations in Fig. 2. Nevertheless, this represents a lower limit to the domain size, as the diffraction peaks are also broadened by molecular disorder and large strains associated with the growth of domains within an amorphous matrix. Another possible contribution to the peak width is the residual 'doping' of the CuPc and PTCBI-rich phases with PTCBI and CuPc, respectively.

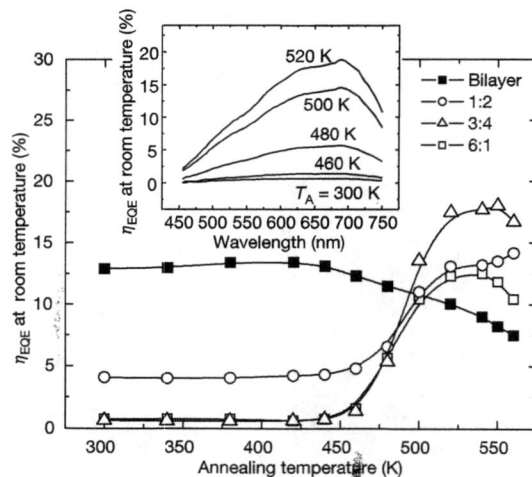

Figure 4 The room-temperature external quantum efficiency after annealing at various temperatures of bilayer and multi-layer devices. Closed squares, bilayer device with layer structure: ITO/400 Å CuPc/400 Å PTCBI/1,000 Å Ag. Open symbols, mixed-layer devices with layer structures: ITO/100 Å CuPc/600 Å CuPc:PTCBI (x: y)/100 Å PTCBI/1,000 Å Ag, where x: y is 1:2 (circles), 3:4 (triangles) and 6:1 (open squares). The cells were subsequently annealed for 2 min at 340 K and 380 K, then every 20 K between 420 K and 540 K, and 550 K and 560 K, each time returning to room temperature between annealing steps to measure η_{EQE}. Inset, Room-temperature η_{EQE} after annealing at various temperatures of a device with layer structure: ITO/100 Å CuPc/600 Å CuPc:PTCBI (3:4)/ 100 Å PTCBI/1,000 Å Ag. The cell was annealed and measured as in the main figure.

Table 1 **Effect of treatments on the properties of a mixed-layer solar cell**

	J_{SC} (μA cm^{-2})	V_{OC} (V)	FF	η_P (%)
As-grown bilayer	340	0.33	0.52	0.75 ± 0.05
As-grown mixed layer	15.5	0.26	0.25	$(1.3 \pm 0.1) \times 10^{-2}$
First anneal ($T_{A1} = 520$ K)	190*	0.10	0.27	$(6.5 \pm 0.4) \times 10^{-1}$
Contact replacement	250	0.30	0.26	0.25 ± 0.2
Second anneal ($T_{A2} = 500$ K)	880	0.44	0.31	1.5 ± 0.1

Shown are the room-temperature performance characteristics of a ITO/400 Å CuPc/400 Å PTCBI/1,000 Å Ag bilayer and ITO/100 Å CuPc/600 Å CuPc:PTCBI (6:1)/100 Å PTCBI/1,000 Å Ag mixed-layer solar cell. The illumination source is a tungsten-halogen lamp with a power density of 7.8 mW cm^{-2}. Here, J_{SC} is the short-circuit current density.
*This result for J_{SC} is in contrast to the results for η_{EQE} of a device with an identical layer structure (Fig. 4), where η_{EQE} of the annealed mixed-layer device approaches that of the as-grown bilayer device. This apparent contradiction is a consequence of the higher optical power levels used during measurements of the current–voltage characteristics as compared to the η_{EQE} measurements.

The temperature-dependent η_{EQE} data also allow us to estimate the maximum operating lifetime (t_{op}) of the devices employing these structurally metastable materials. For example, for $t_{op} = 10$ yr at a cell temperature T_{op}, we obtain $T_{op} = (k\ln(t_{op}/t_{A1})/\Delta E_A + T_{A1}^{-1})^{-1} \approx 53\,^\circ$C where t_{A1} is the annealing time ($\sim$1 min) at $T_{A1} = 540$ K required to effect the observed morphological changes, and $\Delta E_A = (1.1 \pm 0.1)$ eV is the activation energy characteristic of the annealing process estimated from the data in Fig. 4. Phase separation during normal operation of the cells is therefore not expected to contribute significantly to degradation of cell performance.

In Table 1 the room-temperature characteristics of a mixed-layer device are listed as a function of the annealing treatment. For reference, the performance parameters of a bilayer device are also shown. Prior to annealing, the short-circuit current density ($J_{SC}^M = 15.5\,\mu$A cm^{-2}) of the mixed-layer cell is more than an order of magnitude smaller than that of the bilayer ($J_{SC}^B = 340\,\mu$A cm^{-2}), leading to a low $\eta_P = (1.3 \pm 0.1) \times 10^{-2}$%. After annealing at $T_{A1} = 520$ K, $J_{SC}^M = 190\,\mu$A cm^{-2}. The drop in V_{OC} from 0.26 V to 0.10 V partially offsets the gains in J_{SC}, leading to $\eta_P = (6.5 \pm 0.4) \times 10^{-1}$%.

The drop in V_{OC} is attributed to an increased resistance arising from a reduction in disorder[23] at the organic/Ag interface that is due to the annealing process. Hence, improvements in performance are achieved by replacing the contact by peeling off the 'confining' Ag layer and replacing it by deposition of a 120 Å BCP/1,000 Å Ag contact. This contact replacement results in an increased $J_{SC}^M = 250\,\mu$A cm^{-2} and $V_{OC} = 0.30$ V (see Table 1). Annealing this device a second time at $T_{A2} = 500$ K further improves J_{SC}^M to 880 μA cm^{-2}. The open circuit voltage of $V_{OC}^M = 0.44$ V also exceeds that of the bilayer device ($V_{OC}^B = 0.33$ V). The η_P of the twice-annealed mixed-layer device with a replaced contact is $\eta_P = (1.5 \pm 0.1)$%, a twofold improvement over the bilayer with $\eta_P = (0.75 \pm 0.05)$%.

The contact replacement strategy was used to fabricate a solar cell with a high η_P value under standard AM1 illumination conditions at an intensity of 105 mW cm^{-2} (that is, $\sim$1 Sun). The device layer structure: ITO/150 Å CuPc/440 Å CuPc:PTCBI (1:1)/100 Å PTCBI/1,000 Å Ag was first annealed at $T_{A1} = 520$ K for 2 min. The contact was subsequently peeled off and replaced by deposition of a 150 Å BCP/1,000 Å Ag contact. The solar cell performance characteristics after the second anneal are shown in Fig. 5a as a

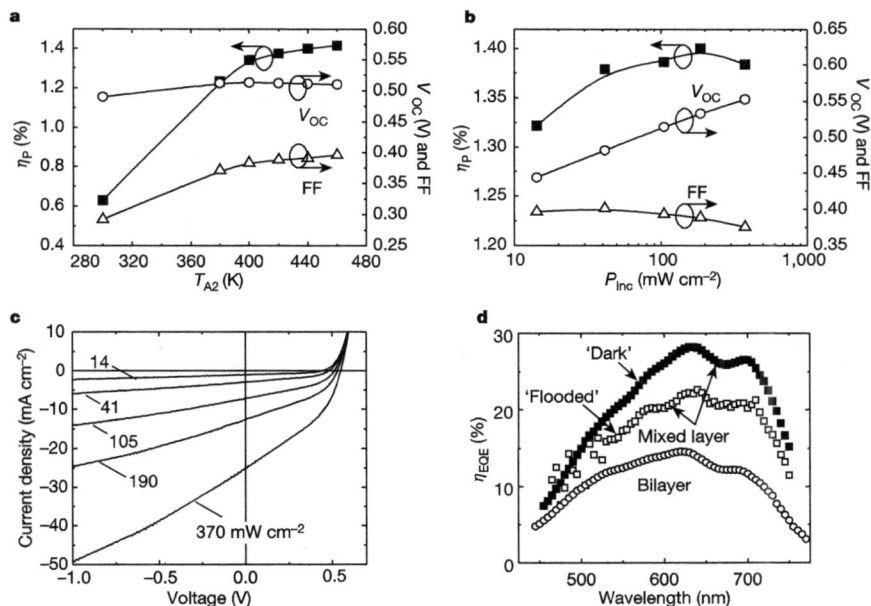

Figure 5 Parameters affecting the room-temperature power conversion efficiency. **a**, η_P, the open-circuit voltage, V_{OC}, and the fill factor, FF, as functions of the second annealing temperature, T_{A2}, for the layer structure: ITO/150 Å CuPc/440 Å CuPc:PTCBI (1:1)/100 Å PTCBI/150 Å BCP/1,000 Å Ag, where the BCP/Ag layers were deposited after the first anneal (at $T_{A1} = 520$ K). The second annealing process is essentially complete at $T_{A2} = 400$ K, so the mechanism leading to cell improvement is different from that of the first annealing step. We infer that the role of the second annealing process is to remove contaminants such as H$_2$O or O$_2$ from the donor–acceptor interfaces, which provide sites for exciton and/or charge recombination. Indeed, a similar increase in η_P is observed when a sample that was exposed to air after the first anneal was annealed a second time. Air exposure causes a rapid decrease in η_P, reducing it to less than 50% of the pre-exposure value. Here, the pre-exposure η_P is recovered after annealing to 400 K.

It is possible that some 'forming' of the donor–acceptor mixed layer/BCP contact also occurs during the second thermal treatment. **b**, Room-temperature η_P, V_{OC} and FF, as functions of the incident optical power intensity P_{inc}, after the second annealing process at $T_{A2} = 460$ K for the same layer structures as in **a**. The photocurrent has a linear dependence on the illumination intensity as shown in **c**, and the increase in V_{OC} with increased illumination intensity offsets the decrease in FF, resulting in η_P being nearly independent of the illumination intensity. **c**, Room-temperature current density–voltage characteristic of the device of **b** at various incident power levels. **d**, External quantum efficiency, η_{EQE}, of the mixed-layer device of **b**, measured with (open squares) and without (closed squares) flooding by 105 mW cm^{-2} AM1 illumination. For comparison, the η_{EQE} of an optimized ITO/200 Å CuPc/200 Å PTCBI/150 Å BCP/Ag bilayer structure is also shown (open circles).

function of T_{A2}. A maximum efficiency is reached for $T_{A2} = 460$ K, with $\eta_P = (1.42 \pm 0.07)\%$ representing the highest efficiency (by ~50%) achieved for CuPc/PTCBI photovoltaic 'Tang' cells over the last 16 yr (ref. 1).

The dependence of the performance characteristics of this device on the incident optical power is shown in Fig. 5b. The slightly lower η_P at 105 mW cm^{-2} as compared to Fig. 5a is due to minor device degradation. In Fig. 5c, the current–voltage characteristics are also shown as a function of intensity. At -1 V bias, the photocurrent density is approximately twice that obtained under short-circuit conditions. This strong dependence of photocurrent on applied bias suggests that carrier collection ultimately limits η_P. Optimizing this η_{CC} may lead to >twofold improvements in J_{SC}, and hence η_P.

In Fig. 5d, η_{EQE} is shown, measured with (open squares) and without (filled squares) flooding by 105 mW cm^{-2} AM1 white-light illumination. For comparison, η_{EQE} of an optimized bilayer device: ITO/200 Å CuPc/200 Å PTCBI/150 Å BCP/Ag (ref. 24) is also shown (open circles). The peak 'dark' $\eta_{EQE}^M = 28\%$ of the annealed mixed-layer device is twice that of the bilayer device $\eta_{EQE}^B = 14\%$. The decrease in η_{EQE} upon flooding with white light is a consequence of the increased carrier concentration under illumination, which increases the recombination probability, and hinders charge transport because of space-charge build-up within the complex folds of the bulk heterojunction structure.

The observed efficiency enhancement relies on the annealing of mixed-layer films in a confined geometry, that is, with a metal contact that prevents stress relief during morphological relaxation that typically occurs in molecular materials at elevated temperatures. Measurements on mixed-layer devices after annealing show dramatic increases in their external quantum efficiencies by a factor of up to ~30. We anticipate that further understanding of the nanoscale structure of confined materials, and the aspects of phase separation of binary mixtures of small-molecular-weight materials observed here, will facilitate the realization of high-performance organic photovoltaic and other optoelectronic devices. □

Methods

The photovoltaic cells were deposited on glass substrates pre-coated with a 1,500-Å-thick, transparent, conducting ITO anode (sheet resistance 40 Ω per square). The substrates were cleaned immediately before transferring them into the vacuum system for film deposition[25]. The organic materials were commercially obtained and purified before deposition using thermal gradient sublimation[18]. The photoactive materials used were CuPc and PTCBI, and bathocuproine (BCP) was used as a contact layer. The organic layers were grown by high vacuum thermal evaporation (base pressure 10^{-7}–10^{-6} torr) from a tungsten 'boat' onto a room-temperature substrate. This was followed by the deposition of the metal cathode through a shadow mask, resulting in contact diameters of 0.3 mm and 1 mm. No dependence of the device characteristics on device area was observed.

After fabrication, the cells were transferred to a vacuum chamber held at 30 mtorr with a heating stage, electrical probes and windows for optical access. The temperature ramp rate of the heating stage was fixed at 15 °C min^{-1}. Electrical characterization was performed during annealing using a semiconductor parameter analyser to obtain the current–voltage characteristics. For in situ photovoltaic power efficiency measurements, the devices were illuminated through the substrate with a 1,000-W solar simulator (Oriel Corp.) equipped with an AM1 filter. To measure η_{EQE}, a monochromatic beam of variable wavelength light chopped at 400 Hz (50% duty cycle) was focused onto a 1-mm diameter device. The photocurrent was measured using a lock-in amplifier referenced to the chopper frequency.

To gain a better understanding of the underlying physical process of the phase separation on the performance of mixed-layer photovoltaic cells, we implemented a model using cellular automata. This approach provides a numerically efficient and at the same time phenomenologically sound method of discretely simulating recrystallization and grain growth[26]. Briefly, a volume is discretized into a three-dimensional array in a simple cubic lattice containing $N_x \times N_y \times N_z = N$ cells. We define the z-direction as the growth direction (that is, perpendicular to the substrate plane). Periodic boundary conditions are applied in the x and y directions. The free energy of a configuration is:

$$E = \frac{1}{2}\sum_{i=1}^{N}\sum_{j=1}^{6} E_{M(i),M(j)} \tag{1}$$

where j sums over all nearest neighbours, $M(i)$ is the material at location i, and $E_{A,B}$ is the free energy associated with the molecular contact between molecules A and B. In this scheme, the cohesive energy per mole of material A is $E_{coh} = 3N_A E_{A,A}$, where N_A is Avogadro's number. E_{coh} is the evaporation enthalpy which can be obtained by thermogravimetry. In our simulations, only two materials CuPc

($E_{coh,CuPc} = 176$ kJ mole^{-1}; ref. 27) and PTCBI are used. Since $E_{coh,PTCBI}$ is unknown, and since most small molecular organic materials used in organic electronic devices have similar E_{coh} values[27], we assume for convenience that $E_{PTCBI} = E_{CuPc}$. Furthermore, we assume that $2E_{CuPc,PTCBI} = E_{CuPc,CuPc} = E_{PTCBI,PTCBI}$ (ref. 28).

The lattice is initialized to mimic the as-grown mixed structures. Subsequently, phase segregation is modelled using a single transformation rule: two neighbouring molecules can exchange positions. Assuming that R_0 is the rate at which molecular exchanges are attempted per cell, the rate of attempts able to overcome the energy barrier, ΔE_A, of exchanging two molecules is a function of temperature: $R(T) = R_0 \exp(-\Delta E_A/kT)$, where k is Boltzmann's constant and T is the absolute temperature. The activation energy associated with the switching of two molecules is prohibitively high because it would require the molecules to deform significantly. The actual process thus involves the presence of vacancies whose activation energy is that responsible for the generation of those vacancies.

Received 25 April; accepted 23 July 2003; doi:10.1038/nature01949.

1. Tang, C. W. A two-layer organic solar cell. *Appl. Phys. Lett.* **48**, 183–185 (1986).
2. Yu, G., Zhang, C. & Heeger, A. J. Dual-function semiconducting polymer devices: light-emitting and photodetecting diodes. *Appl. Phys. Lett.* **64**, 1540–1542 (1994).
3. Halls, J. J. M. *et al.* Efficient photodiodes from interpenetrating networks. *Nature* **376**, 498–500 (1995).
4. Granström, M. *et al.* Laminated fabrication of polymeric photovoltaic diodes. *Nature* **395**, 257 (1998).
5. Yu, G., Gao, J., Hummelen, J., Wudl, F. & Heeger, A. J. Polymer photovoltaic cells—enhanced efficiencies via a network of internal donor–acceptor heterojunctions. *Science* **270**, 1789–1791 (1995).
6. Peumans, P., Yakimov, A. & Forrest, S. R. Small molecular weight organic thin-film photodetectors and solar cells. *J. Appl. Phys.* **93**, 3693–3723 (2003).
7. Padinger, F., Rittberger, R. S. & Sariciftci, N. Effects of postproduction treatment on plastic solar cells. *Adv. Funct. Mater.* **13**, 85–88 (2003).
8. Shaheen, S. E. *et al.* 2.5% efficient organic plastic solar cells. *Appl. Phys. Lett.* **78**, 841–843 (2001).
9. Aernouts, T. *et al.* Extraction of bulk and contact components of the series resistance in organic bulk donor-acceptor-heterojunctions. *Thin Solid Films* **403–404**, 297–303 (2002).
10. Feng, W., Fujii, A., Lee, S., Wu, H. & Yoshino, K. Broad spectral sensitization of organic photovoltaic heterojunction device by perylene and C_{60}. *J. Appl. Phys.* **88**, 7120–7123 (2000).
11. Dittmer, J. J. *et al.* Crystal network formation in organic solar cells. *Sol. Energy Mater. Sol. Cells* **61**, 53–61 (2000).
12. Tsuzuki, T., Shirota, Y., Rostalski, J. & Meissner, D. The effect of fullerene doping on photoelectric conversion using titanyl phthalocyanine and a perylene pigment. *Sol. Energy Mater. Sol. Cells* **61**, 1–8 (2000).
13. Geens, W., Aernouts, T., Poortmans, J. & Hadziioannou, G. Organic co-evaporated films of a PPV-pentamer and C-60: model systems for donor/acceptor polymer blends. *Thin Solid Films* **403–404**, 438–443 (2002).
14. Hiramoto, M., Suemori, K. & Yokoyama, M. Photovoltaic properties of ultramicrostructure-controlled organic co-deposited films. *Jpn. J. Appl. Phys.* **41**, 2763–2766 (2002).
15. Halls, J. J. M. & Friend, R. H. The photovoltaic effect in a poly(p-phenylenevinylene)/perylene heterojunction. *Synth. Metals* **85**, 1307–1308 (1997).
16. Peumans, P. & Forrest, S. R. Very-high-efficiency double-heterostructure copper phthalacyanine/C_{60} photovoltaic cells. *Appl. Phys. Lett.* **79**, 126–128 (2001).
17. Peumans, P., Bulovic, V. & Forrest, S. R. Efficient, high-bandwidth organic multilayer photodetectors. *Appl. Phys. Lett.* **76**, 3855–3857 (2000).
18. Forrest, S. R. Ultrathin organic films grown by organic molecular beam deposition and related techniques. *Chem. Rev.* **97**, 1793–1896 (1997).
19. Kim, J., Lee, J., Han, C., Lee, N. & Chung, I.-J. Effect of thermal annealing on the lifetime of polymer light-emitting diodes. *Appl. Phys. Lett.* **82**, 4238–4240 (2003).
20. Niu, Y., Hou, Q., Yuan, M. & Cao, Y. Effects of thermal annealing on light-emitting devices based on fluorene-copolymers with thiophene and ethylenedioxythienylene. *Synth. Met.* **135–136**, 477–478 (2003).
21. Lee, T.-W. & Park, O. O. Effect of electrical annealing on the luminous efficiency of thermally annealed polymer light-emitting diodes. *Appl. Phys. Lett.* **77**, 3334–3336 (2000).
22. Prabakaran, R., Kesavamoorthy, R., Reddy, G. & Xavier, F. Structural investigation of copper phthalocyanine thin films using X-ray diffraction, Rama scattering and optical absorption measurements. *Phys. Stat. Sol.* **229**, 1175–1186 (2002).
23. Baldo, M. A. & Forrest, S. R. Interface-limited injection in amorphous organic semiconductors. *Phys. Rev. B* **64**, 085201 (2001).
24. Peumans, P., Bulovic, V. & Forrest, S. R. Efficient photon harvesting at high optical intensities in ultrathin double-heterostructure photovoltaic diodes. *Appl. Phys. Lett.* **76**, 2650–2652 (2000).
25. Burrows, P. E. *et al.* Relationship between electroluminescence and current transport in organic heterojunction light-emitting devices. *J. Appl. Phys.* **79**, 7991–8006 (1996).
26. Raabe, D. "Cellular automata in materials science with particular reference to recrystallization simulation". *Annu. Rev. Mater. Res.* **32**, 36–76 (2002).
27. Shtein, M., Gossenberger, H. F., Benziger, J. B. & Forrest, S. R. Material transport regimes and mechanisms for growth of molecular organic thin films using low-pressure organic vapor phase deposition. *J. Appl. Phys.* **89**, 1470–1476 (2001).
28. Israelachvili, J. N. *Intermolecular and Surface Forces* 176–212 (Academic, London, 1992).

Acknowledgements We thank N. Stroustrup for his assistance, and gratefully acknowledge the financial support of the Air Force Office of Scientific Research, the National Renewable Energy Laboratory, and the Global Photonic Energy Corporation. S.U. is currently on leave from Nippon Oil Corporation, Central Technical Research Laboratory, 8 Chidori-cho, Naka-ku, Yokohama 231-0815, Japan.

Competing interests statement The authors declare that they have no competing financial interests.

Correspondence and requests for materials should be addressed to S.R.F. (forrest@princeton.edu).

THERMOELECTRIC CONVERTERS

Complex thermoelectric materials

Thermoelectric materials, which can generate electricity from waste heat or be used as solid-state Peltier coolers, could play an important role in a global sustainable energy solution. Such a development is contingent on identifying materials with higher thermoelectric efficiency than available at present, which is a challenge owing to the conflicting combination of material traits that are required. Nevertheless, because of modern synthesis and characterization techniques, particularly for nanoscale materials, a new era of complex thermoelectric materials is approaching. We review recent advances in the field, highlighting the strategies used to improve the thermopower and reduce the thermal conductivity.

G. JEFFREY SNYDER* AND ERIC S. TOBERER

Materials Science, California Institute of Technology, 1200 East California Boulevard, Pasadena, California 91125, USA

***e-mail: jsnyder@caltech.edu**

The world's demand for energy is causing a dramatic escalation of social and political unrest. Likewise, the environmental impact of global climate change due to the combustion of fossil fuels is becoming increasingly alarming. One way to improve the sustainability of our electricity base is through the scavenging of waste heat with thermoelectric generators (Box 1). Home heating, automotive exhaust, and industrial processes all generate an enormous amount of unused waste heat that could be converted to electricity by using thermoelectrics. As thermoelectric generators are solid-state devices with no moving parts, they are silent, reliable and scalable, making them ideal for small, distributed power generation[1]. Efforts are already underway to replace the alternator in cars with a thermoelectric generator mounted on the exhaust stream, thereby improving fuel efficiency[2]. Advances in thermoelectrics could similarly enable the replacement of compression-based refrigeration with solid-state Peltier coolers[3].

Thermoelectrics have long been too inefficient to be cost-effective in most applications[4]. However, a resurgence of interest in thermoelectrics began in the mid 1990s when theoretical predictions suggested that thermoelectric efficiency could be greatly enhanced through nanostructural engineering, which led to experimental efforts to demonstrate the proof-of-principle and high-efficiency materials[5,6]. At the same time, complex bulk materials (such as skutterudites[7], clathrates[8] and Zintl phases[9]) have been explored and found that high efficiencies could indeed be obtained. Here, we review these recent advances, looking at how disorder and complexity within the unit cell as well as nanostructured materials can lead to enhanced efficiency. This survey allows us to find common traits in these materials, and distill rational design strategies for the discovery of materials with high thermoelectric efficiency. More comprehensive reviews on thermoelectric materials are well covered in several books[1,10–12] and articles[3,5,6,8,13–18].

CONFLICTING THERMOELECTRIC MATERIAL PROPERTIES

Fundamental to the field of thermoelectric materials is the need to optimize a variety of conflicting properties. To maximize the thermoelectric figure of merit (zT) of a material, a large thermopower (absolute value of the Seebeck coefficient), high electrical conductivity, and low thermal conductivity are required. As these transport characteristics depend on interrelated material properties, a number of parameters need to be optimized to maximize zT.

CARRIER CONCENTRATION

To ensure that the Seebeck coefficient is large, there should only be a single type of carrier. Mixed n-type and p-type conduction will lead to both charge carriers moving to the cold end, cancelling out the induced Seebeck voltages. Low carrier concentration insulators and even semiconductors have large Seebeck coefficients; see equation (1). However, low carrier concentration also results in low electrical conductivity; see equation (2). The interrelationship between carrier concentration and Seebeck coefficient can be seen from relatively simple models of electron transport. For metals or degenerate semiconductors (parabolic band, energy-independent scattering approximation[19]) the Seebeck coefficient is given by:

$$\alpha = \frac{8\pi^2 k_B^2}{3eh^2} m^* T \left(\frac{\pi}{3n}\right)^{2/3},$$ (1)

where n is the carrier concentration and m^* is the effective mass of the carrier.

The electrical conductivity (σ) and electrical resistivity (ρ) are related to n through the carrier mobility μ:

$$1/\rho = \sigma = ne\mu.$$ (2)

Figure 1a shows the compromise between large thermopower and high electrical conductivity in thermoelectric materials that must be struck to maximize the figure of merit zT ($\alpha^2\sigma T/\kappa$), where κ is the thermal conductivity. This peak typically occurs at carrier concentrations between 10^{19} and 10^{21} carriers per cm^3 (depending on the material system), which falls in between common metals and semiconductors — that is, concentrations found in heavily doped semiconductors.

EFFECTIVE MASS

The effective mass of the charge carrier provides another conflict as large effective masses produce high thermopower but low electrical conductivity. The m^* in equation (1) refers to the density-of-states effective mass, which increases with flat, narrow bands with high density of states at the Fermi surface. However, as the inertial effective mass is also related to m^*, heavy carriers will move with slower velocities, and therefore small mobilities, which in turn leads to low electrical conductivity (equation (2)). The exact relationship between effective mass and mobility is complex, and depends on electronic structure, scattering mechanisms and anisotropy. In principle, these effective mass terms can be decoupled in anisotropic crystal structures[20].

A balance must be found for the effective mass (or bandwidth) for the dominant charge carrier, forming a compromise between high effective mass and high mobility. High mobility and low effective mass is typically found in materials made from elements with small electronegativity differences, whereas high effective masses and low mobilities are found in materials with narrow bands such as ionic compounds. It is not obvious which effective mass is optimum; good thermoelectric materials can be found within a wide range of effective masses and mobilities: from low-mobility, high-effective-mass polaron conductors (oxides[14], chalcogenides[21]) to high-mobility, low-effective-mass semiconductors (SiGe, GaAs).

ELECTRONIC THERMAL CONDUCTIVITY

Additional materials design conflicts stem from the necessity for low thermal conductivity. Thermal conductivity in thermoelectrics comes from two sources: (1) electrons and holes transporting heat (κ_e) and (2) phonons travelling through the lattice (κ_l). Most of the electronic term (κ_e) is directly related to the electrical conductivity through the Wiedemann–Franz law:

$$\kappa = \kappa_e + \kappa_l \tag{3a}$$

and

$$\kappa_e = L\sigma T = ne\mu LT, \tag{3b}$$

where L is the Lorenz factor, 2.4×10^{-8} J^2 K^{-2} C^{-2} for free electrons.

Box 1 Thermoelectric devices

The thermoelectric effects arise because charge carriers in metals and semiconductors are free to move much like gas molecules, while carrying charge as well as heat. When a temperature gradient is applied to a material, the mobile charge carriers at the hot end tend to diffuse to the cold end. The build-up of charge carriers results in a net charge (negative for electrons, e$^-$, positive for holes, h$^+$) at the cold end, producing an electrostatic potential (voltage). An equilibrium is thus reached between the chemical potential for diffusion and the electrostatic repulsion due to the build-up of charge. This property, known as the Seebeck effect, is the basis of thermoelectric power generation.

Thermoelectric devices contain many thermoelectric couples (Fig. B1, bottom) consisting of n-type (containing free electrons) and p-type (containing free holes) thermoelectric elements wired electrically in series and thermally in parallel (Fig. B1, top). A thermoelectric generator uses heat flow across a temperature gradient to power an electric load through the external circuit. The temperature difference provides the voltage ($V = \alpha\Delta T$) from the Seebeck effect (Seebeck coefficient α) while the heat flow drives the electrical current, which therefore determines the power output. In a Peltier cooler the external circuit is a d.c. power supply, which drives the electric current (I) and heat flow (Q), thereby cooling the top surface due to the Peltier effect ($Q = \alpha TI$). In both devices the heat rejected must be removed through a heat sink.

The maximum efficiency of a thermoelectric material for both power generation and cooling is determined by its figure of merit (zT):

$$zT = \frac{\alpha^2 T}{\rho\kappa}$$

zT depends on α, absolute temperature (T), electrical resistivity (ρ), and thermal conductivity (κ). The best thermoelectrics are semiconductors that are so heavily doped their transport properties resemble metals.

For the past 40 years, thermoelectric generators have reliably provided power in remote terrestrial and extraterrestrial locations most notably on deep space probes such as *Voyager*. Solid-state Peltier coolers provide precise thermal management for optoelectronics and passenger seat cooling in automobiles. In the future, thermoelectric systems could harness waste heat and/or provide efficient electricity through co-generation. One key advantage of thermoelectrics is their scalability — waste heat and co-generation sources can be as small as a home water heater or as large as industrial or geothermal sources.

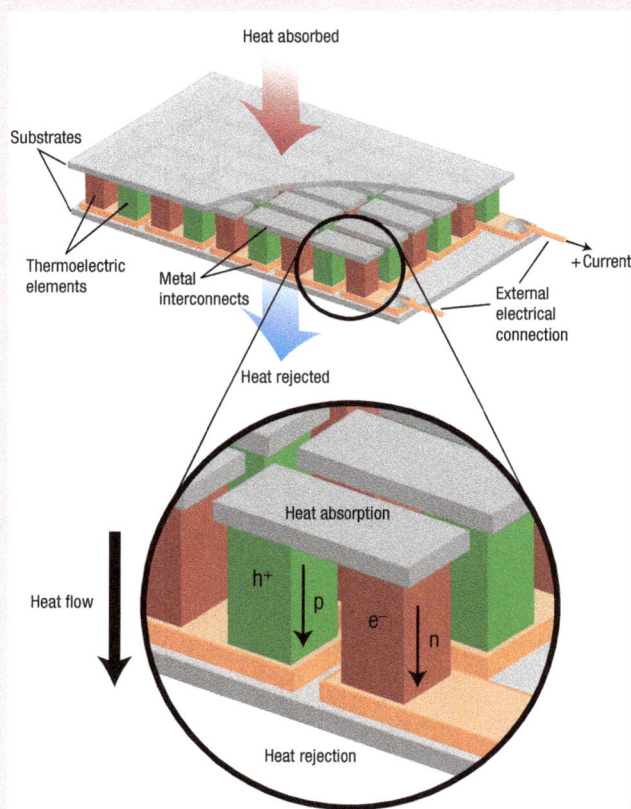

Figure B1 Thermoelectric module showing the direction of charge flow on both cooling and power generation.

The Lorenz factor can vary particularly with carrier concentration. Accurate assessment of κ_e is important, as κ_l is often computed as the difference between κ and κ_e (equation (3)) using the experimental electrical conductivity. A common source of uncertainty in κ_e occurs in low-carrier-concentration materials where the Lorenz factor can be reduced by as much as 20% from the free-electron value. Additional uncertainty in κ_e arises from mixed conduction, which introduces a bipolar term into the thermal conductivity[10]. As this term is not included in the Wiedemann–Franz law, the standard computation of κ_l erroneously includes bipolar thermal conduction. This results in a perceived increase in κ_l at high temperatures for Bi_2Te_3, PbTe and others, as shown in Fig. 2a. The onset of bipolar thermal conduction occurs at nearly the same temperature as the peak in Seebeck and electrical resistivity, which are likewise due to bipolar effects.

As high zT requires high electrical conductivity but low thermal conductivity, the Wiedemann–Franz law reveals an inherent materials conflict for achieving high thermoelectric efficiency. For materials with very high electrical conductivity (metals) or very low κ_l, the Seebeck coefficient alone primarily determines zT, as can be seen in equation (4), where $(\kappa_l/\kappa_e) \ll 1$:

$$zT = \frac{\alpha^2/L}{1 + \dfrac{\kappa_l}{\kappa_e}} \quad . \tag{4}$$

LATTICE THERMAL CONDUCTIVITY

Glasses exhibit some of the lowest lattice thermal conductivities. In a glass, thermal conductivity is viewed as a random walk of energy through a lattice rather than rapid transport via phonons, and leads to the concept of a minimum thermal conductivity[22], κ_{min}. Actual glasses, however, make poor thermoelectrics because they lack the needed 'electron-crystal' properties — compared with crystalline semiconductors they have lower mobility due to increased electron scattering and lower effective masses because of broader bands. Good thermoelectrics are therefore crystalline materials that manage to scatter phonons without significantly disrupting the electrical conductivity. The heat flow is carried by a spectrum of phonons with widely varying wavelengths and mean free paths[23] (from less than 1 nm to greater than 10 μm), creating a need for phonon scattering agents at a variety of length scales.

Thermoelectrics therefore require a rather unusual material: a 'phonon-glass electron-crystal'[24]. The electron-crystal requirement stems from the fact that crystalline semiconductors have been the best at meeting the compromises required from the electronic properties (Seebeck coefficient and electrical conductivity). The phonon-glass requirement stems from the need for as low a lattice thermal conductivity as possible. Traditional thermoelectric materials have used site substitution (alloying) with isoelectronic elements to preserve a crystalline electronic structure while creating large mass contrast to disrupt the phonon path. Much of the recent excitement in the field of thermoelectrics is a result of the successful demonstration of other methods to achieve phonon-glass electron-crystal materials.

ADVANCES IN THERMOELECTRIC MATERIALS

Renewed interest in thermoelectrics is motivated by the realization that complexity at multiple length scales can lead to new mechanisms for high zT in materials. In the mid 1990s, theoretical predictions suggested that the thermoelectric efficiency could be greatly enhanced by quantum confinement of the electron charge carriers[5,25]. The electron energy bands in a quantum-confined structure are progressively narrower as the confinement increases and the dimensionality decreases. These narrow bands should produce high effective masses and therefore large Seebeck coefficients. In addition, similar sized, engineered heterostructures

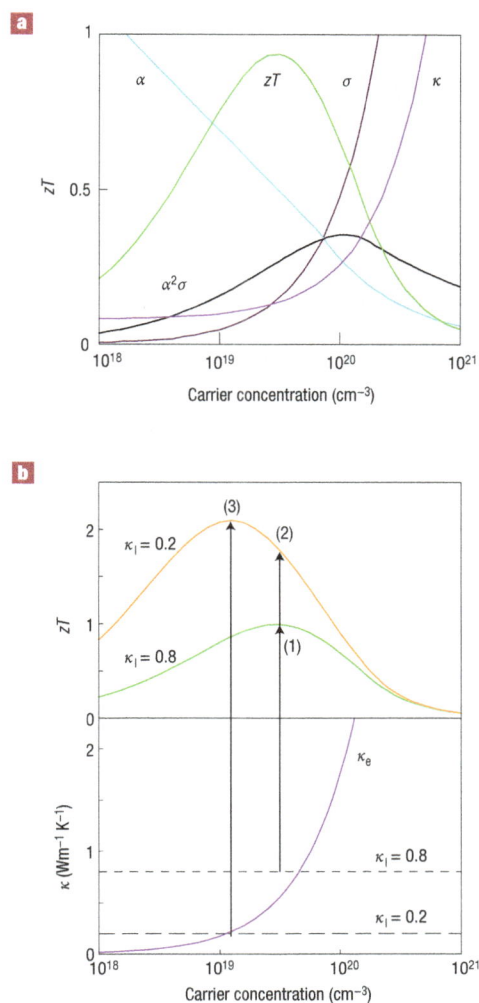

Figure 1 Optimizing zT through carrier concentration tuning. **a**, Maximizing the efficiency (zT) of a thermoelectric involves a compromise of thermal conductivity (κ; plotted on the y axis from 0 to a top value of 10 W m^{-1} K^{-1}) and Seebeck coefficient (α; 0 to 500 μV K^{-1}) with electrical conductivity (σ; 0 to 5,000 Ω^{-1}cm^{-1}). Good thermoelectric materials are typically heavily doped semiconductors with a carrier concentration between 10^{19} and 10^{21} carriers per cm^3. The thermoelectric power factor $\alpha^2\sigma$ maximizes at higher carrier concentration than zT. The difference between the peak in $\alpha^2\sigma$ and zT is greater for the newer lower-κ_l materials. Trends shown were modelled from Bi_2Te_3, based on empirical data in ref. 78. **b**, Reducing the lattice thermal conductivity leads to a two-fold benefit for the thermoelectric figure of merit. An optimized zT of 0.8 is shown at point (1) for a model system (Bi_2Te_3) with a κ_l of 0.8 Wm^{-1} K^{-1} and κ_e that is a function of the carrier concentration (purple). Reducing κ_l to 0.2 Wm^{-1} K^{-1} directly increases the zT to point (2). Additionally, lowering the thermal conductivity allows the carrier concentration to be reoptimized (reduced), leading to both a decrease in κ_e and a larger Seebeck coefficient. The reoptimized zT is shown at point (3).

may decouple the Seebeck coefficient and electrical conductivity due to electron filtering[26] that could result in high zT. Even though a high-ZT device based on these principles has yet to be demonstrated, these predictions have stimulated a new wave of interest in complex thermoelectric materials. Vital to this rebirth has been interdisciplinary collaborations: research in thermoelectrics

Box 2 State-of-the-art high-zT materials

To best assess the recent progress and prospects in thermoelectric materials, the decades of research and development of the established state-of-the-art materials should also be considered. By far the most widely used thermoelectric materials are alloys of Bi_2Te_3 and Sb_2Te_3. For near-room-temperature applications, such as refrigeration and waste heat recovery up to 200 °C, Bi_2Te_3 alloys have been proved to possess the greatest figure of merit for both n- and p-type thermoelectric systems. Bi_2Te_3 was first investigated as a material of great thermoelectric promise in the 1950s[12,16–18,84]. It was quickly realized that alloying with Sb_2Te_3 and Bi_2Se_3 allowed for the fine tuning of the carrier concentration alongside a reduction in lattice thermal conductivity. The most commonly studied p-type compositions are near $(Sb_{0.8}Bi_{0.2})_2Te_3$ whereas n-type compositions are close to $Bi_2(Te_{0.8}Se_{0.2})_3$. The electronic transport properties and detailed defect chemistry (which controls the dopant concentration) of these alloys are now well understood thanks to extensive studies of single crystal and polycrystalline material[85,86]. Peak zT values for these materials are typically in the range of 0.8 to 1.1 with p-type materials achieving the highest values (Fig. B2a,b). By adjusting the carrier concentration zT can be optimized to peak at different temperatures, enabling the tuning of the materials for specific applications such as cooling or power generation[87]. This effect is demonstrated in Fig. B2c for PbTe.

For mid-temperature power generation (500–900 K), materials based on group-IV tellurides are typically used, such as PbTe, GeTe or SnTe[12,17,18,81,88]. The peak zT in optimized n-type material is about 0.8. Again, a tuning of the carrier concentration will alter the temperature where zT peaks. Alloys, particularly with $AgSbTe_2$, have led to several reports of $zT > 1$ for both n-type and p-type materials[73,89,90]. Only the p-type alloy $(GeTe)_{0.85}(AgSbTe_2)_{0.15}$, commonly referred to as TAGS, with a maximum zT greater than 1.2 (ref. 69), has been successfully used in long-life thermoelectric generators. With the advent of modern microstructural and chemical analysis techniques, such materials are being reinvestigated with great promise (see section on nanomaterials).

Successful, high-temperature (>900 K) thermoelectric generators have typically used silicon–germanium alloys for both n- and p-type legs. The zT of these materials is fairly low, particularly for the p-type material (Fig. B2b) because of the relatively high lattice thermal conductivity of the diamond structure.

For cooling below room temperature, alloys of BiSb have been used in the n-type legs, coupled with p-type legs of $(Bi,Sb)_2(Te,Se)_3$ (refs 91,92). The poor mechanical properties of BiSb leave much room for improved low-temperature materials.

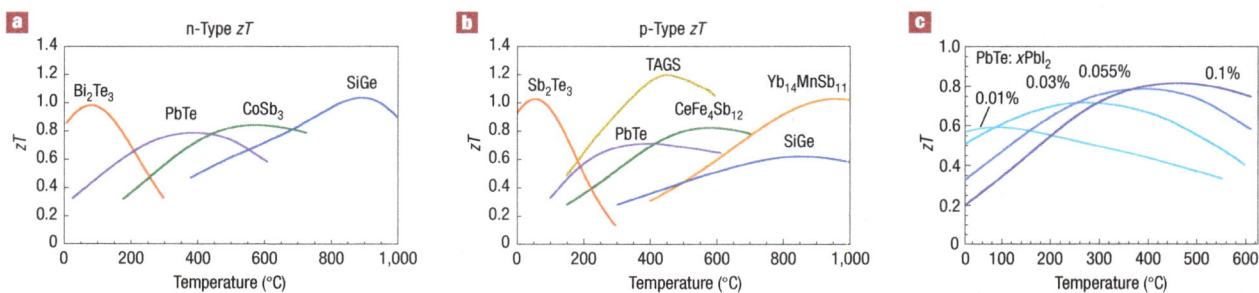

Figure B2 Figure-of-merit zT of state-of-the-art commercial materials and those used or being developed by NASA for thermoelectric power generation. **a**, p-type and **b**, n-type. Most of these materials are complex alloys with dopants; approximate compositions are shown. **c**, Altering the dopant concentration changes not only the peak zT but also the temperature where the peak occurs. As the dopant concentration in n-type PbTe increases (darker blue lines indicate higher doping) the zT peak increases in temperature. Commercial alloys of Bi_2Te_3 and Sb_2Te_3 from Marlow Industries, unpublished data; doped PbTe, ref. 88; skutterudite alloys of $CoSb_3$ and $CeFe_4Sb_{12}$ from JPL, Caltech unpublished data; TAGS, ref. 69; SiGe (doped $Si_{0.8}Ge_{0.2}$), ref. 82; and $Yb_{14}MnSb_{11}$, ref. 45.

requires an understanding of solid-state chemistry, high-temperature electronic and thermal transport measurements, and the underlying solid-state physics. These collaborations have led to a more complete understanding of the origin of good thermoelectric properties.

There are unifying characteristics in recently identified high-zT materials that can provide guidance in the successful search for new materials. One common feature of the thermoelectrics recently discovered with $zT>1$ is that most have lattice thermal conductivities that are lower than the present commercial materials. Thus the general achievement is that we are getting closer to a 'phonon glass' while maintaining the 'electron crystal'. These reduced lattice thermal conductivities are achieved through phonon scattering across various length scales as discussed above. A reduced lattice thermal conductivity directly improves the thermoelectric efficiency, zT, (equation (4)) and additionally allows re-optimization of the carrier concentration for additional zT improvement (Fig. 1b).

There are three general strategies to reduce lattice thermal conductivity that have been successfully used. The first is to scatter phonons within the unit cell by creating rattling structures or point defects such as interstitials, vacancies or by alloying[27]. The second strategy is to use complex crystal structures to separate the electron-crystal from the phonon-glass. Here the goal is to be able to achieve a phonon glass without disrupting the crystallinity of the electron-transport region. A third strategy is to scatter phonons at interfaces, leading to the use of multiphase composites mixed on the nanometre scale[5]. These nanostructured materials can be formed as thin-film superlattices or as intimately mixed composite structures.

COMPLEXITY THROUGH DISORDER IN THE UNIT CELL

There is a long history of using atomic disorder to reduce the lattice thermal conductivity in thermoelectrics (Box 2). Early work by

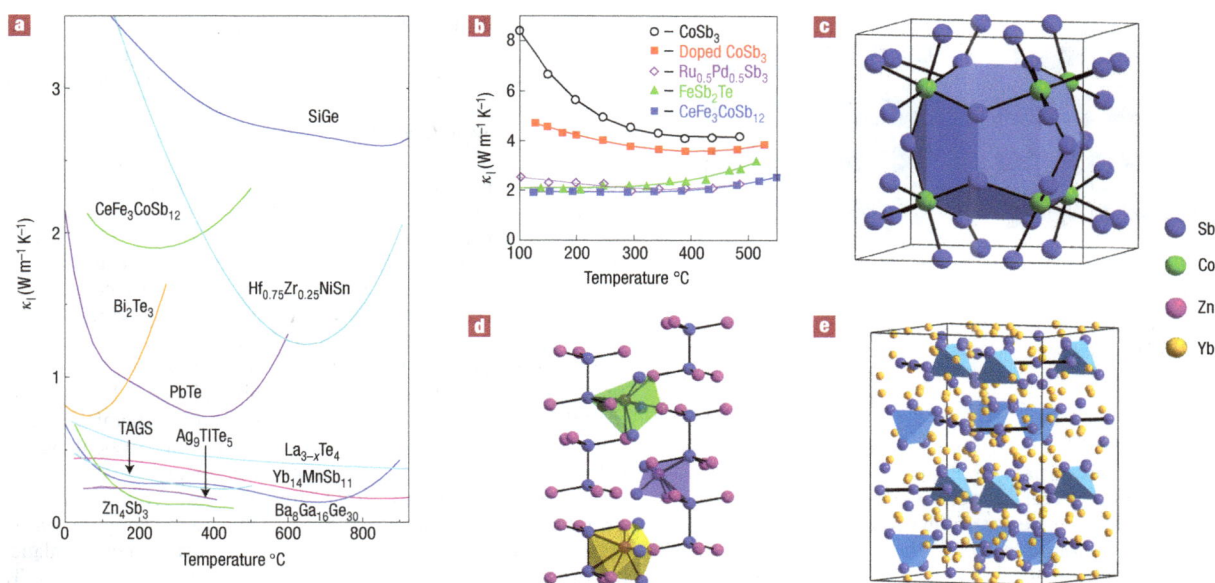

Figure 2 Complex crystal structures that yield low lattice thermal conductivity. **a**, Extremely low thermal conductivities are found in the recently identified complex material systems (such as $Yb_{14}MnSb_{11}$, ref. 45; $CeFe_3CoSb_{12}$, ref. 34; $Ba_8Ga_{16}Ge_{30}$, ref. 79; and Zn_4Sb_3, ref. 80; Ag_9TlTe_5, ref. 40; and $La_{3-x}Te_4$, Caltech unpublished data) compared with most state-of-the-art thermoelectric alloys (Bi_2Te_3, Caltech unpublished data; PbTe, ref. 81; TAGS, ref. 69; SiGe, ref. 82 or the half-Heusler alloy $Hf_{0.75}Zr_{0.25}NiSn$, ref. 83). **b**, The high thermal conductivity of $CoSb_3$ is lowered when the electrical conductivity is optimized by doping (doped $CoSb_3$). The thermal conductivity is further lowered by alloying on the Co ($Ru_{0.5}Pd_{0.5}Sb_3$) or Sb ($FeSb_2Te$) sites or by filling the void spaces ($CeFe_3CoSb_{12}$) (ref. 34). **c**, The skutterudite structure is composed of tilted octahedra of $CoSb_3$ creating large void spaces shown in blue. **d**, The room-temperature structure of Zn_4Sb_3 has a crystalline Sb sublattice (blue) and highly disordered Zn sublattice containing a variety of interstitial sites (in polyhedra) along with the primary sites (purple). **e**, The complexity of the $Yb_{14}MnSb_{11}$ unit cell is illustrated, with $[Sb_3]^{7-}$ trimers, $[MnSb_4]^{9-}$ tetrahedra, and isolated Sb anions. The Zintl formalism describes these units as covalently bound with electrons donated from the ionic Yb^{2+} sublattice (yellow).

Wright discusses how alloying Bi_2Te_3 with other isoelectronic cations and anions does not reduce the electrical conductivity but lowers the thermal conductivity[28]. Alloying the binary tellurides (Bi_2Te_3, Sb_2Te_3, PbTe and GeTe) continues to be an active area of research[29–32]. Many of the recent high-zT thermoelectric materials similarly achieve a reduced lattice thermal conductivity through disorder within the unit cell. This disorder is achieved through interstitial sites, partial occupancies, or rattling atoms in addition to the disorder inherent in the alloying used in the state-of-the-art materials. For example, rare-earth chalcogenides[18] with the Th_3P_4 structure (for example $La_{3-x}Te_4$) have a relatively low lattice thermal conductivity (Fig. 2a) presumably due to the large number of random vacancies (x in $La_{3-x}Te_4$). As phonon scattering by alloying depends on the mass ratio of the alloy constituents, it can be expected that random vacancies are ideal scattering sites.

The potential to reduce thermal conductivity through disorder within the unit cell is particularly large in structures containing void spaces. One class of such materials are clathrates[8], which contain large cages that are filled with rattling atoms. Likewise, skutterudites[7] such as $CoSb_3$, contain corner-sharing $CoSb_6$ octahedra, which can be viewed as a distorted variant of the ReO_3 structure. These tilted octahedra create void spaces that may be filled with rattling atoms, as shown in Fig. 2c with a blue polyhedron[33].

For skutterudites containing elements with low electronegativity differences such as $CoSb_3$ and $IrSb_3$, there is a high degree of covalent bonding, enabling high carrier mobilities and therefore good electron-crystal properties. However, this strong bonding and simple order leads to high lattice thermal conductivities. Thus, the challenge with skutterudites has been the reduction of the lattice thermal conductivity. Doping $CoSb_3$ to carrier concentrations that optimize

zT adds enough carriers to substantially reduce thermal conductivity[34] through electron–phonon interactions (Fig. 2b). Further reductions can be obtained by alloying either on the transition metal or the antimony site.

Filling the large void spaces with rare-earth or other heavy atoms further reduces the lattice thermal conductivity[35]. A clear correlation has been found with the size and vibrational motion of the filling atom and the thermal conductivity leading to zT values as high as 1 (refs 8,13). Partial filling establishes a random alloy mixture of filling atoms and vacancies enabling effective point-defect scattering as discussed previously. In addition, the large space for the filling atom in skutterudites and clathrates can establish soft phonon modes and local or 'rattling' modes that lower lattice thermal conductivity.

Filling these voids with ions adds additional electrons that require compensating cations elsewhere in the structure for charge balance, creating an additional source of lattice disorder. For the case of $CoSb_3$, Fe^{2+} frequently is used to substitute Co^{3+}. An additional benefit of this partial filling is that the free-carrier concentration may be tuned by moving the composition slightly off the charge-balanced composition. Similar charge-balance arguments apply to the clathrates, where filling requires replacing group 14 (Si, Ge) with group 13 (Al, Ga) atoms.

COMPLEX UNIT CELLS

Low thermal conductivity is generally associated with crystals containing large, complex unit cells. The half-Heusler alloys[8] have a simple, cubic structure with high lattice thermal conductivity ($Hf_{0.75}Zr_{0.25}NiSb$ in Fig. 2a) that limits the zT. Thus complex crystal structures are good places to look for improved materials. A good

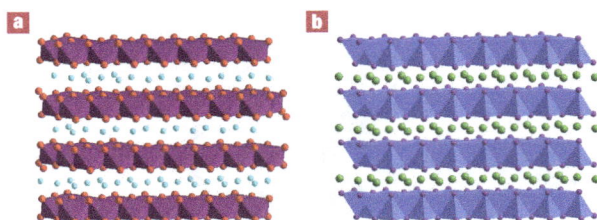

Figure 3 Substructure approach used to separate the electron-crystal and phonon-glass attributes of a thermoelectric. **a**, Na_xCoO_2 and **b**, $Ca_xYb_{1-x}Zn_2Sb_2$ structures both contain ordered layers (polyhedra) separated by disordered cation monolayers, creating electron-crystal phonon-glass structures.

example of a complex variant of Bi_2Te_3 is $CsBi_4Te_6$, which has a somewhat lower lattice thermal conductivity than Bi_2Te_3 that has been ascribed to the added complexity of the Cs layers and the few Bi–Bi bonds in $CsBi_4Te_6$ not found in Bi_2Te_3. These Bi–Bi bonds lower the bandgap compared with Bi_2Te_3, dropping the maximum zT of $CsBi_4Te_6$ below room temperature with a maximum zT of 0.8 (refs 8,36). Like Bi_2Te_3, the layering in $CsBi_4Te_6$ leads to an anisotropic effective mass that can improve the Seebeck coefficient with only minor detriment to the mobility[8]. Many ordered MTe/Bi_2Te_3-type variants (M = Ge, Sn or Pb)[37,38] are known, making up a large homologous series of compounds[39], but to date $zT < 0.6$ is found in most reports. As many of these materials have low lattice thermal conductivities but have not yet been doped to appropriate carrier concentrations, much remains to be done with complex tellurides.

Low lattice thermal conductivities are also seen in the thallium-based thermoelectric materials such as Ag_9TlTe_5 (ref. 40) and Tl_9BiTe_6 (ref. 41). Although these materials do have complex unit cells, there is clearly something unique about the thallium chemistry that leads to low thermal conductivity (0.23 W m^{-1} K^{-1} at room temperature[40]). One possible explanation is extremely soft thallium bonding, which can also be observed in the low elastic modulus these materials exhibit.

The remarkable high zT in Zn_4Sb_3 arises from the exceptionally low, glass-like thermal conductivity (Fig. 2a). In the room-temperature phase, about 20% of the Zn atoms are on three crystallographically distinct interstitial sites as shown in Fig. 2d. These interstitials are accompanied by significant local lattice distortions[42] and are highly dynamic, with Zn diffusion rates almost as high as that of superionic conductors[43]. Pair distribution function (PDF) analysis[44] of X-ray and neutron diffraction data shows that there is local ordering of the Zn interstitials into nanoscale domains. Thus, the low thermal conductivity of Zn_4Sb_3 arises from disorder at multiple length scales, from high levels of interstitials and corresponding local structural distortions and from domains of interstitial ordering. Within the unit cell, Zn interstitials create a phonon glass, whereas the more ordered Sb framework provides the electron-crystal component.

One common characteristic of nearly all good thermoelectric materials is valence balance — charge balance of the chemical valences of all atoms. Whether the bonding is ionic or covalent, valence balance enables the separation of electron energy bands needed to form a bandgap. Complex Zintl compounds have recently emerged as a new class of thermoelectrics[9] because they can form quite complex crystal structures. A Zintl compound contains a valence-balanced combination of both ionically and covalently bonded atoms. The mostly ionic cations donate electrons to the covalently bound anionic species. The covalent bonding allows higher mobility of the charge-carrier species than that found in purely ionic materials. The combination of the bonding types leads to complex structures with the possibility of multiple structural units in the same structure. One example is $Yb_{14}MnSb_{11}$ (refs 45,46), which contains $[MnSb_4]^{9-}$ tetrahedra, polyatomic $[Sb_3]^{7-}$ anions, as well as isolated Sb^{3-} anions and Yb^{2+} cations (Fig. 2e). This structural complexity, despite the crystalline order, enables extremely low lattice thermal conductivity (0.4 W m^{-1} K^{-1} at room temperature; Fig. 2a). Combined with large Seebeck coefficient and high electrical conductivity, $Yb_{14}MnSb_{11}$ results in a zT of ~1.0 at 900 °C. This zT is nearly twice that of p-type SiGe used in NASA spacecraft and has led to rapid acceptance of $Yb_{14}MnSb_{11}$ into NASA programmes for development of future thermoelectric generators. The complexity of Zintl structures also makes them ideal materials for using a substructure approach.

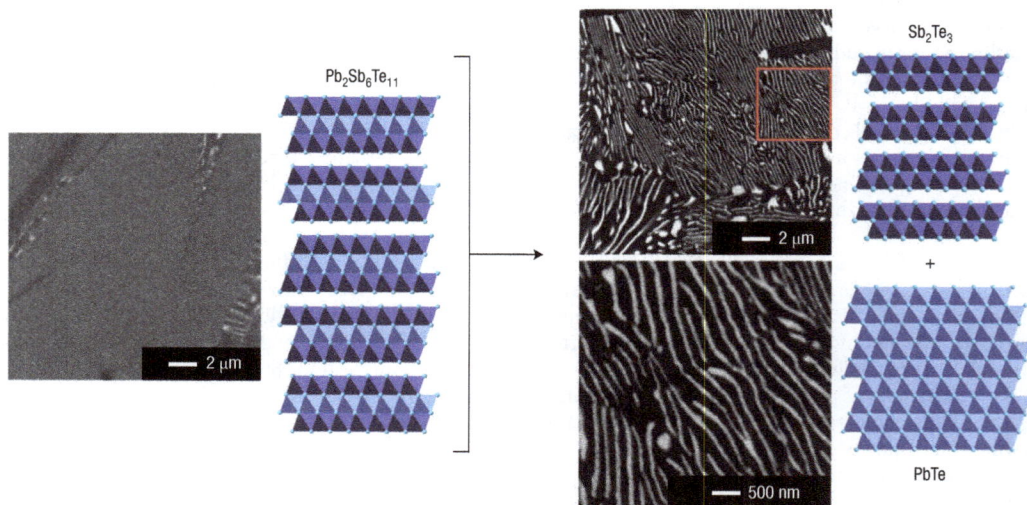

Figure 4 Nanostructured thermoelectrics may be formed by the solid-state partitioning of a precursor phase. The metastable $Pb_2Sb_6Te_{11}$ phase (left) will spontaneously assemble into lamellae of Sb_2Te_3 and PbTe (ref. 76; right). These domains are visible with backscattering scanning electron microscopy, with the dark regions corresponding to Sb_2Te_3 and the light regions to PbTe. Electron backscattering diffraction reveals that the lamellae are oriented with coherent interfaces, shown schematically (right).

Many materials have been reported with $zT > 1.5$ but few have been confirmed by others, and no devices have been assembled that show the efficiency expected from such high-zT materials. This is due to the complexity of fabricating devices, measurement uncertainty and materials complications.

The inherent difficulty in thermoelectrics is that direct efficiency measurements require nearly as much complexity as building an entire device. Thus practical assessment of the thermoelectric figure of merit typically relies on measuring the individual contributing material properties (σ, α and κ). Measurements of the thermoelectric properties are conceptually simple but results can vary considerably, particularly above room temperature where thermal gradients in the measurement system add to systematic inaccuracies. As a typical zT measurement above room temperature requires the measurement of σ, α and κ (from the density, heat capacity, C_p, and thermal diffusivity, D) each with uncertainty of 5% to 20%, the uncertainty in zT from

$$\frac{\Delta z}{z} = 2\frac{\Delta\alpha}{\alpha} + \frac{\Delta\sigma}{\sigma} + \frac{\Delta C_p}{C_p} + \frac{\Delta D}{D}$$

could easily reach 50%. Accuracy is particularly important for the Seebeck coefficient because it is squared in the calculation of zT and there are few standards with which to calibrate systems. In addition, a variety of geometric terms are required in these calculations (density, thickness and coefficient of thermal expansion[79]).

The sensitivity of the materials themselves to impurities and dopant concentrations further complicates measurements. This is because of the strong dependence of conductivity and to a lesser extent Seebeck coefficient on carrier concentration (Fig. 1a). Small inhomogeneities can result in large variations in thermoelectric properties within a sample[71], making repeatability and combining results of different measurements difficult. For example, combining Seebeck and resistivity on one sample or set of contacts and thermal conductivity on another could lead to spurious results. Likewise, reliable property values are particularly difficult to obtain when sublimation, microstructural evolution, electrochemical reactions and phase transitions are present. Thus, even the act of measuring a sample at high temperatures can alter its properties

The hallmarks of trustworthy measurements are slow, physical trends in properties. Typical materials (Box 2) show a linear or concave downward trend in Seebeck coefficient with temperature and only slow variation with chemical doping. Abrupt transitions to high-zT materials (as a function of temperature or composition) are unlikely. Thin-film samples are particularly difficult to measure. Electrical conductivity often depends critically on the perceived thickness of the conducting layer — if the substrate or quantum-well walls become conducting at any time, it can lead to an erroneously high electrical conductivity estimate of the film. Thermal conductivity and the Seebeck effect likewise depend significantly on the assumption that the substrate and insulating superlattice layers do not change with processing, atmosphere, or temperature. One should be encouraged by results of $zT > 1$ but remain wary of the uncertainties involved to avoid pathological optimism.

SUBSTRUCTURE APPROACH

One method for circumventing the inherent materials conflict of a phonon-glass with electron-crystal properties is to imagine a complex material with distinct regions providing different functions. Such a substructure approach would be analogous to the enabling features that led to high-T_c superconductivity in copper oxides. In these materials, the free charge carriers are confined to planar Cu–O sheets that are separated by insulating oxide layers. Precise tuning of the carrier concentration is essential for superconductivity. This is enabled by the insulating layers acting as a 'charge reservoir'[47] that houses dopant atoms that donate charge carriers to the Cu–O sheets. The separation of the doping regions from the conduction regions keeps the charge carriers sufficiently screened from the dopant atoms so as not to trap carriers, which would lead to a low mobility, hopping conduction mechanism rather than superconductivity.

Likewise, the ideal thermoelectric material would have regions of the structure composed of a high-mobility semiconductor that provides the electron-crystal electronic structure, interwoven with a phonon-glass. The phonon-glass region would be ideal for housing dopants and disordered structures without disrupting the carrier mobility in the electron-crystal region, much like the charge-reservoir region in high-T_c superconductors. The electron crystal regions will need to be thin, on the nanometre or ångström scale, so that phonons with a short mean free path are scattered by the phonon-glass region. Such thin, low-dimensional electron-transport regions might also be able to take advantage of quantum confinement and/or electron filtering to enhance the Seebeck coefficient. Skutterudites and clathrates represent a 0-dimensional version of the substructure approach, with isolated rattlers in an electron-crystal matrix.

The thermoelectric cobaltite oxides (Na_xCoO_2 and others such as those based on the Ca–Co–O system) may likewise be described using a substructure approach[14,48,49]. The Co–O layers form metallic layers separated by insulating, disordered layers with partial occupancies (Fig. 3a). Oxides typically have low mobilities and high lattice thermal conductivity, due to the high electronegativity of oxygen and the strong bonding of light atoms, respectively. These properties give oxides a distinct disadvantage for thermoelectric materials. The relatively large Seebeck values obtained in these systems has been attributed to spin-induced entropy[50]. The success of the cobaltite structures as thermoelectric materials may be an early example of how the substructure approach overcomes these disadvantages. The study of oxide thermoelectrics benefits greatly from the variety of structures and synthetic techniques known for oxides, as well as our understanding of oxide structure–property relationships.

Zintl compounds may be more appropriate for substructure-based thermoelectrics due to the nature of their bonding. The covalently bound anion substructures can adopt a variety of topologies — from 0-dimensional isolated single ions, dimers and polyatomic anions to extended one-, two- and three-dimensional chains, planes and nets[9]. This covalently bound substructure enables high carrier mobilities whereas the ionic cation substructure is amenable to doping and site disorder without disrupting the covalent network. The valence-balanced bonding in these materials frequently leads to bandgaps that are of a suitable size for thermoelectric applications.

The substructure approach is clearly seen in the Zintl compound $Ca_xYb_{1-x}Zn_2Sb_2$ (ref. 51), whose structure is similar to Na_xCoO_2, with sheets of disordered cations between layers of covalently bound Zn–Sb (Fig. 3b). $Ca_xYb_{1-x}Zn_2Sb_2$ demonstrates the fine-tuning ability in the ionic layer concomitant with a modest reduction in lattice thermal conductivity due to alloying of Yb and Ca. The Ca^{2+} is slightly more electropositive than Yb^{2+}, which enables a gradual changing of the carrier concentration as the Yb:Ca ratio is changed. This doping produces disorder on the cation substructure

Box 4 Thermoelectric efficiency

The efficiency of a thermoelectric device depends on factors other than the maximum zT of a material. This is primarily due to the temperature dependence of all the materials properties (α, σ, κ) that make up $zT(T)$. For example, even for state-of-the-art Bi_2Te_3, which has a peak zT value of 1.1, the effective device ZT is only about 0.7 based on the overall performance of the device as a cooler or power generator.

Here we use ZT (upper case) to distinguish the device figure of merit from lower-case $zT = \alpha^2\sigma/\kappa$, the material's figure of merit[10]. For a Peltier cooler the device ZT is most easily measured from the maximum temperature drop obtained (ΔT_{max}).

$$\Delta T_{max} = \frac{ZT_c^2}{2} \ .$$

For a generator, the maximum efficiency (η) is used to determine ZT:

$$\eta = \frac{\Delta T}{T_h} \cdot \frac{\sqrt{1 + ZT} - 1}{\sqrt{1 + ZT} + \frac{T_c}{T_h}} \ .$$

Like all heat engines, the maximum power-generation efficiency of a thermoelectric generator is thermodynamically limited by the Carnot efficiency ($\Delta T/T_h$). If temperature is assumed to be independent and n-type and p-type thermoelectric properties are matched (α, σ and κ), (an unrealistic approximation in many cases)

the maximum device efficiency is given by the above equation with $Z = z$.

To maximize efficiency across a large temperature drop, it is imperative to maximize the device ZT, and not just a peak materials zT. One method to achieve this is to tune the material to provide a large average $zT(T)$ in the temperature range of interest. For example, the peak zT in PbTe may be tuned from 300 °C to 600 °C (see Fig. B2c). For large temperature differences (needed to achieve high Carnot efficiency) segmenting with different materials that have peak zT at different temperatures (see Fig. B2a,b) will improve device ZT (ref. 93). Functionally graded materials can also be used to continuously tune zT instead of discrete segmenting[94].

For such large ΔT applications the device ZT can be significantly smaller than even the average zT due to thermoelectric incompatibility. Across a large ΔT, the electrical current required for highest-efficiency operation changes as the materials properties change with temperature or segment[95]. This imposes an additional materials requirement: the thermoelectric compatibility factors ($s = (\sqrt{(1 + zT)} \pm 1)/\alpha T$) with – for power generation[96] or + for cooling[97]) must be similar. For high efficiency, this term needs to be within about a factor of two across the different temperature ranges[95]. A compelling example of the need for compatibility matching is segmenting TAGS with SiGe. The compatibility is so poor between these materials that replacing SiGe with $Yb_{14}MnSb_{11}$ quadruples the device efficiency increase for adding the additional high-temperature segment[45].

but not the conducting anion substructure such that the bandgap and carrier mobility is unchanged. The disorder on the cation substructure does indeed lower the lattice thermal conductivity producing a modest increase in zT. However, the relatively simple structure of $Ca_xYb_{1-x}Zn_2Sb_2$ leads to relatively high lattice thermal conductivities ($\sim$1.5 W m^{-1} K^{-1}) — suggesting further methods to reduce the conductivity, such as nanostructuring, would lead to an improved material.

Given the broad range of phonons involved in heat transport, a substructure approach may only be one component in a high-performance thermoelectric material. Long-wavelength phonons require disorder on longer length-scales, leading to a need for hierarchical complexity. Combining a substructure approach with nanostructuring seems to be the most promising method of achieving a high-Seebeck, high-conductivity material that manages to scatter phonons at all length scales.

COMPLEX NANOSTRUCTURED MATERIALS

Much of the recent interest in thermoelectrics stems from theoretical and experimental evidence of greatly enhanced zT in nanostructured thin-films and wires due to enhanced Seebeck and reduced thermal conductivity[5]. Reduced thermal conductivity in thin-film superlattices was investigated in the 1980s[52], but has only recently been applied to enhanced thermoelectric materials. Recent efforts[53-55] on Bi_2Te_3-Sb_2Te_3 and PbTe–PbSe films and Si nanowires[56,57] have shown how phonon scattering can reduce lattice thermal conductivity to near κ_{min} values[22,58] (0.2–0.5 W m^{-1} K^{-1}). Thin films containing randomly embedded quantum dots likewise achieve exceptionally low lattice thermal conductivities[59,60]. Very high zT values (>2) have been reported in thin films but the difficulty of measurements makes them a challenge to reproduce in independent labs (Box 3). It is clear however that nanostructured thin-films and wires do exhibit lattice

thermal conductivities near (or even below[61]) κ_{min}, which results in higher material zT, but improvements of electrical and thermal contacts to these materials in a device are needed before higher device ZT (Box 4) is achieved.

The use of bulk (mm^3) nanostructured materials would avoid detrimental electrical and thermal losses and use the existing fabrication routes. The challenge for any nanostructured bulk material system is electron scattering at interfaces between randomly oriented grains leading to a concurrent reduction of both the electrical and thermal conductivities[27]. The effect of grain-boundary scattering in a silicon–germanium system has been extensively studied, as the system possesses excellent electron-crystal properties but very high thermal conductivities. In 1981, the synthesis of polycrystalline silicon germanium alloys were described, and the decrease in thermal conductivity with smaller grain size was tracked[62]. Compared with single crystals of SiGe alloys, polycrystalline materials with grains on the order of 1 μm show an enhanced zT. However, later experiments on materials with grains between 1–100 μm found that the increased phonon scattering was offset by the decrease in electrical conductivity[63]. Nevertheless, recent work suggests that truly nanostructured SiGe enhances zT (ref. 5).

The results on epitaxial thin-films suggests that the ideal nanostructured material would have thermodynamically stable, coherent, epitaxy-like, interfaces between the constituent phases to prevent grain-boundary scattering of electrons. Thus, a promising route to nanostructured bulk thermoelectric materials relies on the spontaneous partitioning of a precursor phase into thermodynamically stable phases[64]. The growth and characterization of such composite microstructures have been studied in metals for decades because of their ability to greatly improve mechanical strength. The use of microstructure to reduce thermal conductivity in thermoelectrics, dates to the 1960s[65]. For example, crystals pulled from an InSb–Sb eutectic alloy, which forms rods of Sb as thin as 4 μm in an InSb matrix, shows a clear decrease in thermal conductivity with smaller rod diameters, in both parallel and perpendicular directions[66]. Additionally, no major decrease in electrical conductivity or Seebeck coefficient was observed. In a similar manner, several eutectics form from two thermoelectrics form such as rock salt–tetradymite (for example, $PbTe-Sb_2Te_3$) revealing a variety of layered and dendritic microstructures[64,67]. A fundamental limitation of such an approach is that rapid diffusion in the liquid phase leads to coarse microstructures[67,68].

Microstructural complexity may explain why $(AgSbTe_2)_{0.15}(GeTe)_{0.85}$ (TAGS) and $(AgSbTe_2)_x(PbTe)_{1-x}$ (LAST), first studied in the 1950s, have remained some of the materials having highest known zT. Originally believed to be a true solid solution with the rock-salt structure, recent interest has focused on the nanoscale microstructure and even phase separation that exists in these and related alloy compositions. From early on it was predicted that lattice strain in TAGS could explain the low lattice thermal conductivities of 0.3 W m⁻¹ K⁻¹ (ref. 69). Recent work points to the presence of twin-boundary defects in TAGS as an additional source of phonon scattering[70]. Inhomogeneities on various length scales[69] have been found in LAST alloys, which may be associated with the reports of high zT; however, this also makes reproducibility a challenge. In LAST, Ag–Sb rich nanoparticles 1–10 nm in size as well as larger micrometre-sized features precipitate from the bulk[71–73]. The nanoparticles are oriented within the rock-salt crystal with coherent interfaces, therefore electronic conductivity is not significantly reduced. Conversely, the large density difference between the different regions leads to interfacial scattering of the phonons, reducing the thermal conductivity. Through these mechanisms, thermal conductivities of the order of 0.5 W m⁻¹ K⁻¹ at 700 K have been observed. A variety of other materials have been formed that have oriented nanoparticle inclusions in a PbTe matrix[31,32,74].

As the thermoelectric properties of nanostructured materials should depend on the size and morphology of the microstructural features, the materials science of microstructural engineering should become increasingly important in the development of thermoelectric nanomaterials. Our group has focused on the partitioning of quenched, metastable phases that then transform into two phases during a controlled process[75,76]. By restricting the partitioning to solid-state diffusion at low temperatures, the resulting microstructures are quite fine. Figure 4 shows a sample of $Pb_2Sb_6Te_{11}$ after quenching and the microstructure that results on annealing at 400 °C. A lamellar spacing of 360 nm is observed, corresponding to 80-nm PbTe and 280-nm Sb_2Te_3 layer thicknesses. The lamellar spacing can be controlled from below 200 nm to several micrometres. One appealing aspect of this lamellar growth is that the low lattice mismatch between the PbTe (111) and Sb_2Te_3 (001) planes leads to coherent interfaces between the lamellae. Controlled partitioning of a precursor solid can also be done from a glass, as in the case of the alloy glass $(GeSe_2)_{70}(Sb_2Te_3)_{20}(GeTe)_{10}$, which devitrifies to fine lamellae of $GeSe_2$ and $GeSb_4Te_7$ (ref. 77). The numerous possibilities for controlling such reactions will introduce complexity at multiple length scales to thermoelectric materials engineering.

CONCLUDING REMARKS

The conflicting material properties required to produce a high-efficiency (phonon-glass electron-crystal) thermoelectric material have challenged investigators over the past 50 years. Recently, the field has undergone a renaissance with the discovery of complex high-efficiency materials that manage to decouple these properties. A diverse array of new approaches, from complexity within the unit cell to nanostructured bulk and thin-film materials, have all led to high-efficiency materials. Given the complexity of these systems, all of these approaches benefit from collaborations between chemists,

physicists and materials scientists. The global need for sustainable energy coupled with the recent advances in thermoelectrics inspires a growing excitement in this field.

doi:10.1038/nmat2090

References

1. Rowe, D. M. (ed.) *CRC Handbook of Thermoelectrics* (CRC, Boca Raton, 1995).
2. Matsubara, K. in *International Conference on Thermoelectrics* 418–423 (2002).
3. DiSalvo, F. J. Thermoelectric cooling and power generation. *Science* **285**, 703–706 (1999).
4. Rowe, D. M. (ed.) *CRC Handbook of Thermoelectrics: Macro to Nano* (CRC, Boca Raton, 2005).
5. Dresselhaus, M. S. *et al.* New directions for low-dimensional thermoelectric materials. *Adv. Mater.* **19**, 1043–1053 (2007).
6. Chen, G., Dresselhaus, M. S., Dresselhaus, G., Fleurial, J. P. & Caillat, T. Recent developments in thermoelectric materials. *Int. Mater. Rev.* **48**, 45–66 (2003).
7. Uher, C. in *Thermoelectric Materials Research I* (ed. Tritt, T.) 139–253 (Semiconductors and Semimetals Series 69, Elsevier, 2001).
8. Nolas, G. S., Poon, J. & Kanatzidis, M. Recent developments in bulk thermoelectric materials. *Mater. Res. Soc. Bull.* **31**, 199–205 (2006).
9. Kauzlarich, S. M., Brown, S. R. & Snyder, G. J. Zintl phases for thermoelectric devices. *Dalton Trans.* 2099–2107 (2007).
10. Goldsmid, H. J. *Applications of Thermoelectricity* (Methuen, London, 1960).
11. Tritt, T. M. (ed.) *Recent Trends in Thermoelectric Materials Research* (Academic, San Diego, 2001).
12. Heikes, R. R. & Ure, R. W. *Thermoelectricity: Science and Engineering* (Interscience, New York, 1961).
13. Sales, B. C. Electron crystals and phonon glasses: a new path to improved thermoelectric materials. *Mater. Res. Soc. Bull.* **23**, 15–21 (1998).
14. Koumoto, K., Terasaki, I. & Funahashi, R. Complex oxide materials for potential thermoelectric applications. *Mater. Res. Soc. Bull.* **31**, 206–210 (2006).
15. Mahan, G. D. in *Solid State Physics* Vol. 51 (eds Ehrenreich, H. & Spaepen, F.) 81–157 (Elsevier, 1998).
16. Rosi, F. D. Thermoelectricity and thermoelectric power generation. *Solid-State Electron.* **11**, 833–848 (1968).
17. Rosi, F. D., Hockings, E. F. & Lindenblad, N. E. Semiconducting materials for thermoelectric power generation. *RCA Rev.* **22**, 82–121 (1961).
18. Wood, C. Materials for thermoelectric energy-conversion. *Rep. Prog. Phys.* **51**, 459–539 (1988).
19. Cutler, M., Leavy, J. F. & Fitzpatrick, R. L. Electronic transport in semimetallic cerium sulfide. *Phys. Rev.* **133**, A1143–A1152 (1964).
20. Bhandari, C. M. & Rowe, D. M. in *CRC Handbook of Thermoelectrics* (ed. Rowe, D. M.) Ch. 5, 43–53 (CRC, Boca Raton, 1995).
21. Snyder, G. J., Caillat, T. & Fleurial, J.-P. Thermoelectric transport and magnetic properties of the polaron semiconductor $Fe_xCr_{3-x}Se_4$. *Phys. Rev. B* **62**, 10185 (2000).
22. Slack, G. A. (ed.) *Solid State Physics* (Academic Press, New York, 1979).
23. Dames, C. & Chen, G. in *Thermoelectrics Handbook Macro to Nano* (ed. Rowe, D. M.) Ch. 42 (CRC, Boca Raton, 2006).
24. Slack, G. A. in *CRC Handbook of Thermoelectrics* (ed. Rowe, M.) 407–440 (CRC, Boca Raton, 1995).
25. Hicks, L. D. & Dresselhaus, M. S. Effect of quantum-well structures on the thermoelectric figure of merit. *Phys. Rev. B* **47**, 12727–12731 (1993).
26. Zide, J. M. O. *et al.* Demonstration of electron filtering to increase the Seebeck coefficient in $In_{0.53}Ga_{0.47}As/In_{0.53}Ga_{0.28}Al_{0.19}As$ superlattices. *Phys. Rev. B* **74**, 205335 (2006).
27. Bhandari, C. M. in *CRC Handbook of Thermoelectrics* (ed. Rowe, D. M.) 55–65 (CRC, Boca Raton, 1995).
28. Wright, D. A. Thermoelectric properties of bismuth telluride and its alloys. *Nature* **181**, 834–834 (1958).
29. Kusano, D. & Hori, Y. Thermoelectric properties of p-type $(Bi_2Te_3)(0.2)$ $(Sb_2Te_3)(0.8)$ thermoelectric material doped with PbTe. *J. Jpn Inst. Met.* **66**, 1063–1065 (2002).
30. Zhu, P. W. *et al.* Enhanced thermoelectric properties of PbTe alloyed with Sb_2Te_3. *J. Phys. Condens. Matter* **17**, 7319–7326 (2005).
31. Poudeu, P. F. P. *et al.* Nanostructures versus solid solutions: Low lattice thermal conductivity and enhanced thermoelectric figure of merit in $Pb_{9.6}Sb_{0.2}Te_{10-x}Se_x$ bulk materials. *J. Am. Chem. Soc.* **128**, 14347–14355 (2006).
32. Poudeu, P. F. R. *et al.* High thermoelectric figure of merit and nanostructuring in bulk p-type $Na_{1-x}Pb_mSb_yTe_{m+2}$. *Angew. Chem. Int. Edn* **45**, 3835–3839 (2006).
33. Feldman, J. L., Singh, D. J., Mazin, II, Mandrus, D. & Sales, B. C. Lattice dynamics and reduced thermal conductivity of filled skutterudites. *Phys. Rev. B* **61**, R9209–R9212 (2000).
34. Fleurial, J.-P., Caillat, T. & Borshchevsky, A. in *Proc. ICT'97 16th Int. Conf. Thermoelectrics* 1–11 (IEEE Piscataway, New Jersey, 1997).
35. Sales, B. C., Mandrus, D. & Williams, R. K. Filled skutterudite antimonides: A new class of thermoelectric materials. *Science* **272**, 1325–1328 (1996).
36. Chung, D. Y. *et al.* A new thermoelectric material: $CsBi_4Te_6$. *J. Am. Chem. Soc.* **126**, 6414–6428 (2004).
37. Shelimova, L. E. *et al.* Thermoelectric properties of $PbBi_4Te_7$-based anion-substituted layered solid solutions. *Inorg. Mater.* **40**, 1146–1152 (2004).
38. Shelimova, L. E. *et al.* Crystal structures and thermoelectric properties of layered compounds in the $ATe-Bi_2Te_3$ (A = Ge, Sn, Pb) systems. *Inorg. Mater.* **40**, 451–460 (2004).
39. Kanatzidis, M. G. Structural evolution and phase homologies for "design" and prediction of solid-state compounds. *Acc. Chem. Res.* **38**, 359–368 (2005).
40. Kurosaki, K., Kosuga, A., Muta, H., Uno, M. & Yamanaka, S. Ag_9TlTe_5: A high-performance thermoelectric bulk material with extremely low thermal conductivity. *Appl. Phys. Lett.* **87**, 061919 (2005).
41. Wolfing, B., Kloc, C., Teubner, J. & Bucher, E. High performance thermoelectric Tl_9BiTe_6 with an extremely low thermal conductivity. *Phys. Rev. Lett.* **86**, 4350–4353 (2001).
42. Toberer, E. S., Sasaki, K. A., Chisholm, C. R. I., Haile, S. M. & Snyder, G. J. Local structure of interstitial Zn in β-Zn_4Sb_3. *Phys. Status Solidi* **1**, 253–255 (2007).

43. Chalfin, E., Lu, H. X. & Dieckmann, R. Cation tracer diffusion in the thermoelectric materials $Cu_xMo_6Se_8$ and "beta-Zn_4Sb_3". *Solid State Ionics* **178**, 447–456 (2007).

44. Kim, H. J., Božin, E. S., Haile, S. M., Snyder, G. J. & Billinge, S. J. L. Nanoscale alpha-structural domains in the phonon-glass thermoelectric material beta-Zn_4Sb_3. *Phys. Rev. B* **75**, 134103 (2007).

45. Brown, S. R., Kauzlarich, S. M., Gascoin, F. & Snyder, G. J. $Yb_{14}MnSb_{11}$: New high efficiency thermoelectric material for power generation. *Chem. Mater.* **18**, 1873–1877 (2006).

46. Kauzlarich, S. M., Brown, S. R. & Snyder, G. J. Zintl phases for thermoelectric devices. *Dalton Trans.* 2099–2107 (2007).

47. Cava, R. J. Structural chemistry and the local charge picture of copper-oxide superconductors. *Science* **247**, 656–662 (1990).

48. Terasaki, I., Sasago, Y. & Uchinokura, K. Large thermoelectric power in $NaCo_2O_4$ single crystals. *Phys. Rev.B* **56**, 12685–12687 (1997).

49. Shin, W. & Murayama, N. Thermoelectric properties of (Bi,Pb)-Sr-Co-O oxide. *J. Mater. Res.* **15**, 382–386 (2000).

50. Wang, Y. Y., Rogado, N. S., Cava, R. J. & Ong, N. P. Spin entropy as the likely source of enhanced thermopower in $Na_xCo_2O_4$. *Nature* **423**, 425–428 (2003).

51. Gascoin, F., Ottensmann, S., Stark, D., Haile, S. M. & Snyder, G. J. Zintl phases as thermoelectric materials: Tuned transport properties of the compounds $Ca_xYb_{1-x}Zn_2Sb_2$. *Adv. Funct. Mater.* **15**, 1860–1864 (2005).

52. Yao, T. Thermal-Properties of AlAs/GaAs Superlattices. *Appl. Phys. Lett.* **51**, 1798–1800 (1987).

53. Touzelbaev, M. N., Zhou, P., Venkatasubramanian, R. & Goodson, K. E. Thermal characterization of Bi_2Te_3/Sb_2Te_3 superlattices. *J. Appl. Phys.* **90**, 763–767 (2001).

54. Caylor, J. C., Coonley, K., Stuart, J., Colpitts, T. & Venkatasubramanian, R. Enhanced thermoelectric performance in PbTe-based superlattice structures from reduction of lattice thermal conductivity. *Appl. Phys. Lett.* **87**, 23105 (2005).

55. Beyer, H. *et al.* High thermoelectric figure of merit *ZT* in PbTe and Bi_2Te_3-based superlattices by a reduction of the thermal conductivity. *Physica E* **13**, 965–968 (2002).

56. Hochbaum, A. I. *et al.* Enhanced thermoelectric performance of rough silicon nanowires. *Nature* **451**, 163–167 (2008).

57. Boukai, A. I. *et al.* Silicon nanowires as efficient thermoelectric materials. *Nature* **451**, 168–171 (2008).

58. Cahill, D. G., Watson, S. K. & Pohl, R. O. Lower limit to thermal conductivity of disordered crystals. *Phys. Rev. B* **46**, 6131–40 (1992).

59. Kim, W. *et al.* Cross-plane lattice and electronic thermal conductivities of ErAs: InGaAs/InGaAlAs superlattices. *Appl. Phys. Lett.* **88**, 242107 (2006).

60. Kim, W. *et al.* Thermal conductivity reduction and thermoelectric figure of merit increase by embedding nanoparticles in crystalline semiconductors. *Phys. Rev. Lett.* **96**, 045901 (2006).

61. Chiritescu, C. *et al.* Ultralow thermal conductivity in disordered, layered WSe_2 crystals. *Science* **315**, 351–353 (2007).

62. Rowe, D. M., Shukla, V. S. & Savvides, N. Phonon-scattering at grain-boundaries in heavily doped fine-grained silicon-germanium alloys. *Nature* **290**, 765–766 (1981).

63. Vining, C. B., Laskow, W., Hanson, J. O., Vanderbeck, R. R. & Gorsuch, P. D. Thermoelectric properties of pressure-sintered $Si_{0.8}Ge_{0.2}$ thermoelectric alloys. *J. Appl. Phys.* **69**, 4333–4340 (1991).

64. Jang, K.-W. & Lee, D.-H. in *Fourteenth International Conference on Thermoelectrics* 108 (IEEE, 1995).

65. Goldsmid, H. J. & Penn, A. W. Boundary scattering of phonons in solid solutions. *Phys. Lett. A* **27**, 523–524 (1968).

66. Liebmann, W. K. & Miller, E. A. Preparation phase-boundary energies, and thermoelectric properties of InSb-Sb eutectic alloys with ordered microstructures. *J. Appl. Phys.* **34**, 2653–2659 (1963).

67. Ikeda, T. *et al.* Solidification processing of alloys in the pseudo-binary PbTe-Sb_2Te_3 system. *Acta Mater.* **55**, 1227–1239 (2007).

68. Aliev, M. I., Khalilova, A. A., Arsaly, D. G., Ragimov, R. N. & Tanogly, M. Electrical and thermal properties of the GaSb-$FeGa_{1.3}$ eutectic. *Inorg. Mater.* **40**, 331–335 (2004).

69. Skrabek, E. A. & Trimmer, D. S. in *CRC Handbook of Thermoelectrics* (ed. Rowe, D. M.) 267–275 (CRC, Boca Raton, 1995).

70. Cook, B. A., Kramer, M. J., Wei, X., Harringa, J. L. & Levin, E. M. Nature of the cubic to rhombohedral structural transformation in $(AgSbTe_2)_{15}(GeTe)_{85}$ thermoelectric material. *J. Appl. Phys.* **101**, 053715 (2007).

71. Chen, N. *et al.* Macroscopic thermoelectric inhomogeneities in $(AgSbTe_2)_x(PbTe)_{1-x}$. *App. Phys. Lett.* **87**, 171903 (2005).

72. Androulakis, J. *et al.* Nanostructuring and high thermoelectric efficiency in p-type $Ag(Pb_{1-y}Sn_y)_mSbTe_{2+m}$. *Adv. Mater.* **18**, 1170–1173 (2006).

73. Hsu, K. F. *et al.* Cubic $AgPb_mSbTe_{2+m}$: Bulk thermoelectric materials with high figure of merit. *Science* **303**, 818–821 (2004).

74. Sootsman, J. R., Pcionek, R. J., Kong, H. J., Uher, C. & Kanatzidis, M. G. Strong reduction of thermal conductivity in nanostructured PbTe prepared by matrix encapsulation. *Chem. Mater.* **18**, 4993–4995 (2006).

75. Ikeda, T. *et al.* Solidification processing of alloys in the pseudo-binary PbTe-Sb_2Te_3 system. *Acta Mater.* **55**, 1227–1239 (2007).

76. Ikeda, T. *et al.* Self-assembled nanometer lamellae of thermoelectric PbTe and Sb_2Te_3 with epitaxy-like interfaces. *Chem. Mater.* **19**, 763–767 (2007).

77. Clavagueramora, M. T., Surinach, S., Baro, M. D. & Clavaguera, N. Thermally activated crystallization of $(GeSe_2)_{70}(Sb_2Te_3)_{20}(GeTe)_{10}$ alloy glass - morphological and calorimetric study. *J. Mater. Sci.* **18**, 1381–1388 (1983).

78. Rowe, D. M. & Min, G. Alpha-plot in sigma-plot as a thermoelectric-material performance indicator. *J. Mater. Sci. Lett.* **14**, 617–619 (1995).

79. Toberer, E. S., Christensen, M., Iversen, B. B. & Snyder, G. J. High temperature thermoelectric efficiency in $Ba_8Ga_{16}Ge_{30}$. *Phys. Rev. B* (in the press).

80. Caillat, T., Fleurial, J. P. & Borshchevsky, A. Preparation and thermoelectric properties of semiconducting Zn_4Sb_3. *J. Phys. Chem. Solids* **58**, 1119–1125 (1997).

81. Gelbstein, Y., Dashevsky, Z. & Dariel, M. P. High performance n-type PbTe-based materials for thermoelectric applications. *Physica B* **363**, 196–205 (2005).

82. Vining, C. B., Laskow, W., Hanson, J. O., Vanderbeck, R. R. & Gorsuch, P. D. Thermoelectric properties of pressure-sintered $Si_{0.8}Ge_{0.2}$ thermoelectric alloys. *J. Appl. Phys.* **69**, 4333–4340 (1991).

83. Culp, S. R., Poon, S. J., Hickman, N., Tritt, T. M. & Blumm, J. Effect of substitutions on the thermoelectric figure of merit of half-Heusler phases at 800 °C. *Appl. Phys. Lett.* **88**, 042106 (2006).

84. Goldsmid, H. J. & Douglas, R. W. The use of semiconductors in thermoelectric refrigeration. *Brit. J. Appl. Phys.* **5**, 386–390 (1954).

85. Scherrer, H. & Scherrer, S. in *Thermoelectrics Handbook Macro to Nano* (ed. rowe, D. M.) Ch. 27 (CRC, Boca Raton, 2006).

86. Kutasov, V. A., Lukyanova, L. N. & Vedernikov, M. V. in *Thermoelectrics Handbook Macro to Nano* Ch. 37 (CRC, Boca Raton, 2006).

87. Kuznetsov, V. L., Kuznetsova, L. A., Kaliazin, A. E. & Rowe, D. M. High performance functionally graded and segmented Bi_2Te_3-based materials for thermoelectric power generation. *J. Mater. Sci.* **37**, 2893–2897 (2002).

88. Fritts, R. W. in *Thermoelectric Materials and Devices* (eds. Cadoff, I. B. & Miller, E.) 143–162 (Reinhold, New York, 1960).

89. Fleischmann, H., Luy, H. & Rupprecht, J. Neuere Untersuchungen an Halbleitenden IV VI-I V VI_2 Mischkristallen. *Zeitschrift Für Naturforschung A* **18**, 646–649 (1963).

90. Fleischmann, H. Wärmeleitfähigkeit, Thermokraft und Elektrische Leitfähigkeit von Halbleitenden Mischkristallen Der Form $(A_{x/2}^I Bi_{1-x}^{IV} C_{x/2}^V)D^{VI}$. *Zeitschrift Für Naturforschung A* **16**, 765–780 (1961).

91. Yim, W. M. & Amith, A. Bi-Sb alloys for magneto-thermoelectric and thermomagnetic cooling. *Solid State Electron.* **15**, 1141 (1972).

92. Sidorenko, N. A. & Ivanova, L. D. Bi-Sb solid solutions: Potential materials for high-efficiency thermoelectric cooling to below 180 K. *Inorg. Mater.* **37**, 331–335 (2001).

93. Snyder, G. J. Application of the compatibility factor to the design of segmented and cascaded thermoelectric generators. *Appl. Phys. Lett.* **84**, 2436–2438 (2004).

94. Müller, E., Drasar, C., Schilz, J. & Kaysser, W. A. Functionally graded materials for sensor and energy applications. *Mater. Sci. Eng. A* **362**, 17–39 (2003).

95. Snyder, G. J. in *Thermoelectrics Handbook Macro to Nano* (ed. Rowe, D. M.) Ch. 9 (CRC, Boca Raton, 2006).

96. Snyder, G. J. & Ursell, T. Thermoelectric efficiency and compatibility. *Phys. Rev. Lett.* **91**, 148301 (2003).

97. Ursell, T. S. & Snyder, G. J. in *Twenty-first International Conference on Thermoelectrics* ICT'02 412 (IEEE, Long Beach, California, USA, 2002).

Acknowledgements

We thank Jean-Pierre Fleurial and Thierry Caillat for discussions concerning skutterudites, Marlow Industries, Cronin Vining, Yaniv Gelbstein, Ken Kurosaki for data and discussions and JPL-NASA and the Beckman Institute at Caltech for funding.

Enhanced thermoelectric performance of rough silicon nanowires

Allon I. Hochbaum[1]*, Renkun Chen[2]*, Raul Diaz Delgado[1], Wenjie Liang[1], Erik C. Garnett[1], Mark Najarian[3], Arun Majumdar[2,3,4] & Peidong Yang[1,3,4]

Approximately 90 per cent of the world's power is generated by heat engines that use fossil fuel combustion as a heat source and typically operate at 30–40 per cent efficiency, such that roughly 15 terawatts of heat is lost to the environment. Thermoelectric modules could potentially convert part of this low-grade waste heat to electricity. Their efficiency depends on the thermoelectric figure of merit ZT of their material components, which is a function of the Seebeck coefficient, electrical resistivity, thermal conductivity and absolute temperature. Over the past five decades it has been challenging to increase $ZT > 1$, since the parameters of ZT are generally interdependent[1]. While nanostructured thermoelectric materials can increase $ZT > 1$ (refs 2–4), the materials (Bi, Te, Pb, Sb, and Ag) and processes used are not often easy to scale to practically useful dimensions. Here we report the electrochemical synthesis of large-area, wafer-scale arrays of rough Si nanowires that are 20–300 nm in diameter. These nanowires have Seebeck coefficient and electrical resistivity values that are the same as doped bulk Si, but those with diameters of about 50 nm exhibit 100-fold reduction in thermal conductivity, yielding $ZT = 0.6$ at room temperature. For such nanowires, the lattice contribution to thermal conductivity approaches the amorphous limit for Si, which cannot be explained by current theories. Although bulk Si is a poor thermoelectric material, by greatly reducing thermal conductivity without much affecting the Seebeck coefficient and electrical resistivity, Si nanowire arrays show promise as high-performance, scalable thermoelectric materials.

The most widely used commercial thermoelectric material is bulk Bi_2Te_3 and its alloys with Sb, Se, and so on, which have $ZT = S^2 T/\rho k \approx 1$, where S, ρ, k and T are the Seebeck coefficient, electrical resistivity, thermal conductivity and absolute temperature, respectively. It is difficult to scale bulk Bi_2Te_3 to large-scale energy conversion, but fabricating synthetic nanostructures for this purpose is even more difficult and expensive. Si, on the other hand, is the most abundant and widely used semiconductor, with a large industrial infrastructure for low-cost and high-yield processing. Bulk Si, however, has a high k ($\sim 150\,\mathrm{W\,m^{-1}\,K^{-1}}$ at room temperature)[5], giving $ZT \approx 0.01$ at 300 K (ref. 6). The spectral distribution of phonons contributing to the k of Si at room temperature is quite broad. Because the rate of phonon–phonon Umklapp scattering scales as ω^2, where ω is the phonon frequency, low-frequency (or long-wavelength) acoustic phonons have long mean free paths and contribute significantly to k at high temperatures[7–10]. Thus, by rational incorporation of phonon-scattering elements at several length scales, the k of Si is expected to decrease dramatically. The ideal thermoelectric material is believed to be a phonon glass and an electronic crystal. Here, we show that by using roughened nanowires, we can reduce the thermal conductivity to $\sim 1.6\,\mathrm{W\,m^{-1}\,K^{-1}}$, with the phonon contribution close to the amorphous limit, without significantly modifying the power factor S^2/ρ, such that $ZT \approx 1$ at room temperature. Further reduction of nanowire diameter is likely to increase ZT to >1.

Wafer-scale arrays of Si nanowires were synthesized by an aqueous electroless etching (EE) method[11–13]. The technique is based on the galvanic displacement of Si by $Ag^+ \rightarrow Ag^0$ reduction on the wafer surface. The reaction proceeds in an aqueous solution of $AgNO_3$ and HF acid. Briefly, Ag^+ reduces onto the Si wafer surface by injecting holes into the Si valence band and oxidizing the surrounding lattice, which is subsequently etched by HF. The initial reduction of Ag^+ forms Ag nanoparticles on the wafer surface, thus delimiting the spatial extent of the oxidation and etching process. Further reduction of Ag^+ occurs on the nanoparticles, not the Si wafer, which becomes the active cathode by electron transfer from the underlying wafer. Ag dentritic growth on the arrays can be washed off with deionized water after the synthesis. The arrays were washed in a concentrated nitric acid bath for at least one hour to remove all residual Ag from the nanowire surfaces. After the nitric acid bath, no Ag particles were observed during transmission electron microscopy (TEM) analysis and no Ag peaks appeared in the energy-dispersive X-ray spectra of the nanowires. Furthermore, the reaction proceeds at or near room temperature (295 K), so no diffusion of Ag atoms into a covalent solid lattice—such as Si—should be expected.

Nanowires synthesized by this approach were vertically aligned and consistent throughout batches, and across large areas up to wafer-scale. Figure 1a is a cross-sectional scanning electron microscope (SEM) image of one such array, and the inset shows a one-inch-square nanowire array. Key parameters of the reaction were identified using p-type ⟨100⟩-oriented, nominally 10–20 Ω cm, Si as the etch wafer. Both etching time and $AgNO_3$ concentration controlled nanowire length, roughly linearly, down to 5 μm at short immersion times (<10 min). At longer etching times, nanowire lengths were controllable up to 150 μm, while longer wires were too fragile to preserve the array. Wafers cut to ⟨100⟩, ⟨110⟩ and ⟨111⟩ orientations all yielded nanowire arrays etched normal to the wafer surface over most of the wafer area. Similar results were obtained for EE of both n- and p-type wafers with resistivities varying from 10 to 10^{-2} Ω cm ($\sim 10^{14}$ to $10^{18}\,\mathrm{cm^{-3}}$ dopant concentrations). Because thermoelectric modules consist of complementary p- and n-type materials wired in series, the generality and scalability of this synthesis are promising for fabrication of Si-based devices.

After etching, the fill factor of the nanowires was approximately 30% over the entire wafer surface. The nanowires varied from 20 to 300 nm in diameter with an average diameter of approximately 100 nm, as measured from TEM micrographs (Fig. 1b). The

[1]Department of Chemistry, [2]Department of Mechanical Engineering, [3]Department of Materials Science and Engineering, University of California, Berkeley, California 94720, USA. [4]Materials Sciences Division, Lawrence Berkeley National Laboratory, Berkeley, California 94720, USA.
*These authors contributed equally to this work.

Nature | Vol 451 | 10 January 2008 | doi:10.1038/nature06381

nanowires were single crystalline, as shown by the selected area electron diffraction pattern (top inset) and high-resolution TEM image of the Si lattice of an EE nanowire in Fig. 1c. In contrast to the smooth surfaces of typical vapour–liquid–solid (VLS)-grown, gold-catalysed Si nanowires (Fig. 1d)[14,15], those of the EE Si nanowires are much rougher. The mean roughness height of these nanowires varied from wire to wire, but was typically 1–5 nm with a roughness period of the order of several nanometres. This roughness may be attributed to randomness of the lateral oxidation and etching in the corrosive aqueous solution or slow HF etching and faceting of the lattice during synthesis.

The main advantage of using Si nanowires for thermoelectric applications lies in the large difference in mean free path lengths between electrons and phonons at room temperature: 110 nm for electrons in highly doped samples[16,17] and ~300 nm for phonons[10]. Consequently, incorporating structures with critical dimensions/ spacings below 300 nm in Si should reduce the thermal conductivity without significantly affecting S^2/ρ. The thermal conductivity of these hierarchically structured Si nanowires was characterized using devices consisting of resistive coils supported on parallel, suspended SiN_X membranes[14,18]. This construction allows us to probe thermal transport in individual nanowires. The membranes are thermally connected through a bridging nanowire, with negligible leakage from heat transfer by means other than conduction through the wire. The thermal conductivity was extracted from the thermal

conductance using the dimensions of the nanowire, as determined by SEM. To anchor the nanowire to the membranes and reduce thermal contact resistance, a Pt/C composite was deposited on both ends using a focused electron beam (Fig. 2a, also see Supplementary Information). The contact resistance at the interface between the nanowire and the pad is negligible relative to the nanowire thermal resistance. This condition was verified by measuring the thermal conductivity of a large nanowire (135 nm diameter) after two rounds of thermal anchoring with Pt/C pads. The second thermal anchoring doubled the contact area of the nanowire with the Pt/C pad and the SiN_X membrane, and the measured thermal conductivity of the wire remained unchanged. Hence, the nanowire thermal resistance dominates over that of the contacts (see Supplementary Fig. 2).

Figure 2b shows the measured thermal conductivity of both VLS and EE Si nanowires. It has been shown that the k of VLS Si nanowires is strongly diameter-dependent[14], which is attributed to boundary scattering of phonons. We found that EE Si nanowires exhibit a diameter dependence of k similar to that of VLS-grown wires. The magnitude of k, however, is five- to eightfold lower for EE nanowires of comparable diameters. Because the phonon spectrum is broad and Planck-like, k can be reduced by introducing scattering at additional length scales beyond the nanowire diameter[1–4,19]. In the case of the EE nanowires, the roughness at the nanowire surface behaves like secondary scattering phases. The roughness may contribute to higher rates of diffuse reflection or backscattering of phonons at

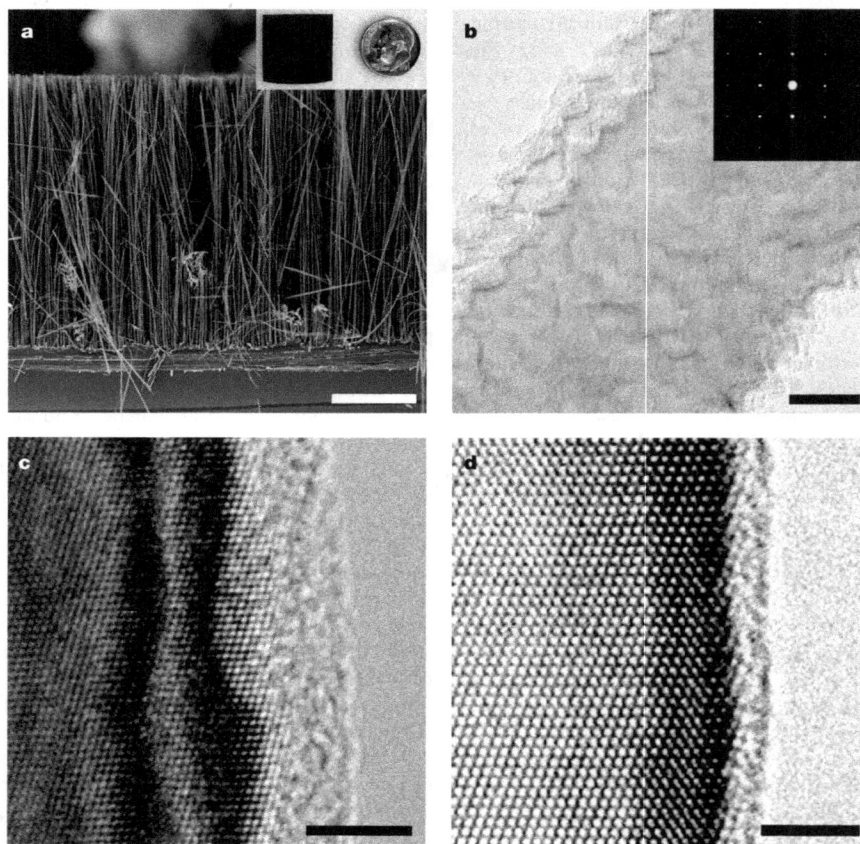

Figure 1 | Structural characterization of the rough silicon nanowires.
a, Cross-sectional SEM of an EE Si nanowire array. Dendritic Ag growth can be seen within the array—a product of Ag^+ reduction onto the wafer during reaction. The Ag is etched in nitric acid after the synthesis, and elemental analysis confirms it is dissolved completely. Inset, an EE Si nanowire array Si wafer chip of the typical size used for the syntheses. Similar results are obtained on entire 4-inch wafers. The chip is dark and non-reflective owing to light scattering by, and absorbing into, the array. b, Bright-field TEM

image of a segment of an EE Si nanowire. The roughness is clearly seen at the surface of the wire. The selected area electron diffraction pattern (inset) indicates that the wire is single crystalline all along its length. c, High-resolution TEM image of an EE Si nanowire. The roughness is evident at the interface between the crystalline Si core and the amorphous native oxide at the surface, and by undulations of the alternating light/dark thickness fringes near the edge. d, High-resolution TEM of a VLS-grown Si nanowire. Scale bars for a–d are 10 μm, 20 nm, 4 nm and 3 nm, respectively.

Nature | Vol 451 | 10 January 2008 | doi:10.1038/nature06381

the interfaces. These processes have been predicted to affect the k values of Si nanowires, but not to the extent observed here[20,21]. The peak k of the EE nanowires is shifted to a much higher temperature than that of VLS nanowires, and both are significantly higher than that of bulk Si, which peaks at around 25 K (ref. 5). This shift suggests that the phonon mean free path is limited by boundary scattering as opposed to intrinsic Umklapp scattering.

While the above wires were etched from high-resistivity wafers, the peak ZT of semiconductor materials is predicted to occur at high dopant concentrations ($\sim 1 \times 10^{19}$ cm^{-3}; ref. 22). To optimize the

ZT of EE nanowires, lower resistivity nanowires were synthesized from $10^{-1}\,\Omega$ cm B-doped p-Si $\langle 111 \rangle$ and $10^{-2}\,\Omega$ cm As-doped n-Si $\langle 100 \rangle$ wafers by the standard method outlined above. Nanowires etched from the $10^{-2}\,\Omega$ cm and less resistive wafers, however, did not produce devices with reproducible electrical contacts, probably owing to greater surface roughness, as observed in TEM analysis. Consequently, more optimally doped nanowires were obtained by post-growth gas-phase B doping of wires etched from $10^{-1}\,\Omega$ cm wafers (see Supplementary Information). The resulting nanowires have an average $\rho = 3 \pm 1.4$ mΩ cm (as compared to $\sim 10\,\Omega$ cm for wires from low-doped wafers).

Figure 2c shows the k of small-diameter nanowires etched from 10, 10^{-1}, and $10^{-2}\,\Omega$ cm wafers. The post-growth doped nanowire (52 nm diameter) etched from a $10^{-1}\,\Omega$ cm wafer has a slightly lower k than the lower-doped wire of the same diameter. This small decrease in k may be attributed to higher rates of phonon-impurity scattering. Studies of doped and isotopically purified bulk Si have revealed a reduction of k as a result of impurity scattering[6,23,24]. Owing to the atomic nature of such defects, they are expected to predominantly scatter short-wavelength phonons. On the other hand, nanowires etched from a $10^{-2}\,\Omega$ cm wafer have a much lower k than the other nanowires, probably as a result of the greater surface roughness.

In the case of the 52 nm nanowire, k is reduced to 1.6 ± 0.13 W m^{-1} K^{-1} at room temperature. For comparison, the temperature-dependent k of amorphous bulk SiO$_2$ (data points used from http://users.mrl.uiuc.edu/cahill/tcdata/tcdata.html agree with measurement in ref. 25) is also plotted in Fig. 2c. As can be seen from the plot, k of these single-crystalline EE Si nanowires is comparable to that of insulating glass. Indeed, k of the 52 nm nanowire approaches the minimum k predicted and measured for Si: ~ 1 W m^{-1} K^{-1} (ref. 26). The resistivity of a single nanowire of comparable diameter (48 nm) was measured (see Supplementary Information) and the electronic contribution to thermal conductivity (k_e) can be estimated from the Wiedemann–Franz law[16]. For measured $\rho = 1.7$ mΩ cm, $k_e = 0.4$ W m^{-1} K^{-1}, meaning that the lattice thermal conductivity ($k_l = k - k_e$) is 1.2 W m^{-1} K^{-1}.

By assuming the mean free path due to boundary scattering $\ell_b = Fd$, where $F > 1$ is a multiplier that accounts for the specularity of phonon scattering at the nanowire surface and d is the nanowire diameter, a model based on Boltzmann transport theory was able to explain[27] the diameter dependence of thermal conductivity in VLS nanowires, as observed in ref. 14. Because the thermal conductivity of EE nanowires is lower and the surface is rougher than that of VLS ones, it is natural to assume $\ell_b = d$ ($F = 1$), which is the smallest mean free path due to boundary scattering. However, this still cannot explain why the phonon thermal conductivity approaches the amorphous limit for nanowires with diameters ~ 50 nm. In fact, theories that consider phonon backscattering, as recently proposed by ref. 21, cannot explain our observations either. The thermal conductivity in amorphous non-metals[26] can be well explained by

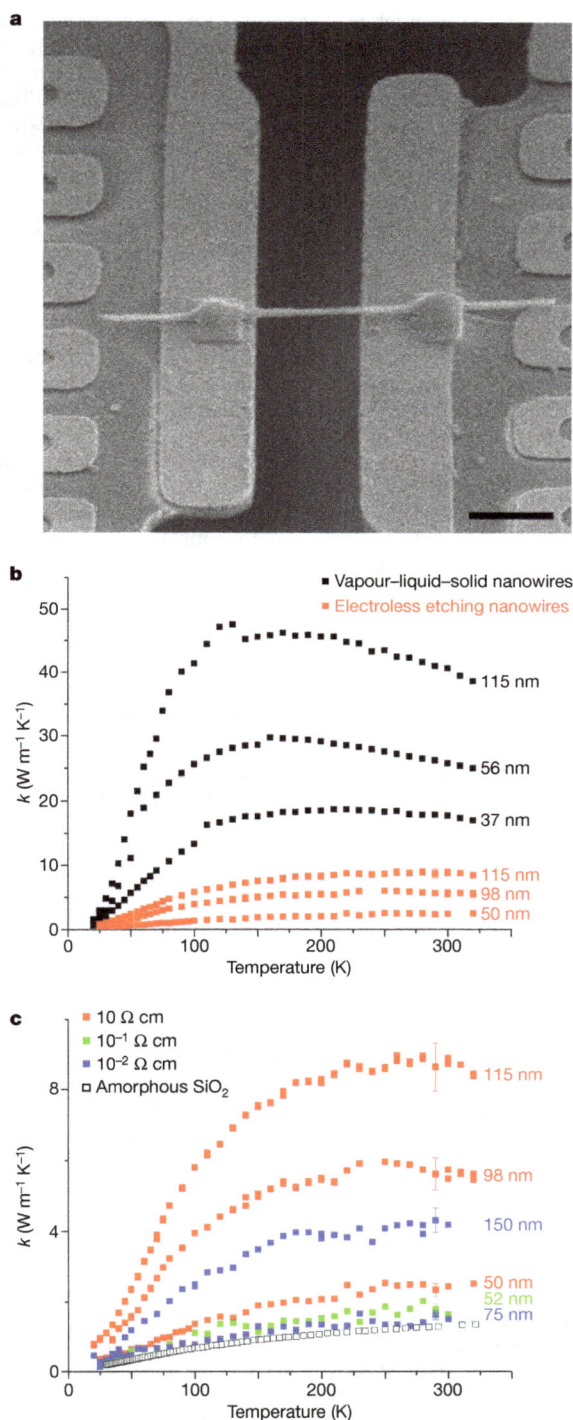

Figure 2 | Thermal conductivity of the rough silicon nanowires. a, An SEM image of a Pt-bonded EE Si nanowire (taken at 52° tilt angle). The Pt thin film loops near both ends of the bridging wire are part of the resistive heating and sensing coils on opposite suspended membranes. Scale bar, 2 μm. **b**, The temperature-dependent k of VLS (black squares; reproduced from ref. 14) and EE nanowires (red squares). The peak k of the VLS nanowires is 175–200 K, while that of the EE nanowires is above 250 K. The data in this graph are from EE nanowires synthesized from low-doped wafers. **c**, Temperature-dependent k of EE Si nanowires etched from wafers of different resistivities: 10 Ω cm (red squares), 10^{-1} Ω cm (green squares; arrays doped post-synthesis to 10^{-3} Ω cm), and 10^{-2} Ω cm (blue squares). For the purpose of comparison, the k of bulk amorphous silica is plotted with open squares. The smaller highly doped EE Si nanowires have a k approaching that of insulating glass, suggesting an extremely short phonon mean free path. Error bars are shown near room temperature, and should decrease with temperature. See Supplementary Information for k measurement calibration and error determination.

assuming that the phonon mean free path $\ell = \lambda/2$, where λ is the phonon wavelength, which invokes a Debye-like short-range coherence in an atomically disordered lattice. However, there seems no justifiable reason to make this assumption for the single-crystal EE Si nanowires, because their diameters are about 100-fold larger than the lattice constant. To the best of our knowledge, there is currently no theory that can explain why a single-crystalline Si nanowire that is ~ 50 nm in diameter should behave like a phonon glass. On the basis of the difference between VLS and EE nanowires, we suspect that the roughness plays a strong role in screening a broad spectrum of

phonons, fundamentally altering phonon transmission through these confined structures. The exact mechanism, however, remains unknown.

To calculate the nanowire ZT, ρ and S measurements were carried out on individual highly doped nanowires. One such measurement on a 48 nm diameter wire is shown in Fig. 3a. Nanowires were measured in a horizontal geometry on 200 nm SiN_X films on Si substrates with a microfabricated heating element, and 2- and 4-point probe electrodes (see Supplementary Fig. 3). The power factor was calculated as $S^2/\rho = 3.3 \times 10^{-3}$ W m^{-1} K^{-2} for the nanowire at 300 K. The ratio of the power factor of optimally doped bulk Si to that of the EE Si nanowire as a function of temperature is plotted in Fig. 3b (with bulk values taken from ref. 6). The nanowire power factor decreases gradually relative to bulk with decreasing temperature, possibly due to a longer electron mean free path. On the other hand, as temperature decreases, the disparity between k of the nanowire and bulk grows. At low temperatures, long-wavelength phonon modes, which contribute strongly to thermal transport in bulk, are efficiently scattered in the roughened nanowires.

Figure 3b charts the ratio of $k_{bulk}:k_{nw}$ for the 52 nm highly doped EE Si nanowire as a function of temperature. Whereas the k_{nw} is two orders of magnitude lower than k_{bulk} at room temperature, this ratio reaches more than four orders of magnitude at low temperature. Also shown is $k_{bulk}:k_{nw}$ for highly doped bulk Si, for which $k_{bulk}:k_{nw}$ is greatly reduced at low temperature. The large disparity persists unchanged, however, near room temperature. As a result, the degradation of the nanowire power factor with decreasing temperature is offset by the significant decrease in k, resulting in a relatively constant ZT enhancement factor for the EE Si nanowire.

ρ and S of the 48 nm nanowire were used for the ZT calculation because the diameter is close to that of the 52 nm wire for which k has been measured. The nanowire ZT is highest near room temperature at 0.6 (Fig. 3c). As compared to optimally doped bulk Si ($\sim 1 \times 10^{19}$ cm^{-3}), the ZT of the EE nanowire is nearly two orders of magnitude greater throughout the temperature range measured[6]. The large increase in ZT is due to the significant decrease of k as compared to bulk while maintaining a high power factor. The hierarchical structuring of the EE Si nanowires allows selective scattering of phonons by dopants, nanoscale surface roughness, and dimensional confinement, while leaving electronic transport largely unaffected.

In conclusion, we have shown that it is possible to achieve $ZT = 0.6$ at room temperature in rough Si nanowires of ~ 50 nm diameter that were processed by a wafer-scale manufacturing technique. With optimized doping, diameter reduction, and roughness control, the ZT is likely to rise even higher. This ZT enhancement can be attributed to efficient scattering throughout the phonon spectrum by the introduction of nanostructures at different length scales (diameter, roughness and point defects). The significant reduction in thermal conductivity observed in this study may be a result of changes in the fundamental physics of heat transport in these quasi-one-dimensional materials. By achieving broadband impedance of phonon transport, we have demonstrated that the EE Si nanowire system is capable of approaching the limits of minimum lattice thermal conductivity in Si. Modules with the performance reported here, and manufactured from such a ubiquitous material as Si, may find wide-ranging applications in waste heat salvaging, power generation, and solid-state refrigeration. Moreover, the phonon scattering techniques developed in this study could significantly augment ZT even further in other materials to produce highly efficient solid-state thermoelectric devices.

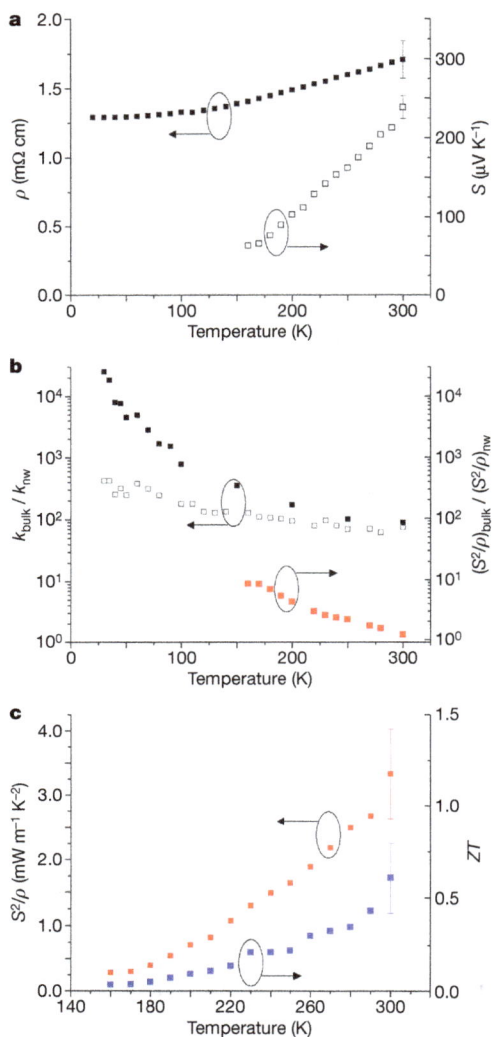

Figure 3 | Thermoelectric properties and ZT calculation for the rough silicon nanowire. a, S (open squares) and ρ (solid squares) of the highly doped EE 48 nm nanowire. See Supplementary Information for error analysis. **b**, Ratio of intrinsic bulk Si k (ref. 5) to that of a highly doped EE nanowire 50 nm in diameter. $k_{bulk}:k_{nw}$ increases dramatically with decreasing temperature, from 100 at 300 K to 25,000 at 25 K (solid squares). As compared to highly doped bulk Si (1.7×10^{19} cm^{-3} As-doped, data adapted from ref. 6), $k_{bulk}:k_{nw}$ increases from 75 at 300 K to 425 at 30 K (open squares). Red squares show the ratio of the power factor of optimally doped bulk Si relative to the nanowire power factor as a function of temperature. **c**, Single nanowire power factor (red squares) of the nanowire and calculated ZT (blue squares) using the measured k of the 52 nm nanowire in Fig. 2c. By propagation of uncertainty from the ρ and S measurements, the error bars are 21% for the power factor and 31% for ZT (assuming negligible temperature uncertainty, which seems valid given that the measurements are stable to better than ± 100 mK).

METHODS SUMMARY

Nanowires were typically etched from B-doped Si wafers of different resistivities in aqueous solutions of 0.02 M AgNO$_3$ and 5 M HF for several hours. Excess Ag was removed in a nitric acid bath for at least one hour. Highly doped nanowires were achieved by annealing arrays at 850 °C for one hour in BCl$_3$ vapour. The structure and microstructure of nanowire arrays and individual nanowires were characterized using SEM and TEM.

For thermal conductivity measurements, nanowires were either drop-cast onto the microfabricated devices from dispersions in isopropanol, or placed directly on the devices by micromanipulation with narrow tungsten probe tips (GGB Industries) mounted on a scanning stage (Marzhauser SM 3.25). The thermal conductivity of individual nanowires was measured by the previously described method[14,18].

The electrical conductivity and Seebeck coefficient of EE Si nanowires were measured by drop-casting isopropanol dispersions of nanowires onto Si wafer substrates coated with a 200-nm-thick silicon nitride film. Metal contact lines and a heating coil were fabricated on the same wafers using standard optical lithography (see Supplementary Fig. 3). 2- and 4-point I–V measurements, and dimensions from SEM images, were used to determine ρ for individual nanowires. S of single nanowires was measured by applying a current to the heating coil and measuring the temperature between the two inner 4-point probe contacts. 4-point measurements of both contact lines, and measured R versus T calibration curves were used to calculate the ΔT between them. S was calculated by $S = \Delta V / \Delta T$.

Full Methods and any associated references are available in the online version of the paper at www.nature.com/nature.

Received 7 June; accepted 9 October 2007.

1. Majumdar, A. Thermoelectricity in semiconductor nanostructures. *Science* **303**, 777–778 (2004).
2. Hsu, K. F. *et al.* Cubic AgPb$_m$SbTe$_{2+m}$: bulk thermoelectric materials with high figure of merit. *Science* **303**, 818–821 (2004).
3. Harman, T. C., Taylor, P. J., Walsh, M. P. & LaForge, B. E. Quantum dot superlattice thermoelectric materials and devices. *Science* **297**, 2229–2232 (2002).
4. Venkatasubramanian, R., Siivola, E., Colpitts, T. & O'Quinn, B. Thin-film thermoelectric devices with high room-temperature figures of merit. *Nature* **413**, 597–602 (2001).
5. Touloukian, Y. S., Powell, R. W., Ho, C. Y. & Klemens, P. G. (eds) *Thermal Conductivity: Metallic Elements and Alloys, Thermophysical Properties of Matter* Vol. 1, 339 (IFI/Plenum, New York, 1970).
6. Weber, L. & Gmelin, E. Transport properties of silicon. *Appl. Phys. A* **53**, 136–140 (1991).
7. Nolas, G. S., Sharp, J. & Goldsmid, H. J. in *Thermoelectrics: Basic Principles and New Materials Development* (eds Nolas, G. S., Sharp, J. & Goldsmid, H. J.) Ch. 3 (Springer, Berlin, 2001).
8. Asheghi, M., Leung, Y. K., Wong, S. S. & Goodson, K. E. Phonon-boundary scattering in thin silicon layers. *Appl. Phys. Lett.* **71**, 1798–1800 (1997).
9. Asheghi, M., Touzelbaev, M. N., Goodson, K. E., Leung, Y. K. & Wong, S. S. Temperature-dependent thermal conductivity of single-crystal silicon layers in SOI substrates. *J. Heat Transf.* **120**, 30–36 (1998).
10. Ju, Y. S. & Goodson, K. E. Phonon scattering in silicon films with thickness of order 100 nm. *Appl. Phys. Lett.* **74**, 3005–3007 (1999).
11. Peng, K. Q., Yan, Y. J., Gao, S. P. & Zhu, J. Synthesis of large-area silicon nanowire arrays via self-assembling nanochemistry. *Adv. Mater.* **14**, 1164–1167 (2002).
12. Peng, K., Yan, Y., Gao, S. & Zhu, J. Dendrite-assisted growth of silicon nanowires in electroless metal deposition. *Adv. Funct. Mater.* **13**, 127–132 (2003).
13. Peng, K. *et al.* Uniform, axial-orientation alignment of one-dimensional single-crystal silicon nanostructure arrays. *Angew. Chem. Intl Edn.* **44**, 2737–2742 (2005).
14. Li, D. *et al.* Thermal conductivity of individual silicon nanowires. *Appl. Phys. Lett.* **83**, 2934–2936 (2003).
15. Hochbaum, A. I., Fan, R., He, R. & Yang, P. Controlled growth of Si nanowire arrays for device integration. *Nano Lett.* **5**, 457–460 (2005).
16. Ashcroft, N. W. & Mermin, N. D. *Solid State Physics* Chs 1, 2 and 13 (Saunders College Publishing, Fort Worth, 1976).
17. Sze, S. M. *Physics of Semiconductor Devices* Ch. 1 (John Wiley & Sons, New York, 1981).
18. Shi, L. *et al.* Measuring thermal and thermoelectric properties of one-dimensional nanostructures using a microfabricated device. *J. Heat Transf.* **125**, 881–888 (2003).
19. Kim, W. *et al.* Thermal conductivity reduction and thermoelectric figure of merit increase by embedding nanoparticles in crystalline semiconductors. *Phys. Rev. Lett.* **96**, 045901 (2006).
20. Zou, J. & Balandin, A. Phonon heat conduction in a semiconductor nanowire. *J. Appl. Phys.* **89**, 2932–2938 (2001).
21. Saha, S., Shi, L. & Prasher, R. Monte Carlo simulation of phonon backscattering in a nanowire. *Proc. ASME Int. Mech. Eng. Congr. Exp. (5–10 November 2006)* art. no. 15668 1–5 (ASME, Chicago, 2006).
22. Rowe, D. M. (ed.) *CRC Handbook of Thermoelectrics* Ch. 5 (CRC Press, Boca Raton, 1995).
23. Brinson, M. E. & Dunstan, W. Thermal conductivity and thermoelectric power of heavily doped n-type silicon. *J. Phys. C* **3**, 483–491 (1970).
24. Ruf, T. *et al.* Thermal conductivity of isotopically enriched silicon. *Solid State Commun.* **115**, 243–247 (2000).
25. Cahill, D. G. & Pohl, R. O. Thermal conductivity of amorphous solids above the plateau. *Phys. Rev. B* **35**, 4067–4073 (1987).
26. Cahill, D. G., Watson, S. K. & Pohl, R. O. Lower limit to the thermal conductivity of disordered crystals. *Phys. Rev. B* **46**, 6131–6140 (1992).
27. Mingo, N., Yang, L., Li, D. & Majumdar, A. Predicting the thermal conductivity of Si and Ge nanowires. *Nano Lett.* **3**, 1713–1716 (2003).

Supplementary Information is linked to the online version of the paper at www.nature.com/nature.

Acknowledgements We thank T.-J. King-Liu and C. Hu for discussions and J. Goldberger for TEM analysis. We acknowledge the support of the Division of Materials Sciences and Engineering, Office of Basic Energy Sciences, DOE. A.I.H. and R.C. thank the NSF-IGERT and ITRI-Taiwan programs, respectively, for fellowship support. We also thank the National Center for Electron Microscopy and the UC Berkeley Microlab for the use of their facilities. R.D.D. thanks the GenCat/Fulbright programme for support.

Author Information Reprints and permissions information is available at www.nature.com/reprints. Correspondence and requests for materials should be addressed to A.M. (majumdar@me.berkeley.edu) and P.Y. (p_yang@berkeley.edu).

Silicon nanowires as efficient thermoelectric materials

Akram I. Boukai[1]†, Yuri Bunimovich[1]†, Jamil Tahir-Kheli[1], Jen-Kan Yu[1], William A. Goddard III[1] & James R. Heath[1]

Thermoelectric materials interconvert thermal gradients and electric fields for power generation or for refrigeration[1,2]. Thermoelectrics currently find only niche applications because of their limited efficiency, which is measured by the dimensionless parameter ZT—a function of the Seebeck coefficient or thermoelectric power, and of the electrical and thermal conductivities. Maximizing ZT is challenging because optimizing one physical parameter often adversely affects another[3]. Several groups have achieved significant improvements in ZT through multi-component nanostructured thermoelectrics[4-6], such as Bi_2Te_3/Sb_2Te_3 thin-film superlattices, or embedded PbSeTe quantum dot superlattices. Here we report efficient thermoelectric performance from the single-component system of silicon nanowires for cross-sectional areas of 10 nm × 20 nm and 20 nm × 20 nm. By varying the nanowire size and impurity doping levels, ZT values representing an approximately 100-fold improvement over bulk Si are achieved over a broad temperature range, including $ZT \approx 1$ at 200 K. Independent measurements of the Seebeck coefficient, the electrical conductivity and the thermal conductivity, combined with theory, indicate that the improved efficiency originates from phonon effects. These results are expected to apply to other classes of semiconductor nanomaterials.

The most efficient thermoelectrics have historically been heavily doped semiconductors because the Pauli principle restricts the heat-carrying electrons to be close to the Fermi energy[1] for metals. The Wiedemann–Franz law, $\kappa_e/\sigma T = \pi^2/3(k/e)^2 = (156\,\mu V\,K^{-1})^2$, where κ_e is the electronic contribution to κ, constrains $ZT = S^2\sigma T/\kappa$, where S is the Seebeck coefficient (or thermoelectric power, measured in $V\,K^{-1}$), and σ and κ are the electrical and thermal conductivities, respectively. Semiconductors have a lower density of carriers, leading to larger S values and a κ value that is dominated by phonons (κ_{ph}), implying that the electrical and thermal conductivities are somewhat decoupled[1]. κ can be reduced by using bulk semiconductors of high atomic weight, which decreases the speed of sound. However, this strategy has not yet produced materials with $ZT > 1$.

For a metal or highly doped semiconductor, S is proportional to the energy derivative of the density of electronic states. In low-dimensional (nanostructured) systems the density of electronic states has sharp peaks[7-9] and, theoretically, a high thermopower. Harnessing this electronic effect to produce high-ZT materials has had only limited success[10,11]. However, optimization of the phonon dynamics and heat transport physics in nanostructured systems has yielded results[4-6]. Nanostructures may be prepared with one or more dimensions smaller than the mean free path of the phonons and yet larger than that of electrons and holes. This potentially reduces κ without decreasing σ (ref. 12). Bulk silicon (Si) is a poor thermoelectric ($ZT_{300\,K} \approx 0.01$; ref. 13), and this phonon physics is important for our Si nanowires, in which the electronic structure remains bulk-like.

Figure 1 shows images of the devices and the Si nanowires we used for these experiments. Details related to the fabrication, calibration and experimental measurements are presented in the Supplementary

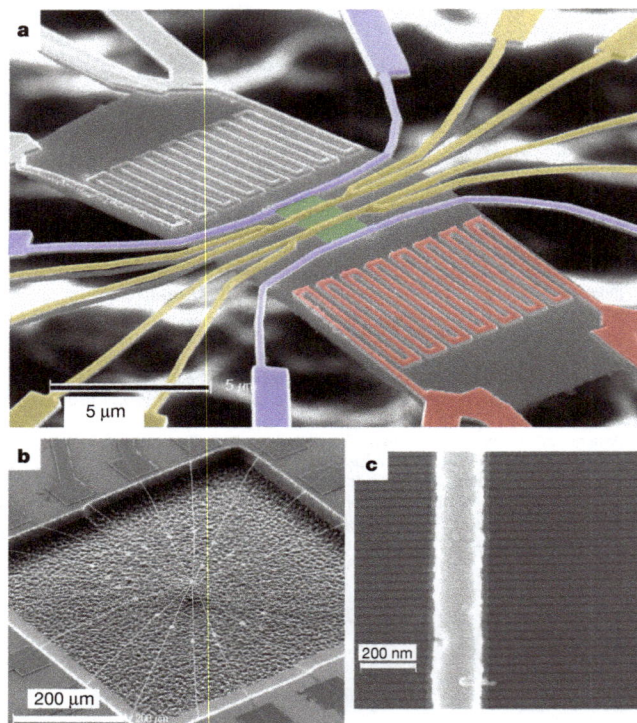

Figure 1 | Scanning electron micrographs of the device used to quantitate the thermopower and electrical and thermal conductivity of Si nanowire arrays. a, This false-colour image of a suspended platform shows all electrical connections. The central green area is the Si nanowire array, which is not resolved at this magnification. The four-lead yellow electrodes are used for thermometry to quantify the temperature difference across the nanowire array. The thermal gradient is established with either of the two Joule heaters (the right-hand heater is coloured red). The yellow and blue electrodes are combined to carry out four-point electrical conductivity measurements on the nanowires. The grey region underlying the nanowires and the electrodes is the 150-nm-thick SiO_2 insulator that is sandwiched between the top Si(100) single-crystal film from which the nanowires are fabricated, and the underlying Si wafer. The underlying Si wafer has been etched back to suspend the measurement platform, placing the background of this image out of focus. **b,** Low-resolution micrograph of the suspended platform. The electrical connections radiate outwards and support the device. **c,** High-resolution image of an array of 20-nm-wide Si nanowires with a Pt electrode.

[1]Division of Chemistry and Chemical Engineering, MC 127-72, 1200 East California Blvd, California Institute of Technology, Pasadena, California 91125, USA.
†These authors contributed equally to this work.

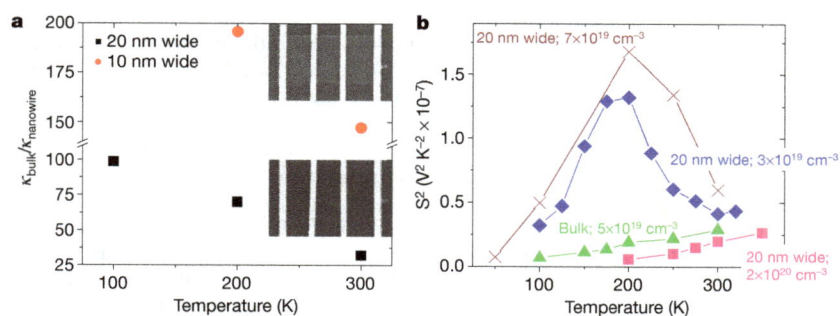

Figure 2 | Factors contributing to ZT for various Si nanowires. All nanowires are 20 nm in height. a, The temperature dependence of the thermal conductivity κ, presented as $\kappa_{bulk}/\kappa_{nanowire}$ to highlight the improvement that the reduction of κ in nanowires lends to ZT. κ_{bulk} values, which are slightly below the true bulk value for Si, are taken from an identically measured 520 nm × 35 nm-sized film. The inset scanning electron microscope micrographs show the region of the device containing the nanowires before (top) and after (bottom) the XeF_2 etch to remove the nanowires. **b**, The temperature dependence of S^2 for 20-nm-wide Si nanowires at various p-type doping concentrations (indicated on the graph). Note that the most highly doped nanowires (pink line) yield a thermopower similar to that of bulk Si doped at a lower level. For nanowires doped at slightly higher and slightly lower concentrations than the bulk, S peaks near 200 K. This is a consequence of the one-dimensional nature of the Si nanowires.

Information. The platform permits four-point measurements of the electrical conductivity of the nanowires, Joule heating to establish a thermal gradient across the nanowires, and four-point thermometry to quantify that gradient[10,14]. The resistance of the four-point thermometry electrodes is typically two orders of magnitude smaller than the resistance of the nanowires. The measurement platform is suspended in vacuum to allow measurement of nanowire thermal conductivity[15,16]. For all measurements, the Si nanowires could be selectively removed using a XeF_2 etch, thus allowing for measurements of the contributions to σ, S and κ from the platform and oxide substrate.

There are several ways to prepare Si nanowires, including materials methods for bulk production[17]. We wanted Si nanowires in which the dimensions, impurity doping levels, crystallographic nature, and so on, were all quantifiable and precisely controlled. We used the super-lattice nanowire pattern transfer (SNAP) method[18], which translates the atomic control over the layer thickness of a superlattice into control over the width and spacing of nanowires. Si nanowires made via SNAP inherit their impurity dopant concentrations directly from the single-crystal Si epilayers of the silicon-on-insulator substrates from which they are fabricated[19]. These epilayers were 20- or 35-nm-thick Si(100) films on 150 nm of SiO_2, and were p-type impurity (boron) doped using diffusion-based doping[19]. Four-point probe conductivity measurements of the silicon-on-insulator films were used to extract dopant concentrations. We prepared nanowire arrays several micrometres long, with lateral width × thickness dimensions of 10 nm × 20 nm, 20 nm × 20 nm and 520 nm × 35 nm. The last approximates the bulk and, in fact, measurements on the sample obtained bulk values for S, σ and κ. Measurements of κ for our nanowires were consistent with literature values for materials grown (round) Si nanowires[16].

All values of S, σ and κ reported here are normalized to individual nanowires, although each experiment used a known number of nanowires ranging from 10 to 400. The Si microwires and nanowires were prepared using the same substrates, doping methods, and so on, but different patterning methods (electron-beam lithography versus SNAP).

Measurements of κ and S^2 for Si nanowires (and microwires) for different nanowire sizes and doping levels are presented in Fig. 2. More complete data sets, electrical conductivity data and a statistical analysis are presented in the Supplementary Information. The nanowire electrical conductivity is between 10 and 90% of the bulk, depending on nanowire dimensions. A reduced σ probably arises from surface scattering of charge carriers[19]. Nevertheless, all nanowires are highly doped and most exhibit metallic-like conductivity (increasing σ with decreasing T).

The temperature dependence of κ for a microwire and 10- and 20-nm-wide nanowires were recorded at modest statistical resolution to establish trends. This data indicated that κ drops sharply with shrinking nanowire cross-section (Fig. 2a) and that the 10-nm-wide nanowires exhibited a κ value (0.76 ± 0.15 W m^{-1} K^{-1}) that was below the theoretical limit of 0.99 W m^{-1} K^{-1} for bulk Si (ref. 20). Thus, very large data sets were collected for 10- and 20-nm-wide nanowires to allow for a more precise determination of κ.

Our observed high ZT for Si nanowires (Fig. 3) occurs because κ is sharply reduced and the phonon drag component of the thermo-power S_{ph} becomes large. Below, we show that S_{ph} increases because of a three-dimensional to one-dimensional crossover of the phonons participating in phonon drag and decreasing κ. However, we first discuss why our measured κ at 300 K for 10-nm-wide Si nanowires is less than κ_{min} (ref. 20).

The derivation of κ_{min} assumes that the minimum path length of wavelength λ phonons is $\lambda/2$ and that the phonons are described by the Debye model using bulk sound speeds with no optical modes. The $\lambda/2$ value is an order-of-magnitude estimate and is difficult to determine precisely, much like the minimum electron mean free path used to calculate the Mott–Ioffe–Regel σ_{min}. Also, κ_{min} is proportional to the transverse and longitudinal acoustic speeds of sound[20]. These are reduced in our nanowires at long wavelengths because the modes

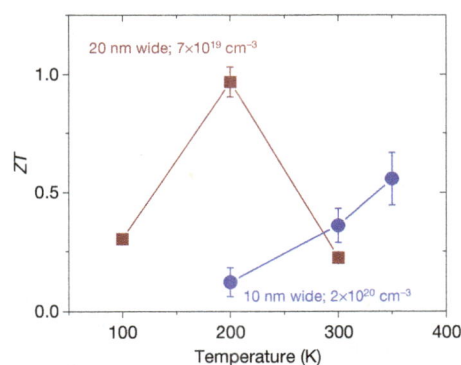

Figure 3 | Temperature dependence of ZT for two different groups of nanowires. The cross-sectional area of the nanowires, and the p-type doping level, are given. The 20-nm-wide nanowires have a thermopower that is dominated by phonon contributions, and a ZT value ~1 is achieved near 200 K. The smaller (10-nm-wide) nanowires have a thermopower that is dominated by electronic contributions. The ZT at 350 K is calculated using the thermal conductivity value for the 10-nm-wide nanowires at 300 K. The error bars represent 95% confidence limits.

become one-dimensional. The ratio of the one-dimensional to two-dimensional longitudinal speeds of sound is $[(1 + v)(1 - 2v)/(1 - v)]^{1/2} = 0.87$ where $v = 0.29$ is the Poisson ratio of Si. The transverse acoustic speed goes to zero at long wavelength because $\omega \propto k^2 d$ where d is the nanowire width[21]. Therefore, the bulk κ_{min} estimate above is invalid for our nanowires and values smaller than κ_{min} are attainable.

For all but the most highly doped nanowires, S peaks near 200 K (Fig. 2b). This peak is unexpected: similarly doped bulk Si exhibits a gradual decrease in S as T is reduced (green trace). For $T < 100$ K, a peaked $S(T)$ is observed for metals and lightly doped semiconductors and is due to phonon drag[13,22,23].

Phonon drag is generally assumed to vanish with decreasing sample dimensions because the phonon path length is limited by the sample size[24–26]. This seems to eliminate phonon drag as the reason for the peak in our nanowires. We show below that the phonon wavelengths participating in drag are of the order of or larger than the wire width. This leads to a three-dimensional to one-dimensional crossover of these modes and removes the cross-sectional wire dimensions from limiting the phonon mean path (see Fig. 4 inset). The nanowire boundaries are incorporated into the one-dimensional mode and are not an obstacle to phonon propagation. Therefore, the limiting size becomes the wire length ($\sim 1 \mu m$) and phonon drag 'reappears' at very small dimensions.

In addition, classical elasticity theory[21] is valid for the phonon wavelengths considered here[27], leading to thermoelastic damping[21,28,29] of sound waves proportional to κ. Thus S_{ph} is further enhanced, owing to the observed reduced thermal conductivity κ. A detailed discussion is in the Supplementary Information.

It might seem that elasticity theory leads to a contradiction because κ is proportional to the mean phonon lifetime. If the phonon lifetimes increase as stated above, then κ should also increase. But

because the elasticity expression is only valid for long-wavelength modes, and κ is the average of all modes, there is no contradiction.

We now consider separately the electronic and phonon contributions to the thermopower—$S = S_e + S_{ph}$—for the nanowire data at $T > 200$ K. Charge carriers dissipate heat to the lattice through a process that first involves momentum conserving (non-dissipative) electron–phonon collisions. The phonons that contribute to phonon drag cannot have a wavelength shorter than λ_{min}, which is determined by the size of the Fermi surface. Phonon drag is observed in metals only at low T because the Fermi surface is large and the heat carrying short-wavelength phonons have short lifetimes. At low T (<20 K), $S_{ph} \propto T^3$ from the phonon specific heat ($\propto T^3$). For $kT \gg \Theta_{Debye}$, the specific heat becomes constant and the number of phonons available for phonon–phonon scattering is $\propto T$, leading to $S_{ph} \propto 1/T$ (ref. 1).

For p-type Si, the holes are near the valence band maximum. The phonon drag modes are acoustic with the largest wavevector $k_{ph} = 2k_{Fermi} = 0.2$ Å^{-1} (for impurity doping of 3×10^{19} cm^{-3}). The shortest wavelength is $\lambda_{ph} = 2\pi/k = 31$ Å.

Umklapp (non-momentum-conserving) phonon–phonon scattering processes determine the rate of phonon heat dissipation. The Debye energy Θ_D sets the energy scale for Umklapp scattering. The number of Umklapp phonons available to dissipate the long-wavelength phonons is given by the Bose–Einstein function:

$$N_U = \frac{1}{e^{\Theta_D/T} - 1}$$

leading to a scattering rate $1/\tau_{ph} \propto N_U$. When $T \gg \Theta_D$, $1/\tau_{ph} \propto T$. Because $\Theta_D = 640$ K for Si, the full Bose–Einstein expression must be applied for $T \lesssim 350$ K. The electronic contribution S_e is estimated from the Mott formula[1]:

$$S_e(T) = \frac{\pi^2 k^2 T}{3e}\left(\frac{\partial \ln \sigma(\varepsilon)}{\partial \varepsilon}\right) \approx (283 \mu V\ K^{-1})(kT/E_F)$$

where the conductivity derivative equals the reciprocal of the energy scale over which it varies (the Fermi energy E_F). Assuming hole doping occurs in the heavier Si valence band (mass 0.49), this leads to $E_F = 0.072$ eV $= 833$ K and $k_F = 0.1$ Å^{-1} for the number density of boron dopant atoms $n = 3 \times 10^{19}$ cm^{-3}. Thus $S_e(T) = aT$ where $a = 0.34 \mu V\ K^{-2}$.

The $T > 200$ K thermopower data of the 20-nm-wide wire (doping $n = 3 \times 10^{19}$ cm^{-3}) fits:

$$S = S_e + S_{ph} = aT + b[\exp(\Theta_D/T) - 1]$$

where a, b and Θ_D are varied to obtain the best fit (Fig. 4). The coefficients are $a = 0.337 \mu V\ K^{-2}$, $b = 22.1 \mu V\ K^{-1}$, and $\Theta_D = 534$ K. The coefficient a is almost identical to our estimate of $0.34 \mu V\ K^{-2}$. Thus phonon drag explains the observed thermopower. Consistent with measurements[24] of phonon drag in bulk Si, S in our nanowires increases significantly at lighter doping. This data, plus a fit of the Fig. 4 data using the experimental $\Theta_D = 640$ K, is presented in the Supplementary Information. Rather than fitting by varying a, b and Θ_D, we fixed Θ_D to its known experimental value (640 K) and allowed only a and b to vary.

The phonon drag contribution to S is of the form[23,30]:

$$S_{ph} \propto \left(\frac{\tau_{ph}}{\mu T}\right)$$

τ_{ph}, the phonon lifetime, is $\propto 1/\kappa$ from elasticity theory. μ is the electron mobility. ZT scales as (neglecting S_e):

$$S_{ph} \propto \frac{1}{\mu T \kappa}$$

$$\sigma \propto n\mu$$

$$ZT \propto \frac{n}{\mu T \kappa^3}$$

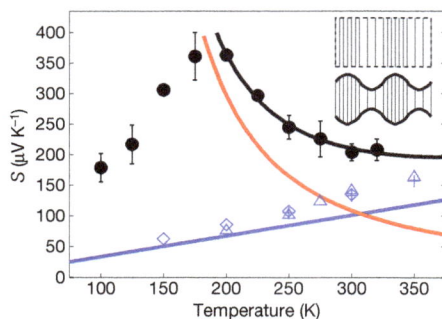

Figure 4 | Thermopower calculation plotted along with experimental data (black points) from a 20-nm-wide Si nanowire p-type doped at 3×10^{19} cm^{-3}. The black curve is the fitted expression for the total thermopower $S_e + S_{ph}$. The red curve is the phonon contribution S_{ph} and the blue line is the electronic term S_e arising from the fit. The fit has maximum error 6.1 $\mu V\ K^{-1}$ and root-mean-square error 1.8 $\mu V\ K^{-1}$. The experimental error bars represent 95% confidence limits and at 150, 200 and 225 K are smaller than the data points. The blue data points are experimental values for bulk wires (doping 2×10^{20} cm^{-3}; crosses), 10-nm-wide nanowires (doping 7×10^{19} cm^{-3}; diamonds), and 20-nm-wide wires (doping 1.3×10^{20} cm^{-3}; triangles) where only a linear-T electronic contribution was found. This data are close to the extracted electronic contribution from the black data points (blue line) and shows that the fitted linear term is reasonable. The drop in S to 0 as $T \to 0$ occurs because the phonon mean free path reaches the sample size and the specific heat tends to 0 according to the third law of thermodynamics. The inset shows the character of a three-dimensional bulk longitudinal acoustic phonon mode (top) and a one-dimensional mode when the wavelength is larger or of the order of the width (bottom). The one-dimensional mode incorporates the existence of the boundary by transverse expansion (or compression) for longitudinal compression (or expansion). The ratio of the transverse strain to the longitudinal strain is the Poisson ratio (0.29 for Si).

Nature | Vol 451 | 10 January 2008 | doi:10.1038/nature06458

leading to increased ZT with decreasing mobility. This is the opposite conclusion reached from when we consider only S_e (ref. 7).

We have combined experiment and theory to demonstrate that semiconductor nanowires can be designed to achieve extremely large enhancements in thermoelectric efficiency, and we have shown that the temperature of maximum efficiency may be tuned by changing the doping and the nanowire size. Theory indicates that similar improvements should be achievable for other semiconductor nanowire systems because of phonon effects. These nanowire thermoelectrics may find applications related to on-chip heat recovery, cooling and power generation. Additional improvements through further optimization of nanowire size, doping and composition should be possible.

METHODS SUMMARY

Single-crystalline Si nanowires were fabricated using the SNAP process[18]. The nanowires were doped p-type using a boron-containing spin-on dopant (see Supplementary Information for details)[19]. Electron-beam lithography was used to create Ti/Pt electrodes for the electrical and heat transport measurements. The entire device was suspended using a XeF_2 etch, leaving the nanowires anchored to a thin SiO island (see Supplementary Information for details). The nanowire electrical conductivity was measured by a Keithley 2400 using a four-point measurement to eliminate contact resistance. For measurement of S and κ, a temperature difference was created across the ends of the nanowires by sourcing a direct current through one of the resistive heaters. The resistance rise of each thermometer was recorded simultaneously using a lock-in measurement (Stanford Research Systems SRS-830) as the temperature was ramped upwards. The resistance of the thermometers was typically two orders of magnitude smaller than the nanowire array. For measurement of S, the thermoelectric voltage, as a response to the temperature difference, was recorded using a Keithley 2182A nanovoltmeter. A difference measurement was used to determine κ, whereby the κ value of the nanowires plus the oxide island was subtracted from the κ value of the oxide island. The thermal conductivity of the oxide island was determined by removing the nanowires with a highly selective XeF_2 etch (see Supplementary Information for details).

Received 15 June; accepted 2 November 2007.

1. MacDonald, D. K. C. *Thermoelectricity: An Introduction to the Principles* (Wiley, New York, 1962).
2. Mahan, G., Sales, B. & Sharp, J. Thermoelectric materials: New approaches to an old problem. *Phys. Today* **50**, 42–47 (1997).
3. Chen, G. *et al.* Recent developments in thermoelectric materials. *Int. Mater. Rev.* **48**, 45–66 (2003).
4. Venkatasubramanian, R. *et al.* Thin-film thermoelectric devices with high room-temperature figures of merit. *Nature* **413**, 597–602 (2001).
5. Harman, T. C. *et al.* Quantum dot superlattice thermoelectric materials and devices. *Science* **297**, 2229–2232 (2002).
6. Hsu, K. F. *et al.* Cubic AgPb$_m$SbTe$_{2+m}$: Bulk thermoelectric materials with high figure of merit. *Science* **303**, 818–821 (2004).
7. Hicks, L. D. & Dresselhaus, M. S. Thermoelectric figure of merit of a one-dimensional conductor. *Phys. Rev. B* **47**, 16631–16634 (1993).
8. Mahan, G. D. & Sofo, J. O. The best thermoelectric. *Proc. Natl Acad. Sci. USA* **93**, 7436–7439 (1996).
9. Humphrey, T. E. & Linke, H. Reversible thermoelectric nanomaterials. *Phys. Rev. Lett.* **94**, 096601 (2005).
10. Boukai, A., Xu, K. & Heath, J. R. Size-dependent transport and thermoelectric properties of individual polycrystalline bismuth nanowires. *Adv. Mater.* **18**, 864–869 (2006).
11. Yu-Ming, L. *et al.* Semimetal-semiconductor transition in Bi$_{1-x}$Sb$_x$ alloy nanowires and their thermoelectric properties. *Appl. Phys. Lett.* **81**, 2403–2405 (2002).
12. Majumdar, A. Enhanced thermoelectricity in semiconductor nanostructures. *Science* **303**, 777–778 (2004).
13. Weber, L. & Gmelin, E. Transport properties of silicon. *Appl. Phys. A* **53**, 136–140 (1991).
14. Small, J. P., Perez, K. M. & Kim, P. Modulation of thermoelectric power of individual carbon nanotubes. *Phys. Rev. Lett.* **91**, 256801 (2003).
15. Li, S. *et al.* Measuring thermal and thermoelectric properties of one-dimensional nanostructures using a microfabricated device. *J. Heat Transf.* **125**, 881–888 (2003).
16. Li, D. *et al.* Thermal conductivity of individual silicon nanowires. *Appl. Phys. Lett.* **83**, 2934–2936 (2003).
17. Morales, A. M. & Lieber, C. M. A laser ablation method for the synthesis of semiconductor crystalline nanowires. *Science* **279**, 208–211 (1998).
18. Melosh, N. A. *et al.* Ultra-high density nanowire lattices and circuits. *Science* **300**, 112–115 (2003).
19. Wang, D., Sheriff, B. A. & Heath, J. R. Complementary symmetry silicon nanowire logic: Power-efficient inverters with gain. *Small* **2**, 1153–1158 (2006).
20. Cahill, D. G., Watson, S. K. & Pohl, R. O. Lower limit to the thermal conductivity of disordered crystals. *Phys. Rev. B* **46**, 6131–6140 (1992).
21. Landau, L. D. & Lifshitz, E. M. in *Theory of Elasticity* 3rd edn, 138 (Butterworth Heinemann, Oxford, 1986).
22. Pearson, W. B. Survey of thermoelectric studies of the Group 1 metals at low temperatures carried out at the National Research Laboratories, Ottawa. *Sov. Phys. Solid State* **3**, 1024–1033 (1961).
23. Herring, C. Theory of the thermoelectric power of semiconductors. *Phys. Rev.* **96**, 1163–1187 (1954).
24. Geballe, T. H. & Hull, G. W. Seebeck effect in silicon. *Phys. Rev.* **98**, 940–947 (1955).
25. Behnen, E. Quantitative examination of the thermoelectric power of n-type Si in the phonon drag regime. *J. Appl. Phys.* **67**, 287–292 (1990).
26. Trzcinksi, R., Gmelin, E. & Queisser, H. J. Quenched phonon drag in silicon microcontacts. *Phys. Rev. Lett.* **56**, 1086–1089 (1986).
27. Maranganti, R. & Sharma, P. Length scales at which classical elasticity breaks down for various materials. *Phys. Rev. Lett.* **98**, 195504 (2007).
28. Lifshitz, R. & Roukes, M. L. Thermoelastic damping in micro- and nanomechanical systems. *Phys. Rev. B* **61**, 5600–5609 (2000).
29. Zener, C. Internal friction in solids. I. Theory of internal friction in reeds. *Phys. Rev.* **52**, 230–235 (1937).
30. Gurevich, L. The thermoelectric properties of conductors. *Zhurnal Eksperimentalnoi I Teoreticheskoi Fiziki* **16**, 193–228 (1946).

Supplementary Information is linked to the online version of the paper at www.nature.com/nature.

Acknowledgements We thank D. Wang for discussions and J. Dionne, M. Roy, K. Kan and T. Lee for fabrication assistance. This work was supported by the Office of Naval Research, the Department of Energy, the National Science Foundation, the Defense Advanced Research Projects Agency, and a subcontract from the MITRE Corporation.

Author Contributions A.I.B., Y.B., J.-K.Y. and J.R.H. contributed primarily to the design and execution of the experiments. J.T.-K. and W.A.G. contributed primarily to the theory.

Author Information Reprints and permissions information is available at www.nature.com/reprints. Correspondence and requests for materials should be addressed to J.R.H. (heath@caltech.edu).

Thin-film thermoelectric devices with high room-temperature figures of merit

Rama Venkatasubramanian, Edward Siivola, Thomas Colpitts & Brooks O'Quinn

Research Triangle Institute, Research Triangle Park, North Carolina 27709, USA

Thermoelectric materials are of interest for applications as heat pumps and power generators. The performance of thermoelectric devices is quantified by a figure of merit, ZT, where Z is a measure of a material's thermoelectric properties and T is the absolute temperature. A material with a figure of merit of around unity was first reported over four decades ago, but since then—despite investigation of various approaches—there has been only modest progress in finding materials with enhanced ZT values at room temperature. Here we report thin-film thermoelectric materials that demonstrate a significant enhancement in ZT at 300 K, compared to state-of-the-art bulk Bi_2Te_3 alloys. This amounts to a maximum observed factor of ~2.4 for our p-type Bi_2Te_3/Sb_2Te_3 superlattice devices. The enhancement is achieved by controlling the transport of phonons and electrons in the superlattices. Preliminary devices exhibit significant cooling (32 K at around room temperature) and the potential to pump a heat flux of up to 700 W cm^{-2}; the localized cooling and heating occurs some 23,000 times faster than in bulk devices. We anticipate that the combination of performance, power density and speed achieved in these materials will lead to diverse technological applications: for example, in thermochemistry-on-a-chip, DNA microarrays, fibre-optic switches and microelectrothermal systems.

The performance of thermoelectric devices depends on the figure of merit (ZT) of the material, given by

$$ZT = (\alpha^2 T / \rho K_T) \qquad (1)$$

where α, T, ρ and K_T are the Seebeck coefficient, absolute temperature, electrical resistivity and total thermal conductivity, respectively. Z, the material coefficient, can be expressed in terms of lattice thermal conductivity (K_L), electronic thermal conductivity (K_e) and carrier mobility (μ), for a given carrier density (p) and the corresponding α, yielding equation (2), below. Here L_0 is the Lorenz number, approximately 1.5×10^{-8} V^2 K^{-2} in non-degenerate semiconductors, q is the electronic charge, and σ is electrical conductivity.

$$Z = \frac{\alpha^2 \sigma}{K_L + K_e} \simeq \frac{\alpha^2}{\left(\dfrac{K_L}{\mu p q}\right) + L_0 T} \qquad (2)$$

State-of-the-art devices utilize alloys, typically p-$Bi_xSb_{2-x}Te_{3-y}Se_y$ ($x \approx 0.5$, $y \approx 0.12$) and n-$Bi_2(Se_yTe_{1-y})_3$ ($y \approx 0.05$) for the 200–400 K temperature range. For certain alloys, K_L can be reduced more strongly than μ, leading[1] to enhanced ZT. A ZT of 0.75 at 300 K in p-type $Bi_xSb_{2-x}Te_3$ ($x \approx 1$) was reported[2] 40 years ago. Since then, there has been modest progress in the ZT of thermoelectric materials near 300 K. The highest ZT in any bulk thermoelectric material at 300 K appears to be ~1.14 for p-type $(Bi_2Te_3)_{0.25}(Sb_2Te_3)_{0.72}(Sb_2Se_3)_{0.03}$ alloy[3], although there has been a report[4] of an estimated $ZT > 2$ at 300 K in $Bi_{0.5}Sb_{1.5}Te_3$ alloy under a hydrostatic pressure of 2 GPa. A ZT value of greater than one in crystals of Bi–Sb alloy under a magnetic field[5] at ~100 K has been reported. There is no apparent thermodynamic upper limit on ZT; in addition to its engineering importance, values of ZT significantly larger than unity are of interest to transport theory[6].

Several possible approaches to enhancing ZT have been investigated. In bulk materials, cage-like structures simulating a phonon glass/electron crystal[7] have been examined, with a view to reducing K_L without a deterioration of the electronic mobilities. A value of $ZT > 1$ in $LaFe_3CoSb_{12}$ was reported[8] at $T > 700$ K, and attributed primarily to reduction of K_L from La-filling[9]. A ZT of ~1.35 was reported for the skutterudite $CeFe_{3.5}Co_{0.5}Sb_{12}$ at ~900 K (ref. 10) where, although a dramatic reduction in K_L of the filled skutterudite was observed near 300 K, there was apparently less role of Ce-filling at higher temperatures where the enhanced ZT was observed. A ZT

of ~0.8 at 225 K was observed[11] in $CsBi_4Te_6$, significantly higher than in $Bi_{0.5}Sb_{1.5}Te_3$, apparently due to the smaller bandgap of the former compound (M.G. Kanatzidis, personal communication). From these results, we observe the following: (1) A ZT significantly greater than 1 has not been demonstrated at ordinary temperatures (300 K). (2) A one-to-one correlation between lower K_L and enhanced ZT has not been established. (3) More importantly, the concept of individually tailoring the phonon properties without producing a deterioration of electronic transport, thereby enhancing ZT, has not been established.

Thin-film thermoelectric materials[12] offer tremendous scope for ZT enhancement. Three generic approaches have been proposed to date. One involves the use of quantum-confinement effects[13] to obtain an enhanced density of states near the Fermi energy. Using such effects, a ZT of 0.9 at 300 K and 2.0 at 550 K, using estimated thermal-conductivity values, has been reported[14] in $PbSe_{0.98}Te_{0.02}/PbTe$ quantum-dot structures. The second approach[15] involves phonon-blocking/electron-transmitting superlattices. These structures utilize the acoustic mismatch between the superlattice components to reduce K_L (refs 16, 17), rather than using the conventional alloying approach, thereby potentially eliminating alloy scattering of carriers. The third thin-film approach is based on thermionic effects in heterostructures[18,19]. Here we report a ZT at 300 K of ~2.4 in p-type Bi_2Te_3/Sb_2Te_3 superlattices and a similar, although less dramatic, $ZT \approx 1.4$ in n-type $Bi_2Te_3/Bi_2Te_{2.83}Se_{0.17}$ superlattices. These ZT values were measured in devices using the Harman technique[20], in which parameters related to ρ, K_T and α (that make up ZT) are measured at the same place, at the same time, with current flowing. This method has been extended with a variable-thickness approach (C.B. Vining, personal communication) to obtain the intrinsic ZT and other parameters.

Potential ideality of superlattices in the Bi_2Te_3 system

High-quality superlattices have been produced[21,22] in the Bi_2Te_3 system, with one of the individual layers as small as 10 Å, using a low-temperature growth process. Such ultra-short-period superlattices offer significantly higher in-plane carrier mobilities (parallel to the superlattice interfaces) than alloys, owing to the near absence of alloy scattering and random interface carrier scattering[23]. We show that the enhanced carrier mobilities in monolayer-range superlattices are effective in the cross-plane direction for certain

When citing this article, please cite the original version as shown on the contents page of this chapter.

superlattices, where we can also obtain reduced K_L and so an enhanced ZT as per equation (2).

We adapted the transmission line model (TLM) technique[23] used for the measurement of specific electrical contact resistivities (ρ_c)[24] to determine the cross-plane electrical transport. This adaptation was feasible with the low ρ_c ($< 10^{-7}\,\Omega\,cm^2$) achievable in these materials. With such ρ_c, the transfer lengths (L_t) are small, and so an effective 'squeezing' of the current occurs. A typical TLM device consists of adjacent metallized areas, spaced a few micrometres apart, serving as top current-injection contacts. The current is confined to the thin-film on the substrate. The film between the contacts is etched anisotropically to create various device thickness underneath the contacts[23]. Measuring the intercept resistances using successive etch-steps gives the electrical resistivity perpendicular to the superlattice interfaces. We observe that the average length of the contact is $\sim 250\,\mu m$, much greater than L_t ($\sim 1\,\mu m$). Thus the top metal contact, the thermally conducting substrate, and the inactive (where current is not flowing) thermoelectric material act like thermal shunts, in addition to exacerbated radiative/convective losses from the close-to-surface L_t region, for any cooling developed across this region. Thus we have performed an isothermal electrical resistivity measurement, confirmed by negligible Peltier voltage (in a Harman method[20] set-up) in such TLM structures. Further, the cross-plane electrical resistivities determined by the TLM method were confirmed by the variable-thickness ZT method in some of the high-ZT samples, as discussed subsequently. This method has been tested on n-Bi_2Te_3 thin films with known theoretical and experimental electrical anisotropy[25]; we measured an anisotropy of 3.8 ± 0.32 and the theoretical anisotropy is ~ 4.1 (ref. 25). The validity was also confirmed in p-type $Bi_xSb_{2-x}Te_3$ ($x \approx 0.63 \pm 0.12$) alloy[26] films.

The anticipated heterostructure band diagram in these short-period/shallow potential superlattices is shown in Fig. 1a, where the valence-band offset is expected to be less than the average thermal energy of carriers. The variation of the mobility anisotropy as a function of superlattice period, for $d_{Bi_2Te_3} = d_{Sb_2Te_3}$, is shown in Fig. 1b. $d_{Bi_2Te_3}$ and $d_{Sb_2Te_3}$ are the thickness of the Bi_2Te_3 and Sb_2Te_3 layers, respectively, that make up a period. The carrier concentration is isotropic, so the ratio of electrical resistivity anisotropy is inversely related to the mobility anisotropy (or $\mu_\perp/\mu_{in\text{-}plane}$, where $\mu_\perp$ is the cross-plane mobility). The mobility anisotropies of the 10Å/10Å, 10Å/20Å, 10Å/50Å, 20Å/20Å, 20Å/30Å and 20Å/40Å structures are shown in Fig. 1c. (Here 10Å/50Å, for example, indicates $d_{Bi_2Te_3} = 10\,Å$ and $d_{Sb_2Te_3} = 50\,Å$ in the superlattice.) The measured cross-plane mobilities in the Bi_2Te_3/Sb_2Te_3 superlattices appear to be consistent with a miniband conduction across the superlattice interfaces, with fully developed or larger valence-band offsets in structures for Bi_2Te_3 thickness $\geq 30\,Å$. The data are qualitatively similar to the transport models developed for AlGaAs/GaAs superlattices[27,28]. Note from Fig. 1b, c that the 10Å/50Å, 20Å/40Å and 10Å/20Å Bi_2Te_3/Sb_2Te_3 superlattices offer cross-plane electrical conductivities comparable to in-plane values. It has been shown[29] that periods of 50–60 Å are desirable for minimizing K_L; thus the 10Å/50Å and 20Å/40Å superlattices are useful for enhanced ZT.

Bi_2Te_3/Sb_2Te_3 superlattices show significantly reduced K_L. A full description of the measurement procedure, the data, and the mechanisms are presented elsewhere[29]. The behaviour of these superlattices are consistent with phonon-transmission experiments in GaAs/AlGaAs superlattices[30], although we have invoked coherent backscattering of phonons at superlattice interfaces[29], similar to photon-localization in highly scattering media[31,32]. Values of K_L of the superlattice structures show a minimum of $\sim 0.22\,W\,m^{-1}\,K^{-1}$ at $\sim 50\,Å$; this is a factor of 2.2 smaller than the K_L of $Bi_{0.5}Sb_{1.5}Te_3$ alloy ($\sim 0.49\,W\,m^{-1}\,K^{-1}$) along the same c axis. We note that a theory for the thermal conductivity minimum has been proposed[33]. K_L for all the 60-Å-period structures, whether it be 30Å/30Å or 10Å/50Å

20Å/40Å, is about $0.25\,W\,m^{-1}\,K^{-1}$. K_L appears to be more dependent on the superlattice period[29], relatively independent of the thickness of the constituents, whereas the electronic transport depends on the period and the relative thickness of the constituents (Fig. 1b, c).

A K_L value of $0.22\,W\,m^{-1}\,K^{-1}$ can be compared with the minimum thermal conductivities (K_{min}) predicted for Bi_2Te_3, using the full-wavelength or half-wavelength models of Slack[34] and Cahill[35] for the phonon mean free path. Table 1 shows these predicted values, along with experimental K_L values of alloys along the trigonal axis (c axis) or perpendicular to it (a-b axis). The K_L values in the superlattices approach the values of K_{min} estimated from the models of Slack and Cahill. When all the phonons have a mean free path equal to the lattice spacing, K_{min} is expected[36] to be between the predictions of the models of Slack and Cahill. The phonon mean free path from kinetic theory is approximately 2.2 Å, using a K_L of $0.22\,W\,m^{-1}\,K^{-1}$ (ref. 29), assuming that all phonons have similar mean free paths—this is justifiable, given that the Debye temperature of Bi_2Te_3 is $\sim 160\,K$, far lower than the temperature of measurement (300 K).

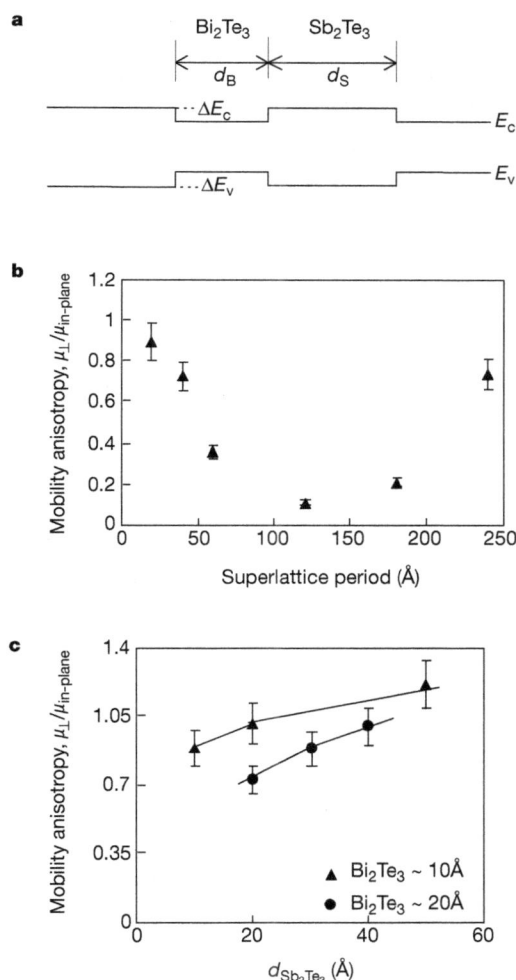

Figure 1 Hole transport across the Bi_2Te_3/Sb_2Te_3 superlattice interface. **a**, The anticipated heterojunction band diagram. $\Delta E_v + \Delta E_c < 2kT$, $\Delta E_v \approx kT$, $d_s + d_B =$ superlattice period (here ΔE_v and ΔE_c represent the valence and conduction band offsets, respectively, in the heterostructure, and $kT \approx 0.0285\,eV$ at 300 K). **b**, The observed hole mobility anisotropy versus superlattice period for the case of $d_B = d_s = (1/2)$ period. **c**, The hole mobility anisotropy in other Bi_2Te_3/Sb_2Te_3 superlattices.

Table 1 Theoretical and experimental lattice thermal conductivities

Material	Thermal conductivity (W m^{-1} K^{-1})
K_{min} of Bi$_2$Te$_3$ (a-b axis), Slack model[34]	0.55
K_{min} of Bi$_2$Te$_3$ (c axis), Slack model[34]	0.28
K_{min} of Bi$_2$Te$_3$ (a-b axis), Cahill model[35]	0.28
K_{min} of Bi$_2$Te$_3$ (c axis), Cahill model[35]	0.14
K_L of Bi$_{2-x}$Sb$_x$Te$_3$ alloy (a-b axis)	0.97
K_L of Bi$_{2-x}$Sb$_x$Te$_3$ alloy (c axis)	0.49
K_L of Bi$_2$Te$_3$/Sb$_2$Te$_3$ superlattice (c axis)	0.22

Lattice thermal conductivity (K_L) of the Bi$_2$Te$_3$/Sb$_2$Te$_3$ superlattice (period ~50 Å) compared with K_L observed in the respective alloys and the theoretical minimum lattice thermal conductivity (K_{min}) from various models.

Variable-thickness *ZT* measurements

ZT of a thermoelectric material can be obtained from a unipolar thermoelement across which a temperature difference is developed by the Peltier effect using a quasi-steady-state current[20]. The unipolar thermoelement consists of a mesa device with two ohmic contacts for current injection. One end of the device is effectively attached to a heat sink at a fixed temperature by being on a thermally conducting substrate. A schematic of the four-probe measurement is shown in Fig. 2a. A current (I) leads to ohmic (V_R) and Peltier (V_0) voltages across the thermoelement, with the total voltage V_T being $V_R + V_0$ at steady state. With current off, V_R decays within the dielectric relaxation time but V_0 decays according to the thermal time constant. V_R includes resistance from the bulk resistance of the mesa and from that of the two contacts (equation (3)).

$$V_R = \rho_\perp (l/a)I + 2(\rho_c/a)I \tag{3}$$

Here, $\rho_\perp$, ρ_c (in Ω cm^2), l and a are respectively the cross-plane electrical resistivity of the superlattice, the average specific resistivity

of the two contacts, and the height and the cross-sectional area of the device. V_0, given by equation (4) below, is affected by any difference in heating at the two contacts. As shown below, the ρ_c in the devices is rather small; hence, differences in the two contacts, causing any relative voltages, is even smaller. Even so, currents were reversed, and an average V_0 was used to account for minor differences.

$$V_0 = (\alpha_\perp^2/K_T)T(l/a)I \tag{4}$$

From equations (3) and (4), we get extrinsic or device figure of merit (ZT_e) as (V_0/V_R) shown in equation (5)[20] below. In thick (large-l) bulk elements, where $2(\rho_c/\rho_\perp l)$ is rather small, ZT_e approaches the intrinsic figure of merit (ZT_i) given by equation (1).

$$ZT_e = V_0/V_R = \{\alpha_\perp^2 T/K_T\rho_\perp\}[1 + 2(\rho_c/\rho_\perp l)]^{-1} \tag{5}$$

We can get $\rho_\perp$ and $(\alpha_\perp^2/K_T)T$, respectively, from the variation of V_R and V_0 with l (or mathematically equivalent, Il/a). Thus ZT_i is:

$$ZT_i = [\partial V_0/\partial(Il/a)]/[\partial V_R/\partial(Il/a)] \tag{6}$$

The data from variable-device-thickness experiments are shown in Fig. 2b and c. We point out that each data point in Fig. 2b and c is an average of 125 transient measurements, stored digitally and averaged out. The variable-thickness method for extracting ZT_i works for medium-to-good ρ_c, that is, medium-to-good device ZT_e shown in Fig. 2b and c, respectively. In Fig. 2b, we obtain ZT_e (given by V_0/V_R) for the largest mesa thickness of ~1.35 μm of ~0.66; even so, the ratio of the two slopes, from equation (6) above, gives a ZT_i of 2.07. The y-intercept of V_R versus Il/a is used to obtain ρ_c, ~1.45 × 10^{-7} Ω cm^2 for the sample in Fig. 2b. This is more than a factor of ten smaller than in bulk technology, but still high for thin-film devices, limiting the extrinsic ZT to ~0.66. In Fig. 2c, however, a combination of thicker 10Å/50Å Bi$_2$Te$_3$/Sb$_2$Te$_3$ superlattice (thickest mesa ~2.67 μm) and an improved ρ_c of

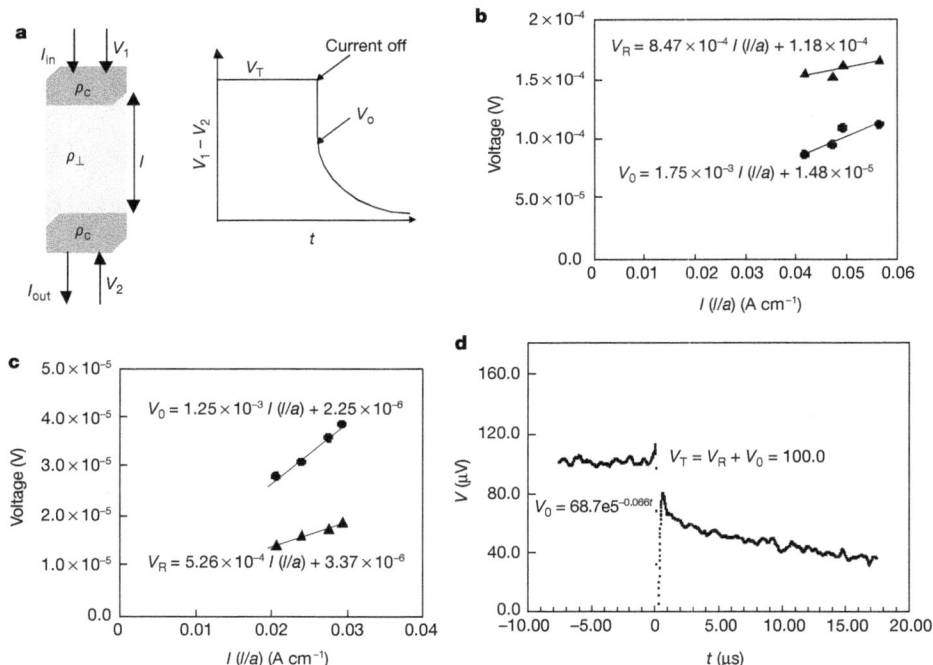

Figure 2 The *ZT* of a thermoelectric device. **a**, Direct measurement by a four-probe Harman method. (See text for definitions of symbols.) **b, c**, Variable-thickness *ZT* measurements. V_R and V_0 obtained from Harman method for each thickness indicate ZT_i ≈ 2.07 and 2.34 in **b** and **c**, respectively. Working conditions: **b**, I = 20.5 mA and area, 4.9 × 10^{-5} cm^2; **c**, I = 20.3 mA and area, 1.8 × 10^{-4} cm^2. Triangles, V_R; circles, V_0. **d**, Harman-method transient on a 5.4-μm-thick 10Å/50Å Bi$_2$Te$_3$/Sb$_2$Te$_3$ superlattice

device, shown for one current direction. The average extrinsic *ZT* for the two current directions is 2.38 ± 0.19. The voltage spike at t ≈ 0, from momentary inductive coupling between the current and voltage circuits, prevents reliable determination of residual V_0 for t < 1 μs; thus extrapolation of data is used from t > 1 μs to get V_0. The transient is an average of 125 consecutive measurements, stored digitally and averaged out, which should lead to minimal error in extrapolation in spite of the spike.

$1.5 \times 10^{-8}\ \Omega\ \text{cm}^2$ leads to a ZT_e value of ~ 2.04 for the thickest device. Based on the ratio of two slopes, a ZT_i of ~ 2.34 is deduced.

For the $10\text{Å}/50\text{Å}$ Bi_2Te_3/Sb_2Te_3 superlattice in Fig. 2b, the measured $\rho_\perp$ is $8.47 \times 10^{-4}\ \Omega\ \text{cm}$. With $\rho_{\text{in-plane}}$ of $9.48 \times 10^{-4}\ \Omega\ \text{cm}$, $\rho_\perp/\rho_{\text{in-plane}}$ or $\mu_\perp/\mu_{\text{in-plane}}$ is ~ 1.12. For the sample in Fig. 2c, $\rho_\perp$ is $5.26 \times 10^{-4}\ \Omega\ \text{cm}$ and $\rho_{\text{in-plane}}$ is $5.5 \times 10^{-4}\ \Omega\ \text{cm}$, thus giving an anisotropy of 1.05. These values are comparable to that indicated in Fig. 1c for the $10\text{Å}/50\text{Å}$ Bi_2Te_3/Sb_2Te_3 superlattice, confirming the ideal electron (hole) transmission characteristics of the structure.

Figure 2d shows the Harman-method transient[20] for a ~ 5.4-μm-thick $10\text{Å}/50\text{Å}$ Bi_2Te_3/Sb_2Te_3 superlattice thermoelement. The average ZT_e for two current directions was 2.38 ± 0.19 at 300 K. The material parameters for this superlattice film indicated a ZT_i of 2.59, and the measured ZT_e of 2.38 translates to a ρ_c of $\sim 1.3 \times 10^{-8}\ \Omega\ \text{cm}^2$. Thus, going from a superlattice device of ~ 1.35 μm to ~ 2.67 μm to ~ 5.4 μm, we see no significant change in ZT_i and so the ZT_e begins to approach ZT_i because of a decreasing role of ρ_c.

As a check, we have measured ZT on 1-mm-thick and thinned-down (5–20 μm) thermoelements made from bulk p-type $Bi_xSb_{2-x}Te_{3-y}Se_y$ ($x \approx 0.63 \pm 0.12$; $y \approx 0.12$) alloys, comparable to those studied in ref. 3. The variable-thickness method indicates a ZT_i of ~ 1.09 at 300 K along the a-b axis in agreement with the results of ref. 3, indicating no unexpected benefits resulting from utilizing a thin-film version of the commercial material. Also, ZT in alloy films (non-superlattice structures) along the c axis were in the range of 0.4 ± 0.13 for carrier levels of $\sim 3 \times 10^{19}\ \text{cm}^{-3}$, consistent with reported values for bulk materials[26].

Using temperature-dependent measurements of ρ, α and K_T of the p-type superlattice—and noting that there is little electrical anisotropy—the estimated ZT_i versus T of $10\text{Å}/50\text{Å}$ p-type Bi_2Te_3/Sb_2Te_3 superlattice is shown in Fig. 3, along with those of other materials. It appears that the superlattices would offer enhanced performance compared to the bulk p-type $Bi_{2-x}Sb_xTe_3$ alloys at lower temperatures. This is not surprising, given that the efficacy of superlattices in reducing K_L improves at lower temperatures as in Si/Ge superlattices[17]. As 'proof of advantage' of enhanced ZT at lower temperatures, we have obtained four times the cooling with Bi_2Te_3/Sb_2Te_3 superlattice devices compared to bulk p-type $Bi_{2-x}Sb_xTe_3$ devices at 210 K, for similar l/a, thermal load and parasitics. (Parasitics are unintentional heat losses from conductive, convective and any radiative processes.) Figure 3 shows that the p-type $10\text{Å}/50\text{Å}$ Bi_2Te_3/Sb_2Te_3 superlattice offers improved ZT compared to $CsBi_4Te_6$ alloy[11] at ~ 210 K.

Phonon-blocking/electron-transmitting structures

The results obtained with the $10\text{Å}/50\text{Å}$ Bi_2Te_3/Sb_2Te_3 superlattices indicate that we can fine-tune the phonon and hole (charge carriers)

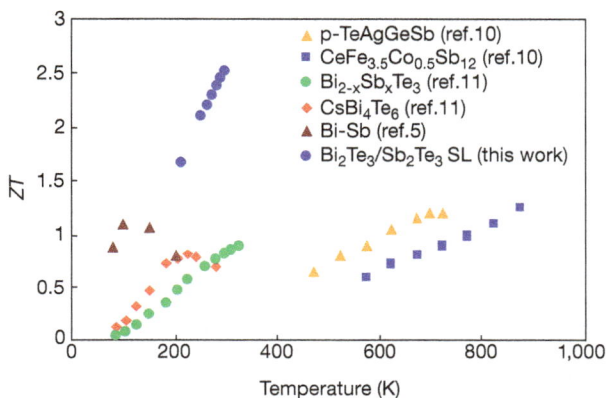

Figure 3 Temperature dependence of ZT of $10\text{Å}/50\text{Å}$ p-type Bi_2Te_3/Sb_2Te_3 superlattice compared to those of several recently reported materials.

transport[37] to improve ZT. Other structures that show $ZT > 1$ at 300 K include $10\text{Å}/40\text{Å}$ and $20\text{Å}/40\text{Å}$ Bi_2Te_3/Sb_2Te_3 superlattices. It is useful to compare the transport characteristics of phonons and holes. The product[38] kl_{mfp}, where k and l_{mfp} are respectively the average wavevector and mean free path, is used as a measure of the amount of blocking (or lack of it) for phonons and holes. For holes, we estimate the thermal velocity (v_{th}), de Broglie wavelength, and the wavevector magnitude as $\sim 2.1 \times 10^7\ \text{cm s}^{-1}$, ~ 114 Å, and $\sim 5.5 \times 10^6\ \text{cm}^{-1}$, respectively. From the relation[39] $(l_{\text{mfp}})_\perp \approx (v_{\text{th}}m^*\mu_\perp/q)$, we obtain $(l_{\text{mfp}})_\perp$ as 136 Å (for the sample with $ZT_i \approx 2.34$, $\mu_\perp$ of $383\ \text{cm}^2\ \text{V}^{-1}\ \text{s}^{-1}$ is deduced from $\rho_\perp$ obtained from variable-thickness ZT). (Here m^* is the effective mass of charge carriers; holes in this case.) For an average phonon wavelength of ~ 30 Å and hole wavelength of ~ 114 Å, we obtain $(kl_{\text{mfp}})_{\text{phonons}} \approx 0.5$ and $(kl_{\text{mfp}})_{\text{holes}} \approx 7.6$. We believe that this comparison probably captures the phonon-blocking/electron-transmitting nature of certain superlattices[37].

n-type superlattices

We have also obtained encouraging results with n-type $10\text{Å}/50\text{Å}$ $Bi_2Te_3/Bi_2Te_{2.83}Se_{0.17}$ superlattices, indicating $ZT_i > 1$ at 300 K. The variable-thickness ZT measurements, similar to those shown in Fig. 2b and c, indicate a ZT_i of ~ 1.46 at 300 K in these superlattices. The best extrinsic ZT_e for an n-type device has been ~ 1.2 at 300 K. The closeness of ZT_i and ZT_e is from obtaining a ρ_c value of $\sim 1.2 \times 10^{-8}\ \Omega\ \text{cm}^2$. The cross-plane (along c axis) electrical resistivity in these n-type superlattices is $\sim 1.23 \times 10^{-3}\ \Omega\ \text{cm}$, not significantly higher than the in-plane (along a-b axis) electrical resistivity of $1.04 \times 10^{-3}\ \Omega\ \text{cm}$. Thus in the $10\text{Å}/50\text{Å}$ n-type super-lattices, apparently due to weak-confinement/near-zero band-offset, there is minimal anisotropy between in-plane and cross-plane electrical resistivities, similar to the $10\text{Å}/50\text{Å}$ and other p-type Bi_2Te_3/Sb_2Te_3 superlattices (Fig. 1c). The lack of electrical anisotropy in both the high-performance p-type and n-type superlattices, comparing properties along the a-b and c crystallographic axes, is in marked contrast to the electrical anisotropy observed in both p-type and n-type bulk materials[26].

The reason for the less-than-impressive ZT_i of 1.46 at 300 K in $Bi_2Te_3/Bi_2Te_{2.83}Se_{0.17}$ superlattices, compared to ~ 2.4 at 300 K in the best p-type Bi_2Te_3/Sb_2Te_3 superlattices, is its higher K_L. From its $\partial V_0/\partial(Il/a)$ and an α of $\sim -238\ \mu\text{V K}^{-1}$, a cross-plane K_T of $\sim 9.45\ \text{mW cm}^{-1}\ \text{K}^{-1}$ is obtained. Using the Weidemann–Franz law and a $\rho_\perp$ value of $1.23 \times 10^{-3}\ \Omega\ \text{cm}$, we obtain an electronic thermal conductivity of $\sim 3.7\ \text{mW cm}^{-1}\ \text{K}^{-1}$. Thus the cross-plane lattice thermal conductivity is $\sim 5.8\ \text{mW cm}^{-1}\ \text{K}^{-1}$. This is more like the c-axis K_L of bulk alloys[26]—and much higher than the value of $\sim 2.5\ \text{mW cm}^{-1}\ \text{K}^{-1}$ observed in 60-Å-period p-type Bi_2Te_3/Sb_2Te_3 superlattices in ref. 29, and higher than that derived from the analysis of data in Figs. 2b, c. We believe that these results indicate near-ideal superlattice interfaces in the Bi_2Te_3/Sb_2Te_3 system[21], where the compositional modifications of either Bi or Sb are accomplished within regions enclosed by the Te–Te van der Waals bond. Thus, it should be possible to obtain mirror-like superlattice interfaces, leading to potential reflection effects for reducing K_L (ref. 29). In contrast, in the $Bi_2Te_3/Bi_2Te_xSe_{3-x}$ system, where both Se and Te are expected to be present at the van der Waals interface, we anticipate much unintended compositional mixing. Thus, the lattice thermal conductivity is likely to be more typical of an alloy. This was further evident in our observations of a lack of any significant improvement in electronic mobilities in n-type $Bi_2Te_3/Bi_2Te_{2.83}Se_{0.17}$ superlattices (compared to n-type $Bi_2Te_{3-x}Se_x$ alloys), in contrast to a marked enhancement of carrier mobilities observed in p-type Bi_2Te_3/Sb_2Te_3 superlattices relative to p-type $Bi_{2-x}Sb_xTe_3$ alloys[21]. Therefore, enhancement of ZT_i to 1.46 in n-type super-lattices (from ~ 1 for bulk n-type alloy) appears to be attributable to the lack of electrical anisotropy and the typical lower lattice thermal conductivity associated with the c axis ($\sim 5.8\ \text{mW cm}^{-1}\ \text{K}^{-1}$

compared to ~10 mW cm^{-1} K^{-1} along the *a-b* axis). Better n-type superlattices with low K_L could be possible in those superlattices that have negligible compositional mixing at the van der Waals interface.

Thermoelectric devices for localized, rapid cooling/heating

Figure 4a shows the sub-ambient cooling obtained in a p-type superlattice device. With an ambient temperature $T_{ambient}$ of about 298 K, we have 32.2 K of cooling (down to −7 °C)—without any forced heat removal by blowing air or running water at the heat sink. Cooling was measured with 12.5-μm thermocouple wires (bead size ~50 μm). Thus the measured cooling is almost always with a load comparable to the thin-film device and in spite of the significant probe-heating on top of the mesa. Under similar testing conditions (same *l/a*, lack of heat removal at heat sink, residual thermal load/heat loss), we measured ~18.4 K of cooling in p-type bulk devices (Fig. 4a). In the linear regime (current < 1 A) where heat-sinking considerations are less, the superlattice device shows a factor of 2.2 cooling compared to bulk. We observed 40 K of cooling for the superlattice device when the heat sink is maintained at 353 K using a large-wattage heater, equivalent to a semi-infinite heat-source. This is analogous to the cooling of heat-sensitive devices in regions adjacent to hotspots in a chip. These results can be compared to recent measurements of cooling of 2.8 K and 6.9 K, measured using micro-thermocouples at 298 K and 373 K, respectively, in SiGeC/Si devices[40] of larger *l/a* ratio (~12.5).

Thin-film thermoelements lead to large cooling power densities (P_D) for similar maximum heat pumping (Q_m)[41] as per equation (7) below, following equation (1). As the area of the device, *a*, is reduced proportional to its thickness, *l*, Q_m is unaffected. However, cooling power density ($P_D \approx Q/a$) is increased. Here $T_{hot\text{-}side}$ and $T_{cold\text{-}side}$ refer to the absolute temperature of the hot and cold junctions, respectively, at either end of the active thermoelement.

$$Q_m = (a/l)\{[0.5\alpha^2 T_{cold\text{-}side}^2/\rho] - [K_T(T_{hot\text{-}side} - T_{cold\text{-}side})]\} \quad (7)$$

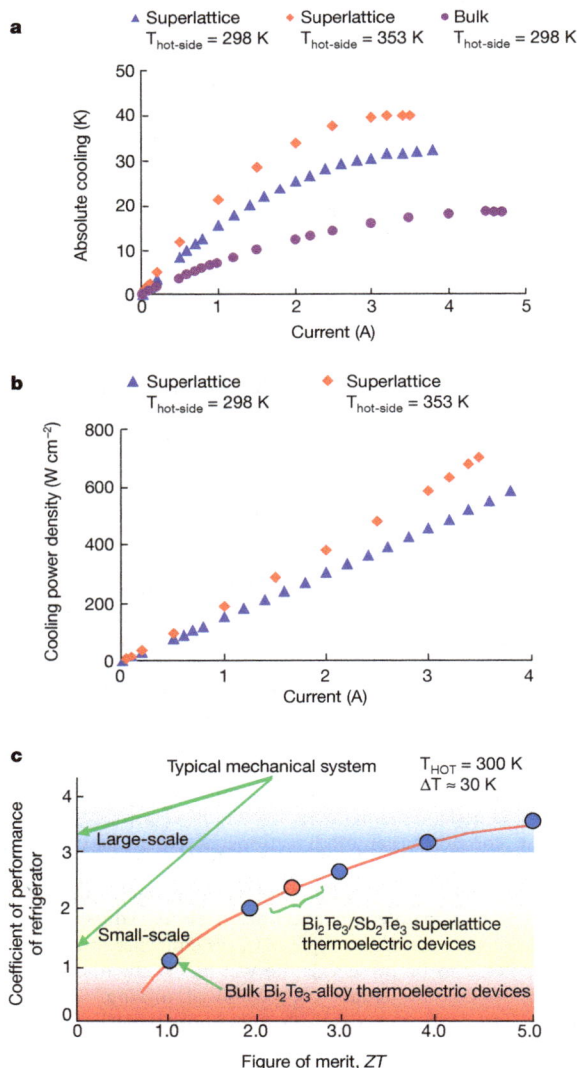

Figure 4 Cooling properties. **a**, Observed cooling as a function of current in a 150-μm circular device made from a p-type superlattice (data shown for hot side at 298 K (blue triangles) and 353 K (red diamonds)) is compared with that of a bulk device (hot side at 298 K, purple circles). **b**, Estimated cooling power density for superlattice devices as a function of current; blue triangles, hot side at 298 K, red diamonds, hot side at 353 K). **c**, Potential COP as a function of *ZT* with various technologies. T_{HOT} refers to the heat-sink temperature.

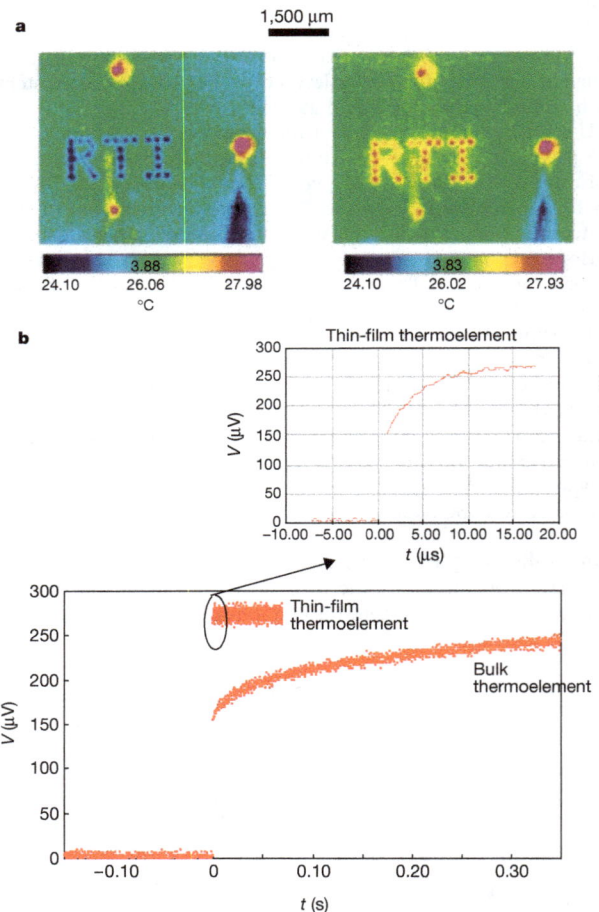

Figure 5 Demonstration of localized and high-speed cooling/heating obtainable with thin-film devices. **a**, Infrared images of spot cooling/heating with thin-film (p-type only) superlattice devices. Note current-injection areas (probes) do not change with current reversal; current through each element, ~65 mA. The infrared images of spots, corresponding to localized cooling or heating, spelling out RTI (Research Triangle Institute) were made using photolithographically processed micro-thermoelements. **b**, Comparison of thermal/cooling time response of thin-film (~5-μm) superlattice device and a bulk device.

In addition to offering low ZT, which translates to low coefficient of performance (COP) in refrigeration, the bulk thermoelectric devices have low P_D ($\sim 1\,W\,cm^{-2}$), another disincentive for high-power electronic/microprocessor applications. Using the values of K_T and α of the superlattices, available P_D as a function of current is estimated in Fig. 4b for the superlattice device. We estimate a value of P_D of $\sim 700\,W\,cm^{-2}$ at 353 K and $\sim 585\,W\,cm^{-2}$ at 298 K at the measured maximum cooling in superlattice devices compared to a value of $\sim 1.9\,W\,cm^{-2}$ in the bulk device of Fig. 4a. The effect of higher ZT on COP is shown in Fig. 4c. The translation of the device ZT to COP of refrigeration would involve efficient heat removal at the heat sink, reduction of thermal resistances at the interface between the active device and the two heat-spreaders, and the fabrication of p–n couples with minimal interconnect resistances. A sub-ambient cooling of 22.5 K at 300 K has been obtained in early p–n couples, currently limited by lower ZT of the n-thermoelement, large interconnect-resistance between the p- and n-thermoelements and absence of forced heat removal.

The thin-film devices also allow the concept of localized cooling by matching the footprint of the refrigeration devices to that of thermal load. This is made feasible by the combination of micro-electronic processing[42] and the ability to place the thermoelectric devices at points of interest; the reversal of current allows spot heating. This flexibility is demonstrated by the infrared images in Fig. 5a. In addition to placing cooling power where required, these thin-film superlattice thermoelements are fast-acting—about 23,000 times faster than bulk devices. This faster response is demonstrated in Fig. 5b. The thin-film device achieves steady-state cooling (indicated by the development of cooling-induced Seebeck voltage) in 15 μs, while the bulk thermoelement takes about 0.35 s. This is a result of the response time associated with the transport of heat through the thin (micrometres) rather than through the millimetres associated with bulk devices. The thermal response time is about $4l^2/\pi^2 D$, where l is the thickness of the thermoelement and D is the thermal diffusivity. This rapid, high-performance cooling and heating, capable of high power densities and with the ability to be locally applied, could have applications in technologies ranging from genomic/proteomic chips to fibre-optic switching[37].

Power conversion with thin-film superlattice devices

We also tested the thin-film microdevices in a power-conversion mode. These superlattice thin-film devices, 5.2 μm thick, can develop a ΔT of 70 K across them, with a corresponding open-circuit voltage indicating an average α of $\sim 243\,\mu V\,K^{-1}$. The ΔT translates into a temperature gradient of $\sim 134,000\,K\,cm^{-1}$. The typical Seebeck coefficient suggests no unusual departure from classical behaviour at these gradients. Due to the higher ZT, modest power-conversion efficiencies can be achieved even for low ΔT. Lightweight, high-power-density devices could be useful for portable power and thermal scavenging.

Thin-film thermoelectric technology could be implemented in a modular fashion: that is, units of thin-film thermoelectric heat-pump or power-conversion devices could be scaled for various capacities. We have reported here several techniques that could be useful in creating such modules: the bonding of thin films to a heat sink, the removal of substrate, photo-lithography to achieve appropriate l/a ratios, the preparation of low-resistivity contacts, and the ability to preserve the properties of the nanometre-scale super-lattices during detailed processing. □

Received 6 June; accepted 24 August 2001.

1. Ioffe, A. F. Semiconductor Thermoelements and Thermoelectric Cooling (Infosearch, London, 1957).
2. Wright, D. A. Thermoelectric properties of bismuth telluride and its alloys. Nature 181, 834 (1958).
3. Ettenberg, M. H., Jesser, W. A. & Rosi, F. D. in Proc. 15th Int. Conf. on Thermoelectrics (ed. Caillat, T.) 52–56 (IEEE, Piscataway, NJ, 1996).
4. Polvani, D. A., Meng, J. F., Chandrashekar, N. V., Sharp. J. & Badding, J. V. Large improvement in thermoelectric properties in pressure-tuned p-type Sb$_{1.5}$Bi$_{0.5}$Te$_3$. Chem. Mater. 13, 2068–2071 (2001).
5. Yim, W. M. & Amith, A. Bi-Sb alloys for magneto-thermoelectric and thermomagnetic cooling. Solid State Electron. 15, 1141–1165 (1972).
6. Vining, C. B. in Proc. 11th Int. Conf. on Thermoelectrics (ed. Rao, K. R.) 276–284 (Univ. Texas at Arlington, 1992).
7. Slack, G. A. & Tsoukala, V. G. Some properties of semiconducting IrSb$_3$. J. Appl. Phys. 76, 1665–1671 (1994).
8. Sales, B. C., Mandrus, D. & Williams, R. K. Filled sketterudite antimonides: a new class of thermoelectric materials. Science 272, 1325–1328 (1996).
9. Mandrus, D. et al. Filled sketterudite antimonides: Validation of the electron-crystal phonon-glass approach to new thermoelectric materials. Mater. Res. Soc. Symp. Proc. 478, 199–209 (1997).
10. Fleurial, J. P. et al. in Proc. 15th Int. Conf. on Thermoelectrics (ed. Caillat, T.) 91–95 (IEEE, Piscataway, NJ, 1996).
11. Chung, D. Y. et al. CsBi$_4$Te$_6$: A high-performance thermoelectric material for low-temperature application. Science 287, 1024–1027 (2000).
12. Tritt, T. M., Kanatzidis, M. G., Lyon, H. B. Jr & Mahan, G. D. Thermoelectric materials—New directions and approaches. Mater. Res. Soc. Proc. 478, 73–84 (1997).
13. Hicks, L. D. & Dresselhaus, M. S. Effect of quantum-well structures on the thermoelectric figure of merit. Phys. Rev. B 47, 12727–12731 (1993).
14. Harman, T. C., Taylor, P. J., Spears, D. L. & Walsh, M. P. in Proc. 18th Int. Conf. on Thermoelectrics (ed. Ehrlich, A.) 280–284 (IEEE, Piscataway, NJ, 1999).
15. Venkatasubramanian, R. et al. in Proc. 1st Natl Thermogenic Cooler Workshop (ed. Horn, S. B.) 196–231 (Center for Night Vision and Electro-Optics, Fort Belvoir, VA, 1992).
16. Venkatasubramanian, R. Thin-film superlattice and quantum-well structures—a new approach to high-performance thermoelectric materials. Naval Res. Rev. 58, 31–40 (1996).
17. Lee, S. M., Cahill, D. G. & Venkatasubramanian, R. Thermal conductivity of Si-Ge superlattices. Appl. Phys. Lett. 70, 2957–2959 (1997).
18. Mahan, G. D. & Woods, L. M. Multilayer thermionic refrigeration. Phys. Rev. Lett. 80, 4016–4019 (1998).
19. Shakouri, A. & Bowers, J. E. Heterostructure integrated thermionic coolers. Appl. Phys. Lett. 71, 1234–1236 (1997).
20. Harman, T. C., Cahn, J. H. & Logan, M. J. Measurement of thermal conductivity by utilization of the Peltier effect. J. Appl. Phys. 30, 1351–1359 (1959).
21. Venkatasubramanian, R. et al. Low-temperature organometallic epitaxy and its application to superlattice structures in thermoelectrics. Appl. Phys. Lett. 75, 1104–1106 (1999).
22. Venkatasubramanian, R. Low temperature chemical vapor deposition and etching apparatus and method. US Patent No. 6071351 (6 June 2000).
23. Venkatasubramanian, R. in Recent Trends in Thermoelectric Materials Research III (ed. Tritt, T. M.) Ch. 4 (Academic, San Diego, 2001).
24. Berger, H. H. Models for contacts to planar devices. Solid State Electron. 15, 145–158 (1972).
25. Drabble, J. R., Groves, R. D. & Wolfe, R. Galvanomagnetic effects in n-type Bi$_2$Te$_3$. Proc. Phys. Soc. 71, 430–434 (1957).
26. Scherrer, H. & Scherrer, S. in CRC Handbook of Thermoelectrics (ed. Rowe, D. M.) 211–237 (CRC Press, Boca Raton, FL, 1995).
27. Palmier, J. F. in Heterojunctions and Semiconductor Superlattices (eds Allan, G., Bastard, G., Boccara, N., Launoo, N. & Voos, M.) 127–145 (Springer, Berlin, 1986).
28. Cappaso, F., Mohammed, K. & Cho, A. Y. Resonant tunneling through double barriers, perpendicular quantum transport phenomena in superlattices, and their device applications. IEEE J. Quant. Electron. QE-22, 1853–1869 (1986).
29. Venkatasubramanian, R. Lattice thermal conductivity reduction and phonon localizationlike behavior in superlattice structures. Phys. Rev. B 61, 3091–3097 (2000).
30. Narayanamurti, V., Störmer, H. L., Chin, M. A., Gossard, A. C. & Weigmann, W. Selective transmission of high-frequency phonons by a superlattice: the dielectric phonon filter. Phys. Rev. Lett. 43, 2012–2015 (1979).
31. John, S. Strong localization of photons in certain disordered dielectric superlattices. Phys. Rev. Lett. 58, 2486–2489 (1987).
32. Wiersma, D. S., Bartolini, P., Lagendik, A. & Righini, R. Localization of light in a disordered medium. Nature 390, 671–673 (1997).
33. Simkin, M. V. & Mahan, G. D. Minimum thermal conductivity of superlattices. Phys. Rev. Lett. 84, 927–930 (2000).
34. Slack, G. in Solid State Physics, 1–71 (eds Ehrenreich, H., Seitz, F. & Turnbull, D.) Series 34 (Academic, New York, 1979).
35. Cahill, D. G., Watson, S. K. & Pohl, R. O. Lower limit to the thermal conductivity of disordered crystals. Phys. Rev. B 46, 6131–6140 (1992).
36. Goldsmid, H. J. in Proc. 18th Int. Conf. on Thermoelectrics (ed. Ehrlich, A.) 531–535 (IEEE Press, Piscataway, NJ, 1999).
37. Venkatasubramanian, R. Thin-film Thermoelectric Cooling and Heating Devices for DNA Genomics/ Proteomics, Thermo-Optical Switching-Circuits, and IR Tags (US Patent Filing, Ser. No. 60/282,185, 2001).
38. Sheng, P. Introduction to Wave Scattering, Localization, and Mesoscopic Phenomena (Academic, London, 1995).
39. Kittel, C. Introduction to Solid State Physics (Wiley, New York, 1976).
40. Fan, X. et al. SiGeC/Si superlattice microcoolers. Appl. Phys. Lett. 78, 1580–1582 (2001).
41. Cadoff, I. B. & Miller, E. Thermoelectric Materials and Devices (Reinhold, New York, 1960).
42. Venkatasubramanian, R. Thin-film thermoelectric device and fabrication method of same. US Patent No. 6300150 (9 October 2001).

Acknowledgements

This work was supported by the US Defense Advanced Research Projects Agency and the US Office of Naval Research.

Correspondence and requests for materials should be addressed to R.V. (e-mail: rama@rti.org).

BATTERIES AND SUPERCAPACITORS

Ionic-liquid materials for the electrochemical challenges of the future

Michel Armand[1], Frank Endres[2], Douglas R. MacFarlane[3], Hiroyuki Ohno[4] and Bruno Scrosati[5]*

Ionic liquids are room-temperature molten salts, composed mostly of organic ions that may undergo almost unlimited structural variations. This review covers the newest aspects of ionic liquids in applications where their ion conductivity is exploited; as electrochemical solvents for metal/semiconductor electrodeposition, and as batteries and fuel cells where conventional media, organic solvents (in batteries) or water (in polymer-electrolyte-membrane fuel cells), fail. Biology and biomimetic processes in ionic liquids are also discussed. In these decidedly different materials, some enzymes show activity that is not exhibited in more traditional systems, creating huge potential for bioinspired catalysis and biofuel cells. Our goal in this review is to survey the recent key developments and issues within ionic-liquid research in these areas. As well as informing materials scientists, we hope to generate interest in the wider community and encourage others to make use of ionic liquids in tackling scientific challenges.

t was with simple triethylammonium nitrate that a pure low-melting salt — an ionic liquid — was first identified more than a century ago. In the 1930s, a patent application described cellulose dissolution using a molten pyridinium salt above 130 °C. It was the need for a sturdy medium for nuclear fuel reprocessing that prompted the study of low-melting-point chloroaluminates. Among the 'onium' cations with positive nitrogen(s), those derived from the imidazolium ring proved to be the best choice in terms of melting points and electrochemical stability.

At roughly the same time, the need for new anions for organic polymer electrolytes based on polyethylene oxide led to the concept of a 'plasticizing anion', that is, an anion having a delocalized charge and multiple conformations differing only marginally in energy. The archetype of such anions is the bis(trifluoromethylsulphonyl)amide $(CF_3SO_2-N-SO_2CF_3)$ ion, also known as NTf$_2$, in which the extremely electron-withdrawing CF_3SO_2- groups are conjugated and linked by flexible S–N–S bonds. When combined with an imidazolium cation, such as the ethylmethylimidazolium cation, this anion produces a fluid ionic liquid (melting point –15 °C) with an ion conductivity comparable to that of the best organic electrolyte solutions; it shows no decomposition or significant vapour pressure up to ~300–400 °C (ref. 1). Surprisingly, it is not miscible with water (~1,000 p.p.m. in equilibrium with liquid H$_2$O), and thus defies the conventional wisdom that states polarity is synonymous with hydrophilicity.

Ionic liquids (Box 1) then developed rapidly, with a reinvestigation of ions, for example quaternary ammonium cations, that had been avoided previously by organic chemists because of unsymmetrical shapes that hindered easy purification through crystallization. The organic chemistry community had earlier engaged in research of media with controllable Lewis acidity (chloroaluminate ionic liquids), but the modern era of ionic liquids has produced numerous 'neutral' ionic liquids, that is, those based on ions which are unreactive towards acids or bases, be they Lewis or Brønsted. As a result, it is now difficult to name an organic reaction that has not been performed successfully in these potentially 'green' solvents, which can be recycled almost indefinitely with no or minimal use of volatile organic compounds. Most products made in ionic liquids can be distilled off, in the case of small molecules, or extracted with water or hydrocarbon solvents, at least one of which is usually immiscible with the ionic liquid.

It is this unique solvent potential that makes ionic liquids key materials for the development of a range of emerging technologies. The advent of ionic liquids has made viable processes that fail, or are even impossible, with conventional solvents. Water-sensitive metals or semiconductors that previously could not be deposited from conventional water baths can now, by turning to ionic liquids, be directly electroplated. Energy devices, such as the polymer-electrolyte-membrane fuel cells, lithium batteries and supercapacitors presently under development to address the challenges of increasing energy costs and global warming, may greatly benefit from a switch to low-vapour-pressure, non-flammable, ionic-liquid-based electrolytes. The significant dissolution power of ionic liquids also extends to macromolecules, as best illustrated by the solutions of up to 25 wt% cellulose that can be obtained; this has great consequences for the textile industry and for the only ethically acceptable source of biofuels, polyosides from wood. Until it was realized that cellulose was soluble in ionic liquids, the only known solvent for cellulose was an explosive organic compound. Ionic liquids also have applications in biology and biomimetic processes, as their enzymes can be remarkably active, opening a vast field for bioinspired catalysis and biofuel cells.

The scientific and technological importance of ionic liquids today spans a wide range of applications. New types of lubricants and seals, and fluids for thermal engines and adsorption refrigeration have been suggested on the basis of the diverse and unique properties of ionic liquids. We have focused this review on some of the most significant examples of recent key developments in ionic-liquid applications made possible by their unique solvent capabilities (Fig. 1 and Box 1). These include electrodeposition, energy management, bioscience and biomechanics. Our main objective is to show how the progress recently achieved in these crucial fields could not have been possible without the advent of ionic liquids.

[1]LRCS CNRS 6007, Université de Picardie Jules Verne, F-80039 Amiens, France, [2]Institute of Particle Technology, Clausthal University of Technology, D-38678 Clausthal-Zellerfeld, Germany, [3]School of Chemistry and ARC Centre of Excellence for Electromaterials Science, Monash University, Clayton, Victoria 3800, Australia, [4]Department of Biotechnology, Tokyo University of Agriculture and Technology, Tokyo 184-8588, Japan, [5]Department of Chemistry, University Sapienza, 00185 Rome, Italy. *e-mail: bruno.scrosati@uniroma1.it

When citing this article, please cite the original version as shown on the contents page of this chapter.

Box 1 | Structure of the archetype ionic liquid.

Ionic liquids are low-temperature molten salts, that is, liquids composed of ions only. The salts are characterized by weak interactions, owing to the combination of a large cation and a charge-delocalized anion. This results in a low tendency to crystallize due to flexibility (anion) and dissymmetry (cation). The archetype of ionic liquids is formed by the combination of a 1-ethyl-3-methylimidazolium (EMI) cation and an N,N-bis(trifluoromethane)sulphonamide (TFSI) anion. This combination gives a fluid with an ion conductivity comparable to many organic electrolyte solutions and an absence of decomposition or significant vapour pressure up to ~300–400 °C. Ionic liquids are basically composed of organic ions that may undergo almost unlimited structural variations because of the easy preparation of a large variety of their components. Thus, various kinds of salts can be used to design the ionic liquid that has the desired properties for a given application. These include, among others, imidazolium, pyrrolidinium and quaternary ammonium salts as cations and bis(trifluoromethanesulphonyl)imide, bis(fluorosulphonyl)imide and hexafluorophosphate as anions.

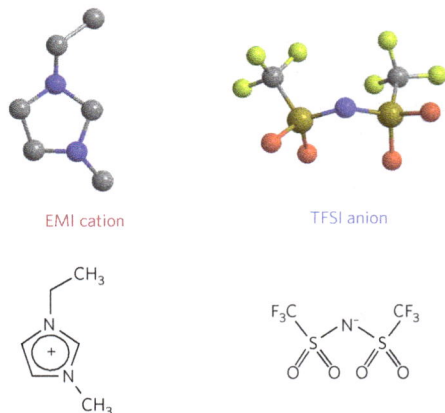

EMI cation   TFSI anion

Box 2 | Designs of ionic liquids for specific applications.

On the basis of their composition, ionic liquids come in different classes that basically include aprotic, protic and zwitterionic types, each one suitable for a specific application.

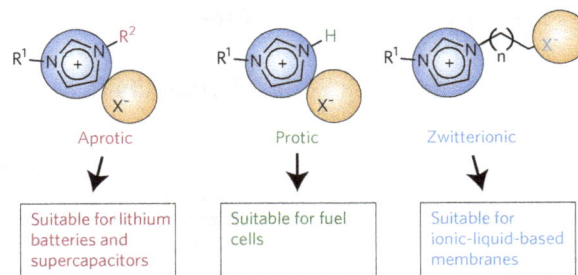

Aprotic   Protic   Zwitterionic

Suitable for lithium batteries and supercapacitors   Suitable for fuel cells   Suitable for ionic-liquid-based membranes

Ionic liquids

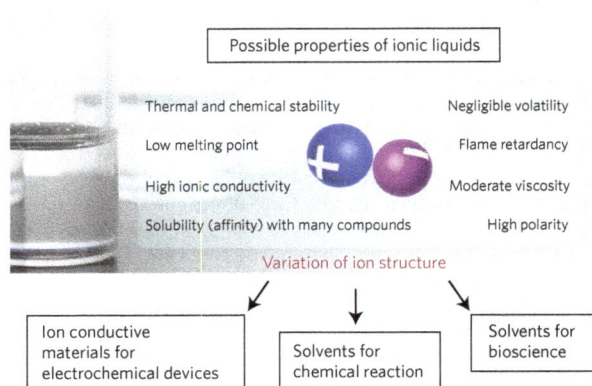

Possible properties of ionic liquids

Thermal and chemical stability   Negligible volatility

Low melting point   Flame retardancy

High ionic conductivity   Moderate viscosity

Solubility (affinity) with many compounds   High polarity

Variation of ion structure

Ion conductive materials for electrochemical devices   Solvents for chemical reaction   Solvents for bioscience

Figure 1 | Ionic liquids, that is, salts with melting points below 100 °C, are potential candidates to be new 'green' reaction media with a number of important properties. They can be exploited in various applications, which range from energy storage and conversion to metal deposition, and can be used as reaction media and in chemistry, biochemistry and even biomechanics.

Electrodeposition

Ionic liquids are superior media for the electrodeposition of metals and semiconductors, and have an unprecedented potential to revolutionize electroplating. Key advantages that enable them to overcome the limits imposed by common aqueous or organic media are their wide electrochemical window, spanning up to 6 V in some cases; extremely low vapour pressures, which allow deposition at temperatures well above 100 °C; and numerous, only partly understood, cation/anion effects that make it possible to influence the morphology and crystal size of deposits[2–4].

Processes that are impossible in water baths become viable if ionic liquids are used. These include the direct electroplating of water-sensitive metals such as aluminium and possibly even magnesium, as well as many other metals having a deposition potential conflicting with water decomposition. Aluminium is an important metal for corrosion protection of steel as, in contrast to industrially used zinc coatings, it is self-passivating in air and is much more sustainable in terms of known ore resources. However, with an electrode potential of −1.7 V relative to the normal hydrogen electrode, it cannot be deposited from aqueous media, and the well-known SIGAL process[5,6] involving alkylaluminium compounds in aromatic solvents is hazardous owing to the high flammability of the aluminium precursors used, requiring the strict exclusion of oxygen. The electrodeposition of aluminium from ionic liquids has

been discussed several times in the literature[7–9] and is sufficiently advanced that it could be introduced to industry within the next five years as a commercial process. Nonetheless, although ionic liquids are air stable and some are hydrophobic, the presence of $AlCl_3$ solute species in the ionic-liquid solution as aluminium precursors still requires the exclusion of moisture, and this is a challenge for the next generation of aluminium-plating baths. AlF_4^-, which is less prone to releasing HF than $AlCl_3$ is to releasing HCl, would be a good choice; however, it disproportionates into AlF_6^{3-} at room temperature. Stabilizing such species in ionic liquids would be very promising.

This process has shown an interesting cation effect: in one ionic liquid, the deposited aluminium is nanocrystalline, with grain sizes of around 50 nm; with a different cation the grain size is, under practically the same conditions, around 5 μm (ref. 10). It is remarkable that just variation of the ionic liquid leads to different morphologies, without the addition of any brightener or grain refiner. This effect results from the solvation layers forming at the interface between the electrode and the ionic liquid; these vary from liquid to liquid and depend upon the metal salt dissolved. Ionic liquids have thus been proven to be excellent media for deposition of a range of other metals including manganese, nickel, tin and copper; these and presumably many others can easily be deposited from liquids with the dicyanamide anion $(CN)_2N$, which combines charge delocalization and stability at reducing potentials with an ability to complex the metal cation and therefore dissolve and stabilize it[11]. A further advantage over aqueous plating baths is that many ionic liquids are benign media without toxicity issues in comparison with the cyanides that are involved in many aqueous systems (for example $[Ag(CN)_2]^-$, which is used in silver plating).

It could be argued that, owing to the very wide electrochemical windows, it should be trivial to deposit reactive elements such as titanium, tantalum, molybdenum and other transition metals that, from the thermodynamic point of view, should be reduced at electrode potentials similar to that of aluminium or even less negative. The main difficulty with these transition elements is the multiplicity of redox states resulting in the electrochemical equivalent of short circuits. From the thermodynamic point of view, the electrochemical window of many ionic liquids is wide enough to allow the deposition of many reactive elements including titanium (from titanium(II)), which lies very close to the deposition potential for lithium. The electrodeposition of thin tantalum layers is possible within a narrow set of experimental parameters, and it has thus been successfully deposited from ionic liquids at elevated temperature, for example to coat biomedical implants[12,13]. On the other hand, attempts to deposit titanium from titanium halides have hitherto been unsuccessful owing to the formation of titanium subvalent halides, which redissolve into the electrolyte, instead of elemental titanium[14].

As a proof of principle, it was shown that nanoparticles of chromium, molybdenum and tungsten could be made from their respective carbonyl compounds as the source of metal in the bath[15]. Once again, a dependence of the particle size on the type of ionic liquid was observed. It seems to be much easier to deposit these elements as alloys; accordingly, the deposition of a ternary aluminium–molybdenum–titanium alloy has recently been described[16].

Ionic liquids also have the potential to be the basis of important approaches in nanoscience. There is great interest in making well-defined metal nanowires, for example using template-assisted techniques. Iron, cobalt and nickel nanowires are, owing to their ferromagnetic properties, interesting for fundamental research. These elements can be obtained in aqueous solution, but as a result of hydrogen coevolution the nanowires are usually of limited quality. This problem does not exist in ionic liquids, as there is no hydrogen evolution, and the template-assisted deposition of silver[17], cobalt[18], germanium and silicon[19] has been reported to deliver compact nanowire mats of these metals; see Fig. 2 for a typical example.

Furthermore, ionic liquids are of importance in semiconductor electrodeposition; recently it was shown that photoluminescent Si_xGe_{1-x} with a bandgap of at least 1.5–3.2 eV can be made by electrodeposition from an ultrapure ionic liquid containing silicon and germanium halides (Fig. 3). During deposition, different colours, from orange to green, are observed in the visible spectrum; these are due to a quantum size effect of the semiconductor particles with sizes between 2 and 20 nm (ref. 20). These results show that the ionic liquid allows deposits so pure that photoluminescence effects can be seen. In addition, the material absorbs visible light and may open the way to a simple electrochemical fabrication of inexpensive solar cells.

However, although the feasibility of metal and semiconductor deposition from ionic liquids has been demonstrated in a variety of cases, there are some fundamental issues that still have to be clarified. The solubility of precursor compounds departs appreciably from that in conventional solvents. The basic electrochemistry of the processes is yet to be fully understood and the structure of the double layer, which, owing to the formation of solvation layers, can obviously not be well described with the models developed for aqueous solutions, must be clarified. The role of the large ions in influencing

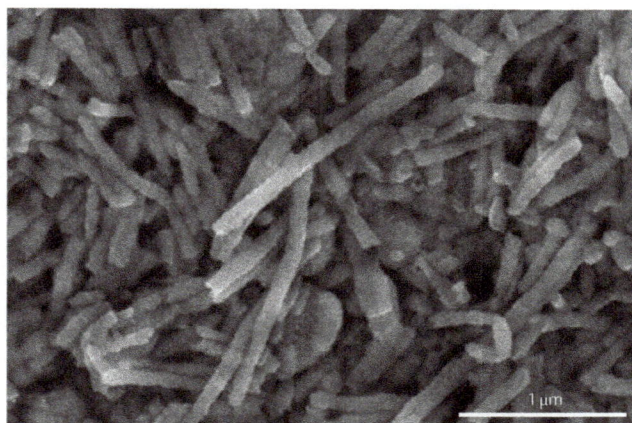

Figure 2 | Processes that are impossible in water baths become viable if an ionic liquid solvent is used. The use of ionic liquid baths also improves the quality and the morphology of the deposits. The figure shows a high-resolution scanning electron microscope image of germanium nanowires made by electrodeposition from ionic liquids. Figure reprinted from ref. 19, © 2008 RSC.

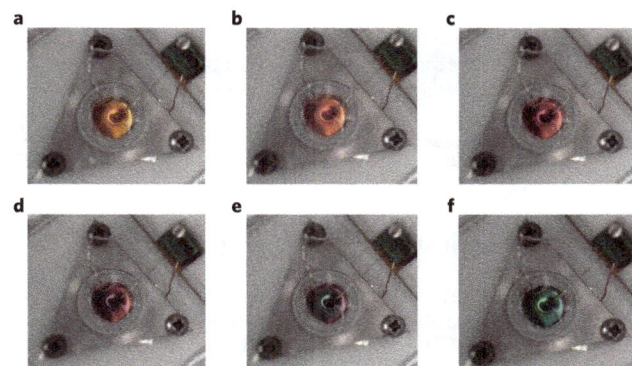

Figure 3 | Deposits of photoluminescent semiconductors, for example Ge_xSi_{1-x} obtained from ionic liquid baths. Different colours are seen in the visible spectrum, owing to quantum size effects of the semiconductor particles. This is a clear demonstration that ionic liquids allow deposits of particle size in the nanometre range that are pure enough to make visible the photoluminescence effect. The panels correspond to potentials of −2.62 V (**a**), −2.68 V (**b**), −2.74 V (**c**), −2.78 V (**d**), −2.83 V (**e**), and −2.85 V (**f**). Figure reprinted from ref. 20, © 2008 RSC.

the diffusion kinetics and the mechanism of deposit nucleation and growth are other aspects requiring further investigation.

Energy management

Ionic liquids are also likely to be important in the field of energy. Global concerns about present energy policy, which relies on fossil fuels and suffers the associated economical and ecological problems, call for the increased use of renewable energy sources in households and industry and for the large-scale replacement of internal-combustion-engine vehicles by those with zero or low emissions. This in turn requires the availability of suitable systems both to store the energy in power stations harnessing intermittent sources (solar or wind) and to power electric or hybrid vehicles. Electrochemical systems, such as batteries, are ideal for this purpose[21]. In particular, lithium batteries, which have the highest energy efficiency of all known electrochemical storage systems and which have reached an advanced state of technological development, seem to be the power source of choice. Lithium batteries can have

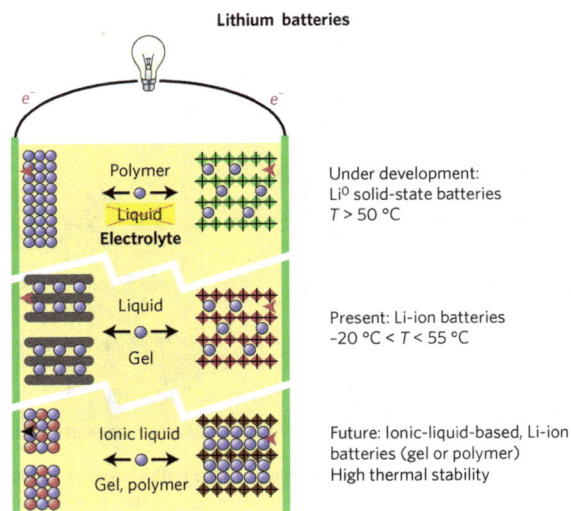

Figure 4 | In their basic structure, lithium batteries are formed by two lithium-exchanging electrodes separated by a lithium-ion-conducting electrolyte. The most common version uses an electrode combination based on a graphite negative electrode with a lithium cobalt oxide positive electrode connected either by a liquid solution of a lithium salt in an organic solvent mixture or by the same solution trapped in a gel. The electrochemical process is the cycling of lithium ions from one electrode to the other. These lithium-ion batteries are commercially available and dominate the consumer-electronics market. However, their use in more demanding applications, such as in sustainable transport, is prevented by a number of drawbacks, among which safety is the main concern. Safety can be improved by excluding liquids, that is, by using solvent-free, polymer electrolytes. The safe operation of this solid-state version, which replaces the graphite with lithium metal at the negative side, has been demonstrated. However, these batteries operate well only at temperatures higher than 50 °C, which limits their range of application. The real breakthrough in the field may come by using ionic-liquid-based gel or polymer electrolytes in combination with revolutionary electrode materials: the high thermal stability and low vapour pressure of the ionic liquids ensure reliability by preventing thermal runaway and pressure build-up, and the innovative electrode structure is expected to enhance the capacity and energy density of the battery.

different configurations (Fig. 4). However, their implementation for energy storage and in vehicle applications has been slowed by safety concerns surrounding the use of large-scale lithium cells. Undesired reactions between the battery components and the liquid organic electrolyte, triggered by unpredictable events such as short circuits or local overheating, lead to an exothermic reaction of the electrolyte with the electrode materials, producing a rapid increase of the battery temperature and, eventually, to fire or explosion.

Again, the unique properties of ionic liquids may help to solve the problem as they are practically non-flammable, which is a significant safety asset. The replacement of the conventional, flammable and volatile, organic solutions with ionic-liquid-based, lithium-ion-conducting electrolytes may greatly reduce, if not prevent, the risk of thermal runaway. This provides the lithium battery with the level of safety that is required for their large-scale application in important and strategic markets. Extensive work in this direction is in progress, and the testing of ionic liquids as new electrolyte media for future, highly safe, lithium batteries is under way in many industrial and academic laboratories[22–25]. However, the results vary widely, as can be expected from the wide variety of ionic liquids investigated. Key aspects to be understood are, among others, the thermodynamic and kinetic stabilities of the ionic liquids with respect to the electrode materials. It would be naive to expect the

organic ionic-liquid cations to be stable at the extremely negative potentials of alkali-metal deposition; in other words, the average electronegativity of an ion based on carbon, hydrogen or nitrogen cannot be as low that of Li^+ or K^+. However, the kinetic metastability of the cation and/or protective, passivating-layer formation on the electrode have been welcome observations[26]. The structure of the interfaces and the nature of the electrode-protecting films are other aspects that need particular attention.

These aspects will be analysed in this review to evaluate their effective influence on the progress of lithium battery technology. The electrolytes that are under study are solutions of a lithium salt in an ionic liquid. The high solubility of lithium salts, even though it requires low-lattice-energy salts (for example $LiBF_4$ or $Li[NTf_2]$), is quite surprising as, in contrast to organic polar solvents, ionic liquids do not possess free electron pairs to provide for a solvation interaction with the cation, other than those on the anion. The resultant structures, probably involving polyanionic species such as $[LiX_n]^{(n-1)}$ still need to be better understood. In principle, ionic liquids are only metastable at the potential of lithium reduction; the cations, despite often being resonance stabilized, are sensitive to electron injection, with the subsequent formation of a radical, or proton elimination. For instance, the archetypal ionic-liquid cation, ethylmethylimidazolium, possesses an acidic C2 proton that limits its stability below 1.3 V relative to Li^+/Li^0 (−1.7 V in comparison with H^+/H_2). Yet more and more reports seem[26–28] to show that in half-cells or full batteries, selected ionic-liquid electrolytes behave well, not only with the conventional graphite negative electrode, but also for Li^0 plating. The latter had been abandoned in organic electrolytes because of uneven plating of the metal during cycling (that is, dendrite formation), with short circuits and dramatic runaway reactions. Graphite was thus selected for 'Li-ion technology' but is still a weak point in terms of safety, as solvents can forcefully co-intercalate and exfoliate the graphite with large release of heat.

This behaviour does not seem to occur with ionic liquids, as the solvation is provided by anions that have no tendency to co-intercalate at such potentials. The very high-capacity Li_xSi ($0 < x < 4.4$) alloying electrode (capacity reaching 2,000 mA h g^{-1}) also behaves quite well in contact with ionic liquids. The NTf_2 electrolytes have been shown to cycle well with Li^0 (ref. 26), offering a dramatic increase in possible energy densities and a simpler technology for high-performance batteries. The chemical manipulation of the cation (quaternary ammonium instead of ethylmethylimidazolium) and of the anion $[(FSO_2)_2N]^-$ instead of $[(CF_3SO_2)_2N]^-$) allows the fine-tuning of the interactions at the interface and passivating-layer formation. Owing to their high anodic stability, ionic liquids function without problem at the positive electrode, even with highly oxidizing materials (for example $LiMn_{1.5}Ni_{0.5}O_4$ and $LiMnPO_4$). The main problem is technological, as the only acceptable (sustainable) current collector in practical batteries is aluminium, which tends to become corroded at such high voltages. The battery electrolytes based on ionic liquids are thus serious contenders in the design of safe batteries for electric and hybrid vehicles. There is also hope, in the longer term, of building lithium batteries using oxygen from air as the oxidizer at the cathode, which would produce a great increase in capacity, even if the end product of oxygen reduction is Li_2O_2, that is, only two electrons per O_2 instead of the four possible[29]. The fact that many of the ionic liquids of interest in this area are hydrophobic is an asset; however, when a lithium salt is dissolved, the electrolyte becomes very water sensitive, and new salts must be designed to counter this problem.

The development of a zero-emission, hydrogen-powered car has been a highly desirable goal for some time. Although the concept is very appealing, the development of hydrogen cars is still restricted to a few models operating in limited geographical regions. The core of the hydrogen car is the hydrogen fuel cell. Fuel cells have been known for more than a century, but some of the issues that have so far hindered their commercial production are still unsolved[30]. These are mainly related to limited functional characteristics of the perfluorosulphonic-acid-membrane electrolyte, the conductivity of which is strongly humidity dependent; consequently, its operating temperature is limited to roughly 100 °C (ref. 31). These issues may be addressed to an extent by switching to ionic-liquid-based polymer membranes. This strategy has been tested by immobilizing ionic liquids, by polymerization of the components or gelification by a neutral macromolecule with an architecture allowing a water-independent proton conductivity associated with a high thermal stability[32-35] (Fig. 5).

In this case, the proton relay molecule, instead of being H_2O, is an imidazole or an amine. Not only is the operating temperature higher, but also the pH increases from 7 to 11. Neutral-to-basic pHs may open new possibilities in terms of catalyst designs, the media being far less aggressive than the highly acidic perfluorosulphonic membrane, with the ultimate goal of replacing platinum. Catalysts made from platinum are considered to be a major barrier to the ultimate widespread uptake of this technology, because of its cost and scarcity. The investigation of ionic-liquid membranes is still in its infancy, but their potential in important, advanced technologies justifies the great attention devoted to them, and this makes it possible to foresee important developments in the near future.

The proper development of electric — or hybrid — vehicle technology requires a complementary component to assist the battery during moments of high power demand such as start-up and acceleration. Supercapacitors — electrochemical devices based on two inert electrodes separated by an electrolyte — are the components of choice[36]. The challenge in supercapacitor technology is to improve the energy density without sacrificing cycle life and safety. Present supercapacitors are limited to a few watt hours per kilogram, which is well below the specific energy required to provide power assists in hybrid vehicles, and they must be used in tandem with batteries. Hence, structure modifications are required to raise this value, for example by enhancing the voltage of the device. This in turn requires the use of electrolytes that can withstand high voltages without undergoing decomposition. Some limited success has been achieved by passing from aqueous to aprotic solvents, for example acetonitrile or propylene carbonate[36]. Ionic liquids, owing to their wide electrochemical stability window, are ideal for this application, as demonstrated recently[37,38]. However, the development of ionic-liquid-based supercapacitors is still in its preliminary stages and further progress is needed to ensure their full exploitation. The challenge is in the choice of ionic liquids that feature wide electrochemical stability windows combined with high ionic conductivity, as well as in the design of electrode structures capable of assuring proper wettability by the ionic liquid.

Bioscience

Bioscience is among the most interesting and important emerging areas where ionic liquids are beginning to play a part. Water has long been believed to be the unique solvent for biomolecules and biosystems. Indeed, almost all studies on enzymatic reactions have been carried out in aqueous media. On the other hand, for long-life biodevices, there are a number of drawbacks to this situation, such as the volatility of water, the limited temperature range and the very narrow pH range for protein stability. Ionic liquids have the potential to overcome these issues. However, to be effective in energy biodevices, such as biofuel cells, ionic liquids must ensure the dissolution of natural biomaterials, as well as producing stable enzymatic reactions.

Dissolution of natural biomaterials, especially of cellulose, is one of the most important targets for energy conversion. Cellulose is produced at a rate of 10^{11} t yr^{-1} in nature and, in our opinion, is the only source of chemicals for energy that is ethically acceptable. Recently, large efforts have been devoted to the dissolution of

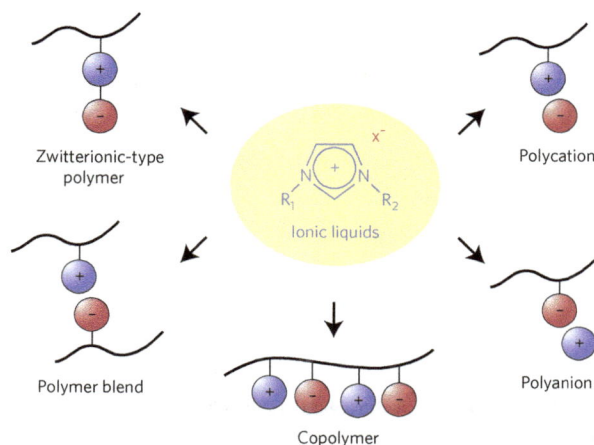

Figure 5 | Although polymer-electrolyte-membrane fuel cells have been known for a long time, they have not yet reached large-scale development as some issues are still unresolved. These are mainly related to limited functional characteristics of the perfluorosulphonic-acid-membrane electrolyte, whose conductivity relies on the humidity level; consequently, its stability is limited to temperatures below ~100 °C. These issues may be addressed to an extent by switching to ionic-liquid-based polymer membranes. Various strategies have been tested for the preparation of these membranes, including polymerization of the components with the formation of polycations, polyanions, copolymers and special, double-ion structures, as well as blends with a neutral macromolecule, typically polyvinylidene fluoride. These revolutionary architectures offer a water-independent proton conductivity associated with a high thermal stability, which allows high operational temperatures, as well as basic pHs. This opens up new possibilities in terms of catalyst design, with the ultimate goal of replacing platinum, which is impractically rare.

cellulose using polar ionic liquids. Several studies on the preparation of polar and low-melting-point ionic liquids have shown that the key parameter for achieving the dissolution of cellulose is a strong hydrogen-bond-accepting ability of the selected ionic liquid[39,40]. Initial success was achieved by using 1-butyl-3-methylimidazolium chloride ([bmim]Cl)[41] and 1-allyl-3-methylimidazolium chloride ([amim]Cl)[42]. The important role of the chloride anion is the formation of ionic liquids that show the required hydrogen-bonding ability[43]. This, however, is counterbalanced by a drawback associated with the melting point of [bmim]Cl (70 °C); hence, extra energy is needed to operate with ionic liquids based on [bmim]Cl. New ionic liquids, in which the phosphate anion derivatives were used instead of the chloride anion, were investigated to solve this issue, and the results demonstrated that such fluids can dissolve cellulose without heating[44]. Thus, owing to these successes, dissolution of cellulose has been achieved; the next step is to preserve its enzymatic reactions in ionic liquids.

The dissolution of enzymes and the maintenance of their effective reactivity require the preservation of their highly ordered structure, a condition that is difficult to achieve even in ordinary ionic liquids, as for the most part they have little ability to solubilize proteins without denaturation. As schematically shown in Fig. 6, there are three different ways to solubilize proteins in ionic liquids. A common method is simply the direct dispersion of the enzymes in ionic liquids. Some lipases are known to be effective in this kind of dispersion. Polyethylene glycol (PEG) modification is the second most generally accepted method for achieving a homogeneous solubilization of proteins in many ionic liquids[45,46]. PEGylation (that is, chemical modification of PEG chains) is applicable to most proteins and PEGylated proteins are thus soluble in most ionic liquids because of the excellent affinity of PEG chains

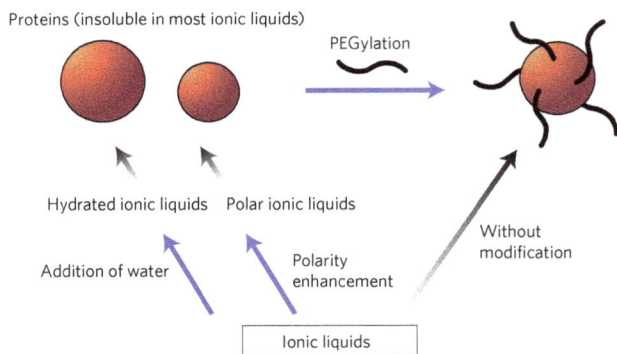

Figure 6 | There are several strategies to solubilize proteins into ionic liquids. A common one is simply the direct dispersion of the enzymes in ionic liquids. Polar ionic liquids are sometimes excellent media in which to solubilize native proteins without any additives or protein modification. Another strategy involves the modification of polyethylene glycol (PEG) on the protein surface, that is, PEGylation. Here, the solubility is improved by the excellent affinity of PEG chains with ions, including those constituting ionic liquids. Finally, mixtures of ionic liquids and a small amount of water, called hydrated ionic liquids, are other potential candidates as solvents for proteins. Because water molecules are strongly bound directly to ions and no free water molecules exist, the solution properties of this hydrated ionic liquid are similar to those of pure ionic liquids.

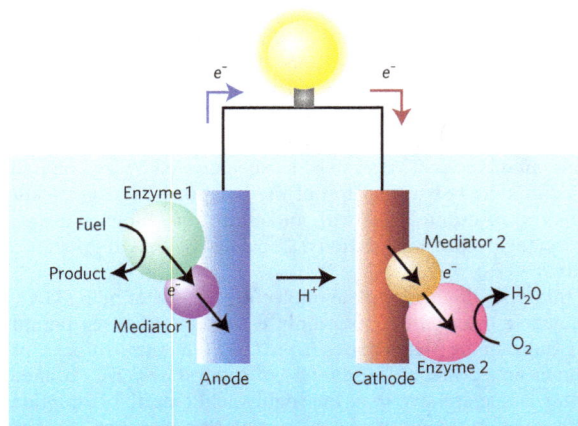

Figure 7 | Sketch for a biofuel cell, that is, a device capable of generating energy with the aid of enzymes. Practical biofuel cells have already been demonstrated using aqueous electrolyte solutions. There is a general opinion that the use of ionic liquids instead of aqueous salt solutions prevents enzymatic activity as a result of the environment that they provide. Considering the large variety of substrates and enzymes available for biofuel cell applications, the use of ionic liquids as electrolytes paves the way for the development of unlimited combinations for novel bio-energy conversion systems, making biofuel cells attractive next-generation energy-conversion devices.

with ions, including those constituting ionic liquids. Crown ethers sometimes show effects similar to PEG[47].

A third, more sophisticated approach is based on a relatively new concept. It is not very easy to design ionic liquids having all the desired properties; therefore, an alternative method can be used to achieve the goal of finely controlling the ionic-liquid property, namely that of mixing it with other ionic liquids or even with molecular liquids. Addition of small amounts of water to ionic liquids may strongly influence the protein solubility while retaining the properties of the selected ionic liquid. It is important to note that the chemical activity of water remains very low because all the water molecules are strongly solvated to the ions. Effectively, Fujita and co-workers reported that a water/ionic-liquid mixture, called a hydrated ionic liquid, is an excellent solvent for proteins[48,49]. Choline dihydrogen phosphate ([Ch][DHP]), composed of a chaotropic cation and a kosmotropic anion, is a typical example. Hydrated (20% water) [Ch][DHP] was shown to be an excellent solvent for cytochrome c and other proteins. In this liquid, cytochrome c was reported to show electron transfer activity even 18 months after storage at room temperature[50]. It appears that the small amount of water present in the ionic liquid provides the required hydrogen-bonding environment for the protein and hence renders it soluble. Such hydrated ionic liquids are likely to be the solvents of choice for natural biomaterials from now on.

Even in water-free conditions, owing to their strong hydrogen-bonding ability, chloride-based ionic liquids can also act as solvents for native proteins under ambient conditions. For instance, cytochrome c was found to be soluble in dry [amim]Cl and [bmim]Cl (ref. 51), showing a quasi-reversible electron transfer reaction at an electrode[52]. Nevertheless, homogeneous dissolution of proteins in ionic liquids remains one of the most critical challenges in this field. In fact, proteins other than cytochrome c were found not to be active in dry [amim]Cl and [bmim]Cl, suggesting a unique stabilization of the haem complex with covalent bonds in the globin domain of cytochrome c. On the other hand, this also strongly reinforces the idea that, by proper tuning, suitable ionic liquids can be designed to be effective solvents for other proteins.

Following the dissolution of cellulose and enzymes in ionic liquids, the remaining goal is the enzymatic depolymerization of

cellulose to generate glucose or other oligosaccharides. Relatively recent reports have described enzymatic hydrolysis in ionic liquid/water mixtures[53], which suggests that the goal can be effectively achieved.

Another important goal in the field is the generation of energy from oligosaccharides, cellobiose, glucose or further derivatives. The use of ionic liquids may be of considerable help here also. The generation of electrical energy with the aid of enzymes in a 'biofuel cell' (Fig. 7) has only been developed in aqueous media. However, the above-mentioned issues associated with the use of aqueous media greatly affect the cell operation. Thus, the replacement of water solutions with ionic liquids is expected to enhance the operational life of the biofuel cell, as well as improving its maintenance-free characteristic[54]. However, in this case also a proper selection of the ionic liquid is necessary to achieve valid results. For instance, ionic liquids are required that are good solvents for the (macro)molecular substrate, but that do not support the activity of any virus, algae, fungus, bacteria or similar. On the other hand, many fundamental studies carried out on water-based biofuel cells can be of great help, as they can provide a criterion to be used for the selection of the most suitable ionic liquid to ensure the desired enzymatic energy conversion. Adequately designed ionic liquids should support high solubility and stability of target enzymes as well as preservation of their activity. It is very likely that ionic-liquid-based biofuel cells will be key energy devices of the future.

Biomechanics

Electromechanical actuators, namely devices that show reversible deformation under the stimulus of an applied voltage (Fig. 8), have attracted attention for their biomimetic capability. Among various configurations, those based either on metal-coated ion-exchange polymers or intrinsically electron-conducting polymers seem to be the most promising. In the first case, the device is based on a membrane, coated with a metal layer (for example gold or silver) on both sides, each of which acts as an electrode. It has been known for some time that these composites can act as electrochemical transducers that behave like artificial muscles for biomechanical and biomimetic applications. Their operation requires humidification because the ion conduction in the polymer typically requires swelling using a solvent, which is usually water. One of the obstacles to the widespread use of such transducers is thus their dependence on the degree of hydration, as evaporation becomes a potential issue. The narrow electrochemical stability window of water also limits the extent of the applied voltage and, thus, the level of strain that the actuators can generate.

There is considerably more scope for production of high levels of strain in the intrinsically conducting polymer variant of these devices. Electrode materials, such as polypyrrole or polyaniline, are used that show a change in volume as they are cycled between their oxidized and reduced states. Aqueous or aprotic solvent electrolytes have conventionally been used, but these limit the performance and lifetime of the devices for the reasons discussed above. The issue may be solved by turning to ionic liquids. The use of ionic liquids in electrochemical actuators was pioneered by Lu and co-workers[55,56], who showed that ionic liquids could indeed provide a huge increase in the performance of these devices. Owing to the use of ionic liquids, an operational life in excess of 10^6 cycles is now an easily achievable target.

Since the beginning of these studies, it has been clear that the origin of the improvement was associated with the high electrochemical stability of the ionic liquids, discussed above, which allows the conducting polymer to undergo its redox cycle with no interference from side reactions involving the electrolyte solvent. This prompted intensive further work in this field over the past five years. For example, Zhou et al.[57] demonstrated the fabrication of a completely solid-state device. This was achieved by gelling the electrolyte by inclusion of a cross-linked polymer component, which was polymerized in situ. An additional benefit of the use of hydrophobic ionic liquids is that they have less tendency to absorb or mix with water in an aqueous environment, thus enabling potential in vivo applications. Xi et al.[58] have shown that the use of an ionic liquid electrolyte allowed a poly(3-methylthiophene)-based actuator to achieve increased isotonic actuation with increased stress. Hence, the work output per cycle increases with increasing stress, a highly welcome feature for a practical device. Lu et al.[59] have described axial-geometry actuators based on polyaniline yarn and hollow fibres; these actuators were able to generate stresses of up to 0.85 MPa, exceeding that of skeletal muscle. The 'yarn-in-fibre' approach has the potential to allow the creation of a multi-fibre device in which the required stress capability can be achieved by simple combination of a sufficient number of fibres.

The other feature of ionic liquids as electrolytes for use in combination with conducting polymers is that the anions typically involved, such as PF_6, BF_4, CF_3SO_3 or $[(CF_3SO_2)_2N]$, are often the same as those used as counter ions in the conducting polymers. The reason for this is related to the similar set of properties that are optimum in both cases: diffuse or shielded charge creating a weakly interacting ion of low basicity[60]. Thus, the active ion is present in the ionic-liquid electrolyte at high concentration, often around 5 mol l^{-1}, and the redox cycling process of such polymer–electrolyte combinations takes place without any diffusion limit in the electrolyte. This feature also allows the polymerization of the conducting polymer to be carried out in the ionic liquid[61]. Carbon-nanotube-based actuators with ionic liquids as electrolytes have recently been described[62,63].

In most actuator applications, it is important that the ionic liquid is present in a gelled or otherwise semi-solid form, while still maintaining its ion conduction properties. A number of approaches have been investigated to achieve this, with some degree of success. Vidal and co-workers have developed ionic-liquid-based interpenetrating network electrolyte systems for use in actuators and shown that these can produce very long lifetimes[64–66]. An alternative approach has involved the use of nitrile rubber impregnated with

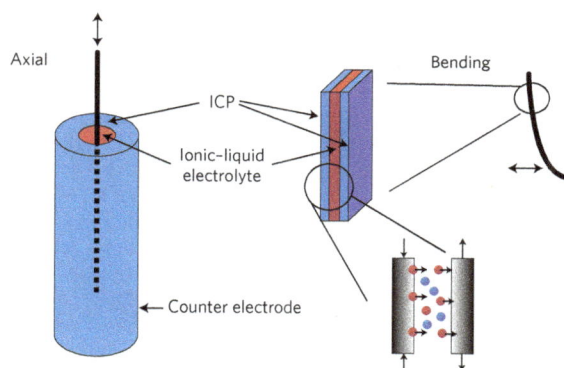

Figure 8 | An electromechanical actuator produces a mechanical bending or axial motion in response to an electrical stimulus. The voltage applied to the two intrinsically conducting polymer (ICP) electrodes causes a small oxidation or reduction reaction to take place, changing the volume of the electrode material. This produces a stress in the material at usefully high strain. Gelled ionic-liquid-based electrolytes provide the ideal combination of high conductivity coupled with high thermal and electrochemical stability for this technology.

an ionic liquid[67,68]. In this case, the actuator displacement was found to vary with anion size, as expected if the change in volume of the actuator is principally the result of incorporation of the anion. A further approach that is gaining some attention is the possible use of organic ionic plastic crystal solid electrolytes[69,70]. Devices utilizing membranes into which ionic liquids have been impregnated have also been reported[71,72]; these show considerably improved stability over aqueous electrolyte analogues.

Overall, the prospects appear to be excellent for practical biomimetic actuator devices having the required strain, lifetime and rate capabilities. For reasons associated with both properties and chemical structure, ionic liquids have become the electrolyte materials of choice and are contributing to significant progress in this field.

Concluding remarks

Lewis-neutral ionic liquids date from just 15 years ago, and have been the basis of a quiet revolution in materials science; their use is increasingly being implemented in devices. In 2007, a conference on the subject drew the attendance of several hundred scientists. We have focused here on the applications of ionic liquids in the field of energy, which is today's greatest challenge because of its implications in climate change, war, peace and humanitarian progress. It is clear that the field of batteries is very close to maturity for transportation purposes, and, as stated, safety is the main hurdle that ionic liquids can help to overcome. Conventional fuel cells have received enormous funding and publicity but have returned disappointing results. Here, a change of direction is needed. This may occur perhaps through the use of a Grotthuss conduction mechanism for the proton but not one based on water. Ultimately, especially if the 'hydrogen economy' proves too difficult to implement, biofuel cells in which enzymes can break down more complex molecules like ethanol will be required. The ability of the ionic liquid to stabilize the complex protein folding that allows the activity of an enzyme is, in itself, a revolution.

In the past two or three years, the tremendous potential offered by ionic liquids as solvents and electrolytes has been realized in a broad swath of materials preparation and device applications[73]. Ionic liquids are thus no longer laboratory curiosities; they have become an important component in the materials science field.

Nonetheless, there remain many aspects of the behaviour of ionic liquids that need urgent investigation if the applications discussed here are to be better understood and developed. Examples are fundamental electrochemical issues such as the structure of the double layer at an electrode in an ionic liquid; the speciation of solute ions such as metal ions; transference numbers and how these are influenced by speciation; dielectric properties and how these merge with conduction properties as a function of frequency; surface chemistry and wetting phenomena of ionic liquids in porous electrodes; thermodynamic quantities such as chemical potentials of solutes in ionic liquids, and how these influence redox potentials; ion association and its effect on thermodynamic and transport properties; and interactions of ionic liquids with solutes and interfaces. The simple dissolution of a lithium salt always results in electrolytes in which only a fraction of the current is actually carried by Li+ ions. It would be interesting to see whether the incorporation of poly(Li salts), that is, lithium salts with a polyanionic chain, through specific interaction with the fixed negative charge of polarizing Li+ ions, could change the flux balance.

All of these fundamental aspects of the physical chemistry and electrochemistry of ionic liquids remain to be thoroughly explained, and they promise to further improve the potential of their various electrochemical applications.

References

1. Earle, M. J. *et al.* The distillation and volatility of ionic liquids. *Nature* **439**, 831–834 (2006).
2. Wasserscheid, P. & Welton, T. (eds). *Ionic Liquids in Synthesis* (Wiley-VCH, 2007).
3. Endres, F., Abbott, A. & MacFarlane, D. R. (eds). *Electrodeposition from Ionic Liquids* (Wiley-VCH, 2008).
4. Abbott, A. P. & McKenzie, K. J. Application of ionic liquids to the electrodeposition of metals. *Phys. Chem. Chem. Phys.* **8**, 4265–4279 (2006).
5. Ziegler, K. & Lehmkuhl, H. Die elektrolytische Abscheidung von Aluminium aus organischen Komplexverbindungen. *Z. Anorg. Allg. Chem.* **283**, 414–424 (1956).
6. Kautek, W. & Birkle, S. Aluminum-electrocrystallization from metal organic electrolytes. *Electrochim. Acta* **34**, 1213–1218 (1989).
7. Caporali, S. *et al.* Aluminium electroplated from ionic liquids as protective coating against steel corrosion. *Corros. Sci.* **50**, 534–539 (2008).
8. Zein El Abedin, S. Coating of mild steel by aluminium in the ionic liquid [EMIm]Tf₂N and its corrosion performance. *Z. Phys. Chem.* **220**, 1293–1308 (2006).
9. Liu, Q. X., Zein El Abedin, S. & Endres, F. Electroplating of mild steel by aluminium in a first generation ionic liquid: a green alternative to commercial Al-plating in organic solvents. *Surf. Coat. Tech.* **201**, 1352–1356 (2006).
10. Zein El Abedin, S., Moustafa, E. M., Hempelmann, R., Natter, H. & Endres, F. Electrodeposition of nano- and microcrystalline aluminium in three different air and water stable ionic liquids. *ChemPhysChem* **7**, 1535–1543 (2006).
11. Deng, M. J. *et al.* Dicyanamide anion based ionic liquids for electrodeposition of metals. *Electrochem. Commun.* **10**, 213–216 (2008).
12. Zein El Abedin, S., Welz-Biermann, S. U. & Endres, F. A study on the electrodeposition of tantalum on NiTi alloy in an ionic liquid and corrosion behaviour of the coated alloy. *Electrochem. Commun.* **7**, 941–946 (2005).
13. Arnould, C. J., Delhalle, J. & Mekhalif, Z. Multifunctional hybrid coating on titanium towards hydroxyapatite growth: electrodeposition of tantalum and its molecular functionalization with organophosphonic acids films. *Electrochim. Acta* **53**, 5632–5638 (2008).
14. Endres, F. *et al.* On the electrodeposition of titanium in ionic liquids. *Phys. Chem. Chem. Phys.* **10**, 2189–2199 (2008).
15. Redel, E., Thomann, R. & Janiak, C. Use of ionic liquids (ILs) for the IL-anion size-dependent formation of Cr, Mo and W nanoparticles from metal carbonyl M(CO)₆ precursors. *Chem. Commun.* **15**, 1789–1791 (2008).
16. Tsuda, T., Arimoto, S., Kuwabata, S. & Hussey, C. L. Electrodeposition of Al–Mo–Ti ternary alloys in the Lewis acidic aluminum chloride-1-ethyl-3-methylimidazolium chloride room-temperature ionic liquid. *J. Electrochem. Soc.* **155**, D256–D262 (2008).
17. Kazeminezhad, I. *et al.* Templated electrodeposition of silver nanowires in a nanoporous polycarbonate membrane from a nonaqueous ionic liquid electrolyte. *Appl. Phys. A* **86**, 373–375 (2007).
18. Yang, P. X., An, M. Z., Su, C. N. & Wang, F. P. Electrodeposition of cobalt nanowires array from an ionic liquid. *Chinese J. Inorg. Chem.* **23**, 1501–1504 (2007).
19. Al-Salman, R. *et al.* Template assisted electrodeposition of germanium and silicon nanowires in an ionic liquid. *Phys. Chem. Chem. Phys.* **10**, 6233–6237 (2008).
20. Al-Salman, R., Zein El Abedin, S. & Endres, F. Electrodeposition of Ge, Si and Si_xGe_{1-x} from an air- and water-stable ionic liquid. *Phys. Chem. Chem. Phys.* **10**, 4650–4657 (2008).
21. Armand, M. & Tarascon, J.-M. Building better batteries. *Nature* **451**, 652–657 (2008).
22. Matsumoto, H., Sakaebe, H. & Tatsumi, K. Preparation of room temperature ionic liquids based on aliphatic onium cations and asymmetric amide anions and their electrochemical properties as a lithium battery electrolyte. *J. Power Sources* **160**, 1308–1313 (2006).
23. Seki, S. *et al.* Lithium secondary batteries using modified-imidazolium room-temperature ionic liquid. *J. Phys. Chem. B* **110**, 10228–10230 (2006).
24. Shin, J.-H., Henderson, W. A. & Passerini, S. PEO-based polymer electrolytes with ionic liquids and their use in lithium metal-polymer electrolyte batteries. *J. Electrochem. Soc.* **152**, A978–A983 (2005).
25. Garcia, B., Lavallée, S., Perron, G., Michot, C. & Armand, M. Room temperature molten salts as lithium battery electrolyte. *Electrochim. Acta* **49**, 4583–4588 (2004).
26. Howlett, P. C., MacFarlane, D. R. & Hollenkamp, A. F. High lithium metal cycling efficiency in a room-temperature ionic liquid. *Electrochem. Solid-State Lett.* **7**, A97–A101 (2004).
27. Fernicola, A. *et al.* LiTFSI-BEPyTFSI as an improved ionic liquid electrolyte for rechargeable lithium batteries. *J. Power Sources* **174**, 342–348 (2007).
28. Seki, S. *et al.* Reversibility of lithium secondary batteries using a room-temperature ionic liquid mixture and lithium metal. *Electrochem. Solid-State Lett.* **8**, A577–A578 (2005).
29. Ogasawara, T. *et al.* Rechargeable Li₂O₂ electrode for lithium batteries. *J. Am. Chem. Soc.* **128**, 1390–1393 (2006).
30. Kordeesch, K. & Simader, G. (eds) *Fuel Cells and Their Applications* (VCH, 1996).
31. Kreur, K. D. *Handbook of Fuel Cells: Fundamental, Technology & Applications* Vol. 3 (eds Vielstich, W., Lamm, A. & Gasteiger, H. A.) (Wiley, 2003).
32. Fuller, J., Breda, A. C. & Carlin, R. T. Ionic liquid-polymer gel electrolytes. *J. Electrochem. Soc.* **144**, L67–L70 (1997).
33. Navarra, M. A., Panero, S. & Scrosati, B. Novel, ionic-liquid-based, gel-type proton membranes. *Electrochem. Solid-State Lett.* **8**, A324–A327 (2005).
34. Susan, M. A. B. H., Kaneko, T., Noda, A. & Watanabe, M. Ion gels prepared by in situ radical polymerization of vinyl monomers in an ionic liquid and their characterization as polymer electrolytes. *J. Am. Chem. Soc.* **127**, 4976–4983 (2005).
35. Winther-Jensen, B., Winther-Jensen, O., Forsyth, M. & MacFarlane, D. R. High rates of oxygen reduction over a vapor phase-polymerized PEDOT electrode. *Science* **321**, 671–674 (2008).
36. Simon, P. & Gogotsi, Y. Materials for electrochemical capacitors. *Nature Mater.* **7**, 845–854 (2008).
37. Sato, T., Masuda, G. & Takagi, K. Electrochemical properties of novel ionic liquids for electric double layer capacitor applications. *Electrochim. Acta* **49**, 3603–3611 (2004).
38. Arbizzani, C. *et al.* Safe high-energy supercapacitors based on solvent-free ionic liquid electrolytes ionic liquid electrolytes. *J. Power Sources* **185**, 1575–1579 (2008).
39. Fukaya, Y., Sugimoto, A. & Ohno, H. Superior solubility of polysaccharides in low viscosity, polar, and halogen-free 1,3-dialkylimidazolium formates. *Biomacromolecules* **7**, 3295–3297 (2006).
40. Hermanutz, F. *et al.* New developments in dissolving and processing of cellulose in ionic liquids. *Macromol. Symp.* **262**, 23–27 (2008).
41. Swatloski, R. P., Spear, S. K., Holbrey, J. D. & Rogers, R. D. Dissolution of cellulose with ionic liquids. *J. Am. Chem. Soc.* **124**, 4974–4975 (2002).
42. Wu, J. *et al.* Homogeneous acetylation of cellulose in a new ionic liquid. *Biomacromolecules* **5**, 266–268 (2004).
43. Anderson, J. L., Ding, J., Welton, T. & Armstrong, D. W. Characterizing ionic liquids on the basis of multiple solvation interactions. *J. Am. Chem. Soc.* **124**, 14247–14254 (2002).
44. Fukaya, Y., Hayashi, K., Wada, M. & Ohno, H. Cellulose dissolution with polar ionic liquids under mild conditions: required factors for anions. *Green Chem.* **10**, 44–46 (2008).
45. Ohno, H., Suzuki, C., Fukumoto, K., Yoshizawa, M. & Fujita, K. Electron transfer process of poly(ethylene oxide) modified cytochrome c in imidazolium type ionic liquid. *Chem. Lett.* **32**, 450–451 (2003).
46. Nakashima, K., Maruyama, T., Kamiya, N. & Goto, M. Comb-shaped poly(ethylene glycol)-modified subtilisin Carlsberg is soluble and highly active in ionic liquids. *Chem. Commun.* 4297–4299 (2005).
47. Shimojo, K., Nakashima, K., Kamiya, N. & Goto, M. Crown ether-mediated extraction and functional conversion of cytochrome *c* in ionic liquids. *Biomacromolecules* **7**, 2–5 (2006).
48. Fujita, K., MacFarlane, D. R. & Forsyth, M. Protein solubilising and stabilising ionic liquids. *Chem. Commun.* 4804–4806 (2005).
49. Fujita, K. *et al.* Unexpected improvement in stability and utility of cytochrome *c* by solution in biocompatible ionic liquids. *Biotechnol. Bioeng.* **94**, 1209–1213 (2006).
50. Fujita, K. *et al.* Solubility and stability of cytochrome c in hydrated ionic liquids: effect of oxo acid residues and kosmotropicity. *Biomacromolecules* **8**, 2080–2086 (2007).

51. Tamura, K. & Ohno, H. Solubility of cytochrome *c* in ionic liquids and redox response at high temperature. *2nd Internat. Congr. Ionic Liquids*, abstr. 2P09-090, 306 (2007).

52. DiCarlo, C. M., Compton, D. L., Evans, K. O. & Laszlo, J. A. Bioelectrocatalysis in ionic liquids. Examining specific cation and anion effects on electrode-immobilized cytochrome *c*. *Bioelectrochemistry* **68,** 134–143 (2006).

53. Turner, M. B. *et al.* Ionic liquid salt-induced inactivation and unfolding of cellulase from Trichoderma reesei. *Green Chem.* **5,** 443–447 (2003).

54. Motoyama, Y., Nakamura, N. & Ohno, H. An ethanol/dioxygen biofuel cell using hydrophobic ionic liquid as electrolyte. *2nd Internat. Congr. Ionic Liquids*, abstr. 1P09-091, 213 (2007).

55. Lu, W. *et al.* Use of ionic liquids for pi-conjugated polymer electrochemical devices. *Science* **297,** 983–987 (2002).

56. Ding, J. *et al.* Use of ionic liquids as electrolytes in electromechanical actuator systems based on inherently conducting polymers. *Chem. Mater.* **15,** 2392–2398 (2003).

57. Zhou, D. Z. *et al.* Solid state actuators based on polypyrrole and polymer-in-ionic liquid electrolytes. *Electrochim. Acta* **48,** 2355–2359 (2003).

58. Xi, B. B. *et al.* Poly(3-methylthiophene) electrochemical actuators showing increased strain and work per cycle at higher operating stresses. *Polymer (Guildf.)* **47,** 7720–7725 (2006).

59. Lu, W., Norris, I. D. & Mattes, B. R. Electrochemical actuator devices based on polyaniline yarns and ionic liquid electrolytes. *Aust. J. Chem.* **58,** 263–269 (2005).

60. MacFarlane, D. R. *et al.* Lewis base ionic liquids. *Chem. Commun.* 1905–1917 (2006).

61. Pringle, J. *et al.* The influence of the monomer and the ionic liquid on the electrochemical preparation of polythiophene. *Polymer (Guildf.)* **47,** 2047–2058 (2005).

62. Lim, H. T., Lef, J. W. & Yoo, Y. T. Actuation behavior of a carbon nanotube/nafion™ IPMC actuator containing an ionic liquid. *J. Korean Phys. Soc.* **49,** 1101–1106 (2006).

63. Barisci, J. N., Wallace, G. G., MacFarlane, D. R. & Baughman, R. H. Investigation of ionic liquids as electrolytes for carbon nanotube electrodes. *Electrochem. Commun.* **6,** 22–27 (2004).

64. Vidal, F., Plesse, C., Teyssie, D. & Chevrot, C. Long-life air working conducting semi-IPN/ionic liquid based actuator. *Synth. Met.* **142,** 287–291 (2004).

65. Vidal, F., Juger, J., Chevrot, C. & Teyssie, D. Interpenetrating polymer networks from polymeric imidazolium-type ionic liquid and polybutadiene. *Polym. Bull.* **57,** 473–480 (2006).

66. Vidal, F. *et al.* Long-life air working semi-IPN/ionic liquid: new precursor of artificial muscles. *Mol. Cryst. Liq. Cryst.* **448,** 95–102 (2006).

67. Cho, M. S. *et al.* A solid state actuator based on the PEDOT/NBR system. *Sens. Actuators B* **119,** 621–624 (2006).

68. Cho, M. S. *et al.* A solid state actuator based on the PEDOT/NBR system: effect of anion size of imidazolium ionic liquid. *Mol. Cryst. Liq. Cryst.* **464,** 633–638 (2007).

69. MacFarlane, D. R. & Forsyth, M. Plastic crystal electrolyte materials: new perspectives on solid state ionics. *Adv. Mater.* **13,** 957–966 (2001).

70. Long, S., MacFarlane, D. R. & Forsyth, M. Fast ion conduction in molecular plastic crystals. *Solid State Ionics* **161,** 105–112 (2003).

71. Bennett, M. D. & Leo, D. J. Ionic liquids as stable solvents for ionic polymer transducers. *Sens. Actuators A* **115,** 79–90 (2004).

72. Wang, J., Xu, C. Y., Taya, M. & Kuga, Y. A. Flemion-based actuator with ionic liquid as solvent. *Smart Mater. Struct.* **16,** S214–S219 (2007).

73. Ohno, H. (ed). *Electrochemical Aspects of Ionic Liquids* (Wiley, 2005).

Acknowledgements

We thank M. Mastragostino for discussions and information concerning supercapacitors.

Materials for electrochemical capacitors

Electrochemical capacitors, also called supercapacitors, store energy using either ion adsorption (electrochemical double layer capacitors) or fast surface redox reactions (pseudo-capacitors). They can complement or replace batteries in electrical energy storage and harvesting applications, when high power delivery or uptake is needed. A notable improvement in performance has been achieved through recent advances in understanding charge storage mechanisms and the development of advanced nanostructured materials. The discovery that ion desolvation occurs in pores smaller than the solvated ions has led to higher capacitance for electrochemical double layer capacitors using carbon electrodes with subnanometre pores, and opened the door to designing high-energy density devices using a variety of electrolytes. Combination of pseudo-capacitive nanomaterials, including oxides, nitrides and polymers, with the latest generation of nanostructured lithium electrodes has brought the energy density of electrochemical capacitors closer to that of batteries. The use of carbon nanotubes has further advanced micro-electrochemical capacitors, enabling flexible and adaptable devices to be made. Mathematical modelling and simulation will be the key to success in designing tomorrow's high-energy and high-power devices.

PATRICE SIMON[1,2] AND YURY GOGOTSI[3]

[1]Université Paul Sabatier, CIRIMAT, UMR-CNRS 5085, 31062 Toulouse Cedex 4, France
[2]Institut Universitaire de France, 103 Boulevard Saint Michel, 75005 Paris, France
[3]Department of Materials Science & Engineering, Drexel University, 3141 Chestnut Street, Philadelphia 19104, USA
e-mail: simon@chimie.ups-tlse.fr; gogotsi@drexel.edu

Climate change and the decreasing availability of fossil fuels require society to move towards sustainable and renewable resources. As a result, we are observing an increase in renewable energy production from sun and wind, as well as the development of electric vehicles or hybrid electric vehicles with low CO_2 emissions. Because the sun does not shine during the night, wind does not blow on demand and we all expect to drive our car with at least a few hours of autonomy, energy storage systems are starting to play a larger part in our lives. At the forefront of these are electrical energy storage systems, such as batteries and electrochemical capacitors (ECs)[1]. However, we need to improve their performance substantially to meet the higher requirements of future systems, ranging from portable electronics to hybrid electric vehicles and large industrial equipment, by developing new materials and advancing our understanding of the electrochemical interfaces at the nanoscale. Figure 1 shows the plot of power against energy density, also called a Ragone plot[2], for the most important energy storage systems.

Lithium-ion batteries were introduced in 1990 by Sony, following pioneering work by Whittingham, Scrosati and Armand (see ref. 3 for a review). These batteries, although costly, are the best in terms of performance, with energy densities that can reach 180 watt hours per kilogram. Although great efforts have gone into developing high-performance Li-ion and other advanced secondary batteries that use nanomaterials or organic redox couples[4-6], ECs have attracted less attention until very recently. Because Li-ion batteries suffer from a somewhat slow power delivery or uptake, faster and higher-power energy storage systems are needed in a number of applications, and this role has been given to the ECs[7]. Also known as supercapacitors or ultracapacitors, ECs are power devices that can be fully charged or discharged in seconds; as a consequence, their energy density (about 5 Wh kg^{-1}) is lower than in batteries, but a much higher power delivery or uptake (10 kW kg^{-1}) can be achieved for shorter times (a few seconds)[1]. They have had an important role in complementing or replacing batteries in the energy storage field, such as for uninterruptible power supplies (back-up supplies used to protect against power disruption) and load-levelling. A more recent example is the use of electrochemical double layer capacitors (EDLCs) in emergency doors (16 per plane) on an Airbus A380, thus proving that in terms of performance, safety and reliability ECs are definitely ready for large-scale implementation. A recent report by the US Department of Energy[8] assigns equal importance to supercapacitors and batteries for future energy storage systems, and articles on supercapacitors appearing in business and popular magazines show increasing interest by the general public in this topic.

Several types of ECs can be distinguished, depending on the charge storage mechanism as well as the active materials used. EDLCs, the most common devices at present, use carbon-based active materials with high surface area (Fig. 2). A second group of ECs, known as pseudo-capacitors or redox supercapacitors, uses fast and reversible surface or near-surface reactions for charge storage. Transition metal oxides as well as electrically conducting polymers are examples of

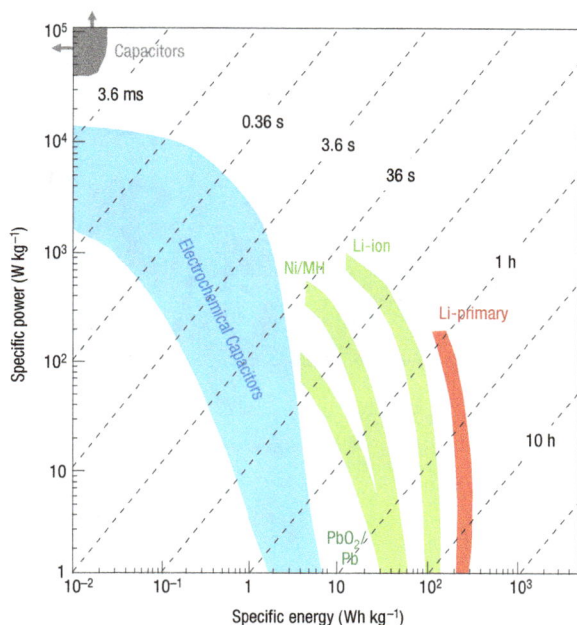

Figure 1 Specific power against specific energy, also called a Ragone plot, for various electrical energy storage devices. If a supercapacitor is used in an electric vehicle, the specific power shows how fast one can go, and the specific energy shows how far one can go on a single charge. Times shown are the time constants of the devices, obtained by dividing the energy density by the power.

pseudo-capacitive active materials. Hybrid capacitors, combining a capacitive or pseudo-capacitive electrode with a battery electrode, are the latest kind of EC, which benefit from both the capacitor and the battery properties.

Electrochemical capacitors currently fill the gap between batteries and conventional solid state and electrolytic capacitors (Fig. 1). They store hundreds or thousands of times more charge (tens to hundreds of farads per gram) than the latter, because of a much larger surface area (1,000–2,000 $m^2 g^{-1}$) available for charge storage in EDLC. However, they have a lower energy density than batteries, and this limits the optimal discharge time to less than a minute, whereas many applications clearly need more[9]. Since the early days of EC development in the late 1950s, there has not been a good strategy for increasing the energy density; only incremental performance improvements were achieved from the 1960s to 1990s. The impressive increase in performance that has been demonstrated in the past couple of years is due to the discovery of new electrode materials and improved understanding of ion behaviour in small pores, as well as the design of new hybrid systems combining faradic and capacitive electrodes. Here we give an overview of past and recent findings as well as an analysis of what the future holds for ECs.

ELECTROCHEMICAL DOUBLE-LAYER CAPACITORS

The first patent describing the concept of an electrochemical capacitor was filed in 1957 by Becker[9], who used carbon with a high specific surface area (SSA) coated on a metallic current collector in a sulphuric acid solution. In 1971, NEC (Japan) developed aqueous-electrolyte capacitors under the energy company SOHIO's licence for power-saving units in electronics, and this application can be considered as the starting point for electrochemical capacitor use in commercial devices[9]. New applications in mobile electronics, transportation

(cars, trucks, trams, trains and buses), renewable energy production and aerospace systems[10] bolstered further research.

MECHANISM OF DOUBLE-LAYER CAPACITANCE

EDLCs are electrochemical capacitors that store the charge electrostatically using reversible adsorption of ions of the electrolyte onto active materials that are electrochemically stable and have high accessible SSA. Charge separation occurs on polarization at the electrode–electrolyte interface, producing what Helmholtz described in 1853 as the double layer capacitance C:

$$C = \frac{\varepsilon_r \varepsilon_0 A}{d} \quad \text{or} \quad C/A = \frac{\varepsilon_r \varepsilon_0}{d} \qquad (1)$$

where ε_r is the electrolyte dielectric constant, ε_0 is the dielectric constant of the vacuum, d is the effective thickness of the double layer (charge separation distance) and A is the electrode surface area.

This capacitance model was later refined by Gouy and Chapman, and Stern and Geary, who suggested the presence of a diffuse layer in the electrolyte due to the accumulation of ions close to the electrode surface. The double layer capacitance is between 5 and 20 $\mu F \, cm^{-2}$ depending on the electrolyte used[11]. Specific capacitance achieved with aqueous alkaline or acid solutions is generally higher than in organic electrolytes[11], but organic electrolytes are more widely used as they can sustain a higher operation voltage (up to 2.7 V in symmetric systems). Because the energy stored is proportional to voltage squared according to

$$E = \frac{1}{2} CV^2 \qquad (2)$$

a three-fold increase in voltage, V, results in about an order of magnitude increase in energy, E, stored at the same capacitance.

As a result of the electrostatic charge storage, there is no faradic (redox) reaction at EDLC electrodes. A supercapacitor electrode must be considered as a blocking electrode from an electrochemical point of view. This major difference from batteries means that there is no limitation by the electrochemical kinetics through a polarization resistance. In addition, this surface storage mechanism allows very fast energy uptake and delivery, and better power performance. The absence of faradic reactions also eliminates the swelling in the active material that batteries show during charge/discharge cycles. EDLCs can sustain millions of cycles whereas batteries survive a few thousand at best. Finally, the solvent of the electrolyte is not involved in the charge storage mechanism, unlike in Li-ion batteries where it contributes to the solid–electrolyte interphase when graphite anodes or high-potential cathodes are used. This does not limit the choice of solvents, and electrolytes with high power performances at low temperatures (down to −40 °C) can be designed for EDLCs. However, as a consequence of the electrostatic surface charging mechanism, these devices suffer from a limited energy density. This explains why today's EDLC research is largely focused on increasing their energy performance and widening the temperature limits into the range where batteries cannot operate[9].

HIGH SURFACE AREA ACTIVE MATERIALS

The key to reaching high capacitance by charging the double layer is in using high SSA blocking and electronically conducting electrodes. Graphitic carbon satisfies all the requirements for this application, including high conductivity, electrochemical stability and open porosity[12]. Activated, templated and carbide-derived carbons[13], carbon fabrics, fibres, nanotubes[14], onions[15] and nanohorns[16] have been tested for EDLC applications[11], and some of these carbons are shown in Fig. 2a–d. Activated carbons are the most widely used materials today, because of their high SSA and moderate cost.

Activated carbons are derived from carbon-rich organic precursors by carbonization (heat treatment) in inert atmosphere

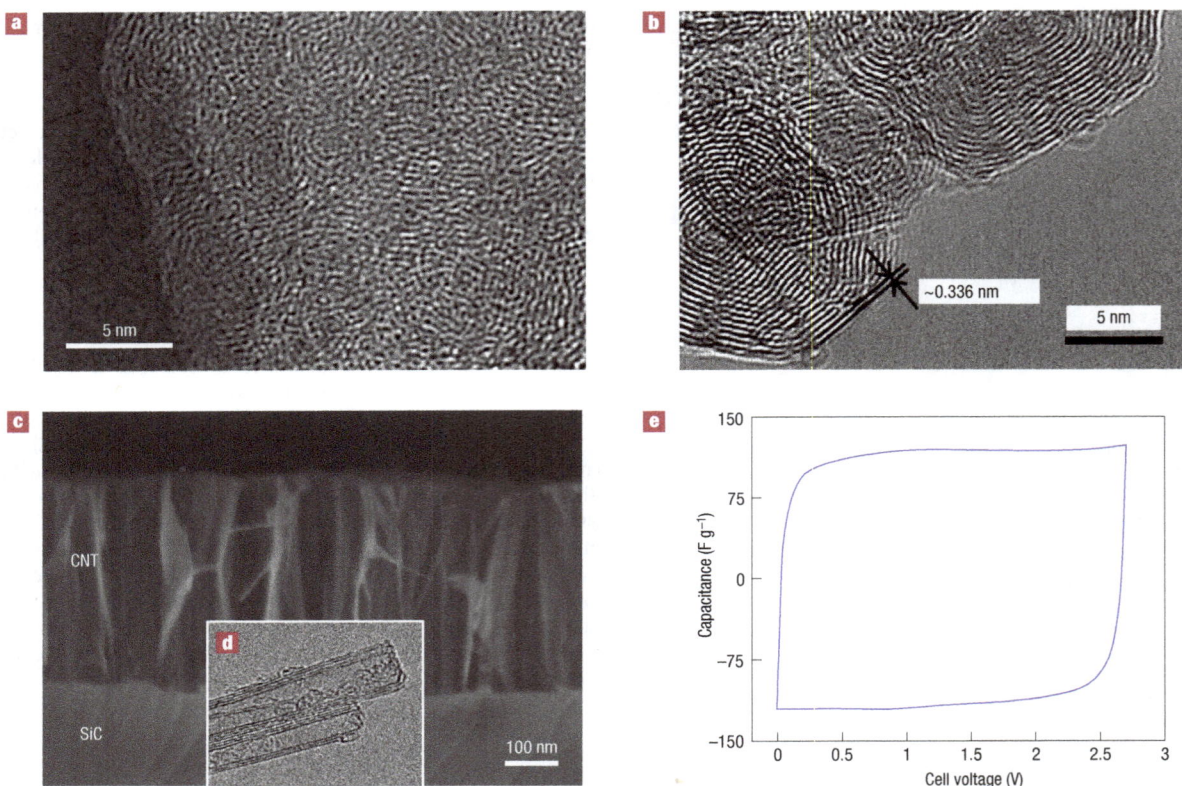

Figure 2 Carbon structures used as active materials for double layer capacitors. **a**, Typical transmission electronic microscopy (TEM) image of a disordered microporous carbon (SiC-derived carbon, 3 hours chlorination at 1,000 °C). **b**, TEM image of onion-like carbon. Reproduced with permission from ref. 80. © 2007 Elsevier. **c**, Scanning electron microscopy image of an array of carbon nanotubes (labelled CNT) on SiC produced by annealing for 6 h at 1,700 °C; inset, **d**, shows a TEM image of the same nanotubes[72]. **e**, Cyclic voltammetry of a two-electrode laboratory EDLC cell in 1.5 M tetraethylammonium tetrafluoroborate NEt_4^+,BF_4^- in acetonitrile-based electrolyte, containing activated carbon powders coated on aluminium current collectors. Cyclic voltammetry was recorded at room temperature and potential scan rate of 20 mV s^{-1}.

with subsequent selective oxidation in CO_2, water vapour or KOH to increase the SSA and pore volume. Natural materials, such as coconut shells, wood, pitch or coal, or synthetic materials, such as polymers, can be used as precursors. A porous network in the bulk of the carbon particles is produced after activation; micropores (<2 nm in size), mesopores (2–50 nm) and macropores (>50 nm) can be created in carbon grains. Accordingly, the porous structure of carbon is characterized by a broad distribution of pore size. Longer activation time or higher temperature leads to larger mean pore size. The double layer capacitance of activated carbon reaches 100–120 F g^{-1} in organic electrolytes; this value can exceed 150–300 F g^{-1} in aqueous electrolytes, but at a lower cell voltage because the electrolyte voltage window is limited by water decomposition. A typical cyclic voltammogram of a two-electrode EDLC laboratory cell is presented in Fig. 2e. Its rectangular shape is characteristic of a pure double layer capacitance mechanism for charge storage according to:

$$I = C \times \frac{dV}{dt} \quad (3)$$

where I is the current, (dV/dt) is the potential scan rate and C is the double layer capacitance. Assuming a constant value for C, for a given scan rate the current I is constant, as can be seen from Fig. 2e, where the cyclic voltammogram has a rectangular shape.

As previously mentioned, many carbons have been tested for EDLC applications and a recent paper[11] provides an overview of

what has been achieved. Untreated carbon nanotubes[17] or nanofibres have a lower capacitance (around 50–80 F g^{-1}) than activated carbon in organic electrolytes. It can be increased up to 100 F g^{-1} or greater by grafting oxygen-rich groups, but these are often detrimental to cyclability. Activated carbon fabrics can reach the same capacitance as activated carbon powders, as they have similar SSA, but the high price limits their use to speciality applications. The carbons used in EDL capacitors are generally pre-treated to remove moisture and most of the surface functional groups present on the carbon surface to improve stability during cycling, both of which can be responsible for capacitance fading during capacitor ageing as demonstrated by Azais et al.[18] using NMR and X-ray photoelectron spectroscopy techniques. Pandolfo et al.[11], in their review article, concluded that the presence of oxygenated groups also contributes to capacitor instability, resulting in an increased series resistance and deterioration of capacitance. Figure 3 presents a schematic of a commercial EDLC, showing the positive and the negative electrodes as well as the separator in rolled design (Fig. 3a,b) and flat design (button cell in Fig. 3c).

CAPACITANCE AND PORE SIZE

Initial research on activated carbon was directed towards increasing the pore volume by developing high SSA and refining the activation process. However, the capacitance increase was limited even for the most porous samples. From a series of activated carbons with different pore sizes in various electrolytes, it was shown that there was no linear relationship between the SSA and the capacitance[19–21]. Some

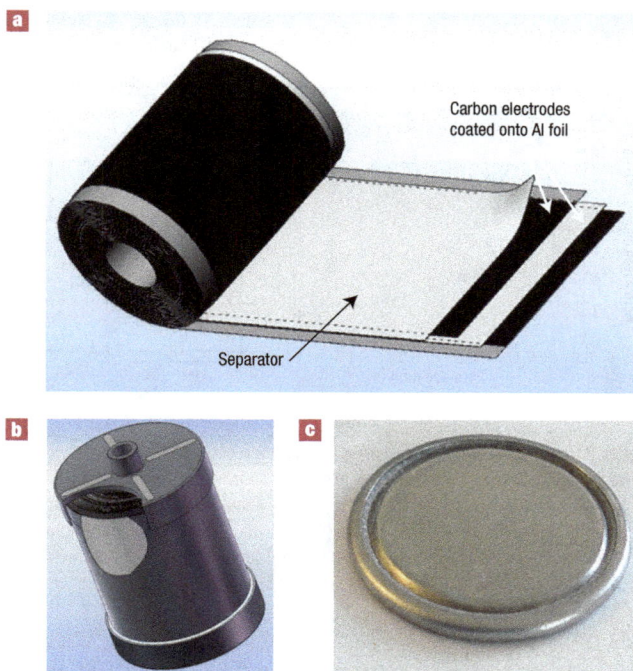

Figure 3 Electrochemical capacitors. **a**, Schematic of a commercial spirally wound double layer capacitor. **b**, Assembled device weighing 500 g and rated for 2,600 F. (Photo courtesy of Batscap, Groupe Bolloré, France.) **c**, A small button cell, which is just 1.6 mm in height and stores 5 F. (Photo courtesy of Y-Carbon, US.) Both devices operate at 2.7 V.

studies suggested that pores smaller than 0.5 nm were not accessible to hydrated ions[20,22] and that even pores under 1 nm might be too small, especially in the case of organic electrolytes, where the size of the solvated ions is larger than 1 nm (ref. 23). These results were consistent with previous work showing that ions carry a dynamic sheath of solvent molecules, the solvation shell[24], and that some hundreds of kilojoules per mole are required to remove it[25] in the case of water molecules. A pore size distribution in the range 2–5 nm, which is larger than the size of two solvated ions, was then identified as a way to improve the energy density and the power capability. Despite all efforts, only a moderate improvement has been made. Gravimetric capacitance in the range of 100–120 F g^{-1} in organic and 150–200 F g^{-1} in aqueous electrolytes has been achieved[26,27] and ascribed to improved ionic mass transport inside mesopores. It was assumed that a well balanced micro- or mesoporosity (according to IUPAC classification, micropores are smaller than 2 nm, whereas mesopores are 2–50 nm in diameter) was needed to maximize capacitance[28].

Although fine-tuned mesoporous carbons failed to achieve high capacitance performance, several studies reported an important capacitive contribution from micropores. From experiments using activated carbon cloth, Salitra et al.[29] suggested that a partial desolvation of ions could occur, allowing access to small pores (<2 nm). High capacitance was observed for a mesoporous carbon containing large numbers of small micropores[30–32], suggesting that partial ion desolvation could lead to an improved capacitance. High capacitances (120 F g^{-1} and 80 F cm^{-3}) were found in organic electrolytes for microporous carbons (<1.5 nm)[33,34], contradicting the solvated ion adsorption theory. Using microporous activated coal-based carbon materials, Raymundo-Piñero et al.[35] observed the same effect and found a maximum capacitance for pore size at 0.7 and

0.8 nm for aqueous and organic electrolytes, respectively. However, the most convincing evidence of capacitance increase in pores smaller than the solvated ion size was provided by experiments using carbide-derived carbons (CDCs)[36–38] as the active material. These are porous carbons obtained by extraction of metals from carbides (TiC, SiC and other) by etching in halogens at elevated temperatures[39]:

$$TiC + 2Cl_2 \rightarrow TiCl_4 + C \tag{4}$$

In this reaction, Ti is leached out from TiC, and carbon atoms self-organize into an amorphous or disordered, mainly sp^2-bonded[40], structure with a pore size that can be fine-tuned by controlling the chlorination temperature and other process parameters. Accordingly, a narrow uni-modal pore size distribution can be achieved in the range 0.6–1.1 nm, and the mean pore size can be controlled with sub-ångström accuracy[41]. These materials were used to understand the charge storage in micropores using 1 M solution of NEt$_4$BF$_4$ in acetonitrile-based electrolyte[42]. The normalized capacitance (μF cm^{-2}) decreased with decreasing pore size until a critical value close to 1 nm was reached (Fig. 4), and then sharply increased when the pore size approached the ion size. As the CDC samples were exclusively microporous, the capacitance increase for subnanometre pores clearly shows the role of micropores. Moreover, the gravimetric and volumetric capacitances achieved by CDC were, respectively, 50% and 80% higher than for conventional activated carbon[19–21]. The capacitance change with the current density was also found to be stable, demonstrating the high power capabilities these materials can achieve[42]. As the solvated ion sizes in this electrolyte were 1.3 and 1.16 nm for the cation and anion[16], respectively, it was proposed that partial or complete removal of their solvation shell was allowing the ions to access the micropores. As a result, the change of capacitance was a linear function of $1/b$ (where b is the pore radius), confirming that the distance between the ion and the carbon surface, d, was shorter for the smaller pores. This dependence published by Chmiola et al.[42] has since been confirmed by other studies, and analysis of literature data is provided in refs 43 and 44.

CHARGE-STORAGE MECHANISM IN SUBNANOMETRE PORES

From a fundamental point of view, there is a clear lack of understanding of the double layer charging in the confined space of micropores, where there is no room for the formation of the Helmholtz layer and diffuse layer expected at a solid–electrolyte interface. To address this issue, a three-electrode cell configuration, which discriminates between anion and cation adsorption, was used[45]. The double layer capacitance in 1.5 M NEt$_4$BF$_4$-acetonitrile electrolyte caused by the anion and cation at the positive and negative electrodes, respectively, had maxima at different pore sizes[45]. The peak in capacitance shifted to smaller pores for the smaller ion (anion). This behaviour cannot be explained by purely electrostatic reasons, because all pores in this study were the same size as or smaller than a single ion with a single associated solvent molecule. It thus confirmed that ions must be at least partially stripped of solvent molecules in order to occupy the carbon pores. These results point to a charge storage mechanism whereby partial or complete removal of the solvation shell and increased confinement of ions lead to increased capacitance.

A theoretical analysis published by Huang et al.[43] proposed splitting the capacitive behaviour in two different parts depending on the pore size. For mesoporous carbons (pores larger than 2 nm), the traditional model describing the charge of the double layer was used[43]:

$$C/A = \frac{\varepsilon_r \varepsilon_0}{b \ln\left(\frac{b}{b-d}\right)} \tag{5}$$

where b is the pore radius and d is the distance of approach of the ion to the carbon surface. Data from Fig. 4 in the mesoporous range

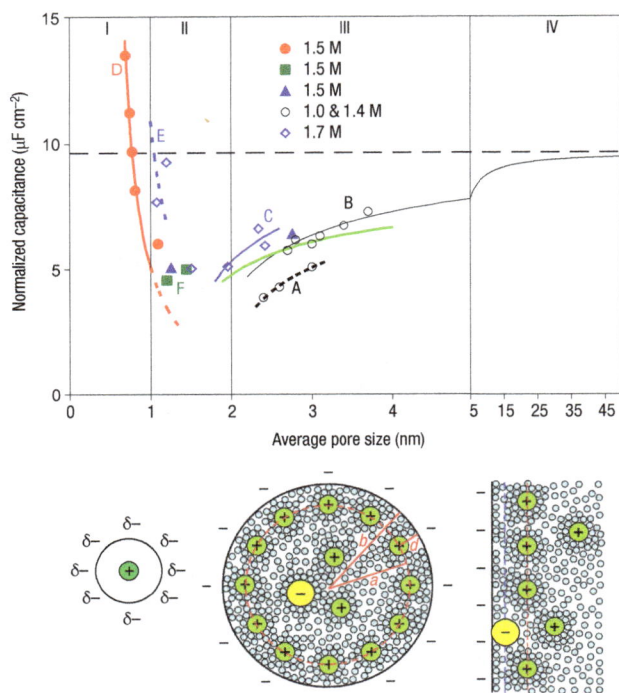

Figure 4 Specific capacitance normalized by SSA as a function of pore size for different carbon samples. All samples were tested in the same electrolyte (NEt$_4^+$,BF$_4^-$ in acetonitrile; concentrations are shown in the key). Symbols show experimental data for CDCs, templated mesoporous carbons and activated carbons, and lines show model fits[43]. A huge normalized capacitance increase is observed for microporous carbons with the smallest pore size in zone I, which would not be expected in the traditional view. The partial or complete loss of the solvation shell explains this anomalous behaviour[42]. As schematics show, zones I and II can be modelled as an electric wire-in-cylinder capacitor, an electric double-cylinder capacitor should be considered for zone III, and the commonly used planar electric double layer capacitor can be considered for larger pores, when the curvature/size effect becomes negligible (zone IV). A mathematical fit in the mesoporous range (zone III) is obtained using equation (5). Equation (6) was used to model the capacitive behaviour in zone I, where confined micropores force ions to desolvate partially or completely[44]. A, B: templated mesoporous carbons; C: activated mesoporous carbon; D, F: microporous CDC; E: microporous activated carbon. Reproduced with permission from ref. 44. © 2008 Wiley.

(zone III) were fitted with equation (5). For micropores (<1 nm), it was assumed that ions enter a cylindrical pore and line up, thus forming the 'electric wire in cylinder' model of a capacitor. Capacitance was calculated from[43]

$$C/A = \frac{\varepsilon_r \varepsilon_0}{b \ln\left(\frac{b}{a_0}\right)} \qquad (6)$$

where a_0 is the effective size of the ion (desolvated). This model perfectly matches with the normalized capacitance change versus pore size (zone I in Fig. 4). Calculations using density functional theory gave consistent values for the size, a_0, for unsolvated NEt$_4^+$ and BF$_4^-$ ions[43].

This work suggests that removal of the solvation shell is required for ions to enter the micropores. Moreover, the ionic radius a_0 found by using equation (6) was close to the bare ion size, suggesting that ions could be fully desolvated. A study carried out with CDCs in a

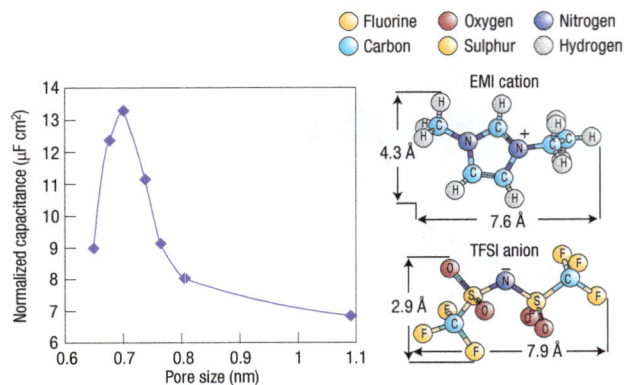

Figure 5 Normalized capacitance change as a function of the pore size of carbon-derived-carbide samples. Samples were prepared at different temperatures in ethyl-methylimidazolium/trifluoro-methane-sulphonylimide (EMI,TFSI) ionic liquid at 60 °C. Inset shows the structure and size of the EMI and TFSI ions. The maximum capacitance is obtained when the pore size is in the same range as the maximum ion dimension. Reproduced with permission from ref. 46. © 2008 ACS.

solvent-free electrolyte ([EMI$^+$,TFSI$^-$] ionic liquid at 60 °C), in which both ions have a maximum size of about 0.7 nm, showed the maximum capacitance for samples with the 0.7-nm pore size[46], demonstrating that a single ion per pore produces the maximum capacitance (Fig. 5). This suggests that ions cannot be adsorbed on both pore surfaces, in contrast with traditional supercapacitor models.

MATERIALS BY DESIGN

The recent findings of the micropore contribution to the capacitive storage highlight the lack of fundamental understanding of the electrochemical interfaces at the nanoscale and the behaviour of ions confined in nanopores. In particular, the results presented above rule out the generally accepted description of the double layer with solvated ions adsorbed on both sides of the pore walls, consistent with the absence of a diffuse layer in subnanometre pores. Although recent studies[45,46] provide some guidance for developing materials with improved capacitance, such as elimination of macro- and mesopores and matching the pore size with the ion size, further material optimization by Edisonian or combinatorial electrochemistry methods may take a very long time. The effects of many parameters, such as carbon bonding (sp versus sp^2 or sp^3), pore shape, defects or adatoms, are difficult to determine experimentally. Clearly, computational tools and atomistic simulation will be needed to help us to understand the charge storage mechanism in subnanometre pores and to propose strategies to design the next generation of high-capacitance materials and material–electrolyte systems[47]. Recasting the theory of double layers in electrochemistry to take into account solvation and desolvation effects could lead to a better understanding of charge storage as well as ion transport in ECs and even open up new opportunities in areas such as biological ion channels and water desalination.

REDOX-BASED ELECTROCHEMICAL CAPACITORS

MECHANISM OF PSEUDO-CAPACITIVE CHARGE STORAGE

Some ECs use fast, reversible redox reactions at the surface of active materials, thus defining what is called the pseudo-capacitive behaviour. Metal oxides such as RuO$_2$, Fe$_3$O$_4$ or MnO$_2$ (refs 48, 49), as well as electronically conducting polymers[50], have been extensively studied in the past decades. The specific pseudo-capacitance exceeds

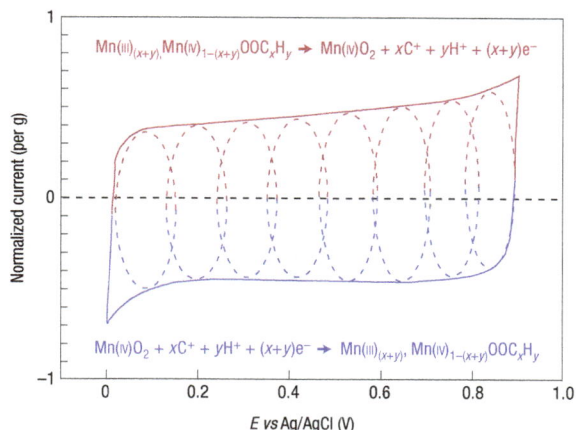

Figure 6 Cyclic voltammetry. This schematic of cyclic voltammetry for a MnO$_2$-electrode cell in mild aqueous electrolyte (0.1 M K$_2$SO$_4$) shows the successive multiple surface redox reactions leading to the pseudo-capacitive charge storage mechanism. The red (upper) part is related to the oxidation from Mn(III) to Mn(IV) and the blue (lower) part refers to the reduction from Mn(IV) to Mn(III).

that of carbon materials using double layer charge storage, justifying interest in these systems. But because redox reactions are used, pseudo-capacitors, like batteries, often suffer from a lack of stability during cycling.

Ruthenium oxide, RuO$_2$, is widely studied because it is conductive and has three distinct oxidation states accessible within 1.2 V. The pseudo-capacitive behaviour of RuO$_2$ in acidic solutions has been the focus of research in the past 30 years[1]. It can be described as a fast, reversible electron transfer together with an electro-adsorption of protons on the surface of RuO$_2$ particles, according to equation (7), where Ru oxidation states can change from (II) up to (IV):

$$RuO_2 + xH^+ + xe^- \leftrightarrow RuO_{2-x}(OH)_x \qquad (7)$$

where $0 \leq x \leq 2$. The continuous change of x during proton insertion or de-insertion occurs over a window of about 1.2 V and leads to a capacitive behaviour with ion adsorption following a Frumkin-type isotherm[1]. Specific capacitance of more than 600 F g^{-1} has been reported[51], but Ru-based aqueous electrochemical capacitors are expensive, and the 1-V voltage window limits their applications to small electronic devices. Organic electrolytes with proton surrogates (for example Li$^+$) must be used to go past 1 V. Less expensive oxides of iron, vanadium, nickel and cobalt have been tested in aqueous electrolytes, but none has been investigated as much as manganese oxide[52]. The charge storage mechanism is based on surface adsorption of electrolyte cations C$^+$ (K$^+$, Na$^+$...) as well as proton incorporation according to the reaction:

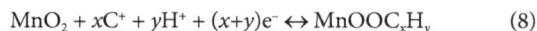

$$MnO_2 + xC^+ + yH^+ + (x+y)e^- \leftrightarrow MnOOC_xH_y \qquad (8)$$

Figure 6 shows a cyclic voltammogram of a single MnO$_2$ electrode in mild aqueous electrolyte; the fast, reversible successive surface redox reactions define the behaviour of the voltammogram, whose shape is close to that of the EDLC. MnO$_2$ micro-powders or micrometre-thick films show a specific capacitance of about 150 F g^{-1} in neutral aqueous electrolytes within a voltage window of <1 V. Accordingly, there is limited interest in MnO$_2$ electrodes for symmetric devices, because there are no oxidation states available at less than 0 V. However, it is suitable for a pseudo-capacitive positive electrode in hybrid systems,

which we will describe below. Other transition metal oxides with various oxidation degrees, such as molybdenum oxides, should also be explored as active materials for pseudo-capacitors.

Many kinds of conducting polymers (polyaniline, polypyrrole, polythiophene and their derivatives) have been tested in EC applications as pseudo-capacitive materials[50,53,54] and have shown high gravimetric and volumetric pseudo-capacitance in various non-aqueous electrolytes at operating voltages of about 3 V. When used as bulk materials, conducting polymers suffer from a limited stability during cycling that reduces the initial performance[9]. Research efforts with conducting polymers for supercapacitor applications are nowadays directed towards hybrid systems.

NANOSTRUCTURING REDOX-ACTIVE MATERIALS TO INCREASE CAPACITANCE

Given that nanomaterials have helped to improve Li-ion batteries[55], it is not surprising that nanostructuring has also affected ECs. Because pseudo-capacitors store charge in the first few nanometres from the surface, decreasing the particle size increases active material usage. Thanks to a thin electrically conducting surface layer of oxide and oxynitride, the charging mechanism of nanocrystalline vanadium nitride (VN) includes a combination of an electric double layer and a faradic reaction (II/IV) at the surface of the nanoparticles, leading to specific capacitance up to 1,200 F g^{-1} at a scan rate of 2 mV s^{-1} (ref. 56). A similar approach can be applied to other nano-sized transition metal nitrides or oxides. In another example, the cycling stability and the specific capacitance of RuO$_2$ nanoparticles were increased by depositing a thin conducting polymer coating that enhanced proton exchange at the surface[57]. The design of specific surface functionalization to improve interfacial exchange could be suggested as a generic approach to other pseudo-redox materials.

MnO$_2$ and RuO$_2$ films have been synthesized at the nanometre scale. Thin MnO$_2$ deposits of tens to hundreds of nanometres have been produced on various substrates such as metal collectors, carbon nanotubes or activated carbons. Specific capacitances as high as 1,300 F g^{-1} have been reported[58], as reaction kinetics were no longer limited by the electrical conductivity of MnO$_2$. In the same way, Sugimoto's group have prepared hydrated RuO$_2$ nano-sheets with capacitance exceeding 1,300 F g^{-1} (ref. 59). The RuO$_2$ specific capacitance also increased sharply when the film thickness was decreased. The deposition of RuO$_2$ thin film onto carbon supports[60,61] both increased the capacitance and decreased the RuO$_2$ consumption. Thin film synthesis or high SSA capacitive material decoration with nano-sized pseudo-capacitive active material, like the examples presented in Fig. 7a and b, offers an opportunity to increase energy density and compete with carbon-based EDLCs. Particular attention must be paid to further processing of nano-sized powders into active films because they tend to re-agglomerate into large-size grains. An alternative way to produce porous films from powders is by growing nanotubes, as has been shown for V$_2$O$_5$ (ref. 62), or nanorods. These allow easy access to the active material, but can only be produced in thin films so far, and the manufacturing cost will probably limit the use of these sophisticated nanostructures to small electronic devices.

HYBRID SYSTEMS TO ACHIEVE HIGH ENERGY DENSITY

Hybrid systems offer an attractive alternative to conventional pseudo-capacitors or EDLCs by combining a battery-like electrode (energy source) with a capacitor-like electrode (power source) in the same cell. An appropriate electrode combination can even increase the cell voltage, further contributing to improvement in energy and power densities. Currently, two different approaches to hybrid systems have emerged: (i) pseudo-capacitive metal oxides with a capacitive carbon electrode, and (ii) lithium-insertion electrodes with a capacitive carbon electrode.

Numerous combinations of positive and negative electrodes have been tested in the past in aqueous or inorganic electrolytes. In most

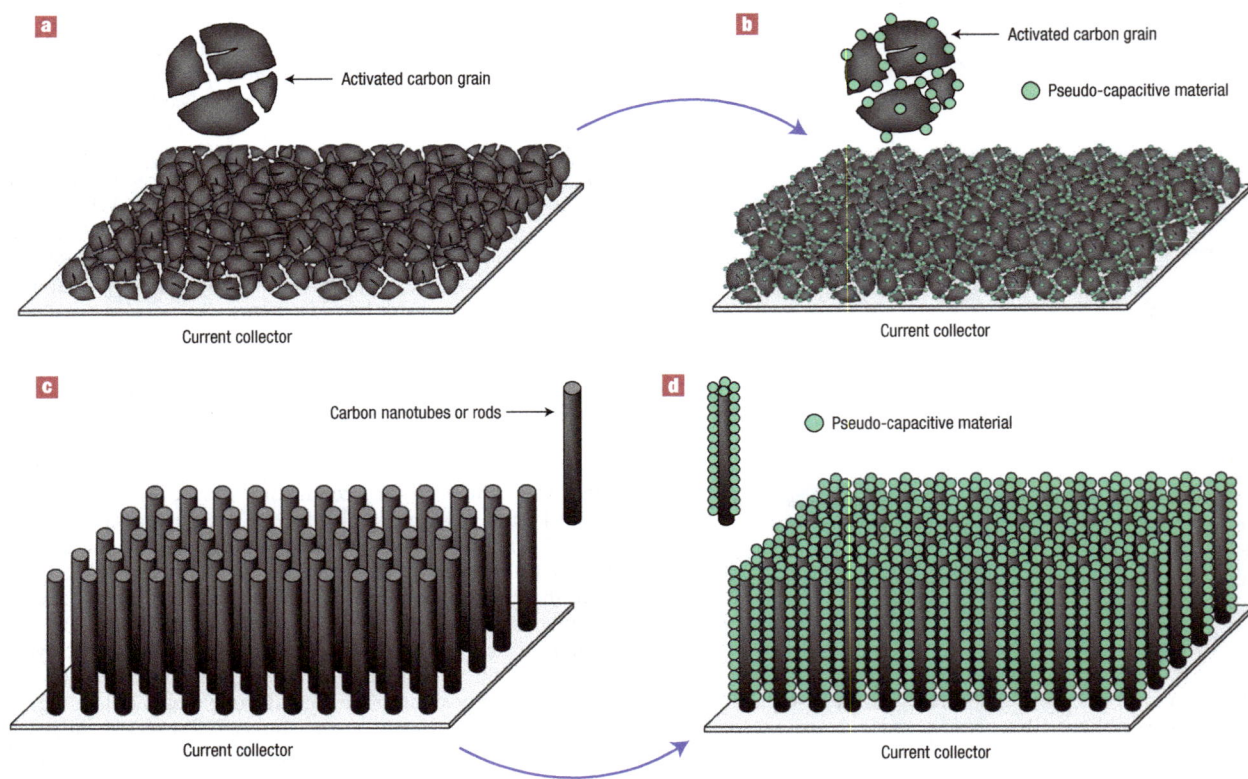

Figure 7 Possible strategies to improve both energy and power densities for electrochemical capacitors. **a**, **b**, Decorating activated carbon grains (**a**) with pseudo-capacitive materials (**b**). **c**, **d**, Achieving conformal deposit of pseudo-capacitive materials (**d**) onto highly ordered high-surface-area carbon nanotubes (**c**).

cases, the faradic electrode led to an increase in the energy density at the cost of cyclability (for balanced positive and negative electrode capacities). This is certainly the main drawback of hybrid devices, as compared with EDLCs, and it is important to avoid transforming a good supercapacitor into a mediocre battery[63].

MnO_2 is one of the most studied materials as a low-cost alternative to RuO_2. Its pseudo-capacitance arises from the III/IV oxidation state change at the surface of MnO_2 particles[58]. The association of a negative EDLC-type electrode with a positive MnO_2 electrode leads to a 2-V cell in aqueous electrolytes thanks to the apparent water decomposition overvoltage on MnO_2 and high-surface-area carbon. The low-cost carbon–MnO_2 hybrid system combines high capacitance in neutral aqueous electrolytes with high cell voltages, making it a green alternative to EDLCs using acetonitrile-based solvents and fluorinated salts. Moreover, the use of MnO_2 nano-powders and nanostructures offers the potential for further improvement in capacitance[64]. Another challenge for this system is to use organic electrolytes to reach higher cell voltage, thus improving the energy density.

A combination of a carbon electrode with a PbO_2 battery-like electrode using H_2SO_4 solution can work at 2.1 V (ref. 65), offering a low-cost EC device for cost-sensitive applications, in which weight of the device is of minor concern.

The hybrid concept originated from the Li-ion batteries field. In 1999, Amatucci's group combined a nanostructured lithium titanate anode $Li_4Ti_5O_{12}$ with an activated carbon positive electrode, designing a 2.8-V system that for the first time exceeded 10 Wh kg^{-1} (ref. 66). The titanate electrode ensured high power capacity and no solid-electrolyte interphase formation, as well as long-life cyclability thanks to low volume change during cycling. Following this pioneering

work, many studies have been conducted on various combinations of a lithium-insertion electrode with a capacitive carbon electrode. The Li-ion capacitor developed by Fuji Heavy Industry is an example of this concept, using a pre-lithiated high SSA carbon anode together with an activated carbon cathode[63,67]. It achieved an energy density of more than 15 Wh kg^{-1} at 3.8 V. Capacity retention was increased by unbalancing the electrode capacities, allowing a low depth of charge/discharge at the anode. Systems with an activated carbon anode and anion intercalation cathode are also under development. The advent of nanomaterials[55] as well as fast advances in the area of Li-ion batteries should lead to the design of high-performance ECs. Combining newly developed high-rate conversion reaction anodes or Li-alloying anodes with a positive supercapacitor electrode could fill the gap between Li-ion batteries and EDLCs. These systems could be of particular interest in applications where high power and medium cycle life are needed.

CURRENT COLLECTORS

Because ECs are power devices, their internal resistance must be kept low. Particular attention must be paid to the contact impedance between the active film and the current collector. ECs designed for organic electrolytes use treated aluminium foil or grid current collectors. Surface treatments have already been shown to decrease ohmic drops at this interface[68], and coatings on aluminium that improve electrochemical stability at high potentials and interface conductivity are of great interest.

The design of nanostructured current collectors with an increased contact area is another way to control the interface between current collector and active material. For example, carbon can be produced in a variety of morphologies[12], including porous films and nanotube brushes that can be grown on various current collectors[69] and that can serve as substrates for further conformal deposition (Fig. 7c and d) of active material. These nano-architectured electrodes could outperform the existing systems by confining a highly pseudo-capacitive material to a thin film with a high SSA, as has been done for Li-ion batteries[70] where, by growing Cu nano-pillars on a planar Cu foil, a six-fold improvement in the energy density over planar electrodes has been achieved[70]. Long's group[64] successfully applied a similar approach to supercapacitors by coating a porous carbon nano-foam with a 20-nm pseudo-capacitive layer of MnO_2. As a result, the area-normalized capacitance doubled to reach 1.5 F cm^{-2}, together with an outstanding volumetric capacitance of 90 F cm^{-3}. Electrophoretic deposition from stable colloidal suspensions of RuO_2 (ref. 71) or other active material can be used for filling the inter-tube space to design high-energy-density devices which are just a few micrometres thick. The nano-architectured electrodes also find applications in micro-systems where micro-ECs can complement micro-batteries for energy harvesting or energy generation. In this specific field, it is often advantageous to grow self-supported, binder-less nano-electrodes directly on semiconductor wafers, such as Si or SiC (ref. 72; Fig. 2c).

An attractive material for current collectors is carbon in the form of a highly conductive nanotube or graphene paper. It does not corrode in aqueous electrolytes and is very flexible. The use of nanotube paper for manufacturing flexible supercapacitors is expected to grow as the cost of small-diameter nanotubes required for making paper decreases. The same thin sheet of nanotubes[14] could potentially act as an active material and current collector at the same time. Thin-film, printable and wearable ECs could find numerous applications.

FROM ORGANIC TO IONIC LIQUID ELECTROLYTES

EC cell voltage is limited by the electrolyte decomposition at high potentials. Accordingly, the larger the electrolyte stability voltage window, the higher the supercapacitor cell voltage. Moving from aqueous to organic electrolytes increased the cell voltage from 0.9 V to 2.5–2.7 V for EDLCs. Because the energy density is proportional to the voltage squared (equation (2)), numerous research efforts have been directed at the design of highly conducting, stable electrolytes with a wider voltage window. Today, the state of the art is the use of organic electrolyte solutions in acetonitrile or propylene carbonate, the latter becoming more popular because of the low flash point and lower toxicity compared with acetonitrile.

Ionic liquids are room-temperature liquid solvent-free electrolytes; their voltage window stability is thus only driven by the electrochemical stability of the ions. A careful choice of both the anion and the cation allows the design of high-voltage supercapacitors, and 3-V, 1,000-F commercial devices are already available[73]. However, the ionic conductivity of these liquids at room temperature is just a few milliSiemens per centimetre, so they are mainly used at higher temperatures. For example, CDC with an EMI/TFSI ionic liquid electrolyte has been shown[46] to have capacitance of 160 F g^{-1} and ~90 F cm^{-3} at 60 °C. In this area, hybrid activated carbon/conducting polymer devices also show an improved performance with cell voltages higher than 3 V (refs 74–76).

For applications in the temperature range –30 °C to +60 °C, where batteries and supercapacitors are mainly used, ionic liquids still fail to satisfy the requirements because of their low ionic conductivity. However, the choice of a huge variety of combinations of anions and cations offers the potential for designing an ionic liquid electrolyte with an ionic conductivity of 40 mS cm^{-1} and a voltage window of >4 V at room temperature[77]. A challenge is, for instance, to find an alternative to the imidazolium cation that, despite high conductivity,

undergoes a reduction reaction at potential <1.5 V versus Li$^+$/Li. Replacing the heavy bis(trifluoromethanesulphonyl)imide (TFSI) anion by a lighter (fluoromethanesulphonyl)imide (FSI) and preparing ionic liquid eutectic mixtures would improve both the cell voltage (because a protecting layer of AlF_3 can be formed on the Al surface, shifting the de-passivation potential of Al above 4 V) and the ionic conductivity[77]. However, FSI shows poor cyclability at elevated temperatures. Supported by the efforts of the Li-ion community to design safer systems using ionic liquids, the research on ionic liquids for ECs is expected to have an important role in the improvement of capacitor performance in the coming years.

APPLICATIONS OF ELECTROCHEMICAL CAPACITORS

ECs are electrochemical energy sources with high power delivery and uptake, with an exceptional cycle life. They are used when high power demands are needed, such as for power buffer and power saving units, but are also of great interest for energy recovery. Recent articles from Miller et al.[7,10] present an overview of the opportunities for ECs in a variety of applications, complementing an earlier review by Kötz et al.[9]. Small devices (a few farads) are widely used for power buffer applications or for memory back-up in toys, cameras, video recorders, mobile phones and so forth. Cordless tools such as screwdrivers and electric cutters using EDLCs are already available on the market. Such systems, using devices of a few tens of farads, can be fully charged or discharged in less than 2 minutes, which is particularly suited to these applications, with the cycle life of EDLC exceeding that of the tool. As mentioned before, the Airbus A380 jumbo jets use banks of EDLCs for emergency door opening. The modules consist of an in series/parallel assembly of 100-F, 2.7-V cells that are directly integrated into the doors to limit the use of heavy copper cables. This application is obviously a niche market, but it is a demonstration that the EDLC technology is mature in terms of performance, reliability and safety.

The main market targeted by EDLC manufacturers for the next years is the transportation market, including hybrid electric vehicles, as well as metro trains and tramways. There continues to be debate about the advantage of using high power Li-ion batteries instead of ECs (or vice versa) for these applications. Most of these discussions have been initiated by Li-ion battery manufacturers who would like their products to cover the whole range of applications. However, ECs and Li-ion batteries should not necessarily be seen as competitors, because their charge storage mechanisms and thus their characteristics are different. The availability of the stored charge will always be faster for a supercapacitor (surface storage) than for a Li-ion battery (bulk storage), with a larger stored energy for the latter. Both devices must be used in their respective time-constant domains (see Fig. 1). Using a Li-ion battery for repeated high power delivery/uptake applications for a short duration (10 s or less) will quickly degrade the cycle life of the system[10]. The only way to avoid this is to oversize the battery, increasing the cost and volume. In the same way, using ECs for power delivery longer than 10 s requires oversizing. However, some applications use ECs as the main power and energy source, benefiting from the fast charge/discharge capability of these systems as well as their outstanding cycle life. Several train manufacturers have clearly identified the tramway/metro market segment as extremely relevant for EC use, to power trains over short distances in big cities, where electric cables are clearly undesirable for aesthetic and other reasons, but also to recover the braking energy of another train on the same line, thanks to the ECs' symmetric high power delivery/uptake characteristics.

For automotive applications, manufacturers are already proposing solutions for electrical power steering, where ECs are used for load-levelling in stop-and-go traffic[78]. The general trend is to increase the hybridization degree of the engines in hybrid electric vehicles, to allow fast acceleration (boost) and braking energy recovery. The on-board energy storage systems will be in higher demand, and a combination of batteries and EDLCs will increase the battery cycle life, explaining why EDLCs are viewed as a partner to Li-ion batteries for this market[78]. Currently, high price limits the use of both Li-ion batteries and EDLC in large-scale applications (for example for load levelling). But the surprisingly high cost of materials used for EDLC is due to a limited number of suppliers rather than intrinsically high cost of porous carbon. Decreasing the price of carbon materials for ECs, including CDC and AC, would remove the main obstacle to their wider use[79].

SUMMARY AND OUTLOOK

The most recent advances in supercapacitor materials include nanoporous carbons with the pore size tuned to fit the size of ions of the electrolyte with ångström accuracy, carbon nanotubes for flexible and printable devices with a short response time, and transition metal oxide and nitride nanoparticles for pseudo-capacitors with a high energy density. An improved understanding of charge storage and ion desolvation in subnanometre pores has helped to overcome a barrier that has been hampering progress in the field for decades. It has also shown how important it is to match the active materials with specific electrolytes and to use a cathode and anode with different pore sizes that match the anion or cation size. Nano-architecture of electrodes has led to further improvements in power delivery. The very large number of possible active materials and electrolytes means that better theoretical guidance is needed for the design of future ECs.

Future generations of ECs are expected to come close to current Li-ion batteries in energy density, maintaining their high power density. This may be achieved by using ionic liquids with a voltage window of more than 4 V, by discovering new materials that combine double-layer capacitance and pseudo-capacitance, and by developing hybrid devices. ECs will have a key role in energy storage and harvesting, decreasing the total energy consumption and minimizing the use of hydrocarbon fuels. Capacitive energy storage leads to a lower energy loss (higher cycle efficiency), than for batteries, compressed air, flywheel or other devices, helping to improve storage economy further. Flexible, printable and wearable ECs are likely to be integrated into smart clothing, sensors, wearable electronics and drug delivery systems. In some instances they will replace batteries, but in many cases they will either complement batteries, increasing their efficiency and lifetime, or serve as energy solutions where an extremely large number of cycles, long lifetime and fast power delivery are required.

doi:10.1038/nmat2297

References

1. Conway, B. E. Electrochemical Supercapacitors: Scientific Fundamentals and Technological Applications (Kluwer, 1999).
2. Service, R. F. New 'supercapacitor' promises to pack more electrical punch. Science 313, 902–905 (2006).
3. Tarascon, J.-M. & Armand, M. Issues and challenges facing rechargeable lithium batteries. Nature 414, 359–367 (2001).
4. Brodd, R. J. et al. Batteries, 1977 to 2002. J. Electrochem. Soc. 151, K1–K11 (2004).
5. Armand, M. & Tarascon, J.-M. Building better batteries. Nature 451, 652–657 (2008).
6. Armand, M. & Johansson, P. Novel weakly coordinating heterocyclic anions for use in lithium batteries. J. Power Sources 178, 821–825 (2008).
7. Miller, J. R. & Simon, P. Electrochemical capacitors for energy management. Science 321, 651–652 (2008).
8. US Department of Energy. Basic Research Needs for Electrical Energy Storage <www.sc.doe.gov/bes/reports/abstracts.html#EES2007> (2007).
9. Kötz, R. & Carlen, M. Principles and applications of electrochemical capacitors. Electrochim. Acta 45, 2483–2498 (2000).
10. Miller, J. R. & Burke, A. F. Electrochemical capacitors: Challenges and opportunities for real-world applications. Electrochem. Soc. Interf. 17, 53–57 (2008).
11. Pandolfo, A. G. & Hollenkamp, A. F. Carbon properties and their role in supercapacitors. J. Power Sources 157, 11–27 (2006).
12. Gogotsi, Y. (ed.) Carbon Nanomaterials (CRC, 2006).
13. Kyotani, T., Chmiola, J. & Gogotsi, Y. in Carbon Materials for Electrochemical Energy Storage Systems (eds Beguin, F. & Frackowiak, E.) Ch. 13 (CRC/Taylor and Francis, in the press).
14. Futaba, D. N. et al. Shape-engineerable and highly densely packed single-walled carbon nanotubes and their application as super-capacitor electrodes. Nature Mater. 5, 987–994 (2006).
15. Portet, C., Chmiola, J., Gogotsi, Y., Park, S. & Lian, K. Electrochemical characterizations of carbon nanomaterials by the cavity microelectrode technique. Electrochim. Acta, 53, 7675–7680 (2008).
16. Yang, C.-M. et al. Nanowindow-regulated specific capacitance of supercapacitor electrodes of single-wall carbon nanohorns. J. Am. Chem. Soc. 129, 20–21 (2007).
17. Niu, C., Sichel, E. K., Hoch, R., Moy, D. & Tennent, H. High power electrochemical capacitors based on carbon nanotube electrodes. Appl. Phys. Lett. 70, 1480 (1997).
18. Azaïs, P. et al. Causes of supercapacitors ageing in organic electrolyte. J. Power Sources 171, 1046–1053 (2007).
19. Gamby, J., Taberna, P. L., Simon, P., Fauvarque, J. F. & Chesneau, M. Studies and characterization of various activated carbons used for carbon/carbon supercapacitors. J. Power Sources 101, 109–116 (2001).
20. Shi, H. Activated carbons and double layer capacitance. Electrochim. Acta 41, 1633–1639 (1995).
21. Qu, D. & Shi, H. Studies of activated carbons used in double-layer capacitors. J. Power Sources 74, 99–107 (1998).
22. Qu, D. Studies of the activated carbons used in double-layer supercapacitors. J. Power Sources 109, 403–411 (2002).
23. Kim, Y. J. et al. Correlation between the pore and solvated ion size on capacitance uptake of PVDC-based carbons. Carbon 42, 1491 (2004).
24. Izutsu, K. Electrochemistry in Nonaqueous Solution (Wiley, 2002).
25. Marcus, Y. Ion Solvation (Wiley, 1985).
26. Jurewicz, K. et al. Capacitance properties of ordered porous carbon materials prepared by a templating procedure. J. Phys. Chem. Solids 65, 287 (2004).
27. Fernández, J. A. et al. Performance of mesoporous carbons derived from poly(vinyl alcohol) in electrochemical capacitors. J. Power Sources 175, 675 (2008).
28. Fuertes, A. B., Lota, G., Centeno, T. A. & Frackowiak, E. Templated mesoporous carbons for supercapacitor application. Electrochim. Acta 50, 2799 (2005).
29. Salitra, G., Soffer, A., Eliad, L., Cohen, Y. & Aurbach, D. Carbon electrodes for double-layer capacitors. I. Relations between ion and pore dimensions. J. Electrochem. Soc. 147, 2486–2493 (2000).
30. Vix-Guterl, C. et al. Electrochemical energy storage in ordered porous carbon materials. Carbon 43, 1293–1302 (2005).
31. Eliad, L., Salitra, G., Soffer, A. & Aurbach, D. On the mechanism of selective electroadsorption of protons in the pores of carbon molecular sieves. Langmuir 21, 3198–3202 (2005).
32. Eliad, L. et al. Assessing optimal pore-to-ion size relations in the design of porous poly(vinylidene chloride) carbons for EDL capacitors. Appl. Phys. A 82, 607–613 (2006).
33. Arulepp, M. et al. The advanced carbide-derived carbon based supercapacitor. J. Power Sources 162, 1460–1466 (2006).
34. Arulepp, M. et al. Influence of the solvent properties on the characteristics of a double layer capacitor. J. Power Sources 133, 320–328 (2004).
35. Raymundo-Pinero, E., Kierzek, K., Machnikowski, J. & Beguin, F. Relationship between the nanoporous texture of activated carbons and their capacitance properties in different electrolytes. Carbon 44, 2498–2507 (2006).
36. Janes, A. & Lust, E. Electrochemical characteristics of nanoporous carbide-derived carbon materials in various nonaqueous electrolyte solutions. J. Electrochem. Soc. 153, A113–A116 (2006).
37. Shanina, B. D. et al. A study of nanoporous carbon obtained from ZC powders (Z = Si, Ti, and B). Carbon 41, 3027–3036 (2003).
38. Chmiola, J., Dash, R., Yushin, G. & Gogotsi, Y. Effect of pore size and surface area of carbide derived carbon on specific capacitance. J. Power Sources 158, 765–772 (2006).
39. Dash, R. et al. Titanium carbide derived nanoporous carbon for energy-related applications. Carbon 44, 2489–2497 (2006).
40. Urbonaite, S. et al. EELS studies of carbide derived carbons. Carbon 45, 2047–2053 (2007).
41. Gogotsi, Y. et al. Nanoporous carbide-derived carbon with tunable pore size. Nature Mater. 2, 591–594 (2003).
42. Chmiola, J. et al. Anomalous increase in carbon capacitance at pore size below 1 nm. Science 313, 1760–1763 (2006).
43. Huang, J. S., Sumpter, B. G. & Meunier, V. Theoretical model for nanoporous carbon supercapacitors. Angew. Chem. Int. Ed. 47, 520–524 (2008).
44. Huang, J., Sumpter, B. G. & Meunier, V. A universal model for nanoporous carbon supercapacitors applicable to diverse pore regimes, carbons, and electrolytes. Chem. Eur. J. 14, 6614–6626 (2008).
45. Chmiola, J., Largeot, C., Taberna, P.-L., Simon, P. & Gogotsi, Y. Desolvation of ions in subnanometer pores, its effect on capacitance and double-layer theory. Angew. Chem. Int. Ed. 47, 3392–3395 (2008).
46. Largeot, C. et al. Relation between the ion size and pore size for an electric double-layer capacitor. J. Am. Chem. Soc. 130, 2730–2731 (2008).
47. Weigand, G., Davenport, J. W., Gogotsi, Y. & Roberto, J. in Scientific Impacts and Opportunities for Computing Ch. 5, 29–35 (DOE Office of Science, 2008).
48. Wu, N.-L. Nanocrystalline oxide supercapacitors. Mater. Chem. Phys. 75, 6–11 (2002).
49. Brousse, T. et al. Crystalline MnO₂ as possible alternatives to amorphous compounds in electrochemical supercapacitors. J. Electrochem. Soc. 153, A2171–A2180 (2006).
50. Rudge, A., Raistrick, I., Gottesfeld, S. & Ferraris, J. P. Conducting polymers as active materials in electrochemical capacitors. J. Power Sources 47, 89–107 (1994).
51. Zheng, J. P. & Jow, T. R. High energy and high power density electrochemical capacitors. J. Power Sources 62, 155–159 (1996).
52. Lee, H. Y. & Goodenough, J. B. Supercapacitor behavior with KCl electrolyte. J. Solid State Chem. 144, 220–223 (1999).
53. Laforgue, A., Simon, P. & Fauvarque, J.-F. Chemical synthesis and characterization of fluorinated polyphenylthiophenes: application to energy storage. Synth. Met. 123, 311–319 (2001).
54. Naoi, K., Suematsu, S. & Manago, A. Electrochemistry of poly(1,5-diaminoanthraquinone) and its application in electrochemical capacitor materials. J. Electrochem. Soc. 147, 420–426 (2000).
55. Arico, A. S., Bruce, P., Scrosati, B., Tarascon, J.-M. & Schalkwijk, W. V. Nanostructured materials for advanced energy conversion and storage devices. Nature Mater. 4, 366–377 (2005).
56. Choi, D., Blomgren, G. E. & Kumta, P. N. Fast and reversible surface redox reaction in nanocrystalline vanadium nitride supercapacitors. Adv. Mater. 18, 1178–1182 (2006).

57. Machida, K., Furuuchi, K., Min, M. & Naoi, K. Mixed proton–electron conducting nanocomposite based on hydrous RuO$_2$ and polyaniline derivatives for supercapacitors. *Electrochemistry* **72**, 402–404 (2004).

58. Toupin, M., Brousse, T. & Belanger, D. Charge storage mechanism of MnO$_2$ electrode used in aqueous electrochemical capacitor. *Chem. Mater.* **16**, 3184–3190 (2004).

59. Sugimoto, W., Iwata, H., Yasunaga, Y., Murakami, Y. & Takasu, Y. Preparation of ruthenic acid nanosheets and utilization of its interlayer surface for electrochemical energy storage. *Angew. Chem. Int. Ed.* **42**, 4092–4096 (2003).

60. Miller, J. M., Dunn, B., Tran, T. D. & Pekala, R. W. Deposition of ruthenium nanoparticles on carbon aerogels for high energy density supercapacitor electrodes. *J. Electrochem. Soc.* **144**, L309–L311 (1997).

61. Min, M., Machida, K., Jang, J. H. & Naoi, K. Hydrous RuO$_2$/carbon black nanocomposites with 3D porous structure by novel incipient wetness method for supercapacitors. *J. Electrochem. Soc.* **153**, A334–A338 (2006).

62. Wang, Y., Takahashi, K., Lee, K. H. & Cao, G. Z. Nanostructured vanadium oxide electrodes for enhanced lithium-ion intercalation. *Adv. Funct. Mater.* **16**, 1133–1144 (2006).

63. Naoi, K. & Simon, P. New materials and new configurations for advanced electrochemical capacitors. *Electrochem. Soc. Interf.* **17**, 34–37 (2008).

64. Fischer, A. E., Pettigrew, K. A., Rolison, D. R., Stroud, R. M. & Long, J. W. Incorporation of homogeneous, nanoscale MnO$_2$ within ultraporous carbon structures via self-limiting electroless deposition: Implications for electrochemical capacitors. *Nano Lett.* **7**, 281–286 (2007).

65. Kazaryan, S. A., Razumov, S. N., Litvinenko, S. V., Kharisov, G. G. & Kogan, V. I. Mathematical model of heterogeneous electrochemical capacitors and calculation of their parameters. *J. Electrochem. Soc.* **153**, A1655–A1671 (2006).

66. Amatucci, G. G., Badway, F. & DuPasquier, A. in *Intercalation Compounds for Battery Materials* (ECS Proc. Vol. 99) 344–359 (Electrochemical Society, 2000).

67. Burke, A. R&D considerations for the performance and application of electrochemical capacitors. *Electrochim. Acta* **53**, 1083–1091 (2007).

68. Portet, C., Taberna, P. L., Simon, P. & Laberty-Robert, C. Modification of Al current collector surface by sol-gel deposit for carbon-carbon supercapacitor applications. *Electrochim. Acta* **49**, 905–912 (2004).

69. Talapatra, S. et al. Direct growth of aligned carbon nanotubes on bulk metals. *Nature Nanotech.* **1**, 112–116 (2006).

70. Taberna, L., Mitra, S., Poizot, P., Simon, P. & Tarascon, J. M. High rate capabilities Fe$_3$O$_4$-based Cu nano-architectured electrodes for lithium-ion battery applications. *Nature Mater.* **5**, 567–573 (2006).

71. Jang, J. H., Machida, K., Kim, Y. & Naoi, K. Electrophoretic deposition (EPD) of hydrous ruthenium oxides with PTFE and their supercapacitor performances. *Electrochim. Acta.* **52**, 1733 (2006).

72. Cambaz, Z. G., Yushin, G., Osswald, S., Mochalin, V. & Gogotsi, Y. Noncatalytic synthesis of carbon nanotubes, graphene and graphite on SiC. *Carbon* **46**, 841–849 (2008).

73. Tsuda, T. & Hussey, C. L. Electrochemical applications of room-temperature ionic liquids. *Electrochem. Soc. Interf.* **16**, 42–49 (2007).

74. Balducci, A. et al. High temperature carbon–carbon supercapacitor using ionic liquid as electrolyte. *J. Power Sources* **165**, 922–927 (2007).

75. Balducci, A. et al. Cycling stability of a hybrid activated carbon//poly(3-methylthiophene) supercapacitor with *N*-butyl-*N*-methylpyrrolidinium bis(trifluoromethanesulfonyl)imide ionic liquid as electrolyte. *Electrochim. Acta* **50**, 2233–2237 (2005).

76. Balducci, A., Soavi, F. & Mastragostino, M. The use of ionic liquids as solvent-free green electrolytes for hybrid supercapacitors. *Appl. Phys. A* **82**, 627–632 (2006).

77. Endres, F., MacFarlane, D. & Abbott, A. (eds) *Electrodeposition from Ionic Liquids* (Wiley-VCH, 2008).

78. Faggioli, E. et al. Supercapacitors for the energy management of electric vehicles. *J. Power Sources* **84**, 261–269 (1999).

79. Chmiola, J. & Gogotsi, Y. Supercapacitors as advanced energy storage devices. *Nanotechnol. Law Bus.* **4**, 577–584 (2007).

80. Portet, C., Yushin, G. & Gogotsi, Y. Electrochemical performance of carbon onions, nanodiamonds, carbon black and multiwalled nanotubes in electrical double layer capacitors. *Carbon* **45**, 2511–2518 (2007).

Acknowledgements

We thank our students and collaborators, including J. Chmiola, C. Portet, R. Dash and G. Yushin (Drexel University), P. L. Taberna and C. Largeot (Université Paul Sabatier), and J. E. Fischer (University of Pennsylvania) for experimental help and discussions, H. Burnside (Drexel University) for editing the manuscript and S. Cassou (Toulouse) for help with illustrations. This work was partially funded through the Department of Energy, Office of Basic Energy Science, grant DE-FG01-05ER05-01, and through the Délégation Générale pour l'Armement.

Nanostructured materials for advanced energy conversion and storage devices

New materials hold the key to fundamental advances in energy conversion and storage, both of which are vital in order to meet the challenge of global warming and the finite nature of fossil fuels. Nanomaterials in particular offer unique properties or combinations of properties as electrodes and electrolytes in a range of energy devices. This review describes some recent developments in the discovery of nanoelectrolytes and nanoelectrodes for lithium batteries, fuel cells and supercapacitors. The advantages and disadvantages of the nanoscale in materials design for such devices are highlighted.

ANTONINO SALVATORE ARICÒ[1],
PETER BRUCE[2], BRUNO SCROSATI[3]*,
JEAN-MARIE TARASCON[4] AND
WALTER VAN SCHALKWIJK[5]

[1]Istituto CNR-ITAE, 98126 S. Lucia, Messina, Italy
[2]School of Chemistry, University of St Andrews, KY16 9ST, Scotland
[3]Dipartimento di Chimica, Università 'La Sapienza', 00186 Rome, Italy
[4]Université de Picardie Jules Verne, LRCS; CNRS UMR-6047, 80039 Amiens, France
[5]EnergyPlex Corporation, 1400 SE 112th Avenue, Suite 210, Bellevue, Washington 98004, USA

*e-mail: scrosati@uniroma1.it

One of the great challenges in the twenty-first century is unquestionably energy storage. In response to the needs of modern society and emerging ecological concerns, it is now essential that new, low-cost and environmentally friendly energy conversion and storage systems are found; hence the rapid development of research in this field. The performance of these devices depends intimately on the properties of their materials. Innovative materials chemistry lies at the heart of the advances that have already been made in energy conversion and storage, for example the introduction of the rechargeable lithium battery. Further breakthroughs in materials, not incremental changes, hold the key to new generations of energy storage and conversion devices.

Nanostructured materials have attracted great interest in recent years because of the unusual mechanical, electrical and optical properties endowed by confining the dimensions of such materials and because of the combination of bulk and surface properties to the overall behaviour. One

need only consider the staggering developments in microelectronics to appreciate the potential of materials with reduced dimensions. Nanostructured materials are becoming increasingly important for electrochemical energy storage[1,2]. Here we address this topic. It is important to appreciate the advantages and disadvantages of nanomaterials for energy conversion and storage, as well as how to control their synthesis and properties. This is a sizeable challenge facing those involved in materials research into energy conversion and storage. It is beyond the scope of this review to give an exhaustive summary of the energy storage and conversion devices that may now or in the future benefit from the use of nanoparticles; rather, we shall limit ourselves to the fields of lithium-based batteries, supercapacitors and fuel cells. Furthermore, from now on we shall refer to nanomaterials composed of particles that are of nanometre dimensions as primary nanomaterials, and those for which the particles are typically of micrometre dimensions but internally consist of nanometre-sized regions or domains as secondary nanomaterials.

LITHIUM BATTERIES

Lithium-ion batteries are one of the great successes of modern materials electrochemistry[3]. Their science and technology have been extensively reported in previous reviews[4] and dedicated books[5,6], to which the reader is referred for more details. A lithium-ion battery consists of a lithium-ion intercalation negative electrode (generally graphite), and a lithium-ion intercalation positive electrode (generally the lithium metal oxide, $LiCoO_2$), these being separated by a lithium-ion conducting electrolyte, for example a solution of $LiPF_6$ in ethylene carbonate-diethylcarbonate. Although such batteries are commercially successful, we are reaching the limits in performance using the current electrode and electrolyte materials. For new generations of rechargeable lithium batteries, not only for applications in consumer electronics but especially for clean energy storage and use in hybrid electric vehicles, further breakthroughs in materials are essential. We must advance the science to advance the technology. When such a situation arises, it is important to open up new avenues. One avenue that is already opening up is that of nanomaterials for lithium-ion batteries

ELECTRODES

There are several potential advantages and disadvantages associated with the development of nanoelectrodes for lithium batteries. Advantages include (i) better accommodation of the strain of lithium insertion/removal, improving cycle life; (ii) new reactions not possible with bulk materials; (iii) higher electrode/electrolyte contact area leading to higher charge/discharge rates; (iv) short path lengths for electronic transport (permitting operation with low electronic conductivity or at higher power); and (v) short path lengths for Li^+ transport (permitting operation with low Li^+ conductivity or higher power). Disadvantages include (i) an increase in undesirable electrode/electrolyte reactions due to high surface area, leading to self-discharge, poor cycling and calendar life; (ii) inferior packing of particles leading to lower volumetric energy densities unless special compaction methods are developed; and (iii) potentially more complex synthesis.

With these advantages and disadvantages in mind, efforts have been devoted to exploring negative and, more recently, positive nanoelectrode materials.

ANODES

Metals that store lithium are among the most appealing and competitive candidates for new types of anodes (negative electrodes) in lithium-ion batteries. Indeed, a number of metals and semiconductors, for example aluminium, tin and silicon, react with lithium to form alloys by electrochemical processes that are partially reversible and of low voltage (relative to lithium), involve a large number of atoms per formula unit, and in particular provide a specific capacity much larger than that offered by conventional graphite[7,8]. For example, the lithium–silicon alloy has, in its fully lithiated composition, $Li_{4.4}Si$, a theoretical specific capacity of 4,200 mA h g[−1] compared with 3,600 mA h g[−1] for metallic lithium and 372 mA h g[−1] for graphite. Unfortunately, the accommodation of so much lithium is accompanied by enormous volume changes in the host metal plus phase transitions. The mechanical strain generated during the alloying/de-alloying processes leads to cracking and crumbling of the metal electrode and a marked loss of capacity to store charge, in the course of a few cycles[8,9].

Although these structural changes are common to alloying reactions, there have been attempts to limit their side effects on the electrode integrity. Among them, the active/inactive nanocomposite concept

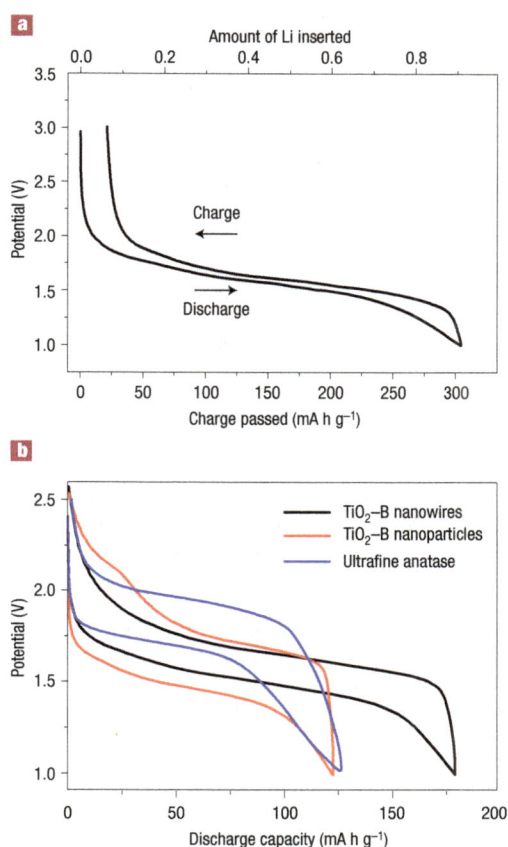

Figure 1 Charge–discharge curves for nanostructured anode materials.
a, Charge–discharge curves for Li_xTiO_2-B nanowires (rate of 10 mA g[−1]). **b**, Comparison of cycling behaviour for TiO_2-B nanowires, TiO_2-B nanoparticles and nanoparticulate anatase, all at 200 mA g[−1].

represents one attractive route. Several authors[7–9] have discussed this approach, which involves intimately mixing two materials, one reacting with lithium wheras the other acts as an inactive confining buffer. Within this composite, the use of nano-size metallic clusters as lithium hosts considerably suppresses the associated strains, and therefore improves the reversibility of the alloying reaction. By applying this concept through different systems such as SnO-based glasses[10], or composites such as Sn–Fe–C (ref. 11), Sn–Mn–C (ref. 12) or Si–C (refs 13–15), several authors have demonstrated that these electrodes show a considerable improvement in their cycling response in lithium cells. The Si–C nanocomposites have attracted considerable interest because they show capacity as high as 1,000 mA h g[−1] for more than 100 cycles[15]. Some of these improvements may arise because the materials avoid cracking, thus maintaining better conduction pathways, or because they incorporate conductive additives such as carbon. Undoubtedly alloy performance can also benefit from nanostructuring. For instance, thin amorphous silicon films deposited on a specially roughened copper foil surface by a sputtering process were shown[16] to have close to 100% reversibility at capacities larger than 3,000 mA h g[−1]. An excellent capacity retention was also noted for silicon electrodes prepared with a nanopillar surface morphology[17] because size confinement alters particle deformation and reduces fracturing.

Table 1 The Gibbs free energy change ΔG of conversion reactions. Values are obtained using thermodynamic data given in the literature[22] for various compounds together with the corresponding electromotive force (e.m.f.) values E, as deduced from the well-known $\Delta G = -nEF$ where F is Faraday's constant.

Compound		ΔG (kg mol^{-1})	E (V) = E_{eq} (volts vs Li$^+$/Li$^\circ$)
CoS$_{0.89}$	(+2)	−265.5	1.73
CoO	(+2)	−347.0	1.79
CoCl$_2$	(+2)	−499.0	2.59
CoF$_2$	(+2)	−528.2	2.74
CoF$_3$	(+3)	−1,023.4	3.54

Perhaps the greatest disadvantage of primary nanoparticles is the possibility of significant side-reactions with the electrolyte, leading to safety concerns (one of the most critical issues for lithium batteries) and poor calendar life. But if the materials fall within the stability window of the electrolyte or at least limit the formation of the solid–electrolyte interface (SEI) layer, then the many advantages of nanoparticles may more easily be exploited. Such an example is $Li_{4+x}Ti_5O_{12}$ ($0 < x < 3$, 160 mA h g^{-1}, 1.6 V versus Li$^+$(1M)/Li). No evidence has been reported for a significant surface layer formation (presumably because the potential is sufficiently high compared with lithium), and this material can be used as a nanoparticulate anode with high rate capability and good capacity retention. Controlling the nanoparticle shape as well as size can offer advantages. This is illustrated by recent results on TiO$_2$-B nanotubes or wires (B designates the form of TiO$_2$ and not boron). Such materials may be synthesized by a simple aqueous route and in high yield, with diameters in the range of 40–60 nm and lengths up to several micrometres. The TiO$_2$-B polymorph is an excellent intercalation host for Li, accommodating up to $Li_{0.91}TiO_2$-B (305 mA h g^{-1}) at 1.5–1.6 V vs Li$^+$(1M)/Li and with excellent capacity retention on cycling (Fig. 1). Interestingly, the rate capability is better than the same phase prepared as nanoparticles of dimension similar to the diameter of the nanowires[18] (Fig. 1). These TiO$_2$-B electrodes are not the only nanotube/wire electrodes. Unsurprisingly, carbon nanotubes have been explored as anodes. Whether the cost of their synthesis is viable remains an open question.

Nanomaterials consisting of nanoparticles or nanoarchitectured materials, as described so far in this review, are not always easy to make because of difficulties in controlling the size and size distribution of the particles or clusters. The potential disadvantage of a high external surface area, leading to excessive side reactions with the electrolyte and hence capacity losses or poor calendar life, has already been mentioned. Such problems may be addressed with internally nanostructured materials (secondary nanomaterials as defined), where the particles are significantly larger than the nanodomains. As well as

reducing side reactions with the electrolyte, this can have the advantage of ensuring higher volumetric energy densities. Of course, as described below, they are not a panacea.

A group of internally nanostructured anodes based on transition metal oxides has recently been described. The full electrochemical reduction of oxides such as CoO, CuO, NiO, Co$_3$O$_4$ and MnO versus lithium, involving two or more electrons per 3d-metal, was shown to lead to composite materials consisting of nanometre-scale metallic clusters dispersed in an amorphous Li$_2$O matrix[19]. Owing to the nanocomposite nature of these electrodes, the reactions, termed 'conversion reactions', are highly reversible, providing large capacities that can be maintained for hundreds of cycles. The prevailing view had been that reversible lithium reaction could occur only in the presence of crystal structures with channels able to transport Li$^+$. The new results, in stark contrast, turn out not to be specific to oxides but can be extended to sulphides, nitrides or fluorides[20]. These findings help to explain the previously reported unusual reactivity of complex oxides including RVO$_4$ (R = In, Fe) or A$_x$MoO$_3$ towards lithium[21,22].

Such conversion reactions offer numerous opportunities to 'tune' the voltage and capacity of the cell[19] owing to the fact that the cell potential is directly linked to the strength of the M–X bonding. Weaker M–X bonding gives larger potentials. The capacity is directly linked to the metal oxidation state, with the highest capacity associated with the highest oxidation states. Thus, by selecting the nature of M and its oxidation state, as well as the nature of the anion, one can obtain reactions with a specific potential within the range 0 to 3.5 V (Table 1), and based on low-cost elements such as Mn or Fe. Fluorides generally yield higher potentials than oxides, sulphides and nitrides.

However, such excitement needs to be tempered because ensuring rapid and reversible nanocomposite reactions is not an easy task. The future of such conversion reactions in real-life applications lies in mastering their kinetics (for example, the chemical diffusion of lithium into the matrix). Great progress has already been achieved with oxides, especially metallic RuO$_2$, which was shown[23] to display a 100% reversible conversion process involving 4e$^-$, and some early but encouraging results have been reported with

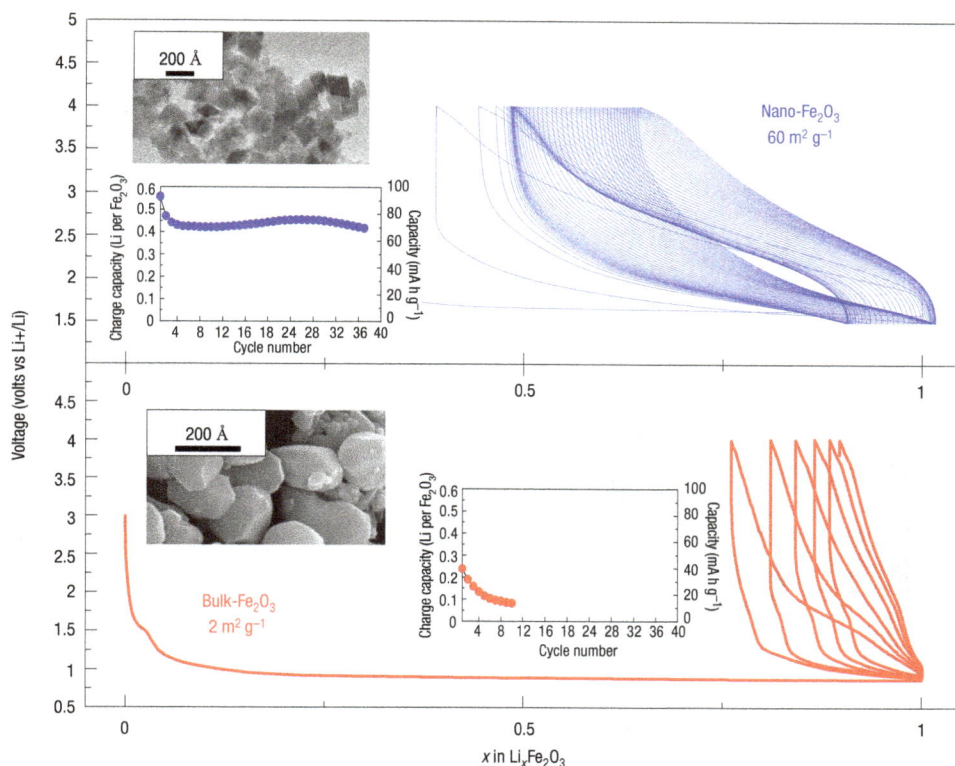

Figure 2 Electrochemical behaviour of bulk and nanostructured α-Fe₂O₃ with voltage–composition curves. The capacity retention and scanning electron micrographs of both samples are shown in the insets.

fluorides[24–26]; but a great deal remains to be done in this area. A further hint that working at the nanoscale may radically change chemical/electrochemical reaction paths of inorganic materials comes from recent studies carried out on the reactivity of macroscopic versus nanoscale haematite (α-Fe₂O₃) particles with Li. With nanometre-scale haematite particles (20 nm), reversible insertion of 0.6 Li per Fe₂O₃ is possible through a single-phase process, whereas large haematite particles (1 to 2 μm) undergo an irreversible phase transformation as soon as ~0.05 Li per Fe₂O₃ are reacted (see Fig. 2)[27]. In this respect, many materials, previously rejected because they did not fulfil the criteria as classical intercalation hosts for lithium, are now worth reconsideration.

CATHODES

This area is much less developed than the nanoanodes. The use of nanoparticulate forms (primary nanomaterials) of the classical cathode materials such as LiCoO₂, LiNiO₂ or their solid solutions can lead to greater reaction with the electrolyte, and ultimately more safety problems, especially at high temperatures, than the use of such materials in the micrometre range. In the case of Li–Mn–O cathodes such as LiMn₂O₄, the use of small particles increases undesirable dissolution of Mn. Coating the particles with a stabilizing surface layer may help to alleviate such problems but can reduce the rate of intercalation, reducing the advantage of the small particles.

A related approach to the formation of silicon nanopillars has been taken for cathode materials. By using a template, for example, porous alumina or a porous polymer, nanopillars of V₂O₅ or LiMn₂O₄ have been grown on a metal substrate[28]. These microfabricated electrode structures have the advantages of the Si example, accommodating volume changes and supporting high rates, although in the case of the manganese oxide, accelerated dissolution is expected. Nanotubes of VO$_x$ have also been prepared and investigated as cathodes[29].

Tuning the electrode material morphology or texture to obtain porous and high-surface-area electrodes constitutes another route to enhance electrode capacities[30]. For V₂O₅, aerogels (disordered mesoporous materials with a high pore volume) were recently reported to have electroactive capacities greater than polycrystalline non-porous V₂O₅ powders[31]. Such aerogels present a large surface area to the electrolyte, and can support high rates, although cyclability can be a problem because of structural changes or very reactive surface groups.

Interesting developments in secondary particle cathodes for lithium batteries have taken place in parallel with the work on secondary particle anodes. By retaining large particles, there is less dissolution than with primary nanoparticulate materials, and high volumetric densities are retained. Conventional wisdom stated that to sustain rapid and reversible electrode reactions in rechargeable lithium batteries, intercalation compounds must be used as the electrodes, and that the intercalation process had to be single-phase, that is, a continuous solid solution on

Figure 3 Transmission electron micrographs of regular and nanostructured spinel. a, Regular $Li_xMn_{2-y}O_4$ spinel. **b,** $Li_xMn_{2-y}O_4$ spinel obtained on cycling layered (O3) $Li_xMn_{1-y}O_2$. The Fourier-filtered image (**b**) highlights the nanodomain structure of average dimensions 50–70 Å. **c,** A schematic representation of the nanodomain structure of $Li_xMn_{2-y}O_4$ spinel derived from layered $Li_xMn_{1-y}O_2$, showing cubic and tetragonal nanodomains.

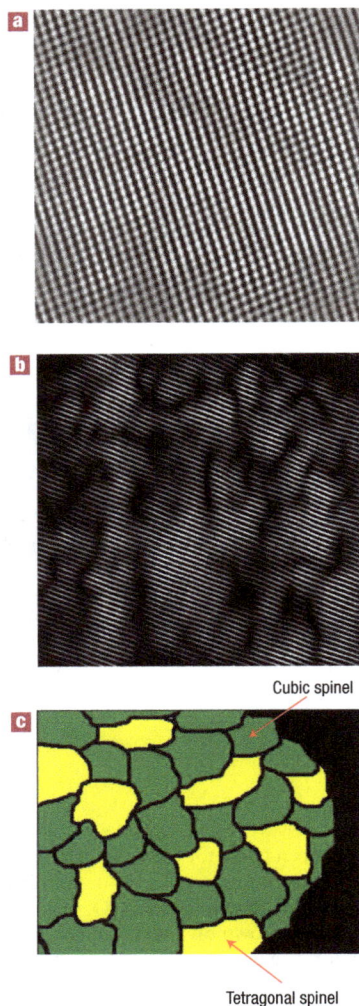

Cubic spinel

Tetragonal spinel

intercalation. However, there are now many examples where lithium intercalation is facile despite undergoing phase transitions, including $LiCoO_2$ and $Li_4Ti_5O_{12}$, especially if there is a strong structural similarity between the end phases (for example, only differences in Li ordering). Such two-phase intercalation reactions are far from being universally reversible. A classic example[32] is the intercalation cathode $Li_xMn_2O_4$, $0 < x < 2$. Cycling is usually confined to the range $0 < x < 1$ to avoid the transformation of cubic $LiMn_2O_4$ to tetragonal $Li_2Mn_2O_4$, which leads to a marked loss of capacity (ability to cycle lithium) because of the 13% anisotropy in the lattice parameters on formation of the tetragonal phase. Layered $LiMnO_2$ with the α-$NaFeO_2$ structure transforms into spinel on cycling but may be cycled with >99.9% capacity retention, despite undergoing the same cubic–tetragonal transformation[33]. The reason is clearly not suppression of the Jahn–Teller driven cubic–tetragonal distortion (something that has been attempted many times with limited success). Instead, the system accommodates the strain associated with the

transformation by developing a nanostructure within the micrometre-sized particles (Fig. 3). The nanodomains of spinel switch between cubic and tetragonal structures, with the strain being accommodated by slippage at the domain wall boundaries[34,35]. The nanodomains form during the layered-to-spinel transformation. Subsequently it has been shown that such a nanostructure can be induced in normal spinel by grinding, with a similar enhancement in cyclability[36]. Interestingly, the layered-to-spinel transformation is very easy, helping to mitigate any ill effects the transformation itself may have on cyclability[37].

A further, somewhat different, example of the benefits of nanoelectrodes within the field of batteries is the optimization of the environmentally benign and low-cost phospho-olivine $LiFePO_4$ phase[38] that displays a theoretical capacity of 170 mA h g^{-1}, as compared with 140 mA h g^{-1} for the $LiCoO_2$ electrode used at present ($LiFePO_4$ operates at a lower voltage). But the insulating character of the olivine means that in practice one could not obtain the full capacity of the material because, as the electrochemical reaction proceeds, 'electronically' isolated areas remain inactive in the bulk electrode. As a result, this material was largely ignored until it was prepared in the form of carbon-coated nanoparticles (Fig. 4) through various chemical and physical means[39–41]. This simultaneously reduces the distance for Li^+ transport, and increases the electronic contact between the particles. Procedures of this kind have led to a greatly improved electrochemical response, and the full capacity of the material is now accessible even under prolonged cycling[39] (see also Fig. 4). This example serves to illustrate some of the advantages of nanoelectrode materials listed at the beginning of this section, and to demonstrate that the search for new electroactive materials is now wider than ever because such materials do not require a particularly high electronic conductivity, nor a high diffusion coefficient for lithium, as had been believed for the past 20 years.

ELECTROLYTES

Progress in lithium batteries relies as much on improvements in the electrolyte as it does on the electrodes. Solid polymer electrolytes represent the ultimate in terms of desirable properties for batteries because they can offer an all-solid-state construction, simplicity of manufacture, a wide variety of shapes and sizes, and a higher energy density (because the constituents of the cell may be more tightly wound). No corrosive or explosive liquids can leak out, and internal short-circuits are less likely, hence greater safety. The most desirable polymer electrolytes are those formed by solvent-free membranes, for example poly(ethylene oxide), PEO, and a lithium salt, LiX, for example $LiPF_6$ or $LiCF_3SO_3$ (ref. 42). The poor ionic conductivity of these materials at room temperature has prevented them from realizing their otherwise high promise. Dispersing nanoscale inorganic fillers in solvent-free, polyether-based electrolytes increases the conductivity several-fold[43–45]. The improvement of the electrolyte transport properties may be explained on the basis of the heterogeneous doping model developed by Maier[46]. Accordingly, nanocomposites can be defined as the distribution of a second (or even third) phase, with particles of nanometric dimension in a matrix that can be amorphous or crystalline. Thus the increase in conductivity may be associated with Lewis acid–base interactions between the surface states of the ceramic nanoparticle with both the polymer chains and the anion of the lithium salt[47]. As in the electrode case, there are of course pros and cons in this particular approach. Indeed, other avenues are being explored to achieve high-conductivity polymer electrolytes. Relevant in this respect are the polymer-in-salt nanostructures[48] and the role of ionic liquids[49].

For 30 years it has been accepted that ionic conductivity in polymer electrolytes occurred exclusively in the amorphous phase, above the glass transition temperature T_g. Crystalline polymer electrolytes were considered to be insulators. But recent studies have shown that this is not the case: the 6:1 crystalline complexes $PEO_6:LiXF_6$; X = P, As, Sb demonstrate ionic conductivity[50,51]. The Li^+ ions reside in tunnels formed by the polymer chains (Fig. 5). Significant increases in ionic conductivity in the crystalline 6:1 complexes are observed on reducing the chain length from 3,000 to 1,000: that is, in the nanometre range[52]. It is evident that in electrolytes, just as in the electrodes described above, control of dimensions on the nanoscale has a profound influence on performance[52]. Crystalline polymer electrolytes represent a radically different type of ionic conductivity in polymers, and illustrate the importance of seeking new scientific directions. The present materials do not support ionic conductivities that are sufficient for applications, but they do offer a fresh approach with much scope for further advances. Recently it has been shown that the conductivity of the crystalline polymer electrolytes may be raised by two orders of magnitude by partial replacement of the XF_6 ions with other mono or divalent anions (ref. 53, and P. G. Bruce, personal communication). Chemical compatibility with selected electrode materials has yet to be evaluated.

SUPERCAPACITORS

Supercapacitors are of key importance in supporting the voltage of a system during increased loads in everything from portable equipment to electric vehicles[54–56]. These devices occupy the area in the Ragone plot (that is, the plot of volumetric against gravimetric energy density) between batteries and dielectric capacitors. Supercapacitors are similar to batteries in design and manufacture (two electrodes, separator and electrolyte), but are designed for high power and long cycle life (>100 times battery life), but at the expense of energy density.

There are two general categories of electrochemical supercapacitors: electric double-layer capacitors (EDLC) and redox supercapacitors. In contrast to batteries, where the cycle life is limited

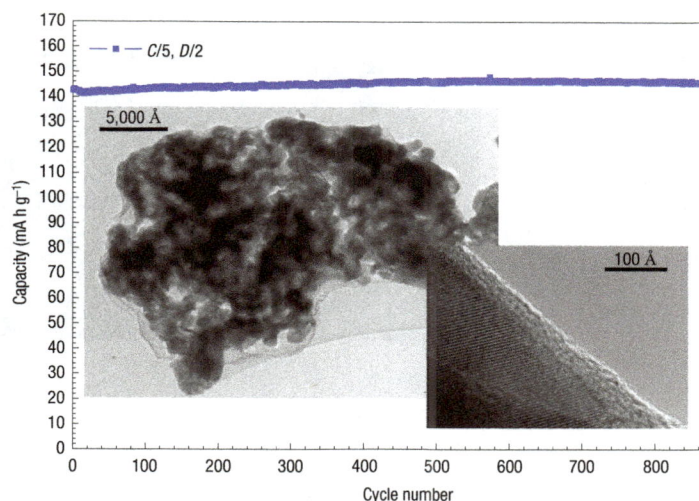

because of the repeated contraction and expansion of the electrode on cycling, EDLC lifetime is in principle infinite, as it operates solely on electrostatic surface-charge accumulation. For redox supercapacitors, some fast faradic charge transfer takes place as in a battery. This gives rise to a large pseudo-capacitance.

Progress in supercapacitor technology can benefit by moving from conventional to nanostructured electrodes. In the case of supercapacitors, the electrode requirements are less demanding than in batteries, at least in terms of electrode compaction, because power prevails over energy density. Thus, the benefits of nanopowders with their high surface area (primary nanoparticles) are potentially more important, hence the staggering interest in nanopowders and their rapid uptake for supercapacitor-based storage sources.

Recent trends in supercapacitors involve the development of high-surface-area activated carbon electrodes to optimize the performance in terms of capacitance and overall conductivity. Attention has been focused on nanostructured carbons, such as aerogels[57], nanotubes[58] and nanotemplates[59]. The advantages of carbon aerogels lie mainly in their low ionic and electronic charging resistance and in their potential use as binderless electrodes. Replacing the standard carbon fibre with carbon aerogel electrodes improves capacitance and cyclability. Because of their unique architecture, carbon nanotubes are now intensively studied as new electrode materials for supercapacitor structures although, as for batteries, cost may be an issue. A critical aspect in nanotechnology for supercapacitors is to reach a compromise between specific surface area (to ensure high capacitance) and pore-size distribution (to permit easy access for the electrolyte).

Nanostructured materials have led to the development of new supercapacitor technologies. Capitalizing on the benefits of nanomaterial electrodes, Telcordia's researchers have, by means of the so-called hybrid supercapacitors (HSCs), proposed a new approach to energy storage[60]. HSCs use one capacitive carbon electrode similar to that of a carbon EDLC; however, the complementary negative electrode consists of a nanostructured lithium

Figure 4 Capacity against cycling number for a lithium coin cell. The experiments used a 1-cm² disc of Bellcore-type plastic laminate made out 4.480 mg of pure $LiFePO_4$ coated with 5% C *in situ*. The overall plastic composition was 69% $LiFePO_4$, 11% C total and 20% Kynar. The cell was cycled at C/5 (0.160 mA cm⁻²), D/2 (−0.400 mA cm⁻²) between 2 and 4.5 V at room temperature. Here, D is discharge rate, C is charge rate. Left inset: enlarged transmission electron micrograph of the $LiFePO_4$ particles used, coated with carbon. The coating used a homemade recipe[103]. This image was recorded with a TECNAI F20 scanning TEM. The quality of the coating — that is, the interface between crystallized $LiFePO_4$ and amorphous carbon — is nicely observed in the high-resolution electron micrograph in the right inset (taken with the same microscope). Courtesy of C. Wurm and C. Masquelier, LRCS Amiens.

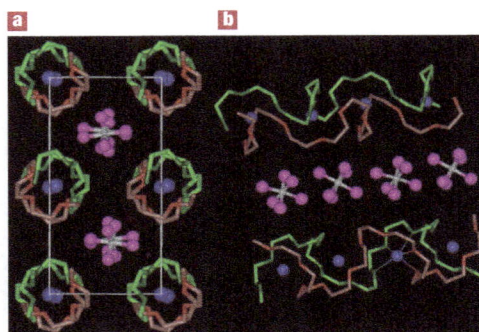

titanate compound that undergoes a reversible faradic intercalation reaction[61]. The key innovation lies in the use of a nanostructured Li$_4$Ti$_5$O$_{12}$ as the negative electrode, resulting in no degradation in performance from charge/discharge-induced strain. Here is a nanomaterial that is being explored for capacitor and battery use because of its excellent cyclability, rate capability and safety. Its use in the hybrid supercapacitor results in a device with cycle life comparable to that of supercapacitors, freeing designers from the lifespan limitations generally associated with batteries. Figure 6 shows some features of these new types of capacitors. Besides Li$_4$Ti$_5$O$_{12}$, other nanostructured oxide materials can be engineered so as to obtain hybrid devices operating over a wide range of voltages. Inorganic nanotubes or nanowires may offer an alternative to nanoparticulate Li$_4$Ti$_5$O$_{12}$, for example the Li$_x$TiO$_2$-B nanowires described in the section on lithium batteries. One can therefore envisage a supercapacitor composed of a carbon nanotube as the positive electrode and an inorganic nanotube capable of lithium intercalation (for example Li$_x$TiO$_2$-B) as the negative electrode.

Fabricating nanopillared electrodes by growing the materials on a substrate was described in the context of lithium batteries. However, a related approach where materials are electro-synthesized in the presence of an electrolyte, which is also a liquid-crystal template, has been used to form nanoarchitectured electrodes, for example Ni or NiOOH, for possible use in supercapacitors[61].

FUEL CELLS

Fuel cell technologies are now approaching commercialization, especially in the fields of portable power sources — distributed and remote generation of electrical energy[62]. Already, nanostructured materials are having an impact on processing methods in the development of low-temperature fuel cells ($T < 200$ °C), the dispersion of precious metal catalysts, the development and dispersion of non-precious catalysts, fuel reformation and hydrogen storage, as well as the fabrication of membrane-electrode assemblies (MEA). Polymer electrolyte membrane fuel cells (PEMFCs) have recently gained momentum for application in transportation and as small portable power sources; whereas phosphoric acid fuel cells (PAFCS), solid oxide fuel cells (SOFCs) and molten carbonates fuel

cells (MCFCs) still offer advantages for stationary applications, and especially for co-generation[62]. Platinum-based catalysts are the most active materials for low-temperature fuel cells fed with hydrogen, reformate or methanol[62]. To reduce the costs, the platinum loading must be decreased (while maintaining or improving MEA performance), and continuous processes for fabricating MEAs in high volume must be developed. A few routes are being actively investigated to improve the electrocatalytic activity of Pt-based catalysts. They consist mainly of alloying Pt with transition metals or tailoring the Pt particle size.

The oxygen reduction reaction (ORR) limits the performance of low-temperature fuel cells. One of the present approaches in order to increase the catalyst dispersion involves the deposition of Pt nanoparticles on a carbon black support. Kinoshita et al. observed that the mass activity and specific activity for oxygen electro-reduction in acid electrolytes varies with the Pt particle size according to the relative fraction of Pt surface atoms on the (111) and (100) faces[63] (Fig. 7). The mass-averaged distribution of the surface atoms on the (111) and (100) planes passes through a maximum (~3 nm) whereas the total fraction of surface atoms at the edge and corner sites decreases rapidly with an increase of the particle size. On the other hand, the surface-averaged distribution for the (111) and (100) planes shows a rapid increase with the particle size, which accounts for the increase of the experimentally determined specific activity with the particle size (Fig. 7). A dual-site reaction is assumed as the rate-determining step:

$$O_2 + Pt \rightarrow Pt\text{–}O_2$$

$$Pt\text{-}O_2 + H^+ + e^- \rightarrow Pt\text{-}HO_2$$

$$Pt\text{–}HO_2 + Pt \rightarrow Pt\text{–}OH + Pt\text{–}O \text{ (r.d.s.)}$$

$$Pt\text{–}OH + Pt\text{–}O + 3H^+ + 3e^- \rightarrow 2Pt + 2H_2O$$

This mechanism accounts for the role of dual sites of proper orientation[63].

Alloying Pt with transition metals also enhances the electrocatalysis of O$_2$ reduction. In low-temperature fuel cells, Pt–Fe, Pt–Cr and Pt–Cr–Co alloy electrocatalysts were observed[62,64,65] to have high specific activities for oxygen reduction as compared with that on platinum. This enhancement in electrocatalytic activity has been ascribed to several factors such as interatomic spacing, preferred orientation or electronic interactions. The state of the art Pt–Co–Cr electrocatalysts have a particle size of 6 nm (ref. 65).

Both CO$_2$ and CO are present in hydrogen streams obtained from reforming. These molecules are known to adsorb on the Pt surface under reducing potentials. Adsorbed CO-like species are also formed on Pt-based anode catalysts in direct methanol fuel cells (DMFCs). Such a poisoning of the Pt surface reduces the electrical efficiency and the power density of the fuel cell[66]. The electrocatalytic activity of Pt against, for example, CO$_2$/CO poisoning is known to be promoted by the presence of a second metal[64,65,67], such as Ru, Sn or Mo. The mechanism by which such synergistic promotion of the H$_2$/CO and methanol oxidation reactions is brought about has been much studied and is still debated. Nevertheless, it turns out that the best performance is obtained from Pt–Ru electrocatalysts with mean particle size 2–3 nm. As in the case of oxygen reduction, the particle size is important for structure-sensitive reactions such as CH_3OH and CO electro-oxidation. The catalytic activity of Pt–Ru surfaces is maximized for the (111) crystallographic plane[68]. According to the bifunctional theory, the role of Ru in these processes is to promote water discharge and removal of strongly adsorbed CO species at low potentials through the following reaction mechanism[66–68]:

$$Ru + H_2O \rightarrow Ru\text{–}OH + H^+ + 1e^-$$

$$Ru\text{–}OH + Pt\text{–}CO \rightarrow Ru + Pt + CO_2 + H^+ + 1e^-$$

A change in the CO binding strength to the surface induced by Ru through a ligand effect on Pt has been reported[66,67].

The synthesis of Pt-based electro-catalysts, either unsupported or supported on high-surface-area carbon, is generally carried out by various colloidal preparation routes. A method recently developed by Bönnemann and co-workers[67] allows fine-tuning of the particle size for the bimetallic Pt–Ru system.

Recent developments in the field of CO-tolerant catalysts include the synthesis of new nanostructures by spontaneous deposition of Pt sub-monolayers on carbon-supported Ru nanoparticles[68,69]. This also seems to be an efficient approach to reduce the Pt loading. Further advances concern a better understanding of the surface chemistry in electrocatalyst nanoparticles and the effects of strong metal–support interactions that influence both the dispersion and electronic nature of platinum sites.

An alternative approach, avoiding the use of carbon blacks, is the fabrication of porous silicon catalyst support structures with a 5-μm pore diameter and a thickness of about 500 μm. These structures have high surface areas, and they are of interest for miniature PEM fuel cells[70]. A finely dispersed, uniform distribution of nanometre-scale catalyst particles deposited on the walls of the silicon pores creates, in contact with the ionomer, an efficient three-phase reaction zone capable of high power generation in DMFCs[70].

As an alternative to platinum, organic transition metal complexes — for example, iron or cobalt organic macrocycles from the families of phenylporphyrins[71] and nanocrystalline transition metal chalcogenides[72] — are being investigated for oxygen reduction, especially in relation to their high selectivity towards the ORR and tolerance to methanol cross-over in DMFCs. The metal–organic macrocycle is supported on high-surface-area carbon, and treated at high temperatures (from 500 to 800 °C) to form a nanostructured compound that possesses a specific geometrical arrangement of atoms. These materials show suitable electrocatalytic activity, even if smaller than that of Pt, without any degradation in performance[71].

As for batteries, the electrolyte is a key component of the fuel cell assembly. Perfluorosulphonic polymer

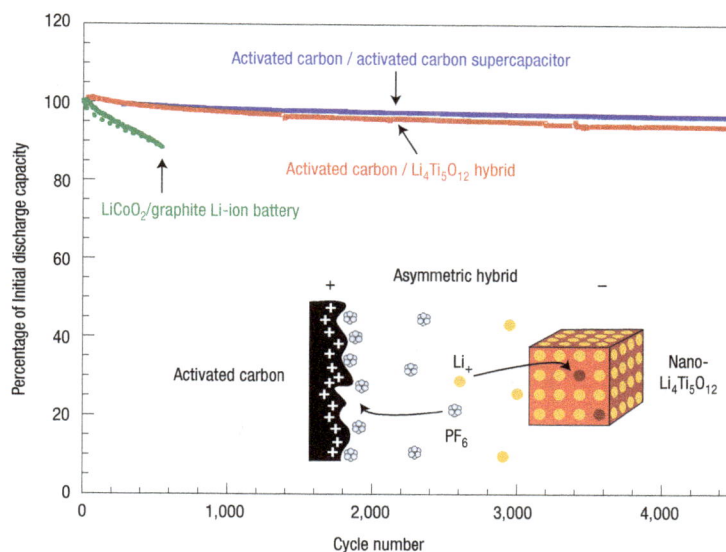

electrolyte membranes (for example Nafion™) are currently used in H_2/air or methanol/air fuel cells because of their excellent conductivity and electrochemical stability[62]. Unfortunately, they suffer from several drawbacks such as methanol cross-over and membrane dehydration. The latter severely hinders the fuel cell operation above 100 °C, which is a prerequisite for the suitable oxidation of small organic molecules involving formation of strongly adsorbed reaction intermediates such as CO-like species[63,66,67]. Alternative membranes based on poly(arylene ether sulphone)[73], sulphonated poly(ether ketone)[74] or block co-polymer ion-channel-forming materials as well as acid-doped polyacrylamid and polybenzoimidazole have been suggested[75–77]. Various relationships between membrane nanostructure and transport characteristics, including conductivity, diffusion, permeation and electro-osmotic drag, have been observed[78]. Interestingly, the presence of less-connected hydrophilic channels and larger separation of sulphonic groups in sulphonated poly(ether ketone) than in Nafion reduces water permeation and electro-osmotic drag whilst maintaining high protonic conductivity[78]. Furthermore, an improvement in thermal and mechanical stability has been shown in nano-separated acid–base polymer blends obtained by combining polymeric N-bases and polymeric sulphonic acids[74]. Considerable efforts have been addressed in the last decade to the development of composite membranes. These include ionomeric membranes modified by dispersing inside their polymeric matrix insoluble acids, oxides, zirconium phosphate and so on; other examples are ionomers or inorganic solid acids with high proton conductivity, embedded in porous non-proton-conducting polymers[75]. Recently, Alberti and Casciola[75] prepared nanocomposite electrolytes by *in situ* formation of insoluble layered Zr phosphonates in ionomeric membranes. Such compounds, for example $Zr(O_3P\text{–}OH)(O_3P\text{–}C_6H_4\text{–}SO_3H)$, show conductivities much higher than the Zr phosphates and comparable to Nafion. In an attempt to reduce

Figure 6 Cycling performance of the new asymmetric hybrid C/nano-$Li_4Ti_5O_{12}$ supercapacitor. Shown for comparison are results for a classical C/C supercapacitor and a commercial lithium-ion battery. The basic concept of this new type of supercapacitor is shown as an inset.

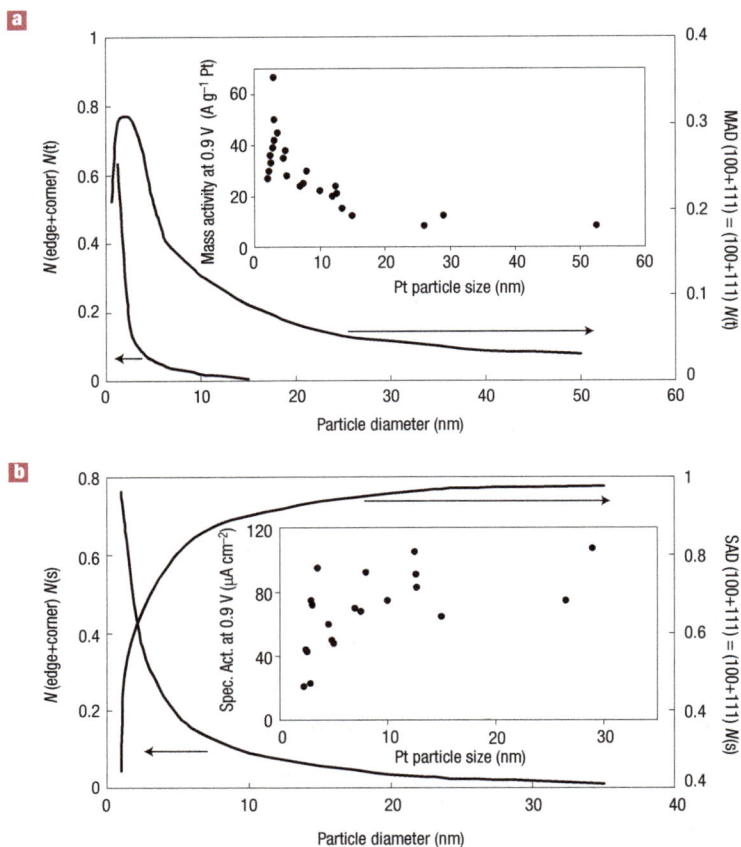

Figure 7 Calculated mass-averaged (a) and surface-averaged distributions (b) as a function of particle size in Pt particles with cubo-octahedral geometry. The values of $N(t)$ and $N(s)$ indicate the total number of atoms and the number of atoms on the surface respectively. The variation with particle size of mass activity (a) and specific activity (b) for oxygen reduction in acid electrolyte is shown in the insets[63].

of composite membranes, under low humidity, needs to be carried out in the next few years to assess these electrolyte materials.

Within the PEMFCs, the Pt/C catalyst is intimately mixed with the electrolyte ionomer to form a composite catalyst layer extending the three-phase reaction zone. This is similar to the composite cathode approach in lithium-ion batteries where the electrode consists of two interpenetrating networks for electron and ion conduction; the benefit of this approach is an enhancement of the interfacial region between catalyst particles and ionomer[62,84]. A reduction in the Pt content to significantly less than 0.5 mg cm^{-2} without degrading the cell performance and life-time has been demonstrated[62,84]. Following this general concept, durable multi-level MEAs are being developed (3M Corporation) using high-speed precision coating technologies and an automated assembly process[85]. Part of the MEA is a nanostructured thin-film catalyst based on platinum-coated nano-whiskers. The approach uses highly oriented, high-aspect-ratio single-crystalline whiskers of an organic pigment material. Typically there are 3×10^9 to 5×10^9 whiskers cm^{-2}. This support permits suitable specific activity of the applied catalysts and aids processing and manufacturing. However, the electrocatalytic activities so obtained are comparable to catalyst–ionomer inks. Platinum-coated nano-whiskers and a cross-section of the MEA are shown in Fig. 8.

The trend towards nanomaterials is not limited to low-temperature fuel cells. Nanostructured electro-ceramic materials are increasingly used in intermediate-temperature solid oxide fuel cells (IT-SOFCs). Although one may start with nanosized particles in the fabrication of SOFCs materials, these are often modified by the temperatures required for cell fabrication (>1,000 °C), thus forming microstructured components with electro-catalytic and ion-conduction properties different from the typical polycrystalline materials[86]. Nanosized YSZ (8% Y_2O_3–ZrO_2) and ceria-based (CGO, SDC, YDC) powders permit a reduction of the firing temperature during the membrane-forming step in the cell fabrication procedure, because their sintering properties differ from those of polycrystalline powders[87]. Furthermore, nanocrystalline ceria, which is characterized by mixed electronic–ionic conduction properties, promotes the charge transfer reactions at the electrode–electrolyte interface[87].

Ionic charge carriers in electro-ceramic materials originate from point defects. In nanostructured systems, the significantly larger area of interface and grain boundaries produces an increase in the density of mobile defects in the space-charge region. This in turn leads to a completely different electrochemical behaviour from that of bulk polycrystalline materials[86,89]. According to Maier[88], these 'trivial' size effects, resulting simply from an increased proportion of the interface, should be distinguished from 'true' size effects occurring when the particle size is smaller than four times the Debye length, where the local properties are changed in terms of ionic and electronic charge-carrier transport. Interestingly, the same space-charge model was

the drawbacks of perfluorosulphonic membranes, nanoceramic fillers have been included in the polymer electrolyte network. Stonehart, Watanabe and co-workers[79] have successfully reduced the humidification constraints in PEMFCs by the inclusion of small amounts of SiO_2 and Pt/TiO_2 (~7 nm) nanoparticles to retain the electrochemically produced water inside the membrane. Similarly modified membranes, containing nanocrystalline ceramic oxide filler, have been demonstrated[80] to operate up to 150 °C. Although it has been hypothesized[75–77] that the inorganic filler induces structural changes in the polymer matrix, the water-retention mechanism appears more likely to be favoured by the presence of acidic functional groups on the surface of nanoparticle fillers[81]. At present, there are no indications that the transport properties are significantly affected by the filler[82]. However, the greater water retention capability of the composite membrane at high temperatures (130 °C–150 °C) and under low humidity[81] should promote the 'vehicular mechanism' of proton conduction as occurs at lower temperatures[83]. Investigation of the lifetime behaviour used by Maier to explain the observed increase in ionic conductivity of dry or hybrid Li-based polymer systems loaded with nano-inorganic fillers (SiO_2, Al_2O_3, TiO_2) as discussed previously.

Notwithstanding the importance of nanostructured materials, the performance of practical fuel cells remains limited by scale-up, stack housing design, gas manifold and sealing. Although progress has been achieved in the direct electrochemical oxidation of alcohol and hydrocarbon fuels[62,90,91], fuel cells are still mostly fed by hydrogen. Much research is now focused on nanostructured hydrides including carbon nanotubes, nanomagnesium-based hydrides, metal-hydride/carbon nanocomposites, and alanates for hydrogen storage[92,93]. Early reports suggested hydrogen adsorption capacities of 14–20 wt% for K- and Li-doped carbon nanotubes[94,95], but today's reliable hydrogen-storage capacities in single-walled carbon nanotubes appear comparable to or even less than those of metal hydrides[96-98], and probably not sufficient to store the amount of hydrogen required for automotive applications, which has been set by the US Department of Energy as 6.5% (ref. 98). Recent developments in this field include modification of Mg-hydrides with transition metals[99,100], and the investigation of boron-nitride nanostructures[101]. Magnesium hydride, MgH_2, is often modified by high-energy ball-milling with alloying elements including[99] Ni, Cu, Ti, Nb and Al so as to obtain, after 20 h of milling, nanoparticles in the range of 20–30 nm providing hydrogen-storage capacities of about 6–11 wt% (refs 99, 100). Besides improving the hydrogen sorption kinetics of MgH_2, partial substitution of Al for Mg in Mg_2Ni hydrogen storage systems seems to increase the life-cycle characteristics, as a consequence of the existence of an Al_2O_3 film on the alloy surface[99]. Further advances resulted from the investigation of MgH_2-V nanocomposites[100] and the addition of very small amounts of nanoparticulate transition metal oxides (Nb_2O_5, WO_3, Cr_2O_3) to Mg-based hydrides[102]. The presence of vanadium as well as the particular nanostructure of the nanocomposite aids the hydrogen penetration into the material. Also, in the case of transition metal oxide promoters, the catalytic effect of compounds such as Cr_2O_3, the reactive mechanical grinding and the nanosized particles seem to produce a substantial improvement in the adsorption and desorption properties of magnesium[102].

CONCLUSION

It is a regrettable feature of important scientific discoveries that they often suffer from hyperbole; perhaps it has always been so, but the speed of today's communications exacerbates the situation. Nanoscience suffers from this disease. Yet there are many genuine scientific advances that fall under the umbrella of nanoscience. This short review demonstrates how moving from bulk materials to the nanoscale can significantly change electrode and electrolyte properties, and consequently their performance in devices for energy storage and conversion. In some cases the effects may be simple

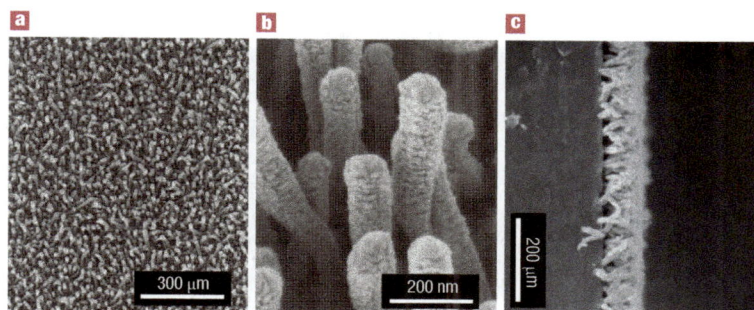

consequences of a reduction in size, for example when nanoparticulate electrodes or electrocatalysts lead to higher electrode/electrolyte contact areas and hence higher rates of electrode reaction. In others the effects may be more subtle, involving internally nanostructured materials or nanostructures with particular morphologies, for example the nanotubes. Space-charge effects at the interface between small particles can result in substantial improvements of properties. There is a profound effect of spatial confinement and contribution of surfaces, due to small particle size, on many of the properties of materials; this challenges us to develop new theory or at least adapt and develop theories that have been established for bulk materials. We also foresee that this subject will bring together the disciplines of materials chemistry and surface science, as both are necessary to understand nanomaterials.

doi:10.1038/nmat1368

Figure 8 Platinum-coated nanostructured whisker supports (0.25 mg cm⁻²). a, Plane view; b, 45° view (higher magnification). The nanostructured film of the MEA (c) shows the Pt-coated nanowhiskers sandwiched between the PEM and the gas-diffusion layer. Courtesy of R. Atanasoski, 3M, St. Paul, Minnesota, USA.

References

1. Nazar, L. F. et al. Nanostructured materials for energy storage. Int. J. Inorg. Mater. **3**, 191–200 (2001).
2. Hirshes, M. Nanoscale materials for energy storage. Mater. Sci. Eng. B **108**, 1 (2004).
3. Scrosati, B. Challenge of portable power. Nature **373**, 557–558 (1995).
4. Tarascon, J.-M. & Armand, M. Issues and challenges facing rechargeable batteries. Nature **414**, 359–367 (2001).
5. Wakihara, W. & Yamamoto, O. (eds) Lithium Ion Batteries—Fundamentals and Performance (Kodansha-Wiley-VCH, Weinheim, 1998).
6. van Schalkwijk, W. & Scrosati, B. (eds) Advances in Lithium-Ion Batteries (Kluwer Academic/Plenum, New York, 2002).
7. Huggins, R. A. in Handbook of Battery Materials (ed. Besenhard, J. O.) Part III, Chapter 4 (Wiley-VCH, Weinheim, 1999).
8. Winter, M. & Besenhard, J. O. Electrochemical lithiation of tin and tin-based intermetallic and composites. Electrochim. Acta **45**, 31–50 (1999).
9. Nazar, L. F. & Crosnier, O. In Lithium Batteries Science and Technology (eds Nazri,G.-A. & Pistoia, G.) 112–143 (Kluwer Academic/Plenum, Boston, 2004).
10. Idota, Y., Kabuto, T., Matsufuji, A., Maekawa, Y. & Miyasaki, T. Tin-based amorphous oxides: a high-capacity lithium-ion storage material. Science **276**, 1395–1397 (1997).
11. Mao, O. & Dahn, J. R. Mechanically alloyed Sn–Fe(–C) powders as anode materials for Li ion batteries. III. Sn₂Fe:SnFe₃C active/inactive composites. J. Electrochem. Soc. **146**, 423–427 (1999).
12. Beaulieu, L. Y. & Dahn, J. R. The reaction of lithium with Sn–Mn–C intermetallics prepared by mechanical alloying. J. Electrochem. Soc. **147**, 3237–3241(2000).
13. Graetz, J., Ahn, C. C., Yazami, R. & Fuetz, B. Highly reversible lithium storage in nanostructured silicon. Electrochem. Solid-State Lett. **6**, A194–197 (2003).
14. Yang, J. et al. Si/C composites for high capacity lithium storage materials. Electrochem. Solid-State Lett. **6**, A154–156 (2003).
15. Novak, P. et al in Int. Meeting Li Batteries IMLB12 Nara, Japan Abstract 9 (2004).
16. Ikeda, H. et al. in Proc. 42nd Battery Symposium, Yokohama, Japan 282 (2001).
17. Green, M., Fielder, E., Scrosati, B., Wachtler, M. & Serra Moreno, J. Structured silicon anodes for lithium battery applications. Electrochem. Solid-State Lett. **6**, A75–79 (2003).
18. Armstrong, A. R., Armstrong, G., Canales, Garcia, J. R. &Bruce, P. G. Lithium intercalation intoTiO₂-B nanowires. Adv. Mater. (in the press).
19. Poizot, P., Laruelle, S., Grugeon, S., Dupont, L. & Tarascon, J.-M. Nano-sized

transition metal oxides as negative electrode material for lithium-ion batteries. *Nature* **407**, 496–499 (2000).

20. Tarascon, J.-M., Grugeon, S., Laruelle, S., Larcher, D. & Poizot, P. In *Lithium Batteries Science and Technology* (eds Nazri, G.-A. & Pistoia, G.) Ch. 7, 220–246 (Kluwer Academic/Plenum, Boston, 2004).

21. Denis, S., Baudrin, E., Touboul, M. & Tarascon, J.-M. Synthesis and electrochemical properties vs. Li of amorphous vanadates of general formula RVO$_4$ (R = In, Cr, Fe, Al, Y) *J. Electrochem. Soc.* **144**, 4099–4109 (1997).

22. Leroux, F., Coward, G. R., Power, W. P. & Nazar, L. F. Understanding the nature of low-potential Li uptake into high volumetric capacity molybdenum oxides. *Electrochem Solid-State Lett.* **1**, 255–258 (1998).

23. Balaya, P., Li, H., Kienle, L. & Maier, J. Fully reversible homogeneous Li storage in RuO$_2$ with high capacity. *Adv. Funct. Mater.* **13**, 621–625 (2003).

24. Badway, F., Cosandey, F., Pereira, N. & Amatucci, G. G. Carbon metal fluoride nanocomposites: high capacity reversible metal fluoride conversion materials as rechargeable positive electrodes for Li batteries. *J. Electrochem. Soc.* **150**, A1318–1327 (2003).

25. Li, H., Ritcher, G. & Maier, J. Reversibile formation and decomposition of LiF clusters using transition metal fluorides as precursors and their application in rechargeable Li batteries. *Adv. Mater.* **15**, 736–739 (2003).

26. Jamnik, J, & Maier, J. Nanocrystallinity effects in lithium battery materials. Aspects of nano-ionics. Part IV. *Phys. Chem. Chem. Phys.* **5**, 5215–5220 (2003).

27. Larcher, D. *et al.* Effect of particle size on lithium intercalation into α-Fe$_2$O$_3$. *J. Electrochem. Soc.* **150**, A133–139 (2003).

28. Sides, C. R., Li, N. C., Patrissi, C. J., Scrosati, B. & Martin, C. R. Nanoscale materials for lithium-ion batteries. *Mater. Res. Bull.* **27**, 604–607 (2002).

29. Nordliner, S., Edström, K. & Gustafsson, T. Electrochemistry of vanadium oxide nanotubes. *Electrochem. Solid State Lett.* **4**, A129 (2001).

30. Le, D. B. *et al.* High surface area V$_2$O$_5$ aerogel intercalation electrodes. *J. Electrochem. Soc.* **143**, 2099–2104 (1996).

31. Dong, W., Rolison, D. R. & Dunn, B. Electrochemical properties of high surface area vanadium oxides aerogels. *Electrochem. Solid State Lett.* **3**, 457–459 (2000).

32. Thackeray, M. M., David, W. I. F., Bruce, P. G. & Goodenough, J. B. Lithium insertion into manganese spinels. *Mater. Res. Bull.* **18**, 461–472 (1983).

33. Robertson, A. D., Armstrong, A. R. & Bruce, P. G. Layered Li$_x$Mn$_y$Co$_{1-y}$O$_2$ intercalation electrodes: influence of ion exchange on capacity and structure upon cycling. *Chem. Mater.* **13**, 2380–2386 (2001).

34. Shao-Horn, Y. *et al.* Structural characterisation of layered LiMnO$_2$ electrodes by electron diffraction and lattice imaging. *J. Electrochem. Soc.* **146**, 2404–2412 (1999).

35. Wang, H. F., Jang, Y. I. & Chiang, Y.-M. Origin of cycling stability in monoclinic and orthorhombic-phase lithium manganese oxide cathodes. *Electrochem. Solid-State Lett.* **2**, 490–493 (1999).

36. Kang, S. H., Goodenough, J. B. & Rabenberg, L. K. Effect of ball-milling on 3 V capacity of lithium manganese oxospinel cathodes. *Chem. Mater.* **13**, 1758–1764 (2001).

37. Reed, J., Ceder, G. & van der Ven, A. Layered-to-spinel phase transformation in Li$_x$MnO$_2$. *Electrochem. Solid-State Lett.* **4**, A78–81 (2001).

38. Padhi, A. K., Nanjundaswamy, K. S., Masquelier, C., Okada, S. & Goodenough, J. B. Effect of structure on the Fe^{3+}/Fe^{2+} redox couple in iron phosphates. *J. Electrochem. Soc.* **144**, 1609–1613 (1997).

39. Ravet, A. *et al.* Electroactivity of natural and synthetic triphylite. *J. Power Sources* **97–98**, 503–507 (2001).

40. Huang, H., Yin, S.-C. & Nazar, L. F. Approaching theoretical capacity of LiFePO$_4$ at room temperature and high rates. *Electrochem. Solid-State Lett.* **4**, A170–172 (2001).

41. Croce, F., Scrosati, B., Sides, B. R. & Martin, C. R. in *206th Meeting Electrochemical Society, Honolulu, Hawaii* Abstract 241(2004).

42. Scrosati. B. & Vincent, C. A. Polymer electrolytes: the key to lithium polymer batteries. *Mater. Res. Soc. Bull.* **25**, 28–30 (2000).

43. Croce, F., Appetecchi, G. B., Persi, L. & Scrosati, B. Nanocomposite polymer electrolytes for lithium batteries. *Nature* **394**, 456–458 (1998).

44. MacFarlane, D. R., Newman, P. J., Nairn, K. M. & Forsyth, M. Lithium-ion conducting ceramic/polyether composites. *Electrochim. Acta* **43**, 1333–1337 (1998).

45. Kumar, B., Rodrigues, S. J. & Scanlon, L. Ionic conductivity of polymer–ceramic composites. *J. Electrochem. Soc.* **148**, A1191–1195 (2001).

46. Maier, J. Ionic conduction in space charge regions. *Prog. Solid State Chem.* **23**, 171–263 (1995).

47. Appetecchi, G. B., Croce, F., Persi, L., Ronci, F. & Scrosati, B. Transport and interfacial properties of composite polymer electrolytes. *Electrochim. Acta* **45**, 1481–1490 (2000).

48. Angell, C. A., Liu, C. & Sanchez, S. Rubbery solid electrolytes with dominant cationic transport and high ambient conductivity. *Nature* **362**, 137–139(1993).

49. Hawett, P. C., MacFarlane, D. R. & Hollenkamp, A. F. High lithium metal cycling efficiency in a room-temperature ionic liquid. *Electrochem. Solid-State Lett.* **7**, A97–101 (2004).

50. MacGlashan, G., Andreev, Y. G. & Bruce, P. G. The structure of poly(ethylene oxide)$_6$:LiAsF$_6$. *Nature* **398**, 792–794 (1999).

51. Gadjourova, Z., Andreev, Y. G., Tunstall D. P. & Bruce, P. G. Ionic conductivity in crystalline polymer electrolytes. *Nature* **412**, 520–523 (2001).

52. Stoeva, Z., Martin-Litas, I., Staunton, E., Andreev Y. G. & Bruce, P. G. Ionic conductivity in the crystalline polymer electrolytes PEO$_6$:LiXF$_6$, X = P, As, Sb. *J. Am. Chem. Soc.* **125**, 4619–2626 (2003).

53. Christie, A. M., Lilley, S. J., Staunton, E, Andreev, Y. G. & Bruce, P. G. Increasing the conductivity of crystalline polymer electrolytes. *Nature* **433**, 50–53 (2005).

54. Conway, B. E. *Electrochemical Supercapacitors.* (Kluwer Academic/Plenum, New York, 1999).

55. Mastragostino, M., Arbizzani, C. & Soavi, F. In *Advances in Lithium-Ion Batteries.* (Van Schalkwijk, W. & Scrosati, B., eds) Ch. 16, 481–505 (Kluwer Academic/Plenum, New York, 2002).

56. Arbizzani, C., Mastragostino, M. & Soavi, S. New trends in electrochemical supercapacitors. *J. Power Sources* **100**, 164–170 (2001).

57. Wang, J. *et al.* Morphological effects on the electrical and electrochemical properties of carbon aerogels. *J. Electrochem. Soc.* **148**, D75–77 (2001).

58. Niu, C., Sichjel, E. K., Hoch, R., Hoi, D. & Tennent, H. High power electrochemical capacitors based on carbon nanotube electrodes. *Appl. Phys. Lett.* **70**, 1480–1482 (1997).

59. Nelson, P. A. & Owen, J. R. A high-performance supercapacitor/battery hybrid incorporating templated mesoporous electrodes. *J. Electrochem. Soc.* **150**, A1313–1317 (2003).

60. Amatucci, G. G., Badway, F., Du Pasquier, A. & Zheng, T. An asymmetric hybrid nonaqueous energy storage cell. *J. Electrochem. Soc.* **148**, A930–939 (2001).

61. Brodd, R. J. *et al.* Batteries 1977 to 2002. *J. Electrochem. Soc.* **151**, K1–11 (2004).

62. Srinivasan, S., Mosdale, R., Stevens, P. & Yang, C. Fuel cells: reaching the era of clean and efficient power generation in the twenty-first century. *Annu. Rev. Energy Environ.* **24**, 281–238 (1999).

63. Giordano, N. *et al.* Analysis of platinum particle size and oxygen reduction in phosphoric acid. *Electrochim. Acta* **36**, 1979–1984 (1991).

64. Freund, A., Lang, J., Lehman, T. & Starz, K. A. Improved Pt alloy catalysts for fuel cells. *Catal. Today* **27**, 279–283 (1996).

65. Mukerjee, S., Srinivasan, S., Soriaga, M. P. & McBreen, J. Role of structural and electronic properties of Pt and Pt alloys on electrocatalysis of oxygen reduction. An *in-situ* XANES and EXAFS investigation. *J. Electrochem. Soc.* **142**, 1409–1422 (1995).

66. Aricò, A. S., Srinivasan, S. & Antonucci, V. DMFCs: From fundamental aspects to technology development. *Fuel Cells* **1**, 133–161 (2001).

67. Schmidt, T. J. *et al.* Electrocatalytic activity of Pt–Ru alloy colloids for CO and CO/H$_2$ electrooxidation: stripping voltammetry and rotating-disk measurements. *Langmuir* **14**, 2591–2595 (1997).

68. Chrzanowski, W. & Wieckowski, A. Enhancement in methanol oxidation by spontaneously deposited ruthenium on low-index platinum-electrodes. *Catal. Lett.* **50**, 69–75 (1998).

69. Brankovic, S. R., Wang, J. X. & Adzic, R. R. Pt submonolayers on Ru nanoparticles–a novel low Pt loading, high CO tolerance fuel-cell electrocatalyst. *Electrochem. Solid-State Lett.* **4**, A217–220 (2001).

70. Hockaday, R. G. *et al.* in *Proc. Fuel Cell Seminar, Portland, Oregon, USA* 791–794 (Courtesy Associates, Washington DC, 2000).

71. Sun, G. R., Wang, J. T.& Savinell, R. F. Iron(III) tetramethoxyphenylporphyrin (Fetmpp) as methanol tolerant electrocatalyst for oxygen reduction in direct methanol fuel-cells. *J. Appl. Electrochem.* **28**, 1087–1093 (1998).

72. Reeve, R. W., Christensen, P. A., Hamnett, A., Haydock, S. A. & Roy, S. C. Methanol tolerant oxygen reduction catalysts based on transition metal sulfides. *J. Electrochem. Soc.* **145**, 3463–3471 (1998).

73. Nolte, R., Ledjeff, K., Bauer, M. & Mulhaupt, R. Partially sulfonated poly(arylene ether sulfone)—a versatile proton conducting membrane material for modern energy conversion technologies. *J. Membrane Sci.* **83**, 211–220 (1993).

74. Kerres, J., Ullrich, A., Meier, F. & Haring, T. Synthesis and characterization of novel acid-base polymer blends for application in membrane fuel cells. *Solid State Ionics* **125**, 243–249 (1999).

75. Alberti, G. & Casciola, M. Composite membranes for medium-temperature PEM fuel cells. *Annu. Rev. Mater. Res.* **33**, 129–154 (2003).

76. Savadogo, O. Emerging membranes for electrochemical systems: (I) solid polymer electrolyte membranes for fuel cell systems. *J. New Mater. Electrochem. Syst.* **1**, 47–66 (1998).

77. Li, Q., He, R., Jensen, J. O. & Bjerrum, N. J. Approaches and recent development of polymer electrolyte membranes for fuel cells operating above 100 °C. *Chem. Mater.* **15**, 4896–4815 (2003).

78. Kreuer, K. D. On the development of proton conducting polymer membranes for hydrogen and methanol fuel cells. *J. Membrane Sci.* **185**, 29–39 (2001).

79. Watanabe, M., Uchida, H., Seki, Y., Emori, M. & Stonehart, P. Self-humidifying polymer electrolyte membranes for fuel cells. *J. Electrochem. Soc.* **143**, 3847–3852 (1996).

80. Aricò, A. S., Creti, P., Antonucci, P. L. & Antonucci, V. Comparison of ethanol and methanol oxidation in a liquid feed solid polymer electrolyte fuel cells at high temperature. *Electrochem. Solid-State Lett.* **1**, 66–68 (1998).

81. Aricò, A. S., Baglio, V., Di Blasi, A. & Antonucci, V. FTIR spectroscopic investigation of inorganic fillers for composite DMFC membranes. *Electrochem. Commun.* **5**, 862–866 (2003).

82. Kreuer, K. D., Paddison, S. J., Spohr, E. & Schuster, M. Transport in proton conductors for fuel cell applications: simulation, elementary reactions and phenomenology. **104**, 4637–4678 (2004).

83. Kreuer, K. D. On the development of proton conducting materials for technological applications. *Solid State Ionics* **97**, 1–15 (1997).

84. Debe, M. K. *Handbook of Fuel Cells—Fundamentals, Technology and Applications* Vol. 3 (eds Vielstich, W., Gasteiger, H. A. & Lamm, A.) Ch. 45, 576–589 (Wiley, Chichester, UK, 2003).

85. Atanasoski, R. in *4th Int. Conf. Applications of Conducting Polymers, ICCP-4, Como, Italy* Abstract 22 (2004).

86. Schoonman, J. Nanoionics. *Solid State Ionics* **157**, 319–326 (2003).

87. Steele, B. C. H. Current status of intermediate temperature fuel cells (It-SOFCs). *European Fuel Cell News* **7**, 16–19 (2000).

88. Maier, J. Defect chemistry and ion transport in nanostructured materials. Part II Aspects of nanoionics. *Solid State Ionics* **157**, 327–334 (2003).

89. Knauth, P & Tuller, H. L. Solid-state ionics: roots, status, and future-prospects. *J. Am. Ceram. Soc.* **85**, 1654–1680 (2002).

90. Perry Murray, E., Tsai, T.& Barnett, S. A. A direct-methane fuel cell with a ceria-based anode. *Nature* **400**, 649–651 (1999).

91. Park, S., Vohs, J. M. & Gorte, R. J. Direct oxidation of hydrocarbons in a solid-oxide fuel cell. *Nature* **404**, 265–267 (2000).

92. Zhou, Y. P., Feng, K., Sun, Y. & Zhou, L. A brief review on the study of hydrogen storage in terms of carbon nanotubes. *Prog. Chem.* **15**, 345–350 (2003).

93. Schlappach, L. & Zuttel, A. Hydrogen-storage materials for mobile applications. *Nature* **414**, 353–358 (2001).

94. Dillon A. C. *et al.* Storage of hydrogen in single-walled carbon nanotubes. *Nature* **386**, 377–379 (1997).

95. Chambers, A., Park, C., Baker, R. T. K. & Rodriguez, N. M. Hydrogen storage in graphite nanofibers. *J. Phys. Chem.* **102**, 4253–4256 (1998).

96. Yang, R. T. Hydrogen storage by alkali-doped carbon nanotubes—revisited. *Carbon* **38**, 623–626 (2000).

97. Dillon, A. C.& Heben, M. J. Hydrogen Storage using carbon adsorbents—past, present and future. *Appl. Phys. A* **72**, 133–142 (2001).

98. Hirscher, M. & Becher, M. Hydrogen storage in carbon nanotubes. *J. Nanosci. Nanotechnol.* **3**, 3–17 (2003).

99. Shang, C. X., Bououdina, M., Song, Y. & Guo, Z. X. Mechanical alloying and electronic simulation of MgH$_2$–M systems (M–Al, Ti, Fe, Ni, Cu, and Nb) for hydrogen storage. *Int. J. Hydrogen Energy* **29**, 73–80 (2004).

100. Bobet, J. L., Grigorova, E., Khrussanova, M., Khristov, M. & Peshev, P. Hydrogen sorption properties of the nanocomposite 90wt.% Mg$_2$Ni–10 wt.% V. *J. Alloys Compounds* **356**, 593–597 (2003).

101. Ma, R. Z. *et al.* Synthesis of boron-nitride nanofibers and measurement of their hydrogen uptake capacity. *Appl. Phys. Lett.* **81**, 5225–5227 (2002).

102. Bobet, J. L., Desmoulinskrawiec, S., Grigorova, E., Cansell, F. & Chevalier, B. Addition of nanosized Cr$_2$O$_3$ to magnesium for improvement of the hydrogen sorption properties. *J. Alloys Compounds* **351**, 217–221 (2003).

103. Audemer, A., Wurm, C., Morcrette, M., Gwizdala, S. & Masquellier, C. Carbon-coated metal bearing powders and process for production thereof. World Patent WO 04/001881 A2 (2004).

Acknowledgements

We thank M. Mastragostino of the university of Bologna for suggestions and help in completing the supercapacitors part of this review. Support from the European Network of Excellence 'ALISTORE' network is acknowledged. P.G.B. is indebted to the Royal Society for financial support.
Correspondence should be addressed to B.S.

Competing financial interests

The authors declare that they have no competing financial interests.

Nanoionics: ion transport and electrochemical storage in confined systems

The past two decades have shown that the exploration of properties on the nanoscale can lead to substantially new insights regarding fundamental issues, but also to novel technological perspectives. Simultaneously it became so fashionable to decorate activities with the prefix 'nano' that it has become devalued through overuse. Regardless of fashion and prejudice, this article shows that the crystallizing field of 'nanoionics' bears the conceptual and technological potential that justifies comparison with the well-acknowledged area of nanoelectronics. Demonstrating this potential implies both emphasizing the indispensability of electrochemical devices that rely on ion transport and complement the world of electronics, and working out the drastic impact of interfaces and size effects on mass transfer, transport and storage. The benefits for technology are expected to lie essentially in the field of room-temperature devices, and in particular in artificial self-sustaining structures to which both nanoelectronics and nanoionics might contribute synergistically.

J. MAIER

is in the Max-Planck-Institut für Festkörperforschung, 70569 Stuttgart, Germany.
e-mail: s.weiglein@fkf.mpg.de

The triumph of electronics in both science and daily life relies, on the one hand, on the availability of solid materials in which atomic constituents exhibit negligible mobilities whereas electronic carriers can be easily excited and transferred; and, on the other, on the necessity of fast and precise information exchange. However, electronic devices cannot satisfy all the technological needs of our society: as far as the transformation of chemical to electrical energy or the chemical storage of electrical energy is concerned — as is possible in batteries and fuel cells, or as far as the transformation of chemical in electrical information is concerned — as verified in (electro)chemical sensors, electrochemical devices are indispensable[1–4]. Solid electrochemical devices use the fact that (at least) one ionic component is immobile whereas another ionic component is highly mobile, and hence combine mechanical durability and electrochemical functionality. Such solids may be solid electrolytes — that is, pure ion-conducting solids — but they may also be mixed conducting solids that exhibit both significant electronic and ionic conductivities and can, for example, be used as powerful electrodes. Mixed conductors can also be used as key materials in related devices such as chemical sensors, chemical filters, chemical reactors and electrochromic windows. Figure 1 provides a schematic overview by using the example of oxide ion conductivity. Beyond that, the study of mixed conductors offers a general phenomenological approach to electrically relevant solids in which the pure ionic and the pure electronic conductor appear as special cases.

In the nanometre regime, interfaces are so closely spaced that their influence on the overall property can become significant if not prevailing. It can even no longer be taken for granted that one can assume semi-infinite boundary conditions; in other words, it cannot be taken for granted that, with increasing distance, the impact of the interface dies off to the bulk value, because the next interface might already be perceptible. In interfacially controlled materials the overall impact of the interfacial regions increases with decreasing size. As long as their local influence is identical to or comparable with that of isolated interfaces we may use the term 'trivial size effects', whereas the term 'true size effects' implies that interfaces perceive each other[5–8]. As long as we can ignore the effect of curvature this distinction is possible.

Nanoelectronics[9–12] is essentially based on the quantum-mechanical confinement effect, a true size effect as a consequence of which the energy of delocalized electrons increases with decreasing size. As far as electronic devices[13] are concerned, trivial size effects are important in varistors (insulating grain boundaries become conductive under bias), PCT ceramics (resistance shows a

When citing this article, please cite the original version as shown on the contents page of this chapter.

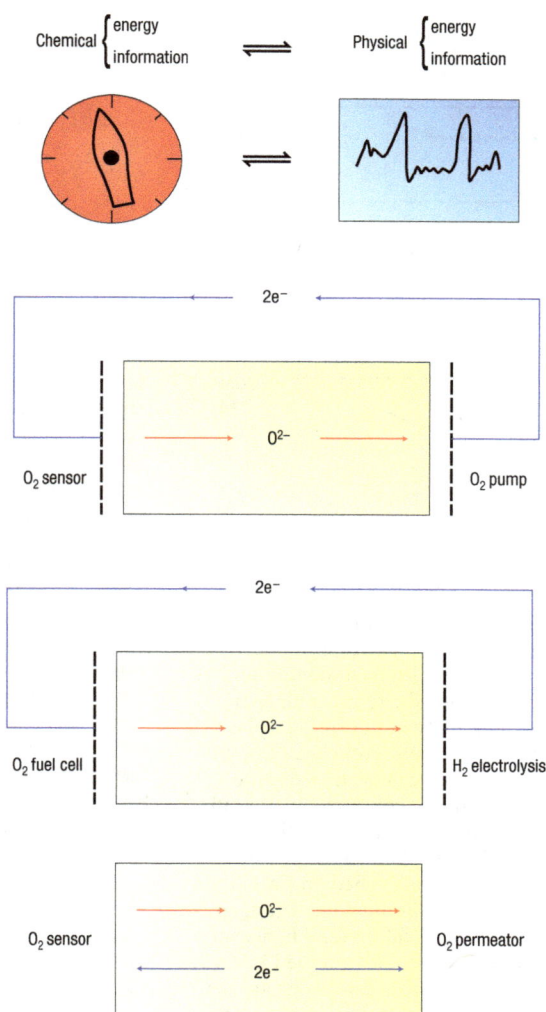

positive temperature coefficient[14]) or Taguchi sensors (adsorbed gases alter the charge-carrier concentration on the surface of semiconductors[15]), in which the interaction between interfacial chemistry and electrical resistance is decisive. Note that in these examples solid-state ionics has a major role.

It has been shown during the past two decades that ionic transport properties are also drastically changed at interfaces, leading to significant trivial and true size effects that at a sufficient number density of interfaces can become prevalent for the whole material[8,16–20]. Not only can an insulator or an electronic conductor be turned into an ionic conductor or an anion conductor into a cation conductor, but artificial conductors can also be generated that are unavailable otherwise. Given this enormous influence of interfaces on ion transport, and given the indispensability of electrochemical devices, it can be expected — and this will be set out in more detail below — that the importance of nanoionics[5–8,16–21] to electrochemistry will be similar to that of nanoelectronics to semiconductor physics. One necessary condition is that the nanostructured arrangement be sufficiently stable. Even though this point is naturally more critical than in nanoelectronics, it has been shown in many cases that metastable interfaces can be durable even up to quite high temperatures.

SIZE EFFECTS IN SOLID-STATE IONICS

Let us first consider the implementation of an interface in a solid system and study its impact on the ionic transport properties. In general this process breaks symmetry by introducing a structural discontinuity. In contrast with the bulk, where electroneutrality must be obeyed, at the interface a narrow charged zone, the so-called space-charge zone, is tolerable and even thermodynamically necessary[13,22–26]. The effects on both electrons and ions can be quantitatively studied, if an abrupt structural change (see Fig. 2) at the interface is assumed[25,26]. Here, the reader may be more interested in a qualitative approach and for this purpose it is advisable to consider the different charge carriers — being strongly coupled by the condition of electroneutrality in the bulk — as independent. Of course, the deviation from charge neutrality results in an electrical field that limits the magnitude and spatial extent of the carrier redistribution. Before considering the impact on charge-carrier concentration more specifically, let us note that because of the different structure of the interfacial core, mobilities change there too. Whether such core effects or space-charge effects are more important for the transport depends on the parameters of the materials and the current direction. If we consider, for example, transport along interfaces, space-charge transport is typically significant if bulk mobilities are high, bulk concentrations are small and carriers are accumulated, whereas in low-mobility materials grain boundaries are likely to offer enhanced mobilities.

Now let us concentrate on the thermodynamically required space-charge effects and consider the simplest case, namely grain boundaries in chemically homogeneous materials. As a result of the difference in structure between boundary and bulk, they are loci of charge separation: in polycrystalline AgCl (Fig. 2d) and polycrystalline CaF_2 the space-charge zone adjacent to the grain boundaries exhibits enhanced vacancy conductivity

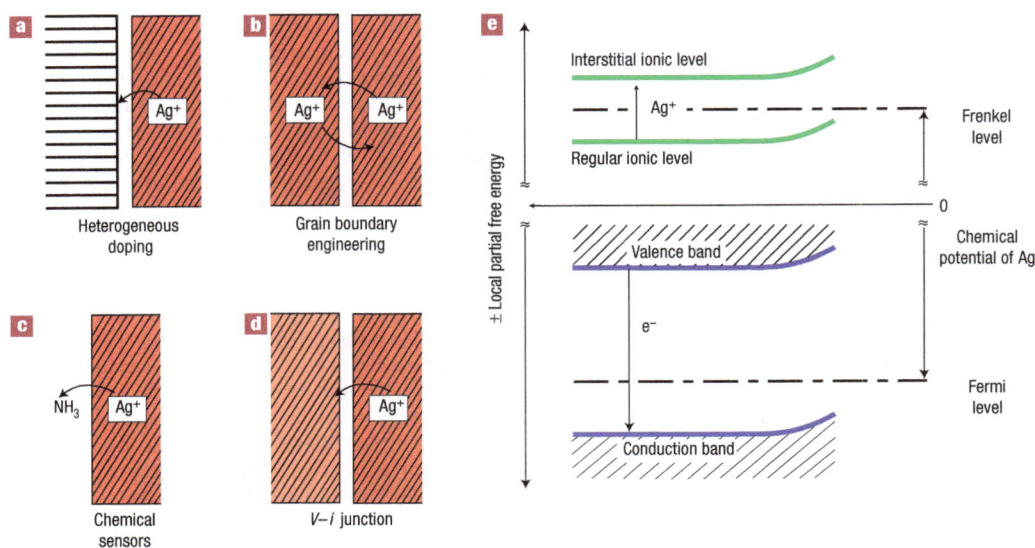

Figure 2 Ion redistribution at equilibrium.
a–d, Four prototype situations involving boundaries to solid ion conductors (using Ag^+ conductor as an example). **a**, Contact to an insulating phase that adsorbs Ag^+; **b**, grain boundary; **c**, contact to an acid–base active gas phase; **d**, contact to a second Ag^+ conductor. **e**, The ionic excitations (Ag^+ transfer from a regular site to an interstitial site, leaving behind a vacancy) can be represented in a (partial free) energy-level diagram, as is commonly used for electronic excitations. At boundaries the charging leads to bending of the levels. The coupling of the ionic (Ag^+, top) and electronic (bottom) pictures is determined by the (Ag) stoichiometry. A given interfacial charge leads to predictable consequences with respect to ionic and electronic carrier concentrations. Reprinted from ref. 92 with kind permission of Springer Science and Business Media.

(silver ion vacancies and fluoride ion vacancies[17]). In nanocrystalline samples especially, this interfacial effect is dominant in overall transport[27]. It can be chemically enforced and enhanced by contamination with NH_3 and SbF_5 (ref. 17), respectively. In $SrTiO_3$ or CeO_2 the grain boundaries show a positive excess charge, which leads to a depression of hole and oxygen vacancy concentrations there and hence to a decrease in ionic and electronic (in the p-regime) conductivities (Fig. 3, top middle)[28–34]. As both carriers are needed for (chemical) oxygen diffusion, the simultaneous depression of both types of conductivity results in a significant chemical resistance exerted by the boundary and observed during an experiment *in situ* in which the oxygen stoichiometry is varied and locally recorded[33]. Whereas the positively charged holes are depleted at such boundaries, the negatively charged conduction electrons have to be accumulated. This phenomenon is clearly seen in (weakly Gd-doped) CeO_2, in which the grain boundaries become n-type conducting whereas the bulk is ionically conducting[28,29,34].

It turned out that a very efficient method of directing or enhancing the space-charge effect is to decorate the grain boundaries with surface-active second-phase particles. Let us specifically consider the addition of fine-grained insulating Al_2O_3 particles to AgCl in which the defect concentration is quite small but in which the defects possess a significant mobility. Then we refer to the situation displayed in Figs 2a and 4d. As the basic oxide acts to adsorb the mobile cation, excess cation vacancies are necessarily formed in the AgCl matrix adjacent to the contact. By this heterogeneous doping process[35] insulators can be transformed into conductors (for example $LiI:Al_2O_3$ refs 36, 37), which was in fact the first composite electrolyte to be studied), anion conductors into cation conductors (for example $TlCl:Al_2O_3$ (ref. 38)), interstitial conductors into vacancy conductors ($AgCl:Al_2O_3$ (ref. 35)), or even superionic transitions can be provoked ($AgI:Al_2O_3$

(ref. 39)). By using mesoporous insulator phases with pore sizes of a few nanometres, this effect can be driven towards very high conductivities[40,41]. Similarly, SiO_2 is active in adsorbing anions; in this way the F^- conductivity in fluorides can be distinctly enhanced[42]. It has recently been shown that heterogeneous doping can even be used to significantly increase the Li^+ conductivity of liquid non-aqueous Li^+ electrolytes[43] (an effect that will be revisited below). The mechanism is the same — in both cases anions are adsorbed; in CaF_2, F^- is removed from the strongly bonded ion arrangement leaving the uncompensated cation (and hence an anion vacancy), whereas in the liquid the Li^+ ions, which are significantly associated with the anions in the bulk, are set free as a consequence of anion adsorption in the contact zones.

A beneficial effect of oxide particles on polymeric Li^+ electrolytes has also been reported[44,45]. As a large fraction of the Li salt is also expected to be undissociated, contributions by a similar mechanism are likely (see also ref. 46).

The inverse effect of a liquid phase on a solid conductor is also known; even gas phases (Fig. 2c) can be used if they are acid–base active, for example NH_3 in the case of AgCl (ref. 47) or BF_3 in the case of CaF_2 (ref. 17), leading to a novel sensor principle.

At the contact of two ionic (mixed) conductors (in addition to a possible charging of the proper interfacial core) a redistribution over both space-charge layers is thermodynamically required[48] (Fig. 2d). This explains the huge conductivity enhancement effects observed in two-phase systems of two silver halides[48,49] or in the CaF_2/BaF_2 heterolayer systems[50,51] (Fig. 3, top left).

The modifications affect not only transport; storage anomalies, too, occur at interfaces[17,20,52,53]. Even if the abrupt structural model is valid and the chemical potential of the element to be stored is spatially invariant within a given phase, a striking effect can occur at the contact[20]. Consider a compound MX that can be stored in neither phase

Figure 3 Size effects classified according to their contribution to the total chemical potential $\tilde{\mu}$ (see the text). Top, from left to right: penetrating accumulation zones in CaF_2/BaF_2 heterolayers, penetrating depletion zones in $SrTiO_3$ nanoceramics, and heterogeneous storage of Li in Li_2O/Ru nanocomposites as examples of electrostatic effects. Middle left: variation of the free formation energy of a localized defect or delocalized electron if confinement becomes perceptible. Bottom left: impact of edges and corners on the surface energy, breakdown of crystal structure. Bottom middle: effects of curvature on the activity of components and the distribution of ions. Bottom right: effects on interaction (interstitials (i) and vacancies (v) cannot separate) and configurational effects (see the text for more details).

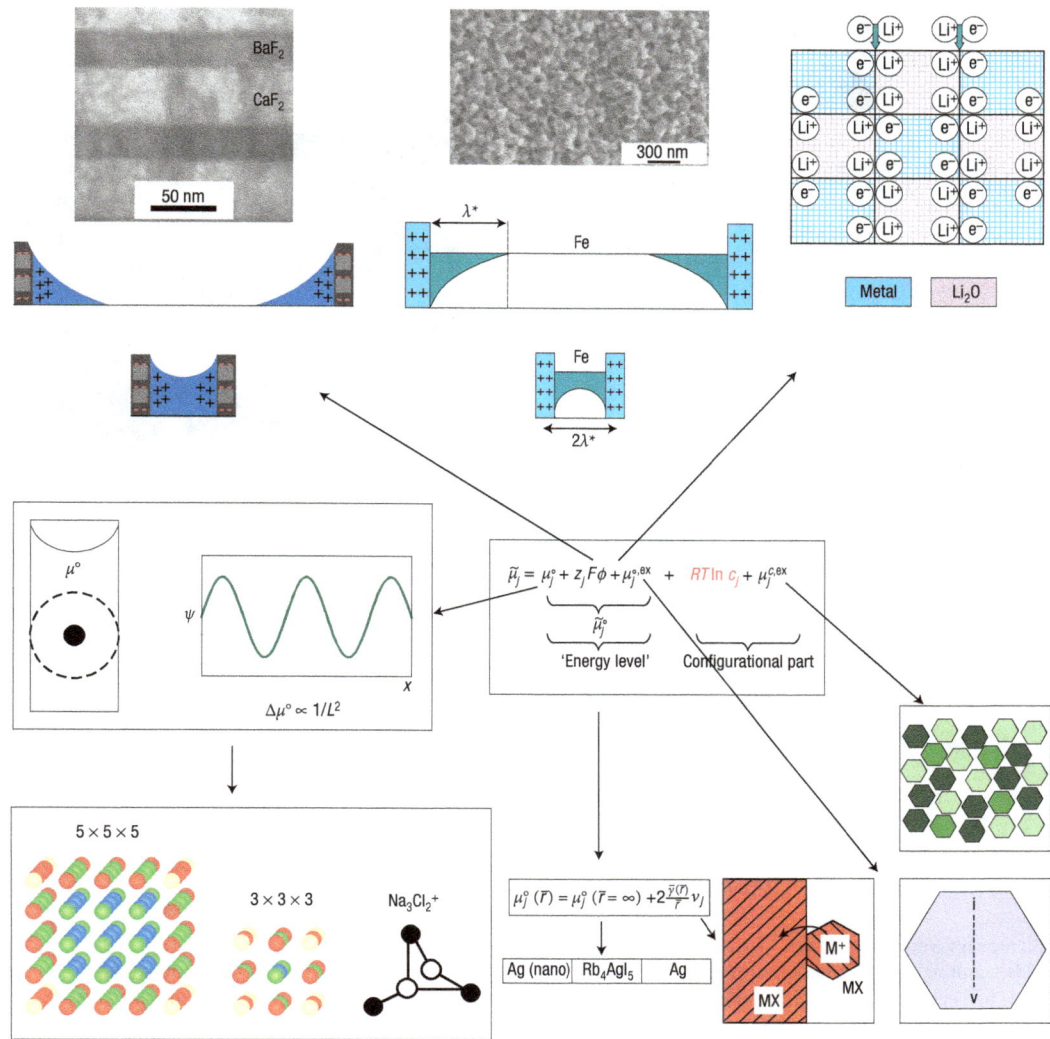

$$\tilde{\mu}_j = \underbrace{\mu_j^\circ + z_j F\phi + \mu_j^{\circ,ex}}_{\tilde{\mu}_j^\circ} + \underbrace{RT \ln c_j + \mu_j^{c,ex}}_{\text{Configurational part}}$$

'Energy level'   Configurational part

$\Delta\mu^\circ \propto 1/L^2$

$5 \times 5 \times 5$ $3 \times 3 \times 3$ $Na_3Cl_2^+$

$$\mu_j^\circ(\bar{r}) = \mu_j^\circ(\bar{r}=\infty) + 2\frac{\bar{\gamma}(\bar{r})}{\bar{r}}v_j$$

Ag (nano) | Rb_4AgI_5 | Ag

A nor phase B; let us assume that in phase A storage is impossible because X^- cannot be accepted (but M^+ can), whereas in phase B the solution of M^+ cations is forbidden (but not that of X^-). Job sharing, however, would allow the composite to store up to (typically) a monolayer of MX at the heterojunction. This storage effect will be considered again in the context of Li batteries (in which $X^- = e^-$; see Fig. 3, top right).

As already mentioned, at interfaces mobilities are altered as well; furthermore, significantly changed local reactivities are expected and have been observed in several investigations (see, for example, ref. 54). A substantial finding in the present context is that point defects act as most efficient acid–base active centres on the surface, as demonstrated by homogeneous and heterogeneous doping of heterogeneous catalysts[55,56].

These examples should convince the reader not only of the richness of the field but also of the powerful additional degree of freedom when leaving single crystals and using interfacially controlled materials.

Now let us briefly consider the thermodynamic consequences of the introduction of point defects (Fig. 3, middle; ref. 16). The generation of a single defect is connected with a severe expenditure of energy that is not ironed out by the alteration of the local (vibrational) entropy; the formation of defects therefore costs locally free energy (see the μ° term in Fig. 3) and would forbid their existence in thermodynamic equilibrium were there not the significant free-energy gain caused by the huge number of microscopically different but energetically equivalent possibilities of distributing these defects (gain in configurational entropy). The different aspects are expressed in Fig. 3 in terms of various contributions to the chemical potential of a defect. Note that the chemical potential of a species measures the increase in the total free energy if the concentration (c) of this species is marginally increased; it is therefore the relevant measure when the concentration that corresponds to the minimum of the free energy is to be discovered. If defects are dilute and randomly distributed, the well-known

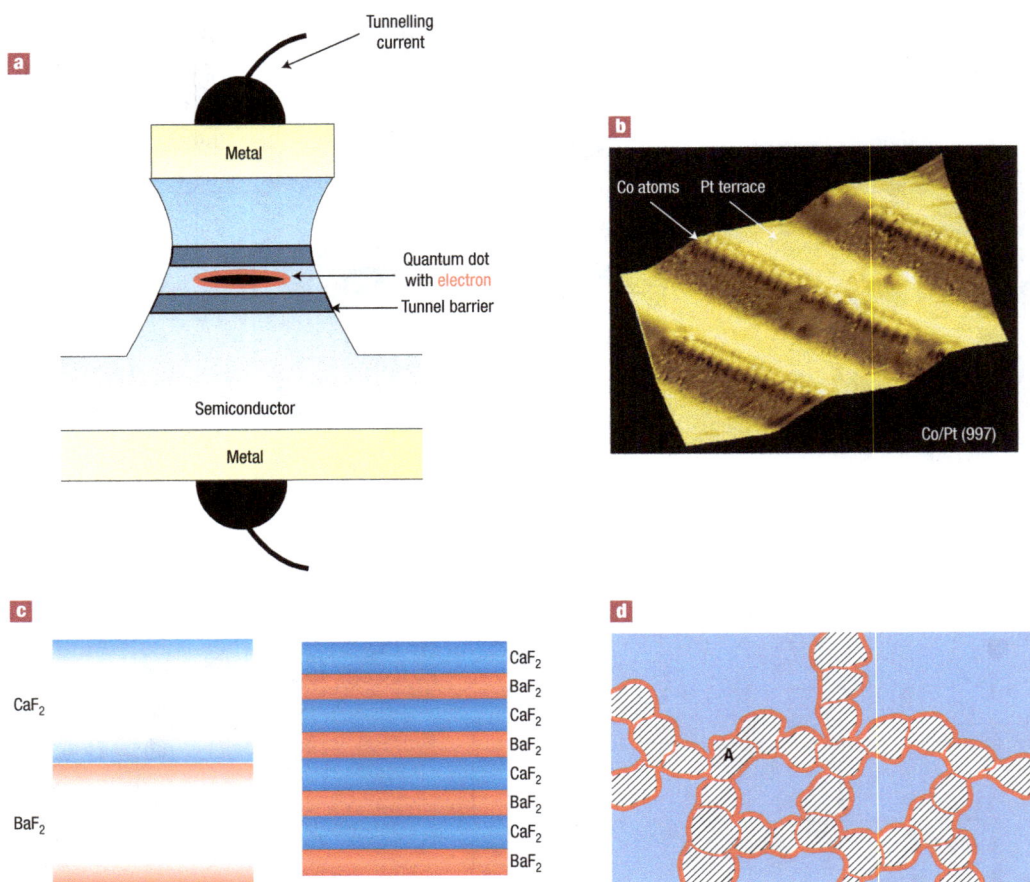

Figure 4 Different dimensionalities of interfacially controlled systems in solid-state electronics and ionics. **a**, Quantum dots in nanoelectronics[11]. **b**, Nanowires as nanomagnets. **c**, Nano-sized ionic heterolayers as artificial ion conductors. **d**, nanocomposite ion conductors consisting of insulator A in ion-conducting matrix MX. Part **a** reprinted from ref. 6. Copyright (2002), with permission from Elsevier. Parts **b** and **c** reprinted from refs 93 and 50, respectively, with permission from *Nature*.

Boltzmann form results ($\mu = \mu^{\circ} + RT\ln c$). Corrections to this occur at higher concentrations where the defects feel each other and/or feel the limited number of available sites (or quantum states). This leads to corrections in terms of energy and entropy (μ^{ex}). (Another correction becomes necessary if concentration gradients are so steep in the interfacial regions that the local properties depend on them and not only on the local value.) In the bulk we meet as many positive as negative charge carriers. This is different at interfaces, with the consequence that electrical field terms ($zF\phi$, where z is charge number, F is Faraday's constant and ϕ is electrical potential) have to be considered explicitly. The total chemical potential, which also contains the electrical effects, is usually called electrochemical potential (and termed $\tilde{\mu}$ in Fig. 3). If the structure does not change abruptly at an interface, the local formation term is a function of the distance. This is particularly important if the interfacial contact results in severe strain, as frequently occurs in epitaxial thin films. The peculiarities of the interfacial situation can even lead to interfacial phase transformations[57,58]. Finally, we also have to take account of the three-dimensional shape of the grains, which is reflected by its (mean) curvature ($1/\bar{r}$, where $\bar{r}$ is mean radius). An increased curvature leads to an enhanced internal pressure that augments the chemical potential by (at least) $2(\bar{\gamma}/\bar{r})v$, where v is the partial molar volume.

It is important in our context that all these contributions (and the list given is not exhaustive) give rise to, or are relevant in terms of, true size effects, the most important of which are addressed in Fig. 3 (for more details see refs 7, 16). Let us consider some of them, and first ignore curvature; that is, let us concentrate on thin films ($1/\bar{r} = 0$). If the spacing of the interfaces is perceptibly less than the effective width of space-charge layers, these layers overlap and the sample is charged throughout. Such effects with regard to accumulation layers (Fig. 3, top left) are discussed in epitaxially grown ionic heterolayers[59] whose thickness has been varied from 1 nm to 1 μm or in AgI nanoplates[60] in which heterolayers appear inherently, which represent the best binary solid electrolyte material at room temperature known so far. The counterpart — namely, penetrating depletion layers (Fig. 3, top middle) — has recently been observed in nanocrystalline $SrTiO_3$ (ref. 61). In the latter case the effective space-charge width is much larger than for the accumulation layers because the screening is much less and as the dielectric constant in $SrTiO_3$ is quite high. The top right panel in Fig. 3 shows the size effect on the storage capacity already mentioned.

In practice, although not a bad approximation in many situations[37,62], the structural change at the interface will not be restricted to one or two atomic layers; in some cases elastic effects can penetrate quite deeply and — if the interfacial spacing is very small — even penetrate the whole sample. Then the μ° term (Fig. 3) will be dependent on position and will be different from the value of the bulk in a large single crystal. However, even if the perfect structure decays in a step-function way, very narrowly spaced interfaces can change the formation energy and entropy of a defect. Let us first consider the energetic part in terms of two phenomena. First, the introduced defect polarizes the neighbourhood and the effective sphere of influence ranges much more widely than the atomic size (Fig. 3, middle left). If this sphere of influence is comparable to or larger than the spacing, this confinement will severely alter the energies[7,16]. This is the classical counterpart of the confinement effect that delocalized electrons experience; they can be much more extended and can perceive the boundaries much earlier. As electrons and ions perceive confinement with different sensitivities, the exploration of size effects of mixed conductors is of fundamental importance. Second, in the same way that electrons and holes cannot separate in tiny crystals[9], this is true for ionic defect pairs or pairs of ionic and electronic defects (Fig. 3, bottom right).

In addition, entropic contributions are affected by the particle size. First, the vibrational properties differ and hence the vibrational entropy. Second, in tiny systems the small number of atoms does not allow one to draw the usual statistical conclusions about the configurational entropy. At best an ensemble of nano-sized particles can be treated as a statistically significant ensemble (Fig. 3, bottom right). The consequence is that the conductivity (or reactivity) may vary significantly from particle to particle and may lead to peculiar effects.

In substantially curved crystals — that is, small ones — a size variation leads to variation in interfacial area, curvature and bulk volume. Like soap bubbles they experience an increased internal pressure (Fig. 3, bottom middle) and hence a greater chemical potential than that in a large single crystal. The most prominent consequence of this is the substantial decrease in the melting point of nanocrystals[63]. The (free) energetic difference shows up directly in the cell voltage of 'symmetrical' electrochemical cells that differ only in the grain size of active electrode components (for example in the all-solid-state cell in Fig. 3, bottom middle, which relates to the measurement of the excess free energy of nanocrystalline silver)[64,65]. For the same reason charging is also predicted if two particles of the same composition and orientation but different sizes come into contact[7]. It has already been mentioned that this excess contribution is proportional to the effective interfacial tension and inversely proportional to the effective radius as long as the surface tension stays constant. At tiny sizes the interfacial tension itself becomes a function of size, for a variety of reasons (such as the importance of edges and corners or the confinement of atoms within interfaces or edges; see Fig. 3, bottom left). At sizes of typically a few nanometres the macroscopic structure finally breaks down (Fig. 3, bottom left). In NaCl, for example, the rock salt structure is well established if the size is of the order of a few lattice constants; however,

below that size completely different structures are thermodynamically stable[66]. Then we leave the regime of solid-state chemistry and enter the regime of cluster chemistry; accordingly, we leave the range of validity of our top-down approach, and a bottom-up approach becomes advisable.

MATERIALS STRATEGIES

So far it has become clear that the deliberate introduction of interfaces ('heterogeneous doping')[25] is a powerful method of modifying the electrochemical properties of materials (see above; ceramics, polymers and liquids, but also glasses[67]) besides the conventional tools (i) of searching for novel ground structures and compositions, (ii) of varying temperature and stoichiometry, or (iii) of introducing impurities. The similarity of 'heterogeneous doping' to conventional, homogeneous doping (dissolution of foreign atoms) is striking because in both cases defects (but of different dimensionality) are irreversibly introduced, resulting in significant and quantifiable effects on charge-carrier density[68].

The benefit of narrowly spaced interfaces that act as fast pathways for ions or components lies not only in the enhanced effective conductivity but also in the possibility of rapid bulk storage resulting from the reduction of the effective diffusion length. Because the equilibration time is proportional to the square of the diffusion length, storage in nanosystems can be enormously fast. A size reduction from 1 mm to 10 nm reduces the equilibration time by 10 orders of magnitude. However, the same effect simultaneously sets a limit on the kinetic stability.

For 10 nm grains not to grow (on a timescale of years), the diffusion coefficient must be smaller than 10^{-20} cm^2 s^{-1}. This constraint does not contradict the material's electrochemical functionality because the diffusion coefficient, which is decisive for morphological stability, relates to the most sluggish component. (Typical values of diffusion coefficients in metals are between 10^{-30} and 10^{-20} cm^2 s^{-1}.) In addition, the danger of growth can be effectively diminished by the presence of second phases: Examples are the hetero-layers already touched upon (see Fig. 4c) or the nanocomposites in the context of Li-battery electrodes (see Fig. 5 and ref. 69).

For atomic-scale structures to exhibit sufficient stability, one has to engineer systems that are in spatial equilibrium (illustrative examples are the β-alumina ion conductors[70] (Fig. 6a), the high-temperature superconductors[71] or the alkali suboxides[72] in which conducting and non-conducting regions coexist on the atomic scale) or to rely on extremely low transport or reaction coefficients (as typically observed in covalent organic systems such as biopolymers).

To benefit from nanomaterials one can make use of confinement in one, two or three dimensions (Fig. 4). The last situation is met in nanodot systems, as represented by isolated grains or grain arrangements as in nanocrystalline ceramics. (Fig. 4a relates to nanoelectronics and displays a quantum dot.) Confinement in two dimensions is met in nanowires, nanotubes or nanoneedles. (The example in Fig. 4b relates to one-dimensional magnets.) Thin films finally offer the possibility of a one-dimensional confinement. In all cases we may refer to single-phase or multiphase systems, such as are realized in nanocomposites (or core–shell structures and nanoporous systems) or nano-sized heterolayer films. Figure 4c, d gives examples relating to nanoionics.

The significance of the possibility of generating systems that rely on the integration of metastable subsystems in the nanometre range can hardly be overestimated. In this regime the distances involved are small enough to permit high storage and rapid exchange of energy and information on an accessible timescale, but the distances are not yet so small that kinetic stability with respect to transport and reaction has to be given up (Fig. 6); it is not accidental that the integration of biological functions (Fig. 6d) is essentially connected with nanostructuring. (Note that, in addition, biomolecules themselves are pronouncedly (meta)stable as a result of the strong covalent bonds.) It is generally expected that on this scale the introduction of higher dimensional defects or the integration or combination of subsystems can lead to non-equilibrium systems with a much higher information content than is possible at equilibrium ('soft materials science'[73]). Artificial nanomaterials of interest may be fabricated by artificial design, as in heterolayers, or be self-assembled, as in dissipative biological structures.

APPLICATIONS

It was stated above that nanosystems reveal their advantages typically at temperatures that are low or moderate. It is no coincidence that the most striking results in the field of electrochemical devices have been obtained in the context of Li batteries that work at room temperature.

In high-performance (secondary) Li batteries a reversible discharge–charge procedure has to be connected with high power density and capacity. The established way to achieve the first is to choose electrodes that accommodate Li without structural change; that is, to use intercalation compounds or — to put it more generally — to work within the homogeneity range of the Li-containing material. It is clear that this restricts the storage capacity severely. So it was a great step forward when Poizot et al.[74] recognized that CoO can almost reversibly take up Li down as far as reduction to the metal or even to the alloy. Although one is dealing with multiphase mixtures, the tiny transport lengths involved make the nanocomposite behave as a semi-fluid phase. The same effect could be verified for a variety of oxides, fluorides and nitrides[75,76]. In RuO_2, four Li per RuO_2 can be reversibly stored with an initial degree of conversion of virtually 100% (Fig. 5)[77].

Another striking phenomenon related to storage can be predicted to occur in such nanocomposites (Fig. 3, top right), which is of fundamental importance and has already been addressed briefly[6]. Let us consider the Li_2O/Ru composites that form in situ during the process of discharging a RuO_2-based Li battery. A significant uptake of Li does not seem possible because neither Ru nor Li_2O can store Li,

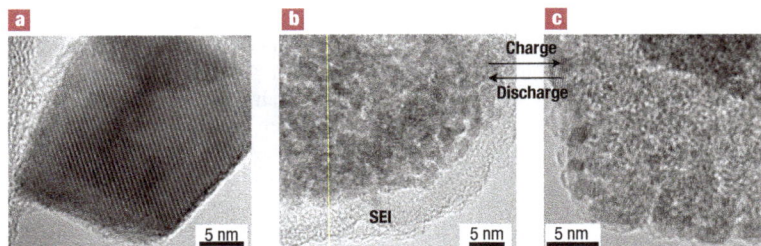

Figure 5 Reversibility of the redox reaction $Li_2O + Ru \leftrightarrow Li + RuO_2$ in the nanocomposite upon Li extraction and incorporation[77]. **a**, Original situation; **b**, lithiated nanocomposite; **c**, delithiated nanocomposite. (SEI: interphase between electrode and electrolyte.) Reprinted with permission from ref. 77.

the first because Li[+] cannot enter the structure (but e[-] can) and in the second the electron is difficult to accommodate (but Li[+] can be). However, the synergistic situation discussed above and shown in Fig. 3 should apply, leading to the possibility of an extra storage of up to about a monolayer of Li[+] per boundary. In nanosystems in which the proportion of interfaces amounts to several tens of per cent by volume, this excess storage is able to deliver an explanation of pseudo-capacitive behaviour observed in many discharge curves close to 0 V with respect to Li. If the grains are so small that the space-charges overlap in Li_2O, this mechanism forms the bridge between a supercapacitor and a battery electrode and may offer a reasonable compromise between speed of discharge and capacity.

The small particle size is also able to explain other apparent anomalies in the discharge curve such as altered absolute voltage values, sloped discharge curves where flat curves would be expected for macro-sized samples, and the occurrence of inhomogeneous distribution of Li within the nanocrystalline electrode[20].

Nano-size effects are also important as far as the electrolyte is concerned. Composites in which LiI is infiltrated in nanoporous Al_2O_3 achieve conductivities as high as 10^{-3} S cm^{-1} at 60 °C, which makes them interesting candidates for battery electrolytes[41]. Oxide additions also proved beneficial in polymer electrolytes[44,45], which because of their high Li[+] conductivities and favourable mechanical properties, are candidates for solid electrolytes in high-performance Li batteries. In these polymer electrolytes, which solvate Li salts, intrinsic structural features extend from the atomistic (functional groups) to the nanometre (side chains, backbone) or even the micrometre (composite of crystalline and amorphous regions) range. In (non-aqueous) liquid electrolytes an important nano-sized heterogeneity is the metastable interface between the Li-rich electrodes and the liquid electrolytes[78], because its kinetic stability is the key criterion for the durability of the cell. Nano-sized heterogeneities can also be generated within the liquid. It has been recently shown (and already mentioned above) that admixtures of tiny SiO_2 particles to non-aqueous liquid electrolytes can substantially improve the battery electrolytes ('soggy sand')[79]. In this way this new class of electrolytes combines high conductivities with the favourable mechanical properties of soft matter (Fig. 6c). Cells using such electrolytes may be operated without a polymer spacer and offer greater safety. The higher

Figure 6 Integration of subfunctions in functional solids. **a**, Equilibrium structure of the ion conductor Na-β-alumina comprising conduction planes and insulating zones on an atomic level[70]. **b, c**, Solid–liquid nanocomposites with non-equilibrium morphology composed of an insulating and a conducting phase and combining mechanical strength and conductivity: **b**, Additive effects in AgCl(s)–AgNO$_3$(aq) composites[98]. **c**, Synergistic effects in 'soggy sand' (in which the solvent is a non-aqueous solvent, the anion is adsorbed counterion and the matrix is SiO$_2$ (ref. 43)) or Nafion (in which the solvent is water, the anion is covalently bonded sulphonic acid, and the matrix is an insulating organic backbone[94,95]). **d**, Lipid bilayer (containing membrane proteins) as the basic element of the plasma membrane of eukaryotic cells regulating the influx and efflux of matter[96]. Parts **a–c** reprinted from refs 97–99, respectively. Copyright (1992, 2002, 2000) with permission from Elsevier. Part **d** reprinted from ref. 96. Copyright (1995) with permission from John Wiley & Sons.

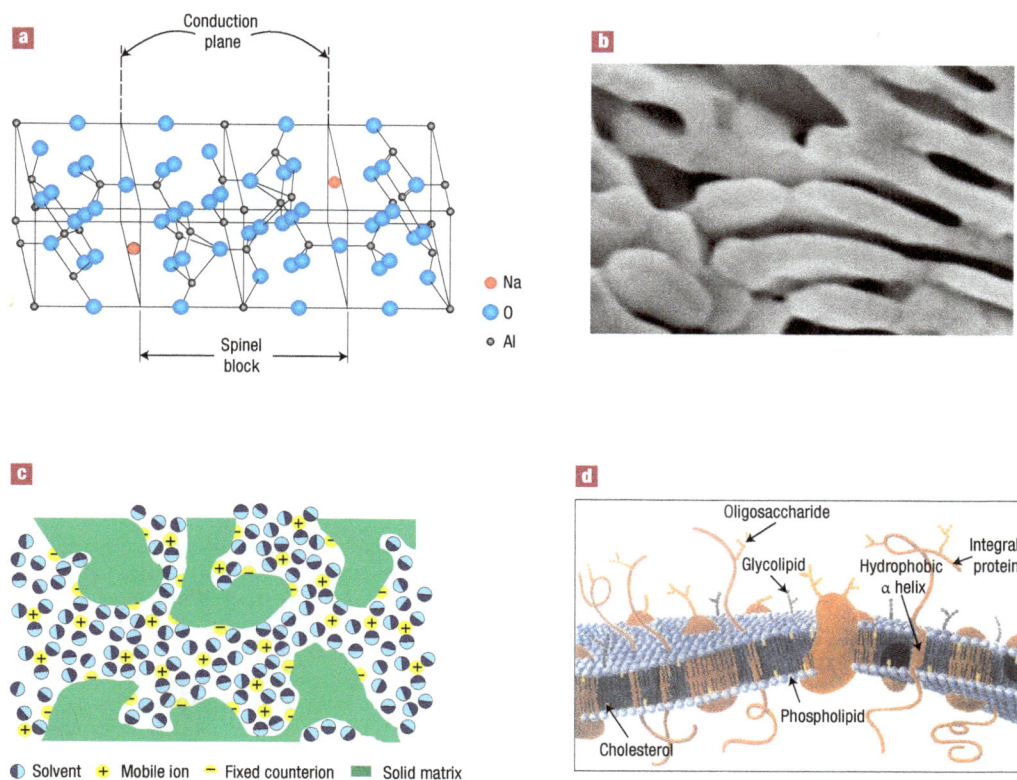

conductivity occurs when percolation between the boundary zones takes place: the particles either touch or are so close that the space–charge zones overlap severely.

In the end, the situation in the 'soggy-sand electrolytes' is quite similar to that of the Nafion electrolytes (used in low-temperature polymer-based fuel cells) except that, in the latter case, the negatively charged counterions are covalently built into the covalent network right from the beginning (Fig. 6c). The protons of these sulphonic acid groups are dissociated away in the tiny water channels interpenetrating the polymer matrix. Clearly, in Nafion the width of these channels is so small that finite size effects might be realized. Variants of this proton electrolyte address modifications of the framework, of the terminal groups as well as of the solvent, for example, using a polysiloxane network instead of the organic backbone, phosphoric acid groups instead of sulphonic acid groups or heterocycles instead of water. The ultimate goals are to prepare fully polymeric solvent-free networks providing high conductivities and to prevent the permeation of molecules such as water or methanol[80,81].

A different issue in the polymer-based fuel cells relates to the electrocatalysts. Here, too, the size is important. It has been shown[54] that metallic nanoparticles can provide a different catalytic activity, and size variation may be a possible way of ending up with a better electrode performance. Generally, enhanced reactivities may be caused by variations in defect concentration, by different fractions of atoms sitting in corners, edges or at the very surface and also by locally varied structure, composition and bonding.

In this context we should also note the attempts to use nanostructured matter for fast hydrogen storage[82–84].

Fuel cells operating at high temperatures have the advantage of improved electrode kinetics such that also hydrocarbons can be electrochemically converted without reforming. As far as the electrolyte is concerned, at these high temperatures the typically stronger activated bulk effects will dominate transport unless extremely tiny particles are used, whose morphology should be difficult to control. Nonetheless nanostructuring might be helpful, for example by trying to block electron conductivity in CeO$_2$. In particular, as regards the electrode a higher rate is feasible simply as a result of an enhanced interfacial area. In addition, however, size effects may be relevant for the electrode reaction itself in terms of modified reactivities. This is particularly relevant if the operation temperature of the fuel cells is to be lowered.

Similar aspects are valid for electrochemical (gas) sensors. Devices relying on complete equilibrium with the gas phase are sensors in which the gas is absorbed: the stoichiometry changes and so does the conductivity (see Fig. 1, bottom cell). The sensor sample itself has to be a mixed conductor. If bulk transport is rate-determining, nanostructuring can be helpful in reducing the transport length, whereas in interfacial control, nanostructuring can be helpful in directly influencing the mechanism. Because in this sensor mode bulk diffusion is decisive for low cross-sensitivity, miniaturization may lead to a selectivity loss, albeit with a gain in response time.

Another important sensor type uses a purely ionically conducting material, namely a solid electrolyte (see Fig. 1, top cell) for which only equilibrium with the ions is achieved and the open-circuit potential is measured. Evidently, in this case the benefits of size effects with respect to the electrode reaction are more important than a potential improvement of the electrolyte conductivity. (The latter is more significant in the amperometric mode.)

Low-temperature (gas) sensors typically relate to the Taguchi principle in which neither is the gas incorporated nor equilibrium with the ions achieved in contact with typically semiconducting oxides. For oxygen, electrons are trapped by the adsorbed gas molecules, an effect that is reflected by the modified surface resistance[15]. Because it is the surface reaction that determines the signal response, nano-size effects are expected to be of significance for the response time. The same is true for drift effects. Although bulk diffusion is usually considered to be sluggish, a recent analysis for SnO_2 has shown that it is a sluggish subsequent step in the surface kinetics that prevents bulk equilibration in nanocrystals[85]. At any rate, nano-size effects are expected to influence performance. An enhanced sensor action with nanocrystalline oxide has indeed been reported for SnO_2 (refs 86, 87). The Taguchi principle can also be generalized to acid–base-active gases; electronic conductors then have to be replaced by ion conductors[88]. For NH_3 sensing, downsizing of the electrode contacts has already been shown to improve the sensor action significantly because of the greater sensitivity to surface properties[89].

Similar remarks apply to devices such as permeators, electrochromic windows, chemical storage devices (see Fig. 1) and photogalvanic cells[90], in which analogous principles are operative. At the end of this article, and as a pertinent outlook, stands the aspect of cellular integration of both electrochemical and electronic 'organs' on the nanometre scale. The integration of tiny sensor, actuator and computer elements — that is, of nano-sized 'hands' and 'legs' in addition to the computer 'brain', the whole ensemble being sustained by an energy-providing 'metabolism' — is expected to characterize future autonomous systems (similar to that sketched in Fig. 7) that could act as adaptive nanomachines. In this context, knowledge from biology (note that biological systems integrate fuel cell, sensor and actuator functions) and nanoscience would converge on learning from living systems how to combine artificial intelligence with artificial sensing and actuating organs into energetically autarchic systems.

Because the implementation of a high density of interfaces can lead to a substantial impact on the overall or even local ionic transport properties of solids and as nanostructured matter can be pronouncedly morphologically stable, the crystallizing field of nanoionics is expected to have a substantial role in future materials research. The span of relevant materials and preparation procedures is enormous,

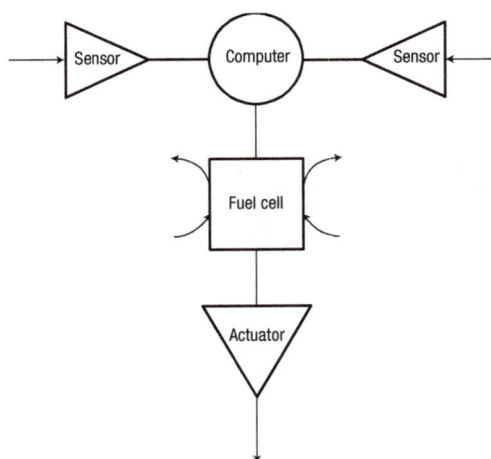

Figure 7 The nano-integration of ionic and electronic 'organs' such as sensors, actuators, computers and fuel cells or batteries results in tiny artificial autonomous systems. Reprinted from ref. 2. Copyright (2004) with permission from John Wiley & Sons Ltd.

ranging from quasi one-dimensional heterolayers to three-dimensional composites on the one hand, and from tailored deposition procedures to self-organization on the other. Not only should new artificial materials come to the fore, but so should systems integrating ionic and electronic functions on the nanometre scale. Much experimental and theoretical research is necessary not only for exploring the full potential of this approach but also for arriving at relevant nanostructures that improve the performance of present devices or enable the design of novel ones. The future will reveal how much remains vision and how much can be realized. At any rate, the exploration of the nanometre regime brings about powerful new degrees of freedom for engineering ionically active materials.

doi:10.1038/nmat1513

References

1. Kudo, T. & Fueki, K. *Solid State Ionics* (Kodansha, Tokyo, 1990).
2. Maier, J. *Physical Chemistry of Ionic Materials. Ions and Electrons in Solids* (Wiley, Chichester, 2004).
3. Bruce, P. G. *Solid State Electrochemistry* (Cambridge Univ. Press, London, 1995).
4. Bouwmeester, H. J. M. & Gellings, P. J. *Solid State Electrochemistry* (CRC Press, New York, 1997).
5. Maier, J. Point defect thermodynamics: Macro- vs. nanocrystals. *Electrochemistry* **68**, 395–402 (2000).
6. Maier, J. Aspects of nano-ionics, part I. Thermodynamic aspects and morphology of nano-structured ion conductors. *Solid State Ionics* **154**, 291–301 (2002); Part II. Defect chemistry and ion transport in nanostructured materials. *Solid State Ionics* **157**, 327–334 (2003); Part III. Nano-sized mixed conductors. *Solid State Ionics* **148**, 367–374 (2002).
7. Maier, J. Nano-ionics: Trivial and non-trivial size effects on ion conduction in solids. *Z. Phys. Chem.* **217**, 415–436 (2003).
8. Maier, J. Space charge regions in solid two phase systems and their conduction contribution. III: Defect chemistry and ionic conductivity in thin films. *Solid State Ionics* **23**, 59–67 (1987).
9. Brus, L. E. Electron–electron and electron–hole interactions in small semiconductor crystallites: The size dependence of the lowest excited electronic state. *J. Chem. Phys.* **80**, 4403–4409 (1984).
10. Reed, M. & Kirle, W. P. *Nanostructure Physics and Fabrication* (Academic, New York, 1989).
11. Haug, R. J. & von Klitzing, K. Prospects for research on quantum dots and single-electron transistors. *FED J.* **6**, 4–12 (1995).
12. Ploog, K. in *Semiconductor Interfaces: Formation and Properties* (eds Lay, G. L., Derrien, J. & Boccara, N.) 10–42 (Springer Proceedings in Physics 22, Berlin, 1987).
13. Sze, S. M. *Semiconductor Devices* (Wiley, New York, 1985).
14. Jonker, G. H. Some aspects of semiconducting barium titanate. *Solid State Electronics* **7**, 895–903 (1964).

15. Seiyama, T., Kato, A., Fujiishi, K. & Nagatani, M. A new detector for gaseous components using semiconductive thin films. *Anal. Chem.* **34**, 1502–1503 (1962).

16. Maier, J. Ionic transport in nano-sized systems. *Solid State Ionics* **175**, 7–12 (2004).

17. Maier, J. Ionic conduction in space charge regions. *Prog. Solid State Chem.* **23**, 171–263 (1995).

18. Tuller, H. L. Ionic conduction in nanocrystalline materials. *Solid State Ionics* **131**, 143–157 (2000).

19. Schoonman, J. Nanostructured materials in solid state ionics. *Solid State Ionics* **135–137**, 5–19 (2005).

20. Jamnik, J. & Maier, J. Nanocrystallinity effects in lithium battery materials. (Aspects of nano-ionics. Part IV). *Phys. Chem. Chem. Phys.* **5**, 5215–5220 (2003).

21. Despotuli, A. L. & Nikolaichik, V. I. A step towards nanoionics. *Solid State Ionics* **60**, 275–278 (1993).

22. Chapman, D. L. A contribution to the theory of electrocapillarity. *Phil. Mag. S. 6* **25**, 475–481 (1913).

23. Kliewer, K. L. & Koehler, J. S. Space charge in ionic crystals. I. General approach with application to NaCl. *Phys. Rev. A* **140**, A1226–A1240 (1965).

24. Poeppel, R. B. & Blakely, J. M. Origin of equilibrium space charge potentials in ionic crystals. *Surf. Sci.* **15**, 507–523 (1969).

25. Maier, J. Defect chemistry and conductivity effects in heterogeneous solid electrolytes. *J. Electrochem. Soc.* **134**, 1524–1535 (1987).

26. Jamnik, J., Maier, J. & Pejovnik, S. Interfaces in solid ionic conductors: Equilibrium and small signal picture. *Solid State Ionics* **75**, 51–58 (1995).

27. Puin, W., Rodewald, S., Ramlau, R., Heitjans, P. & Maier, J. Local and overall ionic conductivity in nanocrystalline CaF$_2$. *Solid State Ionics* **131**, 159–164 (2000).

28. Chiang, Y.-M., Lavik, E., Kosacki, I., Tuller, H. L. & Ying, J. Y. Defect and transport properties of nanocrystalline CeO$_{2-x}$. *Appl. Phys. Lett.* **69**, 185–187 (1996).

29. Tschöpe, A., Sommer, E. & Birringer, R. Grain size-dependent electrical conductivity of polycrystalline cerium oxide. I. Experiments. *Solid State Ionics* **139**, 255–265 (2001).

30. Vollmann, M. & Waser, R. Grain boundary defect chemistry of acceptor-doped titanates: Space charge layer width. *J. Am. Ceram. Soc.* **77**, 235–243 (1994).

31. Denk, I., Claus, J. & Maier, J. Electrochemical investigations of SrTiO$_3$ boundaries. *J. Electrochem. Soc.* **144**, 3526–3536 (1997).

32. Guo, X., Fleig, J. & Maier, J. Separation of electronic and ionic contributions to the grain boundary conductivity in acceptor-doped SrTiO$_3$. *J. Electrochem. Soc.* **148**, J50–J53 (2001).

33. Leonhardt, M., Jamnik, J. & Maier, J. In situ monitoring and quantitative analysis of oxygen diffusion through Schottky-barriers in SrTiO$_3$ bicrystals. *Electrochem. Solid-State Lett.* **2**, 333–335 (1999).

34. Kim, S. & Maier, J. On the conductivity mechanism of nanocrystalline ceria. *J. Electrochem. Soc.* **149**, J73–J83 (2002).

35. Maier, J. Space charge regions in solid two-phase systems and their conduction contribution. I. Conductance enhancement in the system ionic conductor-'inert' phase and application on AgCl·Al$_2$O$_3$ and AgCl·SiO$_2$. *J. Phys. Chem. Solids* **46**, 309–320 (1985).

36. Liang, C. C. Conduction characteristics of lithium iodide aluminium oxide solid electrolytes. *J. Electrochem. Soc.* **120**, 1289–1292 (1973).

37. Chung, R. W. & de Leeuw, S. W. Ionic conduction in LiI-alpha, gamma-alumina: molecular dynamics study. *Solid State Ionics* **175**, 851–855 (2004).

38. Maier, J. & Reichert, B. Ionic transport in heterogeneously and homogeneously doped thallium(I)-chloride. *Ber. Bunsenges. Phys. Chem.* **90**, 666–670 (1986).

39. Lee, J.-S., Adams, S. & Maier, J. Transport and phase transition characteristics in AgI:Al$_2$O$_3$ composite electrolytes. Evidence for a highly conducting 7-layer AgI polytype. *J. Electrochem. Soc.* **147**, 2407–2418 (2000).

40. Yamada, H., Bhattacharyya, A. J. & Maier, J. Extremely high silver ion conductivity in the composites of silver halide (AgBr, AgI) and mesoporous alumina. *Adv. Funct. Mater.* (in the press).

41. Maekawa, H., Tanaka, R., Sato, T., Fujimaki, Y. & Yamamura, T. Size-dependent ionic conductivity observed for ordered mesoporous alumina-LiI composite. *Solid State Ionics* **175**, 281–285 (2004).

42. Hariharan, K. & Maier, J. Enhancement of the fluoride vacancy conduction in PbF$_2$:SiO$_2$ and PbF$_2$:Al$_2$O$_3$ composites. *J. Electrochem. Soc.* **142**, 3469–3473 (1995).

43. Bhattacharyya, A. J. & Maier, J. Second phase effects on the conductivity of non-aqueous salt solutions: 'soggy sand electrolytes'. *Adv. Mater.* **16**, 811–814 (2004).

44. Croce, F., Appetechi, G. B., Persi, L. & Scrosati, B. Nanocomposite polymer electrolytes for lithium batteries. *Nature* **394**, 456–458 (1998).

45. Wieczorek, W., Florjanczyk, Z. & Stevens, R. Composite polyether based solid electrolytes. *Electrochim. Acta* **40**, 2251–2258 (1995).

46. Kasemägi, H., Aabloo, A., Klintenberg, M. K. & Thomas, J. O. Molecular dynamics simulation of the effect of nanoparticle fillers on ion motion in a polymer host. *Solid State Ionics* **168**, 249–254 (2004).

47. Maier, J. & Lauer, U. Ionic contact equilibria in solids — implications for batteries and sensors. *Ber. Bunsenges. Phys. Chem.* **94**, 973–978 (1990).

48. Maier, J. Space charge regions in solid two phase systems and their conduction contribution. II. Contact equilibrium at the interphase of two ionic conductors and the related conductivity effect. *Ber. Bunsenges. Phys. Chem.* **89**, 355–362 (1985).

49. Shahi, K. & Wagner, J. B. Fast ion transport in silver halide solid solutions and multiphase systems. *Appl. Phys. Lett.* **37**, 757–759 (1980).

50. Sata, N., Eberman, K., Eberl, K. & Maier, J. Mesoscopic fast ion conduction in nanometre-scale planar heterostructures. *Nature* **408**, 946–949 (2000).

51. Jin-Phillipp, N. Y. et al. Structures of BaF$_2$-CaF$_2$ heterolayers and their influences on ionic conductivity. *J. Chem. Phys.* **120**, 2375–2381 (2004).

52. Petuskey, W. Interfacial effects on Ag:S nonstoichiometry in silver sulfide/alumina composites. *Solid State Ionics* **21**, 117–129 (1986).

53. Beaulieu, L. Y., Larcher, D., Dunlap, R. A. & Dahn, J. R. Reaction of Li with grain-boundary atoms in nanostructured compounds. *J. Electrochem. Soc.* **147**, 3206–3212 (2000).

54. Brankovic, S. R., Wang, J. X. & Adžić, R. R. A. Pt submomolayers on Ru nanoparticles. a novel low Pt loading, high CO tolerance fuel cell electrocatalyst. *Electrochem. Solid-State Lett.* **4**, A217–A220 (2001).

55. Simkovich, G. & Wagner, C. The role of ionic point defects in the catalytic activity of ionic crystals. *J. Catal.* **1**, 521–525 (1962).

56. Maier, J. & Murugaraj, P. The effect of heterogeneous doping on heterogeneous catalysis: Dehydrohalogenation of tertiary butyl chloride. *Solid State Ionics* **40/41**, 1017–1020 (1990).

57. Lipowsky, R. in *Springer Proceedings in Physics* Vol. 50 (eds Falicov, L. M., Mejia-Lira, F. & Morán-López, J. L.) 158–166 (Springer, Berlin, 1990).

58. Hainovsky, N. & Maier, J. Simple phenomenological approach to premelting and sublattice melting in Frenkel disordered ionic crystals. *Phys. Rev. B* **51**, 15789–15797 (1995).

59. Sata, N., Eberman, K., Eberl, K. & Maier, J. Mesoscopic fast ion conduction in nanometre-scale planar heterostructures. *Nature* **408**, 946–949 (2000).

60. Guo, Y.-G., Lee, J.-S. & Maier, J. AgI nanoplates with mesoscopic superionic conductivity at room temperature. *Adv. Mater.* (in the press).

61. Balaya, P., Jamnik, J. & Maier, J. Mesoscopic size effects in nanocrystalline SrTiO$_3$. *Appl. Phys. Lett.* (submitted).

62. Heifets, E., Kotomin, E. A. & Maier, J. Semi-empirical simulations of surface relaxation for perovskite titanates. *Surf. Sci.* **462**, 19–35 (2000).

63. Buffat, P. & Borel, J.-P. Size effect on the melting temperature of gold particles. *Phys. Rev. A* **13**, 2287–2298 (1976).

64. Knauth, P., Schwitzgebel, G., Tschöpe, A. & Villain, S. Emf measurements on nanocrystalline copper-doped ceria. *J. Solid State Chem.* **140**, 295–299 (1998).

65. Schroeder, A. et al. Excess free enthalpy of nanocrystalline silver, determined in a solid electrolyte cell. *Solid State Ionics* **173**, 95–101 (2004).

66. Martin, T. P. In *Festkörperprobleme, Advances in Solid State Physics* (ed. Grosse, P.) 1–24 (Vieweg, Braunschweig, 1984).

67. Adams, S., Hariharan, K. & Maier, J. Interface effect on the silver ion conductivity during the crystallization of AgI-Ag$_2$O-V$_2$O$_5$ glasses. *Solid State Ionics* **75**, 193–201 (1995).

68. Maier, J. Composite electrolytes. *Mater. Chem. Phys.* **17**, 485–498 (1987).

69. Trifonova, A. et al. Influence of the reductive preparation conditions on the morphology and on the electrochemical performance of Sn/SnSb. *Solid State Ionics* **168**, 51–59 (2004).

70. Beevers, C. A. & Ross, M. A. S. The crystal structure of 'beta alumina' Na$_2$O·11Al$_2$O$_3$. *Z. Krist.* **97**, 59–66 (1937).

71. Philips, J. C. *Physics of High-T$_c$ Superconductors* (Academic, Boston, 1983).

72. Simon, A. Clusters of valence electron poor metals — structure, bonding and properties. *Angew. Chem. Int. Edn Engl.* **27**, 159–183 (1988).

73. Maier, J. Nanoionics and soft materials science. In *Nanocrystalline Metals and Oxides. Selected Properties and Applications* (eds Knauth, P. & Schoonman, J.) 81–110 (*Electronic Materials: Science & Technology* Vol. 7, Kluwer Academic, Boston, Massachusetts, 2002).

74. Poizot, P., Laruelle, S., Grugeon, S., Dupont, L. & Tarascon, J.-M. Nano-sized transition-metal oxides as negative electrode materials for lithium-ion batteries. *Nature* **407**, 496–499 (2000).

75. Li, H., Richter, G. & Maier, J. Reversible formation and decomposition of LiF clusters using transition metal fluorides as precursors and their application in rechargeable Li batteries. *Adv. Mater.* **15**, 736–739 (2003).

76. Badway, F., Cosandey, F., Pereira, N. & Amatucci, G. G. Carbon metal fluoride nanocomposites. high-capacity reversible metal fluoride conversion materials as rechargeable positive electrodes for Li batteries. *J. Electrochem. Soc.* **150**, A1318–A1327 (2003).

77. Balaya, P., Li, H., Kienle, L. & Maier, J. Fully reversible homogeneous and heterogeneous Li storage in RuO$_2$ with high capacity. *Adv. Funct. Mater.* **13**, 621–625 (2003).

78. Aurbach, D. Review of selected electrode–solution interactions which determine the performance of Li and Li ion batteries. *J. Power Sources* **89**, 206–218 (2000).

79. Bhattacharyya, A. J., Dollé, M. & Maier, J. Improved Li-battery electrolytes by heterogeneous doping of nonaqueous Li-salt solutions. *Electrochem. Solid-State Lett.* **7**, A432–A434 (2004).

80. Kreuer, K.-D., Paddison, S. J., Spohr, E. & Schuster, M. Transport in proton conductors for fuel-cell applications: Simulations, elementary reactions, and phenomenology. *Chem. Rev.* **104**, 4637–4678 (2004).

81. Schuster, M. F. H. & Meyer, W. H. Anhydrous proton conducting polymer. *Annu. Rev. Mater. Res.* **33**, 232–261 (2003).

82. Zhou, Y. P., Feng, K., Sun, Y. & Zhou, L. A brief review on the study of hydrogen storage in terms of carbon nanotubes. *Progr. Chem.* **15**, 345–350 (2003).

83. Schlapbach, L. & Züttel, A. Hydrogen-storage materials for mobile applications. *Nature* **414**, 353–358 (2001).
84. Ma, R. Synthesis of boron nitride nanofibers and measurement of their hydrogen uptake capacity. *Appl. Phys. Lett.* **81**, 5225–5227 (2002).
85. Jamnik, J., Kamp, B., Merkle, R. & Maier, J. Space charge influenced oxygen incorporation in oxides: in how far does it contribute to the drift of Taguchi sensors? *Solid State Ionics* **150**, 157–166 (2002).
86. Barsan, N. & Weimar, U. Conduction model of metal oxide gas sensors. *J. Electroceramics* **7**, 143–167 (2001).
87. Pan, X. Q., Fu, L. & Dominguez, J. E. Structure-property relationship of nanocrystalline tin dioxide thin films grown on sapphire. *J. Appl. Phys.* **89**, 6056–6061 (2001).
88. Maier, J. Electrical sensing of complex gaseous species by making use of acid-base properties. *Solid State Ionics* **62**, 105–111 (1993).
89. Holzinger, M., Fleig, J., Maier, J. & Sitte, W. Chemical sensors for acid-base-active gases: Applications to CO_2 and NH_3. *Ber. Bunsenges. Phys. Chem.* **99**, 1427–1432 (1995).
90. Grätzel, M. Photoelectrochemical cells. *Nature* **414**, 338–344 (2001).
91. Maier, J. Ionic and mixed conductors for electrochemical devices. *Radiation Effects Defects Solids* **158**, 1–10 (2003).
92. Maier, J. in *Modern Aspects of Electrochemistry* Vol. 38 (eds Conway, B. E., Vayenas, C. G. & White, R. E.) 1–173 (Springer, New York, 2003).
93. Gambardella, P. *et al.* Ferromagnetism in one-dimensional monatomic metal chains. *Nature* **416**, 301–304 (2002).
94. Kreuer, K.-D., Dippel, T., Meyer, W. & Maier, J. in *Solid State Ionics III* (eds Nazri, G.-A., Tarascon, J.-M. & Armand, M.) 273–282 (*Mat. Res. Soc. Symp. Proc.* Vol. 293, Materials Research Society, Pittsburgh, PA, 1993).
95. Eikerling, M., Kornyshev, A. A., Kuznetsov, A. M., Ulstrup, J. & Walbran, S. Mechanisms of proton conductance in polymer electrolyte membranes. *J. Phys. Chem. B* **105**, 3646–3662.
96. Voet, B. & Voet, J. G. *Biochemistry* (Wiley, New York, 1995).
97. Borg, R. J. & Dienes, G. J. *The Physical Chemistry of Solids* (Academic, San Diego, 1992).
98. Chandra, A. & Maier, J. Properties and morphology of highly conducting inorganic solid-liquid composites based on AgCl. *Solid State Ionics* **145**, 153–158 (2002).
99. Maier, J. Point-defect thermodynamics and size effects. *Solid State Ionics* **131**, 13–22 (2000).

Acknowledgements
The author is indebted to the Max Planck Society and acknowledges support in the framework of the ENERCHEM project.

Competing financial interests
The author declares no competing financial interests.

Issues and challenges facing rechargeable lithium batteries

J.-M. Tarascon* & M. Armand†

**Université de Picardie Jules Verne, Laboratoire de Réactivité et Chimie des Solides, UMR-6007, 33 rue Saint Leu, 80039, Amiens, France*
†*Department of Chemistry, University of Montreal, C.P. 6128 Succ. Centre Ville, Montréal, Quebec H3C 3J7, Canada*

Technological improvements in rechargeable solid-state batteries are being driven by an ever-increasing demand for portable electronic devices. Lithium-ion batteries are the systems of choice, offering high energy density, flexible and lightweight design, and longer lifespan than comparable battery technologies. We present a brief historical review of the development of lithium-based rechargeable batteries, highlight ongoing research strategies, and discuss the challenges that remain regarding the synthesis, characterization, electrochemical performance and safety of these systems

Rechargeable Li-ion cells are key components of the portable, entertainment, computing and telecommunication equipment required by today's information-rich, mobile society. Despite the impressive growth in sales of batteries worldwide, the science underlying battery technology is often criticized for its slow advancement. This is true whatever the technology considered (for example, nickel–cadmium, nickel–metal hydride or Li ion). Certainly, when compared, energy storage cannot keep pace with the rate of progress in the computer industry (Moore's law predicts a doubling of memory capacity every two years), yet the past decade has produced spectacular advances in chemistry and engineering within the emerging technologies of Ni–MeH and Li-ion batteries. These cells are now supplanting the well known Ni–Cd batteries.

A battery is composed of several electrochemical cells that are connected in series and/or in parallel to provide the required voltage and capacity, respectively. Each cell consists of a positive and a negative electrode (both sources of chemical reactions) separated by an electrolyte solution containing dissociated salts, which enable ion transfer between the two electrodes. Once these electrodes are connected externally, the chemical reactions proceed in tandem at both electrodes, thereby liberating electrons and enabling the current to be tapped by the user. The amount of electrical energy, expressed either per unit of weight (W h kg^{-1}) or per unit of volume (W h l^{-1}), that a battery is able to deliver is a function of the cell potential (V) and capacity (A h kg^{-1}), both of which are linked directly to the chemistry of the system. Among the various existing technologies (Fig. 1), Li-based batteries — because of their high energy density and design flexibility — currently outperform other systems, accounting for 63% of worldwide sales values in portable batteries[1]. This explains why they receive most attention at both fundamental and applied levels.

Historical developments in Li-battery research

Before reviewing the present status of research and future challenges for Li-battery technologies, we present a brief historical account of developments over the past 30 years, as personally perceived.

The motivation for using a battery technology based on Li metal as anode relied initially on the fact that Li is the most electropositive (−3.04 V versus standard hydrogen electrode)

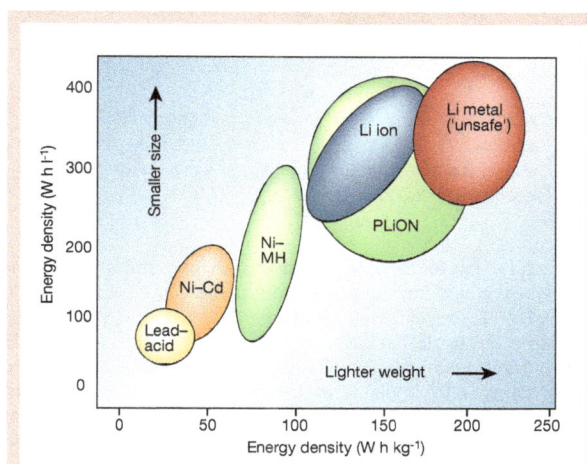

Figure 1 Comparison of the different battery technologies in terms of volumetric and gravimetric energy density. The share of worldwide sales for Ni–Cd, Ni–MeH and Li-ion portable batteries is 23, 14 and 63%, respectively. The use of Pb–acid batteries is restricted mainly to SLI (starting, lighting, ignition) in automobiles or standby applications, whereas Ni–Cd batteries remain the most suitable technologies for high-power applications (for example, power tools).

as well as the lightest (equivalent weight $M = 6.94$ g mol^{-1}, and specific gravity $\rho = 0.53$ g cm^{-3}) metal, thus facilitating the design of storage systems with high energy density. The advantage in using Li metal was first demonstrated in the 1970s with the assembly of primary (for example, non-rechargeable) Li cells[2]. Owing to their high capacity and variable discharge rate, they rapidly found applications as power sources for watches, calculators or for implantable medical devices. Over the same period, numerous inorganic compounds were shown to react with alkali metals in a reversible way. The discovery of such materials, which were later identified as intercalation compounds, was crucial in the development of high-energy rechargeable Li systems. Like most innovations, development of the technology resulted from several contributions. By 1972, the concept of electrochemical intercalation and its potential use were clearly defined[3,4], although the information was not widely disseminated, being reported only in conference proceedings. Before this time, solid-state chemists had been accumulating

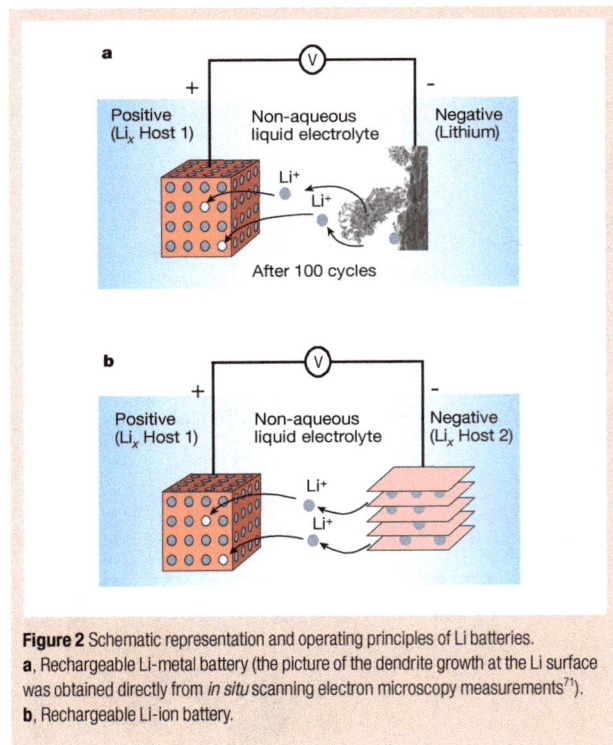

Figure 2 Schematic representation and operating principles of Li batteries. **a**, Rechargeable Li-metal battery (the picture of the dendrite growth at the Li surface was obtained directly from *in situ* scanning electron microscopy measurements[71]). **b**, Rechargeable Li-ion battery.

structural data on the inorganic layered chalcogenides[5,6], and merging between the two communities was immediate and fruitful.

In 1972, Exxon[7,8] embarked on a large project using TiS_2 as the positive electrode, Li metal as the negative electrode and lithium perchlorate in dioxolane as the electrolyte. TiS_2 was the best intercalation compound available at the time, having a very favourable layered-type structure. As the results were published in readily available literature, this work convinced a wider audience. But in spite of the impeccable operation of the positive electrode, the system was not viable. It soon encountered the shortcomings of a Li-metal/liquid electrolyte combination — uneven (dendritic) Li growth as the metal was replated during each subsequent discharge–recharge cycle (Fig. 2a), which led to explosion hazards. Substituting Li metal for an alloy with Al solved the dendrite problem[9] but, as discussed later, alloy electrodes survived only a limited number of cycles owing to extreme changes in volume during operation. In the meantime, significant advances in intercalation materials had occurred with the realization at Bell Labs that oxides, besides their early interest for the heavier chalcogenides[10,11], were giving higher capacities and voltages. Moreover, the previously held belief that only low-dimensional materials

could give sufficient ion diffusion disappeared as a framework structure (V_6O_{13}) proved to function perfectly[12]. Later, Goodenough, with Li_xMO_2 (where M is Co, Ni or Mn)[13,14], would propose the families of compounds that are still used almost exclusively in today's batteries.

To circumvent the safety issues surrounding the use of Li metal, several alternative approaches were pursued in which either the electrolyte or the negative electrode was modified. The first approach[15] involved substituting metallic Li for a second insertion material (Fig. 2b). The concept was first demonstrated in the laboratory by Murphy *et al.*[16] and then by Scrosati *et al.*[17] and led, at the end of the 1980s and early 1990s, to the so-called Li-ion or rocking-chair technology. The principle of rocking-chair batteries had been used previously in Ni–MeH batteries[18,19]. Because of the presence of Li in its ionic rather than metallic state, Li-ion cells solve the dendrite problem and are, in principle, inherently safer than Li-metal cells. To compensate for the increase in potential of the negative electrode, high-potential insertion compounds are needed for the positive electrode, and emphasis shifted from the layered-type transition-metal disulphides to layered- or three-dimensional-type transition-metal oxides[13]. Metal oxides are more oxidizing than disulphides (for example, they have a higher insertion potential) owing to the more pronounced ionic character of 'M–O' bonds compared with 'M–S' bonds. Nevertheless, it took almost ten years to implement the Li-ion concept. Delays were attributed to the lack of suitable materials for the negative electrode (either Li alloys or insertion compounds) and the failure of electrolytes to meet — besides safety measures — the costs and performance requirements for a battery technology to succeed. Finally, capitalizing on earlier findings[20,21], the discovery of the highly reversible, low-voltage Li intercalation–deintercalation process in carbonaceous material[22] (providing that carefully selected electrolytes are used), led to the creation of the $C/LiCoO_2$ rocking-chair cell commercialized by Sony Corporation in June 1991 (ref. 23). This type of Li-ion cell, having a potential exceeding 3.6 V (three times that of alkaline systems) and gravimetric energy densities as high as 120–150 W h kg^{-1} (two to three times those of usual Ni–Cd batteries), is found in most of today's high-performance portable electronic devices.

The second approach[24] involved replacing the liquid electrolyte by a dry polymer electrolyte (Fig. 3a), leading to the so-called Li solid polymer electrolyte (Li-SPE) batteries. But this technology is restricted to large systems (electric traction or backup power) and not to portable devices, as it requires temperatures up to 80 °C. Shortly after this, several groups tried to develop a Li hybrid polymer electrolyte (Li-HPE) battery[25], hoping to benefit from the advantages of polymer electrolyte technology without the hazards associated with the use of Li metal. 'Hybrid' meant that the electrolyte included three components: a polymer matrix (Fig. 3b) swollen with liquid solvent and a salt. Companies such as Valence and Danionics were involved in developing these polymer batteries, but HPE systems never materialized at the industrial scale because Li-metal dendrites were still a safety issue.

With the aim of combining the recent commercial success enjoyed by liquid Li-ion batteries with the manufacturing advantages

Figure 3 Schematic representations of polymer electrolyte networks. **a**, Pure (dry) polymer consisting of entangled chains, through which the Li ions (red points) move assisted by the motion of polymer chains. **b**, A hybrid (gel) network consisting of a semicrystalline polymer, whose amorphous regions are swollen in a liquid electrolyte, while the crystalline regions enhance the mechanical stability. **c**, A poly-olefin membrane (Celgard for instance) in which the liquid electrolyte is held by capillaries.

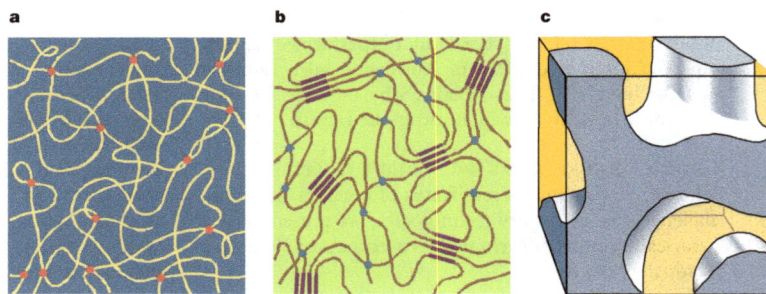

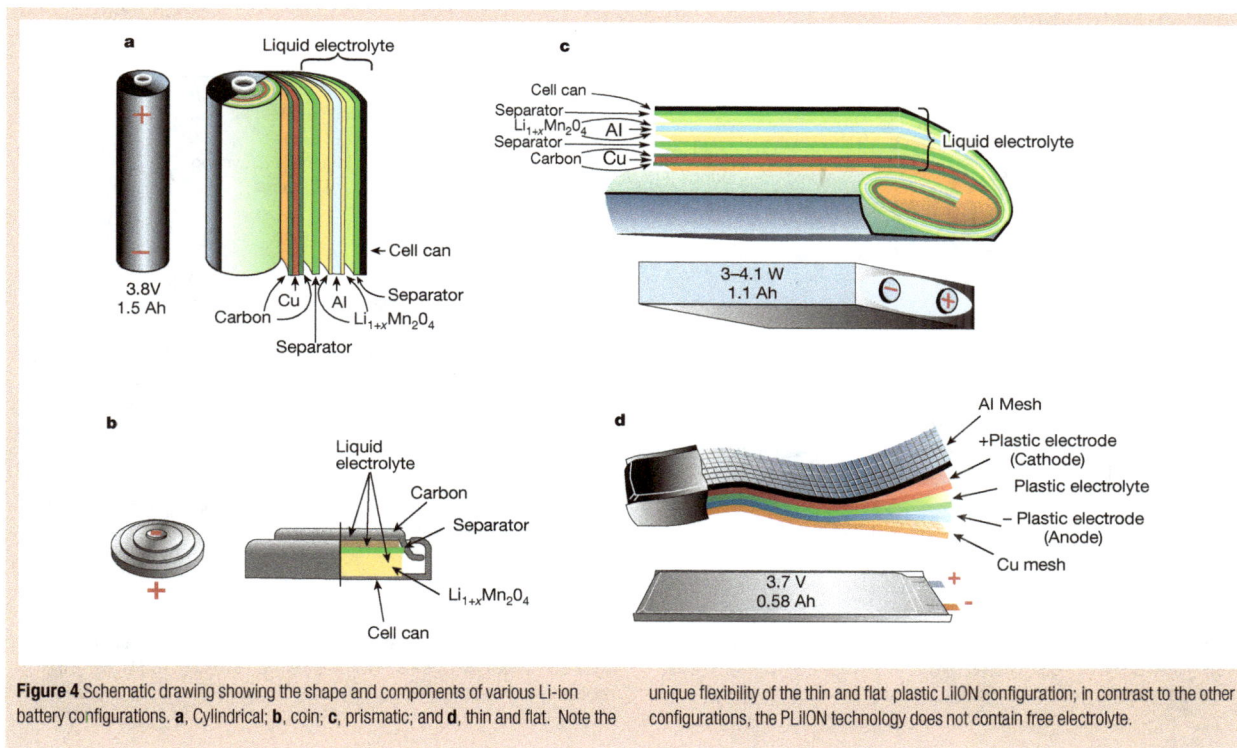

Figure 4 Schematic drawing showing the shape and components of various Li-ion battery configurations. **a**, Cylindrical; **b**, coin; **c**, prismatic; and **d**, thin and flat. Note the unique flexibility of the thin and flat plastic LiION configuration; in contrast to the other configurations, the PLiION technology does not contain free electrolyte.

presented by the polymer technology, Bellcore researchers introduced polymeric electrolytes in a liquid Li-ion system[26]. They developed the first reliable and practical rechargeable Li-ion HPE battery, called plastic Li ion (PLiON), which differs considerably from the usual coin-, cylindrical- or prismatic-type cell configurations (Fig. 4). Such a thin-film battery technology, which offers shape versatility, flexibility and lightness, has been developed commercially since 1999, and has many potential advantages in the continuing trend towards electronic miniaturization. Finally, the 'next generation' of bonded liquid-electrolyte Li-ion cells, derived from the plastic Li-ion concept, are beginning to enter the market place. Confusingly called Li-ion polymer batteries, these new cells use a gel-coated, microporous poly-olefin separator bonded to the electrodes (also gel-laden), rather than the P(VDF-HFP)-based membrane (that is, a copolymer of vinylidene difluoride with hexafluoropropylene) used in the plastic Li-ion cells.

Having retraced almost 30 years of scientific venture leading to the development of the rechargeable Li-ion battery, we now describe some of the significant issues and opportunities provided by the field by highlighting the various areas in need of technological advances.

Present status and remaining challenges

Whatever the considered battery technology, measures of its performance (for example, cell potential, capacity or energy density) are related to the intrinsic property of the materials that form the positive and negative electrodes. The cycle-life and lifetime are dependent on the nature of the interfaces between the electrodes and electrolyte, whereas safety is a function of the stability of the electrode materials and interfaces. Compared with mature batteries technologies, such as lead–acid or Ni–Cd, rechargeable Li-based battery technologies are still in their infancy, leaving much hope for improvement over the next decade. Such improvements should arise from changes in battery chemistry and cell engineering. Advances in active chemistry are left to the solid-state chemists' creativity and innovation in the design and elaboration of new intercalation electrodes. At the same time, they must bear in mind that it is impossible to predict the demands

that might be placed on tomorrow's portable devices, which in turn places different requirements on the active material chemistry. For instance, with respect to the lower operating voltages of emerging electronics, much debate has focused on whether we should develop a low-voltage active chemistry or rely entirely on electronics (d.c.–d.c. converters) and persist in searching for high-voltage active Li chemistry. Finding the best-performing combination of electrode–electrolyte–electrode can be achieved only through the selective use of existing and new materials as negative and positive electrodes, and of the right electrolyte combination, so as to minimize detrimental reactions associated with the electrode–electrolyte interface — the critical phase of any electrochemical system.

Materials for positive electrodes

The choice of the positive electrode depends on whether we are dealing with rechargeable Li-metal or Li-ion batteries (Fig. 5)[27]. For rechargeable Li batteries, owing to the use of metallic Li as the negative electrode, the positive electrode does not need to be lithiated before cell assembly. In contrast, for Li-ion batteries, because the carbon negative electrode is empty (no Li), the positive one must act as a source of Li, thus requiring use of air-stable Li-based intercalation compounds to facilitate the cell assembly. Although rechargeable Li-SPE cells mainly use Li-free V_2O_5 or its derivatives as the positive electrode, $LiCoO_2$ is most widely used in commercial Li-ion batteries, deintercalating and intercalating Li around 4 V.

Initially, the use of layered $LiNiO_2$ was considered[28], as this displayed favourable specific capacity compared with $LiCoO_2$. But expectations were dismissed for safety reasons after exothermic oxidation of the organic electrolyte with the collapsing delithiated Li_xNiO_2 structure. Delithiated Li_xCoO_2 was found to be more thermally stable than its Li_xNiO_2 counterpart. Thus, substitution of Co for Ni in $LiNi_{1-x}Co_xO_2$ was adopted to provide a partial solution to the safety concerns surrounding $LiNiO_2$.

Although the reversible delithiation of $LiCoO_2$ beyond 0.5 Li is feasible, delithiation for commercial applications has been limited to that value for safety reasons (charged cut-off limited to around

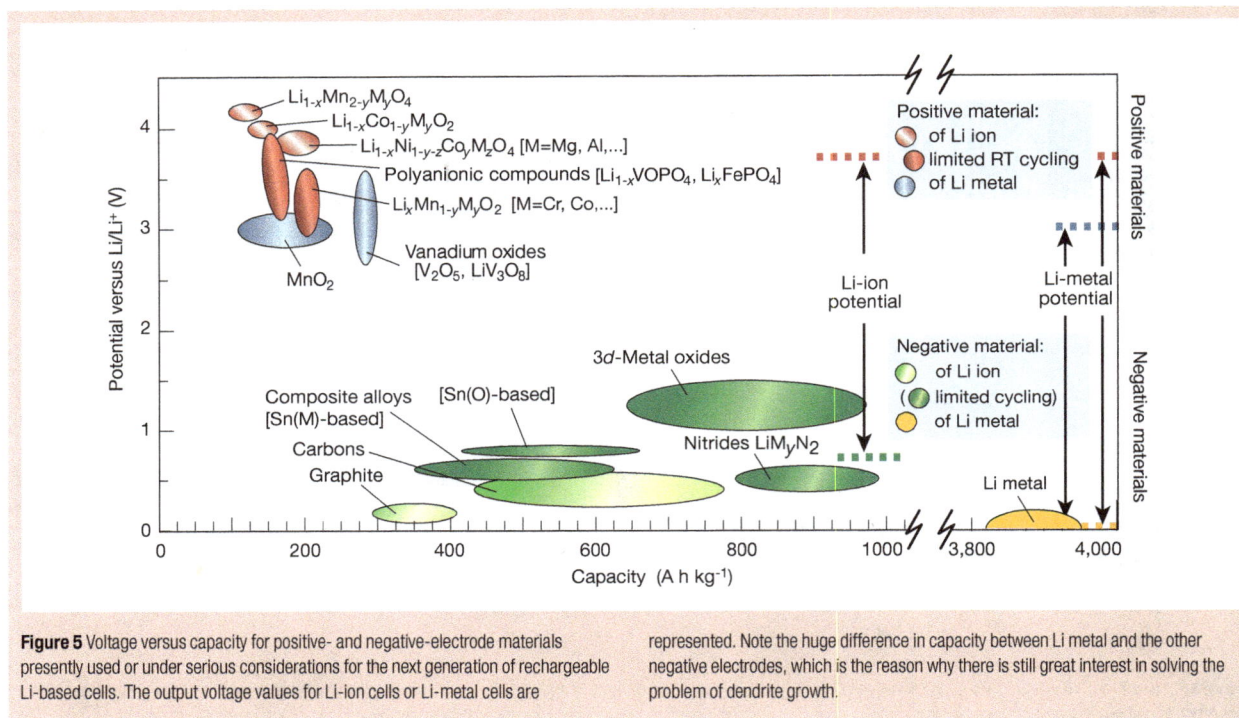

Figure 5 Voltage versus capacity for positive- and negative-electrode materials presently used or under serious considerations for the next generation of rechargeable Li-based cells. The output voltage values for Li-ion cells or Li-metal cells are represented. Note the huge difference in capacity between Li metal and the other negative electrodes, which is the reason why there is still great interest in solving the problem of dendrite growth.

4.2 V). Several routes were investigated to circumvent these safety and capacity issues. Among them was the successful stabilization of the layered structural framework by an electrochemically inert di-, tri- or tetravalent cationic substitute for Ni or Co (Al, Ga, Mg or Ti). This led to $LiNi_{1-x}Ti_{x/2}Mg_{x/2}O_2$ phases[29], which were claimed to be safe and which displayed practical capacities of $180 \, mA \, h \, g^{-1}$ compared to only $140 \, mA \, h \, g^{-1}$ for $LiCoO_2$. Another line of investigation involved the synthesis by *chimie douce* ('soft chemistry') of the layered $LiFeO_2$ and $LiMnO_2$ phases to take advantage of the Fe^{4+}/Fe^{3+} and Mn^{4+}/Mn^{3+} redox couples, respectively. In spite of the numerous and diverse synthesis methods, attempts to prepare electrochemically attractive $LiFeO_2$ phases failed. In contrast, research on $LiMnO_2$ has been more fruitful[30], and the structural instability of the layered phase reversing to the spinel $Li_xMn_2O_4$ upon cycling has recently been diminished through cationic substitution by chromium $(Li_{1+x}Mn_{0.5}Cr_{0.5}O_2)$[31]. These materials exhibit a capacity of $190 \, mA \, h \, g^{-1}$ (larger than that expected from the full oxidation of Mn^{3+} to Mn^{4+}) with little capacity fading upon cycling. It seems that within these materials, the role of Mn is to stabilize the layered structure of the chromium oxide, and that the large capacity is nested in the Cr oxidation state that changes reversibly from +3 to +6. It is therefore unfortunate that Cr presents major toxicity and pricing issues.

The spinel $LiMn_2O_4$, although possessing ≈10% less capacity than $LiCoO_2$, has an advantage in terms of cost and is perceived as being 'green' (that is, non-toxic and from abundant material source). Additionally, it has long been recognized as a potential alternative cathode[14]. Its implementation has been delayed because of limited cycling and storage performances at elevated temperatures, although these hurdles were overcome recently by synthesizing dually substituted $LiMn_{2-x}Al_xO_{4-y}F_y$ spinel phases[32], and by altering their surface chemistry[33].

In the search for improved materials for positive electrodes, it has been recognized recently that NaSICON (a family of Na super-ionic conductors) or olivine (magnesium iron silicate) oxyanion scaffolded structures (Fig. 6), built from corner-sharing MO_6 octahedra (where M is Fe, Ti, V or Nb) and XO_4^{n-} tetrahedral anions (where X is S, P, As, Mo or W), offer interesting possibilities[34]. Polyoxyanionic structures possess M–O–X bonds; altering the nature of X will change (through an inductive effect) the iono-covalent character of the M–O bonding. In this way it is possible to systematically map and tune transition-metal redox potentials. For instance, with the use of the phosphate polyanions PO_4^{3-}, the Fe^{3+}/Fe^{2+} and V^{4+}/V^{3+} redox couples lie at higher potentials than in the oxide form. One of the main drawbacks with using these materials is their poor electronic conductivity, and this limitation had to be overcome[35,36] through various materials processing approaches, including the use of carbon coatings, mechanical grinding or mixing, and low-temperature synthesis routes to obtain tailored particles. $LiFePO_4$, for example, can presently be used at 90% of its theoretical capacity ($165 \, mA \, h \, g^{-1}$) with decent rate capabilities, and thus is a serious candidate for the next generation of Li-ion cells (Fig. 7). As expected in the light of these promising results, polyoxyanionic-type structures having XO_4^{n-} entities (where X is Si, Ge) are now receiving renewed attention with respect to their electrochemical performance as electrode materials.

Although numerous classes of insertion–deinsertion materials were synthesized over the past 20 years, no real gain in capacity was achieved. One possible way to achieve higher capacities is to design materials in which the metal-redox oxidation state can change reversibly by two units (M^{n+2}/M^n), while preserving the framework structure, and having molecular masses similar to those of the presently used $3d$ metal-layered oxides (for example, $LiCoO_2$). Such an approach is feasible with W-, Mo- or Nb-based metal oxides[37], but there is no overall gain in specific energy with these heavier elements. Inserting more than one Li ion per transition metal is also feasible with a few V-based oxides (V^{5+} is reduced to an average state of 3.5 in 'ω-$Li_3V_2O_5$' (ref. 38) or to 3.67 in $Li_5V_3O_8$). In principle, except for coordination number requirements, there is no obvious reason why this should not happen with other early transition metals and, in this respect, the recent finding of the reversible Cr^{6+}/Cr^{3+} redox couple in a $3d$ metal-layered compound provides encouragement.

Tuning the morphology or texture of the electrode material to obtain porous and high-surface-area composite electrodes constitutes another exciting, although less exploited, route to enhance electrode capacities[39]. Indeed, V_2O_5 aerogels, which are

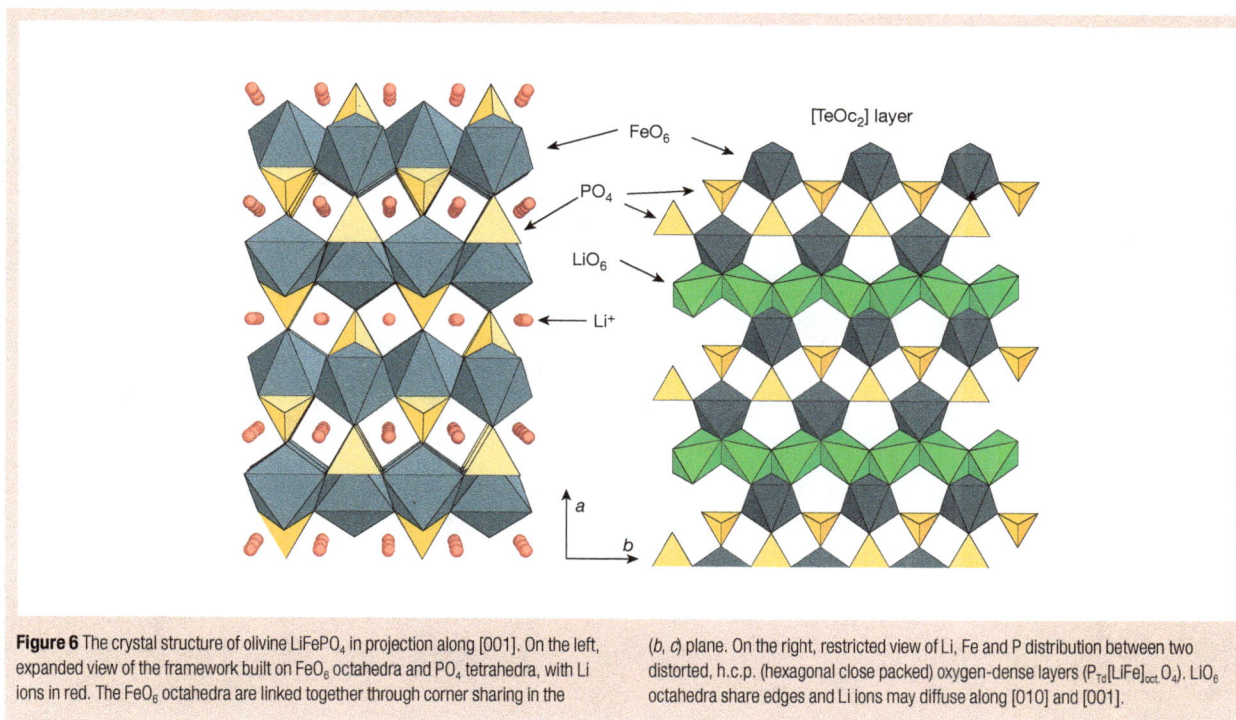

Figure 6 The crystal structure of olivine LiFePO$_4$ in projection along [001]. On the left, expanded view of the framework built on FeO$_6$ octahedra and PO$_4$ tetrahedra, with Li ions in red. The FeO$_6$ octahedra are linked together through corner sharing in the (b, c) plane. On the right, restricted view of Li, Fe and P distribution between two distorted, h.c.p. (hexagonal close packed) oxygen-dense layers (P$_{Td}$[LiFe]$_{oct}$O$_4$). LiO$_6$ octahedra share edges and Li ions may diffuse along [010] and [001].

mesoporous materials in which nanometre-sized domains are networked through a continuous, highly porous volume of free space, were reported recently to have electroactive capacities up to 100% greater than polycrystalline non-porous V$_2$O$_5$ powders and superior power rate capabilities[40] compared to usual V$_2$O$_5$ powders. Such extra capacity apparently derives from the onset of a pure capacitance, associated with the large surface area and high-porosity aerogel matrix, which adds to the existing faradic component. Conductive oxide aerogels such as V$_2$O$_5$ and MnO$_2$ therefore have the potential to boost the field of energy storage once the capacity penalty (in terms of W h l^{-1}) attributable to their poor tap density ($\simeq$0.2 g cm^{-3}) is overcome. Tailor-made nanostructured materials, such as aerogels, create new opportunities not only at the applied level, but also at the fundamental level where some elementary questions, such as the exact mechanism governing these large capacities, remain unanswered.

A radically different approach[41] takes advantage of the facile and reversible redox cleavage of the sulphur–sulphur bond to give lithium thiolate: –SS– + 2Li$^+$ + 2e$^-$ $\Leftrightarrow$ –SLi + LiS–. Depending on the electron, the voltage withdrawing power of the moieties attached to the sulphur can be up to 3 V (sulphur itself works at 2.4 V). Although promising in principle in terms of capacity and cost, these systems presently suffer from the relative low density of the reactants and solubility of the resulting thiolates in the electrolyte, leading to self discharge.

Materials for negative electrodes

As a result of numerous chemical (pyrolitic processing) or physical (mechanical milling) modifications, carbon negative electrodes[42] display electrochemical performances that are improving continuously. Reversible capacities of around 450 mA h g^{-1} are now being reached, compared with a practical value of 350 mA h g^{-1} for graphite (372 mA h g^{-1} for the end compound LiC$_6$). In parallel, ongoing research efforts are focused on searching for carbon alternatives in the hope of finding materials (Fig. 4) with both larger capacities and slightly more positive intercalation voltages compared to Li/Li$^+$, so as to minimize any risks of high-surface-area Li plating at the end of fast recharge, which are associated with safety problems. Such an effort resulted in the emergence of Li transition-metal nitrides as a new potential class of anode materials[43], owing to the large, stable and reversible capacity (600 mA h g^{-1}) displayed by one family member, Li$_{3-x}$Co$_x$N. This result triggered worldwide interest, although performances of the other newly reported Li-based nitrides unfortunately display inferior electrochemical performances compared to the Co phase. Furthermore, use of Li$_{3-x}$Co$_x$N is constrained by the restrictive manufacturing requirements for handling such moisture-sensitive negative electrodes.

Throughout the search for carbon alternatives, much effort has been devoted to the use of Li alloys. The first commercial cell was

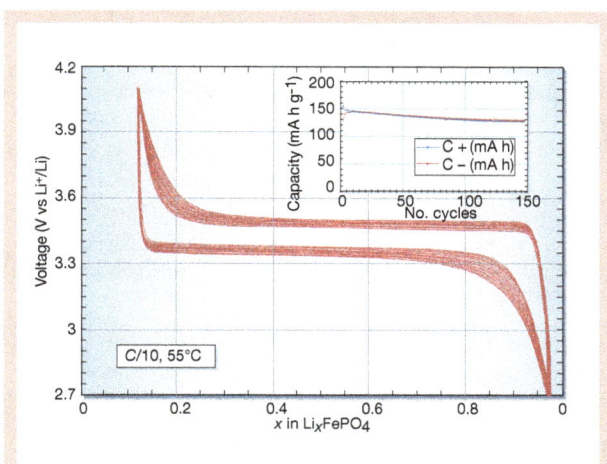

Figure 7 Cycling behaviour at 55 °C of an optimized LiFePO$_4$/C composite electrode (83% of active material) at a scan rate of C/10. Fine particles of LiFePO$_4$ were obtained from annealing at 500 °C a solid intimate mixture resulting from evaporation of an aqueous solution containing Li, Fe(III) and P precursors. The composite electrode was obtained from ball-milling LiFePO$_4$ with carbon SP. From ref. 36.

introduced in the 1980s by Matsushita; this was based on Wood's metal (a low-melting alloy of Bi, Pb, Sn and Cd), whose cycling performances were found to deteriorate with increased depth of discharge. While attractive in terms of gravimetric capacity, Li alloys suffer from cyclability issues resulting from large Li-driven volume swings (up to 200%), which cause disintegration and hence a loss of electrical contacts between particles. Although a reduction in alloy particle size clearly benefits the cyclability by increasing tolerance to stress cracking, so far the gains are not sufficient[44]. However, it became clear that any physical or chemical means of overcoming the problem of reactant expansion should be beneficial, hence the use of composite negative electrodes. The basis behind this concept is the use of a 'buffer matrix' to compensate for the expansion of the reactants, so preserving the electrical pathway[45]. Initially, such a buffer action was achieved by mixing two alloys that reacted at different potentials so that the electrochemically active phase was imbedded in a non-electrochemically active matrix.

A similar approach held considerable promise in 1997, when Fuji announced the commercialization of a new Li-ion technology (STALION) using an amorphous tin composite oxide (ATCO) as negative electrode. This reacts reversibly with Li at about 0.5 V, and has a specific capacity twice that of graphite[46]. *In situ* X-ray diffraction studies of the ATCO electrode led to the conclusion that the Li reactivity mechanism in these composites was based on oxide decomposition by Li through an initial irreversible process to form intimately mixed Li_2O and metallic Sn, followed by a Li alloying reaction to form nanodomains of $Li_{4.4}Sn$ embedded within the Li_2O matrix[47]. However, the STALION cell was never commercialized, in spite of its announcement at the end of 1998. This was most likely due to poor long-term cyclability, the huge and irreversible capacity loss during the first cycle, which was reported by many groups, and the necessity of finding a convenient source for the two initial Li ions needed for the SnO reduction process.

Besides ATCO, other investigations, such as those pursued by Dahn *et al.*[48] on the –Sn–Fe–C system, have also revealed an appealing low-voltage reversible reactivity in composite materials developed as negative electrodes (in spite of initial irreversibility and short-lived capacities). The best experimental proof of the beneficial aspect of the buffer matrix arises from the ability to obtain several hundred cycles on a composite made by precipitating Sn metal at the grain boundaries of electrochemically inactive $SnFe_3C$ grains[49]. However, the cycling performance was improved at the expense of the electrode electrochemical capacity.

A new approach to alleviate the problems of alloy expansion, proposed by Thackeray *et al.*[50], involved selecting intermetallic alloys such as Cu_6Sn_5, InSb and Cu_2Sb that show a strong structural relationship to their lithiated products, Li_2CuSn and Li_3Sb for the Sn and Sb compounds, respectively. InSb and Cu_2Sb electrodes are particularly attractive candidates because they operate through a reversible process of lithium insertion and metal extrusion, with an invariant face-centred-cubic Sb array (that is, this array provides a stable host framework for both the incoming and extruded metal atoms). In the ternary $Li_xIn_{1-y}Sb$ system ($0 < x < 3$, $0 < y < 1$), the Sb array expands and contracts isotropically by only 4%, whereas the overall expansion of the electrode is 46% if the extruded In is taken into account. This expansion is considerably more favourable than the expansion of binary systems such as LiAl, which expand by ~200% during the phase transition of Al to LiAl. InSb and Cu_2Sb electrodes provide reversible capacities between 250 and 300 mA h g^{-1}. Despite the new and elegant concept behind the design of these intermetallic electrodes, they still suffer from poor cyclability, particularly upon the initial cycle; nevertheless, the approach deserves further study.

Based on the peculiar behaviour (that is, large capacity at low potential) of the transition-metal vanadates M–V–O, first proposed by Fuji Co.[51] and later studied by several groups, Poizot *et al.*[52] reinvestigated the reactivity of Li-metal oxide. Surprisingly, they found a Li electrochemical activity for well known oxides, but these

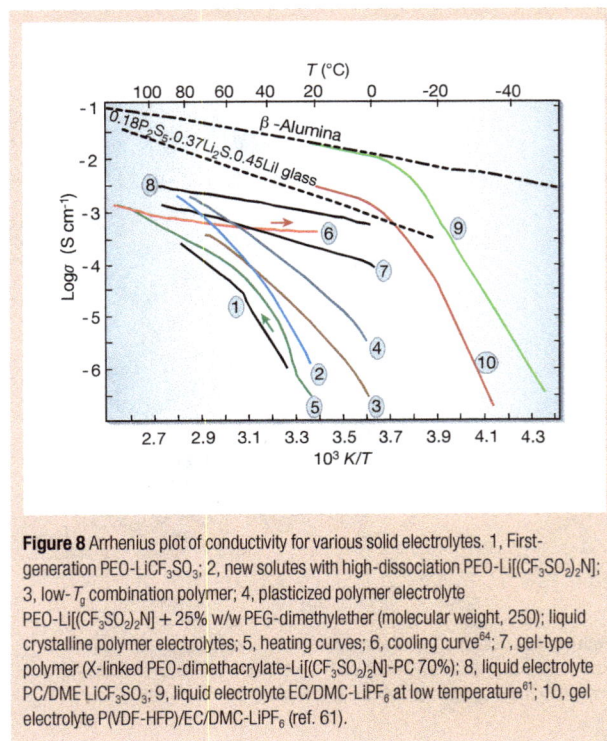

Figure 8 Arrhenius plot of conductivity for various solid electrolytes. 1, First-generation PEO-LiCF$_3$SO$_3$; 2, new solutes with high-dissociation PEO-Li[(CF$_3$SO$_2$)$_2$N]; 3, low-T_g combination polymer; 4, plasticized polymer electrolyte PEO-Li[(CF$_3$SO$_2$)$_2$N] + 25% w/w PEG-dimethylether (molecular weight, 250); liquid crystalline polymer electrolytes; 5, heating curves; 6, cooling curve[64]; 7, gel-type polymer (X-linked PEO-dimethacrylate-Li[(CF$_3$SO$_2$)$_2$N]-PC 70%); 8, liquid electrolyte PC/DME LiCF$_3$SO$_3$; 9, liquid electrolyte EC/DMC-LiPF$_6$ at low temperature[61]; 10, gel electrolyte P(VDF-HFP)/EC/DMC-LiPF$_6$ (ref. 61).

did not react with Li according to the classical processes of Li insertion–deinsertion or Li alloying, the catchwords of the past 20 years. For instance, MO-type compounds (where M is Co, Ni, Fe, Cu or Mn), having a rocksalt structure and containing metal elements (M) that do not alloy with Li, exhibited capacities two to three times those of carbon with 100% capacity retention for up to 100 cycles. The mechanism of Li reactivity in such materials differs from the classical processes, and is nested in the electrochemically driven, *in situ* formation of metal nanoparticles during the first discharge, which enables the formation and decomposition of Li_2O upon subsequent cycling[53]. Remaining issues relate to a problem of surface, with the chemical reactivity being enhanced as the particle size becomes smaller. These findings open new avenues of research aimed at capitalizing on the beneficial effect that particle-size confinement could have within the field of electrochemistry. These and related nanocluster systems under development hold much promise for future developments.

Polymer and liquid electrolytes

Besides the electrodes, the electrolyte, which commonly refers to a solution comprising the salts and solvents, constitutes the third key component of a battery. Although the role of electrolyte is often considered trivial, its choice is actually crucial, and is based on criteria that differ depending on whether we are dealing with polymer or liquid-based Li-ion rechargeable batteries[54]. For instance, working with a highly oxidizing (>4V versus Li/Li$^+$) positive electrode material for Li-ion batteries requires electrolyte combinations that operate well outside their window of thermodynamic stability (3.5 V). This is the reason why early workers in the field ignored very positive cathode materials. But fortunately this electrolyte stability is kinetically controlled, enabling the use of non-aqueous electrolytes at potentials as high as 5.5 V (ref. 55). Similarly, the use of a polymer rather than a liquid electrolyte adds further selection criteria linked to the electrochemical stability of the polymer[56]. There are numerous liquid solvents available, each with different dielectric constants and viscosity, and we can select specific solvents to favour the ionic conductivity of the electrolyte. In contrast, there are only a few Li-based salts or polymers to choose from, the most commonly used ones being based on polyethylene oxide (PEO). The results from research efforts aimed at counterbalancing this deficit have led to the present level of research and development on electrolytes.

Guided by general concepts of viscosity and dielectric constants, optimizing the ionic conductivity of a liquid electrolyte almost becomes a field-trial approach with the hope of finding the key ingredients. For instance, only ethylene carbonate can provide the *ad hoc* protective layer on the surface of graphite that prevents further reaction (continuous electrolyte reduction and self-discharge). Ethylene carbonate is therefore present in almost all commercial compositions, thinned with other solvents owing to its high melting point. Why the homologous propylene carbonate is unsuitable for this protective layer remains an open question, reminding us that chemistry has its secrets.

In contrast, achieving high ionic conductivity in Li-based polymer electrolytes requires a better understanding of the fundamentals of ion dissociation and transport. Both the nature of the polymer–salt interaction and the precise structure of highly concentrated electrolyte solutions have always resisted rationalization. Nevertheless, a principal goal has been to search for new, highly conductive salts with a large electrochemical window, which form a eutectic composition with PEO that melts at the lowest possible temperature[57]. The concept of non-coordinating anions with extensive charge delocalization was achieved with the perfluorosulphonimide $Li^+[CF_3SO_2NSO_2CF_3]^-$ salt (abbreviated as LiTFSI)[58]. Figure 8 shows the marked improvement when passing, with simple PEO, from a 'conventional' $LiCF_3SO_3$ salt (curve 1) to the imide salt (curve 2), where an order of magnitude is gained, not ignoring the larger elastomeric domain towards low temperature. The polymer architecture has a role independently of dissociation. Attaching the side chains of the solvating group to the polymer increases the degrees of freedom as a result of dangling chain ends; this improves conductivity (Fig. 8, curve 3), but compromises the mechanical properties.

Although efforts aimed at enhancing the ionic conductivity of polymer electrolytes have been insufficient to permit operation at room temperature, they have benefited liquid-based electrolyte systems in terms of cost and safety, so that battery manufacturers of Li-ion cells are eager to see the further development of organic anion-based salts able to operate at voltages greater than 4.5 V. LiTFSI is an example of this cross-fertilization. Although extremely resistant to oxidation itself, the electrochemical use of such a salt is limited to 4 V in presence of an Al collector, because a stable and soluble Al salt can be formed as a consequence of the robustness of the anion bonds. With the less stable coordination anions ($LiPF_6$), decomposition occurs immediately and is accompanied by formation of protective AlF_3. However, owing to its high conductivity in any medium, its safety and lack of toxicity, LiTFSI is being used increasingly in Li-ion batteries, the corrosion problem having being solved by simple addition of a passivating coordination-type salt. A wide range of anion-forming systems now exists, especially in the imide family, and these are viewed as candidates for high conductivity and Al passivation.

Having exploited most of the possibilities offered by 'dry' polymers to improve conductivity (ability, amorphous state and lowest possible glass-transition temperature T_g controlling the ion mobility), a remaining option was to use additives, known in polymer science as plasticizers, to act as chain lubricants, so leading to the development of 'hybrid' polymer electrolytes[59]. Indeed, suitable plasticizers are chosen between the same polar solvents as for liquid electrolytes[60], such as propylene carbonate, γ-butyrolactone or polyethylene glycol ethers, or are formed from short-chain PEO (4–25 monomer units). A lightly plasticized material (10–25% additive) improves conductivity by an order of magnitude (Fig. 8, curve 4). Gels, on the other hand, contain 60–95% liquid electrolyte, and are only 2–5 times less conductive than their liquid counterpart[61]

(Fig. 8, curves 7–10). Interestingly, when the gelling agent is a polyether, most of the solvation still takes place through the polymer chains rather than the carbonate solvents, the latter being less prone to donate electron pairs. Understandably, the lightly plasticized systems can be used in a Li-metal configuration, as much of the resilience of the pristine polymer is retained, whereas the much softer gels require a Li-ion configuration.

It is surprising that in spite of the direct link between their ionic conductivity and their degree of amorphicity, very little is known about the structural chemistry of polymer electrolytes. In contrast to the well established dynamic view of ionic conductivity on these materials, Bruce *et al.*[62] recently proposed a structural view, highlighting the importance of aligning or organizing the polymer chains in order to enhance the levels of ionic conductivity. Similarly, Wright and co-workers[63] and Ingram[64] focused on the liquid crystalline state to force the solvating polymer into a conformation that was dictated by the liquid crystal part. The result is a partial decoupling of the conductivity from the glass-transition temperature of the polymer. The conductivity of such liquid crystalline chain polymers is low at room temperature, but reaches liquid-like values at high temperature or when kept under polarization, and remains so upon cooling to room temperature (Fig. 8, curves 5 and 6), without appreciable activation energy[64]. As these new perspectives generate renewed interest in the design of polymer electrolytes, it is hoped that solutions may eventually be found to the problems of ionic conductivity afflicting this class of materials at ambient or subambient temperature.

The addition of nanoparticle fillers (10% w/w), such as Al_2O_3 or TiO_2, to simple PEO compounds increases the conductivity several-fold at 60–80 °C, and prevents crystallization for at least several weeks at room temperature[65]. Two important advantages of these systems are an increase in the apparent Li transport number, from a low of ≈0.3 (common to polymer, liquid and gels) to ≈0.6, and the formation of a stable, low-resistance interface in contact with Li. Because these materials obey different conduction mechanisms, they are presently the focus of many studies, both practical and theoretical[66].

Technologies based on either solid polymer or 'hybrid' polymer electrolytes offer great advantages that will be necessary to meet the flexible, shape-effective requirements dictated by today's electronic miniaturization, while at the same time providing a larger autonomy. Current Li technologies rely on liquid-jellyroll or prismatic-cell configurations. Neither fits well in a multiple-cell configuration. This is in marked contrast with the recent thin, plate-like plastic (PLiON) technology that enables excellent packing efficiency, as multiple plates can be densely packaged in parallel within one cell while preserving the flexibility of the overall package. Future technology improvements should focus on better chemical engineering of the bonded laminates, so as to obtain even thinner cells. Similar attributes can be provided by the solid Li-polymer technology, which in addition exhibits extra capacity and is free of electrolyte leakage. This currently operates at 80 °C. Although warm temperatures may be an advantage for the large batteries required by the transportation sector, problems of conductivity have to be solved for electronic applications, as emphasized earlier.

The electrode–electrolyte interface

The Li-ion cell density can be improved through a selective use of appropriate existing or new materials for negative and positive electrodes. However, optimizing an electrode material is only the first step in the process leading to its implementation in a practical cell. Indeed, while the capacity of a cell is nested in the structural or electronic behaviour of its electrode, poor cell lifetimes are rooted mainly in side reactions occurring at the electrode–electrolyte interface. Thus, mastering the chemical stability of any new electrode material with respect to its operating liquid or polymer electrolyte medium, which requires a control of the electrode–electrolyte interface through surface chemistry, is as important as designing new materials. Tackling interfacial issues is both tedious and complex. We should remember that, despite many years of research devoted to the mechanism by which the solid electrolyte interphase forms on Li or carbonaceous materials, its composition and nature are still the subject of much controversy. In contrast, the positive electrode interface has received little attention over the years, despite its equally crucial role. Its importance is amplified with the Li-ion technology, where high voltages exceed the electrochemical resistance of the electrolyte oxidation, and even favour its catalytically driven decomposition. Thus, it is critical to control the electrode surface so as to modify its catalytic activity towards electrolyte decomposition. The strategy developed to address this issue uses coatings that encapsulate, through chemical or physical means, the electrode grains with either an inorganic or an organic phase. This concept, successfully applied to the spinel $LiMn_2O_4$, is based on minimizing the surface area of the active material in direct contact with the electrolyte[33]. The coating must allow easy diffusion of Li ions and, although insulating in nature, must be thin enough to allow the electrons to tunnel through. Equally relevant is the unexplained role of filler additives in polymer electrolytes[65], which markedly reduce the interfacial impedance in contact with Li.

Thirty years after its initial observation, the key issue of Li dendrite growth, which was thought to be governed mainly by current densities, remains highly topical, especially in light of recent promising results obtained by Aurbach's and Bates' groups. Revisiting Exxon's solvent 1-3-dioxolane, Aurbach and co-workers[67] showed that the use of $LiAsF_6$ salt led to a completely different Li morphology from that obtained from an ethylene carbonate–dimethyl carbonate (EC–DMC) electrolyte. They explained this in terms of the reactivity of dioxolane with lithium, which forms an elastomeric coating endowing the Li surface with plasticity and flexibility, thereby reducing dendrite growth. These findings were implemented in Li/MnO_2 commercial Tadiran cells that, under well defined cycling conditions, are claimed to be safe. The bulk polymerization of the cyclic ether, initiated at the positive electrode on overcharge, acts as a thermal shutdown. Even more spectacular are recent reports by Bates et al.[68] who succeeded in cycling $LiCoO_2$/Li thin-film batteries for more than 50,000 cycles using a glassy electrolyte in ≈ 1-μm-thick films obtained by sputtering techniques. By controlling the uniform Li stripping–plating mechanism, the same authors demonstrated the feasibility of a Li-free, rechargeable, thin-film battery — that is, cells constructed in the discharged state with no Li metal initially present[69]. Such findings, whether resulting from low-current density or the use of solid electrolyte, show that the problem of dendrite growth can be solved, at least with special cell configurations. Visco and co-workers[70] recently showed that a glassy nanometric layer deposited on Li metal completely insulates it from its environment, even in the presence of liquids, and that this coating can be applied at a high production rate. With further work devoted to the implementation of these findings to large-size Li batteries, the development of a Li-free rechargeable battery remains a realistic goal for the future.

The principal challenge for Li-based rechargeable batteries, or indeed for any battery, lies in gaining better understanding and control of the electrode–electrolyte interface in the hope of designing new solid–solid or solid–liquid interfaces. For example, the nature of the secondary reactions occurring at high temperature, which cause cell failure, remains an unanswered question that must be addressed to ensure the practical success of these technologies. In this case, however, the main difficulty stems from a lack of available techniques to probe the evolution of the electrode–electrolyte interface at a local level. We have so far relied (with the exception of X-ray diffraction) on post-mortem rather than in situ studies to determine how the electrodes or interfaces age with time either under cycling or storage conditions, thereby missing key information. But introduction of the plastic Li-ion-type technology has created new opportunities to perform a wide variety of in situ characterization techniques. These include X-ray absorption near-edge structure, nuclear magnetic resonance and Mössbauer spectroscopies, or even scanning electron microscopy observations that allow real-time visualization of dendrite growth at an interface[71]. Efforts aimed at developing new characterization tools must be vigorously pursued so as to create a comprehensive database on the electrode–electrolyte interface.

Conclusion

Consumers are in constant demand for thinner, lighter, space-effective and shape-flexible batteries with larger autonomy. Such demand will continue to generate much research activity towards the development of new cell configurations and new chemistries. In this review we hope to have conveyed the message that the field of energy storage is advancing faster than it perhaps has ever done in the past. The benefits, in terms of weight, size and design flexibility provided by today's state-of-the-art Li-ion configurations, which owe much to the design engineers' striving to develop efficient, economical microtechnologies, are a good illustration. The Li-based battery chemistry is relatively young, and as such is a source of aspirations as well as numerous exciting challenges. The latter are not limited to solid-state chemists. The effort should be highly multidisciplinary with strong roots in the fields of organic and inorganic chemistry, physics, surface science and corrosion. Through materials design we can expect significant improvements in energy density. And although designing new materials can be intuitive or based on chemical concepts, coupling these efforts with those of theorists who are able to perform band-structure calculations on envisioned compounds will prove to be highly beneficial. Of equal importance is a better understanding of the electrode–electrolyte interface to facilitate design of new interfaces. Here the goal is well defined, although we must diverge from the empirical approach used so far, and make full use of the recent progress achieved by in situ characterization. As Li-rechargeable batteries enter their teenage years, scientists and engineers predict an even brighter future lies ahead. □

1. Takeshita, H. Portable Li-ion,worldwide. Proc. Conf. Power 2000, San Diego, 25 September 2000.
2. Ikeda, H., Saito, T. & Tamura, H. in Proc. Manganese Dioxide Symp. Vol. 1 (eds Kozawa, A. & Brodd, R. H.) (IC sample Office, Cleveland, OH, 1975).
3. Steele, B. C. H. in Fast Ion Transport in Solids (ed. Van Gool, W.) 103–109 (North-Holland Amsterdam, 1973).
4. Armand, M. B. in Fast Ion Transport in Solids (ed. Van Gool, W.) 665–673 (North-Holland Amsterdam, 1973).
5. Rouxel, J., Danot, M., & Bichon, M. Les composites intercalaires Na_xTiS_2. Etude gènèrale des phases Na_xTiS_2 et K_xTiS_2. Bull. Soc. Chim. 11, 3930–3936 (1971).
6. Di Salvo, F. J., Schwall, R., Geballe, T. H., Gamble, F. R. & Osieki, J. H. Precursor effects of superconductivity up to 35°K in layered compounds. Phys. Rev. Lett. 27, 310–313 (1971).
7. Whittingham, M. S. Electrochemical energy storage and intercalation chemistry. Science 192, 1226 (1976).
8. Whittingham, M. S. Chalcogenide battery. US Patent 4009052.
9. Rao, B. M. L., Francis, R. W. & Christopher, H. A. Lithium-aluminium electrodes. J. Electrochem. Soc. 124, 1490–1492 (1977).
10. Broahead, J. & Butherus, A.D. Rechargeable non-aqueous battery. US Patent 3791867.
11. Broadhead, J., DiSalvo, F. J. & Trumbore, F. A. Non-aqueous battery using chalcogenide electrode. US Patent 3864167.
12. Murphy, D. W. & Christian, P. A. Solid state electrodes for high energy batteries. Science 205, 651–656 (1979).
13. Mizushima, K., Jones, P. C., Wiseman, P. J. & Goodenough, J. B. Li_xCoO_2 ($0<x\leqslant1$): a new cathode material for batteries of high energy density. Mat. Res. Bull. 15, 783–789 (1980).
14. Thackeray, M. M., David, W. I. F., Bruce, P. G. & Goodenough, J. B. Lithium insertion into manganese spinels. Mat. Res. Bull. 18, 461--472 (1983).
15. Armand, M. B. in Materials for Advanced Batteries (Proc. NATO Symp. Materials Adv. Batteries) (eds Murphy, D. W., Broadhead, J. & Steele, B. C. H.) 145–161 (Plenum, New York, 1980).
16. Murphy, D. W., DiSalvo, F. J., Carides, J. N. & Waszczak, J. V. Topochemical reactions of rutile related structures with lithium. Mat. Res. Bull. 13, 1395–1402 (1978).
17. Lazzari, M. & Scrosati, B. A cyclable lithium organic electrolyte cell based on two intercalation electrodes. J. Electrochem. Soc. 127, 773–774 (1980).
18. Will, F. G. Hermetically sealed secondary battery with lanthanum nickel anode. US patent 3874958 (1975).
19. Percheron-Guegan, A., Achard, J. C., Sarradin, J. & Bronoël, G. Alliages à base de Lanthane et de Nickel et leurs applications électrochimiques. French patent 7516160 (1975).
20. Guérard, D. & Hérold, A. New method for the preparation of lithium insertion compounds in graphite. C.R. Acad. Sci. C 275, 571–572 (1972).
21. Basu, S. Ambient temperature rechargeable battery. US patent 4,423,125 (filing date, 13 September 1982; publication date, 27 Dec 1983).
22. Mohri, M. et al. Rechargeable lithium battery based on pyrolytic carbon as a negative electrode. J. Power Sources 26, 545–551 (1989).
23. Nagaura, T. & Tozawa, K. Lithium ion rechargeable battery. Prog. Batteries Solar Cells 9, 209 (1990).

24. Armand, M., Chabagno, J. M. & Duclot, M. J. in *Fast Ion Transport in Solids Electrodes and Electrolytes* (eds Vashishta, P., Mundy, J.-N. & Shenoy, G. K.) 131–136 (North-Holland, Amsterdam, 1979).

25. Kelly, I. E., Owen, J. R. & Steel, B. H. Poly(ethyleneoxide) electrolytes for operation at near room temperature. *J. Power Sources* 14, 13–21 (1985); *Interfacial Electrochem.* 168, 467 (1984).

26. Tarascon, J.-M., Gozdz, A. S., Schmutz, C., Shokoohi, F. & Warren, P. C. Performance of Bellcore's plastic rechargeable Li-ion batteries. *Solid State Ionics* 86–88, 49–54 (1996).

27. Guyomard, D. in *New Trends in Electrochemical Technology: Energy Storage Systems for Electronics* Vol. 9 (eds Osaka, T. & Datta, M.) 253–350 (Gordon & Breach Science Publishers, 2000).

28. Dahn, J. R., Von Sacken, U., Juzkow, M. W. & Al-Janaby, H. Rechargeable $LiNiO_2$/carbon cells. *J. Electrochem. Soc.* 138, 2207–2211 (1991).

29. Yuan Gao, Yakovleva, M. V. & Ebner, W. B. Novel $LiNi_{1-x}Ti_{x/2}Mg_{x/2}O_2$ compounds as cathode materials for safer lithium-ion batteries. *Electrochem. Solid State Lett.* 1, 117–119 (1998).

30. Armstrong, A. R. & Bruce, P. G. Synthesis of layered $LiMnO_2$ as an electrode for rechargeable lithium batteries. *Nature* 381, 499–500 (1996).

31. Ammundsen, B. *et al.* in *Proc. Int. Symp. Electrochem. Soc.* Vol. 99-24, 57–67 (ECS, Pennington, NJ, 2000).

32. Amatucci, G. G., Pereira, N., Zheng, T. & Tarascon, J.-M. Failure mechanism and improvement of the elevated temperature cycling of $LiMn_2O_4$ compounds through the use of the $LiAl_xMn_{2-x}O_{4-y}F_z$ solid solution. *J. Electrochem. Soc.* 148, A171–A182 (2001).

33. Amatucci, G. G., du Pasquier, A., Blyr, A., Zheng, T. & Tarascon, J.-M. The elevated temperature performance of the $LiMn_2O_4$/C system: failure and solutions. *Electrochem. Acta* 45, 255–271 (1999).

34. Padhi, A. K., Nanjundaswamy, K. S., Masquelier, C., Okada S. & Goodenough, J. B. Effect of structure on the Fe^{3+}/Fe^{2+} redox couple in iron phosphates. *J. Electrochem. Soc.* 144, 1609–1613 (1997).

35. Ravet, N. *et al.* Improved iron-based cathode material. Abstr. No. 127, ECS Fall meeting, Hawaii, 1999.

36. Morcrette, M., Wurm, C., Gaubicher, J. & Masquelier, C. Polyanionic structures as alternative materials for lithium batteries. Abstr. No. 93, Electrode Materials Meeting, Bordeaux Arcachon, 27 May–1 June 2001.

37. Cava, R. J., Murphy, D. W. & Zahurak, S. M. Lithium insertion in Wadsley-Roth phases based on Niobium oxide. *J. Electrochem. Soc.* 30, 2345–2351 (1983).

38. Delmas, C., Brethes, S. & Menetrier, M. $Li_xV_2O_5$-ω, un nouveau matèriau d'électrode pour accumulateur au lithium. *CR Acad. Sci.* 310, 1425–1430 (1990).

39. Le, D. B. *et al.* High surface area V_2O_5 aerogel intercalation electrodes. *J. Electrochem. Soc.* 143, 2099–2104 (1996).

40. Dong, W., Rolison, D. R. & Dunn, B. Electrochemical properties of high surface area vanadium oxides aerogels. *Electrochem. Solid State Lett.* 3, 457–459 (2000).

41. Visco, S. J. & de Jonghe, L. C. in *Handbook of Solid-State Batteries and Capacitors* (ed. Munshi, M. Z. A.) 515 (World Scientific, Singapore, 1995).

42. Dahn, J. R. *et al.* Carbon and graphites as substitutes for the lithium anode. *Industrial Chemistry Library* Vol. 5 (ed. Pistoia, G.) (1994).

43. Shodai, T., Okada, S., Tobishima, S. & Yamabi, I. Study of $Li_{3-x}M_xN$ (M=Co, Ni or Cu) system for use as anode in lithium rechargeable cells. *Solid State Ionics* 86–88, 785–789 (1996).

44. Winter, M. & Besenhard, J. O. Electrochemical lithiation of tin and tin-based intermetallics and composites. *Electrochem. Acta* 45, 31–50 (1999).

45. Anani, A., Crouch-Baker, S. & Huggins, R. A. Kinetics and thermodynamic parameters of several binary alloys negative electrode materials at ambient temperature. *J. Electrochem. Soc.* 134, 3098–3102 (1987).

46. Idota, Y., Kubota, T., Matsufuji, A., Maekawa, Y. & Miyasaka, T. Tin-based amorphous oxide: a high capacity lithium-ion storage material. *Science* 276, 1395–1397 (1997).

47. Courtney, I. A. & Dahn, J. R. Electrochemical and in situ X-ray diffraction studies of the reaction of lithium with tin oxide composites. *J. Electrochem. Soc.* 144, 2045–2052 (1997).

48. Mao, O., Dunlap, R. A. & Dahn, J. R. Mechanically alloyed Sn-Fe(-C) powders as anode materials for Li-ion batteries. I. The Sn_2Fe-C system. *J. Electrochem. Soc.* 146, 405–413 (1999).

49. Beaulieu, L. Y., Larcher, B., Dunlap, R. A. & Dahn, J. R. Reaction of Li with grain-boundary atoms in nano structured compounds. *J. Electrochem. Soc.* 147, 3206–3212 (2000).

50. Kepler, K. D., Vaughey, J. T. & Thackeray, M. M. $Li_xCu_6Sn_5$ (0<x<13): an intermetallic insertion electrode for rechargeable lithium batteries. *Electrochem. Solid State Lett.* 2, 307 (1999).

51. Idota, Y. *et al.* Nonaqueous battery. US Patent No. 5,478,671 (1995).

52. Poizot, P., Laruelle, S., Grugeon, S., Dupont, L. & Tarascon., J.-M. Nano-sized transition-metal oxides as negative-electrode material for lithium-ion batteries. *Nature* 407, 496–499 (2000).

53. Poizot, P., Laruelle, S., Grugeon, S., Dupont, L. & Tarascon., J.-M. Electrochemical reactivity and reversibility of cobalt oxides towards lithium. *C.R. Acad. Sci. II* 681–691 (2000).

54. Dominey, L. A. Current state of the art on lithium battery electrolytes *Industrial Chemistry Library* Vol. 5 (ed. Pistoia, G.) 137–165 (1994).

55. Guyomard, D. & Tarascon, J. M. High voltage stable liquid electrolytes for $Li_{1+x}Mn_2O_4$/carbon rocking-chair lithium batteries. *J. Power Sources* 54, 92–98 (1995).

56. Armand, M. The history of polymer electrolytes. *Solid State Ionics* 69, 309–319 (1994).

57. Fenton, D. E., Parker, J. M. & Wright, P. V. Complexes of alkali metal ions with poly(ethylene oxide). *Polymer* 14, 589 (1973).

58. Armand, M., Gorecki, W. & Andreani, R. in *Second International Meeting on Polymer Electrolytes* (ed. Scrosati, B.) 91–96 (Elsevier, London, 1989).

59. Fauteux, D. in *Polymer Electrolytes Reviews* II (eds MacCallum, J. R. & Vincent, C. A.) 212 (Elsevier, London, 1989).

60. Feuillade, G. & Perche, P. Ion conductive macromodular gels and membranes for solid lithium cells. *J. Appl. Electrochem.* 5, 63–69 (1975).

61. Stallworth, P. E. *et al.* NMR, DSC and high pressure electrical conductivity studies of liquid and hybrid electrolytes. *J. Power Sources* 81, 739–747 (1999).

62. MacGlashan, G. S., Andreev, Y. G. & Bruce, P. Structure of the polymer electrolyte poly(ethylene oxide): $LiAsF_6$. *Nature* 398, 792–793 (1999).

63. Zheng, Y., Chia, F., Ungar, G. & Wright, P. V. Self-tracking in solvent-free, low-dimensional polymer electrolyte blends with lithium salts giving high ambient DC conductivity. *Chem. Commun.* 16, 1459–1460 (2000).

64. Imrie, C. T., Ingram, M. D. & McHattie, G. S. Ion transport in glassy polymer electrolytes. *J. Phys. Chem. B* 103, 4132–4138 (1999).

65. Croce, F., Appetecchi, G. B., Persi, L. & Scrosati, B. Nanocomposite polymer electrolytes for lithium batteries. *Nature* 394, 456–458 (1998).

66. Sata, N., Eberman, K., Eberl, K. & Maier, J. Mesoscopic fast ion conduction in nanometre-scale planar heterostructures. *Nature* 408, 946–949 (2000).

67. Aurbach, D. Review of selected electrode solution interactions which determine the performance of Li and Li-ion batteries. *J. Power Sources* 89, 206–218 (2000).

68. Bates, J. B., Dudney, N. J., Neudecker, B., Ueda, A. & Evans, C. D. Thin film lithium and lithium-ion batteries. *Solid State Ionics* 135, 33–45 (2000).

69. Neudecker, B. J., Dudney, N. J. & Bates, J. B Lithium-free thin film battery with in situ plated Li anode. *J. Electrochem. Soc.* 147, 517–523 (2000).

70. Visco, S. J. The development of reversible lithium metal electrodes for advances Li/S batteries. Conf. Proc. Int. Meeting on Power Sources for Consumer and Industrial Applications, Hawaii, 3-6 September 2001.

71. Orsini, F. *et al.* In situ SEM study of the interfaces in plastic lithium cells. *J. Power Sources* 81–82, 918–921 (1999).

Acknowledgements

The authors thank their colleagues, both in academic institutions and industry, for sharing the gratifying dedication to this field of progress, and P. Rickman for help drawing the figures.

Lithium deintercalation in LiFePO₄ nanoparticles via a domino-cascade model

C. DELMAS[1]*, M. MACCARIO[1], L. CROGUENNEC[1], F. LE CRAS[2] AND F. WEILL[1]

[1]ICMCB-CNRS, site ENSCPB, Université Bordeaux, 87, Av. Dr A. Schweitzer, 33608 Pessac cedex, France
[2]Commissariat à l'Energie Atomique, Laboratoire Composants pour l'Énergie DRT/LITEN/DTNM/LCE, 17, rue des Martyrs—38054 GRENOBLE cedex 9, France
*e-mail: delmas@icmcb-bordeaux.cnrs.fr

Published online: 20 July 2008; doi:10.1038/nmat2230

Lithium iron phosphate is one of the most promising positive-electrode materials for the next generation of lithium-ion batteries that will be used in electric and plug-in hybrid vehicles. Lithium deintercalation (intercalation) proceeds through a two-phase reaction between compositions very close to LiFePO₄ and FePO₄. As both endmember phases are very poor ionic and electronic conductors, it is difficult to understand the intercalation mechanism at the microscopic scale. Here, we report a characterization of electrochemically deintercalated nanomaterials by X-ray diffraction and electron microscopy that shows the coexistence of fully intercalated and fully deintercalated individual particles. This result indicates that the growth reaction is considerably faster than its nucleation. The reaction mechanism is described by a 'domino-cascade model' and is explained by the existence of structural constraints occurring just at the reaction interface: the minimization of the elastic energy enhances the deintercalation (intercalation) process that occurs as a wave moving through the entire crystal. This model opens new perspectives in the search for new electrode materials even with poor ionic and electronic conductivities.

Since the pioneering work of Goodenough and co-workers[1] on the electrochemical behaviour of LiFePO₄, many studies have been devoted to optimizing the material for better electrochemical performances and to trying to understand the lithium intercalation/deintercalation mechanism[2–9]. Armand and co-workers made a significant breakthrough when they showed that a carbon coating formed during LiFePO₄ synthesis simultaneously increases the electronic conductivity of this electrode material and prevents particle growth[10]. During cycling, lithium intercalation/deintercalation occurs through a two-phase reaction[1,6] with, as mentioned for the first time by Yamada and co-workers[11,12], the existence of very narrow solid solutions in the vicinity of the endmembers LiFePO₄ and FePO₄. The extent of these solid-solution domains is strongly related to the particle size, as recently shown by Meethong et al.[13] and has a strong impact on the electrochemical behaviour[14,15].

One of the fascinating peculiarities of LiFePO₄ concerns its ability to be used at very high cycling rates, although both involved phases exhibit very low ionic and electronic conductivities. Therefore, we have to determine why a so-poor electronic and ionic conductor can work at very high rates in the battery.

To try to understand what is going on in lithium deintercalation/intercalation in Li$_x$FePO₄, numerous studies were devoted to trying to establish the relation between the structure and the ionic and electronic transport properties. The theoretical studies done by ab initio calculation by Ceder's group[16] and then by Islam et al.[17] have shown that lithium ions can move easily only in the tunnels parallel to the b direction. Unfortunately, as the solid solutions near both ends of the two-phase domain (Li$_{1-\varepsilon}$FePO₄ and Li$_{\varepsilon}$FePO₄) are very narrow, the number of ionic carriers is very small in comparison with what is observed for other electrode materials. From the electronic point of view, the formation of a very small number of small polarons (Fe²⁺–O–Fe³⁺) in the structure and their very low mobility due to their strong binding to the lithium vacancies in Li$_{1-\varepsilon}$FePO₄ (or to the lithium ions in Li$_{\varepsilon}$FePO₄) prevent high electronic conductivity because both species (Li⁺ and e⁻) have to move simultaneously during the intercalation process[18].

A very interesting electron microscopy study of Li$_{0.50}$FePO₄—which is in fact a mixture of LiFePO₄ and FePO₄—was carried out by Chen et al.[8], who showed that in quite large primary particles, obtained by chemical lithium deintercalation, there are alternating domains of intercalated and deintercalated phases with intermediate zones where defects are concentrated. Moreover, they confirmed the results of Ceder and co-workers[16] concerning the easy diffusion of lithium in the tunnels parallel to the b direction and showed that lithium is extracted at the phase boundary parallel to the bc plane that progresses in the a direction on reaction. More recently, an electron energy-loss spectroscopy (EELS) study carried out by Laffont et al.[9] confirmed that the shrinking core–shell model[1,6] is not relevant to explain lithium deintercalation/intercalation in the olivine structure owing to the strong anisotropy of lithium diffusion. From the EELS results, they showed that there is no solid solution in the interfacial zone between the two endmembers. They proposed an elegant model to describe the deintercalation (intercalation) process, where for partially deintercalated platelet particles lying on the ac plane, the deintercalated phase (FePO₄) remains in the core of the particle and LiFePO₄ on the shell (the particle surface parallel to the b direction). They did not find LiFePO₄ on the surface of the ac plane.

In our laboratory, we try to understand the processes involved in electrochemical cycling starting from LiFePO₄ nanoparticles (with a diameter close to 100 nm). In principle, the chemical reaction and the electrochemical reaction must lead to the same result, but it is well known that in practice some differences can be observed: the chemical reaction requires a double-contact point

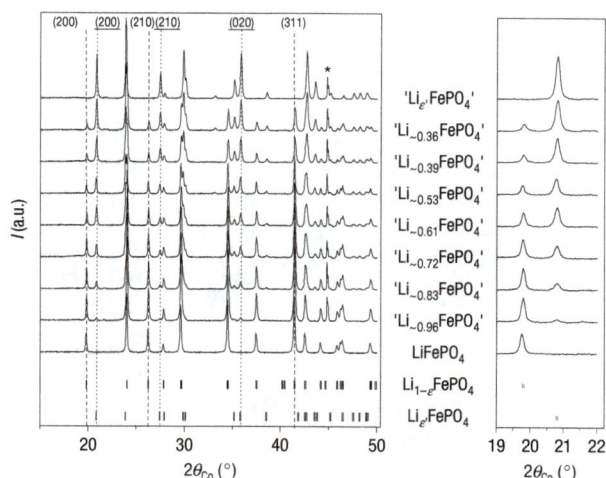

Figure 1 XRD patterns obtained *ex situ* for cast 'Li$_x$FePO$_4$' electrodes recovered at different states of charge during the first cycle of Li ‖ LiFePO$_4$ lithium cells. Miller indices are given in the *Pnma* space group for selected XRD lines for the Li-rich phase (Li$_{1-\varepsilon}$FePO$_4$) and for the Li-deficient phase (Li$_{\varepsilon'}$FePO$_4$). The Miller indices of Li$_{\varepsilon'}$FePO$_4$ are underlined. A magnified view of the (200) diffraction lines is also shown.

In the following, the results obtained by X-ray diffraction (XRD) and electron diffraction on electrochemically deintercalated samples enable us to understand the mechanism that occurs in nanoparticles and to explain why LiFePO$_4$ can be used at very high rates in lithium-ion batteries.

Several Li ‖ LiFePO$_4$ batteries were charged to selected 'Li$_x$FePO$_4$' average compositions (see Supplementary Information, Fig. S1); the materials removed from the cell were characterized by XRD. Figure 1 shows a set of these XRD patterns that illustrates perfectly the two-phase character of the reaction. For 'Li$_x$FePO$_4$' electrodes that consist of mixtures of lithium-rich and lithium-poor phases, the broadening of all the diffraction lines surprisingly remains, in first approximation, almost constant, whatever the overall lithium composition. A structural determination was done for all compositions by the Rietveld method of analysis of the XRD data, with a special emphasis on the determination of the cell parameters and coherence domain lengths $\langle L \rangle$ (Table 1). The cell parameters obtained for the two phases present in the biphased domain are unambiguously different from those of the endmembers (see Supplementary Information, Fig. S2), confirming the existence of solid solutions in the vicinity of the ideal compositions LiFePO$_4$ and FePO$_4$, in good agreement with the results of Yamada *et al.*[12] and Meethong *et al.*[13]. In the following, the Li$_{1-\varepsilon}$FePO$_4$ ($\varepsilon \sim 0$) and Li$_{\varepsilon'}$FePO$_4$ ($\varepsilon' \sim 0$) formulae were considered as the limits of the solid-solution domains. The full-width at half-maximum (FWHM) values determined for selected diffraction lines of the XRD patterns, also reported in Table 1, do not change significantly on deintercalation. Furthermore, there is no continuous change in the coherence domain length values on charge as would have been expected for continuous growth of one phase (Li$_{\varepsilon'}$FePO$_4$) to the detriment of the other (Li$_{1-\varepsilon}$FePO$_4$) by continuous migration of a reaction front in primary particles. A typical result is obtained, as an example, for the 'Li$_{0.83}$FePO$_4$' composition, which corresponds to a 17:83 molar mixture of Li$_{\varepsilon'}$FePO$_4$ and Li$_{1-\varepsilon}$FePO$_4$ phases respectively, but presents very similar coherence domain lengths for both phases. Therefore,

(between the sample and the reagent), whereas the electrochemical reaction requires a triple-contact point (between the sample, the electrolyte and the electronic conductor). Moreover, in the chemical reaction, which can be considered as a chemical short-circuit, the microscopic processes can be slightly different to those involved in the electrochemical reaction that is closer to the thermodynamic equilibrium even at a high rate.

Table 1 Comparison of cell parameters and FWHM values determined for selected diffraction lines of the XRD patterns recorded for 'Li$_x$FePO$_4$' electrodes recovered at different states of charge during the first charge of Li ‖ LiFePO$_4$ lithium cells: (200), (210) and (311) for Li$_{1-\varepsilon}$FePO$_4$ and (200), (210) and (020) for Li$_{\varepsilon'}$FePO$_4$. The coherence domain lengths $\langle L \rangle$ were determined by considering the Thompson–Cox–Hastings function to describe the line profile.

Estimated formula		FWHM (deg)			$a_{hex.}$ (Å)	$b_{hex.}$ (Å)	$c_{hex.}$ (Å)	$\langle L \rangle$ (Å)
	Li$_{1-\varepsilon}$FePO$_4$	(200)	(210)	(311)				
	Li$_{\varepsilon'}$FePO$_4$	(200)	(210)	(020)				
LiFePO$_4$		0.07	0.08	0.09	10.3294	6.0086	4.6948	740±20
'Li$_{\sim0.96}$FePO$_4$'	Li$_{1-\varepsilon}$FePO$_4$	0.06	0.08	0.09	10.3235	6.0044	4.6925	750±60
	Li$_{\varepsilon'}$FePO$_4$	0.07	0.09	—	9.84	5.8	4.79	*
'Li$_{\sim0.83}$FePO$_4$'	Li$_{1-\varepsilon}$FePO$_4$	0.05	0.07	0.08	10.3238	6.0049	4.693	800±30
	Li$_{\varepsilon'}$FePO$_4$	0.10	0.10	0.13	9.827	5.7961	4.7834	730±20
'Li$_{\sim0.72}$FePO$_4$'	Li$_{1-\varepsilon}$FePO$_4$	0.06	0.06	0.08	10.3238	6.005	4.6931	700±20
	Li$_{\varepsilon'}$FePO$_4$	0.10	0.12	0.12	9.826	5.7962	4.7819	600±50
'Li$_{\sim0.61}$FePO$_4$'	Li$_{1-\varepsilon}$FePO$_4$	0.07	0.07	0.08	10.3222	6.004	4.693	730±40
	Li$_{\varepsilon'}$FePO$_4$	0.10	0.12	0.13	9.826	5.7956	4.7814	650±70
'Li$_{\sim0.53}$FePO$_4$'	Li$_{1-\varepsilon}$FePO$_4$	0.07	0.07	0.09	10.3224	6.0043	4.6932	700±20
	Li$_{\varepsilon'}$FePO$_4$	0.10	0.12	0.14	9.828	5.7955	4.7811	750±60
'Li$_{\sim0.39}$FePO$_4$'	Li$_{1-\varepsilon}$FePO$_4$	0.08	0.06	0.08	10.323	6.0041	4.6939	820±100
	Li$_{\varepsilon'}$FePO$_4$	0.09	0.11	0.14	9.828	5.7958	4.7817	930±140
'Li$_{\sim0.36}$FePO$_4$'	Li$_{1-\varepsilon}$FePO$_4$	0.10	0.08	0.10	10.322	6.0039	4.694	650±70
	Li$_{\varepsilon'}$FePO$_4$	0.10	0.11	0.13	9.827	5.7955	4.7805	700±60
'Li$_{\varepsilon'}$FePO$_4$'	Li$_{1-\varepsilon}$FePO$_4$	—	—	—				—
	Li$_{\varepsilon'}$FePO$_4$	0.07	0.09	0.11	9.8207	5.7913	4.7798	760±100

*Calculation of $\langle L \rangle$ was not possible owing to the lack of accuracy for this minor phase.

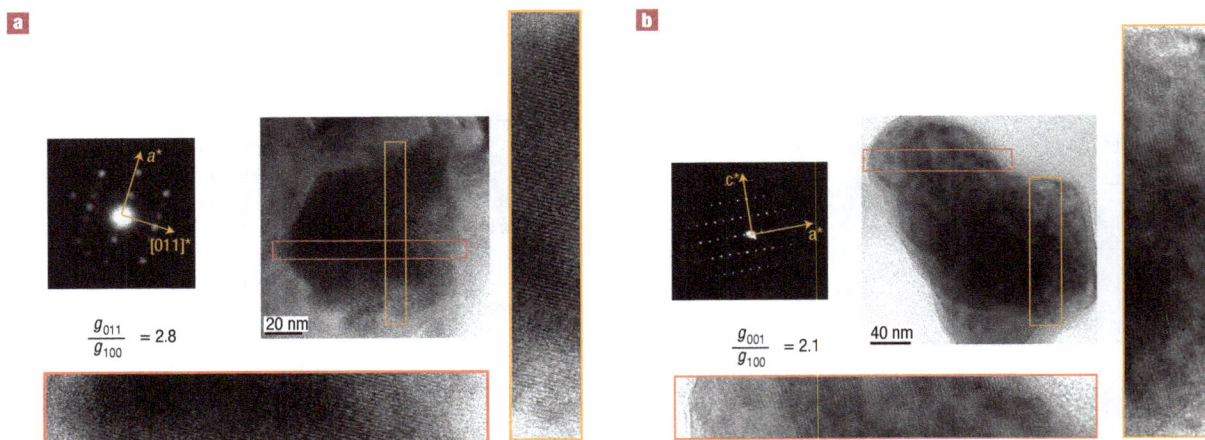

Figure 2 High-resolution images of crystallites with their electron diffraction patterns. Magnified views of two parts of these crystallites are also shown. **a**, $Li_{1-\varepsilon}FePO_4$. **b**, $Li_{\varepsilon'}FePO_4$.

how do we explain the invariance of the coherence domain length whatever the lithium composition of the deintercalated materials? Clearly, this result leads to the assumption that the two-phase mixture corresponds to a mixture of particles that are either fully intercalated or fully deintercalated. This hypothesis can be true only if the reaction front propagation between the intercalated part and the deintercalated part in a given crystal is considerably faster than the nucleation of a $Li_{\varepsilon'}FePO_4$ domain within a $Li_{1-\varepsilon}FePO_4$ crystallite. This result is very important, as it gives a new approach for the lithium deintercalation (intercalation) mechanism in the olivine-type structure. To confirm this result, an electron diffraction study was carried out.

The electron diffraction study was carried out on several batches of electrochemically deintercalated materials with average composition 'Li_xFePO_4' ($x = \sim0.41$, ~0.55 and ~0.61). Two very different charge conditions were chosen to emphasize the relaxation effect on the material microstructure. For the three batches, the electron diffraction study was done on several crystallites with sizes ranging between 80 and 150 nm (28 crystals were studied). The smaller particles were always agglomerated. The study was thus realized either on isolated parts of agglomerated small particles or on isolated larger particles. Very reproducible results were obtained in all cases. They are identical whatever the electrochemical experiment used to obtain the Li_xFePO_4 samples; therefore, only those obtained at the C/20 rate are reported here.

Figure 2 shows typical sets of images. Except in a layer of a few nanometres around the particle, atomic planes can clearly be seen in these images: they are running from one side of the particle to the other without any disordered area. Modification of the contrast can be observed along the atomic planes: it is probably due to a change in the thickness of the particle. Figure 2a shows an image of a whole particle. From the corresponding experimental electron diffraction pattern, it can be deduced that this particle corresponds to $Li_{1-\varepsilon}FePO_4$ oriented along the [$0\bar{1}1$] zone axis. Magnified views of the particle images are also shown. Obviously, this particle is homogeneous and well crystallized everywhere: neither defects nor disordered areas are observed. The Fourier transforms calculated on several parts of this particle always indicate the same R ratio ($R = d_{100}/d_{0kl} = g_{0kl}/g_{100}$, in which g_{hkl} is the reciprocal distance corresponding to the (hkl) plane) between the measured distances along the two orthogonal directions. From these observations, we can affirm that this particle is homogeneous or in other words that there is no trace of the deintercalated $Li_{\varepsilon'}FePO_4$ (see the Supplementary Information). Figure 2b shows another particle observed in the same sample. The electron diffraction pattern proves it is a homogeneous $Li_{\varepsilon'}FePO_4$ particle oriented along the [010] zone axis. The magnified views of the particles confirm the absence of any disordered area. Note that these two particles are representative of all the observed nanocrystals; whatever the composition studied, only two types of crystal were observed: either pure $Li_{1-\varepsilon}FePO_4$ or pure $Li_{\varepsilon'}FePO_4$.

To summarize this part, the high-resolution images show that the nanoparticles are single domains without any defects. The purpose of this microscopic study was to demonstrate the occurrence of only the $Li_{1-\varepsilon}FePO_4$-type phase or only the $Li_{\varepsilon'}FePO_4$-type phase in each crystallite, on the contrary to what was previously reported by other authors for deintercalated Li_xFePO_4 materials obtained by chemical oxidation. These results confirm the XRD results: when lithium deintercalation starts in a given particle, it proceeds very rapidly in the whole crystallite.

To explain the lithium deintercalation (intercalation) behaviour in the olivine system, we have to carefully consider the olivine structure and how Li^+ ions and e^- can move inside the lattice.

The structure is built up of two-dimensional (2D) sheets (Fig. 3a), parallel to the bc plane, with $[FeO_4]_n$ formula made of corner-sharing FeO_6 octahedra (Fig. 3b). These sheets are connected along the a direction by PO_4 tetrahedra to make the 3D skeleton. The very peculiar point of this structure is the presence of one common edge between each PO_4 tetrahedron and each FeO_6 octahedron. In phosphates, owing to the strong covalency of the P–O bond, all P–O and O–O distances are generally fixed and almost invariant within PO_4 tetrahedra (normal P–O bond length is equal to 1.56 Å (ref. 19)). In the olivine structure, this edge-sharing induces strong distortions at the local scale that spread in a cooperative way through all of the crystallite, and therefore considerably reduces the structure flexibility, which is a key point as far as deintercalation (intercalation) is concerned.

During lithium deintercalation, Fe^{2+} ions are oxidized to Fe^{3+} with strong changes in the Fe–O bond lengths and O–O distances in FeO_6 octahedra, leading also to a cooperative structure distortion. The distance modifications are very important as shown in Fig. 3c. This large distortion between the two 3D

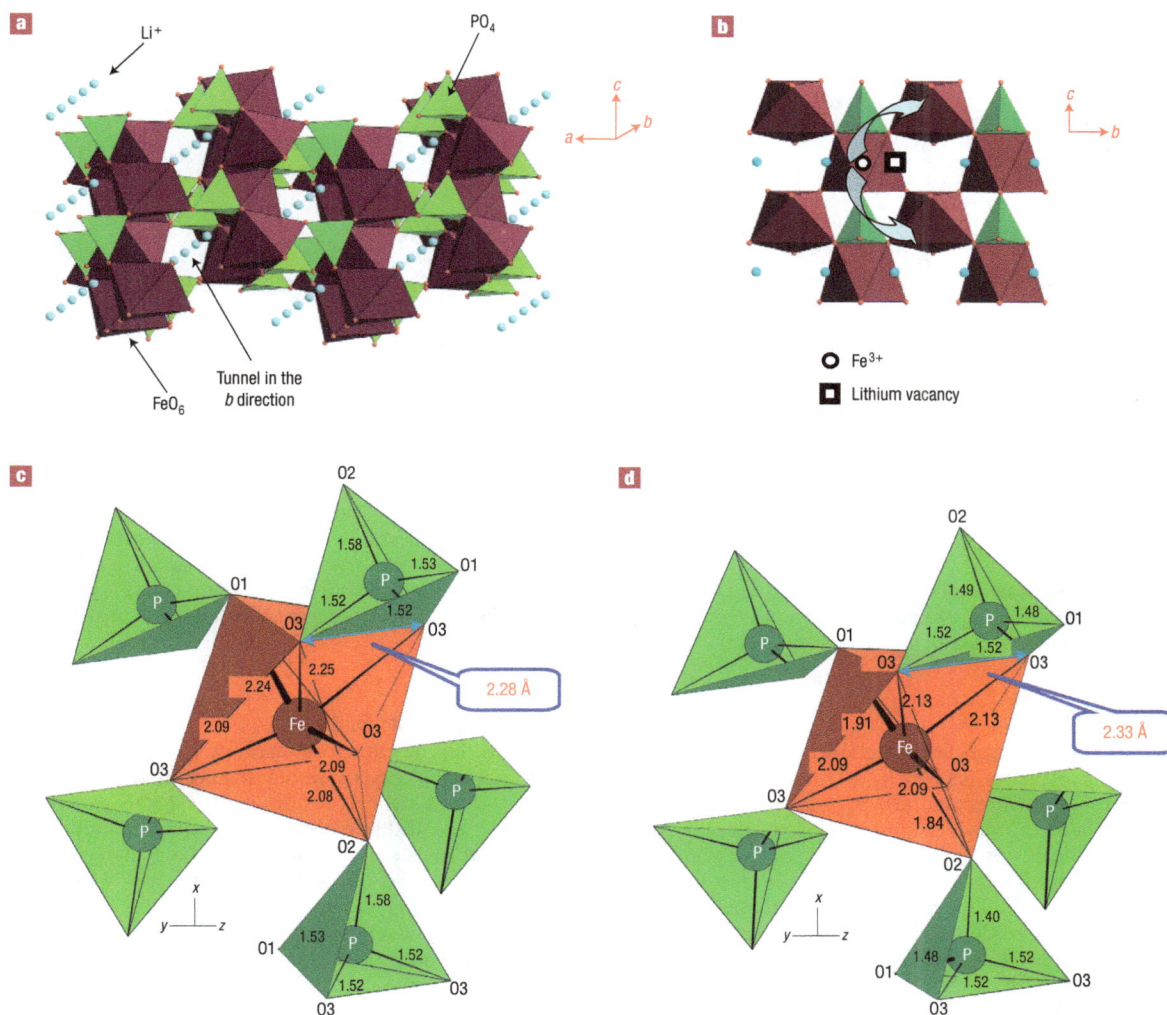

Figure 3 Structure of LiFePO$_4$. a, Perspective view. **b**, Lithium vacancy with an Fe^{3+} ion in its vicinity. **c,d**, Comparison of the interatomic distances in the LiFePO$_4$ (**c**) and FePO$_4$ (**d**) structures to emphasize the very strong distortions.

frameworks also induces a significant change in the O–O distance corresponding to the common edge between the FeO$_6$ octahedron and the PO$_4$ tetrahedron, which evolves from 2.28 to 2.33 Å for LiFePO$_4$ and FePO$_4$, respectively. The differences in both FeO$_6$ and PO$_4$ environments are so important that it is very difficult to introduce a significant amount of Fe^{3+} ions in LiFePO$_4$ and conversely a significant amount of Fe^{2+} ions in FePO$_4$. As a result, a large compositional domain for solid solutions is not allowed. It results in the existence of a biphased domain between two compositions (Li$_{1-\varepsilon}$FePO$_4$ ($\varepsilon \sim 0$) and Li$_{\varepsilon'}$FePO$_4$ ($\varepsilon' \sim 0$)) close to the endmembers LiFePO$_4$ and FePO$_4$, respectively.

These structural distortions also have a huge impact on the electronic conductivity. When an Fe^{3+} ion is localized in the LiFePO$_4$ framework (or an Fe^{2+} ion in the FePO$_4$ framework), a small polaron forms. Ceder and co-workers[18] have studied its migration in the olivine structure from first-principle calculations. They have shown that its intrinsic mobility is quite high if the binding energy with the lithium vacancy is not considered. However, if this binding energy is considered, the activation energy

required for the transfer is significantly increased (from ~200 to ~500 meV). Furthermore, as the solid-solution ranges are very narrow, the charge carrier concentration is very low. Thus, both effects lead to a very low electronic conductivity. As for charge compensation the lithium vacancies must stay near the Fe^{3+} ions during the deintercalation process; the lithium ions and the electrons have to move simultaneously. All experiments and most of the simulations carried out on LiFePO$_4$ have shown that lithium diffusivity in this material is very anisotropic[8,9,16,17]. In particular, Ceder and co-workers[16] showed that the ionic conductivity is quite high in the b direction, whereas it is almost negligible in both other directions. Srinivasan and Newman explain the behaviour of LiFePO$_4$ electrodes using the shrinking core model and thus an isotropic mechanism[20]; nevertheless, their model does not take into account the microscopic process involved during the reaction. Therefore, it can be assumed that their model cannot be applied at the primary particle scale, but at the agglomerate scale.

During electrochemical lithium deintercalation (intercalation) from small particles, we have experimentally found either

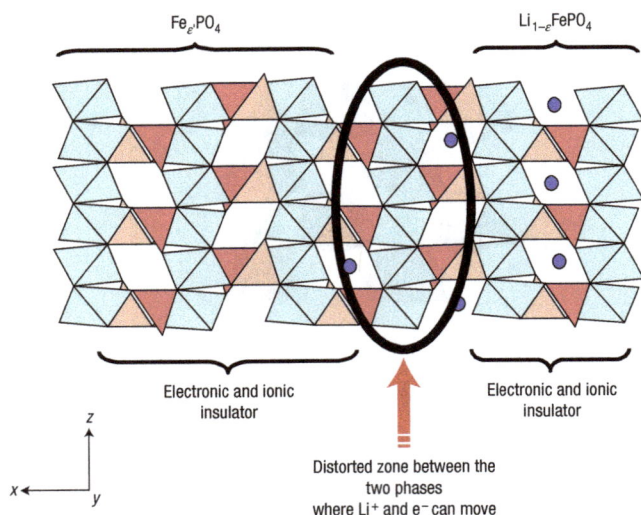

Figure 4 Distorted zone in the *ac* plane between the lithiated (Li$_{1-\varepsilon}$FePO$_4$) and delithiated (Li$_{\varepsilon'}$FePO$_4$) phases during the lithium deintercalation/intercalation process in LiFePO$_4$ olivine-type material.

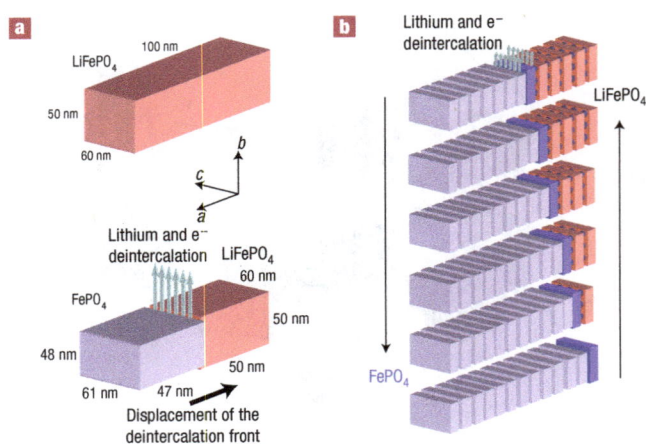

Figure 5 Schematic view of the 'domino-cascade' mechanism for the lithium deintercalation/intercalation mechanism in a LiFePO$_4$ crystallite. Distances are not accurate because studied crystallites are rather isotropic. **a,** Scheme showing a view of the strains occurring during lithium deintercalation. **b,** Layered view of the lithium deintercalation/intercalation mechanism in a LiFePO$_4$ crystallite.

Li$_{1-\varepsilon}$FePO$_4$ or Li$_{\varepsilon'}$FePO$_4$ particles. Therefore, we have to explain how one of the two phases is growing in the other one during lithium deintercalation (intercalation). To understand the whole mechanism, we have to consider simultaneously the ionic and the electronic conductivities, as both processes are coupled at the microscopic scale, as well as the effect of strong local distortions.

As described in Fig. 3b, if we consider a lithium tunnel parallel to the *b* direction where a lithium ion is deintercalated, an electron is extracted from the neighbouring Fe^{2+} ion to form an Fe^{3+} ion. Locally, in this [FeO$_4$]$_n$ layer of corner-sharing FeO$_6$ octahedra, the small polaron can easily move to the next FeO$_6$ octahedron while remaining close to the lithium vacancy (arrows in Fig. 3b). This 2D small polaron mobility in one [FeO$_4$]$_n$ layer was recently discussed by Nazar and co-workers[21] in their study of the Li$_x$FePO$_4$ system at high temperature. As the lithium mobility is high in the *b* tunnel, the ionic and electronic transport processes occur simultaneously and a *b* tunnel starts to be deintercalated. Then, owing to these small polarons, the lithium deintercalation from the tunnels localized in the vicinity of the [FeO$_4$]$_n$ layer becomes easier. Furthermore, as the FeO$_6$ octahedra are not directly connected in the *a* direction, the electronic transfer is very low in that direction. In other words, when a tunnel parallel to the *b* direction starts to be deintercalated, the electronic conductivity that appears locally makes deintercalation in tunnels adjacent to both sides of the distorted [FeO$_4$]$_n$ layer easier. The lithium deintercalation process proceeds very quickly in the vicinity of the [FeO$_4$]$_n$ layer, leading to the formation of a 'FePO$_4$ block' in the crystallite if there are no structural defects to prevent the lithium and electron migrations. This enhancement of the lithium ion mobility at the interface was first postulated by Chen *et al.*[8] and then detailed by Laffont *et al.*[9]. This scheme helps explain why the phase boundary is observed in the *bc* plane[8].

Owing to significant differences in bond lengths and angles between LiFePO$_4$ and the 'FePO$_4$' as-formed block, strong distortions appear locally that destabilize adjacent [FeO$_4$]$_n$ layers. As a result, it is easier to remove lithium ions from the *b* tunnels adjacent to the 'FePO$_4$' block' than to start a new

nucleation of the deintercalated phase elsewhere in the crystal. Nucleation of a new micro-domain of Li$_{\varepsilon'}$FePO$_4$ in another part of the Li$_{1-\varepsilon}$FePO$_4$ crystal would require high activation energy, whereas lithium deintercalation just near the phase boundary consists of the displacement of the phase boundary without any increase in the activation energy. The higher the number of Li$_{1-\varepsilon}$FePO$_4$/Li$_{\varepsilon'}$FePO$_4$ boundaries, the higher the activation energy and the higher the cell polarization during the lithium deintercalation (intercalation) process.

The large mismatch between the intercalated and the deintercalated frameworks induces constraints within the material that make the deintercalation in the vicinity of the boundary (Fig. 4) easier. In this model, when a boundary plane is formed, it moves in the *a* direction through the crystal on lithium deintercalation. This displacement can be considered as a wave going through the crystal without any energy barrier that allows lithium intercalation/deintercalation to proceed at a very high rate. Figure 5 shows a schematic representation of this process within the crystal, which we call the 'domino-cascade model'. This model is also in good agreement with the EELS study reported by Laffont *et al.*[9] that showed that there is no solid solution in the interfacial region between the two limit compositions. In our model, the interface is ideally limited to one 'FePO$_4$ block'. In fact, depending on the size of the crystallites in the *bc* plane, the mismatch between the unit cells can require the presence of several layers in the boundary. The larger the size of the crystallite in the *bc* plane, the larger the mismatch and the higher the number of [FeO$_4$]$_n$ planes involved in the boundary to accommodate the difference in structures. In this case, the local constraints are minimized by energy relaxation, but the driving force for the boundary displacement is also decreased. Furthermore, for large crystallites[8], the local constraints are minimized by energy relaxation with formation of dislocations or cracks. It should also be noted that the size of the homogeneous domains (100 nm range) is very similar to that of our particles. When the cooperative displacement of the phase boundary is stopped, a new nucleation must start elsewhere in the crystal. To summarize, the smaller the particle size in the *bc* plane, the smaller the probability of finding defects, the faster the boundary displacement and the higher the probability of having a mixture of single-phase crystallites.

Nevertheless, note that depending on the deintercalation method (chemical or electrochemical), different phase distributions can be observed in a material of average composition Li_xFePO_4. Certainly, chemical lithium deintercalation must have the tendency to initiate nucleation simultaneously at different points in all the crystallites and thus to lead to the formation of domains in all the crystallites[8,9].

The scheme we propose is also in good agreement with the results recently reported by Allen et al.[22], who carried out a simulation of the intercalation process by using the Avrami–Johnson–Mehl–Eroofev equation. They showed that the nucleation rate is equal to zero and that the phase transformation is controlled by a phase-boundary reaction. Moreover, they determined that the activation energy for the $FePO_4$ to $LiFePO_4$ transformation is very small ($13 kJ mol^{-1}$). Such a result is expected for our model that compares the lithium deintercalation (intercalation) reaction occurring just at the boundary with the displacement of a wave all along the crystal.

Note also that our results are consistent with a study done by Wagemaker et al.[23] on the Li_xTiO_2 system. They have shown that in small particles the coexistence of two phases in the same crystallite, especially when their cell volumes are significantly different, is very difficult owing to the strong system destabilization.

The starting point for the nucleation is unknown, but we can consider that the part of the crystal with the smallest thickness along the b direction is certainly the most favourable for the nucleation. As the growth is faster than the nucleation, the nucleation speed depends on the deintercalation rate. It can be assumed that within the electrode, the particles, better connected to the current collector or having a better carbon coating, will be deintercalated first. As shown by scanning electron microscopy (see Supplementary Information, Fig. S3), these primary particles are agglomerated to form large secondary particles. At the agglomerate scale, a core–shell scheme is likely to occur. The agglomerates during the deintercalation reaction can be constituted of a core of lithiated primary nanoparticles surrounded by a shell of deintercalated primary nanoparticles that would be in agreement with results of simulations[20]. The size and the porosity of these agglomerates would strongly influence the performance of the real batteries, as there are always some inhomogeneities in the electrodes. The domino-cascade mechanism occurs between $Li_{1-\varepsilon}FePO_4$ and $Li_{\varepsilon'}FePO_4$ limit compositions. The extent of these solid-solution domains, which is related to the particle size[13] on both sides of the intercalation domain, must have a strong influence on the electrode behaviour.

Nano $C–LiFePO_4$ obtained by a very short thermal treatment exhibits very high capacity and reversibility in lithium batteries. The XRD and electron microscopy studies show that the biphased material is constituted of a mixture of primary particles of either $Li_{1-\varepsilon}FePO_4$ or $Li_{\varepsilon'}FePO_4$. This result shows that the growth of one phase inside the particle of the other phase is considerably faster than the nucleation. The peculiar olivine structure with edge-sharing between each FeO_6 octahedron and each PO_4 tetrahedron considerably limits the structure flexibility required to allow the formation of a solid solution. Therefore, during the battery charge (discharge), all compositional and structural modifications are localized within the interfacial zone between the two endmember materials. The very high concentration of Li^+/vacancies and Fe^{2+}/Fe^{3+} polarons localized in this interfacial zone enables a very fast reaction. This interfacial zone is destabilized and can move very rapidly inside the crystallite, like a wave going through the particle in the a direction on deintercalation (intercalation). The large structural mismatch between $Li_{1-\varepsilon}FePO_4$ and $Li_{\varepsilon'}FePO_4$ phases is

the driving force of this phenomenon: the elastic energy assists the reaction process. During the reaction, there is a quick displacement of the interfacial reaction front; the activation energy lost on one side of this reaction front is recovered on the other side. This process, which we call 'the domino-cascade model', requires primary nanoparticles (especially in the b and c directions) with a very small concentration of structural defects that would interrupt the displacement of the reaction front.

METHODS

$C–LiFePO_4$ was synthesized by a preliminary mechano-chemical activation of the precursors and then a thermal treatment of this mixture. The reactants, 'iron phosphate—$Fe_{2.84}(PO_4)_2 \cdot \sim 6.5H_2O$' and lithium phosphate, were mixed at a ratio $Li/Fe \sim 1.05$. Cellulose was initially added (10 wt%) with these precursors. This precursors mixture was ball-milled four times for 1 h 30 min each time in a planetary mill using tungsten carbide vessels and agate balls (diameter 10 mm). The resulting mixture was thermal-treated in a tube under argon flow which was introduced in a tubular furnace already at 800 °C. The mixture was maintained at this temperature for 15 min before the sample-containing tube was removed from the furnace and quenched in air.

'Li_xFePO_4' samples ($x = \sim 0.41$, ~ 0.55 and ~ 0.61) were obtained by electrochemical lithium deintercalation from $C–LiFePO_4$. Cast electrodes were prepared from a 500 mg mixture of 80 wt% of $C–LiFePO_4$ active material with 10 wt% of carbon conductive additive and 10 wt% of polyvinylidene fluoride binder, as well as with 0.8 mg of N-methyl pyrrolidone. This mixture was then cast on an aluminium foil and dried overnight at 55 °C. Pellets of 10 mm diameter were finally punched, pressed at $10 T cm^{-2}$ and dried for 48 h at 80 °C under vacuum. Considering their surface and thickness, a mass smaller than 2 mg of active material was cast on each electrode, leading for instance to an average error of $\sim 5\%$ on the reversible capacity. Electrodes were charged up to a given lithium composition in laboratory cells containing a lithium foil as the negative electrode, two separators and a polypropylene foil wetted by the liquid electrolyte (1 M $LiPF_6$ in a mixture of propylene carbonate/ethylene carbonate/dimethyl carbonate (1:1:3)). These lithium cells were assembled in a dry box under argon. Electrochemical tests were carried out with VMP1 apparatus at various rates to check that our results did not come from the electrode technology. In all cases, the recovered capacity agreed with data in the literature ($155 mA h g^{-1}$ at C/20 and $65 mA h g^{-1}$ at 15C), showing that all particles were involved in the electrochemical process. For the 'Li_xFePO_4' ($x \sim 0.55$) sample, the electrochemical lithium deintercalation was carried out at the C/20 rate; the two others ($x \sim 0.41$ and ~ 0.61) were synthesized through a series of continuous galvanostatic charges at decreasing rates (C/100 up to x_3, C/200 up to x_2, C/400 up to x_1, C/800 up to x with $x_3 > x_2 > x_1 > x$) to be very close to the thermodynamic equilibrium. After cycling, the lithium cells were disassembled in a dry box under argon; positive electrodes were rinsed with dimethyl carbonate and dried under vacuum.

Samples for XRD were prepared in a dry box; they were held in a sample holder, which enabled a controlled atmosphere to be maintained around the sample during the experiment. XRD data were collected using a PANalytical X'pert Pro diffractometer (Co $K\alpha$ radiation, iron filter, antiscatter slit of 1° and divergence slit of 1° on the incident beam path). The diffraction patterns were recorded at room temperature in the [15–75]° (2θ) angular range using a 0.0167° (2θ) step and a constant counting time of 11 s. The XRD patterns were refined with the Rietveld method using the Fullprof program[24]. The peaks profile was described with the Thompson–Cox–Hastings function to take into account a possible change in microstructure (size and strains effects) for this material on lithium deintercalation (LaB_6 was used as the standard). Phase quantification, which was possible owing to the fact that there was no significant absorption difference between $Li_{1-\varepsilon}FePO_4$ ($\varepsilon \sim 0$) and $Li_{\varepsilon'}FePO_4$ ($\varepsilon' \sim 0$), enabled us to estimate rather well the percentage of each phase present in the Li_xFePO_4 material and thus to check the lithium composition.

A JEOL 2200FS microscope equipped with a double-tilt specimen stage was used to obtain high-resolution images of the materials. Before observation, a suspension was obtained by grinding the material in an agate mortar with hexane. A droplet of the suspension was deposited on a grid covered with a carbon-supported film. This preparation was done in a glove box, in an argon atmosphere. The grids were brought in the vicinity of the microscope under argon atmosphere and put on the specimen stage in the minimum time needed. A grid was never used twice to prevent any observation of a modified specimen.

Two criteria were used to choose the particles for images: (1) isolated particles and (2) particles oriented along the *a* direction that allow determination of the nature of the phase. Different images were obtained at different magnifications to get a view of the whole particle as well as of different parts of the particle.

Received 6 November 2007; accepted 24 June 2008; published 20 July 2008.

References

1. Padhi, A. K., Nanjundaswamy, K. S. & Goodenough, J. B. Phospho-olivines as positive-electrode materials for rechargeable lithium batteries. *J. Electrochem. Soc.* **144**, 1188–1194 (1997).
2. Chen, Z. H. & Dahn, J. R. Reducing carbon in $LiFePO_4$/C composite electrodes to maximize specific energy, volumetric energy, and tap density. *J. Electrochem. Soc.* **149**, A1184–A1189 (2002).
3. Franger, S., Le Cras, F., Bourbon, C. & Rouault, H. $LiFePO_4$ synthesis routes for enhanced electrochemical performance. *Electrochem. Solid State Lett.* **5**, A231–A233 (2002).
4. Delacourt, C., Wurm, C., Laffont, L., Leriche, J. B. & Masquelier, C. Electrochemical and electrical properties of Nb- and/or C-containing $LiFePO_4$ composites. *Solid State Ion.* **177**, 333–341 (2006).
5. Chen, J. J. & Whittingham, M. S. Hydrothermal synthesis of lithium iron phosphate. *Electrochem. Commun.* **8**, 855–858 (2006).
6. Andersson, A. S., Kalska, B., Haggstrom, L. & Thomas, J. O. Lithium extraction/insertion in $LiFePO_4$: An X-ray diffraction and Mossbauer spectroscopy study. *Solid State Ion.* **130**, 41–52 (2000).
7. Prosini, P. P. Modeling the voltage profile for $LiFePO_4$. *J. Electrochem. Soc.* **152**, A1925–A1929 (2005).
8. Chen, G. Y., Song, X. Y. & Richardson, T. J. Electron microscopy study of the $LiFePO_4$ to $FePO_4$ phase transition. *Electrochem. Solid State Lett.* **9**, A295–A298 (2006).
9. Laffont, L. *et al.* Study of the $LiFePO_4$/$FePO_4$ two-phase system by high-resolution electron energy loss spectroscopy. *Chem. Mater.* **18**, 5520–5529 (2006).
10. Ravet, N. *et al.* Electrode material having improved surface conductivity. CA Patent No. EP1049182 (2000-05-02 2002).
11. Yonemura, M., Yamada, A., Takei, Y., Sonoyama, N. & Kanno, R. Comparative kinetic study of olivine Li_xMPO_4 (M = Fe, Mn). *J. Electrochem. Soc.* **151**, A1352–A1356 (2004).
12. Yamada, A., Koizumi, H., Sonoyama, N. & Kanno, R. Phase change in Li_xFePO_4. *Electrochem. Solid State Lett.* **8**, A409–A413 (2005).
13. Meethong, N., Huang, H. Y. S., Carter, W. C. & Chiang, Y. M. Size-dependent lithium miscibility gap in nanoscale $Li_{1-x}FePO_4$. *Electrochem. Solid State Lett.* **10**, A134–A138 (2007).
14. Yamada, A. *et al.* Room-temperature miscibility gap in Li_xFePO_4. *Nature Mater.* **5**, 357–360 (2006).
15. Meethong, N., Huang, H.-Y. S., Speakman, S. A., Carter, W. C. & Chiang, Y.-M. Strain accommodation during phase transformations in olivine-based cathodes as a materials selection criterion for high-power rechargeable batteries. *Adv. Funct. Mater.* **17**, 1115–1123 (2007).
16. Morgan, D., Van der Ven, A. & Ceder, G. Li conductivity in Li_xMPO_4 (M = Mn, Fe, Co, Ni) olivine materials. *Electrochem. Solid State Lett.* **7**, A30–A32 (2004).
17. Islam, M. S., Driscoll, D. J., Fisher, C. A. J. & Slater, P. R. Atomic-scale investigation of defects, dopants, and lithium transport in the $LiFePO_4$ olivine-type battery material. *Chem. Mater.* **17**, 5085–5092 (2005).
18. Maxisch, T., Zhou, F. & Ceder, G. Ab initio study of the migration of small polarons in olivine Li_xFePO_4 and their association with lithium ions and vacancies. *Phys. Rev. B* **7310**, 104301 (2006) NIL_258–NIL_263.
19. Shannon, R.D. & Prewitt, C.T. Effective ionic radii in oxides and fluorides. *Acta Crystallogr.* **B25**, 925–946 (1969).
20. Srinivasan, V. & Newman, J. Existence of path-dependence in the $LiFePO_4$ electrode. *Electrochem. Solid State Lett.* **9**, A110–A114 (2006).
21. Ellis, B., Perry, L. K., Ryan, D. H. & Nazar, L. F. Small polaron hopping in Li_xFePO_4 solid solutions: Coupled lithium-ion and electron mobility. *J. Am. Chem. Soc.* **128**, 11416–11422 (2006).
22. Allen, J. L., Jow, T. R. & Wolfenstine, J. Kinetic study of the electrochemical $FePO_4$ to $LiFePO_4$ phase transition. *Chem. Mater.* **19**, 2108–2111 (2007).
23. Wagemaker, M., Borghols, W. J. H. & Mulder, F. M. Large impact of particle size on insertion reactions. A case for anatase Li_xTiO_2. *J. Am. Chem. Soc.* **129**, 4323–4327 (2007).
24. Rodriguez-Carvajal, J. Laboratoire Léon Brillouin, <http://www-llb.cea.fr/fullweb/powder.htm> *Laboratoire Léon Brillouin*, (2004).

Supplementary Information accompanies this paper on www.nature.com/naturematerials.

Acknowledgements

The authors wish to thank CEA, ADEME (PVE no. 0366C0072) and Région Aquitaine for financial support.

Author information

Reprints and permission information is available online at http://npg.nature.com/reprintsandpermissions. Correspondence and requests for materials should be addressed to C.D.

High-performance lithium battery anodes using silicon nanowires

CANDACE K. CHAN[1], HAILIN PENG[2], GAO LIU[3], KEVIN McILWRATH[4], XIAO FENG ZHANG[4], ROBERT A. HUGGINS[2] AND YI CUI[2]*

[1]Department of Chemistry, Stanford University, Stanford, California 94305, USA
[2]Department of Materials Science and Engineering, Stanford University, Stanford, California 94305, USA
[3]Environmental Energy Technologies Division, Lawrence Berkeley National Lab, 1 Cyclotron Road, Mail Stop 70R108B, Berkeley, California 94720, USA
[4]Electron Microscope Division, Hitachi High Technologies America, Inc., 5100 Franklin Drive, Pleasanton, California 94588, USA
*e-mail: yicui@stanford.edu

Published online: 16 December 2007; doi:10.1038/nnano.2007.411

There is great interest in developing rechargeable lithium batteries with higher energy capacity and longer cycle life for applications in portable electronic devices, electric vehicles and implantable medical devices[1]. Silicon is an attractive anode material for lithium batteries because it has a low discharge potential and the highest known theoretical charge capacity (4,200 mAh g^{-1}; ref. 2). Although this is more than ten times higher than existing graphite anodes and much larger than various nitride and oxide materials[3,4], silicon anodes have limited applications[5] because silicon's volume changes by 400% upon insertion and extraction of lithium, which results in pulverization and capacity fading[2]. Here, we show that silicon nanowire battery electrodes circumvent these issues as they can accommodate large strain without pulverization, provide good electronic contact and conduction, and display short lithium insertion distances. We achieved the theoretical charge capacity for silicon anodes and maintained a discharge capacity close to 75% of this maximum, with little fading during cycling.

Previous studies in which Si bulk films and micrometre-sized particles were used as electrodes in lithium batteries have shown capacity fading and short battery lifetime due to pulverization and loss of electrical contact between the active material and the current collector (Fig. 1a). The use of sub-micrometre pillars[6] and micro- and nanocomposite anodes[5,7–9] led to only limited improvement. Electrodes made of amorphous Si thin films have a stable capacity over many cycles[5,8], but have insufficient material for a viable battery. The concept of using one-dimensional (1D) nanomaterials has been demonstrated with carbon[10], Co$_3$O$_4$ (refs 11, 12), SnO$_2$ (ref. 13) and TiO$_2$ (ref. 14) anodes, and has shown improvements compared to bulk materials. A schematic of our Si nanowire (NW) anode configuration is shown in Fig. 1b. Nanowires are grown directly on the metallic current collector substrate. This geometry has several advantages and has led to improvements in rate capabilities in metal oxide cathode materials[15]. First, the small NW diameter allows for better accommodation of the large volume changes without the initiation of fracture that can occur in bulk or micron-sized materials (Fig. 1a). This is consistent with previous studies that have suggested a materials-dependent terminal particle size below which particles do not fracture further[16,17]. Second, each Si NW

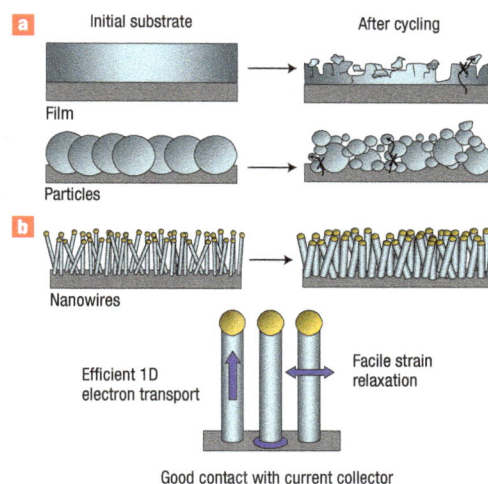

Figure 1 **Schematic of morphological changes that occur in Si during electrochemical cycling. a**, The volume of silicon anodes changes by about 400% during cycling. As a result, Si films and particles tend to pulverize during cycling. Much of the material loses contact with the current collector, resulting in poor transport of electrons, as indicated by the arrow. **b**, NWs grown directly on the current collector do not pulverize or break into smaller particles after cycling. Rather, facile strain relaxation in the NWs allows them to increase in diameter and length without breaking. This NW anode design has each NW connecting with the current collector, allowing for efficient 1D electron transport down the length of every NW.

is electrically connected to the metallic current collector so that all the nanowires contribute to the capacity. Third, the Si NWs have direct 1D electronic pathways allowing for efficient charge transport. In electrode microstructures based on particles, electronic charge carriers must move through small interparticle contact areas. In addition, as every NW is connected to the current-carrying electrode, the need for binders or conducting additives, which add extra weight, is eliminated. Furthermore,

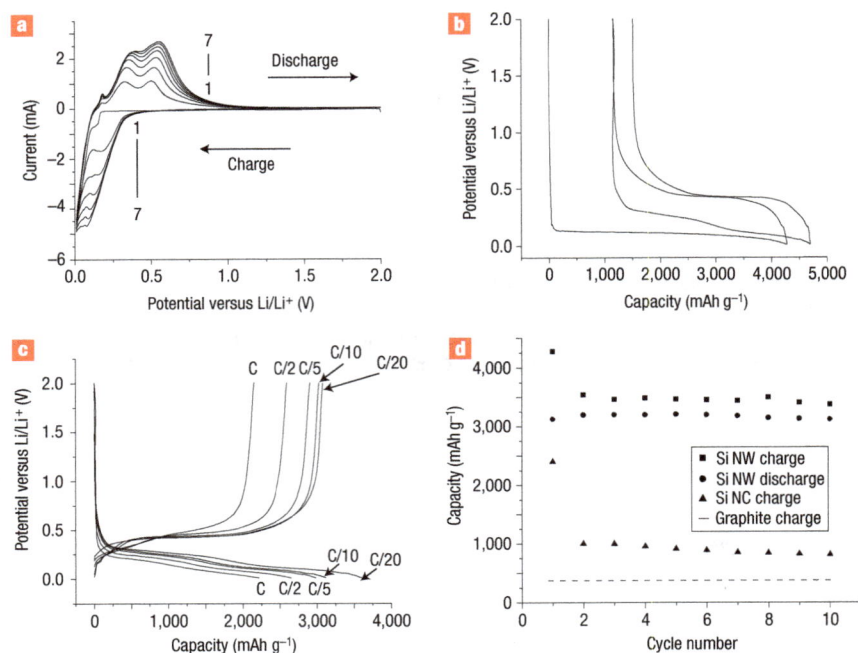

Figure 2 Electrochemical data for Si NW electrodes. a, Cyclic voltammogram for Si NWs from 2.0 V to 0.01 V versus Li/Li+ at 1 mV s⁻¹ scan rate. The first seven cycles are shown. **b**, Voltage profiles for the first and second galvanostatic cycles of the Si NWs at the C/20 rate. The first charge achieved the theoretical capacity of 4,200 mAh g⁻¹ for Li$_{4.4}$Si. **c**, The voltage profiles for the Si NWs cycled at different power rates. The C/20 profile is from the second cycle. **d**, Capacity versus cycle number for the Si NWs at the C/20 rate showing the charge (squares) and discharge capacity (circles). The charge data for Si nanocrystals (triangles) from ref. 8 and the theoretical capacity for lithiated graphite (dashed line) are shown as a comparison to show the improvement when using Si NWs.

our Si NW battery electrode can be easily realized using the vapour–liquid–solid (VLS) or vapour–solid (VS) template-free growth methods[18–23] to produce NWs directly onto stainless steel current collectors (see Methods).

A cyclic voltammogram of the Si NW electrode is shown in Fig. 2a. The charge current associated with the formation of the Li–Si alloy began at a potential of ~330 mV and became quite large below 100 mV. Upon discharge, current peaks appeared at about 370 and 620 mV. The current–potential characteristics were consistent with previous experiments on microstructured Si anodes[6]. The magnitude of the current peaks increased with cycling due to activation of more material to react with Li in each scan[6]. The small peak at 150–180 mV may have been due to reaction of the Li with the gold catalyst, which makes a negligible contribution to the charge capacity (see Supplementary Information, Figs S1 and S2).

Si NWs were found to exhibit a higher capacity than other forms of Si (ref. 5). Figure 2b shows the first and second cycles at the C/20 rate (20 h per half cycle). The voltage profile observed was consistent with previous Si studies, with a long flat plateau during the first charge, during which crystalline Si reacted with Li to form amorphous Li$_x$Si. Subsequent discharge and charge cycles had different voltage profiles characteristic of amorphous Si (refs 24–27). Significantly, the observed capacity during this first charging operation was 4,277 mAh g⁻¹, which is essentially equivalent to the theoretical capacity within experimental error. The first discharge capacity was 3,124 mAh g⁻¹, indicating a coulombic efficiency of 73%. The second charge capacity decreased by 17% to 3,541 mAh g⁻¹, although the second discharge capacity increased slightly to 3,193 mAh g⁻¹, giving a coulombic efficiency of 90%. Both charge and discharge

capacities remained nearly constant for subsequent cycles, with little fading up to 10 cycles (Fig. 2d), which is considerably better than previously reported results[8,9]. As a comparison, our charge and discharge data are shown along with the theoretical capacity (372 mAh g⁻¹) for the graphite currently used in lithium battery anodes, and the charge data reported for thin films containing 12-nm Si nanocrystals[8] (NCs) in Fig. 2d. This improved capacity and cycle life in the Si NWs demonstrates the advantages of our Si NW anode design.

The Si NWs also displayed high capacities at higher currents. Figure 2c shows the charge and discharge curves observed at the C/20, C/10, C/5, C/2 and 1C rates. Even at the 1C rate, the capacities remained >2,100 mAh g⁻¹, which is still five times larger than that of graphite. The cyclability of the Si NWs at the faster rates was also excellent. Using the C/5 rate, the capacity was stable at ~3,500 mAh g⁻¹ for 20 cycles in another device (see Supplementary Information, Fig. S3). Despite the improved performance, the Si NW anode showed an irreversible capacity loss in the first cycle, which has been observed in other work[5]. Although solid electrolyte interphase (SEI) formation has been observed in Si (ref. 28), we do not believe this is the cause of our initial irreversible capacity loss, because there is no appreciable capacity in the voltage range of the SEI layer formation (0.5–0.7 V) during the first charge (Fig. 2b)[8]. Although SEI formation may be occurring, the capacity involved in SEI formation would be very small compared to the high charge capacity we observed. The mechanism of the initial irreversible capacity is not yet understood and requires further investigation.

The structural morphology changes during Li insertion were studied to understand the high capacity and good cyclability of our Si NW electrodes. Pristine, unreacted Si NWs were crystalline

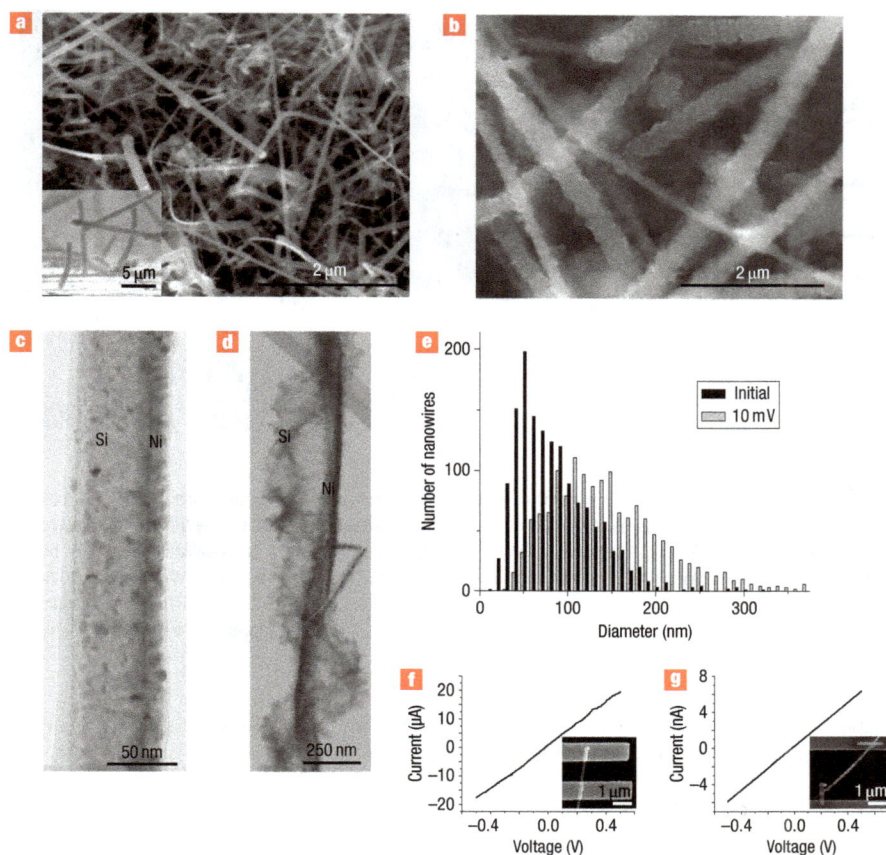

Figure 3 Morphology and electronic changes in Si NWs from reaction with Li. a,b, SEM image of pristine Si NWs before (**a**) and after (**b**) electrochemical cycling (same magnification). The inset in **a** is a cross-sectional image showing that the NWs are directly contacting the stainless steel current collector. **c,d,** TEM image of a pristine Si NW with a partial Ni coating before (**c**) and after (**d**) Li cycling. **e,** Size distribution of NWs before and after charging to 10 mV (bin width 10 nm). The average diameter of the NWs increased from 89 to 141 nm after lithiation. **f,** I−V curve for a single NW device (SEM image, inset) constructed from a pristine Si NW. **g,** I−V curve for a single NW device (SEM image, inset), constructed from a NW that had been charged and discharged once at the C/20 rate.

with smooth sidewalls (Fig. 3a) and had an average diameter of ∼89 nm (s.d., 45 nm) (Fig. 3e). Cross-sectional scanning electron microscopy (SEM) showed that the Si NWs grew off the substrate and had good contact with the stainless steel current collector (Fig. 3a, inset). After charging with Li, the Si NWs had roughly textured sidewalls (Fig. 3b), and the average diameter increased to ∼141 nm (s.d., 64 nm). Despite the large volume change, the Si NWs remained intact and did not break into smaller particles. They also appeared to remain in contact with the current collector, suggesting minimal capacity fade due to electrically disconnected material during cycling.

The Si NWs may also change their length during the change in volume. To evaluate this, 25-nm Ni was evaporated onto as-grown Si NWs using electron beam evaporation. Because of the shadow effect of the Si NWs, the Ni only covered part of the NW surface (Fig. 3c), as confirmed by energy dispersed X-ray spectroscopy (EDS) mapping (see Supplementary Information, Fig. S4). The Ni is inert to Li and acts as a rigid backbone on the Si NWs. After lithiation (Fig. 3d), the Si NWs changed shape and wrapped around the Ni backbone in a three-dimensionally helical manner. This appeared to be due to an expansion in the length of the NW, which caused strain because the NW was attached to the Ni and could not freely expand but rather buckled into a

helical shape. Although the NW length increased after lithiation, the NWs remained continuous and without fractures, maintaining a pathway for electrons all the way from the collector to the NW tips. With both a diameter and length increase, the Si NW volume change after Li insertion appears to be about 400%, consistent with the literature[5].

Efficient electron transport from the current collector to the Si NWs is necessary for good battery cycling. To evaluate this, we conducted electron transport measurements on single Si NWs before and after lithiation (see Methods). The current versus voltage curve on a pristine Si NW was linear, with a 25 kΩ resistance (resistivity of 0.02 Ω-cm) (Fig. 3f). After one cycle, the NWs became amorphous, but still exhibited a current that was linear with voltage with an 8 MΩ resistance (resistivity of 3 Ω-cm) (Fig. 3g). The good conductivity of pristine and cycled NWs ensures efficient electron transport for charge and discharge.

The large volume increase in the Si NWs is driven by the dramatic atomic structure change during lithiation. To understand the structural evolution of NWs, we characterized the NW electrodes at different charging potentials. The X-ray diffraction (XRD) patterns were taken for initial pristine Si NWs, Si NWs charged to 150 mV, 100 mV, 50 mV and 10 mV, as well as after 5 cycles (Fig. 4a). XRD patterns of the as-grown Si NWs

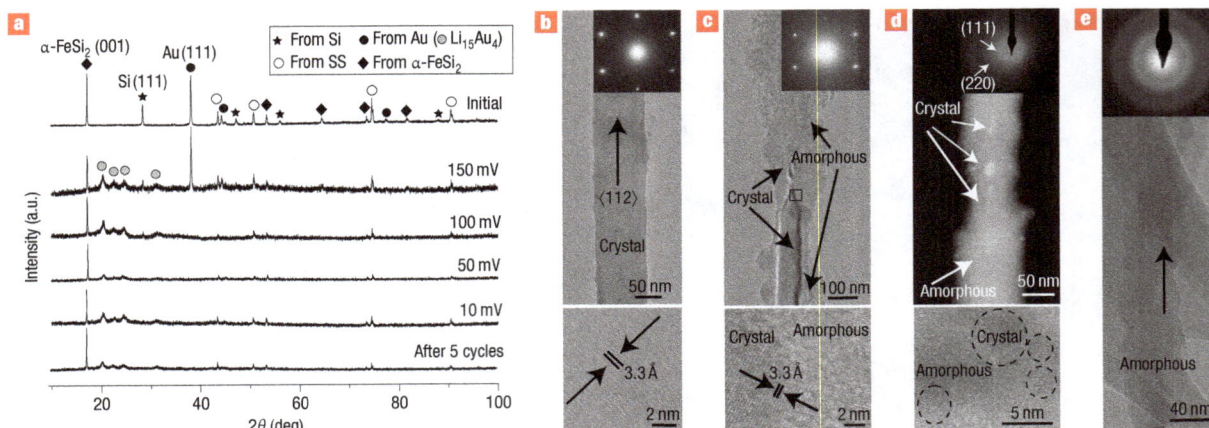

Figure 4 Structural evolution of Si NWs during lithiation. a, XRD patterns of Si NWs before electrochemical cycling (initial), at different potentials during the first charge, and after five cycles. **b–e**, TEM data for Si NWs at different stages of the first charge. **b**, A single-crystalline, pristine Si NW before electrochemical cycling. The SAED spots (inset) and HRTEM lattice fringes (bottom) are from the Si 1/3(224) planes. **c**, NW charged to 100 mV showing a Si crystalline core and the beginning of the formation of a Li_xSi amorphous shell. The HREM (bottom) shows an enlarged view of the region inside the box. **d**, Dark-field image of a NW charged to 50 mV showing an amorphous Li_xSi wire with crystalline Si grains (bright regions) in the core. The spotty rings in the SAED (inset) are from crystalline Si. The HRTEM (bottom) shows some Si crystal grains embedded in the amorphous wire. **e**, A NW charged to 10 mV is completely amorphous $Li_{4.4}Si$. The SAED (top) shows diffuse rings characteristic of an amorphous material.

showed diffraction peaks associated with Si, α-FeSi$_2$, Au (the Si NW catalyst) and stainless steel (SS). The α-FeSi$_2$ forms at the interface between the SS and the Si wires during the high temperature (530 °C) NW growth process. The α-FeSi$_2$ was not found to appreciably react with Li during electrochemical cycling, although a small amount of reaction has been reported[24]. After Si NWs were charged to 150 mV, the higher angle Si peaks disappeared. Only the Si(111) peak was still visible, but its intensity was greatly decreased. This is consistent with the disappearance of the initial crystalline Si and the start of the formation of amorphous Li_xSi. The four broad peaks that appeared in the lower angles are due to the formation of $Li_{15}Au_4$ (see Supplementary Information, Fig. S5). At 100 mV, the pure Au peaks disappeared, indicating that the Au had completely reacted with Li. The Si(111) peak was very weak at 100 mV, and disappeared completely at 50 mV. It appears that Si NWs remain amorphous after the first charge, consistent with the non-flat voltage charging/discharging curve in Fig. 2b. This contrasts, however, with other studies on Si electrodes[25,26], which have reported the formation of crystalline, $Li_{3.75}Si$ at potentials less than 30–60 mV. In situ XRD studies have determined that this crystalline phase only forms at <50 mV for films thicker than ~2 μm (ref. 27). We did not observe this to be the case in our Si NWs, most likely because of their shape and small dimensions.

The local structural features of Si NWs during the first Li insertion were studied with transmission electron microscopy (TEM) and selected area electron diffraction (SAED). The as-grown Si NWs were found to be single-crystalline. Figure 4b shows an example of a typical Si NW with a ⟨112⟩ growth direction[29]. Figure 4c shows a Si NW with a ⟨112⟩ growth direction that was charged to 100 mV. In this case there were two phases present, as expected from the voltage profile. Both crystalline and amorphous phases were clearly seen. The distribution of the two phases was observed both across the diameter (a crystalline core and an amorphous shell) and along the length. The SAED showed the spot pattern for the crystalline phase (Si), but weak diffuse rings from the amorphous phase

(Li_xSi alloy) were also observed. Li ions must diffuse radially into the NW from the electrolyte, resulting in the core–shell phase distribution. The reason for phase distribution along the length is not yet understood. At 50 mV, the Si NW became mostly amorphous with some crystalline Si regions embedded inside the core, as seen from the dark-field image and HRTEM (Fig. 4d). The SAED showed spotty rings representative of a polycrystalline sample and diffuse rings for the amorphous phase. At 10 mV (Fig. 4e), all of the Si had changed to amorphous $Li_{4.4}Si$, as indicated by the amorphous rings in the SAED. These TEM observations were consistent with the XRD results (Fig. 4a) and voltage charging curves (Fig. 2b).

METHODS

Si NWs were synthesized using the VLS process on stainless steel substrates using Au catalyst. The electrochemical properties were evaluated under an argon atmosphere by both cyclic voltammetry and galvanostatic cycling in a three-electrode configuration, with the Si NWs on the stainless steel substrate as the working electrode and Li foil as both reference and counter-electrodes. No binders or conducting carbon were used. The charge capacity referred to here is the total charge inserted into the Si NW, per mass unit, during Li insertion, whereas the discharge capacity is the total charge removed during Li extraction. For electrical characterization, single Si NW devices were contacted with metal electrodes by electron-beam lithography or focused-ion beam deposition. For more detailed descriptions of NW synthesis, TEM and XRD characterization, electrochemical testing, and device fabrication, see the Supplementary Information.

Received 23 July 2007; accepted 14 November 2007; published 16 December 2007.

References

1. Nazri, G.-A. & Pistoia, G. *Lithium Batteries: Science and Technology* (Kluwer Academic/Plenum, Boston, 2004).
2. Boukamp, B. A., Lesh, G. C. & Huggins, R. A. All-solid lithium electrodes with mixed-conductor matrix. *J. Electrochem. Soc.* **128**, 725–729 (1981).
3. Shodai, T., Okada, S., Tobishima, S. & Yamaki, J. Study of $Li_{3-x}M_xN$ (M:Co, Ni or Cu) system for use as anode material in lithium rechargeable cells. *Solid State Ionics* **86–88**, 785–789 (1996).
4. Poizot, P., Laruelle, S., Grugeon, S., Dupont, L. & Tarascon, J.-M. Nano-sized transition-metal oxides as negative-electrode materials for lithium-ion batteries. *Nature* **407**, 496–499 (2000).
5. Kasavajjula, U., Wang, C. & Appleby, A. J. Nano- and bulk-silicon-based insertion anodes for lithium-ion secondary cells. *J. Power Sources* **163**, 1003–1039 (2007).

6. Green, M., Fielder, E., Scrosati, B., Wachtler, M. & Moreno, J. S. Structured silicon anodes for lithium battery applications. *Electrochem. Solid-State Lett.* **6,** A75–A79 (2003).
7. Ryu, J. H., Kim, J. W., Sung, Y.-E. & Oh, S. M. Failure modes of silicon powder negative electrode in lithium secondary batteries. *Electrochem. Solid-State Lett.* **7,** A306–A309 (2004).
8. Graetz, J., Ahn, C. C., Yazami, R. & Fultz, B. Highly reversible lithium storage in nanostructured silicon. *Electrochem. Solid-State Lett.* **6,** A194–A197 (2003).
9. Gao, B., Sinha, S., Fleming, L. & Zhou, O. Alloy formation in nanostructured silicon. *Adv. Mater.* **13,** 816–819 (2001).
10. Che, G., Lakshmi, B. B., Fisher, E. R. & Martin, C. R. Carbon nanotubule membranes for electrochemical energy storage and production. *Nature* **393,** 346–349 (1998).
11. Nam, K. T. *et al.* Virus-enabled synthesis and assembly of nanowires for lithium ion battery electrodes. *Science* **312,** 885–888 (2006).
12. Shaju, K. M., Jiao, F., Debart, A. & Bruce, P. G. Mesoporous and nanowire Co$_3$O$_4$ as negative electrodes for rechargeable lithium batteries. *Phys. Chem. Chem. Phys.* **9,** 1837–1842 (2007).
13. Park, M.-S. *et al.* Preparation and electrochemical properties of SnO$_2$ nanowires for application in lithium-ion batteries. *Angew. Chem. Int. Edn* **46,** 750–753 (2007).
14. Armstrong, G., Armstrong, A. R., Bruce, P. G., Reale, P. & Scrosati, B. TiO$_2$(B) nanowires as an improved anode material for lithium-ion batteries containing LiFePO$_4$ or LiNi$_{0.5}$Mn$_{1.5}$O$_4$ cathodes and a polymer electrolyte. *Adv. Mater.* **18,** 2597–2600 (2006).
15. Li, N., Patrissi, C. J., Che, G. & Martin, C. R. Rate capabilities of nanostructured LiMn$_2$O$_4$ electrodes in aqueous electrolyte. *J. Electrochem. Soc.* **147,** 2044–2049 (2000).
16. Yang, J., Winter, M. & Besenhard, J. O. Small particle size multiphase Li-alloy anodes for lithium-ion batteries. *Solid State Ionics* **90,** 281–287 (1996).
17. Huggins, R. A. & Nix, W. D. Decrepitation model for capacity loss during cycling of alloys in rechargeable electrochemical systems. *Ionics* **6,** 57–63 (2000).
18. Morales, A. M. & Lieber, C. M. A laser ablation method for the synthesis of crystalline semiconductor nanowires. *Science* **279,** 208–211 (1998).
19. Huang, M. H. *et al.* Catalytic growth of zinc oxide nanowires by vapor transport. *Adv. Mater.* **13,** 113–116 (2001).
20. Dick, K. A. *et al.* A new understanding of Au-assisted growth of III-V semiconductor nanowires. *Adv. Funct. Mater.* **15,** 1603–1610 (2005).
21. Pan, Z. W., Dai, Z. R. & Wang, Z. L. Nanobelts of semiconducting oxides. *Science* **291,** 1947–1949 (2001).
22. Wang, Y., Schmidt, V., Senz, S. & Gosele, U. Epitaxial growth of silicon nanowires using an aluminum catalyst. *Nature Nanotech.* **1,** 186–189 (2006).
23. Hannon, J. B., Kodambaka, S., Ross, F. M. & Tromp, R. M. The influence of the surface migration of gold on the growth of silicon nanowires. *Nature* **440,** 69–71 (2006).
24. Netz, A., Huggins, R. A. & Weppner, W. The formation and properties of amorphous silicon as negative electrode reactant in lithium systems. *J. Power Sources* **119–121,** 95–100 (2003).
25. Li, J. & Dahn, J. R. An *in situ* X-ray diffraction study of the reaction of Li with crystalline Si. *J. Electrochem. Soc.* **154,** A156–A161 (2007).
26. Obrovac, M. N. & Krause, L. J. Reversible cycling of crystalline silicon powder. *J. Electrochem. Soc.* **154,** A103–A108 (2007).
27. Hatchard, T. D. & Dahn, J. R. *In situ* XRD and electrochemical study of the reaction of lithium with amorphous silicon. *J. Electrochem. Soc.* **151,** A838–A842 (2004).
28. Lee, Y. M., Lee, J. Y., Shim, H.-T., Lee, J. K. & Park, J.-K. SEI layer formation on amorphous Si thin electrode during precycling. *J. Electrochem. Soc.* **154,** A515–A519 (2007).
29. Wu, Y. *et al.* Controlled growth and structures of molecular-scale silicon nanowires. *Nano Lett.* **4,** 433–436 (2004).

Acknowledgements

We thank Dr Marshall for help with TEM interpretation and Professors Brongersma and Clemens for technical help. Y.C. acknowledges support from the Stanford New Faculty Startup Fund and Global Climate and Energy Projects. C.K.C. acknowledges support from a National Science Foundation Graduate Fellowship and Stanford Graduate Fellowship.
Correspondence and requests for materials should be addressed to Y.C.
Supplementary information accompanies this paper on www.nature.com/naturenanotechnology.

Author contributions

C.K.C. conceived and carried out the experiment and data analysis. H.P., G.L., K.M. and X.F.Z. assisted in experimental work. R.A.H. carried out data analysis. Y.C. conceived the experiment and carried out data analysis. C.K.C., R.A.H. and Y.C. wrote the paper.

The existence of a temperature-driven solid solution in Li$_x$FePO$_4$ for 0 ≤ x ≤ 1

CHARLES DELACOURT, PHILIPPE POIZOT, JEAN-MARIE TARASCON AND CHRISTIAN MASQUELIER*

Laboratoire de Réactivité et de Chimie des Solides, CNRS UMR 6007, Université de Picardie Jules Verne, 33 Rue St. Leu, 80039 Amiens Cedex 9, France
***e-mail: christian.masquelier@sc.u-picardie.fr**

Published online: 20 February 2005; doi:10.1038/nmat1335

Lithium-ion batteries have revolutionized the powering of portable electronics. Electrode reactions in these electrochemical systems are based on reversible insertion/deinsertion of Li$^+$ ions into the host electrode material with a concomitant addition/removal of electrons into the host. If such batteries are to find a wider market such as the automotive industry, less expensive positive electrode materials will be required, among which LiFePO$_4$ is a leading contender. An intriguing fundamental problem is to understand the fast electrochemical response from the poorly electronic conducting two-phase LiFePO$_4$/FePO$_4$ system. In contrast to the well-documented two-phase nature of this system at room temperature, we give the first experimental evidence of a solid solution Li$_x$FePO$_4$ (0 ≤ x ≤ 1) at 450 °C, and two new metastable phases at room temperature with Li$_{0.75}$FePO$_4$ and Li$_{0.5}$FePO$_4$ composition. These experimental findings challenge theorists to improve predictive models commonly used in the field. Our results may also lead to improved performances of these electrodes at elevated temperatures.

A mong the solid-state electrochemistry community, it is still an open question as to whether a single-phase or a two-phase insertion/extraction process is intrinsically advantageous in terms of kinetics as well as of structural stability. The most obvious distinction between the two types is that the equilibrium potential of a single-phase electrode is composition-dependent, whereas that of a two-phase system is constant over the entire composition range. These issues have been rekindled with the arrival of LiFePO$_4$. Eight years after the original work of Padhi[1] was published, phospho-olivines LiMPO$_4$ (M = Fe, Mn, Co) now appear to be potential candidates as positive electrode materials for rechargeable lithium batteries. Owing to smart material processing (carbon coating in particular), Li$^+$ may be easily extracted out of LiFePO$_4$ leading to room-temperature capacities of ~160 mAh g^{-1} (close to the theoretical value of 170 mAh g^{-1})[2–13]. The room-temperature insertion/extraction of lithium proceeds, at 3.45 V versus Li$^+$/Li, in a two-phase reaction between LiFePO$_4$ and FePO$_4$ that crystallize in the same space group[1,14]. The crystal structure of LiFePO$_4$ is well-known, and described in the space group $Pmnb$ with the following unit-cell parameters: a = 6.011(1) Å, b = 10.338(1) Å, c = 4.695(1) Å (refs 15–17). Iron is located in the middle of a slightly distorted FeO$_6$ octahedron, with a Fe–O average bond-length higher than expected for iron in the +2 valence state in octahedral coordination. Lithium is located in a second set of octahedral sites but distributed differently: LiO$_6$ octahedra share edges in order to form LiO$_6$ chains running parallel to [100]$_{Pmnb}$ (Fig. 1), which generates preferential rapid one-dimensional Li$^+$-ion conductivity along that direction[18].

Then come the very intriguing questions of how does Li$^+$ extraction exactly proceed, that is, how small are the values of δ in Li$_{1-\delta}$FePO$_4$ and ε in Li$_\varepsilon$FePO$_4$? On similar concerns, another intriguing issue emerged out of Zhou's 2004 paper[19], in which the authors explicitly report on "a significant failure of the local density approximation (LDA) and the generalized gradient approximation (GGA) to reproduce the phase stability and thermodynamics of mixed-valence Li$_x$FePO$_4$ compounds". We found it quite striking that their calculations converged towards the existence of intermediate Li$_x$FePO$_4$ solid solutions instead of a two-phase mixture, when lithium is extracted from LiFePO$_4$.

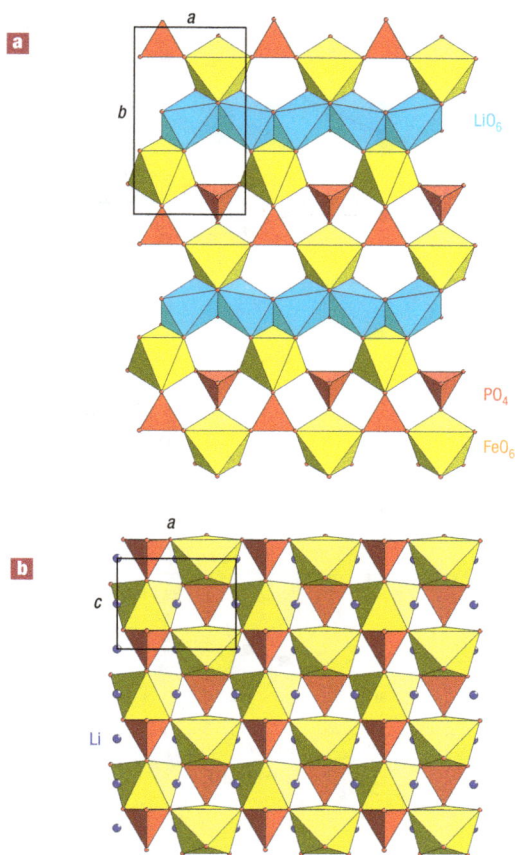

Figure 1 **Projections of the triohylite LiFePO$_4$ structure. a**, representation of a TeOc$_2$ layer and **b**, projection along [010]$_{Pmnb}$. TeOc$_2$ layers consist of LiO$_6$ and FeO$_6$ octahedra (Oc) and PO$_4$ tetrahedra (Te) that occupy 1/2 and 1/8 of the octahedral and tetrahedral interstitials, respectively. LiO$_6$ octahedra form chains isolated from each other by corner-sharing FeO$_6$ octahedral layers perpendicular to [010]. The existence of such Li$^+$ ions channels makes the [100]$_{Pmnb}$ direction the most probable for Li$^+$ ion transport.

Facing such an experimental/theoretical contradiction, we decided to take a closer look at the system in order to add new insight into the intrinsic factors that determine the nature (single versus two-phase process) of the lithium electrochemical insertion process in LiFePO$_4$. It is hoped that such experimental work would help improve theoretical models such as LDA/GGA. Inspired by the previous literature[20] in which a temperature-driven two-phase to one-phase insertion process was demonstrated on Chevrel phases, we decided to investigate the temperature dependence of the LiFePO$_4$/FePO$_4$ phase diagram.

To address the above issues, we investigated the thermal behaviour of a series of x LiFePO$_4$ / $(1-x)$ FePO$_4$ $(0 \leq x \leq 1)$ two-phase mixtures (prepared by either chemical oxidation of LiFePO$_4$ or intimate mixtures of the two end-members). As already described in the literature[21], the chemical oxidation is straightforward, and we obtained well-defined series of x LiFePO$_4$ / $(1-x)$ FePO$_4$ $(0 \leq x \leq 1)$ mixtures, whose compositions were basically governed by the initial amount of NO$_2$BF$_4$ in the acetonitrile solution. As gathered in Fig. 2, the corresponding X-ray diffraction (XRD) patterns display diffraction peaks characteristic of both LiFePO$_4$ and FePO$_4$.

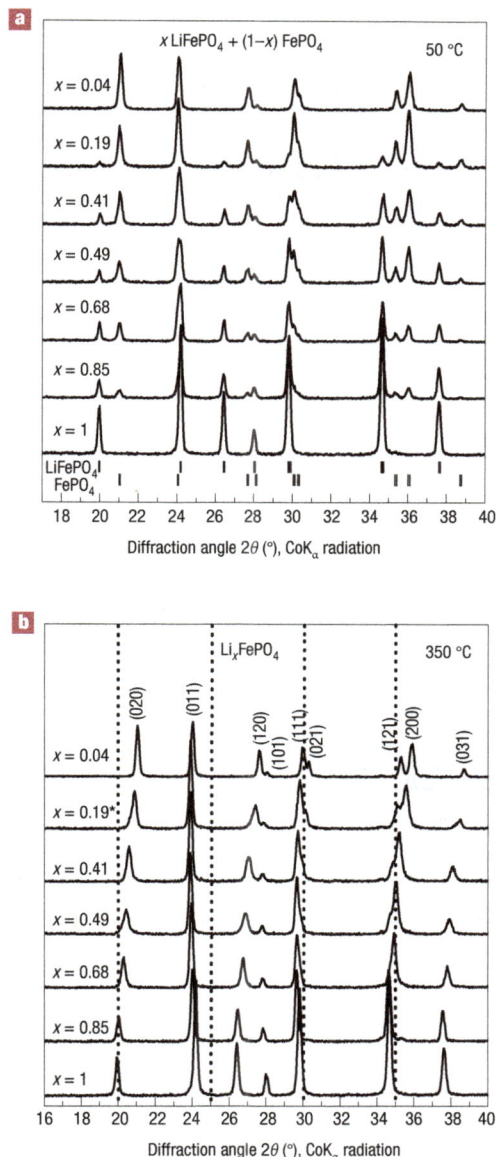

Figure 2 **XRD patterns of Li$_x$FePO$_4$ global compositions at 50 °C and 350 °C.** **a**, x LiFePO$_4$ + $(1-x)$ FePO$_4$ mixtures at 50 °C before heating. **b**, Li$_x$FePO$_4$ single phases with x ranging from 1 to ~0. As indicated by the Bragg positions, pristine samples are composed of the two end-members, LiFePO$_4$ and FePO$_4$. Chemical analysis of the Li/Fe ratio, by atomic absorption spectroscopy as well as Rietveld refinements of the XRD patterns, very satisfactorily converged to the indicated values of x. These samples were prepared either by chemical delithiation ($x = 0.68$, $x = 0.41$, $x = 0.19$ and $x = 0.04$) or simply by mixing the two phases ($x = 0.85$ and $x = 0.49$). At 350 °C (*390 °C for $x = 0.19$), full pattern-matching refinements allow full indexing in the space group $Pmnb$ of the observed Bragg reflections as originated from a single Li$_x$FePO$_4$ olivine-type phase. A continuous shift of (hk0) reflections towards higher 2θ angles when x decreases is clearly depicted.

Temperature-controlled XRD under flowing N$_2$ (to avoid possible oxidation of Fe(II) in Li$_x$FePO$_4$, $0 \leq x \leq 1$) was performed for each of these two-phase mixtures, and is illustrated in Fig. 3 for

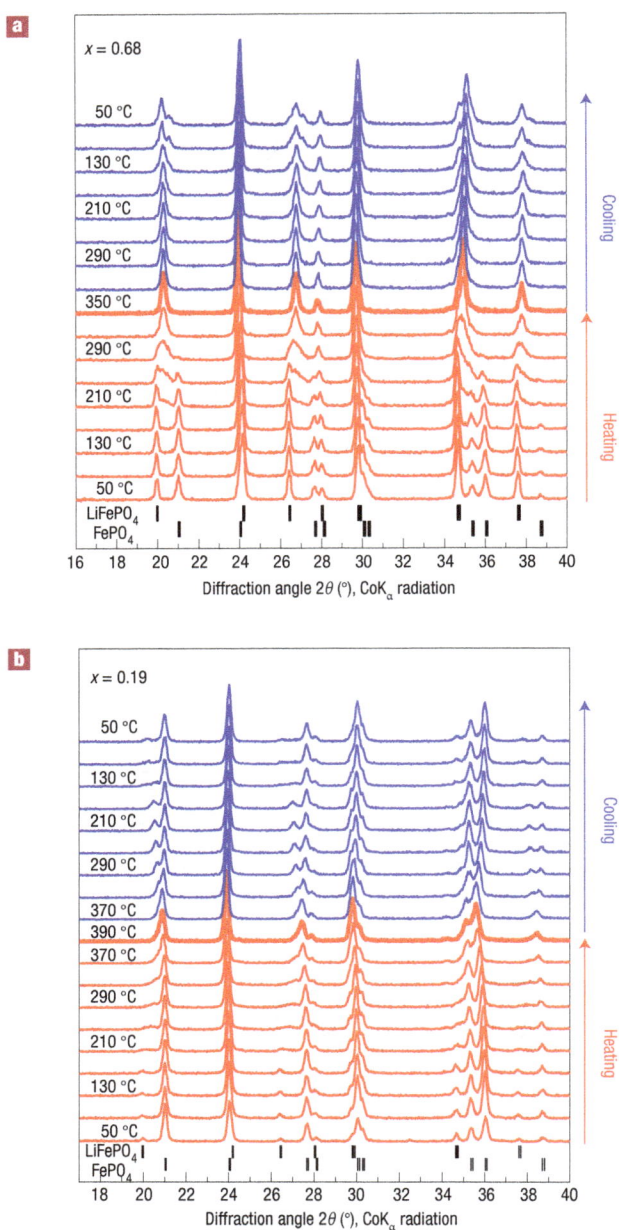

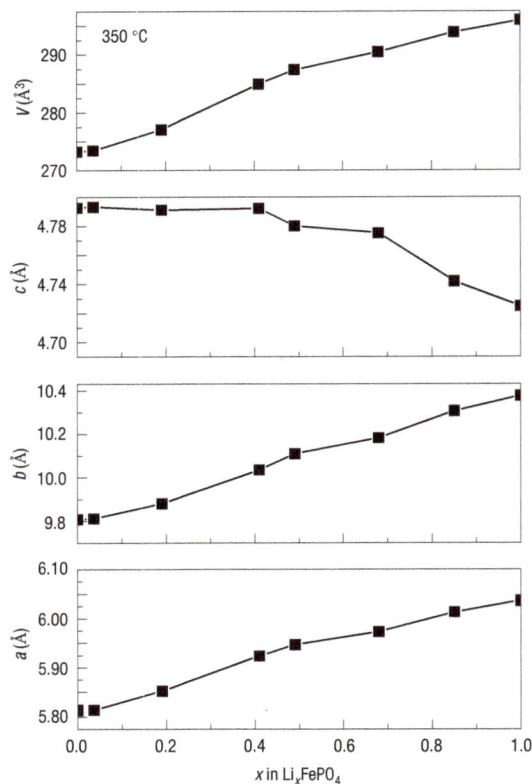

Figure 4 **Unit-cell parameters as a function of x in Li$_x$FePO$_4$ at 350 °C.** These values were refined in the *Pmnb* space group from the data of Fig. 3. Noticeably, from LiFePO$_4$ to FePO$_4$, a and b undergo a linear contraction of 3.7 and 5.5%, respectively, whereas c undergoes a slight expansion of 1.4%.

Figure 3 Temperature-controlled XRD patterns of two selected Li$_x$FePO$_4$ global compositions under N$_2$. a, 'Li$_{0.68}$FePO$_4$' **b,** 'Li$_{0.19}$FePO$_4$'. The transformation of each of the two two-phase samples x LiFePO$_4$ + (1−x) FePO$_4$ into Li$_x$FePO$_4$ single phases is initiated at ~200 °C, and completed at ~350 °C for Li$_{0.68}$FePO$_4$ and ~390 °C for Li$_{0.19}$FePO$_4$. On cooling, these single phases split into complex mixtures of olivine-type phases. Importantly, this behaviour was observed for the whole series of x LiFePO$_4$ + (1−x) FePO$_4$ samples gathered in Fig. 2.

two selected compositions, $x = 0.68$ and $x = 0.19$. As the temperature is raised to around 200 °C, the intensities of the diffraction peaks for both LiFePO$_4$ and FePO$_4$ start to vanish, together with the appearance of broad diffraction peaks intermediate between the reflections of LiFePO$_4$ and FePO$_4$. As the temperature increases,

these diffuse intensities increase, giving rise to well-defined new sets of reflections characteristic of new Li$_x$FePO$_4$ phases. This very broad and continuous temperature range within which the transformation occurs ($\Delta T \sim 150$–200 °C) leads to no thermal signature in differential scanning calorimetry on first heating. Indexing of the XRD reflections of the single phase measured at 350 °C was straightforward using the same space group *Pmnb* as for the two end-members. The unit-cell parameters for all the members of the Li$_x$FePO$_4$ series were obtained from full pattern-matching refinements of XRD data recorded at 350 °C (390 °C for Li$_{0.19}$FePO$_4$). Precise structural descriptions of these new Li$_x$FePO$_4$ compositions with fractional occupancy of the Li site and a mixed-valence Fe^{3+}/Fe^{2+} site, obtained from temperature-controlled neutron diffraction data, are given in another paper (C. Delacourt *et al.*, to be published). Rietveld refinements carried out on neutron diffraction data for Li$_{0.5}$FePO$_4$ solid solutions confirmed the existence of mixed-valence Fe^{3+}/Fe^{2+} single phases. Moreover, valence-bond calculations on the iron site were shown to agree very well with nominal compositions. As shown in Fig. 4, the unit cell constants a and b linearly decrease by −3.7% and −5.5% from $x = 1$ to $x = 0$ in Li$_x$FePO$_4$, respectively, whereas c almost linearly increases by +1.4% from $x = 1$ to $x \approx 0.4$, and remains fairly unchanged from $x \approx 0.4$ to $x = 0$. Interestingly, the emptying of Li sites that are gathered as LiO$_6$ chains running along [100]$_{Pmnb}$ (Fig. 1a) significantly diminishes the average Li–Li site distance within these chains (= $a/2$) and also leads to an even larger contraction of the b parameter, which is characteristic of the interlayer distance between corner-sharing FeO_6 planes (a, c) (Fig. 1b).

On heating at temperatures greater than ~500 °C, Li_xFePO_4 phases start to decompose into mixtures of non-olivine compounds. The nature and extent of each of the formed phosphates and/or oxides depend on the value of x. For Li-rich compositions, Li_xFePO_4 transforms into $LiFePO_4$, monoclinic $Li_3Fe_2(PO_4)_3$ (ref. 22), and iron oxides. As the value of x decreases, the proportion of $LiFePO_4$ decreases and a mixed-valence iron phosphate $Fe^{II}_3Fe^{III}_4(PO_4)_6$ appears[23]. The latter phase was reported[24] as a decomposition product of $x LiFePO_4 + (1-x) FePO_4$. This work, however, did not mention the possible existence of intermediate Li_xFePO_4 phases. As x decreases even more towards Li_0FePO_4, the proportions of $Li_3Fe_2(PO_4)_3$ and $Fe^{II}_3Fe^{III}(PO_4)_6$ significantly decrease while α-$FePO_4$ (ref. 25) is formed. One should note that Li-poor Li_xFePO_4 compositions ($x < 0.10$) transform into non-olivine phases at relatively high temperatures, as previously mentioned[17] for Li_0FePO_4.

Having discovered a whole range of Li_xFePO_4 compositions for ~400 < T < ~500 °C (that is, a new series of mixed-valence Li-deficient Fe^{II}–Fe^{III} phosphates), the next step was to determine whether such compositions could be stabilized at room temperature. Overall, as detailed below, the behaviour of Li_xFePO_4 on cooling is more complicated than on heating, resulting in mixtures of phases depending on x, as well as on the thermal history of the samples. When cooled rather slowly in situ (~1 °C min^{-1}), Li_xFePO_4 solid solution compositions demix into other olivine-type Li_xFePO_4 phases at temperatures that strongly depend on the global Li-content, according to XRD patterns: 370 ± 20 °C for $Li_{0.19}FePO_4$, 150 ± 20 °C for $Li_{0.68}FePO_4$ (Fig. 2). This phase separation occurs in two steps. The first is a phase separation of Li_xFePO_4 into two other olivine-type phases, whose composition depends on the initial value of x and on the temperature. Then, below a temperature of 140 ± 20 °C, this two-phase system turns into a more complex mixture in which the two end-members $LiFePO_4$ and $FePO_4$ coexist with two other olivine-type phases of specific compositions $Li_{x1}FePO_4$ and $Li_{x2}FePO_4$. The cell parameters of these two intermediate phases were consistently determined from the average of a series of refinements of various Li_xFePO_4 starting compositions: $a_1 = 5.95(2)$ Å, $b_1 = 10.20(1)$ Å, $c_1 = 4.73(1)$ Å, and $a_2 = 5.91(2)$ Å, $b_2 = 10.08(1)$ Å and $c_2 = 4.75(2)$ Å (Fig. 5). The values of x_1 and x_2 were interpolated by assuming that a and b follow the Vegard's law, that is, they undergo a linear variation between Li_0FePO_4 and $LiFePO_4$ at room temperature. From interpolation of b, we found with remarkable robustness the values of $x_1 = 0.75 ± 0.02$ and $x_2 = 0.52 ± 0.02$, confirmed by those obtained from interpolation of a, that is, $x_1 = 0.73 ± 0.09$ and $x_2 = 0.55 ± 0.09$.

This result, which has never been previously reported, is of importance as it clearly brings experimental evidence that intermediate olivine-type phases with well-defined compositions can be obtained at room temperature. Another point concerns the values of $x_1 = 3/4$ and $x_2 = 1/2$ suggesting a possible ordering of 3 and 2 Li^+ ions out of 4 within the $LiFePO_4$ unit-cell, respectively. Interestingly, Zhou et al. explored the possible existence of five intermediate compounds ($Li_{0.25}FePO_4$, $Li_{0.5}FePO_4$-a, $Li_{0.5}FePO_4$-b, $Li_{0.5}FePO_4$-c and $Li_{0.75}FePO_4$) from LDA and GGA calculations having lower symmetries (monoclinic or triclinic) than the end members $LiFePO_4$ and $FePO_4$ are predicted[19]. We may wonder though if a Li-ordering within the structure exists (and possibly an induced-charge ordering on the iron sites), as no structural changes such as distortion of the orthorhombic unit-cell or superstructure formation was yet observed. So far, we have only been able to detect, from DSC measurements upon cooling, a broad and small exothermic peak centred at 180 °C for intermediate values of x, whose intensity was maximum for the composition $Li_{0.4}FePO_4$: $\Delta H_{x=0.4} \approx -30$ mJ mol^{-1}. On second heating, a small endothermic peak is observed at

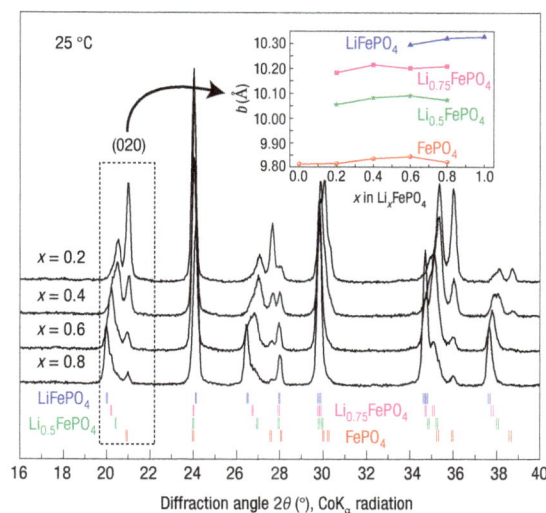

Figure 5 XRD patterns of quenched Li_xFePO_4 (x = 0.8, 0.6, 0.4, 0.2) samples, from 400 °C to 25 °C . Inset: cell parameter b (in $Pmnb$) as a function of x, determined from full pattern-matching for each composition. Quenching of Li_xFePO_4 single phases leads to a complex mixture of three or four olivine-type phases, consisting of the two end-members and in two new metastable phases, namely $Li_{0.75}FePO_4$ and $Li_{0.5}FePO_4$. The proportion of each phase within the mixture depends on the initial composition. XRD patterns were recorded with CoK_α radiation, by steps of 4 s per 0.032°.

T ~220–230 °C, which probably corresponds to the reverse phenomenon.

On aging at room temperature, the dark-green-olive mixture of Li_xFePO_4 phases ($LiFePO_4$, $Li_{0.75}FePO_4$, $Li_{0.5}FePO_4$ and $FePO_4$) slowly transforms into $LiFePO_4$ and $FePO_4$ alone, which is a clear signature of the metastability of both $Li_{0.75}FePO_4$ and $Li_{0.5}FePO_4$ at room temperature. Whatever the quenching protocol (liquid nitrogen or room temperature between stainless steel pieces), mixtures of the four olivine phases were systematically obtained, in proportions depending on the initial Li_xFePO_4 composition. This observation is consistent with the relatively high kinetics of demixing of the intermediate phases at room temperature. From a compilation of all our XRD data collected on cooling and heating, we successfully established a temperature phase-distribution diagram of Li_xFePO_4 (Fig. 6). Note that caution should be exercised in not literally interpreting such a diagram constructed from a still limited number of starting compositions.

Overall, we bring new insights to the thermal behaviour of the binary phase diagram between $LiFePO_4$ and $FePO_4$ as a function of temperature. Previously, only a two-phase electrochemical or chemical process had been reported between these two end-members corresponding to a complete miscibility gap at room temperature. Our findings raise several fundamental questions as to the origin of the driving force enabling the solid solution system to separate into different Li-containing phases by lowering temperature, and furthermore challenge theoretical prediction models commonly used in the field of Li-insertion chemistry. Among the numerous approaches, early stage lattice gas-models based on mean-field theory were used to model the configurational entropy as a function of temperature[20,26]. For instance, simulations with the introduction of a negative term U (indicative of attractive interactions between

a Heating

b Cooling

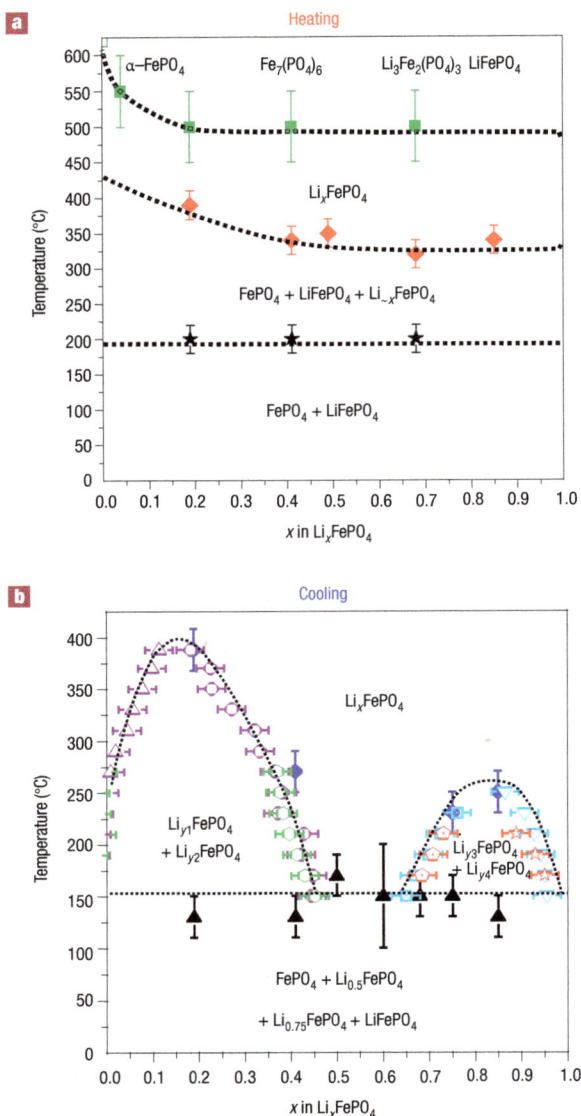

Figure 6 Phase distribution diagrams of Li$_x$FePO$_4$ ($0 \le x \le 1$) established from temperature-controlled XRD data. **a**, On heating and **b**, on cooling. The filled symbols refer to the actual compositions x that were investigated. For a given x composition ($x = 0.85$, $x = 0.75$, $x = 0.68$, $x = 0.6$, $x = 0.50$, $x = 0.41$, $x = 0.19$ and $x = 0.04$) we plotted symbols each time significant changes in the XRD patterns were detected. Open symbols are the result of extensive refinements of the XRD data. On heating, xLiFePO$_4$ + (1−x) FePO$_4$ mixtures transform into a single Li$_x$FePO$_4$ phase through a succession of continuous stages from ~200 to ~350 °C, and then decompose into a mixture of non-olivine phases above ~500 °C. On cooling, Li$_x$FePO$_4$ solid solutions separate into a mixture of two olivine-type phases, whose compositions depend on initial x and T. For instance, the global composition Li$_{0.19}$FePO$_4$ was refined, on cooling at 330 °C, as a mixture of Li$_{0.05}$FePO$_4$ and Li$_{0.27}$FePO$_4$ (from Vegard's law interpolation). Below ~140 °C, a complex mixture of LiFePO$_4$, Li$_{0.75}$FePO$_4$, Li$_{0.25}$FePO$_4$ and FePO$_4$ is obtained, with the proportions of each phase depending on the initial value of x. Vertical error bars correspond to each temperature-controlled XRD data collection. Horizontal error bars are estimated from the standard deviation of the refined cell parameters of each intermediate phase (Li$_{y1}$FePO$_4$, Li$_{y2}$FePO$_4$, Li$_{y3}$FePO$_4$ and Li$_{y4}$FePO$_4$).

Li$^+$-ions) was shown to describe superbly well the temperature-driven phase separation in the Li$_x$Mo$_6$S$_8$ ($0 \le x \le 1$) system for which room-temperature solid solution undergoes a phase separation (for $T < -6$ °C) into a Li-rich and a Li-poor Li$_x$Mo$_6$Se$_8$ phase[20]. Such a term U cannot be determined for LiFePO$_4$ at room temperature, owing to the two-phase nature of the Li-insertion process. In the hope of overcoming such limitation, we are currently performing challenging electrochemical measurements within the temperature range over which the Li$_x$FePO$_4$ system is a solid solution. Furthermore, although it is tempting to compare the Li$_x$FePO$_4$ and Li$_x$Mo$_6$Se$_8$ systems, one might use the elastic modulus as a means to account for the great discrepancy between the temperature domains in which the transition occurs (around room temperature for Li$_x$Mo$_6$Se$_8$ to be compared with ~200–350 °C for Li$_x$FePO$_4$): higher covalency of sulphides and selenides as compared with oxides leads to higher elasticity of the lattices and thus is more favourable to single-phase insertion/extraction processes. Another difference lies in the irreversibility of the transition for the Li$_x$FePO$_4$ system: cooling a Li$_x$FePO$_4$ solid solution reveals Li$_{0.5}$FePO$_4$ and Li$_{0.75}$FePO$_4$ intermediate phases, which were never observed on heating.

More recent models use density functional theory that gives information about the energies of various Li-vacancy configurations in oxide systems. In that respect, classical LDA or GGA have agreed remarkably well with experiments in predicting miscibility and phase stability[27]. Such approaches were also very successful in sorting/predicting phase metastability within the Li$_x$CoO$_2$ system[27], and are taking over from mean-field theory ones. The relative weakness of such methods is that they tend to overly delocalize electrons so that they are unable to properly describe the thermodynamics and the electronic properties of insulating compounds[28]. Classical LDA/GGA[19] led to the prediction that Li$_x$FePO$_4$ should be a solid solution at room temperature over the whole composition range. To counterbalance the discrepancy between theory and experiment, Zhou et al. brought an important correction to LDA and GGA models by explicitly considering coulombic correlation effects through a term U' that corresponds to an 'effective' on-site interaction: they calculated that the threshold value to account for the appearance of the two-phase process was $U' \sim 2.5$ eV.

Lacking theoretical models to fully account for such behaviour, we simply rely on thermodynamics and kinetics from a qualitative point of view. On heating, thermal agitation (entropy) tends to loosen the 'attraction' interactions leading, above 200 °C, to Li$^+$ motion within the structure. Li$^+$–Li$^+$ and Fe^{3+}–Fe^{3+} (or Fe^{2+}–Fe^{2+}) 'attractive' interactions are a consequence of the gain in elastic potential energy arising from phase separation. The peculiar shapes of the XRD peaks observed on progressive heating (intermediate broad signals between those of the two end-members) suggest that Li motion is of a diffusional type, thus leading to a Li gradient between the Li-rich part and the Li-poor part of the grain(s). The lack of any distinguishable thermal effect on first heating suggests a second-order-type phase transition that progressively occurs, without the formation of well-defined intermediate phases. The system evolves through a successive sequence of intermediate states of lower energies (that is, reverse of spinodal decomposition) that results in a continuous lowering of ΔG. On the other hand, by decreasing the temperature (lowering entropy), the single phase Li$_x$FePO$_4$ becomes metastable, as the 'attraction' interactions become prominent. Therefore, phase separation occurs as a succession of intermediate steps as the temperature is progressively decreased. The first one consists of a progressive splitting of Li$_x$FePO$_4$ into two olivine-type phases whose composition depends on x and T. Then below 140 °C, these intermediate phases transform into a mixture of three or four olivine-type phases, according to the initial composition x (LiFePO$_4$, Li$_{0.75}$FePO$_4$, Li$_{0.50}$FePO$_4$ or FePO$_4$) (Fig. 6). On aging at room temperature, Li$_{0.75}$FePO$_4$ and Li$_{0.50}$FePO$_4$ slowly separate into the end-members LiFePO$_4$ and FePO$_4$. Note, Li$_{0.25}$FePO$_4$ was never observed at room temperature, which would suggest that it demixes very quickly. It is legitimate to wonder whether these intermediate phases have a thermodynamic existence in a defined range of temperature, that is, if they belong to the binary diagram.

The onset temperature of a miscibility gap might be influenced by chemical composition such as that demontrated earlier[29] for the Li$_x$Mo$_6$Se$_z$S$_{8-z}$ system: Se substitution for S results in a great lowering of the temperature transition due a smaller value of |U| (ref. 29). Other contributions from Yamada[30–32] and Chiang's group[33] deserve some comments with respect to the Fe^{3+}/Fe^{2+} mixed-valence Li$_x$FePO$_4$ compositions ($0 \leq x \leq 1$) reported here. Yamada's work on the Li$_x$(Mn$_y$Fe$_{1-y}$)PO$_4$ phase diagram at room temperature showed the existence of a solid solution domain when lithium is extracted from Li$_{1-x}$(Mn$^{2+}_y$Fe$^{2+}_{1-y-x}$Fe$^{3+}_x$)PO$_4$ for selected values of y (ref. 31). On the other hand, Chiang and colleagues[33] reported on the possible existence of Li$_{1-a-x}$M$_x$FePO$_4$ or Li$_{1-x}$M$_x$Fe$_{1-b}$PO$_4$ compositions by 'doping' with elements supervalent to Li$^+$ such as Zr or Nb. Our results suggest that Li-deficient compositions may indeed exist but we still question[34] the possible incorporation of Nb or Zr into the triphylite structure and its real implication on the overall electrical conductivity: we have strong evidence, in full accordance with Subramanya-Herle[12], that Fe$_2$P and/or an amorphous (Nb, Fe, C, O, P) wrap around LiFePO$_4$ particles are produced instead, which are very conductive (up to $1.6.10^{-1}$ S cm^{-1}) with an activation energy $\Delta E \sim 0.08$ eV. Instead of using composition as a variable that would lead to single-phase electrochemical domains within the olivine family, our finding also highlights the importance of temperature in creating conditions favourable to single-phase domains containing mixed transition-metal valence states and/or partial occupation of the Li$^+$ sites.

These results open the door to a systematic investigation of the temperature and composition dependence of Li insertion electrodes operating at room temperature through a two-phase process, with the end members having a barely distinguishable structure. Along that line, Li$_4$Ti$_5$O$_{12}$ appears to be among the most suitable candidates to generalize such a finding. Finally, a more basic question to address is whether insertion reactions in insulating compounds can ever proceed as a single-phase process, and thus could solid solution reactions be indicative of an enhancement of intrinsic electronic conductivity. This topic will remain a fruitful area for further investigations.

METHODS

SYNTHESIS

Pristine LiFePO$_4$ was obtained as fine and dispersed particles (~500 nm) from a straightforward aqueous solution route involving a Fe^{III} precursor, as we reported earlier[35]. An aqueous solution containing Li, Fe and P in stoichiometric proportions was first prepared at room temperature by dissolving Fe(NO$_3$)$_3$.9H$_2$O (Sigma Aldrich, 98+%) and LiH$_2$PO$_4$ (Sigma Aldrich, 99%). Slow evaporation under continuous stirring was then followed by a classical two-step thermal treatment at 350 °C and 550 °C under N$_2$/H$_2$ during 10 h. x LiFePO$_4$ / (1−x) FePO$_4$ two-phase mixtures ($0 \leq x \leq 1$)were prepared by chemical oxidation of pristine LiFePO$_4$ with nitronium tetrafluoroborate in acetonitrile (NO$_2$BF$_4$, Acrös Organics, 97%). 1 g of LiFePO$_4$ was immersed in 50 ml of a solution of NO$_2$BF$_4$ in acetonitrile for two days in an environmentally controlled dry box at 20 °C under vigorous magnetic stirring[21]. The depth of delithiation (1−x) was adjusted by the concentration of NO$_2$BF$_4$ in the solution.

CHEMICAL ANALYSIS

Each of the x LiFePO$_4$ / (1−x) FePO$_4$ two-phase mixtures was dissolved in an acidic solution of 10 wt% HCl and 10 wt% HNO$_3$ at ~50 °C under vigorous stirring. Atomic absorption spectroscopy using a Perkin Elmer AAnalyst 300 monitored by the AAWinlab^{TM} software determined the atomic Li/Fe ratios in the obtained solution. Standard solutions of 1.035, 2.07, 2.5, 5 and 6.21 mg l^{-1} were used for the calibration of Fe. Calibration of Li was carried out using standard solutions of 1, 2 and 3 mg l^{-1}.

TEMPERATURE-CONTROLLED XRD

The values of x in x LiFePO$_4$ / (1−x) FePO$_4$ two-phase mixtures (which very satisfactorily agree with those determined by chemical analysis) were determined by refining XRD data collected from a Philips PW 1710 diffractometer (Cu K$_\alpha$ radiation, back monochromator). Instrumental parameters had been previously determined using a LaB$_6$ calibrated powder (NIST) as a standard. Temperature-controlled XRD under N$_2$ was performed on a Bruker D8 Diffractometer (Co K$_\alpha$ radiation, Göbel mirror, radial slits and PSD counter) equipped with an Anton Parr Chamber HTK 1,200 °C. Each pattern (every 20 °C or 50 °C) was recorded for ~30 minutes at constant temperature with a step of 0.032° and an acquisition time of 2 s. Between each fixed temperature, the powder was heated at a rate of 6 °C min^{-1}. The same conditions were used for cooling. All refinements, either full pattern matching or Rietveld refinements were carried out using the Fullprof suite[36].

DIFFERENTIAL SCANNING CALORIMETRY

Differential scanning calorimetry was carried out between room temperature and 400 °C on heating and cooling with a rate of 5° min^{-1} in sealed Al crucibles under Argon using a Mettler DSC 25 apparatus with the TC11 interface. 80–90 mg of x LiFePO$_4$ + (1−x) FePO$_4$ mixtures (with x = 1, 0.8, 0.6, 0.4, 0.2 and 0) were realized by intimate mixing of the powders. To prevent any parasitic thermal effect, FePO$_4$ powders obtained from chemical delithiation of FePO$_4$ were previously heat treated at 400 °C for 24 h.

Received 4 November 2004; accepted 17 January 2005; published 20 February 2005.

References

1. Padhi, A. K., Nanjundaswamy, K. S. & Goodenough, J. B. Phospho-olivines as positive electrode materials for lithium batteries. *J. Electrochem. Soc.* **144**, 1188–1194 (1997).
2. Ravet, N. *et al.* in *196th Electrochemical Society Meeting, Honolulu, Hawaii* Abstract 127 (Electrochemical Society, Pennington, New Jersey, 1999).
3. Dominko, R., Gaberscek, M., Drofenik, J., Bele, M. & Pejovnik, S. A novel coating technology for preparation of cathodes in Li-ion batteries. *Electrochem. Solid State Lett.* **4**, A187–A190 (2001).
4. Huang, H., Yin, S. C. & Nazar, L. F. Approaching theoretical capacity of LiFePO$_4$ at room temperature at high rates. *Electrochem. Solid State Lett.* **4**, A170–A172 (2001).
5. Yamada, A., Chung, S. C. & Hinokuna, K. Optimized LiFePO$_4$ for lithium battery cathodes. *J. Electrochem. Soc.*, **148**, A224–A229 (2001).
6. Franger, S., Le Cras, F., Bourbon, C. & Rouault, H. LiFePO$_4$ synthesis routes for enhanced electrochemical performance. *Electrochem. Solid State Lett.* **5**, A231–A233 (2002).
7. Armand, M., Gauthier, M., Magnan, J. -F. & Ravet, N. Method for synthesis of carbon-coated redox materials with controlled size. World Patent WO 02/27823 A1 (2002).
8. Arnold, G. *et al.* Fine-particle lithium iron phosphate LiFePO$_4$ synthesized by a new low-cost aqueous precipitation technique. *J. Power Sources* **119–121**, 247–251 (2003).
9. Barker, J., Saidi, M. Y. & Swoyer, J. L. Lithium iron(II) phospho-olivines prepared by a novel carbothermal reduction method. *Electrochem. Solid State Lett.* **6**, A53–A55 (2003).
10. Yamada, A. *et al.* Olivine-type cathodes: achievements and problems. *J. Power Sources* **119–121**, 232–238 (2003).
11. Audemer, A., Wurm, C., Morcrette, M., Gwizdala, S. & Masquelier, C., Carbon-coated metal bearing powders and process for production thereof. World Patent WO 2004/001881 A2 (2004).
12. Subramanya Herle, P., Ellis, B., Coombs, N. & Nazar, L. F. Nano-network electronic conduction in iron and nickel olivine phosphates. *Nature Mater.* **3**, 147–152 (2004).
13. Delacourt, C., Wurm, C., Reale, P., Morcrette, M. & Masquelier, C. Low temperature preparation of optimized phosphates for Li-battery applications. *Solid State Ionics* **173**, 113–118(2004).
14. Andersson, A. S., Kalska, B., Häggström, L. & Thomas, J. O. Lithium extraction/insertion in LiFePO$_4$: an X-ray diffraction and Mössbauer spectroscopy study. *Solid State Ionics* **130**, 41–52 (2000).
15. Santoro, R. P. & Newman, R. E. Antiferromagnetism in LiFePO$_4$. *Acta Crystallogr.* **22**, 344–347 (1967).
16. Andersson, A. S. & Thomas, J. O. The source of first-cycle capacity loss in LiFePO$_4$. *J. Power Sources* **97–98**, 498–502 (2001).
17. Rousse, G., Rodriguez-Carvajal, J., Patoux, S. & Masquelier, C. Magnetic structures of the triphylite LiFePO$_4$ and of its delithiated form FePO$_4$. *Chem. Mater.* **15**, 4082–4090 (2003).
18. Morgan, D., Van der Ven, A. & Ceder, G. Li conductivity in Li$_x$MPO$_4$ (M = Mn, Fe, Co, Ni) olivine materials. *Electrochem. Solid State Lett.* **7**, A30–A32 (2004).
19. Zhou, F., Marianetti, C. A., Cococcioni, M., Morgan, D. & Ceder, G. Phase separation in Li$_x$FePO$_4$ induced by correlation effects. *Phys. Rev. B* **69**, 201101 (2004).
20. Dahn, J. R. & Mckinnon W. R. Phase diagram of Li$_x$Mo$_6$Se$_8$ for 0 < x < 1 from *in situ* X-ray studies. *Phys. Rev. B* **32**, 3003–3005 (1985).
21. Wizansky, A. R., Rauch, P. E. & Disalvo, J. F. Powerful oxidizing agents for the oxidative deintercalation of lithium from transition-metal oxides. *J. Solid State Chem.* **81**, 203–207 (1989).
22. Bykov, A. *et al.* Superionic conductors Li$_3$M$_2$(PO$_4$)$_3$ (M=Fe, Sc, Cr): synthesis, structure and electrophysical properties *Solid State Ionics* **38**, 31–52 (1990).
23. Gorbunov, Yu. A. *et al.* Crystal structure of Fe$^{2+}_3$Fe$^{3+}_4$[PO$_4$]$_6$. *Sov. Phys. Dokl.* **25**, 785–787 (1980).
24. Dodd, J., Yazami, R. & Fultz, B. in *206th meeting of the Electrochemical Society, Honolulu, Hawaii* Abstract 425 (Electrochemical Society, Pennington, New Jersey, 2004).
25. Goiffon, A., Dumas, J.- C. & Philippot, E. Phases de type quartz alpha: structure de FePO$_4$ et spectrométrie Mössbauer du Fer-57. *Rev. Chimie Minérale* **23**, 99–110 (1986).
26. Coleman, S. T., Mckinnon W. R. & Dahn, J. R. Lithium intercalation in Li$_x$Mo$_6$Se$_8$: A model mean-field lattice gas. *Phys. Rev. B* **29**, 4147–4149 (1984).
27. Ceder, G. & Van der Ven, A., Phase diagrams of lithium transition metal oxides: investigations from first principles. *Electrochem. Acta* **45**, 131–150 (1999).
28. Anisimov, V. I. in *Advances in Condensed Matter Science* (Gordon & Breach, New York, 2000).
29. Selwyn, L. S., McKinnon, W. R., Dahn, J. R. & Le Page Y. Local environment of Li intercalated in Mo$_6$Se$_8$S$_{8-z}$ as probed using electrochemical methods. *Phys. Rev. B* **33**, 6405–6414 (1986).
30. Yamada, A. & Chung, S. C. Crystal chemistry of the olivine-type Li(Mn$_y$Fe$_{1-y}$)PO$_4$ and (Mn$_y$Fe$_{1-y}$)PO$_4$ as possible 4V cathode materials for lithium batteries. *J. Electrochem. Soc.* **148**, A960–A967 (2001).
31. Yamada, A., Kudo, Y. & Liu, K. -Y. Phase diagram of Li$_x$(Mn$_y$Fe$_{1-y}$)PO$_4$ (0 ≤ x, y ≤ 1). *J. Electrochem. Soc.* **148**, A1153–A1158 (2001).

32. Yamada, A., Kudo, Y. & Liu, K. Y. Reaction mechanism of the olivine-type $Li_x(Mn_{0.6}Fe_{0.4})PO_4$ ($0 \leq x \leq 1$). *J. Electrochem. Soc.* **148**, A747–A754 (2001).

33. Chung, S. Y., Bloking, J. T. & Chiang, Y. M. Electronically conductive phospho-olivines as lithium storage electrodes. *Nature Mater.* **1**, 123–126 (2002).

34. Delacourt, C., Wurm, C., Laffont, L., Leriche, J. B. & Masquelier, C. in *Proc. MRS Fall Meeting 2004* (in the press).

35. Wurm, C., Morcrette, M., Gwizdala, S. & Masquelier, C. Lithium transition-metal phosphate powder for rechargeable batteries. World Patent WO 02/099913 A1 (2002).

36. Rodríguez-Carvajal, J. Recent advances in magnetic structure determination by neutron powder diffraction. *Physica B* **192**, 55–69 (1993).

Acknowledgements

The authors express their sincere gratitude to D. Murphy and M. Touboul for helpful comments and discussions on this manuscript. They also thank UMICORE (Belgium) and C. Wurm for providing the pristine $LiFePO_4$ powder.
Correspondence and requests for materials should be addressed to C.M.

Competing financial interests

The authors declare that they have no competing financial interests.

Nano-network electronic conduction in iron and nickel olivine phosphates

P. SUBRAMANYA HERLE, B. ELLIS, N. COOMBS AND L. F. NAZAR*

University of Waterloo, Department of Chemistry, 200 University Ave. West, Waterloo Ontario Canada N2L 3G1
*e-mail: lfnazar@uwaterloo.ca

Published online: 22 February 2004; doi:10.1038/nmat1063

The provision of efficient electron and ion transport is a critical issue in an exciting new group of materials based on lithium metal phosphates that are important as cathodes for lithium-ion batteries. Much interest centres on olivine-type $LiFePO_4$, the most prominent member of this family[1]. Whereas the one-dimensional lithium-ion mobility in this framework is high[2], the electronically insulating phosphate groups that benefit the voltage also isolate the redox centres within the lattice. The pristine compound is a very poor conductor ($\sigma \sim 10^{-9}\,S\,cm^{-1}$), thus limiting its electrochemical response. One approach to overcome this is to include conductive phases, increasing its capacity to near-theoretical values[3–6]. There have also been attempts to alter the inherent conductivity of the lattice by doping it with a supervalent ion. Compositions were reported to be black p-type semiconductors with conductivities of $\sim 10^{-2}\,S\,cm^{-1}$ arising from minority Fe^{3+} hole carriers[7]. Our results for doped (and undoped) $LiMPO_4$ (M = Fe, Ni) show that a percolating nano-network of metal-rich phosphides are responsible for the enhanced conductivity. We believe our demonstration of non-carbonaceous-network grain-boundary conduction to be the first in these materials, and that it holds promise for other insulating phosphates.

Fascination with the initial report of doped olivine compositions $Li_{0.99}D_{0.01}FePO_4$ (D = Nb, Zr, Mg, Ti) stemmed not only from their enhanced electrochemical performance, but also from the enormous increase in conductivity seen[7] in samples densified at 800 °C. Nonetheless, the description of a highly electronically conductive phosphate challenges conventional wisdom owing to electronic and structural considerations. Small amounts of carbon contaminant that always arise from the organic iron salt precursors[8] have been considered as a contributing factor to conductivity, and factors based on sources of carbon were suspect[9]. This leads to particularly important thermodynamic considerations. Metal phosphates have a propensity to undergo carbothermal reduction at high temperature. The reaction with $LiFePO_4$ and carbon oxidizes the carbon to CO or CO_2, and reduces the neighbouring Fe and P ions in the lattice to Fe_2P and/or Fe_3P. We have detected bulk phosphide formation through X-ray diffraction by deliberate addition of carbon 'gel' to the precursors[10]. The addition of carbon has also been used to form $LiFe^{II}PO_4$ from Fe^{3+} precursors, using careful temperature control[11]. High temperatures result in reduction to Fe_2P, whose presence in the bulk at temperatures above 850 °C has been implicated[7], although it was ruled out as the conductivity source in doped $LiFePO_4$ on the basis of microcontact measurements[12].

We find the carbothermal chemistry can be manipulated through surprisingly subtle variation of the processing conditions and stoichiometry. The nanophase phosphide network that is created in situ within the grain boundaries of the insulating $LiMPO_4$ nanocrystallites forms an efficient electrical conduit that can be directly imaged by transmission electron microscopy (TEM). Furthermore, metal phosphates of more easily reduced elements (Ni) in $LiNiPO_4$ undergo carbothermal reduction at lower temperatures, and can be rendered conductive through this approach. That electronic transport is first and foremost a consequence of the network, rather than a mixed-valent metal $M^{2+/3+}$, state is suggested by these results.

Our first evidence that a second phase could be responsible for enhanced conductivity came from studies of doped compositions similar to those reported previously[7]. We examined a range of Li compositions $Li_xZr_{0.01}FePO_4$, where x varies from 0.99 to 0.87, in addition to studying non-Zr-doped samples. Although it is presumed that $LiFePO_4$ has a very narrow compositional range, evidence from substituted materials $Li_{1-3x}Fe_xNiPO_4$ suggests that Li vacancies can be accommodated in the olivine structure[13]. In addition, the doped $Li_{0.99}D_{0.01}FePO_4$ stoichiometry previously reported[7] for supervalent cations such as Zr^{4+} implies a subvalent state of Fe ($<Fe^{II}$) on the basis of charge balance (that is, for $Li_{0.99}Zr_{0.01}FePO_4$:$Fe^{+1.97}$). Hence, loss of lithium during processing would have to occur to account for Fe^{3+} hole carrier formation. Our deliberate departure from Li stoichiometry should introduce Fe^{3+} 'holes' irrespective of the incorporation of the Zr^{4+} dopant. Following the original procedure[7], $Li_xZr_{0.01}FePO_4$ materials ($x = 0.99 \rightarrow 0.87$) were crystallized at 600 °C, pressed to form a dense pellet, and then sintered in argon at 800 °C. The resultant X-ray diffraction (XRD) patterns for powders and pellets are shown in Fig. 1. The materials prepared at 600 °C (Fig. 1a) are almost entirely triphyllite, $LiFePO_4$. Compositions down to $Li_{0.93}Zr_{0.01}FePO_4$ show reflections due to single-phase $LiFePO_4$. Closer inspection reveals that as the Li content decreases, a slight broadening in the (020) reflection at 30.3° in 2θ becomes evident, which can be resolved as very weak satellite lines attributable to $Fe_2P_2O_7$ (Fig. 1a, inset). On heating the pressed pellets to 800 °C the $Fe_2P_2O_7$ component disappears (Fig. 1b). Mössbauer spectroscopy only revealed the presence of Fe^{2+} (Supplementary Information (Fig. S5). TEM images obtained on microtomed sections of pressed pellets showed a uniform distribution of Zr in the electron energy loss spectra (EELS) map, indicating that the dopant was well dispersed (See Supplementary Information, Fig. S1). Conductivity plots

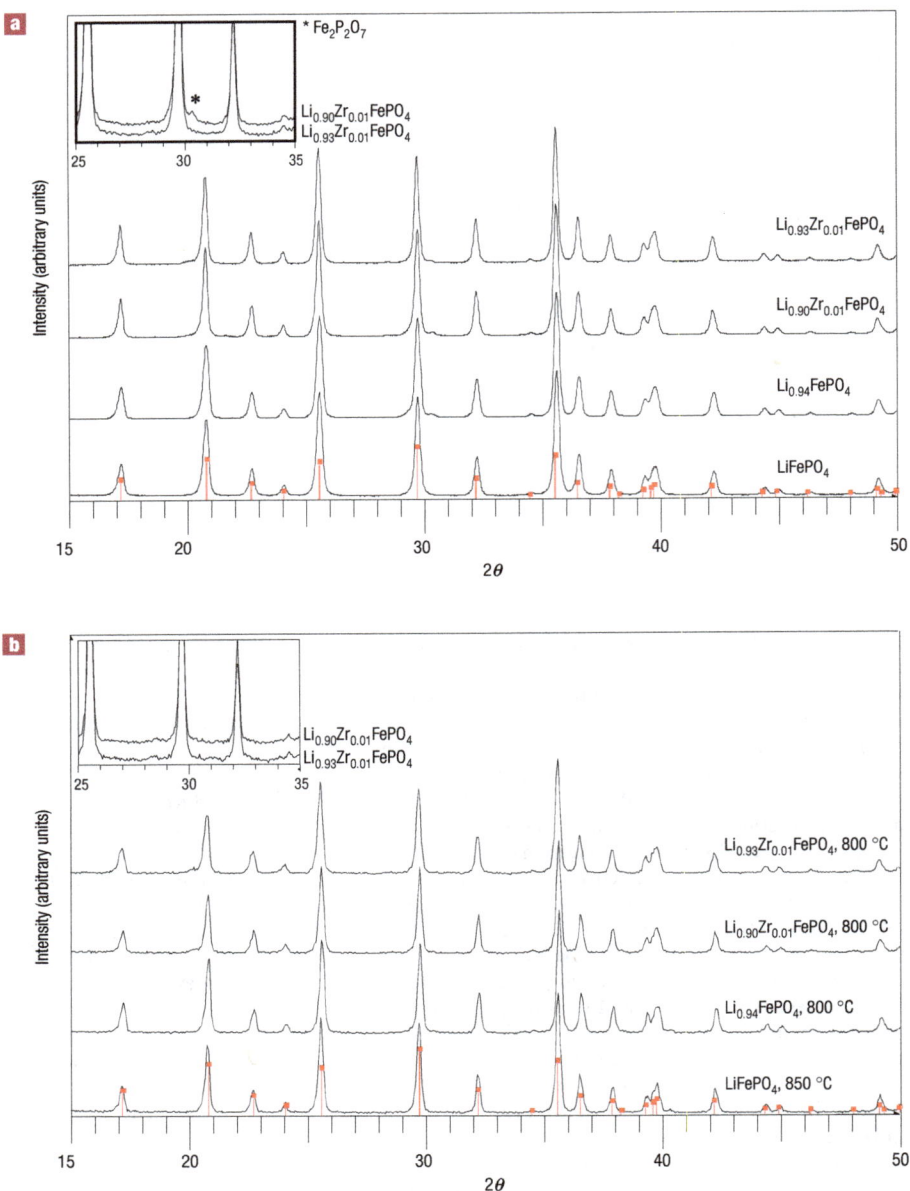

Figure 1 X-ray diffraction patterns of various powders and sintered pellets of Li$_{1-x}$Zr$_y$FePO$_4$ compositions. Near-single-phase triphylite formation occurs at 600 °C in Li substoichiometric samples, although Fe$_2$P$_2$O$_7$ can be detected in the most Li-deficient samples. This component was not detected after sintering the pellets at elevated temperatures in an argon atmosphere. **a**, Li$_{1-x}$Zr$_y$FePO$_4$ powders prepared at 600 °C; **b**, pellets sintered at 800–850 °C. The reflections of LiFePO$_4$ (JCPDS No. 40-1499) are shown on both figures for comparison. We note that a two-phase fit of all of the X-ray data using Rietveld analysis (Supplementary Information, Fig. S2) allowed quantitative assessment of the Fe$_2$P$_2$O$_7$ contribution, showing that it roughly scaled with the substoichiometry as expected. In none of the materials studied, including the Zr-doped samples, was there evidence for solid solution behaviour in Li$_{1-x}$FePO$_4$ in accord with electrochemical measurements, which show two-phase behaviour on Li extraction to form LiFePO$_4$ + FePO$_4$ for all compositions Li$_{1-x}$FePO$_4$.

for all materials (Fig. 2) demonstrate that very high values are achieved for the densified pellets, with conductivities at 300 K in the range of 10^{-2}–10^{-3} S cm^{-1}, and activation energies between 30–40 meV. The values are similar to that reported[7] for Li$_{0.99}$Zr$_{0.01}$FePO$_4$, although the latter did not show appreciable conductivity in our work. The room-temperature value for the composition Li$_{0.96}$Zr$_{0.01}$FePO$_4$ was estimated as 10^{-5} S cm^{-1}. The conductivity generally increases with decreasing

lithium stoichiometry, albeit with some variability due to changes in the processing conditions.

A TEM image of an ultramicrotomed slice taken from a conductive composition, Li$_{0.90}$Zr$_{0.01}$FePO$_4$, is shown in Fig. 3a and corresponding Fe, Li and C maps are displayed in Fig. 3b–d. Together, these show clear evidence of the formation of two distinct nanophases. The regions between the darker LiFePO$_4$ crystallites (20–100 nm in size) in the micrograph appear as black domains in the Fe, Li EELS maps. Particularly startling is the carbon map (Fig. 3d), which reveals that the edges of the $LiFePO_4$ crystallites are coated with a carbon-containing layer between 5 and 10 nm thick, and which forms a network connecting all of the particles and completely filling some of the voids. X-ray fluorescence analysis (EDX) of the regions nearest the grain boundaries gave an Fe/P ratio close to 2:1 (65:35), whereas analysis of the bulk provided an Fe/P ratio of close to 1:1 (45:55), identical to a 'control' sample of non-conducting $LiFePO_4$ densified at 800 °C (Fig. 4, and Supplementary Information, Fig. S3). Thus, Fe_2P and/or iron phosphocarbide, $Fe_{75}P_{15}C_{10}$, is formed by carbothermal reduction of the $LiFePO_4$ with carbon at and within the grain boundaries. Some carbon, arising from both the oxalate precursor and alkoxide in the Zr-dopant, may remain in partially graphitized form. The phosphocarbide is an amorphous, magnetic phase intermediate between Fe_3P and Fe_3C (ref. 14), with a conductivity of about 10^{-1} S cm^{-1}, similar to that of Fe_2P (ref. 15). Their existence gives rise to the observed bulk conductivity in these Zr-doped Li-deficient materials.

Further evidence that these grain-boundary 'nano-networks' can act as efficient electrical conduits comes from our examination of non-Zr doped materials. Samples of Li_xFePO_4 ($x = 0.91$ to 1.00) were prepared using the same process, without the dopant. The values of x were 'valent-equivalent' to the Zr-doped samples described above, where the total valence contribution of $x(Li+) + 0.01Zr$ was kept constant. As in the Zr-doped materials, phase separation to stoichiometric $LiFePO_4$ and a minor amount of $Fe_2P_2O_7$ occurred on crystallization of the precursors at 600 °C, as shown by XRD (for Reitveld ananlysis of phase fraction, see Supplementary Information, Fig. S2). A representative pattern is shown for $Li_{0.94}FePO_4$ in Fig. 1. The fraction of $Fe_2P_2O_7$ correlates with 'x' as expected, and disappears from the pattern at 800 °C (Fig. 1b, inset). Samples pressed at high pressure and heated at 800 °C showed excellent conductivity, achieving the same values as the Zr-doped materials (Fig. 2). Carbon content in $LiFePO_4$ arises from decomposition of the iron oxalate, whereas in the Zr-doped materials, the alkoxide also contributes carbon (12 mol% for the precursor): nonetheless the conductivity of $Li_{0.90}Zr_{0.01}FePO_4$ (800 °C) and $Li_{0.94}FePO_4$ is not substantially different. Samples of Li_xFePO_4 prepared from non-carbon-containing precursors $(Fe_3(PO_4)_2 \cdot 8H_2O + Li_3PO_4)$ were not conductive ($\sigma < 10^{-7}$ S cm^{-1}) at any temperature. Stoichiometric $LiFePO_4$ (iron oxalate precursor) subjected to heat treatment at 800 °C was also essentially non-conductive, with a value less than 10^{-7} S cm^{-1} in accordance with previous reports. However, the conductivity of that sample treated at a slightly higher temperature of 850 °C rose very sharply (Fig. 2), and even exceeded that of the substoichiometric materials to achieve values of about 10^{-2} S cm^{-1}. Carbon from oxalate decomposition is present at both 800 and 850 °C. We conclude that (i) iron phosphide and/or iron phospocarbide at the grain boundaries, not carbon, is primarily responsible for the grain-boundary transport; and (ii) reduction to the phosphide occurs at lower temperatures in substoichiometric $Li_{1-x}FePO_4$ than in its undoped, stoichiometric counterpart. This may be explained by more facile reduction of $Fe_2P_2O_7$ to form iron phosphide, which occurs at a lower temperature than $LiFePO_4$.

These results suggest that the concept can be extended to other phosphates, $LiMPO_4$ (M = Ni, Mn, Co) that have a higher oxidation potential than M = Fe. Samples of Li_xNiPO_4 ($x = 0.94$ to 1.00) were prepared similarly to the iron phosphates, except that the atmosphere and temperature of heat treatment was varied. Unlike $LiFePO_4$, which oxidizes to Fe(III) in air at elevated temperatures, $LiNiPO_4$ is stable in oxygen at high temperature[16], which affords unequivocal confirmation of the carbothermal chemistry. These conditions favour formation of a potentially mixed-valent $Ni^{2+/3+}$ hole-conductor state, but would oxidize conductive nickel phosphides. Preparation of Li_xNiPO_4 by heating carbon-containing precursors at temperatures between 600 and 900 °C in air or oxygen yielded bright-yellow crystalline powders

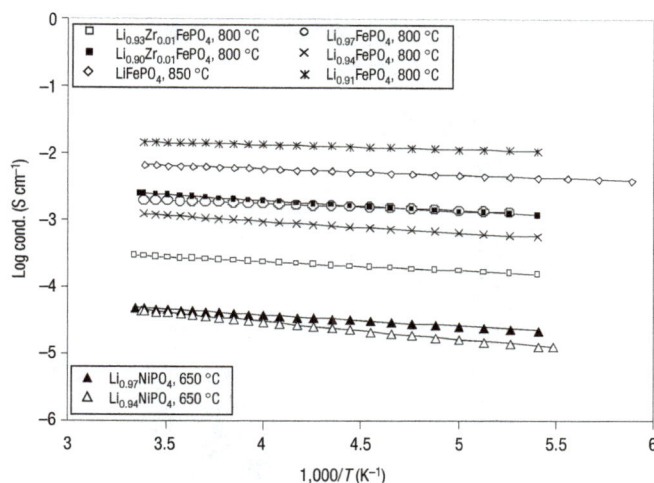

Figure 2 **Conductivity of $Li_{1-x}Zr_yMPO_4$ (M = Fe, Ni) samples.** High conductivity is observed for all samples irrespective of dopant content. All compositions can be made conductive over the temperature range from 300 K to 180 K by choosing suitable sintering temperatures and compositions, showing values in excess of 10^{-3} S cm^{-1} for iron compositions and 10^{-5} S cm^{-1} for the nickel compositions at room temperature. Results are for pellets sintered between 650 °C and 850 °C and measured by four-point probe d.c. methods. For the nickel sample, carbothermal reaction is not initiated below 600 °C, but above this temperature the reaction is sufficiently vigorous that the nucleation of a few larger Ni_3P crystallites ensues. These migrate outside the grain boundaries, resulting in a drop in the overall transport efficiency. We note that heat treatment of Li_xNiPO_4 above 650 °C results in non-conductive light-grey expanded pellets; CO or CO_2 is released by rapid consumption of carbon, resulting in volume expansion and loss of contact between conducting paths. The iron compound, having a lower oxidation potential, provides a wider temperature window for initiation of metal phosphide production before excessive carbothermal reduction takes place (above 925 °C).

for all compositions x. All were non-conductive ($\sigma < 10^{-7}$ S cm^{-1}). For $x = 1$, the diffraction pattern (Supplementary Information, Fig. S4) showed that stoichiometric $LiNiPO_4$ was formed; for $x < 1$, $Li_4P_2O_7$ and $Ni_2P_2O_7$ were minor contaminants. By contrast, heat treatment under argon at 600 °C resulted in deep black materials consisting of $LiNiPO_4$, and minor fractions of Ni_3P (along with $Li_4P_2O_7$ and $Li_2Ni_3P_4O_{14}$). Conductivity measurements of a pressed sintered (650 °C) pellet of these materials showed values of about 10^{-4} S cm^{-1} at room temperature (Fig. 2). That this conductivity is lower than the corresponding iron phosphates is owing to the lower densification temperature, and differences in the energetics of phosphate decomposition. Li_xNiPO_4/C undergoes vigorous carbothermal reaction to form CO/CO_2 at the same temperature at which carbothermal chemistry for Fe just commences. This more facile reduction corresponds to a greater thermodynamic driving force, arising from the easier reduction of Ni^{II} in $LiNiPO_4$ versus Fe^{II} in $LiFePO_4$. The kinetics of these reactions is also affected by the mechanism of phosphide formation. These mechanisms remain to be fully understood for Ni (and Fe) and will be addressed in subsequent studies.

Metal phosphide formation not only explains the conductivity of the $LiFePO_4$ and $LiNiPO_4$ materials at high temperatures, but also the magnetic behaviour of the composite materials. In our iron phosphate materials, conductivity of the composites formed at 800–850 °C was associated with easily detectable ferromagnetism of the samples, despite the antiferromagnetism of $LiFePO_4$ (ref. 17). The ferromagnetic properties of the iron phosphides, $(Fe_2P, Fe_3P$ and $Fe_{75}P_{15}C_{10}$; refs 15, 18

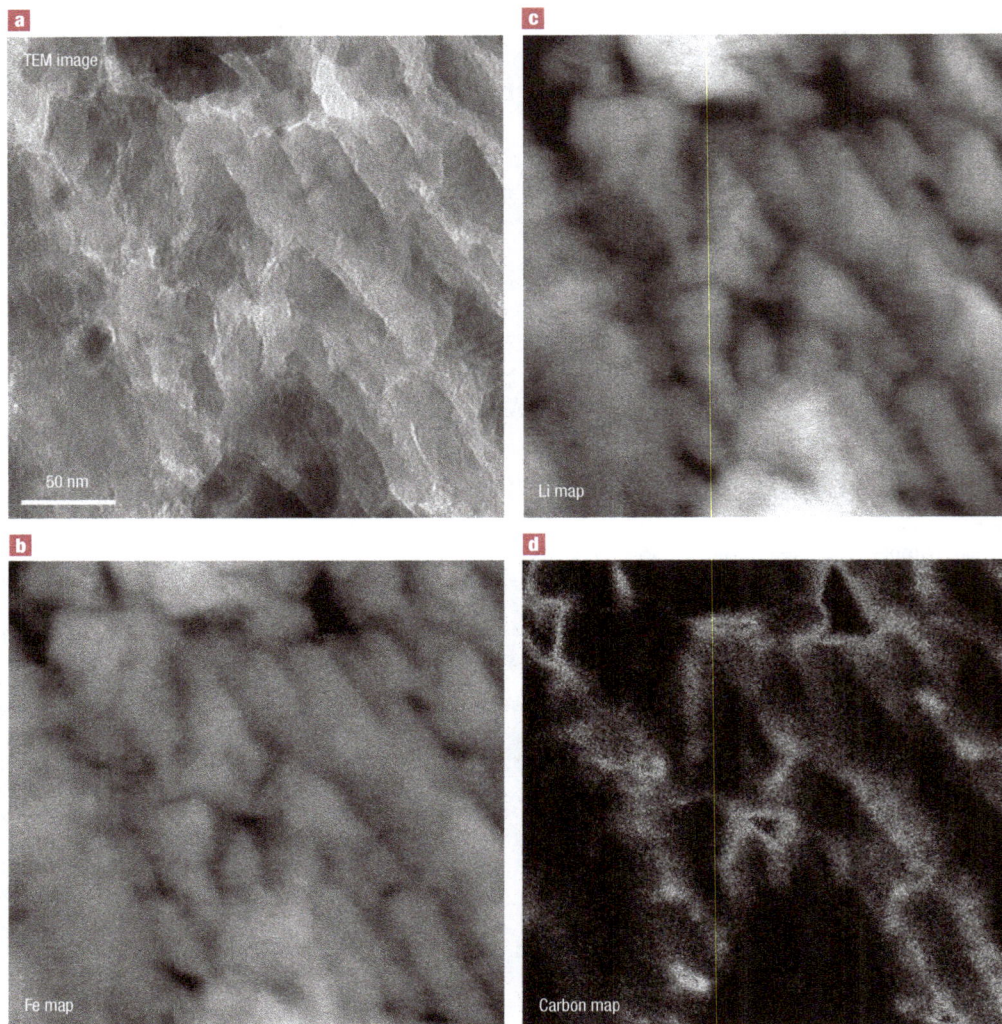

Figure 3 **Elemental mapping of $Li_{0.90}Zr_{0.01}FePO_4$.** **a–d**, The elemental Fe (**b**), Li (**c**) and C (**d**) maps, generated by EELS mapping of the TEM micrograph shown in **a**, illustrate the network formation of carbon and Fe_2P in the interstitial grain-boundary region. In the micrograph, the $LiFePO_4$ crystallites appear as the darker regions, with differing intensity indicating variation in crystallite thickness. The irregularly shaped particles are clearly illustrated by the corresponding bright domains in the Fe and Li EELS maps. The regions between the crystallites (20–100 nm in size) and their edges appear as black domains in the EELS maps and as lighter areas in the micrograph. The scale refers to all the images.

and 14, respectively) account for this behaviour. It also explains the previously reported magnetization of (sintered) $LiFePO_4$ samples noted in Hall measurements[7]. Moreover, although the electronic conductivity of the nickel phosphate is directly associated with nickel phosphide formation, the non-magnetic nature of these materials is consistent with the paramagnetic nature of Ni_3P (ref. 19).

Particularly important is that the grinding procedure used to form these composites gives rise to efficient coating of the carbon precursors responsible for the reduction at the crystallite surface, and beneficially reduces the crystallite size of the particles. We anticipate that the same process should be readily extendable to other olivines such as $LiMnPO_4$ and $LiCoPO_4$, and also to other phosphates. Adroit choices of processing conditions and composition may result in formation of conductive networks at even lower temperatures. Selectively manipulating the reactivity of these materials can also be envisioned to form in situ nano-networks that are applicable not only for energy storage applications, but a range of other useful physical properties that derive

from the combination of insulating and conductive/magnetic materials at the nanoscale level.

METHODS

SYNTHESIS

Powders were synthesized using Li_2CO_3 (99.999%, Alfa Aesar), $FeC_2O_4·2H_2O$ (99.999%, Alfa Aesar) (or $NiC_2O_4·2H_2O$, 99%, Alfa Aesar) and $NH_4H_2PO_4$ (99.998%, Alfa Aesar) as the precursor reagents; where Zr doping was used, it was introduced as $Zr(OC_2H_5)_4·C_2H_5OH$ (Alfa Catalogue Chemicals). Stoichiometries of the reagents were adjusted to prepare the formulations described in the text. The precursors were ball-milled for between 6 and 18 h in silicon nitride media in acetone, removed from the mill and dried, and ground in an argon-filled glove box using an agate mortar and pestle. In a typical preparation of the iron phosphates, 3–4-g batches of the powders were first heated to 350 °C for 10 h in flowing argon (200 ml min⁻¹) using a fine titanium metal getter in the gas stream to control oxygen activity), followed by heat treatment at 600 °C for up to 8 h. Pellets were pressed (5 tons) from ~200 mg of the powder and sintered at either 800 or 850 °C for 10 h. In most cases, pellets were covered with powder of the same composition. A weight loss of about 1.5–2% in the pellets was consistently observed after sintering, which we attributed to evolution of CO/CO_2 and/or Li_2O. Samples of Li_xFePO_4 from non-carbon-containing precursors were made by reacting vivianite with Li_3PO_4 following literature procedures[20],

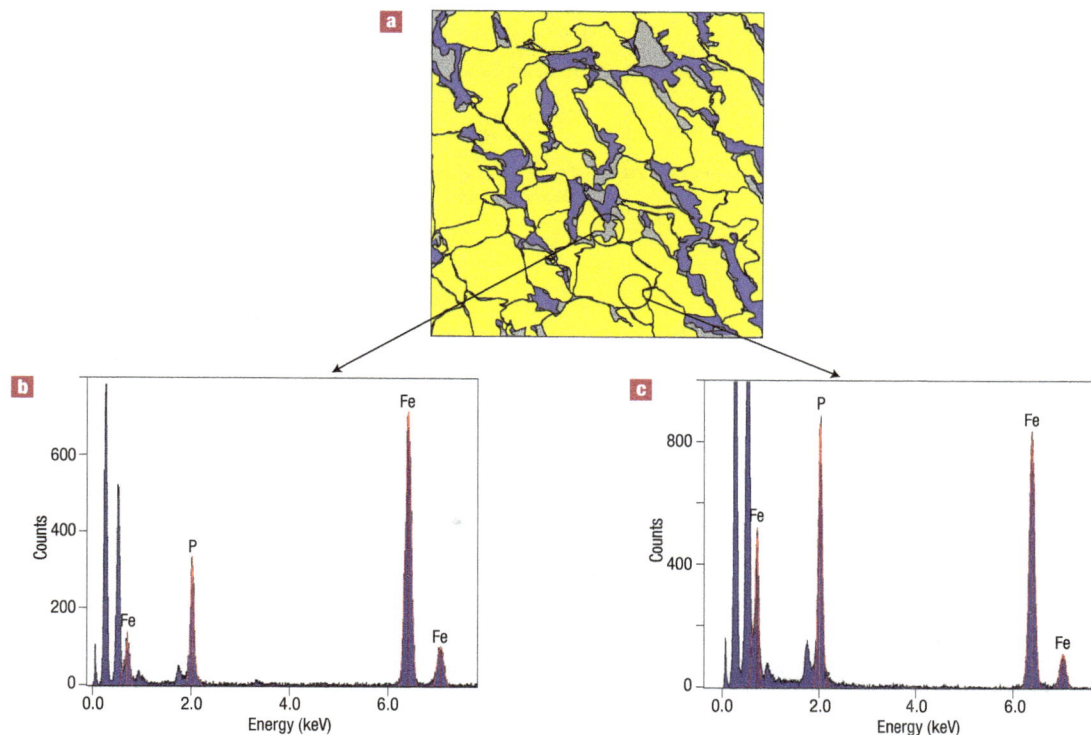

Figure 4 Representation of the Li$_{1-x}$Zr$_{0.01}$FePO$_4$ composite. a, The schematic, generated from the image and superposition of the Fe and C elemental maps depicted in Fig. 3, shows the location of Fe$_2$P,C and carbon in the interstitial grain-boundary region (LiFePO$_4$: yellow; C/Fe$_{75}$P$_{15}$C$_{10}$: blue; Fe$_2$P: grey). The corresponding EDX analysis of the grain-boundary (**b**) and bulk (**c**) regions confirm that the Fe/P ratio of 1:1 is preserved in the bulk, but is close to 2:1 in the grain boundaries. The area fraction of the grain-boundary material is estimated at 5%. The overall semiconducting behaviour observed in Fig. 2 is accounted for by the conductivities of the intergranular materials: (amorphous carbon)[21]; 3.3 x 10^{-1} S cm^{-1} (Fe$_2$P single crystal)[18]; 5.2 × 10^{-2} S cm^{-1} (Fe$_2$P, pressed powders);[10] and ~ 10^{-1} S cm^{-1} (Fe$_{75}$P$_{15}$C$_{10}$)[22].

using stoichiometries (1–0.33x) Li$_3$PO$_4$:(0.33x) NH$_4$H$_2$PO$_4$:Fe$_3$(PO$_4$)$_2$·8H$_2$O. For preparation of the conductive nickel phosphates, the powders were first heated to 350 °C for 10 h in flowing argon, followed by heat treatment at 600 °C for 8 h and sintering of the densified pellet at 650 °C for an additional 10 h. Samples of the non-conductive Li$_x$NiPO$_4$ materials were prepared under the same conditions with the exception of using flowing air as the atmosphere for some experiments.

XRD AND TEM

XRD was performed on a Bruker D8-Advantage powder diffractometer using Cu-Kα radiation (wavelength λ = 1.5405 Å) from 2θ = 10 to 80 degrees at a count rate of 1 s per step of 0.02°. TEM analysis was carried out by embedding a small portion of the sintered pellet in epoxy resin, and slicing the sample with an ultramicrotome. The slice was mounted on thin carbon films supported on a 200-mesh copper grid. TEM imaging and conventional EDX analysis at the grain-boundary region was performed using a Technia 20 instrument (FEI, USA) operating at 200 kV accelerating voltage. EELs maps were generated using a Gatan energy filter (GIF-2000). EDX spot elemental analysis was also performed on the same slice using a Hitachi S5200 operating at 30 kV in STEM mode to confirm the Fe/P ratios (see Supplementary Information, Fig. S3).

CONDUCTIVITY MEASUREMENTS

Pellet surfaces were polished before variable temperature conductivity measurements, which were performed using four-point d.c. methods. Electrode contacts were affixed using silver or gold paste in linear geometry on a thin section of a pellet of approximate dimensions: 1 mm × 1 mm × 5 mm. The activation energies were calculated using the expression $\sigma = \sigma_o e^{-Ea/kT}$.

Received 21 October 2003; accepted 15 December 2003; published 22 February 2004.

References

1. Padhi, A. K., Nanjundaswamy, K. S. & Goodenough, J. B. Phospho-olivines as positive electrode materials for rechargeable lithium batteries. *J. Electrochem. Soc.* **144**, 1188–1194 (1997).
2. Morgan, D., van der Zen, A. & Ceder, G. Li ion conductivity in Li$_x$MPO$_4$ (M = Mn, Fe, Co, Ni) olivine materials. *Electrochem. Solid State Lett.* **7**, A30–A32 (2004).
3. Ravet N. *et al.* Improved iron based cathode material. Abstract No. 127, *Electrochemical Society Fall Meeting*, Honolulu, Hawaii (Electrochemical Society, Pennington, New Jersey, 1999).
4. Li, G., Yamada, A. & Azuma, H. Method for manufacturing active material of positive plate and method for manufacturing non-aqueous electrolyte secondary cell. European Patent EP 1,094,532A1 (2001).
5. Huang, H., Yin, S.-C. & Nazar, L. F. Approaching theoretical capacity of LiFePO$_4$ at room temperature at high rates. *Electrochem. Solid State Lett.* **4**, A170–A172 (2001).
6. Croce, F. *et al.* A novel concept for the synthesis of an improved LiFePO$_4$ lithium battery cathode. *Electrochem. Solid State Lett.* **5**, A47–A50 (2002).
7. Chung, S.-Y., Bloking, J. T. & Chiang, Y.-M. Electronically conductive phospho-olivines as lithium storage electrodes. *Nature Mater.* **1**, 123–128 (2002).
8. Doeff, M. M, Hu, Y., McLarnon, F. & Kostecki, R. Effect of surface carbon structure on the electrochemical performance of LiFePO$_4$. *Electrochem. Solid State Lett.* **6**, A207–A209 (2003).
9. Ravet, N., Abouimrane, A. & Armand, M. Correspondence. *Nature Mater.* **2**, 702 (2003).
10. Ellis, B., Herle, P. S. & Nazar, L. F. LiFePO$_4$ and its doped derivatives. Abstract No. 1074, 203rd *Electrochemical Society Spring Meeting*, Paris (Electrochemical Society, Pennington, New Jersey, 2003).
11. Barker, J., Saidi, M. Y. & Swoyer, J. L. Lithium iron(II) phospho-olivines prepared by a novel carbothermal reduction method. *Electrochem. Solid State Lett.* **6**, A53–A55 (2003).
12. Chung, S.-Y. & Chiang, Y.-M. Microscale measurements of the electrical conductivity of doped LiFePO$_4$. *Electrochem. Solid State Lett.* **6**, A278–A281 (2003).
13. Goni, A., Arriortua, M. I., Barberis, G. E. & Rojo, T. Unexpected substitution in the Li$_{1-3x}$Fe$_x$NiPO$_4$ solid solution. Weak ferromagnetic behaviour. *J. Mater. Chem.* **10**, 423–428 (2000).
14. Berry, B. S. & Pritchet, W. C. Temperature dependence of the ΔE effect in amorphous Fe$_{75}$P$_{15}$C$_{10}$. *Solid State Commun.* **26**, 827–829 (1978).
15. Fujii, H., Hokabe, T., Kamigaichi, T. & Okamoto, T. Magnetic properties of iron phosphide (Fe$_2$P) single crystal. *J. Phys. Soc. Jpn* **43**, 41–46 (1977).
16. Garcia-Moreno, O. *et al.* Influence of the structure on the electrochemical performance of lithium transition metal phosphates as cathodic materials in rechargeable lithium batteries: A new high-pressure form of LiMPO$_4$ (M = Fe and Ni). *Chem. Mater.* **13**, 1570–1576 (2001).
17. Rousse, G., Rodriguez-Carvajal, J., Patoux, S. & Masquelier, C. Magnetic structures of the triphylite LiFePO$_4$ and of its delithiated form, FePO$_4$. *Chem. Mater.* **15**, 4082–4090 (2003).

18. Meyer, A. J. P. & Cadeville, M. C. Magnetic properties of iron-phosphorus compounds. *J. Phys. Soc. Jpn* **17**, 223–225 (1962).

19. Zeppenfeld, K. & Jeitschko, W. Magnetic behaviour of Ni_3P, Ni_2P, NiP_3 and the series $Ln_2Ni_{12}P_7$. *J. Phys. Chem. Solids* **54**, 1527–1531 (1993).

20. Herstedt, M. *et al.* Surface chemistry of carbon-treated $LiFePO_4$ particles for Li-ion battery cathodes studied by PES. *Electrochem. Solid State Lett.* **6**, A202–A206 (2003).

21. Dimitriadis, C. A., Hastas, N. A., Vouroutzis, N., Logothetidis, S. & Panayiotatos, Y. Microstructure and its effect on the conductivity of magnetron sputtered carbon thin films. *J. Appl. Phys.* **89**, 7954–7959 (2001).

22. Axe, J. D., Passell, L. & Tsuei, C. C. Spin waves in an amorphous metallic ferromagnet, $Fe_{75}P_{15}C_{10}$. *AIP Conf. Proc.* **24**, 119–120 (1974).

Acknowledgements

We gratefully acknowledge funding from the National Sciences and Engineering Research Council of Canada (NSERC) through its Discovery Grant Program. We also thank R. A. Dunlap (Physics, University of Dalhousie) for providing the Mössbauer data. We gratefully acknowledge the help of Ian Swainson (Chalk River Neutron Beam Laboratory) in acquiring neutron diffraction data on substoichiometric Li_xFePO_4.

Correspondence and requests for materials should be addressed to L.F.N.

Supplementary Information accompanies the paper on www.nature.com/naturematerials

Competing financial interests

The authors declare that they have no competing financial interests.

Electronically conductive phospho-olivines as lithium storage electrodes

SUNG-YOON CHUNG, JASON T. BLOKING AND YET-MING CHIANG*

Department of Materials Science and Engineering, Massachusetts Institute of Technology, 77 Massachusetts Avenue, Cambridge, Massachusetts 02139, USA
*e-mail: ychiang@mit.edu

Published online: 22 September 2002; doi:10.1038/nmat732

Lithium transition metal phosphates have become of great interest as storage cathodes for rechargeable lithium batteries because of their high energy density, low raw materials cost, environmental friendliness and safety. Their key limitation has been extremely low electronic conductivity, until now believed to be intrinsic to this family of compounds. Here we show that controlled cation non-stoichiometry combined with solid-solution doping by metals supervalent to Li^+ increases the electronic conductivity of $LiFePO_4$ by a factor of ~10^8. The resulting materials show near-theoretical energy density at low charge/discharge rates, and retain significant capacity with little polarization at rates as high as 6,000 mA g^{-1}. In a conventional cell design, they may allow development of lithium batteries with the highest power density yet.

Olivines belong to a general class of 'polyanion' compounds containing compact tetrahedral 'anion' structural units $(XO_4)^{n-}$ (X = P, S, As, Mo or W) with strong covalent bonding, networked to produce higher-coordination sites such as oxygen octahedra that are occupied by other metal ions. In addition to olivine's geophysical importance $(Mg,Fe)_2SiO_4$ olivine is one of the chief constituents of the Earth's mantle), related structures such as Nasicon (sodium super-ionic conductor) have been extensively studied as alkali ion or protonic conductors for use as electrolytes in solid-state electrochemical devices. Now, sparked by work from Goodenough's laboratory[1-3], there is great interest in polyanion compounds as lithium storage electrodes for rechargeable batteries[4-11]. Compounds under scrutiny include Li_xMXO_4 (olivine), $Li_xM_2(XO_4)_3$ (Nasicon structure type) where M = transition metal, $VOPO_4$, $LiFe(P_2O_7)$ and $Fe_4(P_2O_7)_3$, and derivative structures that have additional interstitial metal ions, symmetry-changing displacements, or minor changes in the connectivity of polyhedra. $LiFePO_4$ and $Li(Mn,Fe)PO_4$ olivine (well known as the natural mineral triphylite) are amongst the simplest, most widely studied and potentially most useful of these. $LiFePO_4$ has a high lithium intercalation voltage (~3.5 V relative to lithium metal), high theoretical capacity (170 mA h g^{-1}), low cost, ease of synthesis, and stability when used with common organic electrolyte systems[7,8].

It has, however, become generally accepted[4-6,10,12] that this class of compounds has fundamentally low electronic conductivity, desirable for solid electrolytes but not for applications as ion-storage or fuel-cell electrodes. At the lattice scale, mixed electronic-ionic conductivity is necessary for preservation of overall charge neutrality during lithium ion transport, the chemical (ambipolar) diffusion coefficient being rate-limited by the slower species. At the meso- and microscale, a high electronic conductivity is desirable to prevent the impedance of a typical storage battery electrode, composed of a percolating network of storage particles and a conductive additive such as carbon, from becoming excessive. We know of no previously studied polyanion compound that has been shown to have appreciable electronic conductivity ($>10^{-6}$ S cm^{-1}) at room temperature. Indeed, a recent review[4] on the use of polyanion compounds as storage electrodes comments, "One of the main drawbacks with using these materials is their poor electronic conductivity, and this limitation had to be overcome through various materials processing approaches...". These approaches include coating particles with carbon[9] or co-synthesizing the compounds with carbon[10,11] to surround each

Figure 1 legend:
- + : undoped (700 °C, 2pt)
- ● : 1% Zr (800 °C, 4pt)
- ► : 1% Ti (850 °C, 2pt)
- ▼ : undoped (800 °C, 4pt)
- ▲ : 1% Nb (800 °C, 4pt)
- ⬟ : 0.2% Nb (850 °C, 2pt)
- ■ : 1% Mg (800 °C, 4pt)
- ◄ : 1% Mg (850 °C, 2pt)
- ★ : 1% Nb (850 °C, 2pt)

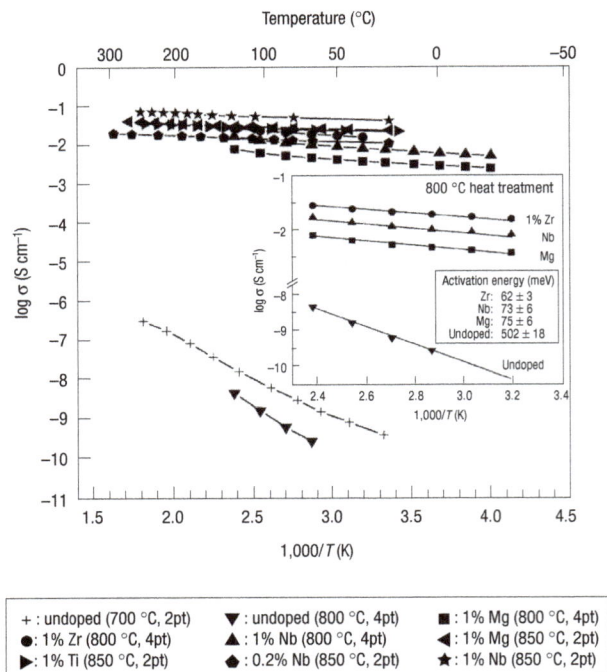

Figure 1 Doped olivines of stoichiometry $Li_{1-x}M_xFePO_4$ show electrical conductivity at room temperature that is a factor of ~10^8 greater than in undoped $LiFePO_4$, and absolute values >10^{-3} S cm^{-1} over the temperature range −20 °C to +150 °C of interest for battery applications. Results are for polycrystals fired at 700–850 °C and measured by two-point d.c. and four-point van der Pauw methods. Inset shows expanded plot for series of dense, single-phase samples fired at 800 °C, showing lower activation energy of the doped compositions (see text).

low-temperature protonic fuel-cell electrodes or membranes for separation of hydrogen gas.

Amongst the various polyanion compounds, the olivine structure[16,17] has the most extensive interconnection of octahedra, with the cations that occupy the M2 sites (Fe site in $LiFePO_4$) forming a corner-sharing network of octahedra in the (010) plane, and the cations on M1 (Li) sites forming edge-sharing chains of octahedra in the [100] direction (Fig. 2). The appreciable electronic conductivity at high temperatures in $(Mg,Fe)_2SiO_4$ (ref. 18) and the existence of other corner-sharing transition metal oxides having high electronic conductivity near room temperature and below (notably the superconducting cuprates, transition-metal-bearing perovskites, and compounds with one-dimensional chains of octahedra such as R_2BaNiO_5 (ref. 19; where R is rare earth) and $La_2Ni_2O_5$ (ref. 20), suggested to us that if multivalent Fe^{2+}/Fe^{3+} solid solutions could be obtained in $LiFePO_4$, substantial room-temperature electronic conductivity might result. But $LiFePO_4$ is known to undergo a first-order phase transition to orthorhombic $FePO_4$ on delithiation. The result is that $Li_{1-x}FePO_4$ compositions actually consist of $LiFePO_4$ and $FePO_4$ phases, both of which seem to be insulating as a result of single iron valency (Fe^{2+} and Fe^{3+} respectively)[1,21]. We theorized that if cation-deficient solid solutions could be retained, good p-type

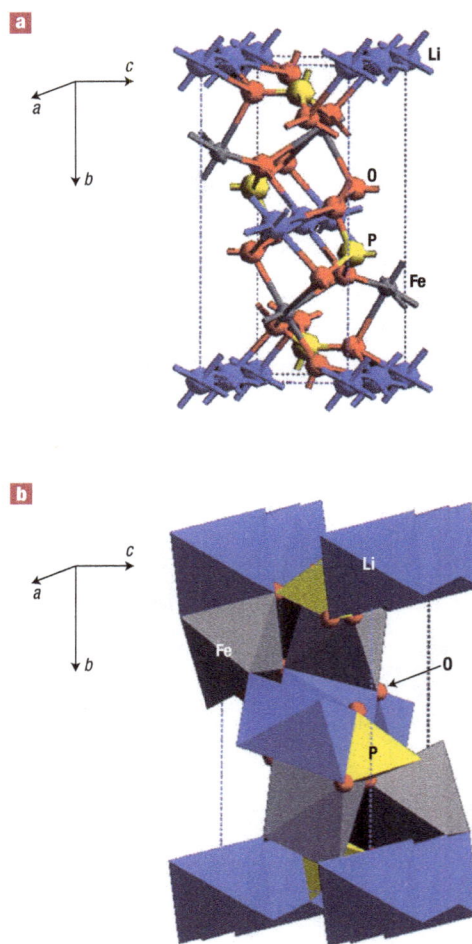

particle with a good electronic conductor. These approaches of course do not increase the lattice electronic conductivity or chemical diffusion coefficient of lithium within the crystal. Additional drawbacks are the need for additional processing, and the loss in energy density due to the electrochemically inert additive, which sometimes constitutes as much as 30 vol% of the resulting composite[10,11].

Here we show that, by selective doping with supervalent cations, the lattice electronic conductivity of $LiFePO_4$ can be increased by a factor of more than 10^8 relative to the pure endmember, reaching values of >10^{-2} S cm^{-1} at room temperature (Fig. 1). The electronic conductivities obtained are well in excess of those in the lithium storage cathodes currently used, such as $LiCoO_2$ (~10^{-3} S cm^{-1}; ref. 13) and $LiMn_2O_4$ (2×10^{-5} to 5×10^{-5} S cm^{-1}; refs 14,15). The resulting doped $LiFePO_4$ materials have lithium storage capacities that are near the theoretical limit of 170 mA h g^{-1} at low charge/discharge rates, and retain significant discharge capacity at rates as high as 6,000 mA g^{-1} of material (40C rate; for an explanation of the C-rate convention, see Methods) in a conventional electrode design. This class of conductive olivines may be of interest for high-power, safe, rechargeable lithium batteries for applications such as hybrid and electric vehicles, back-up power, implantable medical devices and applications that currently use supercapacitor technology. The combination of high electronic and ionic transport at reduced temperatures (room temperature and below) in these compounds also suggests that protonic analogues could be of interest as electrode materials for other electrochemical applications such as

Figure 2 Ordered-olivine structure of $LiFePO_4$: *Pmnb* space group, with Li in M1 site and Fe in M2 site. **a**, Ball–stick model. **b**, Depiction of polyhedral connectivity.

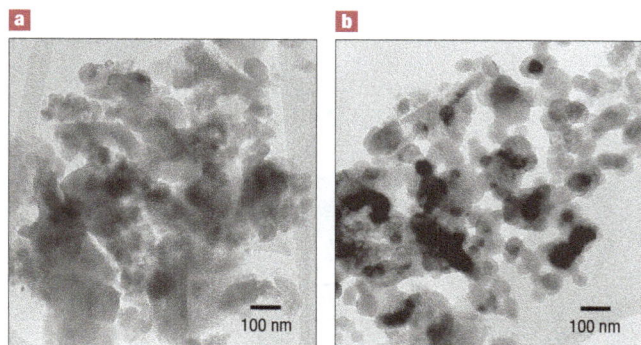

Figure 3 Morphologies of undoped (insulating) and doped (conductive) powders. The morphologies are similar, with 50–200 nm primary crystallites joined in an aggregate structure, as shown here for undoped powder (**a**), and $Li_{0.99}Zr_{0.01}FePO_4$ composition fired at 600 °C, 24 h (**b**).

conductivity should result, by analogy with other cation-deficient transition metal oxides such as $Li_{1-x}CoO_2$ (ref. 13) or the transition metal monoxides, $Fe_{1-x}O$, $Mn_{1-x}O$, and $Ni_{1-x}O$ (see, for example, refs 22–24). On the other hand, substitution of Li^+ or Fe^{2+} with supervalent cations might be expected to result in increased n-type conductivity.

Materials were synthesized as powders and densified compacts were formed using the solid-state reaction of Li_2CO_3, $NH_4H_2PO_4$, and $FeC_2O_4 \cdot 2H_2O$, with the dopant (Mg^{2+}, Al^{3+}, Ti^{4+}, Zr^{4+}, Nb^{5+} or W^{6+}) being added as a metal alkoxide. Final firing was done at 600–850 °C in argon or nitrogen atmosphere. Figure 3 shows typical powder morphologies for undoped and doped powders fired at 600 °C, showing a primary crystallite size of 50–200 nm and an aggregate structure. The surface area measured from the BET isotherm was found to be ~30 $m^2 g^{-1}$ for samples fired at 600 °C, and the compacted density of these powders measured on die-pressed pellets was typically 40–55% of the theoretical density (3.6 $g cm^{-3}$). We used X-ray diffraction (XRD) and transmission electron microscopy (TEM) to make a systematic investigation of the solubility of $LiFePO_4$ for excess Li or Fe, and for a variety of cation dopants. We observed, first, that the undoped compound had a very narrow range of cation non-stoichiometry, with as little as 1% deficiency of the ferrous iron oxalate resulting in a detectable amount of Li_3PO_4 phase. Second, compositions $Li_{1-x}M_xFePO_4$, formulated to allow substitution onto the M1 sites by a cation supervalent to Li^+, showed much higher solubility for the dopants studied (Mg^{2+}, Al^{3+}, Ti^{4+}, Zr^{4+}, Nb^{5+} and W^{6+}) than did compositions $LiFe_{1-x}M_xPO_4$, formulated to induce substitution of the same cations onto the M2 sites. Figure 4 shows an example of the first stoichiometry, $Li_{0.99}Nb_{0.01}FePO_4$, in which elemental mapping shows a uniform distribution of the Nb dopant. Figure 5 compares the X-ray diffraction patterns for several 1 atom% doped powders of each cation stoichiometry; in each case the lithium-deficient stoichiometry (Fig. 5a) has no detectable impurity phases. By contrast, XRD from samples with the same dopants and concentrations in the iron-deficient stoichiometry showed detectable precipitation of Li_3PO_4 (Fig. 5b), and electron microscopy showed impurity phases enriched in the dopant (results not shown). Although the site occupancy of specific dopants has yet to be established, we note that in other olivines the smaller cation tends to occupy the M1 site[16]. Because each of our dopants has an ionic radius in octahedral coordination smaller than that of Fe^{2+} (ref. 25), a preference for the M1 site is expected, although some distribution onto both sites, with some Fe^{2+} being redistributed to the M1 site, would not be surprising.

Electrical conductivity measurements were made on samples sectioned from pellets densified at 700–850 °C. All doped compositions

$Li_{1-x}M_xFePO_4$ (M = Mg, Al, Ti, Nb or W) showed room-temperature conductivities in excess of 10^{-3} S cm^{-1}. Results for several of these dopants are shown in Fig. 1. In contrast, undoped $LiFePO_4$ showed room-temperature conductivity of 10^{-9}–10^{-10} S cm^{-1}, depending on firing temperature. The temperature dependence of conductivity in the doped samples varied somewhat with firing temperature. Samples fired at 700 °C were not dense enough for us to rule out a significant contribution to the d.c. resistivity from the grain boundaries, and samples fired at 850 °C contained a significant fraction of Fe_2P phase. We therefore considered results for dense single-phase samples fired at 800 °C to be most representative of the crystalline conductivity. We assume that the conductivity is predominantly electronic at these high values. The inset in Fig. 1 compares a group of samples processed and measured identically, varying only in the dopant. Samples containing 1 atom% each of Mg^{2+}, Zr^{4+} and Nb^{5+} have conductivities within a factor of three of each other, and show similar activation energies in the range 60–80 meV. The identically processed undoped sample has a much higher activation energy, close to 500 meV. The activation energies may include terms for both defect formation and migration (for example for a polaronic mechanism). Attempts to separate the mobility and carrier concentration using Hall measurements were not successful, owing to magnetization of the samples in the applied field[26,27]. The sign of the thermopower was measured by imposing a temperature gradient across the conductivity samples. Each of the highly conductive doped samples fired at 700 or 800 °C was observed to be p-type, whereas the undoped sample in Fig. 1 (inset) was found to be n-type. First-principles electronic structure calculations[28] confirm that $LiFePO_4$ is a semiconductor with a small indirect gap (~0.3 eV, which is probably an underestimate because of the methods used) and large effective masses of electrons and holes. Thus, the highly conductive doped compositions seem to be extrinsic p-type semiconductors, whereas the undoped stoichiometric $LiFePO_4$ may be an intrinsic semiconductor which is n-type because its electron mobility is higher than its hole mobility.

We also examined the dependence of conductivity on dopant concentration. Taking Nb as an example, it was observed that doping levels of 0.1, 0.5 and 1.0 atom% Nb resulted in room-temperature conductivities of 1.1×10^{-3}, 4.1×10^{-2} and 2.2×10^{-2} S cm^{-1}, respectively, for samples fired at 700 °C. At higher doping levels of 2 and 4 atom%,

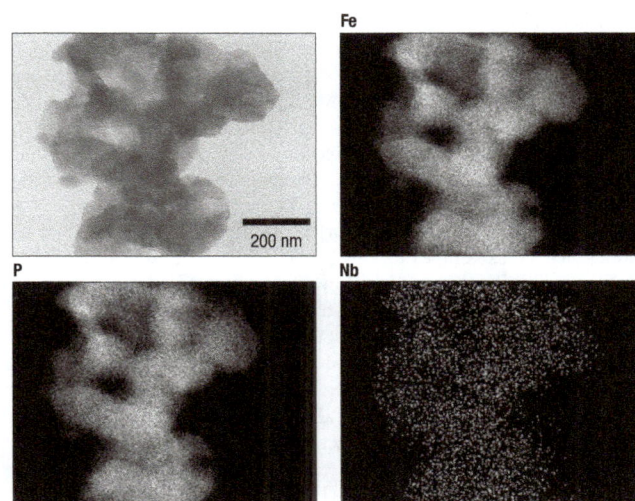

Figure 4 Elemental mapping by scanning TEM (STEM). These maps of a powder of composition $Li_{0.99}Nb_{0.01}FePO_4$ (fired 600 °C, 20 h, in argon) illustrate the uniform dopant solid solution observed in compositions of stoichiometry $Li_{1-x}M_xFePO_4$.

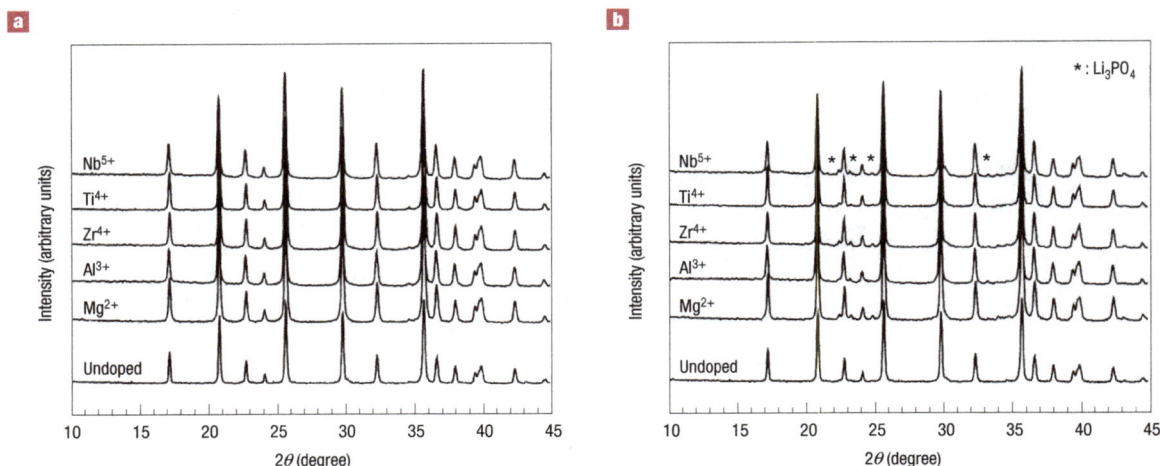

Figure 5 X-ray diffraction of various powders showing the effect of cation stoichiometry on dopant solid-solubility. a, Powders containing 1 atom% dopant in the stoichiometry $Li_{1-x}M_xFePO_4$ fired at 600 °C are single-phase by XRD and TEM/STEM analysis. **b**, Powders containing 1 atom% dopant in the stoichiometry $LiFe_{1-x}M_xPO_4$ show Li_3PO_4 precipitation by XRD, and secondary phases enriched in the dopant by TEM/STEM (not shown).

Nb-enriched impurity phases appeared, and the conductivity decreased to 2.8×10^{-3} and $\sim 10^{-6}$ S cm^{-1}, respectively. The lack of proportionality between dopant level and conductivity within the regime of solid solution is an important consideration for interpretation of the doping mechanism, discussed later.

Because several previous studies have used carbon coatings to increase electronic conductivity, it was important to us to determine whether the increased conductivity was a bulk or surface phase effect. Several additional observations supported an interpretation of the results as a true solid-state doping effect. First, there was a difference in the colour of insulating and conductive samples that allowed us to identify highly conductive samples by appearance alone. Conductive powders and sintered pellets alike were always a deep black, whereas insulating samples, whether they were undoped or doped (that is, being doped but not having the solute in solid solution), were a medium to dark grey. This difference in optical absorption between conductive and insulating compositions cannot be explained by a redistribution of the residual carbon, present at 0.3–1.5 wt% in these samples (see Methods). Second, the increase in electronic conductivity, indirectly observed through electrochemical tests of the powders and directly measured in densified samples, occurs in samples that vary greatly in their specific surface area, from 30 m^2 g^{-1} in the powders fired to 600 °C, to $\sim 3 \times 10^{-4}$ m^2 g^{-1} in the densely sintered samples. It is implausible that a minor surface phase could dominate the conductivity across samples varying by a factor of 10^5 in specific surface area. Finally, TEM of the powders showed no detectable surface films on the crystallites when viewed at lattice-fringe resolution.

The room-temperature conductivity of these doped LiFePO$_4$ materials exceeds that of the well-established intercalation cathodes LiCoO$_2$ and LiMn$_2$O$_4$ in their lithiated (discharged) states[13–15]. Although the lithium ionic conductivity in LiFePO$_4$ has not yet been reliably established, at the present high electronic conductivities the lithium chemical diffusion coefficient is limited by lithium transport (that is, t_{Li} is ~ 0). The structure of our materials (Fig. 3) is almost ideal for providing optimal mixed electronic–ionic transport in a battery system, having a porous aggregate structure in which small primary crystallites of diameter 50–200 nm can be surrounded by the electrolyte, allowing radial lithium diffusion through a small cross-section, while remaining electronically 'wired' together through the sinter necks. At low rates (C/10 to C/30,

see Methods), these materials showed a capacity of ~ 150 mA h g^{-1}, corresponding to $\sim 90\%$ of the theoretical value, when tested as coatings or pressed pellets. At higher rates, remarkable power capabilities were seen when the compound was formulated into a typical lithium battery electrode coating. Figure 6a shows the charge and discharge capacities observed in continuous cycling at rates varying from 15 mA g^{-1} (C/10) to 3,225 mA g^{-1} (21.5C) between the voltage limits of 2.8–4.2 V, at room temperature. For comparison, the behaviour of an electrode prepared using undoped LiFePO$_4$ powder is shown. Figure 6b shows corresponding charge–discharge curves for the doped sample. Other cells tested at rates as high as 6,000 mA g^{-1} (40C) still showed modest polarization, with clear voltage plateaux at ~ 3 V, although charge capacities decreased to ~ 30 mA h g^{-1}. Comparing with published data for this material[1,2,9–11,21], it is clear that the low doping levels used to increase conductivity do not decrease the storage capacity at low rates, but greatly increase the power density that is possible. The low polarization is attributed to the high electronic conductivity at the particle scale, and the decreasing capacity with increasing current rate is most probably due to limited lithium diffusion. At the high rates, it is not clear whether transport within the particles or in the liquid-electrolyte-filled pores is rate-limiting. A lower limit to the solid-state lithium chemical diffusivity can be obtained by assuming depth-independent lithiation through the electrode of spherical primary crystallites of 100 nm diameter. Using this model and the data for capacity versus rate, we calculate a chemical diffusivity of $\sim 5 \times 10^{-15}$ cm^2 s^{-1} at room temperature.

Current lithium ion batteries based on liquid electrolytes, in which laminated electrodes are separated by a microporous film, contain $\sim 30\%$ by weight of cathode storage material, typically LiCoO$_2$. To estimate the performance characteristics of these olivines in a conventional design, we take into account the nearly identical gravimetric capacities of LiFePO$_4$ and LiCoO$_2$, the lower crystalline density of LiFePO$_4$ (3.6 rather than 5.1 g cm^{-3}), and its slightly lower cell voltage when used in conjunction with a carbon electrode (3.25 rather than 3.7 V). Assuming that 15–25% by weight loading of the olivine is possible in an optimized cell, we estimate power and energy densities for a complete cell of 1,300–2,200 W kg^{-1} and 32–53 W h kg^{-1} at a 20C rate, and 2,800–4,670 W kg^{-1} and 18–30 W h kg^{-1} at a 40C rate. Such cells could provide power densities not possible in current nickel metal-hydride (power density 400–1,200 W kg^{-1}, energy density 40–80 W h kg^{-1}) and lithium-ion battery technology (800–2,000 W kg^{-1}, 80–170 W h kg^{-1}), and an energy density several times greater than is possible in supercapacitor technology (4,000–10,000 W kg^{-1}; 5–10 W h kg^{-1}). These capabilities, in a low-cost and ultra-safe storage material, may be especially attractive for hybrid and electric vehicles.

We now address possible doping and defect mechanisms resulting in the observed increase in conductivity. The defect chemistry of LiFePO$_4$ has not been previously studied. We considered numerous models, and rejected interpretations that could not be reconciled with the observations of (1) thermally activated, p-type semiconductivity; (2) absence of a significant dopant concentration dependence of conductivity; (3) similar increases in conductivity for dopants of 2$^+$ to 6$^+$ valence; (4) existence of a two-phase reaction on delithiation, as in undoped LiFeO$_4$; and (5) the apparent retention of high electronic conductivity throughout delithiation, indicated by the combination of high capacity and high rate capability. Note that the substitution of a cation M that is supervalent to Li$^+$ in the composition Li$_{1-x}$M$_x$FePO$_4$ is normally expected to result in donor doping. In oxides, aliovalent solutes can be compensated by electronic or ionic defects. The following point defect reactions (in Kröger–Vink notation[29]) illustrate these mechanisms for an M^{3+} cation that is compensated, respectively, by electrons or by cation vacancies on the M2 site:

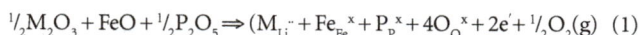

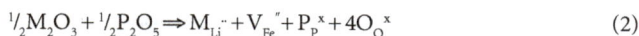

$$\frac{1}{2}M_2O_3 + FeO + \frac{1}{2}P_2O_5 \Rightarrow (M_{Li}^{\cdot\cdot} + Fe_{Fe}^{x} + P_P^{x} + 4O_O^{x} + 2e' + \frac{1}{2}O_2(g) \quad (1)$$

$$\frac{1}{2}M_2O_3 + \frac{1}{2}P_2O_5 \Rightarrow M_{Li}^{\cdot\cdot} + V_{Fe}'' + P_P^{x} + 4O_O^{x} \quad (2)$$

In the first instance, electroneutrality is given by $[M_{Li}^{\cdot\cdot}] = n$, namely the dopant acts directly as a donor species. If the second mechanism is dominant, electroneutrality is given by $[M_{Li}^{\cdot\cdot}] = 2[V_{Fe}'']$, in which case the donor and vacancy charge-compensate one another and no direct effect on the electronic carrier concentration is expected. But it can be shown that also in this instance, secondary defect equilibria should lead to an increase in the n-type conductivity. Clearly, neither of these mechanisms can be dominant in the present materials of high p-type conductivity. An excess of acceptor point defects above and beyond the dopant concentration is necessary. Possible acceptors in the LiFePO$_4$ structure are a high concentration of solutes other than those we detected by ICP chemical analysis, cation vacancies (V_{Li}', V_{Fe}''), or oxygen interstitials (O_i''). The latter defect is unlikely given the nearly hexagonal close-packed

oxygen sublattice in olivine, which should result in a high formation energy for anion vacancies, and the low oxygen activity during firing.

We instead propose a mechanism whereby cation doping on the M1 sites allows the stabilization of solid solutions with a net cation deficiency. That is, the doped olivine endmember has a solid solution of composition Li$_{1-a-x}$M$_x$FePO$_4$ or Li$_{1-x}$M$_x$Fe$_{1-b}$PO$_4$, in which a and b are M1 or M2 vacancy concentrations respectively. If the net charge due to a and b exceeds that due to x, then the material will have a net excess of acceptor defects (Fe^{3+} ions). Taking for example an M^{3+} dopant, the respective valences for a lithium-deficient solid solution are Li$^{1+}_{1-a-x}$M$^{3+}_x$(Fe$^{2+}_{1-a+2x}$Fe$^{3+}_{a-2x}$)(PO$_4$)$^{3-}$. Note that lithium deficiency is particularly likely under high-temperature firing conditions; an excess of lithium is often added in synthesis procedures of intercalation oxides to compensate for lithium volatility. This mechanism is analogous to allowing an extension of the solid solution field for the pure Li-rich endmember phase to cation-deficient solid solutions, Li$_{1-a}$FePO$_4$. We recall that pure LiFePO$_4$ has been observed to decompose immediately to two co-existing phases, LiFePO$_4$ and FePO$_4$, on delithiation, thereby pinning the Li chemical potential and resulting in the flat intercalation voltage as a function of lithium concentration[1,20]. Thus the insulating behaviour of undoped LiFePO$_4$ throughout electrochemical cycling suggests negligible mixed (Fe^{2+}/Fe^{3+}) iron valency in either phase. The retention of either lithium or iron deficiency in the highly lithiated solid solution can therefore result in charge compensation by Fe^{3+} and p-type conductivity, as observed in Li$_{1-x}$CoO$_2$ and numerous other cation-deficient transition metal oxides.

In this two-phase model, the FePO$_4$ endmember phase must also be considered. Our electrochemical data indicate that it also retains high electronic conductivity throughout cycling. The influence of M1 site cation doping is expected to be quite different for this phase. Starting with pure FePO$_4$, in which all iron is trivalent, cation doping will result in the formation of divalent iron: M$^{3+}_x$(Fe$^{2+}_{3x}$Fe$^{3+}_{1-3x}$)PO$_4$. This composition is obtained on delithiation of the solid solution given earlier. The dopant in this instance may be viewed as an 'interstitial' cation donor, occupying normally unoccupied M1 sites, and n-type conductivity should result. The overall picture that emerges is of a two-phase material, one phase p-type and the other n-type, whose relative proportions change with overall lithium concentration. We expect that, macroscopically, a transition from p- to n-type conductivity should be measured for the two-phase material as a whole as delithiation proceeds. Note that these conclusions are qualitatively the same regardless of whether the cation dopant M occupies the M1 site, or preferentially occupies the M2 site and

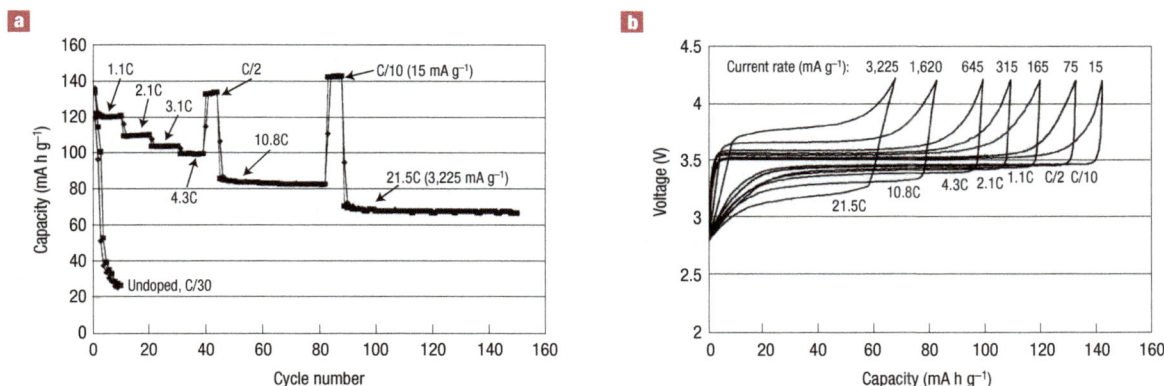

Figure 6 Electrochemical test data for electronically conductive olivine. The material is used in a conventional lithium battery electrode design (78 wt% cathode-active material, 10 wt% Super P(carbon, 12 wt% PVdF binder; 51 μm electrode thickness, 2.5 mg cm^{-2} loading) against a lithium metal counterelectrode. **a,** Cycle testing shows improved reversible capacity in comparison with identically processed undoped LiFePO$_4$ powder. Significant capacity with high coulombic efficiency (>99.5%) is retained at rates as high as 3,225 mA g^{-1} (21.5C). **b,** Charge–discharge curves show little polarization even at the highest current rates, attributed to the high electronic conductivity (Fig. 1).

displaces Fe to the M1 site. We expect the same basic doping mechanism that is effective here to apply to olivines with other transition metals (Mn, Co, Ni) partially or completely substituted for Fe.

METHODS

SYNTHESIS OF POWDERS

Powders were synthesized using Li_2CO_3 (99.999%, Alfa-Aesar, Ward Hill, Massachusetts, USA), $FeC_2O_4\cdot2H_2O$ (99.99%, Aldrich, Milwaukee, Wisconsin, USA), and $NH_4H_2PO_4$ (99.998%, Alfa-Aesar, Ward Hill, Massachusetts, USA) as the source of the main components. The cation dopants were added using the following salts of reagent-grade purity (Alfa-Aesar, Ward Hill, Massachusetts, USA): $MgC_2O_4\cdot2H_2O$, $Al(OC_2H_5)_3$, $Ti(OCH_3)_4(CH_3OH)_2$, $Zr(OC_2H_5)_4$, $Nb(OC_6H_5)_5$, and $W(OC_2H_5)_6$. A typical batch size yielded 2 g of final powder. The results are based on about 50 individual batches of powder synthesized in the course of this study. The starting materials were mixed by ball-milling for 24 h using zirconia milling media, in acetone, followed by drying, then grinding with a mortar and pestle in an Ar-filled glove box before calcining at 350 °C for 10 h in flowing Ar. Final firing for crystallization of the olivine phase was done at 600–800 °C in Ar. Limited experiments were also done using nitrogen gas as the firing atmosphere. No apparent difference in the powder characteristics was detected between argon and nitrogen firing atmospheres.

ANALYSIS OF METAL CONTENT

The metals content of the powders was analysed using inductively coupled plasma emission spectroscopy (ICP), and the carbon concentration by combustion analysis (Luvak, Boylston, Massachusetts, USA). The milling media contributed a low but detectable zirconium concentration of 1.6×10^{-4} to 3.2×10^{-4} mole Zr per mole $LiFePO_4$. Because of concern over a possible role of residual carbon, we carried out synthesis studies using both polypropylene and porcelain milling jars. With the polypropylene jars, even though residual carbon concentrations as high as 3 wt% were observed in the final powders after extended milling, the solid-solubility and conductivity results (that is, the observations of insulating and conductive samples as a function of doping) were as described earlier. The same insulating or conductive behaviour as a function of doping was observed on changing to porcelain milling jars, for which the residual carbon content decreased to ~0.3–1.5 wt% depending on final firing temperature. In none of the samples did we observe a surface coating or continuous carbon phase by high-resolution transmission electron microscopy, and some of the most conductive samples had the lowest analysed carbon concentrations.

XRD AND TEM/STEM

X-ray diffraction was done using Cu K_α radiation and a Rigaku RU-300 instrument (Rigaku Co., Tokyo, Japan). A JEOL-2000FX instrument (JEOL Inc., Tokyo, Japan) operating at 200 kV accelerating voltage was used for TEM analysis. Elemental mapping in Fig. 4 was done using a VG HB603 scanning transmission electron microscope operating at 250 kV accelerating voltage (Vacuum Generators, East Grinstead, UK), equipped with a Link Systems (Oxford Instrument, High Wycombe, UK) energy-dispersive X-ray analyser.

CONDUCTIVITY MEASUREMENTS

Conductivity measurements were made on disc-shaped fired samples by two-point and four-point d.c. methods. For two-point measurements, gold electrodes were sputtered on opposing faces of discs of ~8 mm diameter and ~1 mm thickness. Four-point conductivity measurements were made by the van der Pauw method on discs of ~8 mm diameter lapped to ~300 μm thickness. Gold electrodes were sputtered at the perimeter of the masked discs, and a Bio-Rad HL5500PC instrument (Hercules, California, USA) was used.

ELECTROCHEMICAL TESTS

Electrochemical tests were made on electrodes cast and pressed on aluminium foil, or on thin pressed pellets, prepared from suspensions containing 78 wt% olivine powder (fired at 600 °C), 10 wt% conductive carbon black (Super P™, M.M.M. Carbon, Brussels, Belgium) and 12 wt% polyvinylidine fluoride (PVdF, Alfa Aesar, Ward Hill, Massachusetts, USA), using γ-butyrolactone as the solvent. Typical loadings for coatings 40–100 μm thick were 2.5–5 mg cm^{-2} of cathode powder. The electrodes were assembled in cells constructed of stainless steel and Teflon™, using lithium metal foil as the counterelectrode, a microporous polymer separator (Celgard 2400™, Hoechst Celanese Corporation, Charlotte, North Carolina, USA) and liquid electrolyte mixtures containing 1:1 by weight ethylene carbonate: dimethyl carbonate (EC:DMC) or ethylene carbonate: diethyl carbonate (EC:DEC), and 1 M $LiPF_6$ as the conductive salt. Testing was done using an apparatus constructed from National Instruments (Austin, Texas, USA) modular electronic components, operated by a computer using LabView software. Charge/discharge rates are reported in units of current per mass of phosphate cathode, mA g^{-1}, as well as the C-rate convention, or C/n, where n is the time (h) for complete charge or discharge at the nominal capacity measured at low rates, here taken to be 150 mA h g^{-1} (see Fig. 6b). A 1C rate corresponds to a current rate of 150 mA g^{-1}, which in the ideal case gives complete discharge in 1 h, and a 20C rate corresponds to 3 A g^{-1}, discharging in 3 min.

Received 8 July 2002; accepted 4 September 2002; published 22 September 2002.

References

1. Padhi, A. K., Najundaswamy, K. S. & Goodenough, J. B. Phospho-olivines as positive-electrode materials for rechargeable lithium batteries. *J. Electrochem. Soc.* **144**, 1188–1194 (1997).
2. Padhi, A. K., Najundaswamy, K. S., Masquelier, C., Okada, S. & Goodenough, J. B. Effect of structure on the Fe^{3+}/Fe^{2+} redox couple in iron phosphates. *J. Electrochem. Soc.* **144**, 1609–1613 (1997).
3. Najundaswamy, K. S. *et al.* Synthesis, redox potential evaluation and electrochemical characteristics of NASICON-related-3D framework compounds. *Solid State Ionics* **92**, 1–10 (1996).
4. Tarascon, J.-M. & Armand, M. Issues and challenges facing rechargeable lithium batteries. *Nature* **414**, 359–367 (2001).
5. Gaubicher, J., Le Mercier, T., Chabre, Y., Angenault, J. & Quarton, M. Li/β-VOPO₄: a new 4 V system for lithium batteries. *J. Electrochem. Soc.* **146**, 4375–4379 (1999).
6. Amine, K., Yasuda, H. & Yamachi, M. Olivine $LiCoPO_4$ as 4.8 V electrode material for lithium batteries. *Electrochem. Solid State Lett.* **3**, 178–179 (2000).
7. Yamada, A., Chung, S. C. & Hinokuma, K. Optimized $LiFePO_4$ for lithium battery cathodes, *J. Electrochem. Soc.* **148**, A224–A229 (2001).
8. Andersson, A. S., Thomas, J. O., Kalska, B. & Häggström, L. Thermal stability of $LiFePO_4$-based cathodes. *Electrochem. Solid State Lett.* **3**, 66–68 (2000).
9. Ravet, N. *et al.* Improved iron based cathode material. Abstract No. 127, Electrochemical Society Fall meeting, Honolulu, Hawaii, (1999).
10. Huang, H., Yin, S.-C. & Nazar, L. F. Approaching theoretical capacity of $LiFePO_4$ at room temperature at high rates. *Electrochem. Solid State Lett.* **4**, A170–A172 (2001).
11. Prosini, P. P., Zane, D. & Pasquali, M. Improved electrochemical performance of a $LiFePO_4$-based composite cathode. *Electrochim. Acta* **46**, 3517–3523 (2001).
12. Yang, S., Song, Y., Zavalij, P. Y. & Whittingham, M. S. Reactivity, Stability and electrochemical behavior of lithium iron phosphates. *Electrochem. Comm.* **4**, 239–244 (2002).
13. Molenda, J., Stoklosa, A. & Bak, T. Modifications in the electronic structure of cobalt bronze Li_xCoO_2 and the resulting electrochemical properties. *Solid State Ionics* **36**, 53–58 (1989).
14. Shimakawas, Y., Numata, T. & Tabuchi, J. Verwey-type transition and magnetic properties of the $LiMn_2O_4$ spinels. *J. Solid State Chem.* **131**, 138–143 (1997).
15. Kawaia, H., Nagatab, M., Kageyamac, H., Tukamoto, H. & West, A. R. 5 V lithium cathodes based on spinel solid solutions $Li_2Co_{1+X}Mn_{3-X}O_8$: $-1 \le X \le 1$. *Electrochim. Acta* **45**, 315–327 (1999).
16. Papike, J. J. & Cameron, M. Crystal chemistry of silicate minerals of geophysical interest. *Rev. Geophys. Space Phys.* **14**, 37–80 (1976).
17. Streltsov, V., Belokoneva, E. L., Tsirelson, V. G. & Hansen, N. K. Multipole analysis of the electron density in tryphylite, $LiFePO4$, using X-ray diffraction data. *Acta Crystallogr. B* **49**, 147–153 (1993).
18. Schock, R. N., Duba, A. G. & Shankland, T. J. Electrical conductivity in olivine. *J. Geophys. Res.* **94**, 5829–5839 (1989).
19. Alonso, J. A., Rasines, I., Rodriguez-Carvajal, J. & Torrance, J. B. Hole and electron doping of R_2BaNiO_5 (R = rare earths). *J. Solid State Chem.* **109**, 231–240 (1994).
20. Moriga, T. *et al.* Characterization of oxygen-deficient phases appearing in reduction of the perovskite-type $LaNiO_3$ to $La_2Ni_2O_5$. *Solid State Ionics* **79**, 252–255 (1995).
21. Andersson, A. S., Kalska, B. Häggström, L. & Thomas, J. O. Lithium extraction/insertion in $LiFePO_4$: an X-ray diffraction and Mössbauer spectroscopy study. *Solid State Ionics* **130**, 41–52 (2000).
22. Atkinson, A. in *Advances in Ceramics* Vol. 23 (eds Catlow, C. R. A. & Macrodt, W. C.) 3–26 (American Ceramic Society, Westerville, 1987).
23. Peterson, N. L. Point defects and diffusion mechanisms in the monoxide of the iron-group metals. *Mater. Sci. Forum* **1**, 85–108 (1984).
24. Kofstad, P. *Nonstoichiometry, Diffusion, and Electrical Conductivity in Binary Metal Oxides* 213–264 (Wiley, New York, 1972).
25. Shannon, R. D. Revised effective ionic radii and systematic studies of interatomic distances in halides and chalcogenides. *Acta Crystallogr. A* **32**, 751–767 (1976).
26. Goñi, A. *et al.* Magnetic properties of the $LiMPO_4$ (M = Co, Ni) compounds. *J. Magn. Magn. Mater.* **164**, 251–255 (1996).
27. Kornev, I. *et al.* Magnetoelectric properties of $LiCoPO_4$: microscopic theory. *Physica B* **271**, 304–308 (1999).
28. Xu, Y.-N, Ching, W. Y., Chung, S.-Y., Bloking, J. T. & Chiang, Y.-M. Electronic structure and electronic conductivity of undoped $LiFePO_4$. *Appl. Phys. Lett.* (submitted).
29. Kröger, F. A. *The Chemistry of Imperfect Crystals* (North-Holland, Amsterdam, 1964).

Acknowledgements

We thank W. D. Moorehead, P. Limthongkul and B. P. Nunes for assistance. This research was supported by the US Department of Energy, Basic Energy Sciences, Grant No. DE-FG02-87-ER45307, and used Shared Experimental Facilities at MIT supported by NSF Grant No. 94004-DMR.
Correspondence and requests for materials should be addressed to Y.M.C.

Competing financial interests

The authors declare competing financial interests: details accompany the paper on the *Nature Materials* website (http://www.nature.com/naturematerials).

FUEL CELLS

Advanced anodes for high-temperature fuel cells

Fuel cells will undoubtedly find widespread use in this new millennium in the conversion of chemical to electrical energy, as they offer very high efficiencies and have unique scalability in electricity-generation applications. The solid-oxide fuel cell (SOFC) is one of the most exciting of these energy technologies; it is an all-ceramic device that operates at temperatures in the range 500–1,000 °C. The SOFC offers certain advantages over lower temperature fuel cells, notably its ability to use carbon monoxide as a fuel rather than being poisoned by it, and the availability of high-grade exhaust heat for combined heat and power, or combined cycle gas-turbine applications. Although cost is clearly the most important barrier to widespread SOFC implementation, perhaps the most important technical barriers currently being addressed relate to the electrodes, particularly the fuel electrode or anode. In terms of mitigating global warming, the ability of the SOFC to use commonly available fuels at high efficiency, promises an effective and early reduction in carbon dioxide emissions, and hence is one of the lead new technologies for improving the environment. Here, we discuss recent developments of SOFC fuel electrodes that will enable the better use of readily available fuels.

A. ATKINSON[1], S. BARNETT[2],
R. J. GORTE[3], J. T. S. IRVINE*[4],
A. J. MCEVOY[5], M. MOGENSEN[6],
S. C. SINGHAL[7] AND J. VOHS[3]

[1]Department of Materials, Imperial College, London SW7 2BP, UK
[2]Northwestern University, Department of Materials Science, Evanston, Illinois 60208, USA
[3]University of Pennsylvania Department of Chemical Engineering, Philadelphia, Pennsylvania 19104, USA
[4]School of Chemistry, University of St Andrews, St Andrews, Fife KY16 9ST, UK
[5]ICMB-FSB, Ecole Polytechnique Fédérale, CH-1015 - Lausanne, Switzerland
[6]Risø National Laboratory, Materials Research Department, PO Box 49, Denmark
[7]Pacific Northwest National Laboratory, 902 Battelle Blvd, Richland, Washington 99352, USA
*e-mail: jtsi@st-andrews.ac.uk

Fuel-cell technology has been much heralded in recent years as a keystone of the future energy economy. In association with the hydrogen economy, it has been strongly promoted by the governments of most of the world's leading industrialized nations. There is now a phenomenal commercial interest in fuel-cell technology with new start-up companies being established and major players in the energy market turning their attention to it. The market prospects are certainly in the billions of euros per annum. The degree and extent of market penetration and establishment really only depend on the ability to reduce the cost of these devices while ensuring their long-term stability. The technology is certain to be applied; its scale will depend on the success of researchers in improving performance and cost. It is quite likely that the fuel cell's impact on society will be revolutionary. In the long term, fuel cells will be an essential component of any hydrogen or similar clean-energy economy; in the short-term, they promise greatly enhanced conversion efficiencies of more conventional fuels, and so will deliver large reductions in carbon dioxide emissions.

Fuel cells can be viewed as devices for electrochemically converting chemical fuels into electricity, essentially batteries with external fuel supplies. They offer extremely high chemical-to-electrical conversion efficiencies due to the absence of the Carnot limitation, and further energy gains can be achieved when produced heat is used in combined heat and power, or gas turbine applications. Furthermore, the technology does not produce significant amounts of pollutants such as nitrogen oxides, especially when compared with internal combustion engines.

All fuel cells consist of essentially four components: the electrolyte, the air electrode, the fuel electrode, and the interconnect. In SOFCs, which typically operate at temperatures in the range 800–950 °C, the electrolyte is normally yttria-stabilized zirconia (YSZ), which offers good oxide ion transport while blocking electronic transport. The electrolyte's function is to allow the transport of oxide ions from the electrolyte's interface

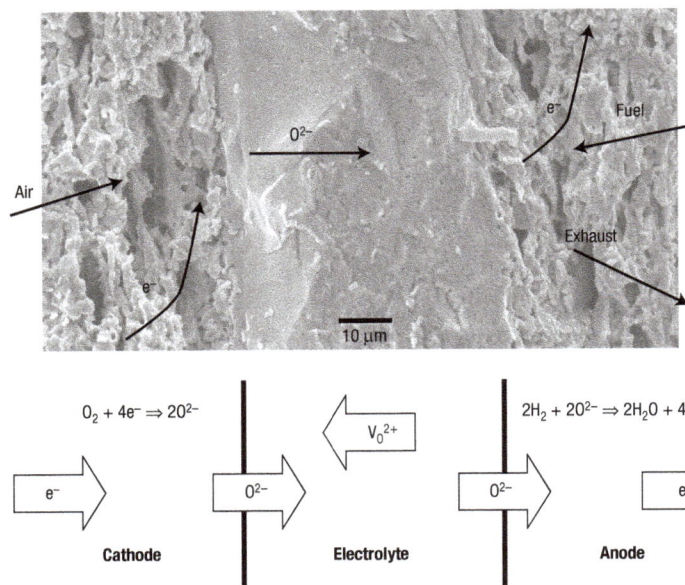

$$O_2 + 4e^- \Rightarrow 2O^{2-}$$

$$2H_2 + 2O^{2-} \Rightarrow 2H_2O + 4e^-$$

V_O^{2+}

e^- O^{2-} O^{2-} e^-

Cathode   **Electrolyte**   **Anode**

with the air electrode to its interface with the fuel electrode. The function of the air electrode, which is typically a composite of lanthanum strontium manganese oxide with YSZ, is to facilitate the reduction of oxygen molecules to oxide ions transporting electrons to the electrode/electrolyte interface, and allowing gas diffusion to and from this interface. The function of the fuel electrode, which is the main subject of this review, is to facilitate the oxidation of the fuel and the transport of electrons from the electrolyte to the fuel/electrode interface. In addition, it must allow diffusion of fuel gas to this interface and exhaust gases away from it. An appropriate microstructure for the SOFC, with particular attention to the electrode/electrolyte interfaces, and a schematic of the electrochemical processes is shown in Fig. 1. The other element is the interconnect, which has traditionally been lanthanum strontium chromite, but in lower temperature variants, attention has been focused on corrosion-resistant metallic alloys. The functions of this component are to transfer electrons between individual cells in the stack, while preventing gas crossover between the fuel and oxidant streams.

SOFCs have a particularly wide range of applications ranging from centralized megawatt-scale generation through to localized applications at greater than 100 kW level for distribution to local domestic generation on the 10 kW scale. There is also a wide range of other applications where cleaner energy is required, such as corrosion protection, uninterruptible power supplies, remote generation and domestic appliances. There are even some SOFC 'palm-power' applications being considered in the 10 W range. The most publicized examples are in the automobile sector, where companies such as Ford, Renault, Delphi and BMW are looking at high-temperature fuel cells for auxiliary power generation, and other companies such as Daimler Chrysler are looking at lower temperature polymer-based fuel cells for complete electrical motive power.

FUELS FOR FUEL CELLS

Each of the many applications for fuel cells in electricity generation has its own characteristic fuel requirements or preferred fuel. Fuels can range from hydrogen to methane to diesel to coal. On the other hand, most fuel cells can work only with fairly pure hydrogen, and those that can use other fuels still typically work best with hydrogen. This mismatch between the desired and/or available fuel and the fuel cell has limited their commercial implementation.

Hydrogen is the fuel normally associated with fuel cells. The first application of fuel cells was for space power — this is well matched with the fuel cell because high-purity fuel (hydrogen) and oxidant (oxygen) are used without regard to cost. Fuel cells are also viewed as a key element in a future hydrogen economy, not only because they can very efficiently convert hydrogen to electricity, but because they may also work well as electrolysers for storing excess energy as hydrogen. However, until hydrogen produced from renewable energy sources is widely available, the hydrogen economy makes little sense as hydrogen is currently produced from hydrocarbons; hence it is more sensible to produce electricity directly from the hydrocarbon fuels. Nonetheless, reversible fuel cell/electrolysers using hydrogen as the storage medium should be viable in the near future for matching distributed-generation electricity supply with widely varying demand levels.

Based on the widespread availability of natural gas — methane with small amounts of other hydrocarbons — most stationary fuel-cell systems have been designed for this fuel. Although direct introduction of either dry natural gas (direct hydrocarbon operation) or steam–natural gas mixtures (internal reforming) to SOFC anodes has been reported, in most cases partial external reforming is used. In this case, the fuel mixture arriving at the fuel-cell anode is primarily methane, hydrogen and carbon monoxide. There is typically a small amount of intentionally added sulphur-containing odorant in natural gas that can degrade anode performance if not removed, and some natural gases intrinsically contain larger amounts. Bio-derived fuels, for example, landfill gas, are often similar to natural gas, but contain carbon dioxide and a range of impurities such as ammonia, hydrogen sulphide and halogens, usually in quite significant concentrations.

Liquid hydrocarbons, for example, propane, are also used for a number of stationary power-generation applications because of their widespread availability. Because of their high energy densities, liquid hydrocarbons are also preferred for portable and transportation applications. Propane and butane are being considered as portable generator fuels because they are readily available, inexpensive, and are the lowest molecular weight (and highest hydrogen-to-carbon ratio) hydrocarbons that can be easily used in liquid form. Their vapour pressures at ambient temperature are low enough so as to not require high-strength tanks, and yet high enough to avoid the need for a fuel pump. Gasoline and diesel are clearly preferred for transportation. JP-8, a kerosene-like fuel, would be the most appropriate one for many military applications. All of these liquid fuels are characterized by a fairly broad range of molecular weights and substantial sulphur-containing impurities, which can cause anode degradation problems including coking and sulphur contamination if not removed.

Alcohols, including methanol and ethanol, have been widely considered for use in portable fuel-cell applications because they are liquid fuels with reasonably high energy densities, are readily available from industrial processes (or renewably from biomass), and have been shown to work with polymer electrolyte membrane fuel cells and SOFCs. Dimethyl ether and ammonia are also industrially produced fuels that have been considered for portable applications. These fuels have the disadvantage of being relatively expensive because of processing costs, but have the advantage of being relatively pure.

There is substantial interest in environmentally friendly use of large coal reserves using fuel cells. Problems are the relatively large amounts of impurities present in many coals, including sulphur and chlorine, and the high degree of carbon dioxide emission from coal relative to hydrocarbons. Also, coal gasification is needed to produce a gaseous fuel that can be introduced into a fuel cell.

Clearly, there is a diverse range of fuels that could be, and are likely to be, used in fuel cells. For high-temperature fuel cells such as the SOFC there is the important possibility of using these fuels without pre-processing, or at least with only partial pre-processing. To achieve this, it is essential to appropriately select and tailor the fuel electrode components and structure.

MATERIALS CHOICE

The materials selection for an SOFC anode is determined by a number of factors. First, the function required of it as the site for the electrochemical oxidation of the fuel associated with charge transfer to a conducting contact. Second, the environment in which it operates, at high temperature in contact not only with the fuel, including possible impurities and increasing concentrations of oxidation products, but also with the other materials, the electrolyte and contact components of the cell, and all this with stability over an adequate commercial lifetime at high efficiency. Third, the processability of the anode, which must be such that an open but well-connected framework can be achieved and retained during the fabrication of the fuel cell. Although in normal operation the ambient oxygen partial pressure is low, it can vary over several orders of magnitude. To accommodate fault conditions, or even just to provide flexibility of operating parameters, the ability to recover even after brief exposure to air at high temperature would be advantageous. A further aspect of this stability is the maintenance of structural integrity over the whole temperature range to which the component is exposed, from the sintering temperature during fabrication through normal operating conditions and then, repeatedly, cycling down to ambient temperature. Compatibility with other cell-component materials implies an absence of solid-state contact reactions, involving interdiffusion of constituent elements of those materials, or formation of reaction product layers that would interfere with anode functionality. It also requires a match of properties, such as shrinkage during sintering, and thermal expansivity

to minimize stresses during temperature variations due to operating procedures, start-up and shut-down.

By definition of its role, it is a requirement that the anode material should be an adequate electronic conductor, and also be electrocatalytically sufficiently active to sustain a high current density with low overpotential loss. However, the catalytic behaviour of anode materials should not extend to the promotion of unwanted side reactions, hydrocarbon pyrolysis followed by deposition of vitreous carbon being an example. An intimate contact between the two solid phases, the electrolyte delivering the oxide ions and the anode on which they are electrically neutralized, is clearly essential, as is access of the fuel and removal of reaction products, these being in the gas phase. On this model, the reaction is therefore sited on a 'three-phase boundary' zone. Low-loss operation implies that the three-phase boundary is not dimensionally limited to a planar interface of solid materials, but is delocalized to provide a 'volumetric' reaction region in three dimensions, porous for gas diffusion and permitting both electron and ion transport. One option is to provide a single-phase electrode with mixed conductivity permitting both oxide ion and electron mobility within the anode material. The alternative is to use a porous composite, as in the nickel-based cermets that have typically been used in SOFCs to date.

NICKEL-CERMET ANODES

After some early investigations using single-phase anodes, nickel–zirconia cermet anodes have been the dominant SOFC anodes for some forty years. Single-phase materials investigated in the earliest SOFC developments included graphite, iron oxide, platinum group, and transition metals[1,2]. Graphite is corroded electrochemically and platinum spalls off in service, presumably due to water-vapour evolution at the metal-oxide interface. As for the transition metals, iron is no longer protected by the reducing activity of the fuel gas once the partial pressures of oxidation products in the anode compartment of an operating cell exceed a critical value, and it then corrodes with formation of a red iron oxide. Cobalt is somewhat more stable, but also more costly. Nickel has a significant thermal expansion mismatch to stabilized zirconia, and at high temperatures the metal aggregates by grain growth, finally obstructing the porosity of the anode and eliminating the three-phase boundaries required for cell operation. As a consequence, all-metal anodes have not found acceptance. This was the context for the introduction of the nickel–zirconia cermet anode by Spacil[3]. His 40-year-old text remains a strikingly modern presentation of the anode specification, making reference to many of the concepts and procedures still current in SOFC technology, and which have made this composite the current standard anode material. As the best transition-metal option, Spacil associated nickel with the stabilized zirconia ceramic material of the electrolyte to confront the two problems mentioned, the difference of thermal expansion that would mechanically stress any nickel anode–electrolyte bond, and the aggregation in service. A minimum metal proportion in the cermet is necessary for continuity of electronic conduction, whereas the zirconia particles

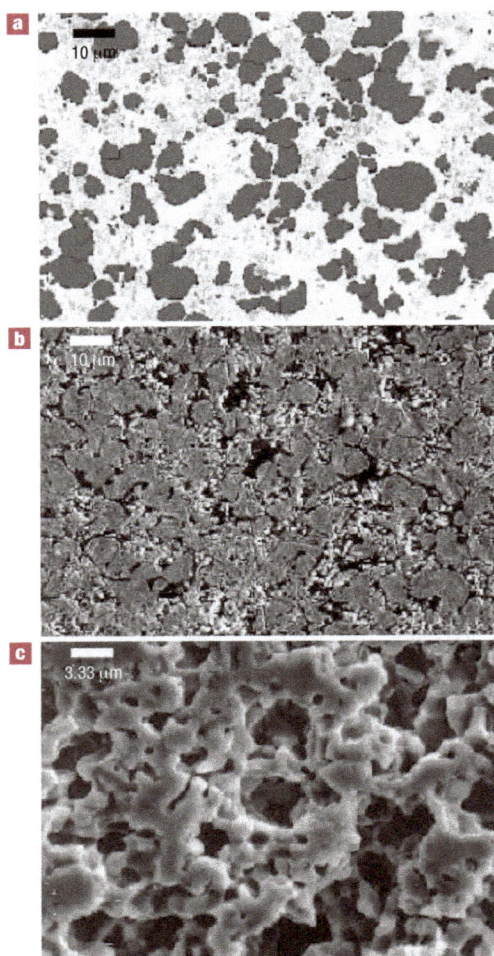

may be non-continuous. Spacil had therefore recognized that the functionality of the ceramic in the composite was essentially structural, to retain the dispersion of the metal particles and the porosity of the anode during long-term operation. Structure and elemental distribution in a typical anode cermet are imaged in Fig. 2 (ref. 4). The provision of oxide-ion mobility complementary to the electronic conductivity and electrocatalytic action of the metal is a very important secondary role of the ceramic, enhancing the electrochemical performance by the delocalization of the electrochemically active zone already mentioned.

The major disadvantage of the nickel-cermet electrode arises from the promotion of the competitive catalytic cracking of hydrocarbon. The rapid deposition of carbon at nickel cermets means that direct oxidation of methane is not technically viable in nickel-containing SOFCs. For natural gas to be used as fuel, it needs to be externally or internally reformed with steam. Impurities in the fuel stream, particularly sulphur, also inhibit anode functionality.

CATALYTIC PROPERTIES

The catalytic properties of the anode in an SOFC are very important in determining the overall cell performance. In the commonly used Ni/YSZ cermet anode, nickel has the dual roles of catalyst for hydrogen oxidation and electrical current conductor. In addition to being an excellent catalyst for the oxidation of hydrogen, nickel is highly active for the steam-reforming of methane. This catalytic property is exploited in so-called internal reforming SOFCs that can operate on fuels composed of mixtures of methane and water. In these systems, the CH_4/H_2O mixture reacts on the nickel anode producing a hydrogen-rich synthesis gas, which then undergoes electrochemical oxidation at the three-phase boundary. Owing to the thermodynamic limitations of the steam-reforming reaction, internal reforming SOFCs generally operate at temperatures above 900 °C. Some application scientists have also termed the type of reforming that occurs within the stack but external to the SOFC, as indirect internal reforming, but this seems misleading and in-stack reforming is a more appropriate term.

Although nickel is an excellent hydrogen oxidation and methane-steam-reforming catalyst, it also catalyses the formation of carbon filaments from hydrocarbons under reducing conditions. The mechanism involves carbon chemisorption on the nickel surface, carbon dissolution into the bulk nickel, and precipitation of graphitic carbon from some facet of the nickel particle after it becomes supersaturated in carbon[5]. Unless sufficient amounts of steam are present along with the hydrocarbon to remove carbon from the nickel surface at a rate faster than that of carbon dissolution and precipitation, the anode will be destroyed. Although thermodynamic calculations are often used to predict the conditions under which carbon formation is favoured[6], carbon fibres can be formed even when carbon is not thermodynamically predicted. This is particularly true in the presence of higher hydrocarbons, as these molecules form carbon on the nickel surface more rapidly than steam can remove it. As a result, even when using methane as the fuel, relatively high steam-to-carbon ratios are needed to suppress this deleterious reaction. Unfortunately, due to the high catalytic activity of nickel for hydrocarbon cracking, this approach does not work for higher hydrocarbons, and it is generally not possible to operate nickel-based anodes on higher hydrocarbon-containing fuels, without pre-reforming with steam or oxygen.

One approach to overcoming the limitations of nickel anodes, which has met with some success, is to augment the oxidation activity of Ni/YSZ cermets through the addition of an oxide-based oxidation catalyst. For example, stable operation on dry methane has been reported at 650 °C in an SOFC using an yttria-doped ceria interlayer between the YSZ electrolyte and the Ni/YSZ cermet anode[7]. Ceria (cerium oxide) is a well-known oxidation catalyst, and might be expected to increase the activity of the anode for the electrochemical oxidation of methane. This approach still requires, however, that the operating temperature be maintained below 700 °C to suppress carbon deposition reactions that take place on nickel.

Most research aimed at overcoming the limitations of nickel-based anodes has focused on the development of alternative anode materials that are catalytically active for the oxidation of methane or higher hydrocarbons, and inactive for cracking reactions that can lead to carbon deposition. This latter criterion is a difficult one, and rules out most transition metals with the possible exception of Cu, Ag and Au. Although these metals make good current collectors, they are not highly active oxidation catalysts. As many metal oxides are excellent oxidation catalysts and are not active for catalytic cracking, most of the work in this area has focused on the development of electronic or mixed electronic–ionic conducting, ceramic anodes. As discussed below, these materials were generally chosen based on their electronic and ionic conductivities rather than their catalytic properties. Although stable performance while operating on hydrocarbons, principally methane, has been achieved in SOFCs using many of these ceramic materials as anodes, it is likely that their catalytic properties are not optimal. The use of composite ceramic anodes has been suggested as one avenue for optimizing both the transport (ionic and electronic) and catalytic properties of ceramic anodes.

Although catalytic anodes that allow for the operation of SOFCs directly on hydrocarbons have been developed, the performance of these anodes (for example, maximum current density) is still not at the level obtained in state-of-the-art SOFCs with Ni/YSZ anodes operating on hydrogen. Thus there is still much room for improvement in the catalytic properties of SOFC anodes designed to use hydrocarbons directly. Understanding the mechanisms of the surface reactions that occur on the anode near the three-phase boundary is likely to be key to further advances in this area. Unfortunately, fundamental insight into these reaction mechanisms is rather limited. For systems that directly use hydrocarbon fuels, the mechanism of the surface oxidation reactions is undoubtedly highly complex, and involves multiple elementary steps and reaction intermediates. At the very least it would be useful to identify the rate-limiting step or steps in the reaction mechanism. Even for Ni/YSZ cermet anodes, the mechanism for hydrogen oxidation reactions is not completely understood. For example, it has been shown that the rate of this reaction is dependent on water partial pressure with higher pressures resulting in higher reaction rates[8]. The mechanism by which water affects this reaction is currently under discussion, with the speculation that the zirconia surface is hydroxylated.

There is clearly a need for fundamental studies of the mechanisms of the catalytic reactions that occur near the three-phase boundary in the anode of an SOFC. Many in situ spectroscopic probes such as Raman and infrared spectroscopy are routinely used in catalysis research to characterize surface intermediates and reaction mechanisms. It is very difficult, however, to apply these techniques to an operating SOFC anode. Nonetheless, some inroads are being made in this area. For example, the application of infrared emission spectroscopy to characterize working SOFC anodes is being pioneered[9]. Although these studies are in the early stages, preliminary results appear very promising. Development of models of anodes that allow other spectroscopic probes, such as photoelectron spectroscopy and detailed kinetic measurements, to be made are also needed in order to advance research in this area.

OTHER CERMETS

To find alternative anodes with lower susceptibility to hydrocarbon cracking, alternative cermets to nickel yttria zirconia have been sought. Interest has largely focused on ceria replacing zirconia and copper replacing nickel. One approach is to alloy copper with nickel; for instance, copper has been added to nickel to reduce the catalytic activity of nickel for hydrocarbon cracking[10]. The influence of adding nickel to $(Ce,Zr)O_2$ or $(Ce,Y,Zr)O_2$ to form a cermet was also studied[11] and cerium oxide is reported as a promising support. In a series of publications, Gorte et al.[12,13] reported that noble metals (Pd, Pt, Rh) had much higher specific rates for water gas shift, steam reforming and carbon dioxide reforming of methane when supported on ceria than when supported on silica or alumina. Gadolinium-doped ceria, $Ce_{0.9}Gd_{0.1}O_{1.95}$ has been studied[14] as an anode material at 900 °C in 5% CH_4 with steam/methane ratios between 0 and 5.5. This material revealed itself to be resistant to carbon deposition, although the reaction rate was controlled by slow methane adsorption. The same catalyst has been studied[15] for methane oxidation, and it was demonstrated that ceria has a low activity for methane oxidation but a high resistance to carbon deposition. It was proposed to add a catalyst (Ni, Rh, Ru) to break the C–H bond more easily.

In contrast to the reported inactivity of acceptor-doped ceria, undoped ceria has been reported to be quite an effective oxidation catalyst, this can be attributed to differing electronic functionalities of the doped and undoped materials. One possibility is that the higher concentration of oxygen vacancies makes the doped material more difficult to reduce and hence a less effective catalyst. Care must be taken, however, in comparing activities in oxidizing conditions as frequently encountered in catalysis experiments, and under fuel conditions that are quite reducing. Ceria is a well-known oxidation catalyst and increases the activity of the anode for the electrochemical oxidation of methane. This approach still requires that the operating temperature be maintained below 700 °C to suppress carbon-deposition reactions that take place on nickel. Another approach being used in the development of cermet anodes that allows the use of fairly dry hydrocarbon, is to use a relatively inert metal such as copper for electrical conductivity and a metal oxide to provide catalytic activity and ionic conductivity. Examples of this type of composite anode include $Cu/CeO_2/YSZ$ (ref. 16) and Cu/YZT (titania-doped yttria zirconia)[17]. The $Cu/CeO_2/YSZ$ anode system is particularly interesting, and has been shown to be effective for the direct use of a variety of hydrocarbon fuels including butane and decane, and highly resistant to deactivation through carbon deposition[18]. The role of copper in these anodes appears to be exclusively current conduction, whereas ceria primarily acts as an oxidation catalyst. It has been shown that SOFCs that have anodes that contain only Cu and YSZ exhibit very poor performance while operating on hydrocarbons. The addition of ceria to the anode, however, produces a marked increase in performance while operating on both hydrogen and hydrocarbons. It is important to note that SOFCs with $Cu/CeO_2/YSZ$ anodes always

Table 1 Summary of properties of some potential oxide anodes at 800 °C under reducing conditions (partial oxygen pressure approximately 10^{-20} atm.).

Material	Electronic conductivity[a] (S cm^{-1})	Ionic conductivity (S cm^{-1})	Oxygen diffusivity (cm^2 s^{-1})	Redox stability[b]	Polarization resistance[c]	CTE (p.p.m. K^{-1})
CeO_2 (ref. 33) 0.5–1	0.1–0.2 (ref. 34)	1×10^{-6} (ref. 34)	XX (ref. 35)	**	12	
$Zr_{1-x-y}Ti_xY_yO_2$ (36)	0.1	1×10^{-2}	?	**	*	10
$La_{0.8}Sr_{0.2}Cr_{0.95}Ru_{0.05}O_3$ (37)	0.6	Low	?	*	**	10
$La_{0.8}Sr_{0.2}Fe_{0.8}Cr_{0.2}O_3$ (ref. 32)	0.5	?	1×10^{-8}	*	**	12
$La_{0.25}Sr_{0.75}Cr_{0.5}Mn_{0.5}O_3$ (ref. 22)	3	?	?	**	***	10
$Sr_{0.86}Y_{0.08}TiO_3$ (ref. 28)	80	Low	?	***	*	11–12
$La_{0.33}Sr_{0.66}TiO_{3.166}$ (refs 29,30)	40	Low	?	***	*, *** d	10
Nb_2TiO_7 (refs 31)	200	Very low	?	X	-	1–2
$Gd_2Ti(Mo,Mn)O_7$ (refs 38,39)	0.1	Reasonable	?	XXX	*	?

[a] Should only be considered as a rough guide as factors such as partial pressure of oxygen, density and microstructure are not standardized.

[b] *** excellent to XXX very poor

[c] These values are strongly dependent on microstructure. and will vary between different laboratories, as an indication values are ,**** < 0.1 Ω cm^2, *** 0.1–0.3 Ω cm^2, ** 0.3–1 Ω cm^2,
* 1–10 Ω cm^2

[d] High performance for composite with ceria

exhibit even higher current densities while operating on hydrogen compared with hydrocarbons. This indicates that catalysis plays a critical role and suggests that enhancements in the catalytic activity of the anode are still needed to optimize performance using hydrocarbon fuels.

ALTERNATIVE ANODE MATERIALS

The basic properties of the 'standard' nickel–zirconia or nickel–ceria cermets are well established, and for these materials the requirement is to optimize their properties through control of microstructure and minor additives. The situation is different for oxide anodes where the requirement is still at the stage of identifying suitable candidates. Oxides are under investigation as single-phase anodes, single-phase anode current collectors, and components of composites for either of these functions. In addition to electrochemical performance, the candidate materials need to display other characteristic properties, some being essential and some desirable. These properties include: electronic conductivity; oxygen diffusivity (ionic conductivity); oxygen surface exchange (reactivity); chemical stability and compatibility; compatible thermal expansion; mechanical strength and dimensional stability under redox cycling. For application as a current collector, electrochemical activity is not required, and therefore oxygen diffusivity and surface exchange properties are not relevant. In this section we briefly review what is

known about these properties for leading candidate materials, and identify areas where further work is required. The materials of interest at this stage are ceria (doped and undoped) and transition metal perovskite and fluorite-related structures. From the point of view of chemical stability under reducing conditions, oxides containing large amounts of Co and Ni are probably not viable, and so the emphasis is on Fe, Mn, Cr and Ti as transition metal ions that can be used to give electronic conductivity. The relevant properties of some typical candidate oxide anode materials are summarized in Table 1.

The target for electronic conductivity for anode materials is often set to be 100 S cm^{-1}, but the actual requirement depends on the cell design, and particularly the length of the current path to the current collection locations. Thus this could be relaxed to as low as 1 S cm^{-1} for a well-distributed current collection. Similarly, if the material were used as a porous support and current collector with thickness of 0.5 mm, it would also need a conductivity greater than approximately 1 S cm^{-1} to maintain losses below 0.1 Ω cm^2. As these are targets for porous structures, the actual requirement for the intrinsic materials properties, that is, in dense form, would need to be about one order of magnitude greater.

A suitable benchmark material is the perovskite $La_{1-x}Sr_xCrO_3$, which has been thoroughly investigated as an interconnect material for SOFCs[19], and is also a potential anode material for SOFCs due to the relatively good stability in both reducing and oxidizing atmospheres at high temperatures[20]. The acceptor doping gives high p-type conductivity in air but, as with all p-type materials, this decreases under reducing conditions. Addition of reducible transition metals such as Ti can introduce significant n-type contribution at low partial pressure of oxygen, but the dilution of Cr on the B-site generally has a greater effect[21]. Substitution of mid-transition metals such as Mn or Fe does not have such a dilution effect, indicating some complementarity of electronic function, and affords an extension of the p-type domain to lower partial oxygen pressures, although conductivities are still below 10 S cm^{-1} under reducing conditions[22]. It is interesting to note that although n-type electronic conductivity would seem the more natural form of electronic defect under fuel conditions, p-type conduction, if it can be retained to low partial oxygen pressures, offers a significant advantage in that conductivity will increase under load.

The reported polarization resistance using chromite perovskites is generally too high for efficient SOFC operation, although significant improvements have been achieved using low-level doping of the B-site. With 3% replacement of Cr by V, methane cracking seems to be avoided, although the polarization resistance is still of the order 10 Ω cm^2 at 900 °C[23]. The introduction of other transition elements into the B-site of La$_{1-x}$Sr$_x$Cr$_{1-y}$M$_y$O$_3$ (M = Mn, Fe, Co, Ni) has been shown to improve the catalytic properties for methane reforming[24]. Of these various dopants, nickel seems to be the most successful, and the lowest polarization resistances have been reported for 10% nickel-substituted lanthanum chromite[25]; however, other workers have found nickel exsolution from 10% nickel-doped lanthanum chromites in fuel conditions[26]. Certainly nickel oxides would not be stable in fuel atmospheres, and although the nickel may be stabilized by the lattice perovskite in higher oxidation state, there will always be the suspicion that the activity of nickel-doped perovskites is due to surface exsolution of nickel metal, and hence raise questions about long-term stability. A particularly successful approach has been to create perovskites with dual B-site occupancy, such as those based on lanthanum chromium manganite[22]. This approach has very successfully combined the good oxidation catalysis properties of lanthanum manganite with the stability and conductivity of the chromite, without compromising any of these good properties by dilution.

Donor doping on the A-site is difficult for 3,3 perovskites, but has been very successful for 2,4 perovskites. Conventional donor doping leaves the A sites filled so that the A:B ratio remains stoichiometric at 1:1. However, a better strategy has been to maintain constant total A-site charge, so that the material is approximately compensated by vacancies on the A-sites (A:B ratio < 1). The Schottky equilibrium between these A-site vacancies and oxygen vacancies, equation (1), means that it becomes easier to create electronic carriers by reduction, equation (2). Conversely, the introduction of oxygen vacancies, for example, for Mg substitution of Ti equation (3), means that increasing oxygen vacancy concentration must make it more difficult to create electronic carriers by reduction at a given partial pressure of oxygen. Thus A-site-deficient strontium

titanates exhibit relatively high conductivity under fuel conditions[27,28]. The standard donor-doping method, replacing Sr by La has also proved very successful, especially in combination with ceria[29]. Effectively these compositions are oxygen hyper-stoichiometric, except when prepared under very reducing conditions, and it has recently been suggested that such lanthanum strontium titanates contain extended defects analogous to the oxygen shear planes observed in lanthanum titanate, La$_2$Ti$_2$O$_7$. Nonetheless, such materials offer very high conductivities[30]. Rutile structures such as NbO$_2$ or Nb$_2$TiO$_6$ offer very high electronic conductivity under fuel conditions, but with very low thermal expansion coefficient[31]. Although initial studies demonstrated good compatibility with zirconia electrolytes[31], later work has shown poor compatibility with ceria electrolytes.

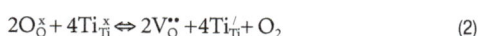

$$O_O^x + A_A^x \Leftrightarrow V_O^{\bullet\bullet} + V_A^{//} + AO \tag{1}$$

$$2O_O^x + 4Ti_{Ti}^x \Leftrightarrow 2V_O^{\bullet\bullet} + 4Ti_{Ti}^{/} + O_2 \tag{2}$$

$$O_O^x + Ti_{Ti}^x + MgO \Leftrightarrow V_O^{\bullet\bullet} + Mg_{Ti}^{//} + TiO_2 \tag{3}$$

The superscript notations in equations (1)–(3) are: x uncharged; • single positive charge, and / single negative charge. All these materials seem to meet the essential requirement of chemical compatibility with YSZ. However, data on other relevant properties are rather sparse. Sr$_{0.86}$Y$_{0.08}$TiO$_3$, La$_{0.75}$Sr$_{0.25}$Cr$_{0.5}$Mn$_{0.5}$O$_3$, La$_{0.8}$Sr$_{0.2}$Fe$_{0.8}$Cr$_{0.2}$O$_3$ and La$_{0.8}$Sr$_{0.2}$Cr$_{0.95}$Ru$_{0.05}$O$_3$ all have acceptable thermal expansion match to YSZ under reducing conditions, but heavier Sr doping for some of these leads to anomalous expansion coefficients and unacceptably large dimensional changes on redox cycling. Better understanding of the factors that contribute to these dimensional changes would be valuable.

Very little is known about the oxygen diffusivity (ionic conductivity) and surface exchange of oxygen under reducing conditions in any of these materials. Such studies have proved valuable in understanding the behaviour of cathodes, and now need to be applied to anode materials. So far only La$_{1-x}$Sr$_x$Fe$_{0.8}$Cr$_{0.2}$O$_3$ has been studied in this way[32].

DIRECT CONVERSION

A subject that is receiving increased attention is the development of new anode materials for direct conversion of hydrocarbon fuels without first reforming those fuels to CO and H$_2$. Elimination of the reforming step would decrease system complexity and avoid the necessity of diluting fuels with steam. For higher hydrocarbons, there are also energy losses associated with the need to partially oxidize the fuel. In principle, an SOFC can operate on any combustible fuel that is capable of reacting with oxide ions coming through the electrolyte. In practice, the high operating temperatures in the presence of hydrocarbons can lead to carbon formation, for example, as discussed above for nickel-based anodes.

Three strategies for direct conversion of hydrocarbon fuels are possible. The first strategy[7] uses conventional nickel-cermet anodes but modifies the operating conditions of the fuel cell. For methane at intermediate temperatures, the rate of carbon

Figure 3 Schematic of Cu/CeO$_2$ YSZ (or SDC) microstructure showing enhanced current collection following tar deposition after initial exposure to butane under operating conditions.

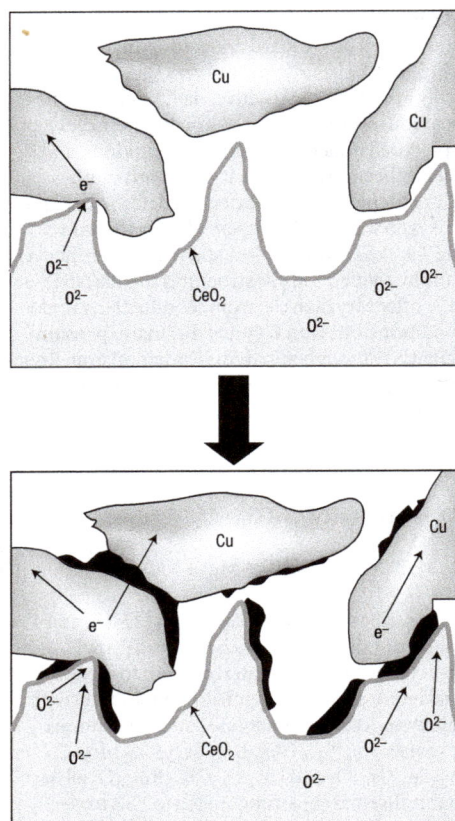

The third strategy for developing SOFCs that can operate directly on hydrocarbon fuels involves anodes made from electronically conductive ceramics, as these are also tolerant towards carbon formation. Reasonable performance can be achieved with direct use of methane using an anode made from La-doped strontium titanate[29,30]; however, the addition of ceria to this doped $SrTiO_3$ enhanced the performance significantly[29]. Reasonable performance in methane can also be obtained with anodes made from Gd-doped ceria[43]. Stable power generation with methane and propane fuels on an anode based on Mn-doped $LaCrO_3$ with YSZ and 5% Ni has also been reported[44]. The addition of nickel, in amounts small enough to avoid carbon formation, was found to enhance the performance of the $LaCrO_3$ anodes. Similar good performance has been achieved using nickel (ref. 24) and Ru (ref. 45) substituted lanthanum chromites, and most recently the split B-site perovskite $(La,Sr)Mn_{0.5}Cr_{0.5}O_3$ that provides both very good performance and good chemical stability[22].

An important issue for direct use of hydrocarbons in all of the strategies is the question of mechanism, as this relates directly to the potential of the electrons that are produced in the cell. Although it has been demonstrated that there is a balance between the production rate for the total oxidation products and the generation of current[16], the cell potentials that have been reported with hydrocarbon fuels are much lower than would be predicted by direct, electrochemical oxidation of the hydrocarbons in a single step. It seems clear that at the very least, the direct-use processes involve an initial activation step, perhaps just involving the breaking of one carbon–hydrogen bond. Indeed, the appearance of tars on Cu/CeO$_2$ composites using hydrocarbons[40] certainly indicates that some pyrolysis reactions may help initiate the hydrocarbon oxidation process in these systems. The efficient oxidation of produced CO or carbon at ceria may well be the key to the effectiveness of these materials in direct use of hydrocarbons, as ceria itself is viewed as being only a moderately good catalyst[16]. Conversely, much of the effort on direct use has focused on reforming activity, especially for transition-metal doping of lanthanum chromite-based materials. It is worthwhile remembering that one of the main oxidation products is steam, so reforming can still occur even with dry fuel sources. This has led to the concept of gradual internal reforming, Fig. 4, which effectively provides a direct utilization mechanism[46]. Preliminary investigations of the activity of the lanthanum strontium chromium manganite perovskite[22] indicate poor reforming and good oxidation activity. This material may therefore act in a similar manner to ceria, and so its catalytic properties seem more closely related to a manganite than a chromite. The various processes that can contribute in combination to effectively achieve direct use are presented in Fig. 5.

MICROSTRUCTURAL COMPROMISE

So far the discussion has related largely to the compositional nature of candidate anodes; however, the microstructure of the electrode is at least as important as its composition. The optimization of durable

deposition may be slow enough such that oxygen anions electrochemically driven through the electrolyte and steam generated by oxidation of methane will remove carbon as it is deposited.

A second strategy to avoid reforming involves replacing nickel cermets with composites containing copper and ceria. Copper is relatively inert towards the carbon-formation reactions that occur on nickel, and stable operation has been observed with even large hydrocarbons over copper-based anodes[16]. Gas-phase pyrolysis reactions can still lead to tar formation on copper cermets; however, the compounds that form on copper tend to be polyaromatics, such as naphthalene and anthracene, rather than graphite. It has been suggested that the polyaromatic compounds enhance anode performance by providing additional electronic conductivity in the anode[40], Fig. 3.

It appears that copper primarily provides electronic conductivity to the anode and is otherwise catalytically inert. This is confirmed by data showing that Au-ceria-SDC (samaria-doped ceria) composites exhibit a similar performance to that of Cu-ceria-SDC anodes, as Au would not be expected to add catalytic activity[41,42]. It is proposed that the function of ceria is primarily that of an oxidation catalyst, although mixed electronic–ionic conductivity could also enhance anode performance. In general, finely dispersed ceria seems more active than doped ceria ceramics, thus the redox oxygen exchange ability of ceria at fuel conditions might be considered as a key factor.

efficient nickel-cermet anodes in recent decades has relied greatly on empirical improvement of materials specifications to control cermet morphology. With modern submicrometre active ceramic powder, the sintering temperature can be significantly decreased, to 1,400 °C or lower, and lower metal contents can be achieved. Associated with the reduced thermal expansivity of the cermet due to the increased ceramic content, stresses during fabrication, reduction and operation are minimized, eliminating microfissuring, which contributes to electrode ageing[47]. Nickel oxide of grain size around 1 μm is now used, whereas the ceramic component may be bidispersed, containing a proportion of coarse powder with grains of 25 μm or larger to form the anode structural skeleton and inhibit the nickel aggregation, mixed with active submicrometre fine powder to promote sintering. These procedures are applied to the conventional electrolyte-supported configuration where the stabilized zirconia substrate, 150 μm or thicker, also provides the overall structural integrity of the cell. Much recent development has extended the function of the material used for the electrochemical anode into becoming also the load- and stress-bearing support for an electrolyte now no thicker than 10 μm. This permits lowering the operating temperature to perhaps 650 °C, while providing adequate cell performance. Under these conditions, materials specifications throughout the system are relaxed, with lower-cost metallic structural and interconnect components, and with diminished thermomechanical stresses and reactions between materials, thereby significantly improving durability. The structural cermet, now up to 1 mm thick, provides not only the anode functionality, but can also serve as a reaction volume for fuel processing, such as internal reforming. However, the lower temperatures significantly diminish the thermal activation of the oxidation reactions, implying increased polarization and giving added importance to considerations of electrocatalysis at the anode[48]. This makes a graded anode structure advisable, with a high porosity, large grain substrate bearing a finer-structured electrocatalytically active anode layer to contact with the electrolyte.

An important distinction between Cu-cermet and nickel-cermet anodes is in the required synthesis conditions. Cu-YSZ cermets prepared by calcining mixtures of CuO and YSZ, the method conventionally used to prepare nickel cermets, are unstable in reducing atmospheres above 650 °C, indeed, Cu seems a remarkably mobile species[49]. It is necessary to calcine the CuO–YSZ mixtures at relatively low temperatures to avoid melting of the CuO, making it less easy to achieve a well connected, YSZ skeleton; however, the easy migration of the Cu phase makes such composites seem untenable with respect to long-term stability. The success of the Cu/CeO$_2$ system[16] is therefore quite extraordinary, and can only be ascribed to a very advantageous microstructure. In this system, a porous skeleton is first obtained by ceramic methods and then the Cu and ceria are added in subsequent, low-temperature steps. The thermal stability is certainly very much improved on that of more conventionally prepared systems with stable operation reported[16] up to 800 °C. Such systems also offer remarkably high

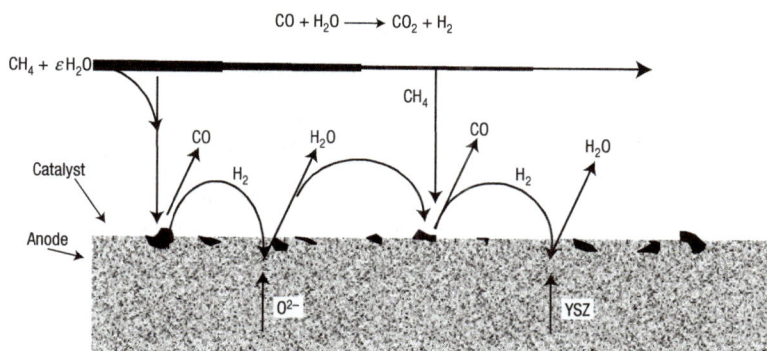

conductivities considering the low copper contents, for example, ~10% by weight. These considerations can only mean that the Cu forms an intimate near-continuous thin network within the retaining skeleton. Some carbon deposition is found to enhance conduction, bridging some of the links between the Cu network[40]. The success of this system ably demonstrates the importance of controlled interfacial and microstructural engineering. Indeed, it must be realized that the anode electrolyte microstructure is the dominant feature governing fuel electrode activity. Therefore care must always be taken in ranking new materials in terms of activity without appropriately considering microstructure.

SUMMARY AND RECOMMENDATIONS FOR RESEARCH ON MATERIALS PROPERTIES

Major advances in SOFC anode development have been achieved in recent years, and there is good encouragement that new fuel electrode formulations for second-generation commercial fuel cells can be found. Such new anodes will offer improved redox tolerance and better resistance stability in hydrocarbon fuels. Other key strategies have been identified, based on oxidation catalysts such as ceria or lanthanum chromium manganite or reforming catalysts such as nickel- or ruthenium-doped lanthanum chromite. Additionally, the function of electronic conductivity in the anode current collector has also been addressed in oxides such as those based on strontium titanate. Although no single material fulfils all the current

Figure 4 The series of processes that lead to the gradual internal methane reforming, where ε is significantly less than 2 (ref. 46).

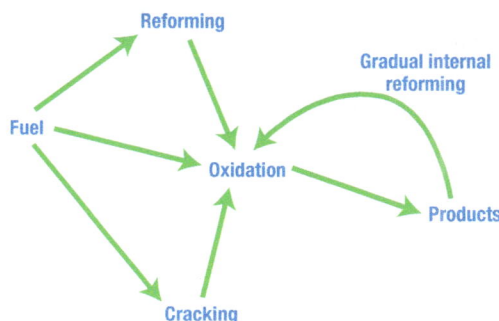

Figure 5 The contributing processes that can give rise to the direct use of hydrocarbon fuels in SOFCs.

collection, electrochemical and catalytic performance indicators required for a supported electrode design to the extent that nickel–zirconia cermets do, a number of systems individually meeting these criteria have been attained. Clearly new composites are possible to fulfil the roles of both the active electrode layer and the current collecting part.

Further research is required to optimize these materials (composition and microstructure); particularly to maximize electronic conductivity without sacrificing essential chemical stability under reducing conditions. Detailed studies are required on the most promising materials to establish and understand dimensional stability on redox cycling and long-term operation, and how this can be improved without compromising electrical conductivity. Studies of oxygen diffusion, ionic conductivity and surface exchange in reducing environments are required on the most promising materials to improve understanding of their electrocatalytic properties. Particular care is necessary to allow reproducibility of results between laboratories, and especially to ensure that meaningful experiments are performed before their results are presented. Understanding of crystal structure, point-defect populations and electronic structure need to be improved at a fundamental level, and related to the key properties mentioned above to guide the search for new materials. More basic work to probe the mechanistic details of fuel utilisation reactions needs to be undertaken. These details are extremely important to rational design of more efficient anodes and thus high-performance SOFCs powered by practical fuels. Practical studies in real fuels need to be expanded, especially addressing issues such as sulphur tolerance and ability to use biofuels. The search for totally new oxide materials with even better properties should continue, because even the best ones only just meet the requirements. Most importantly, this should include the highest possible electronic conductivity, as such a material would also serve as a current collector.

doi:10.1038/nmat1040

References

1. Baur, E. & Preis, H. Uber die Eisenoxyd-Kathode in der Kohle-Luft-Kette. *Z. Elektrochem.* **43**, 727 (1937).
2. Möbius, H.-H. On the history of solid electrolyte fuel cells. *J. Solid State Electrochem.* **1**, 2 (1997).
3. Spacil, H. S. Electrical device including nickel-containing stabilized zirconia electrode. US Patent 3,558,360 (1970).
4. Lee, J. H. *et al.* Quantitative analysis of microstructure and its related electrical property of SOFC anode, Ni-YSZ cermet. *Solid State Ionics* **148**, 1–2 (2002).
5. Toebes, M. L., Bitter, J. H., van Dillen, A. J. & de Jong, K. P. Impact of the structure and reactivity of nickel particles on the catalytic growth of carbon nanofibers. *Catal. Today* **76**, 33–42 (2002).
6. Sasaki, K. & Teraoka, Y. Equilibria in fuel cell gases - I. Equilibrium compositions and reforming conditions. *J. Electrochem. Soc.* **150**, A878–A884 (2003).
7. Murray, E. P., Tsai, T.& Barnett, S. A. A direct-methane fuel cell with a ceria-based anode *Nature* **400**, 649–651 (1999).
8. Mogensen, M., Jensen, K. V., Jorgensen, M. J. & Primdahl, S. Progress in understanding SOFC electrodes. *Solid State Ionics* **150**, 123–129 (2002).
9. Lu, X. Y., Faguy, P. W. & Liu, M.l. In situ potential-dependent FTIR emission spectroscopy - A novel probe for high temperature fuel cell interfaces. *J. Electrochem. Soc.* **149**, A1293–A1298 (2002).
10. Bernardo, C. A., Alstrup, I. & Rostrupnielsen, J. R. Carbon deposition and methane steam reforming on silica-supported Ni-Cu catalysts. *J. Catal.* **96**, 517–534 (1985).
11. Dong, W. S., Roh, H. S., Jun, K. W., Park, S. E. & Oh, Y. S. Methane reforming over Ni/Ce-ZrO2 catalysts: effect of nickel content. *Appl. Catal. A* **226**, 63–72 (2002).
12. Bunluesin, T., Gorte, R. J & Graham, G. W. Studies of the water-gas-shift reaction on ceria-supported Pt, Pd, and Rh: implications for oxygen-storage properties. *Appl. Catal. B* **15**, 107–114 (1998).
13. Sharma, S., Hilaire, S., Vohs, J. M., Gorte, R. J. & Jen, H. W. Evidence for oxidation of ceria by CO₂. *J. Catal.* **190**, 199–204 (2000).
14. Ramirez-Cabrera, E., Atkinson, A. & Chadwick, D. The influence of point defects on the resistance of ceria to carbon deposition in hydrocarbon catalysis. *Solid State Ionics* **136**, 825–831 (2000).
15. Marina, O. A. & Mogensen, M. High-temperature conversion of methane on a composite gadolinia-doped ceria-gold electrode. *Appl. Catal. A* **189**, 117–126 (1999).
16. Park, S. D., Vohs, J. M. & Gorte, R. J. Direct oxidation of hydrocarbons in a solid-oxide fuel cell. *Nature* **404**, 265–267 (2000).
17. Kiratzis, N., Holtappels, P., Hatchwell, C. E., Mogensen, M. & Irvine, J. T. S. Preparation and characterisation of copper/yttria titania zirconia cermets for use as possible solid oxide fuel cell anodes. *Fuel Cells* **1**, 211–218 (2001).
18. Kim, H,. Park, S., Vohs, J. M. & Gorte, R. J. Direct oxidation of liquid fuels in a solid oxide fuel cell. *J. Electrochem. Soc.* **148**, A693–A695 (2001).
19. Minh, N. Q. Ceramic fuel-cells. *J. Am. Ceram. Soc.* **76**, 563–588 (1993).
20. Yokokawa, H., Sakai, N., Kawada, T. & Dokiya, M., Thermodynamic stabilities of perovskite oxides for electrodes and other electrochemical materials. *Solid State Ionics* **52**, 43–56 (1992).
21. Pudmich, G. *et al.* Chromite/titanate based perovskites for application as anodes in solid oxide fuel cells. *Solid State Ionics* **135**, 433–438 (2000).
22. Tao, S.& Irvine, J. T. S. A redox-stable, efficient anode for solid-oxide fuel cells. *Nature Mater.* **2**, 320–323 (2003).
23. Vernoux, P., Guillodo, M., Fouletier, J. & Hammou, A., Alternative anode materials for gradual methane reforming in solid oxide fuel cells. *Solid State Ionics* **135**, 425–431 (2000).
24. Sfeir, J. *et al.* Lanthanum chromite based catalysts for oxidation of methane directly on SOFC anode. *J. Catal.* **202**, 229–244 (2001).
25. Sfeir, J. Van herle, J. & Vasquez, R. in *Proc. 5th European Solid Oxide Fuel Cell Forum* (ed. Huijsmans J.) 570–577 (European SOFC Forum, Switzerland, 2002).
26. Sauvet, A. L. & Irvine, J. T. S. in *Proc. 5th European Solid Oxide Fuel Cell Forum* (ed. Huijsmans J.) 490–498 (European SOFC Forum, Switzerland 2002).
27. Slater, P. R., Fagg, D. P. & Irvine, J. T. S. Synthesis and electrical characterisation of the doped perovskite titanates as potential anode materials for solid oxide fuel cells. *J. Mater. Chem.* **7**, 2495–2498 (1997).
28. Hui, S. Q. & Petric, A., Electrical properties of yttrium-doped strontium titanate under reducing conditions. *J. Electrochem. Soc.* **149**, J1–J10 (2002).
29. Marina, O. A. & Pederson, L. R. in *Proc. 5th European Solid Oxide Fuel Cell Forum* (ed. Huijsmans, J.) 481–489 (European SOFC Forum, Switzerland, 2002).
30. Canales-Vázquez, J., Tao, S. W. & Irvine, J. T. S. Electrical properties in La₂Sr₄Ti₆O₁₉-d: a Potential anode for high temperature fuel cells solid state ionics. **159**, 159–165 (2003).
31. Reich, C., Kaiser, A. & Irvine, J. T. S. Niobia based rutile materials as sofc anodes, fuel cells - from fundamentals to systems. **1**, 249–255 (2001).
32. Ramos, T. & Atkinson, A. in *Ionic and Mixed Conducting Ceramics IV* (eds Ramanarayanan, T. A., Worrell, W. L. & Mogensen, M.) 352–367 (Electrochemical Scociety Proceedings Volume PV2001-28, 2001).
33. Steele, B. C. H. Appraisal of Ce₁₋ᵧGdᵧO₂₋ᵧ/₂ electrolytes for IT-SOFC operation at 500° C. *Solid State Ionics* **129**, 95–110 (2000).
34. Wang, S. R., Kobayashi, T., Dokiya, M. & Hashimoto, T. Electrical and ionic conductivity of Gd-doped ceria. *J. Electrochem. Soc.* **147**, 3606–3609 (2000).
35. Floyd, J. M. Interpretation of transport phenomena in non-stoichiometric ceria. *J. Ind. Technol.* 275–280 (1972).
36. Tao, S. & Irvine, J. T. S. Optimization of mixed conducting properties of Y₂O₃-ZrO₂-TiO₂ and Sc₂O₃-Y₂O₃-ZrO₂-TiO₂ solid solutions as potential SOFC anode materials. *J. Solid State Chem.* **165**, 12–18 (2002).
37. Sauvet, A.-L. & Fouletier, J. Catalytic properties of new anode materials for solid oxide fuel cells operated under methane at intermediary temperature. *J. Power Sources* **101**, 259–266 (2001).
38. Sprague, J. J. & Tuller, H. L. Mixed ionic and electronic conduction in Mn/Mo doped gadolinium titanate. *J. Europ. Ceram. Soc.* **19**, 803–806 (1999).
39. Holtappels, P., Poulsen, F. W. & Mogensen, M. Electrical conductivities and chemical stabilities of mixed conducting pyrochlores for SOFC applications. *Solid State Ionics* **135**, 675–679 (2000).
40. McIntosh, S., Vohs, J. M. & Gorte, R. J. Role of hydrocarbon deposits in the enhanced performance of direct-oxidation SOFCs. *J. Electrochem. Soc.* **150**, A470–A476 (2003).
41. Lu, C., Worrell, W. L., Vohs, J. M. & Gorte, R. J. in *Solid Oxide Fuel Cells VIII* (eds Singhal. S. C. & Dokiya, M.) 773–780 (Electrochemical Society Proceedings

Volume 2003-07, 2003)

42. Lu, C., Worrell, W. L., Gorte, R. J. & Vohs, J. M., SOFCs for direct oxidation of hydrocarbon fuels with samaria-doped ceria electrolyte. *J. Electrochem. Soc.* **150**, A354–A358 (2003).

43. Sfeir, J. *Alternative Anode Materials for Methane Oxidation in Solid Oxide Fuel Cells* Thesis, Ecole Polytechnique Fédérale de Lausanne, Switzerland (2001).

44. Liu, J., Madsen, B. D., Ji, Z. Q. & Barnett, S. A. A fuel-flexible ceramic-based anode for solid oxide fuel cells. *Electrochem. Solid State Lett.* **5**, A122–A124 (2002).

45. Sauvet, A. L. & Fouletier, J. Electrochemical properties of a new type of anode material $La_{1-x}Sr_xCr_{1-y}Ru_yO_{3-\delta}$ for SOFC under hydrogen and methane at intermediate temperatures. *Electrochim. Acta* **47**, 987–995 (2001).

46. Vernoux, P., Guindet, J. & Kleitz, M. Gradual internal methane reforming in intermediate-temperature solid-oxide fuel cells. *J. Electrochem. Soc.* **145**, 3487–3492 (1998).

47. Skarmoutsos, D., Teitz, F. & Nikolopoulos, P. Structure - property relationships of Ni/YSZ and Ni/(YSZ+TiO2) cermets. *Fuel Cells* **1**, 243–248 (2001).

48. Holtappels, P. *Electrocatalysis on Nickel-Cermet E,lectrodes* Jülich Research Centre Report number 3414, and Thesis, Univ. Bonn (1997).

49. Kiratzis, N., Holtappels, P., Hatchwell ,C. E., Mogensen, M. & Irvine, J. T. S. Preparation and characterisation of copper/yttria titania zirconia cermets for use as possible solid oxide fuel cell anodes. *Fuel Cells* **1**, 211–218 (2001).

Acknowledgements

This manuscript is the result of an intense and highly enjoyable workshop organized in Strasbourg, France, December 2002, under the auspices of the European Science Foundation OSSEP programme with support from the US Department of Energy and National Science Foundation. We thank Shanwen Tao (St Andrews) Philippe Stevens (EIER), Jan Van Herle (EPFL) Frank Tietz (Jülich), Jorge Frade (Aveiro), Axel Müller (Karlsruhe), Jan Pieter Ouweltjes (ECN) Anil Virkar (Utah), Olga Marina (PNNL), Truls Norby (Oslo), Tony Petric (McMaster), Elisabeth Siebert (Grenoble), Joseph Sfeir (HTceramix), Wayne Worrell (Pennsylvania), Stuart Adler (Washington), Tatsuya Kawada (Tohoku), Meilin Liu (Georgia), Nguyen Minh (GE) and Anne-Laure Sauvet (Grenoble) for useful discussions.

Competing financial interests

The authors declare that they have no competing financial interests.

Materials for fuel-cell technologies

Brian C. H. Steele* & Angelika Heinzel†‡

*Centre for Ion Conducting Ceramics, Department of Materials, Imperial College, London SW7 2BP, UK (e-mail: b.steele@ic.ac.uk)
†Fraunhofer Institute for Solar Energy Systems, Heidenhofstrasse 2, 79110 Freiburg, Germany
‡Present address: Fachgebiet Energietechnik, Universität Duisburg, Lotharstr. 1-21, 47057 Duisburg, Germany (e-mail: a.heinzel@uni-duisburg.de)

Fuel cells convert chemical energy directly into electrical energy with high efficiency and low emission of pollutants. However, before fuel-cell technology can gain a significant share of the electrical power market, important issues have to be addressed. These issues include optimal choice of fuel, and the development of alternative materials in the fuel-cell stack. Present fuel-cell prototypes often use materials selected more than 25 years ago. Commercialization aspects, including cost and durability, have revealed inadequacies in some of these materials. Here we summarize recent progress in the search and development of innovative alternative materials.

The successful conversion of chemical energy into electrical energy in a primitive fuel cell was first demonstrated[1] over 160 years ago. However, in spite of the attractive system efficiencies and environmental benefits associated with fuel-cell technology, it has proved difficult to develop the early scientific experiments into commercially viable industrial products. These problems have often been associated with the lack of appropriate materials or manufacturing routes that would enable the cost of electricity per kWh to compete with the existing technology, as outlined in a recent survey[2].

The types of fuel cells under active development are summarized in Fig. 1. The alkaline fuel cell (AFC), polymeric-electrolyte-membrane fuel cell (PEMFC) and phosphoric-acid fuel cell (PAFC) stacks essentially require relatively pure hydrogen to be supplied to the anode. Accordingly, the use of hydrocarbon or alcohol fuels requires an external fuel processor to be incorporated into the system. This item not only increases the complexity and cost of the system, but also decreases the overall efficiency as indicated in Fig. 2. In contrast, molten-carbonate fuel cells (MCFCs) and solid-oxide fuel cells (SOFCs) operating at higher temperatures have the advantage that both CO and H_2 can be electrochemically oxidized at the anode. Moreover, the fuel-processing reaction can be accomplished within the stack, which enables innovative thermal integration/management design features to provide excellent system efficiencies (~50%).

Although the introduction of a 'hydrogen economy' might seem an attractive scenario, its implementation is beset with formidable technical and economic difficulties. The cheapest technology for the large-scale production of hydrogen is the steam reforming of natural gas, which produces significant emissions of greenhouse gases[3]. The topic of hydrogen storage is addressed in the accompanying review by Schlapbach and Züttel (see pages 353–358); unless there is a breakthrough in the production of hydrogen and the development of new hydrogen-storage materials, the concept of a 'hydrogen economy' will remain an unlikely scenario. In this article, therefore, we assume that fuel cells have to be designed for operation on hydrocarbon or alcohol fuels to ensure that the technology is able to penetrate the relevant major markets. Otherwise fuel-cell technology will be confined to restricted niche activities where hydrogen might be a commercial option, such as city bus fleets. Clearly the choice of fuel is a further complication in the factors influencing the commercialization of fuel cells.

Constraints on material selection

Materials selection for a commercial product involves an iterative design process that eventually becomes specific to the particular product and application. However, it is possible to make a few general statements about the selection of materials for fuel cells. The combined area-specific resistivity (ASR) of the cell components (electrolyte, anode and cathode) should be below $0.5\,\Omega\,cm^2$ (and ideally approach $0.1\,\Omega\,cm^2$) to ensure high power densities, with targets of $1\,kW\,dm^{-3}$ and $1\,kW\,kg^{-1}$ often mentioned for transport applications. High power densities are also important to reduce costs, as the amount of material per kW is thus minimized. These topics, and considerations of cell efficiencies, are summarized in Box 1.

The need to minimize cell resistivities has a major impact on the selection and processing of the cell components. Cost-effective fabrication of porous electrode structures was achieved for the first time only about 40 years ago. The electrolyte, gaseous reactants, electrocatalyst and current collector have to be brought into close contact within a confined spatial region termed the triple-phase-boundary interface. For the low-temperature systems, the introduction of hydrophobic polytetrafluoroethylene (PTFE or Teflon) greatly simplified the fabrication of porous, liquid-resistant gas-diffusion structures. Metal or carbon powders (or porous carbon papers) provided the electronic pathways, and to further reduce the ASR of the electrode a metallic wire mesh or screen was usually incorporated into the structure. Further improvements in performance were obtained during the 1960s by depositing small crystallites (2–5 nm) of the electrocatalyst (usually platinum or Pt alloys) onto carbon powder or paper. In retrospect, this accomplishment was probably the first manifestation of an engineered nanostructure, and it is not surprising that its implementation more than 40 years ago was so difficult.

High ionic conductivities ($>1\,S\,cm^{-1}$) associated with the liquid KOH, phosphoric acid and molten carbonate electrolytes ensured that, with appropriate design strategies, the ASR values of these components can be small. Although exhibiting lower specific ionic conductivity values, the Nafion membrane used in the PEMFC system can be fabricated relatively easily as a thick film (100 μm) to produce satisfactory ASR values, provided the water content of the film is controlled under the dynamic conditions of cell operation. In contrast, it has been, and continues to be,

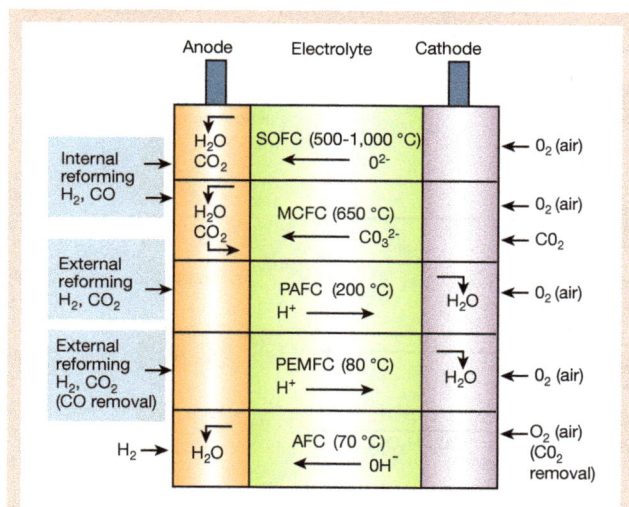

Figure 1 Summary of fuel-cell types. The oxidation reaction takes place at the anode (+) and involves the liberation of electrons (for example, $O^{2-} + H_2 = H_2O + 2e^-$ or $H_2 = 2H^+ + 2e^-$). These electrons travel round the external circuit producing electrical energy by means of the external load, and arrive at the cathode (–) to participate in the reduction reaction (for example, $1/2O_2 + 2e^- = O^{2-}$ or $1/2O_2 + 2H^+ + 2e^- = H_2O$). It should be noted that as well as producing electrical energy and the reaction products (for example, H_2O and CO_2), the fuel-cell reactions also produce heat. The reaction products are formed at the anode for SOFC, MCFC and AFC types, and at the cathode for PAFC and PEMFC types. This difference has implications for the design of the entire fuel-cell system, including pumps and heat exchangers. To maintain the composition of the electrolyte component in the MCFC system, CO_2 has to be recirculated from the anode exhaust to the cathode input. Additionally, the composition of the polymeric-membrane electrolyte has to be carefully controlled during operation by an appropriate 'water management' technology.

difficult to scale-up thick-film technologies to provide cost-effective ceramic solid-oxide electrolyte components with required thicknesses in the range 10–30 μm. Usually the thick-film electrolyte has to be sintered dense at temperatures approaching 1,400 °C; this requires a porous ceramic substrate, which is often the anode or cathode material. The substrate material has to be carefully selected to avoid reaction with the electrolyte, and/or thermal expansion mismatch, during the high-temperature sintering process. The incorporation of a relatively weak, brittle structural component in SOFC stacks is at present restricting the application of SOFC systems to those situations that do not demand rapid temperature fluctuations. In this respect the recent development[2] of sintering procedures below 1,000 °C, which should allow the use of metal substrates, represents a significant advance that will enable the development of more rugged SOFC systems.

Another important component in a fuel-cell stack is the impermeable electronic conducting bipolar plate. This has the dual function of distributing the fuel and air to the anode and cathode, respectively, as well providing the electrical contact between adjacent cells. The corrosive acidic conditions prevailing in the PEMFC and PAFC systems severely restricts the choice of bipolar plate material and at present graphite is usually selected. However, alternative materials or manufacturing methods are mandatory if these systems are ever to attain the target costs. Major research and development (R&D) programmes are examining the behaviour of alternative carbon-based materials produced by injection moulding, or coated stainless steels. For the high-temperature systems (MCFC and SOFC) operating in the temperature range 500–750 °C, appropriate stainless steel compositions can be specified which seem to satisfy the technical and economic constraints. But for SOFCs operating at higher temperatures (800–1000 °C), alternative, more expensive bipolar plate materials have to be specified, which at present incur significant cost penalties.

Additional constraints influencing material selection arise from reliability and durability issues. For transport applications, minimal values of performance degradation (for example, 0.1% over 1,000 h) are required for projected operational lifetimes of 5,000 h. But for stationary applications — for example, distributed CHP (combined heat and power) systems — a similar degradation rate must extend over a period of at least 40,000 h (5 years). These different lifetime targets seem to be introducing problems for PEMFC prototype CHP systems, as the stack components were developed originally for transport applications.

A fuel-cell system also incorporates relevant balance-of-plant items such as pumps, valves, heat exchangers and piping. Although these important components comprise at least half the cost of a fuel-cell system, we will not consider them further here, except to note that for the PAFC system, many of which have now been operated for periods approaching 30,000 h, the main source of system breakdown has been balance-of-plant items. External fuel processors (reformers) are also the subject of intensive development around the world, and a variety of innovative compact reformers using diffusion-bonded printed-circuit components or micro-channel designs[4] also illustrate the impact of materials technology on this aspect of fuel-cell systems.

For more than four decades now, reliable, efficient, fuel-cell systems incorporating AFC stacks have proved their worth in the Apollo spacecraft and space shuttles. Excellent electrode kinetics, when operating on pure hydrogen and oxygen, are an attractive feature of this system. But for terrestrial applications, the additional economic restraints, which include the need to replace hydrogen by cheaper hydrocarbon or alcohol fuels, have provided severe problems for materials selection and the associated fuel-processing technology. After 20 years development, the Elenco consortium abandoned attempts in 1996 to develop a bus powered by an AFC system. Although Zevco have purchased the technical rights, it is anticipated that market penetration of AFC systems will be small, in spite of a recent publication[5] advocating the use of AFC systems with ammonia as the source of hydrogen fuel.

Approximately 200 PAFC co-generation units (International Fuel Cells (IFC) PC25 systems, delivering 200 kW) have been installed around the world, and have exhibited excellent reliability. However, the commercial future of this system is possibly in jeopardy as the manufacturers (IFC and Japanese companies) have been unable to reduce the capital cost sufficiently below US$3,000 per kW_e as originally forecast[6,7]. Most observers[8] believe that for initial market entry the target cost per kW_e must be reduced to around US$1,000 per kW_e, falling to below US$500 per kW_e with volume production. Accordingly, we focus here on materials aspects of the PEMFC, MCFC and SOFC systems, which at present still appear to present opportunities to exploit their potential.

It is important to note that the materials currently being used in PEMFC, MCFC and tubular SOFC prototype demonstration units essentially remain the same as those selected at least 25 years ago[9,10]. Although innovative fabrication and processing routes have improved the attributes (for example, lower cost and lower Pt loadings) of these materials, it is only in the past five years that system engineering and commercialization issues have highlighted the inadequacies of some of the materials originally selected. As indicated in the next two sections, it is these issues that are now driving the development of alternative materials, particularly for the PEMFC and intermediate temperature (IT)-SOFC stacks.

Polymeric-electrolyte-membrane fuel cells

The most important materials under development for PEMFC stacks are construction materials for the cell frames, bipolar plates, electrocatalysts for the fuel and air electrode, and the ion conducting membrane.

Depending on the fuel to be used in the PEM cell, the requirements for these materials are completely different. The simplest case is the operation with pure hydrogen and oxygen or air. Cells with high

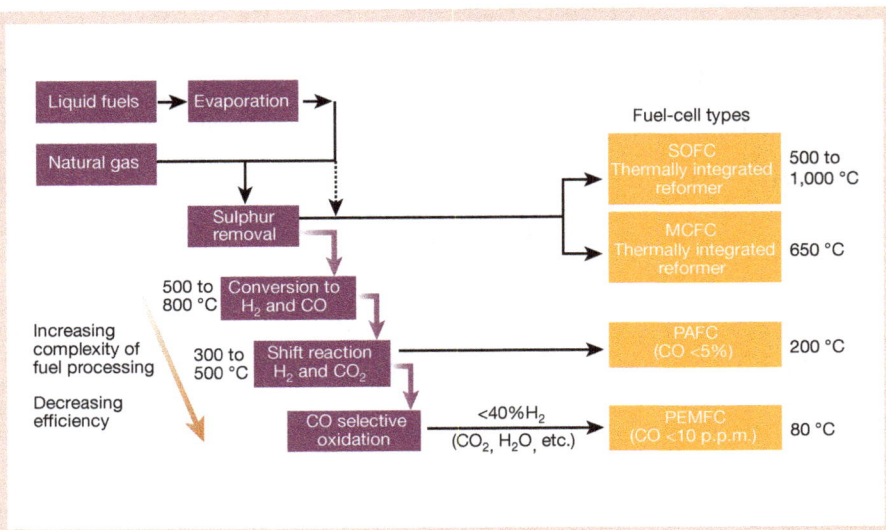

Figure 2 Fuel-cell types and fuel processing. Although selected fuels can be introduced directly into the anode compartment of the high-temperature fuel cells (SOFC and MCFC), better thermal management of the stack can usually be achieved by having separate reformer compartments that are thermally integrated within the stack to produce a mixture of fuel and syngas (H_2 and CO). For the lower-temperature fuel cells (PAFC and PEMFC), external reformers are required. Some of the fuel has to be consumed in these external reformers to maintain the operating temperature. Moreover, dilution of the H_2 fuel reduces performance of the cells, resulting in significant efficiency losses compared with operation on pure H_2. It should be noted that the AFC stack cannot be operated on reformate fuels because of the presence of CO_2 in these gases.

power density and very low degradation are already state of the art. The main requirement for the future is to achieve reduction in the capital cost of the system by attention to materials selection and fabrication, and also by scaling-up the production volume. This target will probably cause a complete review of all the materials used up to now. The second possibility is to operate the PEM cell with a reformate fuel. In that case, inert gases and CO traces are present in the fuel. This is a challenge for the fuel electrode in particular, and a CO-tolerant catalyst is required. The most difficult option is the direct methanol fuel cell (DMFC). The methanol electrode also needs a CO-tolerant catalyst, as adsorbed CO species are formed during the electro-oxidation of methanol. In addition, the small polar methanol molecule behaves in a way similar to water and readily permeates through existing membrane materials. This behaviour leads to a loss in fuel and also to the formation of a mixed potential at the air electrode. The development of innovative membrane materials could remove this disadvantage. As the tolerance to CO by the electrocatalysts is strongly dependent on temperature, an alternative membrane with a better temperature stability is also another

important R&D target. However, an operation temperature above the boiling point of water requires a completely different type of membrane material, as no liquid water will be present under these circumstances for the hydrated protonic-conduction mechanism.

Bipolar plates

Much effort is being expended on the development of cost-effective materials for the bipolar plates. With respect to corrosion resistance, graphite materials are preferred. However, the conductivity of graphite materials is much less than that of metallic materials. Some conductivity values are: C polymers, $\sim 1\ S\ cm^{-1}$; graphite, $10^3\ S\ cm^{-1}$; gold, $45,000 \times 10^3\ S\ cm^{-1}$; Fe alloys, $5,300 \times 10^3\ S\ cm^{-1}$; Ti, $2,400 \times 10^3\ S\ cm^{-1}$. For bipolar plates, polymer/graphite compounds are developed with at least $10\ S\ cm^{-1}$, reducing the resistivity of the bipolar plate well below the resistivity of the membrane. In addition, the fabrication costs of graphite plates incorporating gas-distribution channels are high, making such components too expensive. Moreover, as graphite materials are porous, a binder or resin has to be added to produce the necessary impermeability.

Polymeric materials can be machined more easily and cheaply by hot pressing or injection moulding. Polypropylene, for example, can be mixed with graphite to achieve sufficient electrical conductivity. Higher contents of graphite produce better electrical conductivity values, but the associated mechanical properties become more undesirable as the increased brittleness not only reduces the toughness, but also makes manufacturing more difficult and expensive. Typical carbon contents range between 50 and 80 weight%, and several groups[11–14] are optimizing the machining processes in association with optimization of the material.

Another strategy is to use metallic bipolar plates. The gaseous flow structure can easily be fabricated in thin metal foils by pressing, but only a few metals are sufficiently corrosion-resistant in the acidic environment of the membrane. The most promising materials are stainless steels, as the other candidate metals such as titanium, niobium, tantalum and gold (including gold-plated metals) are too expensive. Stainless steels can provide satisfactory performance for several thousand hours. The steel is protected by a passive layer at the cathode side, but the anode side becomes contaminated by corrosion products[15–18]. Stacks with metallic bipolar plates have been developed by Nuvera and by Siemens.

Electrocatalyst

The second important problem is associated with the electrocatalyst. For operation with pure hydrogen and air, platinum is the most active material. To reduce the cost, nanoparticles of platinum on a carbon

---

Box 1
Cell efficiencies

The overall cell efficiency η is given by the equation

$$\eta = \eta_g\, \eta_v\, \alpha$$

where η_g is the Gibbs efficiency, η_v is the voltage efficiency, and α is the fraction of fuel used.

$$\eta_g = \Delta G/\Delta H = nFE_0/\Delta H$$
$$\eta_v = E/E_0 = (E_0 - IR_c)/E_0$$

where E_0 is the open circuit voltage, and ΔH is the heat of the overall cell reaction. Thus

$$\eta = nF(E_0 - IR_c)\alpha/\Delta H$$

R_c is the area specific resistivity of the cell components (electrolyte, anode and cathode). Note that the cell structure is often termed the membrane–electrode assembly (MEA) for PEMFC systems, and positive–electrolyte–negative (PEN) for SOFC systems.

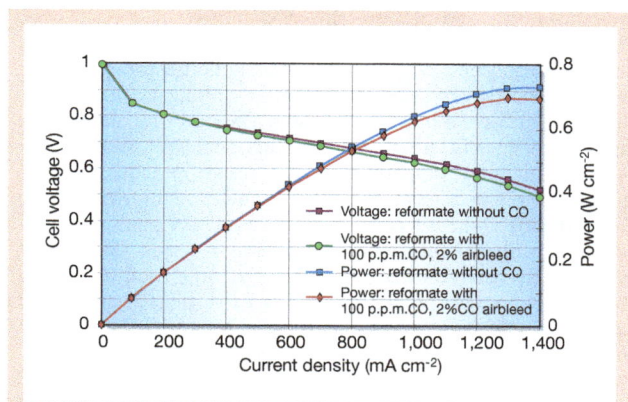

Figure 3 CO tolerance on Pt/Ru anode electrodes. Active cell area, 50 cm^2; precious metal loadings, <0.4 mg cm^{-2}; temperature, 80 °C; pressure, 3 bar; anode fuel reformate, 60% H_2, 25% CO_2, 15% N_2; cathode, air. (From ref. 56.)

support have been developed, and the reduction in noble-metal content without degrading the cell performance has been, and continues to be, an important R&D activity[19]. The platinum loading has been significantly reduced from 2 mg per square centimetre of electrode to values below 0.5 mg cm^{-2} without significant impact on performance and lifetime. In the laboratory, even lower platinum loadings have been examined[20,21].

For fuels containing traces of CO, or methanol in the DMFC, a CO-tolerant catalyst is required. This remains one of the most challenging tasks for the successful development of commercial PEMFC systems[22]. For reformate electrodes as well as for methanol oxidation, the removal of adsorbed CO species is the rate-determining step. The oxidation of adsorbed CO on Pt is slow, and is facilitated by adjacent absorbed OH species. This is the reason why Ru, with its low potential for OH-ads formation, is the most efficient component of the binary catalysts. Pt/Ru and other binary and ternary alloys with these noble metals have been investigated intensively[23], and performance values have increased significantly. The loss in performance is usually expressed in mV for a certain CO content of the fuel, and recent publications show promising results, as indicated, for example, in Fig. 3.

Membranes

Although the US General Electric Company (GE) initiated the development of PEMFCs in the 1950s, it was the introduction of Nafion by DuPont that ensured continuing interest in these systems. Initially Nafion was manufactured[24] for membrane cells used in the production of chlorine (chlor-alkali cells). By 1990, Ballard had overcome many of the engineering problems associated with PEMFC systems, and this had stimulated many groups in the United States and Japan to improve the properties of the original Nafion material[25]. For example, higher ionic conductivities could be attained by selecting perfluorosulphonic acid copolymers with a short pendant group, and it was realized that gas and small-molecule permeability were other important characteristics that had to be improved.

The following properties of polymeric membranes need to be optimized for use in fuel cells: (1) high proton conduction, assured by acid ionic groups (usually SO_3H), depending on sulphonation degree and on thickness of the membrane; (2) good mechanical, chemical and thermal strength requiring the selection of a suitable polymer backbone; mechanical strength for thin membranes can frequently be improved by reinforcement; (3) low gas permeability, which is dependent on material and thickness of the membrane; (4) for DMFC applications low electro-osmotic drag coefficient to reduce methanol crossover.

There is significant interaction between the desired properties of the membrane — high conductivity, low swelling, low gas and

methanol permeability, and stability — and the type of backbone polymer, the degree of sulphonation and the nano-phase separation into hydrophilic and hydrophobic domains (for example, high degrees of sulphonation usually lead to highly conductive membranes, but also to extreme swelling properties). To satisfy these requirements, different approaches have been examined: sulphonated perfluorinated materials with[26] and without[27] microporous support; sulphonated polyhydrocarbons; acid–base complexes and blends with a surplus of acid ionic groups; and inorganic–organic composite materials with improved thermal stability and better water-retaining properties.

Because of their PTFE-like backbone and relatively low equivalent weight, Nafion and related materials are a favoured option and are commonly used in fuel-cell stacks, but the costs remain high. Therefore, much effort is being applied to the development of cheaper, usually fluorine-free, membrane materials. But hydrocarbons often suffer from an insufficient thermal stability, and so more and more aromatic groups have been introduced into the polymer backbone. Polyarylenes[28,29] seem to be the most stable molecules among the hydrocarbons. For example, poly(arylene ether sulphone)-based membranes were prepared by sulphonation of commercially available polymers such as Udel and Victrex[30]. For high ionic conductivity, high sulphonation fractions are desirable, but high sulphonation can lead to extreme swelling even at room temperature. Thus, crosslinking of the polymer chains at the sulphonic acid groups can be included in the synthesis steps to overcome the problems of swelling. However, the long-term stability of these sulphonamide crosslinking bridges remains unproven. Alternative crosslinking methods, such as covalent crosslinking and ionic crosslinking by the introduction of polymeric bases (acid–base blend membranes), have also been examined[31].

For operation at elevated temperatures, which is desirable for high power density DMFC systems and for reformate fuels with CO levels above 100 p.p.m., the conduction mechanism becomes the dominant issue. In the types of membranes described earlier (hydrated polymers), the proton-conduction mechanism is based on the migration of hydrated protons. Above 100 °C, pressurized operation is required to ensure the presence of liquid water. Phosphoric acid and polymers with immobilized heterocycles exhibit a conduction mechanism relying on structure diffusion[32], and can be used at temperatures above the boiling point of water. As phosphoric acid in its liquid form in a porous matrix is well known from PAFC development, new ways to achieve a composite with better properties have been investigated. One polymer material for high-temperature application is polybenzimidazole (PBI), which forms adducts with inorganic acids. Initial publications from Savinell[33] described the use of PBI with phosphoric acid as membrane in fuel-cell experiments. Later developments led to acid–base blend polymers with PBI and sulphonated polyetheretherketones. Reduced methanol permeability and performance data comparable to that of Nafion 112 have been reported[34,35]. Axiva (part of the Hoechst Celanese group) announced the manufacture of PBI-based membrane materials, but only in exclusive cooperation with partners (Plug Power and Honda).

It should be emphasized that if an alternative membrane material does emerge, considerable R&D will still be necessary to optimize and manufacture the new membrane–electrode assembly (MEA). This development has taken many years for Nafion-type MEAs, although some of the expertise gained may be able to be transferred to the new system.

Fuel cells operating at elevated temperatures
Solid-oxide fuel cells

In contrast to other fuel-cell types a SOFC stack can, in principle, be designed to operate within a wide temperature range (500–1,000 °C). It is necessary, therefore, to select the desired temperature of operation. This, in turn, is influenced by the specific application, the type of fuel and the properties of available solid electrolytes. For example, if the SOFC stack is to be integrated with a gas turbine, then system requirements indicate that the temperature of the exhaust gas from the stack should exceed about 850 °C. The steam reformation of fuels such as diesel, petrol and propane to produce H_2 and CO gases for the anode also, at present, requires fuel processors operating in excess of 700 °C. But it is the properties of the solid electrolyte that exert the biggest influence on stack design and materials selection, as indicated earlier.

Figure 4 summarizes how the specific ionic conductivity of a selection of solid electrolytes varies with the reciprocal of temperature. By taking a typical value of 0.15 Ω cm^2 for the maximum ASR value, it is possible to calculate the maximum thickness allowed for a given electrolyte component. For example, a design configuration that specified a self-supported (~150 μm), yttria-stabilized zirconia (YSZ) electrolyte would require a temperature of operation greater than 950 °C. Operation at 950 °C poses major problems for planar (flat) SOFC configurations compared to tubular configurations used, for example, in the Siemens–Westinghouse arrangement. These include stability of the electrode–electrolyte interface, selection of the bipolar-plate material, and the optimal material composition and arrangement for the seals that are necessary in planar SOFC stacks. Economic solutions are still sought today.

Owing to these intrinsic problems with the high-temperature planar configuration, there was early interest in tubular designs that eliminated the high-temperature sealing problem. A variety of tubular configurations was initially examined[36], but the arrangement eventually adopted by Siemens–Westinghouse has so far proved the most successful. This arrangement uses a 1.5-m porous tubular cathode (La(Sr)MnO$_3$, or LSM). After deposition of the La(Sr)CrO$_3$ interconnect strip by plasma spraying, an electrochemical vapour-deposition (EVD) process is used to fabricate an impermeable, thick-film (30–40 μm) electrolyte (YSZ) layer. The cell structure is completed by using a slurry-spray process to deposit the porous Ni-YSZ anode. Although successful from a technical perspective, the EVD process is relatively expensive and efforts are underway to replace this process by an alternative, cheaper fabrication route. However, conventional ceramic routes involving the deposition of YSZ powders and subsequent sintering are constrained by the need to restrict the sintering temperature to below about 1,250 °C to ensure

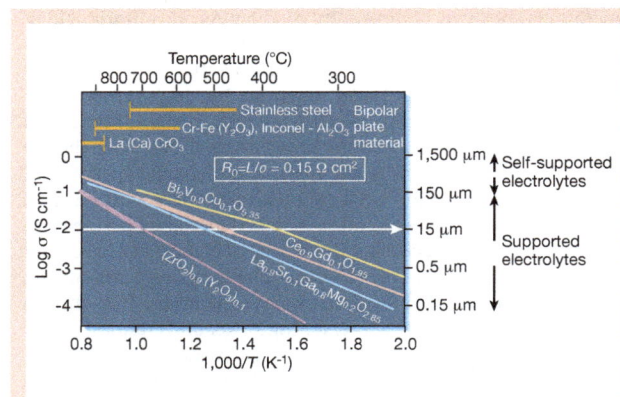

Figure 4 Specific conductivity versus reciprocal temperature for selected solid-oxide electrolytes. To ensure that the total internal resistance (that is, electrolyte + electrodes) of a fuel cell is sufficiently small, the target value for the area specific resistivity (ASR) of the electrolyte is set at 0.15 Ω cm^2. Films of oxide electrolytes can be reliably produced using cheap, conventional ceramic fabrication routes at thicknesses down to ~15 μm. It follows that the specific conductivity of the electrolyte must exceed 10^{-2} S cm^{-1}. This is achieved at 500 °C for the electrolyte Ce$_{0.9}$Gd$_{0.1}$O$_{1.95}$, and at 700 °C for the electrolyte (ZrO$_2$)$_{0.9}$(Y$_2$O$_3$)$_{0.1}$. Although the electrolyte Bi$_2$V$_{0.9}$Cu$_{0.1}$O$_{5.35}$ exhibits higher conductivities, it is not stable in the reducing environment imposed by the fuel in the anode compartment of a fuel cell.

insignificant reaction between the LSM cathode and electrolyte. Selection of the cathode (LSM) and anode (Ni-YSZ) compositions was established during the 1970s by Westinghouse and ABB, after examining a variety of oxide compositions for long- term compatibility with YSZ at elevated temperatures. This empirical conclusion was later explained in terms of available thermochemical data[37].

The Siemens–Westinghouse tubular design remains the most developed SOFC system, and has been evaluated in units generating 25 kW, 100 kW and 200 kW. More recently 200-kW units have been combined with microturbines to provide a system capable of generating electricity around 60% efficiency. Large, multi-megawatt integrated systems are predicted to produce electricity at efficiencies approaching 70%. However, these advances are more concerned with issues of system design rather than the selection and development of new materials.

Intermediate temperature solid-oxide fuel cells

The strategic programmes of large multinational companies (such as Westinghouse, GE and ABB) favoured the development of multi-megawatt, high-temperature SOFC stacks, and these priorities had a strong influence on the development of SOFC designs and materials for two decades from 1970. By 1990, however, it was beginning to be recognized that for smaller SOFC stacks not destined to be integrated with gas turbines, the operating temperature should be lowered as far as possible without compromising the electrode kinetics and internal resistance of the cell. The development of these smaller IT-SOFC stacks for distributed (embedded) CHP units, to produce stand-by power, is also being stimulated by liberalization (deregulation) of electrical supply policies. In addition, many automotive manufacturers are examining whether small SOFC stacks (3–5 kW) can be developed to supply the electrical power for auxiliary functions such as air conditioning in vehicles.

Examples of the most appropriate solid-electrolyte composition for operation at intermediate temperatures (500–750 °C) can be identified from Fig. 4. If we once again assume that the electrolyte component should not contribute more than 0.15 Ω cm^2 to the total cell ASR, then for a thick-film thickness (L) of 15 μm, the associated specific ionic conductivity (σ) of the electrolyte should exceed 10^{-2} S cm^{-1} ($\sigma = L$/ASR = 0.0015/0.15). Examination of Fig. 4 indicates that the ionic conductivity of YSZ attains this target value around 700 °C, and for Ce$_{0.9}$Gd$_{0.1}$O$_{1.95}$ (CGO) the relevant temperature is 500 °C. The use of thinner electrolyte films would allow the operating temperature to be lowered. But at present it seems that the minimum thickness for dense impermeable films that can be reliably mass produced using relatively cheap ceramic fabrication routes is around 10–15 μm. The use of a thick-film electrolyte requires this component to be supported on an appropriate substrate. As the substrate is the principal structural component in these cells, it is necessary to optimize the conflicting requirements of mechanical strength and gaseous permeability.

An IT-SOFC configuration that seeks to retain the specific advantages of both the tubular and planar arrangements is being developed by Rolls-Royce[38]. This integrated planar-stack concept incorporates multi-cell assemblies connected in series and supported by a ceramic substrate, and has many similar features to the original Westinghouse tubular design[39].

Most development work on planar IT-SOFC systems has involved thick-film YSZ electrolytes, and so far most groups have used anode (Ni-YSZ) substrates, which allow the electrolyte powder to be sintered to a dense film around 1,400 °C. One of the problems associated with using porous, composite Ni-YSZ substrates is their relatively poor thermal-expansion compatibility with the YSZ thick film. Accordingly, several groups are examining porous substrates based on Ni-Al$_2$O$_3$ or Ni-TiO$_2$ compositions, with thin interfacial anodic regions incorporating Ni, YSZ and/or doped CeO$_2$. Although replacement of the YSZ can provide better thermal-expansion compatibility, problems still remain over the volume changes associated

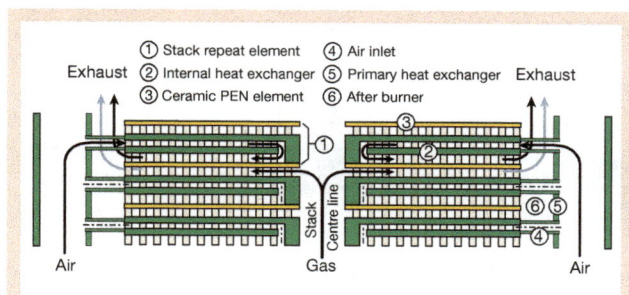

Figure 5 Schematic view of the Sulzer Hexis micro-CHP stack for residential applications. Small (1–5 kW) micro-CHP units are being beta tested for residential applications. The combination of small-scale joint production of electricity and heat is more efficient than the combination of conventional central power stations and decentralized heating equipment. Moreover, security of supply for domestic electrical equipment (for example, central heating pumps and computers) is a major advantage, and part of the emerging trend for distributed power.[57]

with the reduction and oxidation of the Ni component. As the porous substrate–electrolyte films are usually co-fired in air around 1,400 °C, nickel is present as NiO, which has to be carefully reduced by the fuel during the initial heating cycle of the assembled stack. Additionally, operating procedures have to be specified to prevent the nickel re-oxidizing when SOFC stacks are cooled down in the absence of fuel flowing through the anode compartment (of note here is the use of forming gas, N_2/H_2, to protect the Ni-YSZ anode in the Siemens–Westinghouse tubular configuration).

Most IT-SOFC developers are using metallic bipolar plates. Often a ferritic stainless steel is specified because of the low ($12.5 \times 10^{-6} K^{-1}$) thermal-expansion coefficients of these alloys. Moreover, by using compositions stabilized with Nb-Ti, excellent electronic interfacial contacts can be maintained[40] between the cell components for extended periods. Providing appropriate precautions are followed, many R&D laboratories have reported good performance values for IT-SOFC stacks incorporating the following positive–electrolyte–negative (PEN) components: anode-supported thick-film YSZ electrolytes, LSM-YSZ cathodes, and stainless steel bipolar plates.

To minimize sealing requirements, many IT-SOFC stacks have adopted a circular design in which the fuel and air are introduced by means of an appropriate manifold at the centre of the PEN structure. Arrangements are made to distribute the air and fuel gases over the cathode and anode, and the flow rates are adjusted to ensure almost complete conversion of the fuel by the time it reaches the stack periphery. Unreacted fuel and air are then combusted without large temperature changes. These design features minimize sealing problems and allow limited thermal cycling. Examples are provided by the Sulzer Hexis micro-CHP configuration designed for residential accommodation (Fig. 5), and the auxiliary power unit (APU) prototype (Fig. 6) constructed by the Delphi–BMW–Global Thermoelectric consortium for incorporation into vehicles. But at present the heating and cooling rates cannot exceed ~500 °C per hour, owing to the development of stresses associated with thermal-expansion mismatch, and to the brittle glass and ceramic seals. Although this restriction may not be too severe for larger CHP systems (>100 kW), it is not satisfactory for smaller systems (1–10 kW) designed for micro-CHP and APU applications. Further R&D is still required to produce more rugged IT-SOFC stacks. Commercial units can be expected in the next five years once reliability and cost requirements (<US$1,000 per kW) have been effectively demonstrated.

IT-SOFC stacks incorporating alternative components
Although YSZ is still the favoured electrolyte material for SOFC stacks, selection of this material is not without problems and research

continues into the long-term evaluation of scandia-doped ZrO_2, which can exhibit higher ionic conductivities than the traditional YSZ material.

In principle, the use of ceria-based electrolytes such as CGO should allow the cell operating temperature to be lowered to around 500 °C (see Fig. 4). But perceived problems associated with PEN structures incorporating ceria-based electrolytes have restricted investment in this technology. It is well known that, at elevated temperatures, Ce^{4+} ions can be reduced to Ce^{3+} under the fuel-rich conditions prevailing in the anode compartment. The associated electronic conductivity (and deleterious lattice expansion) produces an internal short circuit in the PEN structure, which can significantly degrade the efficiency and performance of cells. However, if the operating temperature is lowered to around 500 °C, then the electronic conductivity is small, and can be neglected under typical operating conditions of the cell[41,42].

Another significant difficulty that has restricted exploitation of the attractive properties of CGO at 500 °C has been the need to develop alternative cathode compositions that function effectively at lower temperatures. Recent developments in this area have been surveyed by Ralph et al.[43], and there are indications[44] that appropriate materials for composite cathodes can be fabricated which exhibit small over-potentials at 500 °C (for example, 0.15 V at 1 A cm^{-2}). Composite anodes such as Ni-CGO also provide adequate performance at 500 °C for simulated syngas fuels, indicating that IT-SOFC stacks at 500 °C are now a viable option.

Operation at intermediate temperatures is stimulating R&D effort into alternative anode compositions to replace the established Ni-YSZ anode. One strategy[45] is to develop electronic conducting oxides that are redox stable, to avoid the use of Ni. This metal can catalyse carbon deposition under certain operating conditions, and also forms NiO, which is accompanied by a deleterious volume expansion, when the anode compartment becomes too oxidizing. Another approach is concerned with the identification of alternative anodes that allow the direct anodic oxidation of hydrocarbons. The claims made in recent publications[46,47] remain controversial[48], but there is little doubt that this topic will remain a fruitful area for further investigations.

Operation at 500 °C allows the use of compliant high-temperature gaskets in place of rigid, brittle glass or ceramic seals, thus permitting greater design flexibility for the stack configuration. At Imperial College, London, researchers have taken advantage of the fact that the thermal-expansion coefficient of CGO and ferritic stainless steel are virtually identical ($12.5 \times 10^{-6} K^{-1}$), so that the thick-film PEN structure can be supported on a porous stainless steel foil. These metal-supported PEN structures are robust, and should withstand the rapid temperature cycles expected during operation of small IT-SOFC stacks.

Another electrolyte, doped $LaGaO_3$ (LSGM), is also attracting much attention for IT-SOFC applications. Although its conductivity is slightly smaller (see Fig. 4) than CGO at 500 °C, its ionic domain is wider and it could be more appropriate to use this electrolyte at temperatures around 600 °C, where the reduction of Ce^{4+} in CGO becomes significant. It has been difficult to fabricate pure single-phase ceramic electrolytes, and second phases such as $SrLaGa_3O_7$ and $La_4Ga_2O_9$ are often detected in the grain boundaries. Whether these phases are responsible for the enhanced reactivity of LSGM, or whether it is an intrinsic property of LSGM, are questions that require urgent. Moreover, the preferred composition, $La_{0.9}Sr_{0.1}Ga_{0.8}Mg_{0.2}O_3$, does not seem to be stable at lower temperatures (P. Majewski, personal communication). Although research continues into the synthesis of alternative oxygen-ion conducting electrolytes, it has proved difficult to prepare alternative materials with an appropriate combination of properties that can displace the traditional fluorite compositions involving ZrO_2 and CeO_2.

Experiments involving single-compartment SOFC fuel cells have been reported. In this configuration a mixture of the fuel and air flows over both electrodes of a ceramic cell. One electrode incorporates an electrocatalyst optimized for the reduction of oxygen, whereas the second electrode is designed to catalyse the fuel oxidation. This configuration eliminates the need for cell sealing and excellent current–voltage (I–V) characteristics have been reported[49]. But the published data need to be interpreted with caution as it seems that the electrodes also catalyse the external oxidation of the fuel. The heat generated by these surface reactions can significantly raise the temperature of small samples, so that the reported data probably refer to I–V values around 100 °C higher than those stated. Parasitic chemical oxidation of the fuel will of course significantly reduce the overall system efficiency, and technological exploitation of this configuration seems unlikely.

Very high performances have been claimed[50] at 500 °C using composite electrolytes incorporating CGO and molten salts. But the long-term stability of such mixtures in a fuel-cell environment must be doubtful and these claims require further independent scrutiny.

Investigations also continue into ceramic proton conductors. The composition $BaZr_{0.9}Y_{0.1}O_{2.95}$, for example, can exhibit a protonic conductivity that approaches the oxygen-ion conductivity of CGO at 500 °C (that is, 10^{-2} S cm^{-1})[51,52]. But solid-state fuel cells incorporating ceramic proton conductors will not be able to electrochemically oxidize CO, and do not, at present, appear to offer any advantages over the oxygen-ion conducting electrolytes in this temperature region. Interesting results have also been reported[53] for the solid acid $CsHSO_4$ at 160 °C. Ceramic protonic conductors are more likely to be exploited in chemical engineering applications requiring the separation and generation of hydrogen.

Molten-carbonate fuel cells

Although the materials (Table 1) used for MCFC stack components have essentially remained unchanged[2] over the past 25 years, significant developments in fabrication technology were introduced during the 1980s. Cost-effective tape-casting techniques now allow the immobilized electrolyte matrix to be manufactured up to a size of 1 m^2. These manufacturing advances were important as the power

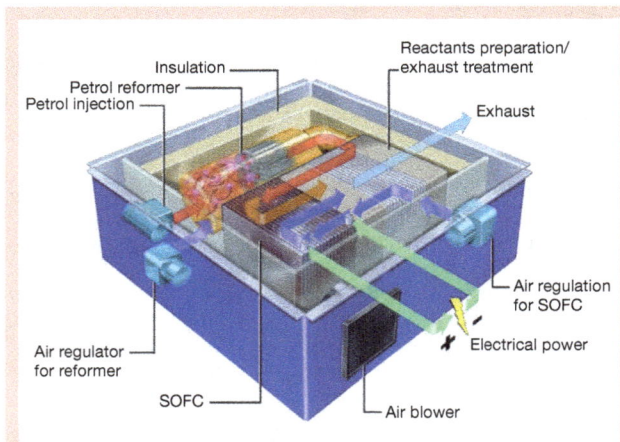

Figure 6 Schematic view of the Delphi–BMW–Global Thermoelectric auxiliary power unit (APU). Depending upon the functions installed in a vehicle, it has been suggested that at least 10% of the output of an internal combustion engine has be used to provide electrical energy. With the introduction of electromagnetic valve trains and smart piezoelectric suspensions this fraction is expected to increase significantly. Furthermore, to maintain air conditioning and refrigeration (in lorries, for example), internal combustion IC engines have to be kept idling even when the vehicle is stationary. Under these low-load conditions the engine is very inefficient and makes significant contributions to environmental pollution. The use of a fuel-cell APU would make a significant contribution to improving air quality and reducing emissions of greenhouse gases.[58]

Table 1 Materials used in the MCFC stack

Anode	Ni-Cr
Cathode	NiO (Li doped)
Electrolyte	Li/Na/KCO$_3$ (60/20/20)
Matrix*	γ-LiAlO$_2$ + Al$_2$O$_3$
Bipolar plate	Stainless steel 310
Wet seal	Aluminized stainless steel

*For electrolyte immobilization.

density of MCFC systems operating at 650 °C is relatively low (~150 mW cm^{-2}), requiring large cell areas to be fabricated. The requirement to recirculate CO_2 from the anode exhaust to the cathode to maintain the composition of the carbonate electrolyte also complicates the balance-of-plant equipment. It follows that the MCFC is only likely to be commercialized only at sizes greater than 100 kW, and originally MCFC plants were conceived as large (multi-megawatt scale), centralized coal-fired power stations.

To develop user confidence it was considered that large plants should be demonstrated as soon as possible, and this led to the construction in the United States of the 1.8-MW Santa Clara system by Energy Research Corporation. Although commissioned in April 1996, evaluation of the plant had to be curtailed owing to problems with the material selected for the stack external manifold. It is now believed that a more appropriate strategy is to develop smaller MCFC systems (~250 kW) for distributed CHP applications using natural gas. An example of this approach is provided by the 300-kW 'Hot Module' unit, developed by MTU Friedrichshafen (a subsidiary of DaimlerChrysler AG), and it is encouraging to note that the first demonstration module has already been operated successfully for more than 1 year (8,000 h) at an electrical efficiency of 47%. Because of the corrosive nature of the molten salt Li/Na/KCO$_3$ electrolyte, lifetime issues have always been of concern for the MCFC system. However, a recent appraisal[54] suggests that most components should attain the target value of 40,000 h, except possibly the coating used to protect the anode structure, which remains the focus of further materials development.

Operating procedures for MCFC systems are influenced by limitations associated with two of the components. In cooling down the hot stack in the absence of a fuel supply, it is necessary to protect the Ni-Cr anode from oxidation by the introduction of an inert gas into the anode compartment. More serious, however, is the inability of the immobilized electrolyte matrix to withstand more than 3 to 5 thermal cycles through the melting point of the molten salt electrolyte. Thus, during stand-by situations the temperature of the MCFC stack must be maintained high enough to prevent solidification of the molten salt electrolyte. Clearly both the MCFC and SOFC systems will initially be competing for the same sub-megawatt CHP market sector. It is expected that both will operate at similar electrical efficiencies, and so the cost of the installed plant will determine which technology eventually has the largest market penetration. It is also interesting to note that both fuel-cell types require procedures to prevent redox reactions damaging the anode, and to restrict thermal cycling of the plant during abnormal operating situations.

Conclusions

Probable applications of fuel cells in the next decade together with a selection of critical materials issues are summarized in Table 2. It is recognized that the capital costs (US$3,000–10,000 per kW) of prototype fuel-cell systems are too high. Although volume production can be expected to reduce these costs, it may be difficult to attain sufficient market share to justify the investment for mass production while competing against established technology. Although significant niche markets exist, such as the PEMFC system for city buses, many observers believe that a more appropriate strategy is to target those sectors of the market (for example, 1–10 kW generation) where the existing technology is inefficient and displays extremely poor

Table 2 Summary of future R&D requirements for fuel-cell materials

Application	Size (kW)	Fuel cell	Fuel	Critical materials issues
Power systems for portable electronic devices	0.001–0.05	PEMFC	H_2	Membranes exhibiting less permeability to CH_3OH, H_2O
		DMFC	CH_3OH	
		SOFC	CH_3OH	Novel PEN structures
Micro-CHP	1–10	PEMFC	CH_4 LPG	CO-tolerant anodes, novel membranes, bipolar plates
		SOFC	CH_4 LPG	More robust thick-film PENs operating at 500–700 °C
APU, UPS, remote locations, scooters	1–10	SOFC	LPG Petrol	More robust thick-film PENs operating at 500–700 °C; rapid start-up
Distributed CHP	50–250	PEMFC	CH_4	CO-tolerant anodes, novel membranes, bipolar plates
		MCFC	CH_4	Better thermal cycling characteristics
		SOFC	CH_4	Cheaper fabrication processes; redox-resistant anodes
City buses	200	PEMFC	H_2	Cheaper components
Large power units	10^3–10^4	SOFC/GT	CH_4	Cheaper fabrication processes for tubular SOFC system

Abbreviations: UPS, uninterruptible power systems; LPG, liquid petroleum gas; GT, gas turbine. Other abbreviations defined in the text.

part-load performance. This, for example, is the strategy adopted recently by the US Department of Energy for the Solid State Energy Conversion Alliance, which aims to mass produce 5-kW SOFC modular stacks with a target cost of US$400 per kW. Once this small-scale technology has demonstrated its reliability, and met cost targets, then larger units based on the same technology can be expected to penetrate other sectors of the stationary power and transport market.

Another area receiving increased attention is the development of fuel-cell power sources for a variety of portable electronic devices. As batteries struggle to keep up with the specific power demands of mobile devices, innovative DMFC designs[55], fabricated using techniques developed for the semiconductor industry, promise energy densities that are between three and five times better than current lithium-ion batteries. The high profit margins and relatively low power demands (~1 W for transmission) associated with cellular telephones, for example, could provide a useful market entry for these small fuel cells, particularly for hybrid systems. □

1. Grove, W. R. On voltaic series and the combination of gases by platinum. *Phil. Mag. Ser. 3* **14**, 127–130 (1839).
2. Steele, B. C. H. Material science and engineering: the enabling technology for the commercialisation of fuel cell systems. *J. Mater. Sci.* **36**, 1053–1068 (2001).
3. Bauen, A. & Hart, J. Assessment of the environmental benefits of transport and stationary fuel cells. *J. Power Sources* **86**, 482–494 (2000).
4. Dicks, A. L. & Larminie, J. in *Proc. Fuel Cell 2000* (ed. Blomen, L.) 357–367 (European Fuel Cell Forum, Oberrohrdorf, Switzerland, 2000).
5. Kordesch, K. *et al.* Alkaline fuel cells applications. *J. Power Sources* **86**, 162–165 (2000).
6. Anahara, R. Total development of fuel cells in Japan. *J. Power Sources* **49**, xi–xiv (1994).
7. Whitaker, J. Investment in volume building: the 'virtuous cycle' in PAFC. *J. Power Sources* **71**, 71–74 (1998).
8. MacKerron, G. Financial considerations of exploiting fuel cell technology. *J. Power Sources* **86**, 28–33 (2000).
9. Kordesch, K. & Simader, G. *Fuel Cells and their Applications* (VCH, Veinheim, Germany, 1996).
10. Larminie, J. & Dicks, A. *Fuel Cell Systems Explained* (Wiley, Bognor Regis, 2000).
11. Borup, R. L. & Vanderborgh, N. E. Design and testing criteria for bipolar plate materials for PEM fuel cell applications. *Mater. Res. Soc. Symp. Proc.* **393**, 151–155 (1995).
12. Barbir, F., Joy, G. C. & Weinberg, D. J. in *Proc. Fuel Cell Seminar 2000* 483–486 (Courtesy Associates, Washington DC, 2000).
13. Scholta, J., Rohland, B., Trapp, V. & Focken, U. Investigations on novel low-cost graphite composite bipolar plates. *J. Power Sources* **84**, 231–234 (1999).
14. Mahlendorf, F., Niemzig, O. & Kreuz, C. in *Proc. Fuel Cell Seminar 2000* 138–140 (Courtesy Associates, Washington DC, 2000).
15. Mallant, R., Koene, F., Verhoeve, C. & Ruiter, A. in *1994 Fuel Cell Seminar* 503–506 (Courtesy Associates, Washington DC, 1994).
16. Zawodzinski, C., Mahlon, S. & Gottesfeld, S. in *1996 Fuel Cell Seminar* 659–662 (Courtesy Associates, Washington DC, 1996).
17. Makkus, R. C., Janssen, A. H. H., de Bruijn, F. A. & Mallant, R. K. A. Use of stainless steel for cost competitive bipolar plates in the SPFC. *J. Power Sources* **86**, 274–282 (2000).
18. Davies, D. P., Adcock, P. L., Turpin, M. & Rowen, S. J. Stainless steel as a bi-polar plate material for solid polymer fuel cells. *J. Power Sources* **86**, 237–242 (2000).
19. Starz, K. A., Auer, A., Lehmann, Th. & Zuber, R. Characterization of platinum-based electrocatalysts for mobile PEMFC applications. *J. Power Sources* **84**, 167–172 (1999).
20. Wilson, M. S., Valerio, J. & Gottesfeld, S. Low platinum loading electrodes for polymer electrolyte fuel cells fabricated using thermoplastic ionomers. *Electrochim. Acta* **3**, 355–363 (1995).
21. Uchida, M., Fukuoka, Y., Sugawara, Y., Ohara, H. & Ohta, A. Improved preparation process of very-low-platinum-loading electrodes for polymer electrolyte fuel cells. *J. Electrochem. Soc.* **145**, 3708–3713 (1998).
22. Gottesfeld, S. *et al.* in *Fuel Cell Seminar 2000* 799–802 (Courtesy Associates, Washington DC, 2000).
23. McNicol, B. D., Rand, D. A. J. & Williams, K. R. Direct methanol-air fuel cells for road transport. *J. Power Sources* **83**, 15–31 (1999).
24. Grot, W., Perfluorinated cation exchange polymers. *Chemie-Ing.-Techn.* MS260/75 (1975).
25. Eisman, G. A. in *Proc. Vol. 86-13* 156–171 (Electrochemical Society, New Jersey, 1986).
26. Kolde, J. A., Bahar, B., Wilson, M. S., Zawodzinski, T. A. & Gottesfeld, S. Advanced composite fuel cell membranes. *J. Electrochem. Soc.* **95**, 193–201 (1995).
27. Wakizoe, M. & Watanabe, A. in *2000 Fuel Cell Seminar* 27–30 (Courtesy Associates, Washington DC, 2000).
28. Huang, R. Y. M. & Kim, J. J. *J. Appl. Polymer Sci.* **89**, 4017, 4029 (1984).
29. Zerfaß, T. Thesis, Univ. Freiburg (1998).
30. Nolte, R., Ledjeff, K., Bauer, M. & Mülhaupt, R. Partially sulphonated poly(arylene ether sulfone)—a versatile proton conducting membrane material for modern energy conversion technologies. *J. Membr. Sci.* **83**, 211–220 (1993).
31. Kerres, J. A. Development of ionomer membranes for fuel cells. *J. Membr. Sci.* **185**, 3–27 (2001).
32. Kreuer, K. D. On the development of proton conducting polymer membranes for hydrogen and methanol fuel cells. *J. Membr. Sci.* **185**, 29–39 (2001).
33. Savinell, R. F. *et al.* A polymer electrolyte for operation at temperatures upto 200C. *J. Electrochem. Soc.* **141**, L46–L48 (1994).
34. Hasiotis, C. *et al.* Development and characterization of acid-doped polybenzimidazole/sulfonated polysulfone blend polymer electrolytes for fuel cells. *J. Electrochem. Soc.* **148**, A513–A519 (2001).
35. Kerres, J., Ullrich, A., Meier, F. & Haring, T. Synthesis and characterization of novel acid-base polymer blends for application in membrane fuel cells. *Solid State Ionics* **125**, 243–249 (1999).
36. Minh, N. Q. & Takahashi, T. *Science and Technology of Ceramic Fuel Cells* (Elsevier, Amsterdam, 1995).
37. Yokokawa, H. Phase diagrams and thermodynamic properties of zirconia based ceramics. *Key Eng. Mater.* **154/155**, 37–74 (1998).
38. Day, M. J. in *4th European SOFC Forum* (ed. McEvoy, A. J.) 133–140 (European Fuel Cell Forum, Oberrohrdorf, Switzerland, 2000).
39. Sverdrup, E. F., Warde, C. J. & Eback, R. L. Design of high temperature solid-electrolyte fuel-cell batteries for maximum power output per unit volume. *Energy Conver.* **13**, 129–141 (1973).
40. Dulieu, D. *et al.* in *3rd European SOFC Forum* (ed. Stevens, P.) 447–458 (European Fuel Cell Forum, Oberrohrdorf, Switzerland, 1998).
41. Steele, B. C. H. Appraisal of $Ce_{1-y}Gd_yO_{2-y/2}$ electrolytes for IT-SOFC operation at 500C. *Solid State Ionics* **129**, 95–110 (2000).
42. Xia, C., Chen, F. & Liu, M. Reduced temperature SOFC fabricated by screen printing. *Electrochem. Solid State Lett.* **4**, A52–A54 (2001).
43. Ralph, J. M., Schoeler, A. C. & Krumpelt, M. Materials for lower temperature SOFC. *J. Mater. Sci.* **36**, 1161–1172 (2001).
44. Doshi, R. *et al.* Development of SOFCs that operate at 500C. *J. Electrochem. Soc.* **146**, 1273–1278 (1999).
45. Irving, J. T. *et al.* in *4th European SOFC Forum* (ed. McEvoy, A. J.) 471–477 (European Fuel Cell Forum, Oberrohrdorf, Switzerland, 2000).
46. Murray, E. P. & Tsai, T. A direct-methane fuel cell with a ceria based anode. *Nature* **400**, 649–651 (1999).
47. Park, S., Craciun, R., Radu, V. & Gorte, R. J. Direct oxidation of hydrocarbons in a SOFC. 1. Methane oxidation. *J. Electrochem. Soc.* **146**, 3603–3606 (1999).
48. Primdahl, S. & Mogensen, M. Exchange current densities in mixed conducting SOFC anodes. (Abstr. BS-PO-24, International Society for Solid State Ionics 2001, 8-13 July 2001, Cairns, Australia.) *Solid State Ionics* (in the press).
49. Hibino, T. *et al.* A low operating temperature SOFC in hydrocarbon-air mixtures. *Science* **288**, 2031–2033 (2000).
50. Zhu, B. Advantages of intermediate temperature SOFC for tractionary applications. *J. Power Sources* **93**, 82–86 (2001).
51. Kreuer, K. D. On the development of proton conducting materials for technological applications. *Solid State Ionics* **97**, 1–15 (1997).
52. Bohn, H. G. & Schober, T. Electrical conductivity of the high temperature proton conductor $BaZr_{0.9}Y_{0.1}O_{2.95}$. *J. Am. Ceram. Soc.* **83**, 768–772 (2000).
53. Haile, S. M., Boysen, D. A., Chisholm, C. R. I. & Merle, R. B. Solid acids as fuel cell electrolytes. *Nature* **410**, 910–913 (2001).
54. Huijsmans, J. P. P. *et al.* An analysis of endurance issues for MCFC. *J. Power Sources* **86**, 117–121 (2000).
55. Hockaday, R. *et al.* in *Proc. Fuel Cell 2000* (ed. Blomen, L.) 37–44 (European Fuel Cell Forum, Oberrohrdorf, Switzerland, 2000).
56. Starz, K. A., Ruth, K, Vogt, M. & Zuber, R. in *Proc. Int. Symp. Fuel Cells for Vehicles* 20-22 November 2000, 210–215 (Nagoya, Japan, 2000).
57. Diethelm, R., Batawi, E. & Honegger, K. in *Proc. Fuel Cell 2000* (ed. Blomen, L.) 203–211 (European Fuel Cell Forum, Oberrohrdorf, Switzerland, 2000).
58. Zizelman, J., Botti, J., Tachtler, J. & Wolfgang, S., SOFC auxiliary power unit: a paradigm shift in electric supply for transportation. *Automotive Eng. Int.* **108** (Delphi Suppl.), 14–20 (2000).

One-dimensional imidazole aggregate in aluminium porous coordination polymers with high proton conductivity

Sareeya Bureekaew[1,2], Satoshi Horike[1,2], Masakazu Higuchi[1], Motohiro Mizuno[3], Takashi Kawamura[1], Daisuke Tanaka[1], Nobuhiro Yanai[1] and Susumu Kitagawa[1,2,4]★

The development of anhydrous proton-conductive materials operating at temperatures above 80 °C is a challenge that needs to be met for practical applications. Herein, we propose the new idea of encapsulation of a proton-carrier molecule—imidazole in this work—in aluminium porous coordination polymers for the creation of a hybridized proton conductor under anhydrous conditions. Tuning of the host–guest interaction can generate a good proton-conducting path at temperatures above 100 °C. The dynamics of the adsorbed imidazole strongly affect the conductivity determined by ^{2}H solid-state NMR. Isotope measurements of conductivity using imidazole-d_4 showed that the proton-hopping mechanism was dominant for the conducting path. This work suggests that the combination of guest molecules and a variety of microporous frameworks would afford highly mobile proton carriers in solids and gives an idea for designing a new type of proton conductor, particularly for high-temperature and anhydrous conditions.

Anhydrous proton-conducting solids that are able to operate at high temperature (~120 °C) are required in fuel-cell technology[1,2]. Heterocyclic organic molecules such as imidazole or benzyl imidazole have attracted considerable attention for this purpose because they are non-volatile molecules with high boiling points and they can exist in two tautomeric forms with respect to a proton that moves between the two nitrogen atoms, which supports a proton-transport pathway[3,4]. The protonic defect may cause local disorder by forming protonated and unprotonated imidazoles. In such materials, the proton transport may occur through structure diffusion that involves proton transfer between the imidazole and the imidazolium ion through the hydrogen-bonded chain, including the molecular reorientation process for subsequent intermolecular proton transfer[1]. Theoretically, the magnitude of the ionic conductivity is given as

$$\sigma(T) = \sum n_i q_i \mu_i$$

where n_i is the number of carriers and q_i and μ_i are the charge and mobility of the carriers, respectively[5]. Both a large amount and a high mobility of ion carriers are required to provide good proton conductivity. Hence, it is important to find suitable materials that meet these requirements. For instance, solid imidazole has a low conductivity (~10^{-8} S cm^{-1}) at ambient temperature[6] although the imidazole density (n_i) is adequately high. This is because the dense packing of imidazole, with strong hydrogen bonding in the solid state, decreases the mobility of each molecule (μ_i).

The main goal of proton-conductor modification should therefore be an improvement of the mobility of proton carriers. It is known that local and translational motion of proton carriers strongly affect the proton-transfer rate[1]. To control the mobility of

proton carriers, further support matrices such as flexible organic polymers or high-porosity solids that afford space in which the carrier can move are considered promising.

Porous coordination polymers (PCPs) or metal organic frameworks constructed from transition-metal ions and organic ligands have received much attention over the past few years because of their promising applications, such as in gas storage[7–11], separation[12–17], catalysis[18–26] and conductivity[27–29]. Recent developments in approaches to combine PCP frameworks and functional guests such as polymers[30,31], metals[22,32–34] or small organic molecules[35–37] on a molecular scale have prompted us to create hybrid materials with unusual performance on the basis of the feature that crystalline nanochannels can afford a unique assembly field for functional guests with specific host–guest interactions. The guests in the nanospace of PCPs show unusual behaviour compared with in the bulk phase because each manner of assembly is heavily dependent on the nature of the PCP channels, such as their size, shape and chemical environment. Hence, we propose the use of PCPs for incorporation with proton-carrier molecules because they can provide desirable working space for carrier molecules, with high mobility and show an appropriate packing structure, for improved proton conductivities at high temperatures and under anhydrous conditions (Fig. 1).

Here, we focused on two types of PCP with one-dimensional (1D) channels and high thermal stability (~400 °C) and imidazole as the guest proton-carrier molecule. Taking the size and shape of imidazole (4.3 × 3.7 Å^2) into consideration, we chose to use the aluminium compounds [Al(μ_2-OH)(1,4-ndc)]$_n$ (**1**; 1,4-ndc : 1,4-naphthalenedicarboxylate; ref. 38) and [Al(μ_2-OH)(1,4-bdc)]$_n$ (**2**; 1,4-bdc : 1,4-benzenedicarboxylate; refs 9, 39), both of which have pore dimensions of about 8 Å, but different pore shapes and

---

[1]Department of Synthetic Chemistry and Biological Chemistry, Graduate School of Engineering, Kyoto University, Katsura, Nishikyo-ku, Kyoto 615-8510, Japan, [2]ERATO Kitagawa Integrated Pores Project, Japan Science and Technology Agency (JST), Kyoto Research Park building #3, Shimogyo-ku, Kyoto 600-8815, Japan, [3]Department of Chemistry, Graduate School of Natural Science and Technology, Kanazawa University, Kanazawa 920-1192, Japan, [4]Institute for Integrated Cell-Material Sciences (iCeMS), Kyoto University, Yoshida, Sakyo-ku, Kyoto 606-850, Japan.
*e-mail: kitagawa@sbchem.kyoto-u.ac.jp.

When citing this article, please cite the original version as shown on the contents page of this chapter.

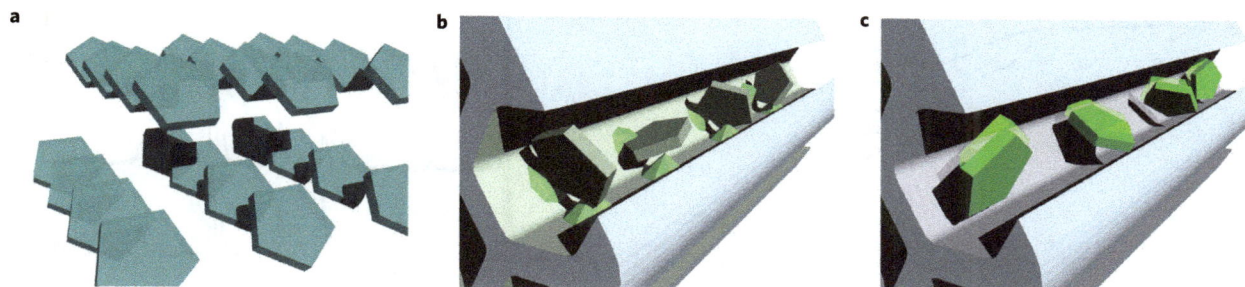

Figure 1 | Imidazole molecules are densely packed with a low mobility that adversely affects the proton-transport process. a, This occurs in the bulk solid. **b**, Imidazole accommodated in a nanochannel containing the active site with a high affinity to imidazole. The strong host–guest interaction retards the mobility of imidazole to afford the low proton conductivity. **c**, Imidazoles are accommodated in a nanochannel without strong host–guest interaction, and therefore, the molecules obtain the high mobility to show high proton conductivity.

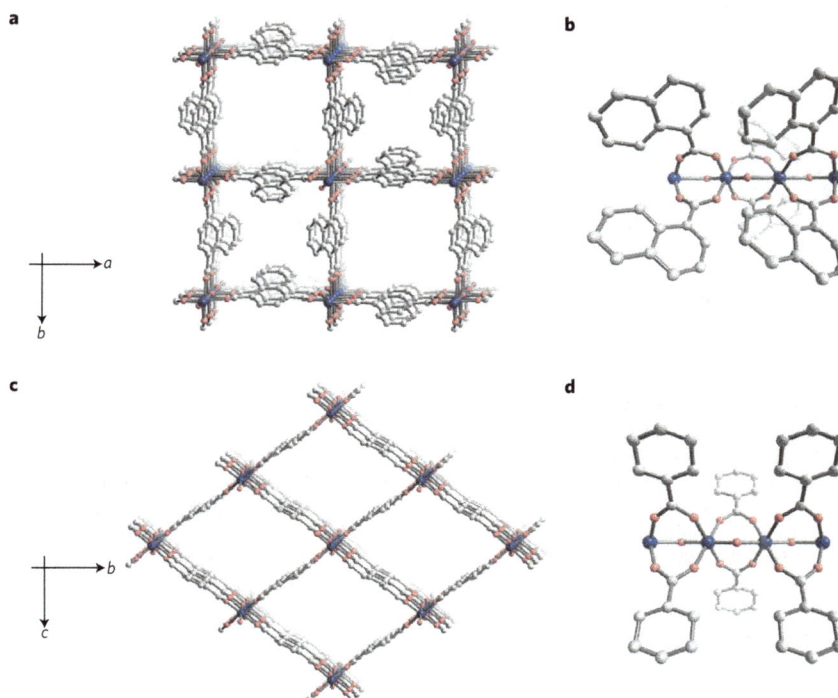

Figure 2 | Aluminium porous frameworks serving as host frameworks for the preparation of proton-conductive materials. a–d, 3D structures of **1** (**a**) and **2** (**c**) and a comparison of the ligand size effect (**b,d**). Al, C and O are represented as blue, grey and red, respectively. H atoms are omitted for clarity.

surface potentials and installed imidazole in each host material. We found that the nanochannels potentiated the different packing of imidazole, compared with the bulk solid imidazole. Conductivity measurements for each of the PCP–imidazole composites at various temperatures and under anhydrous conditions exhibited different profiles because of the different characteristics of the channels of the respective PCPs, resulting in different host–guest interactions, and **1** ⊃ Im had a conductivity of 2.2×10^{-5} S cm^{-1} at 120 °C. We investigated the behaviour of absorbed imidazole in the micropores by means of the solid-state NMR technique and succeeded in determining a good correlation between the features of the host channels, guest mobility and proton conductivity.

Structural information of 1 and 2

The structures of **1** and **2** comprise an infinite number of chains of corner-sharing AlO$_4$(μ_2-OH)$_2$ interconnected by the dicarboxylate ligand, resulting in a 3D framework containing 1D channels. The guest-free structures of **1** and **2** are shown in

Fig. 2. Crystallographic structures show that the pore surfaces of **1** and **2** are composed of hydrophobic (aromatic naphthalene and benzene ring) and hydrophilic (AlO$_4$(μ_2-OH)$_2$) parts. The principal difference between the two aluminium frameworks arises from the difference in ligands 1,4-ndc and 1,4-bdc. As a result of the asymmetric bridging ligand, **1** consists of two kinds of rectangular channel with dimensions of 7.7×7.7 Å^2 and 3.0×3.0 Å^2 running along the c axis (Fig. 2a). This compound shows the property of a rigid framework. Figure 2b shows that steric hindrance arising from the bulky naphthalene ring of the 1,4-ndc ligand of **1** induces restriction of interaction between the polar guest molecule and μ_2-OH and/or the carboxylate group of the framework. As a result of the absence of an accessible hydrophilic pore surface, the hydrophobic character from the aromatic part of the naphthalene ring of the ligand is dominant. In other words, **1** provides two types of microchannel with hydrophobic pore surfaces.

In the case of **2**, the framework exhibits only 1D diamond-shaped channels composed of the smaller benzene moieties of 1,4-bdc, with

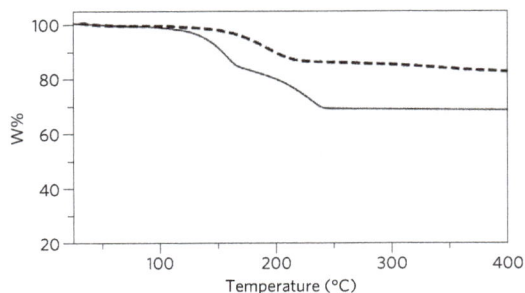

Figure 3 | Thermogravimetric curves for 1 ⊃ Im and 2 ⊃ Im over the temperature range from 25 to 400 °C at a heating rate of 10 °C min⁻¹ under a N₂ atmosphere. Dashed line: 1 ⊃ Im; solid line: 2 ⊃ Im. The guest release of 1 ⊃ Im occurs in one single step that indicates the homogeneous installation of imidazole in 1, whereas that of 2 ⊃ Im occurs in two steps, indicating two types of imidazole (which strongly and weakly interact with 2).

dimensions $8.5 \times 8.5\,\text{Å}^2$, running along the a axis. Eventually, the polar sites on the surface are exposed to guest molecules, which enhances guest-induced structural transformation of **2**, with the aid of the interaction between the imidazoles and μ_2-OH and/or a carboxylate group[40]. Thus, it is intriguing that a simple modification on the organic moiety produces channels with different nature, hydrophobic and hydrophilic for **1** and **2**, respectively[39].

Properties of the inclusion compounds

The thermogravimetric profiles of **1 ⊃ Im** and **2 ⊃ Im** are shown in Fig. 3. Note that **1 ⊃ Im** and **2 ⊃ Im** indicate the imidazole hybrid compound of **1** and **2**, respectively. The existence of imidazole in **1** and **2**, without any reactions/conversion, was confirmed by thermogravimetry/mass spectrometry. The thermogravimetric profile of **1 ⊃ Im** shows 14% weight imidazole loading or 0.6 imidazole/1 Al ion. The release of accommodated imidazole commences at 160 °C and is complete at 225 °C. In the case of **2**, the loaded imidazole amounts to 30% weight or 1.3 imidazole/1 Al ion, which is twice as much as in **1**. The thermogravimetric curve shows that the loss of imidazole molecules in **2 ⊃ Im** occurs in two steps: in the first step, the release of imidazole commences at 130 °C and is completed by approximately 160 °C and in the second step it commences at 160 °C and is completed by 240 °C. The percentage of imidazole loss in the first step is 50%, followed by a residual amount in the second step.

The one single-step mass loss in **1 ⊃ Im** is indicative of uniformly accommodated imidazole molecules. On the other hand, on the basis of the thermogravimetric curve of **2 ⊃ Im**, there are two types of imidazole molecule installed in the channel. According to the amphiphilic nature of the surface of the channel, the imidazoles with a strong interaction with the hydrophilic sites of a μ_2-OH or a carboxylate group are released at a higher temperature (160–240 °C), correlated to the weight loss in the second step, whereas the imidazoles that have less interaction with the pore surface are removed in the first step (below 160 °C).

The crystal structures of **1** and **2** (Fig. 2) reveal the following. Compound **1** consists of two types of 1D channel, namely small channels with dimensions $3.0 \times 3.0\,\text{Å}^2$ in which imidazole is unable to be installed: only large channels with dimensions of $7.7 \times 7.7\,\text{Å}^2$ can install the guest. Compared with **2 ⊃ Im**, half the amount of accommodating imidazole per one Al is reasonable from crystal structures.

Considering that the dispersion of imidazole is uniform in the crystalline channel we could calculate the density of imidazole in the larger channels of **1** ($443\,\text{Å}^3$) and channels of **2** ($750\,\text{Å}^3$). This was calculated from the void space of the guest-free state by using

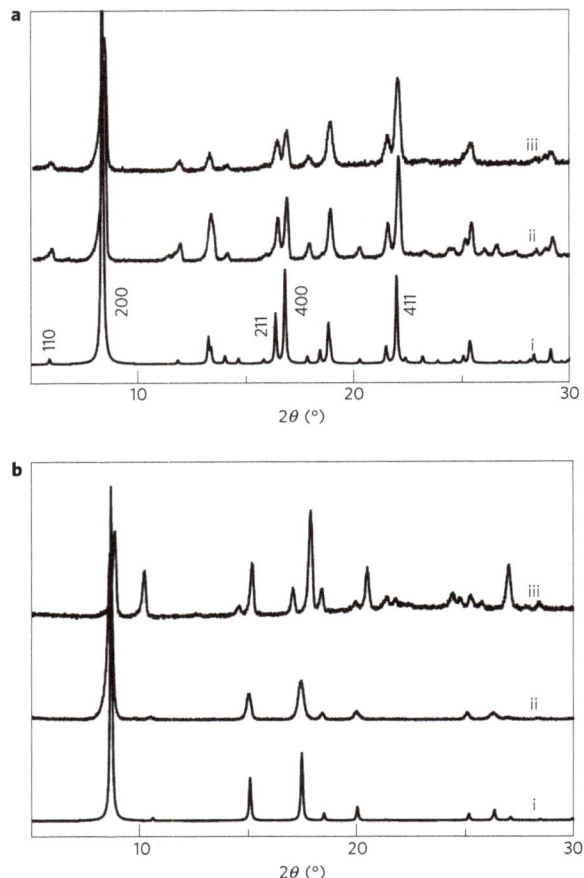

Figure 4 | Difference in flexibility of host frameworks 1 and 2 by imidazole inclusion. a, XRPD patterns of simulated 1·2H₂O (i), 1 (ii) and 1 ⊃ Im (iii). The patterns before and after accommodation of imidazole are identical. b, XRPD patterns of simulated 2 (i), 2 (ii) and 2 ⊃ Im (iii). A change in the XRPD patterns after accommodation of imidazole is observed[39].

the PLATON software package[41]. The values were about 0.15 and $0.19\,\text{g\,cm}^{-3}$, respectively. However, as the structure of **2** changes after accommodation of imidazole, it is difficult to determine the density of imidazole in **2 ⊃ Im**. The densities of imidazoles in **1 ⊃ Im** and **2 ⊃ Im** are much smaller than that of solid bulk imidazole, which is $1.23\,\text{g\,cm}^{-3}$ at ambient temperature[42]. This evidence indicates that the behaviours of imidazoles loaded to the framework considerably differ from bulk imidazole resulting from the space effect.

The X-ray powder diffraction (XRPD) patterns shown in Fig. 4 show that the diffraction pattern of **1 ⊃ Im** is the same as that of apohost **1**, corresponding to the robustness of **1**. Conversely, the peak positions and pattern of **2 ⊃ Im** are different from those of apohost **2** and the shrinkage after installation of imidazole is observed. This is because of strong interaction between the polar imidazole and the hydrophilic pore surface of the flexible **2**.

Conductivity of 1 ⊃ Im and 2 ⊃ Im

We aimed to achieve proton conductivity at temperatures around 100 °C and hence to design composites that are stable at the target temperatures. We already confirmed that the prepared composites **1 ⊃ Im** and **2 ⊃ Im** are stable up to 130 °C without any loss using thermogravimetry. Conductivities of **1 ⊃ Im** and **2 ⊃ Im** were measured by a.c. impedance spectroscopy, which is a versatile

a

Log (σT/S cm^{-1} K)

1,000/T (K^{-1})

b

Z'' ($\Omega \times 10^4$)

Z' ($\Omega \times 10^4$)

c

Z'' ($\Omega \times 10^5$)

Z' ($\Omega \times 10^5$)

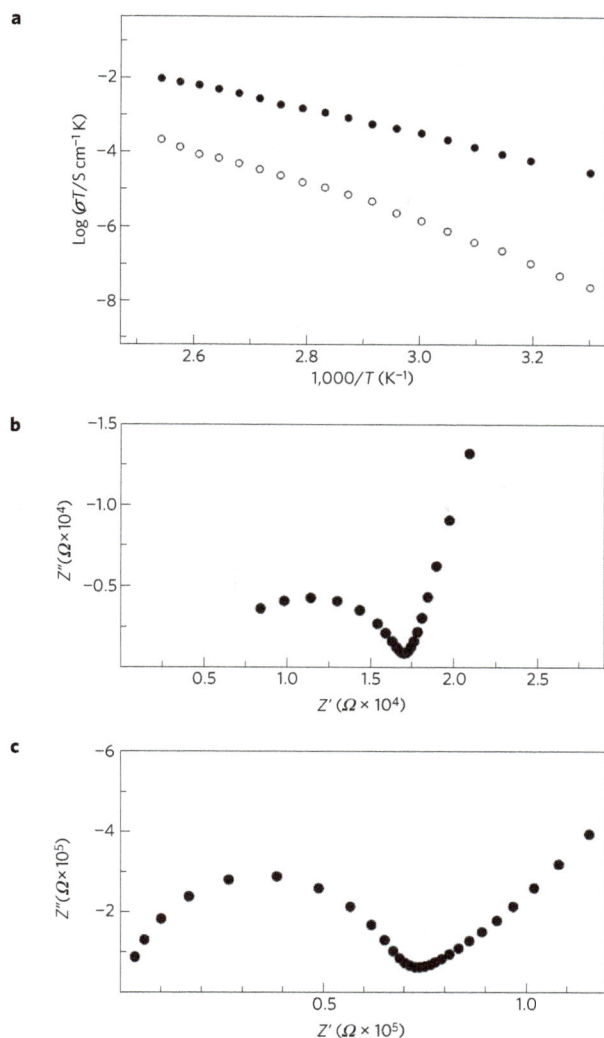

Figure 5 | Temperature dependencies of proton conductivity of 1 and 2 with imidazole molecules. a, Proton conductivity of 1 ⊃ Im (filled circles) and 2 ⊃ Im (open circles) under anhydrous conditions determined by an a.c. impedance analyser. **b,c**, Nyquist diagrams of 1 ⊃ Im (**b**) and 2 ⊃ Im (**c**) at a temperature of 120 °C.

electrochemical tool to characterize intrinsic electrical properties of materials. Figure 5b,c shows Nyquist plots (Z' versus Z'') of the complex impedance measured on 1 ⊃ Im and 2 ⊃ Im under a nitrogen atmosphere at 120 °C. The impedance plots of the two complexes are typical of materials with predominant ionic conductivity. They show one semicircle with a characteristic spur at low frequencies, which indicates blocking of H$^+$ ions at the gold electrodes. The magnitude of Z' decreased with an increase in temperature. The conductivity of the samples was calculated from the impedance value using the following equation

$$\sigma = \frac{L}{Z \cdot A}$$

where σ is the conductivity (S cm^{-1}), L is the measured sample thickness (cm), A is the electrode area (cm^2) and Z is the impedance (Ω).

The temperature dependence of proton conductivities of 1, 1 ⊃ Im and 2 ⊃ Im, measured under anhydrous conditions at temperatures ranging from 25 to 120 °C, are shown in Fig. 5a. Guest-free 1 shows a conductivity lower than 10^{-13} S cm^{-1}, confirmed by d.c. measurement, which is indicative of negligible proton conductivity for this apohost framework. After installation of imidazole, the proton conductivity of 1 ⊃ Im is 5.5 × 10^{-8} S cm^{-1} at room temperature. Although the mole fraction of imidazole in 1 ⊃ Im is much smaller than that of bulk imidazole, the proton conductivity of 1 ⊃ Im is of the same order as that of solid bulk imidazole. This is possibly because of the effect of the nanospace on the dynamic motion of imidazole. The proton conductivity of 1 ⊃ Im improves significantly as the temperature increases: at 120 °C the proton conductivity reaches 2.2 × 10^{-5} S cm^{-1}. Note that bulk imidazole at this temperature is no longer in the solid phase. This increase in the temperature-dependent conductivity of 1 ⊃ Im, compared with the conductivity profile of apohost 1, indicates that a significant improvement in the conductivity arises directly from the accommodated imidazole. Furthermore, the conductivity of 1 ⊃ Im continuously increases with an increase in temperature with the activation energy of 0.6 eV. This result indicates that phase transition does not take place. The mobile imidazole induces high-temperature (>100 °C) proton conductivity in 1 ⊃ Im. We can improve the mobility of imidazole by taking advantage of the isolating effect of PCPs. The evidence from the isotope effect (see Supplementary Fig. S1) shows that the proton conductivity of our systems is mainly contributed by the proton-hopping mechanism.

To improve the proton conductivity we increased the amount of loaded imidazole (the number of charge carriers, n_i) by using 2, which has twice the amount of accessible space for imidazole as the supporting framework. However, the proton conductivity of 2 ⊃ Im at ambient temperature is about 10^{-10} S cm^{-1}, which is lower than that of 1 ⊃ Im. As in the case of 1 ⊃ Im, the conductivity of 2 ⊃ Im increases as the temperature increases with the activation energy of 0.9 eV and it reaches 1.0 × 10^{-7} S cm^{-1} at 120 °C. The proton conductivity of 2 ⊃ Im is about two orders of magnitude lower than that of 1 ⊃ Im, although the amount of loaded imidazole is higher than that of 1 ⊃ Im. This is also possibly because of the difference in dynamic motion of the guest, which is based on the interaction between the guest and the host. Microchannels in compound 1 have a nonpolar potential surface and polar imidazole does not interact strongly with the host framework; therefore, it can move freely in this channel. Nonetheless, in the case of the polar-surface microchannels in 2, the half amount of imidazole interacts strongly with the hydrophilic sites of the host framework. The strong host–guest interactions give rise to the shrinkage of the framework, eventually to a unit cell, resulting in the different environments for imidazole accommodated in 2, compared with in 1. Therefore, because of strong host–guest interaction and dense packing, the imidazoles with strong interaction with the μ_2-OH and/or the carboxylate group of 2 are not allowed to move or rotate freely in the framework. Consequently, the conductivity in 1 ⊃ Im is larger than in 2 ⊃ Im.

Direct observation of dynamics of imidazoles

Solid-state ^{2}H NMR spectroscopy is suitable for examining the dynamics of target molecules selectively[43–45]. We therefore used this analytical technique to determine the mobility and its correlation with the conductivity of adsorbed imidazole in 1 and 2. The ^{2}H NMR powder line shapes are sensitive to local molecular motion and are characterized in terms of both the timescale and the mode of the motion, such as rotation or wobbling behaviour[46,47]. We introduced imidazole-d_4 for each host instead of non-deuterated imidazole and checked that the adsorbed amount for each guest was identical to that in the normal hosts. The ^{2}H NMR spectra of 1 ⊃ Im and 2 ⊃ Im recorded at different temperatures are shown in Fig. 6. In the case of 2 ⊃ Im (Fig. 6, right), at the lowest

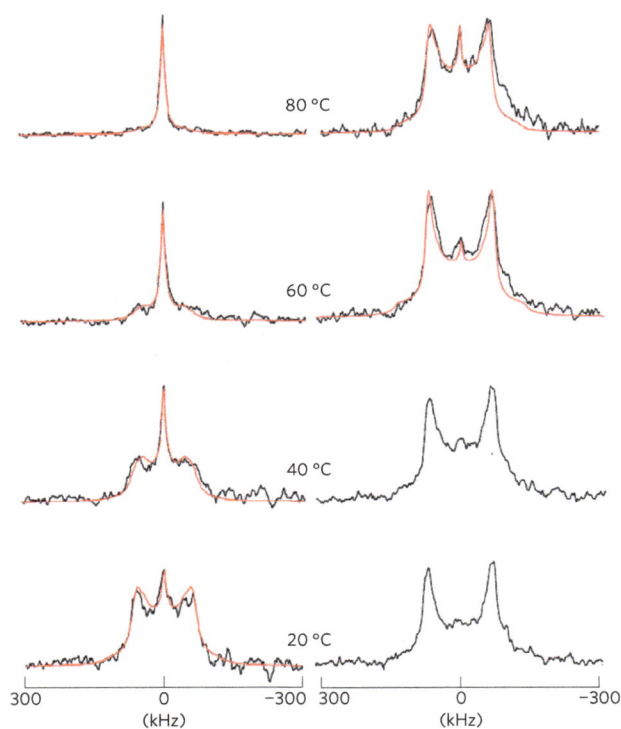

Figure 6 | Adsorbed imidazoles in 1 have a higher mobility than that of 2 in the whole range of temperature. ^{2}H solid-state NMR spectra of 1 ⊃ Im-d_4 (left) and 2 ⊃ Im-d_4 (right). The simulation results are shown as red lines.

measured temperature of 20 °C, we observed a clear Pake-type doublet pattern with a splitting width of 120 kHz, indicating that the adsorbed imidazole-d_4 behaves totally anisotropically. As the temperature increases, a narrow Lorentzian-type peak appears in the middle of the anisotropic powder pattern, corresponding to the emergence of isotropic imidazole by thermal activation. There are two possible explanations for the spectrum: the first is that free motional imidazole with low frequency shows a narrow peak and the second is the simultaneous coexistence of frequencies of slow and fast imidazole. Nonetheless, the spectrum at 40 °C indicates the existence of activated guests in pores and the relative intensity of the activated species increases as the temperature increases to 80 °C.

In the case of 1 ⊃ Im, spectra also show doublet powder patterns at low temperatures (−20 and −60 °C) with the same splitting width as in 2 ⊃ Im at temperatures lower than 30 °C. However, a narrow Lorentzian-type peak starts to appear at 20 °C. This clearly indicates that at ambient temperature the imidazole in 1 can show isotropic motion with a larger frequency than 2. The fraction of isotropic imidazole becomes dominant at 40 °C and at 80 °C we can no longer observe any Pake-doublet pattern at all. This suggests that all adsorbed imidazole within the pores of 1 has a fast isotropic motion. Using the NMR line shapes obtained, we evaluated the motion of the imidazole using a simulation procedure. We used a free rotation model of imidazole molecules with tetrahedral orientations as the main motion because this is associated with the Grotthus mechanism. We succeeded in obtaining theoretical patterns for both samples, at each temperature, on the basis of the tetrahedral free rotational model as shown in Fig. 6 and clearly observed that the rotation frequency of 1 is greater than that of 2. For example, the frequency of 1 at 60 °C is 45 kHz, whereas that of 2 is 10 kHz, and the frequency of 1 at 20 °C

(20 kHz) is still larger than that of 2 at 90 °C (18 kHz) (see Supplementary Fig. S2). Consequently, we are able to conclude that the degree of motional behaviour of the accommodated imidazole of 1 is greater than in the case of 2, which strongly supports the difference in conductivity.

We have presented a new approach to create proton transportation space on the basis of the use of proton-carrier organic molecules to enhance the proton conductivity of solid materials under anhydrous conditions at high temperature. The different values of conductivity of imidazole in compounds 1 and 2 are consistent with the dynamic properties of imidazole adsorbed in the pores. The hydrophilic microporous surface of 2 results in a strong interaction with even half the amount of adsorbed imidazoles and significantly decelerates their mobility, resulting in a poor proton-transfer rate. On the other hand, because of the hydrophobic and flat pore surface of 1, adsorbed imidazole can move more freely than in 2 and than in the bulk phase and we eventually observe higher proton conductivity, which is comparable to that of a conventional organic polymer conductor such as poly(4-vinylimidazole) (ref. 48). PCP can provide an appropriate pore environment and size for the target proton carrier by the fine-tuning of their components. In other words, the optimum mobility and density of proton carriers can be reached by taking advantage of the designability of PCPs. The strategy would be considered significant to prepare hybrid materials with high proton conductivity.

Methods

Materials. Aluminium(III) nitrate nanohydrate Al(NO$_3$)$_3$·9H$_2$O (WAKO, 99.9%); terephthalic acid HO$_2$C–(C$_6$H$_4$)–CO$_2$H (WAKO, 95%); 1,4-naphthalene dicarboxylic acid HO$_2$C–(C$_{10}$H$_8$)–CO$_2$H (WAKO, 95%); imidazole (WAKO, 99%) and imidazole (D-4, CIL, 98%) were used as received. Distilled water was used.

Synthesis of {Al(μ_2-OH)(1,4-ndc)}$_n$ (1). A mixture of Al(NO$_3$)$_3$·9H$_2$O (0.375 g, 1.0 mmol); 1,4-naphthalene dicarboxylic acid (0.108 g, 0.5 mmol) and deionized water (10 ml) was placed in a 23 ml Teflon autoclave and heated at 180 °C for one day. The initial pH of the reaction mixture was 2.5 and the final pH was 2.0. After filtering and washing the crude product with distilled water, a pure, light-yellow powder of 1·2H$_2$O was obtained (yield 80%). The sample was evacuated at 150 °C for 12 h to afford the guest-free compound 1.

Synthesis of {Al(μ_2-OH)(1,4-bdc)}$_n$ (2). The synthesis of 2 was carried out under hydrothermal conditions using Al(NO$_3$)$_3$·9H$_2$O (1.30 g, 3.5 mmol), 1,4-benzenedicarboxylic acid (0.288 g, 2.5 mmol) and distilled water (10 ml). The reaction was carried out in a 23 ml Teflon autoclave. The reaction mixture was heated at 220 °C for three days. After filtering and washing with distilled water, a white powder was obtained. It was identified by powder X-ray diffraction analysis. The excess terephthalic acid in the pores was removed by high-temperature treatment at 330 °C for three days. XRPD analysis revealed that the material was the guest-free compound 2.

Preparation of imidazole-loaded frameworks. Products 1 and 2 were again degassed by heating to 120 °C under reduced pressure for 12 h to remove guest molecules. Imidazole was vapourized into guest-free 1 and 2 at 120 °C, overnight, to yield 1 ⊃ Imi and 2 ⊃ Imi. XRPD patterns of both these compounds confirmed that the frameworks were maintained. The amount of loaded imidazole was determined by thermogravimetric analysis.

Proton conductivity measurement of 1 ⊃ Im and 2 ⊃ Im. Samples for conductivity measurements were prepared by sandwiching the respective powders 1 ⊃ Im and 2 ⊃ Im between two gold-coated electrodes (diameter 3 mm) and then tightly connecting the two electrodes, by means of springs, to ensure good contact between the sample and each electrode. Temperature-dependent conductivities of 1 ⊃ Im and 2 ⊃ Im were determined using a.c. impedance spectroscopy (Solartron SI 1260 Impedance/Gain-Phase analyser), using a homemade cell over the frequency range 1 Hz–10 MHz and with an input voltage amplitude of 100 mV. The measurement cell was filled with nitrogen at atmospheric pressure before recording the measurements. ZView software was used to extrapolate impedance data results by means of an equivalent circuit simulation to complete the Nyquist plot and obtain the resistance values.

^{2}**H solid-state NMR.** Solid-state ^{2}H NMR spectra were recorded using a Varian Chemagnetics CMX-300 spectrometer, at 45.826 MHz and a quadrupole echo pulse sequence. Simulated spectra were produced with FORTRAN programs written by us.

Received 20 November 2008; accepted 30 July 2009; published online 6 September 2009

References

1. Kreuer, K. D., Paddison, S. J., Spohr, E. & Schuster, M. Transport in proton conductors for fuel-cell applications: Simulations, elementary reactions, and phenomenology. *Chem. Rev.* **104,** 4637–4678 (2004).
2. Schuster, M. F. H. & Meyer, W. H. Anhydrous proton-conducting polymers. *Annu. Rev. Mater. Res.* **33,** 233–261 (2003).
3. Jannasch, P. Recent developments in high-temperature proton conducting polymer electrolyte membranes. *Curr. Opin. Colloid Interface Sci.* **8,** 96–102 (2003).
4. Li, S. *et al.* Synthesis and properties of imidazole-grafted hybrid inorganic–organic polymer membranes. *Electrochim. Acta* **51,** 1351–1358 (2006).
5. West, A. R. *Basic Solid State Chemistry* (Wiley, 1999).
6. Kawada, A., McGhie, A. R. & Labes, M. M. Protonic conductivity in imidazole single crystal. *J. Chem. Phys.* **52,** 3121–3125 (1970).
7. Noro, S., Kitagawa, S., Kondo, M. & Seki, K. A new, methane adsorbent, porous coordination polymer. *Angew. Chem. Int. Ed.* **39,** 2081–2084 (2000).
8. Rowsell, J. L. & Yaghi, O. M. Strategies for hydrogen storage in metal-organic frameworks. *Angew. Chem. Int. Ed.* **44,** 4670–4679 (2005).
9. Ferey, G. *et al.* Hydrogen adsorption in the nanoporous metal-benzenedicarboxylate $M(OH)(O_2C-C_6H_4-CO_2)(M = Al^{3+}, Cr^{3+})$, MIL-53. *Chem. Commun.* 2976–2977 (2003).
10. Rosi, N. L. *et al.* Hydrogen storage in microporous metal-organic frameworks. *Science* **300,** 1127–1129 (2003).
11. Li, H., Eddaoudi, M., O'Keeffe, M. & Yaghi, O. M. Design and synthesis of an exceptionally stable and highly porous metal-organic framework. *Nature* **402,** 276–279 (1999).
12. Cychosz, K. A., Wong-Foy, A. G. & Matzger, A. J. Liquid phase adsorption by microporous coordination polymers: Removal of organosulfur compounds. *J. Am. Chem. Soc.* **130,** 6938–6939 (2008).
13. Finsy, V. *et al.* Pore-filling-dependent selectivity effects in the vapor-phase separation of xylene isomers on the metal-organic framework MIL-47. *J. Am. Chem. Soc.* **130,** 7110–7118 (2008).
14. Bradshaw, D. *et al.* Permanent microporosity and enantioselective sorption in a chiral open framework. *J. Am. Chem. Soc.* **126,** 6106–6114 (2004).
15. Dybtsev, D. N. *et al.* Microporous manganese formate: A simple metal-organic porous material with high framework stability and highly selective gas sorption properties. *J. Am. Chem. Soc.* **126,** 32–33 (2004).
16. Min, K. S. & Suh, M. P. Self-assembly and selective guest binding of three-dimensional open-framework solids from a macrocyclic complex as a trifunctional metal building block. *Chem. Eur. J.* **7,** 303–313 (2001).
17. Wang, B. *et al.* Colossal cages in zeolitic imidazolate frameworks as selective carbon dioxide reservoirs. *Nature* **453,** 207–211 (2008).
18. Hasegawa, S. *et al.* Three-dimensional porous coordination polymer functionalized with amide groups based on tridentate ligand: Selective sorption and catalysis. *J. Am. Chem. Soc.* **129,** 2607–2614 (2007).
19. Horike, S., Dinca, M., Tamaki, K. & Long, J. R. Size-selective lewis acid catalysis in a microporous metal-organic framework with exposed Mn^{2+} coordination sites. *J. Am. Chem. Soc.* **130,** 5854–5855 (2008).
20. Cho, S.-H. *et al.* A metal-organic framework material that functions as an enantioselective catalyst for olefin epoxidation. *Chem. Commun.* **24,** 2563–2565 (2006).
21. Horcajada, P. *et al.* Synthesis and catalytic properties of MIL-100(Fe), an iron(III) carboxylate with large pores. *Chem. Commun.* 2820–2822 (2007).
22. Schroder, F. *et al.* Ruthenium nanoparticles inside porous $[Zn_4O(bdc)_3]$ by hydrogenolysis of adsorbed $[Ru(cod)(cot)]$: A solid-state reference system for surfactant-stabilized ruthenium colloids. *J. Am. Chem. Soc.* **130,** 6119–6130 (2008).
23. Ingleson, M. J. *et al.* Generation of a solid Bronsted acid site in a chiral framework. *Chem. Commun.* 1287–1289 (2008).
24. Fujita, M., Kwon, Y. J., Washizu, S. & Ogura, K. Preparation, clathration ability, and catalysis of a two-dimensional square network material composed of cadmium(II) and 4,4′ – bipyridine. *J. Am. Chem. Soc.* **116,** 1151–1152 (1994).
25. Seo, J. S. *et al.* A homochiral metal-organic porous material for enantioselective separation and catalysis. *Nature* **404,** 982–986 (2000).
26. Evans, O. R., Ngo, H. L. & Lin, W. Chiral porous solids based on lamellar lanthanide phosphonates. *J. Am. Chem. Soc.* **123,** 10395–10396 (2001).
27. Ferey, G. *et al.* Mixed-valence Li/Fe-based metal-organic frameworks with both reversible redox and sorption properties. *Angew. Chem. Int. Ed.* **46,** 3259–3263 (2007).
28. Kitagawa, H. *et al.* Highly proton-conductive copper coordination polymer, H2dtoaCu (H2dtoa=dithiooxamide anion). *Inorg. Chem. Commun.* **6,** 346–348 (2003).
29. Sadakiyo, M., Yamada, T. & Kitagawa, H. Rational designs for highly proton-conductive metal-organic frameworks. *J. Am. Chem. Soc.* **131,** 9906–9907 (2009).
30. Uemura, T. *et al.* Radical polymerisation of styrene in porous coordination polymers. *Chem. Commun.* 5968–5970 (2005).
31. Uemura, T. *et al.* Conformation and molecular dynamics of single polystyrene chain confined in coordination nanospace. *J. Am. Chem. Soc.* **130,** 6781–6788 (2008).
32. Mulfort, K. L. & Hupp, J. T. Chemical reduction of metal-organic framework materials as a method to enhance gas uptake and binding. *J. Am. Chem. Soc.* **129,** 9604–9605 (2007).
33. Muller, M. *et al.* Loading of MOF-5 with Cu and ZnO nanoparticles by gas-phase infiltration with organometallic precursors: Properties of Cu/ZnO@MOF-5 as catalyst for methanol synthesis. *Chem. Mater.* **20,** 4576–4587 (2008).
34. Turner, S. *et al.* Direct imaging of loaded metal-organic framework materials (Metal@MOF-5). *Chem. Mater.* **20,** 5622–5627 (2008).
35. Horcajada, P. *et al.* Flexible porous metal-organic frameworks for a controlled drug delivery. *J. Am. Chem. Soc.* **130,** 6774–6780 (2008).
36. Horcajada, P. *et al.* Metal-organic frameworks as efficient materials for drug delivery. *Angew. Chem. Int. Ed.* **45,** 5974–5978 (2006).
37. Tanaka, D. *et al.* Anthracene array-type porous coordination polymer with host–guest charge transfer interactions in excited states. *Chem. Commun.* 3142–3144 (2007).
38. Comotti, A. *et al.* Nanochannels of two distinct cross-sections in a porous Al-based coordination polymer. *J. Am. Chem. Soc.* **130,** 13664–13672 (2008).
39. Loiseau, T. *et al.* A rationale for the large breathing of the porous aluminum terephthalate (MIL-53) upon hydration. *Chem. Eur. J.* **10,** 1373–1382 (2004).
40. Serre, C. *et al.* An explanation for the very large breathing effect of a metal-organic framework during CO_2 adsorption. *Adv. Mater.* **19,** 2246–2251 (2007).
41. Spek, A. L. Single-crystal structure validation with the program PLATON. *J. Appl. Crystallogr.* **36,** 7–13 (2003).
42. Craven, B. M., McMullan, R. K., Bell, J. D. & Freeman, H. C. The crystal structure of imidazole by neutron diffraction at 20 °C and −150 °C. *Acta Crystallogr. B* **33,** 2585–2589 (1977).
43. Horike, S. *et al.* Motion of methanol adsorbed in porous coordination polymer with paramagnetic metal ions. *Chem. Commun.* 2152–2153 (2004).
44. Ueda, T. *et al.* Phase transition and molecular motion of cyclohexane confined in metal-organic framework, IRMOF-1, as studied by 2H NMR. *Chem. Phys. Lett.* **443,** 293–297 (2007).
45. Horike, S. *et al.* Dynamic motion of building blocks in porous coordination polymers. *Angew. Chem. Int. Ed.* **45,** 7226–7230 (2006).
46. Schmidt-Rohr, K. & Spiess, H. W. *Multidimensional Solid-State NMR and Polymers* (Academic, 1994).
47. Abragam, A. *Principles of Nuclear Magnetism* (Oxford Univ. Press, 1961).
48. Bozkurt, A. & Meyer, W. H. Proton conducting blends of poly(4-vinylimidazole) with phosphoricacid. *Solid State Ion.* **138,** 259–265 (2001).

Acknowledgements

This work was supported by Japan Science and Technology Agency (JST).

Author contributions

S.B. and M.H. prepared the aluminium PCPs. S.B. and T.K. measured the ionic conductivity. S.B., N.Y. and D.T. carried out solid-state NMR work and their data analysis was carried out by S.B. and M.M. The work was directed by S.K., and S.B. and S.H. contributed in writing the manuscript.

Additional information

Supplementary information accompanies this paper on www.nature.com/naturematerials. Reprints and permissions information is available online at http://npg.nature.com/reprintsandpermissions. Correspondence and requests for materials should be addressed to S.K.

Parallel cylindrical water nanochannels in Nafion fuel-cell membranes

KLAUS SCHMIDT-ROHR* AND QIANG CHEN

Ames Laboratory and Department of Chemistry, Iowa State University, Ames, Iowa 50011, USA
*e-mail: srohr@iastate.edu

Published online: 9 December 2007; doi:10.1038/nmat2074

The structure of the Nafion ionomer used in proton-exchange membranes of H_2/O_2 fuel cells has long been contentious. Using a recently introduced algorithm, we have quantitatively simulated previously published small-angle scattering data of hydrated Nafion. The characteristic 'ionomer peak' arises from long parallel but otherwise randomly packed water channels surrounded by partially hydrophilic side branches, forming inverted-micelle cylinders. At 20 vol% water, the water channels have diameters of between 1.8 and 3.5 nm, with an average of 2.4 nm. Nafion crystallites ($\sim$10 vol%), which form physical crosslinks that are crucial for the mechanical properties of Nafion films, are elongated and parallel to the water channels, with cross-sections of $\sim$(5 nm)^2. Simulations for various other models of Nafion, including Gierke's cluster and the polymer-bundle model, do not match the scattering data. The new model can explain important features of Nafion, including fast diffusion of water and protons through Nafion and its persistence at low temperatures.

The proton-exchange membrane (PEM) is a central, and often performance-limiting, component of all-solid H_2/O_2 fuel cells. Nafion, the most widely used PEM material, is a perfluorinated polymer that stands out for its high, selective permeability to water and small cations, in particular protons[1-3]. The chemical structure of Nafion combines a hydrophobic Teflon-like backbone with hydrophilic ionic side groups, see Fig. 1a. In hydrated Nafion, these components self-organize to produce a clear peak in small-angle X-ray or neutron scattering (SAXS/SANS) at around $q = 2\pi/(4\,\mathrm{nm})$, where $q = 4\pi/\lambda \sin\theta$ with the scattering angle 2θ and the X-ray or neutron wavelength λ. This 'ionomer peak', see Fig. 1b, has been variously attributed to spherical inverted-micelle water clusters[1,4-9], layered structures[10-13], channel networks[3,14,15] and polymer bundles[16-18].

Although the original Gierke model[1] of a network of spherical water clusters connected by 1-nm-diameter channels is still the most popular, the presence of elongated structures was clearly demonstrated by SAXS studies. First, as pointed out by the Grenoble group, unoriented samples show an $I(q) \sim q^{-1}$ power law typical of long cylinders at small q (refs 15–18), see the log–log plot in Fig. 1c where $I(q) \sim q^{-1}$ is represented by a line of slope -1. Even more importantly, SAXS of oriented Nafion samples does not exhibit a meridional ionomer peak of correlations along the draw direction[19-21]. Progress in the understanding of Nafion and related PEM materials has been seriously hampered by the lack of a convincing model that helps make sense of the bewildering array of data on Nafion and their varied interpretations[2,3].

A quantitative analysis of the published scattering curves[1,16-18] of Nafion should help elucidate the nanometre-scale structure. A few attempts have been made to simulate the SAXS curve[4,11,13,22], but only a limited q range was considered, crystallinity was crucially disregarded[11] and the input parameters did not correspond to realistic three-dimensional density distributions, as shown below. Using a new method of calculating the SAXS curve by numerical Fourier transformation from a given scattering density distribution[23], we will show here that a model featuring long

parallel water channels in cylindrical inverted micelles reproduces all of the SAXS features at hydration levels typical in fuel-cell applications. Such a model was used previously for simulating diffusion through Nafion[24-27], but it was selected *ad hoc* for computational convenience (because only the cross-section of a single pore needs to be considered)[26] and regarded as a 'first-order approximation' of Gierke's model[24]. No independent evidence was given that this model is an accurate representation of the microstructure of Nafion, which was described as 'complex' and 'inhomogeneous'[27], and recent reviews[2,3,28,29] have not acknowledged the parallel-pore structure as one of the accepted models of Nafion. Crystallites, which are crucial for mechanical properties and dominate the scattered intensity at $q < 0.5\,\mathrm{nm}^{-1}$ (refs 30,31), are included in our simulations, and we have simulated $I(q)$ for increasing water content. We will show that a dozen other models[1,3-18,22,32] fail to match the experimental scattering data. Finally, the new model will be shown to explain the excellent transport properties of hydrated Nafion.

SAXS SIMULATION USING THE PARALLEL WATER-CHANNEL MODEL

Figure 2 shows the features of the water-channel model of Nafion, at a hydration level of 20 vol% (11 wt%) water. The structure is made up of $\sim$2.4-nm-diameter water channels (cylinders) lined with hydrophilic side groups, Fig. 2a,b. The channels, which are locally parallel to their neighbours, are stabilized on the outside by the relatively straight helical backbone segments observed by nuclear magnetic resonance[33,34] (NMR). Given a persistence length of between 3 and 5 nm for the Nafion backbone[35,36], the densely packed cylinders with their shell of relatively stiff backbones, see Fig. 2a, can be expected to have a persistence length of tens of nanometres. This backbone stiffness is an important parameter to consider in molecular dynamics simulations of hydrated Nafion. From the scattering density $\rho(\mathbf{x})$, whose cross-section perpendicular to the channel axes is shown in Fig. 2c, we

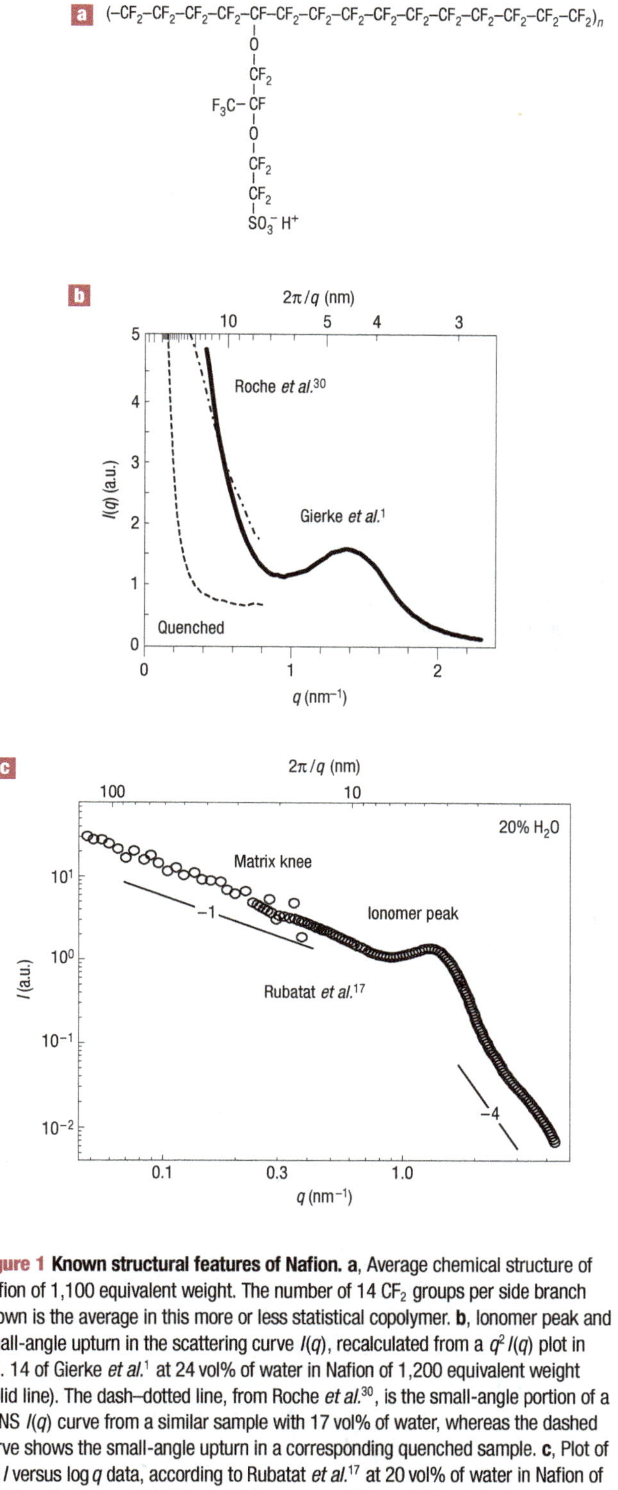

Figure 1 Known structural features of Nafion. a, Average chemical structure of Nafion of 1,100 equivalent weight. The number of 14 CF_2 groups per side branch shown is the average in this more or less statistical copolymer. **b**, Ionomer peak and small-angle upturn in the scattering curve $I(q)$, recalculated from a $q^2 I(q)$ plot in Fig. 14 of Gierke et al.[1] at 24 vol% of water in Nafion of 1,200 equivalent weight (solid line). The dash–dotted line, from Roche et al.[30], is the small-angle portion of a SANS $I(q)$ curve from a similar sample with 17 vol% of water, whereas the dashed curve shows the small-angle upturn in a corresponding quenched sample. **c**, Plot of log I versus log q data, according to Rubatat et al.[17] at 20 vol% of water in Nafion of 1,100 equivalent weight. The scattering pattern of drawn Nafion (not shown here), with the ionomer peak exclusively on the equator[19–21], is another notable structural indicator.

calculate the scattered-intensity curve $I(q)$ using numerical Fourier transformation[23]. The curve for the water-channel model, Fig. 2d, exhibits not only the ionomer peak, but also the experimentally

observed q^{-1} and q^{-4} power laws in the SAXS data. Although the channels were assumed to be straight in this simulation, inclusion of some correlated undulations of the channels does not change the features of the $I(q)$ curve dramatically, as shown in Supplementary Information, Fig. S3.

The water channels have diameters distributed between 1.8 and 3.5 nm, with an average of 2.4 nm. This is smaller than the cluster diameter $d \sim 4$ nm estimated from the SAXS peak position q^* according to $d = 2\pi / q^*$ (ref. 1); indeed, we have shown[23] that for cylinders, this relation does not hold, with d being overestimated by a factor of ~ 1.5. A further shift can occur when the cylinders are surrounded by a polymer layer that increases the diameter of the hard-core repulsion. In our model, the typical minimum thickness of the polymer layer is ~ 0.7 nm (see Supplementary Information, Fig. S1b).

A distribution of channel diameters as used in the simulation is directly proved by the smooth Porod region at 'large' q in the experimental scattering curve; if the size distribution spans less than a factor of 1.5, modulations are seen in the large-q region due to the form factor of the dominant particle size[23]. The diameter may vary along the length of a given channel, within the given distribution of diameters. Simulation results shown in Supplementary Information, Fig. S2 show that without a change in the position of the centre of the channel, this variation in diameters would not affect the scattering curve significantly. However, the variation must be random because no meridional peak of correlations along the draw direction is observed for oriented samples[19,21]. Given that the cylinder diameter relates quite directly to the number of Nafion chains contributing ionic side groups to the cylinder, a major variation in the diameter of a given cylinder seems unlikely on the 20 nm scale. Further investigations of the effects of these details of the model on the scattering curves are desirable.

Crystallites are an important component of the structure of Nafion membranes. They are crucial for mechanical properties, acting as physical crosslinks[31]. The crystallinity at 1,100 equivalent weight has been estimated to range between 5 and 20%, based mostly on wide-angle X-ray diffraction[1]. The intercrystalline repeat length is 10–20 nm, according to the position of the 'matrix knee' in SAXS[1,30]. Whereas it was long believed[1,17] that this corresponds to an interlamellar periodicity along the chain axis, recent careful studies[20] have shown that the repeat is perpendicular to the chain axis, given that the scattering intensity is on the equator for oriented Nafion samples[20,21].

Accordingly, long crystallites of 2–5 nm thickness and x–y (cross-sectional) aspect ratios of 1–1.8 were included in the SAXS simulations producing the simulated curves in Fig. 2d. They suggest that $\sim 3/4$ of the scattered intensity at small q is due to the crystallites, in good agreement with experiments comparing scattering from semicrystalline and non-crystalline Nafion[15,30,31]. As expected, attempts to simulate $I(q)$ with nearly cubic crystalline blocks rather than long particles fit the experimental data less well, producing too little intensity at small q values owing to the q^0 power law for cubes or spheres. Thus, we agree with the recent conclusion of van der Heijden et al.[20] and Kim et al.[15] that the crystallites are elongated and approximately cylindrical. The great length of the crystallites, >50 nm (refs 16–18), is surprising and needs to be critically checked in future work. Although the assignment of the matrix knee to the crystallites is convincing down to $q = 0.25$ nm^{-1}, that is, on a length scale of 25 nm, on the basis of the comparison of data from semicrystalline and non-crystalline Nafion[30,31], the continuing intensity upturn at $q < 0.25$ nm^{-1} could be due to an unrelated large-scale modulation in the electron-density distribution rather than the length of the crystallites. The crystallinity in the simulated

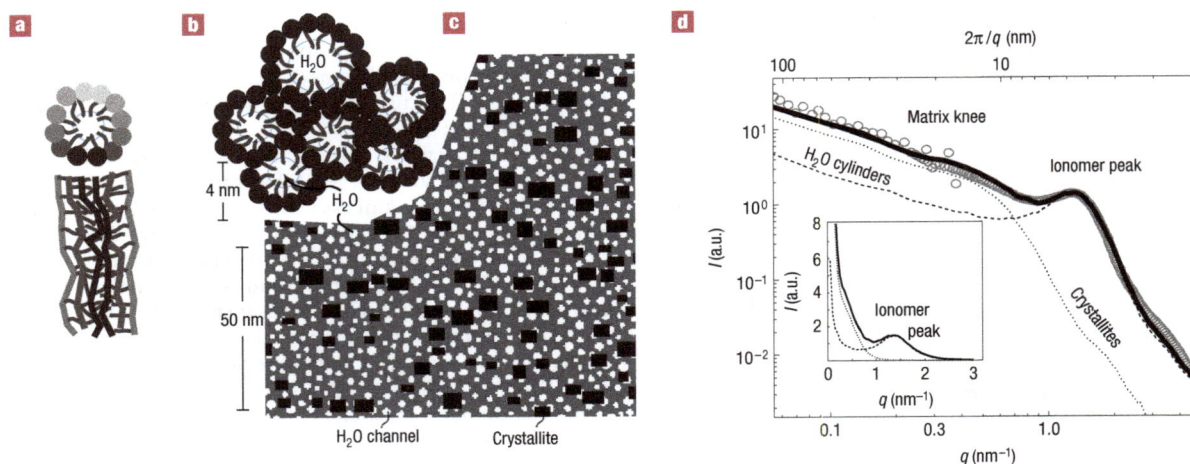

Figure 2 Parallel water-channel (inverted-micelle cylinder) model of Nafion. a, Two views of an inverted-micelle cylinder, with the polymer backbones on the outside and the ionic side groups lining the water channel. Shading is used to distinguish chains in front and in the back. **b,** Schematic diagram of the approximately hexagonal packing of several inverted-micelle cylinders. **c,** Cross-sections through the cylindrical water channels (white) and the Nafion crystallites (black) in the non-crystalline Nafion matrix (dark grey), as used in the simulation of the small-angle scattering curves in **d. d,** Small-angle scattering data (circles) of Rubatat et al.[17] in a log(I) versus log(q) plot for Nafion at 20 vol% of H_2O, and our simulated curve from the model shown in **c** (solid line). The inset shows the ionomer peak in a linear plot of $I(q)$. Simulated scattering curves from the water channels and the crystallites by themselves (in a structureless matrix) are shown dashed and dotted, respectively.

structure was 13% by volume (15% of the dry polymer). Other simulations with crystallinities between 9 and 15% also gave acceptable results. The crystallinity from the straight-cylinder model is probably an overestimate because simulations for undulating channels (see Supplementary Information, Fig. S3) show that correlated undulations of crystallites and water channels reduce the scattered intensity at intermediate and high q values, while leaving the small-angle upturn due to crystallites unchanged.

In scattering experiments on non-crystalline Nafion, produced by quenching or solution-casting[15,30,31,37], the $I(q)$ curve from the hydrated clusters can be observed selectively. It exhibits a broad region of $I(q) \approx$ const. flanked by the ionomer peak and a small-angle upturn that has been proved, by SANS with D_2O/H_2O contrast variation, to be indeed due to the hydrated clusters[37]; the dashed curve in Fig. 1b shows an example. Our scattering curve from the water channels without crystallites (dashed lines in Fig. 2d) reproduces these features within the variation of the experimental curves; the greater[30] or lesser[15] steepness of the experimental small-angle upturn may be attributed to differences in the tortuosity of the water cylinders in the differently prepared samples. In particular, it is to be expected that the small-angle upturn from the water channels is more pronounced in a semicrystalline sample, where the crystallites help align the water cylinders, than in a non-crystalline sample, where cylinders may meander more strongly.

The change in the ionomer peak and other properties of Nafion with water content is pronounced, see Fig. 3a with data from Gierke et al.[1]. The shift and intensity increase of the ionomer peak and small-angle upturn can be reproduced adequately, Fig. 3b, in the water-channel model, with simple swelling of a constant number of water channels in a given volume of polymer; for illustration, Fig. 3c shows matching portions of the scattering density for 10 and 28 vol% water. In contrast, Gierke et al. had to invoke an increase of the number of ionic groups per cluster with water content[1], without specifying the origin of these extra groups. Details regarding the intensity increase in the experimental

and simulated data, which is affected by film-thickness increase and the excess scattering density of the electron-rich sulphonate groups, are discussed in the Supplementary Information.

SAXS SIMULATIONS FOR OTHER MODELS

For comparison, Figs 4–6 show simulations of small-angle scattering for other models of the Nafion nanostructure, namely Gierke's model of spherical clusters on a paracrystalline cubic lattice[1,6,7], the local-order model[4,5], the polymer-bundle model[16–18,20], hydrated bilayers/slabs[10–13] and network models[14,15]. Models without order[9,14,22], which do not produce an ionomer peak, are discussed in the Supplementary Information.

Gierke's popular cluster model[1,6] with spherical clusters on a paracrystalline cubic lattice, Fig. 4a, would produce scattering curves as shown in Fig. 4b. By varying the degree of disorder of the second kind[38] in the paracrystalline lattice, the peaks in the scattering curve and in the radial distribution function $P(r)$ (see insets) can be broadened. These structures can be considered as a valid implementation of the 'local-order model'[4,5] of Nafion. In the original version of this model[4,5], which was an attempt to quantify Gierke's model, an unphysical $P(r)$ with a sharp nearest-neighbour peak separated by a gap down to the baseline from a long-distance plateau without any other maxima was used for quantitative calculations. This violates the Ornstein–Zernike equation relating $P(r)$ to the 'direct' two-particle correlation function[39]. As predicted, the $P(r)$ in the inset of Fig. 4b shows many sharp peaks when the first peak is sharp, whereas the radial distribution functions in Fig. 4b,e,h confirm that a featureless plateau at intermediate distances requires a broadened first maximum that is not separated by a deep gap.

Neither of the simulated scattering curves in Fig. 4b and e matches the features of the experimental data of Fig. 1b. In addition to ionomer peaks that are too sharp, they show a q^0, rather than q^{-1}, power law at small q, which is indeed expected for spheres[38]. The modulations at 'large' q in the log–log plots of Fig. 4 are due to the form factor for a single particle diameter. Note that a wide

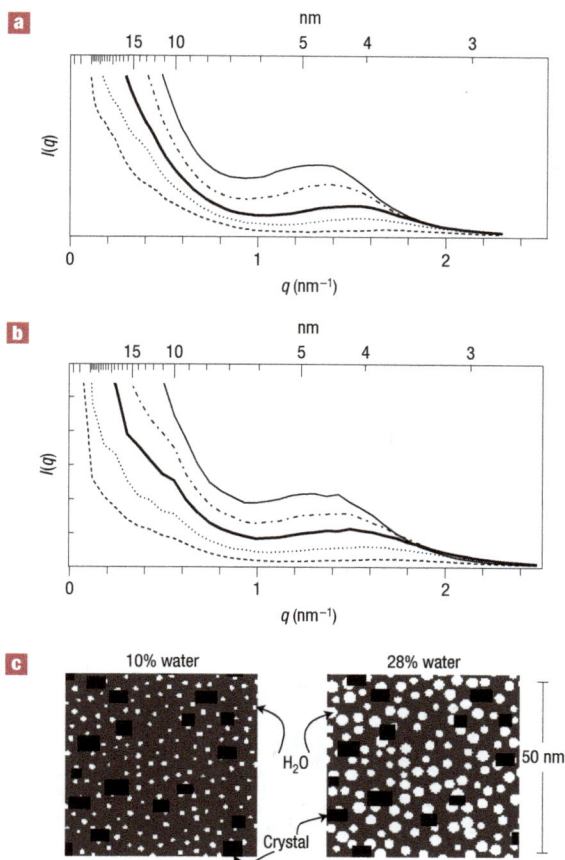

Figure 3 Effect of hydration on the ionomer peak of Nafion and its simulation by the water-channel model. a, Data from Fig. 14 of Gierke *et al.*[1]. 10 vol% water: dashed line; 15%: dotted line; 20%: thick line; 24%: dash–dotted line; 28%: thin solid line. **b,** Simulations using the water-channel model with simple swelling of the channel diameters. Numerical noise is due to the finite number of crystallites. **c,** Corresponding portions of the scattering-density distributions (cross-section perpendicular to the channel axes, with SO_3–H_2O compensation, see Supplementary Information, Fig. S4) for 10 and 28 vol% water. Water channels are white, the non-crystalline matrix is grey and crystallites are black.

size distribution that would remove these modulations is difficult to reconcile with the fairly uniform nearest-neighbour distances in the paracrystalline lattice; larger clusters would coalesce with their neighbours, forming larger aggregates.

Spherical clusters of 3.6 nm diameter that are randomly packed[9] except for hard-core plus polymer-shell repulsion produce a more suitable simulation of the ionomer peak, see Fig. 4g–i, but the experimentally observed power-law decays in the scattering curve are still not reproduced, in particular the small-angle upturn observed even for non-crystalline Nafion[30,31,37]. A q^{-1} dependence at small q could be generated by lining the clusters up along one dimension, but such columns of clusters would produce not only intercolumn equatorial peaks in scattering from oriented samples as observed, but also meridional reflections from the strong correlations within a given column of clusters, inconsistent with experiment[19–21]. Owing to the inability of the spheroidal-cluster model to explain the SAXS data from non-crystalline or oriented samples, it also is not worthwhile to add a large fraction of long

crystallites that would reproduce the small-angle upturn in the $I(q)$ curve for unoriented semicrystalline Nafion.

Polymer bundles in a hydrated matrix were recently proposed as the model for Nafion at both high and low hydration levels, on the basis of a semiquantitative analysis of high-quality scattering data of Nafion[16–18]. Figure 5a shows a cross-section through a model with cylindrical polymer bundles and crystallites of rectangular cross-section, at a water fraction of 20%. The corresponding SAXS curve, Fig. 5b, exhibits relatively pronounced higher-order peaks, inconsistent with the experimental data, which show only a single ionomer peak. The higher-order peaks can be attributed to the sharp peak at the cylinder diameter in the radial distribution function, and they are thus a general feature for arrays of relatively closely packed cylinders (see Supplementary Information, Fig. S5), regardless of the exact details of the larger-scale structure and of the density distribution within the polymer cylinder. Close packing of the polymer bundles is unavoidable here due to the relatively small volume fraction of the water 'matrix'. With a broader distribution of cylinder radii, the first-order peak disappears together with the higher-order peaks (see Supplementary Information, Fig. S6).

Furthermore, it should be noted that in this model, much of the water is occluded in channels surrounded by polymer, see Fig. 5a. In other words, this would essentially be a water-channel model, with channels of awkwardly shaped cross-sections. Their interfacial area, and thus the total energy of the interfaces between the hydrophilic and hydrophobic regions, is significantly higher than in the cylindrical water-channel model for a given water volume fraction near 20%. The irregularly shaped channels could quite easily rearrange into cylindrical channels of smaller interfacial energy, given the significant rotational mobility of chains and side branches[34]. At higher hydration levels (>40 vol% water), for instance produced by swelling Nafion in an autoclave[17], the bundle model is probably correct, but this is not the condition relevant for the membrane in a typical fuel cell.

Ribbons made up of bilayers of polymer chains, with a central bilayer of polymer backbones surrounded by ionic side branches, have been proposed as alternatives to cylinders in the bundles model[17]. Figure 5c,d shows a model with a relatively dense random packing of 3.2-nm-thick and ~6-nm-wide ribbons. The water fraction of 35% (42% without crystallites) is too high and cannot be easily reduced to 20%, for the fundamental reason that a 3.2 nm × 6 nm ribbon surrounded by a water shell of only 0.4 nm thickness has a 71% polymer and 29% water fraction. In addition, the scattering curve obtained from this model contains features not seen in the experimental data, such as a peak around $q = 3$ nm^{-1}. This form-factor peak results from the relatively uniform thickness of the bilayers, which cannot be varied significantly given the fixed dimensions of the stiff helical backbones and well-defined length of the side branches. There are also clear indications of two 'ionomer peaks' in the $I(q)$ curve without crystallites (dashed line): a sharp one at $q = 2\pi/(3.8$ nm$)$ from the short, more strongly correlated dimensions of the ribbon cross-sections, and a broad one at lower $q = 2\pi/(8$ nm$)$, from the long dimension of the cross-sections. Further simulations in Supplementary Information, Fig. S7 with more orientational disorder and wider ribbons corroborate the inadequacies of the ribbon model.

A model with alternating layers of polymer and water[10–13,19] is shown in Fig. 5e. The water layers are distributed on a one-dimensional paracrystalline lattice with an average nearest-neighbour centre-to-centre distance of $a = 4.5$ nm and water-layer widths d ranging between 0.75 and 1.2 nm (23% water fraction) as shown in Fig. 5e. The thickest polymer layers should be considered as crystallites. Even though the polymer layer thicknesses vary so much that some water layers almost merge, the scattering peak in Fig. 5f is narrower than observed experimentally. Note that in most

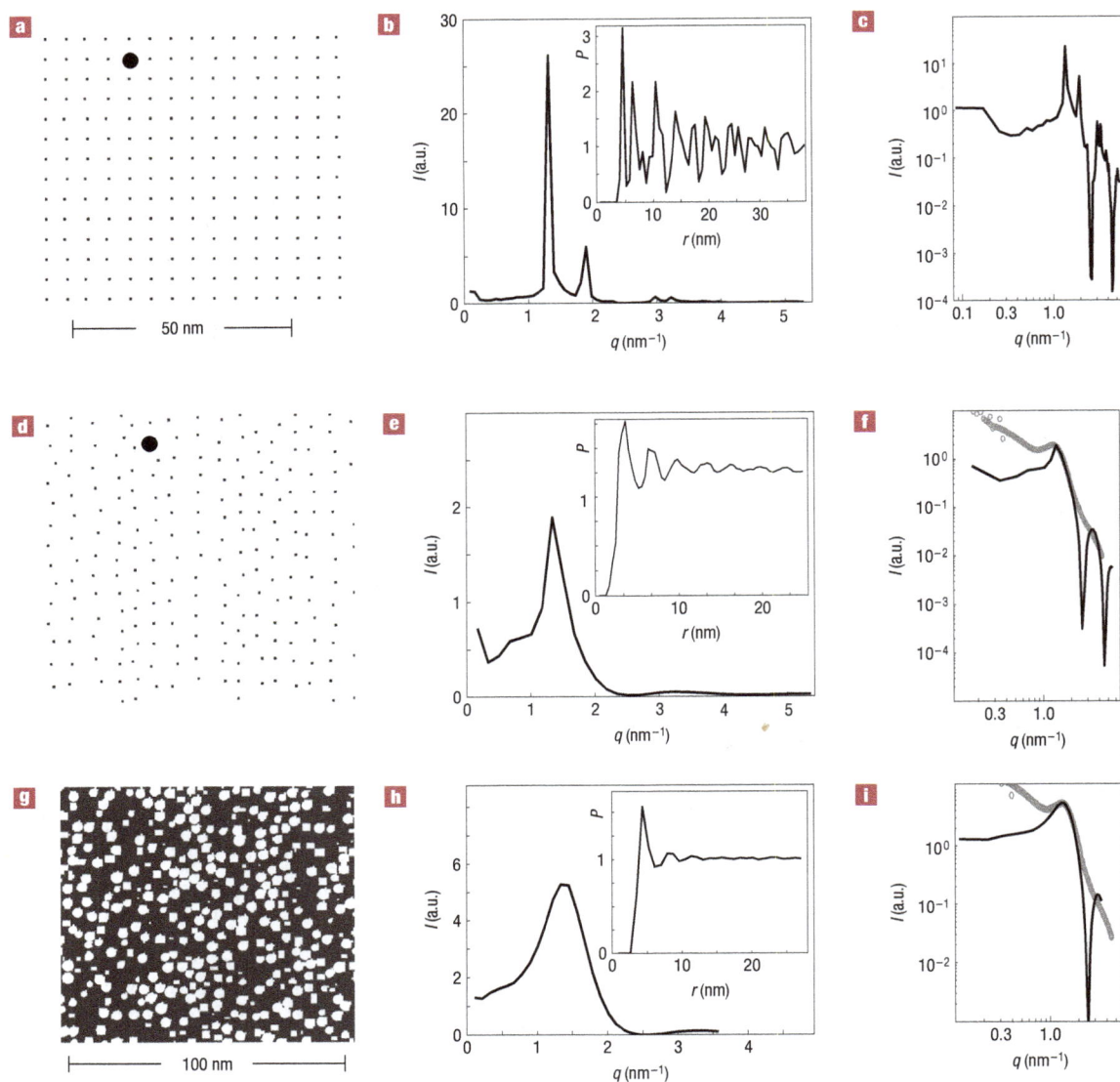

Figure 4 Simulated small-angle scattering curves for spherical-cluster models (Gierke, Yeager and local-order models). The hydration level is 22%. **a**, Paracrystalline cubic lattice, with slight disorder of the second kind. The average lattice spacing was 4.7 nm. The size of one 3.6-nm-diameter spherical cluster is indicated. **b,c**, Linear (**b**) and log–log (**c**) plots of $I(q)$. The radial distribution function $P(r)$ is shown in the inset of **b**. **d**, Same as **a**, but with more pronounced disorder. **e**, Plots of $I(q)$ and $P(r)$ as in **b**. **f**, log(I) versus log(q). **g**, Cross-section through randomly placed (except for hard-core repulsion) spherical water clusters of 3.6 nm diameter. **h,i**, Corresponding scattering curves, and $P(r)$ (inset of **h**). In **f** and **i**, experimental data from Fig. 1c are overlaid in grey.

layered models, the polymer layers are proposed to be bilayers of two Nafion chains packed back-to-back[10–12], which would result in a more regular structure and even sharper scattering peak. Krivandin et al.[13] managed to fit the ionomer peak with a layer model, but only for a water volume fraction of 47%, which is much too high (see the Supplementary Information for details). As expected for layer structures[38], a $I(q) \sim q^{-2}$ power law is seen in the simulated curve of Fig. 5f at small q. It does not match the small-angle upturn in the experimental data. The layers in the simulation were at least 110 nm wide, but the only change for layers of smaller width W would be a transition to a q^{-1} power law for $q < 2\pi/W$ (ref. 23). Note that the small water-layer thickness $d \sim 1$ nm, which derives directly as $d = 0.23\ a = 0.23\ 2\pi/q^*$ from the 23% water volume fraction and the ionomer peak position q^*, would imply

that most of the water is surface-bound, which would slow down water and proton transport dramatically (see below).

Kreuer has proposed a model with a periodic cubic network of cylindrical channels[14]. The main purpose of this model was to identify differences between the structure of different ionomers. For instance, it was concluded that the channels of Nafion are wider and less branched with fewer dead-end channels compared with the channel system in sulphonated polyarylenes. Nafion cylinder segments had to be taken out, which reduced the dimensionality of the lattice, making it more similar to the one-dimensional parallel-channel model. The network model also provides the necessary three-dimensional connections between channels. Figure 6a,c shows examples of two such networks, with node points on a paracrystalline lattice. The corresponding scattering curves, shown

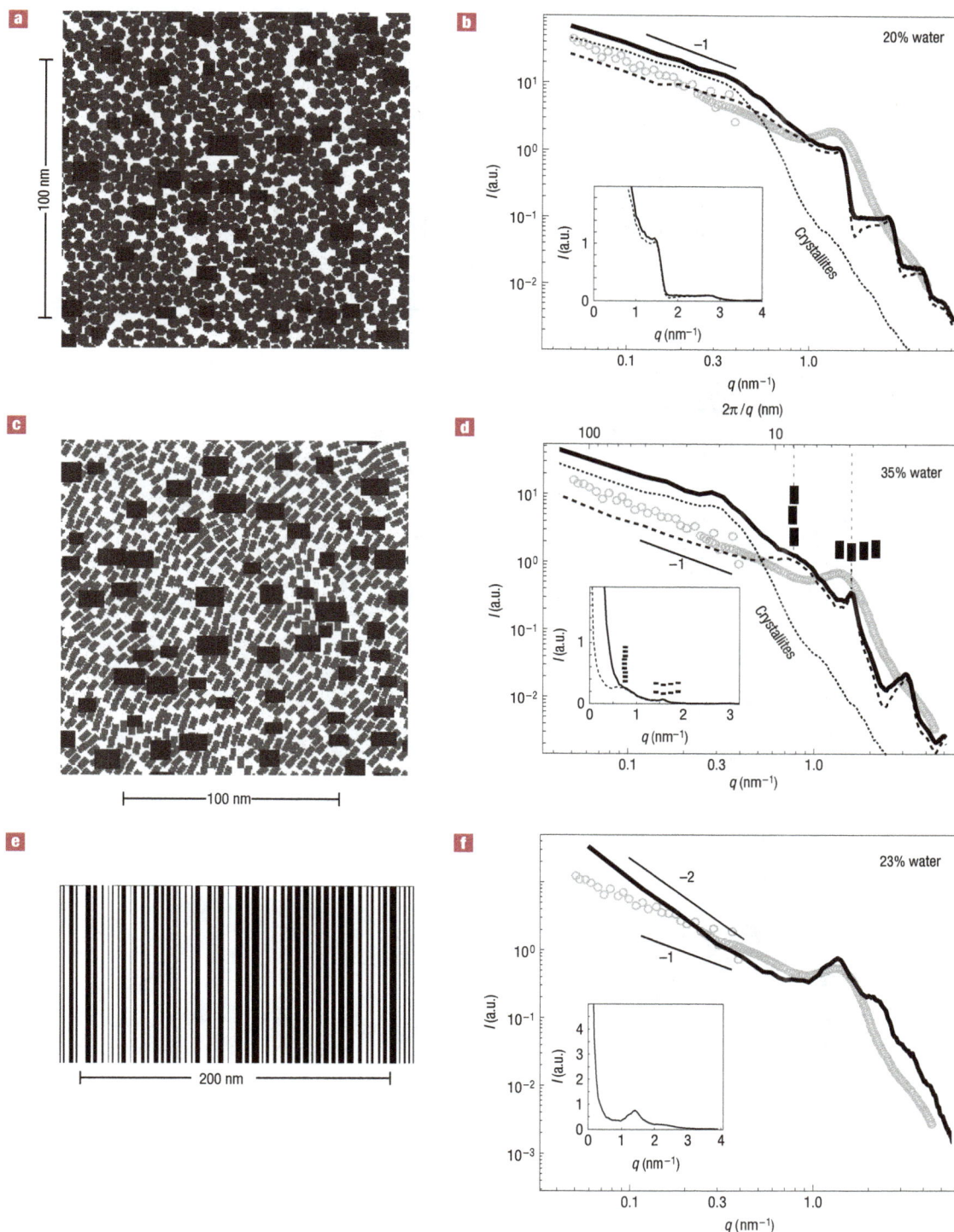

Figure 5 Simulations of small-angle scattering for models with elongated polymer structures. a, Cross-section through a polymer-bundle model at 20 vol% water. The diameters of the parallel polymer bundles (dark grey) ranged between 4 and 5 nm, and 13% crystallites (black rectangles) were included. **b,** Corresponding scattering curve in log–log and linear plots. Simulated scattering curves without crystallites, and for the crystallites by themselves, are shown dashed and dotted, respectively. The experimental data of Fig. 1c are overlaid (grey circles). **c,** Cross-section through a model of 3.2-nm-thick parallel polymer bilayer ribbons (dark grey) and 18% crystallites (larger black rectangles). The water volume fraction is too high, 35%, for the model to be valid, see text. **d,** Corresponding scattering curves in log–log and linear plots analogous to **b**. Stacking of ribbons responsible for the two 'ionomer peaks' is indicated schematically by filled rectangles. **e,** Cross-section through a model of 0.75–1.2-nm-thick water layers (white) on a one-dimensional paracrystalline lattice, separated by polymer layers (black), at 23 vol% water. **f,** Corresponding scattering curves in log–log and linear plots.

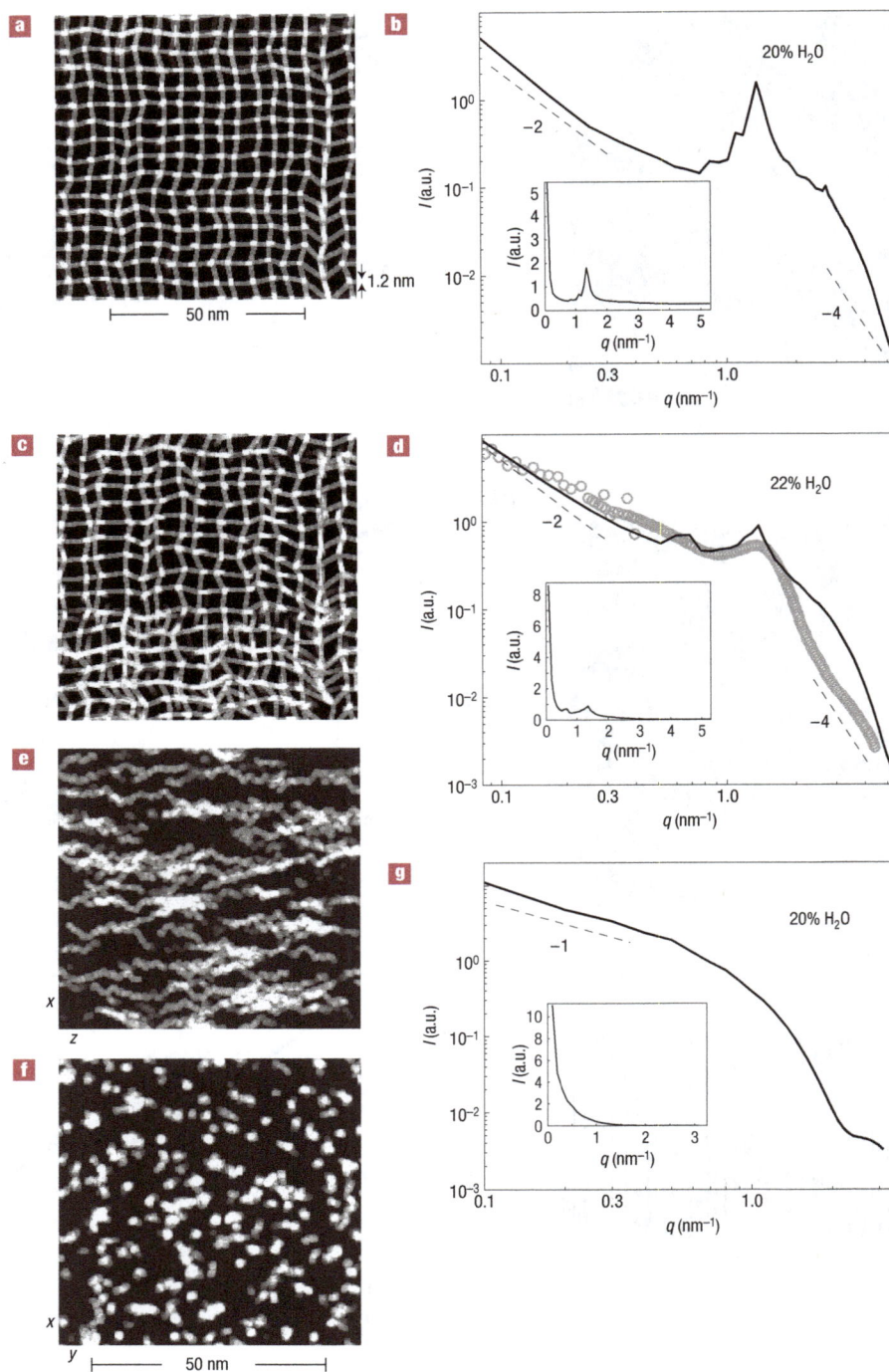

Figure 6 Simulated small-angle scattering curves for network models. a, View of a 5-nm-thick layer of a 'periodic' cubic array of water channels in the hydrophobic matrix. The nodes are on a paracrystalline lattice with 4.6 nm spacing of ± 1.4 nm variation, connected by 1.2-nm-diameter water channels (white or grey). **b**, Corresponding scattering curve in log–log and linear plots. **c**, Same as **a**, but with a larger ± 2.3 nm variation of the lattice spacing. **d**, Corresponding scattering curve in log–log and linear plots. The experimental data of Fig. 1c are overlaid (grey circles). **e**, Side view of a 4-nm-thick layer of 1.5-nm-thick freely merging worm-like water channels. **f**, View along the worm axes. **g**, Corresponding scattering curve in log–log and linear plots.

in Fig. 6b,d, do not match the experimental data particularly well. The 'ionomer peak' at $q^* = 2\pi/(\text{lattice spacing})$ is too sharp, and the slopes in the power-law regions are too steep. A shoulder at $\sim 2q^*$ is observed, as for other paracrystalline models (Figs 4f and 5d), but is absent in the experimental data. Furthermore, it is clear that a tightly crosslinked network of channels as shown in Fig. 6a,c would not deform under uniaxial drawing without changing the spacing in the equatorial direction. It would probably exhibit a meridional reflection, inconsistent with SAXS on drawn samples even at relatively small draw ratios of $\lambda \sim 3$ (refs 19–21).

At first glance, the periodic cubic array of cylindrical channels may seem similar to the parallel-cylinder model presented above. However, not only is the topology different, but also at a given water fraction, the diameter of the cylinders is smaller by more than a factor of two (that is, by more than 100%) because the orthogonal cylinders fill space between the cylinders along the first dimension. In the simulations of Fig. 6, the channel diameter was 1.3 nm, consistent with values given by Kreuer and co-workers[3,14]. In this context, it is interesting to note that if Gierke's model[1] is taken to contain a relatively complete three-dimensional periodic network of 1.2–1.6-nm-diameter channels[40], no water is left to form spherical clusters.

In the model of Kim et al., the hydrated clusters are worm-like micelles that merge freely and thus form a network[15]. Figure 6e and f show two views of such a model perpendicular and along the axes, respectively, of the 1.5-nm-diameter worms. Owing to the absence of correlations between the micelles, no peak is observed in the corresponding scattering curves, Fig. 6g. If significant correlations were included, the frequent merging and free undulation of the micelles would be suppressed, and the model would start to resemble our cylindrical water-channel model.

All models discussed so far attribute the ionomer peak to correlations between hydrated clusters, that is, to the structure factor. In other models of Nafion scattering, the ionomer peak was attributed to the first secondary maximum of the form factor of clusters with a core-shell[32] or parallelepiped[11] structure. Closer inspection shows that the models are inconsistent with the experimental data for Nafion. Most importantly, owing to the main maximum of the form factor at $q = 0$, a strong small-angle upturn is produced by the clusters themselves, not by crystallites, in the form-factor models. This is inconsistent with the limited small-angle upturn observed for non-crystalline Nafion[15,30,31]. It should also be noted that the models require all clusters have the same size and shape (within ±30%), otherwise the secondary maximum identified with the ionomer peak would be strongly broadened. At the same time, the clusters must be allowed to overlap without distortion, otherwise a peak from interparticle correlations would occur. Furthermore, the models with their compact clusters fail to explain the oriented-sample SAXS. Lastly, the models do not provide a path for the observed efficient water diffusion. In summary, these form-factor models are not tenable.

PARALLEL WATER-CHANNEL MODEL AND PROPERTIES OF NAFION

The new, simple water-channel model finally provides a unified view of the structure of Nafion. It explains not only the scattering data of unoriented samples, but also of oriented films or fibres with their exclusively meridional intensity for both the ionomer peak and the small-angle upturn. The stiffness of the helical backbone segments[35,36], which has been confirmed by NMR[33,34], can stabilize the long cylindrical structures. A rationale for the putative regular alternation between clusters and channels in Gierke's model was never given, and we have now shown that the previously elusive channels by themselves fully account for the ionomer peak, without spherical clusters.

The water-channel model naturally accounts for many of the outstanding properties of Nafion, in particular its high proton conductivity and water permeability[27]. In fact, it seems that the water-channel model is the only one that can easily explain why diffusion of water in Nafion at $\phi_{H_2O} = 20\%$ is only one order of magnitude slower than in bulk water[41,42]. The channel diameter in

our model is relatively large, $\sim$2.4 nm, whereas all previous models of Nafion at $\phi_{H_2O} = 20$ vol% contain constrictions of <1.2 nm thickness—the narrow channels in spheroidal-cluster[1,4–7,30] or network models[3,8,14,15], and the thin water layers in sheet-like models[10–13]. These constrictions have a radius of only $\sim$0.6 nm, which equals the thickness of two water molecules and is not much larger than the thickness of a typical bound-water layer[43,44]. They would present significant obstacles to water diffusion, given that diffusion in surface-water layers is orders of magnitude slower than in bulk water[27].

The water self-diffusion coefficient of Nafion is almost an order of magnitude larger than in other sulphonated polymers or ion-exchange resins at a given water volume fraction of $\sim$0.2, which is a previously unexplained observation[45]. The relatively wide, parallel channels that we have identified in Nafion favour a large hydrodynamic component of water/methanol transport (large permeation diffusion coefficient and electro-osmotic drag[46]) as explained in ref. 3, whereas the more random, smaller-diameter structures in other polymers with more flexible backbones and less flexible side chains are less favourable.

The observation that water seems to percolate the polymer at as little as 4 vol%, rather than the 16 vol% in random close-packed structures[17,47,48], can be explained in our model with its non-random water channels that connect different volume elements effectively. Currently, the model does not specify on which length scales beyond 20 nm the channels bend and merge. These important issues require further investigations.

Although about half of the water in Nafion seems to freeze at $\sim$−20 °C (refs 48,49), conductivity and diffusion in Nafion persist below −20 °C down to −50 °C (refs 41,42,48), with a moderately larger activation energy than at higher temperatures. In our model, we can easily explain this in terms of freezing of water in the wider channels but continuing diffusion in the smaller-diameter pores, in which water does not freeze down to −70 °C. In the spheroidal clusters of Gierke's model, which are larger, all water except for the relatively immobile bound surface layer would freeze, which should dramatically slow down diffusion.

Eventually, the water-channel structure of Nafion established here should guide not only molecular dynamics simulations of proton transport in this ionomer[8,29,50], but also the design of ionic polymers, nanocomposites or supramolecular assemblies for PEMs with desirable properties such as lower cost or higher operation temperatures compared with Nafion.

METHODS

Simulations of SAXS curves were carried out by applying our recently introduced algorithm[23], which calculates $I(\mathbf{q}) = |FT(\rho(\mathbf{x}))|^2$ by numerical fast Fourier transformation of the three-dimensional scattering density $\rho(\mathbf{x})$ on a cubic lattice, or of a two-dimensional cross-section through long parallel particles on a square lattice. An exact transformation from discrete to continuous scattering density is followed by calculation of the magnitude-square of $FT(\rho)$ and by spherical averaging of $I(\mathbf{q})$ to yield $I(q)$ (ref. 23). Scattering density distributions $\rho(\mathbf{x})$ were generated to reproduce various models of Nafion, as described in the following for the parallel water-channel model and in the Supplementary Information for others. Generating $\rho(\mathbf{x})$ required significantly more central-processing-unit time (tens of seconds to 20 h) than the subsequent calculation of $I(q)$. The simulations, using the MatLab (MathWorks) programming language, were carried out on a 1.25 GHz PowerPC G4 desktop computer.

For the water-channel simulations, nearly 15,000 approximately round clusters were generated on a square lattice of 1,024² points, corresponding to (530 nm)², representing cross-sections through the long, parallel water cylinders[23]. The cylinder (disc) diameter varied between 1.8 and 3.5 nm, with the distribution shown in Supplementary Information, Fig. S1a. The cylinders (discs) were placed randomly except in the excluded region around the centre of each previously placed cylinder, whose diameter is the sum of the cylinder diameters $+0.9\,\text{nm}$, where the last term provides a polymer shell of around $\sim 0.7\,\text{nm}$ thickness, as seen from the distribution of closest cylinder surface–surface distances plotted in Supplementary Information, Fig. S1b.

Crystallites with thicknesses of between 2 and 5 nm and aspect ratios randomly ranging between 1 and 1.8 were superimposed on the distribution of cylindrical clusters. The centre of every crystallite was placed in a larger-than-average water-free region. The scattering density of the crystallites was set to 1.89, compared with 1.77 for the non-crystalline polymer and 0.99 for the water channels[30]. Double smoothing of the $I(q)$ curves with a three-point filter[23] reduced the effective size along each dimension by 2^2, to 256×256 points or $(130\,\text{nm})^2$, with a 0.52 nm point-to-point resolution. The length of water channels and crystallites is thus $> 100\,\text{nm}$. Technical details of the simulations presented in Figs 3–6 are given in the Supplementary Information.

Received 5 March 2007; accepted 1 November 2007; published 9 December 2007.

References

1. Gierke, T. D., Munn, G. E. & Wilson, F. C. The morphology in Nafion perfluorinated membrane products, as determined by wide- and small-angle X-ray studies. *J. Polym. Sci. Polym. Phys. Edn* **19**, 1687–1704 (1981).
2. Mauritz, K. A. & Moore, R. B. State of understanding of Nafion. *Chem. Rev.* **104**, 4535–4586 (2004).
3. Kreuer, K. D., Paddison, S. J., Spohr, E. & Schuster, M. Transport in proton conductors for fuel-cell applications: Simulations, elementary reactions, and phenomenology. *Chem. Rev.* **104**, 4637–4678 (2004).
4. Dreyfus, B., Gebel, G., Aldebert, P., Pineri, M. & Escoubes, M. Distribution of the 'micelles' in hydrated perfluorinated ionomer membranes from SANS experiments. *J. Phys. France* **51**, 1341–1354 (1990).
5. Gebel, G. & Lambard, J. Small-angle scattering study of water-swollen perfluorinated ionomer membranes. *Macromolecules* **30**, 7914–7920 (1997).
6. Kumar, S. & Pineri, M. Interpretation of small-angle X-ray and neutron scattering data for perfluorosulfonated ionomer membranes. *J. Polym. Sci. B* **24**, 1767–1782 (1986).
7. Ioselevich, A. S., Kornyshev, A. A. & Steinke, J. H. G. Fine morphology of proton-conducting ionomers. *J. Phys. Chem. B* **108**, 11953–11963 (2004).
8. Jang, S. S., Molinero, V., Çagin, T. & Goddard III, W. A. Effect of monomeric sequence on nanostructure and water dynamics in Nafion 117. *Solid State Ion.* **175**, 805–808 (2004).
9. Yeager, H. L. in *Perfluorinate Ionomer Membranes, ACS Symp. Ser.*, Vol. 180 (eds Yeager, H. L. & Eisenberg, A.) 41–64 (American Chemical Society, Washington, 1982).
10. Starkweather, H. W. Jr. Crystallinity in perfluorosulfonic acid ionomers and related polymers. *Macromolecules* **15**, 320–323 (1982).
11. Haubold, H.-G., Vad, T., Jungbluth, H. & Hiller, P. Nanostructure of NAFION: A SAXS study. *Electrochim. Acta* **46**, 1559–1563 (2001).
12. Litt, M. H. A reevaluation of Nafion morphology. *Polym. Preprint* **38**, 80–81 (1997).
13. Krivandin, A. V., Solov'eva, A. B., Glagolev, N. N., Shatalova, O. V. & Kotova, S. L. Structure alterations of perfluorinated sulfocationic membranes under the action of ethylene glycol (SAXS and WAXS studies). *Polymer* **44**, 5789–5796 (2003).
14. Kreuer, K. D. On the development of proton conducting polymer membranes for hydrogen and methanol fuel cells. *J. Membr. Sci.* **185**, 29–39 (2001).
15. Kim, M.-H., Glinka, C. J., Grot, S. A. & Grot, W. G. SANS study of the effects of water vapor sorption on the nanoscale structure of perfluorinated sulfonic acid (NAFION) Membranes. *Macromolecules* **39**, 4775–4787 (2006).
16. Rollet, A.-L., Diat, O. & Gebel, G. A new insight into Nafion structure. *J. Phys. Chem. B* **21**, 3033–3036 (2002).
17. Rubatat, L., Rollet, A.-L., Gebel, G. & Diat, O. Evidence of elongated polymeric aggregates in Nafion. *Macromolecules* **35**, 4050–4055 (2002).
18. Rubatat, L., Gebel, G. & Diat, O. Fibrillar structure of Nafion: Matching Fourier and real space studies of corresponding films and solutions. *Macromolecules* **37**, 7772–7783 (2004).
19. Londono, J. D., Davidson, R. V. & Mazur, S. SAXS study of elongated ionic clusters in poly-perfluorosulfonic acid membranes. *Abstr. Am. Chem. Soc. Polym. Mater. Sci. Eng.* **222**, 342 (2001).
20. van der Heijden, P. C., Rubatat, L. & Diat, O. Orientation of drawn Nafion at molecular and mesoscopic scales. *Macromolecules* **37**, 5327–5336 (2004).
21. Page, K. A., Landis, F. A., Phillips, A. K. & Moore, R. B. SAXS Analysis of the thermal relaxation of anisotropic morphologies in oriented Nafion membranes. *Macromolecules* **39**, 3939–3946 (2006).
22. Elliott, J. A., Hanna, S., Elliott, A. M. S. & Cooley, G. E. Interpretation of the small-angle X-ray scattering from swollen and oriented perfluorinated ionomer membranes. *Macromolecules* **33**, 4161–4171 (2000).
23. Schmidt-Rohr, K. Simulation of small-angle scattering (SAXS or SANS) curves by numerical Fourier transformation. *J. Appl. Cryst.* **40**, 16–25 (2007).
24. Cwirko, E. H. & Carbonell, R. G. Interpretation of transport coefficients in Nafion using a parallel pore model. *J. Membr. Sci.* **67**, 227–247 (1992).
25. Bontha, J. R. & Pintauro, P. N. Water orientation and ion solvation effects during multicomponent salt partitioning in a Nafion cation-exchange membrane. *Chem. Eng. Sci.* **49**, 3835–3851 (1994).
26. Koter, S. Transport of simple electrolyte solutions through ion-exchange membranes-the capillary model. *J. Membr. Sci.* **206**, 201–215 (2002).
27. Choi, P., Jalani, N. H. & Datta, R. Thermodynamics and proton transport in Nafion II. Proton diffusion mechanisms and conductivity. *J. Electrochem. Soc.* **152**, E123–E130 (2005).
28. Paddison, S. J. Proton conduction mechanisms at low degrees of hydration in sulfonic acid-based polymer electrolyte membranes. *Annu. Rev. Mater. Res.* **33**, 289–319 (2003).
29. Blake, N. P., Petersen, M. K., Voth, G. A. & Metiu, H. Structure of hydrated Na-Nafion polymer membranes. *J. Phys. Chem. B* **109**, 24244–24253 (2005).
30. Roche, E. J., Pineri, M., Duplessix, R. & Levelut, A. M. Small-angle scattering studies of Nafion membranes. *J. Polym. Sci. Polym. Phys.* **19**, 1–11 (1981).
31. Moore, R. B. & Martin, C. R. Chemical and morphological properties of solution-cast perfluorosulfonate ionomers. *Macromolecules* **21**, 1334–1339 (1988).
32. Fujimura, M., Hashimoto, T. & Kawai, H. Small-angle X-ray scattering study of perfluorinated ionomer membranes. 2. Models for ionic scattering maximum. *Macromolecules* **15**, 136–144 (1982).
33. Chen, Q. & Schmidt-Rohr, K. ^{19}F and ^{13}C NMR signal assignment and analysis in a perfluorinated ionomer (Nafion) by two-dimensional solid-state NMR. *Macromolecules* **37**, 5995–6003 (2004).
34. Chen, Q. & Schmidt-Rohr, K. Backbone dynamics of the Nafion Ionomer Studied by ^{19}F–^{13}C solid-state NMR. *Macromol. Chem. Phys.* **208**, 2189–2203 (2007).
35. Chu, B., Wu, C. & Buck, W. Light-scattering characterization of poly(tetrafluoroethylene).2. PTFE in perfluorotetracosane-molecular-weight distribution and solution properties. *Macromolecules* **22**, 831–837 (1989).
36. Rosi-Schwartz, B. & Mitchell, G. R. Extracting force fields for disordered polymeric materials from neutron scattering data. *Polymer* **37**, 1857–1870 (1996).
37. Roche, E. J., Pineri, M., Duplessix, R. & Levelut, A. M. Phase separation in perfluorosolfonate ionomer membranes. *J. Polym. Sci. Polym. Phys. Edn* **20**, 107–116 (1982).
38. Guinier, A. *X-Ray Diffraction in Crystals, Imperfect Crystals, and Amorphous Bodies* (Dover, New York, 1994).
39. Klein, R. & D'Aguanno, B. in *Light Scattering, Principles and Development* (ed. Brown, W.) (Clarendon, Oxford, 1996).
40. Capeci, S. W., Pintauro, P. N. & Bennion, D. N. The molecular-level interpretation of salt uptake and anion transport in Nafion membranes. *J. Electrochem. Soc.* **136**, 2876–2882 (1989).
41. Cappadonia, M., Erning, J. W. & Stimming, U. Proton conduction of Nafion-117 membrane between 140 K and room temperature. *J. Electroanal. Chem.* **376**, 189–193 (1994).
42. Saito, M., Hayamizu, K. & Okada, T. Temperature dependence of ion and water transport in perfluorinated ionomer membranes for fuel cells. *J. Phys. Chem. B* **109**, 3112–3119 (2005).
43. Paddison, S. J. & Paul, R. The nature of proton transport in fully hydrated Nafion. *Phys. Chem. Chem. Phys.* **4**, 1158–1163 (2002).
44. Cui, S. T. Molecular self-diffusion in nanoscale cylindrical pores and classical Fick's law predictions. *J. Chem. Phys.* **123**, 054706 (2005).
45. Freger, V. *et al.* Diffusion of water and ethanol in ion-exchange membranes: Limits of the geometric approach. *J. Membr. Sci.* **160**, 213–224 (1999).
46. Ise, M., Kreuer, K. D. & Maier, J. Electroosmotic drag in polymer electrolyte membranes: an electrophoretic NMR study. *Solid State Ion.* **125**, 213–223 (1999).
47. Edmondson, C. A. & Fontanella, J. J. Free volume and percolation in S-SEBS and fluorocarbon proton conducting membranes. *Solid State Ion.* **152–153**, 355–361 (2002).
48. Thompson, E. L., Capehart, T. W., Fuller, T. J. & Jorne, J. Investigation of low-temperature proton transport in Nafion using direct current conductivity and DSC. *J. Electrochem. Soc.* **153**, A2351–A2362 (2006).
49. Tasaka, M., Suzuki, S., Ogawa, Y. & Kamaya, M. Freezing and nonfreezing water in charged membranes. *J. Membr. Sci.* **38**, 175–183 (1988).
50. Cui, S. *et al.* A molecular dynamics study of a Nafion polyelectrolyte membrane and the aqueous phase structure for proton transport. *J. Phys. Chem. B* **111**, 2208–2218 (2007).

Acknowledgements

Work at the Ames Laboratory was supported by the Department of Energy-Basic Energy Sciences (Materials Chemistry and Biomolecular Materials Program) under Contract No. DE-AC02-07CH11358.
Correspondence and requests for materials should be addressed to K.S.-R.
Supplementary Information accompanies this paper on www.nature.com/naturematerials.

Author contributions

K.S.R. ran the simulations and wrote the paper. Q.C. identified various Nafion models in the literature and reprocessed literature data for Fig. 1.

A class of non-precious metal composite catalysts for fuel cells

Rajesh Bashyam[1] & Piotr Zelenay[1]

Fuel cells, as devices for direct conversion of the chemical energy of a fuel into electricity by electrochemical reactions, are among the key enabling technologies for the transition to a hydrogen-based economy[1-3]. Of several different types of fuel cells under development today, polymer electrolyte fuel cells (PEFCs) have been recognized as a potential future power source for zero-emission vehicles[4,5]. However, to become commercially viable, PEFCs have to overcome the barrier of high catalyst cost caused by the exclusive use of platinum and platinum-based catalysts[6-8] in the fuel-cell electrodes. Here we demonstrate a new class of low-cost (non-precious metal)/(heteroatomic polymer) nanocomposite catalysts for the PEFC cathode, capable of combining high oxygen-reduction activity with good performance durability. Without any optimization, the cobalt-polypyrrole composite catalyst enables power densities of about $0.15\,W\,cm^{-2}$ in H_2-O_2 fuel cells and displays no signs of performance degradation for more than 100 hours. The results of this study show that hetero-atomic polymers can be used not only to stabilize the non-precious metal in the acidic environment of the PEFC cathode but also to generate active sites for oxygen reduction reaction.

A schematic representation of the principle of fuel-cell operation is shown in Fig. 1. PEFCs operate with a polymer electrolyte membrane that separates the fuel (hydrogen) from the oxidant (air or oxygen). Precious-metal catalysts, predominantly platinum (Pt) supported on carbon, are used for both the oxidation of the fuel and reduction of the oxygen in a typical temperature range of 80–100°C. Apart from the issue of the high cost of catalyst and other fuel-cell system components (polymer electrolyte membrane, bipolar plates, the rest of the power system, and so on), PEFCs suffer from insufficient performance durability, arising mainly from cathode catalyst oxidation, catalyst migration, loss of electrode active surface area, and corrosion of the carbon support. In a direct methanol fuel cell (DMFC), the Pt cathode also endures a performance loss resulting from the lack of tolerance to methanol diffusing through the membrane from the anode side of the cell[9,10]. Thus, whether using hydrogen or methanol as a fuel, PEFCs are in need of efficient, durable and, most importantly, inexpensive catalysts, as alternatives to Pt and Pt-based materials[11,12]. Although ideally the Pt catalyst should be replaced at both fuel-cell electrodes, the substitution of the cathode catalyst with a non-precious material is likely to result in significantly greater reduction of Pt needed for PEFCs. This is because the slow oxygen reduction reaction (ORR) at the cathode requires much more Pt catalyst than the very fast hydrogen oxidation at the anode[13].

Two approaches are at present gaining momentum to replace Pt, which is scarce (only 37 p.p.b. in the Earth's crust) and expensive ($\sim$US$45\,g^{-1}$, the highest Pt price in 25 years). One approach uses non-Pt catalysts that nonetheless contain precious metals with limited abundance and/or limited world distribution. Such catalysts are typically based on palladium (Pd; ref. 14) or ruthenium (Ru; refs 15, 16). Although Pt is thus avoided, the result is the replacement of one precious metal with another that is on the whole less active than Pt. An alternative approach is to replace Pt with abundant, non-precious materials that are not susceptible to price inflation under high-demand circumstances.

The class of non-precious cathode catalysts that has attracted the most attention over the years is pyrolized metal porphyrins, with cobalt and iron porphyrins viewed as the most promising precursors. The pyrolysis of various such porphyrins, carried out under controlled conditions of temperature, pyrolysis time and metal concentration in the porphyrins, has been shown to produce species of an MeN_x type (Me = Co, Fe), which have been suggested to be the active ORR sites[17-19]. The hypothesis involving Me-N active sites has been supported by the studies correlating fuel-cell performance with nitrogen concentration in the porphyrin catalyst[20]. Most probably owing to the high content of nitrogen, non-pyrolyzed metal porphyrins have been found to exhibit good initial oxygen reduction performance but dismal stability[21].

The three-decade-long search for non-precious-metal catalysts for the PEFC cathode has so far revealed very few materials with

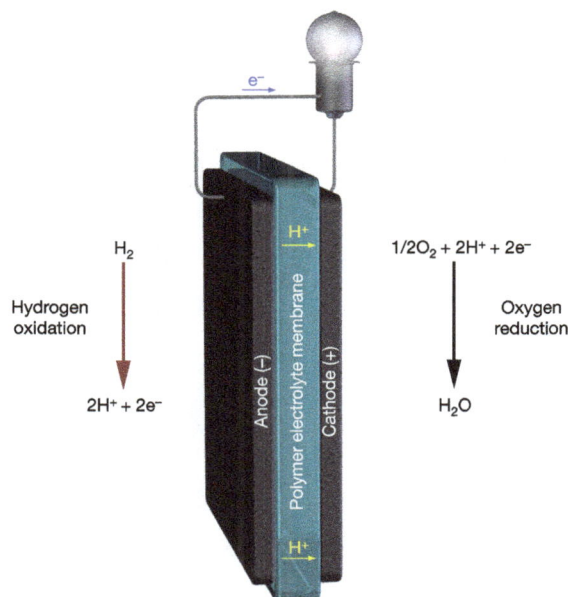

Figure 1 | Polymer electrolyte H_2-O_2 fuel cell.

[1]Los Alamos National Laboratory, Materials Physics and Applications, Los Alamos, New Mexico 87545, USA.

promising ORR activity in either fundamental[22–24] or in fuel-cell studies[25,26], and virtually none with sufficient performance stability. Even the most thoroughly studied pyrolyzed metal-porphyrins continue to lack durability under practical fuel-cell operating conditions, which re-emphasizes the need for more research on alternative non-precious cathode electrocatalysts for polymer electrolyte fuel cells[27].

Here we report major progress towards the development of a new class of inexpensive (non-precious metal)/(heteroatomic polymer) electrocatalysts for oxygen reduction in fuel cells. In particular, we demonstrate a cobalt-polypyrrole-carbon (Co-PPY-C) composite synthesized via a simple chemical method, without resorting to pyrolysis. The obtained catalyst shows good activity with a Co loading of 6.0×10^{-2} mg cm^{-2} and remarkable stability in both H_2-O_2 and H_2-air PEFCs.

Recently, heterocyclic conjugated polymers such as polyaniline, polypyrrole and polythiophene have been the subject of much research owing to their applications in electronics, electrochemistry and electrocatalysis, as well as in biosensors and actuators[28,29]. These applications primarily use the electronic conductivity of conjugated polymers. In this work, however, polypyrrole is solely used as a matrix for entrapping cobalt and generating active ORR sites. The

main reason for introducing polypyrrole into the structure of a composite ORR catalyst was to mimic the atomic configuration occurring in cobalt porphyrins, wherein the Co atom is linked to pyrrole units, thus allowing for the formation of Co-N sites. Consequently, the underlying assumption of this work was to purposely create active Co-N sites without destroying the initial polymer structure. This approach represents a departure from the pyrolysis process, which has been universally used until now to induce ORR activity in the synthesis of non-precious-metal ORR catalysts.

A schematic representation of the polymer structure and presumed configuration of the composite are shown in Fig. 2a and b, respectively. While Fig. 2a depicts the structure of polypyrrole used for entrapping cobalt atoms, the purpose of Fig. 2b is to highlight the importance of a linkage between polypyrrole and Co atoms, established via nitrogen atoms in the pyrrole units. To ensure electronic conductivity in the catalyst phase, the composite shown in Fig. 2b was synthesized on a support made of Vulcan XC 72 carbon (see 'Preparation of polypyrrole-loaded carbon and synthesis of the cobalt-polypyrrole-carbon composite' in the Supplementary Information for details).

The composite was characterized by surface area measurements (Brunauer–Emmett–Teller isotherm method) and thermogravimetric analysis (TGA). The Co loading in the composite was

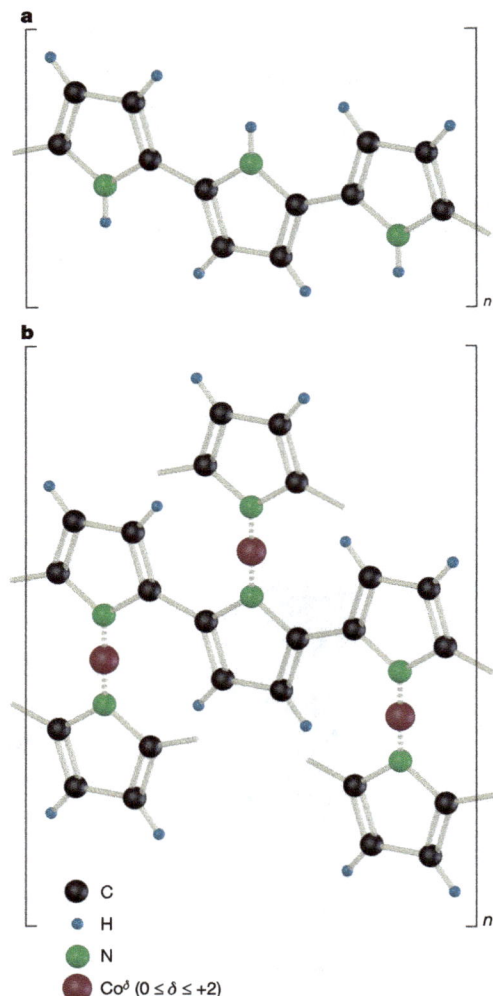

Figure 2 | Schematic representation of the Co-polypyrrole composite catalyst. a, Polypyrrole structure. **b,** Presumed configuration of the Co-PPY catalyst after the entrapment of the cobalt precursor, $Co(NO_3)_2 \cdot 6H_2O$ into polypyrrole and reduction with $NaBH_4$.

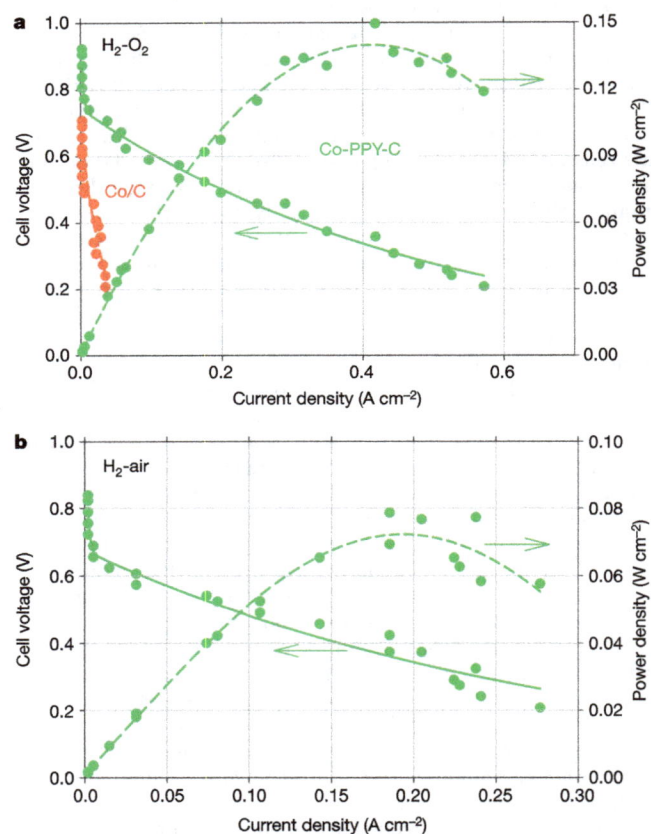

Figure 3 | Fuel-cell performance plots. a, Polarization and power density plots for H_2-O_2 fuel cells with Co-PPY-C composite cathode (green lines) and polarization plot for the cell with Co/C cathode (red line). The Co loading is 6.0×10^{-2} mg cm^{-2}; the cell temperature is 80 °C. **b,** Polarization and power density plots for H_2-air fuel cells with Co-PPY-C composite cathode. The Co loading is 6.0×10^{-2} mg cm^{-2}; the cell temperature is 80 °C. Flow rates of hydrogen and oxygen/air at 5 ml s^{-1} and 9 ml s^{-1} (as referred to the standard conditions), respectively. Anode and cathode gases humidified at 90 °C and 80 °C, respectively. The backpressure of gases is 2.0 atm on both sides of the cell.

determined from TGA. At $124\,m^2\,g^{-1}$, the Brunauer–Emmett–Teller surface area of the Co-PPY-C composite was nearly 50% lower than the surface area of the Vulcan XC 72 support ($240\,m^2\,g^{-1}$), indicating that Vulcan XC 72 properties are modified by the polypyrrole. The complete decomposition of Vulcan XC 72 in the presence of oxygen occurs at 460 °C, whereas that of Co-PPY-C composite occurs at 300 °C. The TGA data suggest that oxidative decomposition of the composite is activated in the presence of cobalt. High-temperature decomposition notwithstanding, the composite catalyst remains stable in the temperature range of PEFC operation: between 80 °C and 100 °C. Based on the TGA, the polypyrrole content in the composite has been found to be nearly 16% by weight.

Membrane electrode assemblies for fuel-cell conditioning and testing of the new catalyst were fabricated with a Co-PPY-C cathode and Pt-Ru black anode (see experimental details in Supplementary Information). The membrane electrode assemblies were subject to conditioning at 0.40 V and at 80 °C under H_2-O_2 fuel-cell operating conditions. The cell performance gradually increased during the conditioning step, reaching a constant level after approximately 30 h (see Supplementary Fig. 1). X-ray absorption near-edge structure (XANES) spectroscopy reveals that during conditioning most of the metallic Co transforms into Co(II) (S. Conradson, R.B. & P.Z., manuscript in preparation). X-ray absorption fine structure (XAFS) data indicate that stability of the Co(II) sites in the composite is achieved thanks to the formation of a coordination bond between Co(II) and N or O (the two elements are not easily distinguishable in XANES/XAFS).

An H_2-O_2 fuel-cell polarization plot recorded after the conditioning step with the cathode made of a newly synthesized Co-PPY-C composite is shown in Fig. 3a (see Supplementary Figs 2 and 3 for catalyst performance reproducibility and standard deviations, respectively). The catalyst generates $\sim$0.2 $A\,cm^{-2}$ at 0.50 V and a maximum power density of $\sim$0.14 $W\,cm^{-2}$. Although this current density is less than that recently reported in ref. 19 for an iron-based catalyst (0.3 $A\,cm^{-2}$ at 0.50 V), it was obtained at a significantly lower absolute pressure of oxygen (2.8 atm versus the 5.0 atm used in ref. 19).

The H_2-O_2 fuel-cell performance of the composite catalyst compares favourably with other non-Pt cathode catalysts, even precious-metal-based materials, such as the recently proposed Pd-Co-Au alloy ($\sim$0.18 $A\,cm^{-2}$ at 0.50 V, 60 °C, 1.5 atm oxygen back pressure) and Pd-Ti alloy (0.1 $A\,cm^{-2}$ at 0.50 V, 60 °C and 9 $mg\,cm^{-2}$)[14]. Furthermore, the loading-independent oxygen reduction turnover rate of 0.83 s^{-1} at 0.50 V and 60 °C (see 'Turnover rates' in the Supplementary Information for details), is 3.5 times higher on the Co-PPY-C composite than on the Pd-Co-Au alloy (0.24 s^{-1}) and 8.3 times higher than on the Pd-Ti alloy (0.10 s^{-1}). The comparison becomes even more favourable for the composite when it is made on the basis of combining the ORR activity with performance durability (see below).

For a reference, a catalyst made by just loading Co onto Vulcan XC 72 was synthesized under the same conditions as the composite catalyst. It is evident from Fig. 3a that activity of the Co/C catalyst is far lower than that of the Co-PPY-C composite. The open cell voltage (that is, the voltage at zero cell current, OCV) of the fuel cell with the Co-PPY-C cathode is 0.22 V higher than that of the fuel cell with the Co/C cathode. In the voltage range between 0.50 V and 0.20 V, current density generated by the Co-PPY-C cell is more than one order of magnitude higher than that delivered by the Co/C cell (Fig. 3a). In spite of having a $\sim$50% catalytic surface area advantage over the Co-PPY-C composite, the peak power density of the Co/C cell is only 0.009 $W\,cm^{-2}$, nearly 16 times lower than that of the Co-PPY-C cell.

The above data provide indirect evidence that electrocatalytic activity of the new catalyst results from the entrapment of the Co sites in the polypyrrole matrix and strong Co–polypyrrole interactions, presumably resulting in the formation of CoN_x active sites. This interaction between Co and polypyrrole is analogous to the interaction occurring between heterocyclic compounds (in particular heterocyclic polymers) placed in contact with metals. Such interaction is known to lead to a charge transfer to the metal, and, consequently, to the formation of a strong heterocycle–metal bonding[30].

An H_2-air fuel-cell polarization plot recorded with a Co-PPY-C composite cathode is shown in Fig. 3b. In spite of the reduced partial pressure of O_2 in air, the cell performance is good, as reflected by the current density of 0.10 $A\,cm^{-2}$ at 0.50 V. The maximum peak power density of the H_2-air cell is $\sim$0.07 $W\,cm^{-2}$, approximately half of that generated with the H_2-O_2 cell (see above).

The most striking property of the new catalyst—or, more likely, of a new class of catalysts represented by the Co-PPY-C catalyst—is demonstrated in Fig. 4. Uncharacteristically for a non-precious catalyst, cell performance with the Co-PPY-C composite cathode is very stable, showing no appreciable drop over 100 h of operation. To the best of our knowledge, this represents the first time that a non-precious cathode catalyst, at a low loading, has shown both a promising activity and stability comparable to that of Pt-based ORR catalysts (compare with the 100-h life test of an E-TEK 20 wt% Pt/C catalyst; Supplementary Fig. 4). The good performance stability of the composite can be seen not only at the relatively low voltage of 0.40 V but also at more practical higher voltages, up to 0.70 V (see Supplementary Figs 5 and 6).

Fundamental electrochemical studies are underway to understand the mechanism of O_2 reduction on the composite materials, such as the Co-PPY-C composite catalyst introduced here. The main goal of future research will be to reduce the overpotential of oxygen reduction at high cathode operating potentials (in the low-current-density region) by at least 0.15–0.20 V and to bring it closer to that measured with pure-Pt and Pt-based cathode catalysts.

To conclude, here we have demonstrated a new non-precious composite catalyst from an entirely new class of (non-precious metal)/(heterocyclic polymer) composite materials, synthesized via a pyrolysis-free process. As shown with the Co-PPY-C composite, such catalysts promise to deliver high ORR activity without any noticeable loss in performance over long fuel-cell operating times. This therefore opens up the possibility of making a variety of other non-precious composite materials from this class of composites for use as catalysts for the PEFC cathode.

Received 7 March; accepted 28 July 2006.

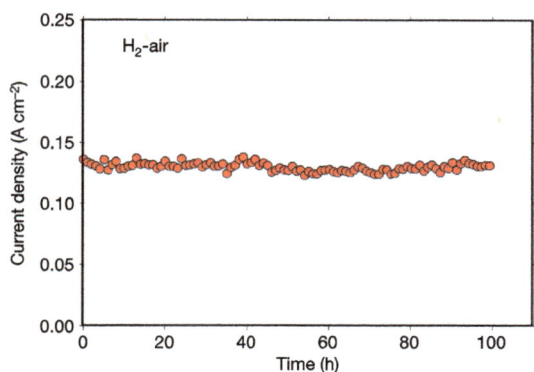

Figure 4 | Durability test of the composite catalyst. Long-term performance of an H_2-air fuel cell with Co-PPY-C composite cathode at 0.40 V. The Co loading is $6.0 \times 10^{-2}\,mg\,cm^{-2}$; the cell temperature is 80 °C. Flow rates of hydrogen and air at 5 $ml\,s^{-1}$ and 9 $ml\,s^{-1}$ (as referred to the standard conditions), respectively. Anode and cathode gases humidified at 90 °C and 80 °C, respectively. The backpressure of gases is 2.0 atm on both sides of the cell.

1. Jacobson, M. Z., Colella, W. G. & Golden, D. M. Atmospheric science: Cleaning the air and improving health with hydrogen fuel-cell vehicles. *Science* **308**, 1901–1905 (2005).
2. Dusastre, V. Materials for clean energy. *Nature* **414**, 331 (2001).
3. Schultz, M. G., Diehl, T., Brasseur, G. P. & Zittel, W. Air pollution and climate-forcing impacts of a global hydrogen economy. *Science* **302**, 624–627 (2003).
4. Steele, B. C. H. & Heinzel, A. Materials for fuel-cell technologies. *Nature* **414**, 345–352 (2001).
5. Brumfiel, G. Hydrogen cars fuel debate on basic research. *Nature* **422**, 104 (2003).
6. Berger, D. J. Fuel cells and precious-metal catalysts. *Science* **286**, 49 (1999).
7. Semelsberger, T. A. & Borup, R. L. Fuel effects on start-up energy and efficiency for automotive PEM fuel cell systems. *Int. J. Hydrogen Energy* **30**, 425–435 (2005).
8. Xie, J., Wood, D. L., More, K. L., Atanassov, P. & Borup, R. L. Microstructural changes of membrane electrode assemblies during PEFC durability testing at high humidity conditions. *J. Electrochem. Soc.* **152**, A1011–A1020 (2005).
9. Zhang, X., Filho, L. P., Torras, C. & Garcia-Valls, R. Experimental and computational study of proton and methanol permeabilities through composite membranes. *J. Power Sources* **145**, 223–230 (2005).
10. Thomas, S. C., Ren, X. M., Gottesfeld, S. & Zelenay, P. Direct methanol fuel cells: Progress in cell performance and cathode research. *Electrochim. Acta* **47**, 3741–3748 (2002).
11. Zhang, J. et al. Platinum monolayer on nonnoble metal-noble metal core-shell nanoparticle electrocatalysts for O_2 reduction. *J. Phys. Chem. B* **109**, 22701–22704 (2005).
12. Zhang, J. et al. Platinum monolayer electrocatalysts for O_2 reduction: Pt monolayer on Pd(111) and on carbon-supported Pd nanoparticles. *J. Phys. Chem. B* **108**, 10955–10964 (2004).
13. Appleby, A. J. Electrocatalysis of aqueous dioxygen reduction. *J. Electroanal. Chem.* **357**, 117–179 (1993).
14. Fernandez, J. L., Raghuveer, V., Manthiram, A. & Bard, A. J. Pd-Ti and Pd-Co-Au electrocatalysts as a replacement for platinum for oxygen reduction in proton exchange membrane fuel cells. *J. Am. Chem. Soc.* **127**, 13100–13101 (2005).
15. Alonso-Vante, N. A. & Tributsch, H. Energy conversion catalysis using semiconducting transition-metal cluster compounds. *Nature* **323**, 431–432 (1986).
16. Alonso-Vante, N., Malakhov, I. V., Nikitenko, S. G., Savinova, E. R. & Kochubey, D. I. The structure analysis of the active centers of Ru-containing electrocatalysts for the oxygen reduction. An in situ EXAFS study. *Electrochim. Acta* **47**, 3807–3814 (2002).
17. Wang, B. Recent development of non-platinum catalysts for oxygen reduction reaction. *J. Power Sources* **152**, 1–15 (2005).
18. Villers, D., Jacques-Bedard, X. & Dodelet, J. P. Fe-based catalysts for oxygen reduction in PEM fuel cells. Pretreatment of the carbon support. *J. Electrochem. Soc.* **151**, A1507–A1515 (2004).
19. Medard, C., Lefevre, M., Dodelet, J. P., Jaouen, F. & Lindbergh, G. Oxygen reduction by Fe-based catalysts in PEM fuel cell conditions: Activity and selectivity of the catalysts obtained with two Fe precursors and various carbon supports. *Electrochim. Acta* **51**, 3202–3213 (2006).
20. Faubert, G. et al. Heat-treated iron and cobalt tetraphenylporphyrins adsorbed on carbon black: physical characterization and catalytic properties of these materials for the reduction of oxygen in polymer electrolyte fuel cells. *Electrochim. Acta* **41**, 1689–1701 (1996).
21. Beck, F. Redox mechanism of chelate-catalyzed oxygen cathode. *J. Appl. Electrochem.* **7**, 239–245 (1977).
22. Jasinski, R. New fuel cell cathode catalyst. *Nature* **201**, 1212 (1964).
23. Atanasoski, R. *Novel Approach to Non-Precious Metal Catalysts*. Report VIIC9 (US DOE, 2005); ⟨http://www.hydrogen.energy.gov/pdfs/progress05/vii_c_9_atanasoski.pdf⟩.
24. Yuasa, M. et al. Modifying carbon particles with polypyrrole for adsorption of cobalt ions as electrocatalytic site for oxygen reduction. *Chem. Mater.* **17**, 4278–4281 (2005).
25. Lalande, G. et al. Physical, chemical and electrochemical characterization of heat-treated tetracarboxylic cobalt phthalocyanine adsorbed on carbon black as electrocatalyst for oxygen reduction in polymer electrolyte fuel cells. *Electrochim. Acta* **40**, 2635–2646 (1995).
26. Lefevre, M. & Dodelet, J. P. Fe-based catalysts for the reduction of oxygen in polymer electrolyte membrane fuel cell conditions: Determination of the amount of peroxide released during electroreduction and its influence on the stability of the catalysts. *Electrochim. Acta* **48**, 2749–2760 (2003).
27. Zelenay, P. *Non-Precious Metal Catalysts*. Report VIIC7 (US DOE, 2005); ⟨http://www.hydrogen.energy.gov/pdfs/progress05/vii_c_7_zelenay.pdf⟩.
28. Lu, W. et al. Use of ionic liquids for pi-conjugated polymer electrochemical devices. *Science* **297**, 983–987 (2002).
29. Hepel, M., Chen, Y. M. & Stephenson, R. Effect of the composition of polypyrrole substrate on the electrode position of copper and nickel. *J. Electrochem. Soc.* **143**, 498–505 (1996).
30. Skotheim, T. A. et al. Highly ordered thin films of polyheterocycles: a synchrotron radiation study of polypyrrole and polythiophene Langmuir-Blodgett films. *Synth. Met.* **28**, C229–C236 (1989).

Supplementary Information is linked to the online version of the paper at www.nature.com/nature.

Acknowledgements We thank the US Department of Energy (Hydrogen, Fuel Cells and Infrastructure Technologies Program) and the Los Alamos National Laboratory for funding this work. We are grateful to E. Brosha, A. Burrell, J. Chlistunoff, F. Garzon, C. Johnston, R. Mukundan and J. Ramsey from the Los Alamos National Laboratory for discussions and technical help.

Author Information Reprints and permissions information is available at www.nature.com/reprints. The authors declare no competing financial interests. Correspondence and requests for materials should be addressed to P.Z. (zelenay@lanl.gov).

Disruption of extended defects in solid oxide fuel cell anodes for methane oxidation

Juan Carlos Ruiz-Morales[1]†, Jesús Canales-Vázquez[1]†, Cristian Savaniu[1], David Marrero-López[2], Wuzong Zhou[1] & John T. S. Irvine[1]

Point defects largely govern the electrochemical properties of oxides: at low defect concentrations, conductivity increases with concentration; however, at higher concentrations, defect–defect interactions start to dominate[1,2]. Thus, in searching for electrochemically active materials for fuel cell anodes, high defect concentration is generally avoided. Here we describe an oxide anode formed from lanthanum-substituted strontium titanate (La-SrTiO₃) in which we control the oxygen stoichiometry in order to break down the extended defect intergrowth regions and create phases with considerable disordered oxygen defects. We substitute Ti in these phases with Ga and Mn to induce redox activity and allow more flexible coordination. The material demonstrates impressive fuel cell performance using wet hydrogen at 950 °C. It is also important for fuel cell technology to achieve efficient electrode operation with different hydrocarbon fuels[3,4], although such fuels are more demanding than pure hydrogen. The best anode materials to date—Ni-YSZ (yttria-stabilized zirconia) cermets[5]—suffer some disadvantages related to low tolerance to sulphur[6], carbon build-up when using hydrocarbon fuels[7] (though device modifications and lower temperature operation can avoid this[8,9]) and volume instability on redox cycling. Our anode material is very active for methane oxidation at high temperatures, with open circuit voltages in excess of 1.2 V. The materials design concept that we use here could lead to devices that enable more-efficient energy extraction from fossil fuels and carbon-neutral fuels.

Over the past few years there has been a growing interest in perovskite-based materials in the search for alternative materials to Ni-YSZ cermets as fuel electrodes in solid oxide fuel cells (SOFCs)[3,10]. Such perovskites are normally oxygen stoichiometric or substoichiometric: here we seek to explore perovskites that are nominally oxygen overstoichiometric as SOFC fuel electrodes. Initially we focus on phases with extended oxygen-rich defects, and then attempt to destabilize these extended defects by reducing the degree of oxygen excess. Titanates with nominal oxygen overstoichiometry are especially interesting owing to their very high electronic conductivity, stability in reducing conditions and resistance to sulphur poisoning[11–14]. Oxides belonging to the La₁₋ₓSrₓTiO₃₊δ system have been previously studied as anode materials showing moderate performance compared to those of the state of the art materials, but better catalytic properties and ionic conductivity are desirable[11,12].

The performance of titanate-based materials was greatly improved by using composites with CeO₂ (ref. 15), based on the catalytic properties of the ceria. These lanthanum strontium titanates are usually treated in the literature as simple cubic perovskites, although the presence of extra oxygen beyond the ABO₃ stoichiometry plays a

critical role in both the structure and the electrochemical properties, as summarized in Fig. 1. The lower members of the La₄Srₙ₋₄TiₙO₃ₙ₊₂ series, n < 7, are layered phases, having oxygen-rich planes in the form of crystallographic shears joining consecutive

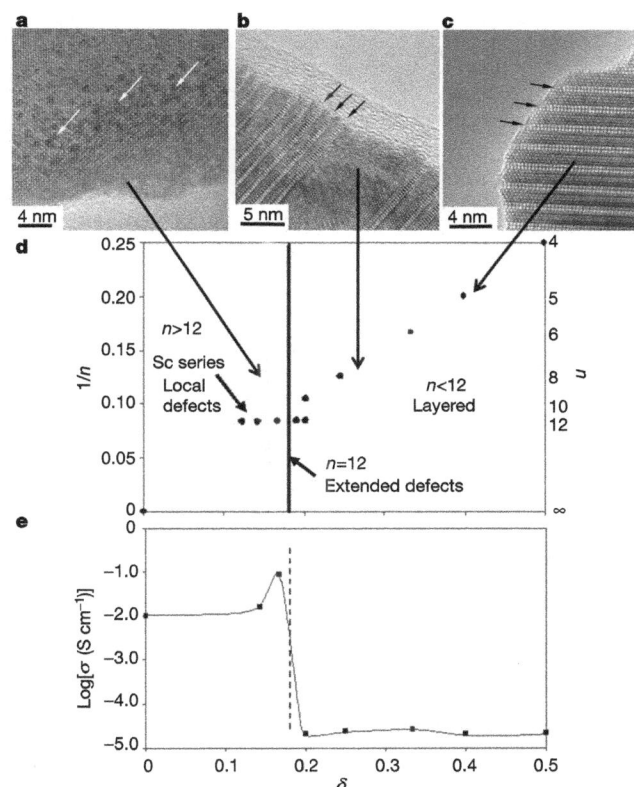

Figure 1 | Relation between microstructure, composition and conductivity of the 'La₄Srₙ₋₄TiₙO₃ₙ₊₂' series. a–c, HRTEM images of samples varying from disordered extended defects (**a**, n = 12) through random layers of extended defects (**b**, n = 8) to ordered extended planar oxygen excess defects (**c**, n = 5). **d,** The location of these phases on the composition map, with 1/n plotted against oxygen excess δ in perovskite unit ABO₃₊δ. **e,** Defect electronic conductivity of grain component as determined by a.c. impedance spectroscopy on samples quenched from 1,300 °C in air, also plotted against δ. 'Sc series' refers to samples in series (La₄Srₙ₋₄)(Tiₙ₋ᵧScᵧ)O₃ₙ₊₂₋ᵧ/₂ (ref. 14).

[1]School of Chemistry, University of St Andrews, St Andrews, Fife KY16 9ST, UK. [2]Department of Inorganic Chemistry, University of La Laguna, 38200, Canary Islands, Spain.
†Present addresses: Department of Inorganic Chemistry, University of La Laguna, 38200, Canary Islands, Spain (J.C.R.-M.); Instituto de Ciencia de los Materiales ICMAB-CSIC, 08193, Barcelona, Spain (J.C.-V.).

When citing this article, please cite the original version as shown on the contents page of this chapter.

Table 1 | Properties of La$_4$Sr$_8$Ti$_{12-x}$M$_x$O$_{38-\delta}$ electrodes

Electrode	R_p (Ω cm^2)		OCV (V)	Representative electrode grain size (μm)
	97.7%H$_2$/2.3%H$_2$O	97.7%CH$_4$/2.3%H$_2$O	97.7%CH$_4$/2.3%H$_2$O	
La$_4$Sr$_8$Ti$_{12}$O$_{38-\delta}$, $x = 0$*	2.97	8.93	0.98	1
M = Sc, ($x = 0.6$)†	0.50	1.20	1.10	0.5–1.0
M = Mn ($x = 1$)‡	0.43	1.14	0.98	0.8
Mn ($x = 0.5$), Ga($x = 0.5$)	0.20	0.57	1.25	0.8

Comparison of open circuit voltage (OCV) and polarization resistance (R_p) measured in indicated gas compositions at 900 °C on doped and undoped La$_4$Sr$_{n-4}$Ti$_n$O$_{3n+2}$.
*Ref. 11; †ref. 14; ‡ref. 23.

blocks. These planes become more sporadic with increasing n (that is, decreasing the oxygen content) until they are no longer crystallographic features, rendering local oxygen-rich defects randomly distributed within a perovskite framework, $n > 11$ (ref. 16). Substitution of Ti^{4+} by Nb^{5+} or Sc^{3+} demonstrates that the oxygen excess parameter (δ) critically determines whether defects are ordered or disordered, with $\delta = 0.167$ being a critical parameter. The presence of such disordered defects appears to strongly affect the redox characteristics of the oxide, as indicated by marked effects on conductivity induced by mild reduction (Fig. 1). Although the materials from this lanthanum strontium titanate oxygen excess series are much easier to reduce, and hence exhibit much higher electronic conductivity than their oxygen stoichiometric analogues, they do not exhibit electrochemical performance comparable to such an effective oxide anode as La$_{0.75}$Sr$_{0.25}$Cr$_{0.5}$Mn$_{0.5}$O$_{3-\delta}$ (ref. 10). This we attribute to the inflexibility of the co-ordination demands of titanium, which strongly prefers octahedral coordination in the perovskite environment.

In order to make the B-site co-ordination more flexible and hence to improve electrocatalytic performance, we have introduced various dopants to replace Ti in La$_4$Sr$_8$Ti$_{12}$O$_{38-z}$ based fuel electrodes, Table 1, and found the most successful to be a combination of Mn and the trivalent Ga ion. Mn supports p-type conduction in oxidizing conditions, and has been previously shown to promote electroreduction under SOFC conditions[17,18]. Furthermore, Mn is known to accept lower coordination numbers in perovskites[19], especially for Mn^{3+}, and thus it may facilitate oxide-ion migration. Similarly Ga is well known to adopt lower co-ordination than octahedral in perovskite-related oxides. The possibility of mixed ionic/electronic conduction is very important, because it would allow the electro-oxidation process to move away from the three-phase electrode/electrolyte/gas interface onto the anode surface, with considerable catalytic enhancement[3].

La$_4$Sr$_8$Ti$_{11}$Mn$_{0.5}$Ga$_{0.5}$O$_{37.5}$ (LSTMG) powders were prepared by solid state reaction from stoichiometric amounts of high purity La$_2$O$_3$, SrCO$_3$, TiO$_2$, Mn$_2$O$_3$ and Ga$_2$O$_3$ fired for 24–48 h. Polarization measurements were performed in a three-electrode arrangement. The electrolyte was a sintered 8 mol% Y$_2$O$_3$ stabilized ZrO$_2$ (YSZ) pellet of 2 mm thickness and 20 mm diameter. La$_{0.8}$Sr$_{0.2}$MnO$_3$ (LSM) and Pt were used as counter-electrode and reference electrode, respectively. The anode was prepared in two configurations, first as a 60-μm-thick layer of 50:50 LSTMG:YSZ and second as an optimized anode with four layers, with graded concentration of YSZ. Each layer was pre-fired at 300 °C and all of them co-fired at 1,200 °C for 2 h. Electrical contacts to the anodes were formed using an Au mesh with small amounts of Au paste to ensure contacting and to avoid any additional catalytic effect. Fuel-cell performance was obtained for these materials as SOFC anodes with 330-μm-thick YSZ(Al$_2$O$_3$) electrolyte and LSM cathode. Platinum paste coated onto LSM and fired at 900 °C was used as the cathode current collector.

La$_4$Sr$_8$Ti$_{11}$Mn$_{0.5}$Ga$_{0.5}$O$_{37.5}$ forms as a single-phase perovskite on firing at 1,400 °C. The structure obtained can be refined as monoclinic, with $a = 5.5287(9)$ Å, $b = 7.8098(5)$ Å, $c = 5.3096(6)$ Å, $\beta = 92.295(7)°$, $V = 229.08(6)$ Å^3. No chemical reactions were observed by X-ray diffraction on firing an intimate mixture of LSTMG and YSZ pressed powder at 1,200 °C in air for 80 h, indicating good chemical compatibility. The phase is stable under fuel conditions at 1,000 °C; the perovskite structure is retained on

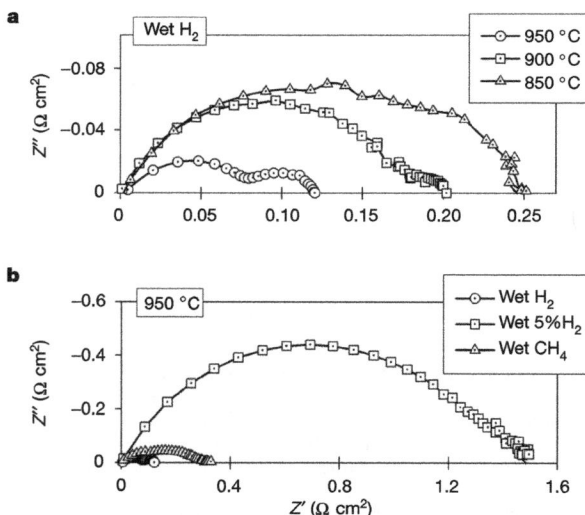

Figure 2 | Polarization measurements on LSTMG/YSZ with varying temperatures and atmospheres. Polarization impedances were measured **a**, under humidified hydrogen at different temperatures, and **b**, under different atmospheres at 950 °C, all humidified at 20 °C on a screen-printed graded La$_4$Sr$_8$Ti$_{11}$Mn$_{0.5}$Ga$_{0.5}$O$_{37.5}$/YSZ working electrode on a 2 mm YSZ electrolyte with La$_{0.8}$Sr$_{0.2}$MnO$_3$ counter and reference electrodes. Z' and Z'' are respectively the real and the imaginary parts of the complex impedance.

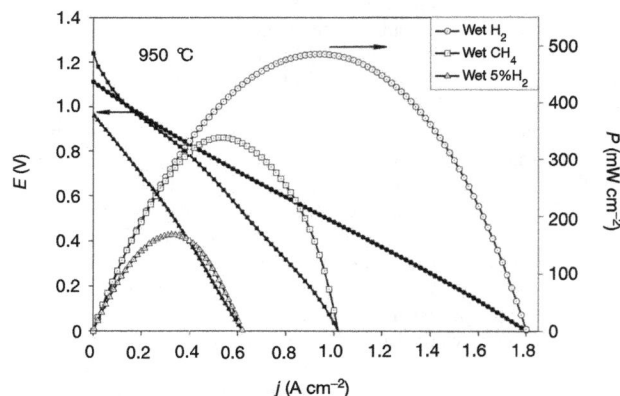

Figure 3 | Performance plots in different atmospheres. Fuel cell performance plots for different fuel gas compositions each containing 2.3% of water at a four-layer optimized LSTMG anode (circle: pure H$_2$, square: pure CH$_4$, triangle: 5%H$_2$) on a 330 μm thin YSZ electrolyte, with LSM cathode in unhumidified oxygen. E is potential difference between electrodes, j is current density and P is power density.

Figure 4 | Electrode interface. Scanning electron microscope image, showing the cross-section of a fuel cell after testing.

firing in 4.9%H_2/2.3%H_2O/92.8%Ar (thereafter termed wet 5%H_2) for 24 h, with the cell changing to orthorhombic $a = 5.5343(6)$ Å, $b = 7.8239(4)$ Å, $c = 5.5322(7)$ Å, $V = 239.54(6)$ Å^3. Detailed high-resolution transmission electron microscopy (HRTEM) studies show exactly the same features that were observed for $La_4Sr_8Ti_{12}O_{38}$ ($n = 12$), that is, this phase is within the localized defect region, as predicted in Fig. 1 for $\delta < 0.167$. On re-oxidation of LSTMG in air, in a thermogravimetric analyser, a weight increase of 0.37% was observed in a step between 400 and 600 °C, corresponding to a change in oxygen content of about 0.5 oxygen atoms per formula unit, consistent with Mn^{II} to Mn^{IV} reoxidation. Some further weight gain of about 0.07% (0.1 O) was observed on holding at 900 °C which may be related to Ti^{III} to Ti^{IV} reoxidation and is fairly typical for these titanates[14].

The total conductivity is 0.004 S cm^{-1} in air at 900 °C and increases up to 0.5 S cm^{-1} at 10^{-15} atm. Activation energies for conduction are 0.16 eV for 200–700 °C, and 0.56 eV for 700–900 °C in wet 5%H_2. The conductivity of $La_4Sr_8Ti_{11}Mn_{0.5}Ga_{0.5}O_{37.5}$ was also measured as a function of oxygen partial pressure at 800 °C and 900 °C. The material shows typical n-type conductivity as the dominant conduction mechanism, but some evidence of p-type conductivity can be found at $p_{O_2} > 0.5$ atm.

The anode polarization resistance was measured using three-electrode geometry as previously described[11]. Results at different temperatures in humidified hydrogen are shown in Fig. 2a and in different atmospheres at 950 °C in Fig. 2b using an optimized anode made of four graded layers with varying LSTMG:YSZ ratios and total thickness 140 μm. The polarization resistances in wet H_2 (97.7%H_2, 2.3%H_2O) were 0.12 Ω cm^2 at 950 °C, 0.20 Ω cm^2 at 900 °C and 0.25 Ω cm^2 at 850 °C. The polarization resistances were 0.12 Ω cm^2 in wet H_2, 1.5 Ω cm^2 in wet 5%H_2 and a remarkably low value, 0.36 Ω cm^2, in wet CH_4 (97.7%CH_4, 2.3%H_2O), at 950 °C. These polarization resistances were attained after about 24 h in fuel conditions; initial polarization resistances were 2–3 times higher. This long time period to achieve equilibration is fairly typical for donor doped strontium titanates that are not cation vacancy compensated, and we attribute this to reorganization of a complex defect structure. The best previous results using metal-free oxide anodes only achieved polarization resistances of twice these values and with much lower OCVs (open circuit voltages) in methane[3]. These polarization resistances are about 15 times less than previously achieved without Mn/Ga addition[11], and less than 50% of the values obtained using just Mn, all for electrodes with similar microstructures (Table 1). Furthermore, these results are comparable with the best cermet electrode performances, and allow both operation in low steam hydrocarbon fuels and good mechanical redox stability.

The OCVs matched the value predicted by the Nernst equation, 0.97 V and 1.13 V at 950 °C, for wet 5%H_2 and wet H_2, respectively. The OCVs in wet CH_4, for a single layer 50:50 YSZ:LSTMG anode, were: 1.39 V at 950 °C, 1.32 V at 900 °C and 1.36 V at 850 °C. These values were reproducible after two days of testing in wet 5%H_2, wet H_2 and wet CH_4. For a four-layer anode, in wet CH_4, the OCVs were: 1.23 V, 1.17 V and 1.16 V at 950 °C, 900 °C and 850 °C, respectively. The high OCVs compared to those typically observed with methane fuels are more significant than, but broadly similar to, that for the addition of precious metal to copper ceria anodes[20]. It seems clear that the thicker, more complex anode structure is less able to fully activate methane at the electrode/electrolyte interface, even though the electrochemical performance of the complex anode is superior in terms of polarization resistance. The OCV trend observed at the thicker electrode is similar to that obtained by Liu and Barnett[21] which they initially attributed to partial oxidation of produced carbon, but later[22] suggested was due to the oxidation of hydrogen, produced by methane reforming with the humidified methane reaching equilibrium at the anode. Here the high OCV we obtained at the thin electrode (1.4 V) implies full oxidation of equilibrated wet CH_4, as this potential is the expected OCV in such an atmosphere.

Figure 3 shows the performance of the LSTMG anode in different atmospheres, at 950 °C, using a two-electrode set-up. The maximum power density in wet H_2 was close to 0.5 W cm^{-2} and the power density in wet CH_4 is two times higher than in 5%H_2, reaching a value of about 0.35 W cm^{-2}. The different slopes in the current–voltage plots under methane seem to suggest different reactions. After running a fuel cell for two days, cycling from 950 °C to 850 °C, in wet 5%H_2, wet H_2 and wet methane, no traces of carbon could be detected visually or by thermal analysis. This anode exhibited a very fine microstructure, Fig. 4, with uniform particle size just less than 1 μm, an estimated 45% porosity and a clean interface with the electrolyte.

This work has demonstrated the concept of inducing functionality through disorder of extended defects. Through partially replacing titanium with some manganese and gallium, an SOFC anode with similar performance in hydrogen to nickel/zirconia cermets has been achieved. In contrast with these cermets, the electrode is remarkably active for the oxidation of CH_4 at high temperatures in the absence of excess steam; moreover, high OCVs reaching stable values between 1.2 V and 1.4 V have been obtained. Optimization of the microstructure allows a very marked improvement in the properties of the anode, but there remains a problem with fairly low lateral conductivity, which (especially for SOFC designs with long current paths) will require an additional current collector for practical application. Thus this material shows considerable promise as an electrochemical anode yielding the lowest polarization results so far reported for an oxide anode and excellent prospects for direct hydrocarbon use.

Received 16 September; accepted 1 November 2005.

1. Kilner, J. A. & Steele, B. C. H. The effect of ion size on the energy of association between oxygen vacancies and dopant cations in oxide solid electrolytes. *J. Electrochem. Soc.* **129**, C143–C148 (1982).
2. Irvine, J. T. S., Feighery, A. J., Fagg, D. P. & García-Martín, S. Structural studies on the optimisation of fast oxide ion transport. *Solid State Ionics* **136/137**, 879–885 (2000).
3. Atkinson, A. et al. Advanced anodes for high-temperature fuel cells. *Nature Mater.* **3**, 17–27 (2004).
4. Primdahl, S., Hansen, J. R., Grahl-Madsen, L. & Larsen, P. H. Sr-doped $LaCrO_3$ anode for solid oxide fuel cells. *J. Electrochem. Soc.* **148**, A74–A81 (2001).
5. Singhal, S. C. & Kendall, K. *High Temperature Solid Oxide Fuel Cells: Fundamentals, Design and Applications* (Elsevier, Oxford, 2004).
6. Matsuzaki, Y. & Yasuda, I. The poisoning effect of sulfur-containing impurity gas on a SOFC anode: Part I. Dependence on temperature, time and impurity concentration. *Solid State Ionics* **132**, 261–269 (2000).
7. Zhu, W. Z. & Deevi, S. C. A review on the status of anode materials for solid oxide fuel cells. *Mater. Sci. Eng. A* **362**, 228–239 (2003).
8. Park, S., Vohs, J. M. & Gorte, R. J. Direct oxidation of hydrocarbons in a solid-oxide fuel cell. *Nature* **404**, 265–267 (2000).

Nature | Vol 439 | 2 February 2006 | doi:10.1038/nature04438

9. Perry, E., Tsai, T. & Barnett, S. A. A direct-methane fuel cell with a ceria-based anode. *Nature* **400**, 649–651 (1999).
10. Tao, S. W. & Irvine, J. T. S. A redox-stable efficient anode for solid-oxide fuel cells. *Nature Mater.* **2**, 320–323 (2003).
11. Canales-Vázquez, J., Tao, S. W. & Irvine, J. T. S. Electrical properties in La$_2$Sr$_4$Ti$_6$O$_{19-\delta}$: a potential anode for high temperature fuel cells. *Solid State Ionics* **159**, 159–165 (2003).
12. Marina, O. A., Canfield, N. L. & Stevenson, J. W. Thermal, electrical, and electrocatalytical properties of lanthanum-doped strontium titanate. *Solid State Ionics* **149**, 21–28 (2002).
13. Mukundan, R., Brosha, E. L. & Garzon, F. H. Sulfur tolerant anodes for SOFCs. *Electrochem. Solid-State Lett.* **7**, A4–A7 (2004).
14. Canales-Vázquez, J., Ruiz-Morales, J. C., Irvine, J. T. S. & Zhou, W. Sc-substituted oxygen excess titanates as fuel electrodes for SOFC. *J. Electrochem. Soc.* **152**, 1458–1465 (2005).
15. Marina, O. A. & Pederson, L. R. in *Proc. 5th European Solid Oxide Fuel Cell Forum* (ed. Huijsmans, J.) 481–489 (European Fuel Cell Forum, Oberrohrdorf, Switzerland, 2002).
16. Canales-Vázquez, J., Smith, M. J., Irvine, J. T. S. & Zhou, W. Studies on the reorganisation of extended defects with increasing n in the perovskite-based La$_4$Sr$_{n-4}$Ti$_n$O$_{3n+2}$ series. *Adv. Funct. Mater.* **15**, 1000–1008 (2005).
17. Holtappels, P., Bradley, J. L., Irvine, J. T. S., Kaiser, A. & Mogensen, M. Electrochemical characterization of ceramic SOFC anodes. *J. Electrochem. Soc.* **148**, A923–A929 (2001).
18. Irvine, J. T. S., *et al.* Optimisation of perovskite titanates and niobates as anode materials for SOFCs. in *Proc. 4th European SOFC Forum* (ed. McEvoy, A. J.) 471–477 (European Fuel Cell Forum, Oberrohrdorf, Switzerland, 2000).
19. Poeppelmeier, K. R., Leonowicz, M. E. & Longo, J. M. CaMnO$_{2.5}$ and Ca$_2$MnO$_{3.5}$ — new oxygen-defect perovskite-type oxides. *J. Solid State Chem.* **44**, 89–98 (1982).
20. McIntosh, S., Vohs, J. M. & Gorte, R. J. Effect of precious-metal dopants on SOFC anodes for direct utilization of hydrocarbons. *Electrochem. Solid-State Lett.* **6**, A240–A243 (2003).
21. Liu, J. & Barnett, S. A. Operation of anode-supported solid oxide fuel cells on methane and natural gas. *Solid State Ionics* **158**, 11–16 (2003).
22. Lin, Y., Zhan, Z., Liu, J. & Barnett, S. A. Direct operation of solid oxide fuel cell with methane fuel. *Solid State Ionics* **176**, 1827–1835 (2005).
23. Ovalle, A., Ruiz-Morales, J. C., Canales-Vázquez, J., Marrero-López, D. & Irvine, J. T. S. Mn-substituted titanates as efficient anodes for direct methane SOFCs. *Solid State Ionics* (submitted).

Acknowledgements This work was funded partly by a EU Marie Curie Fellowship and by EPSRC.

Author Information Reprints and permissions information is available at npg.nature.com/reprintsandpermissions. The authors declare no competing financial interests. Correspondence and requests for materials should be addressed to J.T.S.I. (jtsi@st-and.ac.uk).

A high-performance cathode for the next generation of solid-oxide fuel cells

Zongping Shao & Sossina M. Haile

Materials Science, California Institute of Technology, Pasadena, California 91125, USA

Fuel cells directly and efficiently convert chemical energy to electrical energy[1]. Of the various fuel cell types, solid-oxide fuel cells (SOFCs) combine the benefits of environmentally benign power generation with fuel flexibility. However, the necessity for high operating temperatures (800–1,000 °C) has resulted in high costs and materials compatibility challenges[2]. As a consequence, significant effort has been devoted to the development of inter-mediate-temperature (500–700 °C) SOFCs. A key obstacle to reduced-temperature operation of SOFCs is the poor activity of traditional cathode materials for electrochemical reduction of oxygen in this temperature regime[2]. Here we present $Ba_{0.5}Sr_{0.5}Co_{0.8}Fe_{0.2}O_{3-\delta}$ (BSCF) as a new cathode material for reduced-temperature SOFC operation. BSCF, incorporated into a thin-film doped ceria fuel cell, exhibits high power densities ($1,010\,mW\,cm^{-2}$ and $402\,mW\,cm^{-2}$ at 600 °C and 500 °C, respectively) when operated with humidified hydrogen as the fuel and air as the cathode gas. We further demonstrate that BSCF is ideally suited to 'single-chamber' fuel-cell operation, where anode and cathode reactions take place within the same physical chamber[3]. The high power output of BSCF cathodes results from the high rate of oxygen diffusion through the material. By enabling operation at reduced temperatures, BSCF cathodes may result in widespread practical implementation of SOFCs.

The primary function of the cathode in a fuel cell based on an oxygen-conducting electrolyte is to facilitate the multistep electro-chemical reduction of oxygen[4]:

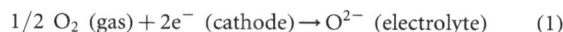

$$1/2\ O_2\ (gas) + 2e^-\ (cathode) \rightarrow O^{2-}\ (electrolyte) \qquad (1)$$

In a single-chamber fuel cell, the cathode must furthermore be inactive towards oxidation of fuel, and operation at reduced temperatures is essential to minimize undesirable gas-phase reactions[2] (Supplementary Fig. 1). In the present work, we use BSCF as a new cathode material and demonstrate excellent performance in both dual-chamber and single-chamber configurations at temperatures lower than 600 °C. This material, a cubic perovskite (Supplementary Fig. 2) in the $BaCoO_{3-\delta}$–$SrCoO_{3-\delta}$ system, was first developed as a high-temperature (>800 °C) oxygen permeation membrane material[5,6]. Unlike typical cathodes, the A-site cation of BSCF perovskite is an alkaline-earth species rather than a rare-earth element.

We first measured the polarization resistance of BSCF by two-electrode impedance methods using symmetric BSCF|electrolyte|BSCF cells, where the electrolyte was ~1-mm-thick 15 mol% samaria-doped ceria (SDC). As is standard in this field, we use the term area specific resistance (ASR) to describe all resistance terms associated with the electrode, whether they occur at the gas–cathode interface, within the bulk of the cathode, or at the cathode–electrolyte interface. Data were collected under a uniform air atmosphere, both with (three-electrode) and without (two-electrode) a reference electrode; the three-electrode configuration allowed direct measurement of the half-cell voltage with and without an applied direct current. The cathode ASR (Fig. 1a) was determined from raw impedance plots as indicated in Fig. 1b, where the high-frequency offset is due primarily to the electrolyte resistance and the radius of the complete arc to the cathode resistance. The ASR, as determined by both techniques, was remarkably low: 0.055–$0.071\,\Omega\,cm^2$ at 600 °C, and 0.51–$0.60\,\Omega\,cm^2$ at 500 °C. Furthermore, the measured value dropped from $\sim0.60\,\Omega\,cm^2$ to $\sim0.27\,\Omega\,cm^2$ at 500 °C under an applied current of $0.14\,A\,cm^{-2}$. Cells fabricated using silver alone as the cathode showed ASR values ~1,000 times larger (Supplementary Fig. 3), demonstrating that silver paste, used to attach silver mesh leads, served only as a current collector. The ASR of the BSCF cathode reported here is substantially lower than that of other single-phase perovskite cathodes measured under similar conditions, and comparable to that of the best composite systems[7–10].

The performance of the BSCF cathode in a conventional, dual-chamber fuel cell was then investigated, again using SDC as the electrolyte. A thin (20 µm) electrolyte layer was supported on a 700-µm-thick Ni + SDC anode, with a 10–20-µm-thick BSCF cathode layer deposited on the opposing side, after first depositing an additional porous interlayer of SDC (<5 µm in thickness). Air was supplied to the cathode chamber and humidified H_2 (3% H_2O) to the anode chamber. Peak power densities of $\sim1,010\,mW\,cm^{-2}$ and $402\,mW\,cm^{-2}$ were obtained at respectively 600 °C and 500 °C (Fig. 2a). These values are more than twice those measured in our laboratory for a similar cell but with SSC + SDC as cathode (Supplementary Fig. 4). In addition to the polarization curves, the cell resistances under open-circuit conditions were measured at various temperatures by impedance spectroscopy. The electrode polarization resistance (the sum of anode and cathode ASRs) is only about $0.021\,\Omega\,cm^2$ at 600 °C, and $0.135\,\Omega\,cm^2$ at 500 °C (Fig. 2), amounting to just 14% and 26% of the resistance of electrolyte at these respective temperatures (Supplementary Fig. 5). It is note-worthy that composite SDC + BSCF cathodes, although still very active for oxygen electroreduction, yielded lower power densities than simple BSCF cathodes.

The trilayer fuel cell was further operated in a single-chamber configuration with a propane + O_2 + He mixture in a 4:9:36 volumetric ratio as the feed gas, and a total flow rate of

490 ml min^{-1}. A peak power density of ~391 mW cm^{-2} was observed at a furnace set temperature of 575 °C, with a current density at short circuit of ~1.9 A cm^{-2}. At 525 °C the respective values were ~ 358 mW cm^{-2} and ~1.7 A cm^{-2}. Upon modifying the BSCF cathode to incorporate 30 wt% SDC, significant improvements in power density were observed in single-chamber mode over simple BSCF. At a furnace set temperature of 500 °C, a peak power density of ~440 mW cm^{-2} was achieved (Fig. 3). Hibino et al.[3] reported a comparably high peak power density of 403 mW cm^{-2} at 500 °C for an electrolyte-supported fuel cell using SSC + SDC as the cathode and ethane as the fuel. However, this cathode was incompatible with propane at temperatures higher than 450 °C (ref. 3). It should be noted that, because of the heat release during partial oxidation at that anode, the real temperature of the single-chamber fuel cell is about 150–245 °C higher than the furnace temperature, depending on the operation conditions (Supplementary Fig. 6). This self-heating in SCFC mode explains the higher power densities obtained from single-chamber than from dual-chamber fuel cells at nominally low temperatures (compare Figs 2a and 3).

The mechanisms responsible for the excellent performance of

BSCF as a fuel-cell cathode were identified by oxygen permeability measurements (Supplementary Figs 7–9) and extensive impedance spectroscopy studies of symmetric cells (Supplementary Figs 10, 11). The oxygen permeation measurements, combined with thermal gravimetric analysis to determine the oxygen vacancy concentration as a function of oxygen partial pressure, revealed that the oxygen vacancy diffusion rate is 7.3×10^{-5} cm^2 s^{-1} at 775 °C and 1.3×10^{-4} cm^2 s^{-1} at 900 °C (temperatures at which such measurements could be accurately performed), from 2 to 200 times greater than that of comparable cathode perovskites[11,12]. In addition, the activation energy for oxygen diffusion in BSCF was found to be less than half that for oxygen surface exchange, 46 ± 2 kJ mol^{-1} versus 113 ± 11 kJ mol^{-1}, suggesting that oxygen surface exchange is the rate-limiting step at low temperatures and that the exceptionally high oxygen diffusivity through BSCF gives it its overall high rate of oxygen electro-oxidation. The oxygen ion conductivity is, in fact, higher than that of SDC. It is for this reason that introduction of a small amount of the electrolyte material decreases cathode performance in dual-chamber configuration, rather than increasing it, as is otherwise observed (for example, in composite LSM + YSZ cathodes[13]).

Impedance spectroscopy of the symmetric cells strongly supported the conclusions of the permeability measurements: specifically, that oxygen diffusion is rapid and surface exchange kinetics are

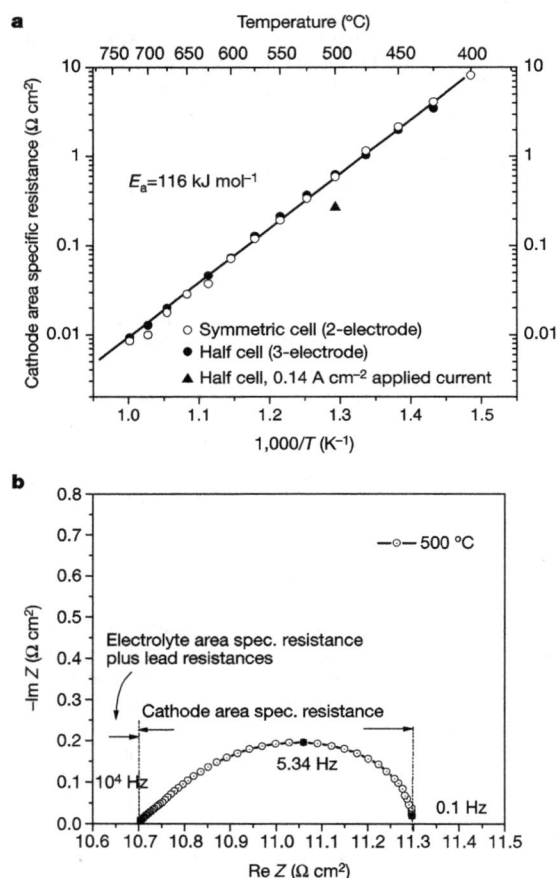

Figure 1 The area specific resistance (ASR) of the cathode material BSCF under air, measured both with (three-electrode, half cell) and without (two-electrode, symmetric cell) a reference electrode. **a**, ASR as a function of temperature, and **b**, a typical impedance spectrum, as obtained from the two-electrode, symmetric cell at 500 °C (raw impedance data have been multiplied by the fuel-cell active area, 0.71 cm^2, so that the cathode ASR corresponds directly to the diameter of the arc associated with the cathode response). The electrolyte, 15 mol% samaria-doped ceria (SDC), is about 1 mm thick, and each electrode about 20 μm thick. The activation energy for the overall cathode resistance is 116 kJ mol^{-1}, comparable to that determined for the surface exchange process, indicating that surface exchange at the cathode–air interface is the rate-limiting step.

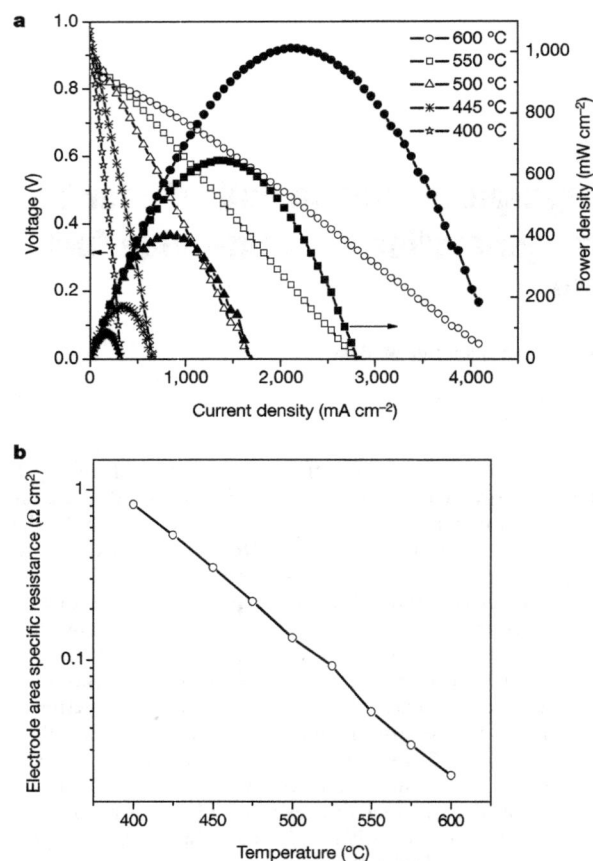

Figure 2 Performance obtained from a $Ba_{0.5}Sr_{0.5}Co_{0.8}Fe_{0.2}O_{3-\delta}$ (~20 μm)|$Sm_{0.15}Ce_{0.85}O_{2-\delta}$ (~20 μm)|Ni + $Sm_{0.15}Ce_{0.85}O_{2-\delta}$ (~700 μm) fuel cell. Air was supplied to the cathode (400 ml min^{-1} at STP) and 3% H_2O-humidified H_2 was supplied to the anode (80 ml min^{-1} at STP). **a**, Cell voltage and power density as functions of current density, and **b**, ASRs of the electrodes (sum of anode and cathode contributions) measured under open circuit conditions. The peak power density reaches 1 W cm^{-2} at 600 °C.

rate-limiting. In particular, (1) good linearity of the cathode ASR versus reciprocal temperature was observed over the temperature range investigated, 400–725 °C, and the derived activation energy, $\sim 116\,kJ\,mol^{-1}$, was almost identical to that determined for the oxygen surface exchange step ($113 \pm 11\,kJ\,mol^{-1}$). (2) At low temperatures, the cathode ASR was sensitive to the presence of CO_2 and H_2O in the atmosphere, gases which could only affect surface and not bulk properties. Carbon dioxide increased the interfacial resistance whereas steam decreased it (Supplementary Fig. 10). (3) An increase in the cathode thickness decreased the ASR (without changing the activation energy) (Supplementary Fig. 11), presumably as a result of increasing the area over which surface exchange could occur. (4) The possibility that interfacial charge transfer could be the rate-limiting step is eliminated by the fact that no arc that could be associated with this step appeared in the impedance data.

The observation that the apparent cathode ASR drops under applied current (without detectable changes in structure as determined by *ex situ* X-ray diffraction), Fig. 1a, is additionally consistent with a picture in which the surface exchange step is rate-limiting. Specifically, it is known that oxygen adsorption and desorption rates can differ significantly, with desorption occurring much more slowly than adsorption[14]. In the absence of an applied current there is no net ion flux, and thus the adsorption and desorption processes contribute equally to the measured resistance. Under an applied current, however, net adsorption occurs at the electrode functioning as the cathode, and the measured rate is dominated by this faster process, resulting in the observed reduction in ASR. Such an effect could not occur for an overall process limited by bulk diffusion.

A key characteristic of a useful cathode for single-chamber fuel-cell applications is a low activity towards fuel oxidation under the oxidant + fuel environment. It is notable that, together with its high activity of oxygen electroreduction, BSCF exhibits the quality of low activity towards propane oxidation. Under stoichiometric conditions ($O_2:C_3H_8 = 5:1$ with 95 vol.% helium) and at 500 °C, the propane conversion rates over BSCF, LSCF and SSC are 5.3%, 35.5% and 16.1%, respectively (Fig. 4). The high activity of SSC

towards propane oxidation is probably responsible for the relatively poor performance of this cathode[3]. Incorporation of electrolyte material into the BSCF cathode under the single-chamber fuel-cell configuration improves performance (unlike the dual-chamber results), and we speculate that the electrolyte is beneficial for limiting the detrimental influence of *in situ* generated CO_2 on the oxygen surface exchange kinetics of BSCF.

For practical applications, in addition to high power density, good fuel-cell stability is essential. Degradation can occur either by reduction/phase decomposition under low oxygen partial pressures or by phase segregation under an oxygen partial pressure gradient. Whereas phase segregation under a steep oxygen concentration gradient has been reported for BSCF[5,15], such conditions are unlikely to exist in the fuel-cell cathode because of rapid oxygen diffusion through BSCF. Furthermore, we found that reduction of BSCF occurs only at oxygen partial pressures of $<10^{-6}$–10^{-8} atm. These partial pressures are equivalent to cathode polarization drops of 0.20–0.23 V at 500–600 °C for a cell exposed to ambient air at the cathode. At lower temperatures, polarization losses increase, as does the tendency towards phase segregation, and thus stability at low temperatures provides a stringent measure of long-term viability. A preliminary examination of stability was carried out using a dual-chamber fuel cell with a BSCF-based cathode, operated at 390 °C under an air/3% humidified H_2 gradient. The cell showed essentially stable performance within the test period of 120 h (Supplementary Fig. 12). Under the fuel-rich, single-chamber conditions employed above, the thermodynamic oxygen partial pressure is as low as 10^{-28} atm at 500 °C, conditions that could well induce the reduction of BSCF. Because such reduction was not observed in our experiments, we conclude that the poor activity of BSCF for propane oxidation results in local oxygen partial pressures in the vicinity of the cathode which are high enough to prevent reduction of cobalt.

The present study demonstrates that very high power densities can be achieved at low temperatures in both dual-chamber and single-chamber fuel cells using $Ba_{0.5}Sr_{0.5}Co_{0.8}Fe_{0.2}O_{3-\delta}$ as a cathode. A significant fraction of the overall cell resistance even at 500 °C ($\sim 74\%$) in dual-chamber configuration is attributable to the 20-μm-thick electrolyte. The SDC used here has a measured conductivity of $\sim 4.7 \times 10^{-3}\,S\,cm^{-1}$ at 500 °C (Supplementary Fig. 5), which is lower than the best reported value for doped ceria, $\sim 9.5 \times 10^{-3}\,S\,cm^{-1}$ (ref. 16). Thus, a decrease in electrolyte resist-

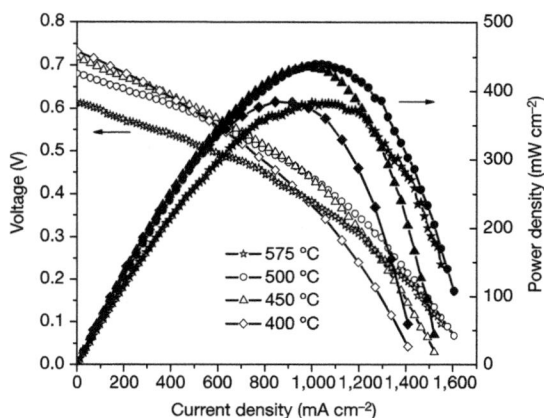

Figure 3 Cell voltage and power density as functions of current density obtained in single-chamber mode from a $Ba_{0.5}Sr_{0.5}Co_{0.8}Fe_{0.2}O_{3-\delta} + Sm_{0.15}Ce_{0.85}O_{2-\delta}$ ($\sim 20\,\mu$m)|$Sm_{0.15}Ce_{0.85}O_{2-\delta}$ ($\sim 20\,\mu$m)|Ni + $Sm_{0.15}Ce_{0.85}O_{2-\delta}$ ($\sim 700\,\mu$m) fuel cell. The BSCF:SDC ratio in the cathode was 70:30 by weight. A mixture of propane, oxygen and helium, flowing (at STP) at 40 ml min^{-1}, 90 ml min^{-1} and 360 ml min^{-1}, respectively, was used as the feed gas. The temperature reported is that of the furnace; local heating due to partial oxidation at the anode can result in considerably higher temperatures ($\Delta T \approx 188$–240 °C, Supplementary Fig. 6). The peak power density is relatively insensitive to furnace temperature, broadly peaking at 500 °C at a value of $\sim 440\,mW\,cm^{-2}$.

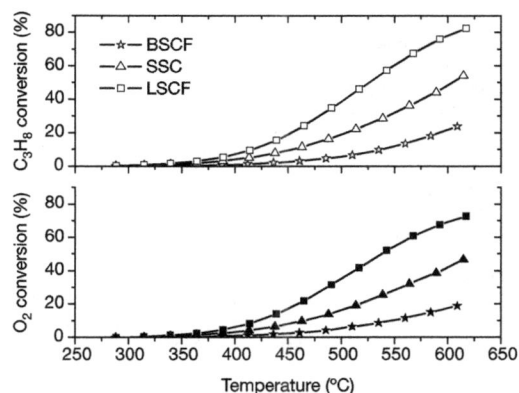

Figure 4 Comparison of the catalytic activities of BSCF ($Ba_{0.5}Sr_{0.5}Co_{0.8}Fe_{0.2}O_{3-\delta}$), SSC ($Sm_{0.5}Sr_{0.5}CoO_{3-\delta}$), and LSCF ($La_{0.6}Sr_{0.4}Co_{0.2}Fe_{0.8}O_{3-\delta}$) oxides towards propane oxidation under stoichiometric conditions. Test samples were exposed to a mixture of propane, oxygen and helium flowing (at STP) at 40 ml min^{-1}, 90 ml min^{-1} and 360 ml min^{-1}, respectively. Powders were calcined for 5 h at appropriate temperatures so as to yield comparable specific surface areas: BSCF at 900 °C (0.62 m^2 g^{-1}), SSC at 1,050 °C (0.66 m^2 g^{-1}) and LSCF at 1,100 °C (0.65 m^2 g^{-1}). Low catalytic activity of BSCF is advantageous for single-chamber fuel-cell operation.

ance by a factor of two could realistically be achieved by optimization of its synthesis and composition, or by reduction of its thickness to ~10 μm (as has already been achieved in the literature[17]), and can be anticipated to result in a peak power density of >600 mW cm^{-2} at 500 °C. Additional increases in power density may be achieved through precise control of the cathode architecture using advanced preparation methods to maximize the surface area over which the oxygen exchange reaction can occur. These steps would probably also yield benefits for operation under single-chamber fuel-cell mode. In addition, a detailed study of possible degradation due to phase segregation induced by the oxygen flux through the cathode, particularly at high current densities, may be necessary for evaluating the long-term viability of BSCF cathodes. □

Methods

Cathode powder preparation

Phase-pure $Ba_{0.5}Sr_{0.5}Co_{0.8}Fe_{0.2}O_{3-\delta}$, $Sm_{0.5}Sr_{0.5}CoO_{3-\delta}$ and $La_{0.6}Sr_{0.4}Co_{0.2}Fe_{0.8}O_{3-\delta}$ powders were synthesized by a sol–gel method in which the appropriate metal nitrates were dissolved in water, and a combination of EDTA and citric acid served as complexing agents. Mild heating induced gelation of the solution, and the resulting gel was held at 250 °C for 12 h to remove organics, and then calcined for 5 h under stagnant air at 900–1,050 °C.

Fuel-cell fabrication

Electrolyte-supported symmetric cells for impedance studies were prepared by nitrogen-borne spray deposition of BSCF powder suspended in isopropyl alcohol onto both sides of an SDC electrolyte disk (~1 mm), followed by calcination at 1,000 °C for 5 h under stagnant air. Silver mesh was attached to the electrode surfaces using silver paste as the current collector. Anode-supported thin-film electrolyte fuel cells (trilayer cells) were fabricated using a dual dry-pressing method. The anode, formed from a 60:40 wt% mixture of NiO and SDC powders was dry-pressed into a pellet, and then an electrolyte powder was distributed on the anode surface and co-pressed with the anode. The resultant bilayer was calcined at 1,350 °C for 5 h in air, and then reduced at 600 °C under flowing hydrogen for 5 h. Either simple BSCF or a 70:30 (by weight) BSCF + SDC powder mixture dispersed in isopropyl alcohol was sprayed on the electrolyte surface. The cell was calcined at 1,000 °C for 5 h under flowing nitrogen. The final fuel cells had the following properties: diameter of 1.33 cm, anode thickness of 0.7 mm, anode porosity of ~46%, electrolyte thickness of ~20 μm, cathode thickness of 10–20 μm, and cathode porosity of ~30%.

Fuel-cell electrochemical characterization

Both symmetric cells and complete trilayer fuel cells were placed in an in-house constructed measurement station, and polarization curves collected using a Keithley 2420 source meter based on the four-probe configuration. In the case of dual-chamber measurements, the trilayer cell was sealed to a quartz tube using silver paste, the cathode side was open to air, and the anode side was exposed to 3% H_2O-humidified H_2 at a flow rate of 80 ml min^{-1} at STP. For the single-chamber fuel-cell test, the whole cell was place in a quartz tube reactor with an inner diameter of 15 mm. Digital mass flow controllers (Aera FC-D980C) were used to introduce propane, oxygen and helium gases to the reactor at various flow rates.

Electrochemical impedance test

The ASR of BSCF cathode was measured in the two-electrode symmetric cell configuration under air. Impedance measurements were performed using a Solartron 1260A frequency analyser under open circuit conditions from 10 mHz to 10^5 Hz. The three-electrode half-cell test was conducted by fixing an additional silver electrode to the middle edge of the SDC as a reference electrode, and then measuring the impedance between the working and reference electrodes measured under both open circuit conditions and current polarization conditions using a combined Solartron 1260A frequency response analyser and a PAR EG&G 273A potentiostat/galvanostat. The impedance of the single cell was also measured under asymmetric atmospheres under open circuit conditions.

Catalytic activity

The catalytic activity of BSCF and other cathode materials towards propane oxidation was tested in a flow-through quartz reactor with a 6 mm inner diameter, containing a fixed bed of 0.3 g catalyst mixed with 1.5 g of silica granules. A gas mixture of He + C_3H_8 + O_2 was passed through the reactor, which was heated to various temperatures using an electrical furnace. The effluent gas was analysed using an in-line Varian CP-4900 Micro GC.

Received 25 February; accepted 25 June 2004; doi:10.1038/nature02863.

1. Steel, B. C. H. & Heinzel, A. Materials for fuel-cell technologies. Nature 414, 345–352 (2001).
2. Brandon, N. P., Skinner, S. & Steele, B. C. H. Recent advances in materials for fuel cells. Annu. Rev. Mater. Res. 33, 183–213 (2003).
3. Hibino, T. et al. A low-operating-temperature solid oxide fuel cell in hydrocarbon-air mixtures. Science 288, 2031–2033 (2000).
4. Fleig, J. Solid oxide fuel cell cathodes: Polarization mechanisms and modeling of the electrochemical performance. Annu. Rev. Mater. Res. 33, 361–382 (2003).
5. Shao, Z. P. et al. Investigation of the permeation behavior and stability of a $Ba_{0.5}Sr_{0.5}Co_{0.8}Fe_{0.2}O_{3-\delta}$ oxygen membrane. J. Membr. Sci. 172, 177–188 (2000).
6. Shao, Z. P., Dong, H., Xiong, G. X., Cong, Y. & Yang, W. S. Performance of a mixed-conducting ceramic membrane reactor with high oxygen permeability for methane conversion. J. Membr. Sci. 183, 181–192 (2001).
7. Huang, K., Wan, J. & Goodenough, J. B. Increasing power density of LSGM-based solid oxide fuel cells using new anode materials. J. Electrochem. Soc. 148, A788–A794 (2001).
8. Ishihara, T., Fukui, S., Nishiguchi, H. & Takita, Y. La-doped $BaCoO_3$ as a cathode for intermediate temperature solid oxide fuel cells using a $LaGaO_3$ based electrolyte. J. Electrochem. Soc. 149, A823–A828 (2002).
9. Ralph, J. M., Rossignol, C. & Kumar, R. Cathode materials for reduced-temperature SOFCs. J. Electrochem. Soc. 150, A1518–A1522 (2003).
10. Xia, C. R. & Liu, M. L. Novel cathodes for low-temperature solid oxide fuel cells. Adv. Mater. 14, 521–523 (2002).
11. Kim, S., Yang, Y. L., Jacobson, A. J. & Abeles, B. Diffusion and surface exchange coefficients in mixed ionic and electronic conducting oxides from the pressure dependence of oxygen permeation. Solid State Ionics 106, 189–195 (1998).
12. Xu, S. J. & Thomson, W. J. Oxygen permeation rates through ion-conducting perovskite membranes. Chem. Eng. Sci 54, 3839–3850 (1999).
13. Tsai, T. & Barnett, S. A. Effect of LSM-YSZ cathode on thin-electrolyte solid oxide fuel cell performance. Solid State Ionics 93, 207–217 (1997).
14. Fukunaga, H., Koyama, M., Takahashi, N., Wen, C. & Yamada, K. Reaction model of dense $Sm_{0.5}Sr_{0.5}CoO_3$ as SOFC cathode. Solid State Ionics 132, 279–285 (2000).
15. van Veen, A. C., Rebeilleau, M., Farrusseng, D. & Mirodatos, C. Studies on the performance stability of mixed conducting BSCFO membranes in medium temperature oxygen permeation. Chem. Commun., 32–33 (2003).
16. Steele, B. C. H. Appraisal of $Ce_{1-y}Gd_yO_{2-y/2}$ electrolytes for IT-SOFC operation at 500 °C. Solid State Ionics 129, 95–110 (2000).
17. Xia, C. R. & Liu, M. L. A simple and cost-effective approach to fabrication of dense ceramic membranes on porous substrates. J. Am. Ceram. Soc. 84, 1903–1905 (2001).

Supplementary Information accompanies the paper on www.nature.com/nature.

Acknowledgements This work was funded by the Defense Advanced Research Projects Agency, Microsystems Technology Office. Additional support was provided by the National Science Foundation through the Caltech Center for the Science and Engineering of Materials. Selected oxygen permeability measurements were carried out in the Laboratory of Reaction Engineering and Energy, Institute of Research on Catalysis, CNRS, France, during the visit of Z.P.S. there, hosted by C. Mirodatos.

Competing interests statement The authors declare that they have no competing financial interests.

Correspondence and requests for materials should be addressed to S.M.H. (smhaile@caltech.edu).

A redox-stable efficient anode for solid-oxide fuel cells

SHANWEN TAO AND JOHN T.S. IRVINE*

School of Chemistry, University of St Andrews, Fife KY16 9ST, Scotland, UK
*e-mail: jtsi@st-and.ac.uk

Published online: 30 March 2003; doi:10.1038/nmat871

S olid-oxide fuel cells (SOFCs) promise high efficiencies in a range of fuels. Unlike lower temperature variants, carbon monoxide is a fuel rather than a poison, and so hydrocarbon fuels can be used directly, through internal reforming or even direct oxidation. This provides a key entry strategy for fuel-cell technology into the current energy economy. Present development is mainly based on the yttria-stabilized zirconia (YSZ) electrolyte[1]. The most commonly used anode materials are Ni/YSZ cermets, which display excellent catalytic properties for fuel oxidation and good current collection, but do exhibit disadvantages, such as low tolerance to sulphur[2] and carbon deposition[3] when using hydrocarbon fuels, and poor redox cycling causing volume instability. Here, we report a nickel-free SOFC anode, $La_{0.75}Sr_{0.25}Cr_{0.5}Mn_{0.5}O_3$, with comparable electrochemical performance to Ni/YSZ cermets. The electrode polarization resistance approaches 0.2 Ω cm^2 at 900 °C in 97% H_2/3% H_2O. Very good performance is achieved for methane oxidation without using excess steam. The anode is stable in both fuel and air conditions, and shows stable electrode performance in methane. Thus both redox stability and operation in low steam hydrocarbons have been demonstrated, overcoming two of the major limitations of the current generation of nickel zirconia cermet SOFC anodes.

Present development of SOFCs is mainly based on the yttria-stabilized zirconia (YSZ) electrolyte, because it exhibits good thermal and chemical stability, high oxide-ion conductivity and mechanical strength at high temperature[1]. The most commonly used anode materials for zirconia-based SOFCs are Ni/ZrO_2 cermets. Because nickel is such a good catalyst for hydrocarbon cracking, these cermets can only be used in hydrocarbon fuels if excess steam is present to ensure complete fuel reforming, diluting fuel and adding to system cost. There are reports that the problem of carbon deposition when using hydrocarbons in a SOFC may be avoided by using a Cu-ceria anode[4], or when an $(Y_2O_3)_{0.15}(CeO_2)_{0.85}$ interface was applied between YSZ and Ni-YSZ cermet anode although only at lower temperatures[5]. Equally problematic, is the poor redox tolerance of nickel cermets that precludes many medium- and small-scale applications. Thus there is considerable interest in finding alternative anode systems.

Different structure types, such as perovskite, fluorite, pyrochlore and tungsten bronze have been investigated as potential anode materials[6–8]. Interesting results have been obtained with lanthanum strontium titanates[9] and especially cerium-doped lanthanum strontium titanate[10];

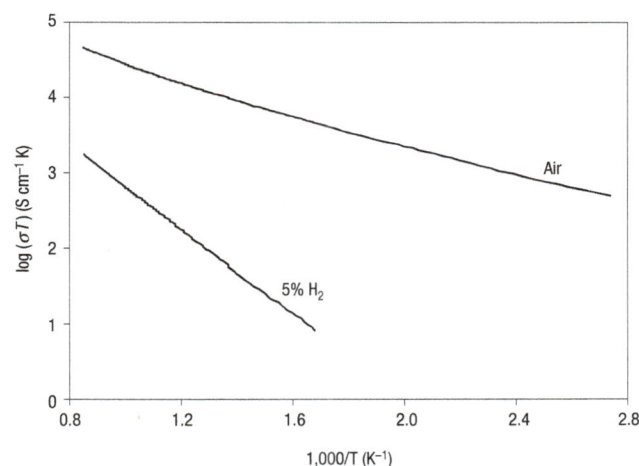

Figure 1 Temperature (*T*) dependence of total conductivity (σ) of $La_{0.75}Sr_{0.25}Cr_{0.5}Mn_{0.5}O_{3-\delta}$ in air and 5% H_2.

however, it is now thought that the cerium-doped anodes are in fact two phase, consisting of a ceria/perovskite assemblage[10]. $LaCrO_3$-based materials have been investigated as interconnect materials for SOFCs[1]; however, they are also potential anode materials for SOFCs due to their relatively good stability in both reducing and oxidizing atmospheres at high temperatures[11]. The reported polarization resistance using these materials is too high for efficient SOFC operation, although significant improvements have been achieved using low-level doping of the B-site. Quite a lot of attention has been focused on 3% replacement of Cr by V, and although methane cracking seems to be avoided[12], the polarization resistance is still of the order[6,12] 10 Ω cm^2. The introduction of other transition elements into the B-site of $La_{1-x}Sr_xCr_{1-y}M_yO_3$ (M = Mn, Fe, Co, Ni) has been shown to improve the catalytic properties for methane reforming[13]. Of the various dopants, nickel seems to be the most successful, and the lowest polarization resistances have been reported for 10% Ni-substituted lanthanum chromite[14]; however, other workers

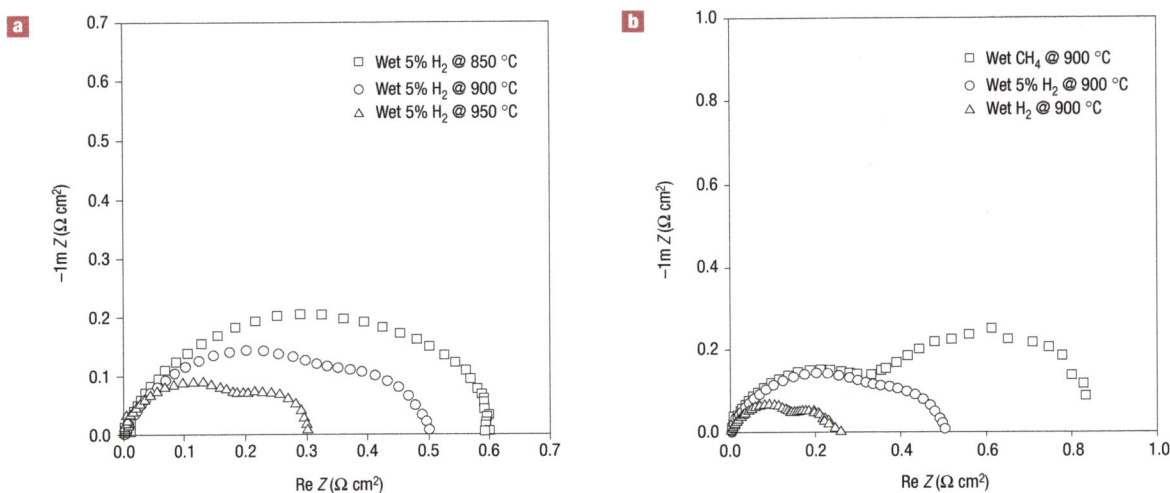

Figure 2 Electrode impedance of an optimized $La_{0.75}Sr_{0.25}Cr_{0.5}Mn_{0.5}O_3$ anode: **a**, in wet 5% H_2, (5% H_2, 3% H_2O, 92% Ar) at 850, 900 and 950 °C; **b**, at 900 °C, in different humidified (3% H_2O) fuel gas compositions at 900 °C. The electrode dispersions have been obtained in a three-electrode set-up. The electrolyte contribution has been subtracted from the overall impedance.

have found nickel exsolution from 10% Ni-doped lanthanum chromites in fuel conditions[15]. Certainly, nickel oxides would not be stable in fuel atmospheres, and although the nickel may be stabilized by the lattice in a higher oxidation state, there will always be the suspicion that the activity of nickel-doped perovskites is due to surface exsolution of nickel metal, and hence poses questions about long-term stability. In a recent report, a composite anode of 5% Ni with a 50/50 mixture of $La_{0.8}Sr_{0.2}Cr_{0.8}Mn_{0.2}O_3$ and $Ce_{0.9}Gd_{0.1}O_{1.95}$ was used for SOFCs with different fuels[16].

No significant weight loss was observed when $LaCrO_3$ was exposed to a reducing atmosphere (partial oxygen pressure, $p_{O_2} = 10^{-21}$ atm at 1,000 °C)[17], this therefore indicates that chromium strongly retains its sixfold coordination. Indeed Cr^{III} is well known to strongly prefer sixfold coordination in its chemistry, thus it is difficult to introduce the oxygen vacancies that are required for oxygen-ion conduction into the $LaCrO_3$ lattice. When the B-sites are doped by other multivalence transition elements that do tolerate reduced oxygen coordination, such as Mn, Fe, Co, Ni and Cu, oxygen vacancies may be generated in a reducing atmosphere at high temperature. Thus a significant degree of B-site doping is required to generate a percolation path for oxygen vacancies to achieve high oxygen-ion conductivity.

In this report, complex perovskites based upon Cr and one or more other transition elements (M) such as V, Mn, Fe, Co, Ni, Cu at the B-sites forming compositions $(La,Sr)_2M_{1-x}Cr_{1+x}O_{6-\delta}$ have been investigated. Previous workers have focused on doped lanthanum chromite, where doping is used in the solid-state chemical sense of up to 20% dopant on the B-site, more usually 5 or 10%. Here we report a complex perovskite where two transition metal species occupy the B-site in excess of the percolation limit (for example, >33%). Such dramatic replacement of an active B-site ion by another element would normally significantly degrade its functionality; however, if the two elements act in a complementary fashion, then a dramatically improved new material may be achieved. Samples containing ~50% Co, Ni or Cu were unstable under fuel conditions, with very significant exsolution of metal. This is not surprising because these oxides are unstable, with reduction to the metal under fuel conditions. The stability limit for FeO is very close to fuel conditions; however, MnO is clearly stable under fuel conditions and so we have focused on M = Mn in the investigations reported herein.

Mn, like Cr, supports p-type conduction in oxidizing conditions, and has previously been shown to promote electroreduction under SOFC conditions[7]. Furthermore, Mn is known to accept lower coordination numbers in perovskites,[18] especially for Mn^{3+} ($3d^4$), and so may enhance oxide-ion migration. Such mixed ionic/electronic conduction would be very important because it allows the electro-oxidation process to move away from the three-phase electrode/electrolyte/gas interface onto the anode surface, with considerable catalytic enhancement.

Two methods were applied for materials preparation. For solid-state reaction La_2O_3, $SrCO_3$, Cr_2O_3 and Mn_2O_3 were mixed after drying to remove the absorbed water and air and fired at 1,000–1,400 °C. The second method used was combustion synthesis from the corresponding nitrates and ethylene glycol. Good fuel-cell performance was obtained for these materials as SOFC anodes with 0.3-mm-thick YSZ (Al_2O_3) electrolyte and $La_{0.8}Sr_{0.2}MnO_3$ (LSM) cathode. Platinum paste coated onto LSM and fired at 900 °C was used as the cathode current collector. When pure $La_{0.75}Sr_{0.25}Cr_{0.5}Mn_{0.5}O_3$ was applied as the anode, polarization resistances of 0.9 and 0.47 Ω cm^2 were observed in wet 5% H_2/Ar and wet H_2 respectively at 925 °C. It has been reported previously that the application of a multilayer LSM cathode can significantly decrease the cathode overpotential[19,20]. Similarly, significant decrease of series and anode polarization resistance was observed when a $La_{0.7}Sr_{0.3}Cr_{0.95}Ru_{0.05}O_3$/YSZ multilayer anode was used[21]. In our experiments, the cathodes were multilayer, first layer: $La_{0.8}Sr_{0.2}MnO_3$, 78 wt%, YSZ 22 wt%; second and the third layer pure $La_{0.8}Sr_{0.2}MnO_3$. The anode with a thickness of about 50 µm was deposited onto the YSZ electrolyte using an isopropanol-based slurry and firing typically 1,000–1,300 °C. Graded anodes were applied to optimize performance, first layer: $La_{0.75}Sr_{0.25}Cr_{0.5}Mn_{0.5}O_3$, 75 wt%, YSZ 25 wt%; second and the third layer are pure $La_{0.75}Sr_{0.25}Cr_{0.5}Mn_{0.5}O_3$. The anode was dried at 300 °C, between each layer application, thus due to particle redistribution, the final anode gradation was only between 5 and 10% YSZ from the top surface to the electrolyte interface, as determined by energy-dispersive spectroscopic analysis (JEOL with OxfordLink EDS) A three-electrode arrangement with thick electrolyte was used to allow the anode processes to be investigated in isolation. The electrolyte was a sintered 8 mol.% Y_2O_3 stabilized ZrO_2 (YSZ) pellet of 2 mm thickness and 20 mm diameter. Pt was painted onto the other

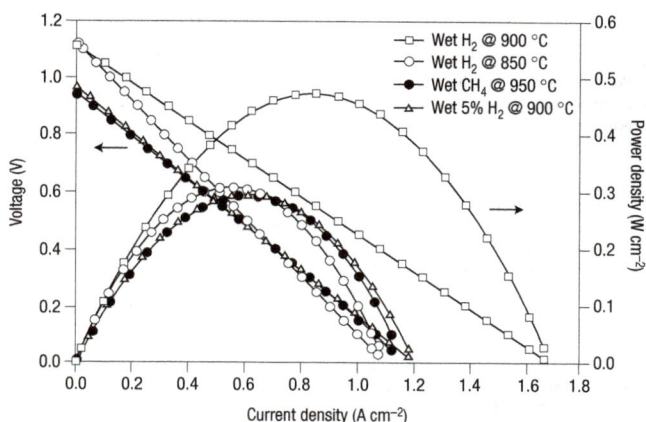

Figure 3 Fuel-cell performance of a ($La_{0.75}Sr_{0.25}$)$_{0.9}$$Cr_{0.5}Mn_{0.5}O_3$ anode in different wet (3% H_2O) atmospheres. Note that the electrolyte resistance, 0.3 Ω at 900 °C, accounts for a polarization loss of 0.15 V at 0.5 A cm^{-2}, and that cathodic losses are also included. The straight-line plots correspond to the voltages (left y axis) and the curved plots the power densities (right y axis)

side of YSZ as counter and reference electrodes. Electronic contacts at anodes were formed using an Au mesh with small amounts of Au paste to ensure contacting and to avoid any extraneous catalytic effect (which could be possible for Pt).

$La_{0.75}Sr_{0.25}Cr_{0.5}Mn_{0.5}O_3$ (LSCM) forms as a single-phase double perovskite on firing at 1,100 °C and above. The structure of $La_{0.75}Sr_{0.25}Cr_{0.5}Mn_{0.5}O_{3-\delta}$ obtained at 1,400 °C may be refined as rhombohedral with $a = 5.4562(3)$ Å, $\alpha = 60.440(9)°$, $V = 116.00(4)$ Å^3. Because Cr and Mn have similar numbers of electrons, cation ordering would not be expected to be observable, even if the phase existed as an ordered double perovskite. No chemical reactions were observed on firing an intimate mixture of LSCM and YSZ at 1,300 °C in air for 80 hours, indicating good chemical compatibility. The phase seems stable under fuel conditions; the perovskite structure is retained on firing in 5% H_2/3% H_2O/92% Ar for 120 hours, with the cell changing to $a = 5.5028(4)$ Å, $b = 5.4823(5)$ Å, $c = 7.767(1)$ Å, but with an increase in unit cell volume of 1%. The perovskite structure tolerates at least 10% A-site deficiency, a feature frequently used to enhance stability of such perovskites in SOFCs. The thermal expansion coefficient is 9.3×10^{-6} K^{-1} in air, and is comparable to that of YSZ, 10.3×10^{-6} K^{-1} in both air and in 5% H_2. On re-oxidation of $La_{0.75}Sr_{0.25}Cr_{0.5}Mn_{0.5}O_{3-\delta}$ in air in a thermogravimetric analyser (TGA), a weight increase of 1.80% was observed, corresponding to a change in oxygen content of 0.25 oxygen atoms per formula unit. Weight changes were reversible on redox cycling, and the perovskite-related structure was retained throughout. The conductivity of $La_{0.75}Sr_{0.25}Cr_{0.5}Mn_{0.5}O_3$ was measured as a function of p_{O_2} at 900 °C. Conductivity is constant at 38 S cm^{-1} p_{O_2} values above 10^{-10} atm; however at lower p_{O_2} values, conductivity decreases, indicating p-type conductivity as the dominant electronic mechanism. The real p_{O_2} at the electrode would be higher under operation due to polarization, which would cause a conductivity increase due to the p-type conduction increasing with effective p_{O_2}. Figure 1 shows the dependence of conductivity on temperature in reducing and oxidizing atmospheres. In both cases the behaviour is indicative of good polaronic semiconductivity, with activation energies for conduction ranging from 0.24 ± 0.02 eV for 300–900 °C, and 0.18 ± 0.01 eV for 20–300 °C in air, and 0.56 ± 0.01 eV in 5% H_2. The change in activation energy on heating in air, may be related to an anticipated rhombohedral to orthorhombic transition

Initial studies of these materials prepared by the combustion method showed good performance when used as anodes in SOFCs. Better performance was achieved using a $Ce_{0.8}Gd_{0.2}O_2$ (CGO) thin film interface prepared by a sol–gel process between the anode and electrolyte. A current of 140 mA cm^{-2} was passed for 4 hours at constant potential without any degradation or observable carbon deposition using LSCM as an anode in 3% H_2O/97% CH_4 at 925 °C with a CGO interface. Such an interface has previously been found to enhance hydrocarbon oxidation and reduce cracking[5]; however on further refinement of the electrode fabrication process, equally good performance was achieved using combustion-synthesis-derived $La_{0.75}Sr_{0.25}Cr_{0.5}Mn_{0.5}O_3$. This anode exhibited a very fine microstructure with uniform particle size of just less than 1 μm and estimated 30% porosity. The anode polarization resistance was measured using three-electrode geometry, and results at different temperatures are shown in Fig. 2a for 5% H_2 and in different atmospheres at 900 °C in Fig. 2b. The polarization resistances in wet 5% H_2 were 0.59 Ω cm^2 at 850 °C, 0.51 Ω cm^2 at 900 °C and 0.30 Ω cm^2 at 950 °C. The series resistance of the electrode is negligible in comparison with that due to the electrolyte, indicating both good lateral and transverse conduction in the electrode. The polarization resistances at 900 °C are: 0.51 Ω cm^2 in wet 5% H_2; 0.26 Ω cm^2 in wet H_2; and 0.87 Ω cm^2 in wet CH_4. The anode polarization in wet H_2 at 900 °C is comparable to a conventional Ni-YSZ cermet anode measured in a similar set-up[22] at 1,000 °C. By tailoring current collection and microstructure, even better Ni/YSZ anodes can be prepared, and similar optimization would be anticipated for this new anode material with future development.

The performance of the A-site deficient variant ($La_{0.75}Sr_{0.25}$)$_{0.9}$$Cr_{0.5}Mn_{0.5}O_3$ anode was also investigated in a conventional two-electrode supported electrolyte cell (Fig. 3). The open circuit voltages (OCVs) for wet 5% H_2 and wet H_2 were close to the value predicted by the Nernst equation, 0.95 and 1.09 V at 900 °C, respectively, and the OCV for wet CH_4 was 0.87 V. The maximum power densities were higher for wet H_2 than wet 5% H_2, with values of 0.47 W cm^{-2} and 0.29 W cm^{-2} respectively at 900 °C. The maximum power density for wet (3% H_2O) CH_4 is about 0.2 Wcm^{-2} at 0.5 V at 900 °C. Similar performance was observed for wet CH_4 at 950 °C as for wet hydrogen at 850 °C. It should be noted that unlike the three-terminal polarization measurements, these two-terminal tests also include polarization losses from both unoptimized electrolyte and cathode, thus the anodic losses are less than 50% of the overall losses. After running a fuel cell in wet methane at 900 °C for 7 hours and cooling down in wet methane, only trace amounts of carbon were detected by TGA-mass spectrometry on the LSCM anode.

LSCM has therefore been demonstrated as a Ni-free single-phase anode with comparable performance in hydrogen to nickel zirconia cermets. In contrast with these cermets, the electrode is active for the electro-oxidation of CH_4 at high temperatures in the absence of excess steam. The relative contributions of direct oxidation and internal reforming have still to be determined. There is evidence that performance is still curtailed by microstructure in the form of second semicircles in the polarization impedance, which might be due to gas concentration or diffusion phenomena, and the steep increase in potential at high currents. These electrodes offer suitable performance for direct incorporation into anode-supported fuel-cell designs, although the absence of very high conductivity in fuel conditions would necessitate the addition of an appropriate current collector in long current path designs such as long tubular.

Received 23 December 2002; accepted 4 March 2003; published 30 March 2003.

References

1. Minh, N. Q. Ceramic fuel-cells. *J. Am. Ceram. Soc.* **76**, 563–588 (1993).
2. Matsuzaki, Y. & Yasuda, I. The poisoning effect of sulfur-containing impurity gas on a SOFC anode: Part I. Dependence on temperature, time and impurity concentration. *Solid State Ionics* **132**, 261–269 (2000).
3. Steele, B. C. H., Kelly, I., Middleton, H. & Rudkin, R. Oxidation of methane in solid-state

electrochemical reactors. *Solid State Ionics* **28–30**, 1547–1552 (1988).

4. Park, S., Vohs, J. M. & Gorte, R. J. Direct oxidation of hydrocarbons in a solid-oxide fuel cell *Nature* **404**, 265–267 (2000).

5. Perry, E., Tsai, T. & Barnett, S. A. A direct-methane fuel cell with a ceria-based anode. *Nature* **400**, 649–651 (1999).

6. Primdahl, S., Hansen, J. R., Grahl-Madsen, L. & Larsen, P. H. Sr-doped LaCrO₃ anode for solid oxide fuel cells. *J. Electrochem. Soc. A* **148**, 74–81 (2001).

7. Holtappels, P., Bradley, J., Irvine, J. T. S., Kaiser, A. & Mogensen, M. Electrochemical characterization of ceramic SOFC anodes. *J. Electrochem. Soc. A* **148**, 923–929 (2001).

8. Pudmich, G. *et al.* Chromite/titanate based perovskites for application as anodes in solid oxide fuel cells. *Solid State Ionics* **135**, 433–438 (2000).

9. Marina, O. A., Canfield, N. L. & Stevenson, J. W. Thermal, electrical, and electrocatalytical properties of lanthanum-doped strontium titanate. *Solid State Ionics* **149**, 21–28 (2002).

10. Marina, O. A. & Pederson, L. R. in *Proc. 5th European Solid Oxide Fuel Cell Forum* (ed. Huijsmans, J.) 481–489 (European SOFC Forum, Switzerland 2002).

11. Yokokawa, H., Sakai, N., Kawada, T. & Dokiya, M. Thermodynamic stabilities of perovskite oxides for electrodes and other electrochemical materials. *Solid State Ionics* **52**, 43–56 (1992).

12. Vernoux, P., Guillodo, M., Fouletier, J. & Hammou, A. Alternative anode materials for gradual methane reforming in solid oxide fuel cells. *Solid State Ionics* **135**, 425–431 (2000).

13. Sfeir, J. *et al.* Lanthanum chromite based catalysts for oxidation of methane directly on SOFC anode. *J. Catal.* **202**, 229–244 (2001).

14. Sfeir, J., Van Herle, J. & Vasquez, R. in *Proc. 5th European Solid Oxide Fuel Cell Forum* (ed. Huijsmans, J.) 570–577 (European SOFC Forum, Switzerland 2002).

15. Sauvet, A. L. & Irvine, J. T. S. in *Proc. 5th European Solid Oxide Fuel Cell Forum* (ed. Huijsmans, J.) 490–498 (European SOFC Forum, Switzerland 2002).

16. Liu, J., Madsen, B. D., Ji, Z. Q. & Barnett, S. A. A fuel-flexible ceramic-based anode for solid oxide fuel cells. *Electrochem. Solid State Lett. A* **5**, 122–124 (2002).

17. Nakamura, T., Petzow, G. & Gauckler, L. J. Stability of the perovskite phase LABO₃ (B = V, Cr, Mn, Fe, Co, Ni) in reducing atmosphere I. Experimental results. *Mater. Res. Bull.* **14**, 649–659 (1979).

18. Poeppelmeier, K. R., Leonowicz, M. E. & Longo, J. M. CaMnO₂.₅ and Ca₂MnO₃.₅ - new oxygen-defect perovskite-type oxides. *J. Solid State Chem.* **44**, 89–98 (1982).

19. Juhl, M., Primdahl, S., Manon, C. & Mogensen, M. Performance/structure correlation for composite SOFC cathodes. *J. Power Sources* **61**, 173–181 (1996).

20. Holtappels, P. & Bagger, C. Fabrication and performance of advanced multi-layer SOFC cathodes. *J. Eur. Ceram. Soc.* **22**, 41–48 (2002).

21. Sauvet A.-L. *Study of Novel Anode Materials for SOFCs Under Methane* PhD Thesis, L'Université Joseph-Fourier de Grenoble, (2001).

22. Brown, M., Primdahl, S. & Mogensen, M. Structure/performance relations for Ni/yttria-stabilized zirconia anodes for solid oxide fuel cells. *J. Electrochem. Soc.* **147**, 475–485 (2000).

Acknowledgements

We thank the Engineering and Physical Sciences Research Council (EPSRC) and the New Energy and Industrial Technology Development Organization (NEDO) (All Perovskite Solid Oxide Fuel Cell UK/Asian SOFC Project) for support.

Correspondence and requests for materials should be addressed to J.I.

Competing financial interests

The authors declare that they have no competing financial interests.

HYDROGEN GENERATION AND STORAGE

Hydrogen-storage materials for mobile applications

Louis Schlapbach[*][†] **& Andreas Züttel**[†]

*EMPA, Swiss Federal Laboratories for Materials Research and Testing, CH-8600 Dübendorf, Switzerland
†University of Fribourg, Physics Department, CH-1700 Fribourg, Switzerland (e-mail: andreas.zuettel@unifr.ch)

Mobility — the transport of people and goods — is a socioeconomic reality that will surely increase in the coming years. It should be safe, economic and reasonably clean. Little energy needs to be expended to overcome potential energy changes, but a great deal is lost through friction (for cars about 10 kWh per 100 km) and low-efficiency energy conversion. Vehicles can be run either by connecting them to a continuous supply of energy or by storing energy on board. Hydrogen would be ideal as a synthetic fuel because it is lightweight, highly abundant and its oxidation product (water) is environmentally benign, but storage remains a problem. Here we present recent developments in the search for innovative materials with high hydrogen-storage capacity.

Energy can be stored in different forms: as mechanical energy (for example, potential energy or rotation energy of a flywheel); in an electric or magnetic field (capacitors and coils, respectively); as chemical energy of reactants and fuels (batteries, petrol or hydrogen); or as nuclear fuel (uranium or deuterium). Chemical and electric energy can be transmitted easily because they both involve electronic Coulomb interaction. Chemical energy is based on the energy of unpaired outer electrons (valence electrons) eager to be stabilized by electrons from other atoms. The hydrogen atom is most attractive because its electron (for charge neutrality) is accompanied by only one proton. Hydrogen thus has the best ratio of valence electrons to protons (and neutrons) of all the periodic table, and the energy gain per electron is very high.

Hydrogen is the most abundant element on Earth, but less than 1% is present as molecular hydrogen gas H_2. The overwhelming majority is chemically bound as H_2O in water and some is bound to liquid or gaseous hydrocarbons. The clean way to produce hydrogen from water is to use sunlight in combination with photovoltaic cells and water electrolysis (see review in this issue by Grätzel, pages 338–344). Other forms of primary energy and other water-splitting processes are also used: the hydrogen consumed today as a chemical raw material (about 5×10^{10} kg per year worldwide) is to a large extent produced using fossil fuels and the reaction of hydrocarbon chains $(-CH_2-)$ with H_2O at high temperatures, which produces H_2 and CO_2. Direct thermal dissociation of H_2O requires temperatures higher than 2,000 °C (>900 °C with a Pt/Ru catalyst).

The chemical energy per mass of hydrogen (142 MJ kg^{-1}) is at least three times larger than that of other chemical fuels (for example, the equivalent value for liquid hydrocarbons is 47 MJ kg^{-1}). Once produced, hydrogen is a clean synthetic fuel: when burnt with oxygen, the only exhaust gas is water vapour, but when burnt with air, lean mixtures have to be used to avoid the formation of nitrogen oxides. Whether hydrogen can be considered a clean form of energy on a global scale depends on the primary energy that is used to split water[1].

The availability of free energy is often unsafe. The mechanical energy of a 1,000-kg car running out of control at 40 km h^{-1} can kill pedestrians. The process of burning hydrogen can be done in an efficient and controlled way to liberate energy at a desirable rate, or in an uncontrolled way with the potential to cause damage. For historical reasons hydrogen has a bad reputation, which is not altogether justified: a more recent analysis[2] of the *Hindenburg* catastrophe shows that the air ship caught fire because of a highly flammable skin material and not because of the hydrogen gas it contained. The safety of hydrogen relies on its high volatility and non-toxicity.

Today, many scientists and engineers, some companies, governmental and non-governmental agencies and even finance institutions are convinced that hydrogen's physical and chemical advantages will make it an important synthetic fuel in the future. After the successful use of hydrogen for space technology, national hydrogen associations were created and joint ventures started. Examples are fuel-cell development projects, the Shell–GfE–Hydro-Québec joint venture[3], the International Hydrogen Energy Association and its technical-economic conferences, and solar hydrogen R&D programmes. But are these ventures aimed at mobility?

There are essentially two ways to run a road vehicle on hydrogen. In the first, hydrogen in an internal combustion engine is burnt rapidly with oxygen from air. The efficiency of the transformation from chemical to mechanical through thermal energy is limited by the Carnot efficiency and is slightly higher for hydrogen–air mixtures (around 25%) than for petrol–air mixtures. When a lean mixture is used, the exhaust gas contains nothing but water vapour; richer mixtures also produce NO_x. In the second method, hydrogen is 'burnt' electrochemically with oxygen from air in a fuel cell, which produces electricity (and heat) and drives an electric engine (see review in this issue by Steele and Heinzel, pages 345–352). Here, the efficiency of the direct process of electron transfer from oxygen to hydrogen is not limited by the Carnot efficiency; it can reach 50–60%, twice as much as the thermal process.

For on-board energy storage, vehicles need compact, light, safe and affordable containment. A modern, commercially available car optimized for mobility and not prestige with a range of 400 km burns about 24 kg of petrol in a combustion engine; to cover the same range, 8 kg hydrogen

Figure 1 Volume of 4 kg of hydrogen compacted in different ways, with size relative to the size of a car. (Image of car courtesy of Toyota press information, 33rd Tokyo Motor Show, 1999.)

are needed for the combustion engine version or 4 kg hydrogen for an electric car with a fuel cell.

Hydrogen is a molecular gas. At room temperature and atmospheric pressure, 4 kg of hydrogen occupies a volume of 45 m³. This corresponds to a balloon of 5 m diameter — hardly a practical solution for a vehicle (Table 1). In the following we will focus on the question of compacting hydrogen, looking at the materials, technology and safety aspects (Fig. 1).

Conventional hydrogen storage

Classical high-pressure tanks made of fairly cheap steel are tested up to 300 bar and regularly filled up to 200 bar in most countries. To store our 4 kg hydrogen still requires an internal volume of 225 litres (about 60 gallons) or 5 tanks of 45 litres each. Novel high-pressure tanks made of carbon-fibre-reinforced composite materials are being developed; these are tested up to 600 bar and filled up to 450 bar for regular use. But they need a special inert inner coating to prevent the high-pressure hydrogen reacting with the polymer. Consequently, another approach is to use hydrogen-inert aluminium tanks and to strengthen them with external carbon-fibre coatings. Spherical containers slightly smaller than 60 cm in diameter would be able to carry our 4 kg, but for practical fabrication a cylindrical shape is preferred.

These high-pressure containers, when full, would contain about 4% hydrogen by mass, but with significant disadvantages: the fuel would be available at a pressure dropping from 450 bar to zero overpressure, so additional pressure control would be essential. High-pressure vessels present a considerable risk — the compression itself is the most dangerous and complicated part. In Japan such vessels are prohibited on the roads in ordinary cars. .

Condensation into liquid or even solid hydrogen is, of course, particularly attractive from the point of view of increasing the mass

per container volume. The density of liquid hydrogen is 70.8 kg m⁻³ (70.6 kg m⁻³ for solid hydrogen). But the condensation temperature of hydrogen at 1 bar is −252 °C and the vaporization enthalpy at the boiling point amounts to 452 kJ kg⁻¹. As the critical temperature of hydrogen is −241 °C (above this temperature hydrogen is gaseous), liquid hydrogen containers are open systems to prevent strong overpressure. Therefore, heat transfer through the container leads directly to the loss of hydrogen. Larger containers have a smaller surface to volume ratio than small containers, so the loss of hydrogen is smaller. The continuously evaporated hydrogen may be catalytically burnt with air in the overpressure safety system of the container or collected again in a metal hydride. (Solid hydrogen is a molecular insulating solid; under high pressure it transforms into metallic, possibly even superconducting hydrogen[4] with T_c of 200–300 °C.)

Cryotechniques for cooling and superinsulated low temperature storage units were developed and proven in space technology. Liquid hydrogen is a fuel in the launching process of the Space Shuttle and in *Ariane*. A Lockheed military-type aircraft and a Tupolev supersonic aircraft have been flown with engines fuelled by liquid hydrogen. BMW have built an automated liquid-hydrogen filling station, and developed and tested several cars running with hydrogen in newly designed vessels to reduce losses by evaporation to below 1.5 mass% per day (BMW, personal communication).

Hydrocarbons with a molecular mass of at least 60 g mol⁻¹ are liquid at room temperature with a density close to 1,000 kg m⁻³. The number of hydrogen atoms per carbon atom can vary in hydrocarbons owing to their ability to change from σ-bonds to π-bonds. Hydrocarbons can be burnt completely by oxidation of carbon into CO_2 and of hydrogen into H_2O; some can also be considered as a liquid storage medium for hydrogen if they can be hydrogenated and dehydrogenated; that is, if their ratio of hydrogen to carbon atoms can be adapted reversibly. Cyclohexane (C_6H_{12}), for example, reversibly desorbs six hydrogen atoms (7.1 mass%) and forms benzene (C_6H_6). Stationary hydrogenation and dehydrogenation under steady-state conditions are managed in numerous chemical plants, but the on-board process under variable conditions is another matter. We do not consider hydrogen storage in NH_3 (5.9 mass%) to be realistic, owing to the corrosive nature of ammonia.

Hydrogen adsorption on solids of large surface area

Hydrogen adsorbs at solid surfaces depending on the applied pressure and the temperature. The variation of attractive surface forces as a function of distance from the surface decides whether van der

Table 1 Physical and chemical properties of hydrogen, methane and petrol

Properties	Hydrogen (H₂)	Methane (CH₄)	Petrol (–CH₂–)
Lower heating value (kWh kg⁻¹)	33.33	13.9	12.4
Self-ignition temperature (°C)	585	540	228–501
Flame temperature (°C)	2,045	1875	2,200
Ignition limits in air (Vol%)	4–75	5.3–15	1.0–7.6
Minimal ignition energy (mW s)	0.02	0.29	0.24
Flame propagation in air (m s⁻¹)	2.65	0.4	0.4
Diffusion coefficient in air (cm² s⁻¹)	0.61	0.16	0.05
Toxicity	No	No	High

Waals-type weak physisorption of molecular hydrogen occurs, or whether dissociation and chemisorption of atomic hydrogen takes place. Owing to the attractive forces, the most stable position for an adsorbed molecule is with its centre at about 1 molecular radius from the surface, and the attractive field rapidly diminishes at greater distances. Once a monolayer of adsorbate molecules or atoms has formed, the gaseous species interacts with the liquid or solid adsorbate. Therefore, the binding energy of the second layer of adsorbates is similar to the latent heat of sublimation or vaporization of the adsorbate. Consequently, adsorption at a temperature at or above the boiling point of the adsorbate at a given pressure leads to the adsorption of a single monolayer. For storage purposes, the adsorption of hydrogen has been studied on carbon species only. Other light and reasonably cheap materials of high surface area may prove to be attractive as well.

The condensation of a monolayer of hydrogen on a solid leads to a maximum of 1.3×10^{-5} mol m^{-2} of adsorbed hydrogen. In the case of a graphene sheet with a specific surface area of 1,315 m^2 g^{-1} as adsorbent, the maximum theoretical concentration is 0.4 H atoms per surface carbon atom or 3.3 mass% hydrogen on sheets with hydrogen atoms on one side. On active carbon with the specific surface area of 1,315 m^{-2} g^{-1}, 2 mass% hydrogen is reversibly adsorbed at a temperature of 77 K (ref. 4). On nanostructured graphitic carbon at 77 K (liquid nitrogen temperature), the reversibly adsorbed quantity of hydrogen correlates with the specific surface area of the sample. It amounts to 1.5 mass% per 1,000 m^2 g^{-1} of specific surface area (Fig. 2), in agreement with the calculation shown in Fig. 3. Nanostructured graphite produced by ball milling for 80 h in a 1-MPa hydrogen atmosphere contains up to 0.95 H atoms per carbon atom or 7.4 mass%. Eighty per cent of this hydrogen desorbs at a temperature of over 600 K. Evidently, moderately high hydrogen absorption can be realized with planar graphitic structures at low temperature[7,8].

Are curved structures more attractive? In microporous solids with capillaries which have a width not exceeding a few molecular diameters (the diameter of H$_2$ is 0.41 nm), the potential fields from opposite walls overlap so that the attractive force acting on hydrogen molecules is greater than that on an open flat surface. The effect of nanotube curvature on the adsorption energy for hydrogen has been investigated theoretically by Stan and Cole[9]. The Feynman (semiclassical) effective potential approximation was used to calculate the adsorption potential and the amount of hydrogen adsorbed on a (13,0) zigzag nanotube. The adsorption potential was found to be 9 kJ mol^{-1} for hydrogen molecules inside the nanotubes at 50 K, about 25% higher than on the flat surface of graphite. The increase arises

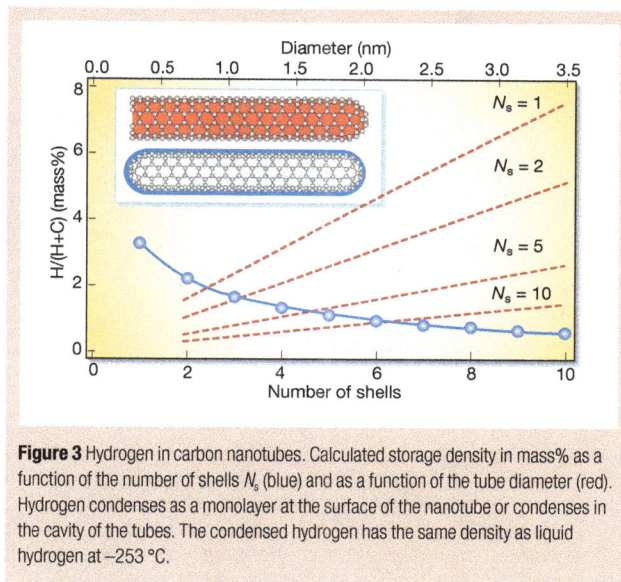

Figure 3 Hydrogen in carbon nanotubes. Calculated storage density in mass% as a function of the number of shells N_s (blue) and as a function of the tube diameter (red). Hydrogen condenses as a monolayer at the surface of the nanotube or condenses in the cavity of the tubes. The condensed hydrogen has the same density as liquid hydrogen at −253 °C.

from the curvature of the surface and the related higher number of carbon atoms interacting with the hydrogen molecule. At low temperatures, far more hydrogen can be adsorbed in the tube than on a flat surface; but the ratio decreases strongly with increasing temperature, from 55 at 50 K to 11 at 77 K.

Conflicting results have been published concerning the reversible storage of hydrogen in carbon nanotubes (see refs 8, 9 for detailed reviews). To a large extent the controversy is caused by insufficient characterization of the carbon material used. It is often a mixture of opened and unopened, single-walled and multi-walled tubes of various diameters and helicities together with other carbonaceous species, just a few of which will be analysed in transmission electron micrographs. To our knowledge, the 'fantastic' results of the Northwestern University group[10] (more than 60 mass% hydrogen in specific carbon fibres) have not been reproduced elsewhere. Heben et al.[11,12] still claim to reach 6–8 mass% reversible storage (cooling at 1 bar down to liquid nitrogen temperature) in high-purity single-walled tubes opened by sonication. But Hirscher et al.[13] revealed that a sonication bar made of titanium alloy results in hydrogen-storing titanium alloy particles in the nanotube material; storage was below 1 mass% when stainless steel sonication bars were used. Zuettel et al.[14] showed by thermal desorption mass spectroscopy that after reaction of carbon nanotubes with hydrogen at elevated temperatures, hydrocarbons are desorbed. The large mass increase observed by Chen et al.[15] when carbon nanotubes filled with alkaline metals were exposed to hydrogen has been attributed to hydroxide formation[16].

Nijkamp et al.[17] have surveyed the storage capacities of a large number of different carbonaceous adsorbents for hydrogen at 77 K and 1 bar. They concluded that microporous adsorbents, for example zeolites and activated carbons, have appreciable sorption capacities. Optimization of sorbent and adsorption conditions is expected to lead to adsorption of 560 ml STP g^{-1}, close to targets set for practical use in vehicles. Zuettel et al.[18] concluded that charging of carbon nanotubes with hydrogen at liquid nitrogen temperature (77 K), or cathodically at ambient conditions, is due to physisorption. The amount of adsorbed hydrogen is proportional to the specific surface area of the carbon nanotube material and limited to 2 mass% for carbon materials (Fig. 2).

Taking all these results together, we consider it to be scientifically interesting and challenging to continue research on the interaction of hydrogen with different and well-characterized carbon nanostructures. Whether a hydrogen-storage material will emerge from it, however, remains an open question.

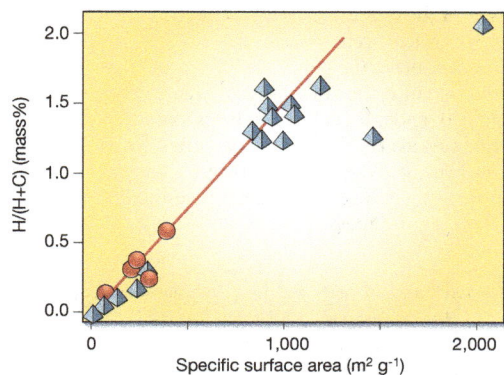

Figure 2 Reversibly stored amount of hydrogen on various carbon materials versus the specific surface area of the samples. Circles represent nanotube samples (best-fit line indicated), triangles represent other nanostructured carbon samples[17].

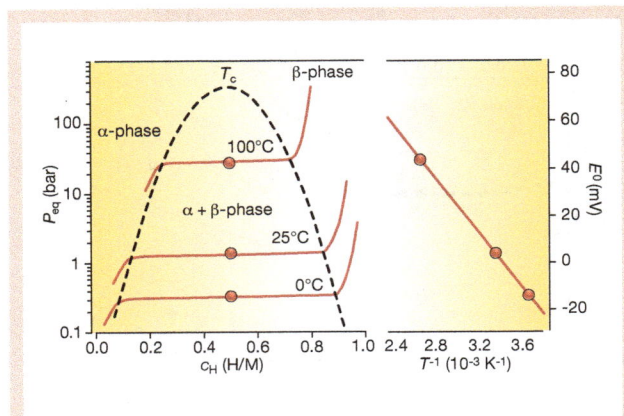

Figure 4 Pressure–concentration–temperature plot and a van't Hoff curve (logarithm of the equilibrium or plateau pressure against the reciprocal temperature); values are for LaNi$_5$. The vertical axes indicate the corresponding hydrogen pressure or the equivalent electrochemical potential. From the slope of the van't Hoff plot, experimental values of the enthalpy of hydride formation ΔH can be evaluated. The plateau pressure $p_{eq}(T)$ as a function of temperature is related to the changes ΔH and ΔS of enthalpy and entropy, respectively, by the van't Hoff equation:
$$\ln(p_{eq}/p_{eq}^0) = (-\Delta H/R)(1/T) + \Delta S/R.$$

In an interesting sideline to work[19] — addressing the potential of metallization and superconductivity of hydrogen, rather than its storage — some hydrogen was reported to be sorbed into graphite intercalation compounds made of alkaline metals between graphene sheets. Intercalation of hydrogen alone between the graphene layers in macroscopic structures has never been reported. The dynamic diameter of a free hydrogen molecule ($d = 0.4059$ nm) is larger than the interlayer distance in graphite ($d = 0.3355$ nm), so a significant size mismatch would have to be accommodated upon entry.

Hydrogen storage by metal hydrides

Many metals and alloys are capable of reversibly absorbing large amounts of hydrogen. Charging can be done using molecular hydrogen gas or hydrogen atoms from an electrolyte. Molecular hydrogen is dissociated at the surface before absorption; two H atoms recombine to H$_2$ in the desorption process. The thermodynamic aspects of hydride formation from gaseous hydrogen are described by pressure–composition isotherms (Fig. 4). The host metal initially dissolves some hydrogen as a solid solution (α-phase). As the hydrogen pressure together with the concentration of H in the metal is increased, interactions between hydrogen atoms become locally important, and we start to see nucleation and growth of the hydride (β) phase. While the two phases coexist, the isotherms show a flat plateau, the length of which determines how much H$_2$ can be stored reversibly with small pressure variations. In the pure β-phase, the H$_2$ pressure rises steeply with the concentration. At higher H$_2$ pressure, further plateaux and further hydride phases may be formed. The two-phase region ends in a critical point T_C, above which the transition from α- to β-phase is continuous. The plateau or equilibrium pressure depends strongly on temperature and is related to the changes ΔH and ΔS of enthalpy and entropy, respectively. As the entropy change corresponds mostly to the change from molecular hydrogen gas to dissolved hydrogen, it is roughly 130 J K^{-1} mol^{-1} for all metal–hydrogen systems under consideration. The enthalpy term characterizes the stability of the metal–hydrogen bond. To reach an equilibrium pressure of 1 bar at 300 K, ΔH should amount to 19.6 kJ mol$_H^{-1}$ (ref. 20). The operating temperature of a metal hydride system is fixed by the plateau pressure in thermodynamic equilibrium and by the overall reaction kinetics[21].

Hydrogen is located in the form of atoms, never molecules, on interstitial sites of the host metal lattice. The lattice expands during hydrogen sorption, often losing some of its high symmetry. As a consequence of the coexistence of the non-expanded α-phase and anisotropically expanded β-phase, lattice defects and internal strain fields are formed, which end in a decrepitation of brittle host metals such as intermetallics. The H atoms vibrate about their equilibrium position, and perform local motions and long-range diffusion.

In terms of electronic structure, the proton acts as an attractive potential to the host metal electrons; electronic bands are lowered in energy and form low-lying bands by hybridization with the hydrogen band, 6–8 eV below the Fermi level. The Fermi level itself is shifted, and various phase transitions (metal–semiconductor, magnetic–non-magnetic, reflecting–transparent, order–disorder) may occur. The metal–hydrogen bond offers the advantage of very high hydrogen density at moderate pressure and desorption of all stored hydrogen at the same pressure.

Which metallic systems are appropriate for hydrogen storage? Many elemental metals form hydrides, for example PdH$_{0.6}$, rare earth REH$_2$, REH$_3$ or MgH$_2$, none of which are in the pressure and temperature range attractive for mobile storage (1–10 bar, 0–100 °C, corresponding to an enthalpy change between 15 and 24 kJ mol$_H^{-1}$). The discovery of hydrogen sorption by intermetallic compounds created great hopes and stimulated R&D efforts worldwide. Well-known compounds and their properties are summarized in Table 2 (ref. 22).

Alloys derived from LaNi$_5$ show some very promising properties, including fast and reversible sorption with small hysteresis, plateau pressure of a few bars at room temperature and good cycling life. The volumetric hydrogen density (crystallographically) of LaNi$_5$H$_{6.5}$ at 2 bar equals that of gaseous molecular hydrogen at 1,800 bar; but advantageously, all the hydrogen desorbs at a pressure of 2 bar. The density for practical purposes is reduced by the packing fraction of LaNi$_5$ powder, but is still above that of liquid hydrogen (Fig. 5). Storage in these intermetallics allows very safe hydrogen handling. But as lanthanum and nickel are large elements, the proportion of hydrogen in LaNi$_5$H$_{6.5}$ remains below 2 mass%. This is attractive for electrochemical hydrogen storage in rechargeable metal hydride electrodes, reaching a capacity of 330 mA h g^{-1}, and produced and sold in more than a billion AB$_5$-type metal hydride batteries per year[23–25].

For gaseous hydrogen fuel tanks to be used in vehicles, however, this is not enough. We need some 4–5 mass% (indeed, 6.5 mass% and 62 kg H$_2$ m^{-3} are the targets of the US Department of Energy). This low mass density is the general weakness of all known metal hydrides working near room temperature (Fig. 6). Of course, many intermetallic compounds and alloys are known that form hydrides with up to 9 mass% hydrogen (such as Li$_3$Be$_2$H$_7$; ref. 26) and 4.5 H atoms per metal atom (BaReH$_9$; ref. 27), but they are not reversible within the required range of temperature and pressure.

When reaction kinetics rather than thermodynamic equilibrium conditions limit the hydride formation and decomposition, various physical and chemical pretreatments can be applied. Ball milling has the benefits of reducing the grain size, increasing the defect concentration and shortening the diffusion path. Fluorination treatments[28] are well suited to render the surface active, apparently by means of the formation of surface nickel precipitates, the role of which is well known in the hydrogen dissociation process.

Many promising new ideas are under investigation with the clear goal of enhancing the mass density. For example, new families of

Table 2 Intermetallic compounds and their hydrogen-storage properties[22]

Type	Metal	Hydride	Structure	mass%	p_{eq}, T
Elemental	Pd	PdH$_{0.6}$	$Fm3m$	0.56	0.020 bar, 298 K
AB$_5$	LaNi$_5$	LaNi$_5$H$_6$	$P6/mmm$	1.37	2 bar, 298 K
AB$_2$	ZrV$_2$	ZrV$_2$H$_{5.5}$	$Fd3m$	3.01	10^{-8} bar, 323 K
AB	FeTi	FeTiH$_2$	$Pm3m$	1.89	5 bar, 303 K
A$_2$B	Mg$_2$Ni	Mg$_2$NiH$_4$	$P6222$	3.59	1 bar, 555 K
Body-centred cubic	TiV$_2$	TiV$_2$H$_4$	b.c.c.	2.6	10 bar, 313 K

alloys are being studied in several Japanese laboratories[29–31] and are included into the national 'Protium' programme. They are based on vanadium, zirconium and titanium as rather electropositive components combined with $3d$ and $4d$ transition metals. Reversible hydrogen-storage capacities approaching 3 mass% around room temperature have been reported.

A higher mass density is reachable only with light elements such as calcium and magnesium. In fact, Mg forms ionic, transparent MgH_2 containing 7.6 mass% hydrogen. But its formation from bulk Mg and gaseous hydrogen is extremely slow, and in thermodynamic equilibrium a plateau pressure of 1 bar requires not room temperature but 300 °C. Different processes have been tried to obtain micro- or nanostructured Mg: precipitations from metal-organic solutions or high-energy ball milling of Mg have proved successful ways to obtain good charging or discharging kinetics at 150 °C, and the thermodynamics evidently are not affected[26,29–31]. Alloying Mg before the hydride formation is another approach: Mg_2Ni forms a ternary complex hydride Mg_2NiH_4, which still contains 3.6 mass% hydrogen. The hydride forms fairly rapidly, probably owing to the presence of Ni as catalyst for the dissociation of molecular hydrogen, but thermodynamically it still requires 280 °C for 1 bar hydrogen. The alloys Mg_2Cu, $Mg_{17}La_2$ and MgAl, and some other known alloys or intermetallic compounds of Mg, react readily (MgAl after ball milling) with hydrogen and decompose into MgH_2 and another compound or hydride. One example is $Mg_2Cu + H_2$ which decomposes to $MgH_2 + MgCu_2$. The reactions are reversible at high temperature.

In addition to the enthalpy change of the hydrogen sorption reaction, we now also have to consider the enthalpy change of the host metal system. Could this make Mg-based storage alloys more attractive? If the transformation of the host metal system that accompanies hydrogen desorption is exothermic itself, the net enthalpy of hydrogen desorption becomes smaller. In the systems studied so far (Mg–Cu, Mg–Al), the temperature for 1 bar equilibrium pressure could be lowered from 300 °C to 280 °C. This is not a large improvement, and it has the penalty that the mass density is thereby reduced as well. Magnesium does not form a binary intermetallic compound with iron, but in the presence of hydrogen it is possible to synthesize the rather stable ternary hydride Mg_2FeH_6 with 5.5 mass% hydrogen[26].

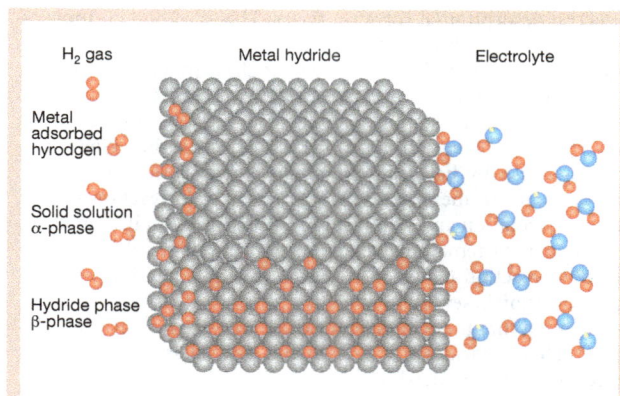

Figure 5 Schematic model of a metal structure with H atoms in the interstices between the metal atoms, and H_2 molecules at the surface. Hydrogen atoms are from physisorbed hydrogen molecules on the left-hand side and from the dissociation of water molecules on the right-hand side.

A different approach is to use composite materials, in which attractive properties of two components are combined to overcome their weaknesses. For instance, magnesium has been ball milled with graphitic carbon or mixed with hydrides showing fast kinetics, such as $LaNi_5$ or Mg_2Ni. It comes as no surprise that the capacities reached values between those of the components[5,26,32].

Alanates and other light hydrides

Some of the lightest elements in the periodic table, for example lithium, boron, sodium and aluminium, form stable and ionic compounds with hydrogen. The hydrogen content reaches values of up to 18 mass% for $LiBH_4$. However, such compounds desorb the hydrogen only at temperatures from 80 °C up to 600 °C, and the reversibility of the reaction is not yet clear for all systems. Bogdanovic and Schwickardi[33] showed in 1996 that the decomposition temperature of $NaAlH_4$ can be lowered by doping the hydride with TiO_2. The same group showed the reversibility of the reaction

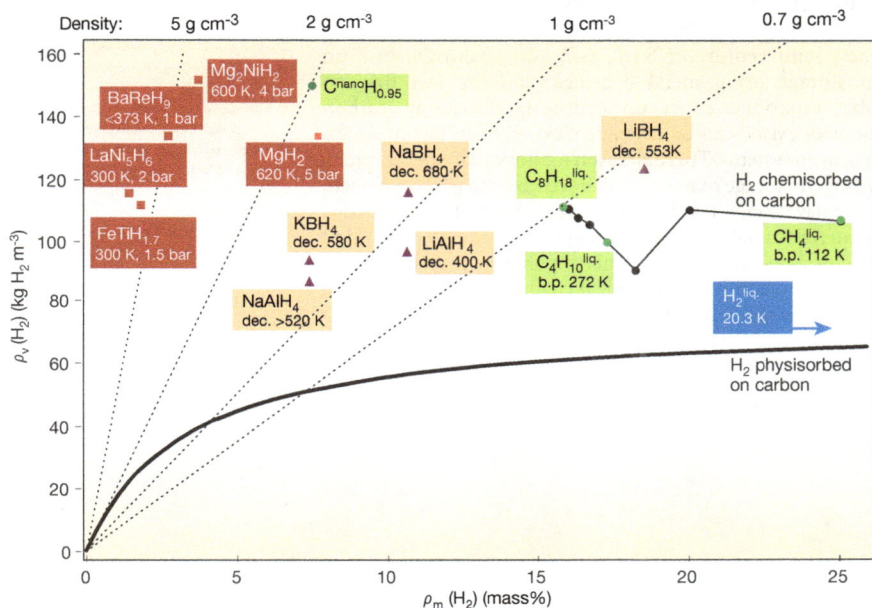

Figure 6 Stored hydrogen per mass and per volume. Comparison of metal hydrides, carbon nanotubes, petrol and other hydrocarbons.

for several desorption/absorption cycles. This is a good example of the potential of such hydrides, which were discovered more than 50 years ago. But several points have to be clarified. First, is the high desorption temperature due to the poor kinetics of the system or due to the thermodynamic stability of the compound? The kinetics can be improved by applying an appropriate catalyst to the system and apparently also by ball milling and the introduction of defects. Second, what are the conditions for a reversible reaction — for example, formation and stabilization of clusters of an intermetallic compound of the remaining metals on desorption? Third, what is the desorption reaction and what are the intermediate reaction products (decomposition in several steps)?

Assuming that these questions can be answered, what would be the role of light hydrides in an on-board fuel cell? The exhaust gas of a fuel cell is water vapour, which could be collected and reused for on-board hydrogen production. The common experiment — shown in many chemistry classes — where a small piece of sodium floating on water produces hydrogen, demonstrates such a process. The sodium is transformed to sodium hydroxide in this reaction. The reaction is not reversible, but the sodium hydroxide could later be removed and reduced in a solar furnace back to sodium. Each sodium atom produces one hydrogen atom, so the corresponding gravimetric hydrogen density of the sodium reaction is slightly more than 4 mass%. Lithium used in the same way would deliver up to 14 mass% of hydrogen. The alkali metals as a hydrogen source are easy to handle, and a car could be refilled within a few minutes. To deliver the necessary 4 kg of hydrogen using the water produced in the fuel cell would take 28 kg of lithium. After using up all the hydrogen the tank would contain 99 kg of lithium hydroxide, ready to be recycled.

Visions for the future

Road traffic is likely to increase even more. Must road vehicles have to satisfy human prestige, and do we need a strong and heavy vehicle cage to protect drivers and passengers from being injured in case of inevitable mistakes? As soon as wireless, electronic control systems are installed to help to avoid crashes, much lighter and more energy-efficient vehicles will be built in countries that do not consider energy availability to be unlimited. In Europe and Japan, there are already small and lightweight cars available with a combustion engine using as little as 3 l of petrol per 100 km (that is, less than half the amount used by conventional, widely used cars). Four kilograms of hydrogen are enough to run 400 km with a combustion engine, and only 2 kg hydrogen for a modern car driven by a fuel cell.

The new joint venture of Shell, GfE and Hydro-Québec[3] on hydrogen storage using metal hydrides, and the fact that no comparable economic effort on hydrogen storage in carbon nanostructures exists, can be taken as clear signs in favour of the metal–hydrogen systems. There is reason for hope that one day much better hydrogen-storage materials will be discovered and developed, rather as we have seen a revolution in high-temperature superconducting materials or hard permanent magnets. We must bear in mind that to develop a sustainable future energy policy requires us to focus not only on the scientific and technical challenge, but also on vital adaptations by the socioeconomic system and a change in attitudes to energy. Sustainability — and humanity — will profit if the price we pay for energy includes costs for long-term production, transport and distribution of energy, materials, and restoration of the damaged environment. □

1. Winter, C. J. & Nitsch, J. Hydrogen as an Energy Carrier: Technologies, Systems, Economy (Springer, 1988).
2. Bain, A. & Van Vorst, W. D. Int. J. Hydrogen Energy 24, 399–403 (1999).
3. Shell Hydrogen, Hydro-Québec (HQ) & Gesellschaft für Elektrometallurgie (GfE). Hydrogen storage joint venture to be established. < http://www.shell.com> Press release (12-07-2001).
4. Nellis, W. J., Louis, A. A. & Ashcroft, N. W. Metallization of fluid hydrogen. Phil. Trans. R. Soc. Lond. A 356, 119–135 (1998).
5. Orimo, S.-I. et al. Hydrogen in the mechanically prepared nanostructured graphite. Appl. Phys. Lett. 75, 3093 (1999).
6. Orimo, S., Matsushima, T., Fujii, H., Fukunaga, T. & Majer, G. Defective carbon for hydrogen storage—thermal desorption property of the mechanically prepared nanostructured graphite. J. Appl. Phys. (in the press).
7. Stan, G. & Cole, M. W. Hydrogen adsorption in nanotubes. J. Low Temp. Phys. 110, 539–544 (1998).
8. Hirscher, M. (ed.) Hydrogen storage in nanoscale carbon and metals. Appl. Phys. A (special issue) 72, 2 (2001).
9. Sholl, C. A. & Gray, E. MacA. (eds) Proc. Int. Symp. Metal Hydrogen Systems—Fundamentals and Applications, Noosa, Australia, 1-6 October 2000. J. Alloys Compounds (in the press).
10. Chambers, A., Park, C., Baker, R. T. K. & Rodriguez, N. M. Hydrogen storage in graphite nanofibers. Phys. Chem. B 102, 4253–4256 (1998).
11. Dillon, A. C. et al. Storage of hydrogen in single-walled carbon nanotubes. Nature 386, 377–379 (1997).
12. Dillon, A. C. et al. Carbon nanotube materials for hydrogen storage. Proc. 2000 DOE/NREL Hydrogen program review, 8-10 May 2000.
13. Hirscher, M. et al. Hydrogen storage in sonicated carbon materials. Appl. Phys. A 72, 129–132 (2001).
14. Züttel A. et al. Hydrogen sorption by carbon nanotubes and other carbon nanostructures. J. Alloys Compounds (in the press).
15. Chen, P., Wu, X., Lin, J. & Tan, K. L. High H$_2$ uptake by alkali-doped carbon nanotubes under ambient pressure and moderate temperatures. Science 285, 91–93 (1999).
16. Hirscher, M. et al. Hydrogen storage in carbon nanostructures. J. Alloys Compounds (in the press).
17. Nijkamp, M. G., Raaymakers, J. E. M. J., Van Dillen, A. J. & De Jong, K. P. Hydrogen storage using physisorption—materials demands. Appl. Phys. A 72, 619–623 (2001).
18. Züttel, A. et al. Hydrogen storage in carbon nanostructures. Int. J. Hydrogen Energy (in the press).
19. Enoki, T., Shindo, K. & Sakamoto, N. Electronic properties of alkali-metal-hydrogen-graphite intercalation compounds. Z. Phys. Chem. 181, 75–82 (1993).
20. Schlapbach, L. (ed.) Hydrogen in Intermetallic Compounds I. Electronic, Thermodynamic, and Crystallographic Properties, Preparation (Topics in Applied Physics Vol. 63) (Springer, 1988).
21. Schlapbach, L. (ed.) Hydrogen in Intermetallic Compounds II. Surface and Dynamic Properties, Applications (Topics in Applied Physics Vol. 67) (Springer, 1992).
22. Sandrock, G. & Thomas, G. The IEA/DOC/SNL on-line hydride databases. Appl. Phys. A 72, 153–155 (2001).
23. Sakai, T., Natsuoka, M. & Iwakura, C. Rare earth intermetallics for metal–hydrogen batteries. Handb. Phys. Chem. Rare Earths 21, 135–180 (1995).
24. Latroche, M., Percheron-Guegan, A. & Chabre, Y. Influence of cobalt content in MmNi$_{(4.3-x)}$Mn$_{0.3}$Al$_{0.4}$Co$_x$ alloy (x = 0.36 and 0.69) on its electrochemical behaviour studied by in situ neutron diffraction. J. Alloys Compounds 295, 637–642 (1999).
25. Schlapbach, L., Felix Meli, F., Züttel, A., Westbrook, J. H. & Fleischer, R. L. (eds) in Intermetallic Compounds: Principles and Practice Vol. 2, Ch. 22 (Wiley, 1994).
26. Zaluska, A., Zaluski, L. & Stroem-Olsen, J. O. Structure, catalysis and atomic reactions on the nano-scale: a systematic approach to metal hydrides for hydrogen storage. Appl. Phys. A 72, 157 (2001).
27. Yvon, K. Complex transition metal hydrides. Chimia 52, 613–619 (1998).
28. Liu, F. J. & Suda, S. A method for improving the long-term storability of hydriding alloys by air water exposure. J. Alloys Compounds 231, 742–750 (1995).
29. Akiba, E. & Iba, H. Hydrogen absorption by Laves phase related BCC solid solution. Intermetallics 6, 461–470 (1998).
30. Kuriiwa, T. et al. New V-based alloys with high protium absorption and desorption capacity. J. Alloys Compounds 295, 433–436 (1999).
31. Tsukahara, M. et al. Hydrogen storage and electrode properties of V-based solid solution type alloys prepared by a thermic process. J. Electrochem. Soc. 147, 2941–2944 (2000).
32. Inoue, H. et al. Effect of ball-milling with Ni and Raney Ni on surface structural characteristics of TiV2.1Ni0.3 alloy. J. Alloys Compounds 325, 299–303 (2001).
33. Bogdanovic, B. & Schwickardi, M. Ti-doped alkali metal aluminium hydrides as potential novel reversible hydrogen storage materials. J. Alloys Compounds 253, 1–9 (1997).

Acknowledgements
We thank the Swiss Federal Office of Energy (BFE), in contract with IEA, the Swiss Federal Office of Education and Science (BBW), and the University of Fribourg and EMPA for support of our hydrogen-storage research projects.

A metal-free polymeric photocatalyst for hydrogen production from water under visible light

Xinchen Wang[1,2]*, Kazuhiko Maeda[3], Arne Thomas[1], Kazuhiro Takanabe[3], Gang Xin[3], Johan M. Carlsson[4], Kazunari Domen[3]* and Markus Antonietti[1]

The production of hydrogen from water using a catalyst and solar energy is an ideal future energy source, independent of fossil reserves. For an economical use of water and solar energy, catalysts that are sufficiently efficient, stable, inexpensive and capable of harvesting light are required. Here, we show that an abundant material, polymeric carbon nitride, can produce hydrogen from water under visible-light irradiation in the presence of a sacrificial donor. Contrary to other conducting polymer semiconductors, carbon nitride is chemically and thermally stable and does not rely on complicated device manufacturing. The results represent an important first step towards photosynthesis in general where artificial conjugated polymer semiconductors can be used as energy transducers.

The search for suitable semiconductors as photocatalysts for the splitting of water into hydrogen gas using solar energy is one of the noble missions of material science. An optimal material would combine an ability to dissociate the water molecules, having a bandgap that absorbs light in the visible range and to remain stable in contact with water. Besides, it should be non-toxic, abundant and easily processable into a desired shape. During the past 30 years, various inorganic semiconductors and molecular assemblies have been developed as catalysts for hydrogen production from water under visible light[1-12]. Semiconductors explored so far are constructed from transition-metal ions with d^0 electronic configuration or post-transition-metal ions of d^{10} configuration, along with group VA or VIA ions as counter-anion components[2,4-12]. For photocatalysis to be chemically productive, precious-metal species[11] such as Pt and RuO_2 must be used in most cases as extra cocatalysts to promote the transfer of photoinduced charge carriers from the bulk to the surface at which water is converted to hydrogen gas. Metal-based complexes (for example, a complex with four manganese ions in photosystem II (ref. 13) and a di-iron centre in hydrogenases[14]) are in natural enzymes the active sites photocatalysing the decomposition of water. Synthetic polymer semiconductors such as polyparaphenylene[15] have also been used for hydrogen production; however, they are active only in the ultraviolet region and have moderate performance. Here, we show that another simple polymer-like semiconductor, made of only carbon and nitrogen, can function as a metal-free photocatalyst for the extraction of hydrogen from water.

Carbon nitrides can exist in several allotropes with diverse properties, but the graphitic phase is regarded as the most stable under ambient conditions. The first synthesis of a polymeric carbon nitride, melon, was already reported by Berzelius and Liebig in 1834 (ref. 16), and it is therefore one of the oldest synthetic polymers

reported. In recent work, we used the thermal polycondensation of common organic monomers to synthesize graphitic carbon nitrides ($g-C_3N_4$) with various architectures[17,18]. The graphitic planes are constructed from tri-s-triazine units connected by planar amino groups (Fig. 1a). The in-plane organization of tri-s-triazine units and the compression of aromatic planes were found to follow the perfection of condensation (see Supplementary Information, Figs S1,S2), enabling the generation of carbon nitride polymers with adjustable electronic properties, while keeping the free shapeability of a polymer made from liquid precursors. The carbon nitrides used in this study were prepared by heating cyanamide (Aldrich, 99% purity) to temperatures between 673 and 873 K (ramp: 2.2 K min^{-1}) for 4 h. For the material condensed at 823 K, an in-planar repeat period of 0.681 nm (for example, the distance of the nitride pores) in the crystal is evident from the X-ray powder diffraction (XRD) pattern (Fig. 1b), which is smaller than one tri-s-triazine unit ($\sim$0.713 nm), presumably owing to the presence of small tilt angularity in the structure. The strongest XRD peak at 27.4°, corresponding to 0.326 nm, is due to the stacking of the conjugated aromatic system, as in graphite (see Supplementary Information, Fig. S2). It should be noted that all of the materials feature some residual amount of hydrogen, which decreases with increasing condensation temperature.

Different thermal condensation enables the finer adjustment of the electronic and optical properties, as indicated by the ultraviolet–visible absorption spectrum for carbon nitrides prepared using different condensation temperatures (see Supplementary Information, Fig. S3). The absorption edge is moved towards longer wavelengths, indicating a decreasing bandgap with increasing condensation temperatures. The bandgap of the condensed graphitic carbon nitride is estimated to be 2.7 eV from its ultraviolet–visible spectrum (Fig. 1c), showing an intrinsic

[1]Max-Planck Institute of Colloids and Interfaces, Department of Colloid Chemistry, Research Campus Golm, 14424 Postdam, Germany, [2]Research Institute of Photocatalysis, State Key Laboratory Breeding Base of Photocatalysis, Fuzhou University, Fuzhou 350002, China, [3]Department of Chemical System Engineering, School of Engineering, The University of Tokyo, Bunkyo-ku, Tokyo 113-8656, Japan, [4]Fritz-Haber-Institute of the Max-Planck-Society, Theory Department, Faradayweg 4-6, D-14195 Berlin, Germany. *e-mail: xcwang@fzu.edu.cn; domen@chemsys.t.u-tokyo.ac.jp.

When citing this article, please cite the original version as shown on the contents page of this chapter.

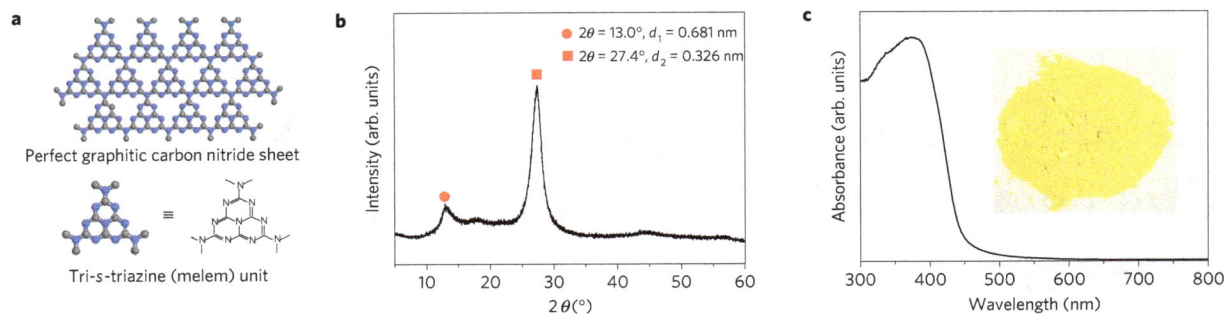

Figure 1 | Crystal structure and optical properties of graphitic carbon nitride. a, Schematic diagram of a perfect graphitic carbon nitride sheet constructed from melem units. **b**, Experimental XRD pattern of the polymeric carbon nitride, revealing a graphitic structure with an interplanar stacking distance of aromatic units of 0.326 nm. **c**, Ultraviolet–visible diffuse reflectance spectrum of the polymeric carbon nitride. Inset: Photograph of the photocatalyst.

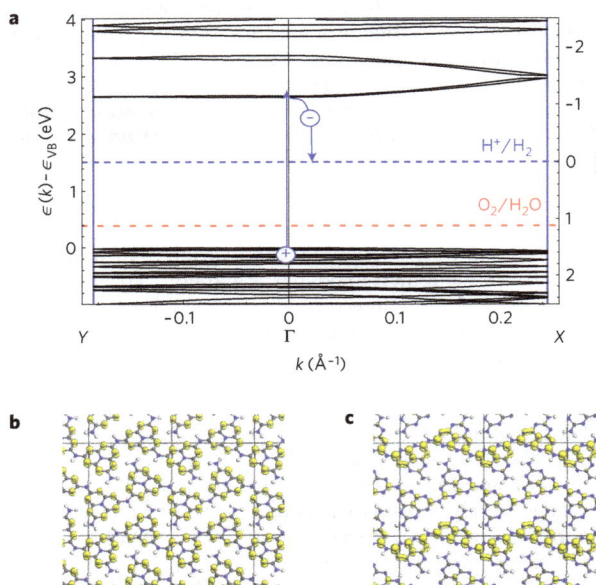

Figure 2 | Electronic structure of polymeric melon.

a, Density-functional-theory band structure for polymeric melon calculated along the chain (Γ-X direction) and perpendicular to the chain (Y-Γ direction). The position of the reduction level for H^+ to H_2 is indicated by the dashed blue line and the oxidation potential of H_2O to O_2 is indicated by the red dashed line just above the valence band. **b**, The Kohn–Sham orbitals for the valence band of polymeric melon. **c**, The corresponding conduction band. The carbon atoms are grey, nitrogen atoms are blue and the hydrogen atoms are white. The isodensity surfaces are drawn for a charge density of $0.01q_e$ Å^{-3}.

semiconductor-like absorption in the blue region of the visible spectrum. This bandgap is sufficiently large to overcome the endothermic character of the water-splitting reaction (requiring 1.23 eV theoretically).

However, in addition to the magnitude of the bandgap, the character of the valence and conduction band and their absolute energies with respect to the reduction and oxidation levels are also important. To analyse the electronic structure, we have carried out density-functional-theory calculations[19] for different reaction intermediates along the condensation path in Supplementary Information, Fig. S1. The calculated bandgap decreases from the highest occupied molecular orbital/lowest unoccupied molecular

orbital (HOMO–LUMO) gap in the melem molecule of 3.5 eV via 2.6 eV for polymeric melon[20] (Fig. 2) down to 2.1 eV for an infinite sheet of a hypothetically, fully condensed g-C$_3$N$_4$. Although the magnitude of the bandgap is underestimated, the trend in our calculations supports our interpretation of the ultraviolet–visible experiments. Figure 2a shows that the band structure of the recently suggested polymeric melon structure[20] has a non-isotropic band structure with a direct bandgap at the Γ point and only dispersion along the Γ–X direction parallel to the chain. The wavefunction of the valence band in Fig. 2b is a combination of the HOMO levels of the melem monomer, which are derived from nitrogen p_z orbitals. The conduction band can similarly be connected to the LUMO of the melem monomer, which consists predominantly of carbon p_z orbitals. Photoexcitation consequently leads to a spatial charge separation between the electron in the conduction band and the hole in the valence band. This suggests that the nitrogen atoms would be the preferred oxidation sites for H_2O to form O_2, whereas the carbon atoms provide the reduction sites for H^+ to H_2. We have finally calculated the absolute value of the reduction and oxidation levels for H_2O to check that they are both located inside the bandgap. The reaction energies were calculated with *ab initio* thermodynamics[21], where the free-energy contribution of the molecules was calculated in the ideal-gas approximation and the solvation energies for H^+ and H_2O were taken from the literature[22,23]. Our value for the reduction level of hydrogen $\mu_e^{H^+/H_2} = -4.18$ eV with respect to the vacuum level corresponds to an absolute value for the normal hydrogen electrode potential (NHE) $V_{NHE} = 4.18$ V. The calculated value for the oxidation level of H_2O, $\mu_h^{O_2/H_2O} = 5.30$ eV with respect to the vacuum level, corresponds to the oxidation potential $V_{O_2/H_2O} - V_{NHE} = 1.12$ V. Both of these values are slightly underestimated compared with experimental values $V_{NHE} = 4.44$ V and $V_{O_2/H_2O} - V_{NHE} = 1.23$ V (ref. 23). Figure 2a shows that the reduction level for H^+ is well positioned in the middle of the bandgap. This ascertains that the reduction process is energetically possible. The oxidation level is located slightly above the top of the valence band, which would permit transfer of holes, but presumably with a low driving force. These calculations provide evidence that carbon nitride has the potential to function as a photocatalyst for hydrogen production.

The photocatalysis experiments were carried out with g-C$_3$N$_4$ in powder form (Fig. 1c, inset) to provide sufficient surface area. It is insoluble in water as well as in acid (HCl, pH=0) or base (NaOH, pH = 14). The as-prepared g-C$_3$N$_4$ achieved steady H_2 production from water containing triethanolamine as a sacrificial electron donor on light illumination ($\lambda > 420$ nm) even in the absence of noble metal catalysts such as Pt, as shown in Fig. 3, curve (i). These results indicate that g-C$_3$N$_4$ functions as a stable 'metal-free' photocatalyst for visible-light-driven H_2 production.

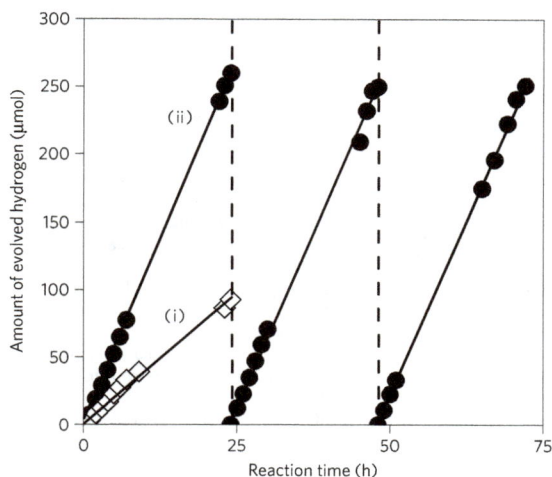

Figure 3 | Stable hydrogen evolution from water by g-C₃N₄. A typical time course of H_2 production from water containing 10 vol% triethanolamine as an electron donor under visible light (of wavelength longer than 420 nm) by (i) unmodified g-C_3N_4 and (ii) 3.0 wt% Pt-deposited g-C_3N_4 photocatalyst. The reaction was continued for 72 h, with evacuation every 24 h (dashed line). Unmodified g-C_3N_4 also photocatalysed steady H_2 production from aqueous methanol solution (10 vol%), as shown in Supplementary Information, Fig. S4.

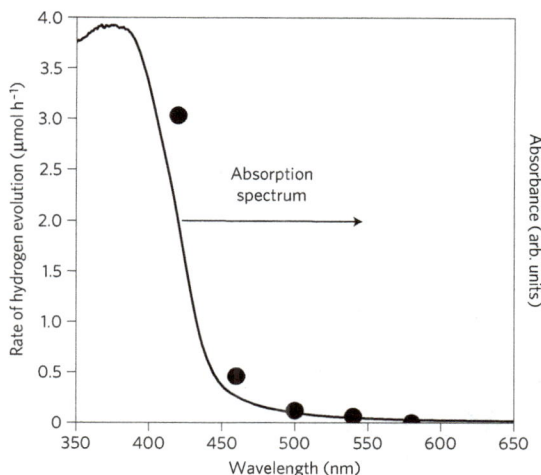

Figure 4 | Wavelength-dependent hydrogen evolution from water by g-C₃N₄. Steady rate of H_2 production from water containing 10 vol% methanol as an electron donor by 0.5 wt% Pt-deposited g-C_3N_4 photocatalyst as a function of wavelength of the incident light. Ultraviolet–visible absorption spectrum of the g-C_3N_4 catalyst is also shown for comparison.

However, the H_2 evolution activity of bare g-C_3N_4 was fluctuant, and exhibited variation from batch to batch (0.1–4 μmol h⁻¹). Modification with a small amount of Pt solved this problem, as described for other systems in the literature[24], improving activity and minimizing the experimental error to lie within 10–15% for different batch samples. This modification violates the 'metal-free' principle, but facilitates electron localization from the conduction band of g-C_3N_4 to the deposited Pt nanoparticles and simplifies H_2 production, here by about a factor of 7. We interpret this finding to be caused by the more favourable hydrogen elimination from Pt–H surface bonds, as compared with the covalent elimination from a hydrogenated g-C_3N_4 surface. Future work should focus on optimizing reaction conditions and improving surface area/surface functionality to catalyse hydrogen elimination even from the bare g-C_3N_4 to avoid the use of noble metals, which was however already shown to be possible.

The H_2 production rate increases with increasing Pt content to a plateau at around 2–4%, beyond which it decreases again. A typical time course of H_2 production using 3 wt% Pt-deposited g-C_3N_4 catalyst is shown in Fig. 3, curve (ii). The fact that the synthesized carbon nitride material contains residual hydrogen requires an evaluation of the origin of the hydrogen to exclude the possibility of the catalyst material as the hydrogen source. To evaluate stability, the reaction was allowed to proceed for a total of 72 h with intermittent evacuation every 24 h under visible-light irradiation ($\lambda > 420$ nm). Continuous H_2 evolution with no noticeable degradation of the carbon nitride was clearly observed from the beginning of the reaction. The total evolution of H_2 after 72 h was 770 μmol, by far exceeding the amount of surface C_6N_8 (tri-s-triazine) units (7.6 μmol) or the loaded Pt nanoparticles (15 μmol). (The number of surface C_6N_8 (tri-s-triazine) units is estimated to be 7.55×10^{-5} mol g⁻¹ of g-C_3N_4. The calculation is based on the specific surface area of g-C_3N_4 (~ 10 m² g⁻¹), which was determined by the Brunauer–Emmett–Teller method at liquid-nitrogen temperature, and the area of C_6N_8 units (~ 0.22 nm⁻²).) To confirm that hydrogen is not coming from g-C_3N_4, we carried out an extra experiment using Pt/g-C_3N_4

under ultraviolet light ($\lambda > 300$ nm). The system steadily produced 4.5 mmol H_2 in 19 h, exceeding the amount of hydrogen (1 mmol) in the catalyst. In most cases where nitrogen-containing materials (for example, (oxy)nitrides) are used as photocatalysts, a low level of N_2 evolution is detected in the initial stage of photocatalytic reaction[6]. This is attributed to the oxidation of N^{3-} species near the material surface to N_2. No N_2 evolution was observed for the present catalyst even after the extended period of irradiation, which may indicate strong binding of N in the covalent carbon nitride. These results indicate excellent stability of the present material for photocatalytic H_2 production. Production of H_2 was also observed when other electron donors including methanol, ethanol and ethylenediaminetetraacetic acid were used instead of triethanolamine, although the hydrogen evolution rates were lower (see Supplementary Information, Fig. S5).

To support that reaction proceeds through light absorption within the carbon nitride polymer, we also examined the dependence of the rate of H_2 evolution on the wavelength of incident light. As shown in Fig. 4, the trend of H_2 evolution rates matches well with that of absorption in the optical spectra. The longest wavelength available for the reaction is 540 nm, which corresponds to the absorption edge of the catalyst. A control experiment showed that indeed no reaction took place in the dark.

It is not the most condensed material (condensed at 873 K) that shows the highest activity, but the more polymeric C_3N_4 material condensed at 823 K. Obviously, some defect structures are crucial, presumably to localize the electron and hole at specific surface termination sites where they can be transferred to the water molecules.

Note that although hydrogen generation from water is an achievement, it constitutes only the minor half of the problem of water splitting. Water oxidation by non-oxide catalysts remains the real challenge of the community. For inorganic sulphides and (oxy)nitrides, the anion components are less stable and, in some cases, more susceptible to oxidation than water. There are some cases where N_2 evolution and sulphur deposition are observed as a result of oxidative decomposition of the photocatalyst[25]. Indeed, our system in the previous experiments did not eliminate molecular oxygen (presumably again because of the covalent character of the

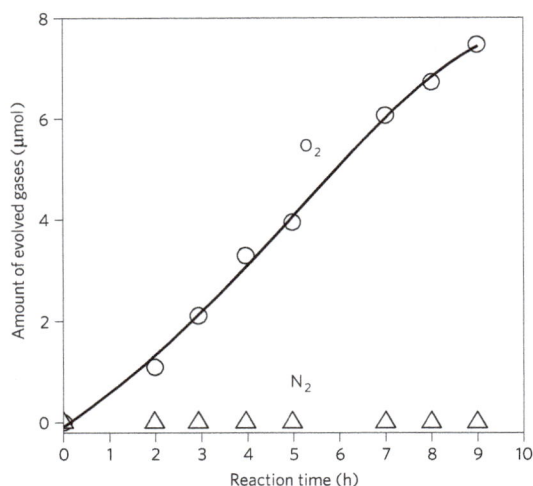

Figure 5 | Oxygen evolution from water by g-C₃N₄. Time courses of O_2 production from water containing 0.01 M silver nitrate as an electron acceptor under visible light (of wavelength longer than 420 nm) by 3.0 wt% RuO_2-loaded g-C₃N₄. La_2O_3 (0.2 g) was used as a buffer (pH 8–9).

oxygen bonds), and a sacrificial reactant to take up the oxygen was used. The bandgap calculations however indicate that in principle oxygen elimination should be thermodynamically possible, as the HOMO was found to lie below the water oxidation potential.

This stimulates further research for constructing a suitable catalytic active site that efficiently promotes O_2 evolution from the g-C₃N₄ surface. As a first trial, we thus attempted to introduce RuO_2, which is well known as a good oxidation catalyst for O_2 evolution[26-28], by an impregnation method[12], using silver nitrate as a sacrificial electron acceptor.

As shown in Fig. 5, the modified catalyst exhibited visible-light activity ($\lambda > 420$ nm) for O_2 evolution with no N_2 evolution, although the activity for O_2 evolution was an order of magnitude lower than that for H_2 evolution (Fig. 3). These results indicate that the loaded RuO_2 is able to take up the photogenerated holes from g-C₃N₄, exhibiting functionality as efficient O_2 evolution sites. The decrease in activity with reaction time is primarily attributable to the deposition of metallic silver on the catalyst surface, which blocks light absorption and obstructs active sites[6,7]. Under ultraviolet irradiation ($\lambda > 300$ nm), the amount of O_2 evolved for 8 h reaction reached 53 µmol (see Supplementary Information, Fig. S6), which is larger than that in the RuO_2 catalyst (about 28 µmol). (The amount of O_2 in the RuO_2 (3.0 wt%)-loaded g-C₃N₄ (0.1 g) is calculated to be about 28 µmol. It should be noted that g-C₃N₄ contains oxygen species as impurities at a rate of about 0.2 wt%.) This fact clearly demonstrates that the observed O_2 evolution is no doubt derived from water oxidation. The low activity for O_2 evolution can be explained by the band-structure calculation, which showed that the thermodynamic driving force of g-C₃N₄ for O_2 evolution is much smaller than that for H_2 evolution (Fig. 3). Although the present preliminary results can be regarded as promising, further optimization for the O_2 evolution system is thus needed.

In summary, we presented here polymeric carbon nitride as a commonly available and simple photocatalyst that is able to generate hydrogen from water even in the absence of noble metals. Although the estimated quantum efficiency of the Pt-modified C_3N_4 catalyst is still rather low (approximately 0.1% with irradiation of 420–460 nm (Fig. 4)), the present result opens new vistas for the search of energy production schemes, using thermally and oxidation-stable polymeric organic semiconductor structures as the functional material that are cheap and commonly available.

Received 21 April 2008; accepted 8 October 2008;
published online 9 November 2008

References

1. Borgarello, E. *et al*. Photochemical cleavage of water by photocatalysis. *Nature* **289**, 158–160 (1981).
2. Kim, Y. I., Salim, S., Huq, M. J. & Mallouk, T. E. Visible-light photolysis of hydrogen iodide using sensitized layered semiconductor particles. *J. Am. Chem. Soc.* **113**, 9561–9563 (1991).
3. Khaselev, O. & Turner, J. A. A monolithic photovoltaic-photoelectrochemical device for hydrogen production via water splitting. *Science* **280**, 425–427 (1998).
4. Sayama, K. *et al*. Stoichiometric water splitting into H_2 and O_2 using a mixture of two different photocatalysts and an IO_3^-/I^- shuttle redox mediator under visible light irradiation. *Chem. Commun.* 2416–2417 (2001).
5. Kato, H. & Kudo, A. Visible-light-response and photocatalytic activities of TiO_2 and $SrTiO_3$ photocatalysts codoped with antimony and chromium. *J. Phys. Chem. B* **106**, 5029–5034 (2002).
6. Hitoki, G. *et al*. An oxynitride, TaON, as an efficient water oxidation photocatalyst under visible light irradiation ($\lambda \leq 500$ nm). *Chem. Commun.* 1698–1699 (2002).
7. Ishikawa, A. *et al*. Oxysulfide $Sm_2Ti_2S_2O_5$ as a stable photocatalyst for water oxidation and reduction under visible light irradiation ($\lambda \leq 650$ nm). *J. Am. Chem. Soc.* **124**, 13547–13553 (2002).
8. Tsuji, I., Kato, H., Kobayashi, H. & Kudo, A. Photocatalytic H_2 evolution reaction from aqueous solutions over band structure-controlled $(AgIn)(x)Zn_2(1-x)S_2$ solid solution photocatalysts with visible-light response and their surface nanostructures. *J. Am. Chem. Soc.* **126**, 13406–13413 (2004).
9. Tsuji, I., Kato, H. & Kudo, A. Visible-light-induced H_2 evolution from an aqueous solution containing sulfide and sulfite over a ZnS–$CuInS_2$–$AgInS_2$ solid-solution photocatalyst. *Angew. Chem. Int. Ed.* **44**, 3565–3568 (2005).
10. Maeda, K. *et al*. GaN:ZnO solid solution as a photocatalyst for visible-light-driven overall water splitting. *J. Am. Chem. Soc.* **127**, 8286–8287 (2005).
11. Maeda, K. *et al*. Photocatalyst releasing hydrogen from water—Enhancing catalytic performance holds promise for hydrogen production by water splitting in sunlight. *Nature* **440**, 295 (2006).
12. Lee, Y. *et al*. Zinc germanium oxynitride as a photocatalyst for overall water splitting under visible light. *J. Phys. Chem. C* **111**, 1042–1048 (2007).
13. Yachandra, V. K. *et al*. Where plants make oxygen—a structural model for the photosynthetic oxygen-evolving manganese cluster. *Science* **260**, 675–679 (1993).
14. de Carcer, I. A., DiPasquale, A., Rheingold, A. L. & Heinekey, D. M. Active-site models for iron hydrogenases: Reduction chemistry of dinuclear iron complexes. *Inorg. Chem.* **45**, 8000–8002 (2006).
15. Yanagida, S., Kabumoto, A., Mizumoto, K., Pac, C. & Yoshino, K. Poly(para)phenylene-catalyzed photoreduction of water to hydrogen. *JCS-Chem. Commun.* **8**, 474–475 (1985).
16. Liebig, J. About some nitrogen compounds. *Ann. Pharm.* **10**, 10 (1834).
17. Groenewolt, M. & Antonietti, M. Synthesis of g-C₃N₄ nanoparticles in mesoporous silica host matrices. *Adv. Mater.* **17**, 1789–1792 (2005).
18. Goettmann, F., Fischer, A., Antonietti, M. & Thomas, A. Chemical synthesis of mesoporous carbon nitrides using hard templates and their use as a metal-free catalyst for Friedel-Crafts reaction of benzene. *Angew. Chem. Int. Ed.* **45**, 4467–4471 (2006).
19. Clark, S. J. *et al*. First principles methods using CASTEP. *Z. f Krist* **220**, 567–570 (2005).
20. Lotsch, B. V. *et al*. Unmasking melon by a complementary approach employing electron diffraction, solid-state NMR spectroscopy, and theoretical calculations-structural characterization of a carbon nitride polymer. *Chem. Eur. J.* **13**, 4969–4980 (2007).
21. Reuter, K. & Scheffler, M. Composition, structure, and stability of $RuO_2(110)$ as a function of oxygen pressure. *Phys. Rev. B* **65**, 035406 (2001).
22. Tissandier, M. D. *et al*. The proton's absolute aqueous enthalpy and Gibbs free energy of solvation from cluster-ion solvation data. *J. Phys. Chem. A* **102**, 7787–7794 (1998).
23. Weast, R. C., Astle, M. J. & Beyer, W. H. *Handbook of Physics and Chemistry* 64th edn, D158 (CRC Press, 1983).
24. Kraeutler, B & Bard, A. J. Heterogeneous photocatalytic preparation of supported catalysts—photodeposition of platinum on TiO_2 powder and other substrates. *J. Am. Chem. Soc.* **100**, 4317–4318 (1978).
25. Maeda, K. & Domen, K. New non-oxide photocatalysts designed for overall water splitting under visible light. *J. Phys. Chem. C* **111**, 7851–7861 (2007).
26. Kalyanasundaram, K. & Grätzel, M. Cyclic cleavage of water into H_2 and O_2 by visible light with coupled redox catalysts. *Angew. Chem. Int. Ed.* **18**, 701–702 (1979).
27. Borgarello, E., Kiwi, J., Pelizzetti, E., Visca, M. & Grätzel, M. Photochemical cleavage of water by photocatalysis. *Nature* **289**, 158–160 (1981).

28. Harriman, A., Pickering, I. J., Thomas, J. M. & Christensen, P. A. Metal oxides as heterogeneous catalysts for oxygen evolution under photochemical conditions. *J. Chem. Soc. Faraday Trans. 1* **84**, 2795–2806 (1988).

Acknowledgements

This work was supported by the Max Planck Society within the framework of the project ENERCHEM, and the Research and Development in a New Interdisciplinary Field Based on Nanotechnology and Materials Science programs of the Ministry of Education, Culture, Sports, Science and Technology (MEXT) of Japan. X.W. is grateful for the financial support from the National Basic Research Program of China (973 program, Grant No. 2007CB613306), NSFC (Grant Nos. 20537010 and 20603007), the New Century Excellent Talents in University of China (NCET-07-0192) and the AvH Foundation. K.M. gratefully acknowledges the support of a Japan Society for the Promotion of Science (JSPS) Fellowship. The authors thank J. Kubota, T. Hisatomi and K. Kamata (Department of Chemical System Engineering, The University of Tokyo) for assistance in the revision of this article.

Additional information

Supplementary Information accompanies this paper on www.nature.com/naturematerials. Reprints and permissions information is available online at http://npg.nature.com/reprintsandpermissions. Correspondence and requests for materials should be addressed to K.D. or X.C.W.

High-capacity hydrogen storage in lithium and sodium amidoboranes

ZHITAO XIONG[1], CHAW KEONG YONG[1], GUOTAO WU[1], PING CHEN[1,2]*, WENDY SHAW[3],
ABHI KARKAMKAR[3], THOMAS AUTREY[3], MARTIN OWEN JONES[4], SIMON R. JOHNSON[4],
PETER P. EDWARDS[4] AND WILLIAM I. F. DAVID[5]

[1]Department of Physics, National University of Singapore, Singapore 117542, Singapore
[2]Department of Chemistry, National University of Singapore, Singapore 117542, Singapore
[3]Pacific Northwest National Laboratories, Richland, Washington 99352, USA
[4]Inorganic Chemistry Laboratory, University of Oxford, South Parks Road, Oxford OX1 3QR, UK
[5]ISIS Facility, Rutherford Appleton Laboratory, Chilton OX11 0QX, UK
*e-mail: phychenp@nus.edu.sg

Published online: 23 December 2007; doi:10.1038/nmat2081

The safe and efficient storage of hydrogen is widely recognized as one of the key technological challenges in the transition towards a hydrogen-based energy economy[1,2]. Whereas hydrogen for transportation applications is currently stored using cryogenics or high pressure, there is substantial research and development activity in the use of novel condensed-phase hydride materials. However, the multiple-target criteria accepted as necessary for the successful implementation of such stores have not yet been met by any single material. Ammonia borane, NH_3BH_3, is one of a number of condensed-phase compounds that have received significant attention because of its reported release of ∼12 wt% hydrogen at moderate temperatures (∼150 °C). However, the hydrogen purity suffers from the release of trace quantities of borazine. Here, we report that the related alkali-metal amidoboranes, $LiNH_2BH_3$ and $NaNH_2BH_3$, release ∼10.9 wt% and ∼7.5 wt% hydrogen, respectively, at significantly lower temperatures (∼90 °C) with no borazine emission. The low-temperature release of a large amount of hydrogen is significant and provides the potential to fulfil many of the principal criteria required for an on-board hydrogen store.

One of the most promising materials suggested as a potential solid-state hydrogen store for a future sustainable hydrogen economy is ammonia borane (NH_3BH_3). This molecular material is a stable solid at room temperature and pressure and is neither flammable nor explosive. Importantly, it contains ∼19.6 wt% hydrogen. Ammonia borane, first synthesized in 1955, is a plastic crystalline solid adopting a tetragonal crystal structure with space group $I\bar{4}mm$ and lattice parameters of $a = b = 5.240\,Å$ and $c = 5.028\,Å$ at room temperature[3]. It thermally decomposes between 70 °C and 112 °C to yield polyaminoborane, $[NH_2BH_2]_n$, and hydrogen[4–8], reaction (1). In turn, polyaminoborane decomposes over a broad temperature range from about 110 °C to approximately 200 °C with further hydrogen loss, forming polyiminoborane, $[NHBH]_n$, and a small fraction of borazine, $[N_3B_3H_6]$, reactions (2) and (3) respectively[5]. The decomposition of $[NHBH]_n$ to BN occurs at temperatures in excess of 500 °C (ref. 5). This final step is not considered practical

for hydrogen storage.

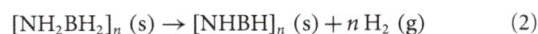

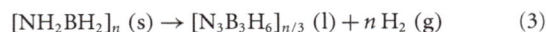

$$n\,NH_3BH_3\,(s) \rightarrow [NH_2BH_2]_n\,(s) + n\,H_2\,(g) \qquad (1)$$

$$[NH_2BH_2]_n\,(s) \rightarrow [NHBH]_n\,(s) + n\,H_2\,(g) \qquad (2)$$

$$[NH_2BH_2]_n\,(s) \rightarrow [N_3B_3H_6]_{n/3}\,(l) + n\,H_2\,(g) \qquad (3)$$

Recent work has investigated various approaches to lower the decomposition temperature of ammonia borane through the use of nanoscaffolds, iridium or base-metal catalyst, carbon cryogel and ionic liquids[9–14] and so on. We have adopted an approach that has been applied in the manipulation of the thermodynamic property of compounds through chemical alteration[2] to modify ammonia borane, that is, through substituting one H in the NH_3 group in BH_3NH_3 with a more electron-donating element. The rationale behind this approach is to alter the polarity and intermolecular interactions (specifically the dihydrogen bonding) of ammonia borane to produce a substantially improved dehydrogenation profile. Here, we report that lithium amidoborane ($LiNH_2BH_3$) and sodium amidoborane ($NaNH_2BH_3$) show substantially different and improved dehydrogenation characteristics with respect to ammonia borane itself. These alkali-metal amidoborane materials are formed through the interactions of alkali-metal hydrides (LiH and NaH) with ammonia borane, which lead to the replacement of a single ammonia borane hydrogen atom by a lithium or sodium atom. We observed that more than 10 wt% and 7 wt% of hydrogen desorbs from $LiNH_2BH_3$ and $NaNH_2BH_3$, respectively, at around 90 °C.

Lithium and sodium amidoboranes were prepared by ball-milling 1:1 molar ratios of ammonia borane (NH_3BH_3) and the corresponding alkali-metal hydrides (see the Supplementary Information). A gradual pressure increase within the ball-mill vessel was observed and the gaseous product was identified as H_2 by mass spectrometry. The amount of hydrogen evolved from the starting chemicals was calculated from this pressure increase

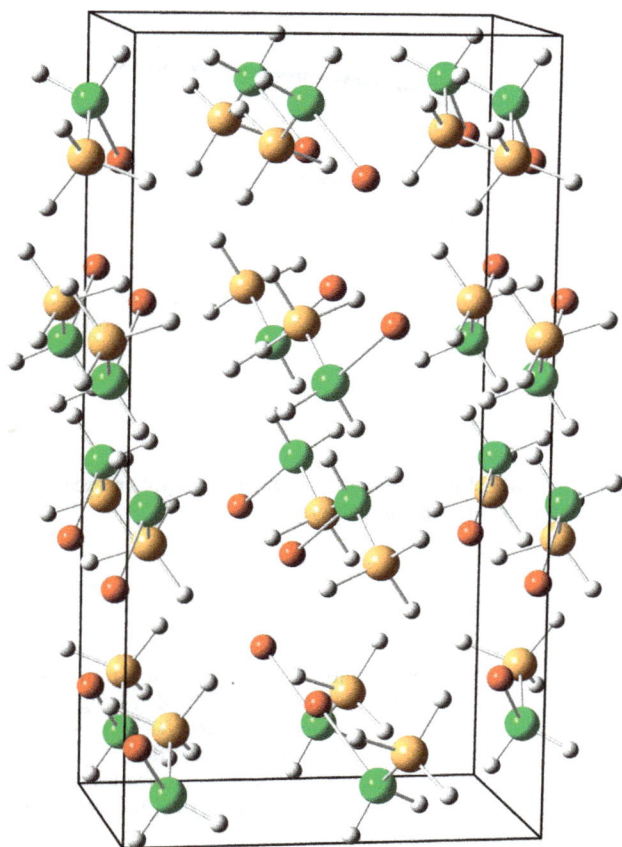

Figure 1 Schematic diagram of the crystal structure of $LiNH_2BH_3$ and $NaNH_2BH_3$ determined from high-resolution X-ray powder diffraction data at room temperature. Boron is represented by orange spheres, nitrogen by green spheres, hydrogen by white spheres and lithium by red spheres.

Figure 2 High-field 289.2 MHz (21.2 T) ^{11}B NMR of $LiNH_2BH_3$ and NH_3BH_3 samples. (i), As-prepared $LiNH_2BH_3$ sample (-19.7 p.p.m.); (ii), untreated NH_3BH_3 (-22.8 p.p.m.); (iii), $LiNH_2BH_3$ sample after dehydrogenation to 140 °C ($+29.8$ p.p.m.); (iv), polyborazylene ($+26.5$ p.p.m.).

and it was observed that about 1.0 molar equivalent of H_2 was released from both $LiH–NH_3BH_3$ and $NaH–NH_3BH_3$ mixtures within 3 h. In contrast, no gaseous products were detected when pure ammonia borane was ball-milled under the same conditions (see the Supplementary Information).

The release of one mole of H_2 per mole of ammonia borane and alkali-metal hydride during the milling process suggests that reactions (4) and (5) have occurred:

$$NH_3BH_3 \text{ (s)} + LiH \text{ (s)} \rightarrow LiNH_2BH_3 \text{ (s)} + H_2 \text{ (g)} \quad (4)$$

$$NH_3BH_3 \text{ (s)} + NaH \text{ (s)} \rightarrow NaNH_2BH_3 \text{ (s)} + H_2 \text{ (g)}, \quad (5)$$

leading to the formation of alkali-metal amidoboranes, $LiNH_2BH_3$ and $NaNH_2BH_3$. It is likely that one of the driving forces for the reactions above may come from the high potential for the $H^{\delta+}$ in NH_3 and $H^{\delta-}$ in alkali-metal hydrides to combine and form H_2 (ref. 2).

The formation of $LiNH_2BH_3$ and $NaNH_2BH_3$ in solution was reported in 1996 (ref. 15) and 1938 (ref. 16), respectively. $LiNH_2BH_3$, termed lithium amidotrihydroborate and used as a reducing agent for organic synthesis, was produced from a reaction between ammonia borane and n-butyllithium in tetrahydrofuran[15]. The identity of the substituted ammonia borane species in solution

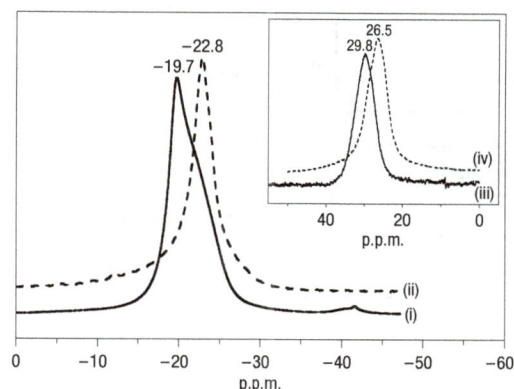

was confirmed by ^{11}B NMR. The reaction of $(CH_3)_2OBH_3$ and Na in liquid ammonia produces $NaNH_2BH_3$ (ref. 16). However, no crystal structure information or decomposition properties of $LiNH_2BH_3$ or $NaNH_2BH_3$ have previously been reported.

Hence, to confirm our assumption that $LiNH_2BH_3$ and $NaNH_2BH_3$ had been formed, high-resolution X-ray powder diffraction data were collected on the post-ball-milled $LiH–NH_3BH_3$ and $NaH–NH_3BH_3$ samples on the high-resolution powder diffractometer ID31 at the ESRF, Grenoble. The structures (Fig. 1) were solved using the computer program DASH[17], which confirmed the stoichiometry to be $LiNH_2BH_3$ and $NaNH_2BH_3$ with Li/Na directly substituting for one of the ammonia hydrogen atoms. $LiNH_2BH_3$ and $NaNH_2BH_3$ are isostructural, crystallizing in the orthorhombic space group $Pbca$. For $LiNH_2BH_3$, the lattice constants are $a = 7.11274(6)$ Å, $b = 13.94877(14)$ Å, $c = 5.15018(6)$ Å, $V = 510.970(15)$ Å^3, whereas for $NaNH_2BH_3$, the lattice constants are $a = 7.46931(7)$ Å, $b = 14.65483(16)$ Å, $c = 5.65280(8)$ Å, $V = 618.764(20)$ Å^3. Both the Li–N (1.98 Å) and Na–N bonds (2.35 Å) are substantially longer than the H–N bond in ammonia borane, which is indicative of substantial $M^+(NH_2BH_3)^-$ ionic character. As expected, the B–N bond length (1.56 Å) is slightly shorter than in ammonia borane, revealing stronger bonding between B and N in the alkali-metal amidoboranes, which is in agreement with the predictions of Armstrong et al.[18].

The substitution of H by a stronger electron-donating alkali metal will induce considerable changes in the electronic state of N with concomitant modification of the chemical bonding between B and N. As a consequence, the chemical bonding of B–N, B–H and N–H will be affected. $LiNH_2BH_3$ was characterized to illustrate these changes. High-field ^{11}B NMR studies revealed a boron species with a chemical shift of -19.7 p.p.m. (Fig. 2), which is consistent with the previously published NMR data for the $LiNH_2BH_3$ species[15]. The subtle downfield shift of the ^{11}B resonance in the $LiNH_2BH_3$ complex compared with that in ammonia borane (-22.8 p.p.m.) suggests that the lithium amido group—$NH_2(Li)$—may form a stronger donor complex with borane, which is consistent with the crystallographically observed shortening of the B–N bond length for $LiNH_2BH_3$.

Dehydrogenation of the $LiNH_2BH_3$ and the $NaNH_2BH_3$ samples was investigated using temperature-programmed-desorption (TPD) and volumetric-release techniques. The post-milled ammonia borane was also tested for comparison. On

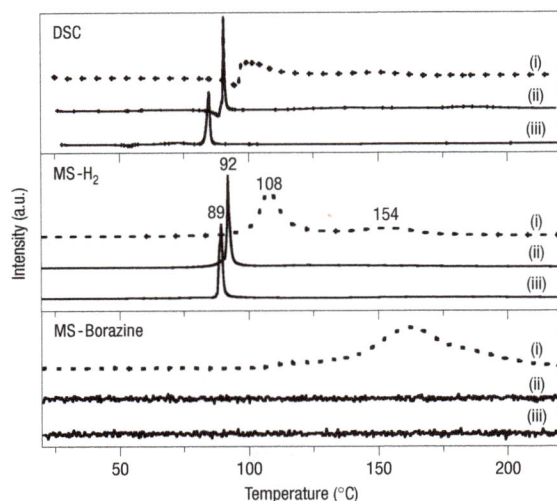

Figure 3 **TPD and DSC spectra.** (i), Post-milled ammonia borane; (ii), Li amidoborane sample; (iii) Na amidoborane sample. MS: mass spectrometry.

Figure 4 **Time dependences of hydrogen desorption from alkali amidoboranes and post-milled BH_3NH_3 samples at about 91 °C.**

decomposition, the milled ammonia borane released hydrogen with a small amount of borazine; maximum decomposition rates occurred at 108 °C and 154 °C, respectively (Fig. 3). In contrast, $LiNH_2BH_3$ decomposed directly to hydrogen on heating, with vigorous hydrogen release at around 92 °C; no borazine was detected (Fig. 3). The thermal dehydrogenation of $NaNH_2BH_3$ resembles that of $LiNH_2BH_3$, but at a slightly lower peak temperature of 89 °C. Volumetric-release measurements showed that about 8 wt% and 6 wt% of hydrogen are released from the $LiNH_2BH_3$ and the $NaNH_2BH_3$ samples, respectively, within the first hour at ~91 °C. Extending the reaction period to 19 h enables about 11 wt% and 7.4 wt% or about 2.0 equivalent moles of H_2 to fully desorb (Fig. 4). In contrast, milled ammonia borane released only about 5.3 wt% or about 0.82 equivalent moles of H_2 under similar conditions. Differential scanning calorimetry (DSC) measurements (Fig. 3) indicated that the heat of hydrogen desorption from $LiNH_2BH_3$ and $NaNH_2BH_3$ is about -3 to -5 kJ mol^{-1}-H_2, which is significantly less exothermic than from the pristine NH_3BH_3 (about -20 kJ mol^{-1}-H_2 (ref. 7)). The nearly thermally neutral dehydrogenation of alkali-metal amidoboranes indicates that rapid near-room-temperature hydrogen release may be feasible if kinetic barriers can be overcome by catalytic modification.

As approximately two molar equivalents of H_2 were released from $LiNH_2BH_3$ and $NaNH_2BH_3$, respectively, the final solid products have the chemical composition of LiNBH and NaNBH, as shown by the dehydrogenation reactions (6) and (7).

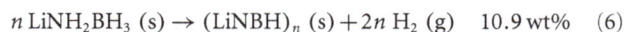

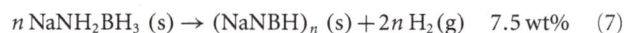

$$n\,LiNH_2BH_3\,(s) \rightarrow (LiNBH)_n\,(s) + 2n\,H_2\,(g) \quad 10.9\,wt\% \quad (6)$$

$$n\,NaNH_2BH_3\,(s) \rightarrow (NaNBH)_n\,(s) + 2n\,H_2\,(g) \quad 7.5\,wt\% \quad (7)$$

High-resolution X-ray powder diffraction data indicated that the post-TPD samples are very poorly crystalline (see the Supplementary Information). A chemical shift of 29.8 p.p.m. was observed in LiNBH in ^{11}B NMR measurements (Fig. 2), consistent with a trigonal planar N–BH–N environment for boron[19], which strongly suggests the final product is a borazine-like or poly-borazine-like compound. Similar phenomena were observed for the NaNBH sample (see the Supplementary Information).

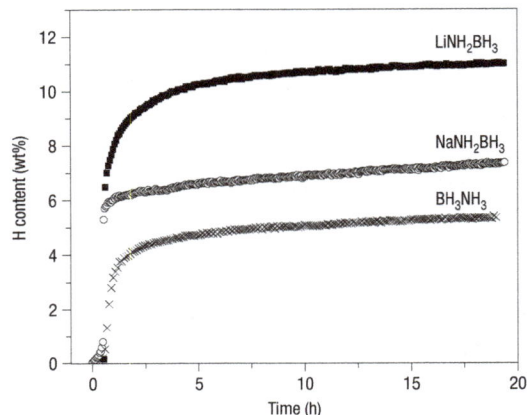

One of the key features of the alkali-metal amidoboranes is the presence of hydrogen atoms with both positive (that is, H in $-NH_2$) and negative (that is, H in $-BH_3$) charges. This is reminiscent of the amide–hydride combination[20,21] and also of borohydride–amide materials[22]. One of the advantages of the dehydrogenation of alkali amidoboranes is that it may not necessarily involve an interface reaction and mass transport through different phases as for the amide–hydride combination[23,24] but rather involve the local combination of $H^{\delta+}$ and $H^{\delta-}$ as in $Li_4BH_4(NH_2)_3$ (ref. 22), which more readily produces H_2.

$LiNH_2BH_3$ and $NaNH_2BH_3$ are the first two examples of a potentially large class of metal amidoboranes. The $[NH_2BH_3]^-$ anion is likely to bond with other metal or metalloid elements to form a variety of corresponding amidoboranes, which will not only open up significant opportunities for hydrogen-storage materials development, but will also bring considerable insights to B–N chemistry.

The search for a safe, economical, hydrogen-rich store is a key materials challenge in the move towards a hydrogen-based energy economy. We have discovered that alkali-metal amidoboranes, $LiNH_2BH_3$ and $NaNH_2BH_3$, satisfy a number of the principal criteria demanded for hydrogen-storage media. $LiNH_2BH_3$ and $NaNH_2BH_3$ provide high storage capacity (10.9 wt% and 7.5 wt%, respectively) of hydrogen at easily accessible dehydrogenation temperatures (about 90 °C) without the unwanted by-product borazine. These materials offer significant advantages over their parent compound, ammonia borane. Alkali-metal amidoboranes are also environmentally harmless, non-flammable, non-explosive and stable solids at room temperature and pressure. At present, the only disadvantageous aspect of these materials is their lack of facile reversibility. However, the almost thermal neutral dehydrogenation of alkali-metal amidoboranes will greatly facilitate their off-board regeneration.

METHODS

SAMPLE PREPARATION

LiH, NaH and NH_3BH_3 were purchased from Sigma-Aldrich with stated purities of 95%, 95% and above 90%, respectively, and were used without further purification. The $LiH–NH_3BH_3$ and $NaH–NH_3BH_3$ mixtures in a hydride/NH_3BH_3 molar ratio of 1:1 and pristine NH_3BH_3 were ball-milled on a Retsch PM400 planetary mill at 200 r.p.m., respectively. Graphite was added to the mixture to enhance ball-milling efficiency. After milling the samples, gases released were identified by a mass spectrometer and the pressure was measured by a pressure gauge. All samples were handled in a MBRAUN glove box filled with purified argon gas.

CHARACTERIZATION

TPD was used over a gas chromatography–mass spectrometry combined system to investigate the thermal desorption properties of the post-milled samples. Detailed operation procedures are described elsewhere[20]. Quantitative measurement of hydrogen desorption at elevated temperatures was carried out on a commercial gas reactor controller provided by the Advanced Materials Corporation. A Netzsch DSC 204 HP unit was applied in detecting the heat flow in the dehydrogenation process.

Structural identifications were carried out on both a Bruker X-ray diffractometer equipped with an *in situ* cell at NUS and the high-resolution synchrotron X-ray powder diffractometer on ID31 at the ESRF, Grenoble.

$^{11}B\{^1H\}$ magic-angle-spinning NMR experiments were carried out at room temperature on a Bruker Advance 400 NMR spectrometer operating at 9.7 T on 128.3 MHz ^{11}B frequency at NUS and on a Varian Unity Inova console operating at 900 MHz ^1H frequency at PNNL. Spectra were obtained using a two-channel custom-built probe with a 3.2 mm rotor, a 4 µs 90° pulse, a 20 s pulse delay and 15 kHz spinning speed. Two hundred to three hundred scans were taken for the ball-milled and post-TPD LiH–NH_3BH_3 samples, whereas model compounds were obtained with 20–40 scans.

Received 5 May 2007; accepted 8 November 2007; published 23 December 2007.

References

1. Schlapbach, L. & Zuttel, A. Hydrogen-storage materials for mobile applications. *Nature* **414**, 353–358 (2001).
2. Grochala, W. & Edwards, P. P. Thermal decomposition of the non-interstitial hydrides for the storage and production of hydrogen. *Chem. Rev.* **104**, 1283–1315 (2004).
3. Shore, S. G. & Parry, R. W. The crystalline compound ammonia-borane, H_3NBH_3. *J. Am. Chem. Soc.* **77**, 6084–6085 (1955).
4. Stowe, A. C., Shaw, W. J., Linehan, J. C., Schmid, B. & Autrey, T. In situ solid state ^{11}B MAS-NMR studies of the thermal decomposition of ammonia borane: Mechanistic studies of the hydrogen release pathways from a solid state hydrogen storage material. *Phys. Chem. Chem. Phys.* **9**, 1831–1836 (2007).
5. Hu, M. G., Geanangel, R. A. & Wendlandt, W. W. The thermal decomposition of ammonia borane. *Thermochim. Acta* **23**, 249–255 (1978).
6. Sit, V., Geanangel, R. A. & Wendlandt, W. W. The thermal dissociation of NH_3BH_3. *Thermochim. Acta* **113**, 379–382 (1987).
7. Wolf, G., Baumann, J., Baitalow, F. & Hoffmann, F. P. Calorimetric process monitoring of thermal decomposition of B–N–H compounds. *Thermochim. Acta* **343**, 19–25 (2000).
8. Baitalow, F., Baumann, J., Wolf, G., Jaenicke-Rößler, K. & Leitner, G. Thermal decomposition of B–N–H compounds investigated by using combined thermoanalytical methods. *Thermochim. Acta* **391**, 159–168 (2002).
9. Gutowska, A. *et al.* Nanoscaffold mediates hydrogen release and the reactivity of ammonia borane. *Angew. Chem. Int. Edn* **44**, 3578–3582 (2005).
10. Denney, M. C., Pons, V., Hebden, T. J., Heinekey, M. & Goldberg, K. I. Efficient catalysis of ammonia borane dehydrogenation. *J. Am. Chem. Soc.* **128**, 12048–12049 (2006).
11. Bluhm, M. E., Bradley, M. G., Butterick, R., Kusari, U. & Sneddon, L. G. Amineborane-based chemical hydrogen storage: Enhanced ammonia borane dehydrogenation in ionic liquids. *J. Am. Chem. Soc.* **128**, 7748–7749 (2006).
12. Keaton, R. J., Blacquiere, J. M. & Baker, R. T. Base metal catalyzed dehydrogenation of ammonia-borane for chemical hydrogen storage. *J. Am. Chem. Soc.* **129**, 1844–1845 (2007).
13. Clark, T. J., Lee, K. & Manners, I. Transition-metal-catalyzed dehydrocoupling: A convenient route to bonds between main-group elements. *Chem. Eur. J.* **12**, 8634–8648 (2006).
14. Feaver, A. *et al.* Coherent carbon cryogel-ammonia borane nanocomposites for H-2 storage. *J. Phys. Chem. B* **111**, 7469–7472 (2007).
15. Myers, A. G., Yang, B. H. & Kopecky, D. J. Lithium amidotrihydroborate, a powerful new reductant. Transformation of tertiary amides to primary alcohols. *Tetrahedron Lett.* **37**, 3623–3626 (1996).
16. Schlesinger, H. I. & Burg, A. B. Hydrides of boron. VIII. The structure of the diammoniate of diborane and its relation to the structure of diborane. *J. Am. Chem. Soc.* **60**, 290–299 (1938).
17. David, W. I. F., Shankland, K. & Shankland, N. Routine determination of molecular crystal structures from powder diffraction data. *Chem. Commun.* **8**, 931–932 (1998).
18. Armstrong, D. R., Perkins, P. G. & Walker, G. T. The electronic structure of the monomers, dimers, a trimer, the oxides and the borane complexes of the lithiated ammonias. *THEOCHEM-J. Mol. Struct.* **122**, 189–203 (1985).
19. Gervais, C. B-11 and N-15 solid state NMR investigation of a boron nitride preceramic polymer prepared by ammonolysis of borazine. *J. Eur. Ceram. Soc.* **25**, 129–135 (2005).
20. Chen, P., Xiong, Z. T., Luo, J. Z., Lin, J. Y. & Tan, K. L. Interaction of hydrogen with metal nitrides and imides. *Nature* **420**, 302–304 (2002).
21. Chen, P., Xiong, Z. T., Luo, J. Z., Lin, J. Y. & Tan, K. L. Interaction between lithium amide and lithium hydride. *J. Phys. Chem. B* **107**, 10967–10970 (2003).
22. Chater, P. A., David, W. I. F., Johnson, S. R., Edwards, P. P. & Anderson, P. A. Synthesis and crystal structure of $Li_4BH_4(NH_2)_3$. *Chem. Commun.* **23**, 2439–2441 (2006).
23. David, W. I. F. *et al.* A Mechanism for non-stoichiometry in the lithium amide/lithium imide hydrogen storage reaction. *J. Am. Chem. Soc.* **129**, 1594–1601 (2007).
24. Chen, P., Xiong, Z. T., Yang, L. F., Wu, G. T. & Luo, W. F. Mechanistic investigations on the heterogeneous solid-state reaction of magnesium amides and lithium hydrides. *J. Phys. Chem. B* **110**, 14221–14225 (2006).

Acknowledgements

P.C., Z.X. and G.W. are grateful for the financial support from A*STAR, Singapore, and helpful discussions with T. Kemmitt and M. Bowden from IRL and L. Sneddon from Univ. Pennsylvania. T.A., W.S. and A.K. would like to thank the DOE Center of Excellence in Chemical Hydrogen Storage for support. M.O.J., S.J. and P.E. would like to thank the EPSRC (SUPERGEN) for financial support. We wish to thank A. Fitch, M. Brunelli and I. Margiolaki for assistance in using the high-resolution beamline ID31 at the ESRF, Grenoble (France). A portion of the research described here was carried out in the Environmental Molecular Sciences Laboratory, a national scientific user facility sponsored by the US Department of Energy's Office of Biological and Environmental Research and located at Pacific Northwest National Laboratory. This work was carried out as a collaboration established by the IPHE project 'Combination of Amine Boranes with MgH_2 & $LiNH_2$ for High Capacity Reversible Hydrogen Storage'.
Correspondence and requests for materials should be addressed to P.C.
Supplementary Information accompanies this paper on www.nature.com/naturematerials.

Computational high-throughput screening of electrocatalytic materials for hydrogen evolution

JEFF GREELEY[1], THOMAS F. JARAMILLO[2], JACOB BONDE[2], IB CHORKENDORFF[2]
AND JENS K. NØRSKOV[1]*

[1]Center for Atomic-scale Materials Design, NanoDTU, Department of Physics, Technical Univ. of Denmark, DK-2800 Kongens Lyngby, Denmark
[2]Center for Individual Nanoparticle Functionality, NanoDTU, Department of Physics, Technical Univ. of Denmark, DK-2800 Kongens Lyngby, Denmark
*e-mail: norskov@fysik.dtu.dk

Published online: 15 October 2006; doi:10.1038/nmat1752

The pace of materials discovery for heterogeneous catalysts and electrocatalysts could, in principle, be accelerated by the development of efficient computational screening methods. This would require an integrated approach, where the catalytic activity and stability of new materials are evaluated and where predictions are benchmarked by careful synthesis and experimental tests. In this contribution, we present a density functional theory-based, high-throughput screening scheme that successfully uses these strategies to identify a new electrocatalyst for the hydrogen evolution reaction (HER). The activity of over 700 binary surface alloys is evaluated theoretically; the stability of each alloy in electrochemical environments is also estimated. BiPt is found to have a predicted activity comparable to, or even better than, pure Pt, the archetypical HER catalyst. This alloy is synthesized and tested experimentally and shows improved HER performance compared with pure Pt, in agreement with the computational screening results.

The *in silico* design of functional materials on the basis of electronic-structure calculations is a longstanding goal of theoretical materials science. Such an effort would require the establishment of a direct link between the macroscopic functionality and the atomic-scale properties of the material, in addition to the development of efficient and accurate methods for solving the electronic-structure problem. The first examples of the use of computational methods to screen for new materials have recently been published[1–10]. Such calculations are computationally demanding, and these studies have generally either used simplified electronic-structure schemes or have considered relatively few material combinations. In the present article, we show that it is possible to carry out moderately large-scale combinatorial screening for alloy catalyst materials using density functional theory (DFT) calculations. We introduce a screening procedure that efficiently combines catalytic activity criteria, detailed stability assessments and a database of DFT calculations on more than 700 binary transition-metal surface alloys. We apply the procedure to the evaluation of alloy catalysts for a fundamental electrochemical process, the hydrogen evolution reaction (HER). One of the most promising candidate materials resulting from the search is a surface alloy of bismuth and platinum. We develop a method to synthesize a BiPt surface alloy and show experimentally that its activity is superior to that of Pt, the archetypical HER catalyst.

The HER, which involves proton reduction and concomitant hydrogen evolution, is important for a variety of electrochemical processes, and technological interest in the HER is spread over applications as diverse as hydrogen fuel cells, electrodeposition and corrosion of metals in acids and storage of energy via H_2 production[11–13]. We choose the HER to illustrate our approach primarily because a simple atomic-scale descriptor for the catalytic activity has previously been established. It has long been known that when the catalytic activity of a material for the HER is plotted as a function of the hydrogen–metal bond strength, a volcano-shaped form is found[14–18]. This behaviour is related to the Sabatier

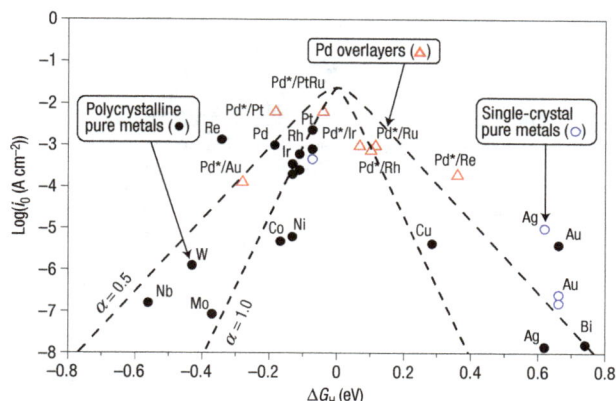

Figure 1 Volcano plot for the HER for various pure metals and metal overlayers. ΔG_H values are calculated at 1 bar of H_2 (298 K) and at a surface hydrogen coverage of either 1/4 or 1/3 ML. Experimental data are compiled in refs 17,20,21. Where available, computational data are taken from the present work; other computational results are taken from the references above. The experimental data have been collected from over 40 years of publications, representing different experimental conditions and surface structures. No corrections for changes in the calculated ΔG_H values with coverage are included, in contrast to the calculated values for Pd overlayers (denoted by Pd*/substrate) presented in ref. 21. The two curved lines correspond to the activity predictions of simple mean-field, microkinetic models, assuming transfer coefficients (α) of 0.5 and 1.0, respectively.

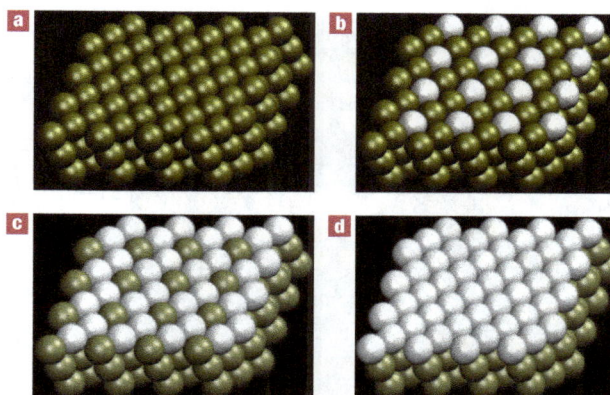

Figure 2 Schematic diagrams of surface alloys at the solute coverages for which calculations are carried out. The solute atoms are shown in white, and the host (substrate) atoms are indicated in dark green. All solute atoms are embedded in the surface layer of the corresponding host metal. **a**, A pure metal. **b**, A surface alloy with solute coverage = 1/3 ML. **c**, A surface alloy with solute coverage = 2/3 ML. **d**, One ML of solute atoms, forming an overlayer.

principle, a general explanatory paradigm in heterogeneous catalysis and electrocatalysis that states that optimal catalytic activity can be achieved on a catalytic surface with intermediate binding energies (or free energies of adsorption) for reactive intermediates[19]. If the intermediates bind too weakly, it is difficult for the surface to activate them, but if they bind too strongly, they will occupy all available surface sites and poison the reaction; intermediate binding energies permit a compromise between these extremes. In the particular case of hydrogen evolution, it turns out that these general principles can be quantified by analysing the free energy of hydrogen adsorption ΔG_H; this quantity is a reasonable descriptor of hydrogen evolution activity for a wide variety of metals and alloys[15,20,21]. As first suggested by Parsons[15], the optimum value should be around $\Delta G_H = 0$. In Fig. 1, we have plotted experimentally measured HER exchange current densities[17,20,21] against ΔG_H values calculated using DFT (see below). Although there is some scatter in the data (the experimental values are from several different authors), the figure seems to suggest that an optimum in the measured HER exchange current densities is found for DFT-derived ΔG_H values very close to (if not identically equal to) zero. We have therefore adopted the approach of using calculated values of $|\Delta G_H|$ to search for new HER catalysts, and we will assume in the following that the closer $|\Delta G_H|$ is to zero, the better the catalyst.

Using periodic $((\sqrt{3} \times \sqrt{3})R30°$ unit cell), self-consistent, DFT calculations (see the Methods section), we evaluate the value of ΔG_H on the 736 distinct binary transition-metal surface alloys that can be formed from the 16 metals Fe, Co, Ni, Cu, As, Ru, Rh, Pd, Ag, Cd, Sb, Re, Ir, Pt, Au and Bi (these elements are simply chosen to give a very broad pool of metallic and semimetallic elements, most of which are thermodynamically resistant to bulk oxide formation in water at 298 K and $U = 0$ V versus the standard hydrogen electrode; SHE). Such alloys, which are composed of a pure metal substrate with a solute element alloyed into the

surface layer (Fig. 2), can exhibit surface properties that are vastly different from the properties of the bulk alloy. In Fig. 3, we show schematically the calculated free energies of hydrogen adsorption on a subset of these surface alloys—those with a 1/3 monolayer (ML) of solute in the surface layer (also see Fig. 2b). The figure clearly demonstrates that a number of binary surface alloys have high predicted activity for the HER.

Our purely computational screening procedure thus identifies a number of interesting candidates for HER catalysts. However, the analysis has so far neglected any consideration of whether or not the indicated alloys will be stable in real electrochemical environments. To estimate the stability of the surface alloys for the HER, we carry out four simple tests for each alloy (see the Supplementary Information). First, we estimate the free-energy change associated with surface segregation events; such events can cause surface solute atoms to segregate into the bulk. Second, we determine the free-energy change associated with intrasurface transformations such as island formation and surface de-alloying. Third, we evaluate the free energy of oxygen adsorption, beginning with splitting of liquid water; facile oxygen adsorption can lead to surface poisoning and/or oxide formation. Finally, we estimate the likelihood that the surface alloys of interest will corrode in acidic environments (pH = 0). For this test, we simply take the free energies of dissolution as reported in the electrochemical series[13].

In Fig. 4, we plot the most pessimistic of the free-energy transformation values determined for each alloy against the absolute magnitude of ΔG_H. The stability considerations immediately eliminate a large number of alloys from consideration; although many alloys have high predicted HER activity, only a small fraction are predicted to be both active and stable in acidic HER environments (Fig. 4), including, among others, surface alloys of BiPt, PtRu, AsPt, SbPt, BiRh, RhRe, PtRe, AsRu, IrRu, RhRu, IrRe and PtRh (note that the last seven of these have solute coverages different from 1/3 ML and therefore are not found in Fig. 3). Thus, these results demonstrate that stability considerations are essential for finding realistic candidate catalysts for hydrogen evolution. Furthermore, the results suggest that more detailed computational analyses and, ultimately, experimental testing of the promising alloys are in order.

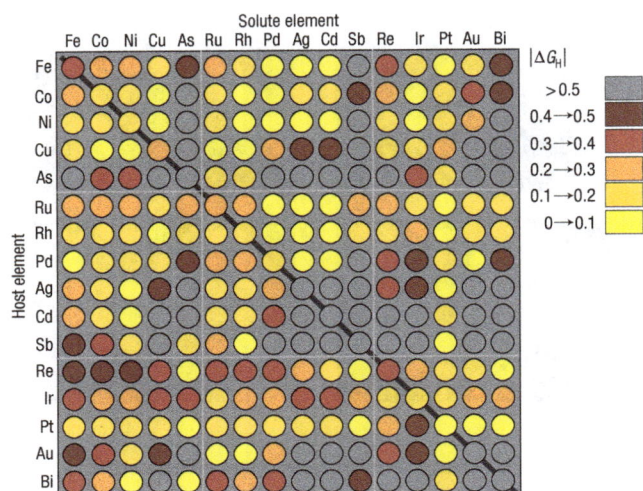

Figure 3 **Computational high-throughput screening for** $|\Delta G_H|$ **on 256 pure metals and surface alloys.** The rows indicate the identity of the pure metal substrates, and the columns indicate the identity of the solute embedded in the surface layer of the substrate. Thus, a point at the intersection of the 'Rh' row with the 'Pd' column, for example, would correspond to a surface alloy with Pd embedded in the (111) surface layer of a pure Rh host. The solute coverage is 1/3 ML in all cases, and the adsorbed hydrogen coverage is also 1/3 ML. The diagonal of the plot corresponds to the hydrogen-adsorption free-energy on the pure metals.

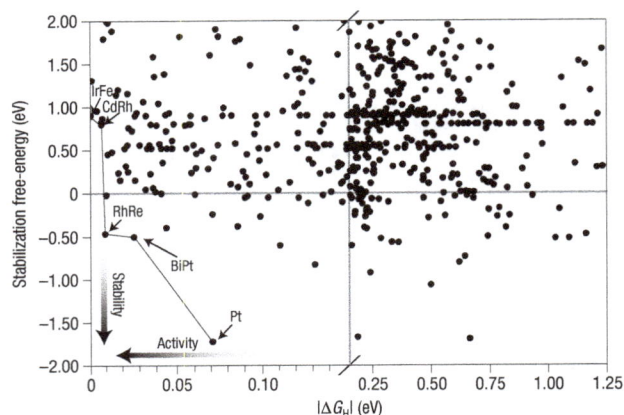

Figure 4 **Pareto-optimal plot of stability and activity of surface alloys for the HER.** The stabilization free-energy can be thought of as a free energy of formation for the surface alloys; the stability of the alloys with respect to various reconstructive/deactivating processes (including surface segregation, island formation, water splitting/oxygen adsorption and metal dissolution) is evaluated for each alloy, and the most pessimistic such energy (that is, the energy that would give the maximum probability that the alloy would destabilize) is plotted. The Pareto-optimal line indicates the best possible compromise between activity and stability, but given the simplicity of our model, other alloys could certainly be worth considering for use as HER catalysts; the alloys in the lower left quadrant, in particular, are promising. For the labelled points, a single element indicates a pure metal. For the bimetallic alloys, the solute is listed first; the solute coverages are 1/3 ML (BiPt), 1 ML (RhRe), 1/3 ML (CdRh) and 2/3 ML (IrFe).

For the present study, we have selected BiPt for further analysis. This surface alloy is particularly interesting for the HER because of the stark contrast between its two constituent elements; pure Pt exhibits high catalytic activity, whereas pure Bi is not active at all (see Fig. 1). A surface alloy formed from these two elements, however, yields a material predicted by the calculations to have an activity comparable to, or even better than, pure Pt. We note, in passing, that this 'counterintuitive' surface alloy might not have been investigated without the aid of a combinatorial screening approach to guide the search for new catalysts.

To study this material in more detail, we first carried out further calculations on more detailed structural models to assess the effect of Bi coverage on our results. These calculations were done at a Bi coverage of 1/4 ML (2 × 2 unit cells—see the Methods section); we note that Bi coverages equal to or greater than 1/2 ML caused very large distortions of the alloy surface and, thus, were not considered in detail. The coverage of adsorbed hydrogen was also 1/4 ML in these analyses. Although the preferred site preference for H on the BiPt surface alloy was different for the 2 × 2 versus $(\sqrt{3} \times \sqrt{3})R30°$ unit cells (face-centred cubic and near-top sites, respectively), the trends in ΔG_H values for the BiPt versus Pt slabs (−0.03 versus −0.07 eV on the $(\sqrt{3} \times \sqrt{3})R30°$ slab and 0.00 versus −0.04 eV on the 2 × 2 slab, respectively) were unchanged. For this more detailed analysis, we also carried out additional stability tests. We estimated the potential versus SHE at which hydroxyl will adsorb on the surface after water splitting; a value of ∼0.4 V was obtained by following the procedure described previously[22]. Thus, although hydroxyl (which preferentially adsorbs on a Bi top site) adsorbs at a potential several tenths of a volt lower than atomic oxygen, even this more reactive species will not be present under typical HER conditions. We also evaluated the adatom formation energy for a BiPt surface alloy in the presence of 1/4 ML of adsorbed hydrogen. Adatom formation was found to be endothermic (∼0.07 eV) in the

presence of adsorbed hydrogen, suggesting that under typical HER conditions, the stable form is that of the surface alloy, and there will only be a relatively small concentration of Bi adatoms present at equilibrium. These results, when combined with the activity and stability calculations described above, paint a picture of BiPt surface alloys as structurally stable systems that exhibit moderately improved hydrogen evolution kinetics compared with pure Pt. It is important to emphasize, however, that the HER model used is quite simple and that the calculated difference in ΔG_H for BiPt and Pt (∼0.04 eV) is small, given the typical accuracy associated with DFT calculations. Thus, it is only possible to conclude from the computational results that BiPt surface alloys might be interesting for the HER. To validate the results of our computational screening, careful experimental tests are clearly necessary.

Mixed Pt and Bi catalysts have been studied experimentally for decades. The two most commonly reported forms are: (1) a Pt surface modified by irreversibly adsorbed Bi (Pt–Bi$_{ir}$) and (2) bulk intermetallic PtBi or PtBi$_2$. Interest in bulk intermetallics has recently increased; they are typically investigated for electro-oxidation reactions (see, for example, ref. 23). Literature on the Pt–Bi$_{ir}$ system—the most commonly studied form of mixed Pt and Bi—extends back to the 1970s; many studies focus on the oxidation of H$_2$, CO and formic acid, in addition to the reduction of O$_2$ (refs 24–33). Perhaps the most pertinent Pt–Bi$_{ir}$ work is that of Gómez et al. who investigated the impact of Bi$_{ir}$ on hydrogen evolution for Pt(100) and Pt(111). They showed that Bi$_{ir}$ severely poisons both Pt surfaces for HER, as evidenced by a significant decrease in current as a function of Bi$_{ir}$ coverage[26,27]. This result seems, at first glance, to contradict our theoretical prediction that mixed Pt and Bi alloys show improved activity for the HER. However, it should be recalled that our calculated BiPt

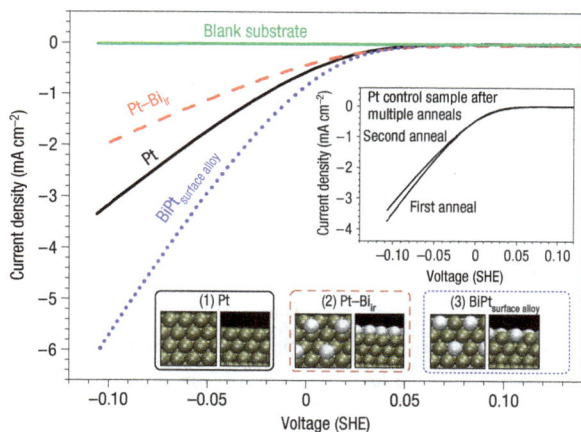

Figure 5 Hydrogen evolution after each stage of BiPt surface alloy synthesis on a fluorine-doped tin-oxide substrate. (1) Pt film after deposition and anneal (2) immediately after Bi UPD (3) after second anneal to form the BiPt surface alloy. The inset represents a control sample—Pt film without Bi—after first and second anneals. Current densities are normalized to the surface area of the initial, pure Pt sample, determined by H UPD.

surface structure is a surface alloy and not Bi_{ir} on Pt. By carefully controlling our electrochemical synthesis procedure (see discussion below), we create a surface alloy, and we show that this specific structure does, in fact, show improved activity for the HER. We note that, to our knowledge, no previous electrochemical studies on the Pt–Bi_{ir} system have attempted to form a surface alloy from the Bi_{ir} submonolayer.

To create the BiPt surface alloy, we used a three-step approach: (1) electrodeposition of an initial Pt film onto an inert support, (2) spontaneous deposition of a submonolayer of Bi_{ir} by Bi underpotential deposition (UPD) and (3) annealing of the Pt–Bi_{ir} precursor to form a BiPt surface alloy. For a given sample, we measured hydrogen evolution after each step of the synthesis. Following the work of Gómez et al., we used the cathodic current density at 0.0 V versus SHE ($i_{0.0\,V}$) as the primary figure of merit for comparison—approximately the exchange current density[26,27,34]. Current–voltage measurements of a representative sample are plotted in Fig. 5. After heating the initial polycrystalline Pt film in a furnace (see the Methods section), the sample demonstrated a current density of $i_{0.0\,V} = 5.6(10)^{-4}\,A\,cm_{Pt}^{-2}$ ($-\log i_{0.0\,V} = 3.25$), consistent with previous literature values for Pt and indicating that the starting material was prepared adequately with negligible contamination[20]. We then adsorbed a submonolayer of Bi_{ir} onto the Pt film by the same Bi UPD method used in previous reports (10 min, 1 mM Bi_2O_3/0.5 M H_2SO_4), resulting in an estimated Bi_{ir} coverage of ~10% by X-ray photoelectron spectroscopy (data not shown). Immediately after Bi UPD, the measured activity of the Pt–Bi_{ir} sample was considerably less than that of the initial Pt film—as expected, the Bi_{ir} blocked Pt sites and poisoned the surface for hydrogen evolution[24]. To complete the synthesis of a BiPt surface alloy, we annealed the Pt–Bi_{ir} precursor to 500 °C in a tube furnace.

Heating the Pt–Bi_{ir} system is the crucial step, as it provides the temperature necessary for Bi to overcome the kinetic barriers associated with alloying into the surface layer of Pt. Evidence of this process was reported by Paffett, Campbell and Taylor who conducted extensive studies of Bi dosed onto a Pt(111) crystal under ultrahigh vacuum conditions[35,36]. They used low-energy electron diffraction and Auger electron spectroscopy to

track morphological changes in a Bi adlayer as a function of coverage and temperature. One of the features they observed was that of a multilayered surface alloy—nominal composition $Pt_{1.00}Bi_{0.95}$—that formed at moderate temperatures (~370 °C) after starting with a multilayer Bi film. These surface-science experiments clearly indicate that Bi can and will form a surface alloy with Pt at elevated temperatures. The synthesis procedure that we have developed shows that surface alloys can also be produced from electrochemically adsorbed species; this is a simple, low-cost and general method that can be applied to numerous electrochemical systems.

Having completed the annealing procedure, we again measured the hydrogen evolution rate. Figure 5 shows significantly improved current–voltage characteristics; quantitatively speaking, $i_{0.0\,V}$ is double that of the Pt–Bi_{ir} precursor and, more importantly, ~50% greater than $i_{0.0\,V}$ of the initial Pt film. The improved performance of this film cannot be attributed to a simple change in morphology or to increased surface area because cyclic voltammetry in the H UPD region after each stage of synthesis indicates, if anything, a decrease in Pt surface area (see the Supplementary Information). To verify that the final anneal did not influence the Pt itself, we conducted several control experiments on similarly prepared pure Pt samples through multiple heat treatments. These data are shown as an inset in Fig. 5 and demonstrate a negligible effect.

The three-step procedure used to synthesize the annealed sample is highly reproducible, and the results provide strong evidence that the final, annealed sample is indeed a BiPt surface alloy, consistent with the ultrahigh vacuum studies of Paffett et al.[35,36]. It is worth noting that Bi itself is a notoriously poor electrocatalyst for the HER[17]; so it is difficult to imagine a morphology involving Bi—other than a surface alloy—that could possibly improve the electrocatalytic activity of Pt for hydrogen evolution. Hence, we have shown that a combination of in silico screening and careful experimental synthesis of promising candidates can lead to improved electrocatalysts, even when the identified metal combinations involve an element that is known to be inert for the HER, or when the metal combinations have previously been studied in different structural forms and have been found to be inactive.

We have thus confirmed experimentally the prediction that the HER activity of BiPt surface alloys should be comparable to that of pure Pt, or perhaps slightly better. This result suggests that our computational screening procedure is a promising technique for use in catalyst searches. The screening procedure can be viewed as a general, systematic, DFT-based method of incorporating both activity and stability criteria into the search for new metal alloy catalysts. As the accuracy and quality of kinetic models and DFT calculations improve, such in silico combinatorial screening procedures should become broadly useful for catalytic materials discovery.

METHODS

The computational analysis is carried out using DACAPO[37], a total energy calculation code. For the high-throughput computational screening, a three-layer slab, periodically repeated in a super-cell geometry with five equivalent layers of vacuum between any two successive metal slabs, is used to determine hydrogen binding energies. Close-packed surfaces are considered in all cases. A $(\sqrt{3} \times \sqrt{3})R30°$ unit cell, corresponding to a hydrogen coverage of 1/3 ML, is used. With this unit cell, and using the 16 elements described in the text, it is possible to create 736 symmetrically distinct pure metal and binary surface alloy slabs—16 pure metals, 240 surface alloys with a pure overlayer of solute metal, 240 with only two solute atoms in the surface layer and an extra 240 with one solute atom in the surface layer. For all slabs, the metal atoms are kept fixed in their bulk-truncated positions, and the hydrogen atoms are allowed to relax until the total force is less than $0.5\,\text{eV}\,\text{Å}^{-1}$. The total energy is then further refined by using the residual forces and estimated harmonic vibrational frequencies for hydrogen ($\sim1,050\,\text{cm}^{-1}$ for three-fold sites and $\sim2,020\,\text{cm}^{-1}$ for top sites) to extrapolate to the bottom of a parabola in energy space. $0.24\,\text{eV}$ is added to the calculated binding energies (with respect to gaseous H_2) to give adsorption free energies[20]. The oxygen binding energies (used for the oxygen adsorption stability criterion) are computed in the same way, except that a slightly higher force cutoff ($0.6\,\text{eV}\,\text{Å}^{-1}$) is used and appropriate vibrational frequencies ($\sim500\,\text{cm}^{-1}$ for three-fold sites and $\sim775\,\text{cm}^{-1}$ for top sites) are used for the energy extrapolations.

For all DFT calculations, adsorption is allowed on only one of the two exposed surfaces of the metal slabs, and the electrostatic potential is adjusted accordingly[38]. Ionic cores are described by ultrasoft pseudopotentials[39], and the Kohn–Sham one-electron valence states are expanded in a basis of plane waves with kinetic energy below $340\,\text{eV}$; a density cutoff of $500\,\text{eV}$ is used. The surface Brillouin zone is sampled with an $18(\sqrt{3}\times\sqrt{3})$ Chadi–Cohen $\mathbf{k}$ point grid. In all cases, convergence of the total energy with respect to the cutoff energies and the $\mathbf{k}$ point set is confirmed. The exchange–correlation energy and potential are described by the generalized gradient approximation[37] (GGA-RPBE98). The self-consistent RPBE98 density is determined by iterative diagonalization of the Kohn–Sham hamiltonian, Fermi population of the Kohn–Sham states ($k_\text{B}T = 0.1\,\text{eV}$) and Pulay mixing of the resulting electronic density[40]. All total energies have been extrapolated to $k_\text{B}T = 0\,\text{eV}$. Zero-point energy effects are assumed to be approximately constant for all metals and alloys considered[20]. Spin-polarization effects are included in the reported results for alloys in which naturally magnetized metals (Ni, Co, Fe) are present. Graphical inserts are produced using visual molecular dynamics[41].

For the more detailed calculations on the BiPt surface alloy, a 2×2 surface unit cell is used, corresponding to a hydrogen coverage of $1/4\,\text{ML}$. A four-layer slab is used for these calculations, and the top two layers of the slab are allowed to relax. The maximum force permitted for any vector component is $0.05\,\text{eV}\,\text{Å}^{-1}$. The $\mathbf{k}$ point grid, in this case, involves $18(1 \times 1)$ Chadi–Cohen points.

Pt was electrodeposited from a $1\,\text{mM}$ H_2PtCl_6 (99.995%, Aldrich) solution onto an inert substrate—fluorine-doped tin-oxide coated onto glass (Hartford Glass). A square wave pulsed deposition was used, referenced to a Ag quasi-reference electrode (AgQRE). The three-step deposition sequence consisted of a $0.5\,\text{s}$ pulse at $-0.9\,\text{V}$ versus AgQRE, followed by a $0.5\,\text{s}$ pulse at $-0.2\,\text{V}$ versus AgQRE, followed by a $1\,\text{s}$ rest interval. The sequence was repeated 120 times, totalling $4\,\text{min}$. The AgQRE was produced by oxidizing a pre-cleaned Ag wire in $1.0\,\text{M}$ HCl at $+1.0\,\text{V}$ versus the saturated calomel electrode for $60\,\text{s}$. Cyclic voltammograms of the deposited and annealed Pt films showed no evidence of contamination, corroborated by X-ray photoelectron spectroscopy measurements. Hydrogen evolution experiments were conducted in N_2-purged H_2SO_4 (pH 0.40). The cyclic voltammogram sweep rate was $5\,\text{mV}\,\text{s}^{-1}$ versus the saturated calomel electrode. A bridge was used during recording of the cyclic voltammograms to prevent Cl^- contamination. All heat treatments were conducted in a tube furnace at $500\,^\circ\text{C}$ in air for $12\,\text{h}$. Samples were transferred in ambient air and protected in plastic containers.

Received 30 June 2006; accepted 5 September 2006; published 15 October 2006.

References

1. Greeley, J. & Mavrikakis, M. Alloy catalysts designed from first principles. *Nature Mater.* **3**, 810–815 (2004).
2. Muller, R. P., Philipp, D. M. & Goddard, W. A. Quantum mechanical-rapid prototyping applied to methane activation. *Top. Catalys.* **23**, 81–98 (2003).
3. Andersson, M. P. *et al.* Towards computational screening in heterogeneous catalysis: Pareto-optimal methanation catalysts. *J. Catalys.* **239**, 501–506 (2006).
4. Greeley, J., Nørskov, J. K. & Mavrikakis, M. Electronic structure and catalysis on metal surfaces. *Annu. Rev. Phys. Chem.* **53**, 319–348 (2002).
5. Toulhoat, H. & Raybaud, P. Kinetic interpretation of catalytic activity patterns based on theoretical chemical descriptors. *J. Catalys.* **216**, 63–72 (2003).
6. Linic, S., Jankowiak, J. & Barteau, M. A. Selectivity driven design of bimetallic ethylene epoxidation catalysts from first principles. *J. Catalys.* **224**, 489–493 (2004).
7. Ceder, G. *et al.* Identification of cathode materials for lithium batteries guided by first-principles calculations. *Nature* **392**, 694–696 (1998).
8. Besenbacher, F. *et al.* Design of a surface alloy catalyst for steam reforming. *Science* **279**, 1913–1915 (1998).
9. Franceschetti, A. & Zunger, A. The inverse hand-structure problem of finding an atomic configuration with given electronic properties. *Nature* **402**, 60–63 (1999).
10. Vitos, L., Korzhavyi, P. A. & Johansson, B. Stainless steel optimization from quantum mechanical calculations. *Nature Mater.* **2**, 25–28 (2003).
11. Jacobson, M. Z., Colella, W. G. & Golden, D. M. Cleaning the air and improving health with hydrogen fuel-cell vehicles. *Science* **308**, 1901–1905 (2005).
12. Hamann, C. H., Hamnett, A. & Vielstich, W. *Electrochemistry* (Wiley-VCH, Weinheim, 1998).
13. Lide, D. R. (ed.) *CRC Handbook of Chemistry and Physics* (CRC Press, New York, 1996).
14. Conway, B. E. & Bockris, J. O. M. Electrolytic hydrogen evolution kinetics and its relation to the electronic and adsorptive properties of the metal. *J. Chem. Phys.* **26**, 532–541 (1957).
15. Parsons, R. The rate of electrolytic hydrogen evolution and the heat of adsorption of hydrogen. *Trans. Faraday Soc.* **54**, 1053–1063 (1958).
16. Gerischer, H. Mechanism of electrolytic discharge of hydrogen and adsorption energy of atomic hydrogen. *Bull. Soc. Chim. Belg.* **67**, 506 (1958).
17. Trasatti, S. Work function, electronegativity, and electrochemical behaviour of metals. III. Electrolytic hydrogen evolution in acid solutions. *Electroanal. Chem. Interfacial Electrochem.* **39**, 163–184 (1972).
18. Krishtalik, L. I. On the influence of hydrogenation of the cathode metal upon the overvoltage of hydrogen. *Electrokhimiya* **2**, 616 (1966).
19. Sabatier, P. Hydrogénations et deshydrogénations par catalyse. *Ber. Deutschen Chem. Gesellschaft* **44**, 1984 (1911).
20. Nørskov, J. K. *et al.* Trends in the exchange current for hydrogen evolution. *J. Electrochem. Soc.* **152**, J23–J26 (2005).
21. Greeley, J., Kibler, L., El-Aziz, A. M., Kolb, D. M. & Nørskov, J. K. Hydrogen evolution over bimetallic systems: Understanding the trends. *ChemPhysChem* **7**, 1032–1035 (2006).
22. Nørskov, J. K. *et al.* Origin of the overpotential for oxygen reduction at a fuel-cell cathode. *J. Phys. Chem. B* **108**, 17886–17892 (2004).
23. Casado-Rivera, E. *et al.* Electrocatalytic activity of ordered intermetallic phases for fuel cell applications. *J. Am. Chem. Soc.* **126**, 4043–4049 (2004).
24. Markovic, N. M. & Ross, P. N. Surface science studies of model fuel cell electrocatalysts. *Surf. Sci. Rep.* **45**, 121–229 (2002).
25. Schmidt, T. J., Stamenkovic, V. R., Lucas, C. A., Markovic, N. M. & Ross, P. N. Surface processes and electrocatalysts on the Pt(hkl)/Bi-solution interface. *Phys. Chem. Chem. Phys.* **3**, 3879–3890 (2001).
26. Gómez, R., Fernández-Vega, A., Feliu, J. M. & Aldaz, A. Hydrogen evolution on Pt single-crystal surfaces—effects of irreversibly adsorbed bismuth and antimony on hydrogen adsorption and evolution on Pt(100). *J. Phys. Chem.* **97**, 4769–4776 (1993).
27. Gómez, R., Feliu, J. M. & Aldaz, A. Effects of irreversibly adsorbed bismuth on hydrogen adsorption and evolution on Pt(111). *Electrochim. Acta* **42**, 1675–1683 (1997).
28. Bowles, B. J. Formation and desorption of monolayers of bismuth on a platinum electrode. *Electrochim. Acta* **15**, 737 (1970).
29. Clavilier, J., Feliu, J. M. & Aldaz, A. An irreversible structure sensitive adsorption step in bismuth underpotential deposition at platinum-electrodes. *J. Electroanal. Chem.* **243**, 419–433 (1988).
30. Hayden, B. E., Murray, A. J., Parsons, R. & Pegg, D. J. UHV and electrochemical transfer studies on Pt(110)-(1x2): The influence of bismuth on hydrogen and oxygen adsorption, and the electro-oxidation of carbon monoxide. *J. Electroanal. Chem.* **409**, 51–63 (1996).
31. Evans, R. W. & Attard, G. A. The redox behavior of compressed bismuth overlayers irreversibly adsorbed on Pt(111). *J. Electroanal. Chem.* **345**, 337–350 (1993).
32. Hamm, U. W., Kramer, D., Zhai, R. S. & Kolb, D. M. On the valence state of bismuth adsorbed on a Pt(111) electrode: an electrochemistry, LEED, and XPS study. *Electrochim. Acta* **43**, 2969–2978 (1998).
33. Kizhakevariam, N. & Stuve, E. M. Coadsorption of bismuth with electrocatalytic molecules—a study of formic-acid oxidation on Pt(100). *J. Vac. Sci. Technol. A* **8**, 2557–2562 (1990).
34. Vetter, K. J. *Electrochemical Kinetics: Theoretical and Experimental Aspects* (Academic, New York, 1967).
35. Paffett, M. T., Campbell, C. T. & Taylor, T. N. The influence of adsorbed Bi on the chemisorption properties of Pt(111)—H_2, CO, and O_2. *J. Vac. Sci. Technol. A* **3**, 812–816 (1985).
36. Paffett, M. T., Campbell, C. T. & Taylor, T. N. Adsorption and growth modes of Bi on Pt(111). *J. Chem. Phys.* **85**, 6176–6185 (1986).
37. Hammer, B., Hansen, L. B. & Nørskov, J. K. Improved adsorption energetics within density-functional theory using revised Perdew-Burke-Ernzerhof functionals. *Phys. Rev. B* **59**, 7413–7421 (1999).
38. Bengtsson, L. Dipole correction for surface supercell calculations. *Phys. Rev. B* **59**, 12301–12304 (1999).
39. Vanderbilt, D. Soft self-consistent pseudopotentials in a generalized eigenvalue formalism. *Phys. Rev. B* **41**, 7892–7895 (1990).
40. Kresse, G. & Furthmuller, J. Efficiency of ab-initio total energy calculations for metals and semiconductors using a plane-wave basis set. *Comput. Mater. Sci.* **6**, 15–50 (1996).
41. Humphrey, W., Dalke, A. & Schulten, K. VMD—visual molecular dynamics. *J. Mol. Graph.* **14**, 33–38 (1996).

Acknowledgements

J.G. and T.F.J. acknowledge H. C. Ørsted Postdoctoral Fellowships from the Technical University of Denmark. J.B. acknowledges support from the Danish Strategic Research Council. The Center for Individual Nanoparticle Functionality is supported by the Danish National Research Foundation. The Center for Atomic-scale Materials Design is supported by the Lundbeck Foundation. We thank the Danish Center for Scientific Computing for computer time. We also thank K. P. Jørgensen and J. Larsen for technical assistance.
Correspondence and requests for materials should be addressed to J.K.N.
Supplementary Information accompanies this paper on www.nature.com/naturematerials.

Competing financial interests

The authors declare that they have no competing financial interests.

Tuning clathrate hydrates for hydrogen storage

Huen Lee[1], Jong-won Lee[1], Do Youn Kim[1], Jeasung Park[1], Yu-Taek Seo[2]*,
Huang Zeng[2], Igor L. Moudrakovski[2], Christopher I. Ratcliffe[2]
& John A. Ripmeester[2]

[1]Department of Chemical and Biomolecular Engineering, Korea Advanced
Institute of Science and Technology (KAIST), Daejeon 305-701, Republic of Korea
[2]Steacie Institute for Molecular Sciences, National Research Council Canada,
Ottawa, Ontario, Canada K1A 0R6

* Present address: Conversion Process Research Center, Korea Institute of Energy Research, PO Box 103,
Jang-dong, Yuseong-gu, Daejeon 305-343, Republic of Korea

The storage of large quantities of hydrogen at safe pressures[1] is a
key factor in establishing a hydrogen-based economy. Previous
strategies—where hydrogen has been bound chemically[2],
adsorbed in materials with permanent void space[3] or stored in
hybrid materials that combine these elements[3]—have problems
arising from either technical considerations or materials cost[2-5].
A recently reported[6-8] clathrate hydrate of hydrogen exhibiting

two different-sized cages does seem to meet the necessary storage
requirements; however, the extreme pressures (~2 kbar)
required to produce the material make it impractical. The
synthesis pressure can be decreased by filling the larger cavity
with tetrahydrofuran (THF) to stabilize the material[9], but the
potential storage capacity of the material is compromised with
this approach. Here we report that hydrogen storage capacities in
THF-containing binary-clathrate hydrates can be increased to
~4 wt% at modest pressures by tuning their composition to
allow the hydrogen guests to enter both the larger and the smaller
cages, while retaining low-pressure stability. The tuning mechanism is quite general and convenient, using water-soluble
hydrate promoters and various small gaseous guests.

The structure II (sII) hydrates constitute a large family of
clathrates with an ideal unit cell $16S \cdot 8L \cdot 136H_2O$, where the large
(L) and small (S) cavities can be filled with guest molecules[10]. The
'solid solution' theory of van der Waals and Platteeuw, the classical
approach to understanding clathrate behaviour, uses the basic
expression:

$$\Delta\mu_w = -kT\Sigma_i \nu_i \ln(1 - \Theta_i) \qquad (1)$$

which relates the free energy difference, $\Delta\mu_w$, between ice and a
hypothetical empty hydrate framework to the minimum occupancy
Θ_i of the hydrate cavities of type i required to render it stable; ν_i are
the number of cages of type i in the unit cell normalized by the
number of water molecules. For sII hydrate, the 'best value' of $\Delta\mu_w$
of 884 J mol^{-1} requires the large cages to be filled to more than 99%
for stability. This is consistent with the observation that many sII
hydrates are known for which the stoichiometry is $L \cdot 17H_2O$ within
experimental error. The small cages can be occupied by a small
guest, leading to a double hydrate, $(2S)_x L \cdot 17H_2O$, generally stable
to higher temperatures, where x is the fractional occupancy of the
small cages, as recently illustrated[9] for THF and H_2 ($x \approx 1$, 1 wt%
H_2). The recently reported H_2 clathrate is also sII with multiple
occupancy of the cages (4 in L, 2 in S)[6-8]. Double hydrates represent
an opportunity to engineer hybrid structures that combine H_2
storage capacity with much less severe synthetic pressures.

We carried out initial studies on materials produced from
5.56 mol% THF solution in water, which gives a clathrate of
composition $THF \cdot 17H_2O$ when cooled below the melting point of
277.3 K. The THF hydrate was then pressurized with H_2 gas at
various pressures up to ~120 bar, and examined for structure, cage
populations and composition by powder X-ray diffraction (PXRD),
Raman and NMR spectroscopy, and direct measurement of the H_2
released on decomposition. From the PXRD results (Supplementary Fig. 1 and Supplementary Table 1), the material was confirmed
to be a sII hydrate according to its phase behaviour (Supplementary
Fig. 2). The Raman spectra (Fig. 1) show four transitions due to
rotational fine structure associated with the high-pressure H_2 gas,
and a broad band that can be identified with H_2 inside hydrate
cages. The hydrate H_2 line is shifted to lower frequency compared to
the free gas[11], as has also been observed for O_2 and N_2 hydrates[12]. ^{1}H
NMR spectroscopy was used to monitor H_2 in the product of the
reaction of H_2 with perdeuterated THF hydrate ($THF-d_8 \cdot 17D_2O$),
so that only H_2 signals and residual protons in water and THF
would be observed. The spectra in Fig. 2 show a broad line at
4.3 p.p.m. that can be attributed to H_2 in the small cavities of the
double hydrate. Full loading of the small cavities of $THF \cdot 17H_2O$
with hydrogen (2H_2 per small cage) will result in a storage capacity
of 2.1 wt% H_2.

Is it possible to obtain higher hydrogen loading, while keeping the
H_2 pressure at a reasonable value? Although double hydrates have
been known for over a century, techniques for hydrate analysis
in terms of guest distribution over the hydrate cage sites were
developed only recently, so little effort has gone into attempts to
tune hydrate compositions. In order to increase the hydrogen
content of the hydrate, the hydrogen guest must also enter the

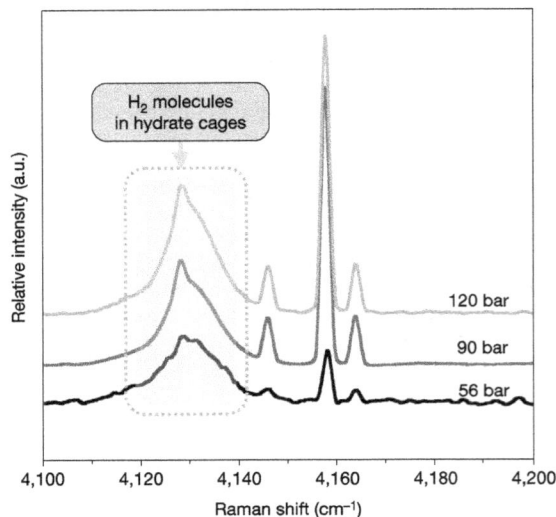

Figure 1 Raman spectra of the THF + H$_2$ double hydrates. Frozen 5.56 mol% THF solutions were stored in a refrigerator at 243 K for at least one day, and then ground to a fine powder (~200 μm). The powdered solid was exposed to hydrogen gas (ultrapure carrier grade, 99.999%) at pressures from 56 to 120 bar and at a temperature of 270 K in a high-pressure Raman cell equipped with two sapphire windows and circular grooves for coolant. The Raman spectrum was obtained using a SPEX 1404p single grating Raman spectrometer, with a CCD detector cooled by liquid nitrogen, using the focused 488 nm line of an Ar-ion laser for excitation. The laser intensity was typically 300 mW. The H–H vibron peak at 4,128 cm^{-1} consists of both the sharp spectral line assigned to hydrogen gas and the broad peak due to hydrogen molecules stored in clathrate hydrate cages. The broad peak becomes higher when hydrate formation pressure increases in the stable region of the pressure–temperature diagram (Supplementary Fig. 2).

Figure 2 Magic angle spinning ^{1}H NMR spectra of the THF + H$_2$ double hydrates formed at 120 bar and 270 K as a function of concentration of THF. The NMR samples were prepared from deuterated water (D$_2$O, 99.9 at.% D) and THF (THF-d$_8$, 99.5 at.% D) by the same procedure as for Raman experiments. The chemical shifts of D$_2$O and THF-d$_8$ were identified from the NMR spectrum of THF-d$_8$ hydrate. Spectra were recorded using a Varian INOVA600 spectrometer, spin rate ~12 kHz, pulse length 5 μs, repetition time 15 s. The samples were transferred to an NMR rotor and analysed at 1 bar and 183 K. As the hydrate is not absolutely stable under these conditions, quantitative spectra cannot be obtained because of some decomposition, although a consistent picture can be constructed from the spectra that illustrate several features. The distribution of hydrogen molecules in both cages depends on temperature and pressure[13].

large cavities of sII. The solid-solution model then requires that the large cage occupancies Θ_L(THF) and Θ_L(H$_2$) become comparable, where the Θ_L are given by $\Theta_{Li} = C_{Li}p_i/(1 + C_{L1}p_1 + C_{L2}p_2)$; the C_{Li} are the Langmuir constants for the large cage, and the p_i are the partial pressures of the guest. As C_L(THF) is much larger than C_L(H$_2$), this requires the partial pressure (concentration) of THF to be lowered considerably in order to place H$_2$ clusters into the large cage.

Hence, a number of experiments were carried out at an H$_2$ pressure of 120 bar where the THF concentration was decreased from 5.56 mol% down to 0.1 mol%. The compositions of the hydrates produced were obtained from measurements of the Raman band intensities. The H$_2$ content of the clathrate produced at each THF loading was calculated from the integral intensity of its Raman band (after appropriate subtraction of intensity from the overlapping gas lines) relative to that of the 5.56 mol% sample and

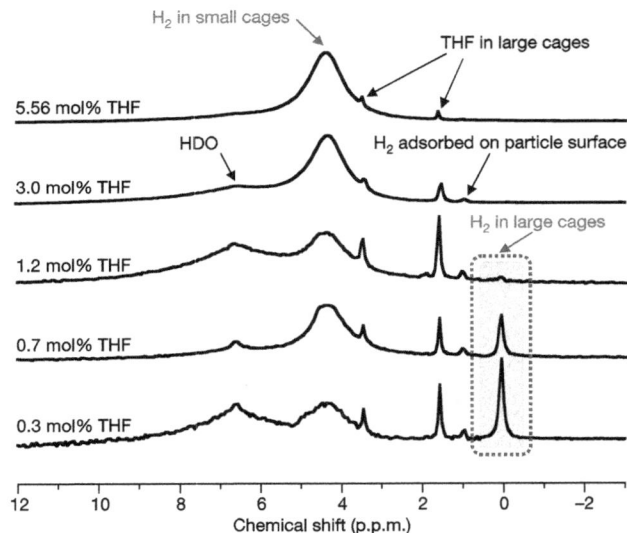

taking into account the different relative volumes of hydrate and ice. Extinction coefficients were assumed to be constant, based on the observation that the ratio of Raman intensity to volumetrically measured H$_2$ content remains constant over the range of THF concentrations (Supplementary Material and Supplementary Fig. 3). The 5.56 mol% sample was taken to be fully loaded, with 2H$_2$ per small cage and 2.09 wt% H$_2$ content. Results are presented in Table 1 as H$_2$:THF mole fractions and wt% H$_2$ content of the hydrate phase (based on the model discussed below). The H$_2$:THF mole ratio of the 'reference' 5.56 mol% sample is 4, whereas for samples with 4.0 and 2.0 mol% THF the mole ratio falls slightly below 4, indicating a small deficit of H$_2$ in the small cages.

What is most interesting, however, is that at 1.0 mol% and down to 0.15 mol% the mole ratio increases above 4 to as high as ~23 for the 0.15 mol% sample, indicating that large cages must also contain H$_2$. The 1 mol% concentration of THF is approximately at the

Table 1 Raman peak areas and H$_2$ content parameters

Mol% THF (starting solution)	Raman peak area (a.u.)	H$_2$:THF mole ratio in hydrate	x	y	Max. wt% H$_2$ in hydrate
10.0	15,796	4.0	0	2	2.09*
5.56	15,406	4.0	0	2	2.09
4.0	9,740	3.44	0	1.719	1.80
2.0	4,896	3.36	0	1.681	1.76
1.5	4,428	4.03	0.003	2	2.09
1.0	3,330	4.52	0.061	2	2.24
0.5	3,085	8.35	0.352	2	3.00
0.2	2,613	17.63	0.630	2	3.80
0.15	2,568	23.10	0.705	2	4.03
0.1	0	0	0	0	0

For $(yH_2)_2 \cdot (4H_2)_x \cdot THF_{(1-x)} \cdot 17H_2O$. a.u., arbitrary units.
*This sample has excess THF (with excess H$_2$ dissolved in the THF). Note that pure hydrogen clathrate $(2H_2)_2 \cdot (4H_2) \cdot 17H_2O$ would have a 5.002 wt% H$_2$ content.

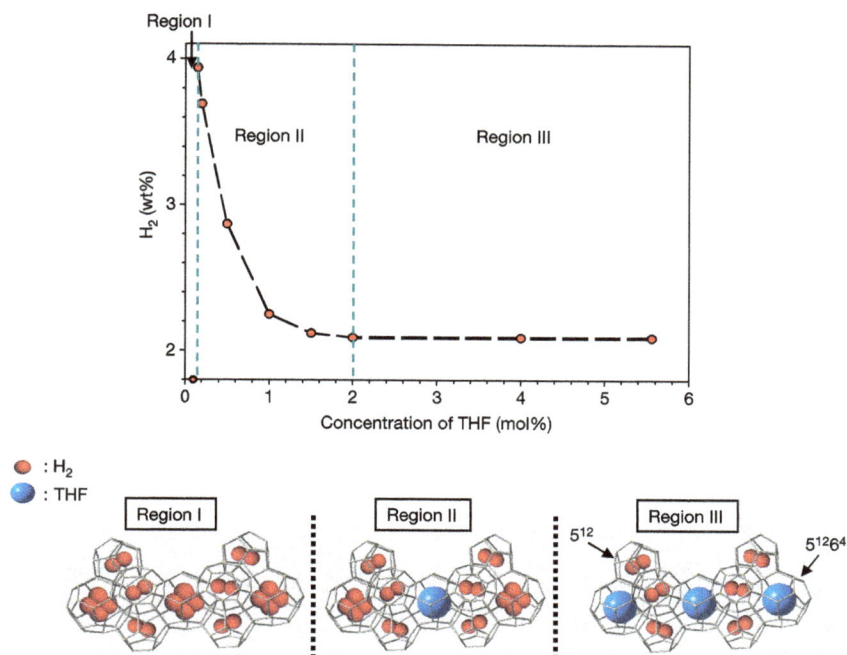

Figure 3 H_2 gas content as a function of THF concentration, and a schematic diagram of H_2 distribution in the cages of THF + H_2 hydrate. (H_2 gas content is calculated from g of H_2 per g of hydrate, and expressed as wt%.) In region III, H_2 molecules are only stored in small cages, while in region II both small and large cages can store H_2 molecules. At the eutectic composition of THF hydrate and below this, solid ice I_h would be the first phase to appear, thus reflecting the fact that the concentration of THF has declined to the point where hydrate no longer forms ($\Theta_L = C_L p/(1 + C_L p) = < 1$). For sII hydrate to be stable relative to ice, the only condition that needs to be satisfied is that the occupancy of the large cage is close to 1, so $\Theta_{THF} + \Theta_{H_2} \approx 1$. This approach represents a general strategy for tuning hydrate compositions—at or below the eutectic composition of water-soluble sII hydrate formers, it is necessary to introduce other guests into the large cages of sII hydrate for stability.

At this stage we can consider the exact nature of the THF hydrate with high H_2 content. We have concluded that some large cages contain THF and some have H_2 clusters, while maintaining the double cage occupancy of the small cages. Thus the formation of double hydrate at low THF concentration (1.0–0.15 mol%) probably progresses as follows:

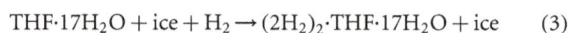

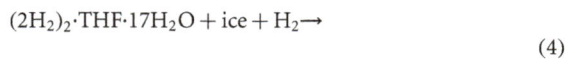

$$\text{THF} + \text{H}_2\text{O} \ (< 1.0 \text{ mol}\%) \rightarrow \text{THF·17H}_2\text{O} + \text{ice} \ (T < 270 \text{ K}) \quad (2)$$

$$\text{THF·17H}_2\text{O} + \text{ice} + \text{H}_2 \rightarrow (2\text{H}_2)_2\text{·THF·17H}_2\text{O} + \text{ice} \quad (3)$$

$$(2\text{H}_2)_2\text{·THF·17H}_2\text{O} + \text{ice} + \text{H}_2 \rightarrow \quad (4)$$
$$(2\text{H}_2)_2\text{·}(4\text{H}_2)_x\text{·THF}_{(1-x)}\text{·17H}_2\text{O}$$

where $(2\text{H}_2)_2$ and $(4\text{H}_2)_x$ represent clusters of two or four H_2 molecules occupying small cages and large cages, respectively (as suggested from very-high-pressure studies of H_2 hydrate[6]). THF hydrate formed in the first process (equation (2)) reacts with H_2 to form a double hydrate (equation (3)), which is the kinetically limited product. Disproportionation of the double hydrate formed in equation (3) with the ice must then occur in order to allow H_2 to occupy large cages and thus produce a thermodynamically stable double hydrate (equation (4)). ^{1}H NMR spectroscopy was used to identify the distribution of guests over the cages in the double

highly dilute THF concentrations of region I, H_2 molecules can still be stored in both cages, but extreme pressures (~2 kbar) are required to form the hydrates. Pure H_2 clathrate $(2\text{H}_2)_2\text{·}(4\text{H}_2)\text{·17H}_2\text{O}$ would have a 5.002 wt% H_2 content.

hydrate with high H_2 content. When the concentration of THF reaches ~1 mol% a new line appears at ~0.15 p.p.m., which continues to grow with decreasing THF concentration. Considering the other experimental evidence for the increased loading of the sample, we assign this peak to H_2 clusters in the large cages (Fig. 2).

We can then calculate the values of x in equation (4) and the effective wt% H_2 for the clathrate formed from each THF concentration. Figure 3 shows a plot of wt% H_2 stored in hydrate as a function of THF concentration—note that these represent a maximum wt% H_2 content, based on the assumptions above. Cross-checks were made with a direct gas release measurement for the 5.56, 0.7 and 0.5 mol% samples, which gave respectively 2.1, 2.4 and 2.7 wt% H_2 compared to 2.1, 2.6 and 3.0 wt% H_2 from the Raman measurements. This can explain the increase in H_2 content in the hydrate, and indeed we observe that the wt% H_2 was 2.24 at 1.0 mol% increasing to 4.03 wt% at 0.15 mol% THF. At even lower THF concentration, 0.1 mol%, a H_2-containing hydrate was no longer formed under the conditions used. We take this to mean that the combination of THF concentration and H_2 pressure was insufficient to fill the large cages in the structure to produce a stable hydrate. When H_2 was applied to frozen 0.2 mol% THF solutions at pressures between 90 and 140 bar, the H_2 content reached a steady state for pressures higher than ~100 bar. If the H_2 in the large cages is present in the form of $(H_2)_4$ clusters, as has been proposed for pure H_2 hydrate, about 60% of the large cages will be occupied by these, with the remainder expected to contain THF. Similar experiments have shown that, upon decreasing the THF concentration below 1.0 mol%, it is possible to introduce other guests such as CH_4 into the large cages of sII hydrate. In recent ^{13}C NMR studies of the sII double hydrate of CH_4 and THF, two CH_4 resonances at −4.5 p.p.m. (small cage) and −8.1 p.p.m. (large cage) were found, but the large cage resonance was only seen in solutions with concentrations below 1.0 mol% THF.

Although the hydrate lattice appears to be tunable for higher

288

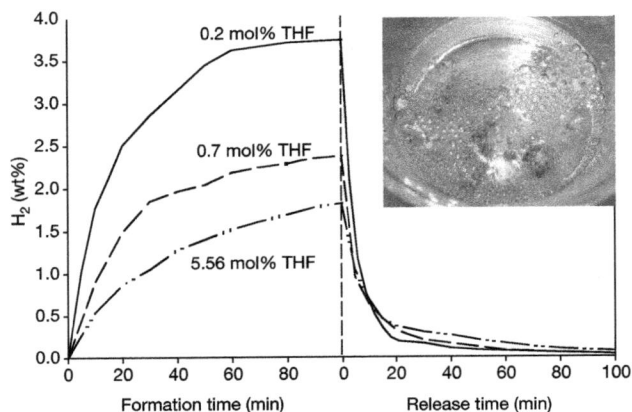

Figure 4 Formation/release kinetics of the H$_2$ + THF double hydrate in the pores of silica beads. At three THF concentrations (5.56, 0.7 and 0.2 mol%) the amount of H$_2$ stored in the cages was measured using an isometric micro-syringe pump. Hydrates were maintained at 1 bar and 270 K for complete release of hydrogen after forming at 120 bar and 270 K. Hydrogen bubbles are observed during hydrate release (inset).

Acknowledgements This work was supported by the Korea Research Foundation and the Brain Korea 21 Project.

Competing interests statement The authors declare that they have no competing financial interests.

Correspondence and requests for materials should be addressed to H.L. (h_lee@kaist.ac.kr) or J.A.R. (John.Ripmeester@nrc-cnrc.gc.ca).

levels of storage, up to ~4.0 wt%, the reaction is so slow as to make the synthetic approach impractical. It is beyond the scope of this Letter to devise an industrial strategy for the efficient synthesis of H$_2$-containing hydrate, but a promising approach is as follows: The reaction of the hydrate/ice mixture with H$_2$ is limited by diffusion through a bulk solid phase. So, if the reacting phase can be dispersed so as to improve the surface area to volume ratio, improved kinetics should be possible. To demonstrate proof of principle, we have dispersed the aqueous phase on a silica bead support, and after exposure to H$_2$ gas, followed the reaction kinetics by measuring the rate at which gas is consumed. Figure 4 shows that the reaction proceeds as expected and that the conversion is now complete in ~1 h; this is a tremendous improvement from the bulk reaction, which took a week or more. We recognize that this is not a practical solution, as the product resides inside the silica gel, and the additional weight takes the material out of the required range as a hydrogen storage material. However, this does show that a dispersed phase will react in a reasonable time, giving a product that is a suitable storage material. □

Received 20 September 2004; accepted 25 January 2005; doi:10.1038/nature03457.

1. Berry, G. D. & Aceves, S. M. Onboard storage alternatives for hydrogen vehicles. *Energy Fuels* **12**, 49–55 (1998).
2. Schlapbach, L. & Zuttel, A. Hydrogen-storage materials for mobile applications. *Nature* **414**, 353–358 (2001).
3. Weitkamp, J., Fritz, M. & Ernst, S. Zeolites as media for hydrogen storage. *Int. J. Hydrogen Energy* **20**, 967–970 (1995).
4. Schimmel, H. G. *et al.* Hydrogen adsorption in carbon nanostructures: Comparison of nanotubes, fibers, and coals. *Chem. Eur. J.* **9**, 4764–4770 (2003).
5. Rosi, N. L. *et al.* Hydrogen storage in microporous metal-organic frameworks. *Science* **300**, 1127–1129 (2003).
6. Mao, W. L. *et al.* Hydrogen clusters in clathrate hydrate. *Science* **297**, 2247–2249 (2002).
7. Patchkovskii, S. & Tse, J. S. Thermodynamic stability of hydrogen clathrates. *Proc. Natl Acad. Sci. USA* **100**, 14645–14650 (2003).
8. Mao, W. L. & Mao, H. Hydrogen storage in molecular compounds. *Proc. Natl Acad. Sci. USA* **101**, 708–710 (2004).
9. Florusse, L. J. *et al.* Stable low-pressure hydrogen clusters stored in a binary clathrate hydrate. *Science* **306**, 469–471 (2004).
10. Jeffrey, G. A. in *Comprehensive Supramolecular Chemistry* Vol. 6 (eds MacNicol, D. D, Toda, F. & Bishop, R.) 757–788 (Pergamon, Oxford, 1996).
11. Nakamoto, K. *Infrared and Raman Spectra of Inorganic and Coordination Compounds* 4th edn (Wiley, New York, 1986).
12. Nakahara, J. *et al.* C. C. Raman spectra of natural clathrates in deep ice cores. *Phil. Mag. B* **3**, 421–430 (1988).
13. Lokshin, K. A. *et al.* Structure and dynamics of hydrogen molecules in the novel clathrate hydrate by high pressure neutron diffraction. *Phys. Rev. Lett.* **93**, 125503 (2004).

Supplementary Information accompanies the paper on **www.nature.com/nature**.

Hydrogen from catalytic reforming of biomass-derived hydrocarbons in liquid water

R. D. Cortright, R. R. Davda & J. A. Dumesic

Department of Chemical Engineering, University of Wisconsin, Madison, Wisconsin 53706, USA

Concerns about the depletion of fossil fuel reserves and the pollution caused by continuously increasing energy demands make hydrogen an attractive alternative energy source. Hydrogen is currently derived from nonrenewable natural gas and petroleum[1], but could in principle be generated from renewable resources such as biomass or water. However, efficient hydrogen production from water remains difficult and technologies for generating hydrogen from biomass, such as enzymatic decomposition of sugars[2-5], steam-reforming of bio-oils[6-8] and gasification[9], suffer from low hydrogen production rates and/or complex processing requirements. Here we demonstrate that hydrogen can be produced from sugars and alcohols at temperatures near 500 K in a single-reactor aqueous-phase reforming process using a platinum-based catalyst. We are able to convert glucose—which makes up the major energy reserves in plants and animals—to hydrogen and gaseous alkanes, with hydrogen constituting 50% of the products. We find that the selectivity for hydrogen production increases when we use molecules that are more reduced than sugars, with ethylene glycol and methanol being almost completely converted into hydrogen and carbon dioxide. These findings suggest that catalytic aqueous-phase reforming might prove useful for the generation of hydrogen-rich fuel gas from carbohydrates extracted from renewable biomass and biomass waste streams.

We consider production of hydrogen by low-temperature reforming (at 500 K) of oxygenated hydrocarbons having a C:O stoichiometry of 1:1. For example, reforming of the sugar–alcohol sorbitol to H_2 and CO_2 occurs according to the following stoichiometric reaction:

$$C_6O_6H_{14}(l) + 6H_2O(l) \rightleftharpoons 13H_2(g) + 6CO_2(g) \qquad (1)$$

The equilibrium constant for reaction (1) per mole of CO_2 is of the order of 10^8 at 500 K, indicating that the conversion of sorbitol in

the presence of water to H_2 and CO_2 is highly favourable. However, the selective generation of hydrogen by this route is difficult because the products H_2 and CO_2 readily react at low temperatures to form alkanes and water. For example, the equilibrium constant at 500 K for the conversion of CO_2 and H_2 to methane (reaction 2) is of the order of 10^{10} per mole of CO_2.

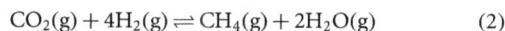

$$CO_2(g) + 4H_2(g) \rightleftharpoons CH_4(g) + 2H_2O(g) \qquad (2)$$

Figure 1 shows a schematic representation of the reaction pathways we believe to be involved in the formation of H_2 and alkanes from oxygenated hydrocarbons over a metal catalyst. The reactant undergoes dehydrogenation steps on the metal surface to give adsorbed intermediates before the cleavage of C–C or C–O bonds. With platinum, the catalyst we use, the activation energy barriers for cleavage of O–H and C–H bonds are similar[10]; however, Pt–C bonds are more stable than Pt–O bonds, so adsorbed species are probably bonded preferentially to the catalyst surface through Pt–C bonds. Subsequent cleavage of C–C bonds leads to the formation of CO and H_2, and CO reacts with water to form CO_2 and H_2 by the water–gas shift reaction (that is, $CO + H_2O \rightleftharpoons CO_2 + H_2$)[11,12].

The further reaction of CO and/or CO_2 with H_2 leads to alkanes and water by methanation and Fischer–Tropsch reactions[13-15]; this H_2 consuming reaction thus represents a series-selectivity challenge. In addition, undesirable alkanes can form on the catalyst surface by cleavage of C–O bonds, followed by hydrogenation of the resulting adsorbed species. This process constitutes a parallel-selectivity challenge. Another pathway that contributes to this parallel-selectivity challenge is cleavage of C–O bonds through dehydration reactions catalysed by acidic sites associated with the catalyst support[16,17] or catalysed by protons in the aqueous solution[18,19], followed by hydrogenation reactions on the catalyst. In addition, organic acids can be formed by dehydrogenation reactions catalysed by the metal, followed by rearrangement reactions[20] that take place in solution or on the catalyst. These organic acids lead to the formation of alkanes from carbon atoms that are not bonded to oxygen atoms.

Table 1 summarizes our experimental results for aqueous-phase reforming of glucose, the compound most relevant to hydrogen production from biomass, as well as for the reforming of sorbitol, glycerol, ethylene glycol and methanol. Reactions were carried out over a Pt/Al_2O_3 catalyst at 498 and 538 K (see Methods for experimental details). The fractions of the feed carbon detected in the effluent gas and liquid streams yield a complete carbon balance for all feed molecules, indicating that negligible amounts of carbon have been deposited on the catalyst. Catalyst performance was stable

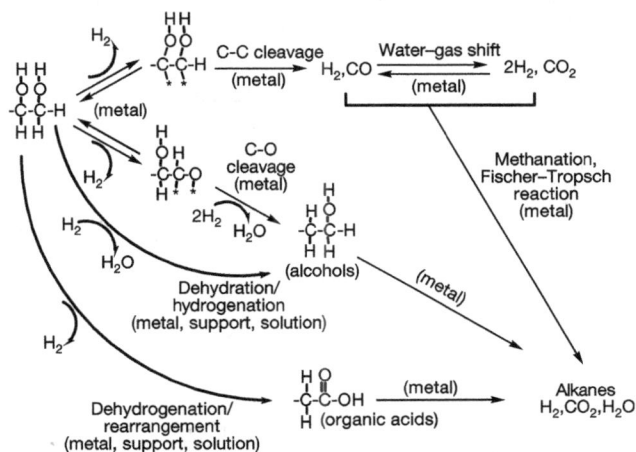

Figure 1 Reaction pathways for production of H_2 by reactions of oxygenated hydrocarbons with water. (Asterisk represents a surface metal site.)

When citing this article, please cite the original version as shown on the contents page of this chapter.

for long periods of time on stream (for example, 1 week). Results from replicate runs agree to within ±3%.

The hydrogen selectivities shown in Table 1 are defined as the number of H_2 molecules detected in the effluent gas, normalized by the number of H_2 molecules that would be present if the carbon atoms detected in the effluent gas molecules had all participated in the reforming reaction. (That is, we infer the amount of converted glucose, sorbitol, glycerol, ethylene glycol and methanol from the carbon-containing gas-phase products and assume that each of the feed molecules would yield 2, 13/6, 7/3, 5/2 or 3 molecules of H_2, respectively.) Alkane selectivity is defined as the number of carbon atoms in the gaseous alkane products normalized by the total number of carbon atoms in the gaseous effluent stream.

Figure 2 illustrates that the selectivity for H_2 production improves in the order glucose < sorbitol < glycerol < ethylene glycol < methanol. Figure 2 also implies that lower operating temperatures result in higher H_2 selectivities, although this trend is in part due to the lower conversions achieved at lower temperatures. The selectivity for alkane production follows a trend with respect to reactant that is opposite to that exhibited by the H_2 selectivity.

The gas streams from aqueous-phase reforming of the oxygenated hydrocarbons were found to contain low levels of CO (that is, less than 300 p.p.m.). For aqueous-phase reforming of glycerol, where analysis of liquid-phase products is more tractable compared to sorbitol and glucose, the major reaction intermediates detected include (in approximate order of decreasing concentration) ethanol, 1,2-propanediol, methanol, 1-propanol, acetic acid, ethylene glycol, acetol, 2-propanol, propionic acid, acetone, propionaldehyde, and lactic acid. Analysis of the gas phase effluent indicates the presence of trace amounts of methanol and ethanol (about 300 p.p.m.).

High hydrogen yields are only obtained when using sorbitol, glycerol and ethylene glycol as feed molecules for aqueous-phase reforming. Although these molecules can be derived from renewable feedstocks[21-24], the reforming of less reduced and more immediately available compounds such as glucose is likely to be more practical; but hydrogen yields for glucose reforming are relatively low (Table 1 and Fig. 2). However, we expect that improvements in catalyst performance, reactor design, and reaction conditions may increase the hydrogen selectivity for the direct aqueous-phase reforming of sugars. For example, the lower H_2 selectivities for the aqueous-phase reforming of glucose, compared to that achieved using the other oxygenated hydrocarbon reactants, are caused at least partially by homogeneous reactions of glucose in the aqueous phase at the temperatures used in this study[19]. Thus, higher selectivities for H_2 from aqueous-phase reforming of glucose might be achieved by reactor designs that maximize the number of catalytically active sites (leading to desirable surface reactions) and

minimize the void volume (leading to undesirable liquid-phase reactions) in the reactor.

Here, using Pt/Al_2O_3 as catalyst, we found that lower glucose concentrations correlated with higher selectivities for hydrogen production, and reforming experiments were therefore conducted at low feed concentrations of 1 wt%, which corresponds to a molar ratio H_2O/C_1 of 165. Processing such dilute solutions is economically not practical, even though reasonably high hydrogen yields are achieved (Table 1). However, undesirable homogeneous reactions, as observed with glucose, pose less of a problem when using sorbitol, glycerol, ethylene glycol and methanol, which makes it possible to generate high yields of hydrogen by the aqueous-phase reforming of more concentrated solutions containing these compounds. For example, we have found that upon increasing the feed concentrations of ethylene glycol or glycerol to 10 wt% (molar ratio $H_2O/C_1 = 15$), it is still possible to achieve high conversions and high selectivities for H_2 production. (A hydrogen selectivity of 97% was achieved with 62% conversion of 10 wt% ethylene glycol; and a hydrogen selectivity of 70% was achieved with 77% conversion of 10 wt% glycerol.) A molar H_2O/C_1 ratio of 15 is still higher than that typically utilized in conventional vapour-phase reforming of hydrocarbons (molar $H_2O/C_1 = 3$ to 5); but our aqueous-phase reforming system has the advantage of not requiring energy-intensive vaporization of water to generate steam.

Operating at higher reactant concentrations and lower conversion levels leads to higher rates of hydrogen production. Rates of hydrogen production are measured as turnover frequencies (that is, rates normalized by the number of surface metal atoms as determined from the irreversible uptake of CO at 300 K). For example, the turnover frequency for production of hydrogen at 498 K from a 1 wt% ethylene glycol solution is 0.08 min^{-1} for the 90% conversion run listed in Table 1 (weight-hourly space velocity, WHSV = 0.008 g of ethylene glycol per g of catalyst per h), but increases to 0.7 min^{-1} for a feed containing 10 wt% ethylene glycol (run at 498 K and at a higher WHSV of 0.12 g of ethylene glycol per g of catalyst per h). The hydrogen selectivity for this latter run is equal to 97% at a conversion of 62%. The turnover frequency for hydrogen production from 10 wt% ethylene glycol at 498 K increased further to 7 min^{-1} when higher space-velocities were used (WHSV = 18 g of ethylene glycol per g of catalyst per h) with a highly dispersed 3 wt% Pt/Al_2O_3 catalyst consisting of smaller alumina particles (63–125 μm) to minimize transport limitations. The hydrogen selectivity for this kinetically controlled run was 99% at a conversion of 3.5%.

The rates of formation of H_2 from aqueous-phase reforming of 10 wt% glucose, sorbitol, glycerol, ethylene glycol and methanol at 498 K over a 3 wt% Pt/Al_2O_3 catalyst under conditions to minimize transport limitations and at low conversions of the reactant have

Table 1 **Experimental data for reforming of oxygenated hydrocarbons**

	Glucose		Sorbitol		Glycerol		Ethylene glycol		Methanol	
Temperature (K)	498	538	498	538	498	538	498	538	498	538
Pressure (bar)	29	56	29	56	29	56	29	56	29	56
% Carbon in liquid-phase effluent	51	15	39	12	17	2.8	11	2.9	6.5	6.4
% Carbon in gas-phase effluent	50	84	61	90	83	99	90	99	94	94
Gas-phase composition										
H_2 (mol. %)	51	46	61	54	64.8	57	70	68.7	74.6	74.8
CO_2 (mol. %)	43	42	35	36	29.7	32	29.1	29	25	24.6
CH_4 (mol. %)	4.0	7.0	2.5	6.0	4.2	8.3	0.8	2.0	0.4	0.6
C_2H_6 (mol. %)	2.0	2.7	0.7	2.3	0.9	2.0	0.1	0.3	0.0	0.0
C_3H_8 (mol. %)	0.0	1.0	0.8	1.0	0.4	0.7	0.0	0.0	0.0	0.0
C_4, C_5, C_6 alkanes (mol. %)	0.0	1.2	0.0	0.6	0.0	0.0	0.0	0.0	0.0	0.0
% H_2 selectivity*	50	36	66	46	75	51	96	88	99	99
% Alkane selectivity†	14	33	15	32	19	31	4	8	1.7	2.7

The catalyst was loaded in a tubular reactor, housed in a furnace, and reduced prior to reaction kinetics studies. The reactor system was pressurized with N_2, and the reforming reaction was carried out at the listed reaction conditions. Each reaction condition was run for 24 h, during which the experimental data were collected. Further experimental details are provided in the Methods.
* % H_2 selectivity = (molecules H_2 produced/C atoms in gas phase)(1/RR) × 100, where RR is the H_2/CO_2 reforming ratio, which depends on the reactant compound. RR values for the compounds are: glucose, 2; sorbitol, 13/6; glycerol, 7/3; ethylene glycol, 5/2; methanol, 3. We note that H_2 and alkane selectivities do not add up to 100%, because they are based independently on H-balances and C-balances, respectively. % Carbon in gas and liquid phase effluents add to 100% for a complete carbon balance. Slight ± deviations from 100% are caused by experimental error.
† % Alkane selectivity = (C atoms in gaseous alkanes/total C atoms in gas phase product) × 100.

been measured to be 0.5, 1.0, 3.5, 7.0 and 7.0 min^{-1}, respectively. These turnover frequencies correspond to hydrogen production rates of 50, 100, 350, 700, and 700 l of H$_2$ (at standard temperature and pressure, 273 K and 1.01 bar) per l of reactor volume per h for each feed molecule, respectively. These rates compare favourably to the maximum rate of hydrogen production from glucose of about 5×10^{-3} l of H$_2$ per l of reactor volume per hour by enzymatic routes[5]. Normalization of the rates by the mass of catalyst used yields rates of about 3×10^3, 6×10^3, 2×10^4, 4×10^4, and 4×10^4 μmol g^{-1} h^{-1} for hydrogen production at 498 K, from 10 wt% glucose, sorbitol, glycerol, ethylene glycol and methanol, respectively. These rates can be compared to the maximum value of 7×10^2 μmol g^{-1} h^{-1} reported for hydrogen production from glucose by enzymatic routes. (For this comparison, we have assumed a typical value of 100 units per mg of protein, where a unit of enzyme activity corresponds to the amount of enzyme which under standard assay conditions converts 1 μmol of substrate per min).

If the hydrogen produced were fed to a fuel cell operating at 50% efficiency, the rate of hydrogen production in our reformer would generate approximately 1 kW of power per l of reactor volume. Electrical power might thus be generated cost-effectively by an integrated fuel-cell/liquid-phase reformer system fed with a low-cost carbohydrate stream derived from waste biomass (for example, corn stover, wheat straw, wood waste). The practical use of aqueous-phase reforming reactions in this manner would depend on efficient feed recycling strategies and on efficient separation of hydrogen from the gaseous effluent stream. The gaseous effluents separated from the main product hydrogen could be combusted to generate the energy necessary for the liquid-phase-reforming reactor. However, some fuel cell applications might not require extensive purification because the main components in the reformer gas effluent other than hydrogen are CO$_2$ and methane, which can act as diluents[25]. Alcohols and organic acids are also present in the gas effluent, but only at trace levels (300 p.p.m. and about 5 p.p.m., respectively) which may not lead to irreversible poisoning[26].

Reforming reactions between hydrocarbons and water to generate hydrogen are endothermic, and conventional steam-reforming of petroleum thus depends on the combustion of additional hydrocarbons to provide the heat needed to drive the reforming reaction. In contrast, the energy required for the aqueous-phase reforming of oxygenated hydrocarbons may be produced internally, by allowing a fraction of the oxygenated compound to form alkanes through exothermic reaction pathways. In this respect, the formation of a mixture of hydrogen and alkanes from aqueous-phase reforming of glucose, as accomplished in the present study, is essentially neutral energetically, and little additional energy is required to drive the reaction. In fact, the energy contained in these alkanes could be used as a feed to an internal combustion engine or suitable fuel cell; this would allow the use of biomass-derived energy to drive the aqueous-phase reforming of glucose (and biomass more generally) with high yields to renewable energy.

While the present findings establish that Pt-based catalysts show high activities and good selectivity for the production of hydrogen from sugars and alcohols by aqueous-phase reforming reactions, improvements are necessary to render the process useful. Highly active catalytic materials that can satisfy the series and parallel selectivity challenges outlined in Fig. 1, but at a lower materials cost than for Pt, are particularly desirable. Moreover, new combinations of catalysts and reactor configurations are needed to obtain higher hydrogen yields from more concentrated solutions of glucose, given that glucose is the only compound we have tested that is directly relevant to biomass utilization. We believe that such improvements are possible, for example, by searching for catalysts that exhibit higher activity at lower temperatures, to minimize the deleterious effects of homogeneous decomposition reactions. □

Methods

Experiments for the aqueous-phase reforming of glucose, sorbitol, glycerol, ethylene glycol and methanol were performed over a 3 wt% Pt catalyst supported on nanofibres of γ-alumina (500 m^2 g^{-1}, Argonide Corp.). The catalyst was prepared by incipient wetness impregnation of alumina with tetraamine platinum nitrate solution, followed by drying at 380 K, calcination at 533 K in flowing oxygen and reduction at 533 K in flowing hydrogen. Chemisorption experiments using carbon monoxide at 300 K showed a CO uptake of 105 μmol per g of catalyst. A stainless steel tubular reactor (having an inner diameter of 5 mm and length of 45 cm) was loaded with 4.5 g of the pelletized Pt/Al$_2$O$_3$ catalyst, which was then reduced under flowing hydrogen at 533 K. The total pressure of the system was then increased by addition of nitrogen to a value slightly higher than the vapour pressure of water at the reaction temperature. The system pressure was controlled by a backpressure regulator. An aqueous solution containing 1 wt% of the oxygenated compound was fed continuously, using a high-performance liquid chromatography (HPLC) pump, at 3.6 ml h^{-1} into the reactor heated to the desired reaction temperature. Under these conditions, the WHSV was 0.008 g of oxygenated compound per g of catalyst per h through the reactor. The effluent from the reactor was water-cooled in a double-pipe heat exchanger to liquefy the condensable vapours. The fluid from this cooler was combined with the nitrogen make-up gas at the top of the cooler, and the gas and liquid were separated in a stainless-steel vessel (about 130 cm^3) maintained at the system pressure. The effluent liquid was drained periodically (every 12 h) for total organic carbon (TOC) analysis and for detection of the primary carbonaceous species using gas chromatography and HPLC. The effluent gas stream passed through the back-pressure regulator and was analysed with several online gas chromatographs. The kinetic data for each reaction condition were typically collected over a 24-h period, after which the reaction conditions were changed. The catalyst performance was stable for times on stream of at least 1 week.

Received 6 February; accepted 23 July 2002; doi:10.1038/nature01009.

1. Rostrup-Nielsen, J. Conversion of hydrocarbons and alcohols for fuel cells. *Phys. Chem. Chem. Phys.* **3**, 283–288 (2001).
2. Kumar, N. & Das, D. Enhancement of hydrogen production by enterobacter cloacae IIT-BT 08. *Process Biochem.* **35**, 589–593 (2000).
3. Woodward, J. *et al.* Enzymatic hydrogen production: Conversion of renewable resources for energy production. *Energy Fuels* **14**, 197–201 (2000).
4. Yokoi, H. *et al.* Microbial hydrogen production from sweet potato starch residue. *J. Biosci. Bioeng.* **91**, 58–63 (2001).
5. Woodward, J., Orr, M., Cordray, K. & Greenbaum, E. Enzymatic production of biohydrogen. *Nature* **405**, 1014 (2000).
6. Garcia, L., French, R., Czernik, S. & Chornet, E. Catalytic steam reforming of bio-oils for the production of hydrogen: Effects of catalyst composition. *Appl. Catal. A* **201**, 225–239 (2000).
7. Amphlett, J. C., Leclerc, S., Mann, R. F., Peppley, B. A. & Roberge, P. R. Fuel cell hydrogen production by catalytic ethanol-steam reforming. *Proc. 33rd Intersoc. Energy Convers. Eng. Conf.* **269**, 1–7 (1998).
8. Marquevich, M., Czernik, S., Chornet, E. & Montane, D. Hydrogen from biomass: Steam reforming of model compounds of fast-pyrolysis oil. *Energy Fuels* **13**, 1160–1166 (1999).
9. Milne, T. A., Elam, C. C. & Evans, R. J. *Hydrogen from Biomass: State of the Art and Research Challenges* 1–82 (National Renewable Energy Laboratory, Golden, CO, 2002).
10. Greeley, J. & Mavrikakis, M. A first-principles study of methanol decomposition on Pt(111). *J. Am. Chem. Soc.* **124**, 7193–7201 (2002).
11. Grenoble, D. C., Estadt, M. M. & Ollis, D. F. The chemistry and catalysis of the water gas shift reaction. 1. The kinetics over supported metal catalysts. *J. Catal.* **67**, 90–102 (1981).
12. Hilaire, S., Wang, X., Luo, T., Gorte, R. J. & Wagner, J. A comparative study of water-gas shift reaction over ceria supported metallic catalysts. *Appl. Catal. A* **215**, 271–278 (2001).
13. Iglesia, E., Soled, S. L. & Fiato, R. A. Fischer-Tropsch synthesis on cobalt and ruthenium. *Metal*

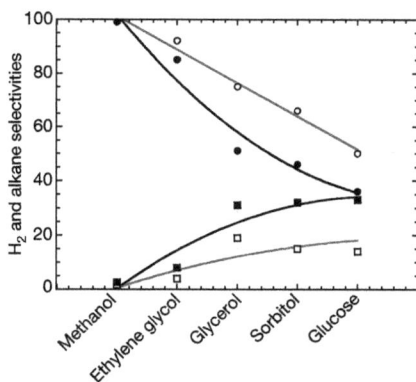

Figure 2 Selectivities (%) versus oxygenated hydrocarbon. H$_2$ selectivity (circles) and alkane selectivity (squares) from aqueous-phase reforming of 1 wt% oxygenated hydrocarbons over 3 wt% Pt/Al$_2$O$_3$ at 498 K (open symbols) and 538 K (filled symbols). The aqueous feed solution was fed to the reactor at a weight-hourly space velocity of 0.008 g of oxygenated hydrocarbon per gram of catalyst per hour. High conversions of the reactant were achieved under these conditions (50–99% conversion to gas-phase carbon, as indicated in Table 1) to provide a rigorous test of the carbon mass balance.

dispersion and support effects on reaction rate and selectivity. *J. Catal.* **137,** 212–224 (1992).

14. Kellner, C. S. & Bell, A. T. The kinetics and mechanism of carbon monoxide hydrogenation over alumina-supported ruthenium. *J. Catal.* **70,** 418–432 (1981).

15. Vannice, M. A. The catalytic synthesis of hydrocarbons from H₂/CO mixtures over the group VIII metals V. The catalytic behaviour of silica-supported metals. *J. Catal.* **50,** 228–236 (1977).

16. Bates, S. P. & Van Santen, R. A. Molecular basis of zerolite catalysis: A review of theoretical simulations. *Adv. Catal.* **42,** 1–114 (1998).

17. Gates, B. *Catalytic Chemistry* (Wiley, New York, 1992).

18. Eggleston, G. & Vercellotti, J. R. Degradation of sucrose, glucose and fructose in concentrated aqueous solutions under constant pH conditions at elevated temperature. *J. Carbohydr. Chem.* **19,** 1305–1318 (2000).

19. Kabyemela, B. M., Adschiri, T., Malaluan, R. M. & Arai, K. Glucose and fructose decomposition in subcritical and supercritical water: Detailed reaction pathway, mechanisms, and kinetics. *Ind. Eng. Chem. Res.* **38,** 2888–2895 (1999).

20. Collins, P. & Ferrier, R. *Monosaccharides: Their Chemistry and Their Roles in Natural Products* (Wiley, West Sussex, England, 1995).

21. Li, H., Wang, W. & Deng, J. F. Glucose hydrogeneration to sorbitol over a skeletal Ni-P amorphous alloy catalyst (Raney Ni-P). *J. Catal.* **191,** 257–260 (2000).

22. Blanc, B., Bourrel, A., Gallezot, P., Haas, T. & Taylor, P. Starch-derived polyols for polymer technologies: Preparation by hydrogenolysis on metal catalysts. *Green Chem.* **2,** 89–91 (2000).

23. Narayan, R., Durrence, G. & Tsao, G. T. Ethylene glycol and other monomeric polyols from biomass. *Biotechnol. Bioeng. Symp.* **14,** 563–571 (1984).

24. Tronconi, E. *et al.* A mathematical model for the catalytic hydrogenolysis of carbohydrates. *Chem. Eng. Sci.* **47,** 2451–2456 (1992).

25. Larminie, J. & Dicks, A. *Fuel Cell Systems Explained* 189 (Wiley, West Sussex, England, 2000).

26. Amphlett, J. C., Mann, R. F. & Peppley, B. A. On board hydrogen purification for steam reformer/PEM fuel cell vehicle power plants. *Hydrogen Energy Prog. X, Proc. 10th World Hydrogen Energy Conf.* **3,** 1681–1690 (1998).

Acknowledgements

We thank K. Allen, J. Shabaker and G. Huber for assistance in reaction kinetics measurements. We also thank G. Huber for help with catalyst preparation/ characterization and for TOC analyses, and M. Sanchez-Castillo for assistance with analysis of reaction products. We thank M. Mavrikakis and researchers at Haldor Topsøe A/S for reviews and discussion. This work was supported by the US Department of Energy (DOE), Office of Basic Energy Sciences, Chemical Science Division.

Competing interests statement

The authors declare competing financial interests: details accompany the paper on *Nature*'s website (http://www.nature.com/nature).

Correspondence and requests for materials should be addressed to J.A.D. (e-mail: dumesic@engr.wisc.edu).

Direct splitting of water under visible light irradiation with an oxide semiconductor photocatalyst

Zhigang Zou*, Jinhua Ye†, Kazuhiro Sayama* & Hironori Arakawa*

* Photoreaction Control Research Center (PCRC), National Institute of Advanced Industrial Science and Technology (AIST), 1-1-1 Higashi, Tsukuba, Ibaraki 305-8565, Japan
† Materials Engineering Laboratory (MEL), National Institute for Materials Science (NIMS), 1-2-1 Sengen, Tsukuba, Ibaraki 305-0047, Japan

....

The photocatalytic splitting of water into hydrogen and oxygen using solar energy is a potentially clean and renewable source for hydrogen fuel. The first photocatalysts suitable for water splitting[1], or for activating hydrogen production from carbohydrate compounds made by plants from water and carbon dioxide[2], were developed several decades ago. But these catalysts operate with ultraviolet light, which accounts for only 4% of the incoming solar energy and thus renders the overall process impractical. For this reason, considerable efforts have been invested in developing photocatalysts capable of using the less energetic but more abundant visible light[3-7], which accounts for about 43% of the incoming solar energy. However, systems that are sufficiently stable and efficient for practical use have not yet been realized. Here we show that doping of indium-tantalum-oxide with nickel yields a series of photocatalysts, $In_{1-x}Ni_xTaO_4$ ($x = 0$–0.2), which induces direct splitting of water into stoichiometric amounts of oxygen and hydrogen under visible light irradiation with a quantum yield of about 0.66%. Our findings suggest that the use of solar energy for photocatalytic water splitting might provide a viable source for 'clean' hydrogen fuel, once the catalytic efficiency of the semiconductor system has been improved by increasing its surface area and suitable modifications of the surface sites.

The photocatalytic splitting of water using oxide semiconductors is initiated by the direct absorption of a photon, which creates separated electrons and holes in the energy band gap of the material[3]. These charges can move to the surface of the semiconductor particle and react with water, provided the potential difference of this reaction exceeds 1.23 eV, dictated by the standard potential E^0 of water being −1.23 (standard hydrogen electrode, SHE, pH = 0).

The oxide semiconductor system we used for this reaction is $InTaO_4$, which crystallizes in the monoclinic wolframite-type structure with space group $P2/a$. The structure (Fig. 1a and b) contains TaO_6 and InO_6 octahedra in a unit cell. The InO_6 octahedra share edges to form zigzag chains along the [100] direction, with individual $[InO_6]_\infty$ octahedra chains linked by TaO_6 octahedra into a three-dimensional network. The volumes of the InO_6 and TaO_6 octahedra in $InTaO_4$ are 13.601 and 10.648 Å^3, respectively[8]. Because the ionic radius of Ni^{2+} (0.78 Å) is much smaller than that of In^{3+} (0.92 Å), substituting In^{3+} by Ni^{2+} should reduce the volume of the InO_6 octahedra, and hence the cell volume in $InTaO_4$. This should in turn affect the direct metal–metal bonding in the crystal, with the In–In distance shortened, resulting in profound alterations of the electronic properties of the compounds[8,9].

Single phases of $In_{1-x}Ni_xTaO_4$ were synthesized by a solid-state reaction at 1,100 °C, using pre-dried In_2O_3, Ta_2O_5 and NiO (99.99% purity) as starting materials. To increase the photocatalytic activity of the materials[10,11], we used impregnation with aqueous $Ni(NO_3)_2$ or $RuCl_3$ solution to load the oxide semiconductor surface with 1.0 wt% partly oxidized nickel or RuO_2; these loaded materials act as electron traps and hydrogen evolution sites. The Ni-loaded photocatalysts were calcined at 350 °C for 1 h in air and reduced in H_2 atmosphere (200 torr) at 500 °C for 2 h, then treated in O_2 atmosphere (100 torr) at 200 °C for 1 h. The reduction–oxidation treatment[11] produced a double-layered structure of metallic Ni and NiO (denoted NiO_y) on the surface of the photocatalyst; this double-layered structure suppresses the backward reaction of water splitting, which is activated by metallic nickel surfaces. The Ru-loaded photocatalysts were calcined at 500 °C for 2 h in air.

The water-splitting experiments were carried out with 0.5 g powdered photocatalyst suspended in 250 ml of pure water in a pyrex glass cell. A 300-W Xe arc lamp was focused through a shutter window and a 420 nm long pass filter was placed on the surface of the cell. The gases evolved were determined by the thermal conductivity detector (TCD) gas chromatograph, which was connected to the glass-made gas circulating line attached to the pyrex glass cell.

The results of the photocatalytic reaction and photophysical parameters are listed in Table 1. The non-doped catalyst, $NiO_y/InTaO_4$, is active, but the activity was significantly enhanced by Ni doping of $InTaO_4$. As a typical example, Fig. 2 shows the evolution of H_2 and O_2 from pure water containing suspensions of $NiO_y/In_{0.9}Ni_{0.1}TaO_4$ and $RuO_2/In_{0.9}Ni_{0.1}TaO_4$ under visible light irradiation ($\lambda > 420$ nm). The rates of H_2 and O_2 evolution were about 16.6 and 8.3 μmol h^{-1}, respectively, and the quantum yield at 402 nm was estimated to be 0.66% by using an interference filter ($\lambda = 402$ nm; half-width, 15.3 nm). For $RuO_2/In_{0.9}Ni_{0.1}TaO_4$, the rates of H_2 and O_2 evolution were about 8.7 and 4.3 μmol h^{-1}, respectively. The gas formation rate of the NiO_y loaded sample is about twice as large as that of the RuO_2-loaded sample. The gas evolution stopped when the light was turned off, showing that the reaction is induced by the absorption of visible light and not by tribological processes, such as the so-called mechano-catalysis.

After evacuating the reaction system and re-running the experiment, almost identical gas production rates were achieved in the second as well as third runs. More than 7,200 μmol gases evolved (H_2, about 4,800 μmol; and O_2, about 2,400 μmol) during the course of a 400-h experiment. The catalyst samples remained unchanged during the course of reaction, suggesting that the

When citing this article, please cite the original version as shown on the contents page of this chapter.

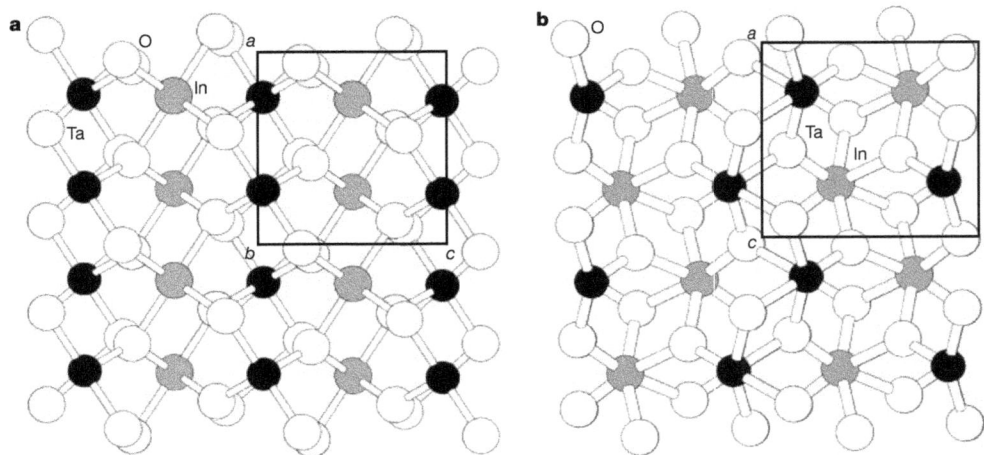

Figure 1 Schematic illustration of the InTaO$_4$ structure. Shown is the InTaO$_4$ structure with monoclinic wolframite-type, with the lattice parameters $a = 5.1552(1)$, $b = 5.7751(1)$, $c = 4.8264(1)$ (Å) and $\beta = 91.373(1)°$ obtained from full-profile structure refinement of XRD data using the Rietveld program REITAN[12]. **a**, The view along the b-axis; **b**, the view along the c axis.

photocatalytic activity is stable under visible light irradiation. The turnover number—the ratio of total amount of gas evolved to catalyst (1,412 μmol in our system)—exceeded 5 after 400 h reaction time. The turnover number in terms of reacted electrons relative to the amount of Ni loaded on the surface of the sample and Ni doped in InTaO$_4$ reached 39 at 400 h reaction time, indicating that the reaction occurs catalytically.

Wavelength dependence of the photocatalytic gas evolution on the photocatalyst was investigated using different cut-off filters. As a comparison, the Pt/TiO$_2$ photocatalyst (P-25) was measured using the same method. Although photocatalytic activity was observed when using In$_{1-x}$Ni$_x$TaO$_4$ and a cut-off filter of $\lambda > 500$ nm, it is much lower than that observed under $\lambda > 420$ nm light irradiation. The rates of H$_2$ and O$_2$ evolution were 1.91 and 0.86 μmol h^{-1} on

In$_{0.9}$Ni$_{0.1}$TaO$_4$, respectively, under $\lambda > 500$ nm light irradiation. The activity disappeared when the wavelength of light irradiation was larger than 550 nm. The results are in good agreement with the observation of diffuse reflectance spectra (see Fig. 3). By contrast, no photocatalytic activity was obtained on Pt/TiO$_2$ under visible light irradiation ($\lambda > 420$ nm).

X-ray diffraction analysis showed that there is no observable structural difference between the samples before and after reaction. Full-profile structure refinement of XRD data was performed using the Rietveld program REITAN[12], showing that all samples crystallize in the same wolframite structure, monoclinic with space group $P2/a$, and that the lattice parameters decrease along all three axes as the Ni content x is increased, as long as $x < 0.15$. Figure 4 shows the change of V/Z with doping content, where V and Z are cell volume and the number of formulas per cell, respectively. The inset shows the c-axis parameter as a function of Ni content, to illustrate the lattice parameter changes induced by doping. There is a linear decrease of V/Z with increasing Ni content as long as $x < 0.15$, while V/Z has a small expansion when the Ni content $x > 0.15$. In compounds with a Ni-doping content greater than 0.2, an impurity

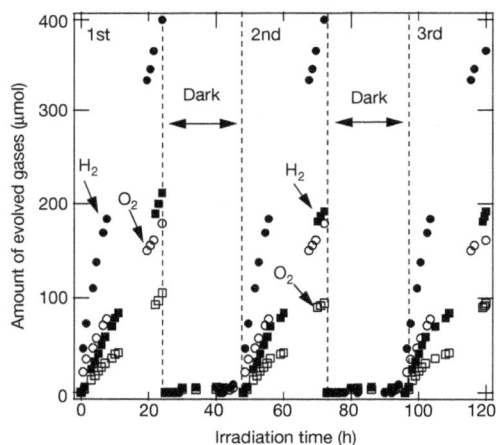

Figure 2 Photocatalytic H$_2$ and O$_2$ generation. Shown are the evolution of H$_2$ and O$_2$ from pure water using as catalyst a suspension of NiO$_y$/In$_{0.9}$Ni$_{0.1}$TaO$_4$ (solid circles, H$_2$; open circles, O$_2$) and RuO$_2$/In$_{0.9}$Ni$_{0.1}$TaO$_4$ (solid squares, H$_2$; open squares, O$_2$). Experiments were done using 0.5 g catalyst powder suspended in 250 ml pure water in a pyrex glass cell under visible light irradiation ($\lambda > 420$ nm). Light source: 300-W Xe lamp. The gases evolved were determined by TCD gas chromatograph. The measurement uncertainties were about 0.05%.

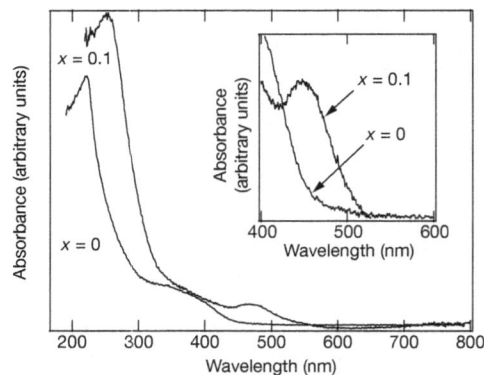

Figure 3 Optical properties of the photocatalyst. The main panel shows the ultraviolet–visible diffuse reflectance spectra of In$_{1-x}$Ni$_x$TaO$_4$ ($x = 0$ and 0.1) at room temperature, with the inset providing an expanded view of the spectra in the wavelength region from 400 to 600 nm.

Table 1 Rates of gas evolution and physical properties of the photocatalysts

Photocatalyst	BET† $(m^2 g^{-1})$	t‡ $(r_A + r_0)/\sqrt{2}(r_B + r_0)$	Rate of evolution ($\mu mol\ h^{-1}$)§			
			NiO$_y$/		RuO$_2$/	
			H$_2$	O$_2$	H$_2$	O$_2$
SiO$_2$*			0.0	0.0	0.0	0.0
InTaO$_4$	1.35	0.793	3.2	1.1	0.75	0.35
In$_{0.95}$Ni$_{0.05}$TaO$_4$	0.87	0.790	4.2	2.1	2.0	1.0
In$_{0.90}$Ni$_{0.10}$TaO$_4$	0.85	0.787	16.6	8.3	8.7	4.3
In$_{0.85}$Ni$_{0.15}$TaO$_4$	0.82	0.784	8.3	4.1	4.8	2.3
In$_{0.80}$Ni$_{0.20}$TaO$_4$	0.81	0.780	4.3	0.9	0.8	0.4

* In order to investigate the effect of NiO$_y$ on photocatalytic activity onto the surface of photocatalysts, the 1.0 wt% NiO$_y$/SiO$_2$ sample was prepared because SiO$_2$ does not have photocatalytic activity under visible light irradiation.
† Their BET surface areas were calculated from N$_2$ isotherms at −196 °C.
‡ t is the tolerance factor for structural stability.
§ NiO$_y$/ is 1.0 wt% NiO$_y$ loading on the surface of photocatalyst; RuO$_2$/ is 1.0 wt% RuO$_2$ loading on the surface of photocatalyst. The rate of gas evolution was calculated for 24 h using 0.5 g catalyst under visible light irradiation $\lambda > 420$ nm.

phase, NiTa$_2$O$_6$, appears and increases its volume fraction significantly with increasing Ni content. It is known[13] that the structural stability of oxides consisting of octahedra such as ABO$_3$, can be estimated by calculating the tolerance factor defined as $t = (r_A + r_0)/\sqrt{2}(r_B + r_0)$, where r_A, r_B and r_0 are the radii of the respective ions. $0.79 < t < 1.1$ would be the ideal cubic structure. In this case, r_A is a maximum for the non-doped InTaO$_4$ compound. As shown in Table 1, the t value decreases with increasing Ni content x, the t value becoming smaller than 0.78 when $x > 0.20$, suggesting that the geometrical arrangement in the oxide governs the structural stability.

Figure 3 illustrates the light absorption properties of In$_{1-x}$Ni$_x$TaO$_4$, showing that the visible absorption spectra of these compounds are characteristic of photocatalysts able to respond to visible light. The band gap of these compounds can be estimated from plots of the square root of Kubelka–Munk functions F(R) versus photon energy[14]. One of the most characteristic features is that the E_g value (E_g is bandgap energy) is narrowed with Ni doping. The bandgap is changed from 2.6 (non-doped) to 2.3 eV (0.1 Ni-doped). This is considered largely to be a consequence of the Ni 3d level[15]. The band structure of NiO is assigned to Ni 3d^8 and Ni 3d^9 (ref. 15).

The bandgap change in Ni-doped compounds can be attributed to internal transitions in a partly filled Ni d shell. This is supported by the appearance in the Ni-doped compounds of an ultraviolet–visible absorption band at 420–520 nm (see Fig. 3 inset), corresponding to an energy range of about 2.9 to 2.3 eV: we note that the correlation splitting of Ni 3d^8 and 3d^9 is 2.4 eV. It is still unknown how a Ni d–d transition could induce electron–hole separation and the generation of hydrogen and oxygen, but it seems clear that the d orbitals play an important role in the photoexcitation and photocatalytic activity observed. In any case, the narrower bandgap will facilitate excitation of an electron from the valence band to the conduction band in the doped oxide semiconductor, thus increasing the photocatalytic activity of the material. □

Received 5 July; accepted 1 October 2001.

1. Honda, K. & Fujishima, A. Electrochemical photolysis of water at a semiconductor electrode. *Nature* **238**, 37–38 (1972).
2. Kawai, T. & Sakata, T. Conversion of carbohydrate into hydrogen fuel by a photocatalytic process. *Nature* **286**, 474–476 (1980).
3. Linsebigler, A. L., Lu, G. & Yates, J. T. Jr Photocatalysis on TiO$_2$ surfaces: principles, mechanisms, and selected results. *Chem. Rev.* **95**, 735–758 (1995).
4. Geoffrey, B. S. & Thomas, E. M. Visible light photolysis hydrogen using sensitized layered metal oxide semiconductors: the role of surface chemical modification in controlling back electron transfer reactions. *J. Phys. Chem. B* **101**, 2508–2513 (1997).
5. Kim, Y. II., Salim, S., Huq, M. J. & Mallouk, T. E. Visible light photolysis of hydrogen iodide using sensitized layered semiconductor particles. *J. Am. Chem. Soc.* **113**, 9561–9563 (1991).
6. Yoshimure, J., Ebina, Y., Kondo, J. & Domen, K. Visible light induced photocatalytic behavior of a layered perovskite type niobate, RbPb$_2$Nb$_3$O$_{10}$. *J. Phys. Chem.* **97**, 1970–1973 (1993).
7. Kudo, A. & Mikami, I. New In$_2$O$_3$(ZnO)$_m$ photocatalysts with laminal structure for visible light induced H$_2$ or O$_2$ evolution from aqueous solutions containing sacrificial reagents. *Chem. Lett.* 1027–1028 (1998).
8. Zou, Z., Ye, J. & Arakawa, H. Structural properties of InNbO$_4$ and InTaO$_4$: correlation with photocatalytic and photophysical properties. *Chem. Phys. Lett.* **332**, 271–277 (2000).
9. Zou, Z., Ye, J. & Arakawa, H. Substitution effects of In^{3+} by Al^{3+} and Ga^{3+} on the photocatalytic and structural properties of the Bi$_2$InNbO$_7$ photocatalyst. *Chem. Mater.* **13**, 1765–1769 (2001).
10. Kim, H. G., Hwang, D. W., Kim, J., Kim, Y. G. & Lee, J. Highly donor-doped (110) layered perovskite materials as novel photocatalysts for overall water splitting. *Chem. Commun.* 1077–1078 (1999).
11. Kudo, K. *et al.* Nickel-loaded K$_4$Nb$_6$O$_{17}$ photocatalyst in the decomposition of H$_2$O into H$_2$ and O$_2$: Structure and reaction mechanism. *J. Catal.* **120**, 337–352 (1989).
12. Izumi, F. J. A software package for the Rietveld analysis of X-ray and neutron diffraction patterns. *Crystallogr. Ass. Jpn* **27**, 23–31 (1985).
13. Machida, M., Yabunaka, J.-i. & Kijima, T. Synthesis and photocatalytic property of layered perovskite tantalates, RbLnTa$_2$O$_7$ (Ln = La, Pr, Nd, and Sm). *Chem. Mater.* **12**, 812–817 (2000).
14. Kim, Y. II., Atherton, S., Brigham, E. S. & Mallouk, T. E. Sensitized layered metal oxide semiconductor particles for photochemical hydrogen evolution from nonsacrificial electron donors. *J. Phys. Chem.* **97**, 11802–11810 (1993).
15. Dare-Edwards, M. P., Goodenough, J. B., Hammett, A. & Nicholson, N. D. Photoelectrochemistry of nickel (II) oxide. *J. Chem. Soc. Faraday Trans.* **77**, 643–661 (1981).

Acknowledgements

This work was supported in part by the COE project, Japan Ministry of Education and Science.

Correspondence and requests for materials should be addressed to Z.Z. (e-mail: z.zou@aist.go.jp) or H.A. (e-mail: h.arakawa@aist.go.jp).

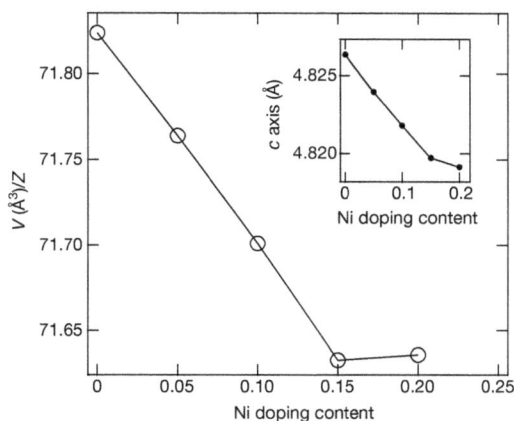

Figure 4 Changes in crystal structure with doping. Shown is the change of V/Z with Ni doping content x, where V and Z are the unit cell volume and the number of formulas per cell, respectively. The inset shows Ni dependence of the c-axis parameter, as an example of the lattice parameter change. All the parameters were obtained from full-profile structure refinements of XRD data using the Rietveld program REITAN[12]. The measurement uncertainties were about 0.002 (Å^3), for V/Z and 0.0001 (Å) for the c-axis parameter.

SUPERCONDUCTORS

Materials science challenges for high-temperature superconducting wire

Twenty years ago in a series of amazing discoveries it was found that a large family of ceramic cuprate materials exhibited superconductivity at temperatures above, and in some cases well above, that of liquid nitrogen. Imaginations were energized by the thought of applications for zero-resistance conductors cooled with an inexpensive and readily available cryogen. Early optimism, however, was soon tempered by the hard realities of these new materials: brittle ceramics are not easily formed into long flexible conductors; high current levels require near-perfect crystallinity; and — the downside of high transition temperature — performance drops rapidly in a magnetic field. Despite these formidable obstacles, thousands of kilometres of high-temperature superconducting wire have now been manufactured for demonstrations of transmission cables, motors and other electrical power components. The question is whether the advantages of superconducting wire, such as efficiency and compactness, can outweigh the disadvantage: cost. The remaining task for materials scientists is to return to the fundamentals and squeeze as much performance as possible from these wonderful and difficult materials.

S. R. FOLTYN[1], L. CIVALE[1], J. L. MACMANUS-DRISCOLL[1,2], Q. X. JIA[1], B. MAIOROV[1], H. WANG[3] AND M. MALEY[1]

[1]Superconductivity Technology Center, Los Alamos National Laboratory, Los Alamos, New Mexico 87545, USA

[2]Department of Materials Science and Mettallurgy, University of Cambridge, Cambridge CB2 3QZ, UK

[3]Department of Electrical and Computer Engineering, Texas A & M University, College Station, Texas 77843-3128, USA

*e-mail: sfoltyn@lanl.gov

Since the 1986 discovery of high-temperature superconducting (HTS) materials[1], the promise of zero-resistance devices operating at liquid-nitrogen temperature has fuelled a worldwide research investment that is now around one billion US dollars. Most research has been in the electric power area for applications such as magnets, motors and power-transmission lines; all power applications share a common requirement that the superconducting material be formed into a long, strong and flexible conductor so that it can be used like the copper wire it is intended to replace. And this is where the problems began, because the HTS materials are ceramics that are more like a piece of chalk than the ductile metal copper.

The first solution to this problem was to pack Bi–Sr–Ca–Cu–O (BSCCO) superconducting powder into a silver tube[2]. Following a series of rolling and heating steps the end product was a 4-mm-wide tape capable of carrying over 100 A at liquid nitrogen temperature and flexible enough to be used in the above-mentioned applications. This so-called first-generation wire is the workhorse of the present-day HTS industry, with many hundreds of kilometres of tape produced for various demonstration projects, primarily power cables and motors. Unfortunately this type of conductor relies heavily on the use of silver, which makes it too expensive for most commercial applications and severely limits prospects for economy-of-scale cost reductions in the future. Further, BSCCO rapidly loses its ability to carry supercurrent in a magnetic field at liquid nitrogen temperature, negating the advantage of using an inexpensive cryogen in applications such as motors and generators.

In the early 1990s an alternative approach was conceived[3–5], which was quickly shown to offer competitive performance[6,7]. Instead of imparting the requisite strength and crystallinity to the superconductor by the powder-in-tube method, 'second-generation' wire uses epitaxial growth of a superconducting coating on a thin metal tape. One advantage is that very little silver is needed, making significant cost reductions possible. Another advantage of this idea is that it allows use of the compound $YBa_2Cu_3O_{7-\delta}$ (YBCO), which retains much higher current-carrying ability in a magnetic field.

As recently as six years ago[8], however, there were serious and legitimate questions about whether second-generation wire could be produced in long lengths with the performance required for commercial applications, and whether it could be done at a competitive price. The first of these questions has now been definitively answered by recent announcements from three companies[9–11] about YBCO tapes that are more than 100 metres long, carrying over 200 amperes per centimetre width at 77 K. One of these companies, SuperPower, has delivered 10 km of wire for a power-distribution cable project in Albany, New York[12]. Another, American Superconductor, has delivered a similar amount for various projects, including two

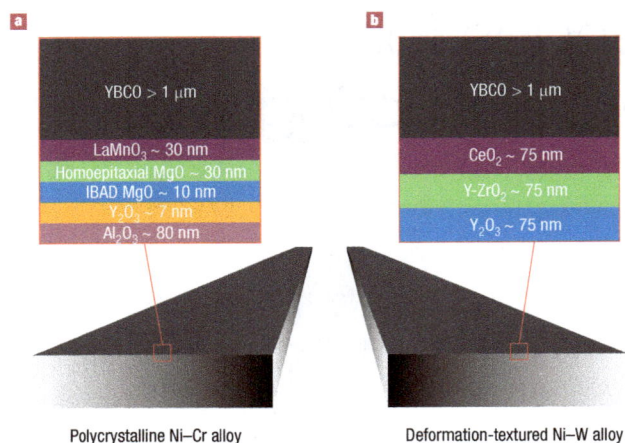

Figure 1 There are two main technologies being used in the manufacture of coated-conductor tape. **a**, The architecture used by SuperPower, which is based on ion-beam-assisted deposition (IBAD). In this case a commercial alloy is used for the substrate and crystalline texture is developed by IBAD of magnesium oxide. The YBCO coating process is metal–organic chemical vapour deposition. A different IBAD material, an 800-nm-thick layer of $Gd_2Zr_2O_7$, is being successfully used in Japan. A 400-nm CeO_2 layer is followed by the YBCO, both laser-deposited. **b**, The textured-metal approach (known as RABiTS, for rolling-assisted biaxially textured substrate) used by American Superconductor. In this case, cubic texture is imparted to the substrate itself through a series of rolling and annealing steps, and YBCO is produced by solution-based metal–organic deposition. In either case it is desirable that the substrate material be strong, oxidation resistant, and non-magnetic. In both technologies, the intermediate oxide layers are epitaxially grown by physical vapour deposition, the YBCO is capped with a protective layer of silver, and the conductor can be clad in copper for stability — finished tapes are typically 50–200 µm thick and 4 mm wide.

and therefore represents the obvious goal of most of the efforts of fundamental research.

From a theoretical perspective, as superconductivity is associated with the pairing of carriers, J_c is limited by the carrier-pair dissociation induced by high values of current density. This limit can be readily estimated from known quantities and is roughly five times higher than the best measured values currently in the literature. It might be tempting to conclude that the difference is simply a materials processing issue; however, the highest present-day J_c values are less than double those reported in the early years of YBCO film fabrication[14,15]. This improvement over the past two decades is directly attributable to working materials issues such as process optimization and substrate preparation, but no such obvious issues can be identified today. It is also worth noting that, after many decades of optimization, commercial Nb-based low-T_c superconducting wires exhibit a similar gap between the state-of-the-art and the depairing limit. Conversely, nothing in our present theoretical knowledge indicates that the gap cannot be closed, and in fact a renewed effort has been proposed[16] to understand the difference between physical limit and reality.

All methods used to deposit YBCO coatings fall into one of two categories, namely *in situ* and *ex situ* (referring to whether the YBCO crystal structure is grown during the film deposition or after). Each method produces characteristic structural features, but even with significant differences in microstructure films from different deposition methods exhibit remarkably similar performance. Figure 2 illustrates how J_c of YBCO films — in this case grown by pulsed laser deposition (PLD) — decreases when the coating is made thicker or when it is subjected to an external magnetic field. The first is important because thicker films are needed to reach higher current in the conductor, and the second is important because many envisioned applications (magnets, motors, generators) require operation of the conductor in a strong magnetic field. A significant distinction between the two phenomena shown in Fig. 2 is that, although there is sound and well-known physics underlying the decrease of J_c with increasing magnetic field, there is no fundamental physical reason for J_c to decrease with film thickness. To understand the differences we need to explore the factors controlling J_c in superconductors.

The most pedestrian factors that limit J_c are material imperfections — such as cracks, voids, or secondary phases — that partially block current. Although these defects presented a serious challenge in the early days of coated conductors, process optimization has largely relegated them to a secondary problem. Another issue, which has proved especially difficult, is the weak-link effect that causes an exponential decrease of J_c with increasing misorientation angle between adjacent grains[17]. In the development of coated conductors this has been the most formidable obstacle of all because it means that YBCO coatings must have a crystalline template for high J_c, yet there is no such thing as a flexible kilometre-long single crystal to use as a template. So perhaps the crowning achievements in the field have been the IBAD and RABiTS technologies (Fig. 1) that enable us to approximate a very long single-crystal substrate. And the approximation is a good one in that IBAD MgO texture has been improved to the point of supporting J_c values equal to those on single-crystal substrates[18], and, through a fortuitous set of circumstances, J_c values are comparable for *ex situ* films on RABiTS[19].

Once the above-mentioned problems have been solved, the property that determines the ability of a coated conductor to carry supercurrent is referred to as 'flux pinning'. Many articles, and indeed books, have been written on this complex subject[20–22], but the basic concept can be explained simply: In the interior of Type II superconductors such as YBCO, magnetic field exists in the form of tubular structures called flux lines or vortices, each of them carrying one unit of magnetic flux, or flux quantum[23]. The application of an electric current to the superconductor generates a lateral force on the vortices known as the Lorentz force, and the resulting vortex

fault-current-limiter demonstrations. At over ten times the price (per kiloamp-metre) of copper wire the question of commercial viability remains to be answered, but the gap is closing as industry gears up for large-scale production. Figure 1 compares the two primary coated-conductor designs being manufactured today[13].

With such remarkable progress it may seem that the role of HTS materials research has come to an end. One could ask 'Isn't the material good enough?' and argue that all that remains to be done is more process engineering and applications development. One answer is that present-day YBCO may have good enough performance to be competitive in some applications. However, it is generally true that higher performance translates to lower cost for a given application, so that a better answer is that in two important areas — performance with increasing thickness or increasing magnetic field — the potential exists to substantially improve YBCO, enabling a lower cost/performance ratio, a larger range of applications, and increased odds of commercial success. In this article, we present the challenges in these two areas and the approaches being taken to address them.

AIMING AT A HIGHER CRITICAL CURRENT

The fundamental property that makes superconductors attractive for power applications is the ability to carry current without losses. This absence of dissipation is however limited to current densities lower than a critical value, J_c, which translates into a critical current I_c for a wire with a defined cross section. Increasing J_c means being able to carry a higher current within the same wire,

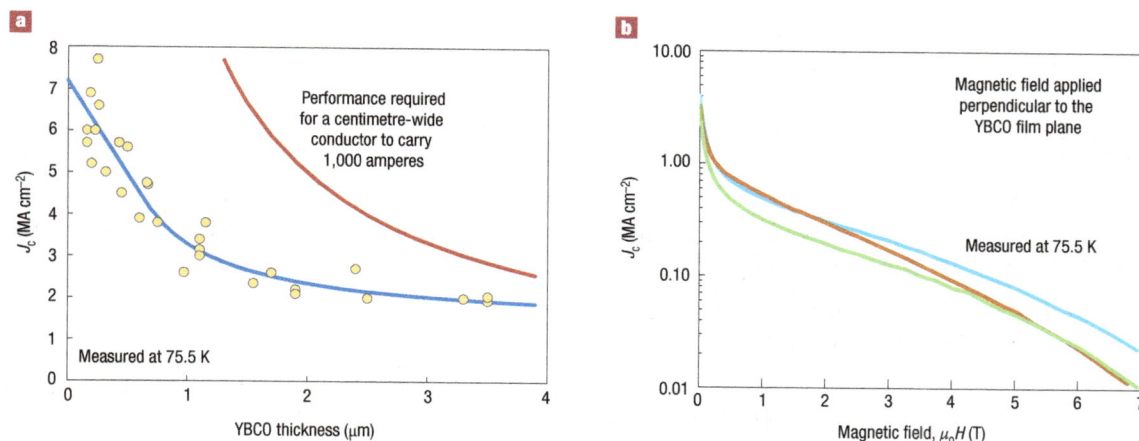

Figure 2 The critical current density of YBCO decreases sharply with film thickness and applied magnetic field — improvements are needed in both areas to increase the commercial viability of HTS coated conductors. **a**, The J_c for very thin films approaches 8 MA cm^{-2}, but it drops rapidly within the first 2 μm of film thickness. Conductors required to carry 1,000 amperes per centimetre of width will need a coating thickness greater than 1 μm. **b**, If the application requires operation in a magnetic field, as most do, performance is further reduced. Measurements for three films of ~1 μm thickness are shown; the variations in field dependence are not unusual. All coatings were fabricated at Los Alamos on single-crystal substrates by pulsed-laser deposition, but similar thickness and field dependence has been observed for films on different substrates and deposited by other processes.

motion dissipates energy and causes electrical resistance to appear in the superconductor. Only if vortices can be immobilized by a counteracting 'pinning force' can a superconductor sustain high current density. The higher the current density, the greater the Lorentz force acting on the vortices — the critical current density is essentially the point at which the Lorentz force begins to exceed the maximum available pinning force.

What is the origin of the pinning force? It arises from the presence of localized material defects or crystalline imperfections that reduce the vortex energy such that vortices tend to remain pinned at the bottom of these potential energy wells. Important properties are the depth of the potential well (pinning energy), as well as the size, shape and density of pinning sites. To maximize J_c we must engineer the optimum vortex-pinning landscape. It has been known since the first epitaxial YBCO films were produced in the late 1980s that their J_c was 10–100 times higher than YBCO crystals because of the large density of defects inherent in thin-film growth. The challenge is uncovering which of these film defects (Box 1 and Fig. 3) are most effective at pinning vortices, and how we can produce films with more of them. Ironically, then, the trick is not to make the material more perfect, but, in very specific ways, less.

ENGINEERING DEFECTS FOR IMPROVED FLUX PINNING

The relationship between material defects and J_c is complex. The pinning force of one individual defect depends on its size and shape, as well as on its composition and structural interaction with the matrix. Things get more complicated when we consider all of the defects in a superconductor, because pinning is not simply additive. In general, increasing the density of pinning sites will initially increase the macroscopic J_c, but beyond some optimum density performance will deteriorate. The situation also depends on the applied magnetic field (H). At low H vortices are far apart, minimizing their mutual interactions; in this 'single vortex pinning' regime, J_c is independent of H. As H increases, vortex–vortex interactions increase and J_c decreases.

Although Box 1 describes a wide variety of defects that are potential pinning centres, the pinning itself can be divided into one

of two categories. 'Correlated' pinning arises from approximately parallel arrays of extended linear or planar defects. Their effect is strongest when the applied magnetic field (which determines the orientation of the vortices) is aligned with them, and it decreases as the misalignment between defects and vortices increases. The signature of correlated pinning is a peak in the angular dependence of J_c. In a broader definition, linear defects with a certain amount of splay can be included in the correlated category. In fact it has been theoretically predicted[24] and experimentally demonstrated[25] that a certain amount of splay improves pinning by keeping double-kink excitations from sliding. 'Random' or 'uncorrelated' pinning is due to randomly distributed localized defects, and in this case the influence is relatively uniform for all field orientations. A defining characteristic of correlated defects, which gives them the potential to provide strong pinning, is that the total pinning force on a fully trapped vortex grows linearly with the vortex length, as does the total Lorenz force. In contrast, the pinning force of random defects usually adds statistically and, moreover, the adjustment of the vortex system to its pinning landscape involves a cost in elastic energy. Because of this, random pinning has sometimes been incorrectly associated with weak pinning. Although this may be the case for point-like defects, randomly distributed nanoparticles are capable of producing very strong pinning.

Figure 4 illustrates the various types of pinning observed in different kinds of films. In angular dependence plots like this, YBCO almost always has a strong peak at 90° (that is, when the applied field is parallel to the YBCO a–b planes). In addition to this intrinsic peak, many vapour-deposited (*in situ*) films have a second peak at 0° (parallel to the YBCO c axis), which is correlated with linear defects related to columnar growth. In *ex situ* films, such as those grown by the solution-based metal organic deposition (MOD), planar defects typically enhance the 90° peak. As discussed in detail later and shown in the figure, addition of nanoparticles to either type of film can drastically alter the angular dependence.

One significant difficulty in increasing J_c in YBCO is that, owing to the size of the vortex cores[22,23], the best pinning defects are nanometre sized, and producing such defects requires special tools. The earliest of these tools was particle irradiation, with which the

Box 1 A close look at defects

There are a number of defect types that can act as pinning centres, which are represented in Fig. B1. Some of these can be observed in the TEM micrograph of Fig. 3.

Precipitates There are several sources for precipitates in YBCO films. The most common is the growth of secondary phases (for example, Y_2O_3 or Ba–Cu–O) caused by deviation of composition from the desired 1:2:3 cation ratio. Precipitates such as $BaZrO_3$ can also be intentionally added to YBCO to improve performance in a magnetic field.

Twin boundaries The superconducting phase of YBCO has an orthorhombic crystal structure. In films with the crystallographic *c*-axis perpendicular to the substrate, domains with perpendicular *a*- and *b*-axis orientations arise. Twin boundaries are the planar defects where these domains meet.

In-plane misorientation All films, especially those grown on coated-conductor substrates that inherently have imperfect crystalline texture, have deviations in the alignment of *a* and *b* axes. Atomic order is disrupted at the boundaries where misaligned grains meet, resulting in strain and dislocations.

Threading dislocations Dislocations between misoriented grains typically thread through the entire film thickness. Even with perfect alignment, as is the case for single-crystal substrates, dislocations form between growth islands because of imperfect lattice matching between YBCO and the substrate.

Surface roughness Nearly all films have surface roughness due to a multitude of effects, including shadowing, columnar growth, porosity, strain and volumetric changes. Roughness usually increases with film thickness. A particular kind of roughness in YBCO is the growth of grains with the *a* (or *b*) axis perpendicular to the substrate.

Antiphase boundaries YBCO is a layered compound, requiring a specific sequence of appropriately oxygenated yttrium, barium and copper layers. In the growth process film domains usually coalesce with the layers matching yttrium-to-yttrium, and so forth. A boundary between imperfectly matched domains is referred to as an antiphase boundary.

Voids In vapour deposition, voids can appear as a consequence of surface roughness when shadowing or reduced mobility prevent vapour from filling valleys. In *ex situ* processes, where the film is deposited and crystallized in separate steps, pores and voids can result from volumetric changes.

Out-of-plane misorientation Rarely, grains form with the *c* axis tilted with respect to the film perpendicular. Possible causes are film roughness, a tilted substrate structure or other substrate defect.

Misfit dislocations The lattice spacing of YBCO is generally different from the substrate materials it is deposited on. To some extent this mismatch can be accommodated by elastic strain, but when energetically favourable, dislocations — in the form of extra (or missing) half-planes of atoms — form to relieve the stress.

Planar defects One type of planar defect is a precipitate of a non-123 phase of YBCO, such as the copper-rich 124 phase. Another type is a stacking fault in which the YBCO layer structure is disrupted by an extra or missing layer of, for example, copper oxide.

Point defects Precipitates are agglomerations of a large number of atoms; in contrast, point defects are atomic-scale disruptions in the crystal structure. The latter can be in the form of vacancies or interstitials (missing or extra atoms), or impurities. In YBCO point defects also result when yttrium and barium change places. The exchange is more likely in compounds where yttrium has been replaced with a rare-earth element such as Nd, Sm or Eu. In this case the ion sizes are more similar to Ba, facilitating the exchange.

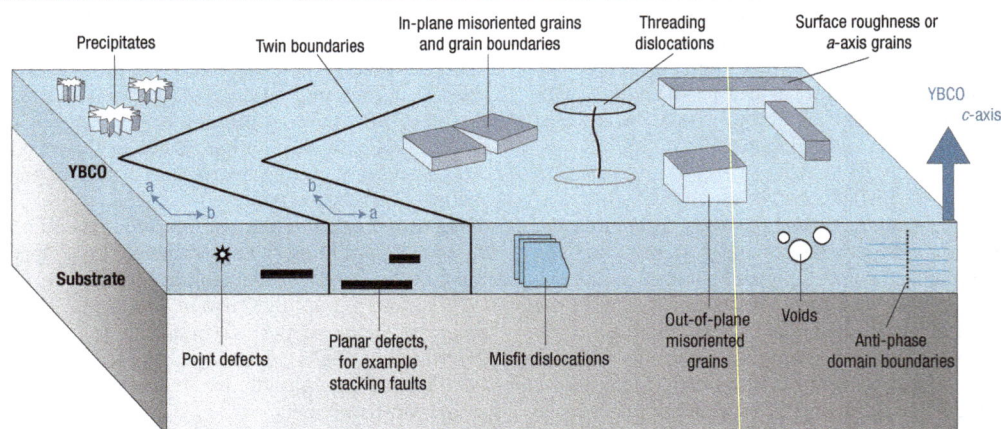

Figure B1 Many thin-film defects have been proposed as flux pinning sites in YBCO; anything that locally disturbs the crystalline perfection over a scale of 0.1–1 nm is a candidate. Strain fields associated with defects may also pin vortices. The challenges in engineering defect structures for the best performance are to determine which ones are beneficial, and to tailor their density to produce the desired effect without obstructing current.

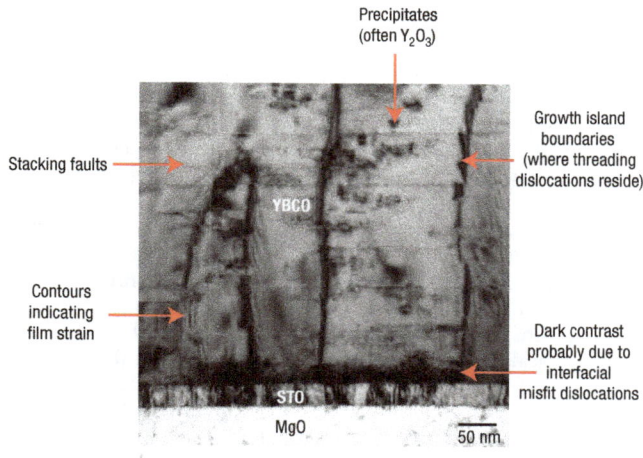

Figure 3 Transmission electron microscope image of a laser-deposited YBCO film on a SrTiO₃-buffered MgO crystal. Films of this type have very good performance, but are also heavily populated with defects like the ones shown in Fig. B1 — the two aspects are related but the exact relationships are frequently difficult to decipher. Absent from this film are defects that are known to block current, such as voids, large-angle grain boundaries and a-axis grains.

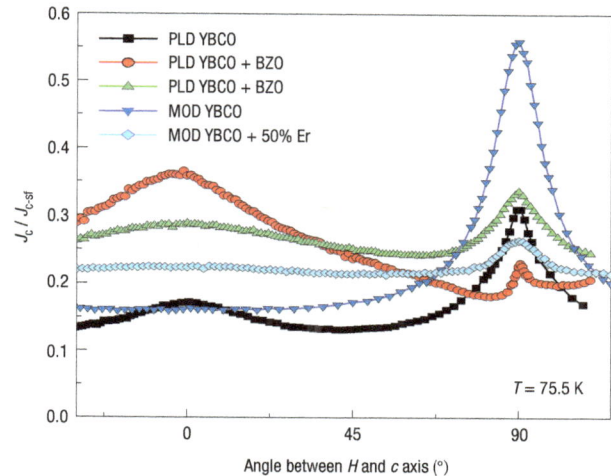

Figure 4 Comparison of the angular dependence of different film types at 1 T shows that the situation is complex and no single orientation can tell the full story. In the figure, zero degrees corresponds to the field orientation perpendicular to the film plane (parallel to the YBCO c-axis), and at 90° the field is parallel to the YBCO a–b planes; J_c is normalized to self-field values ($J_{c\text{-sf}}$) to facilitate the comparison. This type of measurement is essential for elucidating the types of important flux-pinning defects present in films and must ultimately become a routine part of enhancement research.

J_c of YBCO single crystals can be increased by one or two orders of magnitude. Irradiation with electrons[26], protons[27], fast neutrons[28,29] and light ions generates randomly distributed defects either of atomic or nanometre size (uncorrelated pinning), whereas irradiation with high-energy heavy ions produces columnar defects[30] (correlated pinning). It was quickly found, however, that adding irradiation defects to YBCO thin films resulted in a marginal J_c improvement at best[14,31–33], which was not surprising given the high density of pinning centres naturally occurring in coatings. On a more practical side, irradiation of kilometre-long wires did not seem to be commercially feasible. Alternate approaches were clearly needed.

Such alternate approaches can take two distinct paths. We can try to identify which of the naturally present defects in YBCO films are the most effective pinning centres and then try to manipulate their characteristics to improve performance, or we can add extrinsic defects. In either case, clear correlations need to be established between processing, microstructure and properties. To that end pinning improvement must be evaluated by mapping J_c as a function of temperature and the strength and orientation of an external magnetic field, and the material must receive nanoscale scrutiny to catalogue the defects that may act as pinning sites.

An early effort[34] following the first path involved changing the density of a common type of thin-film defect known as a threading dislocation. It was shown that these linear defects — which are parallel to the crystalline c axis of YBCO and are therefore geometrically similar to the columnar defects created by heavy-ion irradiation — occur predominantly at the boundaries between growth islands and that their density can be changed by changing the size of the islands. The density of this particular defect was found to have little effect on J_c in 'self-field', but increasing the density extended the low-field plateau (the single vortex-pinning regime) to higher field. (Self-field is the very weak magnetic field generated by current in the superconductor; it is proportional to the product of J_c and film thickness and is approximately 0.01 T for a 1-μm-thick film with a J_c of 1 MA cm⁻²; ref. 35.) This was the first reported quantitative correlation between a specific, observable defect type and its effect on YBCO film performance.

In the remainder of this article we focus on the issue of flux pinning, specifically the widespread and ongoing efforts aimed at enhancing both the self-field and in-field performance of YBCO through the intentional incorporation of defects beyond those normally present in YBCO films.

INCREASING THICKNESS WITHOUT DECREASING J_c

Before entering into the details of the efforts made to improve the pinning efficiency in an applied magnetic field, we examine the obstacles in enhancing the critical current by simply increasing the thickness of the superconducting wires, when only self-field is present. Looking at Fig. 2a, three possible approaches can be imagined for making a conductor that carries 1,000 amperes in a 1-cm width, which is a US Department of Energy goal for the year 2010[36]. An obvious approach is to make a coating more than 5-μm thick, but this adds considerably to the conductor cost, and is not feasible for every film-production method. Another approach is a thin film with very high J_c, but films less than 1-μm thick would require an improbably high J_c of over 10 MA cm⁻². Only one practical course of action is then left: to make films more than 1-μm thick that retain, as much as possible, the high J_c of much thinner films. Only recently, with the taming of more urgent problems, has this issue begun to attract serious attention from the HTS community.

The starting point is to examine the tantalizingly high J_c values that can be achieved in very thin films, and the reasons why they have not been achieved in thicker films. First it should be pointed out that, although the 7 MA cm⁻² results shown in Fig. 2a are for PLD films, most deposition techniques — representing considerable microstructural diversity — have produced equally high or even higher J_c values[37–39]. In every such case, however, J_c drops rapidly as the films are made thicker. So the issue is a general one and not specific to a single deposition method. Further, confusion arises from the characterization of the phenomenon as a decreasing J_c,

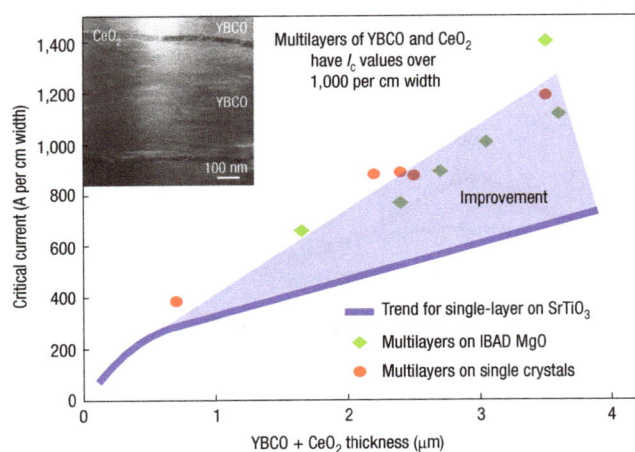

Figure 5 By periodically inserting 30-nm CeO₂ layers every 0.5 μm or so in a thick YBCO film (inset: TEM image), a higher critical current density is maintained. In this plot performance is expressed in terms of the current that could be carried by a one-centimetre-wide conductor with the same J_c and thickness as the test samples. The single-layer trend shown here is derived from the data of Fig. 2a.

because every deposition method is also capable of producing films having very weak or even flat thickness dependence, that is, with little or no decrease[40]. In these cases, however, J_c is not flat at 7 MA cm⁻², but at a lower level, typically half or less. So in reality strong thickness dependence is better characterized as a sharp increase from a relatively flat baseline as the films are made thinner. And we believe it is this increase near the film–substrate interface that holds the key to higher-current coated conductors.

It is easy to imagine that thicker films have lower J_c for the reason proposed when thickness dependence was first reported two decades ago[41]. The quality of an epitaxial coating, and hence J_c, is best near the substrate, and as the film is grown thicker current-blocking defects such as misoriented crystallites, cracks or voids become more prevalent. This is a very compelling argument because the deterioration of crystalline and/or morphological quality has been well-documented in YBCO films made by a variety of deposition methods. The challenge is to establish a quantitative correlation between observed microstructural decay and J_c thickness dependence. We have previously presented one such case in which it was found that YBCO above a certain thickness carried no current whatsoever because of a transition to a very porous microstructure[42]. However, when this problem was solved[18] (by using smoother metal substrates) there was still a thickness dependence similar to that in Fig. 2a. This shows that, even when microstructural decay has been reduced to an insignificant level, strong thickness dependence can persist.

Searching for an explanation, we have looked at how the incremental — or layer-by-layer — critical current density (j_c) varies as a function of distance between the layer and the substrate[43]. This led us to the hypothesis that a dense array of misfit dislocations[44] is raising j_c near the interface. In a successful test of this hypothesis we have recently shown that YBCO on a low mismatch substrate (NdGaO₃), with greatly reduced dislocation density, exhibits a flat thickness dependence[45].

In an earlier test of the interfacial enhancement hypothesis, we created additional interfaces within a thick film by growing alternating layers of YBCO and CeO₂. The result was the highest current ever achieved in a coated conductor. As shown in Fig. 5, a J_c of 4 MA cm⁻² was maintained in multilayers as thick as 3.5 μm,

resulting in conductor samples with the equivalent capacity of up to 1,400 amperes in a 1-cm width[43].

In addition to demonstrating a way to reach very high current levels, another aspect of this work is whether it helps to identify the primary mechanism responsible for enhanced thin-film J_c. In an independent experiment reproducing the multilayer effect, J_c enhancement was attributed to segmenting of vortices by the CeO₂ interlayers[46], based on the argument that thickness dependence is an intrinsic physical phenomenon[47]. And we have shown previously that CeO₂-based multilayers had improved YBCO microstructure[48], suggesting that multilayers could function by overcoming microstructural decay. We now believe that, although microstructure is improved, the main role of the CeO₂ interlayers is to periodically reintroduce interfacial enhancement throughout the entire film thickness. But regardless of the mechanism, the multilayer approach is clearly an effective way to maintain high J_c levels in thick films. Further, the self-field J_c is improved without degrading the field dependence, or, said another way, the increase of self-field J_c in thick films leads to comparable enhancement in an applied magnetic field[45].

ADDING DEFECTS FOR IMPROVED FIELD DEPENDENCE

Although it seems we have limited means for improving self-field J_c, researchers are now finding multiple routes to systematically incorporate defects in YBCO coatings to stem the reduction of J_c in a magnetic field. This work is critically important in two ways: First, being able to add defects in a controlled manner and observe the result is a powerful tool in the quest to develop a general understanding of defect-performance relationships in YBCO films; second, improving field dependence of HTS coated conductors has practical and immediate relevance because most envisioned HTS applications require operation in a magnetic field. All efforts to improve field dependence fall into one of three categories, as described in the following sections.

SUBSTRATE SURFACE DECORATION

Surface decoration, the earliest method used, has been accomplished by depositing metal (Ag, ref. 49; Ir, ref. 50) or oxide (Y₂O₃, refs 51,52) nanoparticles on the substrate before YBCO film growth, or by processing an oxide layer (CeO₂, ref. 53; SrTiO₃, ref. 54) in such a way that nanoscale outgrowths develop naturally on the deposition surface. The particles, ranging in size from 10–100 nm, produce a disruption in the YBCO that can enhance flux pinning. There is, however, no clear and consistent picture regarding the nature of the disruption and the effect on performance. The most likely scenario is that the YBCO lattice planes are buckled or distorted above the nanoparticles, resulting in low-angle grain boundaries or dislocations that may thread to the substrate surface.

In one case Y₂O₃ surface particles produced c-axis-correlated defects, with the expected enhancement of J_c when the field is parallel to these defects[51]. In contrast, J_c enhancement that appears as random pinning has been produced with iridium surface particles[50], but random pinning is typically associated with a homogeneous distribution of point defects distributed throughout the film volume, and not the planar distribution here. The authors report buckling of the YBCO planes above the particles, and the films are thin (100–200 nm)[50]; perhaps strain fields present through the full thickness are creating the appearance of a volumetric particle distribution.

So, interestingly, surface decoration with two different nanoparticle materials has given rise to two different types of pinning. A clue to the difference may be that in the iridium case (and in a second similar case[53]) the nanoparticles have partially reacted with YBCO, whereas the Y_2O_3 nanoparticles are intact. Conceivably the volumetric change associated with the reactions[55] has provided an alternate strain-relieving mechanism, reducing the driving force for dislocation formation, and eliminating correlated pinning enhancement.

IMPURITY ADDITIONS

If nanoparticles on the substrate surface are effective at enhancing field dependence, then couldn't this effect be multiplied by distributing particles throughout the bulk of the YBCO? This question is the basis for some of the more recent attempts to improve performance of YBCO coatings.

In one example, the impurity was a non-superconducting phase of YBCO, Y_2BaCuO_5, controllably inserted into the superconductor matrix using a second laser-deposition target. By alternately depositing thin YBCO layers and discontinuous '211' layers, a pancake-like array of impurity sites were grown into the film, resulting in improved pinning for fields parallel to the c axis[56]. When angular dependence was measured later, improvement was found to be even stronger parallel to the a–b planes[57], as expected for such planar defects.

An alternative to depositing the impurity separately is to simply add it to the YBCO source material. In the first demonstration of this technique, Zr and excess Ba were mixed into the powders used to fabricate a ceramic laser-deposition target, and films were grown under standard YBCO processing conditions[58]. The result was a multitude of ~10 nm $BaZrO_3$ particles that, although distributed randomly throughout the YBCO, produced a significant c-axis-correlated enhancement of J_c. Because of the ease with which nanoparticles can be incorporated — essentially films are produced in the usual way at no additional cost — this technique has been widely duplicated by numerous researchers.

In one variation, yttria-stabilized zirconia (YSZ) was added to the YBCO target, and $BaZrO_3$ formed by reaction[59], presumably leaving a Ba-deficient YBCO film matrix. A fascinating feature that appeared in this work was the self-assembly of vertical arrays of $BaZrO_3$ particles, aptly named "bamboo structure". Similar self-assembly has been observed in many other material systems, such as InAs in GaAs (ref. 60). It is thought to arise when impurity islands preferentially nucleate in the strain field above buried impurity particles[61]. The strain is due in part to the crystalline lattice mismatch between the impurity and host material, and this is probably why 'bamboo' is observed in a high mismatch (9%) system such as YBCO/$BaZrO_3$ whereas it is not observed when the mismatch is lower (2.5%), as with Y_2O_3 nanoparticles[62].

Although interesting, the significance of the bamboo structure is not clear. Certainly columns of non-superconducting particles seem to be ideal for pinning flux perpendicular to the film plane. But in ref. 59 the bamboo structure offers little or no enhancement of J_c over the control samples at technically relevant magnetic field levels. This is at first a difficult interpretation because the researchers[59] have also substituted Gd for Y in the superconductor, but on close inspection it is evident that most of the in-field improvement is due to the Gd substitution, and there is little change when YSZ is added to either YBCO or GdBCO. Further, there is no c-axis enhancement (the claim to the contrary is an artefact of normalization), as would be expected from the observed bamboo columns. A somewhat different story emerges in a later publication by the same authors[63]. For thicker films with a less-dense field of $BaZrO_3$ bamboo, GdBCO still offers improvement over YBCO, but further improvement is exhibited by addition of ZrO_2 and a strong c-axis peak is apparent at 3 T (ref. 63). In another $BaZrO_3$ experiment[64], strong c-axis-correlated pinning enhancement of field dependence was observed. In this case, both bamboo and a random $BaZrO_3$ particle distribution were reported,

along with misfit dislocations surrounding the bamboo columns (no dislocations were reported in ref. 59).

We can only speculate about why a seemingly ideal columnar array of small particles does not always result in correlated enhancement. In the clear case without enhancement the bamboo density is very high, perhaps altering the strain fields to make dislocation formation less energetically favourable. Another possibility is that correlated pinning enhancement is not apparent because current obstruction[65] by the high density of bamboo columns lowers the overall J_c, but the small suppression of self-field J_c suggests otherwise. In the original $BaZrO_3$ work[58], bamboo was not reported, but strong c-axis-correlated enhancement was measured and attributed to linear dislocations observed above the particles. The picture that emerges, then, is that dislocations related to the $BaZrO_3$ nanoparticles produce correlated pinning enhancement, but the particles themselves produce variable effects in YBCO.

To reinforce this, yet another kind of result involving the addition of $BaZrO_3$ has been recently reported — in this case the YBCO is grown by the MOD process. Unlike all of the PLD results described above, the solution process has produced non-epitaxial particles in the film volume surrounded by a variety of crystalline defects. The result is a spectacular level of random-pinning-like J_c enhancement[66].

RARE-EARTH ADDITIONS AND/OR SUBSTITUTIONS

Perhaps the most complex and poorly understood pinning enhancement mechanisms arise from replacement of yttrium in YBCO films with rare-earth (RE) elements. Although it has long been known that such REBCO compounds show different and sometimes enhanced field-dependent J_c behaviour compared with YBCO, the reasons for this are uncertain. However, obvious influences are the increased transition temperature (T_c) values[67], and growth thermodynamics and kinetics that are thought to yield improved film microstructures[68].

In bulk melt processed materials, light rare-earth compositions (LREBa$_2$Cu$_3$O$_{7-\delta}$ with LRE = Nd, Sm, Eu, Gd) fabricated under oxygen-controlled growth have additional flux pinning because of solid solution between the rare-earth atoms and Ba, which produces compositional fluctuations and spatial variations in T_c (ref. 69). Translating bulk results to films is not trivial, however, because controlling LRE–Ba substitution levels relies critically on growth temperature, oxygen pressure and growth rate. As there is only a narrow window of deposition conditions for obtaining high-quality epitaxial films, there is little room to optimize conditions for substitution.

Having said this, film compositions that include the LREs can show improved pinning over YBCO. Interestingly, pinning enhancement is usually random, meaning that the added defects must be point-like in character. The exact nature of these defects is a matter of ongoing speculation. One candidate is cation disorder, such as a rare-earth atom on a Ba site, which results in a point defect surrounded by a strain field. Although cation disorder can be quantified by Raman spectroscopy[70,71], to date no quantitative correlation with pinning enhancement has been attempted.

REPLACEMENT OF YTTRIUM

The simplest substitution approach is to replace yttrium with a rare earth element while maintaining the 123 stoichiometry. As noted previously, GdBa$_2$Cu$_3$O$_{7-\delta}$ shows significant in-field enhancement over YBCO, perhaps due to an increased density of stacking faults[59].

Promising levels of enhancement have also been reported for EuBCO (ref. 72) and SmBCO (ref. 73), although these compositions must be deposited at higher temperature[74-76] and oxygen pressure[77] to control the amount of RE–Ba exchange and to achieve the required crystallinity in compounds with higher melting temperatures. Such modifications are not needed in the heavier

Table 1 A compilation of results for films reported to have enhanced J_c in a magnetic field relative to standard YBCO. Two technologically relevant operating conditions — 1 T at 77 K and 3 T at 65 K — are tabulated; because of the paucity of published data for the latter, J_c values for 3 T, 77 K are also listed. At lower magnetic field strengths, such as for power-transmission cables and transformers, operation at 77 K (the boiling point of liquid nitrogen at sea level) is feasible. Higher-field operation is needed for motors and generators, and the sharp drop in J_c at these field levels can be partially overcome by lowering the temperature to 65 K (the boiling point of liquid nitrogen in vacuum).

Enhancement method	Film deposition process and substrate	Film thickness (µm)	J_c @ 1 T, 77 K (MA cm^{-2})	J_c @ 3 T, 77 K (MA cm^{-2})	J_c @ 3 T, 65 K (MA cm^{-2})	J_c @ self-field, 77 K (MA cm^{-2})	Reference
Substrate surface decoration							
Surface decoration with CeO$_2$	PLD, sapphire	0.20	0.43	0.12	1.5	3.1	53
Surface decoration with Y$_2$O$_3$	PLD, SrTiO$_3$	0.15	0.70	0.30		2.1	51
Surface decoration with Ir	PLD, SrTiO$_3$	0.20	0.50	0.11		6.6	50
Surface decoration with Y$_2$O$_3$	PLD, SrTiO$_3$	0.15	0.60	0.40		2.7	52
Surface decoration with SrTiO$_3$	PLD, IBAD MgO	5.0	0.17*	0.07*		0.9*	54
Impurity addition							
Bulk addition of 5 mol.% BaZrO$_3$	PLD, SrTiO$_3$	0.75	0.19*	0.07*		1.8*	58
Bulk addition of 5 mol.% BaZrO$_3$	PLD, IBAD MgO	0.90	0.43*	0.16*		1.5*	58
Quasi-layers of Y$_2$BaCuO$_5$	PLD, LaAlO$_3$ or SrTiO$_3$	0.26	0.90	0.40		4.5	56
Quasi-layers of Ir (BaIrO$_3$)	PLD, SrTiO$_3$	0.30	0.50	0.13		4.3	99
Bulk addition of 2 vol.% BaZrO$_3$	PLD, RABiTS	0.20	0.60	0.20	1.0	2.7	64
Sectored target with BaSnO$_3$	PLD, LaAlO$_3$	0.30	0.30	0.28	1.0	1.0	99
Bulk addition of 2 vol.% YSZ	PLD, IBAD Gd$_2$Zr$_2$O$_7$	1.16	0.56	0.36		1.7	101
Bulk addition of 2 vol.% BaZrO$_3$	PLD, RABiTS	3.0	0.38	0.11	0.77	1.3	102
<10 mol.% BaZrO$_3$ in YBCO precursor	MOD, SrTiO$_3$	0.20-0.27	2.0	0.7	2.3	6.5	66
Bulk addition of 5 mol.% BaZrO$_3$	PLD, SrTiO$_3$	0.18	0.75*	0.36*			†
Bulk addition of 5 mol.% BaZrO$_3$	PLD, SrTiO$_3$	0.85	0.78*	0.34*			†
Bulk addition of 5 mol.% BaZrO$_3$	PLD, SrTiO$_3$	1.0	0.67*	0.25*			†
RE addition or substitution							
(Nd$_{1/3}$Eu$_{1/3}$Gd$_{1/3}$)BCO mixture	PLD, SrTiO$_3$	0.05	0.28	0.07	0.40	3.0	82
(Nd,Eu,Gd)BCO trilayer	PLD, SrTiO$_3$	0.125	1.0	0.20			103
(Dy$_{1/3}$Ho$_{2/3}$)BCO	PLD, SrTiO$_3$	1.4	0.45*	0.12*		2.3*	83
EuBCO	PLD, SrTiO$_3$	0.16	0.77*	0.19*		5.5*	72
(Y$_{2/3}$Sm$_{1/3}$)BCO	PLD, SrTiO$_3$	0.80	0.37*	0.18*		3.3*	85
(Y$_{2/3}$Sm$_{1/3}$)BCO	PLD, IBAD MgO	1.20	0.45*	0.14*		2.4*	85
30% Y-rich precursor	BaF$_2$, RABiTS	0.70	0.41	0.13	0.60	2.5	88
Sm$_{1.08}$B$_{1.92}$C$_{3O7-\delta}$ with seed layer	PLD, MgO	0.40-0.70	1.1	0.5		3.2	94
GdBCO	PLD, IBAD Gd$_2$Zr$_2$O$_7$	0.25	0.80	0.25		3.0	59
(Y$_{0.9}$Sm$_{0.1}$)BCO	MOCVD, IBAD MgO	1.0	0.21	0.07	0.45	1.9	91
(Gd$_{0.8}$Er$_{0.2}$)BCO	PLD, IBAD Gd$_2$Zr$_2$O$_7$	0.42	0.65	0.20		2.0	81
NdBCO	PLD, RABiTS	0.17	0.40			2.2	86
GdBCO	PLD, IBAD Gd$_2$Zr$_2$O$_7$	3.6	0.17	0.06		1.4	63
20% Y-rich precursor	BaF$_2$, RABiTS	3.0	0.57*	0.18*		1.5*	90
YBCO / (Y + Dy$_{0.5}$)BCO bilayer	MOD, RABiTS	1.75	0.55*	0.17*		3.0	104
(Y + Dy$_{0.5}$)BCO	MOD, RABiTS	0.80	0.75			2.9	92
(Y$_{0.9}$Sm$_{0.1}$)BCO	MOCVD, IBAD MgO	2.1	0.48*			3.0	105
Combined methods							
Sm$_{1+x}$B$_{2-x}$CO films and surface particles	PLD, MgO	0.70	1.30	0.63		3.5	95
GdBCO + 2 vol.% YSZ	PLD, IBAD Gd$_2$Zr$_2$O$_7$	0.25	0.70	0.25		2.0	59
GdBCO + 5 mol % bulk addition of ZrO$_2$	PLD, IBAD Gd$_2$Zr$_2$O$_7$	2.28	0.60	0.21		1.5	63
NdBCO with 2 vol % addition of BaZrO$_3$	PLD, RABiTS	0.17	0.60			2.4	86
(YBCO + 2% BaZrO$_3$)/CeO$_2$ multilayer	PLD, RABiTS	1.0	0.46			2.0	106

*These samples were measured at Los Alamos at the local boiling point temperature of liquid nitrogen, 75.5 K, which gives higher J_c values than measurement at 77 K. Although not exact, based on our experience we have made the following adjustments to compensate for the temperature difference: self-field J_cs have been reduced by 10%; 1 T J_cs have been reduced by 15%; and 3 T J_cs have been reduced by 20%.

†B. Maiorov, unpublished work.

Key: PLD = pulsed laser deposition; IBAD = ion-beam assisted deposition; RABiTS = rolling-assisted biaxially textured substrate; BaF$_2$ = an *ex situ* YBCO process using a fluoride precursor; MOCVD = metal–organic chemical vapour deposition; MOD = a solution-based *ex situ* YBCO process.

rare-earth elements, which are less prone to exchange with Ba and have lower melting points[78]. Ironically, when a way was found to reduce the deposition temperature of EuBCO (by depositing it on a thin YBCO seed layer), high self-field J_c was achieved but the in-field enhancement vanished[79].

MIXTURES

The next level up in complexity is mixtures of rare earth elements and yttrium that retain the 123 composition (RE,Y)Ba$_2$Cu$_3$O$_{7-\delta}$. The promise here is that additional enhancement will result from the strain induced by lattice mismatch between the various 123 components[80]. Reports include (Gd$_{0.8}$Er$_{0.2}$)[81], (Nd$_{1/3}$Gd$_{1/3}$Eu$_{1/3}$)[82], and several Y–Sm mixtures, but the enhancement in all cases is unremarkable except that when present it is random and not correlated. Given that the

possibilities for creating mixed rare-earth compounds are limitless it is desirable to evaluate these mixtures in a systematic way. Two readily quantifiable parameters that can be used to do this are the variance (how different are the ion sizes) and the average ion size.

In the only study to date of variance effects in YBCO (ref. 83) a series of mixtures with two rare-earth elements was prepared with a constant average ion size (similar to yttrium in an effort to minimize RE–Ba exchange). The result is striking: At low fields parallel to the c axis — where J_c drops most rapidly with field — high variance samples (for example, (Nd,Yb)BCO) show no enhancement. Enhancement increases with decreasing variance and for the lowest variance ((Dy,Ho)BCO) J_c is about double that for YBCO. Earlier variance studies in other perovskite oxides have found a correlation between variance and displacement of oxygen ions[84]; such displacements would result in local strain effects that are consistent with the random pinning enhancement observed in YBCO. In a complementary study, the effect of changing RE ion size while maintaining constant variance was examined[85]. No systematic dependence of pinning behaviour on average RE ion size was found, illustrating the difficulty in controlling the defects associated with cation intermixing.

In many LREBCO films, the pronounced c-axis peak typical in PLD YBCO is absent. When an attempt was made to add both kinds of pinning — random enhancement with NdBCO plus correlated enhancement with $BaZrO_3$ — the roles reversed: a strong c-axis peak appeared in the NdBCO and the $BaZrO_3$ added only random enhancement[86].

NON-STOICHIOMETRY

Yet another variation of rare-earth-based enhancement methods involves an intentional deviation from the 123 stoichiometry, usually by increasing the fraction of rare-earth or yttrium, resulting in oxide nanoparticles of the excess material[87–90]. Such particles can also spontaneously appear in nominally stoichiometric films[91]. In either case, these particles generally have not shown significant enhancement in field, although enhancement of the self field J_c from excess Y_2O_3 has been reported[62]. That particles such as yttrium- or rare-earth-oxides often do not produce pinning enhancement at first seems unusual, as $BaZrO_3$ particles frequently lead to enhancement through the formation of dislocations. The difference may be that in the former case the particles have a relatively small lattice mismatch, and the stress due to their formation can be accommodated elastically without dislocations. In the case of ex situ films, however, the story is very different. A recent article reports that in solution-processed $Y(RE_{0.5})BCO$ compositions with RE = Ho, Er or Dy, the RE-oxide particles do result in enhanced pinning in a broad angular range about the c-axis direction[92]. A different ex situ approach using vapour-deposited precursors produced random enhancement in a 20% Y-rich composition[90].

In another non-stoichiometric composition, it has been found that the optimum deposition temperature for $Sm_{1+x}Ba_{2-x}Cu_3O_{7-\delta}$ can be lowered as x is increased from 0 to 0.12 (ref. 93). Building on earlier seed-layer work with EuBCO[79], it was further found that a thin layer deposited at high temperature (830 °C), enables growth of a high-quality film of 0.4 to 0.7 µm thickness (with x = 0.08) at the very low temperature of 740 °C (Yoshida et al.[94]), leading to one of the highest reported J_c values at 1 T, 77 K. This is the first example in which the in-field advantage of the light rare-earth compounds has been realized without needing the high deposition temperatures that could be impractical for the metal substrates of coated conductors. In a further development, Yoshida et al. have added SmBCO nanoparticles to the surface of the seed layer, reaching even higher J_c in field[95]. Angular data are not available to assess the relative contributions of the rare-earth and surface-decoration methods of enhancement.

COMPARING THE QUALITY OF DEFECTS

Literature reports such as those described above typically demonstrate the efficacy of a particular approach by comparison with a standard sample, but difficulty arises because of the wide range of observed behaviour of the control samples. To address this issue and facilitate an objective comparison, in Table 1 we have compiled absolute J_c values obtained with different approaches. In an effort to be concise and to include the greatest possible body of literature we have made the following three simplifications: (1) Two technologically relevant operating conditions — a field of 1 T at a temperature of 77 K, and 3 T at 65 K — have been chosen for comparison. As only a fraction of the reports include 65 K data, we have also listed J_c at 3 T, 77 K. (2) Only one field direction — parallel to the YBCO c-axis — is used in Table 1. We acknowledge that this

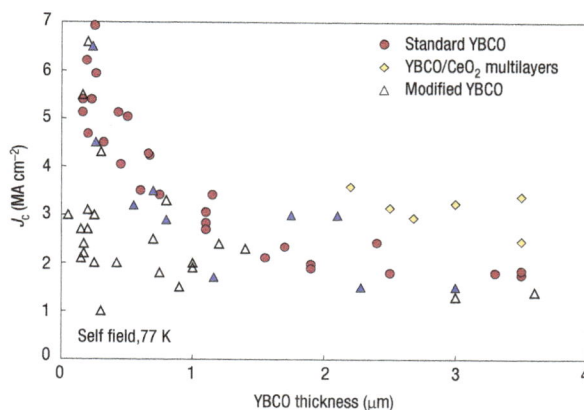

Figure 6 Comparison of the self-field J_c values for enhanced samples from Table 1 with the unmodified PLD films from Fig. 2a. None of the modified samples have self-field J_c values higher than standard YBCO, and many enhancement attempts result in lower self-field values. Samples measured at Los Alamos may have a somewhat higher J_c due to the lower boiling point temperature of nitrogen (75.5 K). The difference varies from sample to sample (it depends on T_c, among other things); to correct we have reduced all Los Alamos-measured J_cs by the average difference of 10%. Filled triangles are the samples that showed the greatest enhancement at 1 T.

is a serious shortcoming because J_c can vary considerably with field orientation, as shown in Fig. 4, and described in detail in ref. 96, but again we are constrained by the limited amount of angle-dependent data in the literature. The c-axis orientation is still informative, however, because this is where enhancement methods typically have the greatest effect. (3) It is not practical to present a comprehensive summary of every attempt to enhance YBCO in field. Instead, we list only those results that show clear improvement relative to the authors' YBCO control samples.

Even with these simplifications, it is still difficult to compare enhancement methods because films of different thickness, which can inherently have different J_c values, are used in the literature. We address this issue by plotting the Table 1 J_c values as a function of film thickness. For the purpose of evaluating the degree of enhancement we have also plotted results for unmodified laser-deposited YBCO films measured under similar conditions.

First we examine self-field results for the modified YBCO films in Table 1, and see that, although some enhanced-film J_cs are comparable to good-quality YBCO, many fall below the benchmark (Fig. 6). In several cases, authors report that modified films have higher self-field J_c than their control samples, suggesting that methods used to attempt enhancement in field may at least partially compensate for lack of pinning in non-optimum YBCO. However, Fig. 6 clearly shows that in no case does modification push J_c above what is observed in optimized, unmodified YBCO. The sole exception is for YBCO/CeO_2 multilayers, which exhibit the higher thick-film J_cs for which they were designed.

Next, in Fig. 7a, we plot J_c for modified films at 1 T, and compare them with 1 T values for unmodified YBCO films measured at our laboratory over the past few years. This comparison removes the uncertainty associated with variable control-sample results, and it enables us to compare films of similar thickness. We see that, on the basis of absolute J_c values, approximately two-thirds of the enhancement attempts yield results that are no better than J_c for optimized YBCO. Figure 7b shows a similar result at 3 T and 77 K.

Interestingly, the films that do exhibit improvement in field come from the full range of enhancement possibilities, meaning

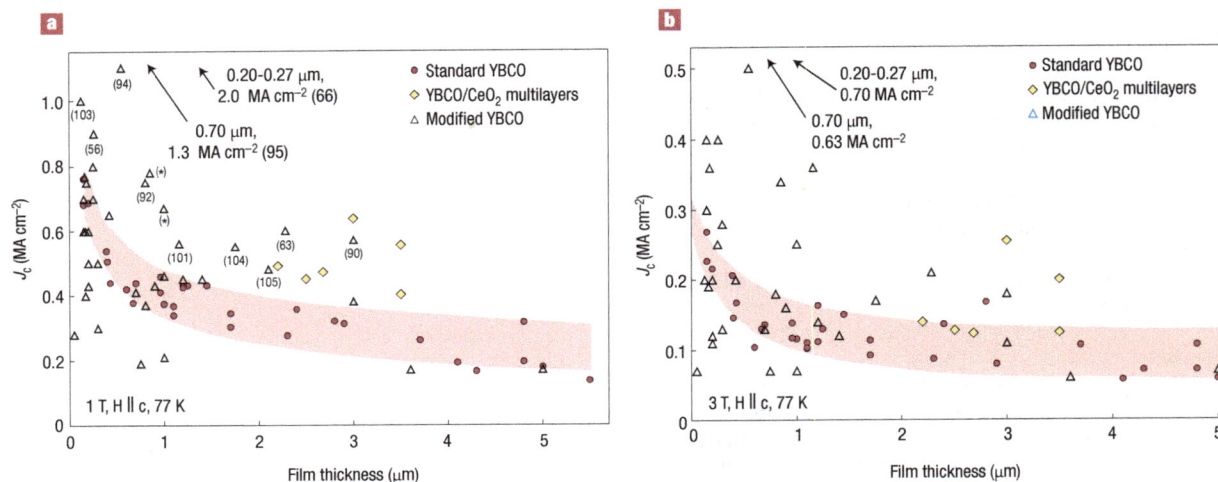

Figure 7 Many attempts to enhance the performance of YBCO in a magnetic field have produced results that are no better than standard YBCO. Films showing the greatest enhancement come from the full range of modification techniques. **a**, The 1-T results of Table 1 are compared with standard PLD YBCO and YBCO/CeO$_2$ multilayers — numbers in parenthesis are references in the text and Table 1 (*B. Maiorov, unpublished data). Two samples, indicated by arrows, are off-scale. (To correct for measurement at 75.5 K, all J_c values obtained at Los Alamos have been reduced by 15%.) **b**, The 3 T, 77 K results of Table 1 are compared with standard PLD YBCO and YBCO/CeO$_2$ multilayers. Two samples, indicated by arrows, are off-scale. (To correct for measurement at 75.5 K, all J_c values obtained at Los Alamos have been reduced by 20%.)

that there are multiple paths for adding beneficial defects to YBCO. This is good news because nearly all of the research in this area is conducted using laser deposition, whereas two of the three major efforts to develop a coated-conductor manufacturing process use different YBCO deposition technologies. Having the flexibility associated with a diverse range of enhancement methods increases the likelihood of finding one that is compatible with any YBCO process. In fact, enhancement methods pioneered with PLD are already showing promise in non-PLD processes[66,92,97].

One way to place the above results into perspective is shown in Fig. 8. This type of log–log plot clearly reveals the important aspects of field dependence: the self-field J_c, the low-field plateau region in which J_c remains relatively constant, and the power-law region in which $J_c \propto H^{-\alpha}$. The curves illustrate different ways to improve J_c in the power-law region, where most HTS applications will operate. Increasing the self-field J_c raises the entire curve, extending the plateau to higher field delays the onset of decay, and reducing the magnitude of alpha lessens the rate of decay. Examples of each type of field-enhancement exist and have been described in this review; however, the predominant effect of defect engineering seems to be the reduction of α. For this reason we would expect a greater proportion of enhancement methods to be successful at higher fields, where lower α produces a stronger effect. We do not observe this difference in Fig. 7b because 3 T is beyond the power-law region at 77 K, but the effect should appear at 65 K.

Except for thick-film multilayers, there are no enhancement examples that function by raising the self-field J_c, and as noted above there are many enhancement attempts with depressed self-field J_c values. Films in the latter category require a very large reduction of α to be competitive at 1 T, and it is evident in Fig. 7 that many of the enhancement attempts did not reduce α enough to overcome low starting J_c values. The films with greatest enhancement, on the other hand, have high self-field J_c values, highlighted as filled symbols in Fig. 6. The important point here is that the benefit of a reduced α can easily be negated by failure to maintain a high self-field J_c.

Conversely, even greater improvement could be realized if one could enhance more than one aspect (self field, plateau, α) of the same

film, but early attempts indicate that this is not straightforward. As described previously, addition of BaZrO$_3$ to REBCO has produced mixed results. Our own attempt to combine multilayers and BaZrO$_3$ additions to simultaneously reap the benefits of high I_c and enhanced field dependence were unsuccessful. We found that the high J_c near the film–substrate interface — a key requirement for multilayers — was suppressed by the BaZrO$_3$ particles[45].

Another example of the unpredictable nature of combining different types of defects involves the angular dependence of YBCO films grown from solution. Such films typically do not have the characteristic c-axis peak found in vapour-deposited ones, but there is enhanced pinning parallel to the film plane due to planar growth defects. When rare-earth-oxide particles are added a c-axis peak appears, but the particles terminate the planar defects lowering the a–b peak[92,98]. This again demonstrates that we cannot arbitrarily add defects to a film and expect a superposition of pinning effects — they do not add in a simple linear way

CONCLUDING REMARKS

It has been a long journey to bring high-temperature superconducting wire to the threshold of commercial production, but a successful one. Despite the technological successes, however, there is still much about YBCO that we do not understand — not even the mechanism responsible for high-temperature superconductivity itself. But the time has come for concerted study of the interrelationships between processing, film defects, and flux pinning, because crossing the threshold of commercial viability may depend on the kind of performance improvements that can be achieved through a better understanding of these relationships. As fruitful as our empirical efforts to discover the ideal defect structure have been, we still have much to learn.

The learning can begin with a careful analysis of what has gone before, and at this point it must be stated that comparison of an enhanced sample with an arbitrary reference standard can cloud the picture and impede progress. Only through the use of absolute standards can we assess the relative merits of different techniques, begin to identify important — and unimportant — defects, and

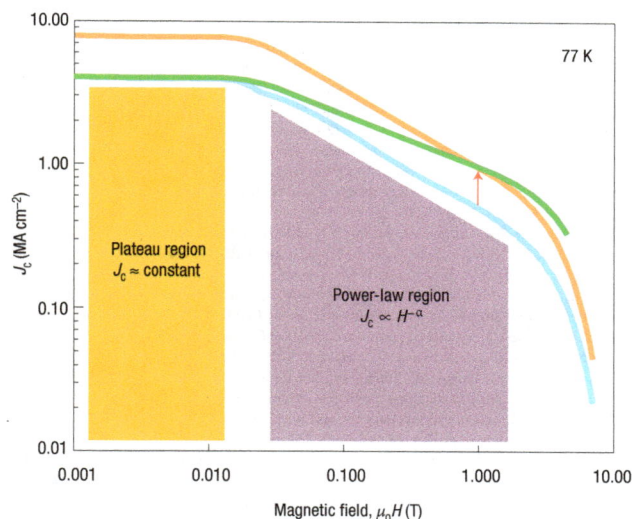

Figure 8 Three factors that determine J_c in a magnetic field are the self-field value, the extent of the plateau, and the rate of decay beyond the plateau. The blue curve is one of the plots from Fig. 2b that has been replotted on the log–log scale. The orange line is a curve with the same plateau and α, but with double the self field J_c — the result is a doubling at 1 T. The green line has a lower α, which is an alternative path to raising J_c at 1 T. A third path (not shown) is to extend the plateau to higher field. Typical α values for PLD YBCO are 0.5–0.6; most enhancement methods function by reducing α to as low as 0.2.

make steady and systematic progress in extracting the best possible performance from YBCO. Instead of being the final answer, Figs 6 and 7 represent an initial effort — limited to a single temperature and field orientation — that we hope will provide a useful starting point for further discussion in the HTS community.

We should also note that Fig. 7 contains some extraordinary results that in thicker films would surely translate to over 1 MA cm^{-2} at 1 T in films 1 μm thick. Although healthy scepticism is always in order until such results are reproduced, we recall many examples in coated-conductor development for which the 'champion' samples of yesterday have become the routine results of today, and we can only assume that the same will be true here. If so, we already have an indication of the untapped potential of YBCO films.

As a step towards reaching that potential, our conclusions at this point can be stated succinctly: (1) Pinning by dislocations has been unambiguously established. Dislocations are operative in all three regimes: Interfacial misfit dislocations are associated with enhanced self-field J_c; inter-island threading dislocations extend the plateau region; and particle-induced dislocations reduce the slope of the power-law region. (2) The precise role of nanoparticles varies with deposition method and the nature of the particles. In vapour-deposited films they function primarily by disrupting the YBCO rather than being pinning sites themselves (at least at 77 K). Particles with large lattice mismatch, for example, often produce extended defects and c-axis-correlated pinning, whereas particles with small lattice mismatch rarely enhance pinning. In *ex situ* films, however, both types of particle can be effective and both random and correlated pinning have been observed. (3) Substituting light rare-earth elements for yttrium generally produces random pinning enhancement associated with point defects, but the source of extra pinning has not been definitively identified. (4) Pinning enhancements are not necessarily additive — self-field and in-field

J_c; random and correlated pinning; a–b and c-axis pinning often seem to be mutually exclusive.

In reaching these conclusions we have reviewed the existing catalogue of process modifications, the defects they produce, and how they impact performance. We have established a methodology for judging the improvement of existing and future efforts in the area. And we have distilled from all of this some fascinating mysteries about what works and does not work when we attempt to manipulate the defect structure in YBCO films. In so doing we hope to sharpen the focus on the perennial materials science issue of structure–property relationships in these fascinating and important materials.

doi:10.1038/nmat1989

References

1. Bednorz, G. & Muller, K. A. Possible high-T_c superconductivity in the Ba-La-Cu-O system. *Z. Phys. B* **64,** 189–193 (1986).
2. Heine, K., Tenbrink, J. & Thoner, M. High field critical current densities in Bi$_2$Sr$_2$CaCu$_2$O$_{8+x}$ / Ag-wires. *Appl. Phys. Lett.* **55,** 2441–2443 (1989).
3. Iijima, Y., Tanabe, N., Kohno, O. & Ikeno, Y. In-plane aligned YBa$_2$Cu$_3$O$_{7-x}$ thin films deposited on polycrystalline metallic substrates. *Appl. Phys. Lett.* **60,** 769–771 (1992).
4. Reade, R. P., Berdahl, P., Russo, R. E. & Garrison, S. M. Laser deposition of biaxially textured yttria-stabilized zirconia buffer layers on polycrystalline metallic alloys for high critical current Y-Ba-Cu-O thin films. *Appl. Phys. Lett.* **61,** 2231–2233 (1992).
5. Doi, T. J., Yuasa, T., Ozawa, T. & Higashiyama, K. in *Proc. 7th Int. Symp. Superconductivity (ISS 1994)* 817–820 (Kitakyushu, Japan, 1994).
6. Wu, X. D. *et al.* Properties of YBa$_2$Cu$_3$O$_{7-δ}$ thick films on flexible buffered metallic substrates. *Appl. Phys. Lett.* **67,** 2397–2399 (1995).
7. Goyal, A. *et al.* High critical current density superconducting tapes by epitaxial deposition of YBa$_2$Cu$_3$O$_x$ thick films on biaxially textured metals. *Appl. Phys. Lett.* **69,** 1795–1797 (1996).
8. Larbalestier, D., Gurevich, A., Feldmann, D. M. & Polyanskii, A. High-T_c superconducting materials for electric power applications. *Nature* **414,** 368–377 (2001).
9. Selvamanickam, V. *et al.* Progress in scale-up of second generation HTS conductor. *Phys. C* doi:10.1016/j.physc.2007.04.236 (2007).
10. Rupich, M. *et al.* in *US Department of Energy Wire Development and Applications Workshop* (Panama City, Florida, 2007); available at http://www.energetics.com/wire07/agenda.html.
11. Ibi, A. *et al.* Development of long YBCO coated conductors by IBAD-PLD method. *Physica C* **445–448,** 525–528 (2006).
12. D'Errico, R. A. SuperPower announces first major wire shipment. *The Business Review* (December 20, 2006); available at <http://albany.bizjournals.com/albany/stories/2006/12/18/daily35.html>.
13. Paranthaman, M. P. & Izumi, T. (eds) High-performance YBCO-coated superconductor wires. *Mater. Res. Soc. Bull.* **29,** 533–589 (2004).
14. Roas, B., Hensel, B., Saemann-Ischenko, G. & Schultz, L. Irradiation-induced enhancement of the critical current density of epitaxial YBa$_2$Cu$_3$O$_{7-x}$ thin films. *Appl. Phys. Lett.* **54,** 1051–1053 (1989).
15. Wu, X. D. *et al.* Large critical current densities in YBa$_2$Cu$_3$O$_{7-x}$ thin films made at high deposition rates. *Appl. Phys. Lett.* **57,** 523–525 (1990).
16. Sarrao, J. & Kwok, W. K. *Basic Research Needs for Superconductivity* (US DOE, Washington, DC, 2006); <http://www.sc.doe.gov/bes/reports/abstracts.html#S>.
17. Dimos, D., Chaudhari, P. & Mannhart, J. Superconducting transport properties of grain boundaries in YBa$_2$Cu$_3$O$_7$ bicrystals. *Phys. Rev. B* **41,** 4038–4049 (1990).
18. Foltyn, S. R. *et al.* Strongly coupled critical current density values achieved in YBa$_2$Cu$_3$O$_{7-δ}$ coated conductors with near-single-crystal texture. *Appl. Phys. Lett.* **82,** 4519–4521 (2003).
19. Feldmann, D. M. *et al.* Grain orientations and grain boundary networks of YBa$_2$Cu$_3$O$_{7-δ}$ films deposited by metalorganic and pulsed laser deposition on biaxially textured Ni-W substrates. *J. Mater. Res.* **21,** 923–934 (2006).
20. Campbell, A. M. & Evetts, J. E. Flux vortices and transport current in type-II superconductors. *Adv. Phys.* **21,** 194–428 (1972).
21. Larkin, A. I. & Ovchinnikov, Yu. N. Pinning in Type II superconductors. *J. Low Temp. Phys.* **34,** 409–428 (1979).
22. Blatter, G., Feigel'man, M. V., Geshkenbein, V. B., Larkin, A. I. & Vinokur, V. M. Vortices in high-temperature superconductors. *Rev. Mod. Phys.* **66,** 1125–1388 (1994).
23. Tinkham, M. *Introduction to Superconductivity* (McGraw Hill, New York, 1975).
24. Hwa, T., Le Doussal, P., Nelson, D. R. & Vinokur, V. M. Flux pinning and forced vortex entanglement with splayed columnar defects. *Phys. Rev. Lett.* **71,** 3545–3548 (1993).
25. Civale, L., *et al.* Reducing vortex motion in YBa$_2$Cu$_3$O$_7$ crystals with splay in columnar defects. *Phys. Rev. B* **50,** 4102–4105 (1994).
26. Giapintzakis, J. *et al.* Production and identification of flux-pinning defects by electron irradiation in YBa$_2$Cu$_3$O$_{7-x}$ single crystals. *Phys. Rev. B* **45,** 10677–10683 (1992).
27. Civale, L. *et al.* Defect independence of the irreversibility line in proton irradiated Y-Ba-Cu-O crystals. *Phys. Rev. Lett.* **65,** 1164–1167 (1990).
28. van Dover, R. B. *et al.* Critical currents near 10^6 Acm^{-2} at 77 K in neutron-irradiated single-crystal YBa$_2$Cu$_3$O$_7$. *Nature* **342,** 55–57 (1989).
29. Sauerzopf, F. M. *et al.* Neutron-irradiation effects on critical current densities in single-crystalline YBa$_2$Cu$_3$O$_{7-δ}$. *Phys. Rev. B* **43,** 3091–3100 (1991).
30. Civale L. *et al.* Vortex confinement by columnar defects in YBa$_2$Cu$_3$O$_7$ crystals: Enhanced pinning at high fields and temperatures. *Phys. Rev. Lett.* **67,** 648–651 (1991).
31. Schindler, W., Roas, B., Saemann-Ischenko, G., Schultz, L. & Gerstenberg, H. Anisotropic enhancement of the critical current density of epitaxial YBa$_2$Cu$_3$O$_{7-x}$ films by fast neutron irradiation. *Physica C* **169,** 117–122 (1989).
32. Roas, B. *et al.* Effects of 173 MeV ^{129}Xe ion irradiation on epitaxial YBa$_2$Cu$_3$O$_{7-x}$ films. *Europhys. Lett.* **11** 669–674 (1990).

33. Christen, D. K. *et al.* Orientation-dependent critical currents in $YBa_2Cu_3O_{7-x}$ epitaxial thin films: evidence for intrinsic flux pinning? *AIP Conf. Proc.* **219,** 336–342 (1991).
34. Dam, B. *et al.* Origin of high critical currents in $YBa_2Cu_3O_{7-\delta}$ superconducting thin films. *Nature* **399,** 439–442 (1999).
35. Babaei Brojeny, A. A. & Clem, J. R. Self-field effects upon the critical current density of flat superconducting strips. *Supercond. Sci. Technol.* **18,** 888–895 (2005).
36. Daley, J. G. in *US Department of Energy Wire Development Workshop* (St. Petersburg, Florida 2006); available at <http://www.energetics.com/meetings/wire05/pdfs/session8/session8c.pdf>.
37. Smith, J. A., Cima, M. J. & Sonnenberg, N. High critical current density thick MOD-derived YBCO films. *IEEE Trans. Appl. Supercond,* **9,** 1531–1534 (1999).
38. Araki, T. & Hirabayashi, I. Review of a chemical approach to $YBa_2Cu_3O_{7-x}$ -coated superconductors – metalorganic deposition using trifluoroacetates. *Supercond. Sci. Technol.* **16,** R71–R94 (2003).
39. Palau, A., Puig, T., Obradors, X., Feenstra, R. & Gapud, A. A. Correlation between grain and grain-boundary critical current densities in *ex situ* coated conductors with variable $YBa_2Cu_3O_{7-\delta}$ layer thickness. *Appl. Phys. Lett.* **88,** 122502 (2006).
40. Kim, S. I. *et al.* Mechanisms of weak thickness dependence of the critical current density in strong-pinning *ex situ* metal-organic-deposition-route $YBa_2Cu_3O_{7-x}$ coated conductors. *Supercond. Sci. Technol.* **19,** 968–979 (2006).
41. Luborsky, F. E. *et al.* Reproducible sputtering and properties of Y-Ba-Cu-O films of various thicknesses. *J. Appl. Phys.* **64,** 6388–6391 (1988).
42. Foltyn, S. R. *et al.* Relationship between film thickness and the critical current of $YBa_2Cu_3O_{7-\delta}$ coated conductors. *Appl. Phys. Lett.* **75,** 3692–3694 (1999).
43. Foltyn, S. R. *et al.* Overcoming the barrier to 1000 A/cm-width superconducting coatings. *Appl. Phys. Lett.* **87,** 162505 (2005).
44. Wang, H., Foltyn, S. R., Arendt, P. N., Jia, Q. X. & Zhang, X. Identification of the misfit dislocations at $YBa_2Cu_3O_{7-\delta}$ / $SrTiO_3$ interface using moiré fringe contrast. *Physica C* **144,** 1–4 (2006).
45. Foltyn, S. R. & Civale, L. in *US Department of Energy Superconductivity for Electric Systems Annual Peer Review* (Arlington, Virginia, 2006); available at <http://www.energetics.com/meetings/supercon06/agenda.html>.
46. Wang, X. & Wu, J. Z. Effect of interlayer magnetic coupling on the J_c of $YBa_2Cu_3O_7$/insulator/ $YBa_2Cu_3O_7$ trilayers. *Appl. Phys. Lett.* **88,** 062513 (2006).
47. Gurevich, A. Thickness dependence of critical currents in thin superconductors. Preprint available at <http://lanl.arxiv.org/abs/cond-mat/0207526> (2002).
48. Jia, Q. X., Foltyn, S. R., Arendt, P. N. & Smith, J. F. High-temperature superconducting thick films with enhanced supercurrent carrying capability. *Appl. Phys. Lett.* **80,** 1601–1603 (2002).
49. Crisan, A., Fujiwara, S., Nie, J. C., Sundaresan, A. & Ihara, H. Sputtered nanodots: a costless method for inducing effective pinning centers in superconducting films. *Appl. Phys. Lett.* **79,** 4547–4549 (2001).
50. Aytug, T. *et al.* Enhancement of flux pinning and critical currents in $YBa_2Cu_3O_{7-\delta}$ films by nanoscale iridium pretreatment of substrate surfaces. *J. Appl. Phys.* **98,** 114309 (2005).
51. Matsumoto, K. *et al.* Enhancement of critical current density of YBCO films by introduction of artificial pinning centers due to the distributed nano-scaled Y_2O_3 islands on substrates. *Physica C* **412–414,** 1267–1271 (2004).
52. Mele, P. *et al.* Critical current enhancement in PLD $YBa_2Cu_3O_{7-x}$ films using artificial pinning centers. *Physica C* **445–448,** 648–651 (2006).
53. Nie, J. C. *et al.* Evidence for c-axis correlated vortex pinning in $YBa_2Cu_3O_{7-\delta}$ films on sapphire buffered with an atomically flat CeO_2 layer having a high density of nanodots. *Supercond. Sci. Technol.* **17,** 845–852 (2004).
54. Maiorov, B. *et al.* Influence of naturally grown nanoparticles at the buffer layer in the flux pinning in $YBa_2Cu_3O_7$ coated conductors. *Supercond. Sci. Technol.* **19,** 891–895 (2006).
55. Holesinger, T. G. *et al.* A comparison of buffer layer architectures on continuously processed YBCO coated conductors based on the IBAD YSZ process. *IEEE Trans. Appl. Supercond.* **11,** 3359–3364 (2001).
56. Haugan, T., Barnes, P. N., Wheeler, R., Meisenkothen, F. & Sumption, M. Addition of nanoparticle dispersions to enhance flux pinning of the $YBa_2Cu_3O_{7-x}$ superconductor. *Nature* **430,** 867–870 (2004).
57. Haugan, T. *et al.* Flux pinning strengths and mechanisms of YBCO with nanoparticle additions. *IEEE Trans. Appl. Supercond.* (in the press).
58. MacManus-Driscoll, J. L. *et al.* Strongly enhanced current densities in superconducting coated conductors of $YBa_2Cu_3O_{7-x}$ + $BaZrO_3$. *Nature Mater.* **3,** 439–443 (2004).
59. Yamada, Y. *et al.* Epitaxial nanostructure and defects effective for pinning in Y(RE)$Ba_2Cu_3O_{7-x}$ coated conductors. *Appl. Phys. Lett.* **87,** 132502 (2005).
60. Goldstein, L., Glas, F., Marzin, J. Y., Charasse, M. N. & Le Roux, G. Growth by molecular beam epitaxy and characterization of InAs/GaAs strained-layer superlattices. *Appl. Phys. Lett.* **47,** 1099–1101 (1985).
61. Shchukin, V. A. & Bimberg, D. Spontaneous ordering of nanostructures on crystal surfaces. *Rev. Mod. Phys.* **71,** 1125–1171 (1999).
62. Wang, H. *et al.* Microstructure and transport properties of Y-rich $YBa_2Cu_3O_{7-\delta}$ thin films. *J. Appl. Phys.* **100,** 053904 (2006).
63. Takahashi, K. *et al.* Investigation of thick PLD-GdBCO and ZrO_2 doped GdBCO coated conductors with high critical current on PLD-CeO_2 capped IBAD-GZO substrate tapes. *Supercond. Sci. Technol.* **19,** 924–929 (2006).
64. Goyal, A. *et al.* Irradiation-free, columnar defects comprised of self-assembled nanodots and nanorods resulting in strongly enhanced flux-pinning in $YBa_2Cu_3O_{7-\delta}$ films. *Supercond. Sci. Technol.* **18,** 1533–1538 (2005).
65. Pan, V. *et al.* Supercurrent transport in $YBa_2Cu_3O_{7-\delta}$ epitaxial thin films in a dc magnetic field. *Phys. Rev. B* **73,** 054508 (2006).
66. Gutierrez, J. *et al.* Strong isotropic flux pinning in $YBa_2Cu_3O_{7-x}$-$BaZrO_3$ films derived from chemical solutions. *Nature Mater.* **6,** 367–373 (2007).
67. Lin, J. G. *et al.* Origin of the R-ion effect on T_c in $RBa_2Cu_3O_7$. *Phys. Rev. B* **51,** 12900–12903 (1995).
68. Kwon, C. *et al.* Improved superconducting properties of $SmBa_2Cu_3O_{7-\delta}$ films using $YBa_2Cu_3O_{7-\delta}$ buffer layers. *Phil. Mag. B* **80,** 45–51 (2000).
69. Pradhan, A. K., Muralidhar, M., Murakami, M. & Koshizuki, N. Studies of flux pinning behaviour in melt processed ternary (Nd-Eu-Gd)$Ba_2Cu_3O_y$ superconductors. *Supercond. Sci. Technol.* **13,** 761–765 (2000).
70. Gibson, G., Cohen, L. F., Humphreys, R. G. & MacManus-Driscoll, J. L. A Raman measurement of cation disorder in $YBa_2Cu_3O_{7-x}$ thin films. *Physica C* **333,** 139–145 (2000).
71. Venkataraman, K., Baurceanu, R. & Maroni, V. A. Characterization of $MBa_2Cu_3O_{7-x}$ thin films by Raman microspectroscopy. *Appl. Spectrosc.* **59,** 639–649 (2005).
72. Jia, Q. X. *et al.* Comparative study of $REBa_2Cu_3O_7$ films for coated conductors. *IEEE Trans. Appl. Supercond.* **15,** 2723–2726 (2005).
73. Kwon, C *et al.* Fabrication and characterization of (rare-earth)-barium-copper-oxide (RE123 with RE = Y, Er, and Sm) films. *IEEE Trans. Appl. Supercond.* **9,** 1575–1578 (1999).
74. Stauble-Pumpin, B. *et al.* Growth mechanisms of coevaporated $SmBa_2Cu_3O_y$ thin films. *Phys. Rev. B* **52,** 7604–7628 (1995).
75. Jia, Q. X., Foltyn, S. R., Coulter, J. Y., Smith, J. F. & Maley, M. P. Characterization of superconducting $SmBa_2Cu_3O_7$ films grown by pulsed laser deposition. *J. Mater. Res.* **17,** 2599–2603 (2002).
76. Sudoh, K., Yoshida, Y. & Takai, Y. Effect of deposition conditions and solid solution on the $Sm_{1+x}Ba_{2-x}Cu_3O_{6+\delta}$ thin films prepared by pulsed laser deposition. *Physica C* **384,** 178–184 (2003).
77. Konishi, M. *et al.* Growth and characterization of Sm-Ba-Cu-O films and other RE-Ba-Cu-O films on PLD-CeO_2/IBAD-GZO/metal substrates. *J. Phys. Conf. Series* **43,** 174–177 (2006).
78. MacManus-Driscoll, J. L., Alonso, J. A., Wang, P. C., Geballe, T. H. & Bravman, J. C. Studies of structural disorder in $ReBa_2Cu_3O_{7-x}$ thin films (Re = rare earth) as a function of rare-earth ionic radius and film deposition condition conditions. *Physica C* **232,** 288–308 (1994).
79. Jia, Q. X. *et al.* The role of a superconducting seed layer in the structural and transport properties of $EuBa_2Cu_3O_{7-x}$ films. *Appl. Phys. Lett.* **83,** 1388–1390 (2003).
80. Radhika Devi, A. *et al.* Enhanced critical current density due to flux pinning from lattice defects in pulsed laser ablated $Y_{1-x}Dy_xBa_2Cu_3O_{7-\delta}$ thin films. *Supercond. Sci. Technol.* **13,** 935–939 (2000).
81. Konishi, M. *et al.* J_c-B characteristics of RE-Ba-Cu-O (RE = Sm, Er, and [Gd,Er]) films on PLD-CeO_2/IBAD-GZO/metal substrates. *Physica C* **445–448,** 633–636 (2006).
82. Cai, B., Holzapfel, B., Hanisch, J., Fernandez, L. & Schultz, L. Magnetotransport and flux pinning characteristics in $RBa_2Cu_3O_{7-\delta}$ (R = Gd, Eu, Nd) and $(Gd_{1/3}Eu_{1/3}Nd_{1/3})Ba_2Cu_3O_{7-\delta}$ high-T_c superconducting thin films on $SrTiO_3$ (100). *Phys. Rev. B* **69,** 104531 (2004).
83. MacManus-Driscoll, J. L. *et al.* Systematic enhancement of in-field critical current density with rare-earth ion size variance in superconducting rare-earth barium cuprate films. *Appl. Phys. Lett.* **84,** 5329–5331 (2004).
84. Rodriguez-Martinez, L. M. & Attfield, J. P. Cation disorder and size effects in magnetorestrictive manganese oxide perovskites. *Phys. Rev. B* **54,** R15622–R15625 (1996).
85. MacManus-Driscoll, J. L. *et al.* Rare earth ion size effects and enhanced critical current densities in $Y_{2/3}Sm_{1/3}Ba_2Cu_3O_{7-x}$ coated conductors. *Appl. Phys. Lett.* **86,** 032505 (2005).
86. Wee, S. H. *et al.* Strong flux-pinning in epitaxial $NdBa_2Cu_3O_{7-\delta}$ films with columnar defects comprised of self-assembled nanodots of $BaZrO_3$. *Supercond. Sci. Technol.* **19,** L42–L45 (2006).
87. Campbell, T. A. *et al.* Flux pinning effects of Y_2O_3 nanoparticulate dispersions in multilayered YBCO thin films. *Physica C* **423,** 1–8 (2005).
88. Gapud, A. A., Feenstra, R., Christen, D. K., Thompson, J. R. & Holesinger, T. G. Temperature and magnetic field dependence of critical currents in YBCO coated conductors with processing-induced variations in pinning properties. *IEEE Trans. Appl. Supercond.* **15,** 2578–2581 (2005).
89. Cui, X. M. *et al.* Enhancement of flux pinning of TFA-MOD YBCO thin films by embedded nanoscale Y_2O_3. *Supercond. Sci. Technol.* **19,** 844–847 (2006).
90. Solovyov, V. F. *et al.* High critical currents by isotropic magnetic-flux-pinning centres in a 3-μm-thick $YBa_2Cu_3O_7$ superconducting coated conductor. *Supercond. Sci. Technol.* **20,** L20-L23 (2007).
91. Song, X. *et al.* Evidence for strong flux pinning by small dense nanoprecipitates in a Sm-doped $YBa_2Cu_3O_{7-\delta}$ coated conductor. *Appl. Phys. Lett.* **88,** 212508 (2006).
92. Zhang, W. *et al.* Control of flux pinning in MOD YBCO coated conductor. *IEEE Trans. Appl. Supercond.* (in the press).
93. Sudoh, K., Yoshida, Y. & Taki, Y. Effect of deposition conditions and solid solution on the $Sm_{1+x}Ba_{2-x}Cu_3O_{6+\delta}$ thin films prepared by pulsed laser deposition. *Physica C* **384,** 178–184 (2003).
94. Yoshida, Y. *et al.* High-critical-current-density epitaxial films of $SmBa_2Cu_3O_{7-x}$ in high fields. *Jpn. J. Appl. Phys.* **44,** L129–L132 (2005).
95. Yoshida, Y. *et al.* High-critical-current-density $SmBa_2Cu_3O_{7-x}$ films induced by surface nanoparticle. *Jpn. J. Appl. Phys.* **44,** L546–L548 (2005).
96. Maiorov, B & Civale, L. in *Flux pinning and ac loss studies on YBCO coated conductors* (eds. Paranthaman, M. P. & Selvamanickam, V.) (Nova Science, in the press).
97. Selvamanickam, V., Xie, Y., Reeves, J. & Chen, Y. MOCVD-based YBCO-coated conductors. *Mater. Res. Soc. Bull.* **29,** 579–582 (2004).
98. Holesinger, T. *et al.* Progress in nano-engineered microstructures for tunable high-current, high temperature superconducting wires. *Adv. Mater.* (in the press).
99. Hanisch, J., Cai, C., Huhne, R., Schultz, L. & Holzapfel, B. Formation of nanosized $BaIrO_3$ precipitates and their contribution to flux pinning in Ir-doped $YBa_2Cu_3O_{7-\delta}$ quasi-multilayers. *Appl. Phys. Lett.* **86,** 122508 (2005).
100. Varanasi, C. V. *et al.* Flux pinning enhancement in $YBa_2Cu_3O_{7-x}$ films with $BaSnO_3$ nanoparticles. *Supercond. Sci. Technol.* **19,** L37–L41 (2006).
101. Kobayashi, H. *et al.* Investigation of magnetic properties of YBCO film with artificial pinning centers on PLD/IBAD metal substrate. *Physica C* **445–448,** 625–627 (2006).
102. Kang, S. *et al.* High-performance high-T_c superconducting wires. *Science* **311,** 1911–1914 (2006).
103. Cai, C., Holzapfel, B., Hanisch, J. & Schultz, L. Direct evidence for tailorable flux-pinning force and its anisotropy in $RBa_2Cu_3O_{7-\delta}$ multilayers. *Phys. Rev. B* **70,** 212501 (2004).
104. Miller, D. J., Holesinger, T. G., Feldmann, D. M. & Rupich, M. W. in *US Department of Energy Superconductivity for Electric Systems Annual Peer Review* (Arlington, Virginia, 2006); available at <http://www.energetics.com/meetings/supercon06/agenda>.
105. Selvamanickam, V, Yie, Y. & Reeves, J. in *US Department of Energy Superconductivity for Electric Systems Annual Peer Review* (Arlington, Virginia, 2006); available at <http://www.energetics.com/meetings/supercon06/agenda>.
106. Kang, S., Leonard, K. J., Martin, P. M., Li, J. & Goyal, A. Strong enhancement of flux pinning in $YBa_2Cu_3O_{7-\delta}$ multilayers with columnar defects comprised of self-assembled $BaZrO_3$ nanodots. *Supercond. Sci. Technol.* **20,** 11–15 (2007).

High-T_c superconducting materials for electric power applications

David Larbalestier, Alex Gurevich, D. Matthew Feldmann & Anatoly Polyanskii

Applied Superconductivity Center, Department of Materials Science and Engineering, Department of Physics, University of Wisconsin, Madison, Wisconsin 53706 USA (e-mail: larbalestier@engr.wisc.edu)

Large-scale superconducting electric devices for power industry depend critically on wires with high critical current densities at temperatures where cryogenic losses are tolerable. This restricts choice to two high-temperature cuprate superconductors, $(Bi,Pb)_2Sr_2Ca_2Cu_3O_x$ and $YBa_2Cu_3O_x$, and possibly to MgB_2, recently discovered to superconduct at 39 K. Crystal structure and material anisotropy place fundamental restrictions on their properties, especially in polycrystalline form. So far, power applications have followed a largely empirical, twin-track approach of conductor development and construction of prototype devices. The feasibility of superconducting power cables, magnetic energy-storage devices, transformers, fault current limiters and motors, largely using $(Bi,Pb)_2Sr_2Ca_2Cu_3O_x$ conductor, is proven. Widespread applications now depend significantly on cost-effective resolution of fundamental materials and fabrication issues, which control the production of low-cost, high-performance conductors of these remarkable compounds.

The potential applications of superconductors — Kamerlingh Onnes's name for materials that lose electric resistance on cooling below a transition temperature, T_c — were apparent to Onnes almost immediately. In 1913, just two years after his discovery, Onnes talked in Chicago about the design of very powerful magnets far exceeding the fields achievable by iron; these would cost as much as a battleship if made from copper and cooled with liquid air, but be affordable if made from superconducting wires. By that time he had already tested a Ni alloy coated with Pb-rich superconducting solder, but this lost superconductivity at fields of less than 50 mT. He ascribed this unexpected setback to bad places in the wire, a problem he anticipated would soon be fixed without difficulty!

In fact, applications had to wait 50 more years, particularly because the physics of superconductivity in magnetic fields were seriously misunderstood[1]. This need not have been so, because London and Shubnikov made important breakthroughs in understanding the magnetic properties of superconductors in the 1930s. By careful alloying experiments, Shubnikov pointed out the vital distinction between type I superconductors, in which currents flow only at the surface and superconductivity is destroyed by weak fields, as in Onnes' 1913 experiment, and a new type of super conductor, now called a type II superconductor, capable of carrying bulk supercurrent at high fields. The key understanding that the behaviour of type II superconductors is due to quantized magnetic vortices was achieved by Abrikosov in the 1950s.

In spite of these theoretical insights, it was not until 1961 that Kunzler *et al.*[2] showed that high-field applications really were possible. Drawing Sn inside a Nb wire and reacting it to form the brittle intermetallic Nb_3Sn, they reached a current density $J > 10^5$ A cm^{-2} at 8.8 T and 4.2 K. This astounding result demonstrated that superconductivity could indeed exist in very high magnetic fields. In fact, Onnes's vision for a 10 T, high-field magnet has been abundantly fulfilled. Laboratory superconducting magnets exist by their thousands, some producing fields exceeding 20 T. Fermilab, Brookhaven, DESY and CERN all have accelerators

composed of kilometres of superconducting bending and focusing magnets (see, for example, the proceedings of the biennial Applied Superconductivity Conferences published in odd-year volumes of *IEEE Trans. Magn.* up to 1989, and *IEEE Trans. Appl. Supercond.* since 1991). The medical technique of magnetic resonance imaging was developed using very homogeneous, persistent-mode superconducting magnets, a business now exceeding US$3 billion per year.

All of these applications are based on two low-temperature superconducting (LTS) materials, Nb-Ti alloy ($T_c = 9$ K) or Nb_3Sn ($T_c = 18$ K). For such applications, superconductivity is a true enabling technology and the use of liquid helium or helium-driven refrigerators is immaterial. But helium cooling tends to be too expensive for replacement of conventional electrotechnology based on Cu and Fe, two cheap and well understood materials. The first period of superconducting applications, from approximately 1962 to 1986, explored various power applications but found them too complex and expensive, thus relegating superconductivity to new high-technology and medical applications where no competitors existed. By 1986, the highest-T_c compounds possessed the A15 structure, the lightest of which, Nb_3Ge, has a T_c of 23 K (ref. 3), 5-K higher than for Nb_3Sn.

Prospects changed dramatically in late 1986, when Bednorz and Muller[4] discovered superconductivity at 30 K in the layered cuprate, $LaBa_2CuO_{4-x}$. In early 1987, superconductivity in $YBa_2Cu_3O_x$ (YBCO) at 92 K was announced[5], well above the boiling point of liquid nitrogen (77 K). Soon T_c rose to more than 130 K (ref. 6) in $Hg_2Ba_2Ca_2Cu_3O_x$. Widespread replacement of Cu and Fe by these new high-temperature superconductors (HTSs) was broadly discussed and significant public and private programmes to build electric utility superconducting devices were put in place in Japan, Europe and the United States[7]. Major components of the generation, transmission (power cables and devices for superconducting magnetic energy storage), distribution (transformers and fault current limiters) and end-use (motor) devices have been built, primarily using the $(Bi,Pb)_2Sr_2Ca_2Cu_3O_x$ (Bi-2223) conductor[7].

Superconductivity can have a significant role in deregulated electricity markets and in lessening CO_2 emissions and other environmental impacts, but significant market penetration of HTS devices requires HTS wires that fully exploit their fundamental current-carrying capability. The announcement by Akimitsu in January 2001 that the binary compound MgB_2 superconducts at up to 39 K (ref. 8) has generated new interest in superconductors for power applications, owing to the cheap abundance of Mg and B and the potential of analogous compounds.

Conductor requirements for power technology

A conductor is more than just the superconductor. Conductors for power applications are multifilamentary wires or tapes in which many superconducting filaments are embedded in a matrix of a normal metal, such as Cu or Ag, which provides protection against magnetic flux jumps and thermal quenching[9] (Fig. 1). Such wires must have sufficient strength to withstand the fabrication process, device winding, cool-down and electromagnetic stresses, and be capable of being made or cabled to sufficient size to carry operating currents from hundreds to thousands of amperes at costs comparable to Cu. Estimates of the acceptable cost range from about US$10 to US$100 per kA m, the scale being set at its lower end by the cost for Cu (~$US15 per kA m) and at the upper end by higher-cost applications where superconductivity has advantages not possessed by present technology[7,10]. Such higher-cost applications include high power density underground power cables in inner cities, environmentally friendly, oil-free HTS transformers, or superconducting magnetic energy-storage devices for power networks where low reactance and instantaneous redistribution of power is vital[11].

These requirements define a parameter set that restricts present choice to Bi-2223 or YBCO. The combination of critical current density J_c, field and operating temperature is summarized in Table 1. Target superconductor current densities must attain 10^4–10^5 A cm^{-2} in fields of 0.1–10 T at temperatures of 20–77 K. The data in Table 1 represent an ongoing dialogue between the R&D and applied communities that tends to push up the operating temperature, current density and field so as to better separate a Cu and Fe technology limited to fields of 1–2 T from a higher-field, superconductor-based technology. Reliable and inexpensive refrigeration is a critical part of the HTS technology and significant advances have been made in this area too[12].

At present, the only HTS conductor in production is Ag-sheathed Bi-2223 with T_c ~108 K (its lower-T_c sibling $Bi_2Sr_2CaCu_2O_x$ has only a small application in the power sector). Figure 1 shows a cross-section of a Bi-2223 conductor of the type being used for the Detroit Edison 100-MVA power cable, in which multiple strands are wound to make a fully transposed 6,000-A cable operating at 70–75 K. The individual strand is 0.2 mm by 4 mm with a critical current I_c of the order of 125 A in zero field at 77 K. Underpinning conductor R&D programmes worldwide is the conviction that such conductors have neither the magnetic field performance for applications at liquid nitrogen temperatures (~65–80 K), nor sufficiently low costs. The potential lower-cost alternative is a biaxially textured YBCO conductor or possibly MgB_2 or an analogous compound[13]. We now explore the underlying materials issues that define these beliefs.

Superconducting properties, crystal structure and anisotropy

Figure 2 presents the important magnetic field–temperature (H–T) phase diagram for the three actual (Nb-Ti, Nb_3Sn and Bi-2223) and

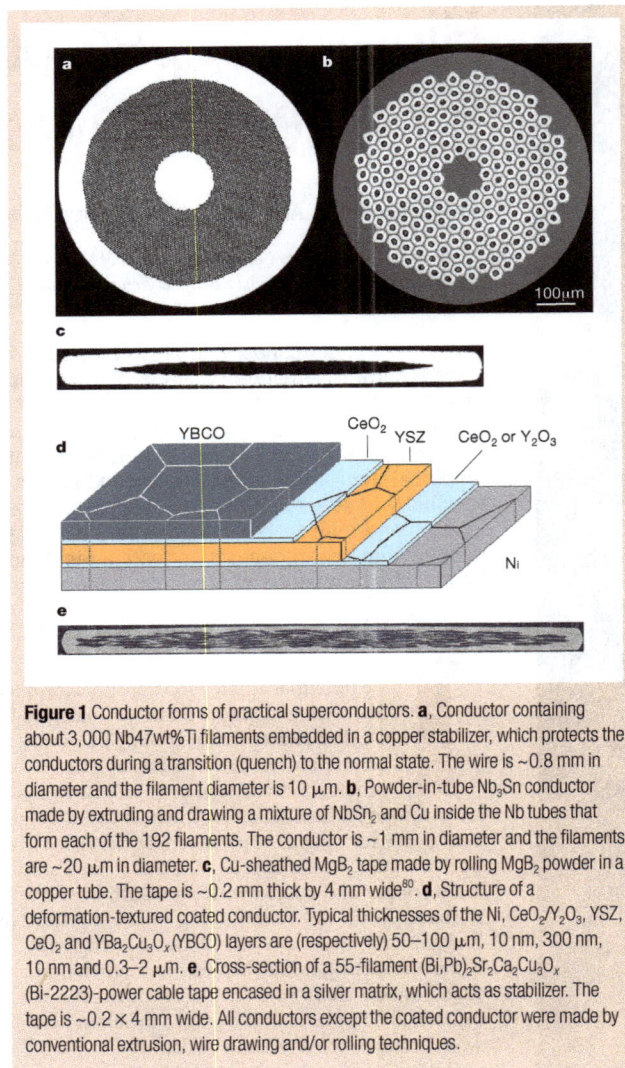

Figure 1 Conductor forms of practical superconductors. **a**, Conductor containing about 3,000 Nb47wt%Ti filaments embedded in a copper stabilizer, which protects the conductors during a transition (quench) to the normal state. The wire is ~0.8 mm in diameter and the filament diameter is 10 μm. **b**, Powder-in-tube Nb_3Sn conductor made by extruding and drawing a mixture of $NbSn_2$ and Cu inside the Nb tubes that form each of the 192 filaments. The conductor is ~1 mm in diameter and the filaments are ~20 μm in diameter. **c**, Cu-sheathed MgB_2 tape made by rolling MgB_2 powder in a copper tube. The tape is ~0.2 mm thick by 4 mm wide[80]. **d**, Structure of a deformation-textured coated conductor. Typical thicknesses of the Ni, CeO_2/Y_2O_3, YSZ, CeO_2 and $YBa_2Cu_3O_x$ (YBCO) layers are (respectively) 50–100 μm, 10 nm, 300 nm, 10 nm and 0.3–2 μm. **e**, Cross-section of a 55-filament $(Bi,Pb)_2Sr_2Ca_2Cu_3O_x$ (Bi-2223)-power cable tape encased in a silver matrix, which acts as stabilizer. The tape is ~0.2 × 4 mm wide. All conductors except the coated conductor were made by conventional extrusion, wire drawing and/or rolling techniques.

two potential (YBCO and MgB_2) conductor materials. Their different phase diagrams result from their distinctly different physical parameters and crystal structures, as shown in Fig. 3 and Table 2. All five are type II superconductors for which bulk superconductivity exists up to an upper critical field $H_{c2}(T)$, which can exceed 100 T for Bi-2223 and YBCO. In fact, applications are limited by a lower characteristic field, the irreversibility field $H^*(T)$ at which J_c vanishes. For the isotropic cubic superconductors, Nb-Ti and Nb_3Sn, $H^*(T)$ is about $0.85H_{c2}(T)$ (refs 14,15). All three higher-T_c compounds have anisotropic layered structures, which result in significant anisotropy of the upper critical field, $\eta = H_{c2}^{\parallel}(T)/H_{c2}^{\perp}(T)$, parallel and perpendicular to the superconducting layers. Here η has values of 1 for Nb-Ti and Nb_3Sn, 2–3 for MgB_2 (refs 16,17), 5–7 for YBCO, and

Table 1 Industry consensus wire performance requirements for various utility device applications

Application	J_c (A cm^{-2})	Field (T)	Temperature (K)	I_c (A)	Wire length (m)	Strain (%)	Bend radius (m)	Cost (US$ per kA m)
Fault current limiter	10^4–10^5	0.1–3	20–77	10^3–10^4	1,000	0.2	0.1	10–100
Large motor	10^5	4-5	20–77	500	1,000	0.2–0.3	0.05	10
Generator	10^5	4–5	20–50	>1,000	1,000	0.2	0.1	10
SMES*	10^5	5–10	20–77	~10^4	1,000	0.2	1	10
Transmission cable	10^4–10^5	<0.2	65–77	100 per strand	100	0.4	2 (cable)	10–100
Transformer	10^5	0.1–0.5	65–77	10^2–10^3	1,000	0.2	1	<10

*SMES, superconducting magnetic-energy storage.
Data supplied by R. Blaugher.

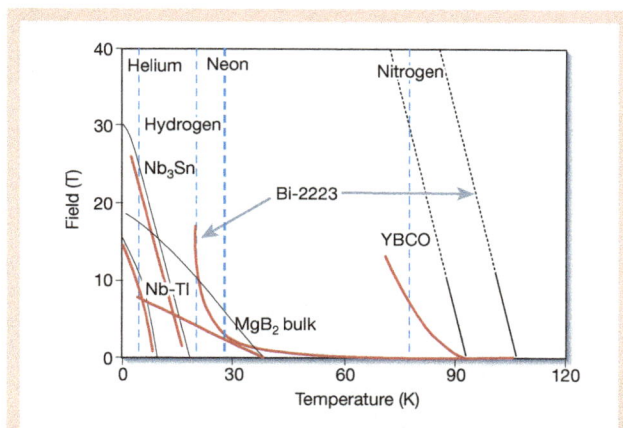

Figure 2 Magnetic field–temperature diagram for Nb47wt%Ti, Nb_3Sn, MgB_2, Bi-2223 and YBCO. The upper critical field H_{c2} at which bulk superconductivity is destroyed is indicated in black, while the irreversibility field H^* at which the bulk critical current density goes to zero is indicated in red. It can be seen that $H^*(T)$ is close to $H_{c2}(T)$ for Nb-Ti and Nb_3Sn, about half of $H_{c2}(T)$ for MgB_2, and much lower than $H^*(T)$ for YBCO and Bi-2223. There is not much scope for using Bi-2223 at 77 K because its irreversibility field (~0.3 T) is so low. As a result, applications at 77 K are restricted mostly to power cables for which the self-field is well below H^*(77K).

50–200 for Bi-2223 (ref. 18). In all cases, H_{c2} and H^* are lower for field applied parallel to the c-axis and perpendicular to the a–b planes. Figure 2 shows that strongly anisotropic Bi-2223 exhibits an enormous suppression of H^*(77K) to the very low value of ~0.2 T, well below H_{c2}(77K), which is of order 50 T for H parallel to the c-axis[19]. This low irreversibility field prevents use of Bi-2223 at 77 K in any significant field (although power cables are possible) and provides one of the key arguments for developing a second-generation HTS technology based on YBCO, for which H^*(77K) ~ 7 T.

The structural origin of these important superconducting anisotropies is illustrated in Fig. 3 and Table 2. Nb-Ti, normally Nb47-50wt%Ti, is a body-centred cubic, solid-solution alloy with short electron mean-free-path, high resistivity and coherence length ξ(4K) = 5 nm. Nb_3Sn has ξ(4K) = 3 nm and possesses the cubic A15 crystal structure, in which three orthogonal chains of Nb atoms separate the Sn atoms lying at cube corners and cube centre. MgB_2 has the hexagonal space group $P6/mmm$ composed of alternating B and Mg sheets, the B being nested in the Mg interstices[8]. It can form as a low-resistivity (ρ(40K) = 0.4 $\mu\Omega$ cm), perhaps perfectly stoichiometric compound[20] for which $\xi(4K)^\perp$ = 6.5 nm and $\xi(4K)^\parallel$ = 2.5 nm, where the perpendicular direction along the c-axis is taken with respect to the B planes in MgB_2 and the CuO_2 planes in HTSs. YBCO is a complex, layered perovskite centred on a Y layer, around which are stacked the CuO_2 plane of strong superconductivity and a double charge-reservoir layer of O-Ba-O and O-Cu. In Bi-2223, the charge-reservoir layer consists of a double (Bi,Pb)-O and its neighbouring Sr-O layers, the superconducting layers being a stack of three CuO_2 layers interleaved by two Ca layers. The anisotropy of Bi-2223 is so

large that H^*(77K) is only 0.2–0.3 T, restricting 77-K applications to low self-field uses such as power cables, even though H_{c2}(77K) exceeds 10 T.

Flux pinning and the critical current density

Under equilibrium conditions, magnetic flux penetrates the bulk of a type II superconductor above the lower critical field H_{c1}, which is of order 10-20 mT for the materials under consideration. Over most of the available H–T space, $H > H_{c1}$, this magnetic flux exists as a lattice of quantized line vortices or fluxons[18,21]. Each fluxon is a tube of radius of the London penetration depth $\lambda(T)$, in which superconducting screening currents circulate around a small non-superconducting core of radius $\xi(T)$, where $\xi(T)$ is the superconducting coherence length. The flux carried by the screening currents in each fluxon equals the flux quantum $\phi_0 = 2 \times 10^{-15}$ Wb. Bulk superconductivity is destroyed when the normal cores overlap at the upper critical field, $H_{c2}(T) = \phi_0/2\pi\mu_0\xi(T)^2$. In isotropic materials such as Nb-Ti and Nb_3Sn, vortex lines are continuous, but the weak superconductivity of the blocking layers of HTS compounds produces a stack of weakly coupled 'pancake' vortices whose circulating screening currents are mostly confined within the superconducting CuO_2 planes[18].

Superconductors can carry bulk current density only if there is a macroscopic fluxon density gradient[22], defined by the Maxwell equation $\nabla \times \mathbf{B} = \mu_0\mathbf{J}$. This gradient can be sustained only by pinning the vortices (flux pinning) at microstructural defects. Increasing T and H weaken the potential wells at which vortices are pinned, thus lessening $H^*(T)$ and $J_c(T,H)$. Flux pinning is determined by spatial perturbations of the free energy of the vortex lines due to local interactions of their normal cores and screening currents with these microstructural imperfections[22]. In addition, the fluxon structure is subjected to the Lorentz force $\mathbf{F}_L = \mathbf{J} \times \mathbf{B}$ of the macroscopic current. The critical current density $J_c(T,H)$ is then defined by the balance of the pinning and Lorentz forces, $F_p = F_L$, where F_p is the volume summation over all microstructural pinning defects in the strongly interacting network of flux lines. Ideally, a type II superconductor can carry any non-dissipative current density J smaller than J_c. When J exceeds J_c, a superconductor switches into a dissipative, vortex-flow state driven by the Lorentz force.

This description of flux pinning immediately suggests tailoring the defect structure of the conductor to maximize J_c. In fact, this does not correspond well to the essentially empirical way in which conductors have been developed, especially those of Bi-2223 and YBCO, for which the important flux-pinning defects are largely unknown. In practice, it is the critical current I_c of a wire of cross-section A that is measured, normally at the finite electric field of 1 μV cm^{-1}. The physics of vortex dynamics and pinning is enormously interesting[18,22], but equally important in practical terms is the fact that the actual cross-section over which transport currents flow in HTS conductors is much less than the whole[23]. What has greatly held back the development of Bi-2223 conductors in particular is uncertainty about whether the long-range utilization of the current-carrying cross-section of state-of-the-art conductors is 5, 25 or 50%. It is certain that it is not 100%, for reasons discussed later. Irrespective of such percolative current-flow effects, the upper limit to the critical current is set by

Table 2 Basic material and critical current density relevant parameters for practical superconductors

Material	Crystal structure	Anisotropy	T_c (K)	H_{c2}	H^*	In-plane coherence length $\xi(0)$ (nm)	In-plane penetration depth $\lambda(0)$ (nm)	Depairing current density (A cm^{-2}), 4.2 K	Critical current density (A cm^{-2})	$\rho(T_c)$ ($\mu\Omega$ cm)
Nb47wt%Ti	Body-centred cubic	Negligible	9	12 T (4 K)	10.5 T (4 K)	4	240	3.6×10^7	4×10^5 (5 T)	60
Nb_3Sn	A15 cubic	Negligible	18	27 T (4 K)	24 T (4 K)	3	65	7.7×10^7	~10^6	5
MgB_2	$P6/mmm$ hexagonal	2–2.7	39	15 T (4 K)	8 T (4 K)	6.5	140	7.7×10^7	~10^6	0.4
YBCO	Orthorhombic layered perovskite	7	92	>100 T (4 K)	5–7 T (77 K)	1.5	150	3×10^8	~10^7	~40–60
Bi-2223	Tetragonal layered perovskite	~50–100	108	>100 T (4 K)	~0.2 T (77 K)	1.5	150	3×10^8	~10^6	~150–800

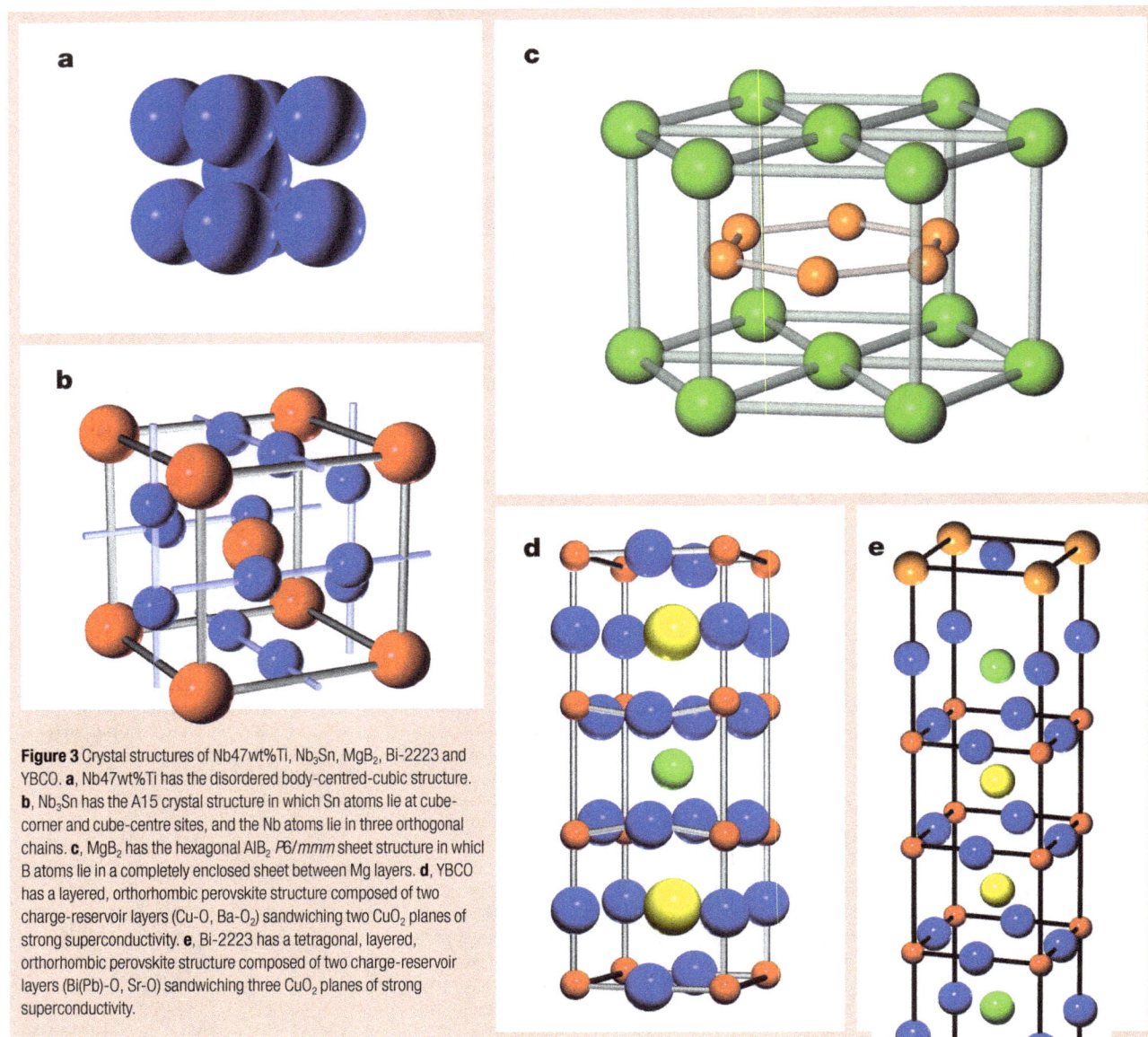

Figure 3 Crystal structures of Nb47wt%Ti, Nb_3Sn, MgB_2, Bi-2223 and YBCO. **a**, Nb47wt%Ti has the disordered body-centred-cubic structure. **b**, Nb_3Sn has the A15 crystal structure in which Sn atoms lie at cube-corner and cube-centre sites, and the Nb atoms lie in three orthogonal chains. **c**, MgB_2 has the hexagonal AlB_2 $P6/mmm$ sheet structure in which B atoms lie in a completely enclosed sheet between Mg layers. **d**, YBCO has a layered, orthorhombic perovskite structure composed of two charge-reservoir layers (Cu-O, Ba-O_2) sandwiching two CuO_2 planes of strong superconductivity. **e**, Bi-2223 has a tetragonal, layered, orthorhombic perovskite structure composed of two charge-reservoir layers (Bi(Pb)-O, Sr-O) sandwiching three CuO_2 planes of strong superconductivity.

the flux-pinning current density, which cannot exceed the depairing current density $J_d = \phi_0/[3/(\sqrt{3})\pi\mu_0\lambda^2(T)\xi(T)]$, the maximum supercurrent density circulating near the vortex cores.

The best information that we possess about the limits to flux pinning in practical superconductors has come from extensive studies of Nb-Ti[24]. Exceptionally strong pinning by a dense, ~20–25-vol% lamellar structure of normal (that is, non-superconducting) α-Ti ribbons, about 1 nm (0.2ξ) thick and aligned parallel to the wire transport current, can produce a J_c that approaches 5–10% of J_d at zero field and 4.2 K. The flux-pinning mechanism is also known for Nb_3Sn, in which J_c is determined by the magnetic interaction of the fluxon currents with grain boundaries[25,26]. In this case, $J_c(T,H)$ increases with decreasing grain size, thus encouraging low-temperature methods of phase formation that produce a fine grain size[27]. There are also indications of grain-boundary pinning in MgB_2 (refs 28–30). All three materials possess zero-field J_c values, which can exceed 1 MA cm^{-2} at 4.2 K.

The important flux-pinning interactions in conductor forms of YBCO and Bi-2223 are largely unknown, partly because the coherence length is so small that even atomic-sized point defects can pin fluxons. The critical temperature of HTSs is extremely sensitive to the carrier (hole) density, which is in turn determined by local oxygen (and other) non-stoichiometries, so that even a weak hole-depletion at crystalline defects can locally drive an HTS into an antiferromagnetic insulator. The proximity of the superconducting state to the metal–insulator transition, the d-wave pairing symmetry, the very short in-plane coherence length ($\xi(0) = 1.5$–2 nm) and the much longer Debye screening length ($l_D \sim \xi(0)$) all combine to produce significant suppression of the superconducting order parameter $\Delta(\mathbf{r})$ near crystal defects such as impurities, dislocations and grain boundaries[31], thus making them effective pinning centres. In fact, recent scanning tunnelling microscopy of Bi-2212 single crystals[32] has revealed significant variations of $\Delta(\mathbf{r})$ near single-atom impurities (Zn and Ni), as well as striking intrinsic variations of $\Delta(\mathbf{r})$ on scales of 1–2 nm even in pure crystals[33].

This extreme sensitivity of HTSs to nanoscale defects, along with the intrinsic spatial fluctuations of $\Delta(\mathbf{r})$ from 25 meV to 60 meV on the scale of ξ or less[33], provide effective core pinning of vortices. There is strong experimental evidence that point defects such as Zn and Ni impurities[32] or oxygen vacancies[34], or line defects such as edge[35] and

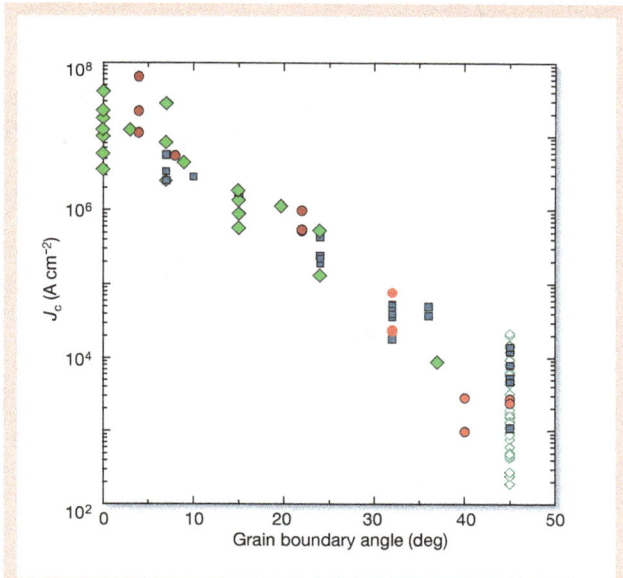

Figure 4 Transport critical current density at 4.2 K measured in thin films of YBCO grown on [001] tilt bicrystal substrates of SrTiO₃ (squares[37] and filled diamonds[44]), Y₂O₃-stabilized ZrO₂ (circles[97]), and bi-epitaxial junctions (open diamonds[98]) of varying misorientation angle θ. Data of refs 97 and 44 were taken at 77 K and have been multiplied by a factor of 10.9 to make them comparable with the data at 4.2 K. Despite a significant scattering, $J_c(\theta)$ exhibits a universal exponential dependence on θ. Data compilation courtesy of J. Mannhart[37].

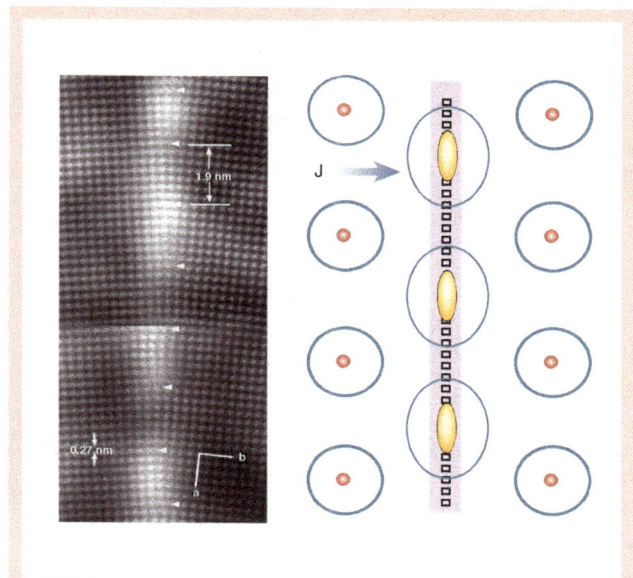

Figure 5 Grain boundary structure and its effect on vortex properties. The left panel is a Fourier-filtered transmission electron micrograph of an 8° [001] tilt Bi₂Sr₂CaCu₂Oₓ bicrystal, showing the edge dislocation structure at the grain boundary and the channels of good conduction that lie between their cores. Stacking faults are indicated by the white regions between alternate pairs of dislocations. Strong strains are found at the core of each dislocation. On the right is a schematic illustration of the consequences of the dislocation structure for vortices at the boundary. Abrikosov vortices are indicated within the grains, but their cores become elongated when they sit at the boundary, owing to the reduced current transparency of the boundary. The grain-boundary dislocations are indicated by the black squares and the carrier-depleted zone by the pink-shaded region. Vortices at the grain boundary acquire a mixed Abrikosov–Josephson character, which strongly reduces their pinning force along the boundary.

screw[36] dislocations, can effectively pin pancake vortices in layered HTSs. Thus, all forms of HTSs possess crystalline defects of different kinds, which provide strong flux pinning at low temperatures. Consistent with this, YBCO and Bi-2223 films[37,38] can exhibit $J_c(4K)$ values of more than 10 MA cm⁻² and single crystals of YBCO (ref. 39) can exhibit values up to 1 MA cm⁻². Although these J_c values are about an order of magnitude lower at 77 K, they are more than sufficient for power applications (Table 1). Moreover, they are still much lower than the depairing limit: $J_d \approx 300$ MA cm⁻² at 4 K for characteristic values $\lambda(0) \approx 150$ nm and $\xi(0) \approx 1.5$ nm. Thus, unlike Nb–Ti where the material is rather close to its performance limit, there is enormous untapped potential for improved HTS material performance.

The main obstacle for HTS conductors is that their current-carrying capability deteriorates rapidly with increasing temperature and magnetic field. As shown in Fig. 2, J_c vanishes at the irreversibility field $H^*(T)$, which is far below $H_{c2}(T)$ at 77 K. An independent, significant limit on J_c is current blockage by grain boundaries, as described in the next section. But the underlying cause of the suppression of $H^*(T)$ at higher temperatures is due to the layered structure of HTSs, which greatly facilitates depinning of stacks of weakly coupled pancake vortices by thermal fluctuations[15,18]. For $H < H^*(T)$, vortex fluctuations cause thermally activated creep of magnetic flux[40], producing measurable dissipation well below J_c defined at 1 μV cm⁻¹. Anisotropy strongly enhances the effect of thermal vortex fluctuations[18], thus suppressing H^* and enhancing flux creep in Bi-2223 much more than in YBCO. Both H^* and J_c can be significantly improved by irradiation, which creates effective columnar pinning tracks[41]. For example, neutron irradiation of U-doped Bi-2223 tapes fissions the U and produces heavy-ion tracks that raise $H^*(77K)$ to over 1 T, while also reducing the anisotropy of J_c (ref. 42).

Grain boundaries

Large devices need kilometres of polycrystalline conductors and the influence of grain boundaries is therefore critical. The most unambiguous evidence that grain boundaries are strong barriers to current

flow in HTSs comes from detailed studies of [001] tilt YBCO bicrystals on SrTiO₃ or Y₂O₃-stabilized ZrO₂ (YSZ) substrates. These have shown that the critical current density J_b across the grain boundary drops exponentially below that of the grains, $J_b = J_0 \exp(-\theta/\theta_c)$, as a function of the misorientation angle θ between the neighbouring crystallites[37,43–45], where $\theta_c \approx 2$–5°, depending on the value of the intragrain J_c (Fig. 4). This extreme sensitivity to misorientation, coupled with the intrinsic anisotropy of the HTS compounds, dictates the need to texture conductors into the tape forms shown in Fig. 1, so as to shift the grain-boundary misorientation distribution to as small a value of θ as possible. For Bi-2223, the rolling deformation used to make the tape produces a marked uniaxial, c-axis texture, whereas for YBCO-coated conductors, the essential goal is to make both in- and out-of-plane texture so that the tape behaves as a quasi-single crystal, even though the YBCO grain size is <1 μm in diameter[46,47].

This behaviour of HTS materials is in strong contrast to metallic LTS and MgB₂, in which grain boundaries are not only transparent to current, but also significantly contribute to flux pinning, increasing the overall J_c as the grain size decreases[26,28,48]. Although Table 2 might suggest that HTS grain boundaries are weak links because of their short coherence length, in fact a value of $\xi(77K)$ of ~4 nm for YBCO is actually longer than for Nb₃Sn at 4 K. Moreover, low carrier density makes grain boundaries in (Ba,Pb)BiO₃ also weak-linked, even though $\xi(4K) \sim 7$ nm for this compound[49,50]. In fact, the weak-link (that is, current-obstructing) behaviour of grain boundaries in HTSs results from the same factors, which otherwise enhance the pinning of vortices by point defects, namely the strong dependence of T_c on hole concentration and low carrier density. A tilt grain boundary is built of a chain of edge dislocations spaced by $d = b/2\sin(\theta/2)$, where

$b \approx 0.4$ nm is the Burgers vector. As the misorientation angle θ increases, the spacing between the insulating dislocation cores decreases, becoming of the order of the coherence length ξ at $\theta \approx 5$–$7°$. For higher angles, a grain boundary becomes a Josephson weak link, because of the local suppression of the order parameter in the current channels between the dislocations, and J_b decreases exponentially with increasing θ[31]. Extensive electron microscopy has established that the hole content is depressed in the vicinity of grain boundaries and that there is only partial occupancy of cations in the structural units that form the boundary[51,52].

Grain boundaries possess extra ionic charge, whose magnitude increases with increasing θ and causes hole depletion in a layer of thickness given by the Thomas-Fermi screening length $l_D \approx \xi(0)$ (Fig. 5). As a result, the grain boundary is in a locally underdoped state, the local order parameter suppression increasing as θ increases[31,52]. That is why local overdoping by partially substituting Ca for Y in YBCO (refs 53,54) is effective. Adding extra carriers to the grain boundary with Ca-doping and reducing local strains can significantly increase J_b (by a factor of eight for a 24° [001] tilt boundary at 4.2 K and 0 T (ref. 53), and by a factor of three for a 7°-bicrystal at 55 K and 3 T (ref. 55)). Although overdoping ameliorates, rather than removes, the problem of current obstruction by grain boundaries in HTSs, it may certainly be useful for conductors[43–45,56,57], because the critical angle θ_c is only 2–5°.

Performance of HTS conductors in a magnetic field is determined by the nature of the vortices at the grain boundaries and their pinning interaction with structural defects, and by their magnetic interactions with strongly pinned vortices within the grains (Fig. 5). Because of the extended core of grain-boundary vortices, they are generally pinned more weakly than vortices in the grains (ref. 58, and Gurevich *et al.*, submitted), and grain boundaries become barriers to current flow because their critical current density $J_b(T,H)$ is smaller than $J_c(T,H)$ in the grains. This behaviour is clearly visible in the magneto-optical images of Fig. 6, which shows preferential flux penetration along a network of coated-conductor grain boundaries[56] having $\theta > 4°$.

The influence of misorientation is less certain for compounds other than YBCO, as all other HTS compounds are harder to grow as good films on bicrystal substrates. To first approximation, thin films of $Bi_2Sr_2CaCu_2O_x$ and Bi-2223 show similar exponential decrease of $J_b(\theta)$ (refs 59,60) to that seen for YBCO, suggesting a common mechanism for suppression of critical current on grain boundaries. Experiments on float-zone $Bi_2Sr_2CaCu_2O_x$ bicrystals indicate much weaker dependence of J_b on θ than for thin films[61], but the intragranular critical current densities are so small that they might not reach the critical current density of the grain boundaries. An extensive magneto-optical study of [001] tilt, thin-film YBCO bicrystals[62] has shown that the temperature dependences of J_c and J_b are significantly different.

Current percolation in polycrystals

Any conductor must be polycrystalline. All conductor materials except Nb-Ti are brittle, making cracks endemic in Nb_3Sn, MgB_2, Bi-2223 and YBCO. In addition, the last three compounds all are prone to porosity. The net result is that current percolates through a polycrystalline network containing many obstructions, some of which partially block the current while others result in a total block[23]. Thus the local fraction of current-carrying cross-section is significantly less than unity. Magneto-optical imaging is an effective method of visualizing non-uniformities of current flow in polycrystalline conductor forms, as shown in Fig. 7 for MgB_2, Bi-2223 and YBCO. These images show the 'roof' pattern of the perpendicular magnetic field, resulting from screening currents induced by an external field. For MgB_2 the finest, visible-scale variation seems to be controlled by ~10% porosity[63] and not by grain boundaries[17,20,64]. For the Bi-2223 tape, a transverse feathering of the image is caused

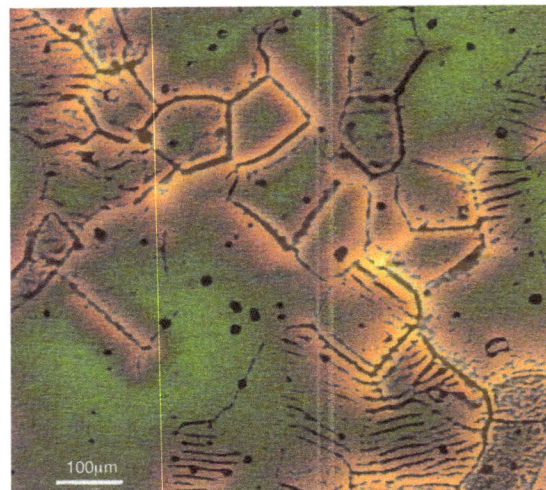

Figure 6 Magneto-optical image of the flux penetrating into a typical deformation-textured YBCO coated conductor overlaid on a light-microscope image of the underlying Ni substrate. Darker (green) areas are regions that are well connected electromagnetically, whereas the lighter (orange) flux network indicates where the local current density is reduced. Some grain boundaries in the Ni substrate (black) do not appear in the magneto-optical image. However, all grain boundaries with misorientations greater than 4° do appear[56].

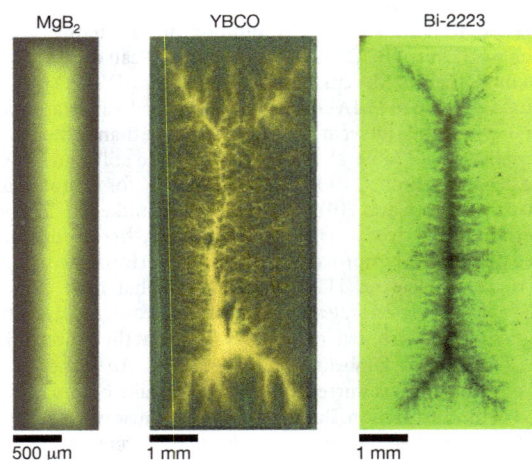

Figure 7 Magneto-optical images of (left to right) a sintered MgB_2 slab[63], a YBCO IBAD-coated conductor, and a Bi-2223 monocore tape[23]. Each tape shows a global roof-top pattern of the perpendicular component of magnetic field above the superconductor surface. This 'rooftop' pattern arises from induced magnetization currents circulating through the entire sample. The MgB_2 slab shows the most uniform current flow, even though it is imaged in an applied field of zero after cooling in 120 mT to 37 K, only 2 K below T_c. Almost no spatial variation of J_c is visible. The coated conductor was imaged after field-cooling in 60 mT to 11 K, well below its T_c of 90 K, then reducing the applied field to zero. The Bi-2223 tape was imaged at 11 K after cooling without field and then applying a filed of 120 mT. Some feathering of the image transverse to the tape length indicates a longitudinal variation of the current, probably associated with rolling defects.

374

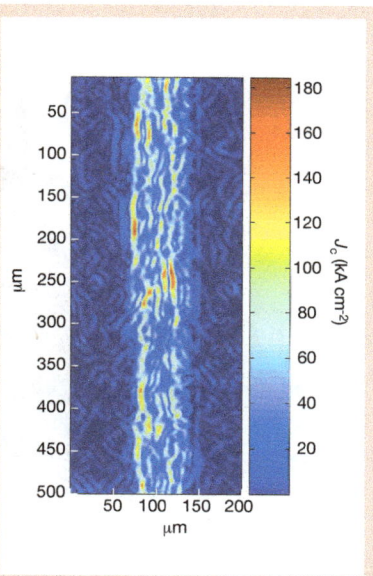

Figure 8 Colour map of the spatial distribution of the local critical current density in the same monocore Bi-2223 tape imaged in Fig. 7. The tape was imaged in a slab geometry produced by sectioning the tape along a centre line parallel to the tape length. The image from which the current reconstruction was performed was made by cooling the slab to 77 K in zero field, then applying a field of 36 mT parallel to the half width of the slab. The transport critical current density J_c(77K,0T) of the tape was 35 kA cm^{-2}, about the median of the distribution seen in the J_c map. Peak values of the J_c exceeded 180 kA cm^{-2}.

principally by quasiperiodic fluctuations of the superconductor thickness and residual rolling damage[23]. For the YBCO-coated conductor, there are multiple lines emanating from the central roof pattern, indicating local, higher current density loops superimposed upon the long-range current. Thus magneto-optical examination shows explicitly that there are multiple scales of current loops flowing in polycrystalline HTS samples. Such images pose the question: by how much is the average current density — defined by $J_c = I_c/A$, where A is the total cross-section, rather than the active cross-section $A_{effective}$ — reduced below the flux-pinning critical current density established within each grain?

To answer such a question requires good understanding of the intragrain critical current densities. In fact, the magnitude of the intragrain J_c of Bi-2223 remains unclear, because few high-quality bulk or thin-film single crystals have been made. Several thin-film results[38,60] suggest that J_c(77K,0T) can achieve 1 MA cm^{-2}, whereas a single-crystal study[65] gave J_c values significantly less than 10^5 A cm^{-2}. At 77 K and 0 T, Bi-2223 conductors can achieve up to ~75 kA cm^{-2} for small current conductors[66] and 40–50 kA cm^{-2} for high current conductors having I_c ~150 A (refs 19,67). Figure 8 shows the results of a recent magneto-optical current reconstruction[23] on a high-J_c monofilament conductor with J_c(77K,0T) = 35 kA cm^{-2}. Remark-

ably, J_c can achieve local values of 180 kA cm^{-2}, up to five times higher than the transport J_c. Although there is a tendency for the highest-J_c regions to be located at the Ag–superconductor interface, in fact very-high-J_c regions are located throughout the tape. Such local variability is not surprising for a seven-component system, because it is hard to control the phase-conversion reaction of $Bi_2Sr_2CaCu_2O_x$ and other constituents to Bi-2223. Images such as Fig. 8 offer the possibility of directly correlating the local microstructure to the local J_c. It is interesting to note that regions of J_c exceeding 100 kA cm^{-2} are 50–100 μm long, several times the ~20-μm grain length. Although earlier work[68] suggested that high-J_c regions were confined mainly to the Ag interface region, more recent results show that better processing is producing much more uniform J_c distribution across Bi-2223/Ag tapes, resulting in significant gains in tape performance.

State-of-the-art Bi-2223 conductors contain grains ~20 μm long by ~1 μm thick, distributed in 30–60 filaments, each of which is of the order of 5–10 μm thick by 200–300 μm wide. The current percolates in a three-dimensional manner through a polycrystalline array, which possesses a significant c-axis texture, of the order of 10–12° full-width at half-maximum (FWHM), but has little in-plane texture. By contrast, YBCO-coated conductors rely on epitaxy and are controlled mostly by the two-dimensional percolation[69,70]. Figure 9 images the local field produced by current flowing along a track in a representative coated conductor. An extended planar array of misoriented grains obstructs the current, forcing it to flow non-uniformly and causing local dissipation, even at $0.5I_c$, where, as usual, I_c is defined at the electric field criterion of 1 μV cm^{-1} established across the whole track. As I increases, the current-obstructing network expands, eventually becoming continuous across the whole conductor. Such images show that I_c is determined locally, with many regions of the sample showing no signs of dissipation even at $I = 1.5I_c$. A similar conductor had a global J_c(77K,0T) of 1.2 MA cm^{-2}, but individual clusters of YBCO grains had J_c values[57] exceeding 5 MA cm^{-2}. Thus, even in the biaxially textured coated conductors for which both in- and out-of-plane texture FWHM values are significantly below 10°, there is abundant evidence for current percolation around a network of obstructing grain boundaries and other defects.

Present J_c(77K,0T) values in 1–2-cm-long coated conductor prototypes[71–74] typically fall in the range 1–3 MA cm^{-2}. Further performance improvement may be impeded by electromagnetic and thermal instabilities, such as flux jumping, hot-spot formation and quench propagation, the same factors which limit the current-carrying capability of low-T_c superconductors[9,75]. Whatever future microstructure improvements occur, current-obstructing obstacles are inevitable in any practical long-length forms. Because the electromagnetic and thermal stability of the current-carrying state is

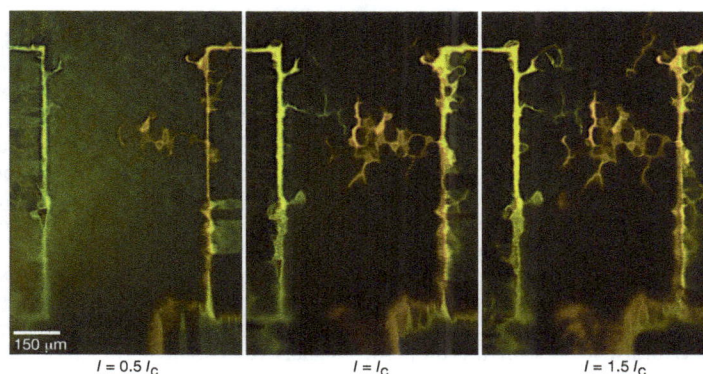

Figure 9 Magneto-optical images at 77 K of the self-field produced by an applied transport current in a RABiTS coated conductor sample for which the full-width J_c(77K,0T) was 0.7 MA cm^{-2}. A laser was used to cut a link restricting the current flow to a region 0.5 × 1.1 mm, which had a J_c(77K,0T) of 0.6 MA cm^{-2}. Images shown are at applied currents for the link of (left to right) $0.5I_c$, I_c, and $1.5I_c$, where I_c is defined at 1 μV cm^{-1}. A cluster of YBCO grain boundaries is visible in all the images, indicating where in the link J_c is limited well before the onset of bulk dissipation. As the current is increased, percolation near the visible cluster of grain boundaries becomes more pronounced, but above and below this cluster current flows more uniformly, and preferentially near the edges of the link. Even at $1.5I_c$, much of the link is supporting a current less than its local J_c. Patterning of bridges has shown that the intragrain J_c can reach 5 MA cm^{-2}, several times the J_c of this bridge. It is clear that current is percolating through a network of grain boundaries, which also result in localized dissipation in the track.

influenced by hot spots near obstacles, the architecture of coated conductors needs to be optimized so as to reduce local dissipation by current redistribution to the Ag stabilizer, and enhance heat transfer from the YBCO to the coolant. An important question is what is the acceptable size and concentration of obstructive hot spots, which provides stable conductor operation.

Because obstructions initiate dissipation, the spatial distribution of the electric field $E(x,y)$ near such obstructions controls the I_c of the conductor. But unlike current distributions, which can be extracted from magneto-optical images, the electric field distribution is much harder to measure. Recently, analytical methods for the calculation of $E(x,y)$, taking account of the highly nonlinear E–J characteristic of a superconductor, have been developed[76,77]. For $J < J_c$, the E–J relation is determined by thermally activated flux creep, which is conventionally approximated as $E = E_c(J/J_c)^n$, where $E_c \sim 1~\mu V~cm^{-1}$, and the exponent $n(T,H)$ varies from 20–30 for $H \ll H^*$ to $n \approx 1$ at $H = H^*$. Calculations for a planar defect perpendicular to current flow show that the nonlinearity of $E(J)$ strongly enhances the electric field in extended regions of length $L_\perp \sim na$ perpendicular to the current, on a scale much larger than the defect size a. This behaviour is in stark contrast to ohmic conductors, in which current flow becomes essentially uniform at a distance $\sim a$ away from the defect. These long-range disturbances of $E(x,y)$ become particularly important in a coated conductor, because even a small defect can produce a large hot spot that spans the entire conductor if $a > d/n$, where d is the sample width. If this rather weak condition is satisfied (typically, $n \sim 20$–30 in HTSs), then even small defects produce significant local excess dissipation. Figure 10 shows the calculated distribution of the electric field near a planar defect of length $a = 0.1d$ and $n = 20$. Even for such a small defect, the disturbance of $E(x,y)$ is much stronger than the applied field E_0 and extends all the way to the opposite side of the film[77].

Materials fabrication considerations

One of our fundamental conclusions is that both Bi-2223 and YBCO fulfil the essential requirements for wide applications of superconductivity in the electric utility network. Conductors of Bi-2223 (and $Bi_2Sr_2CaCu_2O_x$) are available today from several companies for service in 'real-world' utility sites. Cost and performance are not yet as good as are needed for widespread substitution of copper and iron, but significant special applications exist even at much higher costs. A key question for a cost-sensitive market for HTSs is to understand how much residual J_c performance capability there is in the materials themselves. As we have emphasized, the intragrain J_c capability of both Bi-2223 and YBCO is several times higher than realized in today's conductors. Material defects control the overall critical current density of present conductors and these defects are intimately tied to their particular fabrication processes. Although the specifics of particular fabrication processes are beyond the scope of this article, certain general points can be made to emphasize the decisive role that materials engineering now must play in developing this technology.

As Figure 1 shows, four of the five conductors are made by conventional composite metal-working techniques. A merit of conventional metal working is that there is little conceptual barrier to scale-up. Making longer wires is mostly a question of starting with larger pieces and using larger equipment. Both MgB_2 and Bi-2223 wires or tapes use essentially the same technology as that used for Nb_3Sn powder-in-tube composites. In fact, there is now no fundamental barrier to the production of a kilometre-long MgB_2 wire, and many groups have already demonstrated powder-in-tube conductors[30,78–82]. What is less clear is how much of a niche there is for MgB_2 to fill, because even in its high resistivity form which maximizes the upper critical field[17] it seems that the perpendicular irreversibility field is less than for Nb_3Sn. This is likely to restrict MgB_2 to applications of lower field, say 2-3 T, where the ability to operate at ~ 20 K is decisive economically[13]. By contrast, it has proven hard to scale up the

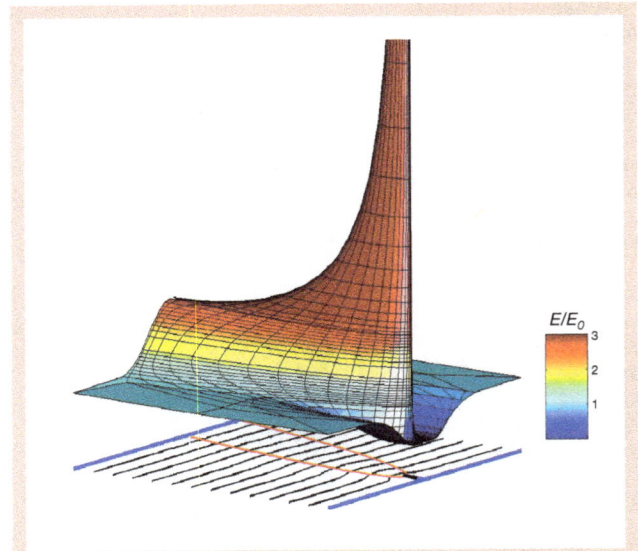

Figure 10 Spatial distribution of the electric field $E(x,y)$ for the power-law E–J characteristic, $E = E_c(J/J_c)^n$, near a planar defect of length a in a film of width d, $n = 20$ and $a = 0.1 d$, as calculated in refs 76,77. The lower part of the figure shows the current streamlines (black) and the boundaries (red) of the hot spot of strongly enhanced electric field lying in the film plane. It is clear that even a small defect produces the extended hot spot that effectively blocks the entire cross-section and can result in thermal instabilities[75].

YBCO-coated conductor to a genuinely continuous process, even after multiple proofs of principle have shown that an appropriate substrate–buffer–YBCO architecture can be made by physical vapour deposition and/or chemical methods[71–74,83]. The key issue here is to understand the defect populations that are characteristic of each fabrication method, and the capability and cost associated with reducing them to tolerable levels.

In assessing progress on HTS conductors, it is particularly relevant to recall that Nb_3Sn is now celebrating 40 years as a conductor and is still improving[84,85]. Bi-2223 has had 10 years of industrial development and is also still developing strongly in attaining higher current density, reducing cost and in becoming more mechanically robust[86–88]. The complexity of its microstructure and the interactions between fabrication and basic materials issues is slowly becoming clear[89], not least because multiple companies worldwide are selling the conductor in competition with each other. As Fig. 8 shows, the key issue is to increase the connectivity of the conductor. The prospect of two- to fivefold improvements in the performance of the present process, coupled to cost reductions triggered by significant scale-up and alternative routes such as overpressure processing[90] or melt processing[91], could accelerate the early penetration of superconducting technology into the utility network. At the same time, this would provide time and rationale for solving the complex issues associated with scale-up of the YBCO coated conductors.

Any manufacturing method for YBCO conductors relies on the epitaxial deposition of YBCO onto a textured template of one or more oxide buffer layers and a normal metal substrate. This template is usually created through introduction of texture either into the buffer layers by ion beam-assisted deposition (IBAD)[47], or into the metal substrate by deformation texturing using the rolling-assisted biaxially textured substrate (RABiTS)[46] approach. The IBAD method enables the use of strong, non-magnetic Ni-superalloy substrates on which an IBAD of aligned YSZ is deposited. This process can achieve a high degree of texture with a layer of YSZ that is $\sim 0.5~\mu m$ thick. But such a process is widely considered to be too slow to be commercial. IBAD of MgO produces good texture within the first 1-2 nm and thus may be rapid enough to be commercial[92], but the degree of texture in present IBAD-MgO-coated conductors is not yet as good as in IBAD-YSZ variants[93]. The inclined substrate deposition (ISD)[94] process is much quicker than IBAD, but the texture is significantly worse[95]. Ion-texturing processes for buffer-layer deposition[96] are receiving renewed attention owing to their simplicity, but their success depends on producing sufficient texture so to compete with the slower and more expensive IBAD process.

All of these methods of producing a textured buffer layer involve physical vapour-deposition processes, which some believe are inherently too expensive for copper and iron replacement[71]. An advantage of the RABiTS approach is that all buffer-layer and YBCO-coating steps can be performed with non-vacuum processes. A strong [100] cube texture is introduced into the substrate by conventional rolling and recrystallization[46]. Although the RABiTS approach has been developed mainly with pure Ni, alloyed substrate materials are being developed to increase the strength and reduce the magnetism of the Ni.

At this stage it is not at all clear which route to a coated conductor will win out. The design of a coated conductor is such that a substrate thickness of ~50 μm is needed to support 1–5 μm of YBCO, giving a superconductor fraction of ~5–10%, compared with the 25–50% for conductors made by metal-working techniques. Short samples of both IBAD and RABiTS conductors can both exceed 2 MA cm^{-2} for YBCO layers up to ~0.5 μm. But thicker YBCO layers tend to exhibit less epitaxy, more porosity and thus smaller J_c. Degradation of the J_c of the YBCO layer as it grows is a serious and poorly understood problem. The principal cause seems to be loss of epitaxy once the layer gets thicker than ~0.5 μm. A proper understanding is required, as virtually all growth methods except liquid-phase epitaxy seem to share this degradation.

Summary

The fundamental crystal and electronic structure of superconductors determines their structural anisotropy and superconducting properties. Generally, the higher the value of T_c, the higher the anisotropy, and the greater the upper critical field and sensitivity to nanoscale variations of the superconducting properties. This variation is positive in the case of point and line defects, because it can lead to strong flux-pinning and high critical current density. The low carrier density, short coherence length, d-wave pairing symmetry and proximity to the metal–insulator transition of HTSs makes interfaces such as grain boundaries significant barriers to current flow, except at very small misorientations. In the higher carrier density LTS materials such as Nb-Ti, Nb$_3$Sn, and MgB$_2$, grain boundaries act as beneficial flux-pinning sites without being barriers to current flow.

Anisotropy exerts two additional, important influences on conductor design, and causes significant suppression of H^* as compared to H_{c2}. MgB$_2$, YBCO and Bi-2223 all exhibit higher critical current density, irreversibility field and upper critical field when the magnetic field is applied parallel to their layered structure. But in practice, either because of the inevitable misalignments that are present in any real wire or because the magnetic field cannot always be arranged to be parallel to the layers, it is the poorer perpendicular properties that generally control performance. An additional burden for YBCO and Bi-2223 is that the grain-to-grain misalignment must be held to values of 5° or less if obstruction of current across the grain boundaries is to be minimized. Thus planar conductor forms, which confer texture, are preferred.

These fundamental materials issues restrict the fabrication technologies available for conductors. So far all three industrially available conductors, Nb-Ti, Nb$_3$Sn and Bi-2223, use composite metal-working techniques, which are relatively easy to scale up from laboratory to factory. It is already clear that this technique will work too for MgB$_2$. But it does not work for YBCO, for which a multilayer epitaxial technique is required. Many techniques using a variety of chemical and physical vapour-deposition techniques can be used to make short samples that demonstrate the concept of the coated conductor. At present the coated conductor R&D community is attempting to work out which fabrication techniques work cost-effectively in continuous processes. Broad and significant applications exist and await only the resolution of the vital materials issues that control development of the cheap, high-performance conductor-fabrication technology that underpins all superconducting applications. □

1. Berlincourt, T. G. Type-II superconductivity: quest for understanding. *IEEE Trans. Magn.* **23**, 403–412 (1987).
2. Kunzler, J. E., Buehler, E., Hsu, L. & Wernick, J. Superconductivity in Nb$_3$Sn at high current density in a magnetic field of 88 kgauss. *Phys. Rev. Lett.* **6**, 89–91 (1961).
3. Gavaler, J. Superconductivity in Nb-Ge films above 22 K. *Appl. Phys. Lett.* **23**, 480–482 (1973).
4. Bednorz, G. & Muller, K. A. Possible high-T$_c$ superconductivity in the Ba-La-Cu-O system. *Z. Phys. B* **64**, 189–193 (1986).
5. Wu, M. K. *et al.* Superconductivity at 93 K in an new mixed-phase Y-Ba-Cu-O compound system at ambient pressure. *Phys. Rev. Lett.* **58**, 908–912 (1987).
6. Schilling, A., Cantoni, M., Guo, J. D. & Ott, H. R. Superconductivity above 130 K in the Hg-Ba-Ca-Cu-O system. *Nature* **363**, 56–58 (1993).
7. Larbalestier, D. C. *et al.* Power Applications of Superconductivity in Japan and Germany (World Technology and Engineering Center, Loyola College, MD, September 1997).
8. Nagamatsu, J., Nakagawa, N., Muranaka, T., Zenitani, Y. & Akimitsu, J. Superconductivity at 39 K in magnesium diboride. *Nature* **410**, 63–64 (2001).
9. Wilson, M. *Superconducting Magnets* (Clarendon, Oxford, 1983).
10. Dew-Hughes, D. *Physics and Materials Science of Vortex States, Flux Pinning and Dynamics* NATO Science Ser. E, Vol. 356 (eds Kossowsky, R., Bose, S., Pan, V. & Durosoy, Z.) 705–730 (Kluwer Academic, Dordrecht, 1999).
11. Hassenzahl, W. V. Superconductivity, an enabling technology for 21st century power systems? *IEEE Trans. Appl. Supercond.* **11**, 1447–1453 (2001).
12. Radebaugh, R. in *Advances in Cryogenic Engineering* Vol. 44 (eds Haruyama, T., Mitsui, T. & Yamafuji, K.) 33–44 (Elsevier Science, 1997).
13. Grant, P. Rehearsals for prime time. *Nature* **411**, 532–533 (2001).
14. Suenaga, M., Ghosh, A. K., Hu, Y. & Welch, D. O. Irreversibility temperatures of Nb$_3$Sn and Nb-Ti. *Phys. Rev. Lett.* **66**, 1777–1780 (1991).
15. Brandt, E. H. The flux line lattice in superconductors. *Rep. Prog. Phys.* **58**, 1465–1594 (1995).
16. Xu, M. *et al.* Single crystal MgB$_2$ with anisotropic superconducting properties. Preprint cond-mat/0105271 at <http://xxx.lanl.gov> (2001).
17. Patnaik, S. *et al.* Electronic anisotropy, magnetic field-temperature phase diagram and their dependence on resistivity in c-axis oriented MgB$_2$ thin films. *Supercond. Sci. Technol.* **14**, 315–319 (2001).
18. Blatter, G., Feigelman, M. V., Geshkenbein, V. B., Larkin, A. I. & Vinokur, V. M. Vortices in high-temperature superconductors. *Rev. Mod. Phys.* **66**, 1125–1388 (1994).
19. Schwartzkopf, L. A., Jiang, J., Cai, X. Y., Apodaca, D. & Larbalestier, D. C. The use of the in-field critical current density, J_c(0.1T), as a better descriptor of (Bi,Pb)$_2$Sr$_2$Ca$_2$Cu$_3$O$_x$/Ag tape performance. *Appl. Phys. Lett.* **75**, 3168–3170 (1999).
20. Canfield, P. C. *et al.* Superconductivity in dense MgB$_2$ wires. *Phys. Rev. Lett.* **86**, 2423–2426 (2001).
21. Tinkham, M. *Introduction to Superconductivity.* (McGraw Hill, New York, 1975).
22. Campbell, A. M. & Evetts, J. E. Flux vortices and transport current in type-II superconductors. *Adv. Phys.* **21**, 194–428 (1972).
23. Polyanskii, A. A. *et al.* Examination of current limiting mechanisms in monocore Ag/BSCCO tape with high critical current density. *IEEE Trans. Appl. Supercond.* **11**, 3269–3270 (2001).
24. Meingast, C. & Larbalestier, D. C. Quantitative description of a very high critical current density Nb-Ti superconductor during its final optimization strain. II. Flux pinning mechanisms. *J. Appl. Phys.* **66**, 5971–5983 (1989).
25. Kramer, E. J. Dynamics of dislocation dipole motion in the flux line lattice of type-II superconductors. *J. Appl. Phys.* **41**, 621–629 (1970).
26. Dew-Hughes, D. The role of grain boundaries in determining J_c in high-field, high current superconductors. *Phil. Mag.* **55**, 459–449 (1987).
27. Cooley, L. D., Lee, P. J. & Larbalestier, D. C. Changes in flux pinning curve shape for flux-line separations comparable to grain size in Nb$_3$Sn wires. *Adv. Cryo. Eng.* (in the press).
28. Eom, C. B. *et al.* High critical current density and enhanced irreversibility field in superconducting MgB$_2$ thin films. *Nature* **411**, 558–560 (2001).
29. Song, X. *et al.* Anisotropic grain morphology, crystallographic texture and their implications for the electromagnetic properties of polycrystalline MgB$_2$ thin films, sintered pellets and filaments. *Supercond. Sci. Technol.* (in the press).
30. Jin, S., Mavoori, H., Bower, C. & van Dover, R. B. High critical current in iron-clad superconducting MgB$_2$ wires. *Nature* **411**, 563–565 (2001).
31. Gurevich, A. & Pashitskii, E. A. Current transport through low-angle grain boundaries in high-temperature superconductors. *Phys. Rev. B* **57**, 13878–13893 (1998).
32. Pan, S. H. *et al.* Imaging the effect of individual zinc impurities on superconductivity in Bi$_2$Sr$_2$CaCu$_2$O$_{8+δ}$. *Nature* **403**, 746–750 (2000).
33. Pan, S. H. *et al.* Microscopic electronic inhomogeneity in the high-T$_c$ superconductor Bi$_2$Sr$_2$CaCu$_2$O$_{8+x}$. *Nature* **413**, 282–285 (2001).
34. Kes, P. Flux pinning and creep in high-temperature superconductors. *Physica C* **185**, 288–291 (1991).
35. Diaz, A., Mechin, L., Berghuis, P. & Evetts, J. E. Evidence for vortex pinning by dislocations in YBa$_2$Cu$_3$O$_7$ low angle grain boundaries. *Phys. Rev. Lett.* **80**, 3855–3858 (1998).
36. Dam, B. *et al.* Origin of high critical currents in YBa$_2$Cu$_3$O$_7$ superconducting thin films. *Nature* **399**, 439–442 (1999).
37. Hilgenkamp, H. & Mannhart, J. Grain boundaries in high-Tc superconductors. *Rev. Mod. Phys.* (in the press).
38. Miu, L. *et al.* Vortex unbinding and layer decoupling in epitaxial Bi$_2$Sr$_2$Ca$_2$Cu$_3$O$_{8+δ}$ films. *Phys. Rev. B* **52**, 4553–4558 (1995).
39. Vargas, J. L. & Larbalestier, D. C. Flux pinning by ordered oxygen deficient phases in nearly

stoichiometric YBa$_2$Cu$_3$O$_{7-\delta}$ single crystals. *Appl. Phys. Lett.* **60**, 1741–1743 (1992).

40. Yeshurun, Y., Malozemoff, A. P. & Shaulov, A. Magnetic relaxation in high-temperature superconductors. *Rev. Mod. Phys.* **68**, 911–949 (1996).

41. Civale, L. Vortex pinning and creep in superconductors with columnar defects. *Supercond. Sci. Technol.* **10**, A11–A28 (1997).

42. Tonies, S. *et al.* On the current transport limitations in Bi-based high temperature superconducting tapes. *Appl. Phys. Lett.* **78**, 3851–3853 (2001).

43. Dimos, D., Chaudhari, P. & Mannhart, J. Superconducting transport properties in YBa$_2$Cu$_3$O$_{7-\delta}$ bicrystals. *Phys. Rev. B* **41**, 4038–4049 (1990).

44. Heinig, N. F., Redwing, R. D., Nordman, J. E. & Larbalestier, D. C. Strong to weak coupling transition in low misorientation angle thin film YBa$_2$Cu$_3$O$_{7-\delta}$ bicrystals. *Phys. Rev. B* **60**, 1409–1417 (1999).

45. Verebelyi, D. *et al.* Low angle grain boundary transport in YBa$_2$Cu$_3$O$_{7-\delta}$ coated conductors. *Appl. Phys. Lett.* **76**, 1755–1757 (2000).

46. Goyal, A. *et al.* Texture formation and grain boundary network in rolling assisted biaxially textured substrates and in epitaxial YBCO films. *Micron* **30**, 463–478 (1999).

47. Iijima, Y., Kakimoto, K., Kimura, M., Takeda, K. & Saitoh T. Reel to reel continuous formation of Y-123 coated conductors by IBAD and PLD method. *IEEE Trans. Appl. Supercond.* **11**, 2816–2821 (2001).

48. Larbalestier, D. C. & West, A. W. New perspective on flux pinning in niobium-titanium composite superconductors. *Acta Metall.* **32**, 1871–1881 (1984).

49. Cooley, L. D., Daeumling, M., Willis, T. W. & Larbalestier, D. C. Weak link behavior in polycrystalline BaPb$_{0.75}$Bi$_{0.25}$O$_3$. *IEEE Trans. Magn.* **25**, 2314–2316 (1989).

50. Takagi, T., Chiang, Y. M. & Roshko, A. Origin of grain-boundary weak links in BaPb$_{1-x}$Bi$_x$O$_3$ superconductor. *J. Appl. Phys.* **68**, 5750–5758 (1990).

51. Babcock, S. E. & Vargas, J. L. The nature of grain boundaries in high-T$_c$ superconductors. *Annu. Rev. Mater. Sci.* **25**, 193–222 (1995).

52. Browning, N. D. *et al.* The atomic origin of reduced critical currents at [001] tilt grain boundaries in YBa$_2$Cu$_3$O$_7$ thin films. *Physica C* **294**, 183–193 (1998).

53. Schmehl, A. *et al.* Doping induced enhancement of the critical currents of grain boundaries in YBa$_2$Cu$_3$O$_7$. *Europhys. Lett.* **47**, 110–115 (1999).

54. Hammerl, G. *et al.* Enhanced supercurrent density in polycrystalline YBa$_2$Cu$_3$O$_7$ at 77K from calcium doping of grain boundaries. *Nature* **407**, 162–164 (2000).

55. Daniels, G. A., Gurevich, A. & Larbalestier, D. C. Improved strong magnetic field performance of low angle grain boundaries of calcium and oxygen overdoped YBa$_2$Cu$_3$O$_7$. *Appl. Phys. Lett.* **77**, 3251–3253 (2000).

56. Feldmann, D. M. *et al.* Influence of nickel substrate grain structure on YBa$_2$Cu$_3$O$_7$ supercurrent connectivity in deformation-textured coated conductors. *Appl. Phys. Lett.* **77**, 2906–2908 (2000).

57. Feldmann, D. M. *et al.* Inter and intra-grain transport measurements in YBa$_2$Cu$_3$O$_{7-x}$ deformation textured coated conductors. *Appl. Phys. Lett.* (in the press).

58. Gurevich, A. & Cooley, L. D. Anisotropic flux pinning in a network of planar defects. *Phys. Rev. B* **50**, 13563–13576 (1994).

59. Amrein, T., Schultz, L., Kabius, B. & Urban, K. Orientation dependence of grain boundary critical current in high-T$_c$ bicrystals. *Phys. Rev. B* **51**, 6792 (1995).

60. Attenberger, A. *Electrische transportmessungen uber definierte Korngrenzen-Strukturen in epitaktischen Bi-2223 Schichten.* Thesis, Technische Universitat Dresden (2000).

61. Tsay, Y. N. *et al.* Transport properties of Bi$_2$Sr$_2$CaCu$_2$O$_{8+\delta}$ bicrystals with [001] tilt grain boundaries. *IEEE Trans. Appl. Supercond.* **9**, 1622–1625 (1999).

62. Polyanskii, A. A. *et al.* Magneto-optical study of flux penetration and critical current densities in [001] tilt YBa$_2$Cu$_3$O$_7$ thin-film bicrystals. *Phys. Rev. B* **53**, 8687–8697 (1996).

63. Polyanskii, A. A. *et al.* Magneto-optical studies of the uniform critical state in bulk MgB$_2$. *Supercond. Sci. Technol.* **14**, 811–815 (2001).

64. Larbalestier, D. C. *et al.* Strongly linked current flow in polycrystalline forms of the superconducting MgB$_2$. *Nature* **410**, 186–189 (2001).

65. Chu, S. Y. & McHenry, M. E. Growth and characterization of (Bi,Pb)$_2$Sr$_2$Ca$_2$Cu$_3$O$_x$ single crystals. *J. Mat. Res.* **13**, 589–595 (1998).

66. Malozemoff, A. P. *et al.* HTS wires at commercial performance levels. *IEEE Trans. Appl. Supercond.* **9**, 2469–2473 (1999).

67. Huang, Y. B. *et al.* Progress in Bi-2223 tape performance. *Adv. Cryo. Eng.* (in the press).

68. Pashitski, A. E., Polyanskii, A., Gurevich, A., Parrell, J. A. & Larbalestier, D. C. Magnetic granularity, percolation and preferential current flow in a silver-sheathed Bi$_{1.8}$Pb$_{0.4}$Sr$_2$Ca$_2$Cu$_3$O$_{8+x}$ tape. *Physica C* **246**, 133 (1995).

69. Holzapfel, B. *et al.* Grain boundary networks in Y123 coated conductors. Formation, properties and simulation. *IEEE Trans. Appl. Supercond.* **11**, 3872–3875 (2001).

70. Rutter, N. & Glowacki, B. Modelling of orientation relations in 2-D percolative systems of buffered metallic substrates for coated conductors. *IEEE Trans. Appl. Supercond.* **11**, 2730–2733 (2001).

71. Rupich, M. W. *et al.* Low cost Y-Ba-Cu-O coated conductors. *IEEE Trans. Appl. Supercond.* **11**, 2927–2930 (2001).

72. Maeda, T. *et al.* YBa$_2$Cu$_3$O$_x$ thick films grown on textured metal substrates by liquid-phase epitaxy

73. Solovyov, V. *et al.* Ex situ post-deposition processing for large area YBa$_2$Cu$_3$O$_x$ films and coated tapes. *IEEE Trans. Appl. Supercond.* **11**, 2939–2943 (2001).

74. Holesinger, T. *et al.* A comparison of buffer layer architectures on continuously processed YBCO coated conductors based on the IBAD YSZ process. *IEEE Trans. Appl. Supercond.* **11**, 3359–3364 (2001).

75. Gurevich, A. Thermal instability near planar defects in superconductors. *Appl. Phys. Lett.* **78**, 1891–1893 (2001).

76. Gurevich, A. & Friesen, M. Nonlinear current flow in superconductors with planar obstacles. *Phys. Rev. B* **62**, 4004–2025 (2000).

77. Friesen, M. & Gurevich, A. Nonlinear current flow in superconductors with restricted geometries. *Phys. Rev. B* **63**, 064521-1-26 (2001).

78. Glowacki, B. A. *et al.* Superconductivity of powder-in-tube MgB$_2$ wires. *Supercond. Sci. Technol.* **14**, 193–199 (2001).

79. Suo, H. L. *et al.* High transport and inductive critical currents in dense Fe- and Ni-clad MgB$_2$ tapes using fine powder. *Adv. Cryo. Eng.* (in the press).

80. Grasso, G. *et al.* Preperation and properties of unsintered MgB$_2$ superconducting tapes. *Adv. Cryo. Eng.* (in the press).

81. Kumakura, H., Matsumoto, A., Fujii, H., Takano, Y. & Togano, K. Fabrication and superconducting properties of powder-in-tube processed MgB$_2$ tapes and wires. *Adv. Cryo. Eng.* (in the press).

82. Grasso, G. *et al.* Large transport critical currents in unsintered MgB$_2$ superconducting tapes. *Appl. Phys. Lett.* **79**, 230–232 (2001).

83. Malozemoff, A. P. *et al.* D. F. Low-cost YBCO coated conductor technology. *Supercond. Sci. Technol.* **13**, 473–476 (2000).

84. Pyon, T. & Gregory, E. Nb$_3$Sn conductors for high energy physics and fusion applications. *IEEE Trans. Appl. Supercond.* **11**, 3688–3691 (2001).

85. Field, M., Hentges, R., Parrell, J., Zhang, Y. & Hong, S. Progress with Nb$_3$Sn conductors at Oxford Instruments–Superconducting Technology. *IEEE Trans. Appl. Supercond.* **11**, 3692–3695 (2001).

86. Wang, W. G. *et al.* Engineering critical current density improvement in Ag-Bi-2223 tapes. *IEEE Trans. Appl. Supercond.* **11**, 2983–2986 (2001).

87. Masur, L. *et al.* Long length manufacturing of high performance BSCCO-2223 tape for the Detroit-Edison Power Cable Project. *IEEE Trans. Appl. Supercond.* **11**, 3256–3260 (2001).

88. Hayashi, K. *et al.* Development of Ag-sheathed Bi2223 superconducting wires and their applications. *IEEE Trans. Appl. Supercond.* **11**, 3281–3284 (2001).

89. Jiang, J. *et al.* Through-process study of factors controlling the critical current density of Ag-sheathed (Bi,Pb)$_2$Sr$_2$Ca$_2$Cu$_3$O$_x$ tapes. *Supercond. Sci. Technol.* **14**, 548–556 (2001).

90. Rikel, M. O. *et al.* Overpressure processing Bi2223/Ag tapes. *IEEE Trans. Appl. Supercond.* **11**, 3026–3029 (2001).

91. Flukiger, R. *et al.* Phase formation in Bi,Pb(2223) tapes. *IEEE Trans. Appl. Supercond.* **11**, 3393–3398 (2001).

92. Wang, C. P., Do, K. B., Beasley, M. R., Geballe, T. H. & Hammond, R. H. Deposition of in-plane textured MgO on amorphous Si$_3$N$_4$ substrates by ion-beam-assisted deposition and comparison with ion-beam-assisted deposition yttria-stabilized-zirconia. *Appl. Phys. Lett.* **71**, 2955–2957 (1997).

93. Groves, J. R. *et al.* High critical current density YBa$_2$Cu$_3$O$_{7-\delta}$ thick films using ion beam assisted deposition MgO bi-axially oriented template layers on nickel-based superalloy substrates. *J. Mater. Res.* **16**, 2175–2178 (2001).

94. Hasegawa, K. *et al.* in *Proc. 9th Int. Symp. Supercond. (IIS'96)* Vol. 2, 745–748 (Springer, Tokyo 1997).

95. Metzger, R. Superconducting tapes using ISD buffer layers produced by evaporation of MgO or reactive evaporation of magnesium. *IEEE Trans. Appl. Supercond.* **11**, 2826–2829 (2001).

96. Reade, R. P., Berdhal, P., Russo, R. E. & Garrison, S. M. Laser deposition of biaxially textured yttria-stabilized zirconia buffer layers on polycrystalline metallic alloys for high critical current Y-Ba-Cu-O thin films. *Appl. Phys. Lett.* **61**, 2231–2233 (1992).

97. Ivanov, Z. G. *et al.* Weak links and dc SQUIDs on artificial nonsymmetric grain boundaries in YBa$_2$Cu$_3$O$_{7-x}$. *Appl. Phys. Lett.* **59**, 3030 (1991).

98. Char, K. *et al.* Bi-epitaxial grain-boundary junctions in YBa$_2$Cu$_3$O$_7$. *Appl. Phys. Lett.* **59**, 733–735 (1991).

Acknowledgements

The authors are grateful to many colleagues for discussions and collaborations. In particular, we thank our present and former Madison colleagues S. Babcock, X. Cai, L. Cooley, G. Daniels, C.-B. Eom, E. Hellstrom, J. Jiang, P. Lee, J. Reeves, M. Rikel and X. Song, and colleagues within the Wire Development Group, especially B. Riley (AMSC), V. Maroni (ANL) and T. Holesinger (LANL). Recent collaborations on coated conductors with T. Peterson and P. Barnes (AFRL), R. Feenstra (ORNL) and D. Verebelyi (AMSC) have been particularly helpful.. R. Blaugher (NREL), G. Grasso (Genoa) and J. Mannhart (Augsburg) supplied material for the tables and figures. Finally we thank the Air Force Office of Scientific Research, the Department of Energy and the National Science Foundation for support.

Template engineering of Co-doped BaFe$_2$As$_2$ single-crystal thin films

S. Lee[1], J. Jiang[2], Y. Zhang[3], C. W. Bark[1], J. D. Weiss[2], C. Tarantini[2], C. T. Nelson[3], H. W. Jang[1], C. M. Folkman[1], S. H. Baek[1], A. Polyanskii[2], D. Abraimov[2], A. Yamamoto[2], J. W. Park[1], X. Q. Pan[3], E. E. Hellstrom[2], D. C. Larbalestier[2] and C. B. Eom[1]*

Understanding new superconductors requires high-quality epitaxial thin films to explore intrinsic electromagnetic properties and evaluate device applications[1-9]. So far, superconducting properties of ferropnictide thin films seem compromised by imperfect epitaxial growth and poor connectivity of the superconducting phase[10-14]. Here we report new template engineering using single-crystal intermediate layers of (001) SrTiO$_3$ and BaTiO$_3$ grown on various perovskite substrates that enables genuine epitaxial films of Co-doped BaFe$_2$As$_2$ with a high transition temperature ($T_{c,\rho=0}$ of 21.5 K, where ρ = resistivity), a small transition width ($\Delta T_c = 1.3$ K), a superior critical current density J_c of 4.5 MA cm^{-2} (4.2 K) and strong c-axis flux pinning. Implementing SrTiO$_3$ or BaTiO$_3$ templates to match the alkaline-earth layer in the Ba-122 with the alkaline-earth/oxygen layer in the templates opens new avenues for epitaxial growth of ferropnictides on multifunctional single-crystal substrates. Beyond superconductors, it provides a framework for growing heteroepitaxial intermetallic compounds on various substrates by matching interfacial layers between templates and thin-film overlayers.

Epitaxial pnictide thin films have so far been hard to produce, especially the F-doped highest-T_c rare-earth 1111 phase[13,14]. As a result of both As and F being volatile at the deposition temperature, it is difficult to control the overall stoichiometry of the deposited film[13]. In contrast, Co, which can be used as the dopant in Ba(Fe, Co)$_2$As$_2$ (Ba-122), has a low vapour pressure under growth conditions. Alkaline-earth (AE) 122 phases have been grown on (001)-oriented (La, Sr)(Al, Ta)O$_3$ (LSAT) and LaAlO$_3$ (LAO) single-crystal substrates by us and other groups[10-12]. However, the quality of the films reported so far is not satisfactory because $T_{c,\rho=0}$ and ΔT_c are 14–17 K and 2–4 K, respectively, values that are much lower and broader than those of bulk single crystals[15]. Furthermore, J_c (5 K, self-field (SF)) of these films is ~10 kA cm^{-2}, one to two orders of magnitude lower than in bulk single crystals[15]. We believe a fundamental reason for the low quality of the AE-122 phase on these substrates is that AE-122 is a metallic system, which does not bond well with oxide single-crystal substrates. In particular, LSAT and LAO substrates contain trivalent cations, whereas the alkaline earths of Ba-122 or Sr-122 are divalent, making the bonding between the substrate and Ba-122 poor, which leads to a non-epitaxial, granular and poorly connected superconducting phase in which the grains are separated by high-angle grain boundaries, wetting grain-boundary phases such as FeAs and/or off-stoichiometric grains[16].

To overcome these problems, we explored the use of a template consisting of thin epitaxial layers of the divalent, AE-containing SrTiO$_3$ (STO) or BaTiO$_3$ (BTO) between various single-crystal perovskite substrates and the Ba-122 film. Building on this common feature, we propose the bonding model described in Fig. 1. Recently, we reported a proof-of-principle of this concept by growing Co-doped Ba-122 epitaxial thin films on (001)-oriented STO bicrystal substrates with intragrain J_c values over 1 MA cm^{-2} (4.2 K, SF), values significantly higher than in previously reported epitaxial thin films[17]. However, the STO substrates became electrically conducting during deposition of the Co-doped Ba-122 thin film at 730 °C in the high vacuum (2.7×10^{-5} Pa), owing to formation of oxygen vacancies in the STO (ref. 18). This provides a parallel, current-sharing path between Ba-122 and STO, which compromises normal-state property studies. Clearly, normal-state behaviour and potential device applications[19] need insulating substrates, especially microwave devices, which need substrates such as LSAT or LAO that have a low dielectric constant.

Here, we report single-crystal Co-doped Ba-122 epitaxial thin films using thin STO or BTO templates deposited on various perovskite single-crystal substrates, which include (001) LSAT, (001) LAO and (110) GdScO$_3$, all of which yield Ba-122 films with superior superconducting properties on insulating substrates. The 50–100 unit cell (u.c.)-thick STO and BTO templates were grown by pulsed laser deposition using a KrF excimer laser (248 nm) with high-oxygen-pressure reflection high-energy electron diffraction (RHEED) for digital control and *in situ* monitoring of the epitaxial growth[20]. Co-doped Ba-122 films ~350 nm thick were grown on both bare single-crystal substrates and substrates with STO and BTO templates. Structural and superconducting properties of these films are listed in Supplementary Table SI. We discuss here only films grown on bare STO, bare LSAT, STO/LSAT (that is, an STO template on LSAT) and BTO/LSAT, while emphasizing that similar properties were obtained on (001) LAO and (110) GdScO$_3$ too (see Supplementary Table SI).

Figure 2a and its inset show an atomic force microscope (AFM) image of a 100 u.c. STO template on LSAT and its RHEED pattern. They show atomically flat terraces and single-unit-cell-high (~4 Å) steps, which confirm that the STO template layer is as good as those of bulk single-crystal substrates of STO. It is found that an STO template up to 100 u.c. is fully coherent with the LSAT substrate. The epitaxial and crystalline quality of the Co-doped Ba-122 thin films was measured by four-circle X-ray diffraction

[1]Department of Materials Science and Engineering, University of Wisconsin-Madison, Madison, Wisconsin 53706, USA, [2]Applied Superconductivity Center, National High Magnetic Field Laboratory, Florida State University, 2031 East Paul Dirac Drive, Tallahassee, Florida 32310, USA, [3]Department of Materials Science and Engineering, The University of Michigan, Ann Arbor, Michigan 48109, USA. *e-mail: eom@engr.wisc.edu.

When citing this article, please cite the original version as shown on the contents page of this chapter.

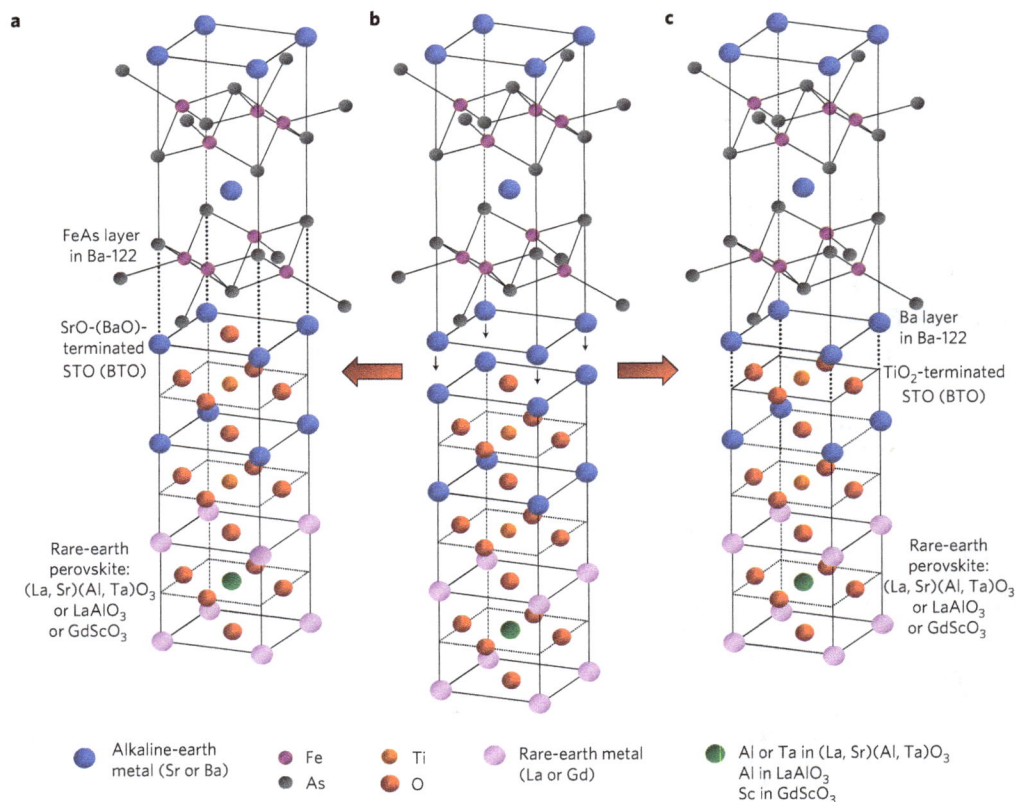

Figure 1 | Schematic model of Ba-122 deposition on a STO (BTO) template grown on various oxide substrates. a–c, The Ba-122 unit cell and the alkaline-earth titanium oxide (AETO) unit cells grown on a rare-earth (RE) perovskite oxide substrate unit cell (**b**). In this example, only two unit cells of AETO and one unit cell of RE perovskite are shown. There are two possible ways the Ba-122 and perovskite lattice bonding can occur. The FeAs layer in Ba-122 bonds strongly to RE-O-terminated RETO as Ba-122 is deposited (the AE-O layer from AETO replaces the Ba layer in Ba-122) (**a**). **c**, The Ba layer in Ba-122 bonds strongly to TiO_2-terminated RETO as Ba-122 is deposited (the Ba layer from Ba-122 replaces AE-O in the AETO) (**c**).

(XRD). Figure 2b shows the out-of-plane θ–2θ scans of the films on bare LSAT and 100 u.c. STO/LSAT. The XRD patterns show that the film 00l reflections dominate, which indicates c-axis growth normal to the template and substrate. An extra peak at $2\theta = \sim 65°$ in the film grown on STO/LSAT is the 002 reflection from Fe (ref. 11), however with an intensity of less than 0.5% of the Ba-122 (004) reflection. The intensity of the Ba-122 00l peaks on STO/LSAT is about two orders of magnitude higher than that on bare LSAT. This indicates that the film on STO/LSAT is highly c-axis oriented, whereas the film on bare LSAT is c-axis textured and polycrystalline. Rocking curves for the 004 reflection were measured to determine the out-of-plane mosaic spread and crystalline quality. As shown in Fig. 2c, the full-width at half-maximum (FWHM) of the 004 reflection rocking curve of the film on STO/LSAT is as narrow as 0.55°, which is the narrowest ever reported for AE-122 thin films, whereas that of the film on bare LSAT is as broad as 3.1°. Remarkably, the FWHM of the film on BTO/LSAT is as narrow as 0.17°, similar to that of a Ba-122 bulk single crystal[21] (see Supplementary Table SI).

In-plane texture and epitaxial quality were determined by azimuthal ϕ scans of the off-axis (112) peak, as shown in Fig. 2d. The film grown on bare LSAT shows broad major peaks ($\Delta\phi = 4.4°$) every 90° with extra broader intermediate-angle peaks, which indicates that the in-plane Ba-122 structure consists of grains with high-angle tilt grain boundaries. In contrast, the film grown on 100 u.c. STO/LSAT shows only sharp, strong peaks ($\Delta\phi = 0.8°$) every 90° characteristic of a truly epitaxial film with perfect in-plane texture. Our Ba-122 thin films are fully strain relaxed. The in-plane

and out-of-plane lattice parameters of all of our Ba-122 films are the same as the bulk single-crystal values.

To verify the crystalline quality, microstructures were studied by transmission electron microscopy (TEM). Figure 3a–d shows the cross-sectional TEM images of the films grown on bare LSAT (Fig. 3a), 100 u.c. STO/LSAT (Fig. 3b), 20 u.c. STO/LSAT (Fig. 3c) and bare STO (Fig. 3d). The insets in the top right corners of Fig. 3a,b,d show the corresponding Ba-122 selected-area electron diffraction patterns, respectively. The Ba-122 films on bare STO and on STO-templated LSAT show epitaxial relationships between the films and the substrate, whereas the film on bare LSAT shows a granular microstructure with many misoriented grains. The inset on the left side of Fig. 3a shows the high-resolution TEM (HRTEM) image of a high-angle grain boundary in the Ba-122 film. The HRTEM image in Fig. 3c shows the structure of interfaces between STO and LSAT and between STO and Ba-122. The localized disorder of atomic arrangement at the STO/Ba-122 interface may be due either to interface reaction or to the ion beam damage during TEM sample preparation. The epitaxial films grown on bare STO and on STO/LSAT show vertically aligned line defects seen in the cross-sectional TEM images of Fig. 3b,d as well as in the top-left side inset in Fig. 3b. The inset in the bottom left corner of Fig. 3b shows a planar-view TEM image of the same film, revealing that the vertically aligned defects are uniformly distributed. Both X-ray energy dispersion spectroscopy and electron energy-loss spectroscopy reveal that the line defects consist of a secondary phase composed mainly of Ba and O and depleted of Fe, Co and As.

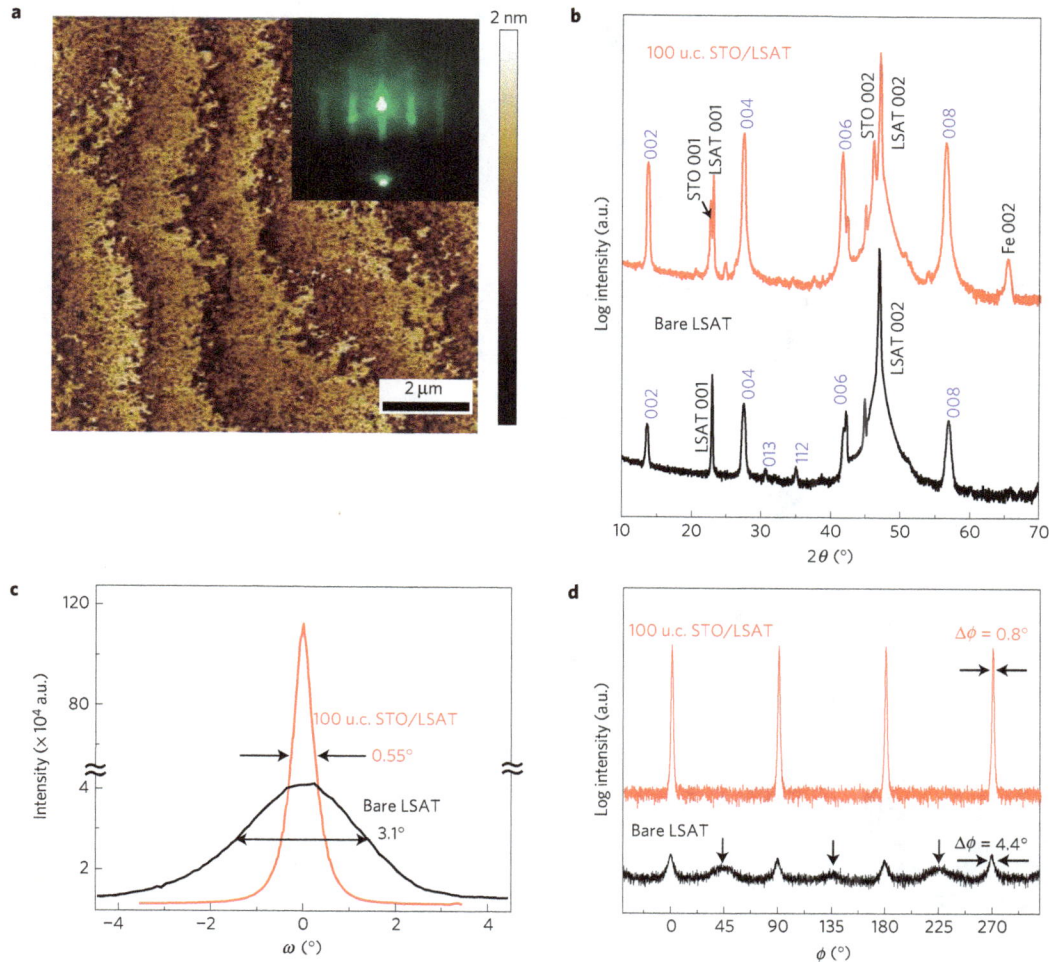

Figure 2 | AFM image and RHEED pattern of the STO template grown on LSAT, and XRD patterns obtained on the Co-doped BaFe$_2$As$_2$ thin films.
a, AFM image of a 100 u.c. STO template grown on LSAT. The RHEED pattern of Ba-122 is shown in the inset. **b**, Out-of plane θ–2θ XRD patterns of the film grown on 100 u.c. STO/LSAT and the film grown on bare LSAT. All non-identified small peaks are reflections of identified peaks resulting from CuKβ and CuKα + β. **c**, Rocking curve and FWHM for (004) reflection of Ba-122. **d**, Azimuthal ϕ scan and $\Delta\phi$ of the off-axis 112 reflection of Ba-122. The vertical arrows above the bare LSAT pattern indicate a peak resulting from misoriented grains separated by high-angle grain boundaries.

To characterize the superconducting transition temperature T_c, resistivity was measured as a function of temperature $\rho(T)$ by a four-point method. As shown in Fig. 4a, the residual resistivity ratio of the Ba-122 film grown on bare STO is as high as 160, a value consistent with oxygen-deficient STO (ref. 18) that we reproduced on bare STO substrates heated in vacuum under the same conditions used for Ba-122 deposition. In contrast, the residual resistivity ratios of films grown on STO/LSAT and BTO/LSAT templates were 1.8 and 2.4, respectively, values characteristic of Co-doped Ba-122 single crystals[22]. Furthermore, we noted that the room-temperature resistivity of a film on bare LSAT is much higher than a film grown on templated LSAT, which we interpret as being due to strong scattering at the high-angle grain boundaries. As shown in Fig. 4b, all films show high $T_{c,\rho=0}$ and narrow ΔT_c except for the film with high-angle grain boundaries grown on bare LSAT. In particular, $T_{c,\rho=0}$ of the film on 100 u.c. STO/LSAT is as high as 21.5 K and ΔT_c is as narrow as 1.3 K, which are the highest and narrowest values ever reported for AE-122 thin films.

Figure 4c shows the zero-field-cooled magnetization T_c transitions. All three epitaxial films have a strong diamagnetic signal, in contrast to the polycrystalline film grown on bare LSAT,

which, as seen in the inset in Fig. 4c, shows a three orders of magnitude smaller diamagnetic signal than the epitaxial films. This is a clear indication of magnetic granularity associated with the inability of screening currents to develop across the high-angle grain boundaries in the film on bare LSAT.

Figure 5a shows J_c as a function of magnetic field for all films determined from vibrating sample magnetometer measurements in fields up to 14 T. Here too all films except that on bare LSAT show high J_c values, over 1 MA cm^{-2} (4.2 K, SF), which are the highest values ever reported for AE-122 thin films and are even better than in bulk single crystals (0.4 MA cm^{-2} at 4.2 K; ref. 15). Remarkably, the J_c of the film on BTO/LSAT is as high as ~4.5 MA cm^{-2}. Furthermore, the J_c values of the two epitaxial films on the templates have only a weak field dependence, indicative of little or no suppression of the J_c by strong fields, as indicated by the STO/LSAT film, which had $J_c = $ ~0.4 MA cm^{-2} even at 14 T. The magneto-optical image of the film on STO/LSAT (inset in Fig. 5a) shows strong flux shielding and a J_c of ~3 MA cm^{-2} (6.6 K, 20 mT), similar to that deduced from the magnetization measurements. Such magneto-optical images confirm that the epitaxial films grown on STO/LSAT or BTO/LSAT are uniform and well connected without any weak links.

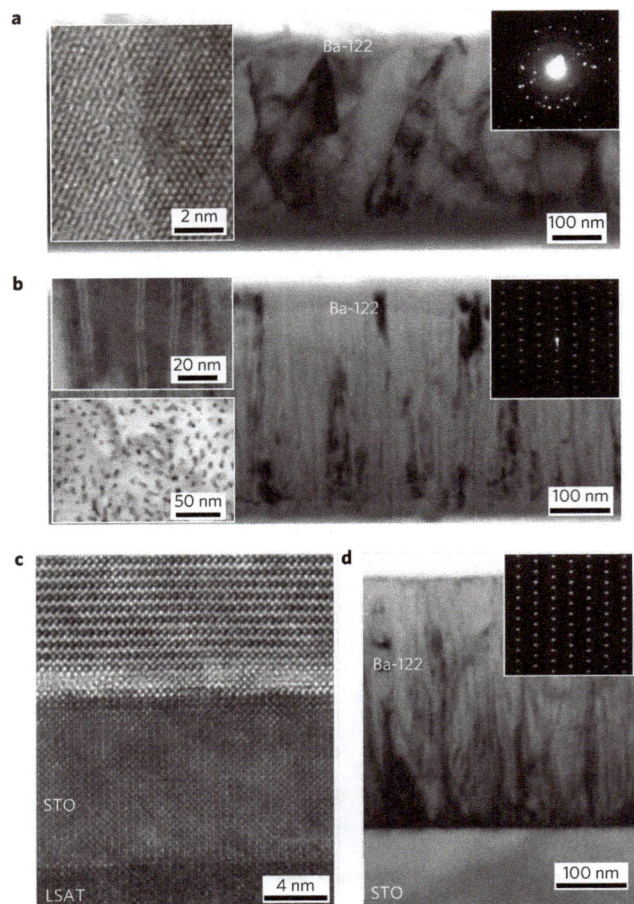

Figure 3 | Microstructures of Co-doped BaFe₂As₂ thin films investigated by TEM. a–d, Cross-sectional TEM micrographs of films on bare LSAT (**a**), 100 u.c. STO/LSAT (**b**), 20 u.c. STO/LSAT (**c**) and bare STO (**d**). They confirm that the films on bare STO or STO-buffered LSAT are epitaxial, whereas the film on bare LSAT is polycrystalline with misoriented grains. Line defects along the c axis exist in the films grown on STO/LSAT and bare STO. The insets on the left side of **b** are planar and cross-sectional views of the same film, showing a uniform distribution of the vertically aligned line defects. The insets on the top right sides of **a,b,d** show the corresponding selected-area electron diffraction patterns. The HRTEM image in **c** shows the interface structures between LSAT and 20 u.c. STO and between 20 u.c. STO and the Ba-122 film.

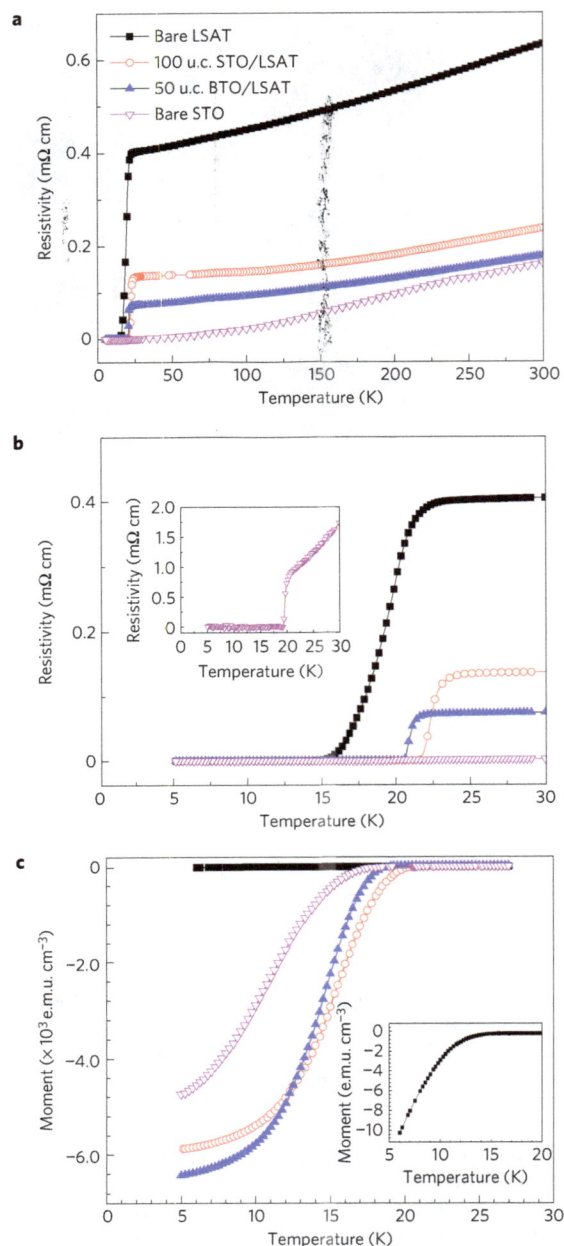

Figure 4 | Resistivity and magnetic moment as a function of temperature. a, $\rho(T)$ from room temperature to below T_c. **b,** Superconducting transition of all films; $T_{c,onset}$ and $T_{c,\rho=0}$ of the film on 100 u.c. STO/LSAT are as high as 22.8 K and 21.5 K, respectively. Inset: Superconducting transition of the film on bare STO. **c,** Magnetic moment as a function of temperature evaluated by warming after zero-field-cooling. A field of 2 mT was applied perpendicular to the plane of the films after cooling to 4.2 K. Inset: An expanded view of the much smaller diamagnetic signal of the film grown on bare LSAT.

We believe that two factors contribute to the high J_c. First, the films on templated LSAT and bare STO have high epitaxial quality with no high-angle tilt grain boundaries, as confirmed by both XRD and TEM analysis. According to our previous report[17], [001] tilt grain boundaries of Ba-122 with $\theta = 6°–24°$ show significant suppression of supercurrent, making it entirely understandable that randomly oriented high-angle tilt grain boundaries would effectively block supercurrent, as suggested by several recent pnictide film reports[10–12,14]. Indeed, the films grown on bare LSAT developed many high-angle grain boundaries, had almost no flux shielding even in low fields (2 mT) and very low J_c. In contrast, films on templated LSAT and bare STO showed high J_c without evidence of weak links. Interestingly, the J_c of the film on STO/LSAT is higher than that of the film on bare STO, which is consistent with our observation of a higher density of line defects in Ba-122 on STO/LSAT than for the film grown on bare STO. We also measured the angular-dependent transport

J_c of a STO/LSAT film, as shown in Fig. 5b. J_c always shows a strong c-axis peak, which opposes the expected electronic anisotropy because the upper critical field H_{c2} is lower for H parallel to the c axis[23–25], making it clear that there is strong pinning along the Ba-122 c axis that is parallel to the vertically aligned defects of the secondary phases. Recently, a similar angular dependence was reported in much lower J_c but granular

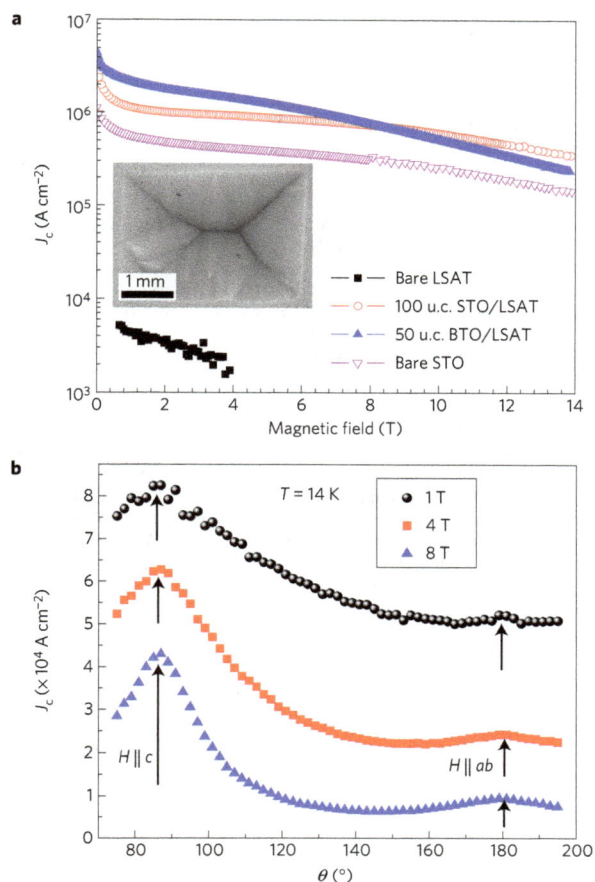

Figure 5 | J_c as a function of magnetic field and its angular dependence.
a, Magnetization J_c as a function of magnetic field at 4.2 K with the field applied perpendicular to the plane of the film. Inset: A magneto-optical image obtained by zero-field-cooling to 6.6 K, then applying a magnetic 20 mT field perpendicular to the plane of the film. **b**, Transport J_c at 14 K and 1, 4 and 8 T as a function of the angle between the applied field and the surface of the Ba-122 films. The magnetic field is always perpendicular to the current-flow direction.

Co-doped Sr-122 films, which was associated with contributions from dilute magnetic pinning[26].

We have also grown Co-doped Ba-122 films on other bare substrates, including (001) LAO and (110) GdScO$_3$ (GSO). In spite of the almost perfect lattice match between Ba-122 and (110)-oriented GSO, the film on bare GSO shows poor quality, just like the films grown on bare LAO (see Supplementary Table SI). Both bare GSO and LAO films show broad peaks from misoriented grains in the ϕ scan. However, when an intermediate layer of STO or BTO was used as a template on the LAO or GSO, the properties of the Ba-122 films were markedly enhanced and misoriented grains were not seen. Remarkably, we could also grow a superior quality Ba-122 film on STO/GSO using template engineering with intermediate STO layers that produced a perfect lattice match with Ba-122. The FWHM of the (004) reflection rocking curve and $\Delta\phi$ of Ba-122 on 50 u.c. STO/GSO were very narrow, 0.24° and 0.3°, respectively (see Supplementary Table SI), values close to those of bulk single crystals[21]. All of the above results show that an STO or BTO template is the key to growing high-quality epitaxial Ba-122 films.

We have developed a versatile new method to turn granular, low-J_c superconducting films into single-crystal, high-J_c and truly

epitaxial films of Ba-122. Our working hypothesis is that we are favouring bonding between the Ba alkaline-earth component of the pnictide phase and the underlying oxide substrate by using template engineering with the intermediate STO or BTO layers. Indeed, the J_c in our high-quality epitaxial films is about 10 times greater than in bulk single crystals[15] and ~400 times greater than in previously reported AE-122 films[10–12]. Template engineering permits truly single-crystal quality (Supplementary Table S1) without any loss of connectivity because of high-angle grain boundaries, and a high density of vertically aligned, secondary phase line defects results in strong c-axis pinning. We believe that this template technique can be applied not only to perovskite single-crystal substrates, but also to other types of oxide substrate or even semiconductor substrate. Indeed, we have demonstrated that this approach yields high-quality Co-doped Ba-122 films on an epitaxial STO template grown on (001) Si substrates[27,28] with $T_{c,\rho=0} = 18\,$K, $\Delta T_c = 1\,$K and no misoriented grains (see Supplementary Information). This approach thus greatly expands substrate choices for high-quality Ba-122 thin films, thus enabling much broader fundamental property investigations of the newly discovered ferropnictide superconductors and parent compounds, as well as exploration of their applications. Furthermore, we expect that epitaxial thin films of other layered intermetallics could be successfully grown on various types of oxide substrate by using similar template-engineering principles.

Methods

Co-doped Ba-122 thin films were grown *in situ* on various single-crystal substrates and STO- or BTO-templated substrates using pulsed laser deposition with a KrF (248 nm) ultraviolet excimer laser in a vacuum of $3 \times 10^{-4}\,$Pa at 730–750 °C. The base pressure before deposition was $3 \times 10^{-5}\,$Pa, and the deposition took place at $3 \times 10^{-4}\,$Pa because of degassing of the substrate heater. The Co-doped Ba-122 target was prepared by solid-state reaction with a nominal composition of Ba/Fe/Co/As = 1:1.84:0.16:2.2. The magnetization of the films that were about 2 mm × 4 mm was measured in a 14 T Oxford vibrating-sample magnetometer at 4.2 K with the applied field perpendicular to the film surface. Magneto-optical imaging was used to examine the uniformity of current flow in the films so as to validate the use of the Bean model for converting the magnetic moment measured in the vibrating sample magnetometer to J_c assuming current circulation across the whole sample without granular effects. For a thin film, $J_c = 15\,\Delta m/(Vr)$, where Δm is the width of the hysteresis loop in e.m.u., V is the film volume in cubic centimetres and r is the radius corresponding to the total area of the sample size, and was calculated from $\pi r^2 = a \times b$ (a and b are the film width and length in centimetres, respectively).

Received 28 September 2009; accepted 5 February 2010; published online 28 February 2010

References

1. Chaudhari, P., Koch, R. H., Laibowitz, R. B., McGuire, T. R. & Gambino, R. J. Critical-current measurements in epitaxial films of YBa$_2$Cu$_3$O$_{7-x}$. *Phys. Rev. Lett.* **58,** 2684–2686 (1987).
2. Oh, B. *et al.* Critical current densities and transport in superconducting YBa$_2$Cu$_3$O$_{7-\delta}$ films made by electron beam coevaporation. *Appl. Phys. Lett.* **51,** 852–854 (1987).
3. Eckstein, J. N. *et al.* Atomically layered heteroepitaxial growth of single-crystal films of superconducting Bi$_2$Sr$_2$Ca$_2$Cu$_3$O$_x$. *Appl. Phys. Lett.* **57,** 931–933 (1990).
4. Zeng, X. *et al.* In situ epitaxial MgB$_2$ thin films for superconducting electronics. *Nature Mater.* **1,** 35–38 (2002).
5. Bu, S. D. *et al.* Synthesis and properties of c-axis oriented epitaxial MgB$_2$ thin films. *Appl. Phys. Lett.* **81,** 1851–1853 (2002).
6. Kamihara, Y., Watanabe, T., Hirano, M. & Hosono, H. Iron-based layered superconductor La[O$_{1-x}$F$_x$]FeAs ($x = 0.05$–0.12) with $T_C = 26\,$K. *J. Am. Chem. Soc.* **130,** 3296–3297 (2008).
7. Chen, X.-H. *et al.* Superconductivity at 43 K in SmFeAsO$_{1-x}$F$_x$. *Nature* **453,** 761–762 (2008).
8. Wang, C. *et al.* Thorium-doping-induced superconductivity up to 56 K in Gd$_{1-x}$Th$_x$FeAsO. *Europhys. Lett.* **83,** 67006 (2008).
9. Rotter, M., Tegel, M. & Johrendt, D. Superconductivity at 38 K in the iron arsenide Ba$_{1-x}$K$_x$Fe$_2$As$_2$. *Phys. Rev. Lett.* **101,** 107006 (2008).
10. Hiramatsu, H., Katase, T., Kamiya, T., Hirano, M. & Hosono, H. Superconductivity in epitaxial films of Co-doped SrFe$_2$As$_2$ with bilayered FeAs structures and their magnetic anisotropy. *Appl. Phys. Express* **1,** 101702 (2008).

11. Katase, T. *et al.* Atomically-flat, chemically-stable, superconducting epitaxial thin film of iron-based superconductor, cobalt-doped $BaFe_2As_2$. *Solid State Commun.* **149**, 2121–2124 (2009).

12. Choi, E. M. *et al. In situ* fabrication of cobalt-doped $SrFe_2As_2$ thin films by using pulsed laser deposition with excimer laser. *Appl. Phys. Lett.* **95**, 062507 (2009).

13. Hiramatsu, H., Katase, T., Kamiya, T., Hirano, M. & Hosono, H. Heteroepitaxial growth and optoelectronic properties of layered iron oxyarsenide, LaFeAsO. *Appl. Phys. Lett.* **93**, 162504 (2008).

14. Haindl, S. *et al.* $LaFeAsO_{1-x}F_x$ thin films: High upper critical fields and evidence of weak link behaviour. Preprint at <http://arxiv.org/abs/0907.2271v1> (2009).

15. Yamamoto, A. *et al.* Small anisotropy, weak thermal fluctuations, and high field superconductivity in Co-doped iron pnictide $Ba(Fe_{1-x}Co_x)_2As_2$. *Appl. Phys. Lett.* **94**, 062511–062513 (2009).

16. Kametani, F. *et al.* Intergrain current flow in a randomly oriented polycrystalline $SmFeAsO_{0.85}$ oxypnictide. *Appl. Phys. Lett.* **95**, 142502 (2009).

17. Lee, S. *et al.* Weak link behaviour of grain boundaries in superconducting $Ba(Fe_{1-x}Co_x)_2As_2$ bicrystals. *Appl. Phys. Lett.* **95**, 212505 (2009).

18. Herranz, G. *et al.* High mobility in $LaAlO_3/SrTiO_3$ heterostructures: Origin, dimensionality, and perspectives. *Phys. Rev. Lett.* **98**, 216803 (2007).

19. Zhang, X. *et al.* Josephson effect between electron-doped and hole-doped iron pnictide single crystals. *Appl. Phys. Lett.* **95**, 062510 (2009).

20. Rijnders, G. J. H. M. *et al. In situ* monitoring during pulsed laser deposition of complex oxides using reflection high energy electron diffraction under high oxygen pressure. *Appl. Phys. Lett.* **70**, 1888–1890 (1997).

21. Wang, X. F. *et al.* Anisotropy in the electrical resistivity and susceptibility of superconducting $BaFe_2As_2$ single crystals. *Phys. Rev. Lett.* **102**, 117005 (2009).

22. Sefat, A. S. *et al.* Superconductivity at 22 K in Co-doped $BaFe_2As_2$ crystals. *Phys. Rev. Lett.* **101**, 117004 (2008).

23. Yamamoto, A. *et al.* Small anisotropy, weak thermal fluctuations, and high field superconductivity in Co-doped iron pnictide $Ba(Fe_{1-x}Co_x)_2As_2$. *Appl. Phys. Lett.* **94**, 062511 (2009).

24. Yuan, H. Q. *et al.* Nearly isotropic superconductivity in $(Ba,K)Fe_2As_2$. *Nature* **457**, 565–568 (2009).

25. Baily, S. A. *et al.* Pseudoisotropic upper critical field in cobalt-doped $SrFe_2As_2$ epitaxial films. *Phys. Rev. Lett.* **102**, 117004 (2009).

26. Maiorov, B. *et al.* Angular and field properties of the critical current and melting line of Co-doped $SrFe_2As_2$ epitaxial films. *Supercond. Sci. Technol.* **22**, 125011 (2009).

27. McKee, R. A., Walker, F. J. & Chisholm, M. F. Crystalline oxides on silicon: The first five monolayers. *Phys. Rev. Lett.* **81**, 3014–3017 (1998).

28. Goncharova, L. V. *et al.* Interface structure and thermal stability of epitaxial $SrTiO_3$ thin films on Si(001). *J. Appl. Phys.* **100**, 014912 (2006).

Acknowledgements

We are grateful to J. Fournelle, M. Putti, A. Xu, F. Kametani, A. Gurevich, P. Li and V. Griffin for discussions and experimental help. Work at the University of Wisconsin was supported by funding from the DOE Office of Basic Energy Sciences under award number DE-FG02-06ER46327, and that at the NHMFL was supported under NSF Cooperative Agreement DMR-0084173, by the State of Florida and by AFOSR under grant FA9550-06-1-0474. A.Y. is supported by a fellowship of the Japan Society for the Promotion of Science. All TEM work was carried out at the University of Michigan and was supported by the Department of Energy under grant DE-FG02-07ER46416.

Author contributions

S.L. fabricated Ba-122 thin-film samples and prepared the manuscript. J.J. carried out electromagnetic characterization and prepared the manuscript. J.D.W. fabricated Ba-122 pulsed laser deposition targets for thin-film deposition. C.W.B., H.W.J., C.M.F. and J.W.P. fabricated thin-film templates on single-crystal substrates. C.T., A.P., D.A. and A.Y. carried out electromagnetic characterizations. S.H.B. and C.W.B. analysed epitaxial arrangement by X-ray diffraction. Y.Z. and C.T.N. carried out TEM measurements. C.B.E., D.C.L., E.E.H. and X.Q.P. supervised the experiments and contributed to manuscript preparation. C.B.E. designed and directed the research. All authors discussed the results and implications and commented on the manuscript at all stages.

Additional information

The authors declare no competing financial interests. Supplementary information accompanies this paper on www.nature.com/naturematerials. Reprints and permissions information is available online at http://npg.nature.com/reprintsandpermissions. Correspondence and requests for materials should be addressed to C.B.E.

Strongly enhanced current densities in superconducting coated conductors of $YBa_2Cu_3O_{7-x}$ + $BaZrO_3$

J. L. MACMANUS-DRISCOLL[1*†], S. R. FOLTYN, Q. X. JIA, H. WANG, A. SERQUIS, L. CIVALE, B. MAIOROV, M. E. HAWLEY, M. P. MALEY AND D. E. PETERSON

[1]Superconductivity Technology Center, Los Alamos National Laboratory, Los Alamos, New Mexico 87545, USA
*Permanent address: Department of Materials Science and Metallurgy, University of Cambridge, Pembroke Street, Cambridge CB2 3QZ, UK
†e-mail: jld35@cam.ac.uk

Published online: 30 May 2004; doi:10.1038/nmat1156

There are numerous potential applications for superconducting tapes based on $YBa_2Cu_3O_{7-x}$ (YBCO) films coated onto metallic substrates[1]. A long-established goal of more than 15 years has been to understand the magnetic-flux pinning mechanisms that allow films to maintain high current densities out to high magnetic fields[2]. In fact, films carry one to two orders of magnitude higher current densities than any other form of the material[3]. For this reason, the idea of further improving pinning has received little attention. Now that

commercialization of YBCO-tape conductors is much closer, an important goal for both better performance and lower fabrication costs is to achieve enhanced pinning in a practical way. In this work, we demonstrate a simple and industrially scaleable route that yields a 1.5–5-fold improvement in the in-magnetic-field current densities of conductors that are already of high quality.

The sources of enhanced pinning in vapour-grown YBCO films are the natural point, line and volume imperfections, probably the most

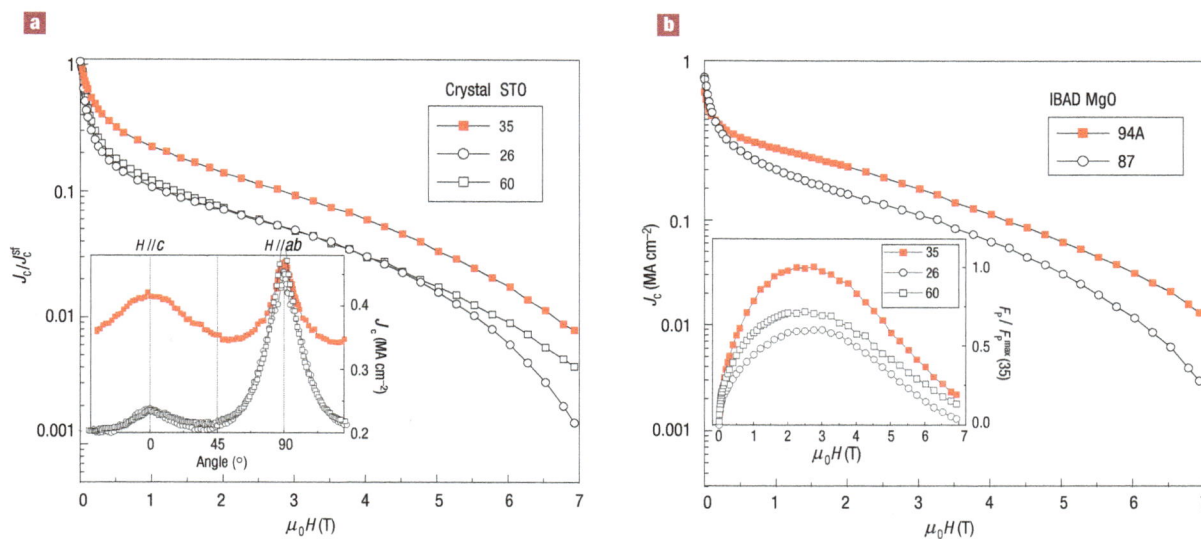

Figure 1 **Critical current density at 75.5 K versus magnetic field applied parallel to the *c* axis for pure YBCO and YBCO+BaZrO₃ films. a,** J_c/J_c^{sf} for single-crystal $SrTiO_3$ substrates. Inset: angular dependence of J_c at 1 T with sample 60 data multiplied by 0.75. **b,** J_c for IBAD-MgO substrates. Open points are for pure YBCO and filled points are for YBCO+BaZrO₃. Inset: $F_p/F_p^{max}(35)$ versus magnetic field applied parallel to the *c* axis for the samples on single-crystal STO.

Table 1 Sample data for reference YBCO films, and for YBCO+BaZrO₃ films on single-crystal STO, MgO-buffered single crystal STO, and STO-buffered-IBAD-MgO.

Sample number	T_c (breadth) (K)	Self-field J_c (75.5 K)	Thickness (µm)	Substrate
Pure YBCO				
26	91.7 (0.7)	2.4	1.55	STO
60	91.5 (0.5)	2.6	1.0	STO
87	92 (3)	2.3	1.2	STO on IBAD
YBCO+BaZrO₃				
12	91 (1)	2.4	0.5	STO
13	92 (2)	2.3	1.7	STO
30	91.5 (1)	> 2.2*†	1.3	STO
35	89.5 (2.5)	2.0	0.75	STO
37	87.8 (0.5)	2.2	1.0	STO
83	90 (1)	>1.8*	1.3	STO on MgO
91	89 (1)	1.5	1.2	STO on MgO
94A	87.5 (1.5)	1.7	0.9	STO on IBAD
94B	88 (1)	2.0	1.0	STO on MgO
95B	88.7 (2)	>2.6*	1.2	STO on MgO

*Current carried in bridge exceeded measurement limit of the current source of 10 A.

†Bridge blew during measurement.

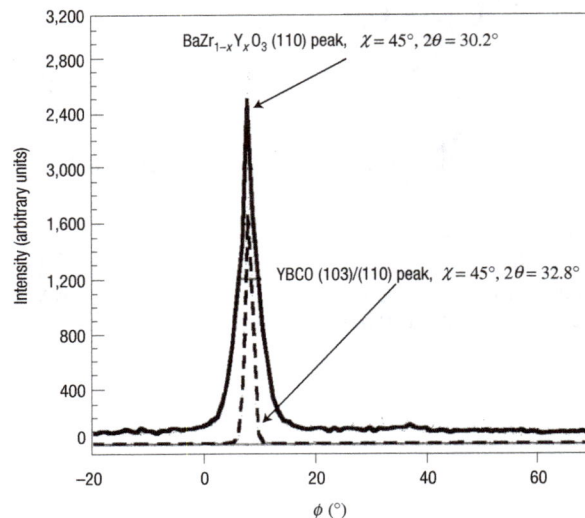

Figure 2 XRD plot showing in-plane alignment of YBCO and BaZrO₃ particles.
ϕ scans at $\chi = 45°$, $2\theta = 30.2°$ corresponding to the BaZrO₃ (110) peak, and $2\theta = 32.8°$ corresponding to the YBCO (103)/(110) peak.

significant of these being the dislocations perpendicular to the substrate plane of film[4]. In terms of dimensionality, dislocations are nearly ideal for pinning magnetic-flux lines. However, the density of dislocations is dominated by the growth island size and their spacing is estimated to be rather large (~100–500 nm)[3,5]. To increase dislocation density, an obvious way would be to decrease the island size that the dislocations bound[6]—for example, by reducing growth temperature—but this is non-trivial because the crystalline quality of the film would be compromised.

Heavy-ion irradiation has been shown to reduce vortex mobility[7,8] but it is impractical for treatment of coated conductors. Other work involving growth of films on mis-cut single-crystal substrates has demonstrated that introduction of columnar growth defects improves the transport critical current density, J_c (at 77 K), by up to 50%, but only at a particular field orientation and magnitude[9]. Other ideas for improving pinning are the introduction of defects by multilayering or addition of particles on the substrate surface[5,10]. When using these methods, some improvements seem possible in thin films on single-crystal substrates.

In this work, prompted by our earlier report that suggested the possibility of enhanced pinning in the presence of epitaxial second phases[11], we study BaZrO₃ additions to YBCO. The main reasons for the choice of BaZrO₃ are: (1) although it can grow heteroepitaxially with YBCO, it has a large lattice mismatch (~9%) so strain between the phases could introduce defects for enhanced pinning; (2) it is a high-melting-temperature phase and so growth kinetics should be slow, leading to small particles; and (3) Zr does not substitute in the YBCO structure[12]. Indeed, single crystals of YBCO are often grown in BaZrO₃-coated crucibles[13]. BaZrO₃ has also been previously investigated[14,15] as a pinning centre in bulk, melt-processed YBCO. However, it was found that the BaZrO₃ agglomerated at the growth fronts of the grains, and, because of heteroepitaxial matching with the YBCO lattice, it acted as a seed to nucleate multiple grains instead of a single domain.

Here we show that nanoparticles of BaZrO₃ grow heteroepitaxially within laser-ablated YBCO films. These particles are easily incorporated in the films from the source target of a ceramic BaZrO₃/YBCO mixture. The particles lead to significant improvement in J_c (at 75.5 K) in films both on single-crystal substrates and on practical, buffered metallic substrates.

Ceramic targets were prepared from: (1) pure YBCO (SCI Inc.), and (2) YBCO + 5 mol.% BaZrO₃. Commercial YBCO powder was used, as well as 99.99% pure powders of Ba(NO)₃ and ZrO₂ (Aldrich). The powders were mixed, ground, pressed and then sintered at 950 °C in flowing oxygen gas. The targets were ablated using pulsed laser deposition with a Lambda Physik KrF excimer laser ($\lambda = 248$ nm), at a repetition rate of 10 Hz. All of the depositions were carried out at the same substrate-to-target distance of 5 cm and an oxygen pressure of 200 mtorr. The substrates used were either single-crystal SrTiO₃ (STO), STO-buffered MgO single crystals, or STO-buffered ion-beam-assisted deposited MgO on Hastelloy[16], hitherto referred to as IBAD-MgO. After deposition at 760–790 °C, samples were cooled to room temperature in O₂ at 300 torr. Ten different YBCO+BaZrO₃ samples were grown with thicknesses in the range 0.5–1.7 µm.

For all the samples, inductive critical temperature (T_c) and J_c measurements in self-field (that is, in no applied magnetic field) at liquid N₂ temperatures were made. Further transport measurements were conducted on some of the samples in liquid N₂, in a magnetic field rotated in a plane perpendicular to the plane of the film but always normal to the current (maximum Lorentz force configuration). Microstructural characterization was carried out by X-ray diffractometry (XRD), atomic force microscopy (AFM) and cross-sectional transmission electron microscopy (TEM).

Table 1 shows the measured data for pure YBCO films and several YBCO+BaZrO₃ films. The T_cs and self-field J_cs of some of the YBCO+BaZrO₃ samples are slightly lower than for the pure YBCO. Figure 1a shows J_c/J_c^{sf} (where J_c^{sf} is self-field J_c) versus magnetic field parallel to the c axis ($H \| c$) for a YBCO+BaZrO₃ film compared with two

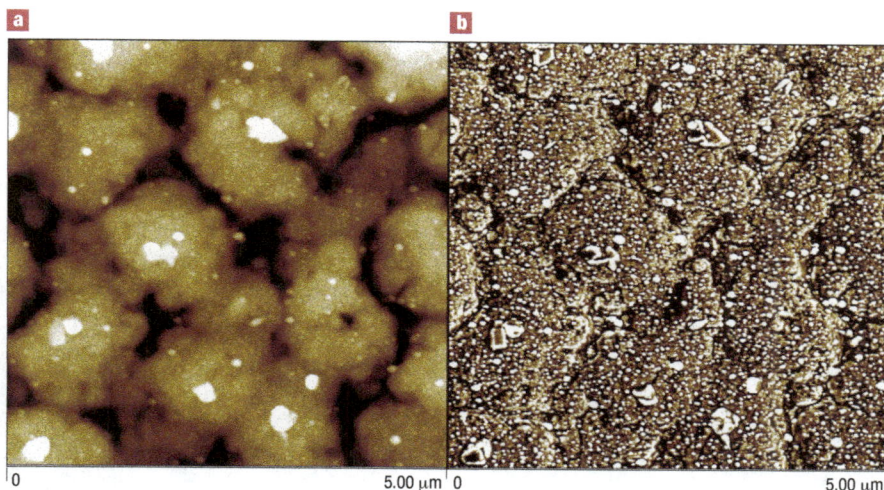

Figure 3 Micrographs of YBCO+BaZrO₃ films grown on single crystal SrTiO₃ showing surface nanoparticles. **a**, Atomic force micrograph, and **b**, phase contrast micrograph.

pure YBCO films of different thickness on STO single crystals. Figure 1b shows J_c versus field ($H\|c$) for a YBCO+BaZrO₃ film compared with a pure YBCO film, both on IBAD-MgO. Figure 1a shows that pure YBCO films of different thickness give very similar forms of the J_c/J_c^{sf} versus field ($H\|c$) curve. We always find this to be the case for samples of different thickness (in the range 0.5–2 μm) if the T_c is the same, and if samples are compared on single crystal or on the same batch of IBAD-MgO (the same batch of IBAD-MgO was used for the samples in Table 1).

The striking result of Fig. 1 is the upwards shift in both normalized J_c, and J_c for the YBCO+BaZrO₃ samples. Identical behaviours were observed for the two other samples measured on single-crystal substrates (13 and 83). Even though the YBCO+BaZrO₃ samples of Fig. 1 have slightly lower T_cs than the pure ones, the in-field J_cs are improved significantly. Hence the BaZrO₃ addition has clearly increased the irreversibility field. On both the single-crystal and IBAD-MgO substrates, accounting for the thickness differences between samples, the J_c values are around a factor of 1.5–2 higher over a wide field range (~1–5 T) and they increase to a factor of around 5 higher at 7 T. In ~1-μm-thick YBCO+BaZrO₃ films on IBAD, the J_cs remain in excess of 0.1 MA cm⁻² at 4.5 T.

The inset of Fig. 1a shows the angular dependence of J_c measured in a field of 1 T for the samples on STO single-crystal. A shift upwards in the relative height of the c-axis angular peak is observed for YBCO+BaZrO₃ compared with the pure YBCO, indicative of strong pinning defects along the c axis in the YBCO+BaZrO₃ film[17]. In fact, the J_c is increased substantially compared with the pure sample across the angular range from 0° to ~80°. The same trend is also observed for samples on IBAD-MgO. The result is important for applications of coated conductors, because the magnetic field will rarely be constrained to a single orientation. Previous measurements of the J_c ($H\|c$) dependencies for pure YBCO films or for the samples on STO, single-crystal STO and IBAD-MgO have been shown to be very similar, suggesting a similar pinning mechanism on different substrates.

The inset of Fig. 1b shows the pinning force ($F_p = \mu_o H J_c$) normalized by the maximum pinning force measured for sample 35 ($F_p^{max}(35)$), versus field for the samples on STO single-crystal. Despite the lower T_c of the YBCO+BaZrO₃ film (35), the pinning force is significantly higher than for the pure YBCO films (26 and 60).

X-ray diffractograms of YBCO+BaZrO₃ films (not shown) show peaks belonging to (00l) YBCO plus an additional, broad, low-intensity peak centred around $2\theta = 42.7°$, which is consistent with (200)BaZrO₃, and/or (200)Ba₂Zr₂₋ₓYₓO₆, and/or (400) Ba₂ZrYO₆ (ref. 18). Surprisingly, the Ba₂ZrYO₆ phase has not been widely reported previously, that is, the assumption has generally been that BaZrO₃ does not react with YBCO. However, the breadth and position of the second phase peak suggests that Ba₂Zr₂₋ₓYₓO₆ particles with a range of x values are present. As shown from X-ray ϕ scans (Fig. 2) of the YBCO (103/110) and BaZrO₃ (110) peaks at $\chi = 45°$ and $2\theta = 32.8°$ and 30.2°, respectively, the particles are in-plane aligned cube-on-cube with the YBCO.

Figure 3 shows AFM micrographs for a YBCO+BaZrO₃ film on STO. Nanoparticles are observed distributed across the film surface. These particles are not observed in pure YBCO films. The phase-contrast image of Fig. 3b indicates that the particles are composed of a phase different from YBCO. Cross-sectional low magnification and high-resolution TEM micrographs (Fig. 4a and b, respectively) confirm the presence of nanoparticles embedded in the YBCO lattice. In addition, c-aligned, misfit edge dislocations of spacing <50 nm are present in the YBCO. Some of these dislocations are indicated by black arrows. The minimum density of columnar defects is at least 400 μm⁻², compared with ~80 μm⁻² previously observed[5] in YBCO films grown by pulsed laser deposition. The dislocations form a family of correlated linear defects that should produce substantial uniaxial pinning along the c direction, consistent with the observed enhancement of the c-axis peak in J_c (inset of Fig. 1a). It seems that the dislocations form as a result of the lattice misfit between the particles and matrix. On the basis of the film and particle lattice parameters of ~3.85 Å and 4.23 Å, respectively, the level of strain is around 9.4% (ref. 19). The high defect density is consistent with the reduced T_cs and self-field J_cs of some of the samples. Although the c-axis dislocations are the source of the enhanced directional pinning, it is possible that some random pinning may also arise directly from the BaZrO₃ particles.

Lattice images (Fig. 4c) and fast Fourier transformations of the particles indicated a cubic structure, with lattice parameter $a \approx 4.23$ Å, consistent with BaZr₁₋ₓYₓO₃. Owing to the small sizes of the particles, it was not possible to determine their composition conclusively. A typical particle-size distribution measured over a ~1 μm² area is shown in

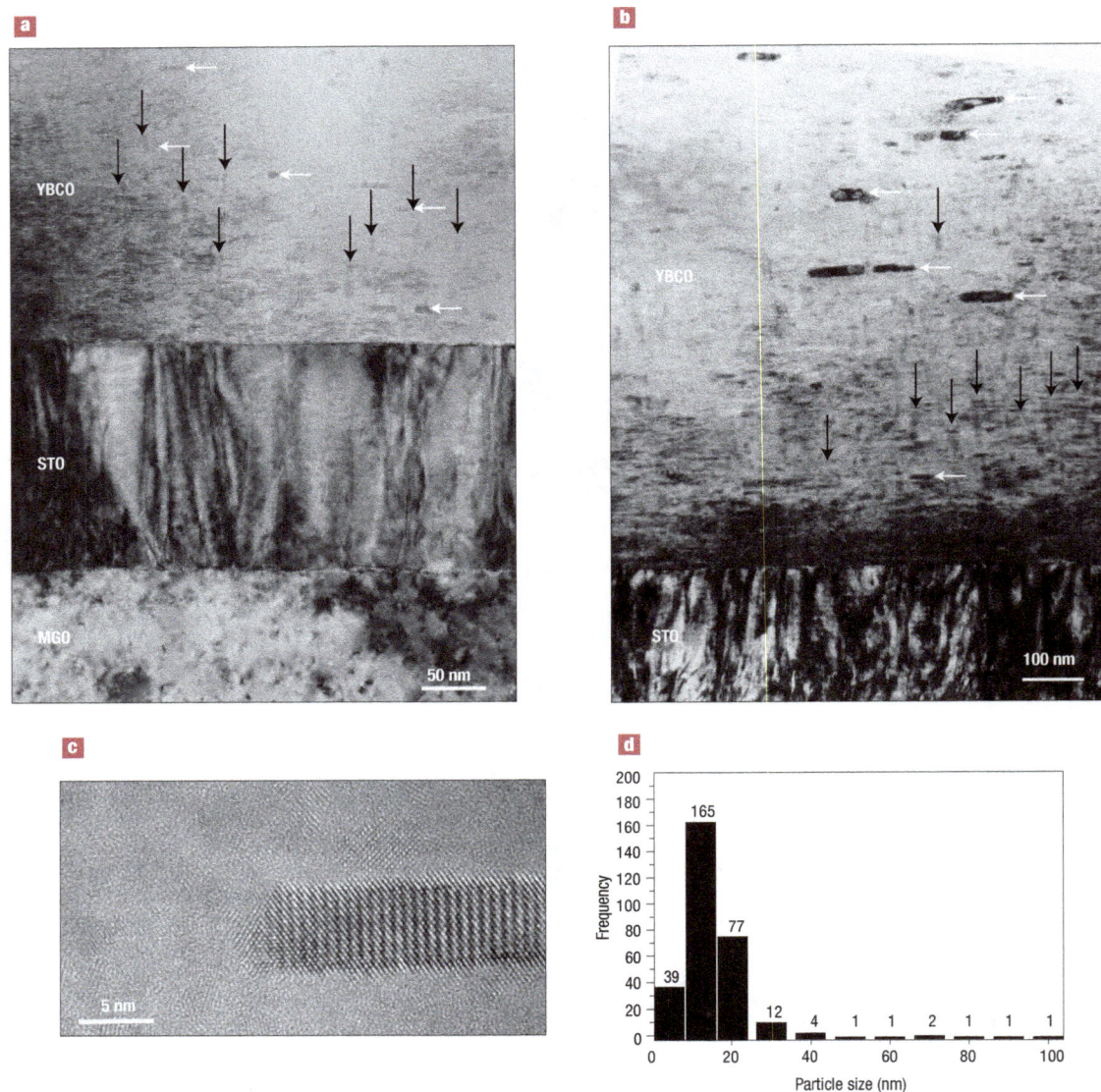

Figure 4 TEM micrographs showing nanoparticles of BaZrO₃ in YBCO+BaZrO₃ films on STO-buffered MgO single-crystals. Some of the nano-particles have been labelled using white arrows and some of the columnar defects by black arrows. **a**, High-magnification micrograph. **b**, Lower-magnification micrograph. **c**, Image of nanoparticle showing lattice fringes consistent with a cubic lattice. **d**, Histogram showing distribution of particle sizes.

Fig. 4d. The nanoparticles ranged in size from 5 nm to 100 nm with a modal particle size of 10 nm.

By implementing a straightforward and inexpensive target compositional modification, J_c enhancements of up to a factor of 5 (depending on field) at liquid N_2 temperatures are achieved in a reproducible way on both single-crystal and buffered metallic substrates. Because only one type of heteroepitaxial addition (BaZrO₃) and only one concentration (5 mol%) were studied, it is likely that other concentrations and different heteroepitaxial second-phase additions will lead to yet greater enhancements in pinning.

Received 13 November 2003; accepted 10 May 2004; published 30 May 2004.

References

1. Selvamanickam, V. *et al.* High temperature superconductors for electric power and high-energy physics. *J. Miner. Met. Mater. Soc.* **50**, 27–30 (1998).

2. Blatter, G., Geshkenbein, V. B. & Koopmann, J. A. G. Weak to strong pinning crossover. *Phys. Rev. Lett.* **92**, 067009 (2004).

3. Dam, B. *et al.* Origin of high critical currents in YBa₂Cu₃O₇₋δ superconducting thin films. *Nature* **399**, 439–442 (1999).

4. Hawley, M., Raistrick, I. D., Beery, J. G. & Houlton, R.J. Growth mechanism of sputtered films of YBa₂Cu₃O₇ studied by scanning tunneling microscopy. *Science* **251**, 1587–1589 (1991).

5. Huijbregtse, J. M. *et al.* Natural strong pinning sites in laser-ablated YBa₂Cu₃O₇₋δ thin films. *Phys. Rev. B* **62**, 1338–1349 (2000).

6. Dam, B., Huijbregtse, J. M. & Rector, J. H. Strong pinning linear defects formed at the coherent growth transition of pulsed-laser-deposited YBa₂Cu₃O₇₋δ films. *Phys. Rev. B* **65**, 064528 (2002).

7. Mazilu, A. *et al.* Vortex dynamics of heavy-ion-irradiated YBa₂Cu₃O₇₋δ: Experimental evidence for a reduced vortex mobility at the matching field. *Phys. Rev. B* **58**, R8909–R8912 (1998).

8. Crescio, E. *et al.* Interplay between as-grown defects and heavy ion induced defects in YBCO films. *Intl J. Mod. Phys. B* **13**, 1177–1182 (1999).

9. Lowndes, D. H. *et al.* Strong, asymmetric flux-pinning by miscut-growth-initiated columnar defects in epitaxial YBa₂Cu₃O₇₋δ films. *Phys. Rev. Lett.* **74**, 2355–2358 (1995).

10. Haugan T. *et al.* Island growth of Y₂BaCuO₅ nanoparticles in (211₋₁.₅ₙₘ/123₋₁₀ₙₘ)×N composite multilayer structures to enhance flux pinning of YBa₂Cu₃O₇₋δ films. *J. Mater. Res.* **18**, 2618–2623 (2003).

11. Berenov, A. *et al.* Microstructural characterization of $YBa_2Cu_3O_{7-\delta}$ thick films grown at very high rates and high temperatures by pulsed laser deposition. *J. Mater. Res.* **18**, 956–964 (2003).

12. Palouse, K. V., Koshy, J. & Damodaran, A. D. Superconductivity in $YBa_2Cu_3O_{7-\delta}$-ZrO_2 systems. *Supercond. Sci. Technol.* **4**, 98–101 (1991).

13. Erb, A., Walker, E., Genoud, J. Y. & Flukiger, R., Improvements in crystal growth and crystal homogeneity and its impact on physics. *Physica C* **282**, 89–92 (1997).

14. Cloots, R. *et al.* Effect of $BaZrO_3$ Additions on the microstructure and physical properties of melt-textured Y-123 superconducting materials. *Mater. Sci. Eng. B* **53**, 154–158 (1998).

15. Luo, Y. Y. *et al.* Effects of precursors with fine $BaZrO_3$ inclusions on the growth and microstructure of textured YBCO. *J. Supercond.* **13**, 575–581 (2000).

16. Groves, R. *et al.* Improvement of IBAD MgO template layers on metallic substrates for YBCO HTS deposition. *IEEE Trans. Appl. Supercond.* **13**, 2651–2654 (2003).

17. Civale, L. *et al.* Angular dependent vortex pinning mechanisms in YBCO coated conductors and thin films. *Appl. Phys. Lett.* **84**, 2121–2123 (2004).

18. Palouse, K. V., Sebastian, M. T., Nair, K., Koshy, J. & Damodaran, A. Synthesis of Ba_2YZrO_6: A new phase in $YBa_2Cu_3O_{7-\delta}$-ZrO_2 system and its suitability as a substrate material for YBCO films. *Solid State. Comm.* **83**, 985–988 (1992).

19. Ramesh, R. *et al.* The atomic structure of growth interfaces in Y-Ba-Cu-O thin-films. *J. Mater. Res.* **6**, 2264–2271 (1991).

Acknowledgements

The work was supported by the Office of Energy Efficiency and Renewable Energy, US Department of Energy. We thank John Durrell, University of Cambridge, UK for helpful discussions. Correspondence and requests for materials should be addressed to J.L.M-D.

Competing financial interests

The authors declare that they have no competing financial interests.

www.ingramcontent.com/pod-product-compliance
Lightning Source LLC
Chambersburg PA
CBHW081047220326
41598CB00038B/7015